MOLECULAR ONCOLOGY OF BREAST CANCER

Edited by

JEFFREY S. ROSS, MD, AND
GABRIEL N. HORTOBAGYI, MD, FACP

JONES AND BARTLETT PUBLISHERS
Sudbury, Massachusetts
BOSTON TORONTO LONDON SINGAPORE

World Headquarters
Jones and Bartlett Publishers
40 Tall Pine Drive
Sudbury, MA 01776
978-443-5000
info@jbpub.com
www.jbpub.com

Jones and Bartlett Publishers Canada
2406 Nikanna Road
Mississauga, ON L5C 2W6
CANADA

Jones and Bartlett Publishers International
Barb House, Barb Mews
London W6 7PA
UK

Production Credits
Chief Executive Officer: Clayton Jones
Chief Operating Officer: Don W. Jones, Jr.
President, Higher Education and Professional Publishing: Robert W. Holland, Jr.
V.P., Sales and Marketing: William J. Kane
V.P., Design and Production: Anne Spencer
V.P., Manufacturing and Inventory Control: Therese Bräuer
Executive Publisher: Christopher Davis
Special Projects Editor: Elizabeth Platt
Editorial Assistant: Kathy Richardson
Director of Marketing: Alisha Weisman
Marketing Manager: Matthew Payne
Manufacturing and Inventory Coordinator: Amy Bacus
Cover Design: Kristin E. Ohlin
Cover Image: © Photos.com
Printing and Binding: Courier Westford
Cover Printing: Courier Westford

Library of Congress Cataloging-in-Publication Data
CIP data not available at time of printing

Printed in the United States of America
08 07 06 05 04 10 9 8 7 6 5 4 3 2 1

CONTENTS

CONTRIBUTORS

Joan Albanell, MD	Department of Medical Oncology, ICMHO & IDIBAPS, Hospital Clinic, Barcelona, Spain
Shahgul Anwar, MD	Department of Pathology and Laboratory Medicine, Albany Medical College, Albany, New York
Mark Ayers, BS	Bristol Myers Squibb Corp., Princeton, NJ
Esteban Ballestar PhD	Cancer Epigenetics Laboratory, Molecular Pathology Program, Spanish National Cancer Center, Spain
Ann Boguniewicz, MD	Department of Pathology and Laboratory Medicine, Albany Medical College, Albany, New York
Melissa Bondy, PhD	Department of Epidemiology, Division of Cancer Prevention, The University of Texas M. D. Anderson Cancer Center, Houston, Texas
Abenaa Brewster, MD, MHS	Department of Epidemiology, Division of Cancer Prevention, The University of Texas M. D. Anderson Cancer Center, Houston, Texas
Fatima Cardoso, MD	Department of Medical Oncology, Jules Bordet Institute, Brussels, Belgium
Lynn Cawkwell, PhD	Division of Cell & Molecular Medicine, University of Hull, Research & Development Building, Castle Hill Hospital, Hull, East Yorkshire, United Kingdom
Carlos Cordon-Cardo, MD, PhD	Division of Molecular Pathology, Memorial Sloan Kettering Cancer Center, New York
Lissandra Dal Lago, MD	Translational Research, Jules Bordet Institute, Free University of Brussels, Belgium
Christine Desmedt, PhD	Translational Research, Jules Bordet Institute, Free University of Brussels, Belgium
Eleftherios Diamandis, MD, PhD	Department of Pathology and Laboratory Medicine, Mount Sinai Hospital and the Department of Laboratory Medicine and Pathobiology, University of Toronto, Ontario, Canada
Virginie Durbecq, PhD	Translational Research, Jules Bordet Institute, Free University of Brussels, Belgium
Manel Esteller MD, PhD	Cancer Epigenetics Laboratory, Molecular Pathology Program, Spanish National Cancer Center, Spain
Jonathan A. Fletcher, MD	Department of Pathology, Brigham & Women's Hospital, Boston, MA
John A. Foekens, PhD	Department of Medical Oncology, Erasmus MC, Rotterdam, The Netherlands
Nadia Harbeck, MD, PhD	Department of Obstetrics and Gynecology, Klinikum rechts der Isar, Technical University of Munich, Munich, Germany
D. Paul Harkin, MD	Cancer Research Centre, Queens University Belfast, University Floor, Belfast City Hospital, Lisburn Road, Belfast, Northern Ireland
Heinz Höfler, MD	Klinikum rechts der Isar, Institut fur Allgemeine Pathologie und Pathologische Anatomie, Technische Universitat München, Munich, Germany
Gabriel N. Hortobagyi, MD	Department of Breast Medical Oncology, The University of Texas M. D. Anderson Cancer Center, Houston, Texas
Sabine Kasimir-Bauer, PhD	Klinik für Frauenheilkunde und Geburtshilfe, Universitätsklinikum Essen, Essen, Germany
Ronald E. Kates, PhD	Department of Obstetrics and Gynecology, Klinikum rechts der Isar, Technical University of Munich, Munich, Germany
Richard D. Kennedy, MD, PhD	Cancer Research Centre, Queens University Belfast, University Floor, Belfast City Hospital, Lisburn Road, Belfast, Northern Ireland
Iraj Khalkhali, MD	Department of Radiology, Harbor UCLA Medical Center, Torrance, California
Marion Kiechle, MD	Department of Obstetrics and Gynecology, Klinikum rechts der Isar, Technical University of Munich, Munich, Germany
Marc Lacroix, PhD	Laboratoire Jean-Claude Heuson de Cancérologie Mammaire, Jules Bordet Institute, Free University of Brussels, Belgium
Gerald P. Linette, MD, PhD	Division of Oncology, Washington University School of Medicine, St. Louis, Missouri and Millennium Pharmaceuticals, Cambridge, MA

Birgit Luber, MD	Klinikum rechts der Isar, Institut fur Allgemeine Pathologie und Pathologische Anatomie, Technische Universitat München, Munich, Germany.
Viktor Magdolen, PhD	Department of Obstetrics and Gynecology, Klinikum rechts der Isar, Technical University of Munich, Munich, Germany
Sabine Maier, PhD	Department of Biomedical Research and Development and Technology Development, Epigenomics AG, Berlin, Germany
Douglas P. Malinowski, PhD	Chief Scientific Officer, Tripath Oncology, Durhsm, North Carolina
John W. M. Martens, PhD	Department of Medical Oncology, Erasmus MC, Rotterdam, The Netherlands
Alfons Meindl, PhD	Department of Obstetrics and Gynecology, Klinikum rechts der Isar, Technical University of Munich, Munich, Germany
Volkmar Müller, MD	Klinik für Frauenheilkunde, Universitätsklinikum Hamburg-Eppendorf, Hamburg, Germany
Andre M. Oliveira, PhD	Department of Pathology, Brigham & Women's Hospital, Boston, MA
Robert Z. Orlowski, MD	Department of Medicine, Division of Hematology/Oncology, and the Lineberger Comprehensive Cancer Center, University of North Carolina at Chapel Hill, Chapel Hill, North Carolina
Klaus Pantel, MD	Institut für Tumorbiologie, Universitätsklinikum Hamburg-Eppendorf, Hamburg, Germany
Martine Piccart, MD, PhD	Translational Research, Jules Bordet Institute, Free University of Brussels, Belgium
Robert Pietrusko, PharmD	Worldwide Regulatory Affairs and Pharmacovigilence, Millennium Pharmaceuticals, Inc., Cambridge, Massachusetts
Lajos Pusztai MD, PhD	Department of Breast Medical Oncology, The University of Texas M. D. Anderson Cancer Center, Houston, Texas
Jennifer E. Quinn, PhD	Cancer Research Centre, Queens University Belfast, University Floor, Belfast City Hospital, Lisburn Road, Belfast
Merrill S. Ross, BS	Department of Pathology and Laboratory Medicine, Albany Medical College, Albany, New York
Jeffrey S. Ross, MD	Department of Pathology and Laboratory Medicine, Albany Medical College, Albany, New York and Division of Molecular Medicine, Millennium Pharmaceuticals, Inc., Cambridge, Massachusetts
Roman Rouzier, MD	Department of Breast Medical Oncology, The University of Texas M. D. Anderson Cancer Center, Houston, Texas
Manfred Schmitt, PhD	Department of Obstetrics and Gynecology, Klinikum rechts der Isar, Technical University of Munich, Munich, Germany
Christos Sotiriou, MD	Translational Research, Jules Bordet Institute, Free University of Brussels, Belgium
James Stec, BS	Millennium Pharmaceuticals Inc., Cambridge, Massachusetts
W. Fraser Symmans, MD	Department of Pathology, The University of Texas M. D. Anderson Cancer Center, Houston, Texas
Christoph Thomssen, MD	Klinik und Poliklinik fur Gynakologie, Universitats-Klinikum Eppendorf, Hamburg, Germany
M. Perla Vargas, MD	Department of Pathology, Harbor UCLA Medical Center, Torrance, California
Hernan I. Vargas, MD	Department of Surgery, Harbor UCLA Medical Center, Torrance, California
Peter Wagner, MD	Department of Breast Medical Oncology, The University of Texas M. D. Anderson Cancer Center, Houston, Texas
Mark Watson, MSc	Division of Cell & Molecular Medicine, University of Hull, Research & Development Building, Castle Hill Hospital, Hull, East Yorkshire, United Kingdom
Xiao-Ming Yin, MD, PhD	Department of Pathology, University of Pittsburgh School of Medicine, Pittsburgh, Pennsylvania

FOREWORD

Molecular Oncology of Breast Cancer comes at just the right time, when the scientific groundwork is being laid for a revolutionary change in the way we practice oncology. In the period prior to sophisticated radiographic imaging, we diagnosed and staged breast cancer based on the physical exam, direct radiographic techniques including mammography, the histopathology of a tumor biopsy, and the findings obtained during surgical treatment with lymph node sampling. Spread of disease as time progressed was determined by radiography and biopsy, performed in response to new signs or symptoms.

With the advent of computerized tomography, in the 1980s we now stage breast cancer and identify metastatic sites with far more accuracy. It was during the 1990s that we began to collect data on abnormalities in genes, gene expression profiles and protein expression in the various types of cancers.

Entry into the 21st century has been accompanied by an avalanche of data on genes and their products that are abnormal in the many types of breast cancer. The technologies utilized to assess breast cancer in clinical departments of pathology, laboratory medicine, and diagnostic imaging have become highly sophisticated. Thousands of recent publications document statistically significant correlations between genetic and molecular abnormalities and the diagnosis, prognosis and response to therapy for patients with breast cancer.

While these correlations have given us increased understanding of "what is wrong" in the breast cancer cell and have helped identify targets for new anticancer therapies, there are only a few examples in which the impact of these scientific findings have changed the way we practice medicine. *HER-2* and the *BRCA-1* and *-2* genes come to mind, in addition to the long-standing usefulness of measuring estrogen receptor levels.

We envision a future when a select panel of genetic and molecular abnormalities will be validated as biological markers that predict risk of breast cancer, detect the presence of cancer at an early time point when only a small tumor burden is present, predict which therapy is likely to be effective in that particular individual's cancer, and assess the response to treatment after days rather than months. For practicing oncologists and researchers interested in the science that will be translated into clinical practice during the next decade, this book offers an invaluable resource.

For investigators interested in the progress that is being made in the genetic and molecular characterization of breast cancer, this book provides a well-organized, up-to-date and exhaustive compendium of discoveries. The authors have taken care to explain methodologies in addition to detailing results of extensive research.

Chapters in *Molecular Oncology of Breast Cancer* present research observations on genes and their products that control cell cycle traversal, a variety of cell functions, cell death, regulation of gene expression, repair of DNA, and cell–cell interactions. They also deal in depth with the various potential applications of diagnostic markers in determining predisposition, screening and early detection, and selection of therapy for breast cancer—in addition to adding to our understanding of the molecular pathogenesis of the disease. I know of no other attempt to accomplish these important goals with this degree of thoroughness.

In summary, this book provides a comprehensive, current assessment of our knowledge of genes and molecular pathways involved in breast cancer, together with practical as well as potential applications to diagnosis, treatment and prevention.

John Mendelsohn, M.D.
President, The University of Texas M. D. Anderson Cancer Center, Houston, Texas

PREFACE

The last three decades witnessed substantial progress in the management of breast cancer. Such progress can be measured in improvements in survival and quality of life. Thus, systematic use of screening mammography was show to decrease tumor size and stage at diagnosis: earlier diagnosis is clearly associated with improved survival. The introduction of several active cytotoxic agents into the management of metastatic and subsequently primary breast cancer demonstrated the chemosensitivity of this malignancy and the incorporation of adjuvant chemotherapy into the multidisciplinary management of primary breast cancer resulted in significant reductions in odds of recurrence and death. Similarly, adjuvant endocrine therapy was shown to reduce odds of recurrence and death for patients with hormone receptor-positive breast cancer. More recently, the development of the first monoclonal antibody targeted to a specific molecular anomaly, HER2-amplification, led to improvements in response rate, time to progression and survival in a subgroup of patients with aggressive and poor prognosis breast cancer.

All these advances were made possible by improved understanding of the natural history and clinical course of breast cancer, research that led to expanded biological knowledge of malignant transformation and progression of tumors and by the utilization of modern clinical trials methodology to test hypotheses in the clinic.

Another important step forward was the demonstration within the past ten years that interfering with estrogenic stimulation, by the use of antiestrogens or ovarian ablation, reduced significantly the incidence of second primary breast cancer, and as shown by formal, prospective randomized trials, also reduced the risk of developing breast cancer in the first place. These studies were the first large-scale demonstration of the efficacy of chemoprevention, based on a modest understanding of the pathophysicology of hormone-dependent breast cancer and of the effects, pharmacokinetics and pharmacodynamics of a group of therapeutic agents, the antiestrogens or selective estrogen receptor modulators.

During the past three decades, the incidence of breast cancer has continued to increase throughout the industrialized world. However, improvements in diagnosis and treatment have kept breast cancer mortality mostly unchanged, with a definite and persistent decline over that past ten years. Thus, not only has there been progress documented in clinical trials, but reduction in mortality has been demonstrated in population-based calculations. Unfortunately, many women still die of breast cancer, and many more develop the disease and have to cope with the anxiety and suffering associated with the diagnosis and currently used treatment approaches. Therefore, much room exists for improvement, and progress in biological understanding might facilitate the next generation of diagnostic, therapeutic and preventive strategies.

The War on Cancer started formally 30 years ago. While the technological and conceptual challenges encountered over this time have been substantial, so has the injection of research funds from government, industry and private philanthropy. Well-funded research expanded enormously our knowledge of basic cell biology, molecular genetics, intra- and intercellular signaling processes and the various molecular differences between normal and malignant breast cells. We have also learned much about the close and mutual interaction between the malignant cells as a community and the surrounding stroma, and the importance of such interactions in the success or failure of the metastatic process. All these areas of new knowledge, when taken to their translational term, lead to the identification of multiple potential targets for risk assessment, diagnosis, prevention or therapeutic interventions.

The editors have followed the burgeoning literature in this exciting field. We have noted the enormous expansion of information and the virtual impossibility of remaining current in all areas relevant to the diagnosis and management of patients with breast cancer, not to mention the intricacy and complexity of biological information on which modern research is based. It became obvious, then, that a detailed compilation of relevant information was necessary, both to serve as a repository of the state of the art, and a source of references for activities related to molecularly based research and management, but also to crystallize current concepts in the field and facilitate further progress. Among many outstanding scientists, we identified a group of individuals considered thought leaders in their chosen area of research interest, to which they have contributed substantial information, and in many cases, led the progress achieved to date.

We hope that this compilation of information will be useful to the readers, whether in laboratory-based or clinical investigation, and will catalyze collaborative efforts to advance the pace of progress. While much progress has been made, much progress remains to be realized, and the progress of the biological revolution as applied to oncology must be rapidly translated into discrete, clinically relevant tools to improve diagnosis and management of this disease, and eventually reduce its incidence by early interventions, before the disease becomes clinically manifest.

Jeffrey S. Ross
Gabriel N. Hortobagyi

CHAPTER 1

Introduction and Background

Gabriel N. Hortobagyi, MD, FACP

Department of Breast Medical Oncology, The University of Texas M. D. Anderson Cancer Center, Houston, Texas

Breast cancer is the most common cancer in women in the United States and most of the industrialized Western World.[1] More than 200,000 women will be diagnosed with breast cancer in 2004 in the United States alone. Worldwide, over 1 million new cases of breast cancer wiill be detected.[2] Breast cancer is the second cause of cancer death in North American women. More than 40,000 women will die this year of metastatic breast cancer in the United States.[1] The mortality rate around the world, especially in developing countries, is much higher, making breast cancer a significant public health problem.[2–4]

Much progress has been made over the past three decades in our understanding of the epidemiology, clinical course, and basic biology of breast cancer.[5] Tangible changes have occurred in areas of early detection, risk assessment, and combined modality treatment that have transformed our ability to manage and, in many cases, cure breast cancer. We review some of these changes below.

Detection and Diagnosis

Over four decades ago, the first successful attempts at imaging the breast were reported.[6] There were daunting challenges to this diagnostic approach, because of the limited contrast provided by various tissue components of the breast and the technical limitations of x-ray imaging. Progress in this area led rapidly to the development of clinical trials to assess the value of systematic attempts at early diagnosis later denominated screening.[7–9] Almost a dozen randomized clinical trials of mammographic screening have been reported.[10,11] The overall results show a 20% to 30% reduction in breast cancer mortality for women randomized to screening mammography and even higher if analysis is restricted to those who were compliant with protocol recommendations.[12,13] Other observations in these trials showed that women undergoing screening had smaller, earlier stage tumors that were highly curable and lent themselves to breast conserving therapy. The strength of the evidence supporting systematic mammographic screening resides in women aged 50 to 65 years. Similar trends occur in women aged 40 to 50 years and in women older than age 65, but these trends do not reach statistical significance because of insufficient numbers of subjects included in controlled trials.[14]

Screening mammography, although clearly effective in reducing mortality, has been criticized for leading to a number of false-positive findings, additional tests, and breast biopsies, thus contributing to cost and morbidity.[14–16] Therefore, there is much room for improvement in diagnostic technology. In recent years, digital mammography was introduced. This technology gives results at least equivalent to film mammography but requires less time and allows computerized manipulation of images, electronic storage, and the use of automated interpretation currently used to triage images for second readings.[17] Another diagnostic approach under evaluation for early diagnosis is magnetic resonance (MR) mammography, especially in women with high-risk familial breast cancers.[18,19] Preliminary results have shown that MR mammography is quite sensitive in young dense breasts, detecting a higher number of benign and malignant lesions but at the cost of decreased specificity. Additional research and standardization are needed to optimize MR imaging for screening purposes.

Thus, imaging provides means for early diagnosis of breast cancer but with low sensitivity and specificity. Physical examination has even lower sensitivity, and breast self-examination has not been shown to contribute to early detection or reduction in breast cancer mortality. There are no blood- or urine-based tests for early detection of breast cancer, although such noninvasive techniques would be of great interest to the breast cancer community.

Metastatic Breast Cancer

The management of metastatic breast cancer has evolved and improved over the past several decades.[5] Breast cancer has a propensity to metastasize to bone, lung pleura, skin, and lymph nodes, among others. The metastatic cascade associated with breast cancer is not a random or mechanical process, but rather there are clear patterns of metastatic preference, with some breast cancers having definite preferences for specific organ sites or tissues. It is common to observe metastatic breast cancers that are limited to osseous structures or the pleura for extended periods of time. Conversely, some breast cancers home to visceral organs, such as liver or brain, early in their clinical course, bypassing completely bone and soft tissue sites. Our ability to predict the metastatic pattern of an individual breast cancer is quite limited, as is our ability to predict prognosis.

Despite our increasing knowledge of the diversity and heterogeneity of breast cancer, we continue to develop and test new diagnostics and therapeutics as if breast cancer were a monolithic entity. However, increasing indications point to genotypic and phenotypic differences that suggest breast cancer is rather a conglomerate of diverse molecular syndromes rather than a single organ-based neoplasm.[20] The first indications leading to this paradigm change occurred over a century ago, when it was observed that some, but certainly not all, breast cancers were highly sensitive to hormonal manipulations.[21] Hormone-sensitive breast cancers also have different metastatic patterns, sensitivity to chemotherapy, compared with hormone-independent tumors. Because of these observations, we approach patients with hormone-sensitive and hormone-insensitive tumors differently. When planning treatment for a patient with newly diagnosed metastasis, we assess the extent of disease spread, the location of metastases, the pace of the disease (slowly vs. rapidly growing), and the estrogen and progesterone receptor content.[5] It is this latter information that will determine to a large extent the initial therapeutic interventions. Thus, patients with hormone receptor-rich breast cancers, especially in the absence of life-threatening disease, are usually offered endocrine therapy as the first and major treatment approach. These patients, who represent almost two thirds of breast cancers, often benefit from several endocrine interventions in sequence, suggesting that although individual tumors might become resistant to specific agents, they do not lose sensitivity to hormonal manipulations.[22]

The estrogen receptor and progesterone receptor assays are well established, generally available, and reasonably reproducible.[23,24] However, they are not standardized, and their major utility is the negative predictive value (for details see Chapter 15). If both types of receptors are completely absent from an individual tumor, the response to hormonal therapy will be less than 5%, probably closer to 0%. Conversely, the presence of either receptor, even in a small percentage of tumor cells, is associated with a 20% to 40% response rate and an additional group of patients in whom lack of tumor progression results in some therapeutic benefit.[24] These statements point to the need for developing better, more sensitive, and more specific predictors of sensitivity to hormonal agents.

The most commonly used endocrine treatments for breast cancer include selective estrogen receptor modulators (SERMs),[25] selective estrogen receptor down-regulators,[26,27] gonadotropin analogues,[28,29] and selective aromatase inhibitors[30] (**Table 1.1**). All these agents are as or more effective than previously used endocrine interventions, and their safety profile is excellent, especially when compared with other therapeutic agents, such as cytotoxic chemotherapy. Today it is considered that selective aromatase inhibitors should be the endocrine agents of choice for postmenopausal women with metastatic breast cancer, followed by SERMs, selective estrogen receptor down-regulators, progestins, and other agents.[22,31] For premenopausal women, the endocrine treatment of choice is SERMs alone or in combination with gonadotropin analogues. An interesting observation in the management of breast cancer is that estrogen receptor–positive tumors develop resistance to specific endocrine interventions but respond to other interventions based on a different mechanism of action. Therefore, resistance to a SERM is not equivalent to hormone resistance. There are no available tests to predict

Table 1.1 Commonly Used Therapeutic Agents for Management of Metastatic Breast Cancer

Endocrine Therapies	Cytotoxic Agents
Selective estrogen receptor modulators (SERMs)	Anthracyclines
Tamoxifen	Doxorubicin
Toremifene	Epirubicin
	Doxil
Selective estrogen receptor down-regulators	Taxanes
Fulvestrant	Docetaxel
	Paclitaxel
Selective aromatase inhibitors	Alkylating agents
Anastrozole	Cyclophosphamide
Exemestane	Melphalan
Letrozole	Nitrosoureas
Gonadotropin analogues	Antimetabolites
Goserelin	5-Fluorouracil
Leuprolide	Methotrexate
Triptorelin	Capecitabine
	Gemcitabine
Progestins	Platinum salts
Medroxyprogesterone acetate	Carboplatin
Megestrol acetate	Cisplatin
Estrogens	Vinca alkaloids
Diethylstilbestrol	Vinblastine
Ethinyl estradiol	Vinorelbine
Androgens	
Fluoxymesterone	

for response or resistance to individual endocrine interventions, nor is there a method to predict when overall endocrine resistance will develop.

Breast cancer is moderately sensitive to cytotoxic chemotherapy, and a large number of agents have been used for its treatment.[5,32] In general, the taxanes and the anthracyclines are considered to have the highest level of activity against breast cancer. However, as single agents, both taxanes and anthracyclines fail to produce tumor regression in about one half of metastatic breast cancers. Members of the other groups of cytotoxics have even more modest antitumor effects. For this reason, combination chemotherapy is preferred when a rapid reduction in tumor volume is considered important. However, objective response to therapy has not always been associated with prolongation of survival, so many oncologists prefer to use, for palliative purposes, sequential as opposed to simultaneous combinations of drugs.[33,34] Achieving a complete clinical response is associated with greater probability of long-term benefit from treatment and with resolution of symptoms.[35]

Achieving a complete response is, in all other areas of oncology the sine qua non for cure. Therefore, progress in the management of metastatic breast cancer is likely to come from the development of combination therapy that produces increasingly high complete remission rates. Although many approaches have been developed and tested over the past three decades to predict response to individual drugs or combinations, not one method has achieved broad acceptance and common usage. Similarly, there is no single method of monitoring response to therapy, and assessment of response to treatment requires combined physical examination, biochemical tests, and imaging studies repeated at periodic intervals. There is therefore much room for improvement. It would be highly desirable to develop easy, reproducible, standardized, and noninvasive or minimally invasive techniques to reliably predict response or resistance to a particular therapeutic agent or combination. There is also a need to develop simple, minimally, or noninvasive methods to determine response to therapy and to monitor such a response over time without having to perform multiple imaging studies. Serum tumor markers, such as carcinoembryonic antigen, CA27.29, and CA15.3, are being used for this latter purpose but are not considered highly reliable.[36,37] Furthermore, many breast cancers do not express these markers; in these tumors, these markers have no utility at all.

Primary Breast Cancer

The most tangible evidence of progress in the management of established breast cancer is the consistent and reproducible improvement in disease-specific and overall survival associated with combined modality therapy of primary breast cancer.[38–41]

Surgery

Over the past three decades, surgical excision has been gradually reduced in extent to the point of preserving most breast tissue for most women with early breast cancer. A wide excision (or lumpectomy) with clear surgical margins is considered the treatment of choice for most women with stages I and II breast cancer.[42,43] Furthermore, the development and rapid adoption of sentinel lymph node mapping and biopsy techniques are in the process of reducing markedly the utilization of axillary lymph node dissection, which is now indicated only for women with lymph node–positive breast cancer.[44–46] Because most primary breast cancers are diagnosed today in a node-negative stage, at least in the United States, Canada, and Western Europe, axillary dissection would probably be indicated for less than one third of all newly diagnosed breast cancers. Recent reports suggest that up to 50% of patients with a positive sentinel lymph node have no further nodal involvement after performance of a full axillary dissection.[47–49] Therefore, there is a need and opportunity to develop predictive markers of lymph node involvement to streamline the process of axillary staging and perhaps eliminate completely invasive axillary procedures from the primary treatment of some patients. Furthermore, the development of more accurate prognostic profiles, perhaps based on genomics or proteomics, would eliminate the need for axillary staging for prognostic purposes and would restrict axillary surgery to those few in whom axillary lymphadenopathy is symptomatic.

Much less progress has been made in the area of prognostic markers.[50,51] Only HER-2 overexpression/amplification (see Chapter 16) has been added to the list of validated and recommended prognostic factors since the introduction of estrogen and progesterone receptor assays over 20 years ago (**Table 1.2**).[37,51–56]

Table 1.2 Validated Prognostic Factors in Patients with Primary Breast Cancer

Factor	Validation Method or Comment
Included in Consensus Recommendations[55,56]	
Nodal status	Multivariate analyses
Tumor size	Multivariate analyses
Estrogen receptor/ progesterone receptor	Multivariate analyses
Histologic type	Multivariate analyses
Histologic grade/nuclear grade	Multivariate analyses
HER-2	Multivariate analyses
Reported to Have Prognostic Value in ≥2 Multivariate Analyses[51,56]	
Thymidine labeling index	Not in general use because of technical complexity
S-phase fraction	Not properly standardized
UPA/uPAR/PAI	Standardized assay requires substantial fresh tissue; not properly standardized for fixed/processed tissues
Not Validated in Multivariate Analyses or Not Properly Standardized[51]	
Ki-67, MIB-1, PCNA, mitosin, Ki-S1	Not validated in multivariate analyses
Cathepsin D	Not validated in multivariate analyses
EGFR	Not properly standardized
P53	Not properly standardized
BCL-2	Not properly standardized

Radiation Therapy

The indications and techniques for radiation therapy for breast cancer have evolved and improved over the past few decades.[38,57–59] Computerized treatment planning and the development of indications based on randomized clinical trials have refined the utilization of these modalities of treatment. There is ongoing interest in identifying patients treated with breast-conserving surgery who might safely forgo radiotherapy; there is also ongoing research to identify those treated with a total mastectomy whose risk of locoregional recurrence is low enough (e.g., less than 10%) to eliminate the need for postmastectomy radiation. There is also a need to identify individual susceptibility to radiation reaction, both to avoid unnecessary and excessive treatment-related toxicity and to determine which patients might need higher than standard doses of radiation therapy for optimal disease control.

Adjuvant Endocrine Therapy

Adjuvant endocrine therapy is undoubtedly the most effective intervention for patients with hormone receptor–positive breast cancer. Tamoxifen is estimated to reduce by about 50% the annual odds of recurrence for estrogen receptor–positive tumors,[40] whereas aromatase inhibitors are estimated to have an effect that is 18% to 43% greater than that of tamoxifen.[60–62] However, not all patients with hormone receptor–positive breast cancer benefit from endocrine therapy, although all are at risk for toxicities related to these treatments. The limitations of the predictive value of estrogen and progesterone receptor assays described earlier certainly apply in this situation too. Therefore, there is a need to develop more sensitive and accurate tests to predict sensitivity and resistance to endocrine therapy in general and perhaps sensitivity and resistance to individual endocrine interventions as well.

Adjuvant Chemotherapy

Adjuvant chemotherapy is another powerful intervention that reduces the annual odds of recurrence and death by 12% to more than 50%, depending on age, menopausal status, and hormone receptor status.[39] However, there are no reliable tests to predict sensitivity or resistance to specific agents or combinations for individual patients. Therefore, chemotherapy decisions for individual patients are made on the basis of clinical trial results or the Oxford meta-analysis[39] and not on a reliable indicator of probable benefit based on the characteristics of the host or the patient's own tumor. There is therefore a need and a major opportunity to develop more accurate predictors of long-term prognosis and, perhaps more importantly, predictors of response to individual cytotoxic agents or combinations that would maximize the probability of cure for individual patients. Similarly, there would be an opportunity to identify patient populations whose probability of recurrence is low enough to preclude the use of adjuvant systemic therapies, or, alternatively, those whose probability of recurrence is moderate or high, but the sensitivity profile makes the use of existing therapies futile. This latter development, in addition to avoiding unnecessary toxicity from inactive drugs, might expedite the development and full assessment of novel potentially more effective interventions.

Over the past couple of decades there has been much interest in the preoperative application of chemotherapy and, more recently, endocrine therapy.[63] Such change in the sequence of utilization of treatments has several potential advantages.[64] First, it results in rapid reduction in tumor extent, facilitating surgical intervention and transforming larger tumors into appropriate candidates for breast-conserving surgery. Second, it provides a crude in vivo assessment of the efficacy of chemotherapy, because the primary tumor can be monitored to assess response to treatment. Third, there are theoretical advantages to the preoperative introduction of systemic therapy based on preclinical observations. Fourth, preoperative chemotherapy establishes an excellent experimental model to assess the biologic effects of treatment, because leaving the primary tumor in place permits repeated sampling of tumor tissue and measurement of various molecular end points. Because of the good correlation of a pathologic complete remission with long-term survival, the preoperative chemotherapy model also provides short-term validation to predictive models based on pretreatment biopsies.[65,66] Although pathologic complete remission is a reasonable intermediate marker, this end point is not reached until 3 to 4 months after initiation of chemotherapy. Therefore, there is much interest in developing and validating predictive markers of pathologic complete remission, on the assumption that pathologic complete remission can be used as a surrogate marker of long-term survival after combined-modality therapy. A variety of histopathologic and biochemical candidates have been evaluated over the years to identify a good predictor of pathologic complete remission (**Table 1.3**)[67]; unfortunately, most have failed to show a high enough predictive value to be of clinical use.[68–95]

Finally, it is important to remember that all currently used treatments for primary breast cancer have the possibility of producing acute side effects and some infrequent but potentially life-threatening delayed toxicities.[96–106] There is therefore the need to develop markers of susceptibility to these effects. One could envision molecular markers associated with increased risk of angiosarcoma after radiation therapy or leukemia or myelodysplastic syndromes after chemotherapy or radiotherapy.[96–106] Predictors of susceptibility to developing endometrial cancer during or after tamoxifen therapy would also be helpful in minimizing this adverse effect, as would identification of susceptible individuals to thromboembolic events.

The development of all the prognostic and predictive markers mentioned above would get us much closer to rational selection of therapies, leaving behind the empiric drug development and treatment selection processes that having served us well throughout history should be supplanted by more sensitive and accurate methods of treatment planning.

The epitome of rational therapy is molecularly targeted therapy (see Chapter 14 of this book).[107–110] The basic hypothesis underlying this strategy is that specific molecular steps in intra- or intercellular signaling under pathologic conditions become dominant drivers of the malignant phenotype. Identification of such molecular anomalies, exemplified by bcr-abl in chronic myelogenous leukemia or HER-2 overexpression in breast cancer, when appropriately validated by correlation with prognosis, is equivalent to the discovery of a therapeutic

Table 1.3 Predictive Factors for Response to Preoperative (Neoadjuvant) Chemotherapy

Author	Ref.	No. of Patients	Chemotherapy Regimens	Predictive Value	Level of Evidence
Nuclear Grade					
Kemeny	68	58	AVCF	Increased response	3
Abu-Farsakh	69	287	FAC	Increased response	3
S-Phase Fraction					
Spyratos	70	35	AVCMF	Increased response	3
Remvikos	71	92	FAC or FTC	Increased response	3
Tubiana-Hulin	72	150	AVCMF	Increased response	3
Chevillard	73;74	87	FAC or FTC	Increased response	3
Apoptotic Index					
Ellis	76;77	27	FEP or AV	Not significant	3
Ellis	78	50	ECF	Not predictive	3
Kandioler	75	67	FEC or paclitaxel	Increased response	3
				Decreased response	3
Symmans	90	11	Paclitaxel	Increased response	4
Buchholz	79	25	FAC or paclitaxel or ATxt	Not predictive	3
HER2					
Resnick	82	40	FAC	Increased response	3
Makris	80	45	MxMeTam_Mi	Decreased response	3/4
Rozan	83	323	FAC + XRT	Not predictive	3
Chang	81	28	MxMe	Not predictive	3/4
Petit	84	79	FEC	Increased response	3
Zhang	86	97	FAC	Not predictive	3
Penault-Llorca	85	115	Doxorubicin, epirubicin or pirarubicin	Increased response	3
p53					
Archer	87	92	Hormone therapy	Not predictive	3
Makris	88	80	MxMeTam_Mi	Not predictive	3/4
Resnick	82	40	FAC	Not predictive	3
Linn	116	51	AC + paclitaxel	Not predictive	4
Formenti	89	35	FU/CIVI/RT	Decreased response	4
Rozan	83	323	FAC + RT	Not predictive	3
Bottini	91	143	Epirubicin or CMF	Decreased response	3
Kandioler	75	67	FEC or	Decreased response	3
			paclitaxel	Increased response	3
BCL-2					
Makris	80	51	MxMeTam_Mi	Not predictive	3/4
Ellis	78	50	ECF	Not predictive	3
Bottini	91	143	Epirubicin or CMF	Not predictive	3
Kandioler	75	67	FEC or	Decreased response	
				Or	
			paclitaxel	Increased response	3
Buchholz	79	25	FAC or paclitaxel or ATxt	Not predictive	3
Pglycoprotein					
Ro	93	48	FAC	Decreased response	3/4
Verrelle	92	17	AVCF	Decreased response	4
Bottini	91	143	Epirubicin or CMF	Decreased response	3
MDR1					
Chevillard	73;74	87	FAC or FTC	Decreased response	3
Transcriptional Profiling					
Chang	94	24	Docetaxel	Increased response	3
Pusztai	95	42	Paclitaxel-FAC	Increased response	3

target. Molecular approaches to inhibit the target are then expected to have a beneficial effect on malignant behavior, resulting in clinical improvement or potentially in cure of the disease. One could also envision the application of such strategies to the prevention of breast cancer, once molecular targets that become apparent early during malignant transformation are identified. A corollary of this concept is that the development of rational therapeutics must be associated with similarly targeted diagnostics.[108] The ability to inhibit a target in the absence of the ability to identify tumors that express that target would dilute our therapeutic efforts. Thus, a molecular intervention without a molecular test is not "targeted therapy."

There are multiple opportunities to use modern diagnostic technology to develop markers of risk, prognosis, and response to therapy (**Table 1.4**). Progress in genomics and proteomics has also led to the identification of a number of potential therapeutic targets. The initial steps in targeted or molecular therapies have had mixed results, with some promising outcomes[111–113] and some disappointments.[114,115] These early results have highlighted the complexity of the biologic systems that drive the malignant phenotype and indicated the need for understanding and validating the role of putative targets in the maintenance of the cancerous process. The development of targeted therapies must go hand in hand with molecular diagnostics to reach the full potential of this strategy. The chapters that follow highlight progress in our understanding of molecular medicine as applied to breast cancer and point to the challenges we face in the application of technologic advances to clinical medicine. They also point out the opportunities we have in harvesting the full benefits of the Human Genome Project and developments in cancer biology in their application to the diagnosis and management of breast cancer.

Table 1.4 Opportunities for Developing Novel or More Effective Molecular Markers in Breast Cancer

Breast cancer prevention
Identify molecular markers of malignant transformation in tissue
Develop molecular markers of malignant transformation in breast nipple aspirate or ductal lavage
Identify genetic markers of risk for carcinogenesis
Identify and validate targets for chemoprevention
Identify and validate intermediate endpoint markers for chemoprevention
Diagnosis of breast cancer
Develop diagnostic molecular test in blood or urine
Develop diagnostic molecular marker in breast nipple aspirate or ductal lavage
Develop molecular markers of breast cancer that lend themselves to direct imaging or labeling for functional imaging (positron emission tomography)
Management of primary breast cancer
Identify potential for invasion and metastasis
Identify markers that predict lymph node invasion
Develop predictors of response to endocrine therapy
Develop predictors of response to individual endocrine agents
Develop predictors of response to chemotherapy
Develop predictors of response to individual cytotoxic agents
Develop predictors of excess radiosensitivity
Develop predictors of radiation resistance
Identify targets for molecular therapy
Management of metastatic breast cancer
Develop markers of disease spread or tissue homing
Develop predictors of response to endocrine therapy
Develop predictors of response to individual endocrine agents
Develop predictors of response to chemotherapy
Develop predictors of response to individual cytotoxic agents
Develop predictors of excess radiosensitivity
Develop predictors of radiation resistance
Develop predictors of susceptibility to specific iatrogenic toxicity (leukemia, thromboembolism, endometrial cancer, etc.)
Identify targets for molecular therapy

References

1. Jemal A, Murray T, Samuels A, et al. Cancer statistics, 2003. Cancer J Clin 2003;53:5–26.
2. Pisani P, Bray F, Parkin DM. Estimates of the worldwide prevalence of cancer for 25 sites in the adult population. Int J Cancer 2002;97:72–81.
3. Collyar DE. Breast cancer: a global perspective. J Clin Oncol 2001;19(18 suppl):101S–105S.
4. Parkin DM, Bray F, Ferlay J, et al. Estimating the world cancer burden: Globocan 2000. Int J Cancer 2001;94:153–156.
5. Hortobagyi GN. Treatment of breast cancer (Review). N Engl J Med 1998;339:974–984.
6. Egan RL. Experience with mammography in a tumor institution. Evaluation of 1,000 studies. Radiology 1960;75:894.
7. Strax P, Venet L, Shapiro S. Mass screening in mammary cancer. Cancer 1969;23:875–878.
8. Moskowitz M, Russell P, Fidler J, et al. Breast cancer screening. Preliminary report of 207 biopsies performed in 4,128 volunteer screenees. Cancer 1975;36:2245–2250.
9. Shapiro S. Evidence on screening for breast cancer from a randomized trial. Cancer 1977;39(6 suppl):2772–2782.

10. Kerlikowske K. Efficacy of screening mammography among women aged 40 to 49 years and 50 to 69 years: comparison of relative and absolute benefit. J Natl Cancer Inst 1997;22:79–86.

11. Kerlikowske K, Grady D, Rubin SM, et al. Efficacy of screening mammography. A meta-analysis. JAMA 1995;273:149–154.

12. Tabar L, Yen MF, Vitak B, et al. Mammography service screening and mortality in breast cancer patients: 20-year follow-up before and after introduction of screening. Lancet 2003;361:1405–1410.

13. Duffy SW, Tabar L, Chen HH, et al. The impact of organized mammography service screening on breast carcinoma mortality in seven Swedish counties. Cancer 2002;95:458–469.

14. Antman K, Shea S. Screening mammography under age 50. JAMA 1999;281:1470–1472.

15. Sox HC. Benefit and harm associated with screening for breast cancer (Editorial). N Engl J Med 1998;338: 1145–1146.

16. Retsky M, Demicheli R, Hrushesky W. Breast cancer screening for women aged 40–49 years: screening may not be the benign process usually thought. J Natl Cancer Inst 2001;93:1572.

17. Pisano ED, Yaffe MJ, Hemminger BM, et al. Current status of full-field digital mammography (Review). Acad Radiol 2000;7:266–280.

18. Tillman GF, Orel SG, Schnall MD, et al. Effect of breast magnetic resonance imaging on the clinical management of women with early-stage breast carcinoma. J Clin Oncol 2002;20:3413-3423.

19. Cecil KM, Schnall MD, Siegelman ES, et al. The evaluation of human breast lesions with magnetic resonance imaging and proton magnetic resonance spectroscopy. Breast Cancer Res Treat 2001;68: 45–54.

20. Perou CM, Sorlie T, Eisen MB, et al. Molecular portraits of human breast tumours. Nature 2000;406: 747–752.

21. Beatson GT. On the treatment of inoperable cases of carcinoma of the mamma: suggestions for a new method of treatment, with illustrative cases. Lancet 1896;2:104–107.

22. Hortobagyi GN. Endocrine treatment of breast cancer. In: Becker KL, ed. Principles and Practice of Endocrinology and Metabolism. Philadelphia: Lippincott Williams &Wilkins, 2001:2039–2046.

23. Palmieri C, Cheng GJ, Saji S, et al. Estrogen receptor beta in breast cancer (Review). Endocr Rel Cancer 2002;9:1–13.

24. Osborne CK. Steroid hormone receptors in breast cancer management (Review). Breast Cancer Res Treat 1998;51:227–238.

25. Osborne CK, Zhao H, Fuqua SA. Selective estrogen receptor modulators: structure, function, and clinical use (Review). J Clin Oncol 2000;18:3172–3186.

26. Osborne CK, Pippen J, Jones SE, et al. Double-blind, randomized trial comparing the efficacy and tolerability of fulvestrant versus anastrozole in postmenopausal women with advanced breast cancer progressing on prior endocrine therapy: results of a North American trial. J Clin Oncol 2002;20: 3386–3395.

27. Vergote I, Faslodex 0020 and 0021 Investigators. Fulvestrant versus anastrozole as second-line treatment of advanced breast cancer in postmenopausal women. Eur J Cancer 2002;38(suppl 6): S57–S58.

28. Pritchard KI. GnRH analogues and ovarian ablation: their integration in the adjuvant strategy (Review). Recent Results Cancer Res 1998;152:285–297.

29. Nicholson RI, Walker KJ. GN-RH agonists in breast and gynaecologic cancer treatment (Review). J Steroid Biochem 1989;33:801–804.

30. Buzdar AU, Robertson JF, Eiermann W, et al. An overview of the pharmacology and pharmacokinetics of the newer generation aromatase inhibitors anastrozole, letrozole, and exemestane (Review). Cancer 2002;95:2006–2016.

31. Hortobagyi GN. Future directions in the endocrine therapy of breast cancer. Breast Cancer Res Treat 2003;80(suppl 1):S37–S39.

32. Hortobagyi GN. The status of breast cancer management: challenges and opportunities. Breast Cancer Res Treat 2002;75(suppl 1):S61–S65.

33. Seidman AD. Monotherapy options in the management of metastatic breast cancer (Review). Semin Oncol 2003;30(2:suppl 3):6–10.

34. Miles D, von Minckwitz G, Seidman AD. Combination versus sequential single-agent therapy in metastatic breast cancer (Review). Oncologist 2002;7(suppl 6):13–19.

35. Greenberg PA, Hortobagyi GN, Smith TL, et al. Long-term follow-up of patients with complete remission following combination chemotherapy for metastatic breast cancer. J Clin Oncol 1996;14:2197–2205.

36. Bast RC Jr, Ravdin P, Hayes DF, et al. 2000 update of recommendations for the use of tumor markers in breast and colorectal cancer: clinical practice guidelines of the American Society of Clinical Oncology [erratum appears in J Clin Oncol 2001;19:4185–4188]. J Clin Oncol 2001;19:1865–1878.

37. Yamauchi H, Stearns V, Hayes DF. When is a tumor marker ready for prime time? A case study of c-erbB-2 as a predictive factor in breast cancer (Review). J Clin Oncol 2001;19:2334–2356.

38. Early Breast Cancer Trialists' Collaborative Group. Favourable and unfavourable effects on long-term survival of radiotherapy for early breast cancer: an overview of the randomised trials. Lancet 2000;355: 1757–1770.

39. Early Breast Cancer Trialists' Collaborative Group. Polychemotherapy for early breast cancer: an overview of the randomised trials. Lancet 1998;352: 930–942.

40. Early Breast Cancer Trialists' Collaborative Group. Tamoxifen for early breast cancer: an overview of the randomised trials. Lancet 1998;351:1451–1467.

41. Early Breast Cancer Trialists' Collaborative Group. Ovarian ablation in early breast cancer: overview of the randomised trials. Lancet 1996;348:1189–1196.

42. Veronesi U, Cascinelli N, Mariani L, et al. Twenty-year follow-up of a randomized study comparing breast-conserving surgery with radical mastectomy for early breast cancer. N Engl J Med 2002;347: 1227–1232.

43. van Dongen JA, Voogd AC, Fentiman IS, et al. Long-term results of a randomized trial comparing breast-conserving therapy with mastectomy: European Organization for Research and Treatment of Cancer 10801 trial. J Natl Cancer Inst 2000;92: 1143–1150.

44. Bonnema J, van de Velde CJ. Sentinel lymph node biopsy in breast cancer (Review). Ann Oncol 2002;13: 1531–1537.

45. Schwartz GF, Giuliano AE, Veronesi U, et al. Proceedings of the consensus conference on the role of sentinel lymph node biopsy in carcinoma of the breast, April 19–22, 2001, Philadelphia, Pennsylvania (Review). Cancer 2002;94:2542–2551.

46. McMasters KM, Tuttle TM, Carlson DJ, et al. Sentinel lymph node biopsy for breast cancer: a suitable alternative to routine axillary dissection in multi-institutional practice when optimal technique is used. J Clin Oncol 2000;18:2560–2566.

47. Hwang RF, Krishnamurthy S, Hunt KK, et al. Clinicopathologic factors predicting involvement of non-sentinel axillary nodes in women with breast cancer. Ann Surg Oncol 2003;10:248–254.

48. Weaver DL, Krag DN, Ashikaga T, et al. Pathologic analysis of sentinel and nonsentinel lymph nodes in breast carcinoma: a multicenter study. Cancer 2000; 88:1099–1107.

49. Veronesi U, Paganelli G, Viale G, et al. Sentinel lymph node biopsy and axillary dissection in breast cancer: results in a large series. J Natl Cancer Inst 1999;91:368–373.

50. Mirza AN, Mirza NQ, Vlastos G, et al. Prognostic factors in node-negative breast cancer: a review of studies with sample size more than 200 and follow-up more than 5 years (Review). Ann Surg 2002;235: 10–26.

51. Ravdin PM. Prognostic factors in breast cancer. In: Bonadonna G, Hortobagyi GN, Gianni AM, eds. Textbook of Breast Cancer—A Clinical Guide to Therapy. St. Louis: Mosby, 1997:35–63.

52. Carr JA, Havstad S, Zarbo RJ, et al. The association of HER-2/neu amplification with breast cancer recurrence. Arch Surg 2000;135:1469–1474.

53. Ferrero-Pous M, Hacene K, Bouchet C, et al. Relationship between c-erbB-2 and other tumor characteristics in breast cancer prognosis. Clin Cancer Res 2000;6:4745–4754.

54. Ross JS, Fletcher JA. HER-2/neu (c-erb-B2) gene and protein in breast cancer (Review). Am J Clin Pathol 1999;112(1 suppl 1):S53–S67.

55. Fitzgibbons PL, Page DL, Weaver D, et al. Prognostic factors in breast cancer. College of American Pathologists Consensus Statement 1999 (Review). Arch Pathol Lab Med 2000;124:966–978.

56. Thor AD, Jeruss JS. Prognostic and predictive markers in breast cancer. In: Bonadonna G, Hortobagyi GN, Gianni AM, eds. Textbook of Breast Cancer—A Clinical Guide to Therapy. London: Martin Dunitz Ltd., 2001:63–84.

57. Nag S, Kuske RR, Vicini FA, et al. Brachytherapy in the treatment of breast cancer (Review). Oncology (Huntingt) 2001;15:195–202.

58. Small W Jr, Lurie RH. Current status of radiation in the treatment of breast cancer (Review). Oncology (Huntingt) 2001;15:469–476.

59. Early Breast Cancer Trialists' Collaborative Group. Effects of radiotherapy and surgery in early breast cancer. An overview of the randomized trials. Early Breast Cancer Trialists' Collaborative Group. N Engl J Med 1995;333:1444–1455.

60. Bianco R, ATAC Trialists' Group. ATAC ("Arimidex", Tamoxifen, Alone or in Combination) trial—anastrozole is superior to tamoxifen as adjuvant treatment in postmenopausal women with early breast cancer. Breast Cancer Res Treat 2003; 76(suppl 1):S155(abst 632).

61. The ATAC Trialists' Group. Anastrozole alone or in combination with tamoxifen versus tamoxifen alone for adjuvant treatment of postmenopausal women with early breast cancer: first results of the ATAC randomised trial. Lancet 2002;359:2131–2139.

62. Goss PE, Ingle JN, Martino S, et al. A randomized trial of letrozole in postmenopausal women after five years of tamoxifen therapy for early-stage breast cancer. N Engl J Med 2003;349:1793–1802.

63. Buchholz TA, Hunt KK, Whitman GJ, et al. Neoadjuvant chemotherapy for breast carcinoma: multidisciplinary considerations of benefits and risks (Review). Cancer 2003;98:1150–1160.

64. Hortobagyi GN, Buzdar AU. Locally advanced breast cancer: a review including the M.D. Anderson experience. In: Ragaz J, Ariel IM, eds. High-risk breast cancer. Berlin: Springer-Verlag, 1991:382–415.

65. Feldman LD, Hortobagyi GN, Buzdar AU, et al. Pathological assessment of response to induction chemotherapy in breast cancer. Cancer Res 1986; 46:2578–2581.

66. Kuerer HM, Newman LA, Smith TL, et al. Clinical course of breast cancer patients with complete pathologic primary tumor and axillary lymph node response to doxorubicin-based neoadjuvant chemotherapy. J Clin Oncol 1999;17:460–469.

67. Hortobagyi GN, Hayes DF, Pusztai L. Integrating newer science into breast cancer prognosis and treatment: a review of current molecular predictors and profiles. Presented at the ASCO Annual Meeting Summaries, 2002, pp. 192–201.

68. Kemeny F, Vadrot J, d'Hubert E, et al. Evaluation Histologique e Radioclinique de L'effet de la Chimiotherapie Premiere sur les Cancers Non Inflammatoires du Sein. Cahiers Cancer 1991;3: 705–714.

69. Abu-Farsakh H, Sneige N, Atkinson EN, et al. Pathologic predictors of tumor response to preoperative chemotherapy in locally advanced breast carcinoma. Breast J 1995;1:96–101.

70. Spyratos F, Briffod M, Tubiana-Hulin M, et al. Sequential cytopunctures during preoperative chemotherapy for primary breast carcinoma. II. DNA flow cytometry changes during chemotherapy, tumor regression, and short-term follow-up. Cancer 1992;69:470–475.

71. Remvikos Y, Jouve M, Beuzeboc P, et al. Cell cycle modifications of breast cancers during neoadjuvant chemotherapy: a flow cytometry study on fine needle aspirates. Eur J Cancer 1993;29A:1843–1848.

72. Tubiana-Hulin M, Malek M, Briffod M, et al. Preoperative chemotherapy of operable breast cancer (stage IIIA). Prognostic factors of distant recurrence. Eur J Cancer 1993;29A(suppl 6):S76.

73. Chevillard S, Lebeau J, Pouillart P, et al. Biological and clinical significance of concurrent p53 gene alterations, MDR1 gene expression, and S-phase fraction analyses in breast cancer patients treated with primary chemotherapy or radiotherapy. Clin Cancer Res 1997;3(12 Pt 1):2471–2478.

74. Chevillard S, Pouillart P, Beldjord C, et al. Sequential assessment of multidrug resistance phenotype and measurement of S-phase fraction as predictive markers of breast cancer response to neoadjuvant chemotherapy. Cancer 1996;77:292–300.

75. Kandioler-Eckersberger D, Ludwig C, Rudas M, et al. TP53 mutation and p53 overexpression for prediction of response to neoadjuvant treatment in breast cancer patients. Clin Cancer Res 2000; 6:50–56.

76. Ellis PA, Smith IE, McCarthy K, et al. Preoperative chemotherapy induces apoptosis in early breast cancer (Letter). Lancet 1997;349:849.

77. Ellis PA, Smith IE, Detre S, et al. Reduced apoptosis and proliferation and increased Bcl-2 in residual breast cancer following preoperative chemotherapy. Breast Cancer Res Treat 1998;48:107–116.

78. Ellis P, Smith I, Ashley S, et al. Clinical prognostic and predictive factors for primary chemotherapy in operable breast cancer. J Clin Oncol 1998;16: 107–114.

79. Buchholz TA, Davis DW, McConkey DJ, et al. Chemotherapy-induced apoptosis and Bcl-2 levels correlate with breast cancer response to chemotherapy. Cancer J 2003;9:33–41.

80. Makris A, Powles TJ, Dowsett M, et al. Prediction of response to neoadjuvant chemoendocrine therapy in primary breast carcinomas. Clin Cancer Res 1997; 3:593–600.

81. Chang J, Powles TJ, Allred DC, et al. Biologic markers as predictors of clinical outcome from systemic therapy for primary operable breast cancer. J Clin Oncol 1999;17:3058–3063.

82. Resnick JM, Sneige N, Kemp BL, et al. p53 and c-erbB-2 expression and response to preoperative chemotherapy in locally advanced breast carcinoma. Breast Dis 1995;8:149–158.

83. Rozan S, Vincent-Salomon A, Zafrani B, et al. No significant predictive value of c-erbB-2 or p53 expression regarding sensitivity to primary chemotherapy or radiotherapy in breast cancer. Int J Cancer 1998; 79:27–33.

84. Petit T, Borel C, Ghnassia JP, et al. Chemotherapy response of breast cancer depends on HER-2 status and anthracycline dose intensity in the neoadjuvant setting. Clin Cancer Res 2001;7:1577–1581.

85. Penault-Llorca F, Cayre A, Bouchet MF, et al. Induction chemotherapy for breast carcinoma: predictive markers and relation with outcome. Int J Oncol 2003;22:1319–1325.

86. Zhang F, Yang Y, Smith T, et al. Correlation between HER-2 expression and response to neoadjuvant chemotherapy with 5-fluorouracil, doxorubicin, and cyclophosphamide in patients with breast carcinoma. Cancer 2003;97:1758–1765.

87. Archer SG, Eliopoulos A, Spandidos D, et al. Expression of ras p21, p53 and c-erbB-2 in advanced breast cancer and response to first line hormonal therapy. Br J Cancer 1995;72:1259–1266.

88. Makris A, Powles TJ, Dowsett M, et al. p53 protein overexpression and chemosensitivity in breast cancer. Lancet 1995;345:1181–1182.

89. Formenti SC, Dunnington G, Uzieli B, et al. Original p53 status predicts for pathological response in locally advanced breast cancer patients treated preoperatively with continuous infusion 5-fluorouracil and radiation therapy. Int J Radiat Oncol Biol Phys 1997;39:1059–1068.

90. Symmans WF, Volm MD, Shapiro RL, et al. Paclitaxel-induced apoptosis and mitotic arrest assessed by serial fine-needle aspiration: implications for early prediction of breast cancer response to neoadjuvant treatment. Clin Cancer Res 2000;6: 4610–4617.

91. Bottini A, Berruti A, Bersiga A, et al. p53 but not bcl–2 immunostaining is predictive of poor clinical complete response to primary chemotherapy in breast cancer patients. Clin Cancer Res 2000;6: 2751–2758.

92. Verrelle P, Meissonnier F, Fonck Y, et al. Clinical relevance of immunohistochemical detection of multidrug resistance P-glycoprotein in breast carcinoma. J Natl Cancer Inst 1991;83:111–116.

93. Ro J, Sahin A, Ro JY, et al. Immunohistochemical analysis of P-glycoprotein expression correlated with chemotherapy resistance in locally advanced breast cancer see comments. Hum Pathol 1990;21: 787–791.

94. Chang JC, Wooten EC, Tsimelzon A, et al. Gene expression profiling for the prediction of therapeutic response to docetaxel in patients with breast cancer. Lancet 2003;362:362–369.

95. Ayers M, Symmans WF, Stec J, et al. Identification of gene expression profiles that predict complete pathologic response to preoperative paclitaxel/FAC chemotherapy in breast cancer. J Clin Oncol 2004; 22:1–10.

96. Keefe DL. Trastuzumab-associated cardiotoxicity (Review). Cancer 2002;95:1592–1600.

97. Michaud LB, Valero V, Hortobagyi G. Risks and benefits of taxanes in breast and ovarian cancer (Review). Drug Safety 2000;23:401–428.

98. Hardenbergh PH, Bentel GC, Prosnitz LR, et al. Postmastectomy radiotherapy: toxicities and techniques to reduce them (Review). Semin Radiat Oncol 1999;9:259–268.

99. Cowen D, Gonzague-Casabianca L, Brenot-Rossi I, et al. Thallium-201 perfusion scintigraphy in the evaluation of late myocardial damage in left-side breast cancer treated with adjuvant radiotherapy (Review). Int J Radiat Oncol Biol Phys 1998;41: 809–815.

100. Nayfield SG, Gorin MB. Tamoxifen-associated eye disease. A review. J Clin Oncol 1996;14:1018–1026.

101. Curtis RE, Boice JD, Moloney WC, et al. Leukemia following chemotherapy for breast cancer. Cancer Res 1990;50:2741–2746.

102. Fisher B, Anderson S, DeCillis A, et al. Further evaluation of intensified and increased total dose of cyclophosphamide for the treatment of primary breast cancer: findings from National Surgical Adjuvant Breast and Bowel Project B-25. J Clin Oncol 1999;17:3374–3388.

103. Smith RE, Bryant J, DeCillis A, et al. Acute myeloid leukemia and myelodysplastic syndrome following doxorubicin-cyclophosphamide adjuvant therapy for operable breast cancer: the NSABP experience. Breast Cancer Res Treat 2001;69:209(abst 2).

104. Mermershtain W, Cohen AD, Koretz M, et al. Cutaneous angiosarcoma of breast after lumpectomy, axillary lymph node dissection, and radiotherapy for primary breast carcinoma: case report and review of the literature. Am J Clin Oncol 2002;25:597–598.

105. Huang J, Mackillop WJ. Increased risk of soft tissue sarcoma after radiotherapy in women with breast carcinoma. Cancer 2001;92:172–180.

106. Marchal C, Weber B, de Lafontan B, et al. Nine breast angiosarcomas after conservative treatment for breast carcinoma: a survey from French comprehensive cancer centers (Review). Int J Radiat Oncol Biol Phys 1999;44:113–119.

107. Hortobagyi GN. New approaches to breast cancer therapy. Breast Cancer Res Treat 2003;80(suppl 1): S1–S2, discussion S13–S18.

108. Mills GB, Kohn E, Lu Y, et al. Linking molecular diagnostics to molecular therapeutics: targeting the PI3K pathway in breast cancer. Semin Oncol 2003;30(5 suppl 16):93–104.

109. Nahta R, Hortobagyi GN, Esteva FJ. Novel pharmacological approaches in the treatment of breast cancer (Review). Expert Opin Invest Drugs 2003;12: 909–921.

110. Cristofanilli M, Hortobagyi GN. Molecular targets in breast cancer: current status and future directions. Endocr Rel Cancer 2002;9:249–266.

111. Esteva FJ, Valero V, Booser D, et al. Phase II study of weekly docetaxel and trastuzumab for patients with HER-2-overexpressing metastatic breast cancer. J Clin Oncol 2002;20:1800–1808.

112. Vogel CL, Cobleigh MA, Tripathy D, et al. Efficacy and safety of trastuzumab as a single agent in first-line treatment of HER2-overexpressing metastatic breast cancer. J Clin Oncol 2002;20:719–726.

113. Slamon DJ, Leyland-Jones B, Shak S, et al. Use of chemotherapy plus a monoclonal antibody against HER2 for metastatic breast cancer that overexpresses HER2. N Engl J Med 2001;344:783–792.

114. Winer E, Cobleigh M, Dickler M, et al. Phase II multicenter study to evaluate the efficacy and safety of Tarceva (erlotinib, OSI-774) in women with previously treated locally advanced or metastatic breast cancer. Breast Cancer Res Treat 2002;76(suppl 1): S115(abst 446).

115. Albain K, Elledge R, Gradishar WJ, et al. Open-label, phase II, multicenter trial of ZD1839 ("Iressa") in patients with advanced breast cancer. Breast Cancer Res Treat 2002;76(suppl 1):S33(abst 20).

116. Linn SC, Pinedo HM, van Ark-Otte J, et al. Expression of drug resistance proteins in breast cancer, in relation to chemotherapy. Int J Cancer 1997;71: 787–795.

CHAPTER 2

Molecular Diagnostic Techniques in Breast Cancer

Jeffrey S. Ross, MD

Albany Medical College, Albany, New York and Millennium Pharmaceuticals, Inc., Cambridge, Massachusetts

The introduction of targeted therapeutics into clinical oncology practice has created major opportunities for further development of molecular diagnostics to serve cancer patients. The approvals of trastuzumab (Herceptin^{TM}) for the treatment of HER-2/neu overexpressing breast cancer and imatinib (Gleevec^{TM}) for the treatment of chronic myelogenous leukemia featuring a bcr/abl translocation and gastrointestinal stromal tumors with selective c-*kit* oncogene activating mutations have expanded the roles of molecular testing in the selection of patients eligible to receive these novel targeted therapies.[1,2] In response, the molecular diagnostics industry is in a state of rapid commercial evolution, focused on technology development and new clinical applications.[3–5] As seen in **Table 2.1**, the in vitro diagnostics industry is now believed to consist of a $33 billion market of which molecular diagnostics, although currently less than 3%, is growing at double digit levels. This modest diagnostics market contrasts with the worldwide pharmaceuticals industry believed to encompass more than 1.1 trillion dollars in total annual sales. Gene-based and molecular diagnostics testing is currently listed at $1 billion in annual worldwide sales but has expanded at a 30–50% rate over the last several years. Of note is that traditional non–molecular-based tests average $26.34 per charged test, whereas molecular diagnostics tests average $113.04. In addition, a variety of enabling technologies, such as laser capture microdissection (LCM) and tissue microarrays (TMA), is accelerating the rate of new biomarker discovery, validation, and the potential introduction of new oncology-related diagnostics into clinical practice at an unprecedented pace. In this chapter we review a series of molecular techniques (**Table 2.2**) and present their application to breast cancer research, current and future treatment, and anti–breast cancer drug discovery.

Table 2.1 The Worldwide In Vitro Diagnostics Industry

	Clinical Pathology	Anatomic Pathology	Molecular Diagnostics
Market size	$26 billion	$6 billion	$1 billion
Growth rate	>5%	5–10%	30–50%
Margins	Low	Medium	High

DNA Targets for Breast Cancer Molecular Diagnostics

The inherent stability of DNA and its resistance to degradation even after tissue fixation and paraffin embedding has made it the most durable target for molecular diagnostic applications. The DNA from human chromosomes has been used to demonstrate the presence of deletions, translocations, gains, or losses of whole chromosomes and a variety of other subtle defects associated with the diagnosis of cancer and disease outcome.[6,7] The introduction and development of the polymerase chain reaction (PCR) has enabled more sophisticated techniques for detecting minute amounts of abnormal DNA in human tumors and relating these findings not only to patient outcome but selection of the therapy.[8]

Metaphase Cytogenetics

Metaphase cytogenetics is arguably the first molecular diagnostic test applied to the field of cancer diagnostics. Ironically, the discovery of the Philadelphia chromosome, the small acrocentric chromosome resulting from the 9:22 (*bcr/abl*) translocation first reported in the early 1950s, also became one of the first gene-based targets in the new era of directed oncology.[2] The approval of the tyrosine kinase inhibitor imatinib (Gleevec^{TM}) as a direct

Table 2.2 Summary of Techniques in the Molecular Oncology Breast Cancer

Target	Methods	Testing Samples	Current Clinical Examples	Future Status
DNA				
	Routine cytogenetics	Blood Bone marrow Fresh tissues	Classification Leukemia Lymphoma Sarcomas	Limited by low sensitivity and resolution
	FISH and CISH	Blood Bone marrow Fresh tissues Paraffin blocks	*bcr/abl* translocation in chronic myelogenous leukemia HER-2/*neu* amplification N-*myc* amplification Sarcoma translocations human papillomavirus genotyping in gynecologic cytology	Selected growth, limited by low resolution
	CGH CGH arrays	Blood Bone marrow Fresh tissues	Detection of chromosomal and single gene gains and losses in tumor tissues vs. normal tissues for research	Predominantly for research use only
	SNPs	Blood	Uncommon familial cancer predisposition Prediction of metabolism and toxicity of anticancer drugs	Continued development, major Bioinformatics challenges
	PCR sequencing	Blood Bone marrow Fresh tissues	Common familial cancer predisposition *BRCA1,2* and other tumor suppressor gene mutations *p53* and other oncogene activating mutations c-*kit* in gastrointestinal stromal tumor Gene rearrangements in lymphoma Minimal residual disease after bone marrow transplant High risk human papillomavirus detection	Continued discovery of new applications, limited by low throughput
	Southern blot	Blood Bone marrow Fresh tissues	Gene rearrangements in lymphoma N-*myc* gene amplification in neuroblastoma	Little further utility due to dilutional effects and lack of automation
	Microsatellite instability	Fresh tissues Urine	Colorectal cancer predisposition Resection margin status Urothelial carcinoma metastasis	Growth potential for selected clinical situations
RNA				
	Transcriptional	Blood Bone marrow	Pharmacogenomic discovery	Substantial growth
	Profiling	Fresh tissues		
	Northern blot	Fresh tissues		
	RT-PCR	Blood Bone marrow Fresh tissues (?paraffin blocks)	Minimal residual disease after bone marrow transplant Micrometastasis detection in sentinel lymph nodes Oncogene activation CpG island hypermethylation to detect prostate cancer TS/DPD levels to predict 5-fluorouracil response	Substantial growth
	ISH, FISH, and CISH	Blood (after cell capture assays) Bone marrow Fresh tissues	Melastatin detection for melanoma prognosis	Limited growth

(*continued*)

Table 2.2 Summary of Techniques in the Molecular Oncology Breast Cancer (*continued*)

Target	Methods	Testing Samples	Current Clinical Examples	Future Status
Protein				
	Western blot	Blood	Serology–immunology tests Infectious agents	Limited
	2D PAGE	Blood Bone marrow Fresh tissues	Discovery of novel proteins and peptides	Research only
	MALDI-TOF	Blood Bone marrow Body fluids	Protein and peptide identification and sequencing SNPs detection	Requires automation to increase throughput
	SELDI	Blood	Detection of early stage ovarian and prostate cancer	Technique undergoing validation; tremendous potential for early cancer detection and therapy monitoring
	IHC	Bone marrow Fresh tissues	Immunophenotyping of lymphoma Solid tumor classification ER/PR status in breast cancer HER-2/*neu* status in breast cancer	Enormous potential if technique can be standardized and reproducibility improved
Bioassays				
	PCR based	Fresh tissues Urine	Telomerase repeat amplification protocol assay	Limited
	Ligand–receptor assays	Fresh tissues	Hormone receptor (ER/PR) Others	Limited
	Truncated protein assays	Blood Bone marrow	Familial polyposis (APC) detection	Limited
General Enabling Techniques				
	Laser capture microdissection	Fresh tissues Paraffin blocks	Biomarker discovery	Research only
	Tissue microarrays	Fresh tissues Paraffin blocks	Biomarker discovery	Research only
	Quantitative digital image analysis	Tissue sections	Biomarker validation Quantification of ISH and IHC	Prognostic and predictive marker assessment
	Microcapillary electrophoresis	Blood Body fluids Tissue-derived fluids	High throughput sequencing Microsatellite instability High throughput proteomics	Substantial potential
	Immunomagnetic cell capture assays	Blood Bone marrow Body fluids	Detection of cancer in blood Pharmacogenomic testing Pharmacodynamic testing	Unknown potential

APC, adenomatosis polyposis coli gene; ER, estrogen receptor; PR, progesterone receptor; TS, thymidylate synthase; DPD, dihydropyrimidine dehydrogenase.

inhibitor of the activated *bcr/abl* transgene catapulted tumor cytogenetics into the modern era of targeted therapeutics. In typical clinical settings, metaphase cytogenetics is mostly used for the classification of leukemia and lymphoma, with significantly less utility for patients suffering from solid tumors. Technical advancements commencing with trypsin digest banding and subsequently with the introduction of fluorescence in situ hybridization (FISH), chromosomal bar coding, multicolor banding, spectral karyotyping, and comparative genomic hybridization (CGH) have contributed important information that has enhanced our understanding of cancer biology and, on occasion, impacted day to day cancer diagnostics.[6] The detailed role of metaphase cytogenetics in the characterization, classification, and management of patients with breast cancer is covered in Chapter 11.

Interphase Cytogenetics: In Situ Hybridization for DNA Targets

FISH

Aneusomies The detection of chromosomal gains and losses or aneusomies using complementary probes for the centromeres of the human chromosomes by FISH during interphase has been used in a variety of research and clinical applications.[9] FISH-based aneusomy detection using urinary cytology samples has outperformed conventional cytology and achieved significant clinical use for the evaluation of patients with recurrent urothelial carcinoma, but this technique is not currently performed routinely on breast cancer specimens.[10]

Translocations Although not sensitive enough to detect small chromosomal changes such as point mutations, the FISH technique has proved to be an excellent method for detecting specific chromosomal translocations. The FISH method for detection of the *bcr/abl* translocation performed on either blood or bone marrow cells to confirm the diagnosis of chronic myelogenous leukemia features greater sensitivity and speed than conventional cytogenetics.[11] FISH has also been used as a method for sub-classifying soft tissue sarcomas, which has led to major reconsideration of the derivation of both round cell and spindle cell lesions.[12]

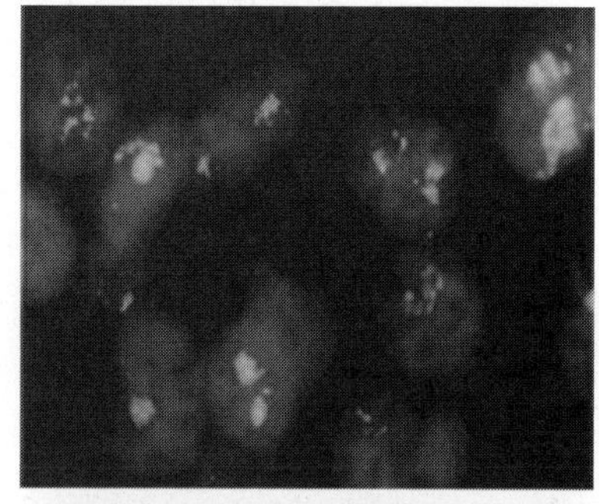

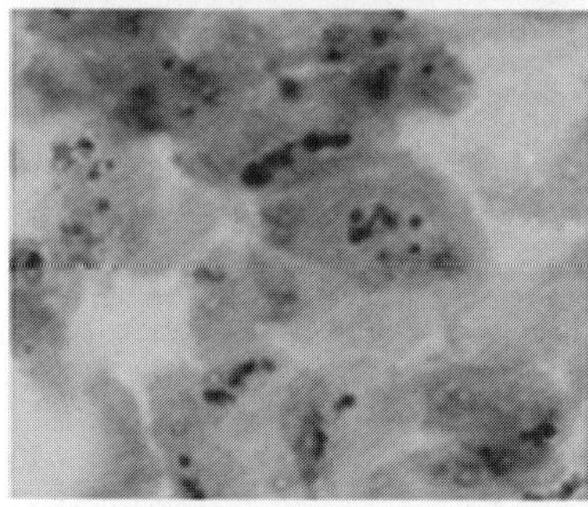

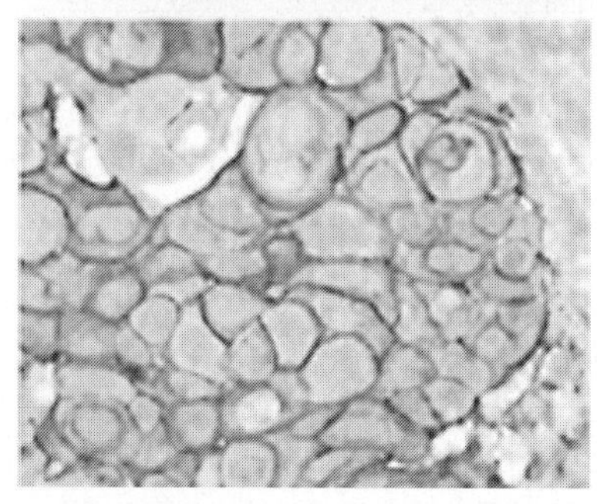

FIGURE 2.1 HER-2/*neu* testing in breast cancer. (Bottom) Immunohistochemistry using Herceptest™ system (Dako Corporation) with continuous membranous 3+ positive immunostaining for HER-2/*neu* protein. (Top) HER-2/*neu* gene amplification detected by FISH (Ventana Inform™ System). (Middle) HER-2/*neu* gene amplification detected by CISH (Zymed System). **See Plate 1 for color image.**

Gene Amplification Detection The most significant application of the FISH technique in cancer diagnostics has been for the detection of amplification of the HER-2/*neu* gene for the selection of breast cancer patients to receive the humanized anti-HER-2/*neu* monoclonal antibody, trastuzumab (Herceptin™).[1,13,14] (HER-2/*neu* testing in breast cancer is covered in depth in Chapter 16.) The FISH test (FIG. 2.1) for gene amplification is performed on 5-μm-thick formalin-fixed paraffin-embedded tissue sections similar to the competing immunohistochemical technique (see below). Approximately, 85% of published studies have linked HER-2/*neu* gene amplification or protein overexpression with adverse outcome in breast cancer and the prediction of therapeutic benefit with trastuzumab.[13,14] The newer chromogenic in situ hybridization test (CISH) is discussed below. Other examples of FISH detection of gene amplifications in clinical practice include the prediction of disease aggressiveness and therapy selection for patients with neuroblastoma featuring amplification of the N-*myc* gene.[15] Additional clinically significant gene amplification tests include the topoisomerase IIαgene for the prediction of anthracycline chemotherapy response[16] and the cyclin D1 gene as a predictor of outcome in breast cancer.[17]

Table 2.3 In Situ Hybridization Comparison

	FISH	CISH	Radiolabeled In Situ Hybridization
Sensitivity	Intermediate	Lowest	Highest
Ease of use	Intermediate	Easiest	Most cumbersome
Automation available	Yes	Yes	No
Slide scoring	Difficult	Easiest	Most difficult
Microscope requirements	Fluorescence	Routine	Routine with dark field condenser
Signal amplification	Limited use	Can increase sensitivity, but complicate slide scoring	Not generally required
Quantification	Excellent for gene amplification scoring	Subjective, may require digital image analysis	Autoradiographic grain counting most accurate
mRNA applications	Limited to none	Yes	Yes
Current clinical applications	Cytogenetics CGH and CGH arrays HER-2/*neu* gene amplification	Viral detection (human papillomavirus, others) Melastatin Light chain restriction in lymphoma/myeloma diagnosis	None

CISH In situ hybridization (ISH) using nonfluorescence probes has also become clinically useful for the management of patient's with cancer and precancer syndromes. In contrast with other radiolabeled and fluorophore-labeled probes (**Table 2.3**), CISH has many advantages, including a relative ease of use, production of permanent slides, simplicity in slide scoring, and potential for automation and high throughput.

HER-2/*neu *Gene Amplification Detection Similar to the FISH, the CISH technique (FIG. 2.1) has the advantages of using a sample with a built-in internal control, a robust DNA target less likely to degrade after tissue processing, and the opportunity for a more objective scoring system when compared with immunohistochemistry (IHC).[18,19] In addition, the CISH method does not require a relatively expensive fluorescence microscope and produces slides that are easily stored and can be reviewed without loss of signal over time. With further refinement possibly by using signal amplification strategies, there is significant potential for the CISH method to eventually become the preferred method of detection of HER-2/*neu* gene amplification for the clinical management of breast cancer patients. CISH has also been used for infectious disease diagnosis, including the identification of high risk types of human papillomavirus in gynecologic cytology specimens.[20–22]

CGH

The FISH technology has been used to perform CGH in which DNA extracted from malignant tumors can be comparatively hybridized against normal or reference DNA samples to discover chromosomal gains and losses of DNA segments in the neoplasms.[23,24] The CGH technology has uncovered numerous genetic defects in tumors that had previously been undetected by conventional cytogenetics.[23,24] Clinical applications of CGH have been limited by the lack of resolution of the original procedure. More recently, a technique of CGH combined with analysis of arrays containing artificial human chromosomes derived from bacteria or yeasts[25–28] has significantly enhanced the precision, sensitivity, and utility of CGH for mapping of tumor-specific genetic alterations (see below).

Single Nucleotide Polymorphism Detection

Greater than one million genetic markers, known as single nucleotide polymorphisms (SNPs), have recently become available for genotyping and phenotyping studies.[29–32] SNPs are the most common type of genetic difference between individuals and provide a tool to survey the genome. Hundreds of thousands of gene-associated SNP candidates have been identified and have uncovered numerous loci that appear to have significant potential to generate clinically useful data for patient management.

Novel genotyping strategies are emerging on a regular basis using a variety of techniques, including oligonucleotide genomic arrays,[29–30] gel and flow cytometry, classic gene sequencing, mass spectroscopy,[31] and microarray or gene chips,[32] all designed to increase the rate of data generation and analysis. In breast cancer, SNP technology has focused on three areas of clinical applications: the detection of predisposition for breast cancer, the prediction of toxicity of chemotherapy related to differences in metabolic rates for patients treated with these drugs, and predicting the efficacy of individual and combinations of anti–breast cancer drugs.

Breast Cancer Predisposition Testing SNP genotyping and gene sequencing (see below) have uncovered a variety of familiar cancer predisposition syndromes based on single and multiple gene variances.[32–34] Genotyping has been widely introduced for the detection of familial cancers of the breast, ovary, colon, melanoma, and multiple endocrine neoplasia (**Table 2.4**) and is covered in depth in the chapter on the predisposition for breast cancer in (Chapter 4).[32–34]

Prediction of Drug Toxicity One of the earliest applications of SNP genotyping in cancer management was the discovery of variations in drug metabolism associated with genomic variations in drug metabolizing enzymes such as the cytochrome P-450 and conjugating enzyme systems.[35] Generally known as *pharmacogenetics,* the wide potential clinical value of germline DNA sequencing will be uncovered as more candidate polymorphisms can be associated with drug toxicities.

Prediction of Drug Efficacy The application of genetic and genomic strategies to predict anticancer drug efficacy and to use the technology in drug discovery, development, and clinical trials encompasses the field of pharmacogenomics.[36–38] Pharmacogenomics platforms can feature testing for either germline SNPs and multigene polymorphisms to predict drug response ("efficacy pharmacogenetics") or the direct profiling of tumors for the expression of genes associated with drug efficacy and resistance (see below).

PCR and Direct Sequencing

The PCR technique is likely the single most important technical advance in the field of cancer molecular diagnostics to date.[39] PCR approaches to DNA have yielded a variety of molecular diagnostic assays recently eclipsed by the applications of the messenger RNA (mRNA) assessment by real-time reverse transcriptase (RT)-PCR (see below).

Tumor Suppressor Gene Mutation Detection PCR-based gene sequencing techniques have greatly enhanced the ability to detect point mutations in tumor suppressor genes such as *p53* (**Table 2.4**).[40] *p53* mutation analysis has been used for the management of urinary bladder cancer, the detection of lung and colorectal cancer, molecular assessment of tumor surgical resection margins, and the monitoring of effectiveness of anti-*p53* directed gene therapies.[41–44] (The *p53* and other tumor suppressor genes of importance in breast cancer are considered in depth in Chapter 23.) PCR technologies are also widely used to assist in the direct sequencing of genomic DNA from family members with familial cancer syndromes, such as detection of the *BRCA1* and *BRCA2* gene mutations associated with the predisposition for breast cancer and other inherited cancer syndromes (see Chapter 4).[45]

Oncogene Activation PCR strategies have similarly been used to detect oncogene activation associated with point mutations in a variety of malignancies including breast cancer (see Chapter 22).[46,47] Recently, PCR-based direct sequencing has been used to detect specific therapy-responsive mutations in the c-*kit* oncogene in imatinib (Gleevec™)-treated gastrointestinal stromal tumors. Although virtually all gastrointestinal stromal tumors overexpress the c-*kit* protein, it has been shown that some point mutations in the c-*kit* gene can produce imatinib-resistant tumors.[47] Recently, resistance or relapse in chronic myelogenous leukemia to imatinib therapy also was shown to involve point mutations in the *abl* oncogene, making the kinase less inhibited by the drug.

Other Applications PCR methods have been developed to detect clonal gene rearrangements characteristic of malignant lymphoid neoplasms and for the detection of recurrent disease after therapy for hematologic malignancies.[48,49] This application has not been used for management of breast cancer. PCR-based testing and the hybrid capture assay have been widely used to detect high risk human papillomavirus genotypes in

Table 2.4 Cancer Predisposition and Tumor Suppressor Genes

Gene Name	Designation	Location	Disease Association
Adenomatous polyposis coli	*APC*	5q21	Familial polyposis coli
Breast cancer 1	*BRCA1*	17q21	Familial breast cancer
Breast cancer 2	*BRCA2*	13q12	Familial breast cancer
E-cadherin	*CDH1*	16q22	Familial stomach cancer Lobular breast cancer Prognosis in epithelial cancers
Cyclin-dependent kinase inhibitor 1C	*CDKN1C*	16q22	Beckwith-Wiedemann Syndrome Wilms Tumor Rhabdomyosarcoma
p16 Cyclin-dependent kinase inhibitor 2A (MTS 1; INK4A)	*CDKN2A*	11p15	Familial melanoma Bladder cancer Other epithelial tumors
Mitogen-activated protein kinase 4	*MAP2K4*	17p11	Pancreatic, breast, colorectal cancer
Multiple Endocrine Neoplasia I (type II associated with the *ret* oncogene)	*MEN1*	11q13	Pituitary adenoma Parathyroid adenoma Islet cell carcinoma Carcinoids
Hereditary nonpolyposis colorectal cancer type I	*hMSH2*	2p22	Colorectal cancer Ovarian cancer Endometrial cancer
Hereditary nonpolyposis colorectal cancer type I	*hMLH1*	3p21	Colorectal cancer Ovarian cancer Endometrial cancer
Neurofibromatosis type 1	*NF1*	17q11	Neurofibromas Gliomas Pheochromocytomas Acute myelogenous leukemia
Neurofibromatosis type 2	*NF2*	22q12	Schwannomas Meningiomas Ependymomas
Phosphatase and tensin homologue	*PTEN*	10q23	Cowden syndrome Gliomas Prostate cancer Endometrial cancer
Retinoblastoma	*RB1*	13q14	Retinoblastoma Osteosarcoma Small cell lung cancer Breast cancer
Serine/threonine kinase 11	*STK11*	19p13	Peutz-Jeghers syndrome
TP 53	*p53*	17p13	Li-Fraumeni syndrome Inactive in >50% of human malignancies
Tuberous sclerosis 1	*TSC1* (KIAA023)	9q34	Hamartomas Renal cell carcinoma Angiomyolipoma
Tuberous sclerosis 2	*TSC2*	16p13	Hamartomas Renal cell carcinoma Angiomyolipoma
von- Hippel-Lindau	*VHL*	3p26	Hemangiomas Renal cell carcinoma Pheochromocytoma
Wilms tumor 1	*WT1*	11p13	Genitourinary dysplasia Familial Wilms tumor

gynecologic cytology specimens but not for breast cancer clinical samples.[50–53]

Southern Blot

The introduction of Southern blot technology to clinical practice represented one of the earliest clinical applications of the emerging field of molecular diagnostics in the 1980s. Southern blotting continues to be used in many laboratories for the detection of clonal gene rearrangements in patients with malignant lymphoid neoplasms[48,49] and for detection of gene amplifications and remains a major protocol test for the clinical management of children with newly diagnosed neuroblastoma.[15] The time-consuming and cumbersome nature of the assay and potential for the target DNA to be diluted by adjacent benign stromal, endothelial, and inflammatory cell DNA have limited the use of Southern blotting for the evaluation of solid tumors such as breast cancer.

Microsatellite Instability Determination

Microsatellite DNA sequences consist of 4–25 tandem repeated nucleotide units that are scattered throughout the human genome.[54–57] The role of microsatellites in cellular homeostasis is unknown. The detection of abnormalities in microsatellites, including point mutations, loss of heterozygosity, and gene methylation, has been associated with a variety of cancer predisposition syndromes[52] and also used for the early detection of malignancy.[58] Mass spectroscopy (see below), including both the matrix assisted desorption and electrospray ionization methods, has recently been introduced to identify microsatellite instability.[59] Microsatellite instability and defects in mismatch repair enzymes such as *hMLH1* and *hMSH2* appear to be relatively uncommon in both hereditary and sporadic breast cancer.[58]

RNA Targets for Breast Cancer Molecular Diagnostics

The measurement of gene expression by detecting absolute and relative mRNA expression levels has become a major approach to cancer molecular diagnostics.[3–5,32,37] The determination of mRNA expression for both novel and known genes by transcriptional profiling using genomic microarrays, RT-PCR, and ISH has found both direct clinical applications and emerging roles in drug discovery and development.[60,61]

Transcriptional Profiling and Genomic Microarrays

The development of printed and spotted genomic microarrays has allowed the rapid screening and accumulation of new information concerning SNPs, gene mutation, and mRNA expression in human malignancies.[62,63] Transcriptional profiling carried out by hybridizing labeled probes derived from disease specimens to arrayed oligo or complementary DNAs (cDNAs) gives a snapshot of the transcriptome where mRNA level of expression of a known and/or novel gene from a breast cancer specimen can be absolutely measured or compared with the gene expression status of a reference sample. Reference samples can include cell lines, normal tissues, and cancer precursor lesions. Biochip and microarray technologies have contributed to the industrialization of the genomic and proteomic discovery process and enable the continued development of personalized medicine and individualized therapies.[32] **Table 2.5** lists the major types of arrays or biochips and their associated fluorescent, radioisotope, or mass spectroscopy reporter systems in current use or in development for oncology molecular diagnostics. Each technique has in common the ability to generate hundreds of thousands of data points requiring sophisticated and complex information systems necessary for accurate and useful data analysis (see Bioinformatics, below).

cDNA Microarrays cDNA microarrays were introduced in the mid-1990s[70] and have achieved widespread use for the expression profiling of human clinical samples and in drug and biomarker target discovery. Using sequence verified cDNA clones, robotic printing, either fluorescent or radioisotope-based signal detection, and usually either glass slide or nylon membrane hybridization surfaces (FIG. 2.2), this technique has generated a wealth of new information in subtyping leukemia/lymphoma, solid tumor classification including breast cancer (see Chapter 6), pharmacogenomics (Chapter 28), and drug and biomarker target discovery.[62,63] In addition, oligonucleotide-based cDNA microarrays produced by companies like Affymetrix, Inc. and Agilent Technologies, Inc. have been used extensively to map the human gene transcriptome. cDNA microarrays have been used to classify

Table 2.5 Types of Microarrays in Oncology Molecular Diagnostics

Feature	cDNA Microarray (spotted arrays)	Oligonucleotide Microarrays	CGH–BAC/YAC Microarrays	Tissue Microarrays
Target type and technology	Double-stranded cDNA Made from tumor mRNA>100 nucleotides PCR products from cDNA clones Competitive hybridization	16–24 bp oligonucleotides In situ synthesis or printing Up to 240,000 oligos/chip Probe redundancy Decreases false positives	DNA fragment arrays Bacterial/yeast artificial chromosomes 100 bp to 100 kb 5000 spots 1 Mb resolution	Paraffin section arrays Frozen section arrays
Types of array surfaces	Glass slides Silicon Nylon membranes	Glass slides Silicon Nylon membranes Gels Beads		Glass slides
Signal detection systems	Fluorophores Radioisotopes Typically dual color	Fluorophores Typically single color	Fluorophores	Fluorophores Chromagens Radioisotopes
Technical limitations	Printing inaccuracies False hybridizations	Greater density but limited by cross-hybridization errors	Limited resolution	Tumor heterogeneity Formalin-based protein and RNA degradation
Automation capability	Robot printing Laser scanning	Robot printing Laser scanning	Fluorescence microscopy Digital imaging	Image analysis
Clinical applications	Leukemia/lymphoma classification Solid tumor classification Carcinoma of unknown primary site Drug selection	Gene re-sequencing SNP discovery Tumor classification SNP prediction of toxicity Drug selection	High resolution cytogenetics for gains and losses Tumor classification New primary tumor vs. metastasis	Currently used for research purposes only
Drug discovery and development applications	Drug target selection Pharmacogenomics discovery Biomarker discovery	Drug target selection Pharmacogenomics discovery Biomarker discovery	Drug target discovery	Drug target validation Pharmacogenomic target validation Biomarker validation

breast cancer[64–66] and to predict the outcome for the disease.[66,67]

Oligonucleotide Microarrays Unlike most cDNA microarray platforms that compare the expression of the unknown tumor sample to reference tissue mRNA expression levels in a two-color system, oligonucleotide arrays typically feature a single color system and do not include a reference mRNA sample that is cohybridized along with the unknown material. Oligonucleotide arrays have been extensively used for transcriptional profiling and detection of SNPs and individual gene point mutations. This method has used photolithography techniques to apply the large number of oligonucleotides on glass and nylon surfaces.[62] Often there is redundancy in the microarray so that a single gene is represented by more than 20 oligonucleotides spanning its entire length, which reduces the rate of false-negative results.[62] This single color approach has been used to profile the relative gene expression differences between samples of human tumors from different patients. Oligonucleotide arrays are also highly regarded for their ability to provide automated high throughput genome wide analysis of SNPs and other clinically relevant DNA polymorphisms and have been used to classify subtypes of breast cancer[68,69] and to predict disease progression[70] and have recently shown substantial promise to assist in the selection of the most efficacious therapy for the disease.[71]

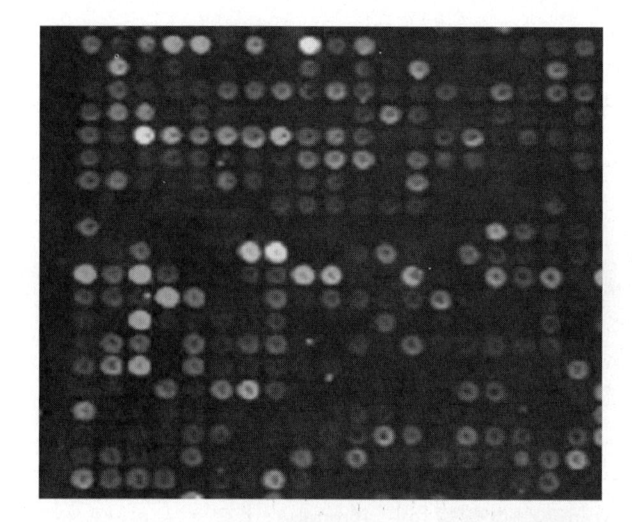

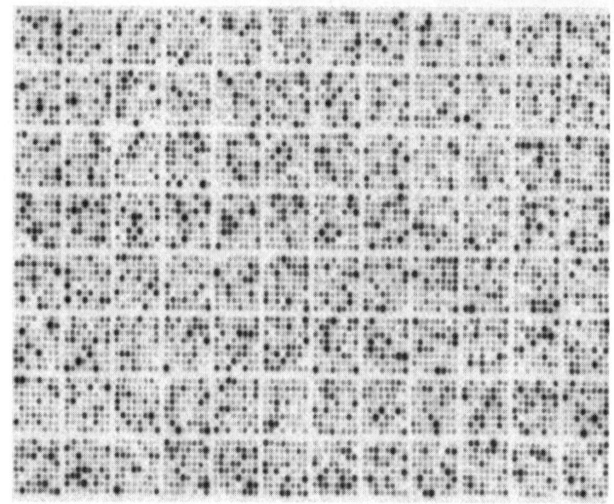

FIGURE 2.2 cDNA Microarrays with each circle representing a distinct cDNA. Probes are generated from isolated RNA(s) by labeling with fluorescent or radioactive nucleotides for detection of specific hybridization. (Top) Fluorophore-based detection comparing Cy3- or Cy5-labeled probes from normal and disease specimens. After hybridization, fluorescent intensities on scanned images are quantified, corrected for background noise, and normalized. (Bottom) Radioisotope detection with a [^{35}S]-nucleotide labeled from a single specimen. The intensity of the densitometric signal is equivalent to the relative abundance. **See Plate 2 for color image.**

CGH Microarrays In DNA array based CGH (array-CGH) the original CGH technique applied to metaphase chromosomes has been replaced by spots of either cloned cDNA or bacterial or yeast artificial human chromosomes (FIG. 2.3).[72] The resolution of array-CGH is significantly greater when compared with metaphase CGH. The technique can be automated to increase throughput and thus opens the door to the further discovery of specific genetic targets in breast cancer that may be exploited by future anticancer drugs.[72,73] Recent studies of breast cancer cell lines and clinical specimens have shown significant correlation between the DNA abnormalities discovered by array-CGH and the RNA expression measured on the same samples when they were profiled on genomic microarrays.[74]

TMAs TMAs allow for a high throughput testing of multiple characterized tumors samples often with known clinical status and follow-up on a single microscopic glass slide.[75–78] TMA can facilitate the discovery of biomarkers and drug targets by allowing for rapid assessment of disease association and drive improved efficiency and productivity by conserving reagents and reducing "hands-on" time. Patient population studies as large as 600 cases can be assessed by ISH and IHC on a single microscopic slide. New digital image analysis hardware and software has been specifically designed to facilitate TMA analysis, allowing for individual core slide scoring to be performed in a semiautomated manner.

Data Analysis and Data Display Each DNA microarray produces tens of thousands of measurements that require significant computer based data analysis to determine biologically relevant patterns. It is critical to have systems in place that efficiently combine association, functional, and gene expression data to assess a gene's potential as a disease marker or drug target. Four techniques have emerged as the predominant method of DNA microarray data analysis: hierarchical clustering, self-organizing maps, multidimensional scaling, and pathway associations[79–80] (discussed in detail in Chapters 6 and 28). Originally described in 1997, the arrangement of hybridization data into a cluster order based on color coded intensities, a measure of comparative gene signals, provided clues as to which genes were signaling in groups.[81–86] Typically, genes whose expression was significantly greater than that of the reference sample are depicted with increasing intensity on the red scale and those with less gene expression than the reference sample displayed in increasing green signals.[81] In hierarchical clustering, the algorithm groups genes in the array based on similarities in their patterns of gene expression. Typically, genes featuring the most similar patterns of expression are grouped next to each other in the vertical axis, and patient samples with similar clinical or disease outcome features are arranged along the horizontal axis as a "heat map" (FIG. 2.4). The grouping of similar gene expression samples with similar disease outcome samples is known as a dendrogram.

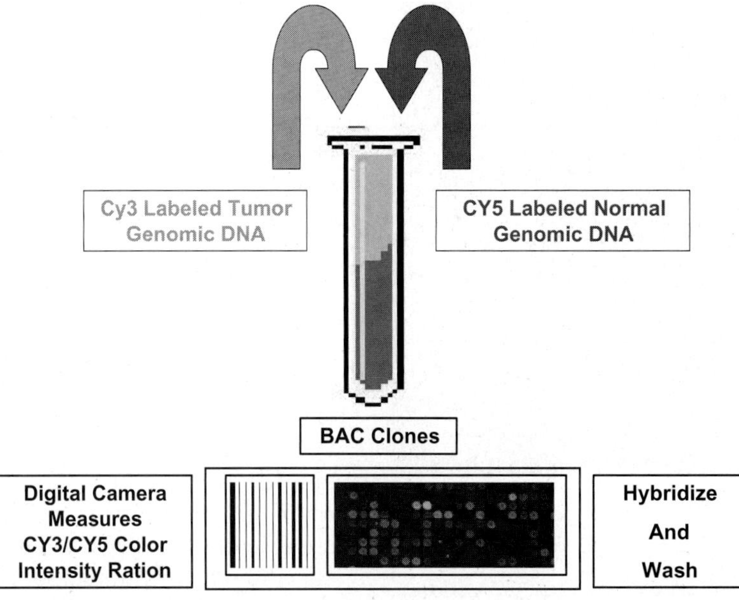

FIGURE 2.3 CGH microarrays. The array-CGH techniques features the labeling of DNA from the unknown tumor sample with the green fluorophore CY3 and the labeling of normal human DNA with the red fluorophore CY5. The DNA from the two sources is then mixed and hybridized to a series of as many as 500 individual clones obtained from either a bacterial (BAC) or yeast (YAC) artificial chromosome set. Portions of the artificial chromosome system that feature increased amounts of DNA (amplifications) in the tumor will reveal green fluorescence, those that feature loss of DNA at a specific locus in the tumor compared with the normal will show red fluorescence, and loci that have an equal amount of DNA in the tumor unknown sample and the normal DNA sample will show a yellow fluorescence. A digital optically coupled camera is used to sum the fluorescence data and submit the relative fluorescence intensities to a high speed computer for analysis. **See Plate 3 for color image.**

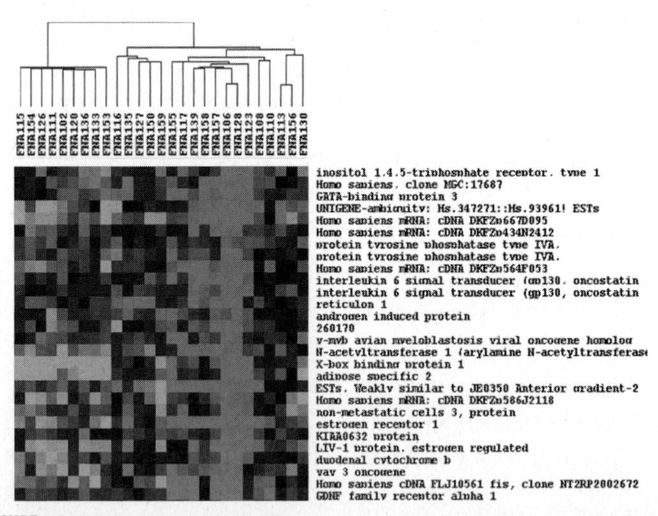

FIGURE 2.4 Breast cancer cDNA microarray results evaluated by hierarchical gene cluster analysis for defining specific gene expression signatures. Hierarchical clustering algorithm allows the clustering of individual tumor profiles on the basis of their similarities to their coexpression with the estrogen receptor alpha gene. Each row represents a tumor, and each column a single gene. A group of specimens featuring a similar gene expression signature is identified by the rectangular box. **See Plate 4 for color image.**

Cluster analysis of genes can be performed in both a *supervised* and *unsupervised* manner. In *supervised analysis,* the bioinformatics system uses machine learning after the computer is provided with an initial batch of categorized data featuring additional classification information such as tumor subtype, grade, stage, and disease outcome. This initial profiling in is known as the "learning or training set." In *unsupervised analysis,* the computer attempts to perform the clustering without exposure to a set of previous training cases.

Transcriptional Profiling and the Prediction of Drug Response Hierarchical clustering of transcriptional profiling data from clinical samples known to respond or resist a single agent or combination of anticancer drugs is a fundamental component of modern pharmacogenomics (see Chapter 28).[35,36,87–90] Initially using a cDNA microarray approach and subsequently employing oligonucleotide microarrays, several groups have now reported on their success at discovering gene expression that can be linked to resistance and responsiveness to standard of care chemotherapy in breast cancer.[85,86] In the next several years, the ability of this approach to personalize the treatment of newly diagnosed cancer patients with individualized selection and dosage of chemotherapeutic agents will be tested on a large scale.

Northern Analysis

Northern blotting was the first mRNA detection method used to measure gene expression patterns in human cancer. Currently, Northern analysis is limited to a research role and is not widely used for clinical assessment of human samples. The technique is cumbersome, slow, and, like Southern blotting, is at risk for the loss of sensitivity due to dilution of the target malignant cell mRNA levels by surrounding nonneoplastic tissues.

Differential Display

Before the development of high throughput microarray technologies, the differential display of mRNA was used. Differential display is a technique in which mRNA expression levels in a cell population are reverse transcribed and then amplified by many separate PCR reactions. This robust and relatively simple procedure allows identification of genes that are differentially expressed in different cell populations and is particularly useful for the discovery of biomarkers.[91] Differential display has previously been used to characterize breast cancer into therapeutic groups[92] but is not currently used for direct clinical testing.

Serial Analysis of Gene Expression

Serial analysis of gene expression is a direct and quantitative measure of gene expression based on the isolation and sequencing of unique sequence serial analysis of gene expression tags.[93] The sequence data are then analyzed to identify the presence and level of gene expression and are combined to form a serial analysis of gene expression library. Similar to differential display, the serial analysis of gene expression method is currently applied mostly to the discovery of biomarkers.

RT-PCR

The introduction of the real-time RT- has allowed a rapid growth in gene expression studies for both hematologic malignancies and solid tumors.[94] In multiplex RT-PCR, housekeeping and gene specific oligonucleotide primers and dye conjugated probes are added to cDNA produced from RNA isolated from clinical samples and a quantitative level of mRNA expression is obtained by normalizing the amplification cycle time for the target gene against that of a housekeeping gene.[94] A variety of commercial closed-system RT-PCR technologies is used for clinical applications, predominantly the TaqMan™ (ABI, Inc.) and the Lightcycler™ (Roche Instruments, Inc.). RT-PCR applications generally focus on the enhanced sensitivity associated with PCR-based strategies due to the ability to detect RNA over a 7-log range.

Leukemia and Lymphoma Diagnosis and Management RT-PCR techniques found their first clinical applications in the field of leukemia and lymphoma management.[95–98] RT-PCR can be used to detect the presence of non-Hodgkin lymphoma in minute tissue samples[96,98] and the recurrence of the disease after bone marrow transplantation.[97] Although the sensitivity of PCR-based methods to detect recurrent lymphoma has caused significant concern as to the specificity for clinically significant versus molecular derived disease relapse, most major cancer centers use the technique on a regular basis to follow their patients, particularly after they have received intensive treatment and bone marrow transplantation.

Solid Tumor Classification The RT-PCR technique has also been extensively used to evolve and revise the classification of solid tumors, including breast cancer. RT-PCR has been used to subclassify breast tumors,[99] to predict prognosis (Chapter 10) and to characterize the mRNA expression of important tumor-specific biomarkers and drug target genes such as *ER*[100] and *HER-2/neu.*[101]

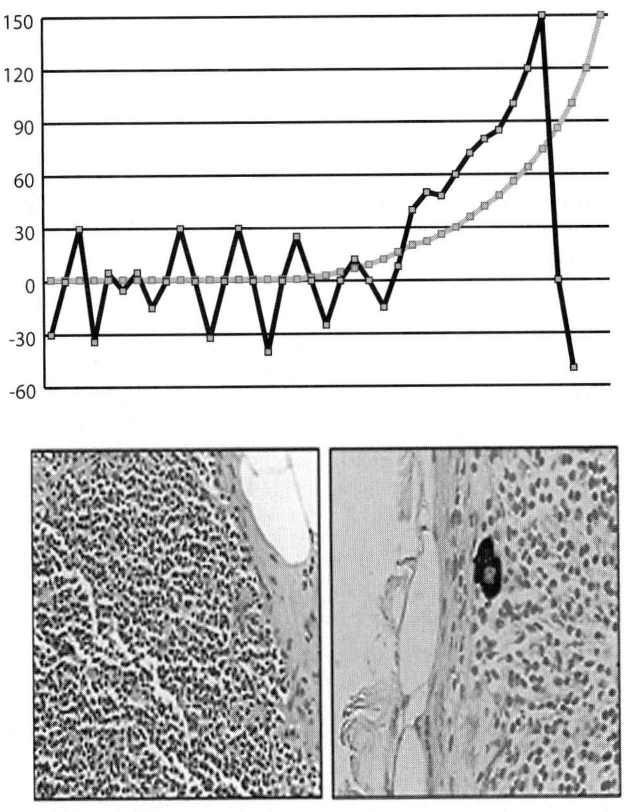

FIGURE 2.5 Micrometastasis detection of breast cancer in a sentinel axillary lymph node biopsy. (Top) Real-time PCR (TaqMan™ detection of cytokeratin 19 mRNA after 25 cycles of amplification (dark curve) normalized to internal control dye (light curve). (Bottom, left) Hematoxylin and eosin stained section of one half of the same lymph node showing no evidence of tumor cells. (Bottom, right) Immunohistochemical stain for cytokeratin 19 demonstrates a single aggregate of two malignant cells beneath the surface in the subcapsular sinusoid. **See Plate 5 for color image.**

Micrometastasis Detection, Sentinel Lymph Nodes, and Resection Margin Status Recently, RT-PCR methods have been used to detect minute amounts of tumor-derived mRNA in lymph node samples and resection margin biopsies from patients suffering from breast cancer[102–104] and a variety of other solid tumors.[105–110] This RT-PCR approach has been compared with serial tissue sections and immunohistochemistry[111,112] for the detection of micrometastases (FIG. 2.5) in sentinel lymph node biopsies. The methods designed to enhance the diagnosis of micrometastasis are currently controversial because it is not certain whether lymph node samples that do not reveal malignant cells on routine microscopy but contain abnormal mRNA expression detected by RT-PCR are clinically different from lymph nodes that do not contain RT-PCR derived aberrant messages.[113] Long-term clinical follow-up and treatment-related follow-up will be needed to determine whether this ultrasensitive method of detecting tumor cells in lymph nodes or at microscopically negative surgical resection margins can guide therapy decisions and improve patient outcomes. Micrometastasis and sentinel lymph node testing is covered in detail in Chapter 9.

Rare Event Detection in Peripheral Blood and Body Fluids Detection of rare circulating tumor cells in peripheral blood and body fluids by RT-PCR has been used as a method for the diagnosis of early stage breast cancer (see Chapter 5), the molecular staging of the disease, and monitoring the therapeutic response.[114–118]

Gene Methylation Status RT-PCR–based measurement of CpG island promoter gene hypermethylation has been extensively studied as an adjunct for the diagnosis of solid tumors.[119–121] Gene silencing by methylation has also been studied in breast cancer (see Chapters 20, 26 and 27).[122–124] The silencing of important disease-related pathways such as *ER* and *BRCA1* have been identified in breast cancer cell lines and clinical samples.[125–128] Finally, methylation enzymes have recently emerged as a target of cancer therapeutics.[129]

Prediction of Drug Response The RT-PCR technique has enabled the introduction of gene expression profiling to predict drug efficacy (pharmacogenomics) into clinical practice. Reflecting that most human tumors are stored after formalin fixation as paraffin blocks, strategies have been designed to amplify mRNA targets extracted from paraffin. Most notable is Taqman™ RT-PCR quantitation of thymidylate synthase and dihydropyrimidine dehydrogenase mRNAs extracted from paraffin-embedded tissues as a predictor of response of cancer to 5-fluorouracil–based therapy.[130] and the recently introduced Oncotype Dx test (Genomic Health Inc.) designed to predict response to single agent tamoxifen treatment in estrogen receptor positive, lymph node negative breast cancer (see Chapter 10).

ISH, FISH, and CISH for mRNA Detection

ISH techniques have been applied to detect cell-specific mRNA levels in tissue sections, aspirates, and smears of various human malignancies and to detect viral DNAs associated with infections and other biomarkers. These approaches have had limited applications in breast cancer.

Protein-Based Targets for Breast Cancer Molecular Diagnostics

The field of proteomics has emerged as a complementary strategy to genomics for oncology drug discovery, molecular diagnostics,[131–135] and specifically for breast cancer.[136–138] Proteomics is a systematic analysis and documentation of the proteins present in a biologic sample. This includes measurements of posttranslational modification, up-/down-regulation of proteins, and absence of expression. The technique requires sophisticated instrumentation for protein separation and identification and documentation in a standardized, easily retrievable, comparative format. Sequencing the human genome had a finite end point; in contrast, the proteome is highly dynamic with almost endless possible variations. Current estimates for the human genome is between 35,000 and 50,000 genes. However, given the fact that many genes exhibit splice variants and that most proteins are posttranslationally modified, it is estimated that at least one million different proteins exist within the proteome. With this degree of complexity, the goal of scientists is not to simply catalog protein components under static conditions but instead to understand how changes in the proteome contributes to homeostasis and normal physiological function and to disease-related proteins and their respective pathways and networks.[134–138] In the clinic, mass spectroscopy proteomic approaches have been developed to analyze blood and body fluids and have been used in clinical trials to detect early stage cancer,[139,140] to predict responses to anticancer drugs,[141] and to discover novel targets for cancer therapeutics.[142]

Western Analysis

Despite its major role in the development of clinical assays and validation of protein targets both for therapeutics and diagnostic purposes, gel electrophoresis and immunoblotting is not commonly used as a specific diagnostic platform for oncology patients. However, immunoblotting is a readily available method to demonstrate the existence of an activated anticancer target and the post-therapy extinguishing or turning off of the targeted pathway by the pharmacologic agent.

Two-Dimensional Gel Electrophoresis

Two-dimensional gel electrophoresis is a technique that has been available for more than two decades and used to characterize complex mixtures of proteins first by charge (isoelectric focusing) and then by molecular size (SDS-gel electrophoresis).[135] The separated proteins can be imaged and quantified after detection with organic dyes (Coomassie Blue), metal reduction (silver stain), fluorophores, metal chelators (Sypro Ruby–fluorescent metal chelator), or radiolabels. Two-dimensional gel electrophoresis has been used both for identification of novel proteins in protein mixtures and as a method for purification of individual proteins before spectroscopy.[135] The technique is challenging, somewhat cumbersome, and requires substantial user expertise. It is predominately done in research laboratories and pharmaceutical companies as "discovery" for the search of novel proteins biomarkers or pathway regulators that will lead to a better understanding of novel disease mechanisms. Two-dimensional gel electrophoresis also serves generally as an enabling tool to better resolve complex protein mixtures to prepare them for identification and sequencing by mass spectroscopy.[135]

Matrix-Assisted Laser Desorption Ionization–Time of Flight (MALDI-TOF)

Mass spectroscopy has evolved as a major method of analysis of the proteome. MALDI-TOF instruments have been developed to characterize proteins and peptides obtained from two-dimensional gel electrophoreses samples and characterization of peptide sequences and, by inference, gene sequences.[143,144] MALDI-TOF is a form of comparative mass spectroscopy in which fragmentation patterns in peptides obtained from an electron bombardment are compared with databases of known amino acid sequences. The MALDI-TOF instruments analyze crystalline protein substrates after laser pulsation followed by energy transfer desorption and gas-phase matrix ion production.[143,144] Both negative and positive molecular ions are detected by the MALDI-TOF system. MALDI-TOF has been used in a variety of applications to breast cancer research and clinical applications, including protein and peptide analysis, nucleic acid sequencing, and drug target discovery.[145–147] The MALDI-TOF technique is also a major method of genotyping patient samples for the detection of clinically important SNPs and polymorphisms.[148]

Surface-Enhanced Laser Desorption and Ionization–Time of Flight (SELDI-TOF) Mass Spectroscopy

SELDI-TOF is a form of mass spectroscopy in which surface-enhanced affinity capture, desorption, and release are performed.[149,150] This technique separates proteins according to their mass and electrical charge generating a spectrum of mass/charge peaks or amplitudes for 15,000 or more proteins or peptides in a single analysis. The height of the peak represents the relative abundance of a protein in the sample. The speed and cost effectiveness of SELDI-TOF is ideal for profiling serum or bodily fluids for disease specific patterns. The most successful application of SELDI is a surface extraction technique in which the extraction, presentation, and structural modification or amplification of the sample is performed using a surface protein chip array.[149,150] SELDI proteomics arrays allow for detection of minute amounts of proteins and peptides in complex protein mixtures such as serum and body fluids. The arrays use covalent linkage of antibodies, receptors, enzymes, ligands, lectins, and DNA and RNA. After protein chip purification on the SELDI surface, TOF mass spectroscopy with laser desorption is performed to characterize the peptides present in the unknown sample. Substantial informatics analysis is then required to evaluate the presence and amount of various peptides. The technique can be highly sensitive for the detection of low concentration peptides, which has caused concern as to possible decreases in specificity. The SELDI technique has been applied to a wide variety of cancers, including carcinoma of the breast, in varying clinical situations.[151] Recently, the SELDI proteomics array method has been combined with a self-learning bioinfomatics system and successfully detected peptide patterns characteristic of early stage ovarian cancer.[152] Although further validation of this approach must be performed on larger groups of patients, the SELDI-TOF technique shows great promise as a potential method of developing new disease markers capable of detecting cancers at early stages.

IHC

IHC has served as a cornerstone of paraffin-based testing of solid tumors for diagnosis, prognosis, prediction of drug response, and validation of new drug and biomarker targets.[153,154] The detection of HER-2/neu protein in breast cancer samples by IHC (FIG. 2.1) linked to use of the drug, trastuzumab (Herceptin™), significantly increased awareness in the limitations of the technique, especially when a semiquantitative imperfect scoring system is used (see Chapter 16).[13,14,155,156] Preanalytic issues, including fixation, tissue processing, and antigen retrieval, can greatly impact the final results of the procedures.[157–160] Image analysis instruments show promise for quantifying results and reducing interobserver and intraobserver variations in slide scoring after staining.[161] Proper standards, controls, and protocols must be adhered to whether IHC is performed manually or on automated instruments.[157–160] Nonetheless, because of the wide availability of antibodies, the familiarity of the procedure in most pathology laboratories, and the adaptability to formalin-fixed paraffin-embedded tissues, it is likely that the role of

Table 2.6 Protein-Based Arrays

Type of Array	Utility	Current Status
cDNA clone arrays	Drug development	Research
Protein microarrays	Drug development	Research
Antibody microarrays	Serum diagnostics Antibody screening	Clinical trials
Aptamer microarrays	Unknown	In development
Microfluidic arrays	Cancer screening	In development

IHC in oncology molecular diagnostics will continue for the foreseeable future.

Protein Arrays

A variety of protein arrays (**Table 2.6**) has recently been designed to facilitate biomarker discovery and clinical molecular diagnostics.[162]

Enzyme Immunoassays

Advances in enzyme-based serum immunodiagnostics have included increased sensitivity, expansion of automated methods, and the discovery of new shed glycoproteins, allowing for closer monitoring of epithelial malignancies.[163,164] Furthermore, a number of inventive technologies have recently been introduced to allow multiplex enzyme-linked immunosorbent assay–based determination of up to 18 analytes in the same assay (i.e., antibody arrays or beads). Multiplexing may allow assessment of the phenotypic and physiologic context in which the target is expressed by comparing with known clinical markers (prostate-specific antigen, CA15.5, CA19.9, CA125, etc.). Correlation with known physiologic parameters will help establish the rules under which a specific target gene or protein operates in normal and disease systems, what the most effective therapy may be, and what happens with therapy.

Truncated Protein Assays

Recently, assays have been designed to detect truncated proteins as indicators of both germline and somatic genetic defects in various forms of cancer.[165,166]

Bioassays

Ligand–Receptor Assays

Ligand binding assays have largely been replaced by direct immunoassays on tissue sections and cytosols and protein extracts of tumor tissues.[167] The advantages of immunolocalization assays as currently used for determining estrogen receptor status in patients with breast cancer must be weighed against the risk of false-positive results when compared with biochemical functional assays, such as the dextran–charcoal radiolabeled estradiol binding technique.[167] The inherent advantages of confirming *both* the measurements of the presence of immunodetectable protein and of the functional status of a cancer-related cell receptor may achieve increasing importance in the era of targeted cancer therapeutics.[168]

PCR-Based Ligand–Receptor Assays

Although a variety of PCR-based bioassays has been developed, the test that has achieved the most widespread use in clinical applications is the telomere repeat amplification protocol procedure for determining telomerase activity.[169] This technique has been used in early detection strategies in several clinical settings, including breast cancer.[170–173] New slide-based direct measurements of telomerase using both ISH for expression of the *hTERT* gene and IHC have been recently introduced but have not had widespread evaluation as to their clinical utility.

General Enabling Technologies

TMAs

TMAs were described previously in the RNA target section.[174–177] TMAs are facilitating drug discovery and development by industrializing the assessment of mRNA and protein target expression in large numbers of clinically defined human and animal samples.[173,174] TMAs are also used to validate disease-specific association of novel tumor biomarkers and prognostic and predictive factors.[178]

Single Cell Analysis and LCM

The discovery of laser microdissection allowed for determining individual cell and tissue specific expression of DNA, RNA, and protein for both clinical and research samples.[179] Before this discovery, the major method of separating complex cell and tissue mixtures for individual cell or group analysis was performed using a fluorescence-based cell sorter (flow cytometry). Laser microdissection includes the techniques of laser-assisted microdissection, laser catapult microdissection, and LCM.[179] Easily applied to both frozen and paraffin sections, LCM features microscopic selection of individual cells or minute tissue areas to be detached from the surrounding sample, or, alternately, the undesired components can be eliminated and the remaining cells of interest obtained for molecular analysis.[180] LCM has enabled the genomic profiling of small tissue biopsies by preventing the dilution of mRNA signals of the desired (i.e., tumor) cells by large numbers of benign stromal and inflammatory cells.[181,182] Subcomponents of individual tissues can be separately selected and the mRNA extracted from this minute material amplified by PCR-based strategies. LCM has had a major impact in proteomics-based discovery where the ability to sample minute tissue areas and profile the proteins by two-dimensional gel electrophoresis and MALDI mass spectroscopy has enhanced novel peptide identification, posttranslational gene modifications, and protein–protein interaction studies.[183–185] Similarly, LCM has been combined with cDNA microarray profiling to elucidate novel gene expression pathways in isolated populations from diseases such as prostate cancer.[186]

Digital Image Analysis

Computer-assisted digital image analysis has been an enabling system for quantitative data accumulation for a wide variety of molecular technologies, including genomic, proteomic, and TMA microarrays. Image analysis has also continued to be applied to the scoring of slides obtained from ISH and IHC procedures.[187,188] Recently, image analysis has been used to improve accuracy and reproducibility of the scoring of slides stained with antibodies to HER-2/neu protein.[189–191] By using cells lines of known HER-2/neu protein expression with digital image analysis, it is anticipated that more reliable assessment of breast cancer specimens can be performed on archival formalin-fixed material and allow an improved selection of patients for potential trastuzumab (Herceptin™)-based therapies.

Microcapillary Electrophoresis

Microcapillary fluidic technologies, including electrophoresis, have enabled the development of high throughput approaches to proteomics.[137–142,192,193] These methods are currently allowing for rapid miniaturized and point of care approaches to molecular diagnostics for cancer and show significant potential to create simplified closed systems for genomic and proteomic profiling.[192]

Immunomagnetic Bead Peripheral Blood Cell Capture

The immunomagnetic bead cell capture system[194,195] has enabled the collection of extremely small numbers of cancer cells from peripheral blood, bone marrow, and body fluids.[196–198] Although still limited by the small numbers of cells captured in most clinical situations and the resulting minute amounts of DNA, RNA, and protein recovered for molecular analysis, this method combined with the latest highly sensitive genomic and proteomic strategies shows promise as a method of performing drug response monitoring and pharmacogenomics to individualize therapy for cancer patients.

Bioinformatics

The wealth of data generated by the emerging high throughput genomic and proteomic biochips is totally dependent on the bioinformatics system chosen to evaluate the data.[199–201] Microarray technologies are having a major impact in cancer disease understanding and biomarker and drug development but continue to struggle with limitations created by a current lack of data comparability.[202] The pathway to clinical use for all new molecular diagnostics–based assays will require an ever more sophisticated and robust bioinformatics platform that can support both the early method validation and the subsequent clinical trials required for regulatory and commercial success.

References

1. Vogel CL, Cobleigh MA, Tripathy D, et al. Efficacy and safety of trastuzumab as a single agent in first-line treatment of HER2-overexpressing metastatic breast cancer. J Clin Oncol 2002;20:719–726.
2. O'Dwyer ME, Druker BJ. STI571: an inhibitor of the BCR-ABL tyrosine kinase for the treatment of chronic myelogenous leukaemia. Lancet Oncol 2000; 1:207–211.
3. Keesee SK. Molecular diagnostics: impact upon cancer detection. Expert Rev Mol Diagn 2002;2:91–92.
4. Poste G. Molecular diagnostics: a powerful new component of the healthcare value chain. Expert Rev Mol Diagn 2001;1:1–5.
5. Leonard DG. The present and future of molecular diagnostics. Mol Diagn 2001;6:71–72.
6. Tonnies H. Modern molecular cytogenetic techniques in genetic diagnostics. Trends Mol Med 2002;8: 246–249.

7. Sandberg AA, Chen Z. Cancer cytogenetics and molecular genetics: detection and therapeutic strategy. In Vivo 1994;8:807–818.
8. Lewis F, Maughan NJ, Smith V, et al. Unlocking the archive—gene expression in paraffin-embedded tissue. J Pathol 2001;195:66–71.
9. Tibiletti MG, Bernasconi B, Dionigi A, et al. The applications of FISH in tumor pathology. Adv Clin Pathol 1999;3:111–118.
10. Skacel M, Liou LS, Pettay JD, et al. Interphase fluorescence in-situ hybridization in the diagnosis of bladder cancer. Front Biosci 2002;7:e27–e32.
11. Pelz F, Kroning H, Franke A, et al. High reliability and sensitivity of the BCR/ABL1. D-FISH test for the detection of BCR/ABL rearrangements. Ann Hematol 2002;81:147–153.
12. Fletcher JA. Cytogenetics and molecular biology of soft tissue tumors. Monogr Pathol 1996;38:37–64.
13. Ross JS, Fletcher JA, Linette GP, et al. The Her-2/neu gene and protein in breast cancer 2003: biomarker and target of therapy. Oncologist 2003;8:307–325.
14. Lohrisch C, Piccart M. HER2/neu as a predictive factor in breast cancer. Clin Breast Cancer 2001;2: 129–135.
15. Bown N. Neuroblastoma tumour genetics: clinical and biological aspects. J Clin Pathol 2001;54:897–910.
16. Sherr CJ. The Pezcoller lecture: cancer cell cycles revisited. Cancer Res 2000;60:3689–3695.
17. Tanner M, Jarvinen P, Isola J. Amplification of HER-2/neu and topoisomerase IIalpha in primary and metastatic breast cancer. Cancer Res 2001;61:5345–5348.
18. Tanner M, Gancberg D, Di Leo A, et al. Chromogenic in situ hybridization: a practical alternative for fluorescence in situ hybridization to detect HER-2/neu oncogene amplification in archival breast cancer samples. Am J Pathol 2000;157:1467–1472.
19. Kumamoto H, Sasano H, Taniguchi T, et al. Chromogenic in situ hybridization analysis of HER-2/neu status in breast carcinoma: application in screening of patients for trastuzumab (Herceptin) therapy. Pathol Int 2001;51:579–584.
20. Roteli-Martins CM, Alves VA, Santos RT, et al. Value of morphological criteria in diagnosing cervical HPV lesions confirmed by in situ hybridization and hybrid capture assay. Pathol Res Pract 2001;197:677–682.
21. Unger ER. In situ diagnosis of human papillomaviruses. Clin Lab Med 2000;20:289–301.
22. Cheung AL, Graf AH, Hauser-Kronberger C, et al. Detection of human papillomavirus in cervical carcinoma: comparison of peroxidase, Nanogold, and catalyzed reporter deposition (CARD)-Nanogold in situ hybridization. Mod Pathol 1999;12:689–696.
23. Houldsworth J, Chaganti RS. Comparative genomic hybridization: an overview. Am J Pathol 1994;145: 1253–1260.
24. Tachdjian G, Aboura A, Lapierre JM, et al. Cytogenetic analysis from DNA by comparative genomic hybridization. Ann Genet 2000;43:147–154.
25. Monni O, Hyman E, Mousses S, et al. From chromosomal alterations to target genes for therapy: integrating cytogenetic and functional genomic views of the breast cancer genome. Semin Cancer Biol 2001; 11:395–401.
26. Forozan F, Karhu R, Kononen J, et al. Genome screening by comparative genomic hybridization. Trends Genet 1997;13:405–409.
27. Veltman JA, Schoenmakers EF, Eussen BH, et al. High-throughput analysis of subtelomeric chromosome rearrangements by use of array-based comparative genomic hybridization. Am J Hum Genet 2002;70:1269–1276.
28. Cai WW, Mao JH, Chow CW, et al. Genome-wide detection of chromosomal imbalances in tumors using BAC microarrays. Nat Biotechnol 2002;20:393–396.
29. Taylor JG, Choi EH, Foster CB, Chanock SJ. Using genetic variation to study human disease. Trends Mol Med 2001;7:507–512.
30. Carlson CS, Newman TL, Nickerson DA. SNPing in the human genome. Curr Opin Chem Biol 2001;5: 78–85.
31. Griffin TJ, Smith LM. Single-nucleotide polymorphism analysis by MALDI-TOF mass spectrometry. Trends Biotechnol 2000;18:77–84.
32. Kallioniemi OP. Biochip technologies in cancer research. Ann Med 2001;33:142–147.
33. Relling MV, Dervieux T. Pharmacogenetics and cancer therapy. Nat Rev Cancer 2001;1:99–108.
34. Borg A. Molecular and pathological characterization of inherited breast cancer. Semin Cancer Biol 2001;11:375–385.
35. Renegar G, Rieser P, Manasco P. Pharmacogenetics: the Rx perspective. Expert Rev Mol Diagn 2001;1: 255–263.
36. Ginsburg GS, McCarthy JJ. Personalized medicine: revolutionizing drug discovery and patient care. Trends Biotechnol 2001;19:491–496.
37. Cordon-Cardo C. Applications of molecular diagnostics: solid tumor genetics can determine clinical treatment protocols. Mod Pathol 2001;14:254–257.
38. Innocenti F, Ratain MJ. Update on pharmacogenetics in cancer chemotherapy. Eur J Cancer 2002;38: 639–644.
39. Davis EG, Chao C, McMasters KM. Polymerase chain reaction in the staging of solid tumors. Cancer J 2002;8:135–143.
40. Weinberg RA. The molecular basis of oncogenes and tumor suppressor genes. Ann N Y Acad Sci 1995;758: 331–338.
41. Sidransky D, Hollstein M. Clinical implications of the p53 gene. Annu Rev Med 1996;47:285–301.
42. Kausch I, Bohle A. Molecular aspects of bladder cancer III. Prognostic markers of bladder cancer. Eur Urol 2002;41:15–29.
43. McCormick F. Cancer gene therapy: fringe or cutting edge? Nat Rev Cancer 2001;1:130–141.
44. Soussi T, Beroud C. Assessing TP53 status in human tumours to evaluate clinical outcome. Nat Rev Cancer 2001;1:233–240.
45. Borg A. Molecular and pathological characterization of inherited breast cancer. Semin Cancer Biol 2001;11:375–385.
46. Apple SK, Hecht JR, Novak JM, et al. Polymerase chain reaction-based K-ras mutation detection of pancreatic adenocarcinoma in routine cytology smears. Am J Clin Pathol 1996;105:321–326.
47. Heinrich MC, Blanke CD, Druker BJ, et al. Inhibition of KIT tyrosine kinase activity: a novel molecular approach to the treatment of KIT-positive malignancies. J Clin Oncol 2002;20:1692–1703.
48. Rubnitz JE, Pui CH. Molecular diagnostics in the treatment of leukemia. Curr Opin Hematol 1999;6: 229–235.
49. Gleissner B, Thiel E. Detection of immunoglobulin heavy chain gene rearrangements in hematologic

malignancies. Expert Rev Mol Diagn 2001;1: 191–200.
50. Unger ER, Duarte-Franco E. Human papillomaviruses: into the new millennium. Obstet Gynecol Clin North Am 2001;28:653–666.
51. Schneider A, Hoyer H, Lotz B, et al. Screening for high-grade cervical intra-epithelial neoplasia and cancer by testing for high-risk HPV, routine cytology or colposcopy. Int J Cancer 2000;89:529–534.
52. Pete I, Szirmai K, Csapo Z, et al. Detection of high-risk HPV (16, 18, 33) in situ cancer of the cervix by PCR technique. Eur J Gynaecol Oncol 2002;23: 74–78.
53. Herbst AL, Pickett KE, Follen M, et al. The management of ASCUS cervical cytologic abnormalities and HPV testing: a cautionary note. Obstet Gynecol 2001;98:849–851.
54. Atkin NB. Microsatellite instability. Cytogenet Cell Genet 2001;92:177–181.
55. Peltomaki P. DNA mismatch repair and cancer. Mutat Res 2001;488:77–85.
56. Saletti P, Edwin ID, Pack K, et al. Microsatellite instability: application in hereditary non-polyposis colorectal cancer. Ann Oncol 2001;12:151–160.
57. Frazier ML, Su LK, Amos CI, et al. Current applications of genetic technology in predisposition testing and microsatellite instability assays. J Clin Oncol 2000;18(21 suppl):70S–74S.
58. Adem C, Soderberg CL, Cunningham JM, et al. Microsatellite instability in hereditary and sporadic breast cancers. Int J Cancer 2003;107: 580–582.
59. van den Boom D, Jurinke C, McGinniss MJ, et al. Microsatellites: perspectives and potentials of mass spectrometric analysis. Expert Rev Mol Diagn 2001;1:383–393.
60. Emilien G, Ponchon M, Caldas C, et al. Impact of genomics on drug discovery and clinical medicine. Q J Med 2000;93:391–423.
61. Herrmann JL, Rastelli L, Burgess CE, et al. Implications of oncogenomics for cancer research and clinical oncology. Cancer J 2001;7:40–51.
62. Harkin DP. Uncovering functionally relevant signaling pathways using microarray based expression profiling. Oncologist 2000;5:501–507.
63. Polyak K, Riggins GJ. Gene discovery using the serial analysis of gene expression technique: implications for cancer research. J Clin Oncol 2001;19: 2948–2958.
64. Dressman MA, Baras A, Malinowski R, et al. Gene expression profiling detects gene amplification and differentiates tumor types in breast cancer. Cancer Res 2003;63:2194–2199.
65. Callagy G, Cattaneo E, Daigo Y, et al. Molecular classification of breast carcinomas using tissue microarrays. Diagn Mol Pathol 2003;12:27–34.
66. Sotiriou C, Neo SY, McShane LM, et al. Breast cancer classification and prognosis based on gene expression profiles from a population-based study. Proc Natl Acad Sci USA 2003;100:10393–10398.
67. Huang E, Cheng SH, Dressman H, et al. Gene expression predictors of breast cancer outcomes. Lancet 2003;361:1590–1596.
68. Pusztai L, Sotiriou C, Buchholz TA, et al. Molecular profiles of invasive mucinous and ductal carcinomas of the breast: a molecular case study. Cancer Genet Cytogenet 2003;141:148–153.
69. Callagy G, Cattaneo E, Daigo Y, et al. Molecular classification of breast carcinomas using tissue microarrays. Diagn Mol Pathol 2003;12:27–34.
70. van de Vijver MJ, He YD, van't Veer LJ, et al. A gene-expression signature as a predictor of survival in breast cancer. N Engl J Med 2002;347:1999–2009.
71. Ramaswamy S, Perou CM. DNA microarrays in breast cancer: the promise of personalised medicine. Lancet 2003;361:1576–1577.
72. Monni O, Hyman E, Mousses S, et al. From chromosomal alterations to target genes for therapy: integrating cytogenetic and functional genomic views of the breast cancer genome. Semin Cancer Biol 2001; 11:395–401.
73. Monni O, Barlund M, Mousses S, et al. Comprehensive copy number and gene expression profiling of the 17q23. amplicon in human breast cancer. Proc Natl Acad Sci USA 2001;98:5711–5716.
74. Pollack JR, Sorlie T, Perou CM, et al. Microarray analysis reveals a major direct role of DNA copy number alteration in the transcriptional program of human breast tumors. Proc Natl Acad Sci USA 2002; 99:12963–12968.
75. Moch H, Kononen T, Kallioniemi OP, et al. Tissue microarrays: what will they bring to molecular and anatomic pathology? Adv Anat Pathol 2001;8:14–20.
76. Bubendorf L, Nocito A, Moch H, et al. Tissue microarray (TMA) technology: miniaturized pathology archives for high-throughput in situ studies. J Pathol 2001;195:72–79.
77. Skacel M, Skilton B, Pettay JD, et al. Tissue microarrays: a powerful tool for high-throughput analysis of clinical specimens: a review of the method with validation data. Appl Immunohistochem Mol Morphol 2002;10:1–6.
78. Kallioniemi OP, Wagner U, Kononen J, et al. Tissue microarray technology for high-throughput molecular profiling of cancer. Hum Mol Genet 2001;10: 657–662.
79. Zhang MQ. Large-scale gene expression data analysis: a new challenge to computational biologists. Genome Res 1999;9:681–688.
80. Werner T. Cluster analysis and promoter modeling as bioinformatics tools for the identification of target genes from expression array data. Pharmacogenomics 2001;2:25–36.
81. Scherf U, Ross DT, Waltham M, et al. A gene expression database for the molecular pharmacology of cancer. Nat Genet 2000;24:236–244.
82. Johnson KF, Lin SM. Critical assessment of microarray data analysis: the 2001 challenge. Bioinformatics 2001;17:857–858.
83. Goryachev AB, Macgregor PF, Edwards AM. Unfolding of microarray data. J Comput Biol 2001;8: 443–461.
84. Davenport RJ. Microarrays. Data standards on the horizon. Science 2001;292:414–415.
85. Ayers M, Symmans WF, Stec J, et al. Gene expression profiling of fine needle aspirations of breast cancer identifies genes associated with complete pathologic response to neoadjuvant taxol/FAC chemotherapy. Breast Cancer Res Treat 2002;76: S64(abst 220).
86. Chang JC, Wooten EC, Tsimelzon A, et al. Gene expression profiling for the prediction of therapeutic response to docetaxel in patients with breast cancer. Lancet 2003;362:362–369.

87. Zanders ED. Gene expression analysis as an aid to the identification of drug targets. Pharmacogenomics 2000;1:375–384.
88. Weinstein JN. Pharmacogenomics—teaching old drugs new tricks. N Engl J Med 2000;343:1408–1409.
89. Innocenti F, Ratain MJ. Update on pharmacogenetics in cancer chemotherapy. Eur J Cancer 2002;38: 639–644.
90. Los G, Yang F, Samimi G, et al. Using mRNA expression profiling to determine anticancer drug efficacy. Cytometry 2002;47:66–71.
91. Zhang JS, Duncan EL, Chang AC, et al. Differential display of mRNA. Mol Biotechnol 1998;10:155–165.
92. Martin KJ, Kritzman BM, Price LM, et al. Linking gene expression patterns to therapeutic groups in breast cancer. Cancer Res 2000;60:2232–2238.
93. Yamamoto M, Wakatsuki T, Hada A, et al. Use of serial analysis of gene expression (SAGE) technology. J Immunol Methods 2001;250:45–66.
94. Jung R, Soondrum K, Neumaier M. Quantitative PCR. Clin Chem Lab Med 2000;38:833–836.
95. Morgan GJ, Pratt G. Modern molecular diagnostics and the management of haematological malignancies. Clin Lab Haematol 1998;20:135–141.
96. Gleissner B, Thiel E. Detection of immunoglobulin heavy chain gene rearrangements in hematologic malignancies. Expert Rev Mol Diagn 2001;1:191–200.
97. Dolken G. Detection of minimal residual disease. Adv Cancer Res 2001;82:133–185.
98. Diaz-Cano SJ, Blanes A, Wolfe HJ. PCR techniques for clonality assays. Diagn Mol Pathol 2001;10: 24–33.
99. Iwao K, Matoba R, Ueno N, et al. Molecular classification of primary breast tumors possessing distinct prognostic properties. Hum Mol Genet 2002;11: 199–206.
100. Hayashi SI, Eguchi H, Tanimoto K, et al. The expression and function of estrogen receptor alpha and beta in human breast cancer and its clinical application. Endocr Relat Cancer 2003;10:193–202.
101. Bieche I, Onody P, Laurendeau I, et al. Real-time reverse transcription-PCR assay for future management of ERBB2-based clinical applications. Clin Chem 1999;45:1148–1156.
102. Turner RR, Giuliano AE, Hoon DS, et al. Pathologic examination of sentinel lymph node for breast carcinoma. World J Surg 2001;25:798–805.
103. Baker M, Gillanders WE, Mikhitarian K, et al. The molecular detection of micrometastatic breast cancer. Am J Surg 2003;186:351–358.
104. Inokuchi M, Ninomiya I, Tsugawa K, et al. Quantitative evaluation of metastases in axillary lymph nodes of breast cancer. Br J Cancer 2003;89:1750–1756.
105. Raj GV, Moreno JG, Gomella LG. Utilization of polymerase chain reaction technology in the detection of solid tumors. Cancer 1998;82:1419–1442.
106. Pantel K, Hosch SB. Molecular profiling of micrometastatic cancer cells. Ann Surg Oncol 2001;8: 18S–21S.
107. von Knebel Doeberitz M, Weitz J, Koch M, et al. Molecular tools in the detection of micrometastatic cancer cells—technical aspects and clinical relevance. Recent Results Cancer Res 2001;158:181–186.
108. Taback B, Morton DL, O'Day SJ, et al. The clinical utility of multimarker RT-PCR in the detection of occult metastasis in patients with melanoma. Recent Results Cancer Res 2001;158:78–92.
109. Jung R, Soondrum K, Kruger W, et al. Detection of micrometastasis through tissue-specific gene expression: its promise and problems. Recent Results Cancer Res 2001;158:32–39.
110. van Diest PJ, Torrenga H, Meijer S, et al. Pathologic analysis of sentinel lymph nodes. Semin Surg Oncol 2001;20:238–245.
111. Noura S, Yamamoto H, Miyake Y, et al. Immunohistochemical assessment of localization and frequency of micrometastases in lymph nodes of colorectal cancer. Clin Cancer Res 2002;8:759–767.
112. Ishida M, Kitamura K, Kinoshita J, et al. Detection of micrometastasis in the sentinel lymph nodes in breast cancer. Surgery 2002;131:S211–S216.
113. Hermanek P. Disseminated tumor cells versus micrometastasis: definitions and problems. Anticancer Res 1999;19:2771–2774.
114. Hu XC, Chow LW. Detection of circulating breast cancer cells by reverse transcriptase polymerase chain reaction (RT-PCR). Eur J Surg Oncol 2000;26: 530–535.
115. Shivers SC, Stall A, Goscin C, et al. Molecular staging for melanoma and breast cancer. Surg Oncol Clin North Am 1999;8:515–526.
116. Baker MK, Mikhitarian K, Osta W, et al. Molecular detection of breast cancer cells in the peripheral blood of advanced-stage breast cancer patients using multimarker real-time reverse transcription-polymerase chain reaction and a novel porous barrier density gradient centrifugation technology. Clin Cancer Res 2003;9:4865–4871.
117. Schroder CP, Ruiters MH, de Jong S, et al. Detection of micrometastatic breast cancer by means of real time quantitative RT-PCR and immunostaining in perioperative blood samples and sentinel nodes. Int J Cancer 2003;106:611–618.
118. Stathopoulou A, Mavroudis D, Perraki M, et al. Molecular detection of cancer cells in the peripheral blood of patients with breast cancer: comparison of CK–19, CEA and maspin as detection markers. Anticancer Res 2003;23:1883–1890.
119. Herman JG, Baylin SB. Promoter-region hypermethylation and gene silencing in human cancer. Curr Top Microbiol Immunol 2000;249:35–54.
120. Esteller M, Herman JG. Cancer as an epigenetic disease: DNA methylation and chromatin alterations in human tumours. J Pathol 2002;196:1–7.
121. Wong IH. Methylation profiling of human cancers in blood: molecular monitoring and prognostication. Int J Oncol 2001;19:1319–1324.
122. Widschwendter M, Jones PA. DNA methylation and breast carcinogenesis. Oncogene 2002;21:5462–5482.
123. Asch BB, Barcellos-Hoff MH. Epigenetics and breast cancer. J Mamm Gland Biol Neoplasia 2001;6:151–152.
124. Yang X, Yan L, Davidson NE. DNA methylation in breast cancer. Endocr Relat Cancer 2001;8:115–127.
125. Hayashi SI, Eguchi H, Tanimoto K, et al. The expression and function of estrogen receptor alpha and beta in human breast cancer and its clinical application. Endocr Relat Cancer 2003;10:193–202.
126. Mueller CR, Roskelley CD. Regulation of BRCA1 expression and its relationship to sporadic breast cancer. Breast Cancer Res 2003;5:45–52.

127. Esteller M, Silva JM, Dominguez G, et al. Promoter hypermethylation and BRCA1 inactivation in sporadic breast and ovarian tumors. J Natl Cancer Inst 2000;92:564–569.
128. Mueller CR, Roskelley CD. Regulation of BRCA1 expression and its relationship to sporadic breast cancer. Breast Cancer Res 2003;5:45–52.
129. Szyf M. Towards a pharmacology of DNA methylation. Trends Pharmacol Sci 2001;22:350–354.
130. Salonga D, Danenberg KD, Johnson M, et al. Colorectal tumors responding to 5-fluorouracil have low gene expression levels of dihydropyrimidine dehydrogenase, thymidylate synthase, and thymidine phosphorylase. Clin Cancer Res 2000;6:1322–1327.
131. Jain KK. Applications of proteomics in oncology. Pharmacogenomics 2000;1:385–393.
132. Lee KH. Proteomics: a technology-driven and technology-limited discovery science. Trends Biotechnol 2001;19:217–222.
133. Simpson RJ, Dorow DS. Cancer proteomics: from signaling networks to tumor markers. Trends Biotechnol 2001;19:S40–S48.
134. Dua K, Williams TM, Beretta L. Translational control of the proteome: relevance to cancer. Proteomics 2001;1:1191–1199.
135. Herrmann PC, Liotta LA, Petricoin EF 3rd. Cancer proteomics: the state of the art. Dis Markers 2001; 17:49–57.
136. Chakravarthy B, Pietenpol JA. Combined modality management of breast cancer: development of predictive markers through proteomics. Semin Oncol 2003;30:23–36.
137. Shin BK, Wang H, Hanash S. Proteomics approaches to uncover the repertoire of circulating biomarkers for breast cancer. J Mamm Gland Biol Neoplasia 2002;7:407–413.
138. Dwek MV, Alaiya AA. Proteome analysis enables separate clustering of normal breast, benign breast and breast cancer tissues. Br J Cancer 2003;89:305–307.
139. Srinivas PR, Srivastava S, Hanash S, et al. Proteomics in early detection of cancer. Clin Chem 2001;47:1901–1911.
140. Bichsel VE, Liotta LA, Petricoin EF 3rd. Cancer proteomics: from biomarker discovery to signal pathway profiling. Cancer J 2001;7:69–78.
141. Hutter G, Sinha P. Proteomics for studying cancer cells and the development of chemoresistance. Proteomics 2001;1:1233–1248.
142. Hanash SM, Madoz-Gurpide J, Misek DE. Identification of novel targets for cancer therapy using expression proteomics. Leukemia 2002;16:478–485.
143. Kolchinsky A, Mirzabekov A. Analysis of SNPs and other genomic variations using gel-based chips. Hum Mutat 2002;19:343–360.
144. Leushner J. MALDI TOF mass spectrometry: an emerging platform for genomics and diagnostics. Expert Rev Mol Diagn 2001;1:11–18.
145. Hondermarck H, Vercoutter-Edouart AS, Revillion F, et al. Proteomics of breast cancer for marker discovery and signal pathway profiling. Proteomics 2001;1: 1216–1232.
146. Wulfkuhle JD, McLean KC, Paweletz CP, et al. New approaches to proteomic analysis of breast cancer. Proteomics 2001;1:1205–1215.
147. Yazidi-Belkoura IE, Adriaenssens E, Vercoutter-Edouart AS, et al. Proteomics of breast cancer: outcomes and prospects. Technol Cancer Res Treat 2002;1:287–296.
148. Griffin TJ, Smith LM. Single-nucleotide polymorphism analysis by MALDI-TOF mass spectrometry. Trends Biotechnol 2000;18:77–84.
149. Weinberger SR, Morris TS, Pawlak M. Recent trends in protein biochip technology. Pharmacogenomics 2000;1:395–416.
150. Merchant M, Weinberger SR. Recent advancements in surface-enhanced laser desorption/ionization-time of flight-mass spectrometry. Electrophoresis 2000;21: 1164–1177.
151. Wulfkuhle JD, McLean KC, Paweletz CP, et al. New approaches to proteomic analysis of breast cancer. Proteomics 2001;1:1205–1215.
152. Petricoin EF, Ardekani AM, Hitt BA, et al. Use of proteomic patterns in serum to identify ovarian cancer. Lancet 2002;359:572–577.
153. Taylor CR, Cote RJ. Immunohistochemical markers of prognostic value in surgical pathology. Histol Histopathol 1997;12:1039–1055.
154. Oertel J, Huhn D. Immunocytochemical methods in haematology and oncology. J Cancer Res Clin Oncol 2000;126:425–440.
155. Hanna W. Testing for HER2. status. Oncology 2001; 61(suppl 2):22–30.
156. Schaller G, Evers K, Papadopoulos S, et al. Current use of HER2. tests. Ann Oncol 2001;12(suppl 1):S97–S100.
157. Taylor CR. The total test approach to standardization of immunohistochemistry. Arch Pathol Lab Med 2000;124:945–951.
158. Werner M, Chott A, Fabiano A, et al. Effect of formalin tissue fixation and processing on immunohistochemistry. Am J Surg Pathol 2000;24:1016–1019.
159. Miller RT, Swanson PE, Wick MR. Fixation and epitope retrieval in diagnostic immunohistochemistry: a concise review with practical considerations. Appl Immunohist Mol Morph 2000;8:228–235.
160. Shi SR, Cote RJ, Taylor CR. Antigen retrieval techniques: current perspectives. J Histochem Cytochem 2001;4:931–937.
161. Esteva FJ, Hortobagyi GN, Sahin AA, et al. Expression of erbB/HER receptors, heregulin and P38 in primary breast cancer using quantitative immunohistochemistry. Pathol Oncol Res 2001;7:171–177.
162. Walter G, Bussow K, Lueking A, et al. High-throughput protein arrays: prospects for molecular diagnosis. Trends Mol Med 2002;8:250–253.
163. Johnson PJ. A framework for the molecular classification of circulating tumor markers. Ann N Y Acad Sci 2001;945:8–21.
164. Thomas CM, Sweep CG. Serum tumor markers: past, state of the art, and future. Int J Biol Markers 2001;16:73–86.
165. Ahnen DJ. The genetic basis of colorectal cancer risk. Adv Intern Med 1996;41:531–552.
166. Traverso G, Shuber A, Levin B, et al. Detection of APC mutations in fecal DNA from patients with colorectal tumors. N Engl J Med 2002;346:311–320.
167. Goussard J. Paraffin section immunocytochemistry and cytosol-based ligand-binding assays for ER and PR detection in breast cancer: the time has come for more objectivity. Cancer Lett 1998;132:61–66.
168. Thorpe R, Wadhwa M, Mire-Sluis A. The use of bioassays for the characterisation and control of

biological therapeutic products produced by biotechnology. Dev Biol Stand 1997;91:79–88.

169. Kim NW, Wu F. Advances in quantification and characterization of telomerase activity by the telomeric repeat amplification protocol (TRAP). Nucleic Acids Res 1997;25:2595–2597.
170. Vasef MA, Ross JS, Cohen MB. Telomerase activity in human solid tumors. Diagnostic utility and clinical applications. Am J Clin Pathol 1999;112:S68–S75.
171. Mokbel K, Williams NJ. Telomerase and breast cancer: from diagnosis to therapy. Int J Surg Invest 2000;2:85–88.
172. Herbert BS, Wright WE, Shay JW. Telomerase and breast cancer. Breast Cancer Res 2001;3:146–149.
173. Orlando C, Gelmini S. Telomerase in endocrine and endocrine-dependent tumors. J Steroid Biochem Mol Biol 2001;78:201–214.
174. Battifora H. The multitumor (sausage) tissue block: novel method for immunohistochemical antibody testing. Lab Invest 1986;55:244–248.
175. Kononen J, Bubendorf L, Kallioniemi A, et al. Tissue microarrays for high-throughput molecular profiling of tumor specimens. Nat Med 1998;4:844–847.
176. Zarrinkar PP, Mainquist JK, Zamora M, et al. Arrays of arrays for high-throughput gene expression profiling. Genome Res 2001;11:1256–1261.
177. Hoos A, Cordon-Cardo C. Tissue microarray profiling of cancer specimens and cell lines: opportunities and limitations. Lab Invest 2001;81:1331–1338.
178. Torhorst J, Bucher C, Kononen J, et al. Tissue microarrays for rapid linking of molecular changes to clinical endpoints. Am J Pathol 2001;159:2249–2256.
179. Todd R, Margolin DH. Challenges of single-cell diagnostics: analysis of gene expression. Trends Mol Med 2002;8:254–257.
180. Curran S, McKay JA, McLeod HL, et al. Laser capture microscopy. Mol Pathol 2000;53:64–68.
181. Best CJ, Emmert-Buck MR. Molecular profiling of tissue samples using laser capture microdissection. Expert Rev Mol Diagn 2001;1:53–60.
182. Maitra A, Wistuba II, Gazdar AF. Microdissection and the study of cancer pathways. Curr Mol Med 2001;1:153–162.
183. Simone NL, Paweletz CP, Charboneau L, et al. Laser capture microdissection: beyond functional genomics to proteomics. Mol Diagn 2000;5:301–307.
184. Craven RA, Banks RE. Laser capture microdissection and proteomics: possibilities and limitation. Proteomics 2001;1:1200–1204.
185. Verma M, Wright GL Jr, Hanash SM, et al. Proteomic approaches within the NCI early detection research network for the discovery and identification of cancer biomarkers. Ann N Y Acad Sci 2001;945:103–115.
186. Rubin MA. Use of laser capture microdissection, cDNA microarrays, and tissue microarrays in advancing our understanding of prostate cancer. J Pathol 2001;195:80–86.
187. Bacus SS, Ruby SG. Application of image analysis to the evaluation of cellular prognostic factors in breast carcinoma. Pathol Annu 1993;28:179–204.
188. Aziz DC, Barathur RB. Quantitation and morphometric analysis of tumors by image analysis. J Cell Biochem 1994;19(suppl):120–125.
189. Esteva FJ, Hortobagyi GN, Sahin AA, et al. Expression of erbB/HER receptors, heregulin and P38. in primary breast cancer using quantitative immunohistochemistry. Pathol Oncol Res 2001;7:171–177.
190. Wang S, Saboorian MH, Frenkel EP, et al. Assessment of HER-2/neu status in breast cancer. Automated Cellular Imaging System (ACIS)-assisted quantitation of immunohistochemical assay achieves high accuracy in comparison with fluorescence in situ hybridization assay as the standard. Am J Clin Pathol 2001;116:495–503.
191. Hanna W. Testing for HER2. status. Oncology 2001; 61:22–30.
192. Bosserhoff AK, Buettner R, Hellerbrand C. Use of capillary electrophoresis for high throughput screening in biomedical applications. A minireview. Comb Chem High Throughput Screen 2000;3: 455–466.
193. Celis JE, Kruhoffer M, Gromova I, et al. Gene expression profiling: monitoring transcription and translation products using DNA microarrays and proteomics. FEBS Lett 2000;480:2–16.
194. Haukanes BI, Kvam C. Application of magnetic beads in bioassays. Biotechnology 1993;11:60–63.
195. Rye PD, Hoifodt HK, Overli GE, et al. Immunobead filtration: a novel approach for the isolation and propagation of tumor cells. Am J Pathol 1997;150: 99–106.
196. Park S, Lee B, Kim I, et al. Immunobead RT-PCR versus regular RT-PCR amplification of CEA mRNA in peripheral blood. J Cancer Res Clin Oncol 2001; 127:489–494.
197. Flatmark K, Bjornland K, Johannessen HO, et al. Immunomagnetic detection of micrometastatic cells in bone marrow of colorectal cancer patients. Clin Cancer Res 2002;8:444–449.
198. Barker SD, Casado E, Gomez-Navarro J, et al. An immunomagnetic-based method for the purification of ovarian cancer cells from patient-derived ascites. Gynecol Oncol 2001;82:57–63.
199. Brazma A, Vilo J. Gene expression data analysis. FEBS Lett 2000;480:17–24.
200. Maughan NJ, Lewis FA, Smith V. An introduction to arrays. J Pathol 2001;195:3–6.
201. Bayat A. Science, medicine, and the future: bioinformatics. BMJ 2002;324:1018–1022.
202. Bustin SA, Dorudi S. The value of microarray techniques for quantitative gene profiling in molecular diagnostics. Tends Mol Med 2002;8:269–272.

CHAPTER 3

Clinical and Molecular Epidemiology of Breast Cancer

Abenaa Brewster, MD, MHS, and Melissa Bondy, PhD

Department of Epidemiology, Division of Cancer Prevention, The University of Texas M. D. Anderson Cancer Center, Houston, Texas

Breast cancer is the most common occurring cancer among women in the United States, representing 32% of all newly diagnosed cancers in women, and is second only to lung cancer as cause of cancer death in women.[1] Breast cancer is currently estimated to affect 215,990 women, and 40,110 will die from the disease in 2004.[1] Female breast cancer incidence rates continue to increase, although the rate of increase has slowed to 0.6% per year since 1986, and in situ breast cancer incidence rates increased by 6.1% per year over the same period.[2] Death rates from breast cancer decreased beginning in the early 1990s, with steeper declines reported among white women (decreases of 2.5% per year) than among black women (decreases of 1.0% per year). Breast cancer death rates among Asian Pacific Islanders, American Indian/Alaskan Native, and Hispanic women were lower than those among black and white women.[2] The incidence and mortality rates by ethnic group are illustrated in FIGURE 3.1: Whites have the highest overall incidence, followed by African Americans.[2]

A number of putative risk factors for breast cancer have been identified (FIG. 3.2). Of those factors examined in epidemiologic studies, aside from age and family history of breast cancer, most risk factors are related to reproductive history and are widely thought to reflect longer lifetime exposures to the endogenous steroid hormones.[3] The hypotheses proposed to explain the role of reproductive risk factors in breast cancer etiology are controversial and are based on the dual effects of estrogens as both promoters of ductal epithelial cell growth and as precursors for mutagenic estrogen metabolites.[4–6] In this chapter, we describe the known and putative breast cancer risk factors, including reproductive and anthropometric factors, breast density, alcohol, smoking, exogenous and endogenous hormonal exposures, obesity, and genetics.

Reproductive Factors

Reproductive factors, such as ages at menarche, parity, first birth, and menopause, determine lifetime exposure to estrogen and are established risk factors for breast cancer.[3] It is estimated that every year of delayed menarche leads to a 4% decrease in risk of breast cancer and every year of delay of menopause causes a 3% increase in risk. Interestingly, pregnancy is associated with a transient increase in breast cancer risk. In a nested case control by Lambe et al.,[7] the increased odds of breast cancer associated with each pregnancy was highest in women aged 30 years and older at the time of their first delivery. In a case-control study nested in the Swedish Fertility Registry, the transient increased risk in breast cancer risk peaked 5 years after delivery (odds ratio, 1.49; 95% confidence interval [CI], 1.01–2.20) in uniparous women compared with nulliparous women.[8]

Similarities in reproductive risk factors between white and black women have been described. Two case-control studies reported a similar effect of the reproductive factors of parity, late age at first birth, and early menopause on risk of breast cancer among white and black women.[9,10] Similar to what has been observed among white women, the nested case-control Black Woman's Health Study reported a dual effect of parity and increasing number of births on breast cancer risk. Compared with having one birth, the relative risk of breast cancer in black women younger than age 45 with four or more births was 2.24 (95% CI,

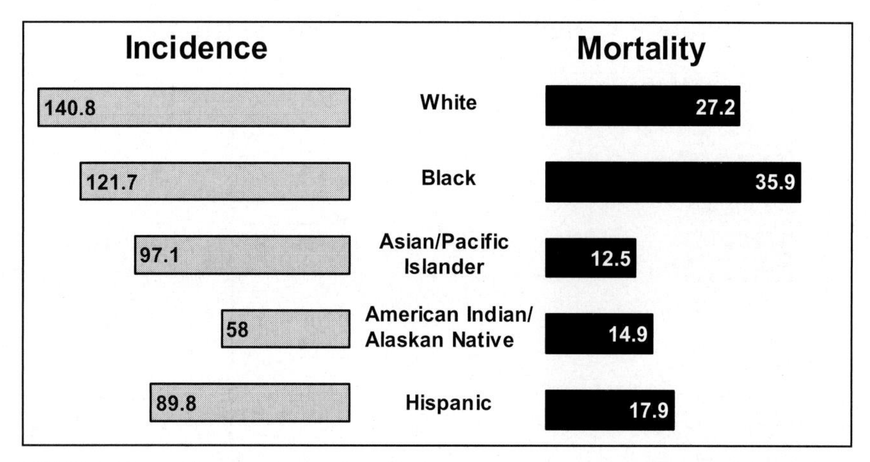

FIGURE 3.1 The incidence and mortality of breast cancer according to ethnicity. (Adapted from Weir et al.[2])

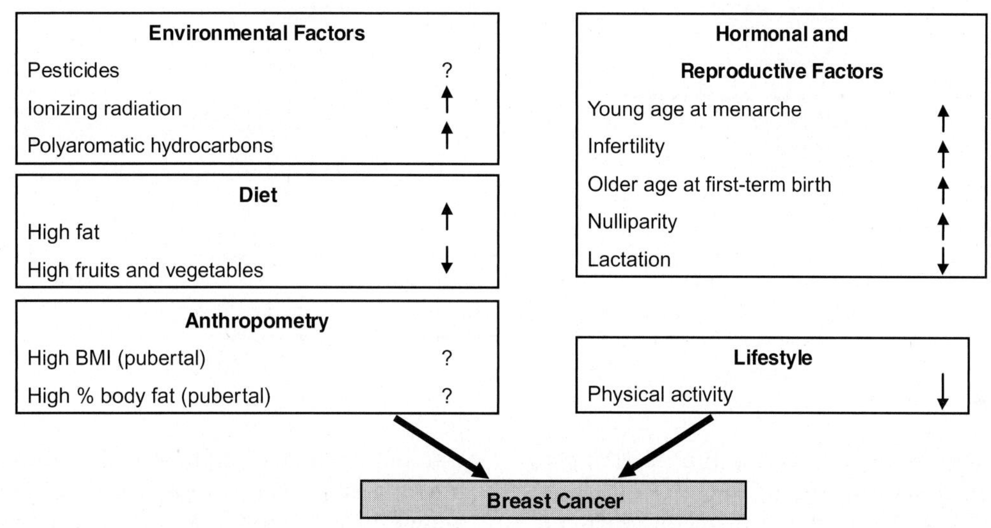

FIGURE 3.2 Environmental, dietary, anthropometric, hormonal, reproductive and lifestyle risk factors for breast cancer.

1.1–5.1). Among black women older than age 45 diagnosed with breast cancer, the relative risk associated with four or more births was 0.50 (95% CI, 0.3–0.9).[11]

Based on the observation that reproductive risk factors correlate with periods of increased breast susceptibility, biologic mechanisms for the relationship between reproductive factors and breast cancer risk have been proposed. For instance, the pubertal and nulliparous breast consists primarily of undifferentiated ductal and lobular structures, and full differentiation of the breast takes place only at full term pregnancy.[12] In animal studies, undifferentiated breast epithelial cells compared with well-differentiated cells were at higher susceptibility to neoplastic transformation by carcinogens.[13] An early full term pregnancy is therefore protective against breast cancer because it induces the terminal differentiation of breast tissue. Identifying the environmental and endogenous exposures that occur during periods of increased breast susceptibility, such as during menarche and nulliparity, may provide important clues to carcinogens influencing breast cancer risk.

Anthropometric Factors

Given the increasing prevalence of obesity worldwide, attention has focused on the influence of anthropometric measures of weight gain, fat distribution, and body mass index (BMI) on breast cancer risk. A pooled analysis indicated an inverse relationship between baseline weight and BMI and breast cancer

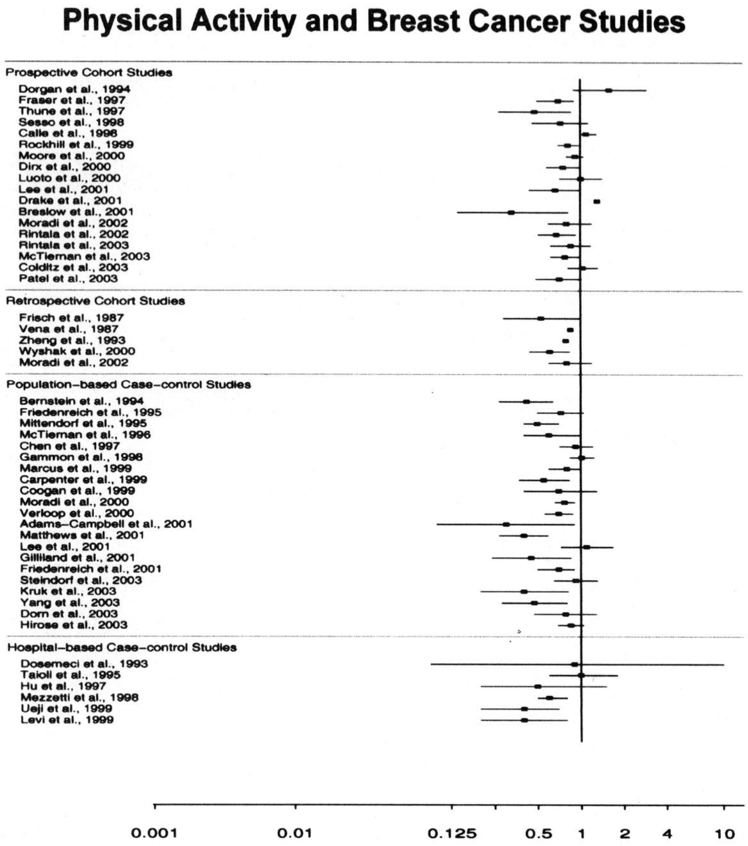

Physical Activity and Breast Cancer Studies

FIGURE 3.3 Relationship between body mass index and relative risk for breast cancer. Graphs show 95% confidence limits. Adapted from van den Brandt et al.[14]

risk among premenopausal women with a positive relationship among postmenopausal women (FIG. 3.3).[14] Based on the accumulation of studies showing a risk for postmenopausal breast cancer with higher weight and BMI, the International Agency for Research on Cancer concluded that weight control is a positive factor for breast cancer prevention in postmenopausal women.[15]

Several investigators have examined the effect of weight gain on breast cancer risk during particular periods of a women's lifetime when endogenous and exogenous exposures may initiate or promote cancer in breast epithelial cells. Weight gain that occurs during the reproductive years of adulthood (often defined as from age 18) has been consistently linked with an increased risk of postmenopausal breast cancer.[16–19] The relationship between excess weight during childhood and breast cancer risk is unclear. Berkey et al.[20] found that although higher childhood BMI was associated with a decreased risk of premenopausal breast cancer, increasing BMI between ages 10 and 20 years was not protective against either pre- or postmenopausal breast cancer. Other studies have shown a protective effect of excess weight in early years on risk of breast cancer.[21–24]

In some studies, central adiposity or the deposition of fat predominantly in the abdominal region has been associated with postmenopausal breast cancer risk independent of BMI[25,26]; other studies did not find such an association.[27] In the prospective Nurse's Health Study, among women who did not use hormone replacement therapy (HRT), the relative risk of breast cancer was 1.88 (95% CI, 1.25–2.85) for women in the highest quartile for waist circumference compared with the lowest quartile. The association between waist-to-hip ratio and breast cancer risk was 1.85 (95% CI, 1.25–2.74).[26] Other epidemiologic studies, including a meta-analysis, found weak or no associations between waist circumference and waist-to-hip ratio and premenopausal breast cancer risk.[25,26,28]

It is proposed that adult obesity influences postmenopausal breast carcinogenesis in part through its effect on increasing circulating levels of endogenous sex hormones from increased aromatase activity in adipose tissue and decreasing concentrations of sex hormone binding globulin. The contribution of endogenous sex hormones to the relationship between BMI and postmenopausal breast cancer risk was examined in a pooled analysis of case-control studies. The relative risk of breast cancer associated with a 5-kg/m^2 increase in BMI decreased from 1.19 (95% CI, 1.05–1.34) to 1.02 (95% CI, 0.89–1.17) after adjusting for free estradiol.[29] These data suggest that an increase in bioavailable estrogen may largely explain the increased breast cancer risk associated with obesity in postmenopausal women.

Breast Density

Mammographic dense breast tissue, characterized by an increase in the proliferation of epithelial cells and stromal fibrosis, has emerged as a significant risk factor for breast cancer. Cohort and nested case-control studies reported a 1.8–6% increase in breast cancer risk associated with high density mammographic parenchymal patterns compared with low density mammographic patterns after adjusting for known breast cancer risk factors.[30–33] With the prevalence of dense breast tissue estimated at 50% for women between the ages of 40 and 49 years and 30% for women between the ages of 70 and 79 years, dense breast tissue may account for a significant number of breast cancer cases.[34] Indeed, the attributable risk for breast cancer for increased breast density was estimated by Boyd et al.[32] to be 30% for women with 50% or greater density.

The biologic basis for the association between mammographic density and risk of breast cancer is unknown. It is hypothesized that breast density serves as an independent marker for factors that influence breast cancer risk, such as premalignant and high risk breast disease, higher levels of endogenous estrogen levels, hereditary susceptibility, and other undetermined components. In the National Breast Screening Study, the relative risk for incident hyperplasia and atypical hyperplasia and/or carcinoma in situ was 13.85 (95% CI, 2.65–72.49) and 9.21 (95% CI, 1.66–51.48), respectively, for women with radiologic density in more than 75% of breast compared with women with no density.[35] Genetic susceptibility has also been reported to contribute to approximately 60% of the variation in breast density.[36] There is interest in understanding the mechanism through which growth hormones and endogenous sex steroid hormones may stimulate epithelial and stromal proliferation, leading to increased breast density and breast carcinogenesis.

Endogenous Estrogen

Breast cancer is a multifactorial disease, with many factors interacting to cause the disease. Prolonged endogenous estrogen exposure resulting from early age at menarche and late age at menopause has been implicated in breast cancer development. Estrogen and its metabolites can act as prooxidants and generate free radicals capable of damaging DNA and inducing mutations.[37] Subsequent stimulation of mutagenic breast epithelial cells by estrogen can then lead to unregulated cell proliferation and differentiation, hallmarks of carcinogenesis.[38] A link between elevated serum estrogen and androgen concentrations and increased breast cancer risk has been reported for postmenopausal women.[39–41] The Endogenous Hormones and Breast Cancer Collaborative Group pooled analyses of nine prospective studies of postmenopausal women and reported a doubling of risk of breast cancer for women in the highest versus lowest quartile of estradiol and free estradiol concentrations.[42] A relationship between high levels of estrogen concentrations at other periods of a woman's life, such as during puberty, and risk of breast cancer has not been defined.

Oral Contraceptive Use

A consensus is growing that oral contraception use is not associated with breast cancer risk. The population-based Women's Contraceptive and Reproductive Experiences (CARE) study interviewed 4682 control subjects and 4575 case subjects (65% white and 35% African American) to determine the risk of breast cancer associated with former and current oral contraceptive use.[43] Seventy-five percent of women in the study had a history of oral contraception use, and the cases were diagnosed between 1994 and 1998. Compared with never-users, the relative risk of breast cancer was 1.0 for current users (95% CI, 0.8–1.3) and 0.9 for former users (95% CI, 0.8–1.0). There was no evidence of increased risk observed with longer duration of use or higher dose of estrogen. Results of the CARE study were similar to those of the older 1986 Cancer and Steroid Hormone population-based case-control study, which also showed no association between oral contraception use and breast cancer risk (relative risk, 1.0; 95% CI, 0.9–1.1).[44] The results of these well-conducted population-based studies contradict those of a meta-analysis of 53 epidemiologic studies showing relative risk for developing breast cancer to be 1.24 for current users of oral contraceptives compared with never-users ($P < 0.001$).[45] Neither the meta-analysis nor the CARE study showed an increased risk of breast cancer associated with oral contraceptive use in women with a family history of breast cancer.

HRT

Over the past 20 years, several cohort studies have examined the association between HRT and breast cancer risk among postmenopausal women. Most cohort studies demonstrated that exposure to HRT increases the risk of breast cancer, although the risk was affected by the type of HRT (single agent vs. combined therapy), the duration of use, and the timing of HRT exposure in relation to diagnosis of breast cancer.[46–51] Only a few cohort studies found no association between postmenopausal HRT exposure and increased breast cancer risk.[52,53]

A meta-analysis of 51 case-control and cohort studies that included 52,705 women with breast cancer and 108,411 women without breast cancer summarized 90% of the worldwide data on the topic. The study observed a 35% increase in breast cancer risk in women using HRT for 5 years or longer, but there was no significant overall increase in the relative risk of breast cancer for those women who stopped HRT 5 or more years before their diagnosis of breast cancer.[54] In the Million Women Study that had a 2.6-year follow-up, breast cancer incidence was significantly increased for current users of estrogen only (relative risk, 1.30; 95% CI, 1.21–1.40; $P < 0.0001$) and estrogen–progesterone users (relative risk, 2.00; 95% CI, 1.88–2.12; $P < 0.0001$) relative to never-users.[55] Other epidemiologic studies observed a stronger association between breast cancer and estrogen plus progesterone therapy compared with estrogen therapy alone.[51]

Given the inherent biases of observational studies, the results of the Women's Health Initiative (WHI) randomized clinical trial, designed primarily to determine whether estrogen or estrogen plus progesterone are effective in the prevention of cardiovascular disease, was awaited with much anticipation. Confirming the results of prior epidemiologic studies, the WHI reported that estrogen plus progesterone use increases the risk of breast cancer. The hazard ratio of breast cancer was 1.24 ($P < 0.001$) for women randomized to receive combined estrogen plus progesterone compared with the placebo group.[56] The hazard ratio of breast cancer increased for every year of estrogen plus progestin use compared with women who received placebo (**Table 3.1**). Contrary to epidemiologic evidence, the WHI estrogen alone arm showed a nonstatistically significant reduction in risk of invasive breast cancer (adjusted hazard ratio, 0.77; 95% CI, 0.59–1.01) compared with placebo.[57] Since the WHI results were published however, all

Table 3.1 Risk of Breast Cancer by Follow-up Year and Randomization in the WHI Study

Years of Follow-up	Estrogen + Progesterone		Placebo		
	No. of Participant-Years	No. of Cases	No. of Participant-Years	No. of Cases	Ratio
Year 1	8435	11	8050	17	0.62
Year 2	8353	26	7980	30	0.83
Year 3	8268	28	7888	23	1.16
Year 4	7926	40	7562	22	1.73
Year 5	5964	34	5566	12	2.64
Year 6 and later	5129	27	4243	20	1.12

forms of HRT use has decreased in the United States and the United Kingdom. It will be several years before it can be determined whether the reduction of exposure to exogenous estrogen will translate into a decrease in incidence of breast cancer in these countries.

Dietary Factors and Breast Cancer Risk

A role for high dietary fat intake in breast cancer risk has been suggested by epidemiologic studies describing higher than average breast cancer incidence in countries with higher consumption of fat.[58] The individual association between nutrition and cancer has been difficult to prove because of difficulty measuring levels of nutritional exposures, assessing critical time periods of exposure, and accounting for multiple nutritional sources. Because the range of nutritional intake is smaller within countries than between countries, intercountry studies are particularly informative. A collaborative pooled analysis of seven prospective cohort studies from four countries with a wide variation in total fat exposure found no significant association of breast cancer risk and intake of saturated, monounsaturated, or polyunsaturated fats.[59] An update of the data, including five additional cohort studies, also found no significant association between types of dietary fat intake and breast cancer risk.[60]

High consumption of fruits and vegetables has been associated with a lower overall incidence of cancer, but evidence for a protective effect on breast cancer remains inconclusive. A pooled analysis of eight cohort studies consisting of 351,625 women with 7377 invasive breast cancer cases showed no significant association between breast cancer risk and dietary fruits and vegetables,[60] but a summary of 26 case-control and cohort studies published from 1982 to 1997 found an inverse relationship of vegetable consumption to breast cancer risk (relative risk, 0.75; 95% CI, 0.66–0.85). No association was observed for fruit consumption.[61]

Vitamin A, carotenoids, and folate have been hypothesized to have biologic properties: carotenoids as antioxidants, vitamin A as inducer of cell differentiation, and folate as an agent that enhances DNA synthesis and repair. Each biologic property was hypothesized to reduce breast cancer risk. Prospective studies comparing dietary levels of vitamin A and carotenoid intake have shown, however, null or weak associations with breast cancer risk.[62–65] A summary of 12 case-control studies reported an inverse association between β-carotene intake and breast cancer risk among postmenopausal women but not among premenopausal women.[66] Four prospective studies have shown a modest inverse relationship between reported dietary folate intake and breast cancer risk, with a statistically significant association observed only among women with high levels of alcohol consumption.[67–69]

Alcohol

Alcohol intake acts as a modest risk factor for breast cancer, influencing risk in a dose-dependent manner. A collaborative analysis of data from 53 epidemiologic studies including 58,515 women with breast cancer and 95,067 control subjects reported a 7.1% increase in the risk of breast cancer for each additional 10 g (1 drink) per day intake of alcohol.[70] Compared with women who did not drink alcohol, the relative risk of breast cancer was 1.03 for women who consumed 5–14 g per day (0.5–1 drink/day) and 1.46 for women who consumed at least 45 g per day (>3.5 drinks/day). Similarly, a pooled analysis of cohort studies found a positive linear relationship between breast cancer risk and alcohol intake (relative risk, 1.09; 95% CI, 1.04–1.13) for an increment of 10 g per day of alcohol.[71]

One of the biologic mechanisms proposed to explain the association between breast cancer development and alcohol intake is alcohol's influence on estrogen concentrations. In an interventional study of postmenopausal women consuming 15 or 30 g of alcohol per day versus placebo for 8 weeks, alcohol intake was associated with increasing levels of estrone sulfate and dehydroepiandrosterone.[72] Other proposed mechanisms for alcohol-induced breast carcinogenesis include the production of free oxygen radicals and toxic metabolites such as acetaldehyde, impairment of DNA repair capacity, increased permeability of cell membranes, and decreased folate absorption and metabolism. Inherited genetic polymorphisms of carcinogen metabolizing genes such as glutathione-S-transferases (GST) M1 and T1 dehydrogenase II may also modulate the association between alcohol intake and increased breast cancer risk.[73]

Smoking

Epidemiologic studies evaluating the association between smoking and breast cancer risk have largely reported null[74,75] or positive associations.[76–78] The contradictory results likely reflect variability of the studies in study design, populations studied, and assessment of smoking exposure. In a collaborative analysis of 53 international epidemiologic studies that included nondrinkers, there was no association with breast cancer for current (relative risk, 0.99; 95% CI, 0.92–1.05) and ever-smokers (relative risk, 1.03; 95% CI, 0.98–1.07) compared with never-smokers.[70] Accounting for variables such as duration of smoking, frequency of smoking, number of pack-years smoked, and timing of smoking exposure has led to a stronger association between smoking and breast cancer risk in some studies[77,79,80] but not in others.[81,82] The relationship between cigarette smoking and breast cancer risk remains unresolved.

Physical Activity

With obesity and physical inactivity poised to surpass tobacco use as the leading preventable causes of death in the United States, attention has refocused on the potentially beneficial aspects of physical activity as a modifiable risk factor for cancer. The preponderance of evidence from case-control[83–85] and cohort studies[86,87] indicates a decreased risk of breast cancer among physically active women, with reductions of risk averaging 20–40% (FIG. 3.4).[88]

Cohort studies have contributed essential information on the benefits of lifetime physical activity and the type, amount, and timing of physical activity needed to influence breast cancer risk. In the multiethnic San Francisco Bay Area Breast Cancer study, lifetime physical activity (recreational, transportation, chores, jobs) was associated with a 25% reduction in breast cancer risk for women in the highest quartile of physical activity compared with the lowest quartile,[89] regardless of menopausal status and ethnic group. The WHI cohort study evaluated the benefits of recreational physical activity among 74,171 postmenopausal women with 1780 newly diagnosed cases of breast cancer. Women with regular strenuous recreational physical activity at least 3 days a week at age 35 years had a 14% decreased risk of breast cancer (95% CI, 0.78–0.95) compared with the less active. Women engaging in 1.25 to 2.5 hours of brisk walking weekly had an 18% decreased risk of breast cancer (relative risk, 0.82; 95% CI, 0.68–0.97) compared with inactive women. Adjustment for daily kilocalorie intake or percentage of calories from fat did not affect the results.[87]

It has been proposed that physical activity lowers pre- and postmenopausal risk of breast cancer through protective mechanisms that result in less biologically available estrogen. These protective mechanisms include delay in menarche, reduction in number of ovulatory cycles, reduced levels of insulin and insulin-like growth factor, reduced ovarian estrogen production, reduction in body fat with

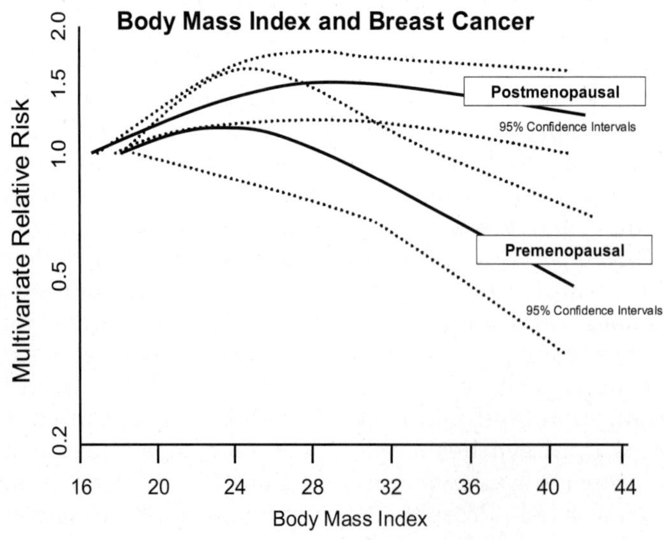

FIGURE 3.4 Summary of studies concerning the impact of physical activity and the risk of breast cancer. From Friedenreich CM.[88]

subsequent decrease in aromatase activity, and increased production of sex hormone binding globulin.[90–93] Randomized clinical trials are needed to determine the efficacy of increased physical activity on breast cancer risk reduction and verify its hypothesized effect on circulating levels of hormone concentrations.[94]

Breast Cancer in Men

Breast cancer in men is infrequent. It is estimated that 1450 men will develop breast cancer and 470 will die from the disease in the year 2004.[1] Although the lifetime risk of being diagnosed with breast cancer is substantially lower in men than women, 0.011% and 13% respectively, men have a higher mortality rate from breast cancer, possibly attributable to the lack of screening and decreased awareness among men of their risk for breast cancer.

Genetic susceptibility is the strongest risk factor for male breast cancer, with the relative risk of developing breast cancer twofold higher in men with sisters and mothers affected with breast cancer compared with men with no family history of breast cancer.[95] Among men with an inherited mutation in the *BRCA2* or *BRCA1* breast cancer susceptibility genes, the lifetime risk of breast cancer is 6.1% and 5.8%, respectively, a risk that is approximately 50 times higher than in the general population.[96,97] Mutations in the *BRCA2* gene account for 15% of prevalent male breast cancer cases, and the frequency ranges from 4% to 21%.[98] The genetic syndrome, Klinefelter disease, which affects 1 in every 1000 men, is estimated to be responsible for another 8% of male breast cancer cases.[99]

Evidently, most male breast cancers may be attributed to other unrecognized genetic and environmental factors. Disorders causing imbalances in the male sex hormones or increasing estrogen concentrations have been proposed to play a role in male breast cancer development. For example, gynecomastia, which may be caused by an overproduction of estrogen, has been associated with an increased risk of developing male breast cancer. Other less common causes of male breast cancer that are similar risk factors for women include radiation exposure, benign breast disease, obesity, and exposure to estrogen medication.[100–102]

Genetic Susceptibility

Having a family history of breast cancer is a significant risk factor for the disease. The Collaborative Group on Hormonal Factors in Breast Cancer analyzed data on 58,209 women with breast cancer and 101,986 control subjects (unselected for family history of breast cancer) and found increasing risk ratios for breast cancer with increasing numbers of affected first-degree relatives. Compared with women who had no affected relative, the risk of breast cancer increased two- to fourfold among women with one or more affected first-degree relatives with breast cancer.[103] Heritable factors are estimated to account for 27% of all breast cancer cases.[104] Mutations in the high penetrance *BRCA1, BRCA2, p53, ATM,* and *pTEN* genes are responsible for only 5–10% of breast cancer cases, with mutations in the *BRCA1* and *BRCA2* genes accounting for about 45% of hereditary breast cancers.[105] The cumulative risk of developing breast cancer in a woman with a mutation in *BRCA1* and *BRCA2* varies depending on the method of ascertainment of the study population. In a clinic-based study of 483 *BRCA1* mutation carriers, the risk of female breast cancer by age 70 was 72.8% (95% CI, 67.9–77.7%) with the average age at diagnosis of female breast cancer 52 years (95% CI, 45–60 years).[97] Linkage studies have shown risks of breast cancer as high as 87% by age 70 in women with a mutation in *BRCA1* and *BRCA2.*[105] Additional cancer risks are conferred by *BRCA1.*[97,106]

Genetic Polymorphisms and Breast Cancer Risk

Because only about 5% to 10% of breast cancers can be attributed to familial cancer predisposition syndromes that involve mutations in single highly penetrant genes, much of the remaining etiologic fraction related to a genetic cause is likely due to common single nucleotide polymorphisms (SNPs) or a combination of SNPs. Some of the pathways associated with breast cancer include the steroid hormone metabolism genes (i.e., *CYP1A1, CYP17, COMT*), carcinogen metabolizing genes (i.e., *SULT1A1, NQO1, GSTs*), cell cycle control genes (*PHB, HER2, CCDN1*), and DNA repair pathway genes (*XRCC1, XRCC3*). Many of these specific SNPs have been reviewed and are briefly summarized below.[107,108] Typically, the breast cancer risk conferred by these SNPs is small to moderate, with odds ratios that reach significance ranging from 1.3 to 2.5.

Steroid Hormone Metabolizing Genes

It is well documented that estrogens play a key role in promoting breast cancer development. Many steroid metabolizing enzymes are active in breast tissue, with aromatase (CYP19) being one of the primary enzymes involved in producing estrogens from androgenic precursors, converting the δ4-androstenedione and testosterone into estrone and bioavailable estradiol.[109] A second important regulatory mechanism for modulating estrogenic activity within the breast tissue is the interconversion of estrone into estradiol by 17β-hydroxysteroid dehydrogenase type 1 and 17β-hydroxysteroid dehydrogenase type 2. Specific genes involved in the steroid hormone pathway encode for ovarian and adrenal synthesis and are expressed in adipose and breast tissue. Examples include *STAR, CYP11A1, HSD3B* family, *HSD18B* family, and *CYP19. STAR* encodes a transport protein that is crucial for translocating cholesterol to mitochondria, where cholesterol is the starting substrate for steroidogenesis. *CYP11A1* is involved in the first step in steroidogenesis that catalyzes and cleaves the cholesterol side chain,[110] and polymorphisms have been found to be associated with excess ovarian androgens.[111] Breast cancer risk has been found to be lower in women with an insertion polymorphism (PROGINS), as well as polymorphic variants of the *ESR1* gene.[112]

In the breast and other extrahepatic tissues, the cytochrome P-450 enzymes, such as CYP1A1 and CYP1B1, catalyze the hydroxylation of E1 and E2. Some molecular epidemiologic studies looking at the role of CYP1A1, CYP1B1, CYP17, and COMT have shown an increased breast cancer risk, whereas others have not.[113] Large-scale studies combining the cohorts are looking at the role of hormone metabolizing genes and their role in breast cancer. By combining these large samples, we will have a better understanding of how these polymorphisms alter estrogen metabolism and modify breast cancer risk.

GST and Carcinogen Metabolizing Genes

Reactive oxygen species may be generated through a number of mechanisms, including metabolism through the E2 pathway. For example, reactive oxygen species are produced through conjugated estrogen mediated redox cycling of quinines and semiquinones.[113] The GSTs are involved in metabolizing or detoxifying products of oxidative damage by catalyzing conjugation of glutathione with reactive oxygen species. The GSTM1, T1, and P1 are known to have enzymatic activity as well as MnSOD. The epidemiologic studies evaluating these genes have also shown mixed results, with some studies showing significant associations and others having none. These mixed results could be a result of small sample size and misclassification of exposures when trying to evaluate a gene–environment interaction. Many of these methodologic issues will be resolved as consortiums combine their studies with common genes and exposures to better understand the genetic heterogeneity of this complex disease.

DNA Repair Genes

Suboptimal repair of DNA damage is a susceptibility factor associated with a variety of cancers, including breast cancer.[114] Case-control studies have reported suboptimal DNA repair capacity among women with breast cancer and women at high risk of breast cancer compared with normal control subjects.[115–119] It is hypothesized that functionally relevant polymorphisms of low penetrance genes involved in DNA repair pathways may contribute to susceptibility to sporadic breast cancer.

Polymorphisms of the *XRCC1* gene, involved in the base excision repair pathway, have been investigated as low penetrance genes that may predispose to breast cancer. The XRCC1*Arg399Gln* polymorphism has been associated with suboptimal DNA repair capacity measured by increased levels of DNA adducts,[120] increased sister chromatid exchange,[121] and hypersensitivity to ionizing radiation.[122] One population-based case-control study found no association between the variant Trp allele and breast cancer in women who participated in the Carolina Breast Cancer Study. The Gln allele was associated with breast cancer in African American but not among white women (odds ratio, 1.7; 95% CI, 1.1–2.4).[123] In the Shanghai Breast Cancer study, there was no association between the XRCC1 Gln allele and breast cancer (odds ratio, 1.20; 95% CI, 0.85–1.69).[124]

The normally encoded BRCA1 and BRCA2 proteins are involved in the DNA repair pathway, and specifically BRCA1 plays a role in the

DNA repair pathways of double stranded break repair and transcription coupled repair.[125–127] Suboptimal repair of double stranded breaks induced by carcinogens such as ionizing radiation may result in breast carcinogenesis. A case-control study of 292 patients from the Anglian Breast Cancer Study of the East Anglian Cancer Registry breast cancer cases was conducted to examine the relationship between polymorphisms of seven genes involved in the double stranded break DNA repair pathway and breast cancer risk.[128] A statistically significant positive association was observed only for two polymorphisms of the *XRCC3* gene (A17893G, C18067T). In a nested case-control study, no association was observed between polymorphisms of *XRCC3* (A4541G, A17893G and C18067T) and breast cancer risk.[129]

Cell Cycle Genes

Polymorphisms of genes involved in cell cycle control are particularly interesting as breast cancer susceptibility alleles. The human epidermal growth factor receptor 2 (HER-2) proto-oncogene is involved in signal transduction and cell proliferation, and overexpression is associated with poorer breast cancer prognosis. An isoleucine to valine polymorphism at codon 655 (I655V) of the *HER-2* gene has been evaluated for breast cancer risk in studies consisting of different populations of women with mixed results. In a case-control study conducted in China, homozygosity for the Val/Val allele was statistically significantly associated with breast cancer risk (odds ratio, 14.1; 95% CI, 1.8–113.4).[130] In three population-based studies of African American, Latina, and European women,[131–133] no association was observed between the HER-2 I655V polymorphisms and breast cancer, whereas in three additional studies, a positive association was seen only among younger women and those with a first-degree family history of breast cancer.[112,134,135]

Future Directions

In this chapter we reviewed the epidemiologic risk factors and described their role in the etiology of breast cancer. As the genetic tools become more sophisticated, we will be able to better define the interactions of genes in combination with risk factors. Research is evolving in understanding risk factors by life cycles during critical periods of a woman's lifetime when the breast is most susceptible to exposures that can initiate or promote cancer in breast epithelial cells (in utero, puberty, pregnancy, and pre- and postmenopausal periods). Identifying exposures that occur during these periods of breast susceptibility may help in our understanding of modifiable risk factors for breast cancer prevention. This will lead to better risk assessment models and improved prevention strategies.

References

1. Jemal A, Tiwari RC, Murray T, et al. American Cancer Society. Cancer statistics 2004. CA Cancer J Clin 2004;54:8–29.
2. Weir HK, Thun MJ, Hankey BF, et al. Annual report to the nation on the status of cancer, 1975–2000, featuring the uses of surveillance data for cancer prevention and control. J Natl Cancer Inst 2003;95: 1276–1299.
3. Kelsey JL, Bernstein L. Epidemiology and prevention of breast cancer. Annu Rev Public Health 1996; 17:47–67.
4. Pike MC, Spicer DV, Dahmoush L, et al. Estrogens, progestogens, normal breast cell proliferation, and breast cancer risk. Epidemiol Rev 1993;15:17–35.
5. Cavalieri EL, Stack DE, Devanesan PD, et al. Molecular origin of cancer: catechol estrogen-3,4-quinones as endogenous tumor initiators. Proc Natl Acad Sci USA 1997;94:10937–10942.
6. Liehr JG. Dual role of oestrogens as hormones and pro-carcinogens: tumour initiation by metabolic activation of oestrogens. Eur J Cancer Prev 1997;6: 3–10.
7. Lambe M, Hsieh C, Trichopoulos D, et al. Transient increase in the risk of breast cancer after giving birth. N Engl J Med 1994;331:5–9.
8. Liu Q, Wuu J, Lambe M, et al. Transient increase in breast cancer risk after giving birth: postpartum period with the highest risk (Sweden). Cancer Causes Control 2002;13:299–305.
9. Schatzkin A, Palmer JR, Rosenberg L, et al. Risk factors for breast cancer in black women. J Natl Cancer Inst 1987;78:213–217.
10. Mayberry RM, Stoddard-Wright C. Breast cancer risk factors among black women and white women: similarities and differences. Am J Epidemiol 1992; 136:1445–1456.
11. Palmer JR, Wise LA. Dual effect of parity on breast cancer risk in African-American women. J Natl Cancer Inst 2003;95:478–483.
12. Russo J, Hu YF, Yang X, et al. Developmental, cellular, and molecular basis of human breast cancer. J Natl Cancer Inst Monogr 2000;27:17–37.
13. Russo J, Wilgus G, Russo IH. Susceptibility of the mammary gland to carcinogenesis. I. Differentiation of the mammary gland as determinant of tumor incidence and type of lesion. Am J Pathol 1979;96: 721–736.
14. van den Brandt P, Spiegelman AD, Yaun SS, et al. Pooled analysis of prospective cohort studies on height, weight, and breast cancer risk. Am J Epidemiol 2000;152:514–527.

15. IARC Handbooks of Cancer Prevention. Lyons, France: International Agency for Research on Cancer, 2002.
16. Ballard-Barbash R, Schatzkin RA, Taylor PR, et al. Association of change in body mass with breast cancer. Cancer Res 1990;50:2152–2155.
17. Barnes-Josiah D, Potter JD, Sellers TA, et al. Early body size and subsequent weight gain as predictors of breast cancer incidence (Iowa, United States). Cancer Causes Control 1995;6:112–118.
18. Huang Z, Hankinson SE, Colditz GA, et al. Dual effects of weight and weight gain on breast cancer risk. JAMA 1997;278:1407–1411.
19. Trentham-Dietz A, Newcomb PA, Storer BE, et al. Weight change and risk of postmenopausal breast cancer (United States). Cancer Causes Control 2000; 11:533–542.
20. Berkey CS, Frazier AL, Gardner JD, et al. Adolescence and breast carcinoma risk. Cancer 1999;85: 2400–2409.
21. Le Marchand L, Kolonel LN, Earle ME, et al. Body size at different periods of life and breast cancer risk. Am J Epidemiol 1988;128:137–152.
22. London SJ, Colditz GA, Stampfer MJ, et al. Prospective study of relative weight, height, and risk of breast cancer. JAMA 1989;262:2853–2858.
23. Magnusson C, Baron J, Persson I, et al. Body size in different periods of life and breast cancer risk in post-menopausal women. Int J Cancer 1988;76: 29–34.
24. Hilakivi-Clarke L, Forsen T, Eriksson JG, et al. Tallness and overweight during childhood have opposing effects on breast cancer risk. Br J Cancer 2001;85: 1680–1684.
25. Kaaks R, Van Noord PA, Den Tonkelaar I, et al. Breast-cancer incidence in relation to height, weight and body-fat distribution in the Dutch "DOM" cohort. Int J Cancer 1998;76:647–651.
26. Huang Z, Willett WC, Colditz GA, et al. Waist circumference, waist:hip ratio, and risk of breast cancer in the Nurses' Health Study. Am J Epidemiol 1999; 150:1316–1324.
27. Morimoto LM, White E, Chen Z, et al. Obesity, body size, and risk of postmenopausal breast cancer: the Women's Health Initiative (United States). Cancer Causes Control 2002;13:741–751.
28. Harvie M, Hooper L, Howell AH. Central obesity and breast cancer risk: a systematic review. Obes Rev 2003;4:157–173.
29. Key TJ, Appleby PN, Reeves GK, et al. Body mass index, serum sex hormones, and breast cancer risk in postmenopausal women. J Natl Cancer Inst 2003;95: 1218–1226.
30. Brisson J, Merletti F, Sadowsky NL, et al. Mammographic features of the breast and breast cancer risk. Am J Epidemiol 1982;115:428–437.
31. Boyd NF, Byng JW, Jong RA, et al. Quantitative classification of mammographic densities and breast cancer risk: results from the Canadian National Breast Screening Study. J Natl Cancer Inst 1995;87: 670–675.
32. Byrne C, Schairer C, Wolfe J, et al. Mammographic features and breast cancer risk: effects with time, age, and menopause status. J Natl Cancer Inst 1995; 87:1622–1629.
33. Maskarinec G, Meng L. A case-control study of mammographic densities in Hawaii. Breast Cancer Res Treat 2000;63:153–161.
34. Harvey JA, Bovbjerg VE. Quantitative assessment of mammographic breast density: relationship with breast cancer risk. Radiology 2004;230:29–41.
35. Boyd NF, Jensen HM, Cooke G, et al. Mammographic densities and the prevalence and incidence of histological types of benign breast disease. Reference Pathologists of the Canadian National Breast Screening Study. Eur J Cancer Prev 2000;9: 15–24.
36. Boyd NF, Dite GS, Stone J, et al. Heritability of mammographic density, a risk factor for breast cancer. N Engl J Med 2002;347:886–894.
37. Bolton JL. Quinoids, quinoid radicals, and phenoxyl radicals formed from estrogens and antiestrogens. Toxicology 2002;177:55–65.
38. Nandi S, Guzman RC, Yang J. Hormones and mammary carcinogenesis in mice, rats, and humans: a unifying hypothesis. Proc Natl Acad Sci USA 1995; 92:3650–3657.
39. Dorgan JF, Longcope C, Stephenson HE, et al. Relation of prediagnostic serum estrogen and androgen levels to breast cancer risk. Cancer Epidemiol Biomarkers Prevent 1996;5:533–539.
40. Hankinson SE, Willett WC, Manson JE, et al. Plasma sex steroid hormone levels and risk of breast cancer in postmenopausal women. J Natl Cancer Inst 1998;90:1292–1299.
41. Toniolo PG, Levitz M, Zeleniuch Jacquotte A, et al. A prospective study of endogenous estrogens and breast cancer in postmenopausal women. J Natl Cancer Inst 1995;87:190–197.
42. The Endogenous Hormones and Breast Cancer Collaborative Group. Endogenous sex hormones and breast cancer in postmenopausal women: reanalysis of nine prospective studies. J Natl Cancer Inst 2002; 94:606–616.
43. Marchbanks PA, McDonald JA, Wilson HG, et al. Oral contraceptives and the risk of breast cancer. N Engl J Med 2002;346:2025–2032.
44. Anonymous. Oral-contraceptive use and the risk of breast cancer. The Cancer and Steroid Hormone Study of the Centers for Disease Control and the National Institute of Child Health and Human Development. N Engl J Med 1986;315:405–411.
45. Anonymous. Breast cancer and hormonal contraceptives: collaborative reanalysis of individual data on 53 297 women with breast cancer and 100 239 women without breast cancer from 54 epidemiological studies. Collaborative Group on Hormonal Factors in Breast Cancer. Lancet 1996;347:1713–1727.
46. Hunt K, Vessey M, McPherson K, et al. Long-term surveillance of mortality and cancer incidence in women receiving hormone replacement therapy. Br J Obstet Gynaecol 1987;94:620–635.
47. Mills PK, Beeson WL, Phillips RL, et al. Prospective study of exogenous hormone use and breast cancer in Seventh-Day Adventists. Cancer 1989;64:591–597.
48. Risch HA, Howe GR. Menopausal hormone usage and breast cancer in Saskatchewan: a record-linkage cohort study. Am J Epidemiol 1994;139:670–683.
49. Persson I, Weiderpass E, Bergkvist L, et al. Risks of breast and endometrial cancer after estrogen and estrogen-progestin replacement. Cancer Causes Control 1999;10:253–260.
50. Colditz GA, Rosner B. Cumulative risk of breast cancer to age 70 years according to risk factor status: data from the Nurses' Health Study. Am J Epidemiol 2000;152:950–964.

51. Schairer C, Lubin J, Troisi, et al. Menopausal estrogen and estrogen-progestin replacement therapy and breast cancer risk. JAMA 2000;283:485–491.

52. Gambrell RD Jr, Maier RC, Sanders BI. Decreased incidence of breast cancer in postmenopausal estrogen-progestogen users. Obstet Gynecol 1993;62: 435–443.

53. Schuurman AG, van den Brandt PA, Goldbohm RA. Exogenous hormone use and the risk of postmenopausal breast cancer: results from The Netherlands Cohort Study. Cancer Causes Control 1995;6:416–424.

54. Anonymous. Breast cancer and hormone replacement therapy: collaborative reanalysis of data from 51 epidemiological studies of 52,705 women with breast cancer and 108,411 women without breast cancer. Collaborative Group on Hormonal Factors in Breast Cancer. Lancet 1997;350:1047–1059.

55. Million Women Study Collaborators. Breast Cancer and hormone replacement therapy in the Million Women Study. Lancet. 2003,362:419–427.

56. Rossouw JE, Anderson GL, Prentice, et al. Risks and benefits of estrogen plus progestin in healthy postmenopausal women: principal results from the Women's Health Initiative randomized controlled trial. JAMA 2002;288:321–333.

57. Anderson GL, Limacher M, Assaf AR, et al. Women's Health Initiative Steering Committee. Effects of conjugated equine estrogen in postmenopausal women with hysterectomy: the Women's Health Initiative randomized controlled trial. JAMA 2004;291:1701–1712.

58. Armstrong B, Doll R. Environmental factors and cancer incidence and mortality in different countries, with special reference to dietary practices. Int J Cancer 1975;15:617–631.

59. Hunter DJ, Spiegelman D, Adami, et al. Cohort studies of fat intake and the risk of breast cancer—a pooled analysis. N Engl J Med 1996;334:356–361.

60. Smith-Warner SA, Spiegelman D, Adami HO. Types of dietary fat and breast cancer: a pooled analysis of cohort studies. Int J Cancer 2001;92:767–774.

61. Gandini S, Merzenich H, Robertson C, et al. Meta-analysis of studies on breast cancer risk and diet: the role of fruit and vegetable consumption and the intake of associated micronutrients. Eur J Cancer 2000;36:636–646.

62. Rohan TE, Howe GR, Friedenreich CM, et al. Dietary fiber, vitamins A, C, and E, and risk of breast cancer: a cohort study. Cancer Causes Control 1993; 4:29–37.

63. Zhang S, Hunter DJ, Forman MR, et al. Dietary carotenoids and vitamins A, C, and E and risk of breast cancer. J Natl Cancer Inst 1999;91:547–556.

64. Terry P, Jain M, Miller AB, et al. Dietary carotenoids and risk of breast cancer. Am J Clin Nutr 2002;76: 883–888.

65. Cho E, Spiegelman D, Hunter DJ, et al. Premenopausal intakes of vitamins A, C, and E, folate, and carotenoids, and risk of breast cancer. Cancer Epidemiol Biomarkers Prev 2003;12:713–720.

66. Howe GR, Hirohata T, Hislop TG, et al. Dietary factors and risk of breast cancer: combined analysis of 12 case-control studies. J Natl Cancer Inst 1990;82: 561–569.

67. Rohan TE, Jain MG, Howe GR, et al. Dietary folate consumption and breast cancer risk. J Natl Cancer Inst 2000;92:266–269.

68. Sellers TA, Kushi LH, Cerhan JR, et al. Dietary folate intake, alcohol, and risk of breast cancer in a prospective study of postmenopausal women. Epidemiology 2001;12:420–428.

69. Zhang SM, Willett WC, Selhub J, et al. Plasma folate, vitamin B6, vitamin B12, homocysteine, and risk of breast cancer. J Natl Cancer Inst 2003;95: 373–380.

70. Hamajima N, Hirose K, Tajima K, et al. Alcohol, tobacco and breast cancer—collaborative reanalysis of individual data from 53 epidemiological studies, including 58,515 women with breast cancer and 95,067 women without the disease. Br J Cancer 2002;87:1234–1245.

71. Smith-Warner SA, Spiegelman D, Yaun SS, et al. Alcohol and breast cancer in women: a pooled analysis of cohort studies. JAMA 1998;279:535–540.

72. Dorgan JF, Baer DJ, Albert PS, et al. Serum hormones and the alcohol-breast cancer association in postmenopausal women. J Natl Cancer Inst 2001;93: 710–715.

73. Park SK, Yoo KY, Lee SJ, et al. Alcohol consumption, glutathione S-transferase M1 and T1 genetic polymorphisms and breast cancer risk. Pharmacogenetics 2000;10:301–309.

74. Millikan RC, Pittman GS, Newman B, et al. Cigarette smoking, N-acetyltransferases 1 and 2, and breast cancer risk. Cancer Epidemiol Biomarkers Prev 1998;7:371–378.

75. Marcus PM, Newman B, Millikan RC, et al. The associations of adolescent cigarette smoking, alcoholic beverage consumption, environmental tobacco smoke, and ionizing radiation with subsequent breast cancer risk (United States). Cancer Causes Control 2000;11:271–278.

76. Morabia A, Bernstein M, Heritier S, et al. Relation of breast cancer with passive and active exposure to tobacco smoke. Am J Epidemiol 1996;143:918–928.

77. Terry PD, Miller AB, Rohan TE, et al. Cigarette smoking and breast cancer risk: a long latency period? Int J Cancer 2002;100:723–728.

78. Reynolds P, Hurley S, Goldberg DE, et al. Active smoking, household passive smoking, and breast cancer: evidence from the California Teachers Study. J Natl Cancer Inst 2004;96:29–37.

79. Calle EE, Miracle-McMahill HL, Thun MJ, et al. Cigarette smoking and risk of fatal breast cancer. Am J Epidemiol 1994;139:1001–1007.

80. Hunter DJ, Hankinson SE, Hough H, et al. A prospective study of NAT2 acetylation genotype, cigarette smoking, and risk of breast cancer. Carcinogenesis 1997;18:2127–2132.

81. Adami HO, Lund E, Bergstrom R, et al. Cigarette smoking, alcohol consumption and risk of breast cancer in young women. Br J Cancer 1988;58:832–837.

82. Egan KM, Stampfer MJ, Hunter DR, et al. Active and passive smoking in breast cancer: prospective results from the Nurses' Health Study. Epidemiology 2002;13:138–145.

83. D'Avanzo B, Nanni O, La Vecchia C, et al. Physical activity and breast cancer risk. Cancer Epidemiol Biomarkers Prev 1996;5:155–160.

84. Shoff SM, Newcomb PA, Trentham-Dietz A, et al. Early-life physical activity and postmenopausal breast cancer: effect of body size and weight change. Cancer Epidemiol Biomarkers Prev 2000;9:591–595.

85. Verloop J, Rookus MA, van der Kooy K, et al. Physical activity and breast cancer risk in women aged 20–54 years. J Natl Cancer Inst 2000;92: 128–135.

86. Moore DB, Folsom AR, Mink PJ, et al. Physical activity and incidence of postmenopausal breast cancer. Epidemiology 2000;11:292–296.

87. McTiernan A, Kooperberg C, White E, et al. Recreational physical activity and the risk of breast cancer in postmenopausal women: the Women's Health Initiative Cohort Study. JAMA 2003;290:1331–1336.

88. Friedenreich CM. Physical activity and cancer prevention: from observational to intervention research. Cancer Epidemiol Biomarkers Prev 2001;10: 287–301.

89. John EM, Horn-Ross PL, Koo J. Lifetime physical activity and breast cancer risk in a multiethnic population: the San Francisco Bay area breast cancer study. Cancer Epidemiol Biomarkers Prev 2003;12: 1143–1152.

90. Toniolo PG. Endogenous estrogens and breast cancer risk: the case for prospective cohort studies. Environ Health Perspect 1997;105(suppl 3):587–592.

91. Hoffman-Goetz L, Apter D, Demark-Wahnefried W, et al. Possible mechanisms mediating an association between physical activity and breast cancer. Cancer 1998;83(3 suppl):621–628.

92. McTiernan A, Ulrich C, Slate S, et al. Physical activity and cancer etiology: associations and mechanisms. Cancer Causes Control 1998;9:487–509.

93. Yu H, Rohan T. Role of the insulin-like growth factor family in cancer development and progression. J Natl Cancer Inst 2000;92:1472–1489.

94. McTiernan A, Schwartz RS, Potter J, et al. Exercise clinical trials in cancer prevention research: a call to action. Cancer Epidemiol Biomarkers Prev 1999;8: 201–207.

95. Rosenblatt KA, Thomas DB, McTiernan A, et al. Breast cancer in men: aspects of familial aggregation. J Natl Cancer Inst 1991;83:849–854.

96. Thompson D, Easton DF. Variation in cancer risks, by mutation position, in BRCA2 mutation carriers. Am J Hum Genet 2001;68:410–419.

97. Brose MS, Rebbeck TR, Calzone KA, et al. Cancer risk estimates for BRCA1 mutation carriers identified in a risk evaluation program. J Natl Cancer Inst 2002;94:1365–1372.

98. Liede A, Karlan BY, Narod SA. Cancer risks for male carriers of germline mutations in BRCA1 or BRCA2: a review of the literature. J Clin Oncol 2004;22: 735–742.

99. Hultborn R, Hanson C, Kopf I, et al. Prevalence of Klinefelter's syndrome in male breast cancer patients. Anticancer Res 1997;17:4293–4297.

100. Sasco AJ, Lowenfels AB, Pasker-de Jong P. Review article: epidemiology of male breast cancer. A meta-analysis of published case-control studies and discussion of selected aetiological factors. Int J Cancer 1993;53:538–549.

101. Hsing AW, McLaughlin JK, Cocco P, et al. Risk factors for male breast cancer (United States). Cancer Causes Control 1998;9:269–275.

102. Lynch HT, Watson P, Narod SA, et al. The genetic epidemiology of male breast carcinoma. Cancer 1999; 86:744–746.

103. Collaborative Group on Hormonal Factors in Breast Cancer. Familial breast cancer: collaborative reanalysis of individual data from 52 epidemiological studies including 58209 women with breast cancer and 101986 women without the disease. Lancet 2001;358:1389–1399.

104. Lichtenstein P, Holm NV, Verkasalo PK, et al. Environmental and heritable factors in the causation of cancer—analyses of cohorts of twins from Sweden, Denmark, and Finland. N Engl J Med 2000;343: 78–85.

105. Ford D, Easton DF, Stratton M, et al. Genetic heterogeneity and penetrance analysis of the BRCA1 and BRCA2 genes in breast cancer families. The Breast Cancer Linkage Consortium. Am J Hum Genet 1998;62:676–689.

106. Thompson D, Easton DF. Cancer incidence in BRCA1 mutation carriers. J Natl Cancer Inst 2002; 94:1358–1365.

107. Dunning AM, Healey CS, Pharoah PD, et al. A systematic review of genetic polymorphisms and breast cancer risk. Cancer Epidemiol Biomarkers Prev 1999;8:843–854.

108. de Jong MM, Nolte IM, te Meerman GJ, et al. Genes other than BRCA1 and BRCA2 involved in breast cancer susceptibility. J Med Genet 2002;39:225–242.

109. Key TJ, Pike M. The role of oestrogens and progestogens in the epidemiology and prevention of breast cancer. Eur J Cancer Clin Oncol 1988;24:29–43.

110. Stocco DM. StAR protein and the regulation of steroid hormone biosynthesis. Annu Rev Physiol 2001;63:193–213.

111. Dunning AM, Ellis PD, McBride S, et al. A transforming growth factorbeta1 signal peptide variant increases secretion in vitro and is associated with increased incidence of invasive breast cancer. Cancer Res 2003;63:2610–2615.

112. Wang-Gohrke S, Chang-Claude J. Population-based, case-control study of HER2 genetic polymorphism and breast cancer risk. J Natl Cancer Inst 2001;93: 1657–1659.

113. Thompson PA, Ambrosone C. Molecular epidemiology of genetic polymorphisms in estrogen metabolizing enzymes in human breast cancer. J Natl Cancer Inst Monogr 2000;27:125–134.

114. Goode EL, Ulrich CM, Potter JD. Polymorphisms in DNA repair genes and associations with cancer risk. Cancer Epidemiol Biomarkers Prev 2002;11: 1513–1530.

115. Helzlsouer K, Harris E, Parshad R, et al. DNA repair proficiency: potential susceptibility factor for breast cancer. J Natl Cancer Inst 1996;88:754–755.

116. Patel RK, Trivedi AH, Arora DC, et al. DNA repair proficiency in breast cancer patients and their first degree relatives. Int J Cancer 1997;73:20–24.

117. Kovacs E, Almendral A. Reduced DNA repair synthesis in healthy women having first degree relatives with breast cancer. Eur J Cancer Clin Oncol 1987; 23:1051–1057.

118. Parshad R, Price FM, Bohr VA, et al. Deficient DNA repair capacity, a predisposing factor in breast cancer. Br J Cancer 1996;74:1–5.

119. Helzlsouer KJ, Harris EL, Parshad R, et al. Familial clustering of breast cancer: possible interaction between DNA repair proficiency and radiation exposure in the development of breast cancer. Int J Cancer 1995;64:14–17.

120. Lunn RM, Langlois RG, Hsieh LL, et al. XRCC1 polymorphisms: effects on aflatoxin B1-DNA adducts and glycophorin A variant frequency. Cancer Res 1999;59:2557–2561.

121. Abdel-Rahman SZ, El Zein RA. The 399Gln polymorphism in the DNA repair gene XRCC1 modulates the genotoxic response induced in human lymphocytes by the tobacco-specific nitrosamine NNK. Cancer Lett 2000;159:63–71.

122. Hu JJ, Smith TR, Miller MS, et al. Amino acid substitution variants of APE1 and XRCC1 genes associated with ionizing radiation sensitivity. Carcinogenesis 2001;22:917–922.

123. Duell EJ, Millikan RC, Pittman GS, et al. Polymorphisms in the DNA repair gene XRCC1 and breast cancer. Cancer Epidemiol Biomarkers Prev 2001;10: 217–222.

124. Shu XO, Cai Q, Gao YT, et al. A population-based case-control study of the Arg399Gln polymorphism in DNA repair gene XRCC1 and risk of breast cancer. Cancer Epidemiol Biomarkers Prev 2003;12:1462–1467.

125. Liu CY, Flesken-Nikitin A, Li S, et al. Inactivation of the mouse Brca1 gene leads to failure in the morphogenesis of the egg cylinder in early post-implantation development. Genes Dev 1996;10:1835–1843.

126. Scully R, Chen J, Plug A, et al. Association of BRCA1 with Rad51 in mitotic and meiotic cells. Cell 1997;88: 265–275.

127. Chen Y, Lee WH, Chew HK. Emerging roles of BRCA1 in transcription regulation and DNA repair. J Cell Physiol 1999;181:385–392.

128. Kuschel B, Auranen A, McBride SA, et al. Variants in DNA double-strand break repair genes and breast cancer susceptibility. Hum Mol Genet 2002;11: 1399–1407.

129. Han J, Hankinson SE, Ranu H, et al. Polymorphisms in DNA double-strand break repair genes and breast cancer risk in the Nurses' Health Study. Carcinogenesis 2004;25:189–195.

130. Xie D, Shu XO, Deng Z, et al. Population-based, case-control study of *HER2* genetic polymorphism and breast cancer risk. J Natl Cancer Inst 2000;92: 412–417.

131. Baxter SW, Campbell IG. Population-based, case-control study of *HER2* genetic polymorphism and breast cancer risk. J Natl Cancer Inst 2001;93: 557–559.

132. Ameyaw M, Thornton N, McLeod HL. Population-based, case-control study of HER2 genetic polymorphism and breast cancer risk. J Natl Cancer Inst 2000;92:1947.

133. Keshava C, McCanlies EC, Keshava N, et al. Distribution of HER2(V655) genotypes in breast cancer cases and controls in the United States. Cancer Lett 2001;173:37–41.

134. Montgomery KG, Gertig DM, Baxter SW, et al. The HER2 I655V polymorphism and risk of breast cancer in women $<$ age 40 years. Cancer Epidemiol Biomarkers Prev 2003;2:1109–1111.

135. Millikan R, Eaton A, Worley K, et al. HER2 codon 655 polymorphism and risk of breast cancer in African Americans and whites. Breast Cancer Res Treat 2003;79:355–364.

CHAPTER
4

Predisposition for Breast Cancer

Marion Kiechle, MD and Alfons Meindl, PhD

Department of Obstetrics and Gynecology, Klinikum rechts der Isar, Technical University of Munich, Munich, Germany

Etiology

Breast cancer is the most frequent carcinoma in women. In more developed countries, the probability of contracting breast cancer between the ages of 20 and 80 is approximately 7.8%, that is, 1 in 13 women.[1] A comparison of familial and sporadic cases shows that familial cancer diagnosed at a younger age is frequently bilateral and frequent in men.[2] In recent years it has been estimated that 5–15% of breast cancer cases involve genetic events. However, careful epidemiologic investigations now indicate that only one third (5%) might be caused by monogenic factors. Most inherited breast cancer cases are now assumed to be polygenic, and most diagnosed breast cancer cases are sporadic without familial clustering (FIG. 4.1).[1,3] In contrast, it can no longer be excluded that at least some of the sporadic cases may be caused through a combination of polygenic variants and exogenous factors.[3]

The penetrance of inherited mutated alleles is modulated by additional low penetrant genetic variants (modifier genes), environmental factors, or both. Factors such as nulliparity, physical exercise, diet, obesity, and birth age influence the age at onset in gene mutation carriers in both sporadic and familial cancer.[4] Indeed, it can be observed in pedigrees from families with multiple occurrences of breast and/or ovarian cancer that the age at onset of the disease is decreasing.[4,5] It is still controversial whether the use of oral contraceptives increases the risk of developing breast cancer in *BRCA* mutation carriers. Some studies calculate a small but statistically significant risk,[6] whereas other studies have demonstrated no greater risk.[7]

The penetrance of predisposing genes for breast cancer is also dependent on age and tissue type. For the two known breast cancer predisposition genes, *BRCA1* and *BRCA2* (see High Penetrance Genes, below), risks have been determined[4] (**Table 4.1**). In the case of mutation carriers in the *BRCA1* or *BRCA2* genes (see below), only breast and ovarian tissues have a relatively high risk to transform into malignant cells. In addition, the probability to change into an aberrant phenotype is age dependent. Likewise, the probability for a *BRCA1* mutation carrier to develop breast cancer at the age of 50 is approximately 29%. However, women with a carrier status and born after 1940 may have higher risks.[4,5] Only slightly increased risks for transformation are described for such tissues as prostate, colon, and pancreas.[8,9]

The known forms of hereditary breast cancer are inherited in an autosomal dominant mode. It should be noted, however, that inactivation of one allele is only a prerequisite for tumor generation and is not sufficient to initiate cancer in normal epithelial breast cells. Indeed, for tumor initiation, the second allele has to be inactivated. This can happen either by loss of heterozygosity in the according regions[10,11] or by epigenetic effects such as methylation (see Chapters 26 and 27).[12] Tumor progression requires the accumulation of additional aberrations in the cellular genome, including the disturbance of TP53 expression that allows cell escape from apoptosis.[13] Based on histologic parameters and gene expression profiling, *BRCA1* tumors can be distinguished from *BRCA2* and sporadic breast cancer samples. Tumor samples from *BRCA1* mutation carriers are normally high grade, infiltrating ductal breast cancers[14] and more often estrogen and HER-2/neu receptor negative than sporadic samples.[14,15] In contrast, *BRCA2* tumors are more difficult to separate from sporadic cases by histologic classification only, but it should be feasible through more validated expression profiling data; for example, tumors from *BRCA2* mutation carriers show an

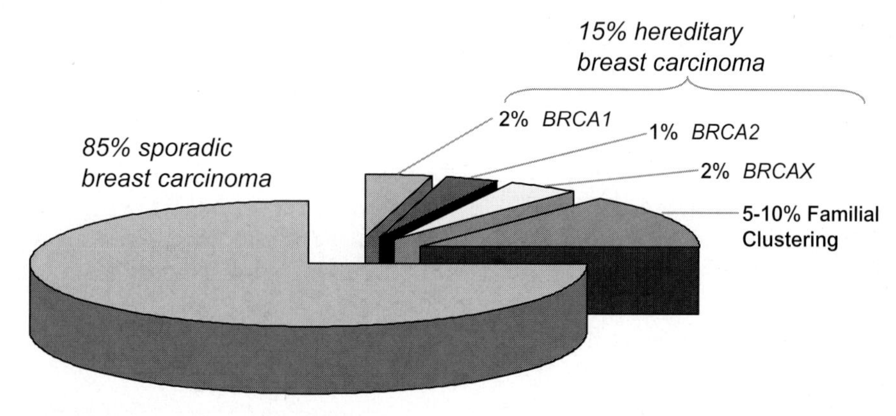

FIGURE 4.1 Origins of breast carcinoma.

Table 4.1 Risks for *BRCA1* or *BRCA2* Mutation Carriers to Develop Breast and/or Ovarian Cancer

	BRCA1		*BRCA2*	
Age (yr)	BCa	OvCa	BCa	OvCa
40	21%	3%	17%	2%
50	39%	21%	34%	4%
60	58%	40%	48%	6%
80	80–90%	60%	80%	30%

expression profile that is different from *BRCA1* but also from sporadic tumors.[16]

Predisposition Markers

It is estimated that 10–15% of all breast cancer cases may be caused by genetic predisposition. The inheritance of a single defective allele is most likely in a family exhibiting clustering of breast cancer cases, especially if it includes cases diagnosed before age 50. Consequently, a dominant monogenic trait can be demonstrated in families with multiple occurrences of breast cancer (high penetrance genes). In contrast, the segregation of multiple deficient alleles is most likely in families with fewer cases of late onset breast cancer (low penetrance genes). The genes associated with hereditary forms of breast cancer are listed in **Table 4.2** and are discussed in the following paragraphs.

High Penetrance Genes

To date, two genes that feature frequent mutations associated with breast cancer predisposition in high risk families have been isolated and validated.[17–19] Other genes also involved in the DNA repair or apoptosis pathways have been characterized but only rarely were shown to be mutated. Some of these genes are restricted to site-specific breast cancer, whereas others are associated with other cancer entities.

BRCA1 The first gene found to be mutated in families with multiple occurrences of breast and/or ovarian cancer was the *BRCA1* gene.[17,19] It has been classified as a classic tumor suppressor gene, and consequently inactivation of the second allele in the resulting tumor could be demonstrated (see above). It is located at 17q21; comprises 24 exons, 22 of them coding; and encodes for a protein consisting of 1863 amino acids. A typical pedigree of a family with a *BRCA1* mutation is given in **FIGURE 4.2**. Currently, more than 2000 different deleterious mutations in the gene are listed in the international Breast Cancer Information Core (BIC) database (http://research.nhgri.nih.gov/bic/). Most of the putative pathogenic mutations found in the gene are either frame-shift or stop mutations that similarly cause the production of truncated proteins.[5,17] Only a few missense mutations, causing single amino acid substitutions, could be demonstrated to be disease causing.[20,21] Recently, a high prevalence of large gene rearrangements in the *BRCA1* gene has been reported for some populations.[22,23] Mutation profiles for different countries have been published,[5,24–26] and it has become apparent that a mosaic of founder- and population-specific mutations exists. Founder mutations are often of Ashkenazi Jewish origin and were shown to be present in different white populations.[5,24–26]

Only a few missense mutations in the *BRCA1* gene can be clearly associated with familial breast cancer. These include amino acid residues that are part of the zinc finger

Table 4.2 List of Genes Associated with Hereditary Breast Cancer

Syndrome	Gene/Region	Primary Carcinoma	Secondary Carcinoma
High Penetrance Genes			
Familial breast/ ovarian carcinoma	*BRCA1* 17q21	Breast and ovaries	Prostate, colon leukemia
Familial breast/ ovarian carcinoma	*BRCA2* 13q13	Breast and ovaries	Prostate, pancreas male breast
Familial breast/ bilateral b.c.	*BACH1* 17q22	Breast	None
Familial breast/ bilateral b.c.	*RAD51* 15q15	Breast	None
Li-Fraumeni syndrome	*TP53* 17p13	Sarcomas, breast	Brain, leukemia
Cowden syndrome	*PTEN* 10q23	Hamartomatous lesions	Polyps
Peutz-Jeghers syndrome	*STK11* 19q	Gastrointestinal polyps	Melanin spots on lips, b.c.
Low Penetrance Genes			
Ataxia telangiectasia	*ATM* 11q22	Lymphomas gliomas	Heterozygous: breast cancer
Li-Fraumeni/ breast cancer	*CHEK2* 13q21	Breast sarcomas	Brain, leukemia

b.c., breast cancer.

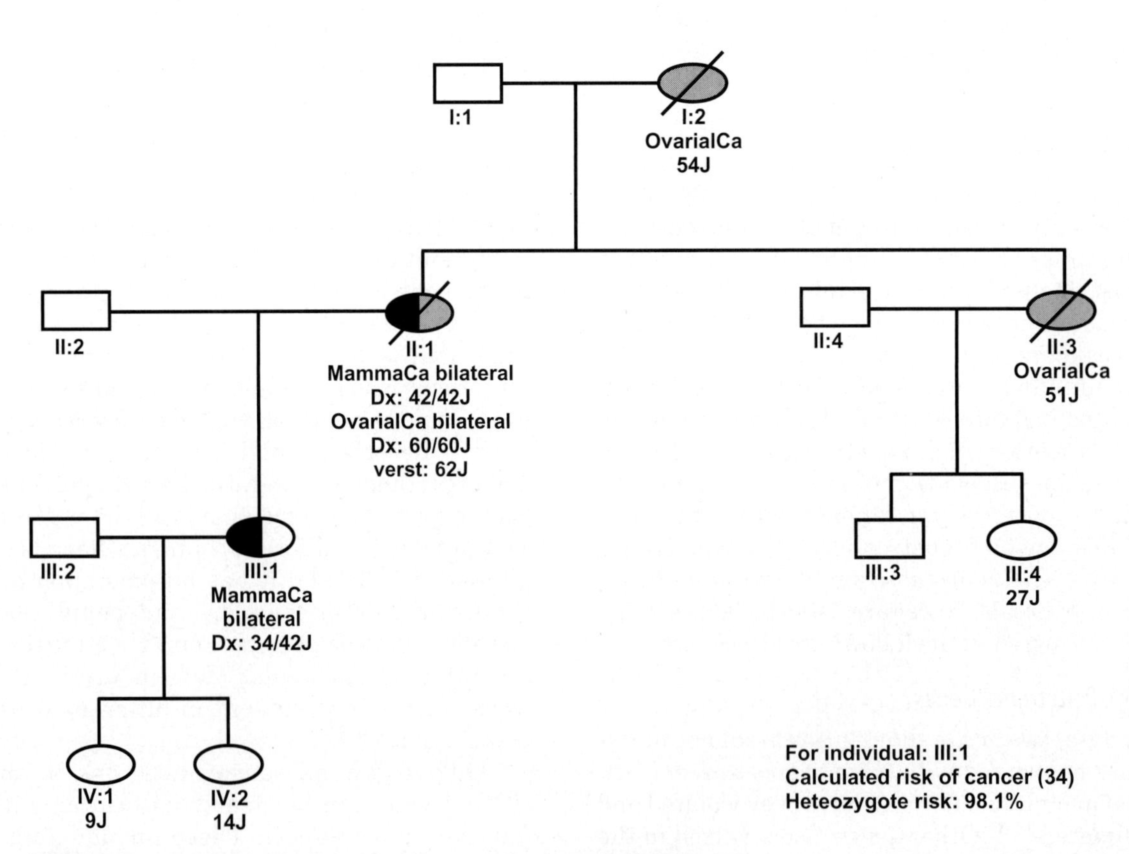

FIGURE 4.2 A mutation in the *BRCA1* gene, causing a truncated protein (1364X).

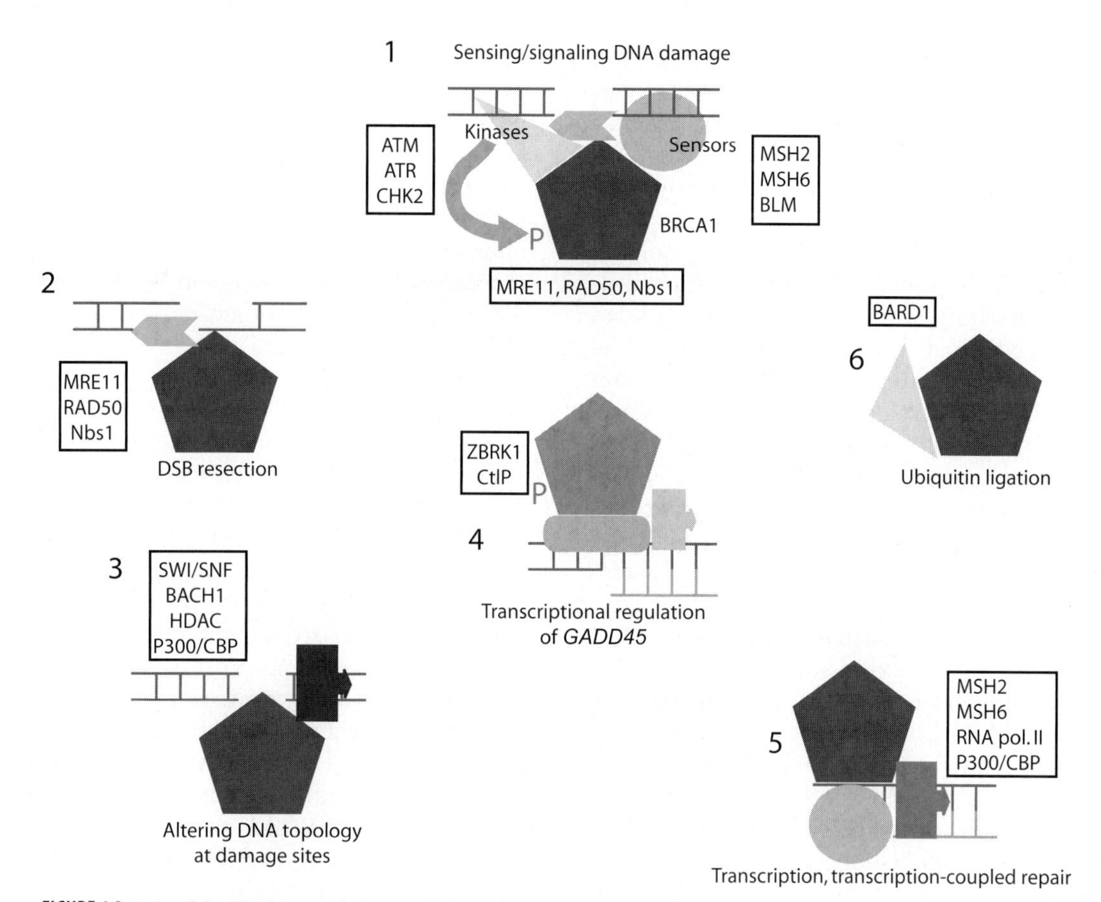

FIGURE 4.3 Role of the BRCA1 protein in the DNA repair process. (From Ref. 28.)

(Cys61Gly, Cys64Y) and amino acids located in the BCRT domain that have been shown to interact with different proteins (e.g., W1837R). Several strategies have been established to evaluate the character of such unclassified variants. These include segregation analysis, evolutionary methods to identify functionally important amino acid sites, loss of heterozygosity analysis in the corresponding tumor samples, and functional assays.[20,21]

A variety of assay approaches has now become available, enabled by the advanced characterization of the BRCA1 protein. Based on the initial observation that the BRCA1 protein is a caretaker of chromosomal stability, several studies demonstrated that it is involved in the repair of double-stranded DNA breaks (for details, see Chapter 25). Such genomic lesions can be repaired by nonhomologous end joining, recombination between homologous DNA sequences, or single-strand annealing.[27,28] In addition, other than its local activities at sites of DNA damage, BRCA1 seems to be involved in processes that are located upstream and downstream of DNA damage responses. On the one hand, BRCA1 is part of protein complexes that apparently have functions intrinsic to the sensing and signaling of different types of DNA lesions; on the other hand, it operates as a transcriptional regulator of genes, whose expression affects downstream biologic responses (FIG. 4.3). The local activities of BRCA1 at sites of DNA damage have been elucidated by several experiments. Sites of DNA damage are rapidly marked by the phosphorylation of the histone species H2A-X, and BRCA1 interacts with the MRE11/RAD50/Nbs1 protein complex to migrate to such sites and resect double-stranded DNA breaks.[28,29] It may be necessary to alter the DNA topology at the damage sites via recruiting the protein complexes SWI/SNF, BACH1, HDAC, and P300/CBP for effective DNA repair as well as the regulation of downstream processes. Included in these downstream processes are the transcriptional regulation of GADD45 and transcription coupled repair (FIG. 4.3). Surprisingly, to date, only the *BACH1* gene has been associated with hereditary breast cancer (see Infrequently

Mutated Genes, below). Upstream processes of these local activities require the phosphorylation of the BRCA1 protein via the kinases ATR, ATM, and CHK2 proteins (FIG. 4.3).[30–32] Interestingly, these two latter kinases could be associated with hereditary breast cancer (see The *ATM* Gene and The *CHK2* Gene, below).

Apart from the breast and ovary, compared with many organs in which there is no cancer predisposition risk, several other organs such as the pancreas, prostate, and colon have a slightly increased risk of cancer development in *BRCA1* mutation carriers.[8,9] This observation indicates that *BRCA1* disruption may have tissue-specific effects that favor malignant transformation. It has been suggested that *BRCA1* functions as an inhibitor of estrogen receptor signaling.[33] Studies have substantiated this observation.[34,35] Evidence that *BRCA1* mediates repression of the estrogen receptor α has been reported, and this process is correlated with the down-regulation of the transcriptional coactivator p300.[34] Further characterization of this pathway may help to explain the tissue specificity of cancer development in *BRCA1*-deficient cancers.

The clinical outcome of patients with germline mutations in *BRCA1* has been evaluated in several studies. Most informative is the comprehensive study of Robson et al.[15] in which two retrospective cohorts of Ashkenazi Jewish women undergoing breast-conserving treatment for invasive cancer were genotyped and the clinical characteristics of the women with and without BRCA founder mutations were compared. In brief, *BRCA1* mutation carriers demonstrated a significantly shorter overall survival. This adverse prognosis could not be linked to tumor size or axillary nodal status. This study also demonstrated that women with *BRCA1* mutations are at substantially increased risk for developing metachronous contralateral breast cancer, whereas no increased risk could be seen for the development of ipsilateral breast cancer when compared with noncarriers.

BRCA2 A second high penetrance gene found to be mutated in families with multiple occurrences of breast cancer is the *BRCA2* gene. It is located at 13q12.3 and consists of 27 exons, 26 of them coding for a protein of 3418 amino acids.[18,19] A typical pedigree with a mutation in the *BRCA2* gene is given in FIGURE 4.4. Similar to the *BRCA1* gene, most of the defined cancer-causing mutations are either frameshift or stop mutations that produce truncated proteins. More than 1000 deleterious mutations in the *BRCA2* gene are listed in the international BIC database. However, compared with *BRCA1*, more missense mutations can be seen in the *BRCA2* gene. Because the BRCA2 protein is not as well characterized as the BRCA1 protein, it is necessary to evaluate the character of the observed unclassified variants carefully. Overall, it appears that most of the listed unclassified variants in the *BRCA2* gene are rare polymorphic variants (unpublished data) rather than pathogenic. Analogous to the *BRCA1* gene, founder mutations are presented frequently in most populations.[36]

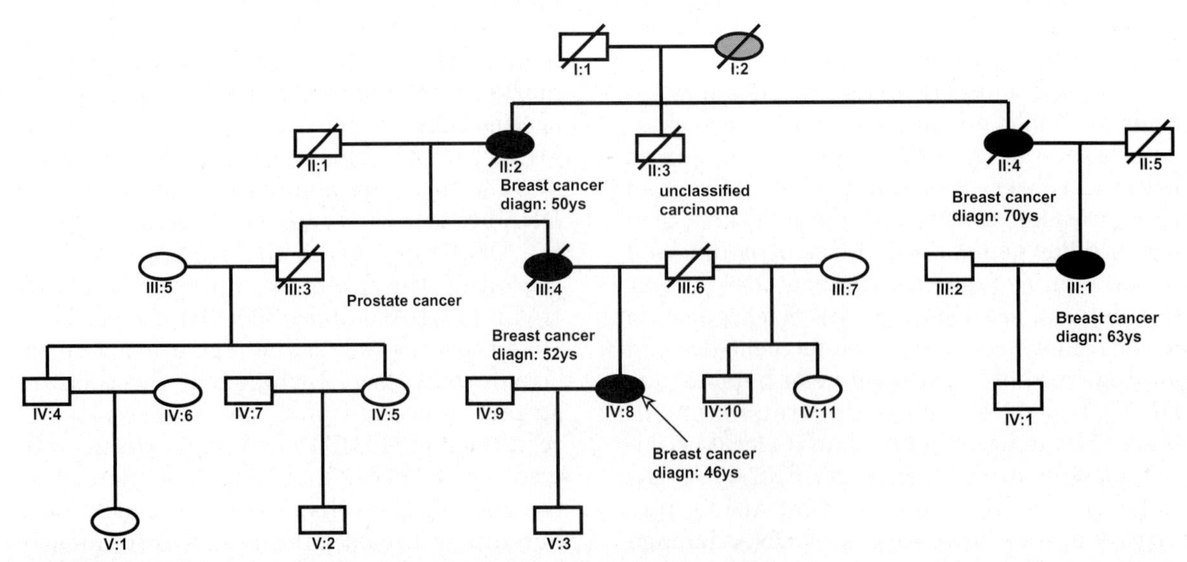

FIGURE 4.4 A mutation in the *BRCA2* gene, causing a truncated protein (S1882X).

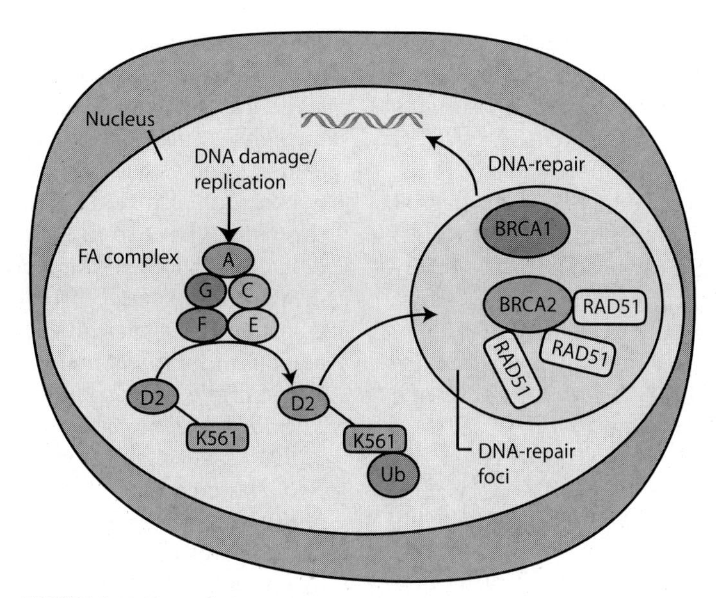

FIGURE 4.5 BRCA2 and FA proteins in DNA repair. (From Ref. 44.)

Mutations in the *BRCA2* gene are normally found in site-specific breast cancer families, including male breast cancer. However, families with a clustering of pancreatic cancer were also shown to harbor *BRCA2* mutations.[37,38] In one study,[37] 26 European families in which at least two first-degree relatives had a histologically confirmed diagnosis of pancreatic ductal adenocarcinoma were screened for mutations in the *BRCA2* gene, and three deleterious and missense aberrations were found. In addition, it could be shown that 2% of men with early onset prostate cancer harbor germline mutations in the *BRCA2* gene. Male *BRCA2* mutation carriers under the age of 60 have a 26-fold risk increase of developing prostate cancer compared with control subjects.[39] However, further studies are required to confirm that *BRCA2* mutations are associated with a higher penetrance in familial cancers of the pancreas and prostate than are *BRCA1* mutations.

Of further interest is the recent finding that biallelic mutations in the *BRCA2* gene have been found in two Fanconi complementation groups.[40] Fanconi anemia (FA) is an autosomal recessive disorder affecting all bone marrow elements and is associated with cardiac, renal, and limb malformations as well as dermal pigmentation changes. Fanconi patients also have an increased cellular sensitivity to DNA cross-linking agents. FA has eight identified complementation groups, and for the complementation groups B and D1, mutations in the *BRCA2* gene can be identified. This finding has now linked the FA pathway to the BRCA1/BRCA2 mediated DNA repair process (**FIG. 4.5**). Several FA proteins, including FANCA, FANCC, FANCE, FANCF, and FANCG, form a constitutive complex in the nucleus of human cells that in response to DNA damage mediates the monoubiquitination of FANCD2. The activated FANCD2 protein is then translocated to chromatin and the DNA-repair foci that contain BRCA1 and BRCA2 (FANCD1). BRCA2 is known to bind directly to RAD51[41] and, similar to BRCA1, is involved in homology-directed DNA repair. Also, similar to the BRCA1 protein, the BRCA2FA pathway can be linked to ATM and to another upstream activator, the MRE11 complex. A study indicates that in response to cellular exposure to ionizing radiation, the ATM protein directly phosphorylates FANCD2.[42] Thus, the FANCD2 protein functions in mediating two signaling pathways by undergoing two different types of posttranslational modifications. A link to the MRE11 complex (NBS-MRE11-RAD50) can be established through the demonstration that the activated form of FANCD2 and NBS1 are colocalized in nuclear foci. Either the disruption of the MRE11-binding carboxy terminal end of NBS1 or the disruption of the FANCD2 monoubiquitination site results in mitomycin hypersensitivity.[43] In summary, both BRCA proteins are involved in DNA repair, a cellular mechanism activated in response to irradiation or chemically induced lesions. As well as the ATM and NBS pathways, respectively, the FA pathway has now linked to this cellular network system that maintains genomic integrity and stability.[44]

As in the case of *BRCA1* mutation carriers, the clinical outcome for *BRCA2* mutation carriers has been evaluated.[15] In contrast to females with predisposing mutations in the *BRCA1* gene, females with mutations in the *BRCA2* gene have no reduced survival after breast cancer therapy. However, similar to *BRCA1* carriers, these patients carry an approximately 30% risk of developing contralateral breast cancer after 10 years of posttreatment follow-up has elapsed. Further follow-up studies are required to confirm these initial data obtained in Ashkenazi Jewish women.

BRCAX Although most of the high risk families with multiple occurrences of site-specific breast cancer or breast and ovarian cancer (six or more cases) map either to 13q13 or 17q21, no further predisposing loci have been determined for most families exhibiting fewer cases of breast cancer. Different models have been developed to determine the genetic transmission modes for the relatively large number of BRCA1/BRCA2 negative families. Based on such models, in most of the families showing familial clustering, polygenic traits may segregate. For only a limited number of families, further high penetrance genes acting in an autosomal dominant form are expected.[45] However, one published study has predicted the involvement of an as yet unknown autosomal recessive gene in a significant portion of such families.[46]

Infrequently Mutated Genes Based on the observation that BRCA1 and BRCA2 are involved in double-stranded DNA break-repair mechanisms, several groups have searched for aberrations in other genes located in this pathway. The *BACH1* gene (*BRCA1*-associated C-terminal helicase-1) codes for a nuclear protein that acts directly with the highly conserved C-terminal BRCT repeats of the BRCA1 protein. The predicted 1249 amino acid long protein, encoded by 20 exons, contains the seven helicase-specific motifs that are conserved among members of the DEAH helicase family. Cantor et al.[47] identified a germline mutation affecting the helicase domain in 2 of 65 patients with either early onset or familial breast cancer. In subsequent experiments the same group demonstrated that these two and a novel third mutation impair the function of the BACH1 protein.[48] This novel protein is both a DNA-dependent ATPase and a 5′ to 3′-DNA helicase. The helicase activity is strictly ATP dependent, and when a Pro47Ala missense mutation is present, this activity is abrogated. Due to another missense mutation (Lys52Arg), ATPase activity is disrupted as the unwinding of DNA and RNA hybrids, an additional function associated with the wild-type. Finally, one missense mutation maintains the ATPase activity but disturbs the unwinding process. Further studies will show whether other variants in the gene will have at least a minor effect on hereditary breast cancer, for example, in combination with CHEK2 (see The *CHK2* Gene, below).

Mouse experiments demonstrated that BRCA2 interacts with RAD51 (see *BRCA2*, above). Recently, Pellegrini et al.[49] showed, on the basis of crystal structures, that the direct binding is mediated by the BRC repeat (BRCA2) and the RecA-homology domain (RAD51). The BRC repeat mimics a motif in RAD51 that serves as an interface for oligomerization between individual RAD51 monomers, thus enabling BRCA2 to control the assembly of the RAD51 nucleoprotein filament that is essential for strand-pairing reactions during DNA recombination. In parallel to the still ongoing functional works on BRCA2 and RAD51, several groups looked for aberrations in the *RAD51* gene. In a study with 20 patients from breast cancer families, Kato et al.[50] identified a missense mutation (Arg150Gln) in two unrelated females, both presenting with bilateral breast cancer. In addition, a single nucleotide polymorphism in the 5′-untranslated region of the *RAD51* gene (135g → c) has been associated with an increase for *BRCA2* mutation carriers.[51] However, further studies, including quantitative reverse transcriptase-polymerase chain reaction, are required to substantiate this finding.

Genes Associated with Cancer Syndromes Breast cancer is also a component of several other cancer syndromes, including Li-Fraumeni syndrome (OMIM 151623), Cowden syndrome (OMIM 158350), and Peutz-Jeghers syndrome (OMIM 175200). All these syndromes are rare, and mutations in the genes causing them contribute only a small fraction of hereditary breast cancer cases.

The *TP53* gene is mutated in Li-Fraumeni syndrome, which is characterized by the occurrence of multiple cancers, including childhood sarcoma and breast cancer (see Chapter 23).[52] Mutations in the *TP53* gene are found in approximately 70% of the Li-Fraumeni families. A higher cancer risk in females than in males was observed by using follow-up studies.[53] Quite recently, mutations in the *CHEK2* gene have also been associated with Li-Fraumeni syndrome (see The *CHK2* Gene, below). Although only a few germline mutations in *TP53* contribute to hereditary breast cancer, somatic inactivation of *TP53* in BRCA mutation related tumors is frequently identified (see above).[54] It is now accepted that inactivation of *TP53* prevents the initiation of apoptosis, prevents the depletion of cells that have accumulated genomic lesions, and thus contributes to the establishment and progression of malignancy.

Multiple hamartomatous lesions, especially of the skin, mucous membranes, breast, and thyroid, and craniomegaly are characteristic lesions of the Cowden syndrome. In 1997, mutations in the *PTEN* (phosphatase and tensin homolog) gene were identified in families with the Cowden syndrome phenotype.[55] Although germline mutations in *PTEN* are rare, *PTEN* is frequently inactivated in sporadic cancers, including glioblastomas and prostate cancer (see Chapter 23).[56] Recent data indicate that *PTEN* is involved in the regulation of apoptosis[56] and angiogenesis.[57] During both processes, the PTEN protein acts through the well-recognized tumor progression associated phosphoinositol-3-kinase and Akt-dependent pathways.

A serine-threonine kinase, the *STK11/LKB1* gene, was found to be mutated in several families with Peutz-Jeghers syndrome.[58] This is an autosomal dominant disorder characterized by melanocytic macules of the lips, multiple gastrointestinal hamartomatous polyps, and an increased risk for various neoplasms, including gastrointestinal and breast cancer. Carrier females with a mutation in the *STK11* gene also have an elevated risk for breast cancer, as has been shown in a British study in which 33 Peutz-Jeghers families were evaluated by follow-up data.[59] Whether the *STK11* gene also plays a role in sporadic breast cancer, as suggested in a recent study,[60] requires validation with larger patient groups.

Low Penetrance Genes

Currently, three genes are considered to act as low penetrance genes for hereditary breast cancer. In contrast to high penetrance genes, such genes act in a concerted way, either in connection with variants in other predisposing genes or in combination with epigenetic or environmental factors. In addition, several modifier genes in breast cancer have been described and are summarized elsewhere.[61]

The *ATM* gene A first gene in which variants have been associated with breast cancer is the *ATM* gene (OMIM 208900). The *ATM* gene is mutated on both alleles in ataxia telangiectasia and codes for a member of the phosphatidylinositol-3 kinase family of proteins. Similar to the two BRCA proteins, ATM plays a key role in monitoring genomic integrity and regulating cell cycle checkpoints, DNA repair, and apoptotic pathways induced by double-stranded DNA breaks. After the occurrence of double-stranded breaks in the cell, ATM interacts with and phosphorylates a number of different targets, including p53, Nibrin, CtIP, BRCA1, and CHEK2 (FIGS. 4.3 and 4.6).[62] Given that there are several confirmed direct functional links between the ATM and BRCA1 proteins, the *ATM* gene also has been implicated in breast carcinoma susceptibility.

In two larger mutation screening reports, variants in the *ATM* gene have been linked to familial breast cancer.[62,63] In both studies, missense mutations in the *ATM* gene were shown to be specific for familial cases but were absent from normal control subjects (**Table 4.2**). However, additional studies are required to confirm that ATM mutations confer increased susceptibility to breast cancer.

The *CHK2* gene Cell cycle checkpoint kinase 2 (CHEK2 [OMIM 604373]) plays an important role in the maintenance of genome integrity and in the regulation of the G2/M cell cycle checkpoint. It has been shown to interact with other proteins involved in DNA repair processes, such as *BRCA1* and *TP53*.[32,64] These findings suggest that *CHK2* could be an attractive candidate gene associated with susceptibility for a variety of cancers (FIG. 4.6). Patients with Li-Fraumeni syndrome were analyzed, and in a small proportion of families a single mutation in exon 10 of *CHK2* was identified.[65] This del1100C variant causes a stop codon after amino acid 380 abrogating the kinase activity. The *CHK2* Breast Cancer Consortium analyzed 718 families with breast and/or ovarian cancer that harbor neither a *BRCA1* nor a *BRCA2* mutation. The 1861delC mutation was identified in 5.1% of the families

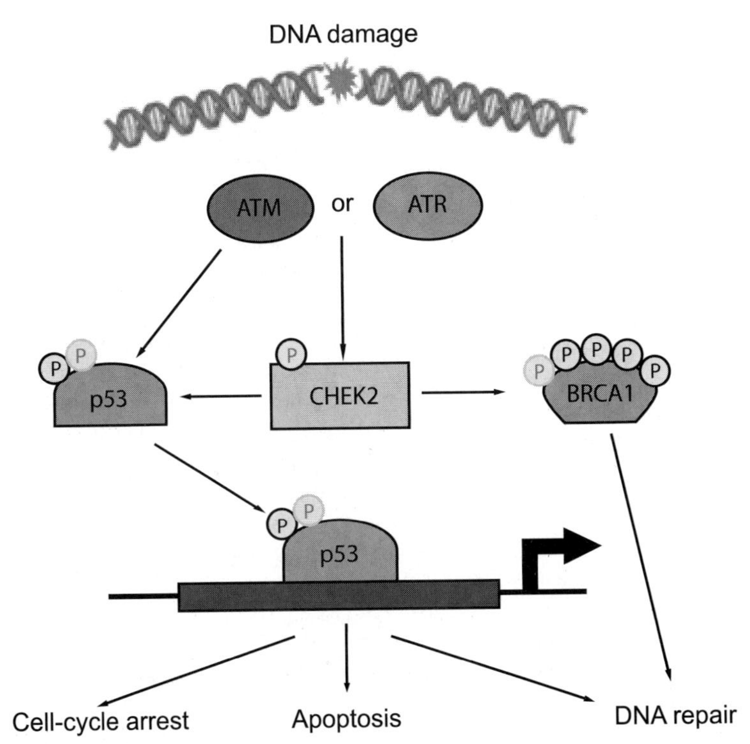

FIGURE 4.6 Possible role of *CHK2* in breast cancer development. (From Ref. 64.)

compared with a prevalence of 1.2% in a healthy control group, indicating a low penetrance of this mutation.[66]

To establish the relevance of *CHK2* mutations in the German population, all 14 coding exons were analyzed in 516 families negative for mutations in both the *BRCA1* and *BRCA2* genes. As in other studies, a statistically significant association for the exon 10 mutations was observed ($P = 0.006$), and carriers had an almost fourfold increased risk for familial breast cancer (relative risk, 3.9). However, the prevalence of the 1100delC mutation in both familial breast cancer patients (1.6%) and control subjects (0.5%) in the German population is significantly lower than reported for other white populations. It shows independent segregation for breast cancer in two of four families analyzed. Furthermore, no increased breast cancer risk was found for the most common missense mutation I157T or the splice site mutation IVS2 + 1G $\rightarrow$ A, which added together for more than half of the variants (12 of 21) observed in patients.[67]

The low prevalence of the exon 10 deletion mutations together with no or an uncertain elevation in risk for other *CHK2* mutations suggests only a limited relevance of mutations in the *CHK2* gene for familial breast cancer. Therefore, at present it appears premature to include *CHK2* in genetic counseling for patients with an unexplained high familial incidence of breast cancer.

Androgen Receptor Gene Mutations in the androgen receptor (*AR*) gene have been linked to different disease entities. Expansions of CAG repeats within the coding region can cause spinal and bulbar muscular dystrophy (Kennedy disease) and different forms of androgen insensitivity.[68] The first link to breast cancer for *AR* was confirmed by the identification of germline mutations in two brothers with breast cancer and Reifenstein syndrome.[69] More recently, a polymorphic CAG repeat in the *AR* gene has been associated with a poorer prognosis for *BRCA1* mutation carriers. Comparison of 165 women with and 139 women without breast cancer revealed that women carrying at least one allele with 29 or 30 CAG repeats manifested the disease earlier.[70]

In summary, an evaluation of genes localized in either apoptosis, DNA repair, or endocrine signaling will provide further low penetrance or modifier genes. Ongoing mapping and cloning studies will show whether a gene segregates recessively or dominantly in high risk families.

Predisposition Testing and Risk Management

Because of the complexity of genetic counseling in hereditary breast cancer, referral to a subspecialized multidisciplinary clinic staffed by practitioners with expertise in this field is generally recommended. In this setting, a new client can be educated by an experienced genetic counselor about the a priori risks associated with breast cancer predisposing germline mutations. At the same visit, the client can also receive information from medical and surgical/gynecologic oncologists concerning screening, diagnosis, and treatment options. Finally, this approach also allows the client to meet with a psychologist/psychiatrist to consider the important psychologic and emotional issues associated with breast cancer predisposition testing and the implications for the client and her family. In such a multidisciplinary counseling setting, the sensitivity and specificity of the applied molecular techniques, the indications for testing, and possible risk managements in *BRCA1/2* mutation carriers and/or *BRCA1/2* negative females from putative *BRCAX* families can be thoroughly considered.

Methods

A variety of laboratory techniques has been used for screening for genetic mutations in hereditary breast cancer. In addition, protein truncation tests have been developed as prescreening methods[71] and can be applied for the rapid identification of the likely presence of underlying pathogenic mutations. However, it is now accepted that screening of index patients in families with either multiple cases or early onset manifestation of breast cancer should be applied by investigating the entire gene to also detect missense mutations and putative splice sites. Two independent methods that are applied currently in routine diagnosis include the direct double-stranded sequencing approach, as performed by the company Myriad Genetics,[19,24] and denaturing high-performance liquid chromatography (DHPLC) analysis of all coding exons with the adjacent intronic sites.[5,36] In families with a significant prior probability of a *BRCA* mutation, testing for large gene rearrangements[22,23] also has to be included. For this approach, a kit based on multiplex ligation-dependent probe amplification is available now.[72] Overall, the sensitivity of this technique is greater than 90%, and when either complete direct double-stranded sequencing or comprehensive DHPLC analysis is combined with multiplex ligation-dependent probe amplification, the specificity is greater than 90%.

Indications for Testing

Genetic testing is recommended only after multidisciplinary genetic counseling. The criterion for testing is a 20% probability of the presence of a mutation.[73–75] Candidates for *BRCA1/2* genetic testing are as follows:

- Two breast cancer cases under age 40 or three under age 50 years;
- Four cases of breast cancer under age 60 years;
- More than four cases of breast cancer at any age;
- Ovarian and breast cancer in a family (breast cancer under age 50 years if only one ovarian and one breast cancer case);
- Early onset female breast cancer at under age 60 years and male breast cancer at any age;
- Single case of breast cancer at under age 40 years if the patient is of Ashkenazi Jewish decent.

Risk Management

Three categories of options for management of women with *BRCA* mutations are available: screening and surveillance, prophylactic surgeries, and chemoprevention. These options should be discussed with women in detail to allow them to make informed decisions regarding their health care.

References

1. Collaborative Group on Hormonal Factors in Breast Cancer. Familial breast cancer: collaborative reanalysis of individual data from 52 epidemiological studies including 58,209 women with breast cancer and 101,986 women without the disease. Lancet 2001;358:1389–1399.
2. Hall JM, Lee MK, Newman B, et al. Linkage of early-onset familial breast cancer to chromosome 17q21. Science 1990;250:1684–1689.
3. Pharoah PD, Antoniou A, Bobrow M, et al. Polygenic susceptibility to breast cancer and implications for prevention. Nat Genet 2002;31:33–36.
4. King MC, Marks JH, Mandell JB. Breast and ovarian cancer risks due to inherited mutations in *BRCA1* and *BRCA2*. Science 2003;302:643–646.
5. Meindl A. German Consortium for Hereditary Breast and Ovarian Cancer. Comprehensive analysis of 989

patients with breast and ovarian cancer provides BRCA1 and BRCA2 mutation profiles and frequencies for the German population. Int J Cancer 2002; 97:472–480.

6. Narod SA, Dube MP, Klijn J, et al. Oral contraceptives and the risk of breast cancer in BRCA1 and BRCA2 mutation carriers. J Natl Cancer Inst 2002; 94:1773–1779.
7. Marchbanks PA, McDonald JA, Wilson HG, et al. Oral contraceptives and the risk of breast cancer. N Engl J Med 2002;346:2025–2032.
8. Thompson D, Easton DF, Breast Cancer Linkage Consortium. Cancer incidence in BRCA1 carriers. J Natl Cancer Inst 2002;94:1358–1365.
9. Breast Cancer Linkage Consortium. Cancer risks in BRCA2 mutation carriers. J Natl Cancer Inst 1999; 91:1310–1316.
10. Smith TM, Easton DF, Evans DG, et al. Allele losses in the region 17q12-21 in familial breast and ovarian cancer involve the wild-type chromosome. Nat Genet 1992;2:128–131.
11. Osorio A, de la Hoya M, Rodriguez-Lopez R, et al. Loss of heterozygosity analysis in tumor samples in patients with familial breast cancer. Int J Cancer 2002;99:305–309.
12. Esteller M. Cancer epigenetics: DNA methylation and chromatin alterations in human cancer. Adv Exp Med Biol 2003;532:39–49.
13. Hakem R, de la Pompa JL, Elia A, et al. Partial rescue of Brca1(5-6) early embryonic lethality by p53 or p21 null mutation. Nat Genet 1997;16:298–302.
14. Lakhani SR, Jacquemier J, Sloane JP, et al. Multifactorial analysis of differences between sporadic breast cancers and cancers involving *BRCA1* and *BRCA2* mutations. J Natl Cancer Inst 1998;90:1138–1145.
15. Robson ME, Chappuis PO, Satagopan J, et al. A combined analysis of outcome following breast cancer: differences in survival based on BRCA1/BRCA2 mutation status and administration of adjuvant treatment. Breast Cancer Res 2004;6:R8–R17.
16. Hedenfalk I, Duggan D, Chen Y, et al. Gene expression profiles in hereditary breast cancer. N Engl J Med 2001;344:539–548.
17. Miki Y, Swensen J, Shattuck-Eidens D, et al. A strong candidate for the breast and ovarian cancer susceptibility gene BRCA1. Science 1994;266:66–71.
18. Wooster R, Bignell G, Lancaster J, et al. Identification of the breast cancer susceptibility gene BRCA2. Nature 1995;378:789–792.
19. Frank TS, Deffenbaugh AM, Reid JE, et al. Clinical characteristics of individuals with germline mutations in BRCA1 and BRCA2: analysis of 10,000 individuals. J Clin Oncol 2002;20:1480–1490.
20. Vallon-Christerson J, Cayanan C, Haraldsson K, et al. Functional analysis of BRCA1 C-terminal missense mutations identified in breast and ovarian cancer families. Hum Mol Genet 2001;10: 353–360.
21. Gorski B, Byrski T, Huzarski T, et al. Founder mutations in the BRCA1 gene in Polish families with breast-ovarian cancer. Am J Hum Genet 2000;66: 1963–1968.
22. Gad S, Caux-Moncoutier V, Pages-Berhouet S, et al. Significant contribution of large BRCA1 gene rearrangements in 120 French breast and ovarian cancer families. Oncogene 2002;21:6841–6847.
23. Petrij-Bosch A, Peelen T, van Vliet M, et al. BRCA1 genomic deletions are major founder mutations in Dutch breast cancer patients. Nat Genet 1997;17: 341–345.
24. Frank TS, Manley SA, Olopade OI, et al. Sequence analysis of BRCA1 and BRCA2: correlation of mutations with family history and ovarian cancer risk. J Clin Oncol 1998;16:2417–2425.
25. Simard J, Tonin P, Durocher F, et al. Common origins of BRCA1 mutations in Canadian breast and ovarian cancer families. Nat Genet 1994;8:392–398.
26. Diez O, Osorio A, Duran M, et al. Analysis of BRCA1 and BRCA2 genes in Spanish breast/ovarian cancer patients: a high proportion of mutations unique to Spain and evidence of founder effects. Hum Mutat 2003;22:301–312.
27. Khanna KK, Jackson SP. DNA double-strand breaks: signalling, repair and the cancer connection. Nat Genet 2001;27:247–254.
28. Venkitaraman AR. Cancer susceptibility and the functions of BRCA1 and BRCA2. Cell 2002;108: 171–182.
29. Wang Y, Cortez D, Yazsi P, et al. BASC, a supercomplex of BRCA1-associated proteins involved in the recognition and repair of aberrant DNA structures. Genes Dev 2000;14:927–939.
30. Tibbetts RS, Cortez D, Brumbaugh KM, et al. Functional interactions between BRCA1 and the checkpoint kinase ATR during genotoxic stress. Genes Dev 2000;14:2989–3002.
31. Cortez D, Wang Y, Qin J, et al. Requirement of ATM-dependant phosphorylation of Brca1 in the DNA damage response to double-strand breaks. Science 1999;286:1162–1166.
32. Zhang J, Willers H, Feng Z et al. Chk2 phosphorylation of BRCA1 regulates DNA double-strand break repair. Mol Cell Biol 2004;24:708–718.
33. Fan S, Wang J, Yuan R, et al. BRCA1 inhibition of estrogen receptor signaling in transfected cells. Science 1999;284:1354–1356.
34. Fan S, Ma YX, Wang C, et al. p300 modulates the BRCA1 inhibition of estrogen receptor activity. Cancer Res 2002;62:141–151.
35. Zheng L, Annab LA, Afshari CA, et al. BRCA1 mediates ligand-independent transcriptional repression of the estrogen receptor. Proc Natl Acad Sci USA 2001; 98:9587–9592.
36. Wagner T, Hirtenlehner K, Shen P, et al. Global sequence diversity of BRCA2: analysis of 71 breast cancer families and 95 control individuals of worldwide populations. Hum Mol Genet 1999;11:413–423.
37. Hahn SA, Greenhalf B, Ellis I, et al. BRCA2 germline mutations in familial pancreatic carcinoma. J Natl Cancer Inst 2003;95:180–181.
38. Naderi A, Couch FJ. BRCA2 and pancreatic cancer. Int J Gastroint Cancer 2002;31:99–106.
39. Edwards SM, Kote-Jarai Z, Meitz J, et al. Two percent of men with early-onset prostate cancer harbor germline mutations in the BRCA2-gene. Am J Hum Genet 2003;72:1–12.
40. Howlett NG, Taniguchi T, Olson S, et al. Biallelic inactivation of BRCA2 in Fanconi anemia. Science 2002;297:606–609.
41. Sharan SK, Morimatsu M, Albrecht U, et al. Embryonic lethality and radiation hypersensitivity mediated by Rad51 in mice lacking Brca2. Nature 1997; 386:804–810.
42. Taniguchi T, Garcia-Higuera I, Xu B, et al. Convergence of the Fanconi anemia and ataxia telangiectasia signaling pathways. Cell 2002;109:459–472.

43. Nakanishi K, Taniguchi T, Ranganathan V, et al. Inter-action of FANCD2 and NBS1 in the DNA damage repair. Nat Cell Biol 2002;4:913–920.
44. D'Andrea AD, Grompe M. The Fanconi anemia/BRCA pathway. Nat Rev Cancer 2003;3:23–34.
45. Antoniou A, Pharoah PD, McMullan G, et al. A comprehensive model for familial breast cancer incorporating BRCA1, BRCA2 and other genes. Br J Cancer 2002;86:76–83.
46. Kaufman DJ, Beaty TH, Struewing JP. Segregation analysis of 231 Ashkenazi Jewish families for evidence of additional breast cancer susceptibility genes. Cancer Epidemiol Biomarkers Prev 2003;12: 1045–1052.
47. Cantor SB, Bell D, Ganesan S, et al. BACH1, a novel helicase-like protein, interacts directly with BRCA1 and contributes to its DNA repair function. Cell 2001;105:140–160.
48. Cantor S, Drapkin R, Zhang F, et al. The BRCA1-associated protein BACH1 is a helicase targeted by clinically relevant inactivating mutations. Proc Natl Acad Sci USA 2004;101:2357–2362.
49. Pellegrini L, Yu DS, Lo T, et al. Insights into DNA recombination from the structure of a RAD51-BRCA2 complex. Nature 2002;420:287–293.
50. Kato M, Yano K, Matsuo F, et al. Identification of a RAD51 alteration in patients with bilateral breast cancer. J Hum Genet 2000;45:133–137.
51. Levy-Lahad E, Lahad A, Eisenberg S, et al. A single nucleotide polymorphism in the RAD51 gene modifies cancer risk in BRCA2 but not BRCA1 carriers. Proc Natl Acad Sci USA 2001;98:3232–3236.
52. Malkin D, Li FP, Strong LC, et al. Germline p53 mutations in a familial syndrome of breast cancer, sarcomas, and other neoplasms. Science 1993;250: 1233–1238.
53. Hwang SJ, Lozano G, Amos CI, et al. Germline p53 mutations in a cohort with childhood sarcoma: sex differences in cancer risk. Am J Hum Genet 2003; 72:975–983.
54. Gasco M, Yulug IG, Crook T. TP53 mutations in familial breast cancer: functional aspects. Hum Mutat 2003;21:301–306.
55. Liaw D, Marsh DJ, Li J, et al. Germline mutations of the PTEN gene in Cowden disease, an inherited breast and thyroid cancer syndrome. Nat Genet 1997;16:64–67.
56. Li J, Yen C, Podsypanina K, et al. PTEN, a putative tyrosine phosphatase gene mutated in human brain, breast and prostate cancer. Science 1997;275:1876–1878.
57. Wen S, Stolarov J, Myers MP, et al. PTEN controls tumor-induced angiogenesis. Proc Natl Acad Sci USA 2001;98:4622–4627.
58. Jenne DE, Reimann H, Nezu J, et al. Peutz-Jeghers syndrome is caused by mutations in a novel serine-threonine kinase. Nat Genet 1998;18:38–43.
59. Lim W, Hearle N, Shah B, et al. Further observations on LKB1/STK11 status and cancer risk in Peutz-Jeghers syndrome. Br J Cancer 2003;89: 308–313.
60. Shen Z, Wen XF, Lan F, et al. The tumor suppressor gene LKB1 is associated with poor prognosis in human breast carcinoma. Clin Cancer Res 2002;8: 2085–2090.
61. De Jong MM, Nolte IM, te Meermann GJ, et al. Genes other than BRCA1 and BRCA2 involved in breast cancer susceptibility. J Med Genet 2002;39: 225–242.
62. Teraoka SN, Maolne KE, Doody DR, et al. Increased frequency of *ATM* mutations in breast carcinoma patients with early onset disease and positive family history. Cancer 2001;93:479–487.
63. Thorstenson YR, Roxas A, Kroiss R, et al. Contributions of ATM mutations to familial breast and ovarian cancer. Cancer Res 2003;63:3325–3333.
64. Bartek J, Falck J, Lukas J. CHK2 kinase—a busy messenger. Nat Rev Mol Cell Biol 2001;2:877–886.
65. Bell DW, Varley JM, Szydlo TE, et al. Heterozygous germ line hCHK2 mutations in Li-Fraumeni syndrome. Science 1999;286:2528–2531.
66. Meijers-Hejboer H, van den Ouweland A, Klijn J, et al. Low-penetrance susceptibility to breast cancer due to CHEK2 1100delC in noncarriers of BRCA1 or BRCA2 mutations. Nat Genet 2002;31:55–59.
67. Dufault MR, Betz B, Wappenschmidt B, et al. Limited relevance of the *CHEK2* gene in hereditary breast cancer. Int J Cancer 2004;110:320–325.
68. Lee HJ, Chang C. Recent advances in androgen receptor action. Cell Mol Life Sci 2003;60:1613–1622.
69. Wooster R, Mangion J, Eeles R, et al. A germline mutation in the androgen receptor gene in two brothers with breast cancer and Reifenstein syndrome. Nat Genet 1992;2:132–134.
70. Rebbeck T, Kantoff PW, Krithivas K, et al. Modification of BRCA1-associated breast cancer risk by the polymorphic androgen-receptor CAG repeat. Am J Hum Genet 1999;64:1371–1377.
71. Hogervorst FB, Cornelis RS, Bout M, et al. Rapid detection of BRCA1 mutations by the protein truncation test. Nat Genet 1995;10:208–212.
72. Schouten JP, McElgunn CJ, Waaijer R, et al. Relative quantification of 40 nucleic acid sequences by multiplex ligation-dependant probe amplification. Nucleic Acids Res 2002;30:e57.
73. Sauven P. Guidelines for the management of women at increased familial risk of breast cancer. Eur J Cancer 2004;40:653–665.
74. Mincey BA. Genetics and management of women at high risk of breast cancer. Oncologist 2003;8:466–473.
75. Schmutzler R, Schlegelberger B, Meindl A, et al. Counselling, genetic testing and prevention in women with hereditary breast and ovarian cancer. Interdisciplinary recommendations of the consortium "Hereditary Breast and Ovarian Cancer" of the German Cancer Aid. Zentralbl Gynakol 2003;125:494–506.

CHAPTER 5

Screening and Early Detection, Serum and Nipple Aspirates, Routine and Molecular Methods Including Reverse Transcriptase-Polymerase Chain Reaction, and Proteomics

Douglas P. Malinowski, PhD

Chief Scientific Officer Tripath Oncology, Durham, NC

Introduction to Breast Cancer Screening

Historically, breast cancer has been detected as a palpable mass either by patient self-examination or with a clinical breast examination conducted by a physician. The disadvantage of this approach is that a clinically detectable palpable mass usually identifies a late stage tumor. Detected at an advanced stage, breast tumors have a high likelihood of a more aggressive invasive phenotype and the potential of metastatic spread to distant organs. As such, the detection of late stage breast cancer is usually associated with a poor prognosis; therapeutic intervention is of limited efficacy and the overall survival rate of the patient is low.

The routine use of x-ray based mammography in the annual screening for women over ages 40–50 years results in the detection of breast carcinoma at an early stage, often before the tumor is of sufficient size to be detected by physical examination. Even though mammography is useful for the routine detection of tumors too small to be detected by physical examination, there are several drawbacks: (1) the presence of dense breast tissue precludes the detection of tumors; (2) lobular lesions (lobular carcinoma in situ and infiltrating lobular carcinoma) are often undetectable by mammography; (3) radiopaque lesions, calcifications, and benign fibroid cysts can often be mistaken for cancerous lesions; and (4) mammography fails to detect breast cancer in 10–15% of patients. To confirm the diagnosis of a suspicious mammography result, a secondary detection method needs to be used to confirm the diagnosis. The use of diagnostic mammography, ultrasonography, and magnetic resonance imaging are all techniques that are used to provide a more definitive identification of carcinoma. A more detailed discussion of mammography and molecular imaging techniques can be found in Chapter 29 by Dr. Varga and colleagues.

The inherent limitations in both sensitivity and specificity for the accurate detection of breast cancer by mammography has been the impetus to identify more accurate, specific, and cost-effective methods to identify breast cancer before the development of overt clinical symptoms. Numerous approaches have been undertaken to identify proteins or nucleic acids either present in the peripheral blood or in aspiration specimens to aid in the timely screening and diagnosis of breast cancer. In this chapter we review the current state of laboratory medicine for the screening of breast cancer using biochemical and molecular diagnostic methods.

Clinical Use of Protein Biomarkers in Disease Detection

Serum Screening for Protein Biomarkers

Breast cancer patients often display elevated levels of proteins in the plasma or serum in response to the presence of disease. These proteins have been shown to display varying degrees of correlation with the presence of primary (operable) breast cancer, recurrent breast disease, prognosis of disease outcome, metastasis, and response to therapy. In this section we review the diagnostic utility of serum protein markers and their clinical relevance to disease management. Although elevated serum levels of certain proteins are associated with primary breast cancer, these are of limited clinical utility in the initial detection of breast cancer due to either limited clinical sensitivity or specificity. As such, there are no biomarkers that are generally recommended for the clinical screening and primary diagnosis and/or detection of breast cancer independent of clinical physical examination or mammography. The guidelines published by the American Society of Clinical Oncology (ASCO) do not recommend the use of any protein biomarkers for the primary screening of breast cancer.[1] However, some of these markers have received a recommendation for the aid in monitoring to detect disease recurrence after therapeutic treatment in advanced breast cancer.[1] Furthermore, the proteins CA15.3 and CA27.29 have received clearance by the US Food and Drug Administration as in vitro diagnostic tests to aid in the monitoring for disease recurrence after breast cancer therapy. Most proteins investigated have shown a limited sensitivity and specificity for the detection of primary breast carcinoma but a good correlation with poor prognosis, disease recurrence, metastasis, and therapeutic effectiveness. These are discussed in more detail below.

CA15.3 CA15.3 (carbohydrate antigen 15.3) is a breast cancer associated antigen that was originally identified through monoclonal antibody reactivity against antigens present in breast carcinoma and human milk-fat globulin membrane fractions. CA15.3 is a member of the mucin protein family, a group of highly glycosylated proteins found in a variety of tissues. Mucin proteins generally function in cell protection, lubrication, and cell adhesion through extracellular matrix interactions.[2] The CA15.3 protein is encoded by the *MUC-1* gene located on chromosome 1q21 and is expressed in a wide variety of different cell types and tissues. The expression of *MUC-1* in breast epithelial cells results in the production of a large protein with a molecular mass of 250,000 to 500,000 Da and contains up to 50% carbohydrate by mass.[2] The protein contains three domains: a large extracellular domain, a transmembrane domain of 31 amino acids, and a small cytoplasmic domain of 69 amino acids. The extracellular domain consists of a core region containing a 20 amino acid repeat. The region is repeated 25–125 times, resulting in a large and variable extracellular domain: $[GVTSAPDTRPAPGSTAPPAH]_n$, where $n = 25–125$.

In normal breast epithelium, the core extracellular domain is extensively glycosylated through serine-linked O-glycosylation.[3] However, in breast tumors, the mucin is underglycosylated, resulting in the exposure of a highly immunogenic core peptide (PDTRP) within the extracellular repeat region. In addition, the repeat peptide sequence of the extracellular domain displays variant sequences creating a wide variety of protein polymorphisms or isoforms of MUC-1 in both normal and cancer breast tissue.[4] The peptide epitope detected by a monoclonal antibody that defines the CA15.3 antigen has been determined to be Thr-Arg-Pro-Ala-Pro-Gly-Ser (TRPAPGS) within the extracellular core repeat region.[5] The MUC-1 protein has been shown to interfere with the function of E-cadherin in cell–cell adherence. As such, overexpression of MUC-1 in carcinoma may help contribute to the metastatic process through this E-cadherin inhibition.[6]

The clinical utility of CA15.3 for breast cancer screening, prognosis, and detection of disease recurrence has been the subject of numerous clinical investigations. CA15.3 is normally detected in the serum of healthy women and in benign breast disease. Elevated serum levels are also associated with breast cancer. Sensitivities for the detection of primary (preoperative) breast cancer range between 13% and 33%.[7,8] As a prognostic marker, high levels of CA15.3 in the serum of postoperative patients have been correlated with poor prognosis and a significantly shorter time to disease progression.[9,10] Furthermore, elevated preoperative levels are associated with early disease recurrence, nodal involvement, and poor prognosis.[11,12] High serum levels of CA15.3 have been shown to correlate in 67% of patients with disease recurrence and in 70–80% of patients with metastatic disease.[8,13] However, 33% of patients with recurrent breast cancer still displayed normal levels of serum CA15.3. CA15.3 is usually elevated in late stage breast cancer (stages 2 or 3) but is not elevated in early breast cancer. The main drawback to the clinical use of CA15.3 is the lack of specificity for breast cancer. CA15.3 is often elevated in cancers of the ovary, lung, and prostate. In addition, it can be elevated in noncancerous conditions such as benign breast and ovarian disease, endometriosis, pelvic inflammatory disease, and hepatitis.

The routine use of CA15.3 for the primary detection of breast cancer is not recommended by the ASCO guidelines because of the non–organ-specific nature of the protein. However, its use in the detection of poor prognosis and for monitoring of disease recurrence is discussed below.

CA27.29 Similar to CA15.3, this biomarker is a glycosylated variant of Mucin 1. As discussed previously, CA27.29 is defined by a specific antibody reactivity that distinguishes it from other mucin proteins. Clinically, it is used to monitor for recurrence in stage 2 and 3 breast cancer. It displays the same specificity issues described for CA15.3 above.[2] The clinical performance of both CA15.3 and CA27.29 in relation to the detection of primary disease, recurrent disease, and metastatic disease is discussed below.

Carcinoembryonic Antigen The carcinoembryonic antigen (CEA) is an 180,000-Da glycoprotein known to be present in the serum of cancer patients, including breast cancer patients. It is a member of the family of cell adhesion molecules within the immunoglobulin superfamily.[14,16] The CEA gene family is located on chromosome 19q13.1 and consists of approximately 28 genes/pseudogenes. The gene encoding CEA was the first gene cloned from this family and the nomenclature is CEACAM5.[15,16] Within the gene family, alternatively spliced transcripts encode a variety of protein isoforms and subsequent posttranslational glycosylation results in a wide variety of CEA antigen proteins. This protein is anchored onto the cell membrane through a glycophosphatidylinositol lipid moiety. A generalized domain structure for CEA is shown in FIGURE 5.1.

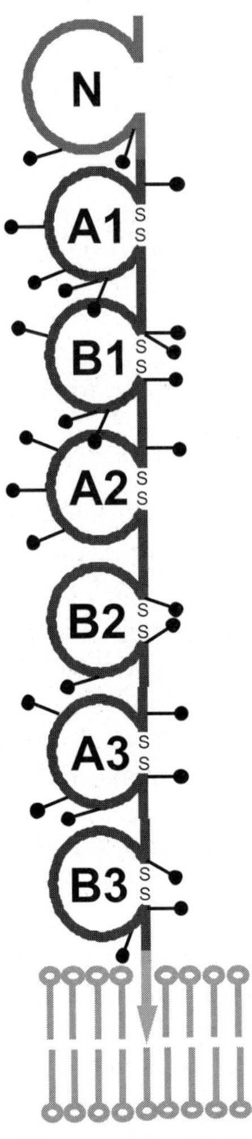

FIGURE 5.1 The domain structure for CEA. The protein contains five immunoglobulin-like domains. The N-terminal domain is an IgV-like domain. The four internal domains are homologous to IgC domains with A and B subtypes. The protein is highly glycosylated through N-linked carbohydrates and is anchored in the outer cell membrane through a GPI moiety. (Courtesy Dr. Wolfgang Zimmermann, Tumor Immunology Laboratory, Department of Urology, University Clinic Grosshadern, Ludwig Maximilians University, Munich, Germany.)

Comparison of CEA, CA15.3, and CA27.29 in Clinical Studies Comparative clinical investigations on the utility of CEA and CA15.3 for the detection and prognosis of breast cancer have been reported. Generally, CEA and CA15.3 demonstrate sensitivities of 13–18% for the detection of primary breast cancer detection.[7,17,18]

Using highly sensitive commercially available diagnostic assays, sensitivities can reach 60%. Specificity for both tumor markers has been reported to be 80–88% in primary disease detection. The serum levels for CEA and CA15.3 have also been studied in the context of disease recurrence. Postoperative levels of CEA that decrease have been shown as strong prognostic factors for disease-free survival.[11] Postsurgical serum levels CEA and CA15.3 are often elevated in patients with disease recurrence and poor disease-free survival; with a sensitivity of 41% for CEA and a sensitivity of 80% for CA 15.5 reported for the detection of metastatic disease.[7,9,11,17,18]

Clinical studies have shown that the incidence of serum elevation for CA15.3 increases with stage of the tumor: 5–30% in stage I, 15–50% in stage II, 60–70% in stage III, and 65–90% in stage IV.[19–21] In these studies, CA27.29 appeared to be more sensitive than CA15.3. Both antigens have been shown to detect recurrence after primary therapy. CA27.29 has been shown to predict disease recurrence with a sensitivity of 57%, a specificity of 98%, and a lead time of 5.3 months from first marker elevation.[22] Based on these data and clinical trials submissions, both CA27.29 and CA15.3 tests have been approved by the US Food and Drug Administration: CA272.9 was approved as an aid in the detection of recurrent breast cancer in women with stages I or II disease. Likewise, CA15.3 was approved as an aid in the management of stage II and II breast cancer. Neither antigen test was approved as a "stand alone" assay and must be used in conjunction with other clinical practices in the monitoring of patients for therapy effectiveness and disease relapse. Although both proteins have been shown to correlate with poor prognosis and disease recurrence, neither CEA nor CA15.3 is recommended for routine clinical screening or monitoring for disease recurrence by the ASCO guidelines.[1] However, in the absence of detectable disease or elevated CA15.3 or CA27.29, an increased level of CEA may be used to identify treatment failure.[1]

α-Fetoprotein α-Fetoprotein (AFP) is the major serum protein in the developing fetus and neonates and represents the fetal equivalent of serum albumin. Late in embryo development, the gene expression for AFP is terminated and the expression of the adjacent gene for serum albumin is initiated. Historically, elevated serum levels of AFP have been associated with certain cancers such as hepatic carcinoma. Elevated serum levels of AFP have been associated with breast cancer, although the sensitivity and specificity do not justify their routine clinical use. Sensitivities for the detection of breast cancer are 22–26%. Specificity for disease was shown to be 80–88%.[18,23] Elevated levels of AFP during pregnancy have been shown to correlate with a decreased risk of breast cancer.[24,25] However, the use of AFP is not used routinely in breast cancer screening because this protein is elevated in a number of other disease states, most notably hepatic carcinoma.

Detection on New Biomarkers in Serum

The identification of circulating proteins (or other biomolecules) in the serum of cancer patients has been the subject of intense investigations in an attempt to find more sensitive and specific tumor antigens than CA15.3 or CA29.27. The use of cancer-specific assays to routinely screen for cancer before the development of overt clinical symptoms is the goal of these research investigations. The identification and subsequent development of such biomarkers would prove useful as an adjunctive technique to mammography in the detection of cancer in asymptomatic women.

Three general approaches have been used to investigate such circulating biomarkers of breast cancer: (1) the detection of breast-specific proteins in the serum of breast cancer patients using enzyme-linked immunosorbent assays, (2) the detection of autoantibodies to circulating breast cancer-specific proteins, or the detection of cancer-associated nucleic acids (either DNA or messenger RNA [mRNA]) present in the circulating cancerous breast epithelial cells in the peripheral blood of cancer patients using highly sensitive polymerase chain reaction–based methods.

HER-2/neu (c-*erb*B2) Protein in Serum Several proteins have been studied in the context of breast cancer-specific expression (or overexpression) and the concomitant presence of these proteins in the blood or serum of the same patients. One such protein that has received considerable attention is the truncated version of HER-2/neu (c-*erb*B2). HER-2/neu (HER-2) is a membrane bound receptor tyrosine kinase with significant sequence homology to epidermal growth factor receptor (EGFR) and plays a significant role in epidermal growth factor signaling in breast tissues. HER-2 participates in epidermal growth factor signal transduction through dimerization with EGFR and other members of the EGFR family. The truncated form of HER-2 represents the extracellular domain of the HER-2 receptor.[26] This extracellular domain is often present in the serum of cancer patients. Clinical utility studies have shown that detection of extracellular domain of HER-2 in the serum is associated with late stage breast cancer; overall the sensitivity is less than 30–38% for the detection of primary breast cancer.[27,28] The presence of shed HER-2 protein in serum has been shown to be associated with tumors expressing high tissue levels of the membrane-bound her2/neu receptor within the carcinoma cells. Generally, serum levels of her2/neu have been measured between 2 and 15 ng/ml in normal and benign breast lesions. Patients with breast cancer have elevated levels between 15 and 75 ng/ml.[29] In addition to the primary detection of cancer in 20–50% of cancer patients, serum HER-2 has been reported to indicate recurrent breast cancer with 50–62% of metastatic disease patients displaying elevated serum levels of HER-2.[28,30] Similar to the observations cited above for CA15.3, serum levels for HER-2 are associated with later stage disease.

Literature reports have shown the following sensitivities as a function of disease stage: 1.7% in stage I, 3% in stage II, 16.7% in stage III, and 36–40% in stage IV.[31] In addition, the presence of HER-2/neu is often associated with disease recurrence or occult/micrometastatic disease before the detection of overt disease symptoms. Elevated levels of HER-2 are often present in 32–50% of metastatic disease patients. These patients usually display a poor response to hormone therapy (estrogen receptor [ER]-positive tumors), a poor response to chemotherapy, and a shorter disease survival.[32,33]

Finally, investigations have shown that patients with elevated levels of her2/neu also develop an immune response to the ectodomain of HER-2 with circulating autoantibodies to the receptor domain. In one study, rising levels of HER-2 in patients with previous HER-2 positive tumors and free of detectable disease are strongly indicative of subsequent clinically detectable disease recurrence.[34] Although preoperative levels of HER-2 are prognostic, they are associated with her2/neu-positive levels in tumors as well as tumor burden; as such elevated serum levels of HER-2 is not an independent prognostic factor.[35] In hormone receptor-positive metastatic breast cancer, the combination of increased serum HER-2 and increased serum CA15.3 was more predictive of poor prognosis and disease outcome than the use of CA15.3 alone.[37] Elevated serum levels of HER-2 have been reported to be associated with the prediction of therapeutic response. These published data have been summarized in a series of reviews on HER-2 and its prognostic and predictive role in managing therapy.[38–40]

Urokinase Plasminogen Activator Receptor Urokinase plasminogen activator receptor (uPAR) is a membrane bound receptor that binds to urokinase plasminogen activator and participates in the activation of the zymogen plasminogen to the active serine protease plasmin. Plasmin participates in a variety of proteolytic processes, including dissolution of the fibrin clot in blood coagulation and the proteolytic degradation of extracellular matrix proteins associated with tissue remodeling. In cancerous cells, the overexpression of these proteins contributes to the invasive behavior of cancer cells. Recent studies have also shown that urokinase plasminogen activator and its subsequent binding to the receptor (uPAR) activates the Ras-ERK and Rho-kinase pathways, leading to cell motility.[41,42] Serum levels of soluble uPAR are detected in normal patients (1.3–6.4 ng/ml) and show an age-related increase in the serum levels in healthy women. Elevated serum levels are detectable in breast cancer patients (1.6–9.2 ng/ml). When the serum levels are adjusted for this age dependent increase in concentration, there is a significant association between increased serum levels of uPAR and disease association. Statistical correlation has been reported for elevated serum levels of uPAR and relapse-free survival as well as overall survival. No correlation is noted with lymph node status or cellular levels of uPAR. Thus, preoperative levels of age-adjusted serum levels of uPAR are associated with poor outcome, independent of lymph node status, tumor size, or ER status.[43]

Table 5.1 summarizes the published observations on the clinical utility of CA15.3, CA27.29, CEA, AFP, soluble HER-2, and uPAR for the diagnosis of breast cancer and prognosis for disease recurrence and metastatic disease.

Additional Serum Proteins **Table 5.2** summarizes a series of serum proteins that have been associated with breast cancer recurrence and the presence of metastatic disease.[44–82] In addition to CEA, CA15.3, and her2/neu,

Table 5.1 Serum Proteins Elevated in Breast Cancer: Most Common Proteins Investigated

Protein	Description	Disease Association	Reference
CEA	Cell adhesion molecule of the Ig superfamily (CEACAM5)	Elevated in breast cancer Serum levels increase with stage of cancer Associated with metastasis and disease recurrence Not specific to breast cancer	1, 36
CA15.3 and CA27.29	Variants of epithelial mucin I (MUC-1)	Elevated in breast cancer Serum levels increase with stage of cancer Associated with metastasis and disease recurrence Not specific to breast cancer	1, 7–13, 17–22
Her2/neu	Extracellular domain of the membrane receptor tyrosine kinase her2/neu (c-*erb*B2)	Elevated in breast cancer Serum levels increase with stage of cancer Associated with metastasis and disease recurrence Associated with her2-positive tumors	27–37
AFP	α-Fetoprotein	Fetal equivalent to serum albumin Expressed in neural tube defects in neonates, hepatic dysfunction, and some cancers	18, 23–25
uPAR	Urokinase plasminogen activator receptor	Present in the serum of healthy women and in benign breast lesions Elevated in breast cancer Circulating levels of uPAR complexed with uPA shown to be lower in breast cancer plasma or serum in comparison with normal patients	43

additional proteins observed in the serum samples of breast cancer patients include several kallikrein proteases, soluble forms of cell adhesion molecules, fragments of cytokeratins, collagen, and additional miscellaneous proteins. The example of angiogenic factors as early markers of breast cancer demonstrate some of the confounding issues associated with the analysis of serum proteins for early disease screening. The angiogenic factors vascular endothelial growth factor (VEGF) and basic fibroblast growth factor (bFGF) are known growth factors responsible for the neovascularization that occurs in early tumor development. Both VEGF and bFGF are often elevated in tumors during angiogenesis. As such, they are obvious candidates for the detection of early stage cancer.

Preliminary reports noted that VEGF was present in the serum of breast cancer patients (**Table 5.2**).[50–52] However, a recently published study, including a case controlled patient study with standardized blood collection and subsequent controlled processing for serum isolation, yielded contradictory conclusions and reported no significant difference in the serum levels of VEGF between breast cancer patients and normal patients.[53] This study pointed out that angiogenic factors including VEGF can be elevated from macrophages that are activated during serum processing. As such, nonstandardized blood collection and specimen processing was attributed to the apparent elevation in these factors independent of disease. This demonstrates the need to carefully control and standardize specimen processing as a critical parameter in the evaluation of these, and presumably other, proteins for the analysis of serum for breast cancer screening and early detection of disease.

In contrast, the same study reported that bFGF was shown to be elevated in the serum of breast cancer patients in this case controlled study even though there was an overlap in serum concentrations between normal and breast cancer patients. At a serum concentration of 1 pg/ml for the cutoff between normal and disease sera, bFGF represents a trace protein analyte requiring highly sensitive and reproducible immunoassays for clinical detection.[53] A summary of the literature reports on the various serum proteins reported to be associated with breast cancer are shown in **Table 5.2**. None of these proteins has been investigated as thoroughly as CEA, CA15.3, and HER-2. As such, they are still the subject of clinical investigations and do not represent validated and acceptable diagnostic aids in the detection and clinical management of breast cancer.

Autoantibodies to Breast Cancer Proteins

Several of the proteins described previously in this chapter are not normal constituents of the

Table 5.2 Miscellaneous Serum Proteins Elevated in Breast Cancer and Associated with Recurrence and Metastatic Disease: Summary of Investigations

Protein Category	Protein Description	Disease Association	Reference
Growth factors	Transforming growth factor-β_1	Associated with recurrent breast cancer	45
	Hepatocyte growth factor	Associated with recurrent breast cancer	46
	Insulin-like binding protein	Increased serum levels of IGFBP1–3 associated with increased risk of breast cancer	47
		Increased levels of IGFBP-2 associated with reduced risk of breast cancer	
	FAS	Prognosis of invasive breast cancer	48, 49
	VEGF	Elevated in metastatic disease	50–52
		Associated with IDC and DCIS in ER-positive tumors	
	Activin A	Secreted in breast cancer; not associated with lymph node metastasis or tumor grade	54
	bFGF	Elevated in breast cancer relative to normal sera	53
Soluble cell adhesion molecules	CD44v	Associated with metastatic disease	55
	E-selectin	Elevated in advanced disease	56, 57
	ICAM-1		
	VCAM-1		
Cytokines	IL-6	Elevated in metastatic disease; none specific to breast cancer	58, 59
	IL-18		60, 61
	Macrophage inhibitory cytokine		62
Proteases	KLK5		63
	KLK8		64
	Cathepsin D		65
	Matrix metalloproteinase-9	Overexpression associated with poor prognosis and in recurrent disease	66
Structural proteins	KI-6 Mucin-like glycoprotein	Sensitivity: 31% in primary tumor 16% stage I 29% stage II 73% recurrent tumor 50% local 89% distant	44
		Useful for surveillance of disease recurrence	
	Collagen		66, 68
	Cytokeratins (fragments of cytokeratin 21.1–CYFRA21.1)	CYFRA21.1 similar to CA15.3	69, 70
	Nuclear matrix proteins		71
Miscellaneous	Mammaglobin	Overexpressed in breast cancer	72
	Osteopontin		73
	Serum prolidase I		74
	Periostin	Marker of bone metastasis of breast cancer	75
	Tissue polypeptide antigen	Similar to CA15.3	76–78
	LDH		79
	Fibrin D dimer	Associated with proteolysis of disease progression; tumor load, metastatic sites and angiogenesis	80
	Chromogranin A		81
	Glutamyl transpeptidase		80
	Superoxide dismutase		80
	Riboflavin carrier protein	Serum levels associated with high tissue levels	82
		Presumably reflects cellular lysis and/or tissue disruption	

DCIS, ductal carcinoma in situ; IDC, invasive ductal carcinoma; LDH, lactate dehydrogenase.

Table 5.3 Proteins Detected with Autoantibodies in Breast Cancer

Protein	Clinical Association	Reference
P53	Autoantibodies to p52 associated with ER-, Ki67-positive breast tumors	83
	No correlation found with her2/neu, CEA, or CA15.3 levels	
	Autoantibodies may be useful to monitor progression of the disease	84
Nucleophosmin	Autoantibodies elevated in recurrent disease and may be useful in predicting recurrence	85
MUC-1	Autoantibodies to MUC-1 shown to correlate with lack of disease progression; may represent a protection against disease progression	86, 87
HER-2/neu	Elevated levels of autoantibodies correlated with serum levels of HER-2/neu, stage of disease, and metastatic disease	88

plasma. As such, these proteins are recognized by the immune system as foreign proteins with a corresponding generation of antibodies. These autoantibodies can be detected in the serum by normal immunoassay methods. Autoantibodies have been reported against the following proteins: p53, nucleophosmin, MUC-1, and HER-2 (**Table 5.3**).[83–88] However, the generation of autoantibodies is expected to be somewhat variable with individual patients with a corresponding variability in clinical sensitivity for disease detection. As such, the detection of autoantibodies is not routinely used in the serum screening for early stage breast cancer.

Screening for Protein Biomarkers in Breast Aspiration Specimens

Protein Analysis of Breast Aspiration Specimens

There are three techniques to sample breast cells for diagnosis independent of excisional biopsy: nipple aspiration fluids, fine-needle aspiration (FNA), and ductal lavage (see Chapter 8 for a discussion of these techniques and the diagnostic issues associated with these different sampling techniques). The application of protein detection chemistry to breast aspiration and nipple specimens has received little attention. This is primarily a reflection of the limited amount of specimen available and the corresponding low levels of proteins associated with these specimens. However, the application of highly sensitive assays based on fluorescence analysis or mass spectrometry has permitted the research investigation of proteins associated in nipple aspirates in cancer patients. For example, the proteases kallikrein 2 and prostate-specific antigen (kallikrein 3, a protein marker routinely detected and associated with prostate cancer) have been shown to be present in normal breast nipple aspirates.[89,90] This observation suggests that these (and perhaps other) proteases may play a significant role in tissue proliferation within the context of hormonal stimulation of mammary glands during the estrogen cycle. This observation has not been expanded to demonstrate a cancer-specific change in these proteins in relation to cancer development.

Recently, the use of microtechnology for enzyme-linked immunosorbent assays has been applied to breast nipple discharges with success. These investigations have shown that many cancer-associated proteases can be detected in breast nipple discharges and that these proteins constitute a disease profile in breast cancer. Common proteins previously associated with breast cancer have been detected in nipple aspirates, including osteopontin, cathepsin D, and soluble her2/neu.[91,92] **Table 5.4** summarizes the protein biomarkers identified in the nipple aspirates of breast cancer patients.[93–100] To date, these interesting experimental observations have not been systematically validated to support their application for routine clinical diagnosis or screening.

Molecular Methods for Screening and Detection of Breast Cancer

Detection of Nucleic Acids in Serum by PCR

PCR has been successfully applied to the detection of a variety of breast epithelial cell-specific nucleic acids in the peripheral blood and plasma from breast cancer patients. Examples of detectable nucleic acids include mitochondrial DNA; mRNA of epithelial-specific genes such as cytokeratins, mammaglobin, and maspin; and a variety of cancer-associated proteins such as CEA. In particular, both mammaglobin and maspin have been reported to be highly specific for the detection of breast cancer based on PCR amplification of peripheral blood

Table 5.4 Summary of Proteins Identified in Breast Cancer Nipple Aspirates

Protein	Clinical Association	Reference
HER-2/neu c-*erb*B2	Overexpressed in nipple aspirate fluid (NAF). Correlation with her2-positive tumors.	92
bFGF	Increased level in NAF for breast cancer patients. Sensitivity of 90%; Specificity 69%.	93
Insulin-like growth factor binding protein-3 (IGFBP-3)	Increased level of IGFBP-3 in NAF of breast cancer. Predictor of breast cancer.	94
Prostate-specific antigen (KLK3)	Decreased levels of prostate-specific antigen in the NAF of breast cancer relative to normal and benign breast disease.	89, 90, 95–97
CEA	Increased levels of CEA in NAF in breast cancer: Sensitivity 32% Specificity 90%	98
Urokinase plasminogen activator (uPA), urokinase plasminogen activator inhibitor (PAI-1), and uPAR	Both uPA and uPAR predictors of cancer presence in NAF.	99
	uPAR independent predictor of advanced disease stage.	99
	PAI-1 associated with breast disease.	100
Osteopontin, cathepsin D		91

specimens.[101,102] Mammaglobin is a member of the lipophilin family of proteolipids that function in the binding of steroid hormones, including androgens. Maspin is a serine protease inhibitor that functions in a variety of cellular processes, including the control of proteolysis, invasion, and metastasis.

Although highly sensitive as an analytical technique, the clinical application of PCR-based detection of nucleic acids in peripheral blood specimens as a primary screening methodology is hampered by a relatively low clinical sensitivity and specificity. These limitations can be ascribed to a number of physiologic parameters or variables: (1) circulating levels of nucleic acids are expected to be a function of tumor burden (i.e., the size of the tumor), (2) the degree of neovascularization and the tissue architecture disruption will influence both cell and cellular component release into the blood, and (3) the half-life of these biomolecules in the peripheral blood is dependent on their clearance through the immune and hepatic systems.

Table 5.5 summarizes the reported investigations on the use of PCR-based detection of breast cancer nucleic acids in the plasma or peripheral blood samples of breast cancer patients.[101–113] To date, these clinical investigations are still preliminary and have not been validated for either routine clinical screening of breast cancer or monitoring for disease recurrence or metastasis. An interesting approach has been undertaken to analyze a number of mRNA transcripts present in breast blood samples using complementary DNA (cDNA) microarray analysis. The basis of this technique, which is gene expression analysis by DNA microarrays, uses a panel of immobilized cDNAs on a microarray to detect mRNA species that are present in peripheral blood samples. Analysis of the peripheral blood samples from breast cancer patients and normal individuals through RNA extraction and subsequent hybridization to the cDNA microarray revealed higher levels of specific mRNA molecules in the peripheral blood sample from a breast cancer patient relative to a normal blood specimen. This approach has been successfully used to identify (and confirm) previous genes known to be overexpressed in breast cancer and present within the peripheral blood of breast cancer patients (FIG. 5.2).[103]

Detection of Nucleic Acids in Breast Aspiration Specimens

Molecular Diagnostics and Breast Aspiration Cytology The use of cytology-based diagnostics for the detection of breast cancer using aspiration specimens is reviewed in Chapter 8. The integration of molecular biomarkers holds significant potential for the improved clinical utility of FNA. For example, the use of HER-2 positive status can be used in conjunction with FNA cytology to distinguish between atypia and carcinoma. Cytology aspiration specimens can lead to indeterminate diagnosis due to compromised cellular morphology and quality. As such, this factor often limits the diagnostic utility of breast aspiration specimens for the detection and confirmation of breast cancer after an abnormal mammogram.

Table 5.5 Summary of Nucleic Acids Detected in Peripheral Blood of Breast Cancer Patients

Nucleic Acid	Clinical Indication	Reference
E-cadherin and p16 DNA methylation	E-cadherin and p16 methylation observed in breast cancer: E-cadherin: 25% tumor and 20% plasma p16: 11% tumor and 8% plasma p16 methylation associated with nodal metastasis	104
Mammaglobin mRNA	Plasma mammaglobin mRNA detected in 54% tumor plasma specimens As an adjunct to CEA and CA15.3, sensitivity and specificity reported to be 81% and 90% More specific plasma marker for breast cancer than either EGFR or cytokeratin-19 mRNA	101, 105
Cytokeratin 19	Detected in plasma: 3.7% normal 14.3% hematological malignancy 32% operable breast cancer 42% metastatic breast cancer Sensitive marker for the detection of operable and metastatic breast cancer	102, 111
CEA	Detected in plasma: 0% normal 3.5% hematological malignancy 10% breast cancer	102
Maspin	Detected in plasma: 0% normal 0% hematological malignancy 9.3% operable breast cancer 14% metastatic breast cancer Maspin mRNA correlated with tumor size in early stage breast cancer	102
Telomerase RNA (hTR) and telomerase reverse transcriptase protein (hTERT) mRNA	hTR was detected in 94% of tumors and 28% of plasma samples hTERT was detected in 94% of tumors and 25% of plasma samples	106, 112
Tumor DNA LOH	Plasma samples shown to display LOH in tumor DNA Microsatellite alterations detected in early stage breast cancer	107 113
Epithelial glycoprotein (EGP-2), CK-19	33% of breast cancer patients had detectable mRNA EGP-2 or CK-19 in peripheral blood	108
mRNA transcriptional panels	Early detection of disseminated breast cancer by gene expression array analysis of peripheral blood	103
MUC-1 mRNA	MUC-1 mRNA detected in peripheral blood specimens: 11% benign breast disease 24% operable breast cancer 45% advanced breast cancer	109
HER-2/neu	HER-2/neu mRNA detected in: 50% stage I breast cancer 83% stage II breast cancer	110

LOH, loss of heterozygosity.

To address these needs, a variety of techniques has been investigated to identify the malignant cells in an aspiration specimen independent of morphologic assessment. These include protein-based detection of specific biomarkers in breast epithelial cells using antibody-based immunohistochemistry (IHC), analysis of chromosomal alterations in breast epithelial cells, and the detection of discrete nucleic acids using PCR-based methods.

Application of Protein-Based Cytology Diagnosis in Breast Aspiration Specimens

IHC assays have been used to identify malignant cells within aspiration cytology specimens. A well-known example is HER-2. The

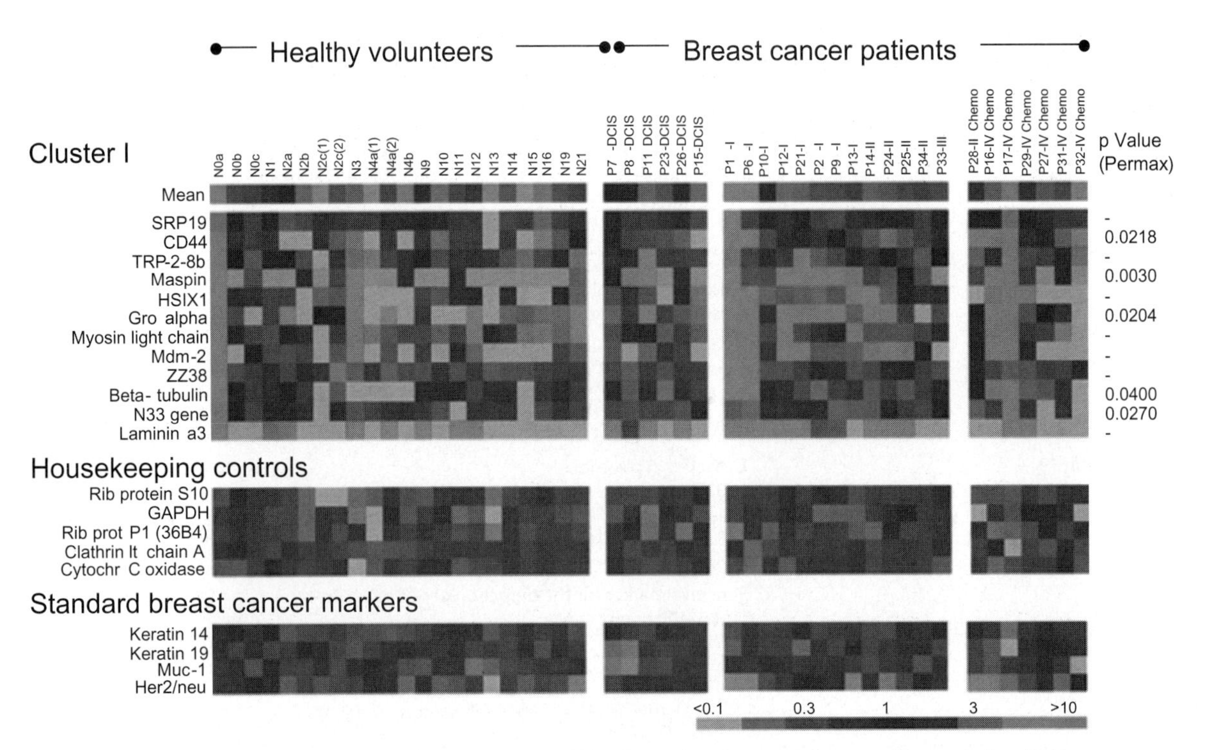

FIGURE 5.2 Microarray analysis of mRNA isolated from peripheral blood samples of 15 normal women (N samples) and 26 breast cancer patients (P samples) with varying degree of disease. Gene expression levels are color coded according to the key. The top panel represents mRNA expression levels between the various samples, including breast tumors of known stage and patients receiving chemotherapy. The middle panel represents standard housekeeping genes with minimal differential expression in breast cancer. The bottom panel represents standard breast cancer markers. (From Ref. 26, with permission.) **See Plate 6 for color image.**

gene encoding HER-2 is located on chromosome 17q12 and is amplified in approximately 20–30% of breast carcinomas. The protein encoded by the amplified *HER-2* gene is also overexpressed in these tumors and can be detected by antibody-based IHC assays.[39,40] The review of HER-2 in breast cancer and treatment options are discussed in Chapter 16. The detection of HER-2 by IHC methods has been successfully used in breast aspiration specimens. In addition to HER-2, specific antibodies are used to detect the presence of the ER and the progesterone receptor (PR) in breast tissue. The detection of ER- and PR-positive tumors (and cytology specimens) is used to identify those breast tumors that will likely benefit from treatment with estrogen hormone antagonists, such as tamoxifen. Collectively, ER/PR and HER-2 positive status are used to identify carcinoma specimens and are further used to guide therapy selection for these hormone responsive tumors.

IHC assays are also used to routinely assess the proliferative state of the malignant cells in breast cytology specimens. Such markers include the proliferation marker, Ki67; overexpression of the tumor suppressor gene *TP53*; and telomerase. An interesting marker (p63) related to myoepithelial cell lineage has been recently reported for the detection of benign cells within aspiration specimens. A summary of the markers used to detect breast carcinoma in aspiration specimens by IHC is shown in **Table 5.6.**[114–125]

Chromosomal Analysis and Comparative Genomic Hybridization in Breast Aspiration Samples The detection of numerical alterations and structural changes[125] in the diploid state of chromosomes in the breast epithelial cells is indicative of precancerous and cancerous malignancies of the breast. There are two commonly used techniques to detect chromosomal alterations associated with breast cancer. The first technique is comparative genomic hybridization (CGH), which compares the ratios of whole genomic hybridization between tumor and normal control DNA on intact chromosomes. This method is sensitive for the detection of chromosomal amplifications and detection events that occur within breast carcinoma. However, the technique requires freshly collected cells and the ability to isolate and prepare chromosomal preparations from

Table 5.6 Summary of Protein Based IHC Detection Biomarkers in Breast Cancer Aspiration Specimens

Protein	Clinical Association	Reference
p63	Detected in myoepithelial cells (MEC) in benign breast disease and in MEC overlaying malignant cell clusters in breast disease	114
HER-2/neu	Detected in FNAs with concordance to histology: 78% with her2/neu FISH 84% with her2/neu IHC	115, 116
Ki67	Ki67 staining correlated with tumor grade in both lymph node–negative and –positive tumors	117, 118
ER, PR	ER and PR overexpression correlated with Ki67 and p53; no correlation with HER-2/neu	119
	Concordance between FNA ICC and histology IHC for the following markers: ER, 92% PR, 76% p53, 75%	120
Bcl-2	Bcl-2 overexpression correlated with favorable prognostic factors: ER, PR, inversely with Ki67 and p53, nuclear grade.	121
nm23	Expression of nm23 protein shows inverse correlation with lymph node status, cytologic grading, and ploidy	122
Vimentin, intermediate filaments	Vimentin detected in 61% of FNA from ductal carcinoma; positively associated with tumor size, lymph node status, tumor grade, and ER-negative status	123
p53	Nuclear accumulation of p53 detected in 28% of carcinoma FNA specimens	124
Pan cytokeratin	Detection of breast epithelial cells within ductal lavage specimens	125

ICC, immunocytochemistry.

cells that are in metaphase, hence enabling the detection of discrete band hybridization events. CGH has been used to identify various chromosomal alterations in breast cancer, including frequent amplifications on chromosome 11q11–11q14.[126,127] However, the heterogeneous nature of chromosomal aneuploidy that occurs within breast cancer complicates the correlation between chromosomal alterations and definitive prognosis, prediction of clinical outcome, or patient management.

A variant of the CGH approach uses cDNAs that span known regions of various chromosomes that are immobilized on microarrays and used to detect specific chromosomal amplifications in breast cancers.[128,129] This approach has been used in breast biopsy specimens to correlate chromosomal alterations with changes in mRNA expression patterns of the encoded genes present on the altered chromosomes. However, it has not been applied to breast cytology specimens. Although a powerful research technique, the use of CGH for routine clinical application and screening has not been widely adopted in routine clinical screening for breast cancer.

Fluorescence In Situ Hybridization Analysis for Chromosomal Amplification in Breast Aspiration Specimens The use of specific chromosomal hybridization probes, labeled with a fluorescent dye, for the specific detection of chromosomal amplifications or deletions based on hybridization to intact chromosomes in a cell preparation has been used in numerous clinical research investigations. This technique, fluorescence in situ hybridization (FISH), permits the detection of both numerical alterations in specific chromosomes using centromere-specific probes and the detection of chromosomal amplifications or deletions using probes for specific chromosome regions. This technique does not require cells in metaphase to perform the analysis; the technique can be applied to nonmitotic interphase cells. Thus, hybridization events are detected as fluorescent spots in a nucleus. As such, the ability to enumerate the number of fluorescent "spots" per nucleus provides an immediate determination of the number of chromosomal regions capable of hybridization to the FISH probes (normally, two spots for each allele in a diploid cell). The use of these interphase FISH probes has been

Table 5.7 Chromosomal Alterations Identified in Breast Aspiration Specimens

Chromosomal Feature	Chromosomal Alteration	Methodology	Clinical Diagnostic	Reference
1q	Amplification of 1q+	FISH on FNA	Invasive breast cancer	126
c-MYC and HER2 oncogenes	Amplification of c-*myc* and her2			
Tumor suppressor gene *p53*	Deletion of *p53*			
Chromosome number	Numerical alterations: 8, 10, and 17			
Aneuploidy	Amplifications and deletions: 1q+, 8q+, 14q−, 16p+, 16q−, 17p−, 17q+, 19q+, 20q+, 21q−, and 22q−	CGH on FNA		127
Chromosome number	Numerical alterations: 1, 8, 11, and/or 17	FISH on ductal lavage		130
Chromosome number	Numerical alterations: 1, 11, and 17	Centromere FISH	Detection of cancer	131
Structural alterations and fusions	Chromosome 16:1 fusion with breakpoints distal to 16q11.2 and proximal to 1q12	FISH	Intracystic papillary breast tumors Detection of low grade cancer	132, 133
Chromosome number	Numerical alterations: 1, 11, and 17	FISH	Differential diagnosis of DCIS and IDC from benign lesions and intraductal papilloma	134
Chromosome number	Numerical alterations: 1, 11, and 17	FISH	Detection of carcinoma within indeterminate ductal lavage	135
Chromosome number	Numerical alterations: 1, 11, and 17	FISH	Detection of carcinoma in FNA specimens	136
Chromosome number	Numerical alterations: 1, 11, and 17	FISH	Grading of breast carcinoma in FNA	137
HER-2/neu	Gene amplification (17q12-q26)	FISH	Detection of Her2-positive cells in NAF	138

DCIS, ductal carcinoma in situ; IDC, invasive ductal carcinoma.

applied successfully to breast cytology specimens (**Table 5.7**).[126,127,130–138] Chromosomal amplifications most often observed in breast aspiration cytology specimens include 1q+, 11q+, and 17q+. Although these techniques are used in clinical investigations, they are of limited clinical utility. They can be used to discriminate suspicious cases of carcinoma but provide little information related to prognosis, risk of progression, or prediction of clinical outcome.

Detection of *HER-2* Amplification in Breast Aspiration Specimens by FISH

The most commonly monitored chromosomal alteration in breast tumors and suspected lesions in breast aspiration specimens is the detection of HER-2/*neu* amplification (HER-2+) on chromosome 17q12. The *HER-2* gene is often amplified in 20–30% of breast carcinomas through amplification of the chromosomal region of 17q12 encoding the gene. Detection of this specific chromosomal amplification is possible through the use of specific FISH probes that span the entire region encoding the c-*erb*B2 gene. FISH permits the specific detection of the chromosomal amplification directly in interphase nuclei of cells, thereby eliminating the need to prepare metaphase chromosomal spreads to detect this amplification. In normal diploid nuclei, there are two alleles for c-*erb*B2 and thus two detectable fluorescent signals per nuclei. Control centromeric probes permit verification that both pairs of chromosome 17 are present per nucleus. In breast cells displaying aneuploidy or amplification at 17q12, the FISH *HER-2* probe will hybridize to these multiple sequences and generate more than two signals per pair of centromeres (FIG. 5.3). This FISH assay is used routinely in the identification of breast cancers with amplified *HER-2*. Clinical performance of FISH in breast cytology specimens has been the subject of multiple

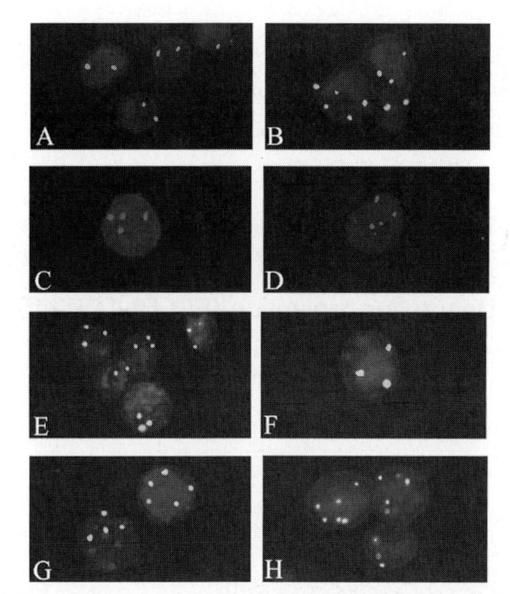

FIGURE 5.3 FISH analysis for chromosomal amplification in breast epithelial cells isolated by ductal lavage (A–D and F–H) and tumor touch prep (E). Chromosomal amplification was detected with centromeric probes for chromosomes 1 (red), 8 (green), 11 (aqua), and 17 (gold). Surgical pathology classification of paired samples: (A) fibroadenoma with florid ductal hyperplasia, (B) ductal carcinoma in situ, (C) invasive ductal carcinoma (IDC), (D) IDC, (E) ductal carcinoma in situ, (F) IDC, (G) IDC, (H) IDC. (From Ref. 118, with permission.) **See Plate 7 for color image.**

investigations. Generally, the concordance of FISH detection of *HER-2* gene amplification in cytology specimens and the corresponding histology sections shows a high degree of concordance (75–92%). In addition, there is an equal level of concordance between FISH and IHC detection of the HER-2 protein in breast cytology specimens. FISH detection of *HER-2* gene amplification has been used to resolve indeterminate results observed in cytology-based diagnosis of aspiration specimens. Generally, FISH demonstrates a sensitivity of 85% and a specificity of 95%. These data have been recently reviewed and summarized.[38]

The specific detection of 17q12 and *HER-2* amplification by FISH has not been routinely performed on breast aspiration samples, because the assay is generally performed on the surgical biopsy specimen from the tumor. However, recent reports have demonstrated that chromosomal aneusomy can be detected in ductal lavage specimens using FISH probes (FIG. 5.3).[125]

Application of PCR Detection of Nucleic Acids in Breast Aspiration Specimens
The use of PCR to detect specific mRNA associated with breast cancer has been successfully applied to aspiration samples. The advantage of PCR is that highly sensitive detection of carcinoma cells is possible, thereby overcoming the limitation in cell adequacy that often accompanies aspiration-based specimens. Examples of successful application of PCR to detect cancer cells present within breast aspirations specimens (FNAs and ductal lavage) are shown in **Table 5.8**.[139–146]

Proteomic Analysis for the Detection of Breast Cancer and Serum Screening

Application of Mass Spectrometry

Recently, mass spectrometry has been applied to the identification of serum proteins that reflect the presence of breast cancer in patients. The concept is based on the hypothesis that breast tumors, once vascularization has occurred, can release proteins into the blood where they can be detected. As mentioned

Table 5.8 Summary of Nucleic Acid Biomarkers Detected in Breast Cancer Aspiration Specimens

Nucleic Acid	Clinical Association	Reference
P53 DNA mutations	Mutation in p53 exons 5–8 detected in 22% breast carcinoma FNA specimens	139
	Mutations detected in 18–24% of FNA specimens	140, 141
Mitochondrial DNA	Mutations detected in D-loop and ND1, ND4, ND5 and cytochrome *b* genes	142
Her2/neu	HER-2 mRNA detected in tumors with lymph node metastasis	143
VEGF	VEGF mRNA expression correlated with degree of angiogenesis	143, 144
Telomerase	73% of breast carcinoma FNA were positive IDC more frequently positive than infiltrating lobular carcinoma	145
	No correlation was detected with tumor grade or lymph node metastasis	
Microsatellite DNA	DNA obtained from FNA was used to detect LOH or MSI by PCR	146

IDC, invasive ductal carcinoma; LOH, loss of heterozygosity; MSI, microsatellite instability.

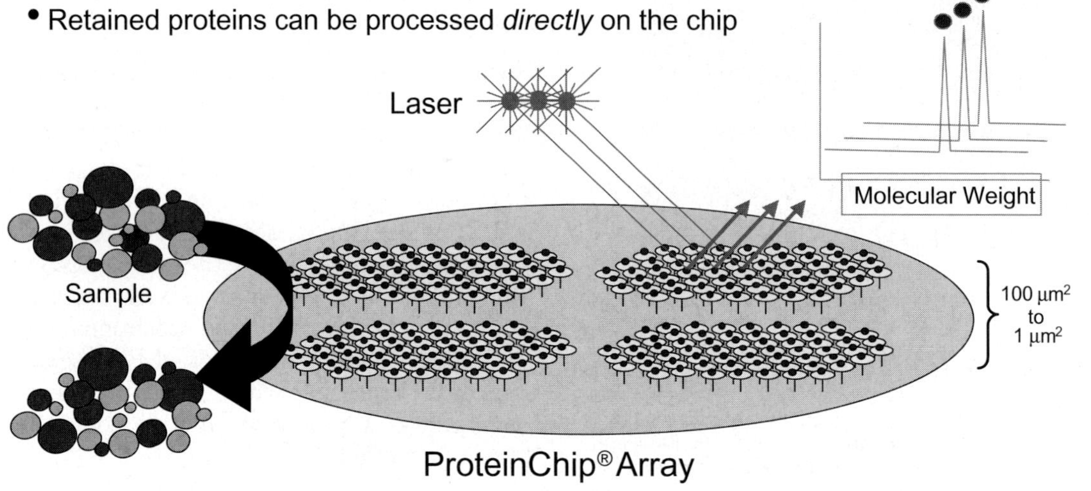

FIGURE 5.4 Diagram of protein enrichment using SELDI compatible Protein BioChip™ arrays. Surface chemistries used include anionic exchange, cationic exchange, hydropholic interactions, immobilized metal chelates, and specific protein immobilizations (i.e., antibodies, lectins, etc.). Enriched proteins are subjected to analysis by SELDI. The ionized protein fragments are detected by time-of-flight mass spectrometry (MS) (see Fig. 5.5). (Courtesy Ciphergen Biosystems, Inc.)

previously for circulating nucleic acids in the blood, the presence of tumor-associated proteins in the serum will be a function of the steady-state concentration of the protein in the serum, which in turn is a function of protein release into the blood system and the corresponding turnover (i.e., degradation) of the protein either through proteolysis, macrophage clearance, and/or hepatic clearance of the protein. Serum levels of these proteins will be low and variable depending on the factors of tumor burden, rate of protein release from the tumor, and subsequent clearance. As such, their detection requires sensitive analytic techniques. One such technology that has received considerable attention recently is mass spectrometry using surface-enhanced laser desorption and ionization (SELDI). The SELDI technique is based on the surface capture of proteins or peptides from a biologic sample using specific chemical interactions on a solid support or surface. These SELDI surfaces use specific chemistries to aid in protein interaction and binding. The most commonly used surface chemistries are ion exchange (anionic and cationic chemistries), metal chelates, hydrophobic interactions, hydrophilic interactions, and functionalized capture surfaces based on covalently immobilized ligands or antibody to provide analyte-specific capture to the surface. Once the protein has been enriched using one or more of the above-mentioned surfaces, the proteins are subjected to ionization using a laser, in which the proteins are ionized into a vacuum chamber where they are detected by a mass spectrometer (FIG. 5.4).

The specific detection of the ionized mass fragments of the proteins is based on time-of-flight (TOF) mass spectrometry. The proteins are detected and the mass of the ionized fragment is determined using the time it takes the ion fragment to reach the detector relative to standard calibrators that are used to calibrate the instrument (FIG. 5.5). The technique is capable of detecting specific proteins in the range of 3000 to 90,000 Da in molecular weight. If the

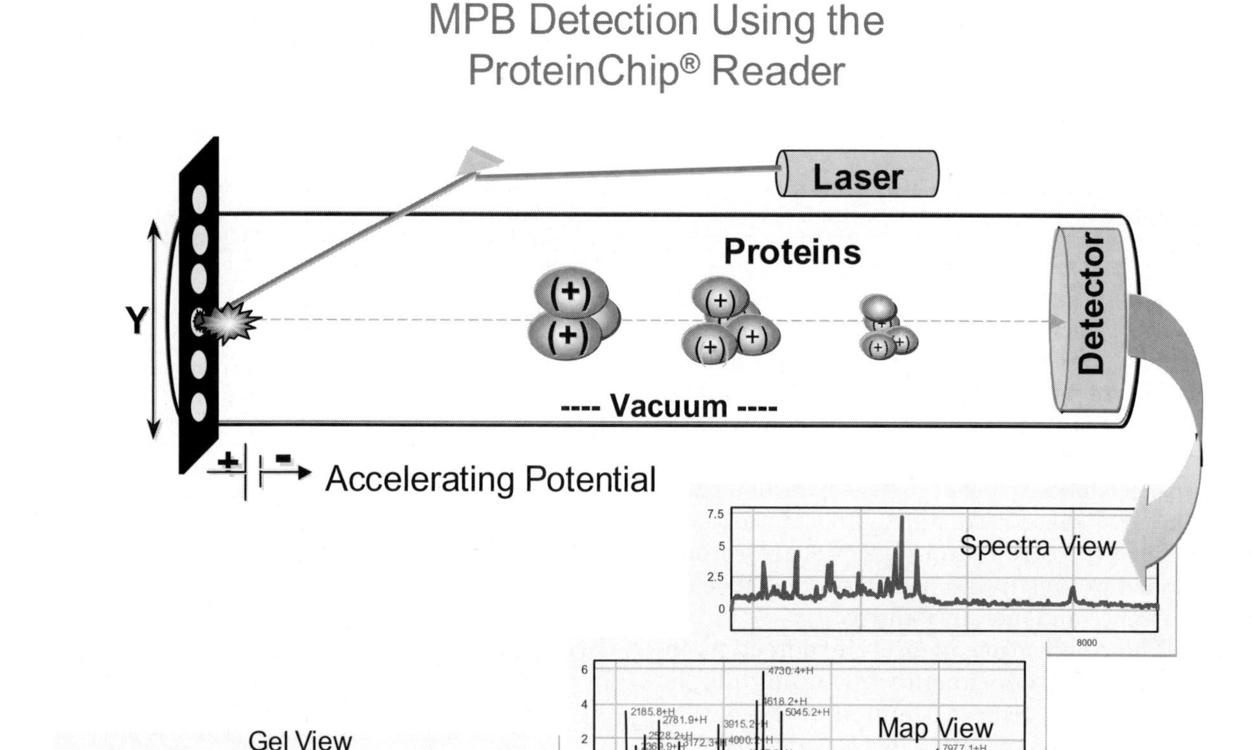

FIGURE 5.5 Diagram of SELDI-based time-of-flight mass spectrometry detection of proteins using the Protein BioChip™ array. The charged protein ions are accelerated through a potential gradient in a vacuum chamber, and the ion fragments are detected through an ion detector. The time of detection is proportional to the mass of the fragment. Calibration of the instrument with protein standards of known molecular mass permits the determination of the relative mass fragments of protein species isolated by the Protein BioChip™ by the mass spectrometer. Data can be visualized as a relative mass unit with a semiquantitative analysis for mass ion abundance. (Courtesy Ciphergen Biosystems, Inc.)

mixture is sufficiently simple, the ionization pattern in the SELDI-TOF analysis can often provide some degree of preliminary protein identification. However, most biologic applications of SELDI-TOF have been focused on complex specimens, such as cancer serum samples, where simple identification of unique proteins is not possible. Rather, these investigations have focused on the identification of protein patterns between normal and disease serum samples.

In the case of breast cancer, initial investigations have focused on the identification of signature profiles of breast cancer serum proteins relative to the serum profile identified in normal patients. The interpretation of these profiles is facilitated through the use of software algorithms to establish the relevant protein signals that constitute a diagnostic profile indicative of cancer. These approaches have been used to report diagnostic profiles indicative of breast cancer. Published studies have reported cross-validated profiles that discriminate between breast cancer and benign disease using serum samples with sensitivities of 80–93% and specificities of 79–91%.[148–149] The application of SELDI-TOF detection of proteomic profiles in breast nipple aspirate samples identified unique proteins of 6500 and 15,940 Da molecular weight that were present in 75–84% of the breast cancer specimens but only observed in 0–9% of normal patient specimens.[150] Furthermore, profiles have been described that are associated specifically with metastatic disease as well as specific for ductal carcinoma in situ in tissue samples, although these approaches have not been used to discriminate tumor types or prognosis within serum or nipple aspirate specimens.[151]

Using this type of proteomics approach, continued research can lead to the identity of the major protein species comprising the proteomic profile. In one instance, this has led to the identification of the protein RS/DJ-1 that has subsequently been verified to be a circulating antigen in 37% of sera from newly diagnosed breast cancer patients.[152]

Summary

Primary screening for the early detection of breast cancer based on the detection of serum proteins is not routinely used because of limited sensitivity and specificity. The most commonly used proteins are CEA, CA15.3, CA27.29, and soluble HER-2/neu. However, the serum presence of these proteins are well documented to be associated with poor prognosis, the detection of early disease recurrence, and metastasis. The use of CA15.3 and CA27.29 are recommended as aids in the monitoring for disease recurrence and therapy failure.

The application of protein detection in breast aspiration samples, including nipple aspirates, has received limited investigation because of low quantity of breast epithelial cells recovered from the aspiration specimens. The detection of circulating nucleic acids in patient peripheral blood through PCR-based methods has yielded interesting results that merit additional clinical validation.

The application of molecular techniques in aspiration specimens, including protein detection by IHC and nucleic acid detection, holds promise to improve the clinical performance of aspiration-based detection of breast cancer. The application of new proteomic approaches based on SELDI mass spectrometry holds promise for the identification of breast cancer based on unique protein signature profiles.

References

1. Bast RC, Ravdin P, Hayes DF, et al. 2000 Update of recommendations for the use of tumor markers in breast and colorectal cancer: clinical practice guidelines of the American Society of Clinical Oncology. J Clin Oncol 2001;19:1865–1878.
2. Duffy MJ. CA15-3 and related mucins as circulating markers in breast cancer. Ann Clin Biochem 1999; 36:579–586.
3. Gendler SJ, Spicer AJ. Epithelial mucin genes. Annu Rev Physiol 1995;57:607–634.
4. Engelmann K, Baldus SE, Hanisch FG. Identification and topology of variant sequences within individual repeat domains of the human epithelial tumor mucin MUC1. J Biol Chem 2001;276:27764–27769.
5. Price MR, Rye PD, Petrakou E, et al. Summary report on the ISOMB TD-4 workshop: analysis of 56 monoclonal antibodies against the MUC1 mucin. J Tumor Biol 19(suppl 1):1–20.
6. Wesseling J, van der Valk SW, Hilkens J. A mechanism for inhibition of E-cadherin-mediated cell-cell adhesion by the membrane-associated mucin episialin/ MUC1. Mol Biol Cell 1996;7:565–577.
7. Molina R, Filella X, Alicarte J, et al. Prospective evaluation of CEA and CA 15.3 in patients with locoregional breast cancer. Anticancer Res 2003;23: 1035–1041.
8. Guadagni F, Ferroni P, Carlini S, et al. A re-evaluation of carcinoembryonic antigen (CEA) as a serum marker for breast cancer: a prospective longitudinal study. Clin Cancer Res 2001;7:2357–2362.
9. Kurebayashi J, Yamamoto Y, Tanaka K, et al. Significance of serum carcinoembryonic antigen and CA 15-3 in monitoring advanced breast cancer patients treated with systemic therapy: a large-scale retrospective study. Breast Cancer 2003;10:38–44.
10. Kumpulainen EJ, Keskikuru RJ, Johansson RT. Serum tumor marker CA 15.3 and stage are the two most powerful predictors of survival in primary breast cancer. Breast Cancer Res Treat 2002;76:95–102.
11. Ebeling FG, Stieber P, Untch M, et al. Serum CEA and CA 15-3 as prognostic factors in primary breast cancer. Br J Cancer 2002;86:1217–1222.
12. Seker D, Kaya O, Adabag A, et al. Role of preoperative plasma CA 15-3 and carcinoembryonic antigen levels in determining histopathologic conventional prognostic factors for breast cancer. World J Surg 2003;27:519–521.
13. Tampellini M, Berruti A, Gorzegno G, et al. Independent factors predict supranormal CA 15-3 serum levels in advanced breast cancer patients at first disease relapse. Tumor Biol 2001;22:367–373.
14. Thompson JA, Grunert F, Zimmermann W. Carcinoembryonic antigen gene family: molecular biology and clinical perspectives. J Clin Lab Anal 1991;5: 344–366.
15. Beauchemin N, Benchimol S, Cournoyer D, et al. Isolation and characterization of full-length functional cDNA clones for human carcinoembryonic antigen. Mol Cell Biol 1987;7:3221–3230.
16. Beauchemin N, Draber P, Dveksler G, et al. Redefined nomenclature for members of the carcinoembryonic antigen family. Exp Cell Res 1999;252: 243–249.
17. Molina R, Filella X, Zanon G, et al. Prospective evaluation of tumor markers (c-erbB-2 oncoprotein, CEA and CA 15.3) in patients with locoregional breast cancer. Anticancer Res 2003;23:1043–1050.
18. Clinton SR, Beason KL, Bryant S, et al. A comparative study of four serological tumor markers for the detection of breast cancer. Biomed Sci Instrum 2003; 39:408–414.
19. Gion M, Mione R, Leon AE, et al. Comparison of the diagnostic accuracy of CA27.29 and CA15.3 in primary breast cancer. Clin Chem 1999;45:630–637.
20. Devine PL, Duroux MA, Quin RJ, et al. CA15-3, CASA, MSA, and TPS as diagnostic serum markers in breast cancer. Breast Cancer Res Treat 1995;34: 245–251.
21. Hayes DF, Bast RC, Desch CE, et al. Tumor marker utility grading system: a framework to evaluate clinical utility of tumor markers. J Natl Cancer Inst 1996;88:1456–1466.

22. Chan DW, Beveridge RA, Muss H, et al. Use of Truquant BR radioimmunoassay for early detection of breast cancer recurrence in patients with stage II and stage III disease. J Clin Oncol 1997;15:2322–2328.
23. Eskelinen M, Kataja V, Hamalainen E, et al. Serum tumour markers CEA, AFP, CA 15-3, TPS and Neu in diagnosis of breast cancer. Anticancer Res 1997;17:1231–1234.
24. Melbye M, Wohlfahrt J, Lei U, et al. Alpha-fetoprotein levels in maternal serum during pregnancy and maternal breast cancer incidence. J Natl Cancer Inst 2000;92:1001–1005.
25. Richardson BE, Hulka BS, Peck JL, et al. Levels of maternal serum alpha-fetoprotein (AFP) in pregnant women and subsequent breast cancer risk. Am J Epidemiol 1998;148:719–727.
26. Schulze G. HER-2/neu gene product in serum—an oncoprotein in the diagnosis and therapy of breast carcinoma. Anticancer Res 2003;23:1007–1010.
27. Carney WP, Neumann R, Lipton A, et al. Potential clinical utility of serum Her2/neu oncoprotein concentrations in patients with breast cancer. Clin Chem 2003;49:1579–1598.
28. Fehm T, Gebauer G, Jager W. Clinical utility of serial serum c-erbB-2 determinations in the follow-up of breast cancer patients. Breast Cancer Res Treat 2002;75:97–106.
29. Meenakshi A, Kumar RS, Kumar NS. ELISA for quantitation of serum C-erbB-2 oncoprotein in breast cancer patients. J Immunoassay Immunochem 2002; 23:293–305.
30. Dittadi R, Zancan M, Perasole A, et al. Evaluation of HER-2/neu in serum and tissue of primary and metastatic breast cancer patients using an automated enzyme immunoassay. Int J Biol Markers 2001;16:255–261.
31. Schwartz MK, Smith C, Schwartz DC, et al. Monitoring therapy by serum HER-2/neu. Int J Biol Markers 2000;15:324–329.
32. Classen S, Kopp R, Possinger K, et al. Clinical relevance of soluble c-erbB-2 for patients with metastatic breast cancer predicting the response to second-line hormone or chemotherapy. Tumor Biol 2002;23:70–75.
33. Lipton A, Ali SM, Leitzel K, et al. Elevated serum Her-2/neu level predicts decreased response to hormone therapy in metastatic breast cancer. J Clin Oncol 2002;20:1467–1472.
34. Molina R, Jo J, Zanon G, et al. Utility of C-erbB-2 in tissue and in serum in the early diagnosis of recurrence in breast cancer patients: comparison with carcinoembryonic antigen and CA 15.3. Br J Cancer 1996;74:1126–1131.
35. Molina R, Jo J, Filella X, et al. c-erbB-2 oncoprotein in the sera and tissue of patients with breast cancer. Utility in prognosis. Anticancer Res 1996;16:2295–2300.
36. Molina R, Zanon G, Filella X, et al. Use of serial carcinoembryonic antigen and CA 15.3 assays in detecting relapses in breast cancer patients. Breast Cancer Res Treat 1995;36:41–48.
37. Ali SM, Leitzel K, Chinchilli VM, et al. Relationship of serum HER-2/neu and serum CA 15-3 in patients with metastatic breast cancer. Clin Chem 2002;48:1314–1320.
38. Ross JS, Fletcher JA, Linette GP, et al. The Her-2/neu gene and protein in breast cancer 2003: biomarker and target of therapy. Oncologist 2003;8:307–325.
39. Ross JS, Fletcher JA. HER-2/neu (c-erb-B2) gene and protein in breast cancer. Am J Clin Pathol 1999;112:S53–S67.
40. Ross JS, Fletcher JA. The HER-2/neu oncogene in breast cancer: prognostic factor, predictive factor, and target for therapy. Oncologist 1998;3:237–252.
41. Jo M, Thomas KS, Wu L, et al. Soluble urokinase-type plasminogen activator receptor inhibits cancer cell growth and invasion by direct urokinase-independent effects on cell signaling. J Biol Chem 2003; 278:46692–46698.
42. Nguyen DH, Hussaini IM, Gonias SL. Binding of urokinase-type plasminogen activator to its receptor in MCF-7 cells activates extracellular signal-regulated kinase 1 and 2 which is required for increased cellular motility. J Biol Chem 1998;273:8502–8507.
43. Riisbro R, Christensen IJ, Piironen T, et al. Prognostic significance of soluble urokinase plasminogen activator receptor in serum and cytosol of tumor tissue from patients with primary breast cancer. Clin Cancer Res 2002;8:1132–1141.
44. Ogawa Y, Shikawa T, Ikeda K, et al. Evaluation of serum KL-6, a mucin-like glycoprotein, as a tumor marker for breast cancer. Clin Cancer Res 2000;6:4069–4072.
45. Sheen-Chen SM, Chen HS, Sheen CW, et al. Serum levels of transforming growth factor beta1 in patients with breast cancer. Arch Surg 2001;136:937–940.
46. Woodbury RL, Varnum SM, Zangar RC. Elevated HGF levels in sera from breast cancer patients detected using a protein microarray ELISA. J Proteome Res 2002;1:233–237.
47. Krajcik RA, Borofsky ND, Massardo S, et al. Insulin-like growth factor I (IGF-I), IGF-binding proteins, and breast cancer. Cancer Epidemiol Biomarkers Prev 2002;11:1566–1573.
48. Sheen-Chen SM, Chen HS, Eng HL, et al. Circulating soluble Fas in patients with breast cancer. World J Surg 2003;27:10–13.
49. Bewick M, Conlon M, Parissenti AM, et al. Soluble Fas (CD95) is a prognostic factor in patients with metastatic breast cancer undergoing high-dose chemotherapy and autologous stem cell transplantation. J Hematother Stem Cell Res 2001;10:759–768.
50. Coskun U, Gunel N, Toruner FB, et al. Serum leptin, prolactin and vascular endothelial growth factor (VEGF) levels in patients with breast cancer. Neoplasm 2003;50:41–46.
51. Lantzsch T, Hefler L, Krause U, et al. The correlation between immunohistochemically-detected markers of angiogenesis and serum vascular endothelial growth factor in patients with breast cancer. Anticancer Res 2002;22:1925–1928.
52. Heer K, Kumar H, Read JR, et al. Serum vascular endothelial growth factor in breast cancer: its relation with cancer type and estrogen receptor status. Clin Cancer Res 2001;7:3491–3494.
53. Granato AM, Nanni O, Falcini F, et al. Basic fibroblast growth factor and vascular endothelial growth factor serum levels in breast cancer patients and healthy women: useful as diagnostic tools? Breast Cancer Res 2003;6:R38–R45.
54. Reis FM, Cobellis L, Tameirao LC, et al. Serum and tissue expression of activin A in postmenopausal

women with breast cancer. J Clin Endocrinol Metab 2002;87:2277–2282.

55. Kopp R, Classen S, Wolf H, et al. Predictive relevance of soluble CD44v6 serum levels for the responsiveness to second line hormone or chemotherapy in patients with metastatic breast cancer. Anticancer Res 2001;21:2995–3000.
56. O'Hanlon DM, Fitzsimons H, Lynch J, et al. Soluble adhesion molecules (E-selectin, ICAM-1 and VCAM-1) in breast carcinoma. Eur J Cancer 2002; 38:2252–2257.
57. Kostler WJ, Tomek S, Brodowicz T, et al. Soluble ICAM-1 in breast cancer: clinical significance and biological implications. Cancer Immunol Immunother 2001;50:483–490.
58. Salgado R, Junius S, Benoy I, et al. Circulating interleukin-6 predicts survival in patients with metastatic breast cancer. Int J Cancer 2003;103: 642–646.
59. Kovacs E. Investigation of interleukin-6 (IL-6), soluble IL-6 receptor (sIL-6R) and soluble gp130 (sgp130) in sera of cancer patients. Biomed Pharmacother 2001;55:391–396.
60. Gunel N, Coskun U, Sancak B, et al. Clinical importance of serum interleukin–18 and nitric oxide activities in breast carcinoma patients. Cancer 2002;95: 663–667.
61. Merendino RA, Gangemi S, Ruello A, et al. Serum levels of interleukin-18 and sICAM-1 in patients affected by breast cancer: preliminary considerations. Int J Biol Markers 2001;16:126–129.
62. Welsh JB, Sapinoso LM, Kern SG, et al. Large-scale delineation of secreted protein biomarkers overexpressed in cancer tissue and serum. Proc Natl Acad Sci USA 2003;100:3410–3415.
63. Yousef GM, Polymeris ME, Grass L, et al. Human kallikrein 5: a potential novel serum biomarker for breast and ovarian cancer. Cancer Res 2003;63: 3958–3965.
64. Kishi T, Grass L, Soosaipillai A, et al. Human kallikrein 8: immunoassay development and identification in tissue extracts and biological fluids. Clin Chem 2003;49:87–96.
65. Berchem G, Glondu M, Gleizes M, et al. Cathepsin-D affects multiple tumor progression steps in vivo: proliferation, angiogenesis and apoptosis. Oncogene 2002;21:5951–5955.
66. Ranuncolo SM, Armanasco E, Cresta C, et al. Plasma MMP-9 (92 kDa-MMP) activity is useful in the follow-up and in the assessment of prognosis in breast cancer patients. Int J Cancer 2003;106:745–751.
67. Jukkola A, Bloigu R, Holli K, et al. Postoperative PINP in serum reflects metastatic potential and poor survival in node-positive breast cancer. Anticancer Res 2001;21:2873–2876.
68. Keskikuru R, Bloigu R, Risteli J, et al. Elevated preoperative serum ICTP is a prognostic factor for overall and disease-free survival in breast cancer. Oncol Rep 2002;9:1323–1327.
69. Sheard MA, Vojtesek B, Simickova M, et al. Release of cytokeratin-18 and -19 fragments (TPS and CYFRA 21-1) into the extracellular space during apoptosis. J Cell Biochem 2002;85:670–677.
70. Giovanella L, Ceriani L, Giardina G, et al. Serum cytokeratin fragment 21.1 (CYFRA 21.1) as tumour marker for breast cancer: comparison with carbohydrate antigen 15.3 (CA 15.3) and carcinoembryonic antigen (CEA). Clin Chem Lab Med 2002;40:298–303.
71. Luftner D, Possinger K. Nuclear matrix proteins as biomarkers for breast cancer. Expert Rev Mol Diagn 2002;2:23–31.
72. Fanger GR, Houghton RL, Retter MW, et al. Detection of mammaglobin in the sera of patients with breast cancer. Tumour Biol 2002;23:212–221.
73. Fedarko NS, Jain A, Karadag A, et al. Elevated serum bone sialoprotein and osteopontin in colon, breast, prostate, and lung cancer. Clin Cancer Res 2001;7:4060–4066.
74. Kir ZO, Oner P, Iyidogan YO, et al. Serum prolidase I activity and some bone metabolic markers in patients with breast cancer: in relation to menopausal status. Clin Biochem 2003;36:289–294.
75. Sasaki H, Yu CY, Dai M, et al. Elevated serum periostin levels in patients with bone metastases from breast but not lung cancer. Breast Cancer Res Treat 2003;77:245–252.
76. Ozyilkan O, Baltali E, Kirazli S. CA 15-3, ceruloplasmin and tissue polypeptide specific antigen as a tumour marker panel in breast cancer. East Afr Med J 2000;77:291–294.
77. D'Alessandro R, Roselli M, Ferroni P, et al. Serum tissue polypeptide specific antigen (TPS): a complementary tumor marker to CA 15-3 in the management of breast cancer. Breast Cancer Res Treat 2001;68:9–19.
78. Sjostrom J, Alfthan H, Joensuu H, et al. Serum tumour markers CA 15-3, TPA, TPS, hCGbeta and TATI in the monitoring of chemotherapy response in metastatic breast cancer. Scand J Clin Lab Invest 2001;61:431–441.
79. Seth LR, Kharb S, Kharb DP. Serum biochemical markers in carcinoma breast. Indian J Med Sci 2003;57:350–354.
80. Dirix LY, Salgado R, Weytjens R, et al. Plasma fibrin D-dimer levels correlate with tumour volume, progression rate and survival in patients with metastatic breast cancer. Br J Cancer 2002;86:389–395.
81. Giovanella L, Marelli M, Ceriani L, et al. Evaluation of chromogranin A expression in serum and tissues of breast cancer patients. Int J Biol Markers 2001;16: 268–272.
82. Karande AA, Sridhar L, Gopinath KS, et al. Riboflavin carrier protein: a serum and tissue marker for breast carcinoma. Int J Cancer 2001;95:277–281.
83. Sangrajrang S, Arpornwirat W, Cheirsilpa A, et al. Serum p53 antibodies in correlation to other biological parameters of breast cancer. Cancer Detect Prevent 2003;27:182–186.
84. Regele S, Vogl FD, Kohler T, et al. p53 autoantibodies can be indicative of the development of breast cancer relapse. Anticancer Res 2003;23:761–764.
85. Brankin B, Skaar TC, Brotzman M, et al. Autoantibodies to the nuclear phosphoprotein nucleophosmin in breast cancer patients. Cancer Epidemiol Biomarker Prev 1998;7:1109–1115.
86. von Mensdorff-Pouilly S, Gourevitch MM, Kenemans P, et al. Humoral immune response to polymorphic epithelial mucin (MUC-1) in patients with benign and malignant breast tumours. Eur J Cancer 1996;32A:1325–1331.
87. Gourevitch MM, von Mensdorff-Pouilly S, Litvinov SV, et al. Polymorphic epithelial mucin (MUC-1)-containing circulating immune complexes in carcinoma patients. Br J Cancer 1995;72:934–938.
88. Visco V, Bei R, Moriconi E, et al. ErbB2 immune response in breast cancer patients with soluble

receptor ectodomain. Am J Pathol 2000;156:1417–1424.

89. Sauter ER, Chervoneva I, Diamandis A, et al. Prostate-specific antigen and insulin-like growth factor binding protein-3 in nipple aspirate fluid are associated with breast cancer. Cancer Detect Prev 2002;26:149–157.

90. Sauter ER, Tichansky DS, Chervoneva I, et al. Circulating testosterone and prostate-specific antigen in nipple aspirate fluid and tissue are associated with breast cancer. Environ Health Perspect 2002; 110:241–246.

91. Varnum SM, Covington CC, Woodbury RL, et al. Proteomic characterization of nipple aspirate fluid: identification of potential biomarkers of breast cancer. Breast Cancer Res Treat 2003;80:87–97.

92. Kuerer HM, Thompson PA, Krishnamurthy S, et al. High and differential expression of HER–2/neu extracellular domain in bilateral ductal fluids from women with unilateral invasive breast cancer. Clin Cancer Res 2003;9:601–605.

93. Hsiung R, Zhu W, Klein G, et al. High basic fibroblast growth factor levels in nipple aspirate fluid are correlated with breast cancer. Cancer J 2002;8:303–310.

94. Sauter ER, Chervoneva I, Diamandis A, et al. Prostate specific antigen and insulin-like growth factor finding protein-3 in nipple aspirate fluid are associated with breast cancer. Cancer Detect Prev 2002;26:149–157.

95. Mitchell G, Sibley PE, Wilson AP, et al. Prostate-specific antigen in nipple aspiration fluid: menstrual cycle variability abd correlation with serum prostate-specific antigen. Tumor Biol 2002;23:287–297.

96. Sauter ER, Welch T, Magklara A, Klein G, et al. Ethnic variation in kallikrein expression in nipple aspirate fluid. Int J Cancer 2002;100:678–682.

97. Black MH, Diamandis EP. The diagnostic and prognostic utility of prostate-specific antigen for diseases of the breast. Breast Cancer Res Treat 2000;59:1–14.

98. Zhao Y, Verselis SJ, Klar N, et al. Nipple fluid carcinoembryonic antigen and prostate-specific antigen in cancer-bearing and tumor-free breasts. J Clin Oncol 2001;19:1462–1467.

99. Qin W, Zhu W, Wagner-Mann C, et al. Nipple aspirate fluid expression of urokinase-type plasminogen activator, plasminogen activator inhibitor-1, and urokinase-type plasminogen activator receptor predicts breast cancer diagnosis and advanced disease. Ann Surg Oncol 2003;10:948–953.

100. Qin W, Zhu W, Wagner-Mann C, et al. Association of uPA, PAT-1, and uPAR in nipple aspirate fluid (NAF) with breast cancer. Cancer J 2003;9:293–301.

101. Grunewald K, Haun M, Urbanek M, et al. Mammaglobin gene expression: a superior marker of breast cancer cells in peripheral blood in comparison to epidermal-growth-factor receptor and cytokeratin-19. Lab Invest 2000;80:1071–1077.

102. Stathopoulou A, Mavroudis D, Perraki M, et al. Molecular detection of cancer cells in the peripheral blood of patients with breast cancer: comparison of CK-19, CEA and maspin as detection markers. Anticancer Res 2003;23:1883–1890.

103. Martin KJ, Graner E, Li Y, et al. High-sensitivity array analysis of gene expression for the early detection of disseminated breast tumor cells in peripheral blood. Proc Natl Acad Sci USA 2001;98:2646–2651.

104. Hu XC, Wong IH, Chow LW. Tumor-derived aberrant methylation in plasma of invasive ductal breast cancer patients: clinical implications Oncol Rep 2003;10: 1811–1815.

105. Lin YC, Wu Chou YH, Liao IC, et al. The expression of mammaglobin mRNA in peripheral blood of metastatic breast cancer patients as an adjunct to serum tumor markers. Cancer Lett 2003;191:93–99.

106. Chen X, Bonnefoi H, Diebold-Berger S, et al. Detecting tumor-related alterations in plasma or serum DNA of patients diagnosed with breast cancer. Clin Cancer Res 1999;5:2297–2303.

107. Chen XQ, Bonnefoi H, Pelte MF, et al. Telomerase RNA as a detection marker in the serum of breast cancer patients. Clin Cancer Res 2000;6:3823–3826.

108. Schroder CP, Ruiters MH, de Jong S, et al. Detection of micrometastatic breast cancer by means of real time quantitative RT-PCR and immunostaining in perioperative blood samples and sentinel nodes. Int J Cancer 2003;106:611–618.

109. de Cremoux P, Extra JM, Denis MG, et al. Detection of MUC1-expressing mammary carcinoma cells in the peripheral blood of breast cancer patients by real-time polymerase chain reaction. Clin Cancer Res 2000;6:3117–3122.

110. Wasserman L, Dreilinger A, Easter D, et al. A seminested RT-PCR assay for HER2/neu: initial validation of a new method for the detection of disseminated breast cancer cells. Mol Diagn 1999;4:21–28.

111. Stathopoulou A, Gizi A, Perraki M, et al. Real-time quantification of CK-19 mRNA-positive cells in peripheral blood of breast cancer patients using the lightcycler system. Clin Cancer Res 2003;9: 5145–5151.

112. Soria JC, Gauthier LR, Raymond E, et al. Molecular detection of telomerase-positive circulating epithelial cells in metastatic breast cancer patients. Clin Cancer Res 1999;5:971–975.

113. Taback B, Giuliano AE, Hansen NM, et al. Microsatellite alterations detected in the serum of early stage breast cancer patients. Ann N Y Acad Sci 2001;945:22–30.

114. Reis-Filho JS, Milanezi F, Amendoeira I, et al. Distribution of p63, a novel myoepithelial marker, in fine-needle aspiration biopsies of the breast: an analysis of 82 samples. Cancer 2003;99:172–179.

115. Nizzoli R, Bozzetti C, Crafa P, et al. Immunocytochemical evaluation of HER–2/neu on fine-needle aspirates from primary breast carcinomas. Diagn Cytopathol 2003;28:142–146.

116. Troncone G, Panico L, Vetrani A, et al. c-erbB-2 expression in FNAB smears and matched surgical specimens of breast cancer. Diagn Cytopathol 1996; 14:135–139.

117. Trihia H, Murray S, Price K, et al. Ki-67 expression in breast carcinoma: its association with grading systems, clinical parameters, and other prognostic factors—a surrogate marker. Cancer 2003;97:1321–1331.

118. Billgren AM, Tani E, Liedberg A, et al. Prognostic significance of tumor cell proliferation analyzed in fine needle aspirates from primary breast cancer. Breast Cancer Res Treat 2002;71:161–170.

119. Klorin G, Keren R. Prognostic evaluation of breast cancer in cytologic specimens. Anal Quant Cytol Histol 2002;24:49–53.

120. Kirjushkina MS, Tchipysheva TA, Ermilova VD, et al. Breast tumor diagnosis in cytologic aspirates using monoclonal antibodies to keratin 8 and 17. Acta Cytol 1997;41:307–312.

121. Bozzetti C, Nizzoli R, Naldi N, et al. Bcl-2 expression on fine-needle aspirates from primary breast carcinoma: correlation with other biologic factors. Cancer 1999;87:224–230.
122. Sauer T, Furu I, Beraki K, et al. nm23 protein expression in fine-needle aspirates from breast carcinoma: inverse correlation with cytologic grading, lymph node status, and ploidy. Cancer 1998;84:109–114.
123. Heatley M, White J, Patterson A, et al. Intermediate filament expression of ductal carcinoma cells in fine needle aspirates of the breast. Acta Cytol 1997;41: 717–720.
124. Alexiev BA. Localization of p53 and proliferating cell nuclear antigen in fine-needle aspirates of benign and primary malignant tumors of the human breast: an immunocytochemical study using supersensitive monoclonal antibodies and the biotin-streptavidin-amplified method. Diagn Cytopathol 1996;15:277–281.
125. King BL, Crisi GM, Tsai SC, et al. Immunocytochemical analysis of breast cells obtained by ductal lavage. Cancer 2002;96:244–249.
126. Heselmeyer-Haddad K, Chaudhri N, Stoltzfus P, et al. Detection of chromosomal aneuploidies and gene copy number changes in fine needle aspirates is a specific, sensitive, and objective genetic test for the diagnosis of breast cancer. Cancer Res 2002;62: 2365–2369.
127. Burki NG, Caduff R, Walt H, et al. Comparative genomic hybridization of fine needle aspirates from breast carcinomas. Int J Cancer 2000;88: 607–613.
128. Holger N, Schulten HJ, Gunawan B, et al. The potential value of comparative genomic hybridization analysis in effusion and fine needle aspiration cytology. Mod Pathol 2002;15:818–825.
129. Pollack JR, Sorlie T, Perou CM, et al. Microarray analysis reveals a direct role of DNA copy number alteration in the transcriptional program of human breast tumors. Proc Natl Acad Sci USA 2002;99: 12963–12968.
130. King BL, Tsai SC, Gryga ME, et al. Detection of chromosomal instability in paired breast surgery and ductal lavage specimens by interphase fluorescence in situ hybridization. Clin Cancer Res 2003;9:1509–1516.
131. Noguchi S, Tsukamoto F, Miyoshi Y, et al. Detection of numerical aberrations in chromosomes by fluorescence in situ hybridization in fine needle aspirates in the preoperative diagnosis of cancer. Gan To Kagaku Ryoho 1999;26:2127–2130.
132. Tsuda H, Takarabe T, Susumu N, et al. Detection of numerical and structural alterations and fusion of chromosomes 16 and 1 in low-grade papillary breast carcinoma by fluorescence in situ hybridization. Am J Pathol 1997;151:1027–1034.
133. Tsuda H, Takarabe T, Shimamura K, et al. Detection of alterations in chromosomes 16 and 1 by fluorescence in situ hybridization in breast tumors cytologically or histologically equivocal for malignancy. Pathobiology 1998;66:268–273.
134. Komoike Y, Motomura K, Inaji H, et al. Diagnosis of ductal carcinoma in situ (DCIS) and intraductal papilloma using fluorescence in situ hybridization (FISH) analysis. Breast Cancer 2000;7:332–336.
135. Yamamoto D, Senzaki H, Nakagawa H, et al. Detection of chromosomal aneusomy by fluorescence in situ hybridization for patients with nipple discharge. Cancer 2003;97:690–694.
136. Ichikawa D, Hashimoto N, Hoshima M, et al. Analysis of numerical aberrations of specific chromosomes by fluorescent in situ hybridization as a diagnostic tool in breast cancer. Cancer 1996;77:2064–2069.
137. McManus DT, Patterson AH, Maxwell P, et al. Interphase cytogenetics of chromosomes 11 and 17 in fine needle aspirates of breast cancer. Hum Pathol 1999; 30:137–144.
138. McManus DT, Patterson AH, Maxwell P, et al. Fluorescence in situ hybridisation detection of erbB2 amplification in breast cancer fine needle aspirates Mol Pathol 1999;52:75–77.
139. Dillon DA, Hipolito E, Zheng K, et al. p53 mutations as tumor markers in fine needle aspirates of palpable breast masses. Acta Cytol 2002;46:841–847.
140. Pollett A, Bedard YC, Li SQ, et al. Correlation of p53 mutations in ThinPrep-processed fine needle breast aspirates with surgically resected breast cancers. Mod Pathol 2002;13:1173–1179.
141. Lavarino C, Corletto V, Mezzelani A, et al. Detection of TP53 mutation, loss of heterozygosity and DNA content in fine-needle aspirates of breast carcinoma. Br J Cancer 1998;77:125–130.
142. Parrella P, Xiao Y, Fliss M, et al. Detection of mitochondrial DNA mutations in primary breast cancer and fine-needle aspirates. Cancer Res 2001;61: 7623–7626.
143. Anan K, Morisaki T, Katano M, et al. Assessment of c-erbB2 and vascular endothelial growth factor mRNA expression in fine-needle aspirates from early breast carcinomas: pre-operative determination of malignant potential. Eur J Surg Oncol 1998;24:28–33.
144. Anan K, Morisaki T, Katano M, et al. Preoperative assessment of tumor angiogenesis by vascular endothelial growth factor mRNA expression in homogenate samples of breast carcinoma: fine-needle aspirates vs. resection samples. J Surg Oncol 1997; 66:257–263.
145. Sugino T, Yoshida K, Bolodeoku J, et al. Telomerase activity in human breast cancer and benign breast lesions: diagnostic applications in clinical specimens, including fine needle aspirates. Int J Cancer 1996; 69:301–306.
146. Zhu W, Qin W, Ehya H, et al. Microsatellite changes in nipple aspirate fluid and breast tissue from women with breast carcinoma or its precursors. Clin Cancer Res 2003;9:3029–3033.
147. Li J, Zhang Z, Rosenzweig J, et al. Proteomics and bioinformatics approaches for identification of serum biomarkers to detect breast cancer. Clin Chem 2002; 48:1296–1304.
148. Vlahou A, Laronga C, Wilson L, et al. A novel approach toward development of a rapid blood test for breast cancer. Clin Breast Cancer 2003;4:203–209.
150. Sauter ER, Zhu W, Fan XJ, et al. Proteomic analysis of nipple aspirate fluid to detect biologic markers of breast cancer. Br J Cancer 2002;86:1440–1443.
151. Wulfkuhle JD, Sgroi DC, Krutzsch H, et al. Proteomics of human breast ductal carcinoma in situ. Cancer Res 2002;62:6740–6749.
152. Le Naour F, Misek DE, Krause MC, et al. Proteomics-based identification of RS/DJ–1 as a novel circulating tumor antigen in breast cancer. Clin Cancer Res 2001;7:3328–3335.

CHAPTER

6

Genomic and Molecular Classification of Breast Cancer

Christos Sotiriou, MD,[1] Christine Desmedt, PhD,[1] Virginie Durbecq, PhD,[1] Lissandra Dal Lago, MD,[1] Marc Lacroix, PhD,[2] Fatima Cardoso, MD,[1] and Martine Piccart, MD, PhD[1]

[1]Translational Research
[2]Laboratoire Jean-Claude Heuson de Cancérologie Mammaire, Jules Bordet Institute, Free University of Brussels, Belgium

Histologic Classification and Clinical Staging

Historically, breast cancer has been divided into smaller groups to standardize comparison of various therapeutic modalities between clinics and to guide treatment. This subdivision is represented by the TNM classification scheme and the clinical stages defined by the American Joint Committee for Cancer Staging, based on pathologic and clinical findings such as tumor size, node involvement, and presence of distant disease.[1] To refine this classification, a variety of pathologic factors, including histologic subtype, nuclear and histologic grade, and lymphatic and/or vascular invasion, has been carefully evaluated as prognostic indicators of clinical course. Although each of these pathologic parameters has been shown to be a useful predictor in expert hands, histologic subclassification has had the greatest applicability. The 30-year survival of women with specific types of invasive carcinomas (tubular, colloid, medullary, lobular, and papillary) is more than 60%, compared with less than 20% for women with other histologic subtypes (ductal carcinomas).[2,3] Unfortunately, these favorable histologies are much less common than invasive ductal carcinoma, which lacks distinct histologic features. Nuclear grade determined by expert breast pathologists has been shown in certain studies to have prognostic power. However, the major criticism related to nuclear grading has been the lack of interobserver consistency.

The recent advances in basic research and genomics have improved our understanding of the biologic basis of tumor development and progression. As a result, a variety of molecular tumor markers characterized in the laboratory has been studied in the clinic for their potential to predict disease outcome or response to therapy in breast cancer patients. Nevertheless, none of these markers appears to provide definitive prognostic or predictive information. Additionally, their prognostic and predictive values are based on patient cohorts and offer little if any individualized information regarding outcome and the need for systemic adjuvant therapy for a given patient. Indeed, patients with the same clinicopathologic parameters can have markedly different treatment responses and clinical courses.

A combination of circumstances, including the advent of array-based technology and progress in the human genome initiative, has provided an ideal opportunity to begin efforts aimed at performing comprehensive molecular and genetic profiling of human cancers, which could lead to a greater understanding of the basic biology of breast cancer. The ability to interrogate tens of thousands of genes simultaneously by using microarray technologies has significantly changed our approach to the analysis of expression profiles. Such comprehensive technologies permit the assessment not only of individual genes, but also of clusters of genes that are coordinately expressed to generate "fingerprints" of biologic states of the cells of origin. Though other techniques analyze differences in gene expression, none matches the ease and the comprehensive nature of the interrogation associated with complementary DNA (cDNA)- or oligonucleotide-based microarray analysis.

Microarray Technique

The concept behind DNA chip or microarray technology is not new: It relies on the accurate binding, or hybridization, of strands of DNA with their precise complementary copies in experimental conditions where one sequence is also bound onto a solid-state substrate.[4] The simultaneous study of thousands of genes, rather than tens to hundreds with traditional techniques, transforms the microarray technique into a powerful whole system analytical tool (FIG. 6.1). Currently, several DNA microarray platforms are used for genome-wide gene expression studies (see Chapter 2). The cDNA microarray represents a popular array type in which double-stranded polymerase chain reaction (PCR) products, amplified from expressed sequence tag clones, are spotted onto glass slides. Theoretically, this approach allows for gene expression profiles to be determined with high validity, reproducibility, and efficiency; however, the construction and use of cDNA microarrays to discriminate among patterns of gene expression presents a number of challenges, including laborious and problematic tracking of cDNA clones and PCR amplicons and cross-hybridization across homologous genes, alternative splice variants, and antisense RNAs.

Many of the technological limitations of microarray applications are addressed by the use of oligonucleotide-based microarrays. There are two major approaches to constructing oligonucleotide arrays. First, microarrays composed of short oligonucleotides (25 bases) may be synthesized directly onto a solid matrix using photolithographic technology (Affymetrix). Second, microarrays composed of long oligonucleotides (55–70 bases) may be either deposited by an ink-jet printing process or spotted by a robotic printing process onto

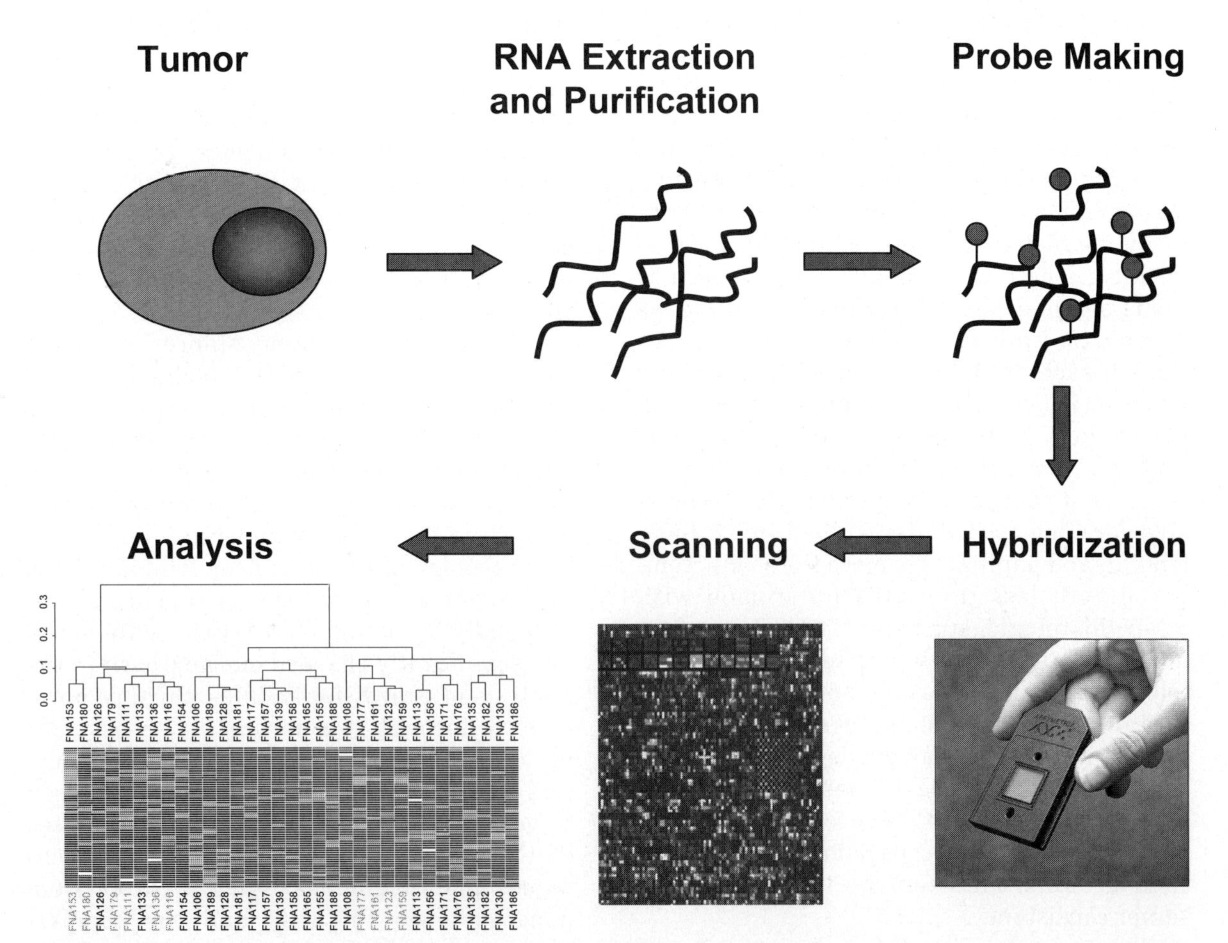

FIGURE 6.1 Steps in microarray analysis (Affymetrix): RNA from tumor to be studied is isolated and purified. The cRNA probes are prepared in the presence of biotin-labeled ribonucleotides (UTP). Hybridization to the Affymetrix Genechip® is performed by resuspending the probe in buffer containing biotinylated control oligos (internal controls). After hybridization, washing and staining steps, the Genechip® is scanned and data are analyzed. **See Plate 8 for color image.**

glass slides (i.e., Agilent, CodeLink). The first approach has been commercially available for several years. A large number of published studies using this system are already available, meaning that it is generally accepted as a method for expression analysis by many laboratories and journals. Although the second approach is less standardized in the literature, the completion of numerous genomic sequences and recent increases in the efficiency of oligonucleotide production have combined to make the use of long oligonucleotide arrays for gene expression studies a very attractive alternative to both short oligonucleotide and full-length spotted cDNA approaches.

Breast Cancer Classification Based on Gene Expression Profiling and Its Prognostic Value

Several reports described the use of microarrays to assess a molecular classification of human breast cancers and investigated the possibility to correlate gene expression profiles with clinical outcome (**Tables 6.1** and **6.2**). Perou et al.[5] from Stanford analyzed variation in gene expression patterns in normal and malignant human breast tissues from 42 individuals. In this population, 20 tumors were sampled twice (before and after a 16-week regimen of doxorubicin chemotherapy), and two tumors were paired with a lymph node metastasis from the same patient. Hierarchical clustering analysis, which organizes tumor samples based on their overall similarity in their gene expression patterns, showed that most tumor specimen pairs clustered together, suggesting that tumor samples from the same patient were still far more similar to each other than to any other patient's tumor specimen. Additionally, these "molecular portraits" revealed by the patterns of gene expression not only uncovered similarities and differences between the tumor samples but identified specific features of physiologic variation.

Another interesting finding was the segregation of the tissue samples into two major subtypes based on the expression levels of 496 cDNA genes ("intrinsic" gene subset), identified as being highly variable in expression across tumors derived from different patients and being minimally variable between tumor samples taken from the same patient. The first subtype was characterized by tumors that were clinically described as estrogen receptor α (ER) positive and the second by tumors that were mostly ER-negative. Interestingly, the ER-positive tumors were distinguished by the relatively high expression of genes normally expressed by breast luminal cells, whereas the ER-negative tumors samples were characterized by the expression of genes expressed by breast basal/myoepithelial cells. The latter group was furthermore subdivided into three different subgroups, the "basal-like," the "*erb*B2-positive," and the "normal breast–like" subgroups with clearly distinct molecular phenotypes.

Such differences in gene expression profiles between clinically ER-positive and ER-negative breast cancers tumors have also been reported by other investigators.[6–8] Gruvberger et al.,[8] using both unsupervised and supervised approaches, identified a list of genes that best discriminated the tumors according to their ER status. Interestingly, they showed that the differences in gene expression profiles between ER-positive and ER-negative tumors could only partly be explained by the activity of a functionally ER pathway, suggesting that these differences might be largely explained on the basis of different cell lineage.

To further address whether these expression signatures segregating according to ER status are due to the molecular/genetic origins of the breast cancer or whether they reflect the cell lineage of the tumor, Desai et al.[9] investigated the profiles of mouse mammary cancers bearing various oncotransgenes using two different promoters targeting different compartments in the mammary gland. They found that the pathways served by the oncogenes determined the ultimate tumor clusters and that the different promoters only altered the substructure and not the major cluster determinants. These results support that both the genetic origins of a tumor and the cell lineage of origin can effectively determine the downstream expression profiles.

Sorlie et al.,[10] from the same Stanford group, refined this classification by analyzing a larger number of breast cancer tumors using a similar microarray cDNA platform. As previously described, they were able to identify three subgroups characterized by low or absent ER gene expression: the basal-like subgroup, characterized by high expression of keratins 5 (*KRT5*) and 17 (*KRT17*) and by laminin and fatty-acid binding protein 7; the *erb*B2-positive subgroup, with high expression of several genes of the *erb*B2 amplicon; and the normal breast–like subgroup. The novel finding in that study was the recognition within the

Table 6.1 Microarray Studies in Breast Cancer

Study	Platform	Patient Cohort	Type of Analysis	Informative Genes	Findings
Perou et al. (2000)[5]	cDNA 8102 features	42 patients with normal or malignant breast tissues: 36 infiltrating ductal carcinomas, 2 lobular, 1 DCIS, 1 fibroadenoma, and 3 normal breasts; 20 pairs of tumors were sampled before and after 16 weeks of doxorubicin chemotherapy	Unsupervised	"Intrinsic" gene subset 496 genes	Identification of four subgroups: ER+/luminal-like, basal-like, *erb*B2+, normal breast-like
Sorlie et al. (2001)[10]	cDNA 8102 features	7 nonmalignant breast samples and 78 carcinomas (71 ductal, 5 lobular, 2 DCIS, including 40 tumors from Ref. 5)	Unsupervised	"Intrinsic" gene subset[5]	(1) Identification of novel luminal-type subclasses: A, B, C (2) Luminal A had the best clinical outcome
Sorlie et al. (2003)[12]	cDNA 8102 features	Population[10] + 38 additional carcinomas	Unsupervised	"Intrinsic" gene subset[5]	Validation of Ref. 10
Gruvberger et al. (2001)[8]	cDNA 6728 features	58 node-negative breast tumors: 28 ER+ and 30 ER–	Supervised	Top 100 ER discriminating genes	Discrimination between ER+ and ER− tumors
West et al. (2001)[6]	Oligonucleotides 5600 genes	49 primary breast tumors: 13 ER+/LN+, 12 ER−/LN+, 12 ER+/LN−, 12 ER−/LN−	Supervised	(1) Top 100 ER discriminating genes (2) Top 100 LN discriminating genes	(1) Discrimination between ER+ and ER− tumors (2) Discrimination between LN+ and LN− tumors
Sotiriou et al. (2003)[11]	cDNA 7650 features	99 invasive ductal carcinomas treated with adjuvant treatment: 53 LN+ and 46 LN−	Unsupervised	706 genes	(1) Identification of six subgroups: luminal 1, 2, and 3; basal-like 1 and 2; *erb*B2+ (2) Luminal 1 had the best clinical outcome
van't Veer et al. (2002)[13]	Oligonucleotides 24479 genes	98 tumors: 78 sporadic tumors LN– from untreated patients (34 metastasis in 5 years, 44 disease free), 18 *BRCA1/2* mutations carriers	Supervised	"Prognosis" signature 70 genes	Prediction of clinical outcome, defined as the presence of metastases at the 5-year mark
van de Vijver et al. (2002)[15]	Oligonucleotides 24,479 genes	295 stage I or II breast cancer patients, both treated and untreated: 151 LN– and 144 LN+	Supervised	"Prognosis" signature[13]	Validation of Ref. 13

DCIS, ductal carcinoma in situ; LN, lymph node.

Table 6.2 Definitions

Term	Definition
Bioinformatics	Field of science in which biology, computer science, and information technology merge to form a single discipline. The ultimate goal of the field is to enable the discovery of new biologic insights as well as to create a global perspective from which unifying principles in biology can be discerned.
Centroid of gene expression	The centroid of gene expression for a class of samples is defined as a multicomponent vector in which each component is the expression of a gene averaged over the sample class.
Expressed sequence tag (EST)	A short strand of DNA that is a part of a cDNA molecule and can act as identifier of a gene.
Gene expression	Transcription of the information contained within the DNA into messenger RNA (mRNA) molecules that are then translated into proteins.
Hierarchical clustering	Bottom up (agglomerative) approach, whereby single expression profiles are successively joined to form nodes, which in turn are then joined further. The process continues until all individual profiles and nodes have been joined to form a single hierarchical tree.
Northern blot	Procedure used mostly to separate and identify RNA fragments, typically via transferring RNA fragments from an agarose gel to a nitrocellulose filter followed by detection with a suitable probe.
Oligonucleotide	Short fragment of a single-stranded DNA.
Polymerase chain reaction (PCR)	Exponential amplification of almost any region of a selected DNA molecule.
RT-PCR	PCR of a reverse transcription product.
Reverse transcription	Transcription from RNA into DNA.
Supervised analysis	Statistical method that makes use of prior knowledge to achieve accurate results. The method is "supervised" or taught on a set of training data for which the outputs are already known. The algorithm attempts to match its predicted outputs to the known outputs.
Unsupervised analysis	Statistical method for microarrays which does not need additional, previously derived, information about the data to be analyzed. The outputs are simply a description of the relationships among the samples or genes. An advantage of this method is that they can detect correlations among genes that would otherwise remain undiscovered.

luminal/ER-positive tumors of at least two different subgroups with distinct molecular signatures: (1) the luminal subgroup A, which was characterized by the highest ER expression and a high expression of GATA binding protein 3, X-box binding protein 1, trefoil factor 3, hepatocyte nuclear factor 3α, and LIV-1, and (2) a second subgroup that could be divided into two smaller units, the luminal subgroups B and C, both showing low to moderate expression of the luminal-specific genes. Moreover, ER-positive tumors could be separated into expression groups that ultimately correlated with the frequency of TP53 mutations. Thus, the luminal subgroup A exhibited a 13% TP53 mutation rate compared with luminal subgroups B and C, which showed a mutation rate of 80% and 40%, respectively.

Interestingly, the authors investigated whether these different subgroups identified by hierarchical clustering analysis were associated with distinct clinical behavior. Although the basal-like and *erb*B2-positive subgroups had the shortest relapse-free and overall survival rates, the luminal subtype A showed the best clinical outcome. The most attractive feature of this analysis was that ER-positive tumors could be subdivided into at least two distinct subgroups, luminal A versus luminal B + C, with different clinical survival rates.

A similar approach was also undertaken by our group, by analyzing RNAs from a cohort of 99 breast cancer patients using 7650-probe element cDNA microarrays.[11] The purpose of this study was to correlate gene expression patterns with known clinicopathologic characteristics and clinical outcome in breast cancer. Our results were significant in their concordance with those of the earlier studies, despite the differences in patient populations, the treatments used, and the technology platforms used. We found that the ER status of the tumor was indeed the most important discriminator of expression subtypes and that tumor grade was a distant second. Other clinical features, namely, lymph node positivity, menopausal status, and tumor size, were not strongly reflected in the expression patterns. Interestingly, in agreement with the previous studies, unsupervised hierarchical clustering analysis using 706-probe elements selected as exhibiting high variability across all tumors segregated the tumors into two main clusters based on their

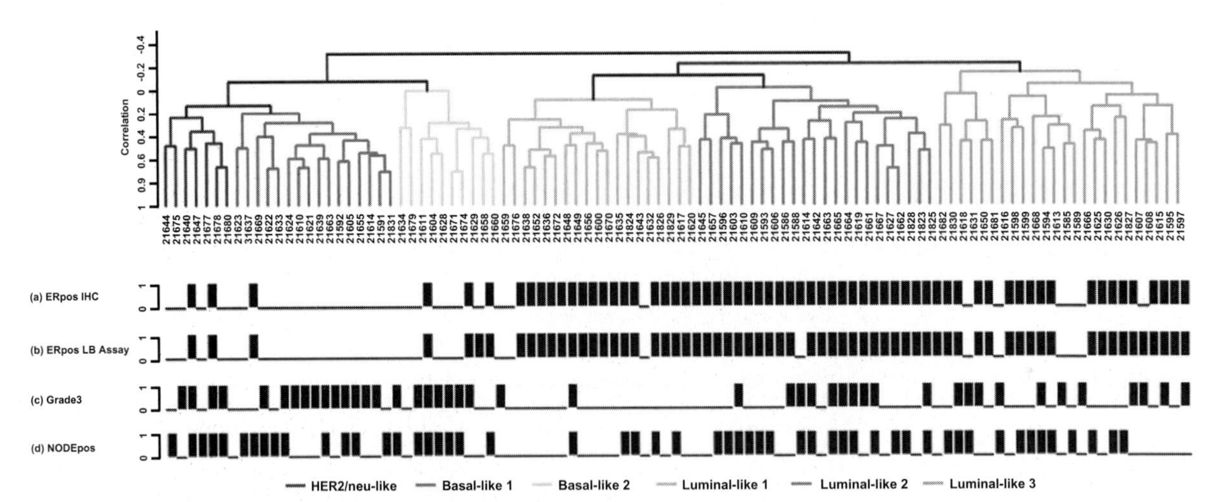

FIGURE 6.2 Dendrogram of 99 breast cancer specimens analyzed by hierarchical clustering analysis. The tumors were separated into two main groups mainly associated with ER status as determined by the ligand-binding assay (LBA) and confirmed by immunohistochemistry (IHC). The dendrogram further branched into smaller subgroups within the ER-positive and ER-negative classes based on their basal and luminal characteristics colored as follows: Her-2/neu subgroup, dark blue; basal-like 1 subgroup, pink; basal-like 2 subgroup, yellow; luminal-like 1 subgroup, light blue; luminal-like 2 subgroup, red; and luminal-like 3 subgroup, green. Black bars represent (a) ER-positive tumors assessed by IHC, (b) ER-positive tumors assessed by LBA, (c) grade 3, and (d) node-positive tumors. **See Plate 9 for color image.**

basal (predominantly ER-negative) and luminal (predominantly ER-positive) characteristics. Furthermore, within each of these clusters we also identified smaller subgroups, namely basal-1, basal-2, Her2-neu, luminal-1, luminal-2, and luminal-3, which were characterized by distinct gene expression signatures reflecting potentially different oncogene-specific pathways (FIG. 6.2). Thus, the ER-negative cluster was characterized by tumors with basal-like expression characteristics as defined by higher gene expression of keratin 5 (*KRT5*), keratin 6 (*KRT6*), metallothionein 1X (*MT1X*), and fatty acid binding protein 7 (*FABP7*). Furthermore, they exhibited higher expression of the secreted frizzled-related protein 1 (*SFRP1*) and the oncogene c-*kit* and lower expression of fibronectin 1 (*FN1*) and mucin-1 (*MUC1*). The basal-1 subgroup was further differentiated by the higher expression of several matrix metalloproteinases and of cell growth–related genes, suggesting a signature for a high proliferate rate. In contrast, the basal-2 subgroup was distinguished by the higher expression of many components of the transcriptional factor AP-1 and by overexpression of activating transcription factor 3 (*ATF3*), caveolin 1 (*CAV1*) and 2 (*CAV2*), hepatocyte growth factor (*HGF*), and transforming growth factor β receptor II (*TGFBR2*). Consistent with the previous studies, a subgroup distinct from the basal-like groups in the ER-negative subset was defined by a high rate of *ERBB2* overexpression. This *erb*B2 subgroup was also further distinguished from the basal-like subgroups by the higher expression of the multidrug resistance protein 1 (*ABCB1*), S100 calcium-binding protein P (*S100P*), fatty acid synthase (*FBXO9*), GTP-binding protein *RAL-B,* member of the *RAS* oncogene family *RAB6A,* fibronectin 1 (*FN1*) and syndecan 1 (*SDC1*), and the lower expression of c-*kit* and c-*myc*. In contrast, the ER-positive luminal clusters showed differential expressions of genes associated with the ER activation pathway and of genes having "luminal" characteristics.

Interestingly, these array-derived tumor subgroups also showed differences in clinical outcome. Thus, the luminal-like subgroup had a significant advantage in both relapse-free and breast cancer survival rates when compared with the basal/*erb*B2 cluster assessed by Kaplan-Meier analysis. Although the three subgroups within the basal-1/basal-2/*erb*B2 (predominantly ER negative) cluster appeared to have similarly poor survival characteristics, the three subgroups within the luminal-like (predominantly ER positive) cluster showed distinct differences in survival. Luminal-1 had the best outcome, with an 80% 10-year relapse-free survival rate. This subgroup also correlated with lower grade tumors and was further characterized by higher expression of c-*kit,* hepatocyte growth factor (*HGF*), insulin-like growth factor-binding protein-3 (*IGFBP3*), *ATF-3,* and components of the *AP-1* transcriptional factor and by lower expression of cell growth-related genes. This was significantly different from the luminal-2 subgroup, which had the worst outcome with a 10-year relapse-free survival rate of 40%. This subgroup was characterized by higher expression of a protein tyrosine phosphatase type IVA member (*PTP4A2*), tumor necrosis factor receptor-associated factor 3 (*TRAF3*), *RAD21,* and BRCA1 associated protein 1 (*BAP1*) and lower expression of *FGFR1, CXCR4, ATF-3,* and *VCAM1*. The luminal-3 subgroup had an intermediate survival outcome of 60% at 10 years. Intriguingly, though some of the ER-positive tumors were *ERBB2* overexpressors, they were dispersed among all the luminal subgroups.

To assess how the "intrinsic" gene list defined by the Stanford group[5,10] could perform in classifying our tumor set, we sought the overlap either between the intrinsic gene list and our cDNA array or between the intrinsic gene list and the 706 variably expressed probe elements in our study. We found 332 (285 unique genes) that overlapped with the full set of probe elements on our cDNA array and 105 (96 unique genes) probe elements that overlapped with our 706-gene list.[11] In both situations, hierarchical cluster analysis segregated our tumors into two distinct subtypes mainly based on their basal (predominantly ER-negative) and luminal (predominantly ER-positive) characteristics. The luminal-like tumors were further segregated into at least two (possibly three) smaller subgroups, which may correspond to the luminal A, B, and C subtypes defined by the Stanford group. Additional analysis revealed that 100% of the specimens in the luminal A subtype in the 332-gene cluster and 77% in the 105-gene cluster were identified as luminal-like 1 tumors in our unsupervised cluster analysis, respectively. In contrast, the specimens in the luminal B and C subgroups were designated either as luminal-like 2 or luminal-like 3 tumors. Although the luminal A subtype showed a favorable clinical outcome when compared with other luminal subtypes, this difference was not statistically significant.

A similar approach was recently carried out by Sorlie et al.[12] In this study, the investigators tested whether the previously proposed subtypes of breast cancer could be generalized by analyzing three independent data sets: an extended Norway/Stanford cohort and two published data sets, those of van't Veer et al.[13] and West et al.[6] For this comparative analysis, the authors correlated the expression centroids, calculated from the core members of each of the five subgroups in the Norway/Stanford data set, with the expression profiles calculated for each sample from the two other data sets. Using hierarchical cluster analysis, they were able to reproduce the five major subgroups characterized by distinct variation in gene expression patterns, as was previously seen. However, the major distinction observed was between the tumors showing high expression of luminal epithelial specific genes, including the *ESR1,* and all other tumors exhibiting expression profile characteristics of either the basal, the *erb*B2, or the luminal B subgroups. These subgroups were also associated with significant differences in outcome. Thus, univariate Kaplan-Meier analysis in the data set from van't Veer et al.[13] showed that the luminal A subgroup tumors exhibited a considerably longer relapse-free survival rate compared with the basal and *erb*B2 subgroups, suggesting that these breast tumor subgroups represent biologically distinct diseases with distinct clinical outcomes.

Genomic Signature of the Metastatic Potential of a Breast Cancer Cell Might Already Be Established in the Primary Tumor Expression Profile

According to a common view, progression from a primary to a metastatic tumor is accompanied by the sequential acquisition of phenotype changes, thus allowing breast cancer cells (BCCs) to invade, disseminate, and colonize distant sites. Based on in vitro data, it has notably been proposed that BCCs in vivo might undergo a transition from the luminal epithelial-like to the mesenchymal-like phenotype. Along the same lines, it has been repeatedly suggested that tumor progression is characterized by a shift from a well-differentiated/low grade to poorly differentiated/high grade category. Nevertheless, most investigations have revealed that progression is not accompanied by major changes in marker expression or grade.[14] These observations corroborate with several recent microarray studies suggesting that the metastatic signature might already be present in the primary breast tumor, thereby challenging the traditional model of metastasis, which holds that most primary tumor cells have low metastatic potential, but rare cells within large primary tumors acquire metastatic capacity through somatic mutations.

From that perspective, van't Veer et al.,[13] applying a supervised approach, sought to identify whether a gene expression signature strongly predictive of a short interval to distant metastases in primary breast cancer tumors exists. For that purpose, the authors investigated a narrow subset of breast cancer patients: T1/T2, N0, and M0, all under 55 years of age treated only with local regional therapies. They found 231 genes significantly associated with disease outcome as defined by the presence of distant metastasis at the 5-year mark. They could then subsequently collapse this list into a core set of 70 prognostic markers. Interestingly, the investigators tested the ability of this array-derived prognostic "expression profile" to correctly identify patients who would need adjuvant chemotherapy and compared it with accepted guidelines for treatment of node-negative breast cancer (National Institutes of Health and St. Gallen consensus guidelines). They found that although the expression profile could correctly identify patients who would need adjuvant chemotherapy, it could effectively reduce the fraction of women not needing adjuvant chemotherapy by about 30%. The same Dutch group applied this signature to a larger test set of 295 node-negative and node-positive breast cancer patients followed for 7 years from the same institution. This study confirmed that the 70-gene prognosis signature could clearly distinguish patients with excellent 10-year survival rates from those with a high mortality rate.[15]

Our group found substantial evidence that the 231 expressed genes reported as separating survival groups in the van't Veer study also had prognostic relevance in our heterogeneous population of node-positive and node-negative patients treated with adjuvant therapy.[11] Of the 231 expressed genes, 93 probe elements (representing 56 unique genes) overlapped with the set of 7650 probe elements represented on our arrays. Hierarchical cluster analysis of this set of 93 probe elements was found to segregate this population into two distinct subgroups with different survival as assessed by univariate Kaplan-Meier analysis. Furthermore, the overlap was investigated between the optimal survival list of 485 probe elements in our population assessed by Cox regression analysis and the 231 genes expressed in the van't Veer prognostic gene set. The intersection consisted of 11 unique genes represented by 14 probe elements. As expected, these 14 elements separated our patients into two major groups, showing a significant difference in survival. Of interest, 5 of the 11 unique genes, *RFC4, MCM6, MAD2L1, BUB1,* and *CKS2,* appear to be involved in DNA replication and chromosomal stability.

In another study West et al.,[6] using Bayesian regression models, were able to identify a gene expression signature discriminating breast cancer samples on the basis of lymph node metastases. Similarly, using a nonlinear statistical analytical approach, Huang and colleagues[16] identified aggregate patterns of gene expression, named mutagenes, that were capable of predicting lymph node metastasis and disease recurrence for individual breast cancer patients. Many of these genes included interferon-induced genes such as *STAT1, MX1, IFT1, ISG115, IFI27,* and *IFI44,* as well as genes involved in T-cell function. Interestingly, the genes implicated in recurrence prediction were distinct from those genes that predict lymph node involvement, suggesting that lymph node and distant metastasis might involve distinct biologic processes.

To explore the molecular differences between primary tumors and metastases, Ramaswamy et al.[17] compared gene expression signatures of 12 adenocarcinoma metastases of different types of tumors and origins with 64 unmatched primary adenocarcinomas. A supervised analytical approach revealed an expression pattern of 21 probes representing 17 unique genes that best discriminated primary and metastatic adenocarcinomas. Intriguingly, when the authors applied this signature to several large gene expression data sets containing molecular profiles from primary solid tumors of diverse types, including breast adenocarcinomas, they found that individuals whose primary tumor bore this metastatic signature had significantly shorter survival times compared with those whose tumors lacked this gene expression program. Interestingly, this type of observation was not seen in lymphomas, supporting the fact that although there is a generic gene expression program related to metastases in a tissue-independent manner, hematopoietic tumors have different mechanisms for navigating the hematologic and lymphoid compartments. Of particular interest was the finding that a considerable proportion of these "metastasis-associated" genes seemed to be derived from the nonepithelial component of the tumor, highlighting the importance of epithelial–mesenchymal interactions to determine tumor cell behavior. This type of tumor–host cell interaction and how different genes may cooperate to fulfill various requirements for the establishment of tissue-specific metastases in breast cancer was examined by other groups as well.

To study the biologic events taking place during hematogenous micrometastasis in bone marrow (BM), Woelfle et al.[18] analyzed gene expression profiles from primary breast tumors from BM-positive and BM-negative breast cancer patients. The investigators were able to identify 86 genes whose differential expression level best discriminate BM-positive from BM-negative breast cancer tumors. Most of these genes, including known metastasis suppressor genes such as *KISS-1* and members of the nm23 metastasis suppressor family *NME3* and *NME4,* were down-regulated in BM-positive tumors, suggesting that the overall transcriptional repression of genes seems important for BM micrometastases. Of interest, many of these genes encode for proteins involved in the extracellular matrix remodeling, cytoskeleton plasticity, and members of the RAS superfamily, pointing to a putative new role of these structural proteins as metastasis suppressors. Another interesting finding was the abnormal up-regulation of genes involved in the JAK/STAT and HIF-1α pathways, underscoring the fact that the dysregulation of these pathways may play a major role in breast cancer BM micrometastasis.

To investigate whether the generated expression profile was specific for BM micrometastases, these investigators, using the same cDNA microarrays, compared expression profiles associated either with lymph node or BM metastases. Interestingly, the specific signature associated with lymphatic dissemination was distinct from the signature associated with BM metastases, supporting the hypothesis that these two routes of dissemination might be governed by different molecular mechanisms and that tumor–host cell interactions may also be important for tumor cell behavior.

A similar approach was used by Kang and colleagues,[19] who investigated the molecular basis for osteolytic bone metastases using the MDA-MB-231 human BCC line as a model. In this study, the authors were able to compare gene expression profiles from various in vivo selected BCC subpopulations with distinct capacities (from low to high) for osteolytic bone metastases when they were inoculated into the left cardiac ventricle of immunodeficient mice. Several conclusions can be drawn from this study. The first is that a gain in bone metastatic activity by MDA-MB-231 subpopulations does not involve an increased expression of the previously reported poor prognosis gene signature by van't Veer et al.[13] Second, it appears that there is a distinct bone metastases gene expression profile that does not overlap with, and is superimposed on, the poor prognostic signature present in the parental MDA-MB-231 BCC with a low capacity for bone metastases. Interestingly, most of the genes in this signature that were overexpressed by more than fourfold encode cell membrane or secretory products that may affect the host environment to favor metastases. They include the bone homing *CXCR4,* the angiogenesis factors *FGF-5* and *CTGF,* the activator of osteoclast differentiation *IL-11,* the matrix metalloproteinase/collagenase *MMP1, folistatin,* the metalloproteinase-disintegrin family member *ADAMTS1,* and *proteoglycan-1.* The third conclusion is that metastasis requires a set of functions beyond those underlying the emergence of the primary tumor.

Summary and Conclusions from the Collective Results of Breast Cancer Microarray-Based Profiling Studies

First, tumors can be grouped according to their composite expression profiles. Such tumor groupings are often related to other molecular or cellular characteristics with known clinical relevance. For example, several studies identified expression signatures related to the well-established major molecular discriminator for breast cancer classification, the ER status. The comprehensive measurement of gene expression in these tissues also provides evidence for many secondary features of potential relevance to clinical practice. For example, *erb*B2 positivity appears to be associated with a distinct expression cluster of ER-negative/basal type tumors. However, although the basal subtype was repeatedly recognized as a distinct group and should be considered as a separate disease with respect to treatment and follow-up, the other subtypes are less clear and require refinement of their molecular definition before they can be reliably defined and diagnosed. Second, these results suggest that molecular profiling might be able to substantially refine cancer prognosis, perhaps well beyond what is possible with other clinical indicators. Third, all these molecular signatures might be generalizable to populations other than those in which they were initially developed and probably across multiple microarray platforms and technologies. These hypotheses, however, need to be validated before implementation of these molecular signatures in the clinic. Finally, the molecular signatures of BCCs captured by microarray technology suggest that the potential of BCCs to metastasize may be genetically established earlier in the disease than previously thought and may be tissue specific.

Molecular Classification of Hereditary and Familial Non-BRCA1/BRCA2 Breast Cancer

The inheritance of mutant *BRCA1* and *BRCA2* genes confers a lifetime risk of breast cancer of 50–85% and a lifetime risk of ovarian cancer of 15–45% (see Chapters 4 and 26). One of the major functions of BRCA1 and BRCA2 proteins is to participate in DNA repair and homologous recombination. Thus, in humans, breast cancer tumors in carriers of mutant *BRCA1* and *BRCA2* genes are characterized by a large number of chromosomal changes, some of which differ depending on the genotype.[20–22]

Tumors with a *BRCA1* mutation are generally negative for both ER and progesterone receptors and are high grade with a high mitotic index, whereas most tumors with a *BRCA2* mutation are ER positive, suggesting that the mutant *BRCA1* and *BRCA2* genes induce the formation of breast tumors through different pathways of tumorigenesis. Additionally, both *BRCA1*- and *BRCA2*-associated breast carcinomas show a low frequency of c-*erb*B2 expression.

To determine whether there are distinct patterns of global gene expression between sporadic and *BRCA1*- and *BRCA2*-related cancers, Hedenfalk et al.[23] compared gene expression profiles derived from breast cancer tissues from patients carrying *BRCA1* and *BRCA2* mutations to breast tissues from sporadic cases. The authors identified 51 genes whose variation in expression among all the experiments best differentiated the three types of cancers, suggesting that gene expression profiles of *BRCA1* and *BRCA2* are generally distinctive from each other as well as from those of sporadic tumors. Moreover, to better decipher differences in gene expression between *BRCA1* mutation– and *BRCA2* mutation–positive tumors, the authors, applying different statistical algorithms, were able to establish a 176-gene signature that best discriminated *BRCA1* mutation– and *BRCA2* mutation–positive tumors. A large set of genes that were up-regulated in *BRCA1* mutation–positive tumors but not in *BRCA2* mutation–positive samples were involved in DNA repair pathways that associated to the cellular response to stress and apoptosis, suggesting that the *BRCA1* mutation might lead to a constitutive stress type state. These results show that inheritable mutations may influence the global patterns of gene expression of a tumor and that although both mutations lead to breast and ovarian cancer, there are significant differences in their gene expression programs. Similar results have been reported by the Amsterdam group.[13]

Although germline mutations in *BRCA1* and *BRCA2* account for most familial breast–ovarian cancer cases, these mutations can only explain a small proportion of familial susceptibility. Very little is known about the genetic basis of non-*BRCA1*/*BRCA2* breast cancers families, namely *BRCAx*. Histopathologic studies have revealed that *BRCAx* tumors are cytologically heterogeneous but are generally of lower grade, with lower mitotic activity than *BRCA1*, *BRCA2*, and sporadic tumors.[24]

Recently, using a complementary strategy of global gene expression profiling, Hedenfalk et al.[25] subclassified *BRCAx* families into genetically more homogeneous groups. In this study, the investigators analyzed gene expression profiles of 16 tumors from eight *BRCAx* families by using cDNA microarrays. Based on the expression profiles and the class discovery method that the authors applied, *BCRAx* tumors could be segregated into two distinct groups represented by 60 significant genes. Many of the genes with increased expression in one of the two groups identified encode for ribosomal proteins, suggesting different capacities for protein synthesis between these groups. Additionally, comparative genomic hybridization analysis in a subset of tumor samples revealed significant differences in copy numbers of 262 cDNA clones between the two groups. Of interest, the localization of these genes suggests the presence of common regions of alterations within the *BRCAx* subgroups, confirming their molecular differences. To further exclude the possibility that this 60 gene expression discriminator was related to unidentified *BRCA1* and *BRCA2* mutations, the authors included in their analysis a number of tumors from known *BRCA1* and *BRCA2* mutation carriers. As expected, neither *BRCA1* nor *BRCA2* tumors were mixed with the *BRCAx* samples when hierarchical clustering and multidimensional scaling analysis was performed, supporting their underlying differences.

These phenotypic differences between *BRCA1* mutation–positive and non-*BRCA1*/*BRCA2*-related breast cancers were furthermore supported by Foulkes et al.[26] In this study, the investigators using immunohistochemistry showed that *BRCA1* mutation–positive tumors were more likely than the non-*BRCA1*/*BRCA2*-related breast cancers to express a basal epithelial phenotype. Among the 292 analyzed breast cancer specimens from Ashkenazi Jewish women, the authors identified 76 that did not overexpress ER or *erb*B2. Of the 72 specimens with sufficient material for testing, 40 expressed stratified epithelial cytokeratin 5 and/or 6, the hallmark of the myoepithelial/basal signature. In a univariate analysis, the expression of cytokeratin 5/6 was associated with *BRCA1*-related breast tumors in a statistically significant way, suggesting its distinctive phenotype. Of interest, these results were furthermore supported by the recent publication by Sorlie et al.,[12] in which all 18 *BRCA1*-related breast tumors had a gene expression profile consistent with a basal-like phenotype.

Similarly, Palacios et al.,[27] using tissue microarrays, showed distinct morphologic and immunohistochemical features in the non-*BRCA1*/*BRCA2* compared with the *BRCA1* tumors as previously reported, whereas *BRCA2* tumors had intermediate characteristics between the two phenotypes. Indeed, non-*BRCA1*/*BRCA2* tumors were more frequently low grade and ER, progesterone receptor, and BCL2 positive, with a low proliferation rate. In contrast, *BRCA2* tumors were distinguished by a higher proliferation rate, frequent normal E-cadherin expression, and a higher frequency of c-*myc* amplification. A comparison with age-matched sporadic cancers showed that non-*BRCA1*/*BRCA2* tumors were less aggressive, less proliferative, and had a lower grade than the sporadic breast tumors matched by age.

In summary, inheritable mutations such as *BRCA1* and *BRCA2* may influence the global patterns of the gene expression of a tumor. Additionally, using large-scale gene expression based class discovery, followed by conventional positional linkage, candidate gene analysis may be an effective approach to finally identify novel breast cancer predisposition genes.

Gene Expression Profiling and the Prediction of Treatment Response in Breast Cancer

One of the most challenging issues of pharmacogenomic research is the identification of markers that can accurately predict patient response to drugs, because it would facilitate the individualization of patient treatment. Such an approach is particularly needed in cancer therapy, where commonly used agents are ineffective in many patients and where side effects are common. Thus, predictions of response have been attempted on the basis of the tumor expression of proliferation and apoptosis markers,[28] endocrine and growth factors, and oncogenes.[29] However, no single tumor marker has been shown to possess a sufficient predictive value to render it clinically useful. The recent development of DNA microarrays raises the possibility of an unbiased genome-wide approach to the genetic basis of drug response (see Chapter 28).

Thus, using oligonucleotide microarrays, Staunton et al.[30] sought to determine whether gene expression signatures in each of the 60 cell lines (originated from different tissue types) in the NCI-60 panel could predict chemosensitivity to several thousands of chemical compounds. Applying this technique, they were able to identify statistically significant gene expression-based classifiers of sensitivity or resistance for 232 compounds based on both drug sensitivity and gene expression data, suggesting the feasibility of individualizing patient treatments based on genetic tumor features.

Similarly, Levenson et al.,[31] using nylon membrane arrays, studied the mechanisms of hormone sensitivity by investigating gene expression profiles after ER activation by four different selective ER modulators and the pure antiestrogen ICI 182,780 in ER-negative MDA-MB-231 BCC line stably transfected with a wild-type ER. The authors found that gene expression patterns of cells after treatment with pure anti-estrogen were not dramatically different from after treatment with selective ER modulators, indicating the activation of a large number of genes through ER-independent mechanisms. Additionally, based on a comprehensive analysis of their array data, they were able to identify "agonistic" and "antagonistic" genes and to propose a selective ER modulator-induced differential signaling pathway. Other groups reported similar results, highlighting the diversity of gene networks and metabolic cell regulatory pathways undertaken by selective ER modulators.[32,33]

However, the prediction of drug sensitivity in the clinic is particularly challenging because drug responses reflect not only properties intrinsic to the target cells, but also interactions between tumor cells and host and host metabolic properties. To date, few studies have applied such a genome-wide approach using clinical material to identify gene expression signatures that could predict drug sensitivity in breast cancer.

In a small pilot study, our group analyzed a set of fine-needle aspiration (FNA) samples from 10 patients before and after primary medical treatment with adriamycin/cyclophosphamide (AC) chemotherapy.[34] Our goal was to determine the feasibility of obtaining representative cDNA expression array profiles from FNA samples performed on breast carcinomas using RNA linear amplification and to correlate such profiles with subsequent clinical response after neoadjuvant chemotherapy. We demonstrated that expression profiles derived from breast tumor FNA samples could be obtained in a reliable and reproducible manner. Similar findings were also obtained by other groups either using FNA samples or core biopsies as tissue material.[7,35–37] Moreover, we provided proof of principle that profiles derived before treatment and changes in profiles shortly after starting treatment may have the potential to predict clinical outcomes with anthracycline-based chemotherapy in individual patients. Using this approach, we were able to identify candidate genes and pathways that may be involved in intrinsic chemoresistance and prove to be useful in identifying targets for the treatment of drug-resistant carcinomas.

With FNA samples, Ayers et al.[38] used a similar approach to evaluate the profile-based prediction of complete pathologic response to neoadjuvant paclitaxel chemotherapy in 24 breast cancer patients. Interestingly, they were able to identify a set of genes that could predict complete pathologic response, whereas individual genes had a limited discriminating power to classify patients into complete responders versus nonresponders.

To investigate whether gene expression profiles could predict a response to docetaxel, Chang et al.[39] analyzed gene expression signatures derived from core biopsies taken before treatment from 24 breast cancer patients. In their study, tumor samples were classified as sensitive or resistant to chemotherapy on the basis of the tumor residual volume at the end of the treatment. Using oligonucleotide-based arrays, the authors identified differential patterns of expression for 92 genes that were significantly correlated with clinical response. In leave-one-out cross-validation analysis, 10 of 11 sensitive tumors (90% specificity) and 11 of 13 resistant tumors (85% sensitivity) were correctly classified, with an accuracy of 88%. Many of the genes highly expressed in the sensitive tumors were involved in cell cycle, cytoskeleton, adhesion, protein transport, transcription, stress, and apoptosis, whereas resistant tumors showed increased expression of genes involved in some transcriptional and transduction pathways. Even though the number of samples was small, the investigators were able to validate their results in an independent set of six patients.

Concluding Remarks and Perspectives

Although microarray analysis of breast cancer has provided valuable information for classifying tumors on a molecular basis and in predicting the clinical outcome and drug response, several significant issues need to be addressed before this powerful tool can be brought into the clinic. Within the next decade, as the cost of conducting microarray experiments is expected to decrease, more academic investigators will include this technology in their arsenal of tools. Microarrays might then be routinely used in the selection, assessment, and quality control of the best drugs for pharmaceutical development as well as for diagnosis and for monitoring desired and adverse outcomes of therapeutic interventions. Nevertheless, despite the potentially enormous benefits of microarrays to public health, challenges must be met to ensure the seamless incorporation of this technology into medical practice. Quality control and quality assurance must be established. The determination of appropriate levels of analytical and biologic validation needed for each medical application of microarrays and their supporting computer-based bioinformatics systems raise new challenges.

These needs are exemplified by a comparison of results generated from experiments done on different microarray platforms (cDNA vs. short oligonucleotides), which showed that although there was a similar pattern of expression for some of the genes, there was a large variation in expression between the platforms tested.[40,41] All this confirms the requirement for standardization of the microarray technique and the need to validate the expression pattern of genes of interest by an alternative RNA quantitative method, such as Northern blot, quantitative reverse transcriptase-PCR, or RNase protection assay, above all when precise quantification is mandatory. Such a multiplatform validation study is currently under investigation in our laboratory with the collaboration of several European institutions involving several distinct breast cancer populations (FIG. 6.3). One of our goals is to establish protocols/classifiers for the molecular profiling of human breast tumors in a multiplatform and in a multipopulation based approach. However, we predict that the implementation of these protocols/classifiers as reliable and tools in daily clinical practice will take some time.

The number of relevant publications in the field of microarrays is increasing exponentially. From 1995 to 1997 fewer than 10 reports featured microarray data. By the time that this review was written, however, approximately 5000 reports had been published in this burgeoning field, with 240 in the breast cancer research area. Thus, to ensure the interpretability of the experimental results generated by the use of microarrays as well as their potential independent validation, there is a crucial need for standardization of data collection. Unfortunately, this is currently lacking in most published microarray reports. Such an initiative has been proposed by the Microarray Gene Expression Data Society, which developed guidelines for submitting microarray data for publication known as "Minimum Information About a Microarray Experiment."[42] These guidelines, which intend to facilitate the interpretation and verification of microarray results, are currently accepted by most journals and must be applied by every investigator involved in microarray studies (for more information see www.mged.org/miame).

An important challenge posed by this technology lies in discerning the biologic meaning of the huge volume of array data. Although no exact rule exists regarding the statistical approaches required for such analysis, various methods are continually being developed. Because DNA microarrays are used for a variety of different purposes (i.e, "class comparison," "class prediction," or "class discovery" studies), the analysis strategy should be determined in light of the overall objectives of the study. Moreover, because it is likely that gene expression profiles will provide information that might affect clinical decision making, such profiling studies must be performed with statistical rigor and must be reported clearly with unbiased statistics, which again are lacking in several microarray studies published so far.[43]

Thus, the validation of all these promising results in larger, independent, and, if possible,

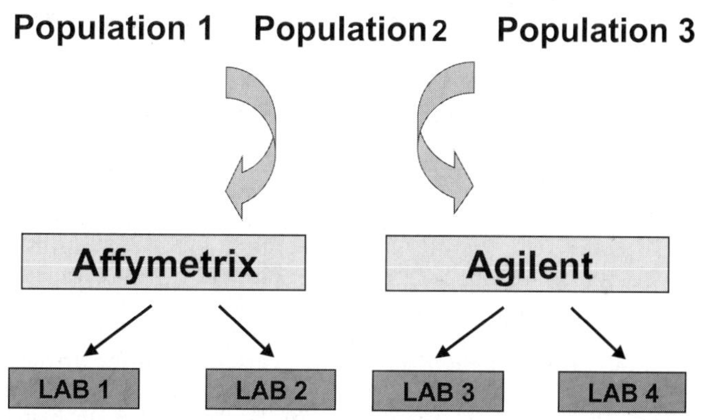

FIGURE 6.3 Example of a multipopulation, multiplatform, and multilaboratory microarray validation/standardization study.

prospective series is urgently needed. With respect to the potential prognostic value of a breast cancer gene signature, the European Organization for Research and Treatment of Cancer Breast Cancer Group (EORTC-BCG) is working to launch such a prospective validation trial with the support of the Breast International Group and a grant from the European Union. The study is designed to compare (in 5000 women with node-negative breast cancer) treatment selection on the basis of classical prognostic/predictive factors (St. Gallen 2003 Guidelines, 2500 patients) or by the expression signature of the 70 genes identified by the Amsterdam group (2500 patients). Although it is anticipated that the clinical outcome will be similar for the two groups, it is thought that the need for adjuvant chemotherapy will be reduced by 10–20% in the group managed according to their tumor gene expression profile. Because ongoing research might show that gene expression arrays are of value in selecting optimal treatment at the time of relapse, our plan is to offer this technology to all women participating in the trial but with a delayed potential use in the "control" group.

Regarding prediction of response to therapy (chemotherapy and endocrine therapy) in breast cancer, our group, with the collaboration of many European centers, launched an ambitious neoadjuvant research program to validate and refine previous results in a prospective manner. Our goal is to establish gene expression signatures associated with response to anthracycline- and endocrine-based therapy. The knowledge we will derive from these studies may substantially improve our ability to provide individualized treatment for breast cancer patients. In oncology we are clearly at a transition point between empirical and molecular medicine, and there is a crucial need to break from tradition in the philosophy and the design of our clinical trials if we want "tailored treatment" to become a reality in a not too distant future.

References

1. Greene FL, Page DL, Fleming ID, et al. (Eds). American Joint Committee for Cancer Staging (AJCC) Staging Manual (6th ed). New York: Springer Co., 2002.
2. Diab SG, Clark GM, Osborne CK, et al. Tumor characteristics and clinical outcome of tubular and mucinous breast carcinomas. J Clin Oncol 1999;17:1442–1448.
3. Fisher ER, Redmond C, Fisher B, et al. Pathologic findings from the National Surgical Adjuvant Breast and Bowel Projects (NSABP). Prognostic discriminants for 8-year survival for node-negative invasive breast cancer patients. Cancer 1990;65:2121–2128.
4. Barrett JC, Kawasaki ES. Microarrays: the use of oligonucleotides and cDNA for the analysis of gene expression. Drug Discov Today 2003;8:134–141.
5. Perou CM, Sorlie T, Eisen MB, et al. Molecular portraits of human breast tumours. Nature 2000;406:747–752.
6. West M, Blanchette C, Dressman H, et al. Predicting the clinical status of human breast cancer by using gene expression profiles. Proc Natl Acad Sci USA 2001;98:11462–11467.
7. Pusztai L, Ayers M, Stec J, et al. Gene expression profiles obtained from fine-needle aspirations of breast cancer reliably identify routine prognostic markers and reveal large-scale molecular differences between estrogen-negative and estrogen-positive tumors. Clin Cancer Res 2003;9:2406–2415.
8. Gruvberger S, Ringner M, Chen Y, et al. Estrogen receptor status in breast cancer is associated with remarkably distinct gene expression patterns. Cancer Res 2001;61:5979–5984.
9. Desai KV, Xiao N, Wang W, et al. Initiating oncogenic event determines gene-expression patterns of human breast cancer models. Proc Natl Acad Sci USA 2002;99:6967–6972.
10. Sorlie T, Perou CM, Tibshirani R, et al. Gene expression patterns of breast carcinomas distinguish tumor subclasses with clinical implications. Proc Natl Acad Sci USA 2001;98:10869–10874.
11. Sotiriou C, Neo SY, McShane LM, et al. Breast cancer classification and prognosis based on gene expression profiles from a population-based study. Proc Natl Acad Sci USA 2003;100:10393–10398.
12. Sorlie T, Tibshirani R, Parker J, et al. Repeated observation of breast tumor subtypes in independent gene expression data sets. Proc Natl Acad Sci USA 2003;100:8418–8423.
13. van't Veer LJ, Dai H, van de Vijver MJ, et al. Gene expression profiling predicts clinical outcome of breast cancer. Nature 2002;415:530–536.
14. Lacroix M, Leclercq G. Relevance of breast cancer cell lines as models for breast tumours: an update. Breast Cancer Res Treat 2004;83:249–289.
15. van de Vijver MJ, He YD, van't Veer LJ, et al. A gene-expression signature as a predictor of survival in breast cancer. N Engl J Med 2002;347:1999–2009.
16. Huang E, Cheng SH, Dressman H, et al. Gene expression predictors of breast cancer outcomes. Lancet 2003;361:1590–1596.
17. Ramaswamy S, Ross KN, Lander ES, et al. A molecular signature of metastasis in primary solid tumors. Nat Genet 2003;33:49–54.
18. Woelfle U, Cloos J, Sauter G, et al. Molecular signature associated with bone marrow micrometastasis in human breast cancer. Cancer Res 2003;63:5679–5684.
19. Kang Y, Siegel PM, Shu W, et al. A multigenic program mediating breast cancer metastasis to bone. Cancer Cell 2003;3:537–549.
20. Ford D, Easton DF, Bishop DT, et al. Risk of cancer in BRCA1-mutation carriers. Lancet 1994;343:692–695.

21. Easton DF, Ford D, Bishop DT. Breast and ovarian cancer incidence in BRCA1-mutation carriers. Am J Hum Genet 1995;56:265–271.
22. Ford D, Easton DF, Stratton M, et al. Genetic heterogeneity and penetrance analysis of the BRCA1 and BRCA2 genes in breast cancer families. Am J Hum Genet 1998;62:676–689.
23. Hedenfalk I, Duggan D, Chen Y, et al. Gene-expression profiles in hereditary breast cancer. N Engl J Med 2001;344:539–548.
24. Lakhani SR, Gusterson BA, Jacquemier J, et al. The pathology of familial breast cancer: histological features of cancers in families not attributable to mutations in BRCA1 or BRCA2. Clin Cancer Res 2000;6: 782–789.
25. Hedenfalk I, Ringner M, Ben-Dor A, et al. Molecular classification of familial non-BRCA1/BRCA2 breast cancer. Proc Natl Acad Sci USA 2003;100:2532–2537.
26. Foulkes WD, Stefansson IM, Chappuis PO, et al. Germline BRCA1 mutations and a basal epithelial phenotype in breast cancer. J Natl Cancer Inst 2003; 95:1482–1485.
27. Palacios J, Honrado E, Osorio A, et al. Immunohistochemical characteristics defined by tissue microarray of hereditary breast cancer not attributable to BRCA1 or BRCA2 mutations: differences from breast carcinomas arising in BRCA1 and BRCA2 mutation carriers. Clin Cancer Res 2003;9(10 Pt 1):3606–3614.
28. Ellis PA, Smith IE, McCarthy K, et al. Preoperative chemotherapy induces apoptosis in early breast cancer. Lancet 1997;349:849.
29. Chang J, Ormerod M, Powles TJ, et al. Apoptosis and proliferation as predictors of chemotherapy response in patients with breast carcinoma. Cancer 2000;89: 2145–2152.
30. Staunton JE, Slonim DK, Coller HA, et al. Chemosensitivity prediction by transcriptional profiling. Proc Natl Acad Sci USA 2001;98:10787–10792.
31. Levenson AS, Svoboda KM, Pease KM, et al. Gene expression profiles with activation of the estrogen receptor alpha-selective estrogen receptor modulator complex in breast cancer cells expressing wild-type estrogen receptor. Cancer Res 2002;62:4419–4426.
32. Frasor J, Danes JM, Komm B, et al. Profiling of estrogen up- and down-regulated gene expression in human breast cancer cells: insights into gene networks and pathways underlying estrogenic control of proliferation and cell phenotype. Endocrinology 2003;144:4562–4574.
33. Coser KR, Chesnes J, Hur J, et al. Global analysis of ligand sensitivity of estrogen inducible and suppressible genes in MCF7/BUS breast cancer cells by DNA microarray. Proc Natl Acad Sci USA 2003;100: 13994–13999.
34. Sotiriou C, Powles TJ, Dowsett M, et al. Gene expression profiles derived from fine needle aspiration correlate with response to systemic chemotherapy in breast cancer. Breast Cancer Res 2002;4:R3.
35. Sotiriou C, Khanna C, Jazaeri AA, et al. Core biopsies can be used to distinguish differences in expression profiling by cDNA microarrays. J Mol Diagn 2002;4:30–36.
36. Assersohn L, Gangi L, Zhao Y, et al. The feasibility of using fine needle aspiration from primary breast cancers for cDNA microarray analyses. Clin Cancer Res 2002;8:794–801.
37. Symmans WF, Ayers M, Clark EA, et al. Total RNA yield and microarray gene expression profiles from fine-needle aspiration biopsy and core-needle biopsy samples of breast carcinoma. Cancer 2003;97: 2960–2971.
38. Ayers M, Symmans WF, Stec J, et al. Gene expression profiles predict complete pathologic response to neoadjuvant paclitaxel/FAC chemotherapy in breast cancer. J Clin Oncol 2004;22:1–10.
39. Chang JC, Wooten EC, Tsimelzon A, et al. Gene expression profiling for the prediction of therapeutic response to docetaxel in patients with breast cancer. Lancet 2003;362:362–369.
40. Kuo WP, Jenssen TK, Butte AJ, et al. Analysis of matched mRNA measurements from two different microarray technologies. Bioinformatics 2002;18: 405–412.
41. Tan PK, Downey TJ, Spitznagel EL Jr, et al. Evaluation of gene expression measurements from commercial microarray platforms. Nucleic Acids Res 2003;31:5676–5684.
42. Brazma A, Hingamp P, Quackenbush J, et al. Minimum information about a microarray experiment (MIAME)-toward standards for microarray data. Nat Genet 2001;29:365–371.
43. Simon R, Radmacher MD, Dobbin K, et al. Pitfalls in the use of DNA microarray data for diagnostic and prognostic classification. J Natl Cancer Inst 2003; 95:14–18.

CHAPTER 7

Cytology: Clinical Sample Procurement and On-Slide Technologies

W. Fraser Symmans, MD

Department of Pathology, University of Texas MD Anderson Cancer Center, Houston, Texas

Cytology is an evolving discipline of anatomic pathology that strives to obtain the most information from the least sample. That is achieved through diagnostic accuracy and the support of molecular pathologic assays that can be applied to these small samples. Cytology provides both a simple test to obtain diagnoses and a flexible tool to obtain cancer cells for translational research studies. The procedures to obtain cytologic samples from the breast are minimally invasive and well tolerated by patients. Cytologic samples can be processed immediately so that microscopic assessment is used to determine adequacy of the biopsy sample and to render a preliminary diagnosis. This provides a powerful clinical tool, because management decisions can be made with the aid of real-time pathologic diagnoses. Furthermore, cytologic samples (particularly from fine-needle aspiration [FNA]) of breast cancer consist of a relatively pure sample of breast cancer cells that are removed directly from the tumor and can be optimally preserved in any medium within seconds. Therefore, cytologic samples have ideal characteristics for many molecular assays with relevance to breast cancer diagnosis and management or translational research that links molecular research to clinical outcomes. Cytology is widely used in breast cancer centers where full advantage is taken of the tolerability of sampling, immediacy of assessment and sample procurement, and flexibility of its application to diagnostic, management, and research algorithms.

Cytology Samples for Pathologic Diagnosis

Whenever a cellular sample can be obtained, a cytologic diagnosis can be rendered. The main issue for all biopsies, including cytology, is the quality of sampling. That is contingent on correct localization of the lesion (clinical or by radiologic guidance) and effective sampling technique. Preparation of the sample for diagnosis is also important. When a cellular sample is obtained by fine needle aspiration, it can be divided onto glass slides to be smeared before fixation and staining (direct smears). Alternatively, samples are mixed into a preservative fluid and then aliquots are placed onto glass slides using an automated device, fixed, and stained (liquid based). The quality of cytologic diagnoses is excellent when methods are correctly applied. Half-hearted attempts at biopsy without proper completion of the procedure or on-site assessment of sample adequacy are a likely source of inadequate sampling and false-negative diagnoses.[1]

Sampling error is the most frequent cause of an incorrect diagnosis for needle biopsies of any type. Sometimes the microscopic appearance of the cells is indeterminate (i.e., both malignant and benign characteristics), and then the diagnosis should be stated as atypical (if really indeterminate) or suspicious for malignancy (if likely to be malignant). As a general guide, less than 10% of diagnoses should be stated as atypical or suspicious, and approximately 30% of atypical diagnoses and 80% of suspicious diagnoses should prove to be malignant.[1–3] Obviously, any diagnosis of insufficient (nondiagnostic), atypical, suspicious, or malignant requires further medical assessment and/or intervention. Confidence in biopsy diagnoses from cells (cytology) or tissue (histology) for patient care in the community depends most on the accuracy of benign diagnoses (measured by the negative predictive value) because no further management may be indicated after a benign result.[1] There must also be confidence in the accuracy of malignant diagnoses (measured by the positive predictive value) because definitive management may follow.[1] Ideally, the negative and positive predictive values for the initial diagnosis of primary breast lesions are the same for FNA biopsy (cytology) and core needle biopsy (tissue cores) of mass lesions.[1]

Core needle biopsies offer a distinct advantage for the diagnosis of primary breast cancer for two main reasons: pathologic assessment of tissue sections from core biopsies allows for confident diagnosis of invasion (error is less than 1%), and preservation of tissue cores in paraffin provides an archival resource of the patient's tumor tissue that may be important if new tests become necessary in the future or if the patient elects to have neoadjuvant (preoperative) treatment and achieves complete pathologic response. Documentation of invasion in the core needle biopsy is used as justification to surgically sample the axillary lymph nodes for staging information (sentinel lymph node biopsy or axillary lymph node dissection) and as justification for prescribing neoadjuvant chemotherapy. In some practices the immediacy of cytologic assessment is used to assess the primary lesion at the time of biopsy, either by scouting FNA of the lesion before core biopsy or by touch imprint cytology of the tissue cores for microscopic assessment of adequacy and the presence of malignant cells.[4,5]

Cytologic samples are commonly obtained for the assessment of possible metastatic disease in regional lymph nodes. In this setting, documentation of invasion is not important, and the goal of sampling is to recognize, or confidently exclude, the presence of breast cancer cells. Clinically positive axillary or supraclavicular lymph nodes are usually sampled by FNA for cytologic diagnosis. At some breast cancer centers, patients with newly diagnosed breast cancer and clinically negative axillae will have an ultrasound evaluation of the axillary nodal basin to look for possible nodal metastasis. If a suspicious (abnormal) lymph node is identified, an ultrasound-guided FNA biopsy of that lymph node is performed.[6,7] Most patients with a macrometastasis will be identified this way.[6,7] Sentinel lymph node biopsy is then only offered to patients who have negative axillary studies. The consequences are that node-positive disease is recognized early so appropriate definitive treatment plans can be decided, and so the frequency of metastatic disease in sentinel lymph node biopsies is decreased, being mostly limited to small metastases and micrometastases.[6,7] Pathologic evaluation of each sentinel lymph node involves fine serial slices (i.e., 2–3 mm thickness) for thorough macroscopic and microscopic study of the nodal tissue. Imprint cytology slides of all cut tissue surfaces are easily obtained for intraoperative assessment without loss of any nodal tissue for the final histopathologic diagnosis.

Imprint slides are stained with a routine stain and reported by the pathologist as an intraoperative consultation so that definitive diagnosis of metastasis will lead to completion axillary dissection during that surgery.[8] The accuracy of intraoperative imprint cytology is equivalent to frozen section but does not consume tissues from the permanent sections.[8–10] The combination of clinical examination, axillary ultrasound, and FNA biopsy during diagnostic workup and the intraoperative evaluation of sentinel lymph nodes with touch imprint cytology will minimize the number of patients who require a second operation for axillary lymph node dissection.

FNA cytology is the most common biopsy technique for diagnosis of possible distant metastases or disease recurrence.[11–14] Staging or follow-up radiologic studies may identify abnormal mass lesions that require pathologic diagnosis. Clinical examination may identify palpable abnormalities that require pathologic diagnosis. In this setting the presence of invasion is not necessary for diagnosis and management because identification of breast cancer cells will define the lesion as metastatic or recurrent. If tumor invasion must be defined in a local recurrence of breast cancer, then core needle biopsy is preferred. Usually, the less invasive FNA biopsy is suitable to obtain a pathologic diagnosis while minimizing procedural risk to the patient. Use of radiologic guidance or palpable biopsy technique to perform the FNA depends on the site and accessibility of the lesion. If the lesion is metastatic breast cancer, the sample is usually highly cellular and sufficient for appropriate molecular pathologic assays and/or translational research protocols.

Pleural, pericardial, peritoneal, cerebrospinal, or other fluids are collected for cytologic diagnosis through a needle and then centrifuged to enrich the cellular component before smearing or centrifuged directly onto glass slides (exfoliative cytology). The diagnostic challenge for the cytologic assessment of distant sites is to recognize breast cancer cells from other cells in the sample that may be normal, reactive, inflammatory, or malignant cells from a different tumor type.

Procurement and Preservation of Clinical Cytology Samples

FNA cytology samples offer the advantage of providing breast cancer cells directly from the tumor onto a slide or into a solution within seconds. However, FNA does provide a limited resource of cancer cells that should be appropriately preserved for specific tests. Advance planning is particularly important if some (or all) of the sample is to be used for specific molecular assays. Advance planning begins with communication and ordering the biopsy procedure. The person obtaining the sample and the person who will assess and process the sample need to know exactly what uses are planned in advance of the procedure. This affects the number of needle biopsies (passes) that are obtained, how each pass will be processed (e.g., number and types of slides prepared, fixatives used, preparation of cell blocks, preservation solutions used, storage temperatures), and even which anesthesia will be used.

Some tests cannot be performed on archival samples if ordered retroactively. A good practice is to indicate on the requisition orders what studies will be needed. If the biopsy is to be performed with radiologic guidance, then this communication has to be with both the radiologist who will perform the biopsy and the laboratory staff who will prepare the sample for each test. An example might be an order sheet that indicates that immunocytochemical studies for estrogen receptor (ER) and progesterone receptor (PR) and fluorescence in situ hybridization (FISH) for HER-2/*neu* gene copy number be performed if the diagnosis of a liver mass is confirmed as metastatic breast cancer. In that example, an immediate interpretation of slides from the first pass could be performed and slides for immunohistochemistry (IHC) and FISH assays would be optimally prepared. If necessary, a dedicated second pass can be obtained specifically for the ancillary tests. Alternatively, a paraffin "cell block" or liquid cell preparation can be prepared in anticipation of future requests for such assays, but those preparations usually require additional passes or cost. The success of most molecular assays depends on optimal sample preparation.

Some of these assays are the sole determinants of whether a patient is eligible for a specific treatment (e.g., endocrine therapies or trastuzumab) and false-negative results from improper sample preparation can be avoided. Under ideal circumstances, the FNA sample of breast cancer is placed in the optimal preservative conditions for each molecular assay within seconds of extraction from the tumor.

Immunocytochemistry and In Situ Hybridization

These are the dominant "on-slide" molecular pathologic assay methods. Immunochemistry is widely used in surgical pathology and cytology to identify specific proteins and determine their expression and localization within cells. A specific antibody to the protein of interest is applied to the sample and then a series of secondary antibody and detection molecules are applied to the sample, so that a visible signal is produced at the site of that protein. Microscopic examination of the slide is then used to identify the frequency and intensity of the signal for expression of that protein and the localization of the protein according to cell type (e.g., cancer cells, fibroblasts, endothelium, lymphocytes) and subcellular distribution (e.g., membrane, cytoplasm, or nucleus). IHC is the application of this method to tissue sections on slides, and immunocytochemistry (ICC) is the application of this same method to cytologic preparations on slides. In situ hybridization is an analogous method that uses nucleotide sequences instead of antibodies and is therefore used to identify specific DNA or RNA sequences. Signal is only visible if there was hybridization of the specific oligonucleotide probe to the targeted DNA or RNA nucleotide sequence and the detection molecular reactions could occur.

There are many applications of immunochemistry in diagnostic pathology. For breast cancer, the IHC expression of many molecules has been tested for possible prognostic relevance. No single IHC assay has achieved the prognostic power sufficient to solely influence patient management decisions. However, immunochemistry is routinely used to identify expression of specific therapeutic targets and so determine eligibility to receive the treatment. IHC for ERs is a standard of care test

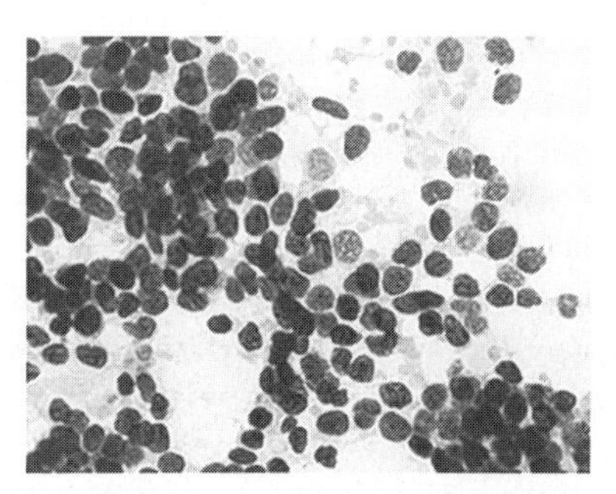

FIGURE 7.1 Immunocytochemical staining for ER demonstrates strong nuclear staining in the tumor cells but not in adjacent lymphocytes. The smear was immediately fixed in Carnoy's solution and stained with Papanicolaou stain. ICC was later performed using antigen retrieval and ER antibody clone 6F11.

for breast cancer (**FIG. 7.1**) and is usually combined with an IHC test for PRs. The results of assays for ER and PR (see Chapter 15) are used to determine whether a patient is eligible for hormonal (endocrine) treatment.[15,16] Both ER and PR assays can also be performed on slides from FNA biopsies and stain ER or PR in cancer cell nuclei. Scattered normal breast epithelial cell nuclei also stain positively for ER or PR by IHC/ICC and should be identified as an internal positive control when interpreting each slide. Failure to identify staining in normal cells may indicate technical failure of the assay and necessitate a repeat study.

Some general points about IHC are outlined with specific points relevant to ICC. There is considerable variability in the quality of different IHC assays. Foremost is the sensitivity and specificity of the primary antibody that detects the protein epitope of interest.[17,18] Extensive study using known control samples is needed before an IHC assay should be applied to any research or clinical use. Each step of the assay must be optimized and then standardized.[19] However, preanalytical variables are often the most problematic during assay development and when comparing samples from different institutions and applying the assay in different laboratories. These factors are the most common source of false-negative staining. The type and duration of fixation is critical to IHC.[20] Fixation is necessary to stabilize proteins and therefore stop the natural degradation of protein epitopes in the sample. Fixation protocols are not standardized, and so IHC methods in different laboratories do vary. Formalin (10% buffered formaldehyde) is the standard fixative solution used and leads to cross-linking of proteins in cells. This cross-linking is a variable and can shield or alter the epitope binding site for a primary antibody in the IHC assay. The extent of protein cross-linking is dependent on the conditions of fixation with formalin.

Most antibodies require an antigen retrieval pretreatment step before staining of formalin-fixed paraffin-embedded tissue sections with IHC to disrupt protein cross-links and expose the protein epitope.[21,22] Antigen retrieval methods usually involve enzymic digestion or heat to disrupt cross-links of proteins. Exact methods of antigen retrieval methods are determined for each antibody specific to each protein target and must be optimized in every laboratory.[23] Preanalytical factors (antigen retrieval methods) are the major source of false-negative IHC results for ER in diagnostic practice.[24–26] Because these methods are not currently standardized in all laboratories, the interpretation of IHC results for clinical care requires pathologists to be critical and vigilant and to make use of any internal positive and negative controls in the sample.

The importance of critical interpretation has been highlighted with IHC for HER-2/*neu*. IHC staining with antibodies to HER-2/*neu* (see Chapter 16) is commonly seen in the cytoplasm of tumor and normal cells, but only circumferential cell membrane localization should be interpreted as true positive staining. Membrane staining is then assessed using a semiquantitative score (1+ to 3+) depending on intensity. This effort to standardize the interpretation methods was necessary to try to identify patients with significant overexpression of HER-2/*neu* in the breast cancer cells so that those patients could be identified as eligible to receive trastuzumab treatment. Normal breast epithelial cells do not express enough HER-2/*neu* on the cell membrane for IHC detection, so interpretation of tumor staining should always be compared with normal breast epithelium as a negative control. The IHC scoring system is semiquantitative and inevitably subjective, so there is known variability in the interpretation of negative versus 1+ staining and for 1+ versus 2+ staining.[27–30] Generally, 3+ staining is accepted as true overexpression for clinical purposes. Although 2+ staining is defined as positive and 1+ staining is defined as negative, many laboratories perform an in situ hybridization assay to assess c-*erb*B2 gene copy number in those cases.[30–32]

Nuclear and cytoplasmic proteins are generally more reliably identified using ICC for cytology specimens than cell membrane proteins. This is partly because the membranes can be torn from cells during the sample acquisition and smearing onto slides. There can also be degeneration and disruption of cell membranes if cells dry on the slide before fixation. ICC assays for cell membrane proteins can therefore be difficult, and ICC for HER-2/*neu* is often not preferred, particularly if there is the capability to perform in situ hybridization for nuclear c-*erb*B2 gene copy number.[33–35]

The main application of in situ hybridization in breast cancer diagnosis is to determine the number of copies of c-*erb*B2 gene that codes for HER-2/neu protein. Overexpression of HER-2/*neu* occurs during the development of 20–30% of invasive breast cancer due to amplification of the c-*erb*B2 gene on the long arm of chromosome 17. IHC is used to detect HER-2/neu protein in the cell membrane, and in situ hybridization is used to detect c-*erb*B2 gene copies in the nucleus. The most common in situ hybridization assay for c-*erb*B2 gene uses a fluorescent-labeled probe that is very sensitive but requires fluorescence microscopy. One commercially available kit also provides a fluorescent-labeled probe that detects a centromeric sequence of chromosome 17 (cep17) and is visualized with a different light filter. Therefore, the FISH assay for c-*erb*B2 is reported as the average number of copies of c-*erb*B2 per cancer cell nucleus (at least four copies defined as amplification) and/or the ratio of the average number of gene copies of c-*erb*B2 compared with the average number of copies of chromosome 17 (more than two defined as amplification) (FIG. 7.2).

There are subtle implications to each method, because the absolute number of copies of c-*erb*B2 gene may determine the protein expression level of HER–2/*neu* in the breast cancer, whereas the ratio of c-*erb*B2:cep17 may correct for aneuploidy of chromosome 17 and select tumors with amplification of c-*erb*B2 gene on each copy of chromosome 17.[36] A chromogenic in situ hybridization test is newly available for detection of c-*erb*B2 gene using a chromagen that is stable and visible with light microscopy.[37,38] The hybridization signal is slightly less distinct than with fluorescence, and it is not possible to compare c-*erb*B2 copy number with a cep17 probe using chromogenic in situ hybridization, but the method does not require a fluorescence microscope for interpretation. Cytologic samples provide FISH signals of excellent quality because the entire nucleus is present on the slide so all cep17 and c-*erb*B2 signals are represented in every cell.[34]

Cytology in Clinical/Translational Research

The clinical utility of cytology has always depended on the ability to obtain a sample for pathologic diagnosis with a minimalist approach that exposes patients to the least discomfort, inconvenience, or risk. This utility is a major advantage for some applications in clinical/translational research of breast cancer, particularly because high resolution imaging methods are now available for safer needle guidance into small lesions in deep body sites and progress in the biomedical sciences has continued toward improved utilization of small samples, from polymerase chain reaction to genomic microarrays, multiplex immunochemistry assays, and even nanotechnology. Sample quality and preparation are critical to the success of these technologies in clinical/translational research and sample quantity is becoming less important.

Cytology is being used to identify patients at risk for development of breast cancer and to

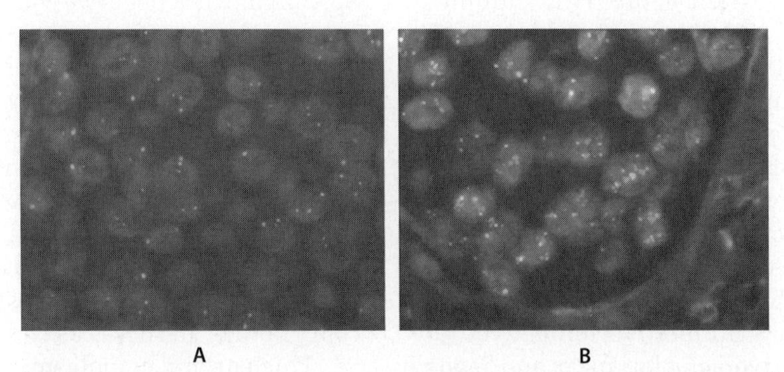

FIGURE 7.2 FISH assay for c-*erb*B2 gene copy number (orange signals) relative to chromosome 17 centromere (cep17) copy number (green signals) on FNA slides of breast cancer: (A) a tumor without amplification and (B) a tumor with amplification. **See Plate 10 for color image.**

assess chemoprevention strategies. FNA of normal breast is used to obtain epithelial cells for cytologic and molecular analyses.[39,40] The epithelial yield is relatively scant but is well suited for assays that are highly sensitive or include amplification. Repeat samples can be obtained for comparison at different times. One example would be to compare changes that occur after administration of a hormonal agent.[41] Cytologic analysis of fluid obtained through the nipple (see Chapter 5) is also being investigated as a method to obtain samples for microscopic and/or molecular studies. Cellular yield is also limited, except in cases of intraductal malignancy, but the cytologic appearance of those cells does indicate underlying pathology in the breast and/or risk of future breast cancer.[42–46] Polymerase chain reaction can be used to amplify DNA or complementary DNA derived from these cells, whereas the accompanying fluid can be used to perform immunochemistry, FISH, and proteomic experiments.[47–51]

There are two approaches to procure fluid from the ductal system of the breast. The simpler method is nipple aspiration for which negative pressure is used to extract fluid from the pretreated nipple. Nipple aspiration fluid volume is much less than ductal lavage fluid, but the procedure is simpler, faster, and more easily tolerated. Ductal lavage requires needle cannulation of selected nipple ducts from each duct orifice at the tip of the nipple. The dependent ductal system is then lavaged with sterile fluid through the cannulated nipple duct. Generally, this is repeated for two or three nipple ducts, but samples are not obtained through all the nipple ducts. This procedure is much more time consuming and costly than nipple aspiration, but it does obtain considerably more cellular and fluid sample and also represents more distal ducts. Epithelial cells and histiocytes are present in ductal fluids in varying numbers, but the fluid component is well suited for proteomic analyses, particularly to study secreted or shed proteins.[49,50] Random FNAs of the breast and nipple aspiration or ductal lavage approaches do not target a specific lesion in the breast or localize sites of abnormality but are important to study global changes within breast tissues or to identify evidence of shed tumor cells in the ductal system.

Serial FNA samples of breast cancer have been used to evaluate cellular or molecular changes from therapy. Neoadjuvant (preoperative) chemotherapy for larger and node-positive breast cancers provides important information about the tumor response to a specific chemotherapy regimen. Breast cancers in individual patients have variable response to a single treatment regimen. Furthermore, the response of a breast cancer to a neoadjuvant chemotherapy regimen is a very strong prognostic indicator of long-term survival. This clinical–pathologic information is unfortunately never available for patients who choose primary surgical management followed by adjuvant (postoperative) chemotherapy, because the tumor is excised before chemotherapy begins. Therefore, neoadjuvant chemotherapy trials provide important opportunities to study the changes that occur in breast cancers during chemotherapy and to learn how to predict likely response to specific treatments.

Pretreatment core needle biopsies are usually obtained for the initial diagnosis of breast cancer and for molecular pathologic assays. FNA has also been used to obtain breast cancer cells at specific times after chemotherapy to compare cellular and molecular responses in the tumor cells. For example, repeated FNAs demonstrated increased ER and PR expression and decreased transforming growth factor-α expression in ER-positive PR-positive breast cancers during tamoxifen treatment.[52,53] Decreased cell proliferation (Ki-67) after 10 days of tamoxifen was shown to be associated with complete or partial clinical response to treatment.[54] The same authors evaluated repeat FNAs at 10 and 21 days after chemoendocrine therapy for breast cancer and described decreased cell proliferation (Ki-67) and ER and increased PR in responsive patients.[55]

Cellular changes in apoptosis and proliferation that are induced by cytotoxic chemotherapy assessed by repeat FNA can predict the overall tumor response after several months of treatment. For example, our pilot study in women receiving paclitaxel for locally advanced breast cancer clearly showed that sustained apoptosis (several days) after the first dose correlated with tumor response after four cycles of treatment (FIG. 7.3).[56,57] Mitotic arrest and early apoptotic response did not correlate with response.[56] Persistent apoptosis (at 1–3 days) correlated with tumor response after different chemotherapy regimens (paclitaxel, docetaxel, anthracycline, mitoxantrone, methotrexate regimens).[58,59] There is also evidence that cellular proliferation and *bcl–2* expression decrease several days after the first chemotherapy dose in more responsive tumors.[58,59] The approach of

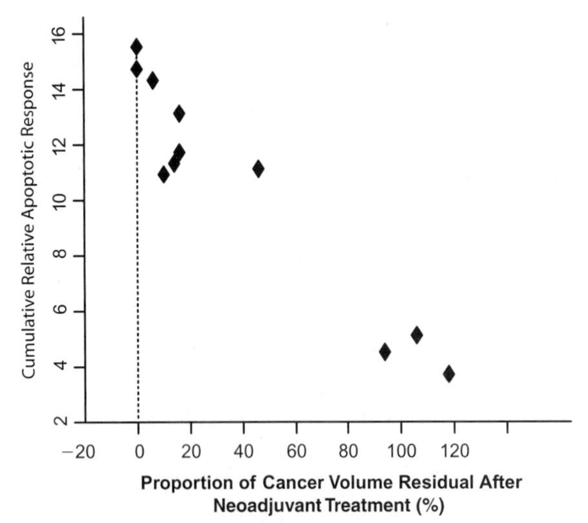

FIGURE 7.3 Cumulative relative apoptotic response during the first 4 days (y-axis) of neoadjuvant paclitaxel chemotherapy (from serial FNAs of 11 breast cancers) is strongly related to pathologic tumor response after four cycles of treatment ($r = -0.97$). Greater initial apoptotic response is predicts less residual cancer volume. (From Ref. 56, with permission.)

serial needle biopsies is not likely to become a standard clinical test, but the information learned from these translational studies is invaluable for understanding which in vivo cellular changes and molecular responses are the most critical to treatment response in our heterogeneous population of humans with breast cancer.

FNA and core needle biopsy contain subtle differences in cellular content. Using on-slide techniques (immunochemistry and in situ hybridization), the microscopic interpretation of staining will take the cell type into account. However, those differences should be understood when planning a translational biopsy study that does not use on-slide technologies. The fundamental difference between FNA and core needle biopsy is that core needle biopsy contains stromal tissues, whereas FNA does not. Using microscopic cell counts we found that FNAs of breast cancers contain a mean of 80% cancer cells, 15% lymphocytes, and 5% nonlymphoid stromal cells, whereas core needle biopsies contain a mean of 50% cancer cells, 20% lymphocytes, and 30% nonlymphoid stromal cells.[60] Most nonlymphoid stromal cells in breast cancer tissue samples are fibroblasts, but there is also increased angiogenesis. When we directly compared transcriptional profiles from microarrays derived from paired FNA and core needle biopsy samples of breast cancer, we found good correlations between samples from the same tumor, indicating that most transcriptional activity

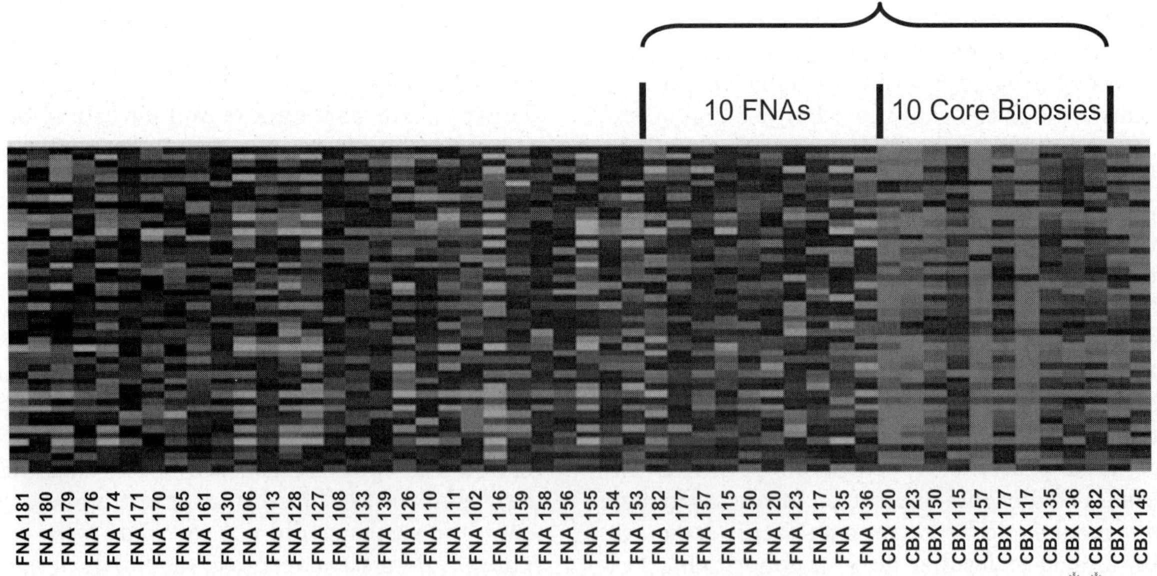

FIGURE 7.4 Differential expression of 51 genes in 10 matched core biopsies and corresponding fine-needle aspirates. These genes are strongly expressed in core needle biopsies, but not FNAs, in the entire study population of 38 FNAs and 12 core needle biopsies. Strong expression is red, and weak expression is green. These were predominantly stromal genes, most commonly endothelial. (From Ref. 60, with permission.) **See Plate 11 for color image.**

is from the cancer cells, but certain genes were differentially expressed according to sample type and are highly expressed in core needle biopsy samples (FIG. 7.4).[60] Although these genes are derived from nonlymphoid stromal cells, their relative expression is not related to the microscopic appearance of the tissue sections or from the microscopic cell counts from those samples.[60] Indeed, most stromal genes that were differentially expressed in core needle biopsy samples are likely to be of endothelial origin.[60]

The message from the study is that different cell types in a biopsy sample do not have identical transcriptional activity and that microscopic appearance or cell counts do not reliably correct for these differences. Most of the transcriptional profile from breast cancer tissues is contributed by the tumor cells, but endothelial cells are more transcriptionally active than other nonlymphoid stromal cells.[60] FNAs enrich for breast cancer cells and contain a minimal component of nonlymphoid stromal cells. Therefore, transcriptional profiles from FNAs of breast cancer are well suited for research of breast cancer gene expression but would be less desirable than core needle biopsy samples if the research question concerns angiogenesis or the tumor–stromal interface.[60] It is likely that there will be similar findings in proteomic studies because extracellular proteins and stromal cellular proteins will be present in tissue samples.

Summary

Cytology techniques offer many attributes for use in molecular oncology. Samples are limited but are increasingly appropriate for sophisticated molecular techniques. FNA biopsy is well tolerated by patients and is a safe and flexible clinical tool that obtains a fresh sample that is enriched with tumor cells. Optimal planning for sample procurement and processing enables molecular assays to be performed to assist clinical management and to advance clinical and translational research.

References

1. Symmans WF, Cangiarella JF, Gottlieb S, et al. What is the role of cytopathologists in stereotaxic needle biopsy diagnosis of nonpalpable mammographic abnormalities? Diagn Cytopathol 2001;24:260–270.
2. Symmans WF, Weg N, Gross J, et al. A prospective comparison of stereotaxic fine-needle aspiration versus stereotaxic core needle biopsy for the diagnosis of mammographic abnormalities. Cancer 1999;85: 1119–1132.
3. Cangiarella JF, Waisman J, Weg N, et al. The use of stereotaxic core biopsy and stereotaxic aspiration biopsy as diagnostic tools in the evaluation of mammary calcification. Breast J 2000;6:366–372.
4. Jacobs TW Silverman JF, Schroeder B, et al. Accuracy of touch imprint cytology of image-directed breast core needle biopsies. Acta Cytol 1999;43:169–174.
5. Green RS, Mathew S. The contribution of cytologic imprints of stereotactically guided core needle biopsies of the breast in the management of patients with mammographic abnormalities. Breast J 2001;7:214–218.
6. Krishnamurthy S, Sneige N, Bedi DG, et al. Role of ultrasound-guided fine-needle aspiration of indeterminate and suspicious axillary lymph nodes in the initial staging of breast carcinoma. Cancer 2002;95: 982–988.
7. Deurloo EE, Tanis PJ, Gilhuijs KG, et al. Reduction in the number of sentinel lymph node procedures by preoperative ultrasonography of the axilla in breast cancer. Eur J Cancer 2003;39:1068–1073.
8. Lee A, Krishnamurthy S, Sahin A, et al. Intraoperative touch imprint of sentinel lymph nodes in breast carcinoma patients. Cancer 2002;96:225–231.
9. Shiver SA, Creager AJ, Geisinger K, et al. Intraoperative analysis of sentinel lymph nodes by imprint cytology for cancer of the breast. Am J Surg 2002;184: 424–427.
10. Henry-Tillman RS, Korourian S, Rubio IT, et al. Intraoperative touch preparation for sentinel lymph node biopsy: a 4-year experience. Ann Surg Oncol 2002;9:333–339.
11. Bizjak-Schwarzbartl M, Fine-needle aspiration biopsy in the diagnosis of metastases in the liver. Diagn Cytopathol 1987;3:278–283.
12. Treaba D, Assad L, Govil H, et al. Diagnostic role of fine-needle aspiration of bone lesions in patients with a previous history of malignancy. Diagn Cytopathol 2002;26:380–383.
13. Ellison E, LaPuerta P, Martin SE. Supraclavicular masses: results of a series of 309 cases biopsied by fine needle aspiration. Head Neck 1999;21:239–246.
14. Gupta RK, Naran S. Fine needle aspiration cytology of cutaneous and subcutaneous metastatic deposits from epithelial malignancies. An analysis of 146 cases. Acta Cytol 1999;43:126–130.
15. Elledge RM, Green S, Pugh R, et al. Estrogen receptor (ER) and progesterone receptor (PgR), by ligand-binding assay compared with ER, PgR and pS2, by immuno-histochemistry in predicting response to tamoxifen in metastatic breast cancer: a Southwest Oncology Group study. Int J Cancer 2000;89: 111–117.
16. Harvey JM, Clark GM, Osborne CK, et al. Estrogen receptor status by immunohistochemistry is superior to the ligand-binding assay for predicting response to adjuvant endocrine therapy in breast cancer. J Clin Oncol 1999;17:1474–1481.
17. FDA. Medical devices; classification/reclassification of immunohistochemistry reagents and kits—FDA. Final rule. Federal Register 1998;63:30132–30142.

18. Taylor CW. FDA issues final rule for classification and reclassification of immunochemistry reagents and kits. Am J Clin Pathol 1999;111:443–444.
19. Shi SR, Cote RJ, Taylor CR. Antigen retrieval immunohistochemistry and molecular morphology in the year 2001. Appl Immunohistochem Mol Morphol 2001;9:107–116.
20. Arber DA. Effect of prolonged formalin fixation on the immunohistochemical reactivity of breast markers. Appl Immunohistochem Mol Morphol 2002;10: 183–186.
21. Taylor CR, Shi SR, Chaiwun B, et al. Strategies for improving the immunohistochemical staining of various intranuclear prognostic markers in formalin-paraffin sections: androgen receptor, estrogen receptor, progesterone receptor, p53 protein, proliferating cell nuclear antigen, and Ki-67 antigen revealed by antigen retrieval techniques. Hum Pathol 1994;25: 263–270.
22. Montero C. The antigen-antibody reaction in immunohistochemistry. J Histochem Cytochem 2003; 51:1–4.
23. Rhodes A, Jasani B, Couturier J, et al. A formalin-fixed, paraffin-processed cell line standard for quality control of immunohistochemical assay of HER-2/neu expression in breast cancer. Am J Clin Pathol 2002; 117:81–89.
24. Rhodes A. Quality assurance in immunohistochemistry. Am J Surg Pathol 2003;27:1284–1285.
25. Rhodes A, Jasani B, Barnes DM, et al. Reliability of immunohistochemical demonstration of oestrogen receptors in routine practice: interlaboratory variance in the sensitivity of detection and evaluation of scoring systems. J Clin Pathol 2000;53:125–130.
26. Rüdiger T, Hofler H, Kreipe HH, et al. Quality assurance in immunohistochemistry: results of an interlaboratory trial involving 172 pathologists. Am J Surg Pathol 2002;26:873–882.
27. Nichols DW, Wolff DJ, Self S, et al. A testing algorithm for determination of HER2 status in patients with breast cancer. Ann Clin Lab Sci 2002;32:3–11.
28. van de Vijver M. Emerging technologies for HER2 testing. Oncology 2002;63(suppl 1):33–38.
29. Hatanaka Y, Hashizume K, Kamihara Y, et al. Quantitative immunohistochemical evaluation of HER2/neu expression with HercepTest™ in breast carcinoma by image analysis. Pathol Int 2001;51: 33–36.
30. Thomson TA, Hayes MM, Spinelli JJ, et al. HER-2/neu in breast cancer: interobserver variability and performance of immunohistochemistry with 4 antibodies compared with fluorescent in situ hybridization. Mod Pathol 2001;14:1079–1086.
31. Tsuda H, Akiyama F, Terasaki H, et al. Detection of HER-2/neu (c-erb B-2) DNA amplification in primary breast carcinoma. Interobserver reproducibility and correlation with immunohistochemical HER-2 overexpression. Cancer 2001;92:2965–2974.
32. Jacobs TW, Gown AM, Yaziji H, et al. Comparison of fluorescence in situ hybridization and immunohistochemistry for the evaluation of HER–2/neu in breast cancer. J Clin Oncol 1999;17:1974–1982.
33. Nizzoli R, Bozzetti C, Crafa P, et al. Immunocytochemical evaluation of HER-2/neu on fine-needle aspirates from primary breast carcinomas. Diagn Cytopathol 2003;28:142–146.
34. Moore JG, To V, Patel SJ, et al. HER-2/neu gene amplification in breast imprint cytology analyzed by fluorescence in situ hybridization: direct comparison with companion tissue sections. Diagn Cytopathol 2000;23:299–302.
35. Sauter G, Feichter G, Torhorst J, et al. Fluorescence in situ hybridization for detecting erbB-2 amplification in breast tumor fine needle aspiration biopsies. Acta Cytol 1996;40:164–173.
36. Wang S, Hossein Saboorian M, Frenkel EP, et al. Aneusomy 17 in breast cancer: its role in HER-2/neu protein expression and implication for clinical assessment of HER-2/neu status. Mod Pathol 2002;15: 137–145.
37. Dandachi N, Dietze O, Hauser-Kronberger C. Chromogenic in situ hybridization: a novel approach to a practical and sensitive method for the detection of HER2 oncogene in archival human breast carcinoma. Lab Invest 2002;82:1007–1014.
38. Zhao J. Wu R, Au A, et al. Determination of HER2 gene amplification by chromogenic in situ hybridization (CISH) in archival breast carcinoma. Mod Pathol 2002;15:657–665.
39. Fabian CJ, Zalles C, Kamel S, et al. Breast cytology and biomarkers obtained by random fine needle aspiration: use in risk assessment and early chemoprevention trials. J Cell Biochem 1997;28–29(suppl): 101–110.
40. Euhus DM, Cler L, Shivapurkar N, et al. Loss of heterozygosity in benign breast epithelium in relation to breast cancer risk. J Natl Cancer Inst 2002;94: 858–860.
41. Conner P, Soderqvist G, Skoog L, et al. Breast cell proliferation in postmenopausal women during HRT evaluated through fine needle aspiration cytology. Breast Cancer Res Treat 2003;78:159–165.
42. Petrakis NL, Ernster VL, King EB, et al. Epithelial dysplasia in nipple aspirates of breast fluid: association with family history and other breast cancer risk factors. J Natl Cancer Inst 1982;68:9–13.
43. Sauter ER, Ross E, Daly M, et al. Nipple aspirate fluid: a promising non-invasive method to identify cellular markers of breast cancer risk. Br J Cancer 1997;76:494–501.
44. Dooley WC, Ljung BM, Veronesi U, et al. Ductal lavage for detection of cellular atypia in women at high risk for breast cancer. J Natl Cancer Inst 2001;93:1624–1632.
45. Wrensch MR, Petrakis NL, Miike R, et al. Breast cancer risk in women with abnormal cytology in nipple aspirates of breast fluid. J Natl Cancer Inst 2001;93:1791–1798.
46. Krishnamurthy S, Sneige N, Thompson PA, et al. Nipple aspirate fluid cytology in breast carcinoma. Cancer 2003;99:97–104.
47. Evron E, Dooley WC, Umbricht CB, et al. Detection of breast cancer cells in ductal lavage fluid by methylation-specific PCR. Lancet 2001;357:1335–1336.
48. Klein P, Glaser E, Grogan L, et al. Biomarker assays in nipple aspirate fluid. Breast J 2001;7:378–387.
49. Kuerer HM, Thompson PA, Krishnamurthy S, et al. High and differential expression of HER-2/neu extracellular domain in bilateral ductal fluids from women with unilateral invasive breast cancer. Clin Cancer Res 2003;9:601–605.
50. Kuerer HM, Thompson PA, Krishnamurthy S, et al. Identification of distinct protein expression patterns in bilateral matched pair breast ductal fluid specimens from women with unilateral invasive breast

carcinoma. High-throughput biomarker discovery. Cancer 2002;95:2276–2282.

51. King BL, Tsai SC, Gryga ME, et al. Detection of chromosomal instability in paired breast surgery and ductal lavage specimens by interphase fluorescence in situ hybridization. Clin Cancer Res 2003;9: 1509–1516.

52. Noguchi S, Motomura K, Inaji H, et al. Up-regulation of estrogen receptor by tamoxifen in human breast cancer. Cancer 1993;71:1266–1272.

53. Noguchi S, Motomura K, Inaji H, et al. Down-regulation of transforming growth factor-alpha by tamoxifen in human breast cancer. Cancer 1993; 72:131–136.

54. Makris A, Powles TJ, Allred DC, et al. Changes in hormone receptors and proliferation markers in tamoxifen treated breast cancer patients and the relationship with response. Breast Cancer Res Treat 1998;48:11–20.

55. Makris A, Powles TJ, Allred DC, et al. Quantitative changes in cytological molecular markers during primary medical treatment of breast cancer: a pilot study. Breast Cancer Res Treat 1999;53:51–59.

56. Symmans WF, Volm MD, Shapiro RL, et al. Paclitaxel-induced apoptosis and mitotic arrest assessed by serial fine-needle aspiration: implications for early prediction of breast cancer response to neoadjuvant treatment. Clin Cancer Res 2000;6:4610–4617.

57. Symmans FW. Breast cancer response to paclitaxel in vivo. Drug Resist Updat 2001;4:297–302.

58. Buchholz TA, Davis DW, McConkey DJ, et al. Chemotherapy-induced apoptosis and Bcl-2 levels correlate with breast cancer response to chemotherapy. Cancer J 2003;9:33–41.

59. Chang J, Ormerod M, Powles TJ, et al. Apoptosis and proliferation as predictors of chemotherapy response in patients with breast carcinoma. Cancer 2000;89: 2145–2152.

60. Symmans WF, Ayers M, Clark EA, et al. Total RNA yield and microarray gene expression profiles from fine-needle aspiration biopsy and core-needle biopsy samples of breast carcinoma. Cancer 2003;97:2960–2971.

CHAPTER
8

Histopathology of Breast Cancer: Correlation with Molecular Markers

W. Fraser Symmans, MD

Department of Pathology, University of Texas MD Anderson Cancer Center, Houston, Texas

Surgical pathology provides a diagnostic and prognostic classification for the diagnosis and management of breast cancer. In this chapter we introduce how molecular markers are currently used in the daily practice of surgical pathology and discuss how molecular markers have influenced our understanding of the pathology of breast cancer. The most important determination in the diagnosis of breast cancer is the presence or absence of invasion. The presence of invasion indicates the potential for metastatic spread and necessitates clinical and pathologic staging of the patient's cancer. Another practical role of surgical pathology is to define the pathologic status of the margins after surgical resection. This is relevant for in situ or invasive breast cancer and contributes to decisions concerning the adequacy of the surgical resection and the requirement for postoperative radiotherapy to reduce the risk of future local recurrence.[1,2] Molecular markers are not used to determine margin status because there is unfortunately no known molecular marker that is generally expressed and is specific to breast cancer cells.

Invasive Breast Carcinoma

Surgical pathology provides a description of the histopathologic appearance of a breast cancer. The description of histologic subtype and grade is a judgment of the inherent biologic aggressiveness of the tumor. The grade of the invasive cancer provides a semiquantitative numerical classification of the tumor differentiation. Poorly differentiated invasive cancers are defined as high grade and generally have more rapid cellular proliferation and tumor growth. Molecular markers for cellular proliferation, such as immunohistochemical stain for Ki-67 or S-phase fraction by flow cytometry, are strongly correlated with tumor grade (see Chapter 10).[3] However, the result of a separate assay for cellular proliferation can be valuable in the assessment of node-negative invasive breast cancer, for which the oncologist must determine the likely risk of future systemic recurrence and decide whether adjuvant chemotherapy will benefit the patient.[4–8]

Histopathologic classification of invasive breast cancer describes the microscopic appearance of the cancer as different histopathologic subtypes. Unfortunately, only some cases (no more than 10%) are of a histopathologic subtype that is considered to have special prognostic significance, such as tubular, mucinous, or papillary carcinoma. Those tumor types characteristically have low grade and low proliferation and are usually estrogen receptor (ER) positive.[9] Medullary carcinoma is an unusual variant of invasive breast cancer that contains poorly differentiated, ER-negative, HER-2/*neu*-negative, highly proliferative carcinoma cells.[10] The tumor is well demarcated from surrounding tissues (not infiltrative) and has syncytial growth and associated lymphoplasmacytic infiltrate. There is some evidence that pure medullary carcinoma has a good prognosis if strict diagnostic criteria are followed.[11–13] True medullary carcinomas are very uncommon (less than 1% of breast cancers), and tumors that fail to reach all the diagnostic criteria (designated as atypical medullary carcinomas) have similar prognosis to usual breast cancer.[11–13] Familial breast cancers are typically high grade and, for the *BRCA1* mutation associated subtype, are most often ER negative.[14–18] It has also been reported that familial breast cancers that have germline loss of *BRCA1* more frequently have atypical medullary appearance (13%) than *BRCA2* associated (3%) or non-*BRCA* associated familial breast cancers (2%) and may have poorer prognosis.[14,19] However, histopathology alone cannot reliably identify familial breast cancer.

Most invasive cancers (at least 90%) are the usual ductal or lobular histopathologic type and have equivalent survival when accounted for tumor stage and grade. Similarities between ductal and lobular subtypes are discussed more fully below. Many molecular markers have been tested for prognostic significance in breast cancer. This is because histopathologic categories have minimal prognostic impact in breast cancer. Molecular markers associated with poor prognosis (reviewed in Chapter 10) are generally associated with higher cellular proliferation (Ki-67, p53, cyclin D1, cyclin E, and cyclin B), growth (HER-2), or invasiveness (uroplasminogen activator, cathepsin D, matrix metalloproteinases), and these are often also associated with high histopathologic grade.[4,20–27] Conversely, molecular markers associated with good prognosis are associated with lower cellular proliferation (p27) and differentiation (ER and progesterone receptor [PR]), and these are often also associated with low histopathologic grade.[28–30] Histopathologic grade is an important prognostic determinant in breast cancer and is generally correlated with the expression of molecular markers for relevant cellular processes, such as proliferation, invasiveness, and differentiation. However, histopathologic grade provides a broad perspective of the tumor biology and does not reliably predict the expression of a molecular marker profile in an individual tumor. In breast cancer, the combined information of tumor stage and grade are strongly prognostic. It is unusual for individual molecular markers to achieve strong prognostic significance in multivariate analyses that include stage and grade. Recently, enzyme-linked immunosorbent assays for the cytosolic ratio of urokinase-like plasminogen activator to inhibitor of plasminogen activator inhibitor type 1 and for tissue inhibitor of metalloproteinase type 1 have been reported to achieve independent prognostic significance (see Chapter 18).[31–34] Identification of a low molecular weight isoform of cyclin E has been described as a molecular marker of poor prognosis (see Chapter 13).[35] However, those assays require fresh tumor sample collection and more sophisticated laboratory techniques than standard immunohistochemistry on a paraffin tissue section. Nevertheless, the enzyme-linked immunosorbent assay for urokinase-like plasminogen activator/plasminogen activator inhibitor type 1 is commonly performed in Europe but has not been widely adopted in the United States. The cyclin E isoform assay is currently performed in a single research laboratory, still needs independent validation, and so is not currently used for routine clinical practice.

Standard practice in breast oncology requires the assessment of ER and PR (see Chapter 15) and HER-2 (Chapter 16) in every invasive breast cancer. Advances in the molecular classification of invasive breast cancer have underlined the important roles that ER and Her-2/*neu* have on the overall gene expression profile in breast cancer. It is interesting to note that these tests were initially used for prognostic assessment but that they gained real clinical utility when targeted molecular therapies were directed at ER and HER-2. In general, 50–70% of newly diagnosed breast cancers are positive for ER and 20–25% are positive for HER-2. Although these phenotypes are not mutually exclusive, most ER-positive breast cancers are HER-2-negative (FIG. 8.1). The converse is also true, although approximately 20–30% of breast cancers are negative for both ER and Her-2/*neu* (FIG. 8.2). In a few examples of well-differentiated breast cancer, such as tubular carcinoma and mucinous carcinoma, and in invasive lobular carcinoma, the pathologist should expect the tumor cells to express ER.[36] Indeed, the ER immunohistochemical study should be repeated if a tumor with that histology is ER negative, especially if the stained tissue section does not contain an internal positive control of normal breast ductal–lobular groups. Otherwise, the histopathologic appearance of the invasive cancer does not reliably predict the tumor status for ER, PR, and HER-2.

Guidelines for interpretation of the clinical, pathologic, and molecular information to decide whether a patient with invasive breast cancer should receive systemic chemotherapy have been outlined from different consensus groups. The National Institutes of Health consensus conference recommendations from the United States are to offer systemic chemotherapy to all patients with node-positive breast cancer or with node-negative invasive breast cancer that measures more than 1 cm.[6] Individualized decisions for women with invasive cancer no more than 1 cm would be influenced by the patient

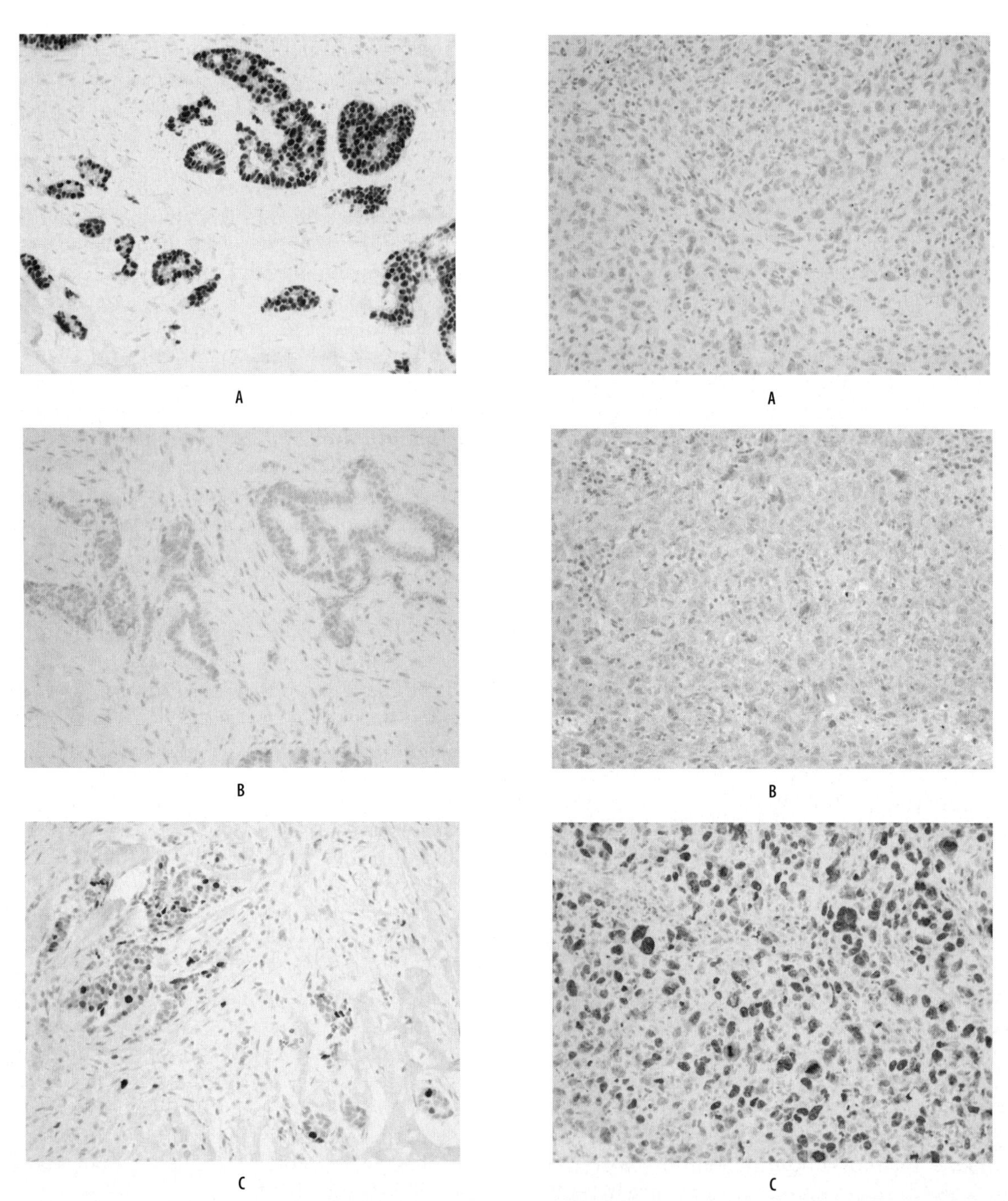

FIGURE 8.1 An example of well-differentiated (low grade) invasive ductal carcinoma that has uniform strong expression of ER (A), no expression of HER-2 (B), and an intermediate proliferation rate with 20% cancer cell nuclei that express Ki-67 (C). **See Plate 12 for color image.**

FIGURE 8.2 An example of poorly differentiated (high grade) invasive ductal carcinoma that has no expression of ER (A), no expression of HER-2 (B), and a high proliferation rate with more than 90% cancer cell nuclei that express Ki-67 (C). **See Plate 13 for color image.**

age and health, histologic type and grade, and molecular markers, such as ER and PR negative, high Ki-67 expression, and possibly HER-2-positive or p53 positive (FIG. 8.3).[6] The St. Gallen conference recommendations from Europe are to offer systemic chemotherapy to women with average to high risk of recurrence, that is, node-positive breast cancer, node-negative invasive breast cancer that measures more than 2 cm or is grade 2 or 3, or is ER and PR negative, or is diagnosed in a woman less than age 35 years.[7,8] Although these stratifications of high and low risk are strongly prognostic for patients with breast cancer, a 70-gene prognostic index derived from microarray gene expression profiles was recently shown to contribute to greater prognostic accuracy when combined with the National Institutes of

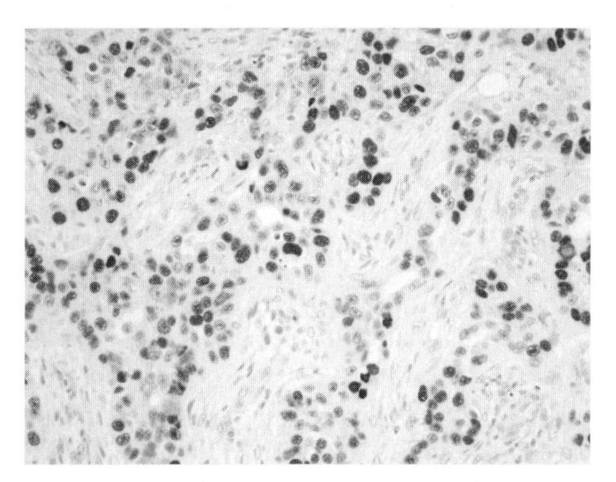

FIGURE 8.3 Immunohistochemical expression of nuclear p53 indicates mutated *p53* gene and dysfunctional p53 protein that accumulates due to decreased clearance. **See Plate 14 for color image.**

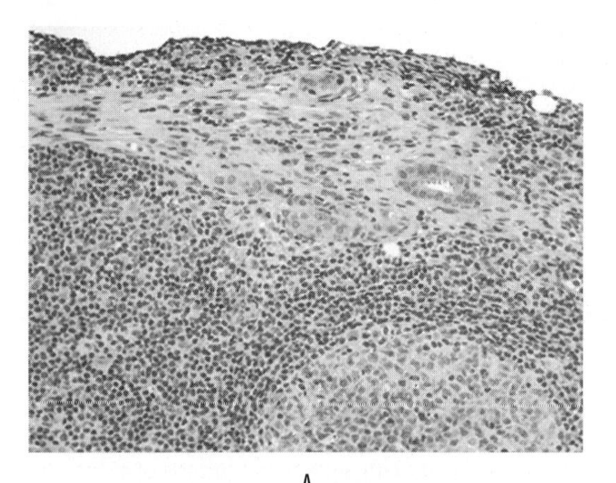

A

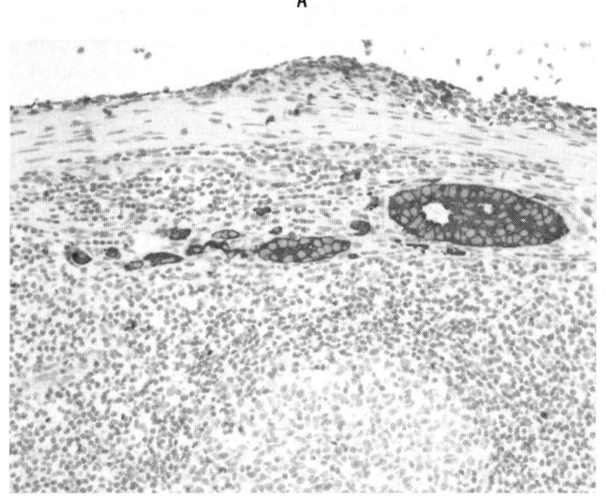

B

FIGURE 8.4 Micrometastasis (<2 mm) in a sentinel lymph node is subtle yet identifiable on the hematoxylin and eosin stain (A) but is readily identified and measurable on the corresponding immunohistochemical stain for a cocktail of cytokeratins (B). **See Plate 15 for color image.**

Health or the St. Gallen assessments of high and low risk.[37]

Surgical pathology also defines pathologic staging of tumor size and lymph node status (T and N stage).[38] There is no role for molecular markers in the definition of pathologic tumor size. However, molecular markers are routinely used in the assessment of sentinel lymph nodes (see Chapters 2 and 5). Immunohistochemical stains for cytokeratins are routinely applied to tissue sections of sentinel lymph nodes to detect occult metastases. Cytokeratins are expressed in epithelial cells, and so this assay is a highly sensitive (but not entirely specific) method to identify carcinoma cells in the lymph node (FIG. 8.4). Routine use of immunohistochemistry for sentinel lymph nodes converts 5–10% of negative sentinel lymph nodes to positive status.[38,39]

In Situ Breast Carcinoma

Molecular pathologic assays are not usually performed as part of the surgical pathologic evaluation of in situ carcinoma. Conceptually, in situ carcinoma is a preinvasive localized phase of breast carcinoma and so is managed by surgical and/or radiotherapy management and does not require systemic therapy. There has been a recent shift in this paradigm because adjuvant hormonal therapy with tamoxifen has been shown to prevent (or delay) the onset of subsequent invasive breast cancer in women with previous ductal carcinoma in situ (DCIS).[40] Tamoxifen is currently being offered to women with DCIS.[41] Therefore, many laboratories have begun to routinely test DCIS for ER and PR (FIG. 8.5), although there is not yet a published study to confirm that the benefit is restricted to hormone receptor–positive DCIS. It is expected that most lobular carcinoma in situ and low grade DCIS should be positive for ER (FIG. 8.5), whereas the ER status of intermediate and high grade DCIS is unpredictable from histopathologic appearance alone.[42]

High grade DCIS has interesting biologic and molecular characteristics. This lesion tends to be locally extensive within the ductal system and may involve single cell (Pagetoid) intraductal spread. Typically, the lesion is histopathologically associated with necrosis, apoptosis, and evidence of cellular proliferation. There is usually high proliferation index (Ki-67 expression), and most high grade DCIS cases have abnormal p53 and/or overexpression of Her-2/*neu*.[43,44] The frequency of p53 and HER-2 abnormalities in high grade DCIS (50–70%) is much higher than in invasive cancer (20–30%).[43,45,46] In concurrent high grade DCIS and invasive cancer, the status of HER-2 is usually the same.[45] High grade DCIS with abnormal p53 expression and high proliferation may be

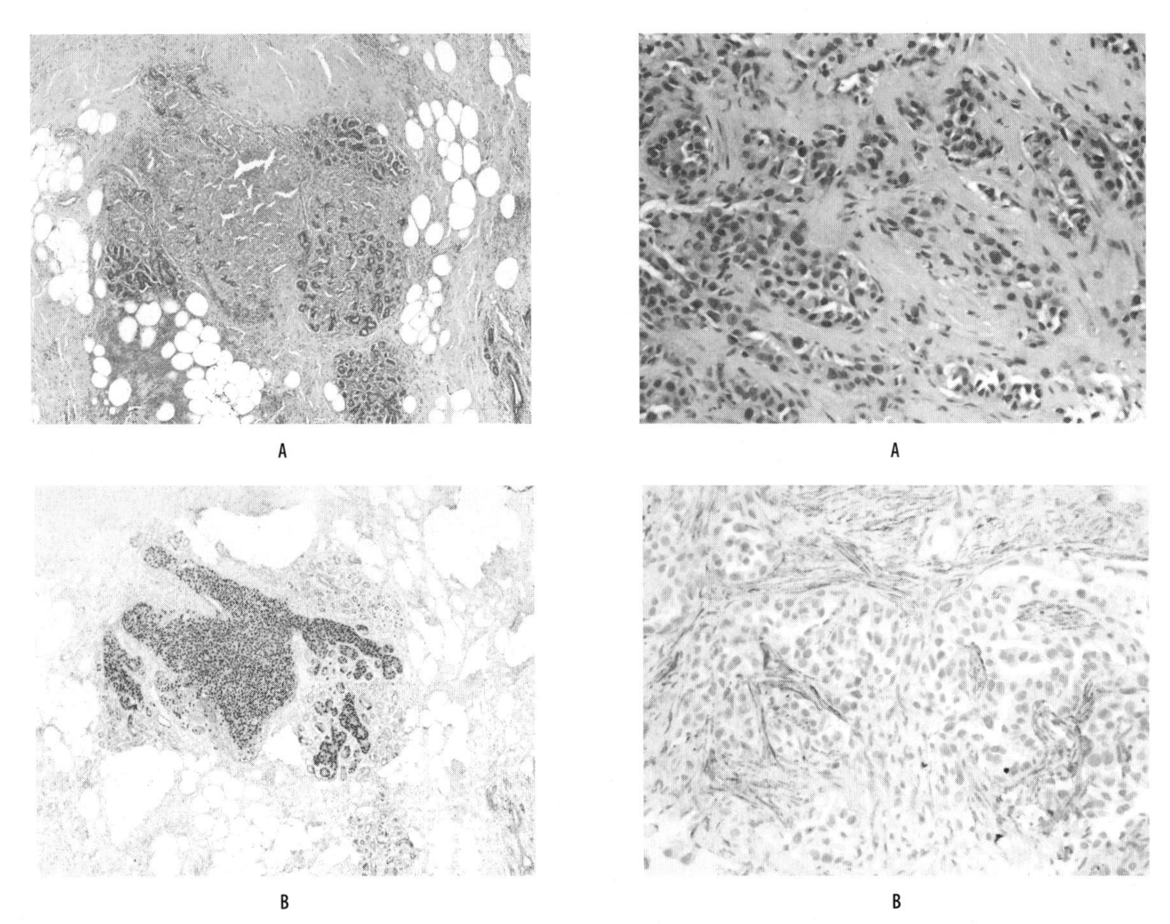

FIGURE 8.5 Intraductal carcinoma (low grade) that partially fills a preexisting lobular structure (A) has strong uniform expression of ER in the intraductal carcinoma cells and scattered less intense expression of ER in the adjacent normal lobular component (B). **See Plate 16 for color image.**

FIGURE 8.6 Invasion can be confirmed using molecular markers for myoepithelial cells or basement membranes. In this example, the epithelial groups within the fibrous stroma of a core biopsy section (A) lack any recognizable surrounding myoepithelial investment when stained for smooth muscle antigen (B). Note that myofibroblasts stain within the desmoplastic stroma (B). **See Plate 17 for color image.**

more likely to develop local recurrence.[47] This suggests that some high grade DCIS lesions are locally aggressive but are delayed in their progression to invasive disease, despite frequent abnormalities of p53 and HER-2.[45]

Molecular Markers to Differentiate Invasion

Invasion is defined by extension of the cancer cells beyond their preexisting ductal and lobular structure, breaching the surrounding myoepithelial cell layer and basement membrane, with invasion into the stroma. This is usually recognizable with microscopy (histopathology) as irregularly infiltrative groups of cancer cells within a characteristic desmoplastic stroma. However, sometimes the histopathologic appearance is indefinite. In such cases, immunohistochemical stains that recognize myoepithelial cells are used to determine whether a myoepithelial cell layer surrounds the groups of cancer cells. These assays detect antigens associated with myofilament organization (e.g., smooth muscle actin and calponin) or nuclear p63 expression.[48–50] Absence of myoepithelium indicates invasion (FIG. 8.6). The interpretation of these stains for myoepithelium in such cases is often challenging because there are increased numbers of myofibroblastic cells in the reactive stroma of invasive or in situ carcinoma (and these share expression of many of the same markers), and preexisting myoepithelial cells in a duct or lobule distended by in situ carcinoma can become attenuated and difficult to identify. Furthermore, these studies are often performed on small needle biopsy samples with known limitations of incomplete sampling and artifacts of tissue compression and disruption. Nonetheless, the implications of determining the presence or absence of invasion for patient management justify the use of these stains in biopsy samples to address this diagnostic important issue.[50]

Ductal Versus Lobular Histology

Surgical pathologists have traditionally classified most invasive breast cancers as ductal type or lobular type. This distinction is based on the histopathologic appearance. Ductal type has nests of invasive cells that may form small ductal structures. The most differentiated form of ductal carcinoma is tubular carcinoma, in which more than 90% of the epithelial structures have ductal appearance and also have lumens. Lobular type is characterized by an invasive pattern of single cells in rows that infiltrate dense fibrotic stroma, encircle preexisting normal structures, and can be identified as microscopic satellite foci variably distant from the main tumor mass. The cells of invasive lobular carcinoma are small, have low nuclear grade, may have an eccentric cytoplasmic mucin vacuole, express ER and PR, and have low proliferative rate.[36] There is evidence that lobular carcinomas are less likely to achieve complete clinical or pathologic response to neoadjuvant (preoperative) chemotherapy, but long-term survival is not different from ductal carcinoma.[51,52] Both findings are probably related to the low grade, low proliferation rate and ER-positive status of invasive lobular carcinoma.[51,52] It has also been demonstrated that invasive ductal and lobular types of carcinoma have similar prognosis when patients are matched for age and tumor stage.[36] Indeed, many invasive breast cancers have shared histopathologic features of ductal and lobular carcinoma.[36] The specific clinical relevance for pure lobular carcinoma lies with its subtle presentation and propensity for missed diagnosis by clinical examination, mammography, and needle biopsy, and its pattern of invasion makes the pathologic assessment of surgical margin status more challenging.[36]

The different pattern of invasive growth seen in lobular carcinoma is largely explained by loss of expression of E-cadherin (FIG. 8.7). The E-cadherin gene encodes an adhesion molecule that is critical for intercellular attachment of epithelial cells (see Chapter 20). In lobular carcinomas there is no expression of E-cadherin protein at the cell membrane due to loss of a gene allele, mutation (encoding a protein that is secreted and not able to form intercellular attachments), and/or silencing of gene expression due to hypermethylation of the promoter site.[53–56] Therefore, E-cadherin is not expressed at the cell membrane, and the carcinoma cells have minimal intercellular attachment. Loss of E-cadherin expression also occurs in lobular carcinoma in situ, explains the characteristic discohesive appearance of the epithelial cells in that lesion, and is related to subsequent development of invasive lobular carcinoma.[57] A recent publication compared the overall gene expression profiles between ductal and lobular types of invasive cancer and identified differential gene expression in very few genes.[58] The most significant difference was reduced expression of E-cadherin in lobular carcinomas.[58] Expression of osteopontin, survivin, and cathepsin B genes was also significantly decreased in invasive lobular cancer.[58] Although complete loss of expression of E-cadherin is usual in invasive lobular carcinoma, reduced expression of E-cadherin does occur in some invasive ductal carcinomas and is related to a more infiltrative growth pattern (FIG. 8.8).[59] In this context, it is

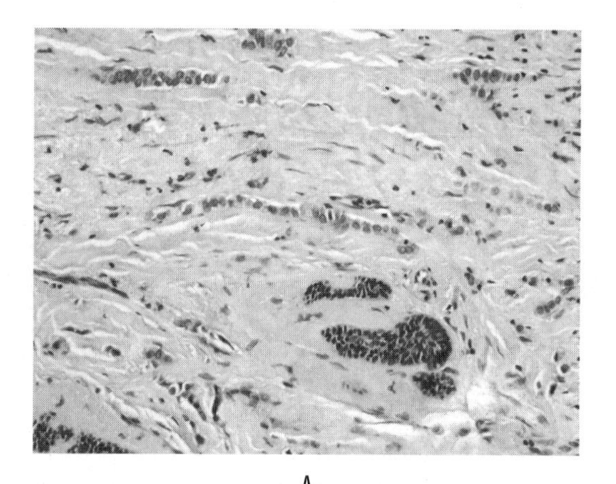

A

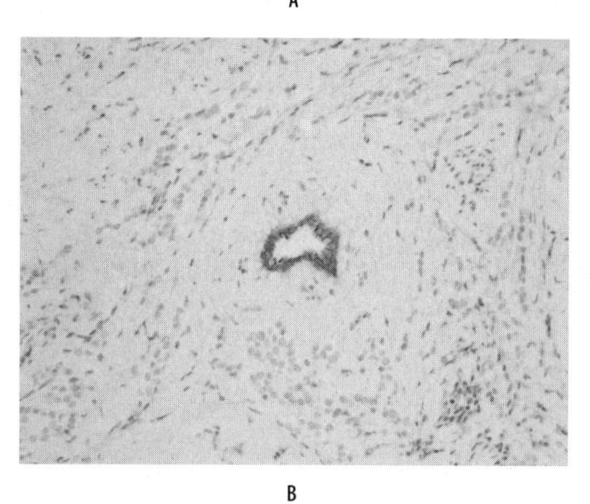

B

FIGURE 8.7 Invasive lobular carcinoma that infiltrates as rows of cancer cells within a dense fibrotic stroma and is adjacent to a preexisting normal ductule (A). An immunohistochemical stain for E-cadherin (B) demonstrates a membranous pattern of strong staining in the epithelium of the normal ductule (center) but complete lack of E-cadherin expression in the invasive lobular carcinoma cells. **See Plate 18 for color image.**

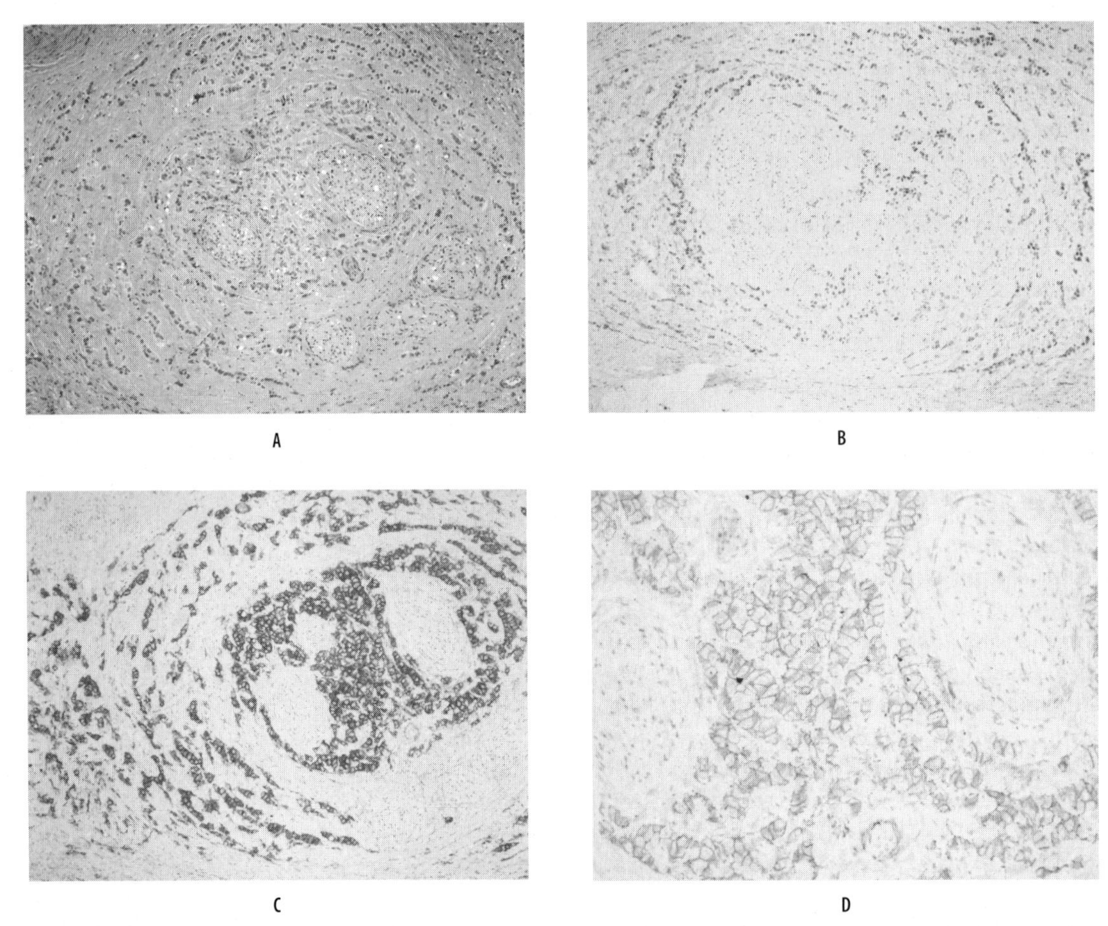

FIGURE 8.8 An invasive carcinoma with rows of cancer cells surrounding nerve fibers and exhibits a pattern of invasion that resembles lobular carcinoma (A). The cancer cells express ER (B); however, they also strongly express HER-2 (C) and have membranous expression of E-cadherin (D). The positive staining for E-cadherin and HER-2 is not typical of invasive lobular carcinoma. The membranous staining for E-cadherin is less intense than in usual ductal carcinoma. **See Plate 19 for color image.**

possible to consider lobular carcinoma of the breast as a variant of invasive carcinoma (rather than a specific histologic type) in which lost expression of a single gene product (E-cadherin) influences intercellular attachment and imparts a characteristic pattern of infiltrative growth.

Molecular Classification of Breast Cancer: Epithelial Differentiation

Gene expression profiles of breast cancer using microarray technology have largely confirmed the concept that ER and Her-2/*neu* are major influences on the biology of breast cancer.[60] ER-negative tumors preferentially express cytokeratins that are usually associated with basal cell types (e.g., CK5, CK6).[60–62] Immunohistochemical studies show that those cytokeratins (e.g., CK5, CK6) are expressed in normal myoepithelial (basal) cells, whereas cytokeratins associated with more differentiated epithelium (e.g., CK8, CK18) are expressed in normal luminal cells (FIG. 8.9).[60,61] Together, these results suggest that both cell types in the normal ductal–lobular structures of the breast can develop into carcinoma and that ER-positive breast cancer may represent luminal cell differentiation, whereas ER-negative breast cancers may represent basal cell differentiation. This concept remains to be proven. However, the results of chemoprevention trials indirectly support this concept by demonstrating that hormonal therapy only prevents hormonally sensitive breast cancer. Tamoxifen therapy largely prevented the development of ER-positive cancers but not ER-negative cancers in the NSABP chemoprevention trial.[63] The implication is that ER-negative cancers develop independently from estrogen even in their preneoplastic development. Also, it has long been

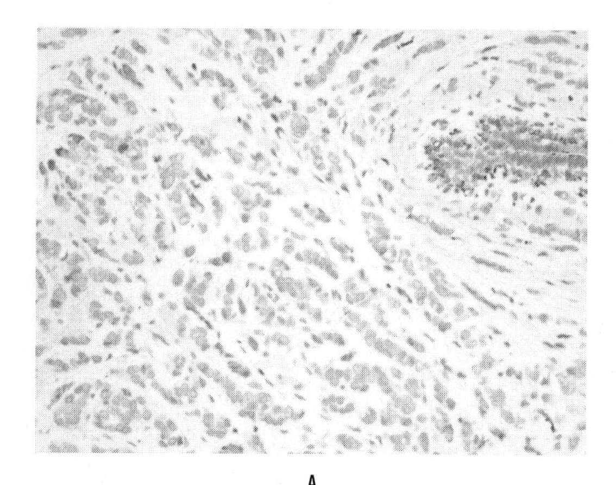

A

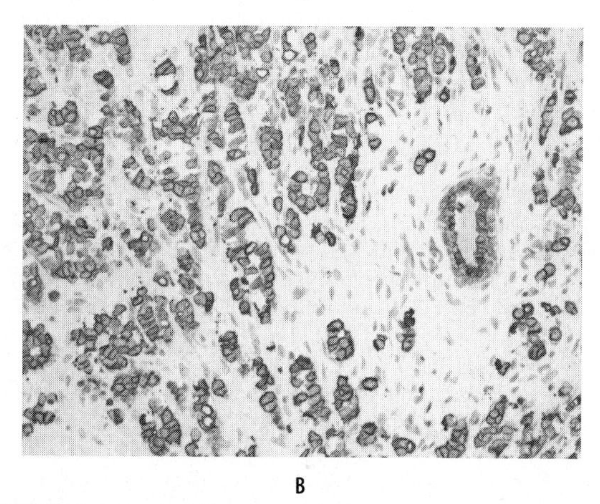

B

FIGURE 8.9 A poorly differentiated (high grade) invasive carcinoma that does not have expression of CK5/6 (A) but has uniform strong expression of CK8/18 (B). Note that CK5 is expressed in the basal epithelial layer of an adjacent normal duct (A), and CK8 is expressed in the luminal epithelial layer of an adjacent normal duct (B). Anti-cytokeratin antibody cocktails to detect CK5 and CK6 (A) and CK8 and CK18 (B) were used for immunohistochemistry. **See Plate 20 for color image.**

acknowledged that ER-positive cancers are generally more differentiated than ER-negative breast cancer, as evidenced by lower overall grade, slower growth rate, and longer time to recurrence. This concept of different stages in epithelial differentiation in different breast cancers forms the basis of a molecular classification that is conceptually parallel to other types of cancer (e.g., non-Hodgkin lymphoma, leukemia, germ cell tumors) and seems more reasonable than the previous concept that breast cancers were anatomically derived from either ducts or lobules.

Epithelial–Mesenchymal Transition

There are rare examples of invasive breast cancer that exhibit mesenchymal characteristics.[64] These can be considered in three categories. First are the poorly differentiated tumors with combined histopathologic elements of sarcoma and carcinoma that have been termed "carcinosarcoma" or "sarcomatoid carcinoma" and are currently classified as "metaplastic carcinoma, high grade."[65] High-grade tumors may be predominantly sarcomatous with focal epithelial differentiation or may have discernible epithelial and mesenchymal components that can be considered to be divergent differentiation within the tumor.[64,65] These lesions may have a metastatic pattern of sarcoma.[65] Next are the low grade spindle cell neoplasms that are essentially mesenchymal but coexpress cytokeratins and therefore contain some epithelial characteristics.[66–69] Those tumors behave like low grade sarcoma, require local surgical control, may be locally recurrent, occasionally metastasize, but are not expected to involve regional lymph nodes.[66,68] Third are the poorly differentiated (high grade) carcinomas, with extensive necrosis, in which immunohistochemical stains for mesenchymal markers (e.g., vimentin or p63) will demonstrate some staining within the tumor cell population.[62,70–72] Those are poorly differentiated breast carcinomas in which some cells have rudimentary epithelial and mesenchymal differentiation that is an indication of their relatively undifferentiated state.

Summary

Advances in the molecular classification of breast cancer are highlighting the importance of cellular differentiation in breast cancer, such as basal or luminal characteristics and expression of ER, in addition to the acquisition of specific molecular aberrations, such as amplification of Her-2/*neu* or silencing of E-cadherin. In this context, most breast cancers can be viewed within a continuum from poorly differentiated biologically aggressive cancers with undifferentiated epithelial and mesenchymal characteristics to the very well-differentiated tubular carcinomas that have slow growth, infrequently metastasize, and express ER. The histopathologic appearance of breast cancer only predicts the profile of expressed molecular markers in a few exceptions (e.g., very well differentiated tumors) but is an unreliable correlate in most breast cancers.

References

1. Silverstein MJ, Lagios MD, Groshen S, et al. The influence of margin width on local control of ductal carcinoma in situ of the breast. N Engl J Med 1999;340:1455–1461.
2. Fisher B, Anderson S, Redmond CK, et al. Reanalysis and results after 12 years of follow-up in a randomized clinical trial comparing total mastectomy with lumpectomy with or without irradiation in the treatment of breast cancer. N Engl J Med 1995;333: 1456–1461.
3. Weidner N, Moore DH 2nd, Vartanian R. Correlation of Ki-67 antigen expression with mitotic figure index and tumor grade in breast carcinomas using the novel "paraffin"-reactive MIB1 antibody. Hum Pathol 1994;25:337–342.
4. Jones S, Clark G, Koleszar S, et al. Low proliferative rate of invasive node-negative breast cancer predicts for a favorable outcome: a prospective evaluation of 669 patients. Clin Breast Cancer 2001;1:310–314, discussion 315–317.
5. Olivotto IA, Bajdik CD, Plenderleith IH, et al. Adjuvant systemic therapy and survival after breast cancer. N Engl J Med 1994;330:805–810.
6. Eifel P, Axelson JA, Costa J, et al. National Institutes of Health Consensus Development Conference Statement: adjuvant therapy for breast cancer, November 1–3, 2000. J Natl Cancer Inst 2001;93:979–989.
7. Goldhirsch A, Glick JH, Gelber RD, et al. Meeting highlights: International Consensus Panel on the Treatment of Primary Breast Cancer. Seventh International Conference on Adjuvant Therapy of Primary Breast Cancer. J Clin Oncol 2001;19: 3817–3827.
8. Goldhirsch A, Wood WC, Gelber RD, et al. Meeting highlights: updated international expert consensus on the primary therapy of early breast cancer. J Clin Oncol 2003;21:3357–3365.
9. Reiner A, Reiner G, Spona J, et al. Histopathologic characterization of human breast cancer in correlation with estrogen receptor status. A comparison of immunocytochemical and biochemical analysis. Cancer 1988;61:1149–1154.
10. Xu R, Feiner H, Li P, et al. Differential amplification and overexpression of HER-2/neu, p53, MIB1, and estrogen receptor/progesterone receptor among medullary carcinoma, atypical medullary carcinoma, and high-grade invasive ductal carcinoma of breast. Arch Pathol Lab Med 2003;127:1458–1464.
11. Rapin V, Contesso G, Mouriesse H, et al. Medullary breast carcinoma. A reevaluation of 95 cases of breast cancer with inflammatory stroma. Cancer 1988;61: 2503–2510.
12. Pedersen L, Holck S, Schiodt T, et al. Medullary carcinoma of the breast, prognostic importance of characteristic histopathological features evaluated in a multivariate Cox analysis. Eur J Cancer 1994;30A: 1792–1797.
13. Pedersen L, Zedeler K, Holck S, et al. Medullary carcinoma of the breast. Prevalence and prognostic importance of classical risk factors in breast cancer. Eur J Cancer 1995;31A:2289–2295.
14. Breast Cancer Linkage Consortium. Pathology of familial breast cancer: differences between breast cancers in carriers of BRCA1 or BRCA2 mutations and sporadic cases. Lancet 1997;349:1505–1510.
15. Lakhani SR. The pathology of familial breast cancer: morphological aspects. Breast Cancer Res 1999;1: 31–35.
16. Lakhani SR, Gusterson BA, Jacquemier J, et al. The pathology of familial breast cancer: histological features of cancers in families not attributable to mutations in BRCA1 or BRCA2. Clin Cancer Res 2000;6: 782–789.
17. Agnarsson BA, Jonasson JG, Bjornsdottir IB, et al. Inherited BRCA2 mutation associated with high grade breast cancer. Breast Cancer Res Treat 1998; 47:121–127.
18. Phillips KA. Immunophenotypic and pathologic differences between BRCA1 and BRCA2 hereditary breast cancers. J Clin Oncol 2000;18(21 suppl): 107S–112S.
19. Stoppa-Lyonnet D, Ansquer Y, Dreyfus H, et al. Familial invasive breast cancers: worse outcome related to BRCA1 mutations. J Clin Oncol 2000;18: 4053–4059.
20. Liu S, Edgerton SM, Moore DH 2nd, et al. Measures of cell turnover (proliferation and apoptosis) and their association with survival in breast cancer. Clin Cancer Res 2001;7:1716–1723.
21. Ferrero JM, Ramaioli A, Formento JL, et al. P53 determination alongside classical prognostic factors in node-negative breast cancer: an evaluation at more than 10-year follow-up. Ann Oncol 2000;11: 393–397.
22. Overgaard J, Yilmaz M, Guldberg P, et al. TP53 mutation is an independent prognostic marker for poor outcome in both node-negative and node-positive breast cancer. Acta Oncol 2000;39:327–333.
23. Cuny M, Kramar A, Courjal F, et al. Relating genotype and phenotype in breast cancer: an analysis of the prognostic significance of amplification at eight different genes or loci and of p53 mutations. Cancer Res 2000;60:1077–1083.
24. Kuhling H, Alm P, Olsson H, et al. Expression of cyclins E, A, and B, and prognosis in lymph node-negative breast cancer. J Pathol 2003;199:424–431.
25. Tsutsui S, Ohno S, Murakami S, et al. Prognostic value of c-erbB2 expression in breast cancer. J Surg Oncol 2002;79:216–223.
26. Reed W, Hannisdal E, Boehler PJ, et al. The prognostic value of p53 and c-erb B-2 immunostaining is overrated for patients with lymph node negative breast carcinoma: a multivariate analysis of prognostic factors in 613 patients with a follow-up of 14–30 years. Cancer 2000;88:804–813.
27. Mirza AN, Mirza NQ, Vlastos G, et al. Prognostic factors in node-negative breast cancer: a review of studies with sample size more than 200 and follow-up more than 5 years. Ann Surg 2002;235:10–26.
28. Barbareschi M, van Tinteren H, Mauri FA, et al. p27(kip1) expression in breast carcinomas: an immunohistochemical study on 512 patients with long-term follow-up. Int J Cancer 2000;89:236–241.
29. Lau R, Grimson R, Sansome C, et al. Low levels of cell cycle inhibitor p27kip1 combined with high levels of Ki–67 predict shortened disease-free survival in T1 and T2 invasive breast carcinomas. Int J Oncol 2001;18:17–23.
30. Tan P, Cady B, Wanner M, et al. The cell cycle inhibitor p27 is an independent prognostic marker in small (T1a,b) invasive breast carcinomas. Cancer Res 1997;57:1259–1263.

31. Look MP, van Putten WL, Duffy MJ, et al. Pooled analysis of prognostic impact of urokinase-type plasminogen activator and its inhibitor PAI-1 in 8377 breast cancer patients. J Natl Cancer Inst 2002;94: 116–128.
32. Harbeck N, Dettmar P, Thomssen C, et al. Risk-group discrimination in node-negative breast cancer using invasion and proliferation markers: 6-year median follow-up. Br J Cancer 1999;80:419–426.
33. Harbeck N, Kates RE, Look MP, et al. Enhanced benefit from adjuvant chemotherapy in breast cancer patients classified high-risk according to urokinase-type plasminogen activator (uPA) and plasminogen activator inhibitor type 1 (n = 3424). Cancer Res 2002;62:4617–4622.
34. Schrohl AS, Holten-Andersen MN, Peters HA, et al. Tumor tissue levels of tissue inhibitor of metalloproteinase-1 as a prognostic marker in primary breast cancer. Clin Cancer Res 2004;10:2289–2298.
35. Keyomarsi K, Tucker SL, Buchholz TA, et al. Cyclin E and survival in patients with breast cancer. N Engl J Med 2002;347:1566–1575.
36. Sastre-Garau X, Jouve M, Asselain B, et al. Infiltrating lobular carcinoma of the breast. Clinicopathologic analysis of 975 cases with reference to data on conservative therapy and metastatic patterns. Cancer 1996;77:113–120.
37. van de Vijver MJ, He YD, van't Veer LJ, et al. A gene-expression signature as a predictor of survival in breast cancer. N Engl J Med 2002;347:1999–2009.
38. Singletary SE, Allred C, Ashley P, et al. Staging system for breast cancer: revisions for the 6th edition of the AJCC Cancer Staging Manual. Surg Clin North Am 2003;83:803–819.
39. Cserni G, Amendoeira I, Apostolikas N, et al. Pathological work-up of sentinel lymph nodes in breast cancer. Review of current data to be considered for the formulation of guidelines. Eur J Cancer 2003; 39:1654–1667.
40. Fisher B, Dignam J, Wolmark N, et al. Tamoxifen in treatment of intraductal breast cancer: National Surgical Adjuvant Breast and Bowel Project B-24 randomised controlled trial. Lancet 1999;353:1993–2000.
41. Yen TW, Hunt KK, Mirza NQ, et al. Physician recommendations regarding tamoxifen and patient utilization of tamoxifen after surgery for ductal carcinoma in situ. Cancer 2004;100:942–949.
42. Baqai T, Shousha S. Oestrogen receptor negativity as a marker for high-grade ductal carcinoma in situ of the breast. Histopathology 2003;42:440–447.
43. Allred DC, Clark GM, Molina R, et al. Overexpression of HER-2/neu and its relationship with other prognostic factors change during the progression of in situ to invasive breast cancer. Hum Pathol 1992; 23:974–979.
44. Bose S, Lesser ML, Norton L, et al. Immunophenotype of intraductal carcinoma. Arch Pathol Lab Med 1996;120:81–85.
45. Latta EK, Tjan S, Parkes RK, et al. The role of HER2/neu overexpression/amplification in the progression of ductal carcinoma in situ to invasive carcinoma of the breast. Mod Pathol 2002;15:1318–1325.
46. Eriksson ET, Schimmelpenning H, Aspenblad U, et al. Immunohistochemical expression of the mutant p53 protein and nuclear DNA content during the transition from benign to malignant breast disease. Hum Pathol 1994;25:1228–1233.
47. Ringberg A, Anagnostaki L, Anderson H, et al. Cell biological factors in ductal carcinoma in situ (DCIS) of the breast-relationship to ipsilateral local recurrence and histopathological characteristics. Eur J Cancer 2001;37:1514–1522.
48. Mukai K, Schollmeyer JV, Rosai J. Immunohistochemical localization of actin: applications in surgical pathology. Am J Surg Pathol 1981;5:91–97.
49. Ribeiro-Silva A, Zamzelli Ramalho LN, Garcia SB, Zucoloto S. Is p63 reliable in detecting microinvasion in ductal carcinoma in situ of the breast? Pathol Oncol Res 2003;9:20–23.
50. Damiani S, Zamzelli Ramalho LN, Garcia SB, et al. Myoepithelial cells and basal lamina in poorly differentiated in situ duct carcinoma of the breast. An immunocytochemical study. Virchows Arch 1999;434: 227–234.
51. Mathieu MC, Rouzier R, Llombart-Cussac A, et al. The poor responsiveness of infiltrating lobular breast carcinomas to neoadjuvant chemotherapy can be explained by their biological profile. Eur J Cancer 2004;40:342–351.
52. Cocquyt VF, Blondeel PN, Depypere HT, et al. Different responses to preoperative chemotherapy for invasive lobular and invasive ductal breast carcinoma. Eur J Surg Oncol 2003;29:361–367.
53. Kanai Y, Oda T, Tsuda H, et al. Point mutation of the E-cadherin gene in invasive lobular carcinoma of the breast. Jpn J Cancer Res 1994;85:1035–1039.
54. Berx G, Cleton-Jansen AM, Strumane K, et al. E-cadherin is inactivated in a majority of invasive human lobular breast cancers by truncation mutations throughout its extracellular domain. Oncogene 1996;13:1919–1925.
55. Droufakou S, Deshmane V, Roylance R, et al. Multiple ways of silencing E-cadherin gene expression in lobular carcinoma of the breast. Int J Cancer 2001; 92:404–408.
56. Sarrio D, Moreno-Bueno G, Hardisson D, et al. Epigenetic and genetic alterations of APC and CDH1 genes in lobular breast cancer: relationships with abnormal E-cadherin and catenin expression and microsatellite instability. Int J Cancer 2003;106(2):208–215.
57. Reis-Filho JS, Cancela Paredes J, Milanezi F, et al. Clinicopathologic implications of E-cadherin reactivity in patients with lobular carcinoma in situ of the breast. Cancer 2002;94(7):2114–2115; author reply 2115–2116.
58. Korkola JE, DeVries S, Fridlyand J, et al. Differentiation of lobular versus ductal breast carcinomas by expression microarray analysis. Cancer Res 2003; 63(21):7167–7175.
59. Goldstein NS. Does the level of E-cadherin expression correlate with the primary breast carcinoma infiltration pattern and type of systemic metastases? Am J Clin Pathol 2002;118(3):425–434.
60. Perou CM, Sorlie T, Eisen MB, et al. Molecular portraits of human breast tumours. Nature 2000; 406(6797):747–752.
61. Korsching E, Packeisen J, Agelopoulos K, et al. Cytogenetic alterations and cytokeratin expression patterns in breast cancer: integrating a new model of breast differentiation into cytogenetic pathways of breast carcinogenesis. Lab Invest 2002;82(11): 1525–1533.
62. Santini D, Ceccarelli C, Taffurelli M, et al. Differentiation pathways in primary invasive breast

carcinoma as suggested by intermediate filament and biopathological marker expression. J Pathol 1996; 179(4):386–391.

63. Fisher B, Costantino JP, Wickerham DL, et al. Tamoxifen for prevention of breast cancer: report of the National Surgical Adjuvant Breast and Bowel Project P-1 Study. J Natl Cancer Inst 1998;90(18): 1371–1388.

64. Al-Nafussi A. Spindle cell tumours of the breast: practical approach to diagnosis. Histopathology 1999;35(1):1–13.

65. Christensen L, Schiodt T, Blichert-Toft M. Sarcomatoid tumours of the breast in Denmark from 1977 to 1987. A clinicopathological and immunohistochemical study of 100 cases. Eur J Cancer 1993;29A(13): 1824–1831.

66. Gobbi H, Simpson JF, Borowsky A, et al. Metaplastic breast tumors with a dominant fibromatosis-like phenotype have a high risk of local recurrence. Cancer 1999;85(10):2170–2182.

67. Gobbi H, Simpson JF, Jensen RA, et al. Metaplastic spindle cell breast tumors arising within papillomas, complex sclerosing lesions, and nipple adenomas. Mod Pathol 2003;16(9):893–901.

68. Sneige N, Yaziji H, Mandavilli SR, et al. Low-grade (fibromatosis-like) spindle cell carcinoma of the breast. Am J Surg Pathol 2001;25(8):1009–1016.

69. Brogi E. Benign and malignant spindle cell lesions of the breast. Semin Diagn Pathol 2004;21(1): 57–64.

70. Fuchs IB, Lichtenegger W, Buehler H, et al. The prognostic significance of epithelial-mesenchymal transition in breast cancer. Anticancer Res 2002; 22(6A):3415–3419.

71. Ribeiro-Silva A, Zambelli Ramalho LN, Britto Garcia S, et al. The relationship between p63 and p53 expression in normal and neoplastic breast tissue. Arch Pathol Lab Med 2003;127(3):336–340.

72. Tsuda H, Takarabe T, Hasegawa T, et al. Myoepithelial differentiation in high-grade invasive ductal carcinomas with large central acellular zones. Hum Pathol 1999;30(10):1134–1139.

CHAPTER
9

Sentinel Lymph Node Dissection and Micrometastasis Detection in Bone Marrow and Lymph Nodes

Volkmar Müller, MD,[1] Sabine Kasimir-Bauer, PhD,[2] and Klaus Pantel, MD, PhD[3]

[1]Klinik für Frauenheilkunde, Universitätsklinikum Hamburg-Eppendorf, Martinistrasse 52, D-20246 Hamburg, Germany
[2]Klinik für Frauenheilkunde und Geburtshilfe, Universitätsklinikum Essen, Hufelandstrasse 55, 45122 Essen, Germany
[3]Institut für Tumorbiologie, Universitätsklinikum Hamburg-Eppendorf, Martinistrasse 52, D-20246 Hamburg, Germany

Biologic Background

The first event in the metastatic cascade of solid tumors like breast cancer is tumor cell dissemination via the regional lymph nodes and/or by tumor circulation in the blood followed by homing in secondary distant organs. Among these distant organs, bone marrow (BM) is a common homing organ for disseminated cancer cells and is therefore a useful indicator organ for the presence of disseminated tumor cells (DTCs) throughout the body.

The current standard of care for surgical management of invasive breast cancer is complete removal of the tumor followed by axillary lymph node dissection (ALND). ALND is not free from complications and short- or long-term sequelae (e.g., scarring, numbness, and lymphedema).[1] An important advance in the evaluation of regional lymph nodes is therefore the development of a more limited dissection, the sentinel lymph node dissection.

According to the Union Internationale Contre le Cancer and the American Joint Commission on Cancer, a "micrometastasis" is defined as a metastatic deposit smaller than 2 mm in a single lymph node, as compared with a metastatic deposit of 2 mm or more in diameter, and/or any metastasis in two or more lymph nodes, which is termed "macrometastasis."[2,3] Since the term micrometastasis also refers to deposits down to a single cell, it has been suggested that a subclassification should distinguish between micrometastasis and single cells. In this regard, the American Joint Commission on Cancer Breast Committee recently defined the classification pN0 (itc) for tumor deposits with a maximum size of 0.2 mm and pN1 (mcm) for deposits larger than 0.2 mm but smaller than 2 mm.[4] The most recent TNM classification for breast cancer[5] does not qualify the presence of single cancer cells in peripheral blood or BM as metastasis (stage M0), but it optionally reports the presence of such cells together with their detection method, for example, M0(i+) for the immunocytochemical detection or M0(mol+) for the detection by molecular methods. In this chapter we discuss the implications of detection of micrometastatic cells in BM and lymph nodes for staging and therapy of breast cancer patients.

Clinical Relevance of Micrometastasis Detection in BM

For patients with primary breast cancer, traditional staging parameters of the tumor (e.g., tumor size, grading, and lymph node status) are determined to estimate the prognosis and to decide whether the patient is in need of systemic adjuvant therapy aimed to prevent metastatic relapse. However, the early seed of metastatic cells in distant organs is missed by traditional tumor staging and can contribute to metastatic relapse. Using sensitive immunocytologic and molecular methods, it is now possible to detect one DTC in the background of one million normal hematopoietic cells. The search for these occult cells in patients with primary

Table 9.1 Examples of Reports Demonstrating the Prognostic Relevance of DTC in BM of Breast Cancer Patients without Overt Metastases (TNM Stage M0)

Reference	Technique	No. of Patients	Detection Rate	Prognostic Value
Landys et al. (13)	IHC	128	19%	DFS,* OS*
Diel et al. (14)	ICC	727	43%	DFS,* OS*
Mansi et al. (15)	ICC	350	25%	DFS, OS
Gebauer et al. (16)	ICC	393	42%	DFS,* OS
Cote et al. (17)	ICC	49	37%	DFS
Braun et al. (18)	ICC	552	36%	DDFS,* OS*
Harbeck et al. (19)	ICC	100	38%	DFS,* OS*
Gerber et al. (10)	ICC	484	31%	DFS,* OS*
Wiedswang et al. (20)	ICC	817	13.2%	DDFS,* OS*
Braun et al. (11)	ICC	150	29%	DDFS,* OS*
Vannucchi et al. (21)	RT-PCR	33	48%	DFS

*Prognostic value supported by multivariate analysis.

DFS, disease-free survival; DDFS, distant disease-free survival; ICC, immunocytochemistry; OS, overall survival.

breast cancer might contribute substantial information for the clinical management of an individual patient.

Prognostic Impact of DTCs in BM

The prognostic importance of the immunocytochemical identification of occult tumor cells in BM of breast cancer patients has been confirmed in various prospective clinical studies comprising more than 3000 patients. Several of these studies also demonstrated that the presence of occult tumor cells in BM is an independent risk factor (**Table 9.1**). However, some studies have not confirmed the prognostic relevance of occult tumor cells,[6–9] probably due to insufficiently sensitive test methods in the older studies.

It is important to mention that the presence of lymph node (micro-) metastases does not correlate with the presence of occult metastatic cells in BM.[10,11] This lack of correlation may be explained by the presence of two independent metastatic pathways in breast cancer, primary lymphogenic or primary hematogenous spread, which seems to be determined by different sets of genes.[12]

Impact of DTCs in BM for Therapy Selection

In breast cancer, the availability of additional factors enabling individual risk assessment is desirable to improve identification of patients at risk for relapse and need of adjuvant treatment (see Chapter 10). One of the opportunities for DTC detection in BM might be the improvement of risk adapted adjuvant treatment strategies. Also, the presence of DTCs in BM of breast cancer cells after adjuvant chemotherapy was demonstrated to be correlated with an impaired prognosis.[22,23] This could identify patients who need additional adjuvant treatment (e.g., antibody-based regimen) after conventional adjuvant chemotherapy with no measurable residual disease.

Clinical Relevance of Sentinel Lymph Node Dissection

The last decade has seen the development of a minimally invasive technique to identify representative nodes, the sentinel nodes (SN), which are the first filter in the regional lymphatic pathway. The SN is indeed the most likely regional node to harbor metastatic carcinoma and seems to be predictive of the histopathology of the remaining lymphatic basin.[24] The SN can be identified by radiocolloid and/or blue dye. A combination of both methods is often applied and thought to be more sensitive.[25,26] However, the application of blue dye alone can miss up to 30–40% of SNs, it is a very time-consuming technique, and the axillary tissue must be dissected blindly until the blue node is located. To overcome this problem, radioactive tracers are injected close to the tumor so that the SN can be identified by lymphoscintigraphy and a gamma ray detecting probe can be used to locate the skin projection of the SN and facilitate the biopsy.

At present, there is no standardization of SN dissection techniques. Probe-directed mapping is complicated by several variables. The first one is the site of injection, which can be peritumoral, intratumoral, subcutaneous, or cutaneous. The second variable is the radiocolloid, which may be filtered sulfur colloid, unfiltered colloid, or radiolabeled albumin. The third one is the definition of SN detection. Using different colloids, SN was defined as (1) radioactivity three times over the background, (2) the node with the highest activity,[27,28] and (3) the node with more than 10 times the radioactivity of adjacent non-SNs.[29]

Intraoperative examination of SNs can be performed by cytologic preparations (imprints, scrapings) and frozen sections. Both methods are comparable, with accuracy and false-negativity rates ranging from 79% to 98% and from 9% to 52%, respectively, for frozen sections and from 77% to 99% and from 5% to 70%, respectively, for intraoperative cytology.[30]

The concept of the SN is not new. Oliver Cope referred to the "Delphian node" in 1963 as the lymph node that "will foretell the nature of the disease process" affecting a nearby organ. Lymphatic mapping in patients with breast cancer actually dates back to 1972 when Haagensen first illustrated the axillary nodal uptake of a vital blue dye during mastectomy. The term "sentinel node" was first used in 1977 by Cabanas et al. to refer to the first involved inguinal node in penile cancer. The validity of the concept was first demonstrated by Morton to select melanoma patients for regional node dissection.[31] Until now, more than 3500 cases of SN biopsy for breast cancer have been described in the literature. The rates for SN identification vary from 71% to 99% and appear to be higher when radioactive tracer techniques are used. The false-negative rate ranges from 1% to 11%.[30,32]

A variety of studies demonstrated that SN biopsy is as accurate as ALND for the detection of micrometastases in patients with T1–T2 tumors. Cserni et al.[30] and the European Working Group for Breast Screening Pathology postulated that remaining non-SN metastases in the axilla after SN biopsy and a negative palpable intraoperative nodal status were rare, even if isolated tumor cells or micrometastases were present in SNs, detected or undetected. These findings were further supported in a review by Noguchi,[24] who summarized four studies, including 446 patients who underwent SN biopsy without ALND. Although the follow-up time was quite short (22–39 months), only one patient with a negative SN biopsy showed an axillary relapse at a median follow-up of 24 months.

DTCs as a Therapeutic Target

Success of standard adjuvant chemotherapy aimed at proliferating cells may be limited because many of the DTCs present after primary resection are nonproliferative or dormant.[33] Cytotoxic chemotherapy regimens might therefore fail to eliminate dormant nonproliferating tumor cells, which may explain metastatic relapse after adjuvant chemotherapy.[23,34] In addition, chemotherapy reduces immunocompetent cells in BM.[35] Thus, complementary biologic therapies are urgently needed for the elimination of DTCs.

Although monoclonal antibody therapy was effective in various experimental systems, the clinical success is limited for patients with advanced-stage solid tumors,[36–38] probably also as a result of the large tumor cell burden and the lack of access that macromolecules have to cells in large tumor masses.[39] Therefore, DTCs should in theory be a much more promising target for antibody-based therapy.[36]

The molecular characterization of DTCs might also open new possibilities for choosing between therapy options. In patients with breast cancer, DTCs in the blood and BM often express the growth factor receptor HER-2/neu,[40–44] which is the target for therapy with the monoclonal antibody trastuzumab (Herceptin®, see Chapter 16). The heterogeneity of solid tumors poses a problem for all types of therapy and limits the chances of complete elimination of all residual tumor cells. For example, expression of the 17-1A antigen (also named Ep-CAM for epithelial cell adhesion molecule), which is also used as therapeutic target,[45–47] is heterogeneous in disseminated mammary carcinoma cells.[48,49]

Clinical Testing Methods for Detection of DTCs in BM

Immunocytochemistry

Using conventional histopathologic techniques at the time of primary diagnosis without clinical signs of metastatic bone disease, the probability to identify DTCs in BM is as low as 4%.[50] These findings imply that histologic evaluation is not sensitive enough for the indicated purpose and lead to the development of sensitive immunocytologic methods. Many studies have used cytokeratins (CKs) as marker antigens; these proteins are stable and expressed in most epithelial tumors.[51,52] Finally, a combination of several antibodies to various CK antigens or broad spectrum anti-CK antibodies have been validated because of the antigenic heterogeneity of tumor cells.[53–56]

The immunocytochemical technique for detection of rare events is very dependent on the reaction parameters. Therefore, both the International Society of Cell Therapy and the National Cancer Institute have recognized the need for standardization of the immunocytochemical assay and for its evaluation in prospective studies.[57,58] On the basis of data available from published methodologic analyses,[18,54,55,57] such a standardized assay may consist of a specificity proven anti-CK monoclonal antibody (i.e., A45-B/B3) and a sufficient sample size (i.e., 2×10^6 mononucleated cells per patient) obtained from two aspiration sites (FIG. 9.1). The use of new automated devices for the microscopic screening of immunostained slides may help to read slides more rapidly and to increase reproducibility of the read-out process by reducing subjectivity and the need for large experience with evaluating cells.[59–64]

To date, most experience with BM screening for occult metastatic breast cancer cells exists for immunocytochemical analyses using Ficoll density gradient centrifugation for tumor cell enrichment. In addition, new enrichment techniques based on improved density gradient methods[65–67] and immunomagnetic procedures[68–70] may increase the sensitivity of the immunocytochemical approaches and might therefore be helpful to obtain more consistent data on the clinical significance of DTCs. However, clinical studies will be required to determine whether the novel enrichment techniques are superior to "standard" methods, such as Ficoll gradients.

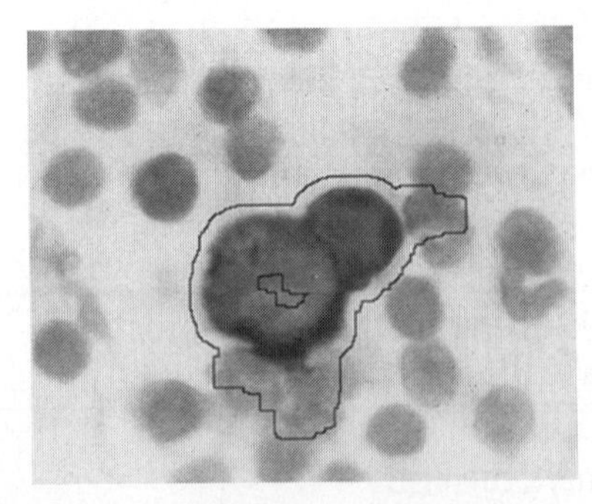

FIGURE 9.1 DTCs from BM of a breast cancer patient stained with an anti-cytokeratin antibody as previously described[18] and detected with the help of an automated microscope.[61] **See Plate 21 for color image.**

Polymerase Chain Reaction

The lack of a unique DNA marker for breast cancer has induced the search for tumor-associated messenger RNA (mRNA) transcripts that can be amplified a million-fold by a reverse transcriptase-polymerase chain reaction (RT-PCR). The major limiting factor for detection of single tumor cells by RT-PCR is illegitimate low level transcription of tumor associated or epithelial-specific genes in normal cells.[71–75] Moreover, deficient expression of the marker gene in DTCs may lower the actual sensitivity of the RT-PCR assay in vivo as compared with experimental model systems that use tumor cell lines for sensitivity estimation.[73] The studies conducted so far applied CK, mucin, carcinoembryonic antigen (CEA), β-human chorionic gonadotropin, mammaglobin, epidermal growth factor receptor, and several other mRNA markers.[21,67,76–82] Quantitative real-time RT-PCR provides interesting prospects for better definition of a cutoff between mRNA expression in normal cells and tumor cells, thereby increasing the specificity of the RT-PCR approach. In addition, exact quantification of the tumor cell load in individual samples might also be an important goal.[80,83–86] However, the amount of marker transcript per tumor cell may vary considerably between tumor cells of an individual patient and also among tumor cells of different patients. Thus, it might be difficult to interpret the PCR results and distinguish between changes in tumor cell numbers or changes in expression levels.

The situation becomes even more complex when the RT-PCR test is used to monitor therapy-induced changes in DTCs because the therapy might affect the expression level of the marker transcript in tumor cells. One possibility to circumvent this serious problem is the use of a multimarker RT-PCR test with independent RNA markers belonging to different gene families.[83,87]

Detection of Micrometastases in Sentinel Lymph Nodes

Conventional routine examination of SNs based on hematoxylin and eosin (H&E) staining of one or two sections through the largest diameter of the lymph node has been shown to underestimate the true incidence of micrometastases. Serial sectioning and CK staining by means of immunohistochemistry (IHC) has significantly increased the rate of occult metastatic disease to the regional nodes up to 35% and led to an "upstaging" of patients (FIG. 9.2).[30,88] Because IHC in addition to conventional staining is expensive, time consuming, and possibly requires more than one pathologist for appropriate analysis, only a few studies analyzed the prognostic significance. Dowlatshahi et al.[89] found a survival disadvantage in patients with occult metastases only identified by means of IHC but undetectable by H&E histology, whereas Millis et al.[90] recently demonstrated that occult axillary lymph node metastases are not of prognostic significance in breast cancer.

Besides the heterogeneity of methods used, one important aspect is the number of sections examined and the size of the sectioning interval. Nodes were sectioned from 5-μm intervals up to 500-μm intervals, indicating

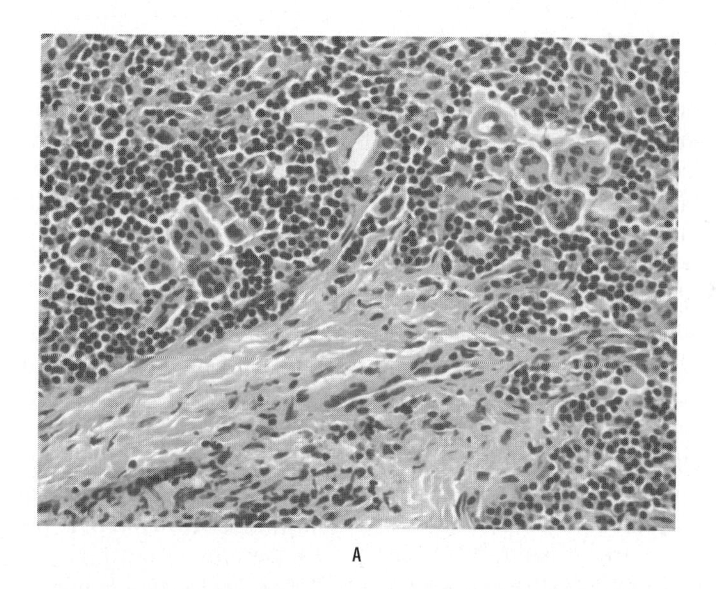

A

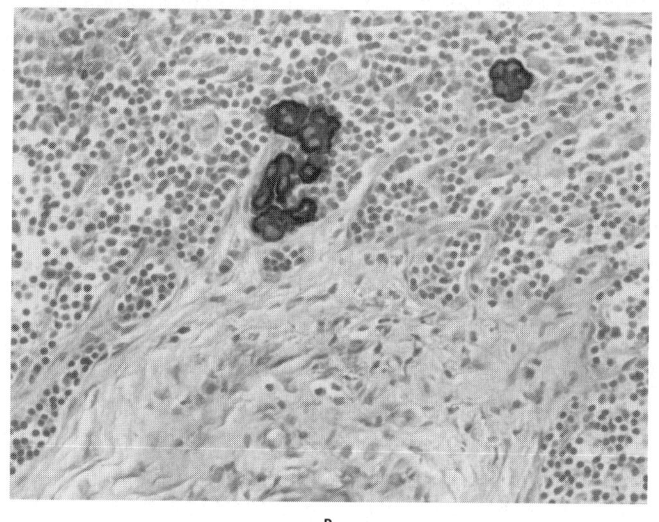

B

FIGURE 9.2 DTC in a lymph node of a breast cancer patient detected with an antibody against cytokeratin. (A) Conventional staining; (B) immunocytochemical staining. (Courtesy of Jeffrey S. Ross, M.D., Albany Medical College, Albany, NY.) **See Plate 22 for color image.**

that the detection of micrometastases correlated with the number of sections examined and the sectioned interval. Viale et al.[91] compared the results of SNs sectioned completely at 50-μm intervals with three to six sections of non-SNs at 100- to 500-μm intervals. Increasing of the sectioning intervals resulted in missing 18% of micrometastatic SNs.

Several groups applied RT-PCR to improve the detection of micrometastases in SN beyond that achievable by IHC (see Chapter 2 and FIG.2.5) and serial sectioning.[92–94] Based on amplification of mRNA tumor markers expressed by breast cancer cells, RT-PCR increased the detection of occult metastases in lymph nodes from 6% to 55% using CEA, CK-19, and mucin-1 as mRNA markers.[95,96] Although RT-PCR is more sensitive than IHC, the method suffers from a lack of specific markers, morphological controls, and the occurrence of false-positive tests because noncancer cells in blood and lymph nodes have also been shown to express these markers.[72,97,98] Furthermore, most studies have relied on single markers so that the use of multiple markers is one approach to improve the sensitivity as well as the specificity of RT-PCR for the detection of micrometastases in SN.

Min et al. evaluated seven potential breast cancer tumor markers in human breast cancer adenocarcinoma lines and in normal nodes from noncancer patients. Mammaglobin transcripts were found in 100% and CEA transcripts in 71% of breast cell lines but were absent in normal nodes.[99] In addition, Kataoka et al.[92] demonstrated that CEA was expressed in 25% of 48 patients with H&E-negative SNs and mammaglobin in 21%. Both groups concluded that RT-PCR, using a multimarker panel, is a powerful and sensitive method of SN diagnosis. In contrast, recent studies suggested that up to 20% of breast cancers do not express mammaglobin at all and that the prognostic significance of micrometastases as detected by RT-PCR remains unclear.

Bostick et al.[102] showed that a multiple-marker RT-PCR and Southern blot analysis, using c-Met, βI → 4GalNAc-T, and P97 mRNA transcripts in 57 SN of 41 patients, improved the detection of occult metastases in the SN when compared with conventional H&E and IHC analysis. The expression of all three tumor mRNA markers in the SN correlated with poor prognostic clinicopathologic factors compared with the expression of 0-2 markers. Ishida et al.[103] postulated that SN biopsy is recommended to detect micrometastases by means of H&E, IHC, and RT-PCR. If SN were only analyzed by H&E staining and/or IHC without PCR, omitting ALND referred by SN biopsy should be avoided. Applying a real-time quantitative RT-PCR for the detection of transcripts for epithelial glycoprotein-2 and CK-19 in SNs and blood samples, Schroder et al.[104] postulated that IHC appears to be more sensitive and specific than H&E and RT-PCR for the detection of micrometastases in SNs.

Future Tests for the Detection and Characterization of DTCs

With an immunocytochemical approach it is possible to characterize DTC by multiple staining or the combination of immunostaining and other analysis. As stated above, this might help to identify antigens that are potential targets for systemic therapies aimed to eliminate minimal residual disease in breast cancer patients. Using a combination of immunocytochemistry and fluorescence in situ hybridization, several groups reported numerical chromosomal aberrations in CK-positive cancer cells in BM, indicating the malignant origin of these disseminated cells. The availability of new protocols for the amplification of the whole genome even of single cells[108–110] has enabled a more detailed analysis of DTCs. Using single cell comparative genomic hybridization, Klein and coworkers were able to demonstrate that CK-positive cells in BM of breast cancer patients without clinical signs of overt metastases are genetically heterogeneous.[109,111,112] A different strategy includes the assessment of the viability and proliferative potential of disseminated cancer cells in conjunction with molecular single cell analysis.[107] The proliferative capacity of CK-positive tumor cells in BM of cancer patients can be assessed using special culture conditions, and the proliferative potential of these cells determines the clinical outcome.[107] As stated above, therapeutic targets like the HER-2/neu could be identified on these cells.[40–44,113]

Current Clinical Utility for the Detection of Micrometastases in BM and Lymph Nodes

In a recently published summary on the current status of micrometastases detection, the National Cancer Institute consensus panel concluded the following: "In summary, the existing evidence warrants the consideration of new clinical trials designed to provide definitive information about the prevalence of DTCs, the relationship to prognosis, and the clinical utility of these findings in patients with solid tumors."[114] However, clinical decisions should not be based on the detection outside clinical trials.

Future Clinical Potential for the Detection of DTCs in BM of Breast Cancer Patients

Success or failure of adjuvant treatment for patients with primary breast cancer can only be assessed after an observation period of several years. The availability of a surrogate marker for monitoring the effectiveness of systemic treatment should make the evaluation and development of new adjuvant therapies possible in a shorter period. Examination of BM and peripheral blood during therapy could indicate whether the therapeutic approach that was used has been effective. Monitoring procedures of this type would be of considerable value. In therapeutic studies, long-term observations are still required to establish whether the therapy-associated reduction in individual disseminated cells[23] is associated with improved prognosis.

In conclusion, there is increasing evidence that the detection and characterization of tumor cells present in BM and lymph nodes of breast cancer patients can provide clinically important information. Examination for occult metastases should be incorporated into future clinical trials that evaluate cancer treatments. In the future, therapy specifically tailored to the disease in subgroups of patients or individual patients with residual disease may represent substantial progress.

Can we really postulate that only colonies of cells are clinically significant and single cells are generally not? Results from BM studies indicate that hematogenous spread of tumor cells takes place in the absence of lymphatic spread. Thus, we suggest BM screening might also be included to better identify patients at risk, previously classified as N0M0.

Current Clinical Utility of the Sentinel Lymph Node Dissection

Because of variations in the SN dissection technique, the rate of SN identification ranges from 79% to 98% and the false-negative rate from 0 to 29%. These data underline the need for quality control, training, and total commitment from the surgeon. Furthermore, lymphatic drainage of the breast can have multiple pathways depending on positioning and local pressure so that more than one SN may be involved. Because IHC and RT-PCR lead to an emerging subgroup of "unofficial" node-positive cases, the question arises whether patients with small tumors and a low risk of relapse, for whom chemotherapy has not been recommended so far, now should be considered as candidates for treatment.

Future Clinical Potential for the Sentinel Lymph Node Biopsy in Breast Cancer Patients

Although the issues discussed above have to be solved and should be taken into account in the formulation of guidelines, SN biopsy is a promising procedure that may play a major role in the management of patients with invasive breast cancer. Further randomized trials comparing SN biopsy alone with ALND should answer whether micrometastases, including single cells, are clinically relevant and which patients with SN micrometatases would benefit from ALND.

References

1. Hoe AL, Iven D, Royle GT, et al. Incidence of arm swelling following axillary clearance for breast cancer. Br J Surg 1992;79:261–262.
2. Hermanek P, Hutter RV, Sobin LH, et al. International Union Against Cancer. Classification of isolated tumor cells and micrometastasis. Cancer 1999;86:2668–2673.
3. Hermanek P. Disseminated tumor cells versus micrometastasis: definitions and problems. Anticancer Res 1999;19:2771–2774.
4. Tjan-Heijnen VC, Buit P, de Widt-Evert LM, et al. Micrometastases in axillary lymph nodes: an increasing classification and treatment dilemma in breast cancer due to the introduction of the sentinel lymph node procedure. Breast Cancer Res Treat 2001;70:81–88.
5. Singletary SE, Allred C, Ashley P, et al. Revision of the American Joint Committee on Cancer staging system for breast cancer. J Clin Oncol 2002;20:3628–3636.
6. Funke I, Schraut W. Meta-analyses of studies on bone marrow micrometastases: an independent prognostic impact remains to be substantiated. J Clin Oncol 1998;16:557–566.
7. Porro G, Menard S, Tagliabue E, et al. Monoclonal antibody detection of carcinoma cells in bone marrow biopsy specimens from breast cancer patients. Cancer 1988;61:2407–2411.

8. Mathieu MC, Friedman S, Bosq J, et al. Immunohistochemical staining of bone marrow biopsies for detection of occult metastasis in breast cancer. Breast Cancer Res Treat 1990;15:21–26.
9. Courtemanche DJ, Worth AJ, Coupland RW, et al. Detection of micrometastases from primary breast cancer. Can J Surg 1991;34:15–19.
10. Gerber B, Krause A, Muller H, et al. Simultaneous immunohistochemical detection of tumor cells in lymph nodes and bone marrow aspirates in breast cancer and its correlation with other prognostic factors. J Clin Oncol 2001;19:960–971.
11. Braun S, Cevatli BS, Assemi C, et al. Comparative analysis of micrometastasis to the bone marrow and lymph nodes of node-negative breast cancer patients receiving no adjuvant therapy. J Clin Oncol 2001; 19:1468–1475.
12. Woelfle U, Cloos J, Sauter G, et al. Molecular signature associated with bone marrow micrometastasis in human breast cancer. Cancer Res 2003;63:5679–5684.
13. Landys K, Persson S, Kovarik J, et al. Prognostic value of bone marrow biopsy in operable breast cancer patients at the time of initial diagnosis: results of a 20-year median follow-up. Breast Cancer Res Treat 1998;49:27–33.
14. Diel IJ, Kaufmann M, Costa SD, et al. Micrometastatic breast cancer cells in bone marrow at primary surgery: prognostic value in comparison with nodal status. J Natl Cancer Inst 1996;88:1652–1658.
15. Mansi JL, Gogas H, Bliss JM, et al. Outcome of primary-breast-cancer patients with micrometastases: a long-term follow-up study. Lancet 1999;354: 197–202.
16. Gebauer G, Fehm T, Merkle E, et al. Epithelial cells in bone marrow of breast cancer patients at time of primary surgery: clinical outcome during long-term follow-up. J Clin Oncol 2001;19:3669–3674.
17. Cote RJ, Rosen PP, Lesser ML, et al. Prediction of early relapse in patients with operable breast cancer by detection of occult bone marrow micrometastases. J Clin Oncol 1991;9:1749–1756.
18. Braun S, Pantel K, Muller P, et al. Cytokeratin-positive cells in the bone marrow and survival of patients with stage I, II, or III breast cancer. N Engl J Med 2000;342:525–533.
19. Harbeck N, Untch M, Pache L, et al. Tumour cell detection in the bone marrow of breast cancer patients at primary therapy: results of a 3-year median follow-up. Br J Cancer 1994;69:566–571.
20. Wiedswang G, Borgen E, Karesen R, et al. Detection of isolated tumor cells in bone marrow is an independent prognostic factor in breast cancer. J Clin Oncol 2003;21:3469–3478.
21. Vannucchi AM, Bosi A, Glinz S, et al. Evaluation of breast tumour cell contamination in the bone marrow and leukapheresis collections by RT-PCR for cytokeratin-19 mRNA. Br J Haematol 1998;103: 610–617.
22. Janni W, Hepp F, Rjosk D, et al. The fate and prognostic value of occult metastatic cells in the bone marrow of patients with breast carcinoma between primary treatment and recurrence. Cancer 2001; 92:46–53.
23. Braun S, Kentenich C, Janni W, et al. Lack of effect of adjuvant chemotherapy on the elimination of single dormant tumor cells in bone marrow of high-risk breast cancer patients. J Clin Oncol 2000; 18:80–86.
24. Noguchi M. Therapeutic relevance of breast cancer micrometastases in sentinel lymph nodes. Br J Surg 2002;89:1505–1515.
25. O'Hea BJ, Hill AD, El-Shirbiny AM, et al. Sentinel lymph node biopsy in breast cancer: initial experience at Memorial Sloan-Kettering Cancer Center. J Am Coll Surg 1998;186:423–427.
26. Bedrosian I, Reynolds C, Mick R, et al. Accuracy of sentinel lymph node biopsy in patients with large primary breast tumors. Cancer 2000;88:2540–2545.
27. Veronesi U, Paganelli G, Galimberti V, et al. Sentinel-node biopsy to avoid axillary dissection in breast cancer with clinically negative lymph-nodes. Lancet 1997;349:1864–1867.
28. Veronesi U, Paganelli G, Viale G, et al. Sentinel lymph node biopsy and axillary dissection in breast cancer: results in a large series. J Natl Cancer Inst 1999;91:368–373.
29. Albertini JJ, Lyman GH, Cox C, et al. Lymphatic mapping and sentinel node biopsy in the patient with breast cancer. JAMA 1996;276:1818–1822.
30. Cserni G, Amendoeira I, Apostolikas N, et al. Pathological work-up of sentinel lymph nodes in breast cancer. Review of current data to be considered for the formulation of guidelines. Eur J Cancer 2003; 39:1654–1667.
31. Cochran AJ, Wen DR, Morton DL. Occult tumor cells in the lymph nodes of patients with pathological stage I malignant melanoma. An immunohistological study. Am J Surg Pathol 1988;12:612–618.
32. Krag D, Weaver D, Ashikaga T, et al. The sentinel node in breast cancer—a multicenter validation study. N Engl J Med 1998;339:941–946.
33. Pantel K, Schlimok G, Braun S, et al. Differential expression of proliferation-associated molecules in individual micrometastatic carcinoma cells. J Natl Cancer Inst 1993;85:1419–1424.
34. Kruger WH, Kroger N, Togel F, et al. Disseminated breast cancer cells prior to and after high-dose therapy. J Hematother Stem Cell Res 2001;10: 681–689.
35. Solomayer EF, Feuerer M, Bai L, et al. Influence of adjuvant hormone therapy and chemotherapy on the immune system analysed in the bone marrow of patients with breast cancer. Clin Cancer Res 2003; 9:174–180.
36. Scott AM and Welt S. Antibody-based immunological therapies. Curr Opin Immunol 1997;9:717–722.
37. van den Eertwegh AJ, Baars A, Pinedo HM. Adjuvant treatment of colorectal cancer: towards tumor-specific immunotherapies. Cancer Metastasis Rev 2001;20(1–2):101–108.
38. Vogel CL, Cobleigh MA, Tripathy D, et al. Efficacy and safety of trastuzumab as a single agent in first-line treatment of HER2-overexpressing metastatic breast cancer. J Clin Oncol 2002;20: 719–726.
39. Jain RK. Physiological barriers to delivery of monoclonal antibodies and other macromolecules in tumors. Cancer Res 1990;50(3 suppl):814s–819s.
40. Brandt B, Roetger A, Heidl S, et al. Isolation of blood-borne epithelium-derived c-erbB-2 oncoprotein-positive clustered cells from the peripheral blood of breast cancer patients. Int J Cancer 1998;76: 824–828.

41. Engel H, Kleespies C, Friedrich J, et al. Detection of circulating tumour cells in patients with breast or ovarian cancer by molecular cytogenetics. Br J Cancer 1999;81:1165–1173.
42. Faivre S, Vago N, Morat L, et al. HER-2 expression in circulating tumor cells (CTC) from patients with metastatic breast cancer. Proc AACR 2002;43:abst. 3639.
43. Braun S, Schlimok G, Heumos I, et al. ErbB2 overexpression on occult metastatic cells in bone marrow predicts poor clinical outcome of stage I-III breast cancer patients. Cancer Res 2001;61:1890–1895.
44. Hayes DF, Walker TM, Singh B, et al. Monitoring expression of HER–2 on circulating epithelial cells in patients with advanced breast cancer. Int J Oncol 2002;21:1111–1117.
45. Hempel P, Muller P, Oruzio D, et al. Combination of high-dose chemotherapy and monoclonal antibody in breast-cancer patients: a pilot trial to monitor treatment effects on disseminated tumor cells. Cytotherapy 2000;2:287–296.
46. Balzar M, Winter MJ, de Boer CJ, et al. The biology of the 17-1A antigen (Ep-CAM). J Mol Med 1999;77: 699–712.
47. Riethmuller G, Holz E, Schlimok G, et al. Monoclonal antibody therapy for resected Dukes' C colorectal cancer: seven-year outcome of a multicenter randomized trial. J Clin Oncol 1998;16:1788–1794.
48. Braun S, Hepp F, Sommer HL, et al. Tumor-antigen heterogeneity of disseminated breast cancer cells: implications for immunotherapy of minimal residual disease. Int J Cancer 1999;84:1–5.
49. Thurm H, Ebel S, Kentenich C, et al. Rare expression of epithelial cell adhesion molecule on residual micrometastatic breast cancer cells after adjuvant chemotherapy. Clin Cancer Res 2003;9:2598–2604.
50. Ridell B, Landys K. Incidence and histopathology of metastases of mammary carcinoma in biopsies from the posterior iliac crest. Cancer 1979;44: 1782–1788.
51. Moll R, Franke WW, Schiller DL, et al. The catalog of human cytokeratins: patterns of expression in normal epithelia, tumors and cultured cells. Cell 1982; 31:11–24.
52. Fuchs E, Cleveland DW. A structural scaffolding of intermediate filaments in health and disease. Science 1998;279:514–519.
53. Osborne MP, Wong GY, Asina S, et al. Sensitivity of immunocytochemical detection of breast cancer cells in human bone marrow. Cancer Res 1991;51: 2706–2709.
54. Pantel K, Schlimok G, Angstwurm M, et al. Methodological analysis of immunocytochemical screening for disseminated epithelial tumor cells in bone marrow. J Hematother 1994;3:165–173.
55. Braun S, Muller M, Hepp F, et al. Micrometastatic breast cancer cells in bone marrow at primary surgery: prognostic value in comparison with nodal status. J Natl Cancer Inst 1998;90:1099–1101.
56. Schlimok G, Funke I, Holzmann B, et al. Micrometastatic cancer cells in bone marrow: in vitro detection with anti-cytokeratin and in vivo labeling with anti-17-1A monoclonal antibodies. Proc Natl Acad Sci USA 1987;84:8672–8676.
57. Borgen E, Naume B, Nesland JM, et al. Standardization of the immunocytochemical detection of cancer cells in BM and blood. I. Establishment of objective criteria for the evaluation of immunostained cells. Cytometry 1999;1:377–388.
58. Pantel K, Cote RJ, Fodstad O. Detection and clinical importance of micrometastatic disease. J Natl Cancer Inst 1999;91:1113–1124.
59. Borgen E, Naume B, Nesland JM, et al. Use of automated microscopy for the detection of disseminated tumor cells in bone marrow samples. Cytometry 2001;46:215–221.
60. Kraeft SK, Sutherland R, Gravelin L, et al. Detection and analysis of cancer cells in blood and bone marrow using a rare event imaging system. Clin Cancer Res 2000;6:434–442.
61. Bauer KD, de la Torre-Bueno J, Diel IJ, et al. Reliable and sensitive analysis of occult bone marrow metastases using automated cellular imaging. Clin Cancer Res 2000;6:3552–3559.
62. Mehes G, Luegmayr A, Ambros IM, et al. Combined automatic immunological and molecular cytogenetic analysis allows exact identification and quantification of tumor cells in the bone marrow. Clin Cancer Res 2001;7:1969–1975.
63. Mehes G, Luegmayr A, Hattinger CM, et al. Automatic detection and genetic profiling of disseminated neuroblastoma cells. Med Pediatr Oncol 2001;36: 205–209.
64. Witzig TE, Bossy B, Kimlinger T, et al. Detection of circulating cytokeratin-positive cells in the blood of breast cancer patients using immunomagnetic enrichment and digital microscopy. Clin Cancer Res 2002;8:1085–1091.
65. Rosenberg R, Gertler R, Friederichs J, et al. Comparison of two density gradient centrifugation systems for the enrichment of disseminated tumor cells in blood. Cytometry 2002;49:150–158.
66. Mitas M, Hoover L, Silvestri G, et al. Lunx is a superior molecular marker for detection of non-small lung cell cancer in peripheral blood. J Mol Diagn 2003;5:237–242.
67. Baker MK, Mikhitarian K, Osta W, et al. Molecular detection of breast cancer cells in the peripheral blood of advanced-stage breast cancer patients using multimarker real-time reverse transcription-polymerase chain reaction and a novel porous barrier density gradient centrifugation technology. Clin Cancer Res 2003;9:4865–4871.
68. Bilkenroth U, Taubert H, Riemann D, et al. Detection and enrichment of disseminated renal carcinoma cells from peripheral blood by immunomagnetic cell separation. Int J Cancer 2001;92:577–582.
69. Flatmark K, Bjornland K, Johannessen HO, et al. Immunomagnetic detection of micrometastatic cells in bone marrow of colorectal cancer patients. Clin Cancer Res 2002;8:444–449.
70. Naume B, Borgen E, Kvalheim G, et al. Detection of isolated tumor cells in bone marrow in early-stage breast carcinoma patients: comparison with preoperative clinical parameters and primary tumor characteristics. Clin Cancer Res 2001;7:4122–4129.
71. Zippelius A, Kufer P, Honold G, et al. Limitations of reverse-transcriptase polymerase chain reaction analyses for detection of micrometastatic epithelial cancer cells in bone marrow. J Clin Oncol 1997; 15:2701–2708.
72. Bostick PJ, Chatterjee S, Chi DD, et al. Limitations of specific reverse-transcriptase polymerase chain reaction markers in the detection of metastases in

the lymph nodes and blood of breast cancer patients. J Clin Oncol 1998;16:2632–2640.

73. Zippelius A, Lutterbuse R, Riethmuller G, et al. Analytical variables of reverse transcription-polymerase chain reaction-based detection of disseminated prostate cancer cells. Clin Cancer Res 2000;6:2741–2750.
74. Jung R, Petersen K, Kruger W, et al. Detection of micrometastasis by cytokeratin 20 RT-PCR is limited due to stable background transcription in granulocytes. Br J Cancer 1999;81:870–873.
75. Silva AL, Diamond J, Silva MR, et al. Cytokeratin 20 is not a reliable molecular marker for occult breast cancer cell detection in hematological tissues. Breast Cancer Res Treat 2001;66:59–66.
76. Kruger W, Krzizanowski C, Holweg M, et al. Reverse transcriptase/polymerase chain reaction detection of cytokeratin-19 mRNA in bone marrow and blood of breast cancer patients. J Cancer Res Clin Oncol 1996;122:679–686.
77. Datta YH, Adams PT, Drobyski WR, et al. Sensitive detection of occult breast cancer by the reverse-transcriptase polymerase chain reaction. J Clin Oncol 1994;12:475–482.
78. Hoon DS, Sarantou T, Doi F, et al. Detection of metastatic breast cancer by beta-hCG polymerase chain reaction. Int J Cancer 1996;69:369–374.
79. Fields KK, Elfenbein GJ, Trudeau WL, et al. Clinical significance of bone marrow metastases as detected using the polymerase chain reaction in patients with breast cancer undergoing high-dose chemotherapy and autologous bone marrow transplantation. J Clin Oncol 1996;14:1868–1876.
80. Slade MJ, Smith BM, Sinnett HD, et al. Quantitative polymerase chain reaction for the detection of micrometastases in patients with breast cancer. J Clin Oncol 1999;17:870–879.
81. Zach O, Kasparu H, Krieger O, et al. Detection of circulating mammary carcinoma cells in the peripheral blood of breast cancer patients via a nested reverse transcriptase polymerase chain reaction assay for mammaglobin mRNA. J Clin Oncol 1999;17:2015–2019.
82. Gerhard M, Juhl H, Kalthoff H, et al. Specific detection of carcinoembryonic antigen-expressing tumor cells in bone marrow aspirates by polymerase chain reaction. J Clin Oncol 1994;12:725–729.
83. Bosma AJ, Weigelt B, Lambrechts AC, et al. Detection of circulating breast tumor cells by differential expression of marker genes. Clin Cancer Res 2002; 8:1871–1877.
84. Guller U, Zajac P, Schnider A, et al. Disseminated single tumor cells as detected by real-time quantitative polymerase chain reaction represent a prognostic factor in patients undergoing surgery for colorectal cancer. Ann Surg 2002;236:768–775, discussion 775–776.
85. Becker M, Nitsche A, Neumann C, et al. Sensitive PCR method for the detection and real-time quantification of human cells in xenotransplantation systems. Br J Cancer 2002;87:1328–1335.
86. Suchy B, Austrup F, Driesel G, et al. Detection of mammaglobin expressing cells in blood of breast cancer patients. Cancer Lett 2000;158:171–178.
87. Taback B, Chan AD, Kuo CT, et al. Detection of occult metastatic breast cancer cells in blood by a multimolecular marker assay: correlation with clinical stage of disease. Cancer Res 2001;61:8845–8850.
88. Pendas S, Dauway E, Cox CE, et al. Sentinel node biopsy and cytokeratin staining for the accurate staging of 478 breast cancer patients. Am Surg 1999;65:500–505, discussion 505–506.
89. Dowlatshahi K, Fan M, Snider HC, et al. Lymph node micrometastases from breast carcinoma: reviewing the dilemma. Cancer 1997;80:1188–1197.
90. Millis RR, Springall R, Lee AH, et al. Occult axillary lymph node metastases are of no prognostic significance in breast cancer. Br J Cancer 2002;86: 396–401.
91. Viale G, Maiorano E, Mazzarol G, et al. Histologic detection and clinical implications of micrometastases in axillary sentinel lymph nodes for patients with breast carcinoma. Cancer 2001;92:1378–1384.
92. Kataoka A, Mori M, Sadanaga N, et al. RT-PCR detection of breast cancer cells in sentinel lymph nodes. Int J Oncol 2000;16:1147–1152.
93. Wascher RA, Bostick PJ, Huynh KT, et al. Detection of MAGE-A3 in breast cancer patients' sentinel lymph nodes. Br J Cancer 2001;85:1340–1346.
94. Branagan G, Hughes D, Jeffrey M, et al. Detection of micrometastases in lymph nodes from patients with breast cancer. Br J Surg 2002;89:86–89.
95. Mori M, Mimori K, Inoue H, et al. Detection of cancer micrometastases in lymph nodes by reverse transcriptase-polymerase chain reaction. Cancer Res 1995;55:3417–3420.
96. Noguchi S, Aihara T, Motomura K, et al. Detection of breast cancer micrometastases in axillary lymph nodes by means of reverse transcriptase-polymerase chain reaction. Comparison between MUC1 mRNA and keratin 19 mRNA amplification. Am J Pathol 1996;148:649–656.
97. Traweek ST, Liu J, Battifora H. Keratin gene expression in non-epithelial tissues. Detection with polymerase chain reaction. Am J Pathol 1993;142: 1111–1118.
98. Hoon DS, Doi F, Giuliano AE, et al. The detection of breast carcinoma micrometastases in axillary lymph nodes by means of reverse transcriptase-polymerase chain reaction. Cancer 1995;76:533–535.
99. Min CJ, Tafra L, Verbanac KM. Identification of superior markers for polymerase chain reaction detection of breast cancer metastases in sentinel lymph nodes. Cancer Res. 1998;58:4581–4584.
100. Leygue E, Snell L, Dotzlaw H, et al. Mammaglobin, a potential marker of breast cancer nodal metastasis. J Pathol 1999;189:28–33.
101. Watson MA, Dintzis S, Darrow CM, et al. Mammaglobin expression in primary, metastatic, and occult breast cancer. Cancer Res 1999;59:3028–3031.
102. Bostick PJ, Huynh KT, Sarantou T, et al. Detection of metastases in sentinel lymph nodes of breast cancer patients by multiple-marker RT-PCR. Int J Cancer 1998;79:645–651.
103. Ishida M, Kitamura K, Kinoshita J, et al. Detection of micrometastasis in the sentinel lymph nodes in breast cancer. Surgery 2002;131(1 suppl): S211–S216.
104. Schroder CP, Ruiters MH, de Jong S, et al. Detection of micrometastatic breast cancer by means of real time quantitative RT-PCR and immunostaining in perioperative blood samples and sentinel nodes. Int J Cancer 2003;106:611–618.
105. Mueller P, Carroll P, Bowers E, et al. Low frequency epithelial cells in bone marrow aspirates from

prostate carcinoma patients are cytogenetically aberrant. Cancer 1998;83:538–546.

106. Fehm T, Sagalowsky A, Clifford E, et al. Cytogenetic evidence that circulating epithelial cells in patients with carcinoma are malignant. Clin Cancer Res 2002;8:2073–2084.

107. Solakoglu O, Maierhofer C, Lahr G, et al. Heterogeneous proliferative potential of occult metastatic cells in bone marrow of patients with solid epithelial tumors. Proc Natl Acad Sci USA 2002;99: 2246–2251.

108. Dietmaier W, Hartmann A, Wallinger S, et al. Multiple mutation analyses in single tumor cells with improved whole genome amplification. Am J Pathol 1999;154:83–95.

109. Klein CA, Seidl S, Petat-Dutter K, et al. Combined transcriptome and genome analysis of single micrometastatic cells. Nat Biotechnol 2002;20: 387–392.

110. Klein CA, Schmidt-Kittler O, Schardt JA, et al. Comparative genomic hybridization, loss of heterozygosity, and DNA sequence analysis of single cells. Proc Natl Acad Sci USA 1999;96:4494–4499.

111. Klein CA, Blankenstein TJ, Schmidt-Kittler O, et al. Genetic heterogeneity of single disseminated tumour cells in minimal residual cancer. Lancet 2002;360: 683–689.

112. Schmidt-Kittler O, Ragg T, Daskalakis A, et al. From latent disseminated cells to overt metastasis: genetic analysis of systemic breast cancer progression. Proc Natl Acad Sci USA 2003;100:7737–7742.

113. Walker M, Hayes D, Gross S, et al. Detection of HER–2/neu cell membrane receptor on circulating tumor cells in patients with advanced breast cancer. Proc AACR 2001;42:abst. 60.

114. Lugo TG, Braun S, Cote RJ, et al. Detection and measurement of occult disease for the prognosis of solid tumors. J Clin Oncol 2003;21:2609–2615.

CHAPTER 10

Prognostic and Predictive Factors Overview

Jeffrey S. Ross, MD[1,2] and Nadia Harbeck, MD, PhD[3]

[1]Department of Pathology and Laboratory Medicine, Albany Medical College, Albany, New York
[2]Division of Oncology Molecular Medicine, Millennium Pharmaceuticals, Inc., Cambridge, Massachusetts
[3]Fauenklinik, Technical University of Munich, Munich, Germany

Based on current incidence rates, it is estimated that an American woman has a one in nine chance of developing breast cancer at some time during her life.[1] This chapter provides a brief overview of both established and emerging biomarkers and molecular diagnostics that have been used to determine breast cancer prognosis and predict the response of the disease to treatment with both conventional and emerging targeted therapies.

Prognostic Versus Predictive Factors

Prognostic Factors

Prognostic factors in clinical use in patients diagnosed with breast cancer are designed to forecast the most likely clinical outcome of the disease without regard to the nature and intensity of the selected treatment. Classic factors widely used in this fashion include the tumor type (infiltrating ductal vs. lobular, ductal subtypes such as medullary, tubular, papillary, and mucinous), tumor grade, tumor size, lymph node status, extent of the intraductal component, and presence of vascular space invasion. Ancillary tests in common use and recommended by both the College of American Pathologists and the American Society of Clinical Oncologists include hormone receptor status and HER-2/*neu* gene/protein status (HER-2). The key features of a clinically useful prognostic factor include ease and reliability of the assay, confirmation that the prognostic significance is not confounded by the type of treatment used, and that the factor provides disease outcome information that is independent from the status of other classic factors. Further criteria and details concerning the mathematic issues and clinical uses of prognostic and predictive factors in breast cancer management can be found in Chapter 30, which discusses the translation of the use of these factors to clinical practice.

Predictive Factors

Biomarkers, genetic mutations and protein abnormalities that are associated with the development of breast cancer are covered in depth in Chapter 4. In this chapter the term "predictive factors" are defined as factors that specifically predict whether a newly diagnosed or relapsed case of breast cancer will or will not respond to a specific single or combination of therapies. For example, the HER-2 status in a newly diagnosed breast cancer can serve both as a stand-alone prognostic factor and as a predictive factor for response to trastuzumab. The estrogen receptor (ER) test, described below, is also an example of a proven predictive factor for the response to hormonal therapy, but ER status is a much weaker general prognostic factor for forecasting therapy-independent disease outcome.

Traditional Morphology-Based Prognostic Factors

Tumor Type

Pure infiltrating ductal carcinoma is by far the most common form of breast cancer (**Table 10.1**)

Table 10.1 Classification of Invasive Carcinomas of the Breast

Breast Cancer Subtype	Incidence	5-Year Overall Survival
Infiltrating ductal carcinoma	70%	80%
Infiltrating lobular carcinoma	10%	85%
Infiltrating mixed ductal, lobular, and unspecified carcinomas	13%	80%
Medullary carcinoma	3%	85%
Mucinous (colloid) carcinoma	2%	95%
Papillary carcinoma	1%	95%
Tubular carcinoma	1%	95%

and accounts for approximately 70% of invasive malignancies.[2] Infiltrating lobular carcinoma is the second most common subtype and encompasses approximately 10% of cases. Combined ductal and lobular carcinomas and carcinomas with features intermediate between ductal and lobular differentiation account for approximately 13% of cases.[2–5] When DNA microarrays are used to classify the gene expression pattern of breast cancers, the lobular versus ductal morphologic classification is replaced by the luminal versus basal differentiation scheme (see Chapter 6).[6–8] In general, invasive ductal carcinomas have a slightly worse prognosis than lobular cancers. Additional subtypes of invasive ductal carcinoma include mucinous, medullary, tubular, and papillary carcinomas. These lesions generally have a more favorable prognosis than infiltrating ductal or infiltrating lobular carcinoma with the exception of medullary carcinoma, whose prognosis is approximately the same as for infiltrating lobular cancer.[2–5]

Tumor Grade

The Bloom-Richardson (Nottingham) histologic grading system for breast cancer (**Table 10.2**) is the most widely used system.[9] Grading has generally predicted prognosis in breast cancer, with grade 1 lesions averaging a 95% 5-year overall survival rate compared with 75% for grade 2 lesions and 50% for grade 3 tumors.[2–5] In most large studies, histologic grading has achieved prognostic significance and in many has retained independent significance on multivariate analysis. However, tumor grading is a subjective analysis, and interobserver variation can confound and limit its utility in disease outcome studies.[10] Nonetheless, tumor grading is a standard and widely respected component of the pathology report on a newly diagnosed breast cancer specimen and is relied on by oncologists when discussing treatment options with patients.

Table 10.2 Histologic Grading for Breast Cancer

	Score	5-Year Overall Survival
Tubule formation (% of carcinoma composed of tubular structures)		
>75%	1	
10–75%	2	
Less than 10%	3	
Nuclear pleomorphism		
Small uniform cells	1	
Moderate increase in size and variation	2	
Marked variation	3	
Mitotic count (per 10 high power fields)		
Up to 7	1	
8–14	2	
15 or more	3	
Final histologic grade		
Grade 1 (scores 3–5)		90%
Grade 2 (scores 6 and 7)		75%
Grade 3 (Scores 8 and 9)		50%

Tumor Size

The measurement of tumor size is also a widely accepted major factor in prognosis assessment in breast cancer and widely used to plan therapy.[2–5] Tumor size is a major component of the breast cancer staging system and, despite issues involving measurement techniques, shrinkage after fixation, and tumors with large stromal components, remains a cornerstone of the surgical pathology report for the disease. It should be noted that the prognostic significance of tumor size appears to be declining as breast cancer screening programs involving mammography and self-examination continue to grow throughout the world. Screening programs lead to earlier diagnosis with decreasing size of the primary tumor at diagnosis. Thus, the size range of newly diagnosed cases in the United States has compressed toward lesions that are well less than 2 cm in greatest diameter when diagnosed. Smaller tumor sizes have also widely impacted ancillary studies by limited the ability to retain tissue for ancillary studies that is not processed by formalin fixation and paraffin embedding (see below).

Lymph Node Status

Most direct studies of breast cancer and meta-analysis reviews of these studies have continually found that the axillary lymph node status in patients with breast cancer remains the single most significant prognostic factor for the disease.[2–5,11–14] The sentinel lymph node biopsy technique has changed the approach to determining axillary lymph node status, although the prognostic significance of micro-metastatic disease detection is not fully established (see Chapter 9).[15,16] Moreover, the clinical significance of micrometastasis detection also remains controversial especially as it relates to therapy selection (see Chapter 9).

Tumor Stage

The widely used TNM staging system for breast cancer (**Table 10.3**) is a fundamental part of the initial management of the disease.[2–5] For patients who present with local and regional (non-stage IV) disease, the tumor size and axillary lymph node status are the major drivers of the pathologic staging for the disease. As seen in **Table 10.3**, the 5-year overall survival rate decreases progressively as patients are classified into higher stage disease with more than 80% of patients with stage IV disease at presentation dying of their disease.

Other Morphology-Based Observations

Extent of Ductal Carcinoma In Situ The extent of the ductal carcinoma in situ component in a resected infiltrating ductal carcinoma specimen has been linked to a higher local disease recurrence rate but not to a significant difference in overall survival.[2–5]

Resection Margin Status Similar to the extent of the ductal carcinoma in situ component, the presence of positive margins (either ductal carcinoma in situ or invasive carcinoma) is associated with an increased rate of local recurrence. Patients with both of these conditions are generally considered for possible additional local surgery or the addition of external beam radiation therapy.[2–5]

Lymphovascular Invasion In most studies, lymphovascular invasion and intralymphatic tumor embolization are adverse prognostic factors.[2–5] It may be difficult to distinguish true lymphovascular invasion from retraction artifact associated with specimen processing. Immunohistochemical (IHC) staining for endothelial-specific markers may improve the accuracy of determining lymphovascular invasion.[2] Compared with lymphatic invasion, tumor invasion of blood vessels may harbor an even worse prognosis, although the distinction between lymphatic and blood vessel invasion may be extremely difficult to make on both biopsy and resection specimens.[2]

Tumor Angiogenesis Tumor angiogenesis measured using IHC to determine microvessel density was first related to breast cancer outcome by Weidner and colleagues in 1991 using an antibody against factor-8 related antigen as an endothelial marker.[17] Subsequently, a number of studies were performed that both agreed and disagreed with the original finding of an independent prognostic significance for microvessel density.[18–20] The lack of correlation of microvessel density counts between studies may relate to differences in the definition of microvessel density and where and how the slides were scored. To date, microvessel density or other measures of angiogenesis in breast cancer have not been directly related to the response of disease to antiangiogenesis therapies such as in response to the anti-vascular endothelial growth factor antibody,

Table 10.3 Pathologic Stage and Survival for Breast Cancer

Stage	Definition	5-Year Survival	7-Year Survival
I	Tumor 2 cm or less in greatest diameter and without evidence of regional (nodal) or distant spread	96%	92%
II	Tumor more than 2 cm but not more than 5 cm in greatest dimension, with regional lymph node involvement but without distant metastases, *or* a tumor of more than 5 cm in diameter without regional (nodal) and distant spread	81%	71%
III	Tumors of any size with possible skin involvement, pectoral and chest wall fixation, and axillary or internal mammary nodal involvement, fixed, but without distant metastases	52%	39%
IV	Tumor of any size with or without regional spread but with evidence of distant metastases	18%	11%

bevacizumab, or in antiangiogenesis small molecule therapeutics in early phase clinical trials (see Chapter 19).

Tumor Infiltrating Lymphocytes The association of tumor infiltrating lymphocytes with a favorable prognosis originally stemmed from the observation that medullary carcinoma in which tumor infiltrating lymphocytes are a standard feature has a better prognosis than standard infiltrating ductal carcinoma. However, despite some reports that found a significant association of tumor infiltrating lymphocyte counts with outcome in routine ductal cancer,[21] the difficulties in performing standardized and reproducible tumor infiltrating lymphocyte counts and the lack of association of tumor infiltrating lymphocyte status with prognosis in many studies have prevented the adoption of this biomarker in most centers.

Skin Involvement and Inflammatory Carcinoma It is widely accepted that direct invasion of the overlying skin and dermal intralymphatic permeation by tumor producing the inflammatory carcinoma phenotype are adverse findings in newly diagnosed breast cancers.[2–5] Although generally believed to have a fatal outcome, recent advances in the therapeutic strategy for inflammatory carcinoma has resulted in improvements in overall survival.[22]

Paget Disease Extension of the intraductal component of an infiltrating ductal carcinoma into the nipple epidermis, the hallmark of Paget disease of the nipple, occurs in approximately 1% of breast cancers.[2–5] The prognosis for patients with Paget disease, which is currently treated with both chemotherapy and radiotherapy, is generally similar to the prognosis for all infiltrating ductal carcinoma as a whole.[23,24] However, given the very high association of Paget disease with HER-2/*neu* amplified infiltrating ductal carcinoma (see Chapter 16), these patients may actually have a reduced overall survival compared with HER-2/*neu* unamplified cases with no involvement of the nipple.[25,26]

Summary of Morphologic Prognostic Factors

In summary, a number of morphology-based prognostic factors (FIG. 10.1), including tumor type, grade, size, node status, and pathologic stage, are well-established standard components of the breast cancer surgical pathology report.[14] The ancillary prognostic and predictive tests described below are all judged by their ability to add meaningful additional information to this list of morphology-driven tests.

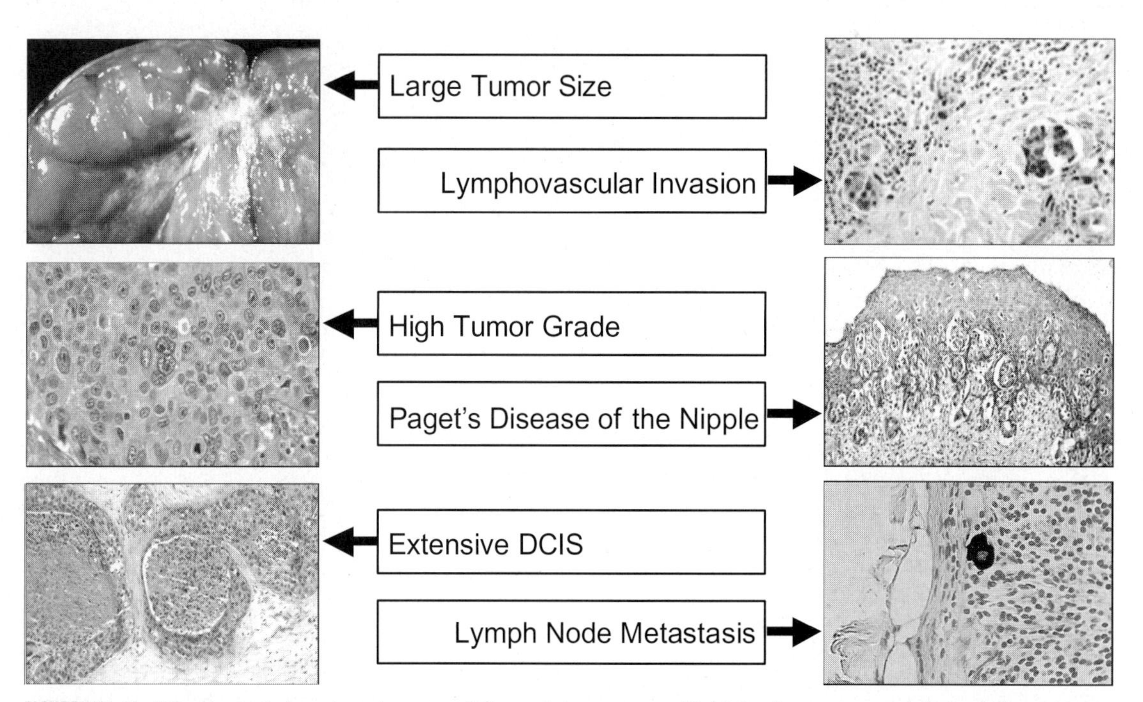

FIGURE 10.1 Traditional morphology-based prognostic factors in breast cancer. Multiple adverse prognostic factors in breast cancer are demonstrated, including large primary tumor size, high grade microscopic appearance, an extensive ductal carcinoma in situ component, intralymphatic invasion, extension to the nipple as Paget disease, and involvement of the axillary lymph nodes (shown as an immunohistochemically identified micrometastasis in a sentinel node biopsy specimen). **See Plate 23 for color image.**

Ancillary Prognostic and Predictive Factors

Commonly Used Tests

FIGURE 10.2 demonstrates the commonly used ancillary tests for breast cancer prognosis.

Hormone Receptor Status The role of ER and progesterone receptor(PR) testing as markers of prognosis and predictors of response to antiestrogen therapy is established as a standard of care for patients with breast cancer (see Chapter 15 for details).[27,28] Positive ER and PR assays are associated with well-differentiated histology, negative lymph node status, diploid DNA content, low cell proliferation rate, and tendency for a relatively indolent clinical course.[27–29] ER/PR-negative tumors are often associated with aggressive disease, including amplification of the HER-2, c-*myc,* and *int-2* oncogenes; mutation of the *p53* gene; and up-regulation of invasion- and metastasis-associated growth factors, growth factor receptors, and proteases.[27,28] The determination of ER/PR status in newly diagnosed breast cancer is required for selection of patients to receive hormonal therapy, and the ER/PR has also been widely used to predict risk for progressive disease.[29] Originally determined on fresh tumor protein extracts and cytosols using a quantitative biochemical competitive binding assay with dextran coated charcoal, the small size of newly diagnosed primary tumors has required a shift to on-slide IHC methods.[30] Despite its limitations, including the lack of standardization, IHC is currently the standard method to determine ER and PR status in breast cancer; in addition, it remains a cornerstone of planning therapy for the disease and appears likely to be used clinically in this fashion for the foreseeable future.

HER-2 Status Amplification and overexpression of the HER-2/*neu* (c-*erb*B2) gene and protein have been identified in 10–34% of invasive breast cancers (see Chapter 16 for details).[31] Most of these studies have linked either *HER-2* gene amplification or HER-2 protein overexpression with adverse prognosis in either node-negative or node-positive disease.[31] In general, when specimens have been carefully fixed, processed, and embedded, there has been excellent correlation between gene copy status determined by fluorescence in situ hybridization (FISH) and protein expression levels determined by IHC.[31] The main use of either method in current clinical practice is focused on the prediction of response to the anti-HER-2 targeted therapy with trastuzumab.[31] For this reason, the American Society of Clinical Oncology and the College of American Pathologists both consider HER-2 testing to be part of the standard workup for a newly diagnosed breast cancer specimen.[32,33]

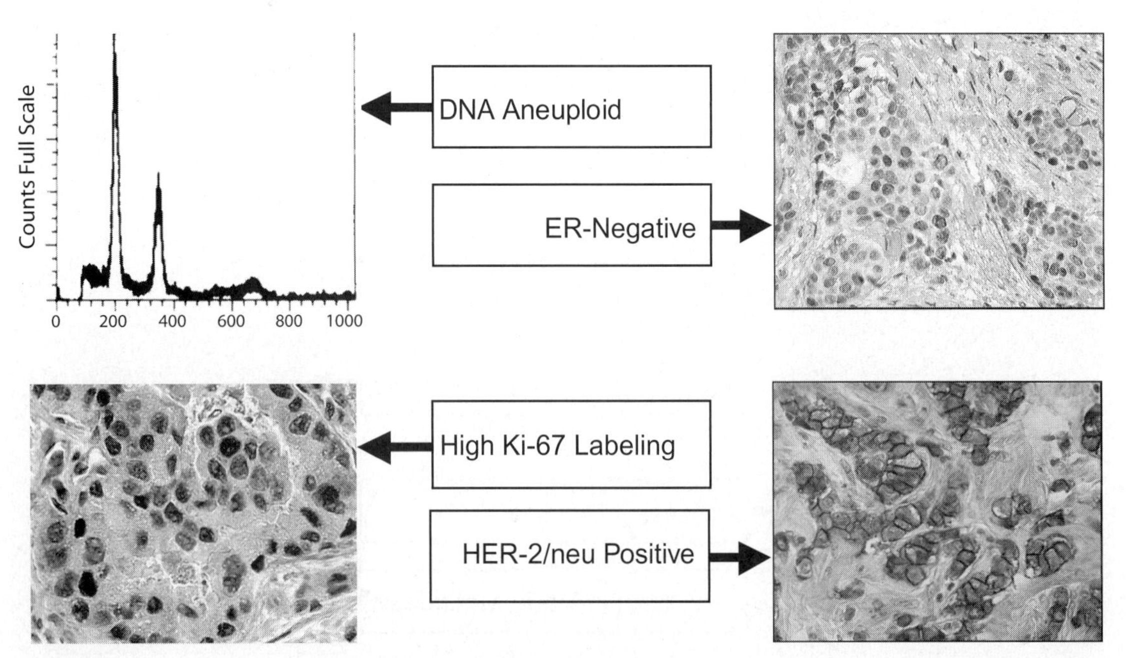

FIGURE 10.2 Ancillary tests for breast cancer prognosis. Multiple assays in common use are demonstrated that are associated with an adverse prognosis, including aneuploid DNA content, a high Ki-67 labeling index, a negative IHC assay for ER protein, and 3+ overexpression of the HER-2/neu protein. **See Plate 24 for color image.**

DNA Ploidy and S Phase Studies on the prognostic significance of DNA content analysis (DNA ploidy) and S phase status have varied greatly, with some investigators finding significant prediction of disease-free and overall survival rates on both univariate and multivariate analysis and others finding no impact on disease outcome (see Chapter 12 for details).[34] The S phase calculation by flow cytometry has generally outperformed ploidy status as a prognostic factor in breast cancer and is advocated by some investigators as a useful clinical parameter. Despite their continuing clinical use in many institutions, neither the American Society of Clinical Oncologists[32] nor College of American Pathologists[33] include ploidy and S phase measurements in their lists of recommended prognostic factors. The lack of a standardized approach to performing this test and interpreting its result is the major reason S phase fraction is not accepted as a standard prognostic marker.

Ki-67 Labeling Cell proliferation labeling measured by Ki-67 immunostaining correlates with the S phase levels calculated by flow cytometry but is generally higher, reflecting the fact that the Ki-67 antigen is also expressed in late G1 as well early G2/M phases of the cell cycle (see Chapter 13 for details).[35] Ki-67 staining has achieved a more consistent significant correlation with breast cancer outcome both on univariate and multivariate analysis than DNA ploidy alone. However, despite being easily performed on formalin-fixed paraffin-embedded tissue sections, this test suffers from the lack of standardization, including the general lack of use of cell line controls of known proliferative index.

Less Common and Research-Based Tests

Cell Cycle Markers Amplification or overexpression of cyclin D1 (*PRAD1*; bcl-1) localized to chromosome 11q13 has also been identified in 20% of clinical breast cancers[36] and has been linked to the expression of the ER[37] and the transition from in situ to invasive ductal breast cancer (see Chapter 13 for details).[38] In a recent study, high levels of the low-molecular-weight isoforms of cyclin E, measured by Western blotting, correlated strongly with decreased disease-specific survival[39]; moreover, levels of total cyclin E also were highly correlative with poor outcome consistent with prior studies performed by IHC.[40] The p21 protein (p21/WAF1/Cip1) is an inhibitor of cyclin-dependent kinases and serves as a critical downstream effector in the p53-specific pathway of cell growth control.[41] Some studies have linked altered expression of p21 with adverse outcome in breast cancer,[42,43] whereas others have not.[44] p27 (kip1) is a cell cycle regulator that acts by binding and inactivating cyclin-dependent kinases.[44] Low p27 expression has been correlated with poor prognosis in many (but not all) studies of patients, especially those with small primary tumors.[45–48]

Oncogenes The measurement of oncogene activity (discussed in detail in Chapter 22) has not played a major role in the clinical assessment of breast cancer specimens to date. The c-*myc* gene is amplified in approximately 16% of breast cancer cases[49] and in most outcome-based studies is associated with decreased disease-free patient survival.[49,50] The H-*ras* gene has been consistently associated with breast cancer progression.[51] Measurements of the c-*fos* (chromosome14q21) and c-*jun* (chromosome 22q13) regulators of the activating protein-1 complex and c-*myb* (chromosome 6q21) have successfully predicted breast cancer recurrence, response to hormonal therapy, and survival.[52]

***p53* and Tumor Suppressor Genes** The prognostic significance of *p53* status in breast cancer has been impacted by the accuracy of IHC versus molecular methods such as single strand conformation polymorphism testing (SSCP), direct sequencing methods, and the yeast colony functional assay (see Chapter 23 for details).[53,54] The *p53* mutation rate is lower in breast cancer than in other carcinomas and has been associated with progressive disease and reduced overall survival.[55] In general, breast carcinomas with *p53* mutations are associated with high histologic grade, high mitotic index, high cell proliferation rate, aneuploid DNA content, negative assays for ER and PR (see Chapter 23 for details),[56–58] and variable association with amplification of oncogenes such as *HER-2, c-myc, ras,* and *int-2.*[59] Some but not all studies have implicated *p53* mutation with resistance to hormonal, adjuvant and neoadjuvant chemotherapy, and combination chemotherapy for metastatic disease encompassing a variety of agents, including anthracyclines and taxanes.[60–67] Currently, determination of *p53* status is not included as a part of the standard of practice for the management of breast cancer. Other tumor suppressor genes such as *Rb* have not been widely applied to breast cancer and are not currently used to stratify patients as to disease relapse risk.

Cell Adhesion Molecules Cell adhesion molecule expression has been extensively studied in breast cancer as a biomarker of tumor development, differentiation, progression, and metastasis (see Chapter 20 for details).[68,69] The E-cadherin–catenin complex has been related to disease outcome in a variety of malignant diseases, including breast cancer.[70] Most published studies have linked loss of expression of E-cadherin with adverse outcome in breast cancer,[71–73] although there have been reports of retained expression indicating disease progression.[74] The most consistent observation concerning the loss of E-cadherin expression in breast cancer has been the association with the infiltrating lobular pattern versus infiltrating ductal pattern of invasive carcinoma (see Chapter 8).[75–77] E-cadherin status has not been widely used to predict the response of breast cancer to therapy. CD44 expression has been associated with the development and progression of breast cancer.[78] Abnormal expression of the standard form of CD44 has been linked to prognosis.[79] Overexpression of the CD44 splice variant v6 has been linked to adverse outcome in some studies[80–82] but not in others.[83] The integrin group and laminin receptor group have been widely studied in breast cancer.[84] Laminin receptor expression has been independently associated with disease outcome in some studies[85,86] but not in others.[87] Altered expression of integrins αv[88] and α6[89,90] have been linked to breast cancer prognosis.

Bcl-2 and Apoptosis (see Chapter 21) In breast cancer, most studies have linked an increased rate of cellular apoptosis with an adverse outcome for the disease.[91–94] Expression of the antiapoptosis-associated gene, Bcl-2, correlates with ER/PR-positive status and has been associated with improved patient survival.[95–97] However, primary tumor bcl-2 expression levels have not been predictive for response to systemic chemotherapy given after relapse.[98] Expression of the proapoptosis gene Bax expression has not been clearly linked to outcome.[99] In addition, activated caspases can act as both initiators and effectors of the apoptotic pathway, and there is evidence that caspases 3, 6, and 8 are associated with higher levels of apoptosis, histologic grade, and tumor aggressiveness in breast cancer.[100] Caspase expression in breast cancer has been linked to overall survival[101] and chemoresistance.[102]

Invasion Associated Proteases (see Chapter 18) Numerous studies in the early 1990s using an immunoassay approach on fresh breast tumor cytosolic preparations showed that elevated cathepsin D levels are an independent predictor of survival in breast cancer.[103–105] Attempts to convert the assay to an IHC-based format have not been successful.[106–108] The serine proteases studied in breast cancer invasion have focused on urokinase plasminogen activator (uPA), uPA receptor, and plasminogen activator inhibitor type 1 (PAI-1). When evaluated on fresh tissue extracts and tumor cytosols, high uPA and PAI-1 levels were consistently associated with disease recurrence and overall patient survival in breast cancer.[109–112] Plasminogen protease levels have also been successfully used as predictors of chemotherapy response.[112] Translation of the uPA/PAI-1 immunoassay to an on-slide IHC format has not, to date, been successful, which has limited widespread use. The matrix metalloproteases (MMPs) are a group of at least 19 zinc metalloenzymes secreted as proenzymes with substantial sequence similarities that are inhibited by metallochelators and specific tissue inhibitors known as tissue inhibitors of metalloproteases (TIMPs).[113] The MMPs include the interstitial collagenases, gelatinases, stromelysins, and membrane-type MMPs and are involved in breast cancer initiation, invasion, and metastasis.[113] High levels of at least three MMPs (MMP-2, MMP-9, and MMP-11) have been found to correlate with poor disease outcome in breast cancer.[114–116]

Emerging Prognostic and Predictive Factors

Oncotype Dx The Oncotype Dx is a reverse transcriptase-polymerase chain reaction (RT-PCR) multiplex assay using a 21 gene probe set and messenger RNA extracted from paraffin blocks of stored breast cancer tissues.[117] Using a cohort of 688 lymph node–negative ER-positive tumors obtained from patients enrolled in two NSABP clinical trials treated with tamoxifen alone, the 21 gene assay predicted disease recurrence to a high level of significance ($P < 0.00001$).[117] This assay has recently become available for new patients.

Transcriptional Profiling and DNA Microarrays (see Chapters 6 and 28) Whole genome transcriptional profiling has been used as a technique for determining prognosis in breast cancer.[6–8,118–120] Gene expression profiles can define cellular functions, biochemical pathways, proliferative activity, and regulatory mechanisms. In a recent DNA microarray analysis on primary breast tumors of 117 node-negative young patients using a supervised classification to identify a poor prognosis gene expression signature, aberrant expression of genes regulating cell cycle, invasion, metastasis, and angiogenesis strongly predicted a short interval to distant metastases.[120] In a follow-up study, the poor prognosis gene expression profile outperformed all currently used clinical parameters in predicting disease outcome, including lymph node status, with an estimated hazard ratio for distant metastases of 5.1 (95% confidence interval, 2.9 to 9.0; $P < 0.001$).[120] DNA microarrays addressing cancer outcomes show variable prognostic performance. Larger studies with appropriate clinical design, adjustment for known predictors, and proper validation are essential for this highly promising technology.[121]

Proteomics Matrix-assisted laser desorption ionization and surface-enhanced laser desorption and ionization mass spectrometry and other proteomics strategies have shown preliminary success for the early detection of ovarian cancer[122] and have recently been applied to breast cancer for the discovery of new and better biomarkers both in serum and nipple aspirate specimens.[123,124] Although further testing of this approach must be performed on larger groups of patients, the surface-enhanced laser desorption and ionization time-of-flight technique shows great promise as a potential method of developing new disease markers capable of detecting cancers at early stages.

Practical Issues Related to Ancillary Tests on Breast Cancer Specimens

A variety of issues impacts the application of ancillary tests and procedures on newly diagnosed breast cancer specimens. These issues impact the ability to standardize these tests, make them reproducible, and enable a thorough evaluation of their true clinical utility. The type of sample required for the analysis is critical for its ultimate validity and clinical utility. The "on-slide" techniques, including IHC and FISH, can be impacted by a variety of important preanalytic issues, including the actual sample procurement and the specimen storage conditions. These issues are reviewed in Chapter 2 and covered in detail in Chapter 16. When IHC is the testing technique, the use of unstained slides versus paraffin blocks can be a significant issue. It is well known that antigens are better preserved in the paraffin block than when they are allowed to oxidize and denature when stored as a precut unstained slide.[125] Another major issue for IHC is the type and intensity of the antigen retrieval method used before the actual staining procedure. This procedure can greatly impact the final results by causing false-positive staining (see Chapter 16 on HER-2/neu protein overexpression detection). Ultimately, the standardization of test procedures must be achieved if the widely used and recently introduced companion diagnostics are going to achieve their goals of improving the response to therapy and overall survival for patients. This will require the use of appropriate controls, strict adherence to test protocols, and the establishment of reproducible intraobserver and interobserver correlation. For IHC-based procedures, improvement in the slide scoring protocol (see Chapters 2 and 16), and for microarray methods, the development of standardized data interpretation (see Chapters 2, 6, and 28) are among the most critical issues. Ultimately, the wide acceptance of new diagnostic procedures for breast cancer will be driven by their overall cost effectiveness. Emerging molecular diagnostics will likely achieve their best adoption when they are predictive of single agent or multiple agent therapy response. Stand-alone prognostic tests may find more resistance for their acceptance in the marketplace unless they can achieve outstanding negative predictive value and safely allow the withholding of cytotoxic therapy for newly diagnosed patients in the adjuvant setting.

Summary and Conclusion

The search for more accurate prognostic tests and tests designed to predict response to therapy for breast cancer will continue to challenge scientists and clinicians. A variety of emerging molecular diagnostic tests is continually being introduced for breast cancer management (**Table 10.4**).[126–129] To date, the most useful prognostic/predictive factors are

Table 10.4 Review of Major Ancillary/Molecular Prognostic Factors in Breast Cancer

Biomarker	Assay	Target of Therapy	Therapeutic	Current Status	Future Prospects
ER/PR	IHC binding assay	Yes	Tamoxifen SERMs Aromatase inhibitors	Standard of care	Improved IHC with antibodies that are negative when ERα is truncated to reduce false positives
HER-2	IHC FISH	Yes	Trastuzumab Other antibodies Gene therapy	Standard of care	CISH assay may replace both IHC and FISH
DNA ploidy	Cytometry	No	—	Common use	Decreased use
S phase	Cytometry	No	—	Common use	Maintained use
Thymidine labeling index	Radioactive ^{3}H thymidine incorporation during DNA synthesis	No	—	Rarely used	Decreased use due to methodologic barrier Has yielded to the Ki-67 labeling index (below)
Ki-67 labeling index	IHC	No	—	Widely used	Continued expansion as replacement of the S phase measurement by flow cytometry
Cyclin D	IHC	Possible	Flavopyridol translocation targets	Clinical trials	May select new drug use such as proteasome inhibitors
Cyclin E	IHC Western	No	—	RUO	Prognostic significance must be validated
EGFR	IHC FISH	Yes	Gefitinib Erlotinib Cetuximab	Increasing use Clinical trials	Targeting the anti-EGFR drugs likely combined with pharmacogenomics
Vascular endothelial growth factor	IHC	Yes	Bevacizumab Small molecules	Increasing use Clinical trials	Increasing use for prognosis; initial targeted therapy disappointing
p53	IHC SSCP Sequencing	Yes	Gene therapy	Increasing use Clinical trials	Targeted therapies disappointing to date
E-cadherin	IHC Methylation-PCR	Yes	5-Azacytidine Demethylation	Increasing use Clinical trials	Diagnosis of pleomorphic lobular carcinoma
CD-44 v6	IHC	No	—	RUO	Predictive significance of v6 splice variant requires validation
Cathepsin D	Immunoassay	No	—	Common use in Europe	IHC studies disappointing; will continue to fade from view
uPA/PAI-1	Immunoassay	Yes	Small molecules (e.g., WX-UK1)	Common use in Europe	Targeted therapies in early stages; IHC assays not validated to date restricting use in the USA
MMP2, 9, 11	IHC	Yes	Marimastat	Clinical trials RUO	Early results of targeted therapy disappointing

Table 10.4 Review of Major Ancillary/Molecular Prognostic Factors in Breast Cancer (*continued*)

Biomarker	Assay	Target of Therapy	Therapeutic	Current Status	Future Prospects
MDR	IHC	Yes	Small molecules	Clinical trials RUO	Continued use
Bcl-2	IHC	Yes	G-3135 proteasome inhibitors	Increasing use Clinical trials	Initial results of targeted therapies disappointing
Telomerase	TRAP IHC ISH	Yes	Small molecules	RUO	Increased use if slide-based assays are successful prognostic factors
NFκB	IHC Western	Yes	Proteasome inhibitors	RUO	Will be used if targeted therapies are successful alone or in combination with cytotoxic drugs
Oncotype Dx	RT-PCR (Paraffin)	No	—	RUO	Recent study of 668 node-negative ER-positive cases treated with tamoxifen only showed 21 gene RT-PCR expression assay could predict risk of disease recurrence at $P < 0.00001$
Transcriptional profiling	cDNA array Oligonucleotide array	No	—	RUO	Continued major expansion of use Predictive marker sets will require multiple cross-validation; could become standard if initial results are confirmed

CISH, chromogenic in situ hybridization; EGFR, epidermal growth factor receptor; MDR, multiple drug resistance factor; RUO, research use only; SERM, selective estrogen receptor modulator; TRAP, telomere repeat amplification protocol.

hormone receptor status and HER-2/neu status and will likely remain the cornerstones of ancillary testing of invasive breast cancer specimens for the foreseeable future. Gene expression profiling performed by either RT-PCR or DNA microarray methods will likely compete with proteomic strategies in this continued effort to develop both prognostic and predictive tests designed to individualize treatment and further fulfill the promise of a truly personalized medicine for breast cancer patients.

References

1. Lacey JV Jr, Devesa SS, Brinton LA. Recent trends in breast cancer incidence and mortality. Environ Mol Mutagen 2002;39:82–88.
2. Rosen PP. Breast Pathology (2nd ed). Philadelphia: Lippincott Williams & Wilkins, 2001:275–293.
3. Elston CW, Ellis IO. Pathological prognostic factors in breast cancer. I. The value of histological grade in breast cancer: experience from a large study with long-term follow-up. Histopathology 1991;19:403–410.
4. Henson DE, Ries LA, Carriaga MT. Conditional survival of 56,268 patients with breast cancer. Cancer 1995;76:237–242.
5. Berg JW, Hutter RVP. Breast cancer. Cancer 1995; 75:257–269.
6. Sorlie T, Perou CM, Tibshirani R, et al. Gene expression patterns of breast carcinomas distinguish tumor subclasses with clinical implications. Proc Natl Acad Sci USA 2001;98:10869–10874.
7. Sotiriou C, Neo SY, McShane LM, et al. Breast cancer classification and prognosis based on gene expression profiles from a population-based study. Proc Natl Acad Sci USA 2003;100:10393–10398.
8. Sorlie T, Tibshirani R, Parker J, et al. Repeated observation of breast tumor subtypes in independent gene expression data sets. Proc Natl Acad Sci USA 2003;100:8418–8423.
9. Bloom HJG, Richardson WW. Histologic grading and prognosis in breast cancer. Br J Cancer 1957;11: 359–377.
10. Robbins P, Pinder S, deKlerk N, et al. Histological grading of breast carcinomas: a study of interobserver variation. Hum Pathol 1995;26:873–879.
11. Merkel DE, Osborne CK. Prognostic factors in breast cancer. Hematol Oncol Clin North Am 1989;3:641–652.
12. Donegan WL. Prognostic factors. Stage and receptor status in breast cancer. Cancer 1992;70(6 suppl): 1755–1764.

13. Clare SE, Sener SF, Wilkens W, et al. Prognostic significance of occult lymph node metastases in node-negative breast cancer. Ann Surg Oncol 1997;4: 447–451.
14. Fitzgibbons PL, Page DL, Weaver D, et al. Prognostic factors in breast cancer. College of American Pathologists Consensus Statement 1999. Arch Pathol Lab Med 2000;124:966–978.
15. Wilke LG, Giuliano A. Sentinel lymph node biopsy in patients with early-stage breast cancer: status of the National Clinical Trials. Surg Clin North Am 2003;83:901–910.
16. Nieweg OE, Estourgie SH, Deurloo EE, et al. Status of lymph node staging. Scand J Surg 2002;91:263–267.
17. Weidner N, Semple JP, Welch WR, et al. Tumor angiogenesis and metastasis—correlation in invasive breast carcinoma. N Engl J Med 1991;324:1–8.
18. Toi M, Bando H, Kuroi K. The predictive value of angiogenesis for adjuvant therapy in breast cancer. Breast Cancer 2000;7:311–314.
19. Gasparini G. Clinical significance of determination of surrogate markers of angiogenesis in breast cancer. Crit Rev Oncol Hematol 2001;37:97–114.
20. Sauer G, Deissler H. Angiogenesis: prognostic and therapeutic implications in gynecologic and breast malignancies. Curr Opin Obstet Gynecol 2003;15: 45–49.
21. Marrogi AJ, Munshi A, Merogi AJ, et al. Study of tumor infiltrating lymphocytes prognostic factors in breast carcinoma. Int J Cancer 1997;74:492–501.
22. Giordano SH, Hortobagyi GN. Inflammatory breast cancer: clinical progress and the main problems that must be addressed. Breast Cancer Res 2003;5: 284–288.
23. Ascenso AC, Marques MS, Capitao-Mor M. Paget's disease of the nipple. Clinical and pathological review of 109 female patients. Dermatologica 1985; 170:170–179.
24. Marshall JK, Griffith KA, Haffty BG, et al. Conservative management of Paget disease of the breast with radiotherapy: 10- and 15-year results. Cancer 2003;97:2142–2149.
25. Wolber RA, DuPuis BA, Wick MR. Expression of c-*erb* B2 oncoprotein in mammary and extramammary Paget's disease. Am J Clin Pathol 1991;96:243–247.
26. Fu W, Lobocki CA, Silberberg BK. Molecular markers in Paget disease of the breast. J Surg Oncol 2001; 77:171–178.
27. Osborne CK. Steroid hormone receptors in breast cancer management. Breast Cancer Res Treat 1998; 51:227–238.
28. Locker GY. Hormonal therapy of breast cancer. Cancer Treat Rev 1998;24:221–240.
29. Bertucci F, Houlgatte R, Benziane A, et al. Gene expression profiling of primary breast carcinomas using arrays of candidate genes. Hum Mol Genet 2000;9:2981–2991.
30. Masood S. Prediction of recurrence for advanced breast cancer. Traditional and contemporary pathologic and molecular markers. Surg Oncol Clin North Am 1995;4:601–632.
31. Ross JS, Fletcher JA, Linette GP, et al. The Her-2/neu gene and protein in breast cancer 2003: biomarker and target of therapy. Oncologist 2003;8: 307–325.
32. Bast RC Jr, Ravdin P, Hayes DF, et al. 2000 update of recommendations for the use of tumor markers in breast and colorectal cancer: clinical practice guidelines of the American Society of Clinical Oncology. J Clin Oncol 2001;19:1865–1878.
33. Hammond ME, Fitzgibbons PL, Compton CC, et al. College of American Pathologists Conference XXXV: solid tumor prognostic factors—which, how and so what? Summary document and recommendations for implementation. Cancer Committee and Conference Participants. Arch Pathol Lab Med 2000;124: 958–965.
34. Ross, JS. DNA Ploidy and Cell Cycle Analysis in Pathology. New York: Igaku-Shoin, 1996:54–55.
35. MacGrogan G, Jollet I, Huet S, et al. Comparison of quantitative and semiquantitative methods of assessing MIB-1 with the S-phase fraction in breast carcinoma. Mod Pathol 1997;10:769–776.
36. Wolman SR, Pauley RJ, Mohamed AN, et al. Genetic markers as prognostic indicators in breast cancer. Cancer 1992;70:1765–1774.
37. Steeg PS, Zhou Q. Cyclins and breast cancer. Breast Cancer Res Treat 1998;52:17–28.
38. Weinstat-Saslow D, Merino MJ, Manrow RE, et al. Overexpression of cyclin D mRNA distinguishes invasive and in situ breast carcinomas from nonmalignant lesions. Nat Med 1995;1:1257–1260.
39. Keyomarsi K, Tucker SL, Buchholz TA, et al. Cyclin E and survival in patients with breast cancer. N Engl J Med 2002;347:1566–1575.
40. Keyomarsi K, O'Leary N, Molnar G, et al. Cyclin E, a potential prognostic marker for breast cancer. Cancer Res 1994;54:380–385.
41. Caffo O, Doglioni C, Veronese S, et al. Prognostic value of p21(WAF1) and p53 expression in breast carcinoma: an immunohistochemical study in 261 patients with long-term follow-up. Clin Cancer Res 1996;2:1591–1599.
42. Oh YL, Choi JS, Song SY, et al. Expression of p21Waf1, p27Kip1 and cyclin D1 proteins in breast ductal carcinoma in situ: Relation with clinicopathologic characteristics and with p53 expression and estrogen receptor status. Pathol Int 2001;51:94–99.
43. Gohring UJ, Bersch A, Becker M, et al. p21(waf) correlates with DNA replication but not with prognosis in invasive breast cancer. J Clin Pathol 2001;54: 866–870.
44. Lau R, Grimson R, Sansome C, et al. Low levels of cell cycle inhibitor p27kip1 combined with high levels of Ki-67 predict shortened disease-free survival in T1 and T2 invasive breast carcinomas. Int J Oncol 2001;18:17–23.
45. Barbareschi M. p27 expression, a cyclin dependent kinase inhibitor in breast carcinoma. Adv Clin Pathol 1999;3:119–127.
46. Barbareschi M, van Tinteren H, Mauri FA, et al. p27(kip1) expression in breast carcinomas: an immunohistochemical study on 512 patients with long-term follow-up. Int J Cancer 2000;89:236–241.
47. Leivonen M, Nordling S, Lundin J, et al. p27 expression correlates with short-term, but not with long-term prognosis in breast cancer. Breast Cancer Res Treat 2001;6:15–22.
48. Nohara T, Ryo T, Iwamoto S, et al. Expression of cell-cycle regulator p27 is correlated to the prognosis and ER expression in breast carcinoma patients. Oncology 2001;60:94–100.
49. Deming SL, Nass SJ, Dickson RB, et al. C-myc amplification in breast cancer: a meta-analysis of its

occurrence and prognostic relevance. Br J Cancer 2000;83:1688–1695.

50. Mizukami Y, Nonomura A, Takizawa T, et al. N-myc protein expression in human breast carcinoma: prognostic implications. Anticancer Res 1995;15: 2899–2905.

51. Rochlitz CF, Scott GK, Dodson JM, et al. Incidence of activating ras oncogene mutations associated with primary and metastatic human breast cancer. Cancer Res 1989;49:357–360.

52. Bland KI, Konstadoulakis MM, Vezeridis MP, et al. Oncogene protein coexpression. Value of HA*ras*, c-*myc*, c-*fos*, and p53 as prognostic discriminants for breast carcinoma. Ann Surg 1995;221:706–720.

53. Gee JM, Barroso AF, Ellis IO, et al. Biological and clinical associations of c-jun activation in human breast cancer. Int J Cancer 2000;89:177–186.

54. Liu MC, Gelmann EP. P53 gene mutations: case study of a clinical marker for solid tumors. Semin Oncol 2002;29:246–257.

55. Gasco M, Shami S, Crook T. The p53 pathway in breast cancer. Breast Cancer Res 2002;4:70–76.

56. Borresen-Dale AL. TP53 and breast cancer. Hum Mutat 2003;21:292–300.

57. Bhargava V, Thor A, Deng G, et al. The association of p53 immunopositivity with tumor proliferation and other prognostic indicators in breast cancer. Mod Pathol 1994;7:361–368.

58. Rosanelli GP, Steindorfer P, Wirnsberger GH, et al. Mutant p53 expression and DNA analysis in human breast cancer. Comparison with conventional clinicopathological parameters. Anticancer Res 1995;15: 581–586.

59. Pelosi G, Bresaola E, Rodella S, et al. Expression of proliferating cell nuclear antigen, Ki-67 antigen, estrogen receptor protein, and tumor suppressor p53 gene in cytologic samples of breast cancer: An immunochemical study with clinical, pathobiological, and histologic correlations. Diagn Cytopathol 1994; 11:131–140.

60. Beck T, Weller EE, Weikel W, et al. Usefulness of immunohistochemical staining for p53 in the prognosis of breast carcinomas: correlation with established prognosis parameters and with the proliferation marker, MIB-1. Gynecol Oncol 1995;57:96–104.

61. Daidone MG, Veneroni S, Benini E, et al. Biological markers as indicators of response to primary and adjuvant chemotherapy in breast cancer. Int J Cancer 1999;84:580–586.

62. Kandioler-Eckersberger D, Ludwig C, Rudas M, et al. TP53 mutation and p53 overexpression for prediction of response to neoadjuvant treatment in breast cancer patients. Clin Cancer Res 2000;6:50–56.

63. Bertheau P, Plassa F, Espie M, et al. Effect of mutated TP53 on response of advanced breast cancers to high-dose chemotherapy. Lancet 2002;360: 852–854.

64. Sjostrom J, Blomqvist C, Heikkila P, et al. Predictive value of p53, mdm-2, p21, and mib-1 for chemotherapy response in advanced breast cancer. Clin Cancer Res 2000;6:3103–3110.

65. Van Poznak C, Tan L, Panageas KS, et al. Assessment of molecular markers of clinical sensitivity to single-agent taxane therapy for metastatic breast cancer. J Clin Oncol 2002;20:2319–2326.

66. Hamilton A, Larsimont D, Paridaens R, et al. A study of the value of p53, HER2, and Bcl-2 in the prediction of response to doxorubicin and paclitaxel as single agents in metastatic breast cancer: a companion study to EORTC 10923. Clin Breast Cancer 2000;1:233–240.

67. Knoop AS, Bentzen SM, Nielsen MM, et al. Value of epidermal growth factor receptor, HER2, p53, and steroid receptors in predicting the efficacy of tamoxifen in high-risk postmenopausal breast cancer patients. J Clin Oncol 2001;19:3376–3384.

68. Ohene-Abuakwa Y, Pignatelli M. Adhesion molecules in cancer biology. Adv Exp Med Biol 2000;465: 115–126.

69. Skubitz AP. Adhesion molecules. Cancer Treat Res 2002;107:305–329.

70. Beavon IR. The E-cadherin-catenin complex in tumour metastasis: structure, function and regulation. Eur J Cancer 2000;36:1607–1620.

71. Charpin C, Garcia S, Bonnier P, et al. Reduced E-cadherin immunohistochemical expression in node-negative breast carcinomas correlates with 10-year survival. Am J Clin Pathol 1998;109:431–438.

72. Parker C, Rampaul RS, Pinder SE, et al. E-cadherin as a prognostic indicator in primary breast cancer. Br J Cancer 2001;85:1958–1963.

73. Yoshida R, Kimura N, Harada Y, et al. The loss of E-cadherin, alpha- and beta-catenin expression is associated with metastasis and poor prognosis in invasive breast cancer. Int J Oncol 2001;18:513–520.

74. Gillett CE, Miles DW, Ryder K, et al. Retention of the expression of E-cadherin and catenins is associated with shorter survival in grade III ductal carcinoma of the breast. J Pathol 2001;193:433–441.

75. Reis-Filho JS, Cancela Paredes J, Milanezi F, et al. Clinicopathologic implications of E-cadherin reactivity in patients with lobular carcinoma in situ of the breast. Cancer 2002;94:2114–2115.

76. Chan JK, Wong CS. Loss of E-cadherin is the fundamental defect in diffuse-type gastric carcinoma and infiltrating lobular carcinoma of the breast. Adv Anat Pathol 2001;8:165–172.

77. Kleer CG, van Golen KL, Braun T, et al. Persistent E-cadherin expression in inflammatory breast cancer. Mod Pathol 2001;14:458–464.

78. Burguignon LY. CD44-mediated oncogenic signaling and cytoskeleton activation during mammary tumor progression. J Mamm Gland Biol Neoplasia 2001;6: 287–297.

79. Joensuu H, Klemi PJ, Toikkanen S, et al. Glycoprotein CD44 expression and its association with survival in breast cancer. Am J Pathol 1993;143: 866–874.

80. Guriec N, Gairard B, Marcellin L, et al. CD44 isoforms with exon v6 and metastasis of primary N0M0 breast carcinomas. Breast Cancer Res Treat 1997;44:261–268.

81. Schumacher U, Horny HP, Horst HA, et al. A CD44 variant exon 6 epitope as a prognostic indicator in breast cancer. Eur J Surg Oncol 1996;22: 259–261.

82. Morris SF, O'Hanlon DM, McLaughlin R, et al. The prognostic significance of CD44s and CD44v6 expression in stage two breast carcinoma: an immunohistochemical study. Eur J Surg Oncol 2001;27: 527–531.

83. Jansen RH, Joosten-Achjanie SR, Arends JW, et al. CD44v6 is not a prognostic factor in primary breast cancer. Ann Oncol 1998;9:109–111.

84. Ivaska J, Heino J. Adhesion receptors and cell invasion: mechanisms of integrin-guided degradation of extracellular matrix. Cell Mol Life Sci 2000;57:16–24.
85. Marques LA, Franco ELF, Tortoni H, et al. Independent prognostic value on laminin receptor expression in breast cancer survival. Cancer Res 1990;50:1479–1483.
86. D'Errico A, Garbisa S, Liotta LA, et al. Augmentation of type IV collagenase laminin receptor, and ki67 proliferation antigen associated with human colon, gastric, and breast carcinoma progression. Mod Pathol 1991;4:239–246.
87. Daidone MG, Silvestrini R, D'Errico A, et al. Laminin receptors, collagenase IV and prognosis in node-negative breast cancers. Int J Cancer 1991;48:529–532.
88. D'Errico A, Garbisa S, Liotta LA, et al. Augmentation of type IV collagenase laminin receptor, and ki67 proliferation antigen associated with human colon, gastric, and breast carcinoma progression. Mod Pathol 1991;4:239–246.
89. Gasparini G, Brooks PC, Biganzoli E, et al. Vascular integrin alpha(v)beta3: a new prognostic indicator in breast cancer. Clin Cancer Res 1998;4:2625–2634.
90. Tagliabue E, Ghirelli C, Squicciarini P, et al. Prognostic value of alpha 6 beta 4 integrin expression in breast carcinomas is affected by laminin production from tumor cells. Clin Cancer Res 1998;4:407–410.
91. Parton M, Dowsett M, Smith I. Studies of apoptosis in breast cancer. BMJ 2001;322:1528–1532.
92. Berardo MD, Elledge RM, de Moor C, et al. bcl-2 and apoptosis in lymph node positive breast carcinoma. Cancer 1998;82:1296–1302.
93. Zhang GJ, Kimijima I, Abe R, et al. Apoptotic index correlates to bcl-2 and p53 protein expression, histological grade and prognosis in invasive breast cancers. Anticancer Res 1998;18:1989–1998.
94. De Jong JS, van Diest PJ, Baak JP. Number of apoptotic cells as a prognostic marker in invasive breast cancer. Br J Cancer 2000;82:368–373.
95. Gonzalez-Campora R, Galera Ruiz MR, Vazquez Ramirez F, et al. Apoptosis in breast carcinoma. Pathol Res Pract 2000;196:167–174.
96. Krajewski S, Krajewska M, Turner BC, et al. Prognostic significance of apoptosis regulators in breast cancer. Endocr Relat Cancer 1999;6:29–40.
97. Silvestrini R, Veneroni S, Daidone MG, et al. The Bcl-2 protein: a prognostic indicator strongly related to p53 protein in lymph node-negative breast cancer patients. J Natl Cancer Inst 1994;86:499–504.
98. Yang Q, Sakurai T, Yoshimura G, et al. Prognostic value of Bcl-2 in invasive breast cancer receiving chemotherapy and endocrine therapy. Oncol Rep 2003;10:121–125.
99. Sjostrom J, Blomqvist C, von Boguslawski K, et al. The predictive value of bcl-2, bax, bcl-xL, bag-1, fas, and fasL for chemotherapy response in advanced breast cancer. Clin Cancer Res 2002;8:811–816.
100. Vakkala M, Paakko P, Soini Y. Expression of caspases 3, 6 and 8 is increased in parallel with apoptosis and histological aggressiveness of the breast lesion. Br J Cancer 1999;81:592–599.
101. Nakopoulou L, Alexandrou P, Stefanaki K, et al. Immunohistochemical expression of caspase-3 as an adverse indicator of the clinical outcome in human breast cancer. Pathobiology 2001;69:266–273.
102. Devarajan E, Sahin AA, Chen JS, et al. Down-regulation of caspase 3 in breast cancer: a possible mechanism for chemoresistance. Oncogene 2002;21:8843–8851.
103. Rochefort H, Chalbos D, Cunat S, et al. Estrogen regulated proteases and antiproteases in ovarian and breast cancer cells. J Steroid Biochem Mol Biol 2001;76:119–124.
104. Thorpe SM, Rocheford H, Garcia M, et al. Association between high concentration of M52,000 cathepsin D and poor prognosis in primary breast cancer. Cancer Res 1989;49:6008–6014.
105. Tandon AK, Clark GM, Chamness GC, et al. Cathepsin D and prognosis in breast cancer. N Engl J Med 1990;322:297–302.
106. Kute TE, Shao ZM, Sugg NK, et al. Cathepsin D as a prognostic indicator for node-negative breast cancer patients using both immunoassays and enzymatic assays. Cancer Res 1992;52:5198–5203.
107. Visscher DW, Sarkar F, LoRusso P, et al. Immunohistologic evaluation on invasion-associated proteases in breast carcinoma. Mod Pathol 1993;6:302–306.
108. Mokbel K, Elkak A. Recent advances in breast cancer (the 37th ASCO meeting, May 2001). Curr Med Res Opin 2001;17:116–122.
109. Harbeck N, Schmitt M, Kates RE, et al. Clinical utility of urokinase-type plasminogen activator and plasminogen activator inhibitor-1 determination in primary breast cancer tissue for individualized therapy concepts. Clin Breast Cancer 2002;3:196–200.
110. Harbeck N, Kates RE, Look MP, et al. Enhanced benefit from adjuvant chemotherapy in breast cancer patients classified high-risk according to urokinase-type plasminogen activator (uPA) and plasminogen activator inhibitor type 1 (n = 3424). Cancer Res 2002;62:4617–4622.
111. Duffy MJ. Urokinase plasminogen activator and its inhibitor, PAI-1, as prognostic markers in breast cancer: from pilot to level 1 evidence studies. Clin Chem 2002;48:1194–1197.
112. Harbeck N, Kates RE, Schmitt M. Clinical relevance of invasion factors urokinase-type plasminogen activator and plasminogen activator inhibitor type 1 for individualized therapy decisions in primary breast cancer is greatest when used in combination. J Clin Oncol 2002;20:1000–1007.
113. Egeblad M, Werb Z. New functions for the matrix metalloproteinases in cancer progression. Nat Rev Cancer 2002;2:161–174.
114. Brinckerhoff CE, Matrisian LM. Matrix metalloproteinases: a tail of a frog that became a prince. Nat Rev Mol Cell Biol 2002;3:207–214.
115. McCawley LJ, Matrisian LM. Matrix metalloproteinases: multifunctional contributors to tumor progression. Mol Med Today 2000;6:149–156.
116. Benaud C, Dickson RB, Thompson EW. Roles of the matrix metalloproteinases in mammary gland development and cancer. Breast Cancer Res Treat 1998;50:97–116.
117. Paik S, Shak S, Tang G, et al. Multi-gene RT-PCR assay for predicting recurrence in node negative breast cancer patients—NSABP studies B-20 and B-14. Presented at the 26th Annual San Antonio Breast Cancer Symposium, San Antonio, Texas, December 3–6, 2003, abst. 16.
118. Bertucci F, Houlgatte R, Benziane A, et al. Gene expression profiling of primary breast carcinomas

using arrays of candidate genes. Hum Mol Genet 2000;9:2981–2991.
119. van't Veer LJ, Dai H, van de Vijver MJ, et al. Gene expression profiling predicts clinical outcome of breast cancer. Nature 2002;415:530–536.
120. van de Vijver MJ, He YD, van't Veer LJ, et al. A gene-expression signature as a predictor of survival in breast cancer. N Engl J Med 2002;347:1999–2009.
121. Ntzani EE, Ioannidis JP. Predictive ability of DNA microarrays for cancer outcomes and correlates: an empirical assessment. Lancet 2003;362:1439–1444.
122. Petricoin EF, Ardekani AM, Hitt BA, et al. Use of proteomic patterns in serum to identify ovarian cancer. Lancet 2002;359:572–577.
123. Li J, Zhang Z, Rosenzweig J, et al. Proteomics and bioinformatics approaches for identification of serum biomarkers to detect breast cancer. Clin Chem 2002;48:1296–1304.
124. Paweletz CP, Trock B, Pennanen M, et al. Proteomic patterns of nipple aspirate fluids obtained by SELDI-TOF: potential for new biomarkers to aid in the diagnosis of breast cancer. Dis Markers 2001;17: 301–307.
125. Jacobs TW, Prioleau JE, Stillman IE, et al. Loss of tumor marker-immunostaining intensity on stored paraffin slides of breast cancer. J Natl Cancer Inst 1996;88:1054–1059.
126. Di Leo A, Cardoso F, Durbecq V, et al. Predictive molecular markers in the adjuvant therapy of breast cancer: state of the art in the year 2002. Int J Clin Oncol 2002;7:245–253.
127. Fukutomi T, Akashi-Tanaka S. Prognostic and predictive factors in the adjuvant treatment of breast cancer. Breast Cancer 2002;9:95–99.
128. Mori I, Yang Q, Kakudo K. Predictive and prognostic markers for invasive breast cancer. Pathol Int 2002; 52:186–194.
129. Ross JS, Linette GP, Stec J, et al. Breast cancer biomarkers and molecular medicine. Expert Rev Mol Diagn 2003;3:573–585.

CHAPTER 11

Breast Cancer Cytogenetics, Chromosomal Abnormalities, and Comparative Genomic Hybridization

Mark Watson, MSc and Lynn Cawkwell, PhD

Division of Cell & Molecular Medicine, University of Hull, Research & Development Building, Castle Hill Hospital, Hull, East Yorkshire, United Kingdom

Giemsa banding (G banding) or variations thereof has been used for over 30 years in the cytogenetic analysis of hematologic malignancies and in prenatal diagnosis. However, its relevance to the cytogenetics of solid malignancies such as breast cancer is limited because of difficulties in obtaining suitable material for analysis (see Production of Suitable Material, below). Giemsa staining produces a banding pattern that is unique to each chromosome but requires chromosomes of sufficient length to allow definition of individual bands.

In recent years the study of breast cancer cytogenetics has improved immensely with the advent of new analysis techniques such as fluorescence in situ hybridization (FISH), which is an interphase based technique, in addition to metaphase based techniques such as multicolor fluorescence in situ hybridization (MFISH), spectral karyotyping (SKY), and comparative genomic hybridization (CGH). These techniques use advanced fluorescent technology to identify chromosomal aberrations, changes in whole chromosome number, and chromosomal locus copy number.

Both metaphase and interphase cytogenetic analyses have specific advantages and disadvantages associated with the type of data obtained and the technical aspects of performing each technique. Analysis of metaphase chromosomes allows changes in ploidy (chromosome number), chromosome translocations, and losses and gains of chromosome material to be identified in one experiment without prior knowledge of these aberrations. However, the techniques can be technically demanding, and the production of tumor metaphases for MFISH and SKY analysis is a major problem. Interphase cytogenetic analysis, on the other hand, is less technically demanding because the probes are less complex and interphase cells can easily be produced in large numbers. This technique can also be used on a wider range of material, including archival tissue samples. However, FISH probes can only target known regions of DNA and not the whole chromosome complement, meaning that a prior global analysis must be performed to identify target regions of interest.

Global Analysis–Metaphase Cytogenetics

Production of Suitable Material

Global cytogenetic techniques, that is, those that allow analysis of the whole chromosome complement in a single experiment, require the production of metaphase chromosomes. In the case of MFISH and SKY, metaphase spreads are produced from the test sample and the fluorescent probes are hybridized to the tumor chromosomes. For CGH analysis, DNA from the tumor sample is hybridized to normal metaphase spreads produced from human lymphocytes (FIG. 11.1). Either way, metaphase chromosome preparations are produced according to the same principles.

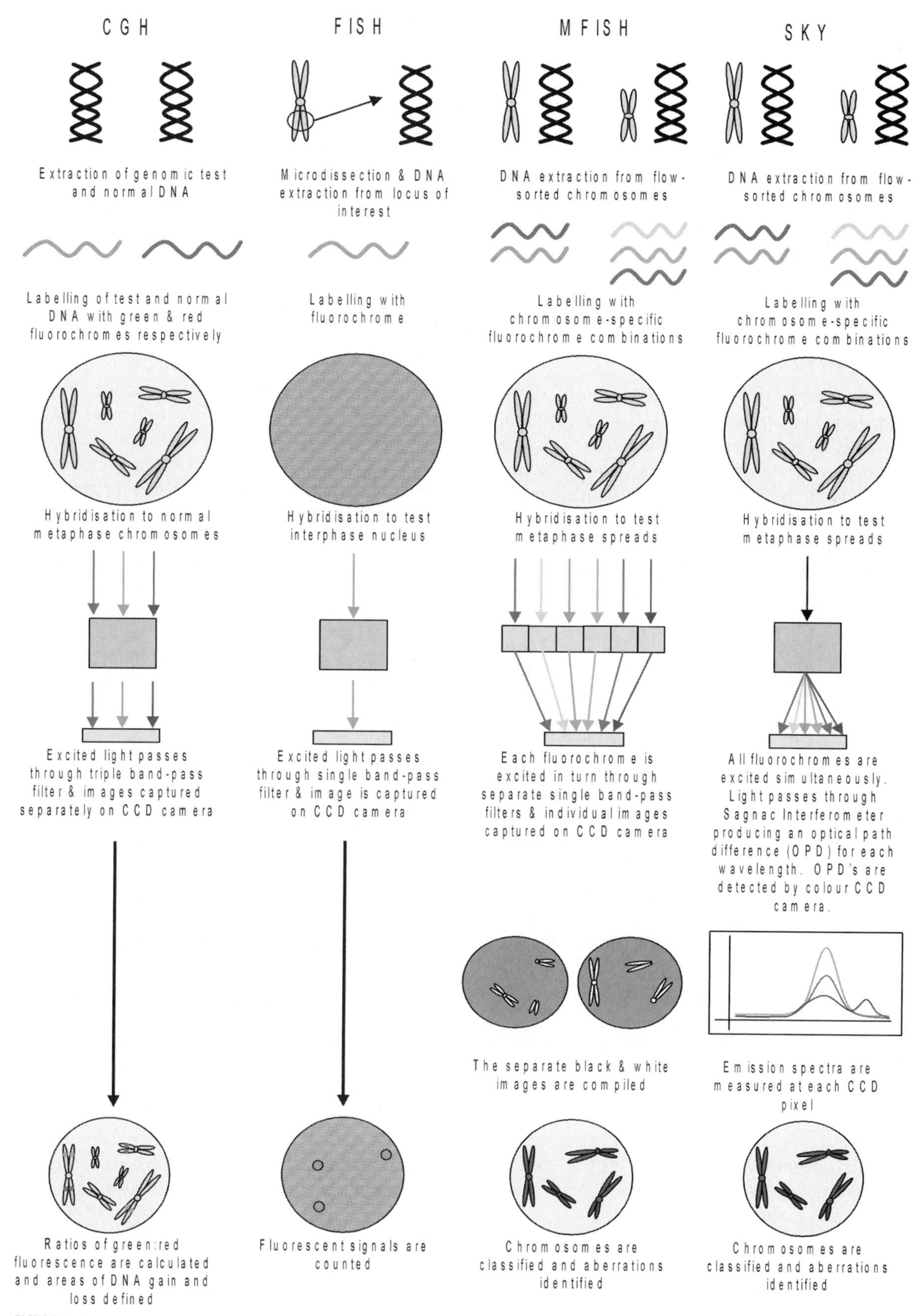

FIGURE 11.1 Comparison of the methodology of fluorescent cytogenetic techniques.

Table 11.1 Resources for Cytogenetic Research Using FISH, MFISH, SKY, and CGH

Resource	Website	Comments
Abbott Diagnostics	http://abbott.com	Supplier of FISH, MFISH, and CGH probes and the PathVysion™ HER-2 status testing kit
Applied Imaging	http://www.appliedimagingcorp.com	Supplier of Quips™, SpectraVysion™, and Karyotyping™ analysis software
The NCBI Spectral Karyotyping & Comparative Genomic Hybridisation Database	http://www.ncbi.nlm.nih.gov/sky/	SKY and CGH resource that allows researchers to view current SKY or CGH data or to submit their own

Production of Tumor Chromosomes

In solid tumor MFISH and SKY analysis, the sample is macerated with either scalpels or scissors, digested with a collagenase enzyme, and cultured. As the cells reach the exponential growth phase, a mitotic inhibitor, for example colcemid, is added to the culture, effectively trapping any dividing cells in metaphase and thus increasing the yield of metaphase spreads. After mitotic inhibitor treatment, the cells are harvested using standard procedures such as adding the enzyme trypsin. The cells are resuspended in a hypotonic solution such as potassium chloride (KCl), causing them to swell (to aid in the spreading of the chromosomes), and finally fixed in a solution of methanol and acetic acid. Metaphase spreads are produced by dropping the swollen cells onto a clean glass microscope slide and allowing to dry.[1] The ideal chromosomes for cytogenetic analysis (especially G banding) are long and thin, allowing high resolution analysis along the length of the chromosome. High quality chromosomes such as these are produced by limiting the mitotic inhibitor treatment time. However, cells cultured from solid tumors such as breast have very long division times (i.e., low mitotic rate), meaning that to produce a sufficient number of metaphase spreads for analysis, the mitotic inhibitor treatment time must be extended. This leads to the production of shorter fatter chromosomes (which are not ideal) for G-banding analysis and for high resolution fluorescent cytogenetic analysis.

MFISH

MFISH is used to analyze an entire karyotype (whole chromosome complement) in a single experiment and thus requires at least 23 unique combinatorial colors to enable identification of each chromosome. This is achieved using five probe fluorophores: red, far red, gold, green, and aqua.[2] Each chromosome is labeled with up to three colors, thus requiring up to three separate probes per chromosome (FIG. 11.1). Metaphase images are captured using a fluorescent microscope equipped with a set of six fluorophore-specific filters and a cooled charged coupled device (CCD) camera. Each metaphase spread is captured six times through the six specific filters, once for each of the five probe fluorophores and once for a 4,6-diamidino-2-phenylindole (DAPI)-banded image. The DAPI image is captured for the software to automatically define chromosome boundaries and aid in their identification. The analysis software (e.g., SpectraVysion™) (**Table 11.1**) compiles the monochrome images and assigns pseudo-colors to each one (according to the filter the image was captured through) (FIG. 11.2).

SKY

SKY allows the simultaneous hybridization and visualization of all 24 human chromosomes in much the same way as MFISH. However, the unique difference is the way in which the metaphase images are captured and analyzed.[3] MFISH involves the capture of five (six including the DAPI image) separate gray-level images corresponding to the five fluorochromes and the subsequent assignment of pseudo-colors to each image. In SKY, all fluorochromes are excited and captured simultaneously through a multiband pass filter. After passing through a series of lenses, the light enters a device called a Sagnac interferometer. Here subtle changes in the optical path of the light from each fluorochrome are produced (known as the optical path difference).[4] A cooled CCD camera is able to detect these

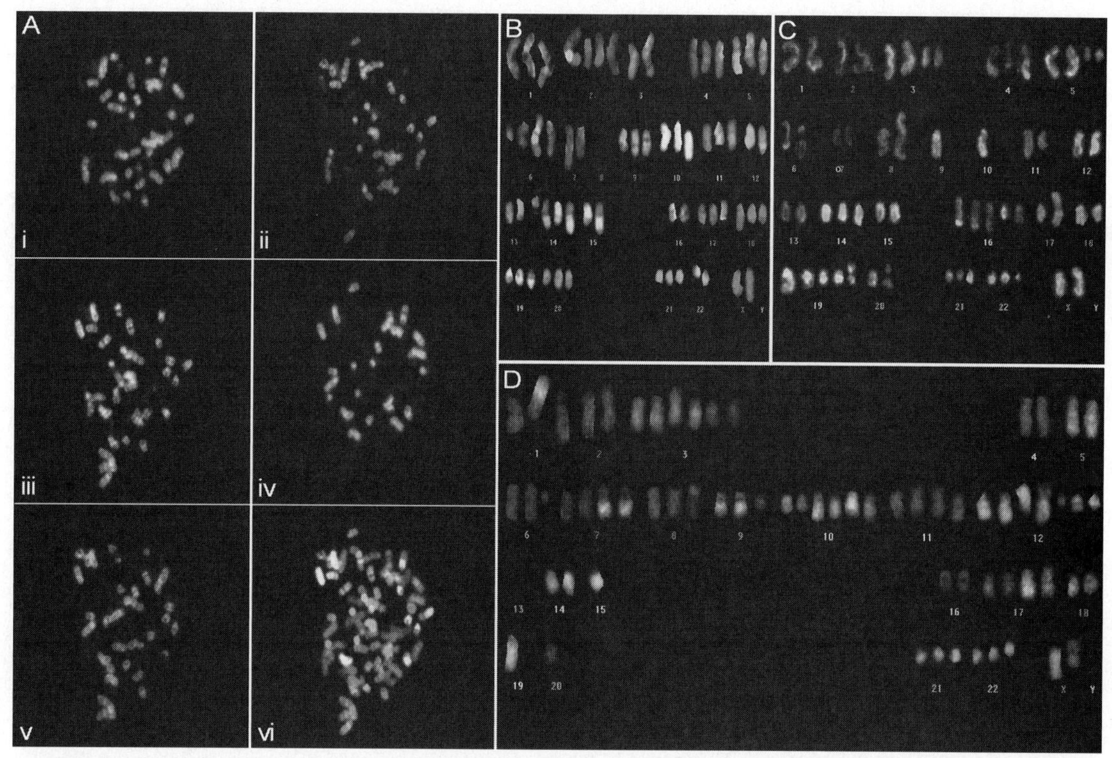

FIGURE 11.2 MFISH. Each of the five color planes, red, green, gold, aqua, and far red (A, i–v, respectively), and the complete composite image (A, vi). Typical analyzed images from the MDA-MB-231 established breast cancer cell line (B) and a short-term culture derived from a primary breast tumor (C) and from the T47D breast cancer cell line (D). **See Plate 25 for color image.**

minute differences in emission spectra at each individual pixel. Thus, for each pixel an interferogram is produced, which is then analyzed by Fourier transformation, the process by which individual spectra of light can be defined from a group of spectra.[5] It is the specific profile of the emission spectra at each pixel that defines the chromosome represented at that point. By comparing each observed emission spectra profile with the known spectral profiles of the chromosomes, classification colors can be assigned to each chromosome.

CGH

CGH analyzes genome-wide losses and gains of chromosomal material and works on the principle of competitive binding between two populations of DNA (one test DNA, e.g., tumor, one reference DNA, e.g., normal), which are simultaneously hybridized to normal metaphase spreads.[6–8] Each DNA population is labeled by nick translation with a detectable fluorophore, usually green fluorescein dUTP for the tumor DNA and red rhodamine dUTP for the normal DNA. The labeled DNA is mixed and hybridized to commercially available metaphase spreads ready prepared from normal human peripheral blood lymphocytes. To inhibit binding to repetitive heterochromatic regions of the chromosomes, an excess of human Cot-1 DNA is added to the hybridization. Fluorescent microscopy and specialized analysis software (Quips™ Karyotyping™, **Table 11.1**) are used to determine the relative ratios of green and red fluorescence at any given chromosomal region. Gray-level images are taken using a CCD camera and a multiband pass excitation filter for each of the two fluorochromes and for the DAPI image (FIG. 11.1). The software compiles the three images and displays them in computer-generated pseudocolors (FIG. 11.3A, i–iv). Because of the nature of the competitive binding, amplification of any area of tumor DNA increases the amount of tumor DNA binding at that region, thus shifting the red–green fluorescence signal in favor of the green. The reverse is true for a deletion of a chromosomal region in the tumor. The proportion of normal DNA to tumor DNA is higher, thus increasing the amount of normal DNA binding at that site, resulting in a red fluorescent signal (FIG. 11.3B). The analysis software uses the DAPI-banded image to classify the chromosomes and then calculates the ratio of

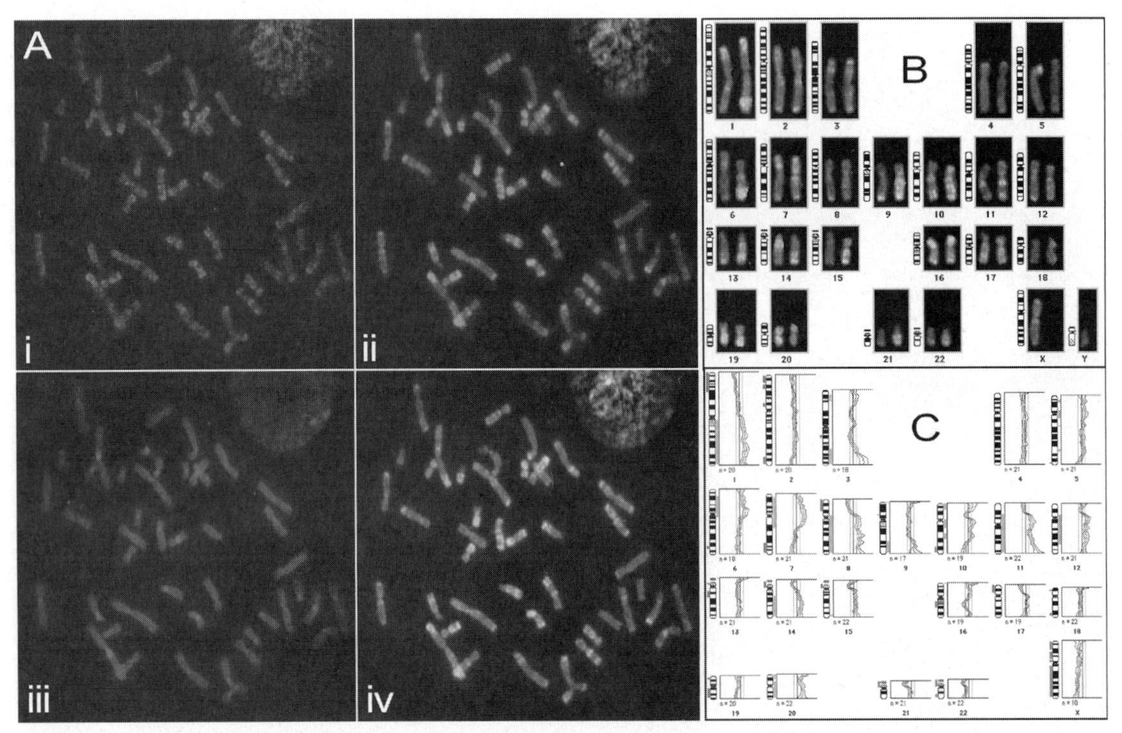

FIGURE 11.3 CGH. The three color planes, red, green, and DAPI (A, i–iii, respectively) and the complete composite image (A, iv). Chromosomes are classified according to the DAPI banded image and red-to-green fluorescence ratios calculated along the length of each chromosome that denote areas of DNA loss and gain, respectively (B). A complete profile compiled from several metaphase spreads (C). **See Plate 26 for color image.**

red to green signals along the length of each chromosome to produce a karyogram of copy number gains and losses with a maximum resolution of about 2Mb (FIG. 11.3C).

Array CGH

Recently, CGH has been applied in the growing field of microarray technology. CGH arrays can comprise thousands of whole genome or chromosome-specific DNA probes produced using plasmid artificial chromosomes or bacterial artificial chromosome clones, which are immobilized onto glass microscope slides.[9] The array CGH technique is presented in detail in Chapter 2. As with standard CGH, test and reference DNAs are fluorescently labeled by nick translation and hybridized (along with Cot-1 blocking DNA) onto the microarray slide.[10] CGH arrays are analyzed using specialized equipment with high resolution CCD cameras that allow color ratios to be calculated at each probe spot to produce a high resolution map of gains and losses. Depending on the size of the array targets and total size of the microarray (genome-wide or chromosome specific), maximum resolutions of around 40 kb can be achieved.[10]

CGH arrays can be applied in combination with complementary DNA arrays that are usually hybridized with complementary DNA transcribed from extracted messenger RNA and reveal patterns of gene expression rather than copy number. Thus, copy number as identified by array CGH can be compared directly with gene expression.[11,12] Recently, this approach identified good correlations between copy number and gene expression in primary breast tumors and established breast cancer cell lines, with 62% of highly amplified genes showing parallel increases of elevated expression and an average 2-fold DNA copy number change associated with approximately a 1.5-fold change in gene expression.[11] Array CGH also has the potential as a tool for the high throughput screening of panels of prognostic or other clinical markers, for example, HER-2 status.[13]

Nonglobal Analysis–Interphase Cytogenetics

Production of Suitable Material

As the title suggests, FISH requires only interphase cells and can therefore be used on cell preparations prepared by means other than live cell culture. These include "touch" preparations whereby the cut surface of the tumor specimen is pressed onto a clean microscope slide, leaving a layer of cells on the surface of the slide. FISH can also be used on formalin-fixed and frozen cryosectioned tissue preparations.

FISH

FISH involves the hybridization of fluorescently-labeled probes to interphase tumor chromosomes and visualization using fluorescent microscopy. FISH-labeled nuclei are analyzed by the counting of fluorescent signals in a large number (more than 100) of individual interphase nuclei.[14] Aneuploidy of a specific chromosome or specific chromosome region is directly related to the number of fluorescent signals. FISH can also identify simple translocations by analysis with special fusion probes.[15]

Technical Limitations of FISH, MFISH, SKY, and CGH

There are several key limitations associated with the use of FISH techniques in the study of breast cancer cytogenetics. In the case of MFISH, where metaphases of a good quality are required, the difficulties in culturing and the low rate of cell division of solid tumors makes this very difficult. MFISH is sometimes complemented with a standard banding technique that enables identification of translocation breakpoints. However, this is rarely achievable with solid malignancies where the chromosome preparations are often of too poor quality for accurate banding analysis. FISH is an excellent tool for the study of breast cancer cytogenetics because it only requires interphase cells. However, because FISH does not produce a complete karyotypic analysis, the investigator must first know what aberration (aneuploidy, translocation, etc.) they are looking for. Most often this requires prior use of another technique such as CGH or MFISH to identify the region or translocation of interest. Other limitations of both techniques include the high cost of consumables and probes and the intensive technical requirements.

There are several limitations associated with the use of CGH. First, as a stand-alone technique, CGH is unable to detect aberrations such as inversions or balanced translocations. Therefore, CGH is complemented with either G banding, MFISH, or SKY analysis. Second, CGH requires at least 50% of the test sample DNA to be tumor in origin. This is important because tumor samples contain varying degrees of normal tissue. As the percentage of normal tissue increases, the reliability of detection of ratio changes decreases.[8] The third limitation of CGH is its limits of resolution, which depend largely on copy number. Low copy number gains are detectable at about 10 Mb and high copy number gains at 2 Mb.[16] Regions of these sizes are large enough to harbor many genes, all of them potential cancer-related genes. Thus, CGH can only precede higher resolution techniques by narrowing down potential regions of interest. Finally, other limitations include the high technical demands and the need to interpret telomeric regions cautiously due to the low fluorescent intensity of these points.

Current Clinical Utility: FISH

Of the fluorescent cytogenetic techniques described herein, only FISH is in current clinical use. The treatment of metastatic breast cancer has advanced significantly since the development of the drug trastuzumab (Herceptin®).[17] Trastuzumab is a humanized monoclonal anti-HER-2 antibody, directed against the growth factor coreceptor HER-2/neu, which is up-regulated in up to 30% of breast cancer patients.[17] Trastuzumab is only effective in patients who have up-regulated HER-2/neu expression because in these patients tumor growth is partly controlled by this receptor. As such, the status of the HER-2/neu protein and/or gene expression must be determined. Because HER-2/neu is a membrane bound receptor, its presence is readily detectable using immunohistochemistry on fixed tumor specimens with anti-HER-2/neu antibodies. However, immunohistochemical analysis requires subjective assessment of protein expression levels based on comparisons with positive and negative controls to produce a score of either 3+ (strongly positive), 2+ (weakly positive), or 1+/0 (negative).[18]

In weakly positive (2+) cases, a more accurate determination of HER-2/neu status is often necessary if an appropriate therapy is to be decided. Results also depend on the specificity of the antibody and the sensitivity of the detection method.[19] HER-2/neu is encoded by the c-*erb*B2 gene, located at chromosome 17q21.1 and an area of frequent amplification in breast cancer.[20] More recently, FISH has been used in the direct detection of amplifications in the c-*erb*B2 gene copy number. The US Food and Drug Administration approved the Abbot PathVision™ kit (**Table 11.1**), a commercially available c-*erb*B2 testing kit comprised of two FISH probes. One probe (Spectrum-Orange™) maps to chromosome region 17q11.2-q12, the location of the c-*erb*B2 gene, and the other probe (SpectrumGreen™) maps to the centromere of chromosome 17. Analysis is based on direct counting of fluorescent signals, with a ratio of orange to green signals of more than two per cell, indicating a c-*erb*B2 gene amplification.

Concordance between immunohistochemistry and FISH results of HER-2/neu status is generally very good, and many research groups have tested the efficacy and advantages of each test for use in routine practice.[21,22] Frequent conclusions drawn from these studies indicate the initial use of the less expensive and less technically demanding immunohistochemistry combined with the more expensive and more technically demanding FISH only for cases of 2+ immunohistochemistry positivity.[23–25] The biologic and clinical significance of HER-2/neu testing in breast cancer is covered in depth in Chapter 16.

Research So Far and Potential Clinical Utility

FISH, MFISH, and SKY

FISH, MFISH, and SKY have been used extensively in the field of breast cancer research. To date, more than 20 established breast cancer cell lines have been completely cytogenetically characterized by SKY[26,27] and MFISH[28] as well astumor-derived short-term cultures by FISH[29] (**Table 11.2**). Among the more significant findings of FISH and SKY using established breast cancer cell lines are amplifications of the c-*myc* oncogene (8q24)[30] and identification of the copy number and translocational instability of chromosome 8 as a whole[26] using SKY alone. Recently, chromosome 8 copy number has been correlated with histologic grade using FISH providing the possibility of a genetic classification system of breast cancer.[31] MFISH analysis of primary breast carcinomas has confirmed the translocational instability of chromosome 8 and also identified a unique translocation involving chromosomes 9 and 17[32] (**Table 11.2**). The frequent translocational involvement (but not the translocation partners) of chromosome 17 in primary breast carcinomas has also been identified using FISH.[29]

CGH

Since its inception, CGH has been used extensively in the analysis of breast cancer

Table 11.2 Summary of Current Cytogenetic Analysis of Breast Cancer Using FISH, MFISH, and SKY

Principal Finding	Tissue Used	Technique Used	References
Most commonly involved chromosome is 8 and most common translocations are 8;11, 8;17, 1;4, and 1;10	19 established breast cancer cell lines and one tumor-derived short-term culture	SKY	26
Most commonly involved chromosomes are 8, 1, 17, 16, and 20 and most common translocations are 8;11, 12;16, 1;16, and 15;17	15 established breast cancer cell lines	SKY	27
Translocations included X;3, 1;5, 2;9, 3;9, 4;20, 4;15, 5;7, 11;20, and 10;12	1 established breast cancer cell line	MFISH	28
Chromosome 17 is most commonly involved in chromosomal rearrangements	12 primary breast carcinomas	FISH	29
High frequency of chromosome 8 structural abnormalities (chromosome 11 being the most frequent translocation partner) Increases in c-*myc* oncogene copy number	16 established breast cancer cell lines	FISH SKY	30
Identification of 9;17 in two breast carcinomas (the sole translocation in one) and one established breast cancer cell line	2 primary breast carcinomas and 5 established breast cancer cell lines	MFISH	32

cytogenetics in both established cell lines and fresh, frozen, and paraffin-embedded tumor tissue. Numerous published studies have identified frequent DNA losses and gains recurring repeatedly in breast cancer cell lines[30,33–35] and tumor tissue[36–48] (**Table 11.3**), the most frequent of these being gains of 1q, 8q, and 17q and losses of 8p, 16q, and 17p. Many of these aberrations have been correlated with known oncogenes, tumor suppressor genes, and other genes associated with the pathogenesis and progression of breast cancer (**Table 11.4**). Aside from simply cataloging these aberrations, CGH has the potential to correlate specific aberrations with disease progression, outcome, and grade. The pathogenesis and progression of breast cancer are still not well understood, but CGH has been used successfully to aid in the study of how breast cancers arise and progress from benign to invasive types. CGH has been used to differentiate between grade I and grade III tumors, and these data suggest that most grade III tumors do not evolve from grade I tumors, they have separate genetic origins.[43] In particular, 65% of grade I tumors were found to have lost chromosome 16q compared with only 16% of grade III tumors.[43]

The progression of premalignant ductal carcinoma in situ to malignant invasive ductal carcinoma is a commonly accepted pathway in breast cancer, but the specific genetic aberrations that link the progression of one to the other are unknown. CGH has enabled low resolution genetic classification of these lesions and identified many common genetic aberrations between them. These include gains on 6p, 10q, 14q, and 15q and losses on 9p[49] and gains of 1q, 3q, 8q, 17q, and 20q and losses of 8p, 13q, and 16q.[50]

The role of CGH in routine clinical use is limited because of its relatively low resolution. The global approach to analysis of this technique makes it more suitable to low resolution identification of chromosomal regions of interest, which may then be targeted with higher

Table 11.3 Summary of Current CGH Studies on Breast Cancer

Gains	Losses	Tissue	References
8q	**8p**	Cell lines	30
8q, **1q**, 7q, 3q, 7p	Xp, **8p**, 18q, Xq	Cell lines	33
8q, **1q**, 20q, 7p, and high level gains of **8q23**, 20q13, 3q25-q26	**8p**, 18q, 1p, Xp, Xq	Cell lines	34
8q22-qter, **8q24**.1, 20q12, **1q41**	Xq11-q12, 13q21, **8p12-pter**	Primary tumors and cell lines	35
1q, **17q**, 19q, 20p, 20q	13q, 14q, **17p**, **16q**, 22q	Primary tumors	36
8q, **1q**, Xq, 5q, 4q, 3q	19p, 1p, **17p**, 22q, 4q, **8p**	Node-negative primary tumors	37
1q, **8q**, high level gains of **8q12-24**, 11q11-13, 20q13-qter	1p, 5q, 6q, 9p, 11q, 13q, **16q**	Primary tumors	38
11q13, **1q**, 19, 16p	14, 13, 18, 5p, 5q, 2q, 6q, 9p	Hypodiploid primary tumors	39
1q, **8q**	**8p**, **16q**, 13q	Primary tumors	40
1q, **8q**, 20q	18q, 13q, 3p	Primary tumors	41
Increased gains of 20q in ERBB2-positive tumors	Increased losses of 18q in ERBB2-positive tumors		
1q, 7, 20, 16, 3, 6, 6q, 8	**16q**, X, 9, 22, 16p, 4, 17, 13	Primary tumors	42
1q, 16p, **8q**–grade I tumors and **1q**, **8q**, **17q**, 20q, 16p–grade II tumors	16q–grade I tumors	Primary tumors	43
1q, **8q**, 16p, **17q**, 19q, 20q	14q, **16q**, **17p**, 21q, 22q	Primary tumors (fine-needle aspirates)	44
1q, **8q23**, **8q24**	**17p**, **8p**	Primary tumors	45
1q, **8q**, 20q	**17p**, **16q**	Primary tumors	46
1q, 5q, **8q**, X	1p33-p36, **16q**, **17p**, 19, 22q	Primary tumors	47
1q, **8q**, **17q**, 20q, 3q, 1p, 5p, 15q	**8p**, 11q, **16q**, **17p**, 18q	Primary tumors (formalin fixed paraffin-embedded)	48

Common aberrations are highlighted in bold.

Table 11.4 Chromosomal Loci of Interest Identified by CGH and Candidate Cancer Genes Mapped to Those Regions

Locus	Finding with CGH	Proposed Genes	Function
1q	Gain	Not known	
8q	Gain	*c-myc* (8q24)	Encodes transcription factor that regulates various growth-related genes
		RB1CC1 (8q11)	RB1CC1 protein regulates the tumor suppressor gene *Rb1*
		NSE2 (8q24.13)	Encodes a breast cancer-associated membrane protein whose function is unknown
8p	Loss	*N33* (8p22)	Putative prostate cancer tumor suppressor gene
16q	Loss	*BCAR1* (16q23.1)	Breast cancer antiestrogen resistance 1 gene
17q	Gain	c-*erb*B2 (17q21.1)	Encodes the HER-2 growth factor receptor protein
		BRCA1 (17q21)	Familial breast cancer susceptibility gene
17p	Loss	*P53* (17p13.1)	Tumor suppressor gene

resolution techniques. However, CGH has been used as a supportive research tool in the field of diagnostic pathology by differentiating between metastases from a primary breast tumor and the development of a secondary primary tumor.[9] In this study, CGH was applied to archival samples from patients who had presented with further tumors several years after an initial presentation and subsequent treatment. In one particular case, a woman presented with two metastases 2 years apart after an initial presentation with a breast tumor and an ovarian tumor 9 years later. CGH analysis revealed that the two metastases shared a large proportion of aberrations with each other and with the initial breast cancer (including gain of 4q, 8p, 16q, and Xp and loss of 3q and 8q), hinting at their shared origin. The ovarian cancer shared no aberrations with the metastases, leading to the conclusion that it was most likely a second primary tumor. An application such as this could enable a more accurate analysis of the course of a patient's disease and potentially provide more accurate prognostic information.

Advances in CGH techniques now means that it can be used on smaller tissue samples, such as those obtained from fine-needle aspirations, biopsies, and from archival material, providing a potential role for CGH in preclinical tumor analysis and/or diagnosis. To this effect, CGH has been used in the analysis of small numbers of cells (10^5 to 10^6 cells) from fine-needle aspirates obtained from breast carcinomas.[44] More recently, CGH has been applied to both fine-needle aspirates and malignant effusions to determine the efficacy of this technique in the analysis of the aforementioned sample types.[51] Both of these studies have not only highlighted the use of CGH in the analysis of aspirates and effusions, they have both demonstrated the supportive role of CGH combined with standard cytologic diagnostic techniques.

Summary

Chromosomal aberrations, losses of tumor suppressor genes, and amplifications of oncogenes such as HER-2 and c-*myc* are key to the pathogenesis and progression of breast cancer and are potentially indicative of patient diagnosis, prognosis, and response to treatment. The development of standardized fluorescent cytogenetic techniques has enabled scientists and clinicians to identify and study complex translocations, aneuploidy, and losses and gains of DNA throughout the chromosome complement. Few of these techniques (FISH being the exception) are in routine clinical use, most likely due to technical requirements, cost, and low specificity of the results obtained. However, these techniques are very useful for identifying unstable regions of the genome suitable for further analysis with techniques with a higher resolution. The advent of high resolution/high throughput florescent techniques such as array CGH may one day mean the potential of routine screening of whole panels of markers of diagnosis, prognosis, and response to treatment on one clinical sample. This could potentially lead to completely personalized diagnosis and treatment systems for a multitude of diseases, including breast cancer.

References

1. Polito P, Cin PD, Debiec-Rychter M, et al. Human solid tumours: cytogenetic techniques. In: Swansbury J, ed. *Cancer Cytogenetics: Methods & Protocols.* Totowa, NJ: Humana Press, 2003;135–150.
2. Speicher MR, Ballard SG, Ward DC. Karyotyping human chromosomes by combinatorial multi-fluor FISH. Nat Genet 1996;12:368–375.
3. Schröck E, du Manoir S, Veldman T, et al. Multicolour spectral karyotyping of human chromosomes. Science 1996;273:494–497.
4. Garini Y, Macville M, du Manoir S, et al. Spectral karyotyping. Bioimaging 1996;4:65–72.
5. Malik Z, Cabib D, Buckwald RA, et al. Fourier transform multipixel spectroscopy for quantitative cytology. J Microsc 1996;182:133–140.
6. Kallioniemi A, Kallioneimi OP, Sudar D, et al. Comparative genomic hybridisation for molecular cytogenetic analysis of solid tumours. Science 1992;258: 818–821.
7. Kallioneimi OP, Kallioneimi A, Sudar D, et al. Comparative genomic hybridisation: a rapid new method for detecting and mapping DNA amplification in tumours. Semin Cancer Biol 1993;4:42–46.
8. Kallioneimi OP, Kallioneimi A, Piper J, et al. Optimising comparative genomic hybridisation for analysis of DNA sequence copy number changes in solid tumours. Genes Chromosomes Cancer 1994;10: 231–243.
9. Weiss MM, Kuipers EJ, Meuwissen SGM, et al. Comparative genomic hybridisation as a supportive tool in diagnostic pathology. J Clin Pathol 2003;56:522–527.
10. Pinkel D, Segraves R, Sudar D, et al. High resolution analysis of DNA copy number variation using comparative genomic hybridisation to microarrays. Nat Genet 1998;20:207–211.
11. Pollack JR, Sørlie T, Perou CM, et al. Microarray analysis reveals a major role of DNA copy number alteration in the transcriptional program of human breast tumours. Proc Natl Acad Sci USA 2002;99: 12963–12968.
12. Hyman E, Kauraniemi P, Hautaniemi S, et al. Impact of DNA amplification on gene expression patterns in breast cancer. Cancer Res 2002;62:6240–6245.
13. Albertson DG. Profiling breast cancer by array CGH. Breast Cancer Res Treat 2003;78:289–298.
14. Netten H, van Vliet LJ, Vrolijk H, et al. Fluorescent dot counting in interphase cell nuclei. Bioimaging 1996;4:93–106.
15. Dubus P, Young P, Beylot-Barry M, et al. Value of interphase FISH for the diagnosis of t(11;14)(q13;q32) on skin lesions of mantle cell lymphoma. Am J Clin Pathol 2002;118:832–841.
16. Lichter P, Joos S, Bentz M, et al. Comparative genomic hybridisation: uses and limitations. Semin Haematol 2000;37:348–357.
17. Gralow J. Herceptin: new treatment modality. The Digital Breast Clinic. http://www.digitalclinic.com/herceptin.htm, 2000.
18. Ellis IO, Dowsett M, Bartlett J, et al. Recommendations for HER2 testing in the UK. J Clin Pathol 2000; 53:890–892.
19. Schaller G, Evers K, Papadopoulos S, et al. Current use of HER2 tests. Ann Oncol 2001;12(suppl 1): S97–S100.
20. Willis S, Hutchins AM, Hammet F, et al. Detailed gene copy number and RNA expression analysis of the 17q12-23 region in primary breast cancers. Genes Chromosomes Cancer 2003;36:382–392.
21. Larsimont D, Di Leo A, Rouas G, et al. HER-2/neu evaluation by immunohistochemistry and fluorescence in situ hybridisation for daily laboratory practice. Anticancer Res 2002;22:2485–2490.
22. Lebeau A, Deimling D, Kaltz C, et al. HER-2/neu analysis in archival tissue samples of human breast cancer: comparison of immunohistochemistry and fluorescence in situ hybridisation. J Clin Oncol 2001; 19:354–363.
23. Dowsett M, Bartlett J, Ellis IO, et al. Correlation between immunohistochemistry (HercepTest) and fluorescence in situ hybridisation (FISH) for HER–2 in 426 breast carcinomas from 37 centres. J Pathol 2003;199:418–423.
24. Cianciulli AM, Botti C, Coletta AM, et al. Contribution of fluorescence in situ hybridisation to immunohistochemistry for the evaluation of HER-2 in breast cancer. Cancer Genet Cytogenet 2002;133: 66–71.
25. Kakar S, Puangsuvan N, Stevens JM, et al. HER-2/neu assessment in breast cancer by immunohistochemistry and fluorescence in situ hybridisation: comparison of results and correlation with survival. Mol Diagn 2000;5:199–207.
26. Davidson JM, Gorringe KL, Chin SF, et al. Molecular cytogenetic analysis of breast cancer cell lines. Br J Cancer 2000;83:1309–1317.
27. Kytölä S, Rummukainen J, Nordgren A, et al. Chromosomal alterations in 15 breast cancer cell lines by comparative genomic hybridisation and spectral karyotyping. Genes Chromosomes Cancer 2000;28: 308–317.
28. Micci F, Teixeira M, Heim S. Complete cytogenetic characterisation of the human breast cancer cell line MA11 combining G-banding, comparative genomic hybridisation, multicolour fluorescence in situ hybridisation, RxFISH, and chromosome-specific painting. Cancer Genet Cytogenet 2001;131:25–30.
29. Anamthawat-Jónsson K, Eyfjörd JE, Ögmundsdóttir HM, et al. Instability of chromosomes 1:3, 16, and 17 in primary breast carcinomas inferred by fluorescence in situ hybridisation. Cancer Genet Cytogenet 1996;88:1–7.
30. Rummukainen J, Kytölä S, Karhu R, et al. Aberrations of chromosome 8 in 16 breast cancer cell lines by comparative genomic hybridisation, fluorescence in situ hybridisation, and spectral karyotyping. Cancer Genet Cytogenet 2001;126:1–7.
31. Bofin AM, Ytterhus B, Fjøsne HE, et al. Abnormal chromosome 8 copy number in cytological smears from breast carcinomas detected by means of fluorescence in situ hybridisation (FISH). Cytopathology 2003;14:5–11.
32. Watson MB, Bahia H, Ashman JNE, et al. Chromosomal alterations in breast cancer revealed by multicolour fluorescence in situ hybridisation. Int J Oncol 2004;25:277–283.
33. Forozan F, Veldman R, Ammerman CA, et al. Molecular cytogenetic analysis of 11 new breast cancer cell lines. Br J Cancer 1999;81:1328–1334.
34. Forozan F, Mahlamaki EH, Monni O, et al. Comparative genomic hybridization analysis of 38 breast cancer cell lines: a basis for interpreting complementary DNA microarray data. Cancer Res 2000;60:4519–4525.
35. Larramendy ML, Lushnikova T, Bjorkqvist AM, et al. Comparative genomic hybridization reveals complex genetic changes in primary breast cancer tumours

and their cell lines. Cancer Genet Cytogenet 2000; 119:132–138.

36. James LA, Mitchel EL, Menasce L, et al. Comparative genomic hybridisation of ductal carcinoma in situ of the breast: identification of regions of DNA amplification and deletion in common with invasive breast cancer. Oncogene 1997;14:1059–1065.

37. Hermsen MA, Baak JP, Meijer GA, et al. Genetic analysis of 53 lymph node-negative breast carcinomas by CGH and relation to clinical, pathological, morphometric, and DNA cytometric prognostic factors. J Pathol 1998;186:356–362.

38. Schwendel A, Richard F, Langreck H, et al. Chromosome alterations in breast carcinomas: frequent involvement of DNA losses including chromosomes 4q and 21q. Br J Cancer 1998;78:806–811.

39. Tanner MM, Karhu RA, Nupponen NN, et al. Genetic aberrations in hypodiploid breast cancer: frequent loss of chromosome 4 and amplification of cyclin D1 oncogene. Am J Pathol 1998;153:191–199.

40. Tirkkonen M, Tanner M, Karhu R, et al. Molecular cytogenetics of primary breast cancer by CGH. Genes Chromosomes Cancer 1998;21:177–184.

41. Isola J, Chu L, DeVries S, et al. Genetic alterations in ERBB2-amplified breast carcinomas. Clin Cancer Res 1999;5:4140–4145.

42. Persson K, Pandis N, Mertens F, et al. Chromosomal aberrations in breast cancer: a comparison between cytogenetics and comparative genomic hybridisation. Genes Chromosomes Cancer 1999;25:115–122.

43. Roylance R, Gorman P, Harris W, et al. Comparative genomic hybridisation of breast tumours stratified by histological grade reveals new insights into the biological progression of breast cancer. Cancer Res 1999;59:1433–1436.

44. Burki NG, Caduff R, Walt H, et al. Comparative genomic hybridisation of fine needle aspirates from breast carcinomas. Int J Cancer 2000;88: 607–613.

45. Gebhart E, Liehr T. Patterns of genomic imbalances in human solid tumours [Review]. Int J Oncol 2000; 16:383–399.

46. Cingoz S, Altungoz O, Canda T, et al. DNA copy number changes detected by comparative genomic hybridisation and their association with clinicopathologic parameters in breast tumours. Cancer Genet Cytogenet 2003;145:108–114.

47. Fazeny-Dörner B, Piribauer M, Wenzel C, et al. Cytogenetic and comparative genomic hybridisation findings in four cases of breast cancer after neoadjuvant chemotherapy. Cancer Genet Cytogenet 2003;146: 161–166.

48. Weber-Mangal S, Sinn HP, Popp S, et al. Breast cancer in young women (< or = 35 years): Genomic aberrations detected by comparative genomic hybridization. Int J Cancer 2003;107:583–592.

49. Aubele M, Mattis A, Zitzelsberger H, et al. Extensive ductal carcinoma in situ with small foci of invasive ductal carcinoma: evidence of genetic resemblance by CGH. Int J Cancer 2000;85:82–86.

50. Buerger H, Otterbach F, Simon R, et al. Different pathways in the evolution of invasive breast cancer are associated with distinct morphological subtypes. J Pathol 1999;189:521–526.

51. Nagel H, Schulten HJ, Gunawan B, et al. The potential value of comparative genomic hybridisation analysis in effusion- and fine needle aspiration cytology. Mod Pathol 2002;15:818–825.

CHAPTER 12

DNA Ploidy and S Phase Analysis in Breast Cancer

Jeffrey S. Ross, MD,[1,2] Merrill S. Ross, BS,[1] Shahgul Anwar, MD,[1] and Ann Boguniewicz, MD[1]

[1]Department of Pathology and Laboratory Medicine, Albany Medical College, Albany, New York
[2]Division of Molecular Medicine, Millennium Pharmaceuticals, Inc., Cambridge, Massachusetts

During the past decade there has been considerable interest in the application of new technologies to identify human malignancy and predict disease outcome. Markers of cell proliferation and the technologies of flow, laser scanning, and image cytometric analysis for the determination of total DNA content in human tumor cells have been among the most frequently used ancillary methods for the detection of neoplasms and the prediction of cancer outcome. In this chapter we consider the biologic background of DNA content measurement in breast cancer along with the traditional and emerging techniques used to assess clinical samples for their DNA ploidy status. We also present the background and utlity of machine calculated S phase results and contrast the utility of these determinations with those based on ploidy status alone. Individual biomarkers of the cell cycle and their biologic basis and clinical utility in breast cancer are covered in Chapter 13.

Biologic Background

Normal Cell Cycle and DNA Ploidy

Human cancer cells actively synthesizing DNA replicate using a cell cycle similar to that of normal cells (FIG. 12.1).[1,2] Normal cells in the resting diploid state (G0 phase) contain 7.14 pg of DNA and enter the cell cycle as the gap 1 (G1) cells. During the DNA synthesis phase (S phase), cells increase their DNA content continuously from 7.14 to 14.28 pg/cell when they reach the tetraploid state with twice the diploid DNA content. The second gap (G2 phase) refers to this tetraploid, or premitotic fraction of cells that undergo mitosis in the M phase to generate two diploid G0 cells, which may reenter the cell cycle or persist in the resting state. A DNA index of 1.0 corresponds to a 2N or 46 chromosome number (7.14 pg) characteristic of G0 and G1 cells. The G2 and M cells feature a 2.0 DNA index that corresponds to a 4N chromosome number of 92 (14.28 pg).

The distribution of a population of cells within the cell cycle generates a pattern known as a DNA ploidy histogram.[2] A DNA ploidy histogram is defined as DNA diploid when the distribution of the total DNA content in the predominant or G0/G1 peak of the normal, unknown, or tumor cell population is equal to that of the total DNA content of the G0/G1 peak of a known to be diploid reference cell population and the S and G2/M phases of the cell population are relatively low. In normal tissues and in most low grade or slowly proliferating neoplasms, approximately 85% of the cell population forms the G0/G1 peak and 15% of the cells comprise the S and G2/M phases.

DNA aneuploidy, also known as nondiploidy, is defined as a DNA content of the G0/G1 peak of a cell population that varies significantly from the G0/G1 peak of the known diploid reference cell population (**Table 12.1**). The DNA index of an aneuploid cell population may rarely be less than 1.0 (hypodiploid) and commonly is far greater than 1.0 (hyperdiploid). Aneuploid tumors with DNA indices located between the 1.0 and

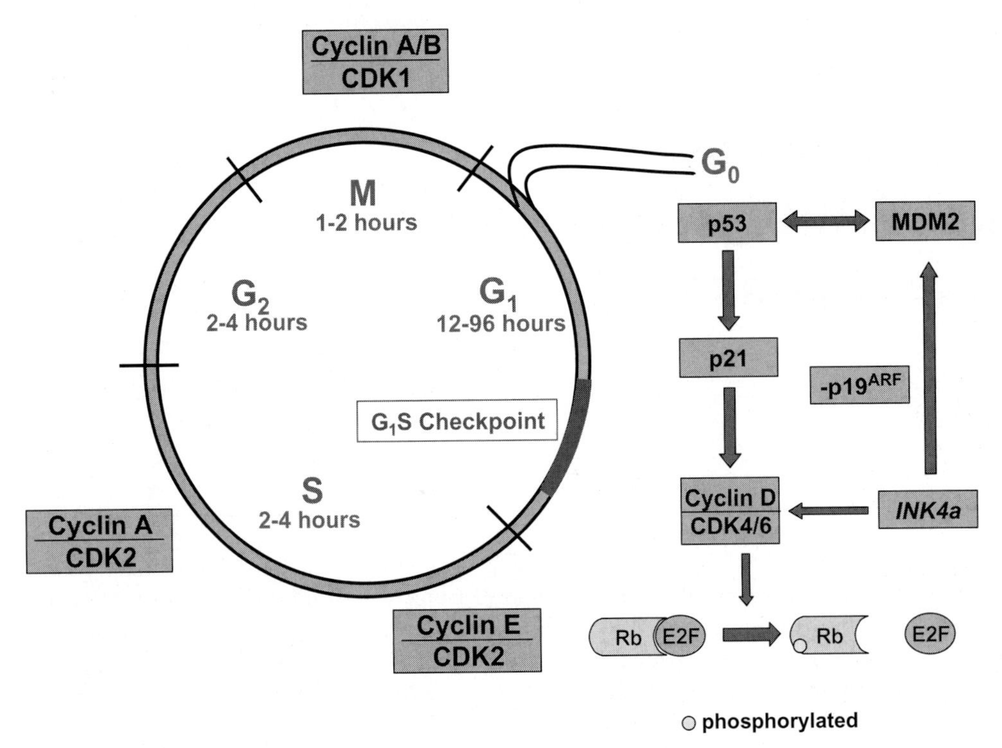

FIGURE 12.1 The normal cell cycle.

Table 12.1 DNA Ploidy Definitions

G0/G1 Peak DNA Index	DNA Ploidy Definition
=1.00 (1.00 ± 0.15)	Diploid
<0.85 or >1.15 (excluding tetraploid)	Aneuploid
<0.85	Hypodiploid
1.15–1.9	Hyperdiploid, hypotetraploid, aneuploid
1.9–2.1	G2/M population tetraploid, G0/G1 tetraploid aneuploid
>2.1	Hypertetraploid aneuploid
Multiple peaks at 1.0, 2.0, 3.0, 4.0, etc.	Multiploid or polyploid

2.0 range are known as hyperdiploid, hypotetraploid, aneuploid tumors. Nondiploid cell populations featuring a DNA index of the G0/G1 main peak at or near 2.0 must be differentiated from diploid tumors with significantly increased G2/M phases. Some investigators refer to these nondiploid tumors with DNA indices near 2.0 as tetraploid tumors. Aneuploid tumors with DNA indices substantially greater than 2.0 are known as hypertertraploid aneuploid tumors. FIGURE 12.2 illustrates the four most frequent types of DNA histograms.

Aneusomy and Aneuploidy

Aneusomy can best be defined as a single cell with an abnormal number of chromosomes. Loss of a single chromosome is termed monosomy; gain of a single chromosome is termed trisomy. Tumors that feature both gains and losses of whole chromosomes or parts of chromosomes are most frequently poorly differentiated neoplasms. In contrast, aneuploidy is defined as a condition of a population of cells when the average DNA content per cell in the cells comprising the G0/G1 phase is significantly different from the normal diploid content (7.14 pg). Aneusomy can be detected by classic cytogenetics or, more recently, in interphase nuclei of smears and tissue sections by using specific probes for chromosomal centromeres and fluorescence in situ hybridization.[3] Although aneusomy and aneuploidy are often detected in the same tumor, it has been reported that fluorescence in situ hybridization based detection of aneusomy is more sensitive than flow or image cytometric detection of DNA aneuploidy for

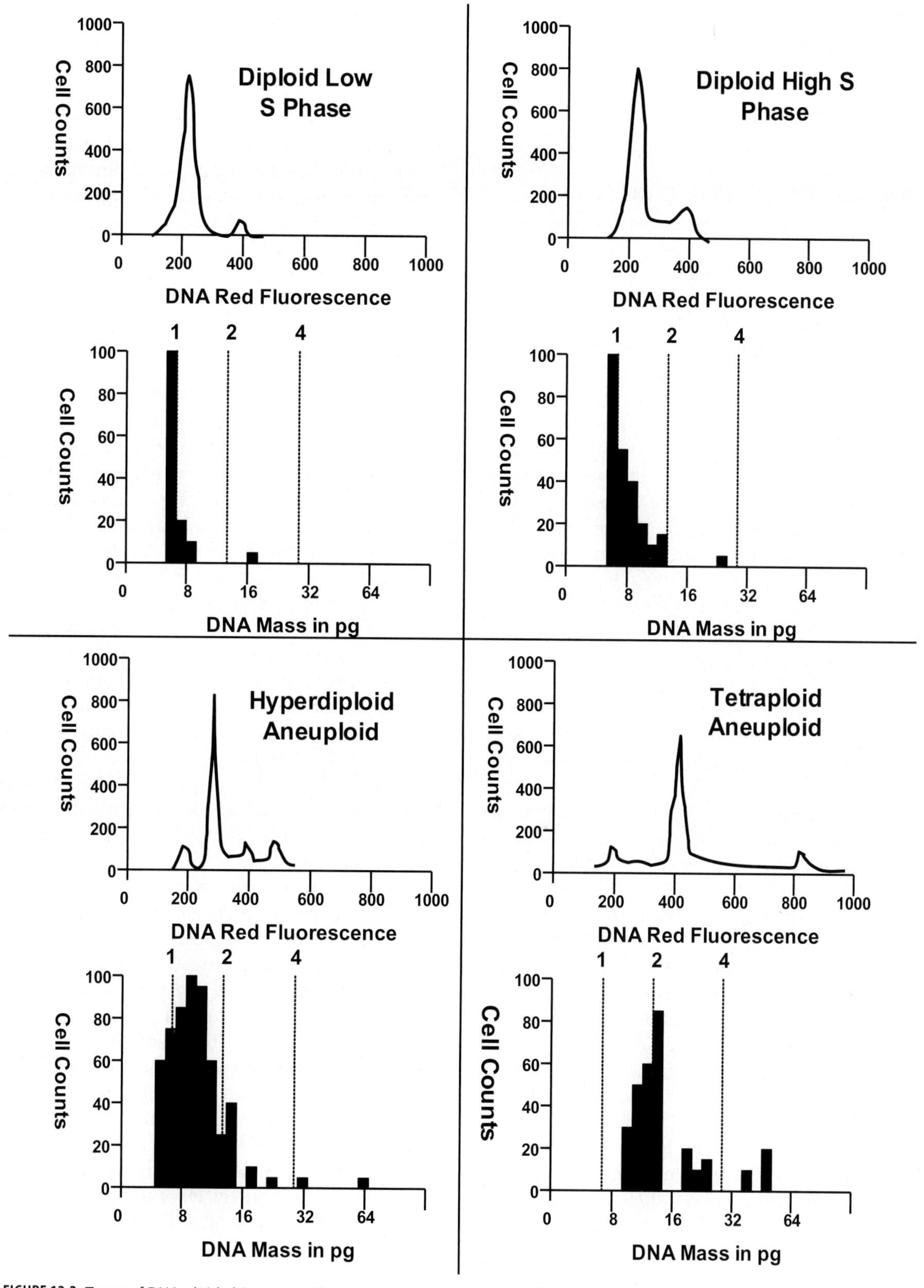

FIGURE 12.2 Types of DNA ploidy histogram. These paired flow cytometry and image analysis histograms show examples of diploid pattern with low S phase (upper left pair), diploid pattern with high S phase (upper right pair), hyperdiploid hypotetraploid aneuploid (lower left pair), and tetraploid aneuploid (lower right pair).

the identification of abnormal DNA content in human tumors.[4,5]

Clinical Testing Methods

Techniques for Measuring DNA Content

Current techniques used to demonstrate DNA content include flow cytometry, image analysis, and laser scanning cytometry.

Flow Cytometry The flow cytometer uses an optical to electronic coupling system and records how a cell interacts with a focused laser beam in terms of scattering of incident light and the cell's ability to emit fluorescence. The photons of light scattered and emitted by a cell after its encounter with the laser beam are separated into various wavelengths by a series of filters and mirrors. Detectors then generate electrical impulses that are converted into digital signals, which are then accumulated in a frequency distribution, or histogram. DNA content is most commonly studied by staining cells with propidium iodide, a DNA binding dye, which can be excited with a standard argon laser. Combined DNA and RNA measurements can be made with the metachromatic fluorochrome acridine orange.[6]

A technique has also been described to simultaneously study DNA and RNA by measuring DNA staining with Hoechst 33342 and RNA staining by Pyronin Y.[7] Flow cytometric DNA ploidy determination features high speed results, allows multiparameter analysis correlating DNA content with antigen expression, and provides the sensitivity for detecting near diploid aneuploid peaks. However, because flow cytometry requires a disaggregation of tissue, there is no simultaneous morphologic comparison.[8] The presence of aneuploid populations may be masked by the inclusion of numerous benign nontumor cells.

Flow cytometry also has the advantage of allowing retrospective studies from paraffin-embedded tissues. Although the best histograms are obtained by fresh or frozen tissues, Hedley and coworkers[9] developed a technique that allowed flow cytometry to be performed on formalin-fixed paraffin-embedded tissue sections. Fixed sections are cut from tissue blocks, dewaxed in xylene, and rehydrated through alcohol solutions. Single cell or nuclear suspensions are obtained after incubation with a 0.5% pepsin/saline solution. Cells are counted, washed, and stained with 4,6-diamidino-2-phenylindole.[10] Modifications of the Hedley technique have allowed good correlations to be obtained from paraffin-embedded specimens with histograms from fresh specimens,[11] although histograms obtained from paraffin-embedded tissues have wider coefficients of variation (CV) and are therefore less precise.[12] Techniques have also been described for performing chromosome analysis by flow cytometry.[13]

Measuring the S Phase by Flow Cytometry S phase measurement is usually obtained through a computer program that estimates the proliferation fraction by using a variety of models to fit the raw data. Those models used to analyze asynchronous populations of specimens often model the G0/G1 and G2/M phases by gaussian distributions, whereas the S phase region is modeled by a polynomial equation, several small gaussians, a trapezoid, or a broadened rectangle.[14–17] A rectangular model is most often used for calculating asynchronous populations with a relatively low S phase.[18] The mathematical models for measuring S phase may be problematic. The intercalating dyes provide static measurements of the percentages of cells having an S phase amount of DNA. This does not indicate which cells are cycling or which are noncycling.

Problems exist in estimating S phase fractions in tumors with overlapping populations.[14] These include cases in which a diploid G2/M overlaps with an aneuploid S phase, S phase overlaps from a near diploid population, or multiple aneuploid populations with close DNA content. The most common problem occurs when a diploid G2/M population overlaps with an aneuploid S phase, as is often seen in breast cancer and in other solid tumors with a DNA index in the range of 1.3 to 1.8.[14] Several modeling systems, such as those described by Dressler and coworkers[19] and McDivitt and colleagues,[20] attempt to estimate S phase in overlapping populations. There can also be difficulty in interpreting the G2/M area of a diploid histogram caused by two simultaneous cells, termed doublets, being interpreted as one cell with tetraploid DNA. Software has become available to allow discrimination of doublets from true G2/M cells.

Image Analysis Computer-based image analysis applies digital technology to quantitative measurements in cytopathology and histopathology. The most widespread uses of this technology are in the quantitation of DNA content and in the quantitation of immunofluorescence and immunohistochemical staining. The availability of image processing instruments allows the pathologist to directly visualize the cell population being studied for DNA measurement, providing simultaneous correlation between DNA studies and cell or tissue morphology. The image analysis system consists of a modified optical microscope, a video camera with digitizer, and a high-speed computer. The image is scanned at intervals known as pixels. At each pixel, a photometric device measures the amount of light absorbed or transmitted. These measurements are recorded by an electronic or mechanical device. A photoelectric cell is used to transform the optical values of light absorbed or transmitted into electronic signals of various amplitude or intensity. The computer assigns a number between 1 and 255 to each pixel in the image, depending on its optical density. Once the image is digitized, a computer reconstruction of the image can be made. DNA content can be measured using Feulgen-stained cells from cytologic preparations or in tissue sections.[21] The dye binds stoichiometrically to the double-stranded DNA molecule.

The computer-based image analysis system uses the summation of the optical density of each Feulgen-stained nucleus to calculate the amount of DNA present based on the Beer-Lambert law. The instrument calculates the DNA index from peaks on the histogram selected by the operator and determines the CV of the G0/G1 peak. This method requires strict control of the staining procedure and rigorous preparation of normal control standards.[22] Because image-based DNA ploidy measurements allow the visual recognition of abnormal cells, this technique facilitates an increased level of sensitivity when low numbers of tumor cells are present and mixed with large numbers of normal cells. Image analysis may therefore be more sensitive in detecting aneuploid tumor populations than flow cytometry when small numbers of tumor cells are present.

In addition, image analysis can allow DNA studies on small specimens, such as biopsies and aspirates, which might not contain enough cells for flow cytometric analysis. However, the method is slow and requires a skilled morphologist. In addition, this method may be insensitive for an aneuploid peak that is near the diploid cutoff. Most comparative studies have shown an excellent correlation between analysis of DNA content by flow cytometry and the computer-based image method.[26] Flow cytometric studies may fail to detect an aneuploid population in 10% of cases in which such a population can be detected by the image method.[23] However, 3% of aneuploid cases detected by flow cytometry, particularly those in the near-diploid region, will not be detected by image analysis.

Quantitative immunohistochemistry, initially designed for the analysis of tumor estrogen and progesterone receptor content, can be performed on the computer-based image analysis system, allowing for an automated assessment of the degree of reactivity of an immunologic marker based on the summation of optical density measurements with a particular chromagen.[24,25] A threshold level for the instrument is set to eliminate nonspecific background staining, and true immunoreactivity can thus be analyzed. The image analysis method can be used to quantitate immunohistochemical measurements of cell cycle regulatory proteins (see Chapter 13).

Laser Scanning Cytometer The laser scanning cytometer has recently been developed to take advantage of the sensitivity of fluorescence-based assays and the specificity of on-slide measurements. The system uses fluorochromes such as propidium iodide applied to cell smears and tissue sections that are analyzed by a high speed computer. The operator selects the cells for study in a fashion similar to an image analysis system or the entire cell population on the slide can be measured automatically. Given its versatility and the ability to study small samples, the laser scanning cytometer shows considerable potential to become a major instrument for the future cytometry laboratory.[26–28]

Major reviews of DNA content analysis highlighted the relative advantages and

Table 12.2 Comparison of Flow and Image Cytometry for the Determination of DNA Ploidy

Method	Advantages	Disadvantages
Flow cytometry	High speed results	There is no simultaneous morphologic comparison of cells during DNA ploidy analysis
	High number of cells counted	Aneuploid peaks of tumor cells low in number may be masked by large numbers of benign cells, stromal cells, and inflammatory cells
	Multiparameter analysis can be performed to subdivide cells for DNA analysis into epithelial and nonepithelial, lymphoid, etc.	Requires disaggregation
	High cell count enhances sensitivity for near diploid aneuploid peaks	
Image analysis	Simultaneous morphologic evaluation of cells during DNA ploidy determination	Relatively slow tedious work and requires skilled morphologist
	Small samples may be used, such as biopsies, small aspirates, curettings, and needle core	May be insensitive for near-diploid aneuploid cell populations
	Does not require disaggregation	Percent CV of G0/G1 peaks may be high for small samples
	Pathologist directed histogram increases sensitivity	

disadvantages of all these technologies.[8,29–31] These comparative studies have highlighted an excellent overall performance of all techniques, with approximately 95% of samples showing similar diploid or aneuploid histograms. A comparison of flow cytometry and image analysis for the determination of DNA ploidy in human neoplasms and the relative advantages and disadvantages of each method[24–25,32] are provided in **Table 12.2**.

Technical Issues

A significant number of technical issues have been considered for their potential impact on the validity and ability to interpret the results of DNA ploidy determinations in breast cancer. In many cases, the contrasting methodologies of flow cytometry and image analysis are similarly affected by these technical issues, whereas in other cases one technique is more affected than the other. In the following section, various important technical issues are considered as to their impact in the preparation and interpretation of a clinically relevant DNA histogram. A list of technical issues that can have an impact on DNA ploidy measurements is shown in **Table 12.3**. In particular, when solid tumors are disaggregated into single cell suspensions for flow or image cytometry,[8,24,29,30] both false-negative and false-positive (aneuploidy) results may be obtained (**Table 12.4**).

Table 12.3 Technical Issues That Have an Impact on DNA Ploidy Analysis

Technical Issues
Specimen volume
Specimen storage
Fixatives
Disaggregation
Stains
Tissue section techniques
Diploid controls and standards
Operator expertise
Quality assurance programs
Tumor heterogeneity

Table 12.4 Benign Neoplasms and Non-Neoplastic Conditions Associated with Aneuploid DNA Content

Conditions
Leiomyoma
Colonic adenoma
Melanocytic nevi
Thyroid adenomas
Pituitary adenomas
Parathyroid adenomas
Adrenal cortical adenomas
Schwannoma
Fibromatosis
Foreign body granulomas
Normal seminal vessicle
Benign tumors with high lymphocyte content

Comparisons of flow and image cytometry in the determination of DNA ploidy status in breast cancer are also shown in FIGURE 12.3.

Specimen Volume The volume of specimens greatly influences the choice of technique and potential statistical significance of DNA ploidy results. Small tissue biopsies may often lack a sufficient number of tumor cells for a flow cytometric determination that usually requires 200,000 events for statistical significance and desirable low CV of the G0/G1 diploid peak. Image analysis DNA ploidy determination can often be done on small specimens by touch preparation, tissue smear, and tissue section. For fine-needle aspiration biopsies, cellularity may be insufficient for flow cytometric DNA ploidy determination and is usually sufficient for smear preparation followed by image analysis. Major issues occur when a small biopsy specimen must be used both for surgical pathologic diagnosis and DNA analysis, as in the needle core biopsy of breast cancer. Insufficient tissue is usually available for disaggregation and preparation of a suspension for flow cytometric DNA determination. Tissue section image analysis, however, can be done on deeper levels of the surgical block.

Specimen Storage Although not a significant issue for prospective and fresh tissue DNA ploidy determination by flow cytometry or image analysis, retrospective studies using processed paraffin blocks may be affected by the nature of specimen storage. Fresh tissues are ideally prepared and analyzed without prolonged storage. For archival specimens, the original temperature of the embedding medium, the original fixative, the length of storage of the paraffin blocks, and the ambient conditions of the paraffin block storage area can all affect the accuracy, reliability, and reproducibility of DNA ploidy determination.

Fixatives Although fresh tissue aspirates and touch preparations are clearly the preferred specimens for DNA ploidy analysis, many retrospective studies and current research studies are based on fixed and processed tissues. Several recent studies have considered the effects of various fixatives on DNA ploidy analysis by flow cytometry[33,34] and image analysis.[35] The lowest percentage of CV and most reproducible histograms by either technique is obtained using 10% (v/v) neutral buffered formalin. Zenker's and B-5 fixatives produce poor percentages of CVs and create substantial artifacts when used for flow cytometry; Bouins and Omni fixatives produce higher percentages of CVs with shifting of the location of G0/G1 peaks. Thus, neutral buffered formalin is an excellent fixative that will cross-link and stabilize DNA for short and intermediate storage of freshly obtained specimens for DNA ploidy before analysis by either flow cytometry or image analysis.

Preparation of Fresh Tissue and Disaggregation The major source of data concerning DNA ploidy in human tumors over the past decade has used archival, paraffin-embedded, stored tissues that have been disaggregated before analysis. The original methods of Hedley et al.[9] established the standard method for disaggregating stored paraffin blocks for flow cytometric DNA ploidy determination as well as for smear preparations of disaggregated tumor cells for image analysis. A significant focus on the disaggregation technique has led to refinements that have greatly improved the accuracy, reliability, and comparability of retrospective DNA ploidy determinations performed in different laboratories. The use of nonenzymatic disaggregation methods has also improved the results of retrospective DNA ploidy studies performed on fixed tissues.

Stains Although propidium iodide, ethidium bromide, and 4,6-diamidino-2-phenylindole are all capable of producing flow cytometric DNA ploidy histograms, propidium iodide is the predominant method of staining nuclei for flow cytometry DNA analysis.[8–10] RNAase pretreatment is necessary to prevent binding of these fluorochromes to RNA. Tissue preparation must be meticulous in that increased propidium iodide binding to damaged or necrotic nuclei can occur and interfere with histogram interpretation. The Feulgen reaction is the standardized staining method for preparation of tissues for image analysis calculation of total nuclear DNA content. Feulgen dye binds to DNA in a stoichiometric manner and does not saturate in optical density as DNA content increases. Although hematoxylin can be used to determine DNA content by image analysis, DNA binding of hematoxylin becomes saturated at increased nuclear DNA contents and does not increase in a linear fashion.

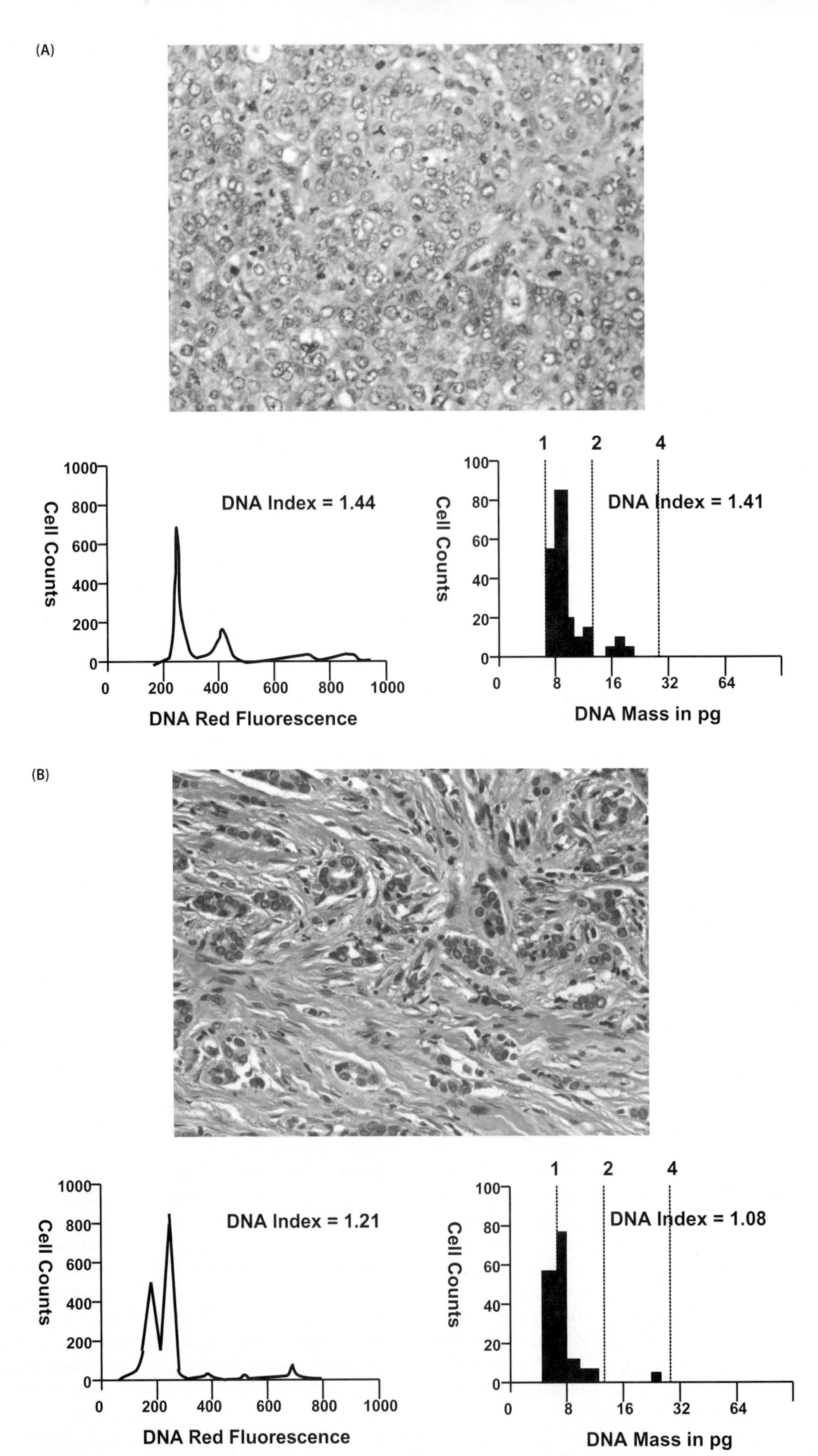

FIGURE 12.3 *(continued)*

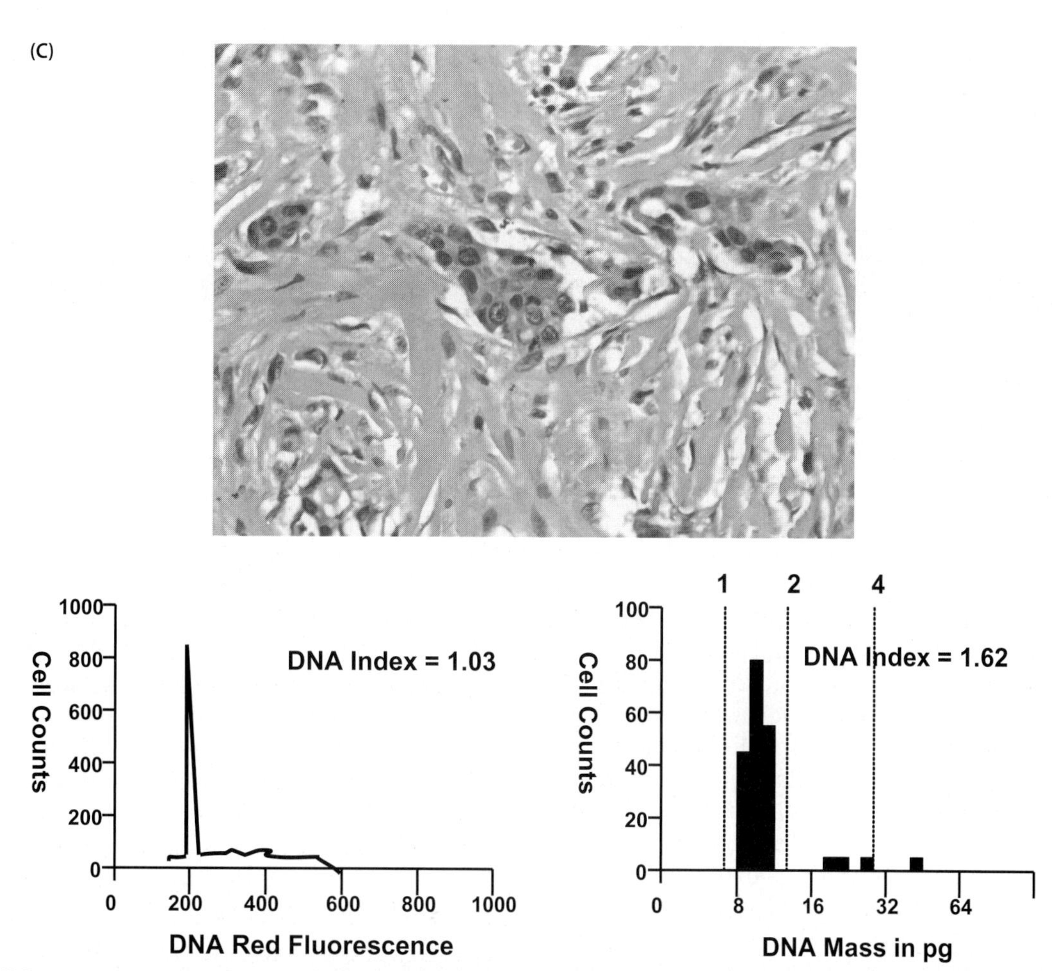

FIGURE 12.3 (A) This specimen features a poorly differentiated infiltrating ductal breast adenocarcinoma in which the image analysis histogram (lower right) and flow cytometric histogram (lower left) are essentially identical. Both instruments indicate a hyperdiploid, hypotetraploid, aneuploid tumor. In addition, the flow cytometric histogram shows diploid stromal and inflammatory cells with separate G0/G1 and G2/M peaks at fluorescence channels 200 and 400, respectively. The aneuploid tumor cell G0/G1 population peak is at the 300 fluorescence channel (DNA index 1.44) and its corresponding G2/M is at 600. The image analysis histogram shows only the tumor cell population, with DNA index of the aneuploid peak similarly at 1.41 (hematoxylin and eosin, ×50). (B) In this infiltrating ductal breast adenocarcinoma, the image analysis histogram (lower left) is diploid with a DNA index of 1.08, whereas the flow cytometric histogram (lower right) features a split G0/G1 peak; a near-diploid aneuploid population is identified with a DNA index of 1.21. This tumor illustrates the increased sensitivity of flow cytometry for near diploid aneuploidy, as the instrument can separate the diploid stromal, benign epithelial, and inflammatory cell populations from the near diploid aneuploid tumor cells (hematoxylin and eosin, ×50). (C) This breast cancer reveals an aneuploid histogram when evaluated by image analysis of tumor cells only with a DNA index of 1.62. The cell suspension obtained by fresh fine-needle aspiration of the resected tumor was not examined by routine microscopy before analysis. On flow cytometric analysis, a diploid pattern was obtained with DNA index of 1.03 representing the predominant benign epithelial cell, inflammatory cell, and stromal cell components. A second aspirate from this specimen was examined microscopically and contained predominantly stromal and inflammatory cells with apparently insufficient numbers of tumor cells for resolution of an aneuploid peak. The relative insensitivity of flow cytometry for identification of aneuploid cell populations when tumor cell concentration in the suspension is low can be improved by microscope study of the suspension before evaluation and by two-color labeling using an epithelial marker antibody to selectively gate for epithelial cells and correspondingly enhance the sensitivity for aneuploidy (hematoxylin and eosin, ×50).

Tissue Section DNA Ploidy The major advantages of tissue section based DNA ploidy determination include the direct simultaneous morphologic analysis of tumor tissue during the histogram preparation, the retrospective nature that allows for rapid accumulation of follow-up data, and the ability to determine DNA content in small needle biopsies, curettings, and cell blocks. The technique, however, is time consuming, tedious, and produces histograms with substantially increased CVs of the G0/G1 peak. Five-micron tissue sections, although creating a significant improvement in cell separation for histogram production, introduce potential errors of nuclear sectioning. Most tissue section methods include the production of a histogram of diploid internal control cells ideally representing benign tissues adjacent to the malignant tissue of similar type. Alternatively, inflammatory, vascular, or stromal cells can be used to create the internal diploid control population to which the DNA content of the malignant cell population can be compared. This comparison assumes that the distribution of sections of both malignant and benign cells will represent the optical density of the intact nuclei of both populations uniformly. Despite these technical considerations, tissue section DNA ploidy analysis has proved to be generally reliable and results in histograms comparable with that obtained from the same tissue block disaggregated and analyzed by flow cytometry or image analysis.

Diploid Controls and Standards A fundamental issue in DNA ploidy determination by flow cytometry and image analysis has been the appropriate use of standards and control cells to which the unknown specimen is compared. In flow cytometry, both human benign cells such as lymphocytes and chicken or trout erythrocytes have been used. The choice of normal tissue as the diploid standard markedly influences the ploidy level when tumors such as breast carcinomas are evaluated by image analysis.[36] It has been recommended that the ideal image analysis histogram can be obtained when the values of DNA content for the reference diploid tissue does not exceed by 15% those of the first G0/G1 peak of the unknown case to be analyzed.[36]

Operator Expertise Regardless of the method used, the expertise of the operator is of paramount importance in the generation of DNA histograms from human tumors. Significant interobserver variability has been reported, for example, in the selection of cells for DNA image cytometry.[37] The sensitive instruments require daily maintenance and standardization. Direct observation as to variations in position of the G0/G1 peak, the percentage of CV, presence of artifacts, and extraneous peaks all must be carefully monitored by the instrument operator. For image analysis, the operator should have a strong background in cytology to allow for the advantages of simultaneous morphologic assessment and DNA analysis that this method can achieve.

Histogram Interpretation Studies have stressed the importance of careful histogram interpretation for image analysis and flow cytometric determination of DNA content in human neoplasms.[7–12] Although separate cell cycle analysis from histograms generated by a flow cytometry or image analysis have included semiobjective mathematic models to partition the cell population into the various cell cycle phases, the identification of DNA aneuploidy has generally been a subjective conclusion determined by visual inspection of the generated histogram. A series of major issues associated with accurate DNA ploidy histogram interpretation is shown in **Table 12.5**. It should be recognized that all tumors and their resulting histograms will not be clearly diploid or aneuploid. A small percentage of histograms will have G0/G1 peaks with DNA index in the 1.12 to 1.19 range that will be "near diploid." It may be best to consider these neoplasms indeterminate for aneuploidy.

Table 12.5 Major Issues in DNA Ploidy Histogram Determination
Exact position of G0/G1 peak
Percent CV of G0/G1 peaks when DNA index in near diploid position
Definition of aneuploidy for system used (the normal diploid D.I. range)
Identifying near diploid aneuploid cell populations
Distinguishing aneuploid cells with DNA indices near 2.0 from the tetraploid component of the diploid cell population (the near tetraploid aneuploidy)

Quality Assurance in DNA Ploidy Determination Given the numerous potential technical issues in histogram generation and interpretation, studies concerned with the overall quality assurance issues in DNA analysis have appeared.[38–40] Considerations of instrument variability, specimen processing variability, staining and histogram generation, histogram interpretation, and clinical application of the results have only recently been subjected to review for potential standardization. Interlaboratory networks for flow cytometry have been developed, and check samples for laboratory cell self-assessment have been issued. Although originally oriented toward flow cytometric DNA analysis of transitional cells from bladder urine, expansion of programs, including proficiency testing to other tumor types and inclusion of image analysis laboratories, has now begun.

Tumor Heterogeneity Tumor cell heterogeneity, the phenomenon of varying diploid and aneuploid stem cell populations in different sites within the same tumor, has been described in a variety of primary neoplasms, including breast carcinoma.[41] Thus, the presence of a diploid histogram obtained from a small sample of a large neoplasm must be interpreted with care, recognizing that multiple samples for either flow cytometry or image analysis can increase the yield of aneuploid results. Multiple in vitro fine-needle aspirations from fresh tumors to obtain cells for flow cytometry from various sites within the resected tumor and multiple touch preparations covering a large surface area for image analysis may decrease the potential impact of tumor heterogeneity on the determination of aneuploidy. Finally, tumor heterogeneity can also be seen in cases of discordance between DNA ploidy pattern of a primary neoplasm and from secondary metastatic deposits. Both diploid primary tumors with aneuploid metastases and aneuploid primary tumors with diploid metastases have been well described.

Flow Versus Image Cytometry

The comparison of techniques of flow cytometric and image analytic determination of human tumoral DNA content, including the relative advantages and disadvantages of both techniques, is summarized in **Table 12.2**. Comparative studies have highlighted an excellent overall comparison of the two methodologies with approximately 95% of samples showing similar diploid or aneuploid histograms by image analysis and flow cytometry.[42] It has been demonstrated that discordance between image analysis and flow cytometry is probably not due to cell loss during flow cytometry cell processing or sampling area differences, but may be due to differences in assessing DNA ploidy in the interpretation of image analysis histograms and/or dilution of aneuploid cells by normal diploid cells in flow cytometry.[43] Although flow cytometry has been the more predominant method of use in previous retrospective studies, current published reports appear evenly divided between image analysis and flow cytometry. Hospitals and laboratories providing DNA ploidy analysis as a routine clinical service also appear approximately equally divided between the flow cytometry and image analysis technologies.

Clinical Relevance and Current Clinical Testing Utility

DNA Ploidy in Breast Cancer

Based on current incidence rates, it is estimated that an American woman has a one in nine chance of developing breast cancer at some time during her life.[1] As would be expected for such a major disease, there is great interest in the discovery, development, and clinical testing of biomarkers and prognostic factors that can be used to guide the selection of therapy for the disease.[44–49]

A large number of prognosis-related studies using flow cytometry and static cell image analysis have been performed on breast cancer specimens to measure total DNA content and determine the S phase fraction.[50–83] The first major retrospective flow cytometric study of DNA content as a predictor of prognosis in breast cancer by Auer et al.[84] divided histograms into four groups ranging from pure diploid to pure aneuploid, the former with an excellent prognosis and the latter with a high relapse and mortality rate. The results of a variety of studies on the prognostic significance of ploidy and S phase status have been notably variable, with some investigators finding significant prediction of disease-free and overall survival on both univariate and multivariate analysis and others finding no impact on disease outcome (**Table 12.6**).[50–83] Of the 34 selected studies highlighted in **Table 12.6**, 16 (47%) found that aneuploid DNA content in primary breast cancer was a significant adverse prognostic factor on multivariate analysis, 6 studies (18%) found ploidy status predictive on univariate analysis only, and 12 studies (35%) found no prognostic significance.

Despite numerous large studies indicating that DNA ploidy status is an independent and powerful prognostic indicator in breast cancer,[50,51,53,55,58,61,66,71,72,74,76–79,83] many investigators have reported that the association of both ploidy and S phase fraction status with prognosis is variable and not conclusive.[52,59,62–65,68,70,73,75,80] Thus, the clinical use of DNA content measurement in breast cancer patient management has been controversial. Other investigations and reviews have questioned the independent prognostic value of DNA ploidy in breast carcinoma.[74,85,86] In large retrospective studies using disaggregation of paraffin blocks and flow cytometry, aneuploidy is identified in approximately 50% of breast adenocarcinomas.[60,63] DNA analysis has been separately correlated with prognosis in both node-negative and node-positive breast cancer patients.[52,60,74] In one report of node-negative patients, multivariate analysis showed that neither the DNA index nor S phase fraction calculation were of predictive value alone but achieved independent status as a predictor of survival when used

Table 12.6 DNA Ploidy and S Phase Measurements in Breast Cancer

Year	Author	Reference No.	Specimen Type	Method	Disease Outcome Statistical Correlation: Univariate Significance	Disease Outcome Statistical Correlation: Multivariate Independent Significance	Comment
1988	Kallionemi	50	Paraffin	FCR, D	Yes	Yes	Ploidy combined with S phase
1988	Fallenius	51	FNA	FCR	Yes	Yes	Average histogram classification; no correlation with node status
1989	Muss	52	Frozen	FCR, D	No	No	DNA index not associated with time to progression or survival
1989	Clark	53	Frozen	FCR, D	Yes	Yes	S phase did not correlate
1989	Christov	54	Fresh	FCR, D	No	No	Correlation with grade
1990	Lewis	55	Paraffin	FC, R, D	Yes	Yes	Only ploidy predicts survival
1990	Toikkanen	56	Paraffin	FCR, D	Yes	No	55% aneuploid rate
1990	Kute	57	Paraffin	FCR, D	Yes	No	Ploidy discriminates estrogen receptor–negative patients
1990	Joensuu	58	Paraffin	FCR, D	Yes	Yes	Independent when combined with S phase
1991	Fisher	59	Paraffin	FCR, D	No	No	1665 cases; S phase correlated
1991	Witzig	60	Paraffin	FCR, D	Yes	No	Separate analysis of node negative and positive
1991	Sharma	61	Paraffin	FCR, D	Yes	Yes	Identify patients for adjuvant treatment
1992	Arnerlov	62	Paraffin	FCR, D	No	No	Ploidy not a significant predictor
1992	Bosari	63	Paraffin	FCR, D	No	No	All multiploid and hypertetraploid patients had recurrence
1994	Leivonen	64	Paraffin	FCR, D	No	No	Study of patients with metastatic or recurrent diseases
1994	Witzig	65	Paraffin	FCR, D	No	No	S phase remained significant
1995	Camplejohn	66	Paraffin	FCR, D	Yes	Yes	Both ploidy and S phase status were independent predictors of outcome
1995	Dieterich	67	Fresh	IAP, D	Yes	No	DNA ploidy status had prognostic significance but was not independent of tumor grade
1995	Pfisterer	68	Fresh	FCR, D	No	No	Neither ploidy or S phase fraction was a significant prognostic factor
1996	Romero	69	Paraffin	FCR, D	Yes	Yes	S phase fraction was an independent predictor in DNA diploid tumors
1997	Bergers	70	Fresh	FCR, D	No	No	Ploidy had no prognostic significance in a study of 1301 patients
1999	Shiao	71	Paraffin	FCR, D	Yes	Yes	Ploidy was independent predictor of survival in white but not African American women
1999	Midulla	72	Paraffin	FCR, D	Yes	Yes	Ploidy was an independent predictor of survival
1999	Wong	73	Paraffin	FCR, D	No	No	Ploidy and S phase did not predict survival
1999	Pinto	74	Fresh	FCR, D	Yes	Yes	Ploidy independently predicted disease-free survival; S phase predicted overall survival

Table 12.6 DNA Ploidy and S Phase Measurements in Breast Cancer (*continued*)

					Disease Outcome Statistical Correlation		
Year	Author	Reference No.	Specimen Type	Method	Univariate Significance	Multivariate Independent Significance	Comment
2000	Mandard	75	Paraffin	FC, IAR, D	No	No	Mitotic index outperformed S phase calculation; ploidy not significant
2001	Bracko	76	Fresh	FCR, D	Yes	Yes	Ploidy independently predicted disease-free and cancer-specific survival
2001	Pinto	77	Fresh	FCP, D	Yes	Yes	S phase outperformed Ki-67 labeling index
2001	Bagwell	78	Paraffin	FCR, D	Yes	Yes	Using 10 adjustments to reclassify histograms, both ploidy and S phase were independent predictors in node-negative patients
2001	Tsutsui	79	Fresh	FC	Yes	Yes	Ploidy was an independent predictor in node-negative patients
2001	Chessevent	80	Paraffin	FCR, D	No	No	Ploidy not significant; S phase was independent predictor in node-negative patients
2002	Chavez-Uribe	81	Paraffin	FC	Yes	No	Ploidy status a significant predictor
2003	Chang	82	Paraffin	FC	Yes	No	On 346 patients, ploidy status was significant on univariate analysis, whereas S phase fraction was significant on both univariate and multivariate analyses
2003	Tsutsui	83	Paraffin	FC	Yes	Yes	DNA ploidy independent predictor using 998 patients

D, disaggregated specimen; FC, flow cytometry; IA, image analysis; P, prospective; R, retrospective.

as a combination variable.[52] In another study of node-positive breast cancer patients, DNA ploidy was a significant prognostic factor on univariate analysis but failed to achieve prognostic significance in multivariate analysis.[60]

A potential clinical role for DNA ploidy analysis in breast cancer has been as a potential factor to aid in the selection of patients with lymph node–negative disease for adjuvant chemotherapy. In a large study including 1665 patients from the National Surgical Adjuvant Breast and Bowel Project, analysis of the outcome data did not confirm the ability of DNA index and percent S phase fraction to independently predict overall survival and discouraged the use of DNA content measurement for the selection of patient-specific therapies.[59] Although a promising study, the National Surgical Adjuvant Breast and Bowel Project concluded that DNA analysis could not detect patients with such a good prognosis as to preclude their receiving adjuvant chemotherapy. A study from India, however, concluded that ploidy status was an independent prognostic factor predicting for recurrence and could be used to identify a subset of node-negative breast cancer patients who could be selected to bypass adjuvant treatment.[61] Interestingly, in patients with recurrent disease studied by flow cytometry who had received either hormonal and or cytotoxic therapy, DNA analysis was unsuccessful in predicting future clinical course, which suggested that treatment could conceivably cause divergent results and diminish the ability of DNA ploidy status to predict subsequent outcome in treated patients.[61] Finally, for lobular breast carcinoma, ploidy and cell cycle analysis have not successfully predicted disease outcome.[87]

Summary and Future Clinical Potential

Although both image analysis and flow cytometry have been used to determine breast carcinoma DNA content, some investigators believe that image analysis may have advantages, including its lower cost, the ability to analyze small specimens, and the capability of detecting rare aneuploid cell populations and characterize them according to a specific morphologic appearance without the potential impact of enzymatic or mechanical tissue disaggregation.[88,89] DNA ploidy and S phase fraction status have been generally correlated with high tumor grade, negative hormone receptor status, and amplification of certain molecular markers, indicating unfavorable prognosis such as the HER-2/*neu* and c-*myc* genes.[90] Although currently of uncertain use as a stand-alone prognostic marker, at many hospitals and cancer centers ploidy status is often included as a component of a panel of prognostic markers for patients with breast cancer. In general, the S phase calculation by flow cytometry has generally outperformed the DNA ploidy status as a prognostic factor in breast cancer. However, neither the American Society of Clinical Oncologists[91] nor the College of American Pathologists[92] includes ploidy and S phase measurements in their respective lists of recommended prognostic factors to be used for breast cancer patient management.

References

1. Gillett CE, Barnes DM. Cell cycle. J Clin Pathol 1998;51:310–316.
2. Ross JS. DNA ploidy and cell cycle analysis in pathology. New York: Igaku-Shoin Medical Publishers, 1996:40–42.
3. Wolman SR. Fluorescence in situ hybridization: a new tool for the pathologist. Hum Pathol 1994;25:586–590.
4. Waters JJ, Barlow AL, Gould CP. FISH. J Clin Pathol Mol Pathol 1998;51:62–70.
5. Lipson B, Pizzolo JG, Elhosseiny AA, et al. Comparison of fluorescence in situ hybridization and flow cytometric DNA ploidy analysis in paraffin-embedded prostatic adenocarcinoma specimens. Anal Quant Histol Cytol 1995;17:93–99.
6. Traganos F, Darzynkiewicz Z, Sharpless T, et al. Simultaneous staining of ribonucleic and deoxyribonucleic acids in unfixed cells using acridine orange in a flow cytofluorometric system. J Histochem Cytochem 1977;25:46–56.
7. Ross JS. DNA ploidy and cell cycle analysis in cancer diagnosis and prognosis assessment. Oncology 1996;10:867–887.
8. Koss LG, Czerniak B, Herz F, et al. Flow cytometric measurements of DNA and other cell components in human tumors: a critical appraisal. Hum Pathol 1989;20:528–548.
9. Hedley DW, Friedlander ML, Taylor IW, et al. Method for analysis of cellular DNA content in paraffin-embedded pathological material using flow cytometry. J Histochem Cytochem 1983;31:1333–1335.
10. Shapiro HM. Flow cytometry of DNA content and other indicators of proliferative activity. Arch Pathol Lab Med 1989;113:591–597.
11. Krause JR, Blank MK. DNA content in fresh versus paraffin-embedded tissue. Anal Quant Cytol Histol 1992;14:89–95.
12. Wersto RP, Libit RL, Koss LG. Flow cytometric DNA analysis of human solid tumors: a review of the interpretation of DNA histograms. Hum Pathol 1991;22:1085–1098.
13. McConnell TS, Cram LS, Baczek N, et al. The clinical usefulness of chromosome analysis by flow cytometry. Semin Diagn Pathol 1989;6:91–107.
14. Dressler LG, Bartow SA. DNA flow cytometry in solid tumors: practical aspects and clinical applications. Semin Diagn Pathol 1989;6:55–82.
15. Baisch H, Beck H-P, Christensen JJ, et al. A comparison of mathematical models for the analysis of DNA histograms by flow cytometry. Cell Tissue Kinet 1992;15:235–249.
16. Dean PN, Jett JH. Mathematical analysis of DNA distributions derived from flow microfluorometry. J Cell Biol 1980;60:525–527.
17. Jett JH, Gurley LR. An improved sum normals technique for cell cycle distribution analysis of flow cytometric DNA histograms. Cell Tissue Kinet 1981;14:413–423.
18. Rich PS, Shackney SE, Schuette WH, et al. A practical graphical method for estimating the fraction of cells in S phase in DNA histograms from clinical tumor samples containing aneuploid cell populations. Cytometry 1983;5:66–74.
19. Dressler LG, Seamer L, Owens MA, et al. S phase estimation in breast cancer by flow cytometry: evaluation of a modeling system. Cancer Res 1987;47:5294–5302.
20. McDivitt RW, Stone KR, Craig RB, et al. A comparison of human breast cell kinetics measured by flow cytometry and thymidine labeling. Lab Invest 1985;52:287–291.
21. Suit PF, Bauer TW. DNA quantitation by image cytometry of touch preparations from fresh and frozen tissue. Am J Clin Pathol 1990;94:49–53.
22. Weid GL, Bartels PH, Bibbo M, et al. Image analysis in quantitative cytopathology and histopathology. Hum Pathol 1989;20:549–570.
23. Claud DR, Weinstein RS, Howeedy A, Straus AK, Coon JS. Comparison of image analysis of imprints with flow cytometry for DNA analysis of solid tumors. Mod Pathol 1989;2:463–466.
24. Bauer TW, Tubbs RR, Edinger MG, et al. A prospective comparison of DNA quantitation by image analysis and flow cytometry. Am J Clin Pathol 1990;93:322–326.
25. Bacus S, Flowers JL, Press MF, et al. The evaluation of estrogen receptor in primary breast carcinomas by computer-assisted image analysis. Am J Clin Pathol 1988;90:233–239.
26. Linder J. Overview of digital imaging in pathology. Pathol Patterns 1990;94(suppl 1):S30–S34.
27. Martin-Reay DG, Kamentsky LA, Weinberg DS. Evaluation of a new slide-based laser scanning cytometer for DNA analysis of tumors. Am J Clin Pathol 1994;102:432–438.
28. Clatch RJ, Walloch JL, Foreman JR, et al. Multiparameter analysis of DNA content and cytokeratin expression in breast carcinoma by laser

scanning cytometry. Arch Pathol Lab Med 1997; 121:585–592.

29. Coon JS, Landay AL, Weinstein RS. Biology of disease: advances in flow cytometry for diagnostic pathology. Lab Invest 1987;57:453–479.
30. Lovett III EJ, Schnitzer B, Keren DF, et al. Application of flow cytometry to diagnostic pathology. Lab Invest 1984;50:115–140.
31. Herman CJ. Cytometric DNA analysis in the management of cancer. Cancer 1992;69:1553–1556.
32. Faranda A, Costa A, Canova S, et al. Image and flow cytometric analyses of DNA content in human solid tumors. A comparative study. Analyt Quant Cytol Histol 1997;19:338–344.
33. Esteban JM, Sheibani K, Owens M, et al. Effects of various fixatives and fixation conditions on DNA ploidy analysis. Am J Clin Pathol 1991;95: 460–466.
34. Herbert DJ, Nishyama RH, Bagwell CB, et al. Effects of several commonly used fixatives on DNA and total nuclear protein analysis by flow cytometry. Am J Clin Pathol 1989;91:535–541.
35. Becker RL, Mikel UV. Interelation of formalin fixation, chromatin compactness and DNA values as measured by flow and image cytometry. Anal Quant Cytol Histol 1990;12:333–341.
36. Kiss R, Gasperin P, Verhest A, et al. Modification of tumor ploidy level via the choice of tissue taken as diploid reference in the digital cell image analysis of Feulgen-stained nuclei. Mod Pathol 1992;5:655–660.
37. Carey FA, Gray E, Salto-Tellez M, et al. Interobserver variation in cell selection for DNA image cytometry. J Clin Pathol 1995;48:616–619.
38. McCarthy RC, Fetterhoff TJ. Issues for quality assurance in clinical flow cytometry. Arch Pathol Lab Med 1989;113:658–666.
39. Bocking A, Chatelain R, Homge M, et al. Representativity and reproducibility of DNA malignancy grading in different carcinomas. Anal Quant Cytol Histol 1989;11:81–86.
40. Coon JS, Deitsch AD, White RW, et al. Check samples for laboratory self assessment in DNA flow cytometry. Cancer 1989;63:1592–1599.
41. Fuhr JE, Frye A, Cattine AA, et al. Flow cytometric determination of breast tumor heterogeneity. Cancer 1991;67:1401–1405.
42. Danque POV, Chen HB, Patil J, et al. Image analysis versus flow cytometry for DNA ploidy quantitation of solid tumors: a comparison of six methods of sample preparation. Mod Pathol 1993;6:270–275.
43. Elsheikh TM, Silverman JF, McCool JW, et al. Comparative DNA analysis of solid tumors by flow cytometric and image analyses of touch imprints and flow cell suspensions. Am J Clin Pathol 1992;98:296–304.
44. Lacey JV Jr, Devesa SS, Brinton LA. Recent trends in breast cancer incidence and mortality. Environ Mol Mutag 2002;39:82–88.
45. Ince TA, Weinberg RA. Functional genomics and the breast cancer problem. Cancer Cell 2002;1:15–17.
46. Baselga J, Norton L. Focus on breast cancer. Cancer Cell 2002;1:319–322.
47. Fukutomi T, Akashi-Tanaka S. Prognostic and predictive factors in the adjuvant treatment of breast cancer. Breast Cancer 2002;9:95–99.
48. Esteva FJ, Sahin AA, Cristofanilli M, et al. Molecular prognostic factors for breast cancer metastasis and survival. Semin Radiat Oncol 2002;12:319–328.
49. Morabito A, Magnani E, Gion M, et al. Prognostic and predictive indicators in operable breast cancer. Clin Breast Cancer 2003;3:381–390.
50. Kallioniemi O, Blanco G, Alavaikko M, et al. Improving the prognostic value of DNA flow cytometry in breast cancer by combining DNA index and S-phase fraction. Cancer 1988;62:2183–2190.
51. Fallenius AG, Franzen SA, Auer GU. Predictive value of nuclear DNA content in breast cancer in relation to clinical and morphologic factors. Cancer 1988;62:521–530.
52. Muss HB, Kute TE, Case LD, et al. The relation of flow cytometry to clinical and biologic characteristics in women with node negative primary breast cancer. Cancer 1989;64:1894–1900.
53. Clark GM, Dressler LG, Owens MA, et al. Prediction of relapse or survival in patients with node negative breast cancer by DNA flow cytometry. N Engl J Med 1989;320:627–633.
54. Christov K, Milev A, Todorov V. DNA aneuploidy and cell proliferation in breast tumors. Cancer 1989;64: 673–679.
55. Lewis WE. Prognostic significance of flow cytometric DNA analysis in node-negative breast cancer patients. Cancer 1990;65:2315–2320.
56. Toikkanen S, Joensuu H, Klemi P. Nuclear DNA content as a prognostic factor in T1-2N0 breast cancer. Am J Clin Pathol 1990;93:471–479.
57. Kute TE, Muss HB, Cooper MR, et al. The use of flow cytometry for the prognosis of stage II adjuvant treated breast cancer patients. Cancer 1990;66: 1810–1816.
58. Joensuu H, Toikkanen S, Klemi PJ. DNA index and S-phases fraction and their combination as prognostic factors in operable ductal breast carcinoma. Cancer 1990;66:331–340.
59. Fisher B, Gunduz N, Constantino J, et al. DNA flow cytometric analysis of primary operable breast cancer. Cancer 1991;68:1465–1475.
60. Witzig TE, Gonchoroff NJ, Therneau T, et al. DNA content flow cytometry as a prognostic factor for node-positive breast cancer. Cancer 1991;68:1781–1788.
61. Sharma S, Mishra MC, Kapur ML, et al. The prognostic significance of ploidy analysis in operable breast cancer. Cancer 1991;68:2612–2616.
62. Arnerlov C, Emdin SO, Lundgren B, et al. Mammographic growth rate, DNA ploidy, and S-phase fraction analysis in breast carcinoma. Cancer 1992;70: 1935–1942.
63. Bosari S, Lee ARKC, Tahan SR, et al. DNA flow cytometric analysis and prognosis of axillary lymph node-negative breast carcinoma. Cancer 1992;70: 1943–1950.
64. Leivonen M, Krogerus L, Nordling S. DNA analysis in advanced breast cancer patients. Cancer Detect Prevent 1994;18:87–96.
65. Witzig TE, Ingle JN, Cha SS, et al. DNA ploidy and the percentage of cells in S-phase as prognostic factors for women with lymph node negative breast cancer. Cancer 1994;74:1752–1760.
66. Camplejohn RS, Ash CM, Gillett CE, et al. The prognostic significance of DNA flow cytometry in breast cancer: results from 881 patients treated in a single centre. Br J Cancer 1995;71:140–145.
67. Dieterich B, Albe X, Vassilakos P, et al. The prognostic value of DNA ploidy and S-phase estimate in

primary breast cancer: a prospective study. Int J Cancer 1995;63:49–54.

68. Pfisterer J, Kommoss F, Sauerbrei W, et al. DNA flow cytometry in node-positive breast cancer. Prognostic value and correlation with morphologic and clinical factors. Anal Quant Cytol Histol 1995;17: 406–412.

69. Romero H, Schneider J, Burgos J, et al. S-phase fraction identifies high-risk subgroups among DNA-diploid breast cancers. Breast Cancer Res Treat 1996;38:265–275.

70. Bergers E, Baak JP, van Diest PJ, et al. Prognostic value of DNA ploidy using flow cytometry in 1301 breast cancer patients: results of the prospective Multicenter Morphometric Mammary Carcinoma Project. Mod Pathol 1997;10:762–768.

71. Shiao YH, Chen VW, Lehmann HP, et al. Patterns of DNA ploidy and S-phase fraction associated with breast cancer survival in blacks and whites. Clin Cancer Res 1997;3:587–592.

72. Midulla C, Cenci M, De Iorio P, et al. DNA ploidy and TLI in association with other prognostic parameters in breast cancer. Anticancer Res 1999;19:381–384.

73. Wong SW, Rangan AM, Bilous AM, et al. The value of S-phase and DNA ploidy analysis as prognostic markers for node-negative breast cancer in the Australian setting. Pathology 1999;31:90–94.

74. Pinto AE, Andre S, Soares J. Short-term significance of DNA ploidy and cell proliferation in breast carcinoma: a multivariate analysis of prognostic markers in a series of 308 patients. J Clin Pathol 1999;52: 604–611.

75. Mandard AM, Denoux Y, Herlin P, et al. Prognostic value of DNA cytometry in 281 premenopausal patients with lymph node negative breast carcinoma randomized in a control trial: multivariate analysis with Ki-67 index, mitotic count, and microvessel density. Cancer 2000;89:1748–1757.

76. Bracko M, Us-Krasovec M, Cufer T, et al. Prognostic significance of DNA ploidy determined by high-resolution flow cytometry in breast carcinoma. Anal Quant Cytol Histol 2001;23:56–66.

77. Pinto AE, Andre S, Pereira T, et al. Prognostic comparative study of S-phase fraction and Ki-67 index in breast carcinoma. J Clin Pathol 2001;54:543–549.

78. Bagwell CB, Clark GM, Spyratos F, et al. Optimizing flow cytometric DNA ploidy and S-phase fraction as independent prognostic markers for node-negative breast cancer specimens. Cytometry 2001;46: 121–135.

79. Tsutsui S, Ohno S, Murakami S, et al. Prognostic value of DNA ploidy in 653 Japanese women with node-negative breast cancer. Int J Clin Oncol 2001;6: 177–182.

80. Chassevent A, Jourdan ML, Romain S, et al. S-phase fraction and DNA ploidy in 633 T1T2 breast cancers: a standardized flow cytometric study. Clin Cancer Res 2001;7:909–917.

81. Chavez-Uribe EM, Vinuela JE, Cameselle-Teijeiro J, et al. DNA ploidy and cytonuclear area of peritumoral and paratumoral samples of mastectomy specimens: a useful prognostic marker? Eur J Surg 2002;168:37–41.

82. Chang J, Clark GM, Allred DC, et al. Survival of patients with metastatic breast carcinoma: importance of prognostic markers of the primary tumor. Cancer 2003;97:545–553.

83. Tsutsui S, Ohno S, Murakami S, et al. Prognostic significance of the combination of biological parameters in breast cancer. Surg Today 2003;33:151–154.

84. Auer GU, Caspersson TO, Wallgren AS. DNA content and survival in mammary carcinoma. Anal Quant Cytol Histol 1980;2:161–165.

85. Frierson HF. Ploidy analysis and S-phase fraction determination by flow cytometry of invasive adenocarcinomas of the breast. Am J Surg Pathol 1991;15: 358–367.

86. Bergers E, van Diest PJ, Baack JPA. Cell cycle analysis of 932 flow cytometric DNA histograms of fresh frozen breast carcinoma material. Cancer 1996;77: 2258–2266.

87. Frost AR, Karcher DS, Terahata S, et al. DNA analysis and s-phase fraction determination by flow cytometric analysis of infiltrating lobular carcinoma of the breast. Mod Pathol 1996;9:930–937.

88. Azua J, Romeo P, Serrano M et al. Prognostic value from DNA quantification by static cytometry in breast cancer. Ann Quant Cytol Histol 1997;19:80–86.

89. Bosari S, Wiley BD, Hamilton WM, et al. DNA measurement by image analysis of paraffin-embedded breast carcinoma tissue. A comparative investigation. Am J Clin Pathol 1991;96:698–703.

90. Bagwell CB, Clark GM, Spyratos F, et al. DNA and cell cycle analysis as prognostic indicators in breast tumors revisited. Clin Lab Med 2001;21:875–895.

91. Bast RC Jr, Ravdin P, Hayes DF, et al. 2000 Update of recommendations for the use of tumor markers in breast and colorectal cancer: clinical practice guidelines of the American Society of Clinical Oncology. J Clin Oncol 2001;19:1865–1878.

92. Hammond ME, Fitzgibbons PL, Compton CC, et al. College of American Pathologists Conference XXXV: solid tumor prognostic factors—which, how and so what? Summary document and recommendations for implementation. Cancer Committee and Conference Participants. Arch Pathol Lab Med 2000;124: 958–965.

CHAPTER

13

Cell Proliferation Markers and Cell Cycle Regulators in Breast Cancer

Jeffrey S. Ross, MD[1] and Carlos Cordon-Cardo[2]

[1]*Department of Pathology and Laboratory Medicine, Albany Medical College, Albany, New York and Division of Molecular Medicine, Millennium Pharmaceuticals, Inc., Cambridge, Massachusetts*

[2]*Division of Molecular Pathology, Memorial Sloan Kettering Cancer Center, New York, New York*

Biologic Background

The cell cycle (FIG. 13.1) is a complex process that responds to the needs of certain cells and tissues to proliferate and balances the production of new cells (cell division) with the facilitation of the death of existing cells (apoptosis).[1,2] Loss of control or dysregulation of the cell cycle is widely believed to be one of the major steps in the development of cancer.[3,4] As eukaryotic cells enter the cell cycle two processes alternate: (1) the doubling of its DNA content (genome) from 7.14 to 14.28 pg in the S phase (synthesis phase) of the cell cycle and (2) the halving of the 14.28 pg of DNA content during the mitotic (M) phase. The cell rest interval between the M and S phases is called Gap 1 (G1), and the interval between the S and M phases is the Gap 2 (G2). The period of the cell cycle when a cell is in any phase other than mitosis is termed the interphase.

The passage of a cell through the cell cycle is regulated by a series of enzymes and proteins.[1–4] The most important cell cycle regulatory proteins are the cyclins and cyclin-dependent kinases (CDKs). The cyclins can be divided into three groups: G1 cyclins, G1/S and S phase cyclins, and M phase cyclins. As cells progress through the cell cycle, the synthesis and degradation of cyclins mirrors their progress. The CDKs consist of three types: the G1 CDKs, a single CDK that regulates the expression of both the G1/S and S cyclins, and an M phase CDK. The expression levels of the CDKs remain relatively constant throughout the cell cycle. However, the actual activation of an individual CDK requires its binding to a cyclin that then triggers the CDK mediated phosphorylation of various protein substrates that further regulate the cell cycle progression. The preparation of the chromosomes for replication is heralded by rising levels of G1 cyclins bound to their respective CDKs. The cell is then prepared for entry into the S phase by the S phase promoting factor, which contains both G1/S and S cyclins bound to a shared CDK. DNA replication begins and continues through the S phase, the G1/S phase cyclins (cyclin E) are degraded, and the level of M phase cyclins begins to rise as the cell enters G2. At this point, the M phase promoting factor (the complex of M phase cyclins with M phase CDK) initiates the following processes: assembly of the mitotic spindle, breakdown of the nuclear envelope, and condensation of the chromosomes. As the cell enters the metaphase of mitosis, the M phase promoting factor activates the anaphase promoting complex, which facilitates the separation and movement of the sister chromatids at the metaphase plate, creating the anaphase and completion of mitosis. The anaphase promoting complex also degrades the M phase cyclins by ubiquitin conjugation and proteolysis via the proteasome,[5,6] triggers the events that allow the sister chromatids to separate by mediating the destruction of the cohesions, turns on synthesis of G1 cyclins to initiate the next rotation of the cycle, and degrades geminin, a protein that has kept the freshly synthesized DNA in S phase from being re-replicated before mitosis. The cell cycle and

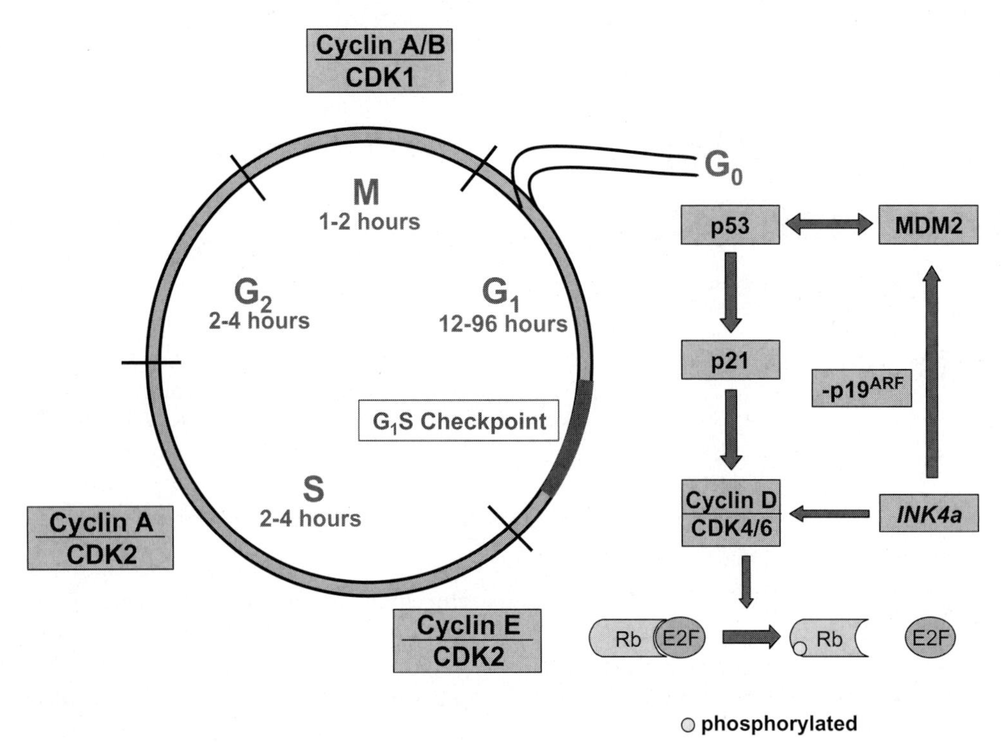

FIGURE 13.1 The cell cycle.

Table 13.1 Techniques for Measurement of Cell Proliferation

Marker	Cell Cycle Compartment Detected
Mitosis counting	M
^{3}H thymidine uptake	S
5-Bromodeoxyuridine labeling	S
Flow cytometry	S (calculated)
Image analysis	S (estimated)
Ki-67 (fresh-frozen tissue)	G_1, S, G_2, M (part)
Ki-67 MIB-1 (paraffin)	G_1, S, G_2, M (part)
PCNA (paraffin)	G_1, S, G_2
p105 (paraffin)	G_1, S, G_2, M
AgNOR (paraffin)	S, G_2, M (part)
Cyclin A	S, G_2, M
Cyclin B	G_2, M (part)
Cyclin D	G_1 (part)
Cyclin E	G_1 (part), S (part)

cell proliferation kinetics (**Table 13.1**) have been extensively studied in breast cancer to further elucidate the disease biology, develop prognostic, and predictive markers of clinical disease and to develop new potential targets of anticancer therapy.[6–10]

Clinical Testing Methods

Mitotic Figure Counting

Although the counting of mitotic figures is the easiest method of assessing cell proliferation, this technique lacks standardization and suffers from lack of interobserver reproducibility.[11] Most often reported as number of mitotic figures per number of high power fields, this method relies on the accurate identification of a mitotic figure. This is a subjective assessment, which can vary with a number of factors, including the thickness of the tissue sections and delay in tissue fixation.[12,13] In addition, many tumors are heterogeneous in that the number of mitotic figures may vary from field to field. Mitotic figure counting is a cornerstone of the most prevalent methods of breast cancer tumor grading (the Nottingham or Bloom-Richardson systems) and is thus performed as part of the standard of practice for all breast cancer specimens.[14]

Tritiated Thymidine Uptake

Other morphologic assessments measure variable components of the cell cycle. Tritiated thymidine (3H-TdR) labeling is a measurement of active S phase proliferating cells.[11,15]

This method requires viable cells. Fresh tissue is cultured with tritiated thymidine for 1 to 2 hours and then processed routinely. A hyperbaric environment and blockade of thymidylate synthetase by 5-fluorouridine or 2′-deoxy-5-fluorouridine potentiates the uptake of thymidine.[16] Autoradiography is performed after 1 week. The thymidine labeling index is reported as the percentage of positive tumor cells per total number of tumor cells counted. This method is also subjective, with the potential for interobserver variation.[17] In addition, it requires radioisotopic labeling, necessitates the use of fresh tissue, and has a prolonged incubation. A study has indicated that tritiated thymidine labeling may be useful in predicting prognosis in gastric cancer.[18]

BUDR Labeling

The thymidine analogue 5-bromodeoxyuridine (BUDR) assay is a monoclonal antibody technique that also measures cells in S phase.[19,20] This method is also potentiated by the use of a hyperbaric environment and by blockade of thymidylate synthetase. However, this technique does not require the use of radioisotopes. Monoclonal antibodies detecting bromodeoxyuridine can be used in immunohistochemical, immunofluorescence, or flow cytometric measurements.[21,22] Studies comparing thymidine and 5-bromodeoxyuridine labeling have noted similar results.[23] Technical factors, including the concentration of bromodeoxyuridine used, can affect the labeling index. Studies have shown that the 5-bromodeoxyuridine labeling index correlates with tumor grade and mitosis counts in a wide variety of tumors but is not used clinically for the diagnosis, prognosis assessment, or management of breast cancer patients.

Ki-67 Immunostaining

Beyond BUDR staining, other antibody techniques using immunohistochemistry (IHC) to measure cell proliferation assess different components of the cell cycle. These techniques do not measure an actual rate of proliferation but are static measurements of cell proliferation associated biomarkers. The antibody Ki-67 was raised against a Hodgkin disease cell line and detects an antigen in the nucleus associated with cell proliferation.[24] The original Ki-67 antigen cannot be detected by IHC methods in formalin-fixed paraffin-embedded tissues and therefore requires the use of fresh tissue with either cryostat sections or touch imprints.[25,26] The antigen detected by Ki-67 begins to be expressed in the mid-G1 phase and is present through the M phase. Noncycling cells and early G1 cells do not stain positively for Ki-67. Several studies have shown good correlation between immunohistochemical staining for Ki-67 and thymidine and bromodeoxyuridine labeling indices[27] as well as with flow cytometric determinations of S phase faction.[28] Ki-67 labeling has been shown to correlate with tumor grade in cancers of the breast[29–33] as well as in a wide variety of other forms of cancer. Cell proliferation labeling measured by Ki-67 immunostaining correlates with the S phase levels calculated by flow cytometry but is generally higher, reflecting the fact that the Ki-67 antigen is also expressed in late G1 as well early G2/M phases of the cell cycle.[33] A comparison of Ki-67 immunostaining with S phase measurements in breast cancer is shown in **Table 13.2**.

Table 13.2 Comparison of Ki-67 Immunostaining with S Phase Measurement in Breast Cancer: One Laboratory's Experience*

Method	Low	Medium	High
S phase flow cytometry (calculated)	<6–8%	8–14%	>14%
Ki-67 immunohistochemistry (% cells staining)	<10–12%	12–18%	>18%

*Unpublished results from the Albany Medical Center internal review of cell proliferation markers in breast cancer. Not intended to be a guide for cutoff points for other laboratories.

Ki-67 MIB-1 Immunostaining Problems with reproducibility in staining of formalin-fixed paraffin-embedded tissues with the original Ki-67 antibody led to the discovery of the MIB-1 clone capable of measuring cell proliferation in processed tissues (FIG. 13.2).[31,32] The MIB-1 clone has proved to be a reliable marker of proliferating cells in a large variety of human neoplasms, including cancers of the lung, breast, gastrointestinal tract, kidney, prostate, and soft tissue sarcomas.[33] Consistent results with this marker have been reported, and it is currently the predominant method of directly measuring cell proliferation in the diagnostic pathology laboratory.

Ki-67 and Prognosis in Breast Cancer A selected series of 20 studies that evaluated the prognostic significance of Ki-67 expression in

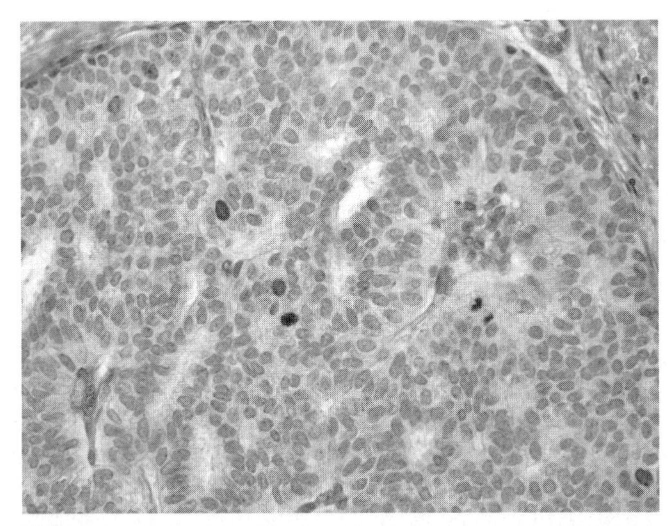

FIGURE 13.2 Ki-67 labeling and cell proliferation in breast cancer. High magnification photomicrograph of a well differentiated infiltrating ductal breast cancer demonstrated rare positively staining nuclei indicating a relatively low Ki-67 staining percentage or labeling index (peroxidase-anti-peroxidase using the MIB-1 clone with hematoxylin counterstain, ×400). **See Plate 27 for color image.**

breast cancer using IHC on processed tissues is shown in **Table 13.3**.[34–53] Of the 20 studies listed in **Table 13.3** involving more than 3000 patients, 17 (85%) reported a significant association of Ki-67 expression with prognosis on univariate analysis. Of the 19 studies that performed multivariate analysis, only 7 (37%) found Ki-67 expression status to be an independent prognostic variable. In summary, cell proliferation measurements appear to feature a greater incidence of statistically significant correlation with breast cancer outcome both on univariate and multivariate analysis than DNA ploidy alone. Although neither the American Society of Clinical Oncology[54] nor College of American Pathologists[55] guidelines include Ki-67 labeling, many laboratories provide this testing at the request of oncologists for use in the treatment of breast cancer.

Perhaps it would be appropriate to note here that although all these markers of proliferation have support in the literature as potential prognostic indicators, there are essentially two "surviving" techniques used routinely by many laboratories: S phase fraction by flow cytometry and Ki-67 by IHC. These two have survived because of the ease with which they can be performed in most laboratories in the community. Although ploidy is reported whenever S phase fraction is, clinicians tend to use the information about proliferation much more than the information about ploidy.

Proliferating Cell Nuclear Antigen

Proliferating cell nuclear antigen (PCNA), or cyclin, is a nonhistone nuclear protein that is a cofactor for DNA polymerase delta. PCNA expression is also detectable in early G1 phase.[56] The levels increase through S phase, level off in G2, and begin to decrease from G2/M to G1. Although there have been variable correlations between PCNA counts and flow cytometric studies, in general, the percentage of PCNA labeled tumor cells appears to be greater than that in S phase, as measured by flow cytometry, probably as a result of an admixture of nontumor cells with tumor in the latter method.[56] Although PCNA appears to be a more specific S phase marker than Ki-67 labeling, it is a less sensitive proliferation marker.[57] Although PCNA IHC results are more easily impacted by preanalytic factors such as the nature of the antigen retrieval system used, a number of studies have indicated that PCNA immunostaining is a useful marker for the study of cell proliferation in a variety of tumors, including infiltrating ductal carcinoma of the breast.[58–62] PCNA staining is not used routinely for the clinical management of breast cancer, however.

DNA Polymerase Alpha

DNA polymerase alpha is a cell cycle related enzyme expressed in all phases of cycling cells but not in resting G0 cells.[63] IHC detection

Table 13.3 Ki-67 Measurements by Immunohistochemistry and Prognosis in Breast Cancer

Year and Study No.	Author	Ref. No.	Specimen Type	Cases	Disease Outcome Statistical Correlation: Univariate Significance	Disease Outcome Statistical Correlation: Multivariate Independent Significance	Comment
1991 (1)	Wintzer	34	Paraffin	63	Yes	No	Ki-67 labeling was independent when a higher cutoff was used in the statistical model
1992 (2)	Sahin	35	Paraffin	42	Yes	—	Ki-67 labeling correlates with adverse prognostic variables
1992 (3)	Locker	36	Paraffin	67	Yes	Yes	Ki-67 levels correlated with overall and disease-free survival
1995 (4)	Pinder	37	Paraffin	177	Yes	Yes	Grade, size, node status, and MIB-1 labeling were all independent predictors
1996	Beck	38	Paraffin	462	Yes	Yes	Ki-67 levels also correlated with p53 status
1997 (5)	Dettmar	39	Paraffin	90	Yes	No	Only S phase measurement by flow cytometry was an independent predictor
1997 (6)	Imamura	40	Paraffin	181	Yes	Yes	MIB-1 LI in DCIS predicted prognosis of invasive carcinoma
1998 (7)	Harbeck	41	Paraffin	100	Yes	No	Only uPA levels were significant predictors of survival on MVA
1998 (8)	Jacquemier	42	Paraffin	162	Yes	Yes	Microvessel density and Ki-67 LI were both independent variables for survival
1999 (9)	Midulla	43	Paraffin	71	Yes	Yes	Ki-67 was only variable predictive of recurrence
2000 (10)	Chang	44	Paraffin	35	Yes	No	Only ER status predicted tamoxifen response rates
2000 (11)	McCready	45	Paraffin	156	No	No	Only size, grade, and node status were predictive
2000 (12)	Vazques-Ramirez	46	Paraffin	72	Yes	No	Only grade and metallothionein levels were independent
2001 (13)	Villar	47	Paraffin	116	Yes	No	Only apoptotic counts were independent predictors
2001 (14)	Schneider	48	Paraffin	52	Yes	Yes	Ki-67 LI of >20% was an independent predictor
2002 (15)	Jones	49	Paraffin	669	Yes	No	Only tumor size was an independent predictor
2002 (16)	Patla	50	Paraffin	48	No	No	Only ER status was predictive
2003 (17)	Schneider	51	Paraffin	212	Yes	No	Only uPA and progesterone receptor status independently predicted lymph node status
2003 (18)	Penault-Lorca	52	Paraffin	115	Yes	No	Ki-67 predicted chemotherapy response on univariate analysis. Only HER-2/*neu* status was predictive on multivariate analysis
2003 (19)	Han	53	Paraffin	175	Yes	No	Ki-67 levels correlated with cyclin E levels; only cyclin E was independent prognostic variable
2003 (20)	Lumachi	54	Paraffin	387	No	No	Only age, tumor size, and ER status correlated with survival

DCIS, ductal carcinoma in situ; LI, labeling index; MVA, multivariate analysis; uPA, urokinase type plasminogen activator.

requires fresh-frozen tissue sections and the marker is not clinically useful at this time.

p105

Antibodies against p105 detect a nuclear antigenic epitope involved with RNA synthesis that increases rapidly in late S phase and then in M phase, although it is also present in all phases of the cycling cells.[64] Although IHC staining for this marker has been used in some neoplasms, there is little evidence of use or utility for breast cancer patients.

Silver Nucleolar Organizer Regions

Nucleolar organizer regions (NORs) are loops of DNA that encode ribosomal RNA production.[65] These regions are believed to be responsible for transcriptional activity. NORs are located on the short arms of five pairs of the acrocentric chromosomes 13, 14, 15, 21, and 22. In interphase they are most commonly located in nucleoli. These regions are associated with proteins that can be detected using silver staining techniques because of their argyrophilia. Normal resting diploid cells have one or two of the argyrophilic nucleolar organizer region-associated proteins (AgNORs) discernible per nucleus.

The number of AgNOR dots is a good index of cell proliferation, with higher number of AgNOR dots associated with a shorter cell cycle.[66] Most studies of breast cancers have shown that AgNOR counts correlate with proliferative activity as measured by flow cytometry or by Ki-67 antibody staining.[67,68] AgNOR counting is not routinely performed in breast cancer specimens for patient care decisions.

Cell Cycle Regulators

Cyclins and their regulatory proteins, the CDKs, serve to control the rate of passage of cells through the cell cycle.[69] The cell cycle checkpoints are controlled by the CDKs, which are composed of the cyclin structural protein and the kinase enzyme. In mammals, a succession of kinases (CDK4, CDK2, and CDC2) are expressed along with a succession of cyclins (D, E, A, and B) as cells go from G1 to S to G2 to M (FIG. 13.1).[70] When the cell detects a problem with a continuation of the cycle, the activation of the cyclin–CDK complex is not completed. If no defect is detected, the cyclin–CDK complex is activated by phosphorylation, and this leads to the activation of a transcription factor by the removal of an inhibitor of the transcription factor. The transcription factor then turns on the transcription of specific genes necessary for the next cell cycle step, including the transcription of the next cyclin and kinase genes. The rapid degradation of cyclins by the proteasome facilitates the transient nature of the cell cycle control. The regulation of cyclins and their respective CDKs in breast cancer has been of keen interest to scientists and oncologists, both for the potential to develop new prognostic categories for the disease and the development of new antiproliferative chemotherapies.[71]

Cyclins

Cyclin D The G1 cyclins and their respective CDKs regulate the passage of cells through the G1 and S phases of the cell cycle. Cyclins D1, D2, D3, and E are the G1 cyclins. The expression of cyclin D1 is rapidly induced as breast cancer cells begin to divide, features a very short half-life, and is rapidly modulated in response to changes in extracellular environment.[69–71] Cyclin D1 expression can be induced by c-*Myc,* AP-1, and NF-κB transcription factors.[72] Regulation of the CDKs Cdk 4 and Cdk 6 is achieved by the expression of the INK4a group of CDK inhibitors (CKIs) ($p15^{INK4a}$, $p16^{INK4a}$, $p18^{INK4a}$, and $p19^{INK4a}$), which do not complex directly with cyclin D1, and the Cip/Kip family of cdk inhibitors ($p21^{Cip/Kip}$, $p27^{Cip/Kip}$, and $p57^{Cip/Kip}$), which inhibit the S phase but not G1 phase of the cell cycle by directly stabilizing the cyclin D1/Cdk4/6 complex.[72] The short half-life of cyclin D1 is believed to be achieved by rapid ubiquitination and degradation by the proteasome pathway (see Chapter 24).[73]

Cyclin D1 (*PRAD1, bcl-1*) is localized to chromosome 11q13 and was originally identified in parathyroid adenomas and believed to be an oncogene.[74] Cyclin D1 gene amplification (FIG. 13.3) was originally described in 10–20% of

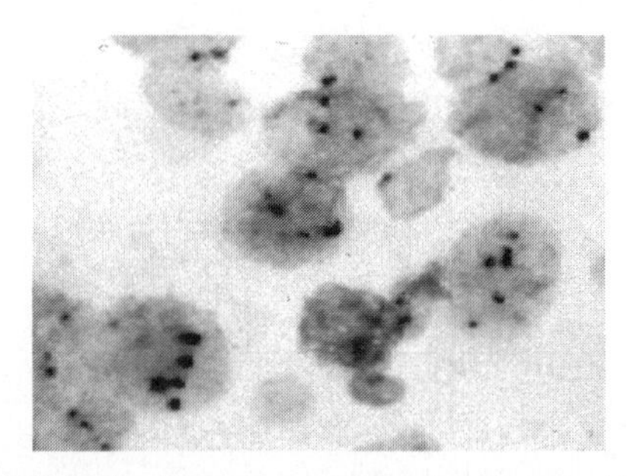

FIGURE 13.3 Cyclin D1 gene amplification detected by chromogenic in situ hybridization. High magnification photomicrograph demonstrating numerous intranuclear copies of the cyclin D1 gene in a case of infiltrating ductal breast cancer (anticyclin D1 probe with immunoperoxidase signal detection with diaminobenzidine and hematoxylin counterstain, ×1000).

mammary carcinomas,[75–77] and subsequent IHC studies found that the cyclin D1 protein was overexpressed in most human breast cancers.[78] Overexpression of cyclin D1 has been linked to the transition to overt carcinoma from premalignant breast lesions,[79] and transgenic mice engineered to overexpress cyclin D1 in the breast tissue germline are prone to develop breast cancer.[80] Cyclin D1 amplification is associated with enhanced cell proliferation via complex formation of the cyclin protein subunits with the cdk4 kinase that stimulates cell cycle progression during the G1 phase.[69–71] Cyclin D1 expression has been linked to the expression of the estrogen receptor[81] (see Chapter 15). A series of 13 studies involving 3041 patients that examined the prognostic significance of cyclin D1 expression in breast cancer[82–94] is listed in **Table 13.4**. From this table it can be seen that the results are mixed: Four of 10 studies (40%) that performed multivariate statistical analysis

Table 13.4 Selected Studies of Cyclin D1 Status and Prognosis in Breast Cancer

					Disease Outcome Statistical Correlation		
Year and Study No.	Author	Ref. No.	Specimen Type	Cases	Univariate Significance	Multivariate Independent Significance	Comment
1995 (1)	McIntosh	82	Paraffin IHC	93	—	—	Cyclin D1 or D3 expression predicts adverse prognosis if coexpressed with EGFR or pRB
1996 (2)	Pelosio	83	Paraffin IHC	180	Yes	Yes	Overexpression of cyclin D1 was an independent favorable prognostic factor
1997 (3)	van Diest	84	Paraffin IHC	148	No	No	Cyclin D1 expression correlated with grade and mitotic index but not with prognosis
1998 (4)	Dublin	85	Paraffin IHC	192	No	No	Cyclin D1 and pRB expression correlated with each other but not with disease outcome
1996 (5)	Seshadri	86	Paraffin Slot Blot	1014	No	No	Cyclin D1 gene amplification status did not correlate with prognosis
1999 (6)	Fenny	87	Paraffin RT-PCR	253	Yes	Yes	Increased Cyclin D1 mRNA correlates with adverse prognosis in ER-positive patients
2000 (7)	Utsumi	88	Paraffin RT-PCR	97	Yes	Yes	Low cyclin D1 mRNA levels independently predicted survival High cyclin D mRNA levels correlated with ER mRNA levels
2000 (8)	Turner	89	Paraffin IHC	98	No	No	Cyclin D1 expression predicted early local relapse after radiation but not late relapse
2002 (9)	Naidu	90	Paraffin PCR + IHC	440	Yes	No	Cyclin D1 gene amplification correlated with adverse outcome, whereas cyclin D1 protein expression was favorable
2002 (10)	Umekita	91	Paraffin IHC	173	Yes	Yes	Cyclin D1 overexpression independently predicted outcome in ER-negative group
2003 (11)	Lim	92	Paraffin IHC	128	Yes	—	Cyclin D1 expression predicted shortened survival and correlated with β-catenin expression
2003 (12)	Hwang	93	Paraffin IHC	175	Yes	—	Cyclin D1 expression was a favorable prognostic factor
2003 (13)	Delozier	94	Paraffin IHC	50	Yes	No	Only a Ki-67 plus cyclin A and mitotic index were significant on MVA

EGFR, epidermal growth factor receptor; MVA, multivariate analysis; RT-PCR, reverse transcriptase-polymerase chain reaction.

concluded that cyclin D1 expression was an independent prognostic variable, whereas the remaining 6 studies (60%) found that it was not an independent predictor.

The regulation of cyclin D2 expression in breast cancer[95] has been related to the silencing of gene expression by promoter methylation.[96] Cyclin D2 expression has not been studied as a prognostic factor for the disease. Cyclin D3 expression has been associated with proliferation and differentiation in both normal tissues and cancer.[97–99] Intense cyclin D3 expression is seen in steroid hormone secreting endocrine cells and endocrine neoplasms.[97–99] Cyclin D3 deregulation has also been linked to the pathogenesis of malignant lymphoma.[98,99]

Cyclin E The CDK2 interacting cyclins, including cyclins E1, E2, A1, and A2, are associated with DNA replication and cellular proliferation. Up-regulation of cyclin E (E1) messenger RNA (mRNA) can result from either gene amplification or transcriptional dysregulation.[100] Cyclin E plays a critical role in the G1/S transition. Overexpression of cyclin E protein is associated with cell proliferation and chromosomal instability.[100–103] Cyclin E1 has been characterized as an adverse prognostic factor in breast cancer (see below). Although cyclin E2 is overexpressed in some breast cancers, its role as a prognostic marker is currently unknown.[100]

Cyclin E1, another G1/S phase regulator, has been associated with tumor biology in a wide range of malignancies, including breast cancer.[71,101] IHC studies initially demonstrated that both cyclin A and cyclin E expression was intense in a subset of breast cancer cases.[88] This was followed by claims that cyclin E expression was a superior prognostic marker when compared with cyclin D1.[103,104] Similar to cyclin D3, cyclin E expression has been associated with estrogen receptor (ER) and progesterone receptor status in breast cancer.[105] Cyclin E status has also been linked to precancerous ductal hyperplasia[106,107] and chromosomal instability[108] in breast cancer. When combined with p27^{Kip1} cdk inhibitor status, cyclin E was found to have independent prognostic significance.[109] Some recent studies have demonstrated strong prognostic significance for cyclin E expression and breast cancer outcome,[110–113] including several that found the marker to be an independent predictor on multivariate analysis.[111–113] In a recent study, both high levels of the low molecular weight isoforms of cyclin E and total cyclin E measured by Western blotting independently correlated with decreased disease-specific survival.[112] Other studies, however, have failed to link cyclin E status with prognosis[114] or found it to predict tamoxifen response but not overall outcome.[115] Continued study of cyclin E as a prognostic marker in primary breast cancer appear warranted before the marker can achieve clinical utility for the management of the disease and be validated as a potential target for anticancer therapy.[116]

Cyclin A The A-type cyclins (A1 and A2) have been linked to the pathogenesis of cancer.[100] Overexpression of cyclin A2 has been associated with cell proliferation and adverse prognosis in a variety of malignant processes.[100] Cyclin A1 is a tissue-specific cyclin that is highly expressed in acute myeloid leukemia and in testicular cancer.[100] The A-type cyclins activate two different CDKs and function in both S phase and mitosis. In S phase, phosphorylation of components of the DNA replication machinery such as CDC6 by cyclin A-CDK is believed to be important for initiation of DNA replication and to restrict the initiation to only once per cell cycle.[117] Cyclin A levels rise during S phase and precipitously fall before metaphase. The synthesis of cyclin A is predominantly regulated by the E2F transcription factor and similarly influenced by the ER pathway.[104] Removal of cyclin A is carried out by ubiquitination and the proteasome, possibly via the same or similar anaphase-promoting complex/cyclosome targeting subunits that are used as for cyclin B.[117] Altered expression of cyclin A has been identified in a variety of human tumors.[117–119]

Studies of cyclin A expression in breast cancer are currently limited in their scope. In one IHC-based report on the expression of cyclins A, D1, D3, and E in 170 breast carcinomas, positive immunoreactivity for cyclin A correlated with early relapse ($P = 0.001$ univariate analysis, $P = 0.006$ multivariate analysis) as well as disease specific survival ($P < 0.0001$).[104] Interestingly, this study found no prognostic association for cyclin D1 or cyclin E.[120] In another study adjusted for lymph node status, age, performance status, T classification, grade, prior surgery, ER status, and tamoxifen use, only overexpression of cyclin A and HER-2/neu protein achieved prognostic significance.[121] Another report found that only p21 IHC-based expression levels combined with S phase determined by flow cytometry or with cyclin A and lymph node status were salient survival prognostic factors.[122] In addition, it has been reported that cyclin A expression levels override the prognostic effects of p53 status in breast cancer.[123] However, some studies have not found cyclin A levels to have prognostic significance.[114]

Cyclin B The cyclin B complex with its specific CDK cdc2 regulates the onset of mitosis via several cell cycle checkpoints that ensure that chromosome segregation does not occur if unreplicated or damaged DNA or misaligned chromosomes are present.[124] The entry into mitosis is also regulated by p53, which blocks cells at the G2 checkpoint via inhibition of Cdc2.[125] The binding of Cdc2 to cyclin B1 is required for its activity, and the direct repression of the cyclin B1 gene by p53 also contributes to blocking entry into mitosis.[125] p53 mediated transcriptional induction of the CKI $p21^{CIP1}$ also regulates entry into mitosis.[126] Studies of the prognostic significance of cyclin B in breast cancer are limited. In one study, the cytoplasmic localization of $p21^{CIP1}$ and high cyclin B levels were significant predictors of poor prognosis.[126] In another study featuring multivariate analysis, both cyclin E (relative risk 2.01, $P = 0.021$) and cyclin B (relative risk 1.85, $P = 0.033$) were identified as independent predictors of metastasis-free survival when the Ki-67 cell proliferation index was omitted, but only cyclin E expression was associated with disease-specific survival (relative risk 2.56, $P = 0.006$).[114] In a follow-on study of additional patients by this group, after inclusion of Ki-67 in the model, cyclin E expression lost its prognostic significance, whereas cyclin B remained the only independent prognostic factor with a hazard ratio of 4.5 ($P = 0.026$) for tumor-related death.[127]

CDKs

DNA replication and mitosis are dependent on the activity of CDK enzymes, which are heterodimers of a catalytic subunit with a corresponding cyclin subunit.[128] The catalytic subunit of CDK is only active when bound to a regulatory cyclin protein, and CDK activity can be inhibited by the binding of CKIs. Whereas the level of CDKs does not change significantly during the cell cycle, the abundance of cyclins and CKIs is tightly controlled, both temporally and spatially, providing precise regulation of cell cycle events.[129] Cell cycle progression involves changing rates of cyclin transcription or proteolysis, with consequent changes in the substrates for the CDKs through the cell cycle.[128] Cyclin accumulation is particularly important in terminating the G1 phase, when it raises CDK activity and starts events leading to DNA replication. For example, the CDK, Cdc2 ($p34^{cdc2}$), and cyclin B form a stoichiometric complex that is necessary for the Cdc2 subunit to gain its protein kinase activity.[130] Inactivation of CDKs can be achieved by several mechanisms: by association with CKIs, including $p16^{INK4a}$, $p15^{INK4b}$, $p21^{Cip1}$, $p27^{Kip1}$, and $p57^{Kip2}$; by disassociation from their cyclin regulatory unit; by dephosphorylation of a conserved threonine residue in the CDK T loop; and by adding an inhibitory phosphate molecule.[129]

$p34^{cdc2}$

As seen above, the CDKs regulate the conversion of quiescent cells to a state of active proliferation, commitment to DNA replication, initiation of DNA replication, and entry into and exit from mitosis. Changes in the activity of the CDKs throughout the cell cycle depend not only on their association with their respective cyclins and on posttranslational phosphorylation–dephosphorylation modifications, but also on specific protein inhibitors and on protein degradation.[131,132] A major CDK is the $p34^{cdc2}$ kinase, a highly regulated serine-threonine kinase that when complexed with cyclins A and B as the M phase promoting factor controls cell entry into mitosis. The $p34^{cdc2}$ kinase interacts with a variety of other kinases and phosphatases,[133] including protein kinase C.[134] Studies of $p34^{cdc2}$ as a prognostic factor in breast cancer are limited. In one study of 88 cases of invasive breast cancer using immunoblotting techniques, the incidence of lymph node metastasis was significantly increased in the $p34^{cdc2}$/cyclin D1 double positive group and low in the double negative group. Relapse-free survival times of $p34^{cdc2}$-positive cases were significantly shorter than those of $p34^{cdc2}$-negative cases.[135] In another report of primary invasive breast cancer tissues from a combined group of 135 lymph node–negative and –positive patients, $p34^{cdc2}$ overexpression significantly correlated with disease related death.[136] In patients who received adjuvant therapy, $p34^{cdc2}$ protein overexpression did not influence disease-free or overall survival when compared with the combined group of treated and untreated patients.[136]

The mechanism by which the mitotic spindle checkpoint couples to the cell survival machinery remained elusive until it was discovered that microtubule stabilization engenders a survival pathway that depends on elevated activity of $p34^{cdc2}$ and increased expression of the apoptosis inhibitor and mitotic regulator, survivin. Pharmacologic, genetic, or molecular ablation of $p34^{cdc2}$ kinase after microtubule stabilization is associated with apoptosis independent of p53 and suppression of tumor growth.[137] Thus, inhibitors of $p34^{cdc2}$ kinase currently in early stages of development have the potential to improve the efficacy of microtubule-stabilizing agents such as the taxanes used to treat malignancies such as breast cancer.[137,138]

Drugs Targeting CDKs

As critical regulators of the cell cycle and with overwhelming evidence suggesting that most human neoplasms result at least in part from cdk hyperactivation and subsequent abrogation of the Rb pathway, novel direct and indirect cdk inhibitors are in various stages of development as anticancer drugs.[139–143] Because CDK deregulation, either through direct or indirect means, is found in most cancer cells, pharmacologic CDK inhibition has become an attractive strategy toward mechanism-based, targeted, and nongenotoxic therapies in oncology. Disrupting the cell cycle by inhibiting CDKs has become an important therapeutic strategy for many pharmaceutical companies in their approaches to the treatment of cancer. Flavopiridol is the first CKI to be tested in clinical trials. It has been shown to cause cell cycle arrest, induce apoptosis, inhibit angiogenesis, and potentiate the effects of cytotoxic chemotherapy.[139,140] Flavopiridol is currently undergoing phase II testing as monotherapy and phase I and/or II evaluation in combination with traditional chemotherapy agents in breast cancer.[139,140] The assessment of CDK inhibition as evidence of flavopiridol's targeted effect in serial biopsies of tumor and surrogate tissues is also under investigation in these protocols.[139,140] Rapamycin and its analogues inhibit the G1/S boundary, target the downstream regulator mTOR, prevent CDK activation, inhibit retinoblastoma protein phosphorylation, and accelerate the turnover of cyclin D1, leading to a deficiency of active CDK4/cyclin D1 complexes.[144]

In addition to flavopiridol, another cdk inhibitor, UCN-01, has demonstrated promising results, with some evidence of antitumor activity achieved when plasma concentrations effective to inhibit cdk-related functions are reached.[139,140] Other indirect CDK inhibitors include proteasome inhibitors (bortezomib, PS-341) and perifosine.[143] The best schedule of administration, combination with standard chemotherapeutic agents, the development of predictive biomarkers, and demonstration of CDK modulation from tumor samples from patients in these trials are important issues that need to be answered to obtain the best possible results with these novel agents.[143] Finally, recent studies have identified important cross-talk between the cell cycle regulatory apparatus and proteins regulating histone acetylation.[145] Histone deacetylase plays an important role in pRB tumor suppression function and transcriptional repression. The recent evidence for cross-talk between the CDKs and histone gene expression on the one hand and cyclin-dependent regulation of histone acetylases on the other suggests chemotherapeutics targeting histone acetylation may have complex and possibly complementary effects with agents targeting CDKs.[145]

CKIs

CDK activity is controlled by cyclin abundance and subcellular location and by the activity of two families of inhibitors, the CKIs. Many hormones, cytokines, growth factors, and enzymes influence cell growth through signal transduction pathways that modify the activity of the cyclins by impacting on the expression of the CKIs.[146,147] The NF-κB family of transcription factors and regulators appears to play a major role in cycle regulation by controlling CKI expression.[147] CKI expression is also regulated by ubiquitination and degradation through the proteasome pathway.[147,148] The major CKIs, p21/WAF1/Cip1 and p27 (kip1), have been widely studied for their prognostic significance in human neoplasms.[149]

p21

The p21 protein (p21/WAF1/Cip1) is an inhibitor of CDKs and serves as a critical downstream effector in the p53-specific pathway of cell growth control.[149,150] A number of IHC-based studies examined the potential prognostic significance of p21 expression in breast cancer.[151–164] **Table 13.5** includes a review of 14 published studies involving 2710 patients who were evaluated for the prognostic significance of p21 expression in node-negative, node-positive, and combined cases of breast cancer. The studies have generally conflicted with significant differences in their conclusions reported by different investigators. Although all the reports used the IHC technique, their methods varied widely, including differences in specimen preparation and fixation, antigen retrieval, automated versus manual methods, antibodies used, and, particularly, in their slide scoring system and data analysis.

From **Table 13.5**, 5 of 13 studies (38%) that featured multivariate analysis found p21 expression status to be an independent predictor of breast cancer outcome.[152–154,158,163] Several studies specifically linked p21 expression to the p53 pathway.[151–153,159,162] In one report of 106 patients, loss of p21 expression was associated with low histologic grade, p53 overexpression, and high mdm-2 expression.[154] In this

Table 13.5 Selected Studies of $p21^{Cip1}$ Status and Prognosis in Breast Cancer

					Disease Outcome Statistical Correlation		
Year and Study No.	Author	Ref. No.	Method	Cases	Univariate Significance	Multivariate Independent Significance	Comment
1996 (1)	Barbareschi	151	IHC	88	Y	—	p21 expression correlated with p53 status
1996 (2)	Caffo	152	IHC	26	Y	Y	Correlated with p53 overexpression
1997 (3)	Wakasugi	153	IHC	104	Y	Y	Correlated with p53 overexpression
1997 (4)	Jiang	154	IHC	106	Y	Y	p21 expression correlated with metastasis
1999 (5)	Reed	155	IHC	77	N	N	LN− patients only; no correlation with outcome
2000 (6)	Mathoulin-Portier	156	IHC	162	N	N	p21 correlated with Ki-67 expression but not with survival
2000 (7)	Bankfalvi	157	IHC	141	N	N	p21 expression associated with p53 and mdm2 status only
2000 (8)	Thor	158	IHC	798	N (LN−) Y (LN+)	N (LN−) Y (LN+)	p21 was an independent predictor in LN+ cases only
2001 (9)	Domagala	159	Immunoblot	222	N	N	p21/p53 combination status predicted adjuvant therapy response but not survival
2001 (10)	Goring	160	IHC	307	N	N	p21 status correlated with DNA replication only
2002 (11)	O'Hanlon	161	IHC	105	N	N	65% of LN+ had p21 overexpression; no correlation with disease outcome
2003 (12)	Pellikainen	162	IHC	420	N	N	p53 status was an independent predictor; p21 status was not
2003 (13)	Michels	163	IHC flow cytometry	104	Y	Y	Using multiple traditional markers, p21 was the most significant predictive factor
2003 (14)	Chasle	164	IHC	50	N	N	Only cyclin A and Ki-67 were significant predictors.

LN−, LN+, lymph node negative and positive, respectively; N, no; Y, yes.

study, low levels of p21 and higher mdm-2 levels directly correlated with the onset of lymph node metastases and shortened patient survival.[154] In another study of patients treated with systemic adjuvant therapy, bivariate analysis of the combined p21 and p53 phenotypes showed that p21+/p53+ tumors were associated with long disease-free and overall survival, whereas p21−/p53+ tumors had the worst prognosis.[150] These authors concluded that p21/p53 heterogeneous expression may be of clinical relevance for the therapeutic response to chemotherapy and hormonal therapy.[150] In a large cooperative group-based study of nearly 800 patients, p21 negativity inversely correlated with p53 immunopositivity and featured an improved outcome if they were node positive but not if they were node negative.[158] Although other studies maintained expression of p21 was a favorable prognostic finding,[151–154,163] these authors hypothesized that their results supported the concept that the loss of p21 expression in the node-positive group was a favorable finding that reflected an enhanced sensitivity of these tumors to chemotherapy.[158] However, another group published findings that conflicted with this opinion, finding that maintained rather than degraded p21 immunopositivity predicted an enhanced response to chemotherapy.[159]

Table 13.5 also includes 8 of 14 studies (57%) that failed to identify prognostic significance for p21 expression in breast cancer on univariate analysis.[155–157,159–162,164] Several reports found p21 expression loss to correlate with other adverse prognostic factors such as tumor grade, size, mitotic index, Ki-67 labeling index, and ER status.[151,155–157] In summary, the prognostic significance of p21 expression in breast cancer continues to be controversial with no clear consensus as to its current or future role for the management of the disease.

p27

Protein p27 was initially discovered in cells arrested by transforming growth factor-α, by contact inhibition, or by lovastatin.[165] The members of the kinase inhibitor protein (KIP) family ($p21^{CIP1}$, $p27^{Kip1}$, and $p57^{Kip2}$) have opposite effects on the function of different CDKs. p27 and p21 have a negative effect on the activities of cyclin E/CDK2 and cyclin A/CDK2 and positive effects on cyclin D/CDK complexes.[165] Three major mechanisms are believed to be responsible for these observations: (1) enhancement of the assembly of cyclin D/CDKs complexes, (2) nuclear localization of these complexes, and (3) increased stability of D type cyclins.[165] In proliferating cells p27 binds to cyclin D/CDKs, whereas in nonproliferating cells p27 is found in complexes with cyclin E/CDK2. The competition for p27 between cyclin D/CDKs and cyclin E/CDK2 complexes seems to be crucial for cell cycle progression.[166] p27 mRNA levels are constant throughout the cell cycle, whereas p27 protein levels are high in quiescent cells and decrease during G1 phase, reaching the lowest point in S phase.[167]

The decrease in p27 protein levels observed in the passage from G1 to S phase is due to a decrease in the p27 half-life, which corresponds to an increased degradation via the ubiquitin–proteasome pathway. During ubiquitination and subsequent degradation, the interaction between a ubiquitin-conjugating enzyme and a substrate is mediated by the action of a ubiquitin protein ligase. Ubiquitin ligases regulating the G1 phase are called SCF complexes. The SCF ligases are each formed by at least four basic subunits: Skp1, a Cullin subunit (Cul1 in metazoans), an F-box protein, and the Roc1/Rbx1 protein.[168,169] Each SCF ligase joins an ubiquitin-conjugating enzyme (Ubc3, Ubc4, or Ubc5) to specific substrates that are recruited by different F-box proteins. Many F-box proteins have been described for various substrates, and studies have demonstrated that Skp2 is the F-box protein involved in p27 degradation (see below).[168,169] Cell cycle levels of Skp2 are inversely correlated with p27 levels, with Skp2 expression low in early/mid-G1 and increasing in late G1, in coordination with the decrease in p27 protein.[166–168] Most nonneoplastic epithelial tissues from a variety of organs express high levels of nuclear p27 protein, but p27 is virtually undetectable in proliferating cells, such as the basal layer of epithelia or lymphocytes in germinal centers.[166,167] Variable loss of p27 protein detection has been identified in a wide variety of human neoplasms.

In contrast with the studies of p21, most reports have indicated that p27 expression loss is a significant adverse prognostic factor in breast cancer (FIG. 13.4).[109,155,169–180] In the series of 14 selected disease outcome studies

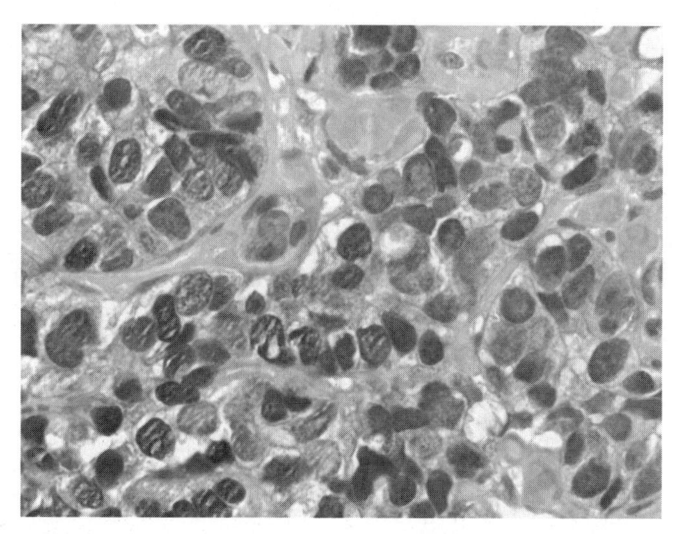

FIGURE 13.4 P27 Expression in breast cancer. Most nuclei stain positively for p27 in this case of moderately differentiated infiltrating ductal breast cancer. This patient was diagnosed with node positive disease, but remained free of recurrent disease at 5 years from the time of diagnosis after receiving adjuvant chemotherapy (peroxidase-anti-peroxidase, ×400). **See Plate 28 for color image.**

listed in **Table 13.6** involving 3502 patients, 11 studies (79%)[109,169–173,175,177,178,180] found p27 expression to be a significant prognostic marker. However, in the 13 studies that used multivariate statistical analysis, only 7 studies (54%)[109,169–171,173,177,178] found the p27 status to be an independent predictive variable when other biomarkers and traditional prognostic factors were included in the model. Most initial studies tested the prognostic value of p27 in breast tumors in comparison with already well-established parameters such as tumor stage, grade, nodal status, hormone receptor levels, and S phase fraction. In one study that assayed both p27 and cyclin E levels in 246 primary breast cancers of women under age 45 years, tumors with both low p27 and elevated cyclin E protein expression had the highest mortality, with both markers achieving independent predictive value for overall survival on multivariate analysis.[109] In another study of 202 patients with breast cancers less than 1 cm in size, p27 expression loss, when compared with clinicopathologic features and p53, HER-2/*neu*, Ki-67, cdc25B, and microvessel density, was an independent risk factor associated with a 3.4-fold increased risk of death, particularly in node-negative tumors.[169] The studies cited in **Table 13.6** generally confirm that high p27 levels detected by IHC are usually associated with low histologic grade, positive ER status, high cyclin D1 expression, and low S phase fraction. p27 protein levels have also inversely correlated with cyclin E dependent kinase activity.[166] The discrepancies seen in the p27 studies are similar to those seen for p21 and include the same issues associated with the IHC techniques and slide scoring systems.

A number of different oncogenic stimuli have been implicated in the inactivation of p27, including amplification of Ras, cyclin D, loss of PTEN, increased RTK activity (HER-2/*neu*, insulin-like growth factor receptor, epidermal growth factor), increased Akt activity, and increased cyclin E or Skp2 (FIG. 13.5).[181] The potential prognostic value for p27 may reflect the fact that it is the readout of multiple different pathways involved in the development of tumors. It also highlights the potential overlap of these pathways. For example, increased Akt activity might increase Skp2 and cyclin D1 levels and directly phosphorylate p27, all of which may collaborate to promote p27 mislocalization and/or degradation.[181] However, the importance of post-translational modifications of p27 and how these modifications impact the various signaling pathways cannot be overemphasized. In summary, in addition to being a promising prognostic marker for breast cancer, the multiple pathways impacted by the p27 CKI

Table 13.6 Selected Studies of $p27^{Kip1}$ Status and Prognosis in Breast Cancer

					Disease Outcome Statistical Correlation		
Year and Study No.	Author	Ref. No.	Specimen Type	Cases	Univariate Significance	Multivariate Independent Significance	Comment
1997 (1)	Porter	109	IHC	246	Y	Y	Both p27 and cyclin E were independent predictors in women under age 45 years
1997 (2)	Tan	169	IHC	202	Y	Y	Both p27 and LN status predicted outcome
1999 (3)	Wu	170	IHC	181	Y	Y	p27 independently predicted overall and disease-free survival
1999 (4)	Tsuchiya	171	IHC	102	Y	Y	p27 status was the only independent prognostic factor
1999 (5)	Gillett	172	IHC	189	Y	N	p27 significant on univariate analysis but reduced to nonsignificance by traditional biomarkers
1999 (6)	Chu	173	IHC	169	Y	Y	p27 was an independent predictor and correlated with tumor grade and ER status
1999 (7)	Reed	155	IHC	77	N	N	p27 was not a significant prognostic factor in this study restricted to LN− patients
2000 (8)	Leong	174	IHC	148	N	N	p27 was not predictive in this study of low grade cancers and correlated with Ki-67 status
2000 (9)	Barbareschi	175	IHC	512	Y	—	p27 predicted a negative outcome for LN− patients
2000 (10)	Volpi	176	IHC	286	N	N	p27 was not a significant prognostic factor; only tumor size and thymidine labeling index were predictive
2001 (11)	Lau	177	IHC	147	Y	Y	p27 was an independent predictor and was more predictive if combined with Ki-67 status
2001 (12)	Nohara	178	IHC	216	Y	Y	Among factors in multivariate analysis, p27 was the most significant individual factor
2001 (13)	Leivonen	179	IHC	197	N	N	p27 status was significant on univariate analysis at 5 years but not at 7 years
2003 (14)	Barnes	180	IHC	830	Y	N	Increased p27 levels were associated with prolonged overall survival on univariate analysis only

LN−, LN+, lymph node negative and positive, respectively; N, no; Y, yes.

suggest that this cell cycle regulator may evolve into an important target for novel anticancer agents and combination chemotherapy research.

Skp2

As seen above, the S phase kinase associated protein Skp2 is required for the ubiquitin mediated degradation of various proteins, including the cdk inhibitor p27.[166–168] A recent study of Skp2 in breast cancer suggests an important role for Skp2 overexpression (FIG. 13.6) in the pathogenesis of ER-negative HER-2/*neu*–negative breast carcinoma,[182] consistent with the proposal that Skp2 can serve as a proto-oncogene. Although limited data are available concerning Skp2 expression as a prognostic marker in breast cancer, this p27 regulator has recently emerged as a favored target of novel anticancer drug development for the disease.[182]

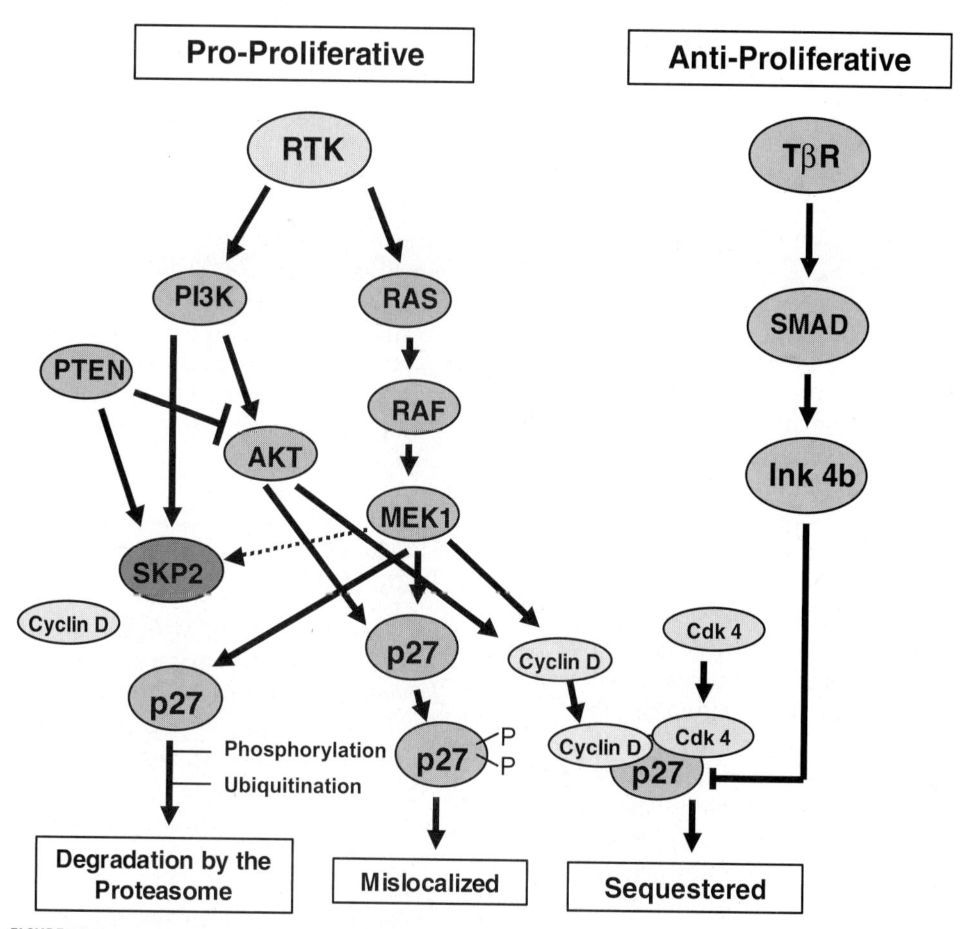

FIGURE 13.5 Targeting the p27 pathway. The p27 pathway has become a significant target for the development of new anticancer drugs. In this diagram, three potential approaches to p27 targeting are shown. *Decreasing tumor cell proliferation* as an anticancer effect by (1) reducing the degradation of p27 by inhibiting Skp2 and *increasing tumor cell proliferation* to overcome the cytotoxic chemotherapy resistance of noncycling tumor cells by (2) inactivating p27 by mislocalizing it in the cytoplasm and (3) stabilizing cyclin D1, which sequesters p27 and prevents the slowing of the cell cycle. (From Ref. 181.)

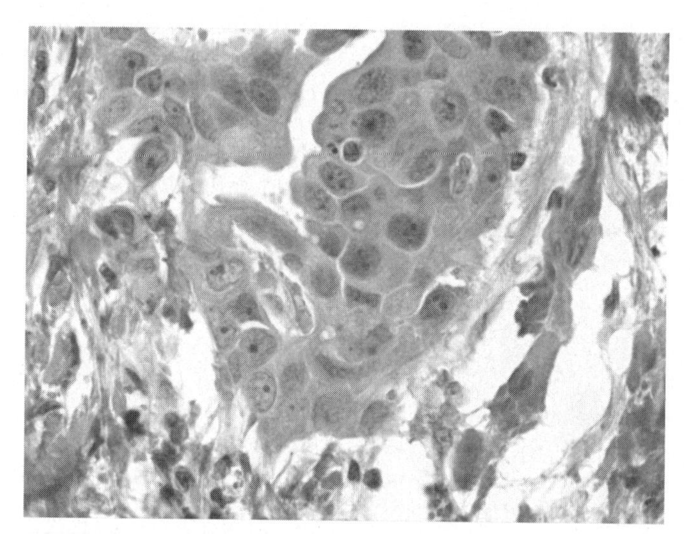

FIGURE 13.6 Skp-2 expression in breast cancer. This photomicrograph shows rare nuclear staining for Skp-2 in the same patient in Figure 4 demonstrating the well-described inverse relationship with p27 expression (peroxidase-anti-peroxidase ×400). **See Plate 29 for color image.**

Tumor Suppressor Proteins Functioning as Cell Cycle Regulators

Tumor suppressor genes are considered in depth in Chapter 23. In this section a brief overview of the potential role(s) of tumor suppressors on the regulation of the cell cycle is presented.

The *p53* Gene

p53 is a tumor suppressor gene localized to chromosome 17p that codes for a multifunctional DNA binding protein involved in cell cycle arrest, DNA repair, differentiation, and apoptosis.[183] The *p53* mutation rate is lower in breast than in other epithelial cancers and has been associated with more aggressive disease and worse overall survival.[184] The prognostic significance of *p53* status in breast cancer has been impacted by the accuracy of IHC versus molecular methods (SSCP, direct sequencing, and the yeast colony functional assay).[183,184] In general, breast carcinomas with *p53* mutations are consistently associated with high tumor grade, high mitotic index, high cell proliferation rate, aneuploid DNA content, negative assays for estrogen and progesterone receptor,[185–187] and variable association with amplification of oncogenes such as HER-2/*neu*, c-*myc*, *ras*, and *int-2*.[188,189] A number of studies have implicated *p53* mutation with resistance to hormonal, adjuvant, and neoadjuvant chemotherapy and combination chemotherapy for metastatic disease encompassing a variety of agents, including anthracyclines and taxanes.[190–195]

The *MDM2* Gene

The *MDM2* gene encodes a protein that binds to the *p53* gene, thereby reducing its cell cycle progression inhibitory role.[196] *MDM2* amplification appears to be a rare event in breast cancer.[196–197] Multiple reports have linked *MDM2* overexpression to adverse outcome in patients with node-negative and node-positive breast cancer[198–200]; however, one report found no correlation.[201]

The Retinoblastoma Gene

Abnormal expression of the retinoblastoma tumor suppressor gene and protein occurs in 10% to 20% of primary breast cancer tissues.[202–203] Retinoblastoma gene alterations have been associated with smaller node-negative tumors but have not been predictive of relapse-free and overall survival.[204]

p16

The *p16* (*INK4A*) tumor suppressor gene is a CKI that inhibits cell growth at the G1/S checkpoint of the cell cycle in concert with Rb, p14, and p15.[205,206] *p16* expression may be lost by a process of mutation (less common) or CpG island hypermethylation of the gene promoter (most common).[205–207] Studies of the *p16* gene, mRNA, and protein in breast cancer have conflicted with some studies linking overexpression with adverse outcome, some studies linking loss of expression with adverse outcome, and some studies finding no association with outcome.[208–211]

Clinical Relevance and Current Clinical Testing Utility

At the current time, widespread clinical use of cell proliferation markers for the management of women with breast cancer is limited to the Ki-67 biomarker, which is typically reported using the MIB-1 antibody as a percent cell staining or labeling index. Although this test is not listed as a recommended or required "category I" test for breast cancer by either the American Society of Clinical Oncology or the College of American Pathologists, it is extremely popular in many regions of the United States and performed on a regular basis in many diagnostic laboratories. Well-defined cutoffs agreed upon by a panel of experts for this marker have not been published, and **Table 13.2** is intended only to show the differences between S phase measured by flow cytometry and direct Ki-67 staining. A number of laboratories have converted from reporting DNA ploidy status and S phase calculation measured by flow or image cytometry to reporting the Ki-67 staining results as the only direct measure of cell proliferation for clinical breast cancer specimens. Although there is not a uniform consensus in the United States or in Europe that Ki-67 staining is clinically useful, a significant number of key opinion leaders are currently favoring that this marker can add important additional information upon which treatment decisions can be made, such as the decision to use or not to add cytotoxic adjuvant therapy to the treatment plan for a patient with a newly diagnosed lymph node negative tumor featuring a small size, well to moderately differentiated tumor grade, and an ER-positive HER-2/*neu*–negative biomarker status.

Summary and Future Clinical Potential

The cell cycle regulators and proliferation markers considered in this chapter are summarized in **Table 13.7**. As seen above, to date only the Ki-67 assay is in widespread clinical use in the United States. The future use of Ki-67 measurements as useful guides to breast cancer management will require the ultimate standardization of the IHC testing method and slide evaluation procedures needed to make the procedure reliable and reproducible in all providing laboratories. By this approach only would the Ki-67 test become a recommended test by the American Society of Clinical Oncology and the College of American Pathologists. Of the other cell cycle regulators, early interest in the potential of cyclin D status appears to have waned and been replaced with keen interest in cyclin E. The recent studies that have shown the predictive power of cyclin E[112] have encouraged expanded studies to confirm on a larger scale whether this marker could be used to plan breast cancer therapy. Similarly, there is an emerging interest in the use of determining the p27 status of newly diagnosed cases and using this information to gauge disease severity.[181]

By far the most significant progress of the cell cycle regulator clinical research in breast cancer resides in the development of therapeutics that target the cell cycle.[141–143] The main thrust of the anticancer drug preclinical research and early clinical trial has focused on the inhibition of CDKs with drugs like flavopiridol.[139,140] Other agents have attempted to restore the expression of naturally occurring CKIs by reducing their degradation. An example of this strategy is the use of proteasome inhibitors.[148] The p27 pathway has been highlighted as a therapy target both as a potential way to slow cell proliferation by inhibiting Skp2 and restoring p27 expression and, alternatively, by directly inhibiting p27 to prevent cancer cells from leaving the cell cycle and maintaining their sensitivity to other agents in a combination therapy strategy.[181]

Table 13.7 Summary of Selected Cell Cycle Regulators in Breast Cancer

Regulatory Protein	Cell Cycle Role/Function	Prognostic Significance	Current Target of Therapy
Ki-67	Unknown	Strong	No
Cyclin D	Promoter	Moderate	No
Cyclin E	Promoter	Strong	No
Cyclin A	Promoter	Moderate	No
Cyclin B	Promoter	Unknown	No
$p34^{cdc2}$	Promoter	Moderate	No
p21	Inhibitor	Weak	No
p27	Inhibitor	Strong	Yes
p15	Inhibitor	Unknown	No
p16	Inhibitor	Weak	Yes
p53	Inhibitor	Strong	Yes
pRb	Inhibitor	Weak	No

References

1. Sandal T. Molecular aspects of the mammalian cell cycle and cancer. Oncologist 2002;7:73–81.
2. Lundberg AS, Weinberg RA. Control of the cell cycle and apoptosis. Eur J Cancer 1999;35:1886–1894.
3. Weinberg RA. How cancer arises. Sci Am 1996;275: 62–70.
4. Sherr CJ. The Pezcoller lecture: cancer cell cycles revisited. Cancer Res 2000;60:3689–3695.
5. Elliott PJ, Ross JS. The proteasome: a new target for novel drug therapies. Am J Clin Pathol 2001;116: 637–646.
6. Adams J. Development of the proteasome inhibitor PS-341. Oncologist 2002;7:9–16.
7. Bundred NJ. Prognostic and predictive factors in breast cancer. Cancer Treat Rev 2001;27:137–142.
8. Zafonte BT, Hulit J, Amanatullah DF, et al. Cell-cycle dysregulation in breast cancer: breast cancer therapies targeting the cell cycle. Front Biosci 2000;5:D938–D961.
9. Daidone MG, Silvestrini R. Prognostic and predictive role of proliferation indices in adjuvant therapy of breast cancer. J Natl Cancer Inst Monogr 2001; 30:27–35.
10. Bagwell CB, Clark GM, Spyratos F, et al. DNA and cell cycle analysis as prognostic indicators in breast tumors revisited. Clin Lab Med 2001;21:875–895.
11. Linden MD, Torres FX, Kubus J, et al. Clinical application of morphologic and immunocytochemical assessments of cell proliferation. Am J Clin Pathol 1992;979(suppl 1):S4–S13.
12. van Diest PJ, Baak JPA, Matze-Cok P, et al. Reproducibility of mitosis counting in 2,469 breast cancer specimens: results from the multicenter morphometric mammary carcinoma project. Hum Pathol 1992; 23:603–607.
13. Gerdes J, Dallenbach F, Lennert K. Growth fractions in malignant non-Hodgkin's lymphomas (NHL) as determined in-situ with the monoclonal antibody Ki-67. Hematol Oncol 1984;2:365–371.
14. Parham DM. Mitotic activity and histological grading of breast cancer. Pathol Annu 1995;30:189–207.
15. Meyer JS, Prey MU, Babcock DS, et al. Breast carcinoma cell kinetics, morphology, stage, and host characteristics. A thymidine labeling study. Lab Invest 1989;54:41–51.

16. Meyer JS. Cell kinetic measurements of human tumors. Hum Pathol 1982;13:874–877.
17. Smallwood JA, Cooper A, Taylor I. The errors of thymidine labelling in breast cancer. Clin Oncol 1983;9:331–335.
18. Amadori D, Bonaguri C, Volpi A, et al. Cell kinetics and prognosis in gastric cancer. Cancer 1993;71:1–4.
19. Gratzner HG. Monoclonal antibody to 5-bromo- and 5-iodo-deoxyuridine: a new reagent for detection of DNA replication. Science 1982;218:474–475.
20. Lloveras B, Edgerton S, Thor AD. Evaluation of in vitro bromo-deoxyuridine labeling of breast carcinomas with the use of a commercial kit. Am J Clin Pathol 1991;95:41–47.
21. Ellwart J, Dormer P. Effect of 5-flouro-2′deoxyuridine (FdUrd) on 5-bromo-2′deoxyuridine (BrdUrd) incorporation into DNA measured with a monoclonal BrdUrd antibody and by the BrdUrd/Hoechst quenching effect. Cytometry 1985;6:513–520.
22. Gratzner HG. Monoclonal antibody to 5-bromo- and 5-iododeoyyuridine: a new reagent for detection of DNA replication. Science 1982;218:474–475.
23. Meyer JS, Nauert J, Koehm S, et al. Cell kinetics of human tumors by in vitro bromodeoxyuridine labeling. J Histochem Cytochem 1989;37:1449–1454.
24. Gerdes J, Schwab U, Lemke H, et al. Production of a mouse monoclonal antibody reactive with human nuclear antigen associated with cell proliferation. Int J Cancer 1983;31:13–20.
25. Gerdes J, Lemke H, Baisch H, et al. Cell cycle analysis of a cell proliferation-associated human nuclear defined by the monoclonal antibody Ki-67. Immunol 1984;133:1710–1715.
26. Schwarting R, Gerdes J, Niehus J, et al. Determination of the growth fraction in cell suspensions by flow cytometry using the monoclonal antibody Ki67. J Immunol Methods 1986;90:65–70.
27. Kamel OW, Franklin WA, Ringus JC, et al. Thymidine labeling index and Ki-67 growth fraction in lesions of the breast. Am J Pathol 1988;134:107–113.
28. Isola JJ, Helin HJ, Hella MJ, et al. Evaluation of cell proliferation in breast carcinoma. Comparison of Ki-67 immunohistochemical study. DNA flow cytometric analysis, and mitotic count. Cancer 1990; 65:1180–1184.
29. Kennedy JC, El-Badawy N, DeRose PB, et al. Comparison of cell proliferation in breast carcinoma using image analysis (Ki-67) and flow cytometric systems. Anal Quant Cytol Histol 1992;14: 304–310.
30. Marchetti E, Querzoli P, Marzola A, et al. Assessment of proliferative rate of breast cancer by Ki-67 monoclonal antibody. Mod Pathol 1990;3:31–34.
31. Key G, Becker MHG, Baron B, et al. New Ki67 equivalent murine monoclonal antibodies (MIB-1-3) generated against bacterially expressed parts of the Ki67 cDNA containing three 62 base pair repetitive elements encoding for the Ki67 epitope. Lab Invest 1993;68:629–636.
32. Catoretti G, Becker MHG, Key G, et al. Monoclonal antibodies against recombinant parts of the Ki67 antigen (MIB-1 and MIB-3) detect proliferating cells in microwave-processed formalin-fixed paraffin sections. J Pathol 1995;177:285–293.
33. MacGrogan G, Jollet I, Huet S, et al. Comparison of quantitative and semiquantitative methods of assessing MIB-1 with the s-phase fraction in breast carcinoma. Mod Pathol 1997;10:769–776.
34. Wintzer HO, Zipfel I, Schulte-Monting J, et al. Ki-67 immunostaining in human breast tumors and its relationship to progress. Cancer 1991;67:421–428.
35. Sahin AA, Ro J, Ro JY, et al. Ki-67 immunostaining in node-negative stage I/II breast carcinoma: significant correlation with prognosis. Cancer 1991;68: 549–557.
36. Pinder SE, Wencyk P, Sibbering DM, et al. Assessment of the new proliferation marker MIB1 in breast carcinoma using image analysis: associations with other prognostic factors and survival. Br J Cancer 1995;71:146–149.
37. Locker AP, Birrell K, Bell JA, et al. Ki67 immunoreactivity in breast carcinoma: relationships to prognostic variables and short term survival. Eur J Surg Oncol 1992;18:224–229.
38. Beck T, Weller EE, Weikel W, et al. Usefulness of immunohistochemical staining for p53 in the prognosis of breast carcinomas: correlations with established prognosis parameters and with the proliferation marker, MIB-1. Gynecol Oncol 1995;57:96–104.
39. Dettmar P, Harbeck N, Thomssen C. Prognostic impact of proliferation-associated factors MIB1 (Ki-67) and S-phase in node-negative breast cancer. Br J Cancer 1997;75:1525–1533.
40. Imamura H, Haga S, Shimizu T, et al. MIB1-determined proliferative activity in intraductal components and prognosis of invasive ductal breast carcinoma. Jpn J Cancer Res 1997;88:1017–1023.
41. Harbeck N, Dettmar P, Thomssen C, et al. Prognostic impact of tumor biological factors on survival in node-negative breast cancer. Anticancer Res 1998; 18:2187–2197.
42. Jacquemier JD, Penault-Llorca FM, Bertucci F, et al. Angiogenesis as a prognostic marker in breast carcinoma with conventional adjuvant chemotherapy: a multiparametric and immunohistochemical analysis. J Pathol 1998;184:130–135.
43. Midulla C, De Iorio P, Nagar C, et al. Immunohistochemical expression of p53, nm23-HI, Ki67 and DNA ploidy: correlation with lymph node status and other clinical pathologic parameters in breast cancer. Anticancer Res 1999;19:4033–4037.
44. Chang J, Powles TJ, Allred D, et al. Prediction of clinical outcome from primary tamoxifen by expression of biologic markers in breast cancer patients. Clin Cancer Res 2000;6:616–621.
45. McCready DR, Chapman JA, Hanna WM, et al. Factors affecting distant disease-free survival for primary invasive breast cancer: use of a log-normal survival model. Ann Surg Oncol 2000;7:416–426.
46. Vazquez-Ramirez FJ, Gonzalez-Campora JJ, Hevia-Alvarez E, et al. P-glycoprotein, metallothionein and NM23 protein expressions in breast carcinoma. Pathol Res Pract 2000;196:553–559.
47. Villar E, Redondo M, Rodrigo I, et al. bcl-2 Expression and apoptosis in primary and metastatic breast carcinomas. Tumour Biol 2001;22:137–145.
48. Schneider J, Gonzalez-Roces S, Pollan M, et al. Expression of LRP and MDR1 in locally advanced breast cancer predicts axillary node invasion at the time of rescue mastectomy after induction chemotherapy. Breast Cancer Res 2001;3:183–191.
49. Jones S, Clark G, Koleszar S, et al. Low proliferative rate of invasive node-negative breast cancer predicts for a favorable outcome: a prospective evaluation of 669 patients. Clin Breast Cancer 2001; 1:310–314.

50. Patla A, Rudnicka-Sosin L, Pawlega J, et al. Prognostic significance of selected immunohistochemical parameters in patients with invasive breast carcinoma concomitant with ductal carcinoma in situ. Pol J Pathol 2002;53:25–27.
51. Schneider J, Pollan M, Tejerina A, et al. Accumulation of uPA-PAI-1 complexes inside the tumour cells is associated with axillary nodal invasion in progesterone-receptor-positive early breast cancer. Br J Cancer 2003;88:96–101.
52. Penault-Llorca F, Cayre A, Bouchet Mishellany F, et al. Induction chemotherapy for breast carcinoma: predictive markers and relation with outcome. Int J Oncol 2003;22:1319–1325.
53. Han S, Park K, Bae BN, et al. Prognostic implication of cyclin E expression and its relationship with cyclin D1 and p27Kip1 expression on tissue microarrays of node negative breast cancer. J Surg Oncol 2003;83:241–247.
54. Lumachi F, Ermani M, Brandes AA, et al. Predictive value of different prognostic factors in breast cancer recurrences: multivariate analysis using a logistic regression model. Anticancer Res 2001;21:4105–4108.
55. Bast RC Jr, Ravdin P, Hayes DF, et al. 2000 Update of recommendations for the use of tumor markers in breast and colorectal cancer: clinical practice guidelines of the American Society of Clinical Oncology. J Clin Oncol 2001;19:1865–1878.
56. Hammond ME, Fitzgibbons PL, Compton CC, et al. College of American Pathologists Conference XXXV: solid tumor prognostic factors—which, how and so what? Summary document and recommendations for implementation. Cancer Committee and Conference Participants. Arch Pathol Lab Med 2000;124: 958–965.
57. Garcia RL, Coltrera MD, Gown AM. Analysis of proliferative grade using anti PCNA/Cyclin monoclonal antibodies in fixed, embedded tissues. Comparison with flow cytometric analysis. Am J Pathol 1989; 134:733–738.
58. Louis DN, Edgerton S, Thor AD, et al. Proliferating cell nuclear antigen and Ki–67 immunohistochemistry in brain tumors: a comparative study. Acta Neuropathol 1991;81:675–679.
59. Frierson HF Jr. Immunohistochemical analysis of proliferating cell nuclear antigen (PCNA) in infiltrating ductal carcinomas: comparison with clinical and pathologic variables. Modern Pathol 1993;6:290–294.
60. Sitonen SM, Isola JJ, Rantala IS, et al. Intratumor variation in cell proliferation in breast carcinoma as determined by antiproliferating cell nuclear antigen monoclonal antibody and automated image analysis. Am J Clin Pathol 1993;99:226–231.
61. Ioachim E, Kamina S, Athanassiadou S, et al. The prognostic significance of epidermal growth factor receptor (EGFR), C-erbB-2, Ki-67 and PCNA expression in breast cancer. Anticancer Res 1996;16: 3141–3147.
62. Sheen-Chen SM, Eng HL, Chou FF, et al. The prognostic significance of proliferating cell nuclear antigen in patients with lymph node-positive breast cancer. Arch Surg 1997;132:264–267.
63. Takashima T, Onoda N, Ishikawa T, et al. Proliferating cell nuclear antigen labeling index and p53 expression predict outcome for breast cancer patients with four or more lymph node metastases. Int J Mol Med 2001;8:159–163.
64. Kawakita N, Seki S, Yanai A, et al. Immunocytochemical identification of proliferating hepatocytes using anti-proliferating cell nuclear antigen (PCNA/Cyclin) monoclonal antibody and comparison with immunocytochemical staining using anti-DNA polymerase alpha monoclonal antibody in serial sections. Am J Clin Pathol 1992;97:S14–S20.
65. Clevenger CV, Epstein AL, Bauer KD. Modulation of the nuclear antigen p105 as a function of cell cycle progression. J Cell Physiol 1987;130:336–343.
66. Deschens J, Weidner N. Nucleolar organizer regions (NOR) in hyperplastic and neoplastic prostate disease. Am J Surg Pathol 1990;14:1148–1155.
67. Trere D, Pession A, Derenzini M. The silver-stained proteins of interphasic nucleolar organizer regions as a parameter of cell duplication rate. Exp Cell Res 1989;184:131–138
68. Mourad WA, Erkman-Balis B, Livingston S, et al. Argyrophilic nucleolar organizer regions in breast carcinoma. Cancer 1992;69:1739–1744.
69. Dervan PA, Gilmartin LG, Loftus BM, et al. Breast carcinoma kinetics: argyrophilic nucleolar organizer region counts correlate with Ki67 scores. Am J Clin Pathol 1989;92:401.
70. Sherr CJ. Mammalian G1 cyclins. Cell 1993;73: 1059–1065.
71. Hartwell LH, Kastan MB. Cell cycle control and cancer. Science 1994;266:1821–1828.
72. Keyomarsi K, Pardee AB. Redundant cyclin overexpression and gene amplification in breast cancer cells. Proc Natl Acad Sci USA 1993;90: 1112–1116.
73. Sandal T. Molecular aspects of the mammalian cell cycle and cancer. Oncologist 2002;7:73–81.
74. Weissman AM. Themes and variations on ubiquitination. Nat Rev Mol Cell Biol 2001;2:161–178.
75. Motokura T, Bloom T, Kim HG, et al. A novel cyclin encoded by a bcl1-linked candidate oncogene. Nature 1991;350:512–515.
76. Wolman SR, Pauley RJ, Mohamed AN, et al. Genetic markers as prognostic indicators in breast cancer. Cancer 1992;70:1765–1774.
77. Hunter T, Pines J. Cyclins and cancer. Cell 1991; 66:1071–1074.
78. Steeg PS, Zhou Q. Cyclins and breast cancer. Breast Cancer Res Treat 1998;52:17–28.
79. Bartkova J, Lukas J, Strauss M, et al. Cell cycle-related variation and tissue-restricted expression of human cyclin D1 protein. J Pathol 1994;172: 237–245.
80. Weinstat-Saslow D, Merino MJ, Manrow RE, et al. Overexpression of cyclin D mRNA distinguishes invasive and in situ breast carcinomas from non-malignant lesions. Nat Med 1995;1:1257–1260.
81. Wang TC, Cardiff RD, Zukerberg L, et al. Mammary hyperplasia and carcinoma in MMTV-cyclin D1 transgenic mice. Nature 1994;369:669–671.
82. Bartkova J, Lukas J, Muller H, et al. Cyclin D1 protein expression and function in human breast cancer. Int J Cancer 1994;57:353–361.
83. McIntosh GG, Anderson JJ, Milton I, et al. Determination of the prognostic value of cyclin D1 overexpression in breast cancer. Oncogene 1995;11: 885–891.
84. Pelosio P, Barbareschi M, Bonoldi E, et al. Clinical significance of cyclin D1 expression in patients with node-positive breast carcinoma treated with adjuvant therapy. Ann Oncol 1996;7:695–703.
85. van Diest PJ, Michalides RJ, Jannink L, et al. Cyclin D1 expression in invasive breast cancer. Correlations

and prognostic value. Am J Pathol 1997;150: 705–711.

86. Dublin EA, Patel NK, Gillett CE, et al. Retinoblastoma and p16 proteins in mammary carcinoma: their relationship to cyclin D1 and histopathological parameters. Int J Cancer 1998;79:71–75.
87. Seshadri R, Lee CS, Hui R, et al. Cyclin DI amplification is not associated with reduced overall survival in primary breast cancer but may predict early relapse in patients with features of good prognosis. Clin Cancer Res 1996;2:1177–1184.
88. Kenny FS, Hui R, Musgrove EA, et al. Overexpression of cyclin D1 messenger RNA predicts for poor prognosis in estrogen receptor-positive breast cancer. Clin Cancer Res 1999;5:2069–2076.
89. Utsumi T, Yoshimura N, Maruta M, et al. Correlation of cyclin D1 mRNA levels with clinico-pathological parameters and clinical outcome in human breast carcinomas. Int J Cancer 2000;89:39–43.
90. Turner BC, Gumbs AA, Carter D, et al. Cyclin D1 expression and early breast cancer recurrence following lumpectomy and radiation. Int J Radiat Oncol Biol Phys 2000;47:1169–1176.
91. Naidu R, Wahab NA, Yadav MM, et al. Expression and amplification of cyclin D1 in primary breast carcinomas: relationship with histopathological types and clinico-pathological parameters. Oncol Rep 2002;9:409–416.
92. Umekita Y, Ohi Y, Sagara Y, et al. Overexpression of cyclin D1 predicts for poor prognosis in estrogen receptor-negative breast cancer patients. Int J Cancer 2002;98:415–418.
93. Lim SC, Lee MS. Significance of E-cadherin/beta-catenin complex and cyclin D1 in breast cancer. Oncol Rep 2002;9:915–928.
94. Hwang TS, Han HS, Hong YC, et al. Prognostic value of combined analysis of cyclin D1 and estrogen receptor status in breast cancer patients. Pathol Int 2003;53:74–80.
95. Chasle J, Delozier T, Denoux Y, et al. Immunohistochemical study of cell cycle regulatory proteins in intraductal breast carcinomas—a preliminary study. Eur J Cancer 2003;39:1363–1369.
96. Lukas J, Bartkova J, Welcker M, et al. Cyclin D2 is a moderately oscillating nucleoprotein required for G1 phase progression in specific cell types. Oncogene 1995;10:2125–2134.
97. Evron E, Umbricht CB, Korz D, et al. Loss of cyclin D2 expression in the majority of breast cancers is associated with promoter hypermethylation. Cancer Res 2001;61:2782–2787.
98. Doglioni C, Chiarelli C, Macri E, et al. Cyclin D3 expression in normal, reactive and neoplastic tissues. J Pathol 1998;185:159–166.
99. Bartkova J, Lukas J, Strauss M, et al. Cyclin D3: requirement for G1/S transition and high abundance in quiescent tissues suggest a dual role in proliferation and differentiation. Oncogene 1998;17:1027–1037.
100. Russell A, Thompson MA, Hendley J, et al. Cyclin D1 and D3 associate with the SCF complex and are coordinately elevated in breast cancer. Oncogene 1999;18:1983–1991.
101. Yasmeen A, Berdel WE, Serve H, et al. E- and A-type cyclins as markers for cancer diagnosis and prognosis. Expert Rev Mol Diagn 2003;3:617–633.
102. Keyomarsi K, O'Leary N, Molnar G, et al. Cyclin E, a potential prognostic marker for breast cancer. Cancer Res 1994;54:380–385.
103. Dutta A, Chandra R, Leiter LM, et al. Cyclins as markers of tumor proliferation: immunocytochemical studies in breast cancer. Proc Natl Acad Sci USA 1995;92:5386–5390.
104. Keyomarsi K, Conte D Jr, Toyofuku W, et al. Deregulation of cyclin E in breast cancer. Oncogene 1995; 11:941–950.
105. Dou QP, Pardee AB, Keyomarsi K. Cyclin E—a better prognostic marker for breast cancer than cyclin D? Nat Med 1996;2:254.
106. Nielsen NH, Arnerlov C, Emdin SO, et al. Cyclin E overexpression, a negative prognostic factor in breast cancer with strong correlation to oestrogen receptor status. Br J Cancer 1996;74:874–880.
107. Scott KA, Walker RA. Lack of cyclin E immunoreactivity in non-malignant breast and association with proliferation in breast cancer. Br J Cancer 1997;76: 1288–1292.
108. Nielsen NH, Arnerlov C, Cajander S, et al. Cyclin E expression and proliferation in breast cancer. Anal Cell Pathol 1998;17:177–188.
109. Spruck CH, Won KA, Reed SI. Deregulated cyclin E induces chromosome instability. Nature 1999;401: 297–300.
110. Porter PL, Malone KE, Heagerty PJ, et al. Expression of cell-cycle regulators p27Kip1 and cyclin E, alone and in combination, correlate with survival in young breast cancer patients. Nat Med 1997;3: 222–225.
111. Donnellan R, Kleinschmidt I, Chetty R. Cyclin E immunoexpression in breast ductal carcinoma: pathologic correlations and prognostic implications. Hum Pathol 2001;32:89–94.
112. Kim HK, Park IA, Heo DS, et al. Cyclin E overexpression as an independent risk factor of visceral relapse in breast cancer. Eur J Surg Oncol 2001;27: 464–471.
113. Keyomarsi K, Tucker SL, Buchholz TA, et al. Cyclin E and survival in patients with breast cancer. N Engl J Med 2002;347:1566–1575.
114. Han S, Park K, Bae BN, et al. Prognostic implication of cyclin E expression and its relationship with cyclin D1 and p27Kip1 expression on tissue microarrays of node negative breast cancer. J Surg Oncol 2003;83:241–247.
115. Kuhling H, Alm P, Olsson H, et al. Expression of cyclins E, A, and B, and prognosis in lymph node-negative breast cancer. J Pathol 2003;199:424–431.
116. Span PN, Tjan-Heijnen VC, Manders P, et al. Cyclin-E is a strong predictor of endocrine therapy failure in human breast cancer. Oncogene 2003;22:4898–4904.
117. Akli S, Keyomarsi K. Cyclin E and its low molecular weight forms in human cancer and as targets for cancer therapy. Cancer Biol Ther 2003;2:S38–S47.
118. Yam CH, Fung TK, Poon RY. Cyclin A in cell cycle control and cancer. Cell Mol Life Sci 2002;59:1317–1326.
119. Bindels EM, Lallemand F, Balkenende A, et al. Involvement of G1/S cyclins in estrogen-independent proliferation of estrogen receptor-positive breast cancer cells. Oncogene 2002;21:8158–8165.
120. Woo RA, Poon RY. Cyclin-dependent kinases and S phase control in mammalian cells. Cell Cycle 2003; 2:316–324.
121. Bukholm IR, Bukholm G, Nesland JM. Over-expression of cyclin A is highly associated with early

relapse and reduced survival in patients with primary breast carcinomas. Int J Cancer 2001;93: 283–287.
122. Michalides R, van Tinteren H, Balkenende AJ, et al. Cyclin A is a prognostic indicator in early stage breast cancer with and without tamoxifen treatment. Br J Cancer 2002;86:402–408.
123. Michels JJ, Duigou F, Marnay J, et al. Flow cytometry and quantitative immunohistochemical study of cell cycle regulation proteins in invasive breast carcinoma: prognostic significance. Cancer 2003;97: 1376–1386.
124. Bukholm IR, Husdal A, Nesland JM, et al. Overexpression of cyclin A overrides the effect of p53 alterations in breast cancer patients with long follow-up time. Breast Cancer Res Treat 2003;80:199–206.
125. Smits VA, Medema RH. Checking out the G(2)/M transition. Biochim Biophys Acta 2001;1519:1–12.
126. Taylor WR, Stark GR. Regulation of the G2/M transition by p53. Oncogene 2001;20:1803–1815.
127. Winters ZE, Hunt NC, Bradburn MJ, et al. Subcellular localisation of cyclin B, Cdc2 and p21(WAF1/CIP1) in breast cancer. Association with prognosis. Eur J Cancer 2001;37:2405–2412.
128. Rudolph P, Kuhling H, Alm P, et al. Differential prognostic impact of the cyclins E and B in premenopausal and postmenopausal women with lymph node-negative breast cancer. Int J Cancer 2003;105:674–680.
129. John PC, Mews M, Moore R. Cyclin/Cdk complexes: their involvement in cell cycle progression and mitotic division. Protoplasma 2001;216:119–142.
130. Lee MH, Yang HY. Negative regulators of cyclin-dependent kinases and their roles in cancers. Cell Mol Life Sci 2001;58:1907–1922.
131. Doree M, Hunt T. From Cdc2 to Cdk1: when did the cell cycle kinase join its cyclin partner? J Cell Sci 2002;115:2461–2464.
132. Doree M, Galas S. The cyclin-dependent protein kinases and the control of cell division. FASEB J 1994;8:1114–1121.
133. Solomon MJ. Activation of the various cyclin/cdc2 protein kinases. Curr Opin Cell Biol 1993;5: 180–186.
134. Coleman TR, Dunphy WG. Cdc2 regulatory factors. Curr Opin Cell Biol 1994;6:877–882.
135. Livneh E, Fishman DD. Linking protein kinase C to cell-cycle control. Eur J Biochem 1997;248:1–9.
136. Ohta T, Fukuda M, Arima K, et al. Analysis of Cdc2 and cyclin D1 expression in breast cancer by immunoblotting. Breast Cancer 1997;4:17–24.
137. Depowski PL, Brien TP, Sheehan CE, et al. Prognostic significance of p34cdc2 cyclin-dependent kinase and MIB1 overexpression, and HER–2/neu gene amplification detected by fluorescence in situ hybridization in breast cancer. Am J Clin Pathol 1999;112: 459–469.
138. O'Connor DS, Wall NR, Porter AC, et al. A p34(cdc2) survival checkpoint in cancer. Cancer Cell 2002; 2:43–54.
139. Chen JG, Yang CP, Cammer M, et al. Gene expression and mitotic exit induced by microtubule-stabilizing drugs. Cancer Res 2003;63:7891–7899.
140. Tan AR, Swain SM. Review of flavopiridol, a cyclin-dependent kinase inhibitor, as breast cancer therapy. Semin Oncol 2002;29:77–85.
141. Sausville EA. Cyclin-dependent kinase modulators studied at the NCI: pre-clinical and clinical studies. Curr Med Chem Anticancer Agents 2003; 3:47–56.
142. Vermeulen K, Van Bockstaele DR, Berneman ZN. The cell cycle: a review of regulation, deregulation and therapeutic targets in cancer. Cell Prolif 2003; 36:131–149.
143. Fischer PM, Gianella-Borradori A. CDK inhibitors in clinical development for the treatment of cancer. Expert Opin Invest Drugs 2003;12:955–970.
144. Senderowicz AM. Cyclin-dependent kinases as targets for cancer therapy. Cancer Chemother Biol Response Modif 2002;20:169–196.
145. Mita MM, Mita A, Rowinsky EK. Mammalian target of rapamycin: a new molecular target for breast cancer. Clin Breast Cancer 2003;4:126–137.
146. Wang C, Fu M, Mani S, et al. Histone acetylation and the cell-cycle in cancer. Front Biosci 2001;6: D610–D629.
147. Li A, Blow JJ. The origin of CDK regulation. Nat Cell Biol 2001;3:182–184.
148. Joyce D, Albanese C, Steer J, et al. NF-kappaB and cell-cycle regulation: the cyclin connection. Cytokine Growth Factor Rev 2001;12:73–90.
149. Elliott PJ, Ross JS. The proteasome: a new target for novel drug therapies. Am J Clin Pathol 2001;116: 637–646.
150. Tsihlias J, Kapusta L, Slingerland J. The prognostic significance of altered cyclin-dependent kinase inhibitors in human cancer. Annu Rev Med 1999;50: 401–423.
151. Oh YL, Choi JS, Song SY, et al. Expression of p21Waf1, p27Kip1 and cyclin D1 proteins in breast ductal carcinoma in situ: relation with clinicopathologic characteristics and with p53 expression and estrogen receptor status. Pathol Int 2001;51:94–99.
152. Barbareschi M, Caffo O, Doglioni C, et al. p21WAF1 immunohistochemical expression in breast carcinoma: correlations with clinicopathological data, oestrogen receptor status, MIB1 expression, p53 gene and protein alterations and relapse-free survival. Br J Cancer 1996;74:208–215.
153. Caffo O, Doglioni C, Veronese S, et al. Prognostic value of p21(WAF1) and p53 expression in breast carcinoma: an immunohistochemical study in 261 patients with long-term follow-up. Clin Cancer Res 1996;2:1591–1599.
154. Wakasugi E, Kobayashi T, Tamaki Y, et al. p21(Waf1/Cip1) and p53 protein expression in breast cancer. Am J Clin Pathol 1997;107:684–691.
155. Jiang M, Shao ZM, Wu J, et al. p21/waf1/cip1 and mdm-2 expression in breast carcinoma patients as related to prognosis. Int J Cancer 1997;74:529–534.
156. Reed W, Florems VA, Holm R, et al. Elevated levels of p27, p21 and cyclin D1 correlate with positive oestrogen and progesterone receptor status in node-negative breast carcinoma patients. Virchows Arch 1999;435:116–124.
157. Mathoulin-Portier MP, Viens P, Cowen D, et al. Prognostic value of simultaneous expression of p21 and mdm2 in breast carcinomas treated by adjuvant chemotherapy with antracyclin. Oncol Rep 2000;7: 675–680.
158. Bankfalvi A, Tory K, Kemper M, et al. Clinical relevance of immunohistochemical expression of p53-targeted gene products mdm-2, p21 and bcl-2 in breast carcinoma. Pathol Res Pract 2000;196:489–501.
159. Thor AD, Liu S, Moore DH 2nd, et al. p(21WAF1/CIP1) expression in breast cancers: associations with

p53 and outcome. Breast Cancer Res Treat 2000;61: 33–43.

160. Domagala W, Welcker M, Chosia M, et al. p21/WAF1/Cip1 expression in invasive ductal breast carcinoma: relationship to p53, proliferation rate, and survival at 5 years. Virchows Arch 2001;439:132–140.
161. Gohring UJ, Bersch A, Becker M, et al. p21(waf) correlates with DNA replication but not with prognosis in invasive breast cancer. J Clin Pathol 2001;54: 866–870.
162. O'Hanlon DM, Kiely M, MacConmara M, et al. An immunohistochemical study of p21 and p53 expression in primary node-positive breast carcinoma. Eur J Surg Oncol 2002;28:103–107.
163. Pellikainen MJ, Pekola TT, Ropponen KM, et al. p21WAF1 expression in invasive breast cancer and its association with p53, AP-2, cell proliferation, and prognosis. J Clin Pathol 2003;56:214–220.
164. Michels JJ, Duigou F, Marnay J, et al. Flow cytometry and quantitative immunohistochemical study of cell cycle regulation proteins in invasive breast carcinoma: prognostic significance. Cancer 2003;97: 1376–1386.
165. Chasle J, Delozier T, Denoux Y, et al. Immunohistochemical study of cell cycle regulatory proteins in intraductal breast carcinomas—a preliminary study. Eur J Cancer 2003;39:1363–1369.
166. Sherr CJ, Roberts JM. CDK inhibitors: positive and negative regulators of G1-phase progression. Genes Dev 1999;13:1501–1512.
167. Chiarle R, Pagano M, Inghirami G. The cyclin dependent kinase inhibitor p27 and its prognostic role in breast cancer. Breast Cancer Res 2001;3:91–94.
168. Hengst L, Reed SI. Translational control of $p27^{Kip1}$ accumulation during the cell cycle. Science 1996; 271:1861–1864.
169. Pagano M, Tam SW, Theodoras AM, et al. Role of the ubiquitin-proteasome pathway in regulating abundance of the cyclin-dependent kinase inhibitor p27. Science 1995;269:682–685.
170. Tan P, Cady B, Wanner M, et al. The cell cycle inhibitor p27 is an independent prognostic marker in small (T1a,b) invasive breast carcinomas. Cancer Res 1997;57:1259–1263.
171. Wu J, Shen ZZ, Lu JS, et al. Prognostic role of p27Kip1 and apoptosis in human breast cancer. Br J Cancer 1999;79:1572–1578.
172. Tsuchiya A, Zhang GJ, Kanno M. Prognostic impact of cyclin-dependent kinase inhibitor p27kip1 in node-positive breast cancer. J Surg Oncol 1999;70: 230–234.
173. Gillett CE, Smith P, Peters G, et al. Cyclin-dependent kinase inhibitor p27Kip1 expression and interaction with other cell cycle-associated proteins in mammary carcinoma. J Pathol 1999;187:200–206.
174. Chu JS, Huang CS, Chang KJ. p27 expression as a prognostic factor of breast cancer in Taiwan. Cancer Lett 1999;141:123–130.
175. Leong AC, Hanby AM, Potts HW, et al. Cell cycle proteins do not predict outcome in grade I infiltrating ductal carcinoma of the breast. Int J Cancer 2000;8: 26–31.
176. Barbareschi M, van Tinteren H, Mauri FA, et al. p27(kip1) expression in breast carcinomas: an immunohistochemical study on 512 patients with long-term follow-up. Int J Cancer 2000;89:236–241.
177. Volpi A, De Paola F, Nanni O, et al. Prognostic significance of biologic markers in node-negative breast cancer patients: a prospective study. Breast Cancer Res Treat 2000;63:181–192.
178. Lau R, Grimson R, Sansome C, et al. Low levels of cell cycle inhibitor p27kip1 combined with high levels of Ki-67 predict shortened disease-free survival in T1 and T2 invasive breast carcinomas. Int J Oncol 2001;18:17–23.
179. Nohara T, Ryo T, Iwamoto S, et al. Expression of cell-cycle regulator p27 is correlated to the prognosis and ER expression in breast carcinoma patients. Oncology 2001;60:94–100.
180. Leivonen M, Nordling S, Lundin J, et al. p27 expression correlates with short-term, but not with long-term prognosis in breast cancer. Breast Cancer Res Treat 2001;67:15–22.
181. Barnes A, Pinder S, Bell J, et al. Expression of p27kip1 in breast cancer and its prognostic significance. J Pathol 2003;201:451–459.
182. Blain SW, Scher HI, Cordon-Cardo C, et al. p27 as a target for cancer therapeutics. Cancer Cell 2003;3: 111–115.
183. Signoretti S, Di Marcotullio L, Richardson A, et al. Oncogenic role of the ubiquitin ligase subunit Skp2 in human breast cancer. J Clin Invest 2002;110: 633–641.
184. Liu MC, Gelmann EP. P53 gene mutations: case study of a clinical marker for solid tumors. Semin Oncol 2002;29:246–257.
185. Gasco M, Shami S, Crook T. The p53 pathway in breast cancer. Breast Cancer Res 2002;4:70–76.
186. Bhargava V, Thor A, Deng G, et al. The association of p53 immunopositivity with tumor proliferation and other prognostic indicators in breast cancer. Mod Pathol 1994;7:361–368.
187. Rosanelli GP, Steindorfer P, Wirnsberger GH, et al. Mutant p53 expression and DNA analysis in human breast cancer. Comparison with conventional clinicopathological parameters. Anticancer Res 1995;15: 581–586.
188. Pelosi G, Bresaola E, Rodella S, et al. Expression of proliferating cell nuclear antigen, Ki-67 antigen, estrogen receptor protein, and tumor suppressor p53 gene in cytologic samples of breast cancer: an immunochemical study with clinical, pathobiological, and histologic correlations. Diagn Cytopathol 1994; 11:131–140.
189. Beck T, Weller EE, Weikel W, et al. Usefulness of immunohistochemical staining for p53 in the prognosis of breast carcinomas: correlation with established prognosis parameters and with the proliferation marker, MIB-1. Gynecol Oncol 1995;57:96–104.
190. Peyrat J-P, Bonneterre J, Lubin R, et al. Prognostic significance of circulating P53 antibodies in patients undergoing surgery for local regional breast cancer. Lancet 1995;345:621–622.
191. Ferrero JM, Ramaioli A, Formento JL, et al. P53 determination alongside classical prognostic factors in node-negative breast cancer: an evaluation at more than 10-year follow-up. Ann Oncol 2000;11: 393–397.
192. Geisler S, Lonning PE, Aas T, et al. Influence of TP53 gene alterations and c-erbB-2 expression on the response to treatment with doxorubicin in locally advanced breast cancer. Cancer Res 2001;61:2505–2512.
193. Montero S, Guzman C, Vargas C, et al. Prognostic value of cytosolic p53 protein in breast cancer. Tumour Biol 2001;22:337–344.

194. Daidone MG, Veneroni S, Benini E, et al. Biological markers as indicators of response to primary and adjuvant chemotherapy in breast cancer. Int J Cancer 1999;84:580–586.

195. Kandioler-Eckersberger D, Ludwig C, et al. TP53 mutation and p53 overexpression for prediction of response to neoadjuvant treatment in breast cancer patients. Clin Cancer Res 2000;6:50–56.

196. Bertheau P, Plassa F, Espie M, et al. Effect of mutated TP53 on response of advanced breast cancers to high-dose chemotherapy. Lancet 2002;360: 852–854.

197. Quesnel B, Preudhomme C, Fournier J, et al. MDM2 gene amplification in human breast cancer. Eur J Cancer 1994;30A:982–984.

198. McCann AH, Kirley A, Carney DN, et al. Amplification of the MDM2 gene in human breast cancer and its association with MDMp53 protein status. Br J Cancer 1995;71:981–985.

199. Jiang M, Shao ZM, Wu J, et al. p21/waf1/cip1 and mdm-2 expression in breast carcinoma patients as related to prognosis. Int J Cancer 1997;74:529–534.

200. Mathoulin-Portier MP, Viens P, Cowen D, et al. Prognostic value of simultaneous expression of p21 and mdm2 in breast carcinomas treated by adjuvant chemotherapy with anthracycline. Oncol Rep 2000; 7:675–680.

201. Cuny M, Kramar A, Courjal F, et al. Relating genotype and phenotype in breast cancer: an analysis of the prognostic significance of amplification at eight different genes or loci and of p53 mutations. Cancer Res 2000;60:1077–1083.

202. Bankfalvi A, Tory K, Kemper M, et al. Clinical relevance of immunohistochemical expression of p53-targeted gene products mdm-2, p21 and bcl-2 in breast carcinoma. Pathol Res Pract 2000;196: 489–501.

203. Benedict WF, Xu H-J, Takahashi R. The retinoblastoma gene: Its role in human malignancies. Cancer Invest 1990;8:535–540.

204. Varley JM, Armour J, Swallow JE, et al. The retinoblastoma gene is frequently altered leading to loss of expression in primary breast tumors. Oncogene 1989;4:725–729.

205. Berns EM, de Klein A, van Putten WL, et al. Association between RB-1 gene alterations and factors of favourable prognosis in human breast cancer, without effect on survival. Int J Cancer 1995;64: 140–145.

206. Lee MH, Yang HY. Negative regulators of cyclin-dependent kinases and their roles in cancers. Cell Mol Life Sci 2001;58:1907–1922.

207. Sherr CJ. The INK4a/ARF network in tumour suppression. Nat Rev Mol Cell Biol 2001;2:731–737.

208. Esteller M. CpG island hypermethylation and tumor suppressor genes: a booming present, a brighter future. Oncogene 2002;21:5427–5440.

209. Hui R, Macmillan RD, Kenny FS, et al. INK4a gene expression and methylation in primary breast cancer: overexpression of p16INK4a messenger RNA is a marker of poor prognosis. Clin Cancer Res 2000;6: 2777–2787.

210. Han S, Ahn SH, Park K, et al. P16INK4a protein expression is associated with poor survival of the breast cancer patients after CMF chemotherapy. Breast Cancer Res Treat 2001;70:205–212.

211. Milde-Langosch K, Bamberger AM, Rieck G, et al. Overexpression of the p16 cell cycle inhibitor in breast cancer is associated with a more malignant phenotype. Breast Cancer Res Treat 2001;67:61–70.

212. Ito Y, Kobayashi T, Takeda T, et al. Expression of p16 and cyclin-dependent kinase 4 proteins in primary breast carcinomas. Oncology 1997;54:508–515.

CHAPTER 14

Targeted Therapy for Cancer: Integrating Diagnostics and Therapeutics

Jeffrey S. Ross, MD

Albany Medical College, Albany, New York and Millennium Pharmaceuticals, Inc., Cambridge, Massachusetts

The recent regulatory approval in the United States and Europe of imatinib mesylate (Gleevec®) for patients with bcr/abl translocation positive chronic myelogenous leukemia (FIG. 14.1) and the subsequent approval for gastrointestinal stromal tumors featuring an activating c-*kit* growth factor receptor mutation infused the oncology drug development community with enthusiasm for anticancer targeted therapy.[1,2] Recent major news magazines have headlined the considerable public interest in new anticancer drugs that exploit disease-specific genetic defects as the target of their mechanism of action.[3,4] Many scientists, clinicians, and pharmaceutical company executives now believe that in the next 5 to 10 years, the integration of molecular oncology and molecular diagnostics will further revolutionize oncology drug discovery and development; customize the selection, dosing, and route of administration of both previously approved traditional agents and new therapeutics in clinical trials; and truly personalize medical care for the cancer patient.[5–8]

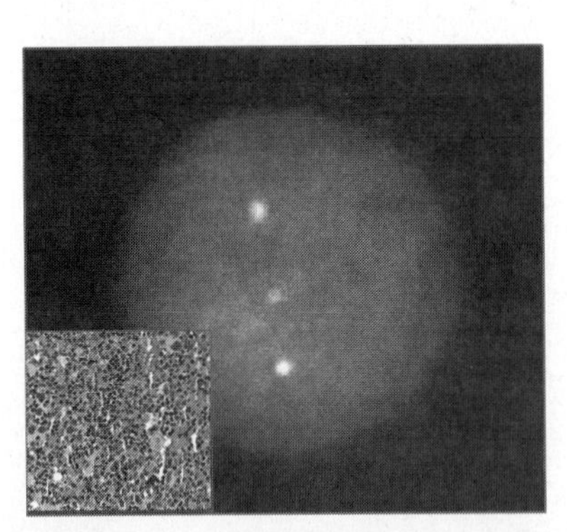

FIGURE 14.1 Chronic myelogenous leukemia and imatinib (Gleevec®) therapy. Photomicrograph demonstrates a *bcr/abl* translocation detected by FISH in a patient with a packed bone marrow biopsy diagnostic of chronic myelogenous leukemia (inset). Note the yellow "fusion" gene product, indicating the apposition of one green (chromosome 22) and one red (chromosome 9) resulting from the translocation. This patient was treated with single agent imatinib (Gleevec®) and achieved complete remission of bone marrow histology and absence of *bcr/abl* by routine cytogenetics and FISH assessment. **See Plate 30 for color image.**

Targeted Therapies for Cancer: Definitions

During the last several years, multiple definitions for the term "targeted therapy" have emerged. From the regulatory perspective, targeted therapy has been defined as a drug in whose approval label there is a specific reference to a simultaneously or previously approved diagnostic test that must be performed before the patient can be considered eligible to receive the drug. The classic example of this definition for targeted therapy is the coapprovals of the anti-breast cancer antibody trastuzumab (Herceptin®) and the eligibility tests (Herceptest®, Pathway®, and Pathvysion®) (see Chapter 16 and below). For many scientists and oncologists, targeted therapy is defined as a drug with a focused mechanism that specifically acts on a well-defined target or biologic pathway that, when inactivated, causes regression or destruction of the malignant process. Examples of this type of targeted therapy include hormonal-based therapies (see Chapter 15), inhibitors of the epidermal growth factor receptor (EGFR) pathway (see Chapter 17), blockers of invasion and metastasis enabling proteins and enzymes (see Chapter 18), antiangiogenesis agents (see Chapter 19), proapoptotic drugs (see Chapter 21), and proteasome inhibitors (see Chapter 24). In addition, most scientists and oncologists consider anticancer antibody therapeutics that seek out and kill malignant cells bearing the target antigen as another type of targeted therapy.

The Ideal Target

The ideal cancer target (**Table 14.1**) can be defined as a macromolecule that is crucial to the malignant phenotype and is not significantly expressed in vital organs and tissues; that has biologic relevance that can be reproducibly measured in readily obtained clinical samples; that is definably correlated with clinical outcome; and that interruption, interference, or inhibition of such a macromolecule yields a clinical response in a significant proportion of patients whose tumors express the target with minimal to absent responses in patients whose tumors do not express the target. For antibody therapeutics, additional important criteria include the use of cell surface targets that when complexed with the therapeutic naked or conjugated antibody, internalize the antigen–antibody complex by reverse pinocytosis, thus facilitating tumor cell killing.

The Original Targeted Therapy: Antiestrogens for Breast Cancer

Arguably the first type of targeted therapy in oncology was the development of antiestrogen therapies for patients with breast cancer that expressed the estrogen receptor (ER) protein (see Chapter 15) (FIG. 14.2).[9] Originally developed as a competitive binding bioassay performed on fresh tumor protein extracts and used to select for hormone production ablation by surgery (oophorectomy, adrenalectomy, and hypophysectomy), the ER and progesterone receptor test format converted to an immunohistochemistry (IHC) platform when the decreased size of primary tumors enabled by mass screening programs produced insufficient material for creating enough fresh tissue to perform the assay.[10] The drug tamoxifen (Nolvadex®), which has both hormonal and nonhormonal mechanisms of action, has been the most widely prescribed antiestrogen for the treatment of metastatic breast cancer and chemoprevention of the disease in high risk women.[11,12] Although, ER and progesterone receptor testing is the front line for predicting tamoxifen response, additional biomarkers, including HER-2/neu (HER-2) and cathepsin D testing, have been used to further refine therapy selection.[13] The introductions of specific estrogen response modulators and aromatase inhibitors such as anastrozole (Arimidex®), letrozole (Femara®), and the combination chemotherapeutic, estramustine (Emcyt®)[14–18] have added new strategies for evaluating tumors for hormonal therapy.

Table 14.1 Features of the Ideal Anticancer Target

- Crucial to the malignant phenotype
- Not significantly expressed in vital organs and tissues
- A biologically relevant molecular feature
- Reproducibly measurable in readily obtained clinical samples
- Correlated with clinical outcome
- When interrupted, interfered with, or inhibited, the result is a clinical response in a significant proportion of patients whose tumors express the target
- Responses in patients whose tumors do not express the target are minimal

Leukemia and Lymphoma Lead the Way

The introduction of immunophenotyping for leukemia and lymphoma was followed by the first applications of DNA based assays, the polymerase chain reaction, and RNA based molecular technologies in these diseases that complemented continuing advances in tumor cytogenetics.[19,20] In addition to the imatinib (Gleevec®) targeted therapy for chronic myelogenous leukemia, other molecular targeted therapy in hematologic malignancies includes the use of all-*trans*-retinoic acid (ATRA) for the treatment of acute promyelocytic leukemia[21]; anti-CD20 antibody therapeutics targeting non-Hodgkin lymphomas, including rituximab (Rituxan®)[22]; and the emerging Flt-3 target for a subset of acute myelogenous leukemia patients (see below).[23]

Thirty Years Later: HER-2 and Trastuzumab (Herceptin®)

After the introduction of hormone receptor testing, some 30 years then elapsed before the next major targeted cancer chemotherapy program for a solid tumor was developed. In the mid-1980s, the discovery of the *HER-2* (c-*erb*B2) gene and protein and subsequent association with an adverse outcome in breast cancer provided clinicians with a new biomarker that could be used to guide adjuvant chemotherapy.[24] The development of trastuzumab (Herceptin®), a humanized monoclonal

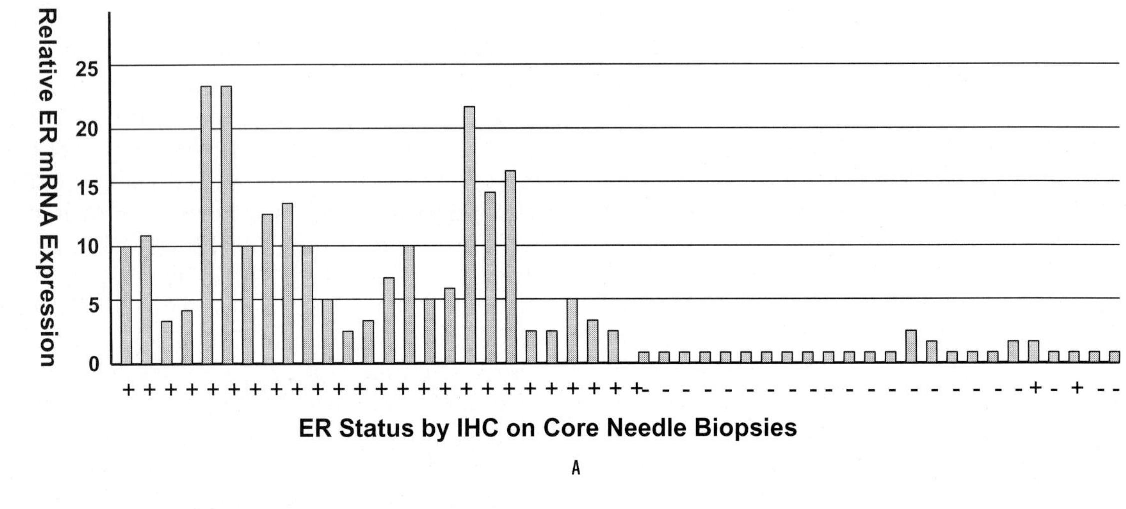

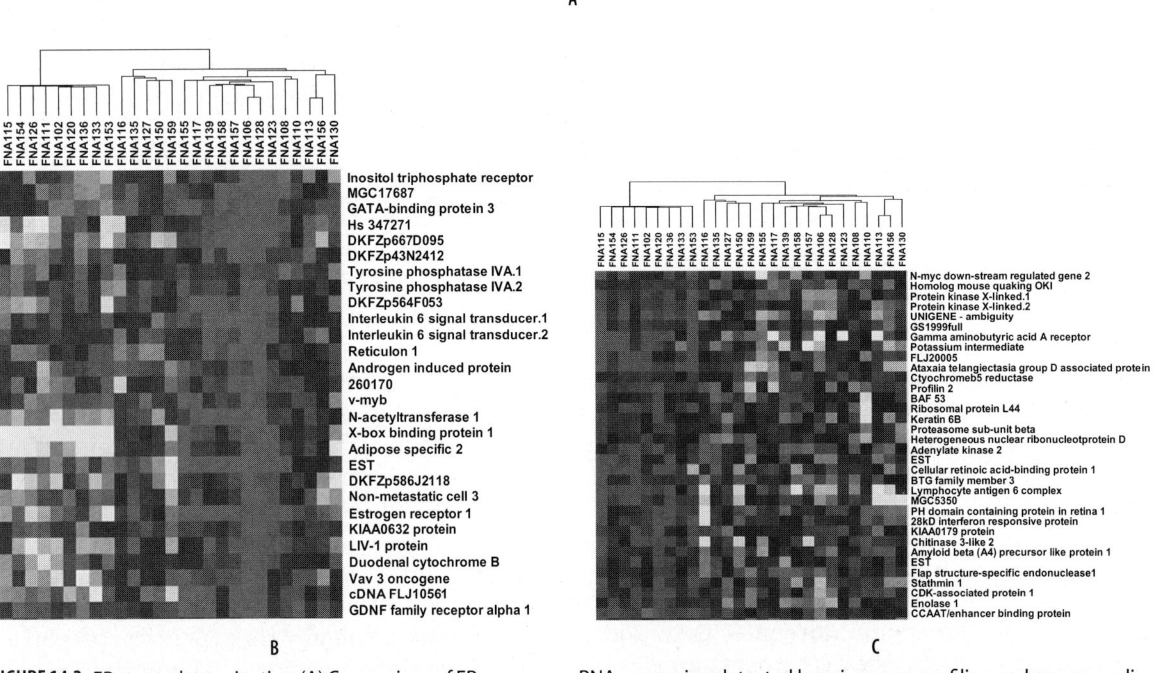

FIGURE 14.2 ER status determination. (A) Comparison of ER messenger RNA expression detected by microarray profiling and corresponding ER protein expression measured by IHC. The concordance between ER levels determined by IHC and ER levels determined by gene expression profiling was about 95%. (B) Genes expressed in ER-positive cases. (C) Genes expressed in ER-negative cases. **See Plate 31 for color image.**

antibody designed to treat advanced metastatic breast cancer that had failed first- and second-line chemotherapy, caused a rapid wide adoption of HER-2 testing of the patients' primary tumors.[25] However, soon after its approval, widespread confusion concerning the most appropriate diagnostic test to determine HER-2 status in formalin-fixed paraffin-embedded breast cancer tissues (see Chapter 16) substantially impacted trastuzumab use.[26–33] Since its launch in 1998, trastuzumab has become an important therapeutic option for patients with HER-2–positive breast cancer.[34–37]

Reports that fluorescence in situ hybridization (FISH) could out-perform IHC in predicting trastuzumab response[38] and the well-documented lower response rates of intermediate (2+) IHC staining versus intense (3+) staining tumors[39] resulted in a variety of approaches, including IHC as a primary screen with follow-on FISH testing of either 1+ cases, 2+ cases, or both 1+ and 2+ cases or primary FISH based testing. In a recently published study where trastuzumab was used as a single agent, the response rates in 111 assessable patients with 3+ IHC staining was 35% and the response rates for 2+ cases was 0%; the response rates in patients with and without *HER-2* gene amplification detected by FISH were 34% and 7%, respectively.[39] In another study of HER-2–positive breast cancer treated with trastuzumab plus paclitaxel, overall response rates ranged from 67% to 81% compared with 41% to 46% in patients with normal expression of HER-2.[40] Interestingly, although FISH based testing is more expensive to perform and is not as widely available as IHC, a recent published review from New York and Italy suggested that FISH was actually the most cost-effective option.[41] With trastuzumab achieving excellent results in the treatment of HER-2–positive advanced metastatic disease and under extensive evaluation in major clinical trials for its potential efficacy when used in earlier stages, the potential role(s) for HER-2 testing as a predictor(s) of responses to other therapies being resolved by large prospective clinical outcome studies and the more convenient gene-based chromogenic in situ hybridization technique "waiting in the wings," the story of HER-2 testing in breast cancer will continue to unfold over the next several years.

Targeted Anticancer Therapies Using Antibodies

An unprecedented number and variety of targeted small molecule and antibody based therapeutics are currently in early development and clinical trials for the treatment of cancer. Therapeutic antibodies have become a major strategy in clinical oncology because of their ability to specifically bind to primary and metastatic cancer cells with high affinity and create antitumor effects by complement-mediated cytolysis and antibody-dependent cell-mediated cytotoxicity (naked antibodies) or by the focused delivery of radiation or cellular toxins (conjugated antibodies).[42–47] Currently, there are six anticancer therapeutic antibodies approved by the US Food and Drug Administration (FDA) for sale in the United States (**Table 14.2**). Therapeutic monoclonal antibodies are typically of the IgG class containing two heavy and two light chains. The heavy chains form a fused "Y" structure with two light chains running in parallel to the open portion of the heavy chain. The tips of the heavy–light chain pairs form the antigen binding sites, with the primary antigen recognition regions known as the complementarity determining regions.

The early promise of mouse monoclonal antibodies for the treatment of human cancers was not realized because (1) unfocused target selection led to the identification of target antigens that were not critical for cancer cell survival and progression, (2) there was a low overall potency of naked mouse antibodies as anticancer drugs, (3) antibodies penetrated tumor cells poorly, (4) there was limited success in producing radioisotope and toxin conjugates, and (5) the development of human antimouse antibodies (HAMA) prevented the use of multiple dosing schedules.[48]

The next advance in antibody therapeutics began in the early 1980s when recombinant DNA technology was applied to antibody design to reduce the antigenicity of murine and other rodent-derived monoclonal antibodies. Chimeric antibodies were developed where the constant domains of the human IgG molecule were combined with the murine variable regions by transgenic fusion of the immunoglobulin genes; the chimeric monoclonal antibodies were produced from engineered hybridomas and CHO cells.[49,50] The use of chimeric antibodies significantly reduced the HAMA responses but did not completely eliminate them.[51,52] Although several chimeric antibodies achieved regulatory approval, certain targets required humanized antibodies to achieve appropriate dosing. Partially humanized antibodies were then developed where the six complementarity determining regions of the heavy and light chains and a limited number of structural amino acids of the murine monoclonal antibody were grafted by recombinant technology to the complementarity determining region depleted human IgG scaffold.[53] Although this process further reduced or eliminated the HAMA responses, in many cases significant further antibody design procedures were needed to reestablish the required specificity and affinity of the original murine antibody.[54,55]

A second approach to reducing the immunogenicity of monoclonal antibodies has been to replace immunogenic epitopes in the murine variable domains with benign amino acid sequences, resulting in a deimmunized variable domain. The deimmunized variable domains are genetically linked to human IgG constant domains to yield a deimmunized antibody (Biovation, Aberdeen, Scotland). Additionally, primatized antibodies were subsequently developed that featured a chimeric antibody structure of human and monkey that, as a near exact copy of a human antibody, further reduced immunogenicity and enabled the capability for continuous repeat dosing and chronic therapy.[56] Finally, fully human antibodies have now been developed using murine sources and transgenic techniques.[56]

Table 14.2 Targeted Anticancer Antibody Therapeutics

Name	FDA Approval	Source and Partners	Type	Target	Indication(s) (both approved and investigational)
Alemtuzumab (Campath®)	05/01	BTG ILEX Oncology Schering AG	Monoclonal antibody, humanized Anticancer, immunologic Multiple sclerosis treatment Immunosuppressant	CD52	Cancer, leukemia, chronic lymphocytic Cancer, leukemia, chronic myelogenous Multiple sclerosis, chronic progressive
Daclizumab (Zenapax®)	03/02	Protein Design Labs Hoffmann-La Roche	Monoclonal IgG_1 Chimeric Immunosuppressant Antipsoriasis Antidiabetic Ophthalmological Multiple sclerosis treatment		Transplant rejection, general Transplant rejection, bone marrow Uveitis Multiple sclerosis, relapsing-remitting Multiple sclerosis, chronic progressive Cancer, leukemia, general Psoriasis Diabetes, type I Asthma Colitis, ulcerative
Rituximab (Rituxan®)	11/97	IDEC Genentech Hoffmann-La Roche Zenyaku Kogyo	Monoclonal IgG_1 Chimeric Anticancer, immunologic Antiarthritic, immunologic Immunosuppressant	CD20	Cancer, lymphoma, non-Hodgkin Cancer, lymphoma, B cell Arthritis, rheumatoid Cancer, leukemia, chronic lymphocytic Thrombocytopenic purpura
Trastuzumab (Herceptin®)	09/98	Genentech Hoffmann-La Roche ImmunoGen	Monoclonal IgG_1 Humanized Anticancer, immunologic	p185neu	Cancer, breast Cancer, lung, non–small cell Cancer, pancreatic
Gemtuzumab (Mylotarg®)	05/00	Wyeth/AHP	Monoclonal IgG_4 Humanized	CD33/ coleacheamycin	Cancer, leukemia, AML (patients older than 60 years)
Ibritumomab (Zevalin®)	02/02	IDEC	Monoclonal IgG_1 Murine Anticancer	CD20/^{90}Y	Cancer, lymphoma, low grade, follicular, transformed non-Hodgkin (relapsed or refractory)
Edrecolomab (Panorex™)	01/95*	Glaxo-Smith-Kline	Monoclonal IgG_{2A} Murine Anticancer	Epithelial cell adhesion molecule (Ep-CAM)	Cancer, colorectal
Tositumomab (Bexxar®)	06/03	Corixa	Anti-CD 20 Murine Monocolonal antibody with ^{131}I conjugation	CD20	Cancer, lymphoma, non-Hodgkin

AML, acute myelogenous leukemia.

Using modern antibody design and deimmunization technologies, scientists and clinicians have attempted to improve the efficacy and reduce the toxicity of anticancer antibody therapeutics.[42,48,57–59] The bacteriophage antibody design system has facilitated the development of high affinity antibodies by increasing antigen binding rates and reducing corresponding detachment rates.[59] Increased antigen binding is also achieved in bivalent antibodies with multiple attachment sites, a feature known as avidity. Modern antibody design has endeavored to create small antibodies that can penetrate to cancerous sites but maintain their affinity and avidity. A variety of approaches has been used to increase antibody efficacy.[42] Clinical trials have recently combined anticancer antibodies with conventional cytotoxic drugs, yielding promising results.[42–45] The application of radioisotope, a small molecule cytotoxic drug, and protein toxin conjugation has resulted in promising results in clinical trials and achieved regulatory approval for several drugs now on the market (see below). Antibodies have also been designed to increase their enhancement of effector functions of antibody-dependent cellular cytotoxicity. Another cause of toxicity of conjugated antibodies has been the limitations of the conjugation technology, which can restrict the ratio of the number of toxin molecules per antibody molecule.[42,43,55] Methods designed to overcome the toxicity of conjugated antibodies include the use of antibody targeted liposomal small molecule drug conjugates and the use of antibody conjugates with drugs in nanoparticle formats to enhance bonding strength that enable controlled release of the cytotoxic agent. Another technique that uses site selective prodrug activation to reduce bystander tissue toxicity is the antibody directed enzyme prodrug therapy. An antibody bound enzyme is targeted to tumor cells. This allows for selective activation of a nontoxic prodrug to a cytotoxic agent at the tumor site for cancer therapy.

A variety of factors can reduce antibody efficacy[48]: (1) limited penetration of the antibody into a large solid tumor or into vital regions such as the brain, (2) reduced extravasations of antibodies into target sites due to decreased vascular permeability, (3) cross-reactivity and nonspecific binding of antibody to normal tissues reduces targeting effect, (4) heterogeneous tumor uptake results in untreated zones, (5) increased metabolism of injected antibodies reduces therapeutic effects, and (6) HAMA and human antihuman antibodies form rapidly and inactivate the therapeutic antibody.

Toxicity has been a major obstacle in the development of therapeutic antibodies for cancer.[42–45] Cross-reactivity with normal tissues can cause significant side effects for unconjugated (naked) antibodies, which can be enhanced when the antibodies are conjugated with toxins or radioisotopes. Immune mediated complications can include dyspnea from pulmonary toxicity, occasional central and peripheral nervous system complications, and decreased liver and renal function. On occasion, unexpected toxic complications can be seen, such as the cardiotoxicity associated with the HER-2/*neu* targeting antibody trastuzumab. Radioimmunotherapy with isotopic-conjugated antibodies can also cause bone marrow suppression (see below).

Unconjugated or naked antibodies include a variety of targeting molecules both on the market and in early and late clinical development. A variety of mechanisms has been cited to explain the therapeutic benefit of these drugs, including enhanced immune effector functions and direct inactivation of the targeted pathways as seen in the antibodies directed at surface receptors such as HER-1 (EGFR) and HER-2.[2–5] Surface receptor targeting can reduce intracellular signaling, resulting in decreased cell growth and increased apoptosis.[56]

As seen in **Table 14.2**, of the seven anticancer antibodies on the market, one is conjugated with a radioisotope Y^{90}-ibritumomab tiuxetan (Zevalin®) and one is conjugated to a complex natural product toxin gemtuzumab ozogamicin (Mylotarg®). Conjugation procedures have been designed to improve antibody therapy efficacy and have used a variety of methods to complex the isotope, toxin, or cytotoxic agent to the antibody.[42,43] Cytotoxic small molecule drug conjugates have been widely tested, but enthusiasm for this approach has been limited by the relatively low potency of these compounds.[42] Fungal derived potent toxins have yielded greater success with the calicheamicin conjugated anti-CD33 antibody gemtuzumab ozogamicin approved for the treatment of acute myelogenous leukemia and a variety of antibodies conjugated with the fungal toxin maytansanoid (DM-1) in preclinical development and early clinical trials. The interest in radioimmunotherapy increased significantly in 2001 with the FDA approvals of the ^{90}Y-conjugated anti-CD20 antibody Y^{90}-ibritumomab tiuxetan and the ^{131}I-conjugated anti-CD20 antibody I^{131}-tositumomab. A variety of isotopes is under investigation in addition to ^{90}Y as potential conjugates for anticancer antibodies.[43] Radioimmunotherapy features the phenomenon of the bystander effect, in which if antigen expression is heterogeneous, extensive tumor cell killing can still take place, even on nonexpressing cells, but can also lead to significant toxicity when the neighboring cells are vital non-neoplastic tissues such as the bone marrow and liver.

Antibody Therapeutics for Hematologic Malignancies

The earliest and most successful clinical use of antibodies in oncology has been for the treatment of hematologic malignancies.[42–45,56,60–63]

By taking advantage of improved recombinant technologies generating more specific and higher affinity monoclonal antibodies with reduced immunogenicity after humanization or deimmunization and the emerging conjugation capabilities, antibody therapeutics have become a major weapon in the treatment of leukemias and lymphomas.[60–63]

Approved in 1997, rituximab (Rituxan®) is arguably the most commercially successful anticancer drug of any type since the introduction of taxanes. Rituximab sales exceeded $700 million in sales in the United States in 2001.[44] Targeting the CD20 surface receptor common to many B cell non-Hodgkin lymphoma subtypes, rituximab is a chimeric monoclonal IgG_1 antibody that induces apoptosis, antibody-dependent cell cytotoxicity, and complement-mediated cytotoxicity[56] and has achieved significantly improved disease-free survival rates compared with patients receiving cytotoxic agents alone.[64–67]

Y^{90}-ibritumomab tiuxetan (Zevalin®) consists of the murine version of the anti-CD20 chimeric monoclonal antibody, rituximab, which has been covalently linked to the metal chelator, MD-DTPA, permitting stable binding of ^{111}In when used for radionucleotide tumor imaging and ^{90}Y when used to produce enhanced targeted cytotoxicity.[68–71] In early 2002, Y^{90}-ibritumomab tiuxetan became the first radioconjugated antibody therapeutic for cancer approved by the FDA. Since its FDA approval, numerous patients who have received Y^{90}-ibritumomab tiuxetan after becoming refractory to a rituximab-based regimen have achieved significant responses.[69,70]

The approval of gemtuzumab ozogamicin (Mylotarg®) by the FDA in 2000 marked the first introduction of a plant toxin conjugated antibody therapeutic.[72–76] Gemtuzumab ozogamicin is targeted against CD33, a surface marker expressed by 90% of myeloid leukemic blasts but absent from stem cells, armed with calicheamicin, a potent cytotoxic antibiotic that inhibits DNA synthesis and induces apoptosis.[72] The current indication for use of gemtuzumab ozogamicin is in acute myelogenous leukemia patients older than 60 years with the recommendation that before the initiation of therapy, the leukemic blast count is below 30,000/mL.[73–75]

Alemtuzumab (Campath®), a humanized monoclonal antibody, was approved in mid-2001 for the treatment of B-cell chronic lymphocytic leukemia in patients who have been treated with alkylating agents and who have failed fludarabine therapy.[77,78] Daclizumab (Zenapax®) is a chimeric monoclonal antibody that targets the interleukin-2 receptor. This antibody is primarily used to prevent and treat patients with organ transplant rejection but has also been used in a wide variety of chronic inflammatory conditions, including psoriasis, multiple sclerosis, ulcerative colitis, asthma, type I diabetes mellitus, uveitis, and also in a variety of leukemias.[79,80] I^{131}-tositumomab (Bexxar®) is a radiolabeled anti-CD20 murine monoclonal antibody approved in 2003 for the treatment of relapsed and refractory follicular/ low grade and transformed non-Hodgkin lymphoma.[81,82]

Antibody Therapeutics for Solid Tumors

Interest in the development of antibody therapeutics for solid tumors among many commercial organizations and universities has been significantly impacted by the technologic advances in antibody engineering and the approval and recent clinical and commercial success of trastuzumab, the only therapeutic antibody approved by the FDA for the treatment of solid tumors (edrecolomab is approved in Germany, but not in the United States). Trastuzumab has been described extensively above.

During the 4 years since the FDA approval of trastuzumab, there have been no additional antibodies approved for the treatment of solid tumors. Nonetheless, significant progress has been made in this field, and a number of both late stage and early stage products show substantial promise.

Cetuximab (Erbitux®)

The EGFR (HER-1) is the target of a variety of small molecule drugs and the late stage antibody cetuximab.[83] Cetuximab, a chimeric monoclonal antibody, binds to the EGFR with high affinity, blocking growth factor binding, receptor activation, and subsequent signal transduction events and leading to cell proliferation.[84] Cetuximab enhanced the antitumor effects of chemotherapy and radiotherapy in preclinical models by inhibiting cell proliferation, angiogenesis, and metastasis and by promoting apoptosis.[84] Cetuximab has been evaluated both alone and in combination with radiotherapy and various cytotoxic chemotherapeutic agents in a series of phase II/III studies that primarily treated patients with either head and neck or colorectal cancer.[84,85] Breast cancer trials are also underway.[86] Although the FDA approval process for cetuximab was initially slowed because of concerns over clinical trial design and outcome data management,[87] the antibody was approved for use in the treatment of advanced metastatic colorectal cancer in February 2004. Similar to trastuzumab, the development of cetuximab also included an immunohistochemical test for determining EGFR overexpression to define patient eligibility to receive the antibody. Thus, cetuximab has joined trastuzumab as an FDA-approved targeted therapy featuring an unconjugated antibody. Although anti-EGFR small molecule drugs (see below) are under continuing clinical trial evaluation in breast cancer,[88] cetuximab is not currently projected as a potential therapeutic for this disease.

Bevacizumab (Avastin®)

Bevacizumab (rhuMAb-VEGF) is a humanized murine monoclonal antibody targeting the vascular endothelial growth factor (VEGF) ligand. VEGF regulates both vascular proliferation and permeability and functions as an antiapoptotic factor for newly formed blood vessels.[89,91] Patients treated with bevacizumab alone have shown significant tumor responses. Patients treated with bevacizumab in combination with conventional chemotherapy have had greater responses.[89,91] In clinical trials for advanced metastatic breast cancer, the initial results of the combination treatment of bevacizumab and paclitaxel showed antitumor activity,[92] but follow-on studies were not convincing that the targeting of VEGF in this clinical setting would be effective. Bevacizumab has also been combined with trastuzumab in a two antibody therapeutic strategy for HER-2 overexpressing breast cancer.[93] The phase II study evaluating bevacizumab in metastatic renal cell carcinoma reached its prespecified efficacy end point earlier than expected. Finally, late-stage clinical trials using bevacizumab with 5-fluorouracil, leucovorin, and CPT-11 (Irinotecan®) in advanced colorectal cancer are currently underway,[94] with recently released results indicating a major prolongation in the time to disease progression and overall survival in patients who received the angiogenesis inhibitor versus those who did not.[95] These data led to an approval by the FDA of bevacizumab for the treatment of metastatic colorectal cancer in February 2004.

Unlike cetuximab, the development of bevacizumab has not included a diagnostic eligibility test. Neither direct measurement of VEGF expression or assessment of tumor microvessel density has been incorporated into the clinical trials or linked to the response rates to the antibody. Thus, bevacizumab cannot be considered a true targeted therapy, and further focusing of this agent for colorectal, breast, lung, and other cancers will likely be inhibited by the inability to individually select patients with a diagnostic test who will be more likely to benefit from its use, either alone or in combination with other known and novel drugs.

Edrecolomab (Panorex®)

Edrecolomab is a murine IgG_{2A} monoclonal antibody that targets the human tumor-associated antigen Ep-CAM (17-1A). Edrecolomab has been approved in Europe (Germany) since 1995, but to date has not been approved by the FDA. In a study of 189 patients with resected stage III colorectal cancer, treatment with edrecolomab resulted in a 32% increase in overall survival compared with no treatment ($P < 0.01$).[96] Edrecolomab's antitumor effects are mediated through antibody-dependent cellular cytotoxicity, complement-mediated cytolysis, and the induction of an anti-idiotypic network.[97] Edrecolomab is also currently being tested in large multicenter adjuvant phase III studies in stage II/III rectal cancer and stage II colon cancer. Edrecolomab was well tolerated when used as monotherapy and added little to chemotherapy-related side effects when used in combination. Sequential treatment of patients with metastatic breast cancer with edrecolomab after adjuvant chemotherapy reduced levels of disseminated tumor cells in the bone marrow and eliminated Ep-CAM–positive micrometastases.[98]

huJ-591 (Anti-$PSMA_{EXT}$)

Prostate-specific membrane antigen (PSMA) is a membrane-bound glycoprotein restricted to normal prostatic epithelial cells, prostate cancer, and the endothelium of the neovasculature of a wide variety of nonprostatic carcinomas and other solid tumors.[99–101] PSMA expression per cell progressively increases in primary prostate cancer, metastatic hormone sensitive prostate cancer, and hormone refractory metastatic disease.[99–101] PSMA expression is increased further in association with clinically advanced prostatic cancer, particularly in hormone refractory disease, and is an ideal sentinel molecule for use in targeting prostatic cancer cells. Increasing expression levels of PSMA in resected primary prostate cancer is associated with increased rates of subsequent disease recurrence.[100] In addition, significant PSMA expression has been identified in the tumor vasculature of a variety of nonprostate tumors, including breast cancer (FIG. 14.3).[101] Humanized and fully human antibodies specific for the extracellular domain of PSMA have been developed. A phase I clinical trial of one of these antibodies, huJ-591 conjugated with ^{90}Y, has yielded promising results.[102] Programs using toxin conjugates with anti-PSMA antibodies have completed preclinical development[103] and are currently in early stage clinical trials for hormone-refractory advanced metastatic prostate cancer. Finally, antibodies to PSMA have been used as diagnostic imaging agents (FIG. 14.3), including the commercially available Prostascint®.[104]

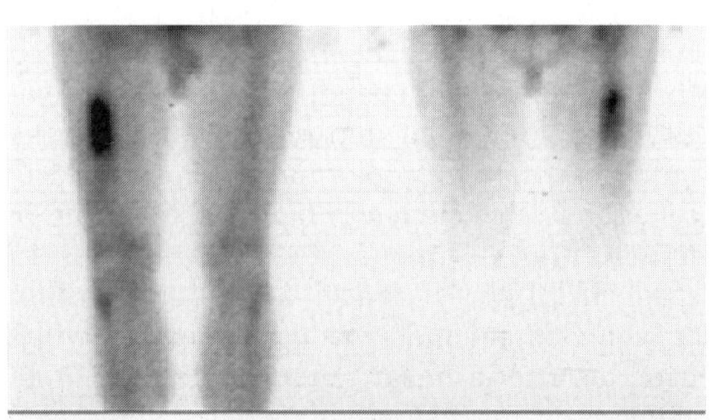

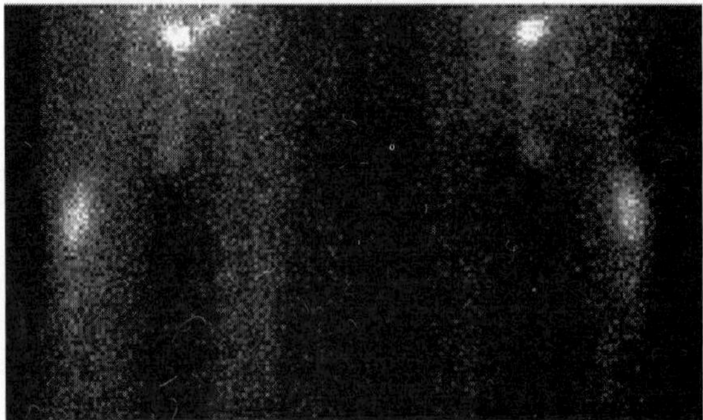

FIGURE 14.3 PSMA expression in non-prostate cancer. (Top) Traditional bone scan demonstrating bilateral activity in the femur indirectly indicating the presence of metastatic renal cell carcinoma. (Bottom) ^{111}I-huJ591$_{EXT}$ diagnostic immunoscintiscan of the same patient showing direct localization of the anti-PSMA antibody conjugate to the sites of metastatic renal cell carcinoma that feature PSMA expression in the tumor neovasculature.

Selected Targeted Anticancer Therapies Using Small Molecules

Table 14.3 lists selected small molecule drugs designed to target specific genetic events and biologic pathways critical to cancer growth and progression.

ATRA

Arguably the first truly targeted therapy after the development of hormonal therapy for breast cancer was the development of ATRA for the treatment of acute promyelocytic leukemia, a subset of acute nonlymphocytic leukemia featuring a disease-defining retinoic acid receptor activating t(15:17) reciprocal translocation.[105,106] For these selected patients, direct targeting of the retinoic acid receptor with ATRA has resulted in very high response rates, delay in disease progression, and long-term cures for these patients.[105,106]

Imatinib (Gleevec®)

The development of imatinib for patients with chronic myelogenous leukemia in 2001 ushered in a new excitement both in the scientific and public communities for targeted anticancer therapy. Imatinib received fast-track approval by the FDA as an ATP-competitive selective inhibitor of *bcr-abl* and has unprecedented efficacy for the treatment of early stage chronic myelogenous leukemia typically achieving durable complete hematologic and complete cytogenetic remissions, with minimal toxicity.[107–109] Imatinib is a true targeted therapy for leukemia in that a test for the bcr/abl translocation must be performed before a patient will be considered as eligible to receive the drug.

Imatinib has also achieved regulatory approval for the treatment of relapsed and metastatic gastrointestinal stromal tumors (GISTs), which characteristically feature an activating point mutation in the c-*kit* receptor tyrosine kinase gene.[110] For GISTs, the response to imatinib treatment appears to be predictable based on the location of the c-*kit* mutation.[111] The use of imatinib in GIST is also an example of targeted therapy as a measurement of c-*kit* expression usually performed by IHC, required to confirm the diagnosis and render the patient eligible for treatment. Interestingly, most commercially available antibodies for c-*kit* recognize the

Table 14.3 Selected Small Molecule Drugs Designed to Target Specific Genetic Events and Biologic Pathways Critical to Cancer Growth and Progression

Target	Drug	Source	Clinical Development Status	Comment
PML-RAR-α in PML	ATRA	Promega	Approved	First true targeted therapy since the introduction of ER testing and hormonal therapy for breast cancer
bcr/abl in CML	Imatinib	Novartis	Approved	Has emerged as standard of care for early stage CML
c-kit in GIST PDGF-α	Imatinib	Novartis	Approved	Responses in relapsed/metastatic GIST can be predicted by the location of the activating *c-kit* mutation
Flt-3 in AML	SU5416 PKC412 MLN-518	Pfizer Novartis Millennium	Early stage clinical trials	Small molecule drugs that target the *flt-3* internal tandem duplication seen in 30% of AML
EGFR in NSCLC	Gefitinib	Astra Zeneca	Approved	Preclinical activity in breast cancer; clinical trials are ongoing
EGFR in glioblastoma	Erlotinib	Genentech/OSI	Pending	In late stage trials in NSCLC and recently (ASCO 2003) shown efficacy in treatment of high grade malignant gliomas
Antiangiogenesis	Thalidomide SU 5416 ZD6474 Endostatin Marimastat Others	Celgene Pfizer/Sugen Astra Zeneca Entremed British Biotech Others	Pending	Thalidomide is approved for treatment of leprosy and widely used to treat multiple myeloma; other agents are in early and mid-stage clinical trials
Bcl-2	G3139	Genta	Pending	Antisense oligonucleotide targets the antiapoptotic gene, *bcl-2*
Proteasome in multiple myeloma	Bortezomib	Millennium	Approved	Proteasome inhibition effective in hematologic malignancies, but of uncertain potential for the treatment of solid tumors

AML, acute myelogenous leukemia; CML, chronic myelogenous leukemia; NSCLC, non–small cell lung cancer; PDGF, platelet derived growth factor; PML, acute promyelocytic leukemia; RAR, retinoic acid receptor.

total c-*kit* and do not distinguish the activated or phosphorylated version, which is the actual target of imatinib.

Currently, the high treatment failure rate is directly linked to the test used to characterize the patients. It is anticipated that either the use of specific antibodies designed to identify the activated c-*kit* gene or directed sequencing of the c-*kit* gene may be required before imatinib is prescribed for patients with recurrent or metastatic GIST. Although c-*kit* activation has not been linked to breast cancer, the platelet derived growth factor α, another target of imatinib, has been associated with breast cancer biology.[112]

FLT-3 Targeted Therapy

In approximately 30% of cases of acute myelogenous leukemia and less frequently in other forms of leukemia, a *flt-3* gene mutation creates an internal tandem duplication that creates an abnormal FLT-3 receptor that promotes the growth and survival of the leukemic cells.[113–115] Three small molecule compounds are in clinical trials for the treatment of acute myelogenous leukemia by targeting the *flt-3* internal tandem duplication (ITD). These drugs are also examples of potential true targeted therapies in that a test for detecting an internal tandem duplication that causes the *flt-3* gene activation will likely be required and incorporated into the FDA drug approval label should these agents be successful in future clinical trials.

Gefitinib (Iressa®)

Gefitinib was approved by the FDA in 2003 as monotherapy for the treatment of patients with locally advanced or metastatic non–small cell lung cancer after failure of both platinum based and docetaxel chemotherapies.[116,117]

Gefitinib is a small molecule drug that targets the EGFR. In contrast with the approval of trastuzumab, the approval of gefitinib did not include an eligibility requirement reference to a specific tumor diagnostic test designed to select patients that were more likely to respond to the drug. Overexpression of EGFR typically identified by IHC is extremely common in both lung and breast cancers,[116–118] but in contrast with HER-2 overexpression, which is virtually limited to cases with gene amplification, multiple mechanisms of dysregulation of EGFR and associated activation of signaling pathways have been described for both of these tumors (see Chapter 17).[116–118] Thus, it has been difficult to develop this drug for expanded indications or combination therapies in the absence of a well-defined efficacy test. However, more recently, two independent groups reported their similar discovery of a specific activating mutation in the tyrosine kinase domain of the EGFR receptor that was associated with a high response of patients with non-small cell lung cancer to gefitinib.[119–120] The clinical significance of this important finding for the development of gefitinib in breast cancer is currently unknown. Preclinical studies of gefitinib have demonstrated activity ion breast cancer cell lines,[118] and clinical trials using the drug in combination with standard cytotoxic drugs are underway (see Chapter 17).

Erlotinib (Tarceva®)

Erlotinib is another targeted inhibitor of EGFR currently in late stage clinical trials for the treatment of non–small cell lung cancer and pancreatic cancer.[121,122] Clinical studies of erlotinib in breast cancer are ongoing (see Chapter 17). Erlotinib has shown efficacy in preclinical models of brain tumors[123] and has recently shown promising results in the treatment of high grade malignant gliomas. To date, similar to gefitinib, the clinical trials for erlotinib have not included an assessment of the EGFR status or other diagnostic test for eligibility to receive the drug.

Antiangiogenesis (SU5416, Thalidomide, Endostatin/Angiostatin, and Marimastat)

A variety of small molecule drugs is currently in clinical trials for breast cancer and other malignancies that target the establishment and growth of tumor blood vessels.[124–127] Additional compounds that target matrix metalloproteases, such as the drug marimastat, are also considered to be angiogenesis inhibitors.[128,129] To date, none of these compounds has linked a diagnostic test such as tumor microvessel density or the expression of an angiogenesis promoting gene or protein in their clinical development plans. Matrix metalloprotease inhibitors are discussed in Chapter 18, and antiangiogenesis therapies are discussed in detail in Chapter 19.

G3139 (Genasense®)

Another emerging strategy in anticancer therapy is the targeting of chemotherapy resistance by overcoming the antiapoptosis mechanisms of cancer cells. An example of this approach is the novel antisense oligonucleotide G3139, which targets the antiapoptotic gene *bcl-2*.[130,131] This agent has been tested mostly in hematologic malignancies and has not been used widely in either preclinical experiments or clinical trials in breast cancer. Therapeutic strategies targeting apoptosis are covered in depth in Chapter 21.

Bortezomib (Velcade®)

Recently, drugs targeting the proteasome have been developed that are designed to impact downstream pathways regulating angiogenesis, tumor growth, adhesion, and resistance to apoptosis.[132,133] One of these agents, bortezomib (PS-341), has recently been approved for the treatment of advanced refractory multiple myeloma.[134] Bortezomib has shown preclinical activity in breast cancer as a single agent and is being tested in combination therapy strategies in multiple clinical trials.[135] Proteasome biology and proteasome inhibitors are discussed in detail in Chapter 24.

The Future Is Now: Pharmacogenomics and Personalized Medicine

Targeted therapy in oncology has been a major stimulus for the evolving field of pharmacogenomics (see Chapter 28). In its broadest definition, pharmacogenomics can encompass both germline and somatic (disease) gene and protein measurements used to predict the likelihood that a patient will respond to a specific single or multiagent chemotherapy regimen and to predict the risk of toxic side effects.[136,137] During the next several years, the field of oncology drug development will see numerous products pass through the approval process and enter the market accompanied by diagnostic tests designed to "personalize" their use, dosage, route of administration, and length of treatment for each patient, one at a time. Only time will tell whether this new approach to anticancer pharmaceuticals will yield breakthrough results, reducing morbidity and mortality and improving outcomes for all who will be afflicted with the disease.

References

1. O'Dwyer ME, Druker BJ. Chronic myelogenous leukaemia—new therapeutic principles. J Intern Med 2001;250:3–9.
2. Mauro MJ, Druker BJ. STI571: targeting BCR-ABL as therapy for CML. Oncologist 2001;6:233–238.
3. Lemonick MD, Park A. New hope for cancer. Time 2001;157:62–69.
4. Brown E, Lewis PH, Nocera J. In search of the silver bullet. Fortune 2001;143:166–170.
5. Amos J, Patnaik M. Commercial molecular diagnostics in the U.S.: the Human Genome Project to the clinical laboratory. Hum Mutat 2002;19: 324–333.
6. Bottles K. A revolution in genetics: changing medicine, changing lives. Physician Exec 2001;27:58–63.
7. Evans WE, McLeod HL. Pharmacogenomics—drug disposition, drug targets, and side effects. N Engl J Med 2003;348:538–549.
8. Weinshilboum R. Inheritance and drug response. N Engl J Med 2003;348:529–537.
9. Osborne CK. Steroid hormone receptors in breast cancer management. Breast Cancer Res Treat 1998; 51:227–238.
10. Wilbur DC, Willis J, Mooney RA, et al. Estrogen and progesterone detection in archival formalin-fixed paraffin embedded tissue from breast carcinoma: a comparison of immunocytochemistry with dextran coated charcoal assay. Mod Pathol 1992;5:79–84.
11. Ciocca DR, Elledge R. Molecular markers for predicting response to tamoxifen in breast cancer patients. Endocrine 2000;13:1–10.
12. Jordan VC. Tamoxifen: a most unlikely pioneering medicine. Nat Rev Drug Discov 2003;2:205–213.
13. Locker GY. Hormonal therapy of breast cancer. Cancer Treat Rev 1998;24:221–240.
14. Ibrahim NK, Hortobagyi GN. The evolving role of specific estrogen receptor modulators (SERMs) Surg Oncol 1999;8:103–123.
15. Miller WR, Anderson TJ, Dixon JM. Anti-tumor effects of letrozole. Cancer Invest 2002;20:15–21.
16. Buzdar AU, Robertson JF, Eiermann W, et al. An overview of the pharmacology and pharmacokinetics of the newer generation aromatase inhibitors anastrozole, letrozole, and exemestane. Cancer 2002;95: 2006–2016.
17. Jordan VC. Antiestrogens and selective estrogen receptor modulators as multifunctional medicines. 1. Receptor interactions. J Med Chem 2003;46: 883–908.
18. Jordan VC. Antiestrogens and selective estrogen receptor modulators as multifunctional medicines. 2. Clinical considerations and new agents. J Med Chem 2003;46:1081–1111.
19. Rubnitz JE, Pui CH. Molecular diagnostics in the treatment of leukemia. Curr Opin Hematol 1999;6: 229–235.
20. Gleissner B, Thiel E. Detection of immunoglobulin heavy chain gene rearrangements in hematologic malignancies. Expert Rev Mol Diagn 2001;1: 191–200.
21. Grimwade D, Lo Coco F. Acute promyelocytic leukemia: a model for the role of molecular diagnosis and residual disease monitoring in directing treatment approach in acute myeloid leukemia. Leukemia 2002;16:1959–1973.
22. Reichert JM. Monoclonal antibodies in the clinic. Nat Biotechnol 2001;19:819–822.
23. Kiyoi H, Naoe T. FLT3 in human hematologic malignancies. Leuk Lymph 2002;43:1541–1547.
24. Ross JS, Fletcher JA, Linette GP, et al. The Her-2/neu gene and protein in breast cancer 2003: biomarker and target of therapy. Oncologist 2003; 8:307–325.
25. Slamon DJ, Leyland-Jones B, Shak S, et al. Use of chemotherapy plus a monoclonal antibody against HER2 for metastatic breast cancer that overexpresses HER2. N Engl J Med 2001;344: 783–792.
26. Schnitt SJ, Jacobs TW. Current status of HER2 testing: caught between a rock and a hard place. Am J Clin Pathol 2001;116:806–810.
27. Hayes DF, Thor AD. c-erbB-2 in breast cancer: development of a clinically useful marker. Semin Oncol 2002;29:231–245.
28. Masood S, Bui MM. Prognostic and predictive value of HER2/neu oncogene in breast cancer. Microsc Res Tech 2002;59:102–108.
29. Paik S, Tan-chui E, Bryan J, et al. Successful quality assurance program for HER-2 testing in the NSAPB trial for Herceptin. Breast Cancer Res Treat 2002; 76(suppl 1):S31.
30. Wang S, Saboorian MH, Frenkel EP, et al. Assessment of HER-2/neu status in breast cancer. Automated Cellular Imaging System (ACIS)-assisted quantitation of immunohistochemical assay achieves high accuracy in comparison with fluorescence in situ hybridization assay as the standard. Am J Clin Pathol 2001;116:495–503.
31. Wang S, Saboorian MH, Frenkel E, et al. Laboratory assessment of the status of Her-2/neu protein and oncogene in breast cancer specimens: comparison of immunohistochemistry assay with fluorescence in situ hybridization assays. J Clin Pathol 2000;53: 374–381.
32. Tanner M, Gancberg D, Di Leo A, et al. Chromogenic in situ hybridization: a practical alternative for fluorescence in situ hybridization to detect HER–2/neu oncogene amplification in archival breast cancer samples. Am J Pathol 2000;157:1467–1472.
33. Zhao J, Wu R, Au A, et al. Determination of HER2 gene amplification by chromogenic in situ hybridization

(CISH) in archival breast carcinoma. Mod Pathol 2002;15:657–665.
34. Hortobagyi GN. Overview of treatment results with trastuzumab (Herceptin) in metastatic breast cancer. Semin Oncol 2001;28:43–47.
35. McKeage K, Perry CM. Trastuzumab: a review of its use in the treatment of metastatic breast cancer overexpressing HER2. Drugs 2002;62:209–243.
36. Shawver LK, Slamon D, Ullrich A. Smart drugs: tyrosine kinase inhibitors in cancer therapy. Cancer Cell 2002;1:117–123.
37. Ligibel JA, Winer EP. Trastuzumab/chemotherapy combinations in metastatic breast cancer. Semin Oncol 2002;29:38–43.
38. Mass RD, Press MF, Anderson S, et al. Improved survival benefit from Herceptin (trastuzumab) in patients selected by fluorescence in situ hybridization (FISH). Proc Am Soc Clin Oncol 2001;85.
39. Vogel CL, Cobleigh MA, Tripathy D, et al. Efficacy and safety of trastuzumab as a single agent in first-line treatment of HER2-overexpressing metastatic breast cancer. J Clin Oncol 2002;20:719–726.
40. Seidman AD, Fornier MN, Esteva FJ, et al. Weekly trastuzumab and paclitaxel therapy for metastatic breast cancer with analysis of efficacy by HER2 immunophenotype and gene amplification. J Clin Oncol 2001;19:2587–2595.
41. Fornier M, Risio M, Van Poznak C, et al. HER-2 testing and correlation with efficacy in trastuzumab therapy. Oncology 2003;16:1340–1358.
42. Carter P. Improving the efficacy of antibody-based cancer therapies. Nat Rev Cancer 2001;1:118–129.
43. Goldenberg DM. Targeted therapy of cancer with radiolabeled antibodies. J Nucl Med 2002;43:693–713.
44. Reichert JM. Therapeutic monoclonal antibodies: trends in development and approval in the US. Curr Opin Mol Ther 2002;4:110–118.
45. Ross JS, Gray K, Gray GS, et al. Anticancer antibodies. Am J Clin Pathol 2003;119:472–485.
46. Hemminki A. From molecular changes to customised therapy. Eur J Cancer 2002;38:333–338.
47. Milenic DE. Monoclonal antibody-based therapy strategies: providing options for the cancer patient. Curr Pharm Des 2002;8:1749–1764.
48. Reilly RM, Sandhu J, Alvarez-Diez TM, et al. Problems of delivery of monoclonal antibodies. Pharmaceutical and pharmacokinetic solutions. Clin Pharmacokinet 1995;28:126–142.
49. Winter G, Harris WJ. Humanized antibodies. Immunol Today 1993;14:243–246.
50. Merluzzi S, Figini M, Colombatti A, et al. Humanized antibodies as potential drugs for therapeutic use. Adv Clin Path 2000;4:77–85.
51. Kuus-Reichel K, Grauer LS, Karavodin LM, et al. Will immunogenicity limit the use, efficacy, and future development of therapeutic monoclonal antibodies? Clin Diagn Lab Immunol 1994;1:365–372.
52. Pimm MV. Possible consequences of human antibody responses on the biodistribution of fragments of human, humanized or chimeric monoclonal antibodies: a note of caution. Life Sci 1994;55:PL45–PL49.
53. Jones PT, Dear PH, Foote J, et al. Replacing the complementarity-determining regions in a human antibody with those from a mouse. Nature 1986;321:522–525.
54. Isaacs JD. From bench to bedside: discovering rules for antibody design, and improving serotherapy with monoclonal antibodies. Rheumatology 2001;40:724–738.
55. Watkins NA, Ouwehand WH. Introduction to antibody engineering and phage display. Vox Sang 2000;78:72–79.
56. Reff ME, Hariharan K, Braslawsky G. Future of monoclonal antibodies in the treatment of hematologic malignancies. Cancer Control 2002;9:152–166.
57. Chester KA, Hawkins RE. Clinical issues in antibody design. Trends Biotechnol 1995;13:294–300.
58. Reff ME, Heard C. A review of modifications to recombinant antibodies: attempt to increase efficacy in oncology applications. Crit Rev Oncol Hematol 2001;40:25–35.
59. Nielsen UB, Marks JD. Internalizing antibodies and targeted cancer therapy: direct selection from phage display libraries. PSTT 2000;3:282–291.
60. Burke JM, Jurcic JG, Scheinberg DA. Radioimmunotherapy for acute leukemia. Cancer Control 2002;9:106–113.
61. Linenberger ML, Maloney DG, Bernstein ID. Antibody-directed therapies for hematological malignancies. Trends Mol Med 2002;8:69–76.
62. Stevenson GT, Anderson VA, Leong WS. Engineered antibody for treating lymphoma. Recent Results Cancer Res 2002;159:104–112.
63. Wiseman GA, Gordon LI, Multani PS, et al. Ibritumomab tiuxetan radioimmunotherapy for patients with relapsed or refractory non-Hodgkin lymphoma and mild thrombocytopenia: a phase II multicenter trial. Blood 2002;99:4336–4342.
64. Dillman RO. Monoclonal antibody therapy for lymphoma: an update. Cancer Pract 2001;9:71–80.
65. Grillo-Lopez AJ, Hedrick E, Rashford M, et al. Rituximab: ongoing and future clinical development. Semin Oncol 2002;29(1 suppl 2):105–112.
66. Coiffier B. Rituximab in the treatment of diffuse large B-cell lymphomas. Semin Oncol 2002;29(1 suppl 2):30–35.
67. Grillo-Lopez AJ. AntiCD20 mAbs: modifying therapeutic strategies and outcomes in the treatment of lymphoma patients. Expert Rev Anticancer Ther 2002;2:323–329.
68. Krasner C, Joyce RM. Zevalin: 90yttrium labeled anti-CD20 (ibritumomab tiuxetan), a new treatment for non-Hodgkin's lymphoma. Curr Pharm Biotechnol 2001;2:341–349.
69. Gordon LI, Witzig TE, Wiseman GA, et al. Yttrium 90 ibritumomab tiuxetan radioimmunotherapy for relapsed or refractory low-grade non-Hodgkin's lymphoma. Semin Oncol 2002;29(1 suppl 2):87–92.
70. Wagner HN Jr, Wiseman GA, Marcus CS, et al. Administration guidelines for radioimmunotherapy of non-Hodgkin's lymphoma with $^{(90)}$Y-labeled anti CD20 monoclonal antibody. J Nucl Med 2002;43:267–272.
71. Dillman RO. Radiolabeled anti-CD20 monoclonal antibodies for the treatment of B-cell lymphoma. J Clin Oncol 2002;20:3545–3557.
72. Stadtmauer EA. Trials with gemtuzumab ozogamicin (Mylotarg®) combined with chemotherapy regimens in acute myeloid leukemia. Clin Lymphoma 2002;2(suppl 1):S24–S28.
73. Nabhan C, Tallman MS. Early phase I/II trials with gemtuzumab ozogamicin (Mylotarg®) in acute myeloid leukemia. Clin Lymph 2002;2(suppl 1):S19–S23.

74. Sievers EL, Linenberger M. Mylotarg: antibody-targeted chemotherapy comes of age. Curr Opin Oncol 2001;13:522–527.
75. Bross PF, Beitz J, Chen G, et al. Approval summary: gemtuzumab ozogamicin in relapsed acute myeloid leukemia. Clin Cancer Res 2001;7:1490–1496.
76. Larson RA, Boogaerts M, Estey E, et al. Antibody-targeted chemotherapy of older patients with acute myeloid leukemia in first relapse using Mylotarg (gemtuzumab ozogamicin). Leukemia 2002;16:1627–1636.
77. Dumont FJ. CAMPATH (alemtuzumab) for the treatment of chronic lymphocytic leukemia and beyond. Expert Rev Anticancer Ther 2002;2:23–35.
78. Pangalis GA, Dimopoulou MN, Angelopoulou MK, et al. Campath-1H (anti-CD52) monoclonal antibody therapy in lymphoproliferative disorders. Med Oncol 2001;18:99–107.
79. Carswell CI, Plosker GL, Wagstaff AJ. Daclizumab: a review of its use in the management of organ transplantation. BioDrugs 2001;15:745–773.
80. Kreitman RJ, Chaudhary VK, Kozak RW, et al. Recombinant toxins containing the variable domains of the anti-Tac monoclonal antibody to the interleukin-2 receptor kill malignant cells from patients with chronic lymphocytic leukemia. Blood 1992;80:2344–2352.
81. Cheson BD. Bexxar (Corixa/GlaxoSmithKline). Curr Opin Investig Drugs 2002;3:165–170.
82. Zelenetz AD. A clinical and scientific overview of tositumomab and iodine I 131 tositumomab. Semin Oncol 2003;30(2 suppl 4):22–30.
83. Mendelsohn J, Baselga J. The EGF receptor family as targets for cancer therapy. Oncogene 2000;19:6550–6565.
84. Baselga J. The EGFR as a target for anticancer therapy—focus on cetuximab. Eur J Cancer 2001;37(suppl 4):S16–S22.
85. Herbst RS, Langer CJ. Epidermal growth factor receptors as a target for cancer treatment: the emerging role of IMC-C225 in the treatment of lung and head and neck cancers. Semin Oncol 2002;29(1 suppl 4):27–36.
86. Leonard DS, Hill AD, Kelly L, et al. Anti-human epidermal growth factor receptor 2 monoclonal antibody therapy for breast cancer. Br J Surg 2002;89:262–271.
87. Reynolds T. Biotech firm faces challenges from FDA, falling stock prices. J Natl Cancer Inst 2002;94:326–328.
88. Morris C. The role of EGFR-directed therapy in the treatment of breast cancer. Breast Cancer Res Treat 2002;75(suppl 1):S51–S55.
89. Rosen LS. Angiogenesis inhibition in solid tumors. Cancer J 2001;7(suppl 3):S120–S128.
90. Rosen LS. Clinical experience with angiogenesis signaling inhibitors: focus on vascular endothelial growth factor (VEGF) blockers. Cancer Control 2002;9(2 suppl):36–44.
91. Chen HX, Gore-Langton RE, Cheson BD. Clinical trials referral resource: current clinical trials of the anti-VEGF monoclonal antibody bevacizumab. Oncology 2000;15:1017, 1020, 1023–1026.
92. Cobleigh MA, Langmuir VK, Sledge GW, et al. A phase I/II dose-escalation trial of bevacizumab in previously treated metastatic breast cancer. Semin Oncol 2003;30(5 suppl 16):117–124.
93. Pegram MD, Reese DM. Combined biological therapy of breast cancer using monoclonal antibodies directed against HER2/neu protein and vascular endothelial growth factor. Semin Oncol 2002;29(3 suppl 11):29–37.
94. Berlin JD. Targeting vascular endothelial growth factor in colorectal cancer. Oncology 2002;16(8 suppl 7):13–15.
95. McCarthy M. Antiangiogenesis drug promising for metastatic colorectal cancer. Lancet 2003;361:1959.
96. Schwartzberg LS. Clinical experience with edrecolomab: a monoclonal antibody therapy for colorectal carcinoma. Crit Rev Oncol Hematol 2001;40:17–24.
97. Haller DG. Update of clinical trials with edrecolomab: a monoclonal antibody therapy for colorectal cancer. Semin Oncol 2001;28(1 suppl 1):25–30.
98. Kirchner EM, Gerhards R, Voigtmann R. Sequential immunochemotherapy and edrecolomab in the adjuvant therapy of breast cancer: reduction of 17–1A-positive disseminated tumour cells. Ann Oncol 2002;13:1044–1048.
99. Israeli RS, Powell CT, Corr JG, et al. Expression of the prostate-specific membrane antigen. Cancer Res 1994;54:1807–1811.
100. Ross JS, Gray K, Mosher R, et al. Prostate specific antigen membrane (PSMA): target validation and prognostic significance in prostate cancer. Mod Pathol 2003;16:167A.
101. Liu H, Moy P, Kim S, et al. Monoclonal antibodies to the extracellular domain of prostate-specific membrane antigen also react with tumor vascular endothelium. Cancer Res 1997;57:3629–3634.
102. Bander NH, Nanus D, Goldsmith S, et al. Phase I trial of humanized monoclonal antibody (mAb) to prostate specific membrane antigen/extracellular domain (PSMAext). Proc ASCO 2002;722.
103. Fracasso G, Bellisola G, Cingarlini S, et al. Anti-tumor effects of toxins targeted to the prostate specific membrane antigen. Prostate 2002;53:9–23.
104. Freeman LM, Krynyckyi BR, Li Y, et al. National Prostascint study group. The role of $^{(111)}$In Capromab Pendetide (Prosta-ScintR) immunoscintigraphy in the management of prostate cancer. Q J Nucl Med 2002;46:131–137.
105. Parmar S, Tallman MS. Acute promyelocytic leukaemia: a review. Expert Opin Pharmacother 2003;4:1379–1392.
106. Fang J, Chen SJ, Tong JH, et al. Treatment of acute promyelocytic leukemia with ATRA and As2O3: a model of molecular target-based cancer therapy. Cancer Biol Ther 2002;1:614–620.
107. Druker BJ. Imatinib alone and in combination for chronic myeloid leukemia. Semin Hematol 2003;40:50–58.
108. O'Brien SG, Guilhot F, Larson RA, et al. Imatinib compared with interferon and low-dose cytarabine for newly diagnosed chronic-phase chronic myeloid leukemia. N Engl J Med 2003;348:994–1004.
109. Goldman JM, Melo JV. Chronic myeloid leukemia—advances in biology and new approaches to treatment. N Engl J Med 2003;349:1451–1464.
110. von Mehren M. Gastrointestinal stromal tumors: a paradigm for molecularly targeted therapy. Cancer Invest 2003;21:553–563.
111. Verweij J, van Oosterom A, Blay JY, et al Imatinib mesylate (STI-571 Glivec, Gleevec) is an active agent

for gastrointestinal stromal tumours, but does not yield responses in other soft-tissue sarcomas that are unselected for a molecular target. Results from an EORTC Soft Tissue and Bone Sarcoma Group phase II study. Eur J Cancer 2003;39:2006–2011.

112. de Jong JS, van Diest PJ, van der Valk P, et al. Expression of growth factors, growth-inhibiting factors, and their receptors in invasive breast cancer. II. Correlations with proliferation and angiogenesis. J Pathol 1998;184:53–57.

113. Gilliland DG, Griffin JD. The roles of FLT3 in hematopoiesis and leukemia. Blood 2002;100:1532–1542.

114. Sawyers CL. Finding the next Gleevec: FLT3 targeted kinase inhibitor therapy for acute myeloid leukemia. Cancer Cell 2002;1:413–415.

115. Kelly LM, Yu JC, Boulton CL, et al. CT53518, a novel selective FLT3 antagonist for the treatment of acute myelogenous leukemia (AML). Cancer Cell 2002;1:421–432.

116. Schiller JH. New directions for ZD1839 in the treatment of solid tumors. Semin Oncol 2003;30(1 suppl 1):49–55.

117. Ranson M. ZD1839 (Iressa): for more than just non-small cell lung cancer. Oncologist 2002;7(suppl 4):16–24.

118. Campiglio M, Locatelli A, Olgiati C, et al. Inhibition of proliferation and induction of apoptosis in breast cancer cells by the epidermal growth factor receptor (EGFR) tyrosine kinase inhibitor ZD1839 ("Iressa") is independent of EGFR expression level. J Cell Physiol 2004;198:259–268.

119. Paez JG, Janne PA, Lee JC, et al. EGFR mutations in lung cancer: Correlation with clinical response to gefitinib therapy. Science 2004;304:1497–1500.

120. Lynch TJ, Bell DW, Sordella R et al. Activating mutations in the epidermal growth factor receptor underlying responsiveness of non-small-cell lung cancer to Gefitinib. N Engl J Med 2004;350:2129–2139.

121. Bonomi P. Erlotinib: a new therapeutic approach for non-small cell lung cancer. Expert Opin Invest Drugs 2003;12:1395–1401.

122. Khalil MY, Grandis JR, Shin DM. Targeting epidermal growth factor receptor: novel therapeutics in the management of cancer. Expert Rev Anticancer Ther 2003;3:367–380.

123. Newton HB. Molecular neuro-oncology and development of targeted therapeutic strategies for brain tumors. Part 1. Growth factor and Ras signaling pathways. Expert Rev Anticancer Ther 2003;3:595–614.

124. Zogakis TG, Libutti SK. General aspects of anti-angiogenesis and cancer therapy. Expert Opin Biol Ther 2001;1:253–275.

125. Thomas DA, Kantarjian HM. Current role of thalidomide in cancer treatment. Curr Opin Oncol 2000;12:564–573.

126. Mendel DB, Laird AD, Smolich BD, et al. Development of SU5416, a selective small molecule inhibitor of VEGF receptor tyrosine kinase activity, as an anti-angiogenesis agent. Anticancer Drug Des 2000;15:29–41.

127. Dell'Eva R, Pfeffer U, Indraccolo S, et al. Inhibition of tumor angiogenesis by angiostatin: from recombinant protein to gene therapy. Endothelium 2002;9:3–10.

128. Brown PD. Ongoing trials with matrix metalloproteinase inhibitors. Expert Opin Investig Drugs 2000;9:2167–2177.

129. Miller KD, Gradishar W, Schuchter L, et al. A randomized phase II pilot trial of adjuvant marimastat in patients with early-stage breast cancer. Ann Oncol 2002;13:1220–1224.

130. Tolcher AW. Regulators of apoptosis as anticancer targets. Hematol Oncol Clin North Am 2002;16:1255–1267.

131. Tamm I, Dorken B, Hartmann G. Antisense therapy in oncology: new hope for an old idea? Lancet 2001;358:489–497.

132. Adams J. Development of the proteasome inhibitor PS-341. Oncologist 2002;7:9–16.

133. Elliott PJ, Ross JS. The proteasome: a new target for novel drug therapies. Am J Clin Pathol 2001;116:637–646.

134. Richardson PG, Barlogie B, Berenson J, et al. A phase 2 study of bortezomib in relapsed, refractory myeloma. N Engl J Med 2003;348:2609–2617.

135. Orlowski RZ, Dees EC. The role of the ubiquitination-proteasome pathway in breast cancer: applying drugs that affect the ubiquitin-proteasome pathway to the therapy of breast cancer. Breast Cancer Res 2003;5:1–7.

136. Weinstein JN. Pharmacogenomics—teaching old drugs new tricks. N Engl J Med 2000;343:1408–1409.

137. Slonim DK. Transcriptional profiling in cancer: the path to clinical pharmacogenomics. Pharmacogenomics 2001;2:123–136.

CHAPTER

15

Hormone Receptors and Hormonal-Based Therapy for Breast Cancer

Jeffrey S. Ross, MD[1] *and Gabriel N. Hortobagyi, MD*[2]

[1]*Albany Medical College, Albany, New York and Millennium Pharmaceuticals, Inc., Cambridge, Massachusetts*
[2]*The University of Texas M.D. Anderson Cancer Center, Houston, Texas*

The finding that removal of the ovaries from premenopausal women improved prognosis in patients with metastatic breast cancer was first described in the late 19th century.[1] Soon thereafter it was noted that only one third of breast cancers responded to ovarian ablation and responses lasted, on average, for only 1–2 years.[2] Subsequently, hormonal-based therapy became a standard of care in the treatment of breast cancer. Treatment approaches included oophorectomy, ovarian irradiation, adrenalectomy, and hypophysectomy.[3] Given the significant morbidity and mortality associated with these treatments, it became a major thrust of breast cancer research to discover a predictive marker that could select the appropriate patients likely to respond to the selected procedure. The identification of the estrogen receptor (ER) and the development of a test designed to predict the outcome of antihormonal therapy in breast cancer heralded the first true targeted therapy in modern cancer treatment. The ER also became a target for the development of new drugs for the treatment and prevention of breast cancer. With the development of tamoxifen for the treatment of ER-positive breast cancers, a new generation of hormonal interventions was launched. Tamoxifen was successively shown to be at least as effective and better tolerated as any other hormonal intervention available at the time for the management of metastatic breast cancer, regardless of age or gender, and soon became the most effective intervention in the management of ER-positive primary breast cancer.[4] The administration of adjuvant tamoxifen for 5 years after diagnosis led to a major improvement in the survival of breast cancer patients.[3] Tamoxifen was also shown to reduce the incidence of breast cancer by about 50% in women at high risk for developing this disease.[5]

Although very well tolerated when compared with other anticancer agents, tamoxifen can produce side effects and toxicities. As a mixed estrogen agonist and antagonist, tamoxifen acts as an antiestrogen in the breast, whereas it has estrogen agonist activity in bone, cardiac lipids, and the endometrium. This latter effect is associated with a two- to fourfold increase in the incidence of endometrial cancer in patients treated with tamoxifen. More recently, the discovery of other selective ER modulators (SERMs) that could feature a stimulatory action in bones and lower circulating cholesterol but have inhibitory actions in breast and uterine tissues has resulted in the development of multifunctional drugs designed to prevent a variety of diseases, including breast and uterine cancer, osteoporosis, and coronary heart disease.[3]

Significance in Breast Cancer

Based on current incidence rates, it is estimated that an American woman has a one in nine chance of developing breast cancer at some time during her life.[6] According to the American Cancer Society, 215,990 new cases of breast cancer in women will be reported and 40,110 women will die of the disease in the United States in 2004 alone.[7] Arguably the first type of targeted therapy in oncology was the development of antiestrogen therapies for patients with breast cancer that expressed the ER protein.[8] Originally developed as a competitive binding bioassay performed on fresh tumor protein extracts and used to select for hormone production ablation by surgery (oophorectomy, adrenalectomy, and hypophysectomy), the ER/progesterone receptor (PR) test format converted to an immunohistochemistry (IHC) platform when the decreased size of primary tumors enabled by mass screening programs produced insufficient material for creating enough fresh tissue to perform the assay.[9]

The drug tamoxifen (Nolvadex®), which has both hormonal and nonhormonal mechanisms of action, has been the most widely prescribed antiestrogen for the treatment of metastatic breast cancer, primary breast cancer, and chemoprevention of the disease in high risk women.[10,11] Although, ER/PR testing is the front line for predicting tamoxifen response, the assay is most useful for its negative predictive value: Tumors with negative ER and PR do not respond to tamoxifen (or any other hormonal therapy), whereas one third of tumors with positive assays for ER or PR will have an objective response to tamoxifen. Thus, additional biomarkers, including pS2, HER-2/neu, and cathepsin D testing, have been used in an attempt to further refine therapy selection.[12] The introductions of specific estrogen response modulators (toremifene, raloxifene, arzoxifene) and selective aromatase inhibitors such as anastrozole (Arimidex®), letrozole (Femara®), and exemestane (Aromasin®); the selective ER down-regulator (SERD) fulvestrant (Faslodex®); and the combination chemotherapeutic, estramustine (Emcyt®)[13–17] have added new strategies for evaluating tumors for hormonal therapy. In this chapter, we present the basic biology of hormone receptors, the techniques for assessing ER and PR status, and a review of the significance of ER and PR status as stand-alone prognostic factors for breast cancer. The uses of the hormone receptor measurements for the prediction of response to hormonal-based and nonhormonal-based therapies are also considered.

Biologic Background of Hormone Receptors

The association of early menarche, late menopause, and nulliparity with an increased risk of breast cancer development has supported the concept that cycling estrogen and progesterone exposure may play a role in the pathogenesis of the disease.[18] Indeed, in patients who undergo oophorectomy at an early age, the incidence of breast cancer is significantly reduced.[19] In postmenopausal women, the major source of estrogen is androgenic precursors derived from the adrenal glands that are converted into estrogen by the aromatase enzyme in peripheral (adipose, muscle) tissues. Moreover, in postmenopausal women with increased body fat, the enzymatic conversion of androgens produced by the adrenal glands is associated with an increased rate of breast cancer development.[20]

ERs

The ER proteins, ER-α and ER-β, are members of the nuclear hormone receptor family. Upon ligand binding, ER dissociates from its inactive binding to HSP90, activates, undergoes a conformational change, dimerizes, and autophosphorylates through intrinsic tyrosine kinases (FIG. 15.1).[18] Activated ER dimers recruit coactivators and/or corepressors and eventually bind to recognition sequences termed estrogen response elements, which regulate the promoter regions of a variety of genes and can activate the mitogen activated protein kinase pathway, resulting in the activation of the AP-1 proteins, fos and jun.[21,22] The determination of ER-α status in breast cancer is considered to be a component of the standard of care for newly diagnosed breast cancer patients given its critical role in defining response to endocrine therapy (see below). Thus, there is great interest in determining the mechanisms that regulate its expression. The loss of ER expression in ER-negative tumors is generally not the result of mutations, deletions, loss of heterozygosity, or polymorphisms within the gene.[18] The main mechanisms by which the expression of ER protein is decreased or eliminated include two reversible epigenetic modifications: DNA methylation and histone deacetylation. The presence or absence of acetyl groups on histone tails (primarily H_3 and H_4) can govern chromatin structure and gene transcription. However, histone protein modification is not the only method of gene silencing but likely interacts with a second mechanism, DNA methylation.

Methylation of the ER Promoter

The main epigenetic modification of human genes is methylation of the nucleotide cytosine (C) when it precedes a guanine (G), forming the dinucleotide CpG. Approximately 6–8% of all cytosines are methylated in normal human DNA. The distribution of CpG dinucleotides in the human genome is not uniform. CpG-rich regions denominate CpG islands. These are usually unmethylated in all normal tissues

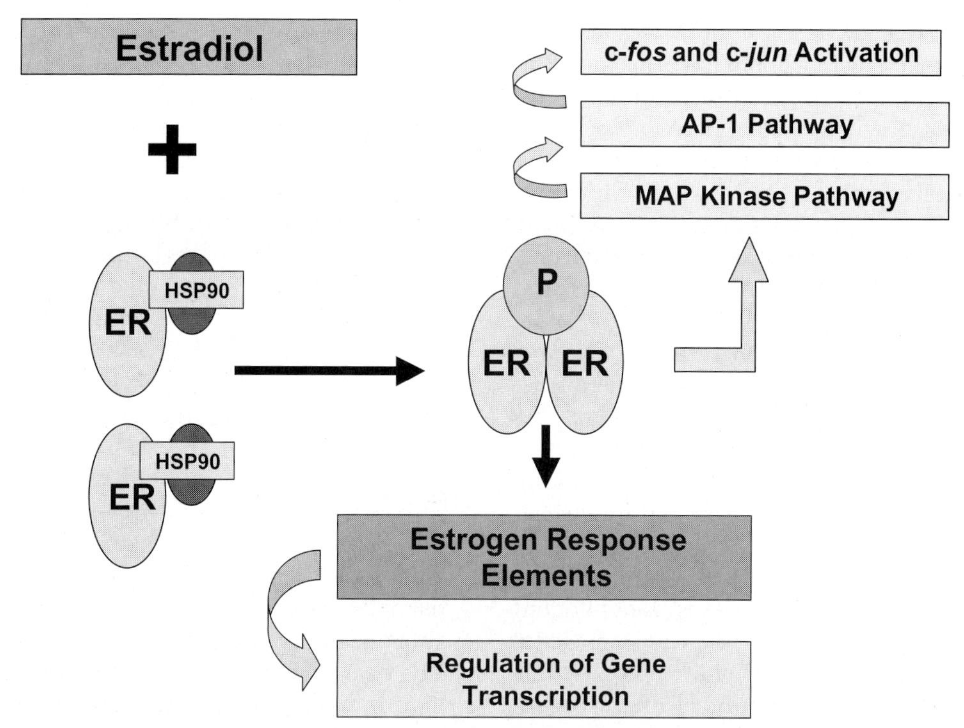

FIGURE 15.1 Activation of ER by estradiol and downstream signaling. Upon ligand binding, ER dissociates from its inactive binding to HSP90, activates, undergoes a conformational change, dimerizes, and autophosphorylates through intrinsic tyrosine kinases. Activated ER dimers bind to recognition sequences, termed estrogen response elements, that regulate the promoter regions of a variety of genes and can activate the mitogen activated protein (MAP) kinase pathway, resulting in the activation of the AP-1 proteins, fos and jun.

and span the 5′ end of genes in the regulatory regions that determine gene expression. When transcription factors are available and the island remains in an unmethylated state with open chromatin configuration associated with hyperacetylated histones, transcription can occur (see Chapter 26).[23] CpG dinucleotides are methylated by the DNA methyltransferase proteins. DNA methylation within the promoter and first exon of genes is correlated with gene silencing and a lack of gene expression.[23] The ER-α promoter region contains a CpG island that is methylated in ER-negative human breast carcinoma cell lines. In addition, histones isolated from these cell lines are deacetylated, suggesting a dual mechanism of methylation and histone deacetylation for ER silencing in these cells (see below).[24]

In ER-negative cell lines, treatment with pharmacologic agents that demethylate the ER gene CpG island, such as 5-Aza-2′-deoxycytidine, results in the restoration of a functional ER protein.[25] In primary breast cancer tissues, ER is hypermethylated in 25–45% of the cases with consistent correlation with the loss of gene expression and clinical progression.[26,27] The PR gene (see below) also undergoes methylation-associated silencing in breast carcinomas with a similar frequency to that of ER.[24] However, because PR expression is induced by ER, the implications of PR gene silencing are more complex than for ER. In summary, although a variety of possible mechanisms might account for the loss of ER and PR gene expression, including structural changes within each gene such as deletions or polymorphisms, it is promoter methylation that is predominantly responsible.[28,29]

ER Silencing by Histone Acetylation

It was noted previously that methylation-related silencing of ER is lost if an inhibitor of histone deacetylase (HDAC) is added to the use of a demethylating drug.[30] The acetyl groups on histone tails (primarily H_3 and H_4) can govern chromatin structure and gene transcription. The treatment of cells with epigenetically silenced genes with HDAC inhibitors can restore gene expression of such antitumor genes as tumor suppressors, cell cycle inhibitors, and hormone receptors.[31–33]

ER, Histone Acetylation, and Metastasis Associated Genes

ER induces changes in gene expression both via its influence on direct gene activation and through the modification of biologic functions of target loci. ER-positive breast cancer cells exhibit activation of the heregulin/HER-2/*neu* pathway that leads to disruption of estradiol signaling and progressively more invasive tumor phenotypes. This process is effected by heregulin-β_1 mediated overexpression of metastasis-associated protein 1 (MTA1), a component of HDAC and nucleosome-remodeling complexes.[34] In breast cancer cell lines, MTA1 and HDACs can function as potent corepressors of estrogen response element driven transcription. The MTA1 protein may serve multiple functions in cellular signaling, chromosome remodeling, and transcription processes that are important in the progression, invasion, and growth of metastatic epithelial cells (see Chapter 26).[35]

The *MTA3* gene is an estrogen-dependent component of the Mi-2/NuRD transcriptional corepressor in breast epithelial cells.[36] *MTA3* constitutes a key component of an estrogen-dependent pathway regulating growth and differentiation. The absence of ER or of *MTA3* leads to aberrant expression of the transcriptional repressor Snail, a master regulator of epithelial to mesenchymal transitions. Aberrant Snail expression results in loss of expression of the cell adhesion molecule E-cadherin (see Chapters 20 and 26), an event associated with changes in epithelial architecture and invasive growth. These results establish mechanistic links between the ER status and invasive growth of breast cancers.[37]

Hormone Receptor Coactivators and Corepressors

It has been shown that a number of different coregulatory factors can modify the signaling of nuclear hormone receptors such as ER.[38] These coactivators and corepressors can stimulate or inhibit ER-mediated target gene activation. Coactivators of ER enhance transcription and include SRC-1, which enhance ER signaling when ER is bound to tamoxifen,[39] SRC-2, which activates ER only when it is bound to estrogens, and SRC-3, which is the only coactivator overexpressed in primary breast cancer clinical samples.[40] Other important steroid receptor coactivators include E6-AP, TIF-2, CBP, P/CAF, SRA, and AIB1.[41–44] Interestingly, the 26S proteasome is required for ER-α and coactivator turnover and for efficient ER-α transactivation (see Chapter 24).[45]

Several inhibitors of ER-mediated signaling known as ER corepressors have been described and include NCoR (SMRT), which can maintain estrogen-stimulated gene expression while reducing the responsiveness to tamoxifen.[46] Reduced expression of another ER corepressor, REA, has been found in clinical samples from patients with advanced disease who have become resistant to antiestrogen therapy.[47] Up-regulation of metastatic tumor antigen 1 is associated with the invasiveness and metastatic potential of several human cancers and acts as a corepressor of nuclear ER-α.[48]

ER-α and ER-β

The ER-α gene (FIG. 15.2) was initially thought to be the only ER gene, but a second ER gene, termed ER-β, was cloned first from the rat and subsequently from the human in 1996.[49,50] ER-α and ER-β are structurally similar with high homology within the DNA and hormone-binding domains. This domain-specific homology suggests that ER-α and ER-β would share similar DNA and ligand-binding function, but the low overall homology between them may explain a variety of differing global effects. ER-β is expressed in a variety of tissues that also feature ER-α expression, including breast, uterus, ovary, prostate, epididymis, testis, pituitary, kidney, thymus, bone, and central nervous system.[51] At present, relatively little is known about the role of ER-β in breast carcinoma. ER-β expression is found by reverse transcriptase-polymerase chain reaction (RT-PCR) techniques in both ductal carcinoma in situ (DCIS) and lobular carcinoma in situ and is lost in late stage invasive lesions.[52] Reliable antibodies against ER-β for IHC are not widely available or validated in clinical material. Thus, the potential role of ER-β in breast cancer remains to be defined.

PRs

Like ER, the PR gene is a member of the nuclear hormone receptor superfamily. Two isoforms of the PR, PR-A and PR-B, are encoded by the same gene, using two distinct transcriptional start sites and yielding proteins that differ in their biologic activities.[53] Although both PR-A and PR-B are highly expressed in normal tissues, PR-B protein concentrations reportedly are elevated in breast carcinoma. The PR-A/PR-B ratio is believed to be an important parameter for

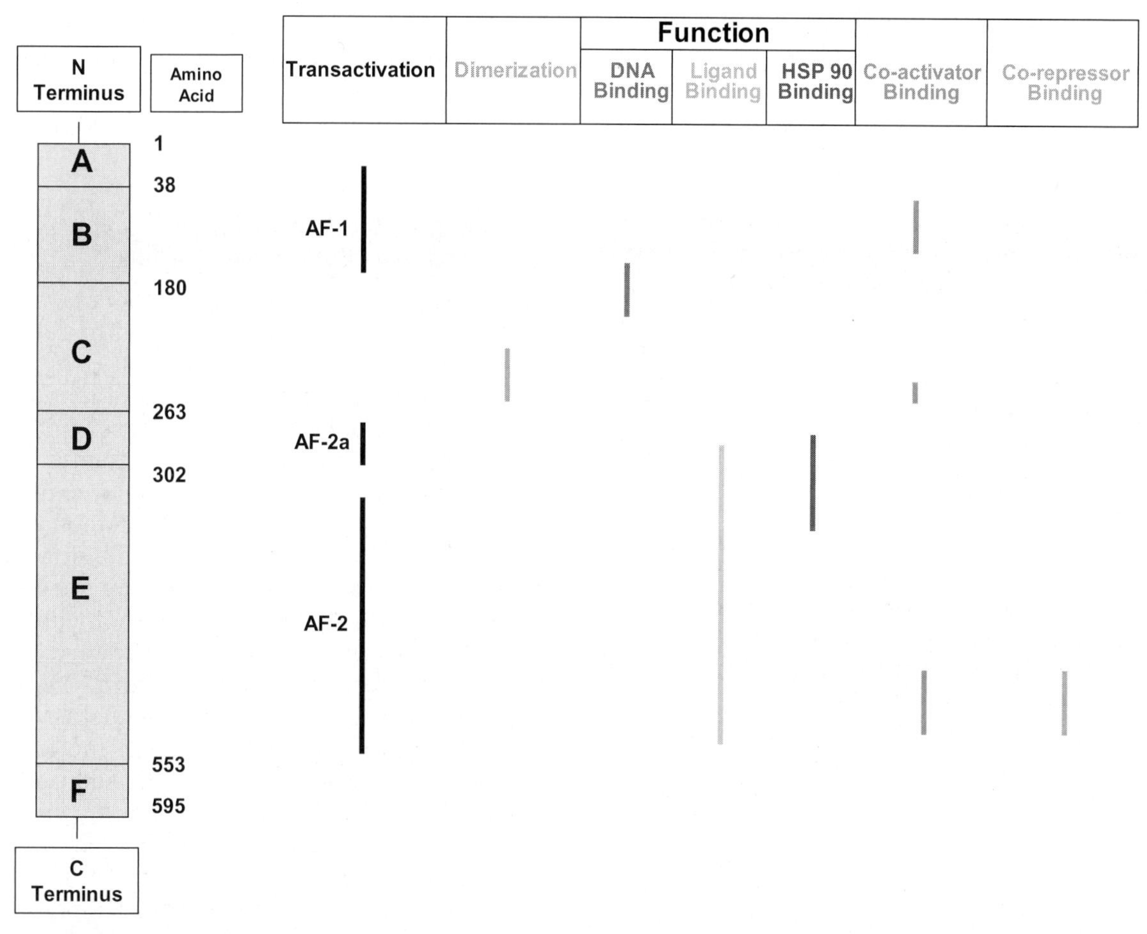

FIGURE 15.2 Structure of the ER-α gene. The ER-α gene encodes a 595 amino acid nuclear hormone receptor protein containing structural and functional domains. The six structural domains are labeled A through F. The seven functional domains are listed on the top and the structural domains that they encompass are indicated by the respective colored lines. (Modified from Ref. 38.) **See Plate 32 for color image.**

progesterone-mediated functions and clinical impact.[54] About 60% of ER-positive tumors also are PR positive, and approximately 75% of these ER/PR-positive tumors respond positively to endocrine therapy (see below).[55] It has been suggested that PR may be necessary for responsiveness to hormonal therapy in that ER-positive/PR-negative tumors are generally less responsive than ER-positive/PR-positive tumors, possibly indicating that the estrogen response pathway may be deactivated in the ER-positive/PR-negative cases.[49] ER-negative/PR-positive tumors are rare and appear to show an intermediate response to endocrine therapy.[55,56] PR expression also is regulated by methylation and/or acetylation of a CpG island within the PR promoter region. Apparently, in the study of breast cancer cell lines, demethylation alone using 5-Aza-2′-deoxycytidine is not sufficient for reactivation of PR.[57] Chromatin reorganization, however, using both HDAC inhibitors and demethylating agents results in the reexpression of PR.[30]

Androgen Receptor

Although the androgen receptor-3 is often coexpressed with the ER and PR in human breast tumors, its role in breast cancer is poorly understood.[58] Variations in testosterone levels and androgen receptor activity may play a role in the etiology of breast cancer.[59] Results from observational studies that have examined polymorphisms in the androgen receptor suggest that the low activity androgen receptor increases breast cancer risk.[59] Testing for androgen receptor in clinical breast cancer samples has not yielded significant prognostic information and has not been recommended for use as a method for selecting patients for hormonal-based therapies.

Hormone Receptors as Prognostic Factors in Breast Cancer

Approximately 70–80% of all invasive primary breast cancers express ER-α protein to some degree and are termed ER positive. ER-positive tumors tend to grow more slowly, are lower grade, are diagnosed in an earlier stage, are DNA diploid, and are generally associated with a slightly better overall prognosis.[55] The role of ER and PR testing as markers of prognosis and predictors of response to antiestrogen therapy is established as a standard of care for patients with breast cancer.[60,61] Since the 1970s, numerous studies involving tens of thousands of patients have evaluated the expression of ER and PR with clinical outcome in breast cancer (**Table 15.1**).[62–119] In reviewing the selected series of prognosis studies included in **Table 15.1**, several basic conclusions emerge:

1. Most studies find both ER and PR status to be a predictor of prognosis in breast cancer at least on univariate analysis.
2. When multivariate analysis is used, a number of studies find the prognostic significance of ER/PR status to be reduced or lost when such factors as tumor size, grade, nodal status, or certain other biomarkers such as HER-2/*neu* status are included in the model.
3. PR status more often retains prognostic significance in multivariate analysis than ER status and thus may be a more powerful prognostic factor.
4. The method used to measure ER/PR status (see below) does not have a major influence on the prognostic significance.
5. The prognostic significance of both ER and PR status appears to be more significant during the first several years after diagnosis and diminishes over time for long-term surviving patients.
6. Subset analysis of breast cancer patients by size, grade, or node status, for example, may create different relative prognostic significance for ER/PR status in each subgroup.
7. The clinical utility of determining ER and PR in breast cancer tissues is more powerful as a predictor of response to hormonal based therapy than as a stand-alone general prognostic factor.

Hormone Receptors in Breast Cancer: Association with Other Prognosis Variables

As stated above, positive ER and PR assays are associated with well-differentiated histology, negative lymph node status, diploid DNA content, low cell proliferation rate, and tendency for a relatively indolent clinical course.[8,12,55,60,61] More than 90% of lobular carcinomas are ER positive, as are most pure tubular carcinomas. Medullary carcinomas and inflammatory carcinomas of the breast are predominantly ER negative. ER/PR-negative tumors are often associated with aggressive disease, including amplification of the HER-2/*neu*, c-*myc*, and *int-2* oncogenes; mutation of the *p53* gene; and up-regulation of invasion- and metastasis-associated growth factors, growth factor receptors, and proteases.[8,12,55,60,61] In addition to these relationships, several proteins are particularly impacted by ER status and are discussed below. The prognostic significance of ER tends to change with time. Patients with ER-positive breast cancer tend to have fewer recurrences early in the course of the disease, although the recurrence rate becomes higher during later years, as compared with patients with ER-negative disease.[120] This crossover of hazard ratios reduces the long-term prognostic value of ER.

Hormone Receptor Status in Hereditary Breast Cancer

Approximately 80% of hereditary breast cancers are accounted for by mutations in the breast cancer susceptibility genes, *BRCA1* (40–45%) and *BRCA2* (35–40%), with the remaining 20% of familial breast cancer linked to mutations in other tumor suppressor genes such as *p53*, *PTEN*, and *ATM*.[121] A link between *BRCA1* and estrogen regulated signaling pathways has been described based on the fact that *BRCA1* can inhibit signaling by the ligand activated ER and suggests that *BRCA1* may be a negative regulator of ER mediated proliferation.[122] It has also been suggested that these data can be used to explain, at least in part, why *BRCA1* mutated tumors manifest specifically in

Table 15.1 Selected Studies of the Hormone Receptor Status and Prognosis in Breast Cancer

					Disease Outcome Statistical Correlation		
Year	Author	Ref. No.	No. of Cases	Method	Univariate Significance	Multivariate Significance	Comment
1979	Cooke	62	286	Biochem	Yes	Yes	ER status independently predicted early recurrence
1979	Hahnel	63	335	Biochem	Yes	Yes	ER independent predictor of disease-free interval
1980	Gapinski	64	274	Biochem	Yes	—	ER predicted recurrence and survival rates
1980	Westerberg	65	270	Biochem	Yes	Yes	ER predicts disease-free survival
1981	Samaan	66	217	Biochem	Yes	—	ER-negative tumors more often featured visceral metastases
1981	Godolphin	67	583	Biochem	Yes	Yes	ER independently predicts recurrence-free survival
1982	Von Maillot	68	222	Biochem	Yes	—	ER predicts disease-free survival
1982	Neifeld	69	132	Biochem	Yes	—	ER positive has better overall prognosis
1983	Brooks	70	155	Biochem	Yes	—	ER positive has favorable prognosis
1983	Mason	71	374	Biochem	Yes	—	ER status provides useful prognostic information. PR status is less useful
1983	Stewart	72	447	Biochem	Yes	Yes	ER predicts survival but not disease-free interval
1984	Mercer	73	2006	Biochem	Yes	—	ER predicts prognosis in both pre- and postmenopausal women
1984	Aamdal	74	233	Biochem	Yes/no	—	ER status predicts short-term but not long-term survival in breast cancer
1984	Alanko	75	286	Biochem	No	—	No correlation of ER status with disease-free interval
1984	Parl	76	121	Biochem	Yes	Yes	ER independent selector of high risk subgroup
1985	Adami	77	170	Biochem	Yes/no	—	ER status predicts disease-free interval in short term only. ER does not predict survival
1985	Butler	78	556	Biochem	No	No	ER status did not predict recurrence or survival
1986	Vollenweider-Zerargui	79	547	Biochem	Yes	—	ER and PR were most predictive in stage II patients
1986	Angus	80	136	IHC	No	—	No prognostic significance for ER when measured by IHC
1987	Clark	81	1015	Biochem	Yes	Yes	ER adds independent prognostic information to tumor size and nodal status
1987	Hawkins	82	372	Biochem	Yes	—	Size, nodal status, and ER were significant prognostic factors
1987	Shek	83	457	Biochem	Yes	Yes	ER was most significant predictor of recurrence
1988	Lampertico	84	614	IHC	Yes	Yes	ER was an independent predictor of survival
1988	Chevallier	85	645	Biochem	Yes	Yes	Both ER and PR were independent predictors of survival

(continued)

Table 15.1 Selected Studies of the Hormone Receptor Status and Prognosis in Breast Cancer (*continued*)

					Disease Outcome Statistical Correlation		
Year	Author	Ref. No.	No. of Cases	Method	Univariate Significance	Multivariate Significance	Comment
1989	Kinsel	86	257	IHC	Yes	—	ER status more predictive of disease-free interval than for survival
1989	Syratos	87	1262	Biochem	No	No	ER was not a significant prognostic factor, but PR status was
1990	McCarty	88	152	IHC	Yes	Yes	ER status was the sole independent prognostic factor
1990	Sigurdsson	89	367	Biochem	Yes	Yes	Both ER and PR were independent prognostic factors
1991	Mathiesen	90	960	Biochem	Yes	No	PR was a better prognostic factor than ER
1991	Muir	91	483	IHC	Yes	Yes	ER combined with PR had the most significant prognostic value
1991	Winstanley	92	767	Biochem	No	No	Size and node status were the only significant factors
1991	Aaltomaa	93	281	IHC	Yes	Yes	ER and PR status both were independent survival predictors
1992	Mouridsen	94	7315	Biochem	Yes	No	ER/PR significant on univariate analysis only
1993	Ewers	95	516	Biochem	Yes	Yes	ER was most significant predictor in multivariate model
1994	Robertson	96	60	IHC	Yes	—	ER and PR status predicted prognosis in patients with locally advanced breast cancer
1994	Beck	97	789	Biochem IHC	Yes	—	PR was more significant than ER when IHC was used
1994	Kommoss	98	241	IHC	Yes	—	Both ER and PR were associated with decreased short-term disease-free survival
1994	Barbareschi	99	233	Biochem	Yes	No	ER and PR significant on univariate analysis only
1995	Stierer	100	288	IHC	Yes	No	ER predicted short term disease-free survival on univariate analysis only
1996	Robertson	101	99	EIA	Yes	Yes	ER determined by enzyme immunoassay predicted survival and response to antiestrogen therapy
1996	Layfield	102	236	IHC	Yes	—	ER quantified by image analysis correlated with survival
1996	Pichon	103	2257	Biochem	Yes	Yes	ER but not PR was an independent prognostic factor. ER status loses significance after 8 years follow-up
1996	Hawkins	104	215	IHC	Yes	No	Only tumor grade was an independent predictor of local recurrence
1996	Collett	105	977	Biochem	Yes	—	ER and PR significant only for the first 5 years after resection
1997	Younes	106	300	Biochem	Yes	Yes	ER status was most predictive when 9 fmol/mg protein was used as the cutoff

Table 15.1 Selected Studies of the Hormone Receptor Status and Prognosis in Breast Cancer (*continued*)

					Disease Outcome Statistical Correlation		
Year	Author	Ref. No.	No. of Cases	Method	Univariate Significance	Multivariate Significance	Comment
1997	Eissa	107	100	EIA	No	No	HER-2/neu and p53 status were the only significant predictors on multivariate analysis
1997	Molino	108	405	Biochem IHC	Yes	Yes	ER and PR were significant prognostic factors and Biochemical and IHC methods closely agreed on ER/PR status
1998	Raabe	109	1335	Biochem	No	No	Only tumor size and node status were significant; ER status was defined as a weak predictor
1998	Yaghan	110	256	IHC	Yes	Yes	ER status independently predicted locoregional recurrence
1999	Chang	111	158	IHC	Yes	Yes	ER status independently predicted survival and response to systemic chemotherapy
1999	Insa	112	439	Biochem IHC	Yes	Yes	ER status independently predicted median survival
2000	Elledge	113	205	Biochem IHC	Yes	Yes	ER status by either method predicted tamoxifen response on univariate analysis, but only IHC determined ER status was an independent predictor
2001	Fisher	114	1259	Biochem IHC	Yes	—	ER was a significant predictor of relapse-free survival
2002	Costa	115	670	Biochem	Yes	Yes/no	Both ER and PR significant on univariate analysis, but only PR was an independent predictor of prognosis
2002	Pinto	116	392	IHC	Yes	Yes	ER and S phase fraction determined by flow cytometry were independent prognostic factors
2003	Chang	117	346	IHC	Yes	Yes/no	Only PR was an independent predictor of survival
2003	Cocquyt	118	135	IHC	No	—	ER was not a significant predictor of complete pathologic response to neoadjuvant chemotherapy
2004	Haffty	119	113	IHC	Yes	Yes/no	Only PR status was an independent predictor of survival in patients with locoregional disease recurrence

Biochem, dextran-coated charcoal competitive binding assay; EIA, enzyme immunoassay.

the breast and ovary rather than in other sites. Finally, it should be noted that *BRCA1* mutation related hereditary breast cancer is typically ER-negative compared with *BRCA2* related tumors, which are usually ER positive. This latter fact has been cited to explain the apparent greater success of chemoprevention of breast cancer using tamoxifen in *BRCA2* patients than in *BRCA1* patients.

ER Related Biomarkers of Significance in Breast Cancer

pS2 Protein

The pS2 is an estrogen inducible small trefoil protein associated with positive PgR and may be a marker of functioning ER, irrespective of ER status.[123,124] pS2 expression also has been linked to the expression of cathepsin D.[121] pS2 expression has consistently been associated with hormone responsive breast cancer featuring a favorable prognosis.[123–131] Use of pS2 measurement continues to show promise as a predictor of indolent disease and clinical hormonal responsiveness in breast cancer. However, the lack of a standardized assay has prevented the incorporation of this marker into standard practice.

Heat Shock Proteins

The heat shock or stress response proteins, HSP27, HSP70, and HSP90, are a family of highly conserved proteins associated with tissue responses to heat, toxins, heavy metals, abnormal pH, certain hormones, drugs, and anoxia.[132] HSP27 and HSP70 expression have been found to be predictors of disease recurrence and outcome in a number of studies of breast cancer.[132–136] HSP overexpression has not predicted hormonal response in one study,[137] correlated with favorable cytotoxic therapy response in a single early study,[138] and inversely correlated with drug response in another.[139] At this time, further studies on the prognostic significance and impact on therapy response for heat shock protein expression in breast cancer is required before these markers can achieve clinical utility. HSP90 is a natural target of herbimycin, ansamycins, and several other compounds. Because HSP90 is involved in the degradation of various proteins, including HER-2, it has become a valuable target for therapeutic intervention. Thus, the anticancer drug geldanamycin and its derivative, 17-AAG, are in clinical trials to determine their antitumor efficacy.

Cathepsin D

Cathepsin D is an estrogen regulated lysosomal aspartyl protease localized to chromosome 11 and believed to facilitate cancer cell migration and promote stromal invasion by the digestion of basement membrane, matrix, and connective tissue.[140] Numerous studies in the early 1990s using an immunoassay approach on fresh breast tumor cytosolic preparations showed that elevated cathepsin D levels are an independent predictor of survival in breast cancer.[141–143] Attempts to convert the assay to an IHC based format have not been successful.[144,145] Thus, given the limitations of the current assay format, interest in using cathepsin D assessment as a prognostic factor and guide to therapy in breast cancer has all but disappeared in the United States, although it is still performed in some centers in Europe.

Clinical Testing Methods

The introduction of testing for ER status in patients with metastatic breast cancer in the early 1970s marked the introduction of targeted therapy for cancer and the first widely accepted integration of a diagnostic test into the selection of therapy for the disease. In this section the main methods of measuring ER and PR in breast cancer clinical specimens (**Table 15.2**) are summarized.

Biochemical Assay

The initial finding that estrogen-related cancer tissues, such as breast carcinoma, overexpressed ERs but nonmalignant tissues did not inspired the search for a method to predict the endocrine responsiveness of breast cancer. It was widely known in the 1950s and 1960s that when surgical ablation was used to reduce estrogen synthesis, only one in three women responded and two women were undergoing surgery and intensive medical care without any hope of a favorable outcome. It was then reasoned that if the ER was necessary for estrogen-stimulated growth, then determination of ER in a tumor specimen could be used to select the appropriate patients for antiestrogen therapy.[146] In 1971, Jensen and colleagues[146] published their studies designed to determine the ER content of tumor cytosols by the size of the 8S ER-estradiol peak and correlated the results with the outcome of adrenalectomy and other ablative therapy. The biochemical competitive radiolabeled estradiol using dextran-coated charcoal (DCC) was rapidly introduced to diagnostic laboratories across the United States and subsequently in Europe and the rest of the world (FIG. 15.3). In 1974 a National Institutes of Health Consensus Conference reported that for the patients evaluated by extramural review, only 8% of ER-negative tumors responded to additive or ablative therapy, whereas approximately 60% of ER-positive patients had an objective response to endocrine therapy.[147] Although this report further expanded the use of ER testing, it was the discovery of the antiestrogen drug, tamoxifen, in the early 1970s and the determination that only patients with ER-positive breast cancer

Table 15.2 Comparison of ER Determination Methods

Technique	Type of Sample	Parameter Measured	Advantages	Disadvantages
Dextran-coated charcoal biochemical assay	Cytosol	Competitive binding of radiolabeled estradiol	Functional assay quantitative Not impacted by truncated ER	Requires large sample Radioactivity Cumbersome Signal contamination by benign elements
IHC	Tissue section	ER antigen	Small sample ready Works after fixation Tumor cell specific Archival samples can be used	Not standardized Cutoffs vary Not quantitative IHC issues include antigen retrieval
EIA	Cytosol	ER antigen	Fresh protein used Quantitative	Requires large sample
RT-PCR	RNA extract	mRNA	Quantitative	Cumbersome Signal dilution by benign elements
Genomic microarray	RNA extract	mRNA	Quantitative Correlation with other pathway gene expression	Not standardized Cumbersome Expensive Signal dilution by benign elements

EIA, enzyme immunoassay.

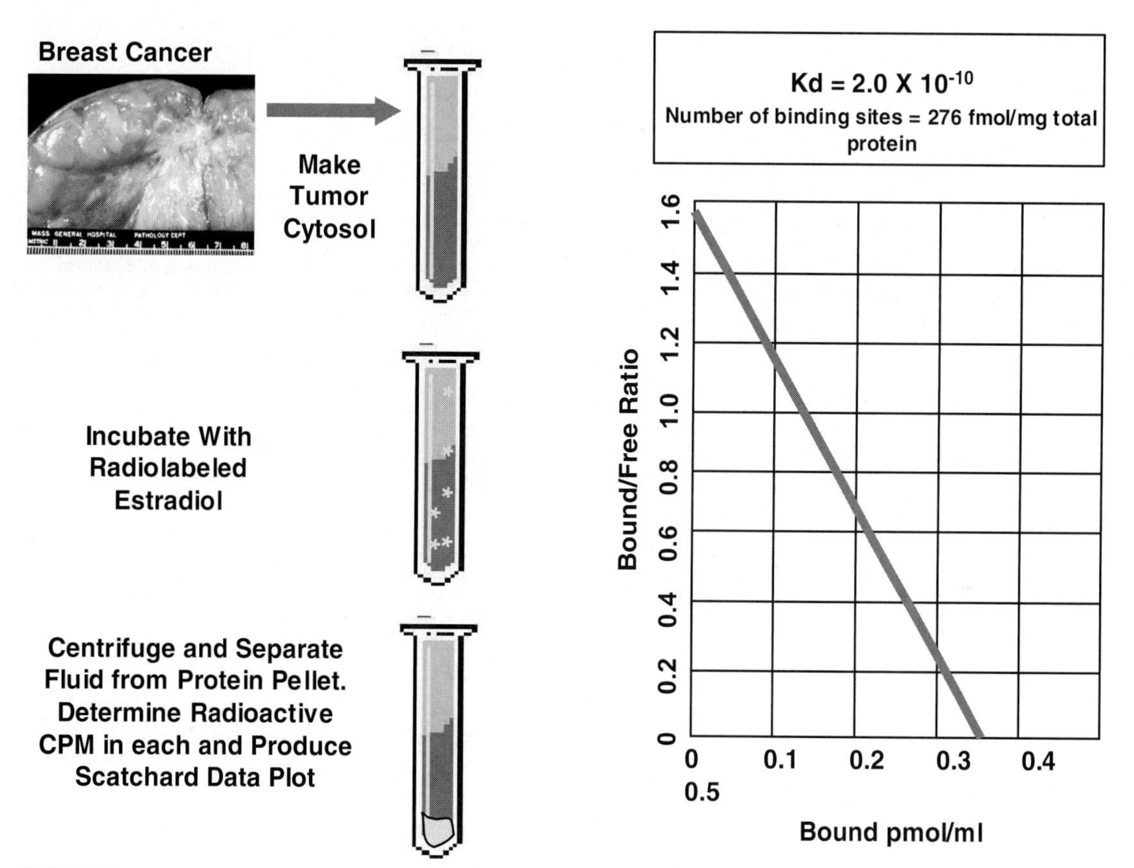

FIGURE 15.3 The dextran-coated charcoal radiolabeled estradiol competitive binding assay for determination of ER status in breast cancer. As shown in the diagram, a fresh breast cancer tumor is digested and a protein-rich tumor cytosol is produced. Radiolabeled estradiol is added to the cytosol and allowed to incubate. If the original tumor contained estrogen receptors, the radiolabeled estradiol will become bound to the cytosol-based protein extract. After centrifugation, most radioactivity will be in the protein pellet rather than in the liquid supernatant as "bound" ligand. If the tumor is ER negative, most radioactivity will remain in the liquid supernatant. On the right, a typical Scatchard data plot is shown for an ER-positive tumor. **See Plate 33 for color image.**

benefited from tamoxifen therapy that expanded the testing until it was extended to virtually all newly diagnosed cases of breast cancer.[148,149] The broad application of tamoxifen for treatment and prevention of breast cancer envisioned in the 1970s reached ultimate success some 25 years later. The translational research strategy was based on the proven preclinical activity of tamoxifen as an antiestrogen, a relatively tolerable incidence of side effects, and ultimately by the targeting of specific patient populations known to be likely responders to the treatment.[3]

The main advantages of the biochemical ligand binding assay revolve around its functional aspects. A positive test result requires an intact and functioning ER. However, to preserve potential ER binding activity, a fresh tumor cytosol must be prepared and either used immediately or frozen and stored for later use. Also, because the assay is blinded to the actual cells binding the radiolabeled estradiol, concern has persisted that benign stromal and epithelial cells whose proteome contributed to the cytosol could cause a false-positive result when mixed with ER-negative breast cancer cells. This nonmorphologic assay has similarly been questioned for its ability to accurately assess ER status in a heterogeneous tumor with foci of ER-positive and ER-negative disease "blended" together. Many investigators questioned whether the 25–35% of patients who tested ER positive with the DCC assay and failed to respond to antiestrogen therapy had actually false-positive ER assay results and their tumors were actually ER negative. To date, this issue has not been resolved. However, as mass screening programs, including self-examination and mammography, took hold in the 1980s, the size of a newly diagnosed breast cancer became smaller and the ability to perform the biochemical assay was reduced to fewer and fewer cases because of the lack of sufficient tumor tissue for production of a cytosol.

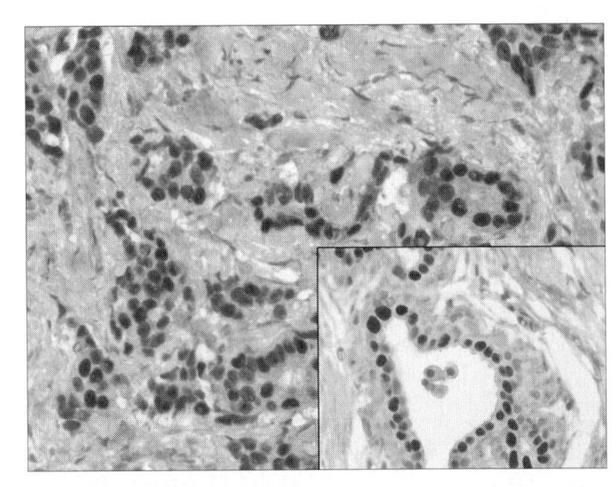

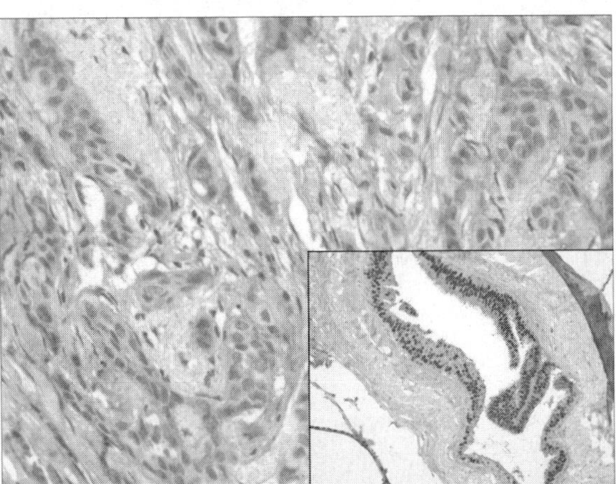

FIGURE 15.4 The immunohistochemical assay for determination of ER status in breast cancer. (Top) The photomicrograph is derived from a strong ER-positive tumor with 100% of the tumor cell nuclei staining intensely for the ER antigen. The positive staining on benign epithelial cells serving as a control is shown in the inset. (Bottom) The tumor shows a negative result with virtually all tumor cells failing to demonstrate nuclear immunostaining. The positive staining on benign epithelial cells again serving as a control is shown in the inset. Although image analysis-based slide scoring of ER and PR immunostaining is advocated by many investigators, it is not widely performed, especially in smaller community hospital pathology laboratories (peroxidase-antiperoxidase, ×200). **See Plate 34 for color image.**

IHC

To address the need to develop an ER test for small samples, Greene and colleagues[150] pursued and developed an in situ assay for ER status using antiestrogen antibodies and IHC (FIG. 15.4). The IHC method had the advantage that it was morphology driven and could report the ER/PR status specifically for the tissue compartment of interest only (e.g., invasive vs. noninvasive neoplasia). Thus, the issues of false-negative results from potential tumor cell dilution by ER-negative benign elements, false-positive results from contaminating ER-positive benign elements, and the issue of tumor heterogeneity could all be addressed with this new test. The advantages of the IHC approach have included its ready application to small biopsies, resections, and fine-needle aspirates and its ability to be used on archival formalin-fixed paraffin-embedded tissues. However, the IHC technique is not standardized, and many laboratories use U.S. Food and Drug Administration–approved reagents in different ways and disagree in the slide scoring system to be used to generate the final results. One method of interpretation that is increasingly used in the community incorporates the percentage of cancer cells that stain and the intensity of the stain into an overall score.[151] Image analysis approaches have further standardized the IHC procedures in some laboratories, but they are not widely used, and most pathology departments continue to use manual nondigital approaches.[152]

Initial and subsequent comparative studies of the IHC and DCC methods showed they were highly concordant.[108,153,154] Similarly, the response rates to antiestrogen therapy have been similar for patients whose tumors were tested by either the DCC or IHC methods, although, on occasion, one investigator has found one technique to be superior.[97] In general, patients with positive ER assays by either biochemical or IHC methods will respond to hormonal therapy in proportion to the receptor protein content (see below). Conversely, the total absence of ER/PR expression is strongly associated with lack of benefit from hormonal therapy. On occasion, when the IHC method is used, some patients may fail to show response to hormonal therapy despite strong staining for ER protein, which may be functionally defective. Data suggest that these tumors might feature a positive immunostain but have an abnormal or truncated ER protein produced by a mutated ER receptor gene that reacts with the anti-ER antibody but fails to bind estrogen.[155,156] However, despite these potential limitations, IHC is currently the standard method to determine ER and PR status in breast cancer, it remains a cornerstone of planning of therapy for the disease and appears likely to be used clinically in this fashion for the foreseeable future.

Enzyme Immunoassay

This method also requires a tumor cytosol produced from fresh tumor tissue.[101,107] Thus, it features the same disadvantages as the DCC test. Although it is not a functional assay, the use of fresh protein eliminates the need for antigen retrieval that is widely used in the IHC procedures and can confound IHC results. This test is quantitative and precise but, similar to other cytosol based tests, is dependent on the "quality" of the tissue that was used to make the cytosol and its relative tumor cell versus stromal, benign epithelial, and inflammatory cell components. This approach is more widely used in Europe where larger primary tumors facilitate the production of protein extracts and cytosols and the HER-2/neu protein and urokinase-related plasminogen proteases can also be measured on aliquots from the same material.

Real-Time RT-PCR

The RT-PCR method has not been widely used to detect ER and PR messenger RNA (mRNA) levels in clinical material and has been predominantly used for research studies only.[157] To date, the ability of ER/PR mRNA levels measured by RT-PCR to predict hormone therapy response has not been tested.

Genomic Microarray and Gene Expression Profiling

A number of recent studies revealed that the DNA microarray approach showed a series of genes expressed in breast cancer according to the tumoral ER status[158–162] (FIG. 15.5). These studies have shown significant promise. For example, one study demonstrated that comprehensive expression profiling could be performed on fine-needle aspirate biopsies and reliably measure ER status with high concordance with IHC results on the same specimens (FIG. 15.6).[160] This study also found that a complex pattern of gene expression, but not including ER itself, could also be used to predict the IHC-based clinical ER status.[160] Thus, microarray-based tests on clinical breast cancer samples could capture conventional prognostic information such as ER/PR status but could also generate additional gene expression information that cannot currently be measured with other methods that could be used to further customize therapy (see Chapter 28).

In addition to the use of high density commercial genomic microarrays, custom microarrays have now been produced by investigators that have yielded comparable results with those obtained from large-scale microarray analysis, can categorize tumor cells into early or late hormonal therapy response types, and may provide new clues in the elucidation of the estrogen-dependent growth mechanisms of cancer.[162] One of the most striking findings in gene expression profiling is that ER expression is the major determinant of the two largest molecularly defined subgroups of breast cancer.[163] Tumors with ER expression also express a large number of ER-dependent genes, whereas those without ER expression present with an entirely different gene expression profile.

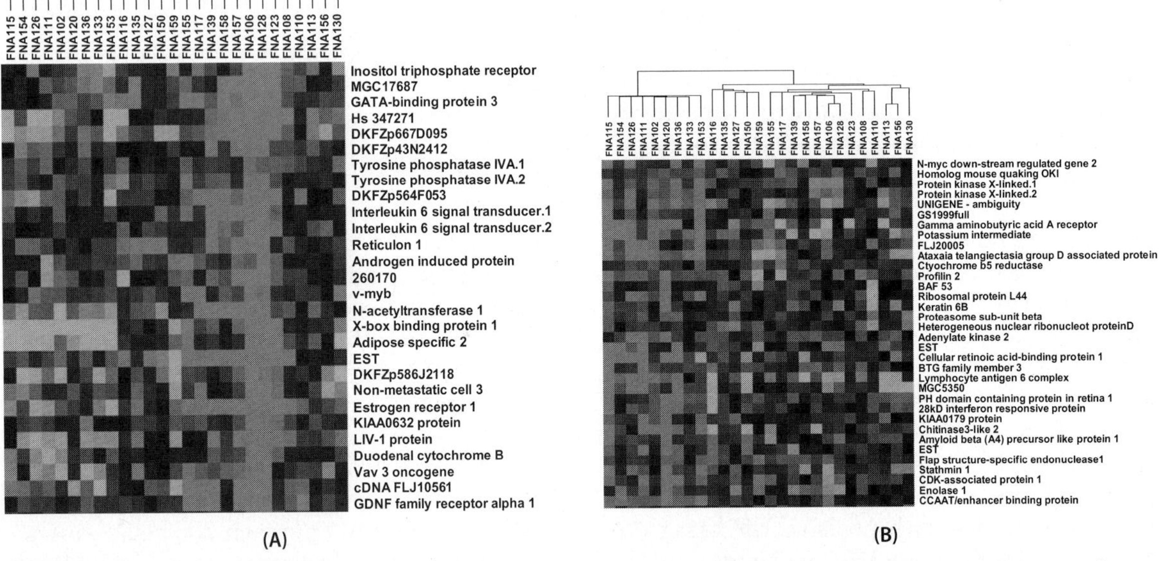

FIGURE 15.5 Breast cancer complementary DNA microarray results evaluated by hierarchical gene cluster analysis for defining specific gene expression signatures. Hierarchical clustering algorithm allows the clustering of individual tumor profiles on the basis of their similarities to their coexpression with the ER-α gene. Each row represents a tumor and each column a single gene. Red indicates up-regulation, green down-regulation, and black no change in relative gene expression. (A) The gene regulation associated with ER-positive status is shown. (B) The up- and down-regulated genes associated with ER-negative status are shown. (From Ref. 151.) **See Plate 31 (Fig. 14.2) for color image.**

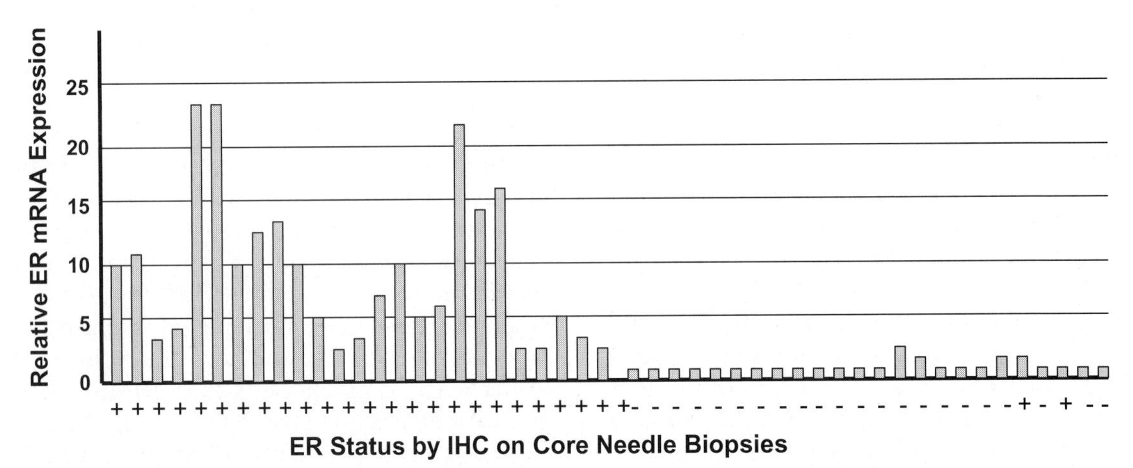

FIGURE 15.6 Comparison of ER mRNA expression detected by microarray profiling and corresponding ER protein expression measured by IHC. The concordance between ER levels determined by IHC and ER levels determined by gene expression profiling was approximately 95%. All cases with high mRNA expression levels were ER positive by IHC. Two cases with low mRNA levels by microarray profiling were ER positive by IHC. From these data, it appears possible that a single test using gene expression profiling combined using either fine-needle aspiration or tissue biopsies from breast cancer patients can capture clinically relevant markers and guide therapy for the disease. (From Ref. 151.)

Clinical Relevance and Current Clinical Testing Utility

Prognostic Factor

As described in detail above, determination of the ER/PR status in breast cancer provides significant prognostic information separate from its ability to predict hormonal therapy response. Investigators continue to disagree as to the actual prognostic power of hormone receptor status, whether it is a dependent or independent factor, whether PR is truly superior to ER in this regard, and for how long after diagnosis the ER/PR status retains prognostic significance. However, the ER/PR status is used in a variety of ways for therapy selection that depend on the prognostic rather than predictive capabilities of the test. For example, in a patient with a lymph node-negative, less than 1 cm, primary breast cancer of low or intermediate grade featuring ER-positive/PR-positive status, most oncologists would recommend the use of hormonal therapy in a chemoprevention strategy rather than cytotoxic drugs but might feel differently if the hormone receptor status was ER negative/PR negative. Thus, it is considered an integral part of the evaluation of all primary breast cancers, regardless of histology, nodal status, or age, to perform ER and PR assays.

Predictive Factor

As described in Chapter 1, hormone-sensitive breast cancers have different metastatic patterns and sensitivity to chemotherapy compared with hormone-independent tumors. Patients with hormone-sensitive and hormone-insensitive tumors are approached differently. When planning treatment for a patient with newly diagnosed metastatic disease, the extent of disease spread, the location of metastases, the pace of the disease (slowly vs. rapidly growing), and the ER and PR content are the critical data that will drive decision making.[164] In the two thirds of patients with hormone receptor–rich breast cancers, especially in the absence of life-threatening disease, endocrine therapy is used as the first and major treatment approach. These patients often benefit from several endocrine interventions in sequence, suggesting that although individual tumors might become resistant to specific agents, they do not lose sensitivity to hormonal manipulations.[165] In general, given the lack of standardization of ER/PR assays as currently used, the major clinical use lies in the negative predictive value of the tests. If an individual tumor is ER negative/PR negative, the response to hormonal therapy is usually less than 5%, probably closer to 0%; conversely, in contrast to an ER-positive/PR-positive tumor, the presence of either receptor, even in a small percentage of tumor cells, is still associated with a 20% to 40% clinical benefit rate.[8] Thus, there is still a need for improving and standardizing the ER/PR assays, reducing false-positive and -negative results that confound therapy response prediction, and more accurately customizing the therapy selection of an individual patient.

Not only is it considered indispensable to determine ER and PR status for all invasive breast cancers, but recent reports suggest that even in patients with noninvasive breast cancer, ER status provides critical information to determine the components of optimal multidisciplinary treatment. Allred and colleagues[166] recently reported the results of the reanalysis of a subset of patients with DCIS treated on NSABP protocol B-24. In this study, patients with fully excised DCIS were randomly assigned to receive tamoxifen or not after completion of surgical therapy. Although the overall results indicated that those who received tamoxifen had a lower rate of local–regional recurrences, the analysis by ER status indicated that the benefit was restricted to patients with ER-positive DCIS. Those with ER-negative DCIS appeared to have little or no benefit from treatment with tamoxifen. In recent years there has been increasing interest in the potential cross-talk between steroid hormone receptor signaling and the various other growth factor induced signaling pathways.

Endocrine Therapy of Breast Cancer

Metastatic Breast Cancer

Metastatic breast cancer is a heterogeneous disease with protean manifestations and great variation in clinical behavior. Patients with ER-positive breast cancer tend to have recurrent disease that targets skin, lymph nodes and other soft tissue, and bone. Visceral metastases (to liver, lung, and brain) are relatively less common, although they occur too. ER-positive breast cancer tends to be moderately well to well differentiated and is associated with lower proliferative activity. Recurrences and metastases tend to occur somewhat later than with ER-negative breast cancer, and overall survival after recurrence is on average 1 or 2 years longer for ER-positive disease.

Endocrine therapy is the most important component of the therapeutic armamentarium of metastatic breast cancer. The most commonly used strategy is to initiate therapy with endocrine agents, exhaust all hormonal therapeutic alternatives, and then turn to chemotherapy for additional palliation.[167] An interesting observation in the management of breast cancer is that ER-positive tumors develop resistance to specific endocrine interventions but respond to other endocrine treatments based on a different mechanism of action. Therefore, resistance to a SERM is not equivalent to hormone resistance. At present, there are no available tests to predict for response or resistance to individual endocrine interventions, nor is there a method to predict when overall endocrine resistance will develop. In fact, palliation and preservation of quality of life are the most important objectives of treatment of metastatic breast cancer. There are multiple endocrine agents with demonstrated activity in metastatic breast cancer. There are also multiple active cytotoxic agents, although the following discussion is restricted to endocrine therapy.

The most commonly used endocrine treatments for breast cancer include SERMs, SERDs, gonadotropin analogues, and selective aromatase inhibitors[168–173] (for details see Chapter 1). These agents are effective, and their safety profile is excellent compared with other therapeutic agents, such as cytotoxic chemotherapy.

In general terms, endocrine agents are used in sequence, with one agent administered continuously until the detection of resistance to that agent. At that time a second endocrine agent is substituted, and so on. The selection of endocrine agents is based on menopausal status and the presence of specific comorbid conditions. Today it is considered that selective aromatase inhibitors should be the endocrine agents of choice for postmenopausal women with metastatic breast cancer. Anastrozole, letrozole, and exemestane have all been compared with megestrol acetate in second-line therapy, demonstrating superiority in therapeutic index. Some of these agents have also been compared with aminoglutethimide, a first-generation nonselective aromatase inhibitor; again, the third-generation selective aromatase inhibitors were shown to be superior to aminoglutethimide. More recent randomized trials compared anastrozole and letrozole with tamoxifen in first-line therapy. In large, randomized, phase III trials, anastrozole and letrozole were shown to be at least as effective and better tolerated as tamoxifen; for some end points considered, the aromatase inhibitors were found to be superior.[174,175] Exemestane, a steroidal aromatase inhibitor, was also found to be superior to tamoxifen in a small, randomized, phase II trial; a phase III trial is ongoing.

There is information that suggests incomplete cross-resistance between the nonsteroidal aromatase inhibitors (anastrozole, letrozole) and exemestane.[176] Therefore, if progressive disease develops in a postmenopausal woman with ER-positive metastatic breast cancer while taking a nonsteroidal aromatase inhibitor, a change in therapy to exemestane is likely to get an additional period of clinical benefit. Other second-line options would include a SERM (tamoxifen or toremifene) or, more recently a SERD (fulvestrant).[177,178] For tamoxifen-resistant metastatic breast cancer, treatment with a nonsteroidal aromatase inhibitor is therapeutically equivalent to treatment with fulvestrant. However, the optimal choice of endocrine therapy after an aromatase inhibitor has not been defined, nor has the best sequence of administration of the various endocrine agents available to treat these patients. The role of progestins, androgens, or estrogens and other agents is even less well determined.[179]

There are interesting hypotheses emerging from laboratory investigations. Osipo et al.[180] and Liu et al.[181] suggested that whereas tamoxifen (and other SERMs) produces long-term suppression of ER-positive breast cancers in vitro and in vivo, some tamoxifen cell lines become resistant to and dependent on continued administration of the SERM. Removal of the SERM and estrogen deprivation is the treatment of choice in this setting. Conversely, ER-positive cell lines treated chronically with estrogen deprivation become exquisitely sensitive to physiologic doses of estrogen, which induce apoptosis in these cells. Taken together, these experiments suggest that alternating estrogen agonist therapy with periods of estrogen deprivation might be the optimal sequence strategy for these various endocrine regimens. This hypothesis, although eminently plausible, needs to be tested in prospective clinical trials.

For premenopausal women, the endocrine treatment of choice is ovarian ablation alone, a SERM alone, or in combination with gonadotropin analogues (or other forms of ovarian suppression/ablation). If ovarian suppression was used first, a SERM would be an appropriate choice upon development of progressive disease; conversely, if a SERM was used first, ovarian suppression is an appropriate second-line intervention. Several small randomized trials and a recent meta-analysis of these trials indicated that the simultaneous use of ovarian suppression and a SERM might improve treatment outcomes.[182] The combination is clearly associated with higher overall response rates and longer duration of response and time to progression. The meta-analysis also suggested a significant survival advantage for the combination. These results, especially in relation to the survival advantage, have been criticized, because none of the trials included a crossover from one single agent to the next; therefore, it remains uncertain whether a similar survival duration could be obtained with the two interventions administered in sequence or simultaneously.

Beyond SERMs and ovarian suppression, all interventions available for postmenopausal women are available and effective for premenopausal women too. If ovarian suppression was permanent (as in the case of surgical oophorectomy or radiation-induced ovarian suppression), a selective aromatase inhibitor, followed upon development of resistance by a SERD or an aromatase inhibitor of a different class, or a progestin would be appropriate choices. For patients with indolent disease and continued clinical benefit from several endocrine interventions, androgen and estrogen therapies are still available and useful.

Although specific tests are yet unavailable, the understanding of the molecular events and pathway biology of these newer hormonal therapies is showing definite progress. It has been shown that tamoxifen, raloxifene, and fulvestrant can reduce the expression of cell cycle proteins, including cyclin D1 and cyclin E, and inhibit the phosphorylation of the retinoblastoma (*Rb*) gene, a major target of the cyclin-associated kinases that are critical in cell cycle progression and cellular proliferation.[183,184] Apoptosis-associated pathways are impacted by SERM treatment, estrogen withdrawal, or aromatase inhibition.[185] SERM treatment also causes reductions in tumor size and S phase fraction.[185] Ongoing research continues to study the mechanism of action of hormonal therapies and has used molecular techniques to uncover new candidate biomarkers that may prove useful in improving the accuracy of pretherapy diagnostics for appropriate classification of patients and the drugs they are most likely to respond to.[186] Although ER/PR testing by biochemical or IHC methods remains the standard approach for predicting response to tamoxifen, additional biomarkers, including HER-2/neu and cathepsin D, have been proposed to further refine therapy selection.[186,187] The introduction of aromatase inhibitors such as anastrozole, letrozole, and exemestane[173] has added new strategies for evaluating tumors for hormonal therapy. For example, it has been reported that ER-positive HER-2–positive tumors may be resistant to tamoxifen but may respond to an aromatase inhibitor.[188,189]

Adjuvant Endocrine Therapy

For patients with primary, ER-positive, and/or PR-positive breast cancer, adjuvant endocrine therapy represents the major and indispensable part of combined modality curative treatment. As for patients with metastatic breast cancer, the selection of optimal therapy is made on the basis of menopausal status and the presence of specific comorbid conditions or risk factors for complications. In contrast to metastatic breast cancer, only some of the several endocrine interventions have been adequately tested in randomized trials, so the choices are far fewer.

Over the past 30 years, multiple randomized clinical trials and several meta-analyses of all clinical trials of adjuvant therapy have demonstrated the significant benefit of adjuvant tamoxifen.[190] When administered for 5 years to patients with hormone receptor–positive breast cancer, tamoxifen is associated with about a 50% reduction in annual odds of recurrence and about a 33% reduction in annual odds of death. The safety profile of this agent, based on several million women-years of experience, is extremely favorable, although rare side effects (thromboembolic events, endometrial cancer) reduce the therapeutic index in postmenopausal women with low risk breast cancer. Although tamoxifen had become the standard of care for all women with hormone receptor–positive primary breast cancer, recent results of large clinical trials are rapidly changing this standard. For postmenopausal women, a selective aromatase inhibitor is the treatment of choice. Because the only trial reported to date compared an aromatase inhibitor with tamoxifen was based on anastrozole, the recommendation is to use this aromatase inhibitor for newly diagnosed patients.[190]

Three additional randomized trials have continued to erode the tamoxifen standard in other areas. Thus, a large clinical trial indicated that adding 5 years of letrozole after 5 years of tamoxifen had been completed produced an incremental reduction in risk of recurrence and death.[191] Two more recent reports suggested that crossing over to an aromatase inhibitor (anastrozole or exemestane) after completing just 2 or 3 years of tamoxifen produced robust additional improvements in outcome.[192,193] Progress in general leads to more questions than answers. At this time it is apparent that aromatase inhibitors have a critical role to play in the management of postmenopausal women with primary breast cancer. However, it is as yet uncertain whether the optimal strategy should be an aromatase inhibitor up front, crossover to an aromatase inhibitor after some treatment with tamoxifen, or an aromatase inhibitor up front followed after some time by tamoxifen or some other endocrine intervention. Treatment with an aromatase inhibitor introduces other concerns, mainly related to the very effective reduction in endogenous estrogen production. Thus, aromatase inhibitor treated patients must be carefully monitored for changes in bone density and cardiac lipid profile.

For premenopausal women with steroid hormone receptor–positive breast cancer, the standard endocrine intervention continues to be tamoxifen administered for 5 years. There is much interest in determining whether the combination of ovarian suppression and tamoxifen would benefit all or at least some premenopausal women, especially in combination with chemotherapy. There are several randomized trials ongoing to determine the indications for total estrogen blockade. There is, however, uncertainty about the optimal duration of temporary ovarian suppression and about the optimal sequence of utilization of the various systemic treatments for premenopausal women.

The exciting results obtained with aromatase inhibitors in postmenopausal women generated obvious questions about the use of these agents for younger women. Because aromatase inhibitors are not known to be effective in women with intact ovarian function, ongoing trials are exploring the use of aromatase inhibitors in combination with ovarian suppression. Although this is a reasonable research strategy, there are no data about the effectiveness or long-term safety of such an approach. A few small statistically underpowered trials have been reported with aminoglutethimide and megestrol acetate.[194–196] The results are inconclusive, so these agents have no established role in the management of primary breast cancer.

Breast Cancer Chemoprevention and Hormonal Therapy

Large prospective clinical trials are evaluating the potential of SERMs and selective aromatase inhibitors to prevent the development of breast cancer in women with a high risk of the disease.[197,198] One such study has shown that 5 years of tamoxifen-based chemoprevention can reduce the risk of breast cancer by up to 50% in high risk women.[197] Four prospective randomized trials of chemoprevention with tamoxifen have been reported to date. In aggregate, these four trials included over 23,000 women at risk for breast cancer.[5,199–204] Although the eligibility criteria were different in each trial and the approach to chemoprevention varied somewhat, half of these patients received tamoxifen, whereas the other half did not. Two of the four trials demonstrated a highly significant reduction in the incidence of breast cancer among tamoxifen treated patients, and a meta-analysis of all four trials confirmed a 38% reduction in risk of recurrence. No survival differences have been documented in these studies. A fifth trial included a different SERM, raloxifene.[205] This study too demonstrated a significant reduction in risk of developing breast cancer in the group treated with the SERM. This trial provided the necessary incentive for a large, prospective, randomized trial (NSABP P-2) comparing tamoxifen to raloxifene in the chemoprevention setting. Accrual to this trial should be completed within the next year or two. Tamoxifen-based chemoprevention appears to be effective in reducing the development of ER-positive but not ER-negative tumors.

On the basis of the exciting results of the Arimidex and Tamoxifen: Alone or in Combination (ATAC) trial, two new chemoprevention trials were initiated. The International Cancer Intervention Study (IBIS-II) trial compares anastrozole administered for 5 years with placebo for women at high risk for developing breast cancer. The second trial, NSABP B-35, compares anastrozole to tamoxifen in the management of women with a completely treated DCIS or lobular carcinoma in situ. These studies should provide additional information about the relative efficacy and safety of these endocrine agents used for prevention of breast cancer.

Summary and Future Clinical Potential

Hormone receptor testing is an established cornerstone for the initial assessment and planning of therapy for patients newly diagnosed with breast cancer. Although the current tests for ER and PR status are not standardized and provide variable positive predictive value, they provide extremely important negative predictive value for the selection of hormone-based therapies at all stages of the disease. Recent molecular and microarray-based studies of the ER gene pathway show the ER gene to be a powerful driver of the expression profile of a large number of downstream genes potentially capable of determining the biologic character of an individual tumor. With the continued development of more specific antiestrogens, ER modulators, and down-regulators, the focus of translational research will also be on the development of more robust and accurate clinical assays that guide the use of these agents in the adjuvant setting, in the treatment of metastatic disease, and in the discovery of new chemoprevention strategies for the high risk patient.

References

1. Beatson GT. On the treatment of inoperable cases of carcinoma of the mamma: suggestions for a new method of treatment with illustrative cases. Lancet 1896;2:104–107.
2. Boyd S. On oophorectomy in cancer of the breast. BMJ 1900;2:1161–1167.
3. Jensen EV, Jordan VC. The estrogen receptor: a model for molecular medicine. Clin Cancer Res 2003;9:1980–1989.
4. Osborne CK. Tamoxifen in the treatment of breast cancer. N Engl J Med 1998;339:1609–1618.
5. Fisher B, Costantino JP, Wickerham DL, et al. Tamoxifen for prevention of breast cancer: report of the National Surgical Adjuvant Breast and Bowel Project P-1 study. J Natl Cancer Inst 1998;90: 1371–1388.
6. Lacey JV Jr, Devesa SS, Brinton LA. Recent trends in breast cancer incidence and mortality. Environ Mol Mutag 2002;39:82–88.
7. Jemal A, Tiwari RC, Murray T, et al. Cancer statistics 2004. CA Cancer J Clin 2004;54:8–29.
8. Osborne CK. Steroid hormone receptors in breast cancer management. Breast Cancer Res Treat 1998; 51:227–238.
9. Wilbur DC, Willis J, Mooney RA, et al. Estrogen and progesterone detection in archival formalin-fixed paraffin embedded tissue from breast carcinoma: a comparison of immunocytochemistry with dextran coated charcoal assay. Mod Pathol 1992;5: 79–84.
10. Ciocca DR, Elledge R. Molecular markers for predicting response to tamoxifen in breast cancer patients. Endocrine 2000;13:1–10.
11. Jordan VC. Tamoxifen: a most unlikely pioneering medicine. Nat Rev Drug Discov 2003;2:205–213.
12. Locker GY. Hormonal therapy of breast cancer. Cancer Treat Rev 1998;24:221–240.
13. Ibrahim NK, Hortobagyi GN. The evolving role of specific estrogen receptor modulators (SERMs) Surg Oncol 1999;8:103–123.
14. Miller WR, Anderson TJ, Dixon JM. Anti-tumor effects of letrozole. Cancer Invest 2002;20:15–21.
15. Buzdar AU, Robertson JF, Eiermann W, et al. An overview of the pharmacology and pharmacokinetics of the newer generation aromatase inhibitors anastrozole, letrozole, and exemestane. Cancer 2002;95: 2006–2016.
16. Jordan VC. Antiestrogens and selective estrogen receptor modulators as multifunctional medicines. 1. Receptor interactions. J Med Chem 2003;46: 883–908.
17. Jordan VC. Antiestrogens and selective estrogen receptor modulators as multifunctional medicines. 2. Clinical considerations and new agents. J Med Chem 2003;46:1081–1111.
18. Keen JC, Davidson NE. The biology of breast carcinoma. Cancer 2003;97:825–833.
19. Hulka BS, Moorman P. Breast cancer: hormones and other risk factors. Maturitas 2001;38:103–116.
20. Trichopoulos D, MacMahon B, Cole P. Menopause and breast cancer risk. J Natl Cancer Inst 1972;48: 605–613.
21. Paech K, Webb P, Kuiper GGJM, et al. Differential ligand activation of estrogen receptors ERα and ERβ at AP1 sites. Science 1997;277:1508–1510.
22. Maruyama S, Fujimoto N, Asano K, et al. Suppression by estrogen receptor beta of AP–1 mediated transactivation through estrogen receptor alpha. J Steroid Biochem Mol Biol 2001;78: 177–184.
23. Esteller M. CpG island hypermethylation and tumor suppressor genes: a booming present, a brighter future. Oncogene 2002;21:5427–5440.
24. Yang X, Phillips DL, Ferguson AT, et al. Synergistic activation of functional estrogen receptor (ER)-α by

DNA methyltransferase and histone deacetylase inhibition in human ER-α-negative breast cancer cells. Cancer Res 2001;61:7025–7029.

25. Ferguson AT, Lapidus RG, Baylin SB, et al. Demethylation of the estrogen receptor gene in estrogen receptor-negative breast cancer cells can reactivate estrogen receptor gene expression. Cancer Res 1995; 55:2279–2283.
26. Lapidus RG, Ferguson AT, Ottaviano YL, et al. Methylation of estrogen and progesterone receptor gene 5′ CpG islands correlates with lack of estrogen and progesterone receptor gene expression in breast tumors. Clin Cancer Res 1996;2:805–810.
27. Nass SJ, Herman JG, Gabrielson E, et al. Aberrant methylation of the estrogen receptor and E-cadherin 5′ CpG islands increases with malignant progression in human breast cancer. Cancer Res 2000;60:4346–4348.
28. Ferguson AT, Lapidus RG, Davidson NE. The regulation of estrogen receptor expression and function in human breast cancer. Cancer Treat Res 1998;94: 255–278.
29. Esteller M. CpG island hypermethylation and tumor suppressor genes: a booming present, a brighter future. Oncogene 2002;21:5427–5440.
30. Yang X, Phillips DL, Ferguson AT, et al. Synergistic activation of functional estrogen receptor (ER)-alpha by DNA methyltransferase and histone deacetylase inhibition in human ER-alpha-negative breast cancer cells. Cancer Res 2001;61:7025–7029.
31. Keen JC, Yan L, Mack KM, et al. A novel histone deacetylase inhibitor, scriptaid, enhances expression of functional estrogen receptor alpha (ER) in ER negative human breast cancer cells in combination with 5-aza 2′-deoxycytidine. Breast Cancer Res Treat 2003;81:177–186.
32. Momparler RL. Cancer epigenetics. Oncogene 2003; 22:6479–6483.
33. Carrozza MJ, Utley RT, Workman JL, et al. The diverse functions of histone acetyltransferase complexes. Trends Genet 2003;19:321–329.
34. Mazumdar A, Wang RA, Mishra SK, et al. Transcriptional repression of oestrogen receptor by metastasis-associated protein 1 corepressor. Nat Cell Biol 2001;3:30–37.
35. Nicolson GL, Nawa A, Toh Y, et al. Tumor metastasis-associated human MTA1 gene and its MTA1 protein product: role in epithelial cancer cell invasion, proliferation and nuclear regulation. Clin Exp Metast 2003;20:19–24.
36. Fujita N, Jaye DL, Kajita M, et al. MTA3, a Mi-2/NuRD complex subunit, regulates an invasive growth pathway in breast cancer. Cell 2003;113: 207–219.
37. Fearon ER. Connecting estrogen receptor function, transcriptional repression, and E-cadherin expression in breast cancer. Cancer Cell 2003;3:307–310.
38. Sommer S, Fuqua SA. Estrogen receptor and breast cancer. Semin Cancer Biol 2001;11:339–352.
39. Smith CL, Nawaz Z, O'Malley BW. Coactivator and corepressor regulation of the agonist/antagonist activity of the mixed antiestrogen, 4-hydroxytamoxifen. Mol Endocrinol 1997;11:657–666.
40. Anzick SL, Kononen J, Walker RL, et al. AIB1, a steroid receptor coactivator amplified in breast and ovarian cancer. Science 1997;277:965–968.
41. Kurebayashi J, Otsuki T, Kunisue H, et al. Expression levels of estrogen receptor-alpha, estrogen receptor-beta, coactivators, and corepressors in breast cancer. Clin Cancer Res 2000;6:512–518.
42. Bouras T, Southey MC, Venter DJ. Overexpression of the steroid receptor coactivator AIB1 in breast cancer correlates with the absence of estrogen and progesterone receptors and positivity for p53 and HER2/neu. Cancer Res 2001;61:903–907.
43. Liu H, Lee ES, Deb Los RA, et al. Silencing and reactivation of the selective estrogen receptor modulator-estrogen receptor alpha complex. Cancer Res 2001;61:3632–3639.
44. Fan S, Ma YX, Wang C, et al. p300 modulates the BRCA1 inhibition of estrogen receptor activity. Cancer Res 2002;62:141–151.
45. Lonard DM, Nawaz Z, Smith CL, et al. The 26S proteasome is required for estrogen receptor-alpha and coactivator turnover and for efficient estrogen receptor-alpha transactivation. Mol Cell 2000;5: 939–948.
46. Lavinsky RM, Jepsen K, Heinzel T, et al. Diverse signaling pathways modulate nuclear receptor recruitment of N-CoR and SMRT complexes. Proc Natl Acad Sci USA 1998;95:2920–2925.
47. Simon SL, Parkes A, Leygue E, et al. Expression of a repressor of estrogen receptor activity in human breast tumors: relationship to some known prognostic markers. Cancer Res 2000;60:2796–2799.
48. Kumar R, Wang RA, Mazumdar A, et al. A naturally occurring MTA1 variant sequesters oestrogen receptor-alpha in the cytoplasm. Nature 2002;418: 654–657.
49. Kuiper GGJM, Enmark E, Pelto-Huikko M, et al. Cloning of a novel estrogen receptor expressed in rat prostate and ovary. Proc Natl Acad Sci USA 1996; 93:5925–5930.
50. Mosselman S, Polman J, Dijkema R. ER β: identification and characterization of a novel human estrogen receptor. FEBS Lett 1996;392:49–53.
51. Kuiper GGJM, Carlsson B, Grandien K, et al. Comparison of the ligand binding specificity and transcript tissue distribution of estrogen receptors α and β. Endocrinology 1997;138:863–870.
52. Fuqua SA, Schiff R, Parra I, et al. Expression of wild-type estrogen receptor β and variant forms in human breast cancer. Cancer Res 1999;59:5425–5428.
53. Zhang X, Liu Y, Lee M, et al. A specific defect in the retinoic acid receptor associated with human lung cancer cell lines. Cancer Res 1994;54:5663–5669.
54. Bamberger A-M, Milde-Langosch K, Schulte H, et al. Progesterone receptor isoforms, PR-B and PR-A, in breast cancer: correlations with clinicopathologic tumor parameters and expression of AP-1 factors. Horm Res 2000;54:32–37.
55. Elledge RM, Fuqua SA. Estrogen and progesterone receptors. In: Harris JR, ed. Diseases of the Breast (vol 2). Philadelphia: Lippincott Williams & Wilkins, 2000:471–488.
56. Graham JD, Roman SD, McGowan E, et al. Preferential stimulation of human progesterone receptor B expression by estrogen in T-47D human breast cancer cells. J Biol Chem 1995;270:30,693–30,700.
57. Ferguson AT, Lapidus R, Davidson NE. Demethylation of the progesterone receptor CpG island is not required for progesterone receptor gene expression. Oncogene 1998;17:577–583.
58. Birrell SN, Hall RE, Tilley WD. Role of the androgen receptor in human breast cancer. J Mamm Gland Biol Neoplasia 1998;3:95–103.

59. Lillie EO, Bernstein L, Ursin G. The role of androgens and polymorphisms in the androgen receptor in the epidemiology of breast cancer. Breast Cancer Res 2003;5:164–173.
60. Goldhirsch A, Colleoni M, Gelber RD. Endocrine therapy of breast cancer. Ann Oncol 2002;13(suppl 4):61–68.
61. Shao W, Brown M. Advances in estrogen receptor biology: prospects for improvements in targeted breast cancer therapy. Breast Cancer Res 2004;6:39–52.
62. Cooke T, George D, Shields R, et al. Oestrogen receptors and prognosis in early breast cancer. Lancet 1979;1:995–997.
63. Hahnel R, Woodings T, Vivian AB. Prognostic value of estrogen receptors in primary breast cancer. Cancer 1979;44:671–675.
64. Gapinski PV, Donegan WL. Estrogen receptors and breast cancer: prognostic and therapeutic implications. Surgery 1980;88:386–393.
65. Westerberg H, Gustafson SA, Nordenskjold B, et al. Estrogen receptor level and other factors in early recurrence of breast cancer. Int J Cancer 1980;26: 429–433.
66. Samaan NA, Buzdar AU, Aldinger KA, et al. Estrogen receptor: a prognostic factor in breast cancer. Cancer 1981;47:554–560.
67. Godolphin W, Elwood JM, Spinelli JJ. Estrogen receptor quantitation and staging as complementary prognostic indicators in breast cancer: a study of 583 patients. Int J Cancer 1981;28:677–683.
68. von Maillot K, Horke W, Prestele H. Prognostic significance of the steroid receptor content in primary breast cancer. Arch Gynecol 1982;231:185–190.
69. Neifeld JP, Lawrence W Jr, Brown PW, et al. Estrogen receptors in primary breast cancer. Arch Surg 1982;117:753–757.
70. Brooks JL, Ryan JA Jr, Bauermeister DE. Prognostic significance of hormone receptors in early recurrence of breast cancer. Am J Surg 1983;145:599–603.
71. Mason BH, Holdaway IM, Mullins PR, et al. Progesterone and estrogen receptors as prognostic variables in breast cancer. Cancer Res 1983;43:2985–2990.
72. Stewart JF, Rubens RD, Millis RR, et al. Steroid receptors and prognosis in operable (stage I and II) breast cancer. Eur J Cancer Clin Oncol 1983;19: 1381–1387.
73. Mercer RJ, Bryan RM, Bennett RC, et al. The prognostic value of oestrogen receptors in breast cancer. Aust N Z J Surg 1984;54:7–10.
74. Aamdal S, Bormer O, Jorgensen O, et al. Estrogen receptors and long-term prognosis in breast cancer. Cancer 1984;53:2525–2529.
75. Alanko A, Heinonen E, Scheinin TM, et al. Oestrogen and progesterone receptors and disease-free interval in primary breast cancer. Br J Cancer 1984; 50:667–672.
76. Parl FF, Schmidt BP, Dupont WD, et al. Prognostic significance of estrogen receptor status in breast cancer in relation to tumor stage, axillary node metastasis, and histopathologic grading. Cancer 1984;54:2237–2242.
77. Adami HO, Graffman S, Lindgren A, et al. Prognostic implication of estrogen receptor content in breast cancer. Breast Cancer Res Treat 1985;5: 293–300.
78. Butler JA, Bretsky S, Menendez-Botet C, et al. Estrogen receptor protein of breast cancer as a predictor of recurrence. Cancer 1985;55:1178–1181.
79. Vollenweider-Zerargui L, Barrelet L, Wong Y, et al. The predictive value of estrogen and progesterone receptors' concentrations on the clinical behavior of breast cancer in women. Clinical correlation on 547 patients. Cancer 1986;57:1171–1180.
80. Angus B, Napier J, Purvis J, et al. Survival in breast cancer related to tumour oestrogen receptor status and immunohistochemical staining for NCRC 11. J Pathol 1986;149:301–306.
81. Clark GM, Sledge GW Jr, Osborne CK, et al. Survival from first recurrence: relative importance of prognostic factors in 1,015 breast cancer patients. J Clin Oncol 1987;5:55–61.
82. Hawkins RA, White G, Bundred NJ, et al. Prognostic significance of oestrogen and progestogen receptor activities in breast cancer. Br J Surg 1987;74: 1009–1013.
83. Shek LL, Godolphin W, Spinelli JJ. Oestrogen receptors, nodes and stage as predictors of post-recurrence survival in 457 breast cancer patients. Br J Cancer 1987;56:825–829.
84. Lampertico P, Stagni F, Crosignani P. Prognostic value of histochemical evaluation of steroid binding sites in breast tumors. Int J Biol Markers 1988;3: 197–202.
85. Chevallier B, Heintzmann F, Mosseri V, et al. Prognostic value of estrogen and progesterone receptors in operable breast cancer. Results of a univariate and multivariate analysis. Cancer 1988;62:2517–2524.
86. Kinsel LB, Szabo E, Greene GL, et al. Immunocytochemical analysis of estrogen receptors as a predictor of prognosis in breast cancer patients: comparison with quantitative biochemical methods. Cancer Res 1989;49:1052–1056.
87. Spyratos F, Hacene K, Tubiana-Hulin M, et al. Prognostic value of estrogen and progesterone receptors in primary infiltrating ductal breast cancer. A sequential multivariate analysis of 1262 patients. Eur J Cancer Clin Oncol 1989;25:1233–1240.
88. McCarty KS Jr, Kinsel LB, Georgiade G, et al. Long-term prognostic implications of sex-steroid receptors in human cancer. Progr Clin Biol Res 1990;322: 279–293.
89. Sigurdsson H, Baldetorp B, Borg A, et al. Indicators of prognosis in node-negative breast cancer. N Engl J Med 1990;322:1045–1053.
90. Mathiesen O, Bonderup O, Carl J, et al. The prognostic value of estrogen and progesterone receptors in female breast cancer. A single center study. Acta Oncol 1991;30:691–695.
91. Muir IM, Reed RG, Stacker SA, et al. The prognostic value of immunoperoxidase staining with monoclonal antibodies NCRC-11 and 3E1.2 in breast cancer. Br J Cancer 1991;64:124–127.
92. Winstanley J, Cooke T, George WD, et al. The long term prognostic significance of oestrogen receptor analysis in early carcinoma of the breast. Br J Cancer 1991;64:99–101.
93. Aaltomaa S, Lipponen P, Eskelinen M, et al. Hormone receptors as prognostic factors in female breast cancer. Ann Med 1991;23:643–648.
94. Mouridsen HT, Andersen J, Andersen KW, et al. Classical prognostic factors in node-negative breast cancer: the DBCG experience. J Natl Cancer Inst Monogr 1992;11:163–166.
95. Ewers SB, Attewell R, Baldetorp B, et al. Prognostic significance of flow cytometric DNA analysis and

estrogen receptor content in breast carcinomas—a 10 year survival study. Breast Cancer Res Treat 1993; 24:115–126.

96. Robertson JF, Ellis IO, Pearson D, et al. Biological factors of prognostic significance in locally advanced breast cancer. Breast Cancer Res Treat 1994;29: 259–264.
97. Beck T, Weikel W, Brumm C, et al. Immunohistochemical detection of hormone receptors in breast carcinomas (ER-ICA, PgR-ICA): prognostic usefulness and comparison with the biochemical radioactive-ligand-binding assay (DCC). Gynecol Oncol 1994;53:220–227.
98. Kommoss F, Pfisterer J, Idris T, et al. Steroid receptors in carcinoma of the breast. Results of immunocytochemical and biochemical determination and their effects on short-term prognosis. Anal Quant Cytol Histol 1994;16:203–210.
99. Barbareschi M, Dalla Palma P, Bevilacqua P, et al. Invasive node-negative breast carcinoma: multivariate analysis of the prognostic value of peritumoral vessel invasion compared with that of conventional clinico-pathologic features. Anticancer Res 1994; 14:2229–2235.
100. Stierer M, Rosen H, Weber R, et al. A prospective analysis of immunohistochemically determined hormone receptors and nuclear features as predictors of early recurrence in primary breast cancer. Breast Cancer Res Treat 1995;36:11–21.
101. Robertson JF, Cannon PM, Nicholson RI, et al. Oestrogen and progesterone receptors as prognostic variables in hormonally treated breast cancer. Int J Biol Markers 1996;11:29–35.
102. Layfield LJ, Saria EA, Conlon DH, et al. Estrogen and progesterone receptor status determined by the Ventana ES 320 automated immunohistochemical stainer and the CAS 200 image analyzer in 236 early-stage breast carcinomas: prognostic significance. J Surg Oncol 1996;61:177–184.
103. Pichon MF, Broet P, Magdelenat H, et al. Prognostic value of steroid receptors after long-term follow-up of 2257 operable breast cancers. Br J Cancer 1996; 73:1545–1551.
104. Hawkins RA, Tesdale AL, Killen ME, et al. Prospective evaluation of prognostic factors in operable breast cancer. Br J Cancer 1996;74:1469–1478.
105. Collett K, Hartveit F, Skjaerven R, et al. Prognostic role of oestrogen and progesterone receptors in patients with breast cancer: relation to age and lymph node status. J Clin Pathol 1996;49:920–925.
106. Younes M, Lane M, Miller CC, et al. Stratified multivariate analysis of prognostic markers in breast cancer: a preliminary report. Anticancer Res 1997;17: 1383–1390.
107. Eissa S, Khalifa A, el-Gharib A, et al. Multivariate analysis of DNA ploidy, p53, c-erbB-2 proteins, EGFR, and steroid hormone receptors for prediction of poor short term prognosis in breast cancer. Anticancer Res 1997;17:1417–1423.
108. Molino A, Micciolo R, Turazza M, et al. Prognostic significance of estrogen receptors in 405 primary breast cancers: a comparison of immunohistochemical and biochemical methods. Breast Cancer Res Treat 1997;45:241–249.
109. Raabe NK, Hagen S, Haug E, et al. Hormone receptor measurements and survival in 1335 consecutive patients with primary invasive breast carcinoma. Int J Oncol 1998;12:1091–1096.
110. Yaghan R, Stanton PD, Robertson KW, et al. Oestrogen receptor status predicts local recurrence following breast conservation surgery for early breast cancer. Eur J Surg Oncol 1998;24:424–426.
111. Chang J, Powles TJ, Allred DC, et al. Biologic markers as predictors of clinical outcome from systemic therapy for primary operable breast cancer. J Clin Oncol 1999;17:3058–3063.
112. Insa A, Lluch A, Prosper F, et al. Prognostic factors predicting survival from first recurrence in patients with metastatic breast cancer: analysis of 439 patients. Breast Cancer Res Treat 1999;56:67–78.
113. Elledge RM, Green S, Pugh R, et al. Estrogen receptor (ER) and progesterone receptor (PgR), by ligand-binding assay compared with ER, PgR and pS2, by immunohistochemistry in predicting response to tamoxifen in metastatic breast cancer: a Southwest Oncology Group Study. Int J Cancer 2000;89: 111–117.
114. Fisher B, Dignam J, Tan-Chiu E, et al. Prognosis and treatment of patients with breast tumors of one centimeter or less and negative axillary lymph nodes. J Natl Cancer Inst 2001;93:112–120.
115. Costa SD, Lange S, Klinga K, et al. Factors influencing the prognostic role of oestrogen and progesterone receptor levels in breast cancer—results of the analysis of 670 patients with 11 years of follow-up. Eur J Cancer 2002;38:1329–1334.
116. Pinto AE, Mendonca E, Andre S, et al. Independent prognostic value of hormone receptor expression and S-phase fraction in advanced breast cancer. Anal Quant Cytol Histol 2002;24:345–354.
117. Chang J, Clark GM, Allred DC, et al. Survival of patients with metastatic breast carcinoma: importance of prognostic markers of the primary tumor. Cancer 2003;97:545–553.
118. Cocquyt VF, Schelfhout VR, Blondeel PN, et al. The role of biological markers as predictors of response to preoperative chemotherapy in large primary breast cancer. Med Oncol 2003;20:221–231.
119. Haffty BG, Hauser A, Choi DH, et al. Molecular markers for prognosis after isolated postmastectomy chest wall recurrence. Cancer 2004;100: 252–263.
120. Hess KR, Pusztai L, Buzdar AU, et al. Estrogen receptors and distinct patterns of breast cancer relapse. Breast Cancer Res Treat 2003;78:105–118.
121. Rosen EM, Fan RG, Pestell RG, Goldberg ID. BRCA1 gene in breast cancer. J Cell Physiol 2003;196:19–41.
122. Fan S, Ma YX, Wang C, et al. p300 modulates the BRCA1 inhibition of estrogen receptor activity. Cancer Res 2003;62:141–151.
123. Ardavanis A, Gerakini F, Amanatidou A, et al. Relationships between cathepsin-D, pS2 protein and hormonal receptors in breast cancer cytosols: inconsistency with their established prognostic significance. Anticancer Res 1997;17:3665–3669.
124. Dittadi R, Biganzoli E, Boracchi P, et al. Impact of steroid receptors, pS2 and cathepsin D on the outcome of N+ postmenopausal breast cancer patients treated with tamoxifen. Int J Biol Markers 1998; 13:30–41.
125. Foekens JA, Rio M-C, Seguin P, et al. Prediction of relapse and survival in breast cancer patients by pS2 protein status. Cancer Res 1990;50:3832–3837.
126. Predine J, Spyriatos P, Prud'homme JF, et al. Enzyme-linked immunosorbent assay of prognosis

and adjuvant hormone therapy. Cancer 1992;69: 2116–2123.

127. Allred DC, Clark GM, Tandon AK, et al. Prognostic significance of immunocytochemistry determined pS2 in axillary node negative breast carcinoma. Breast Cancer Res Treat 1990;16:182.

128. Schwartz LH, Koerner FC, Edgerton SM, et al. pS2 expression and response to hormonal therapy in patients with advanced breast cancer. Cancer Res 1991;51:624–628.

129. Jansen RL, Hupperets PS, Arends JW, et al. pS2 is an independent prognostic factor for post-relapse survival in primary breast cancer. Anticancer Res 1998;18:577–582.

130. Thompson AM, Elton RA, Hawkins RA, et al. PS2 mRNA expression adds prognostic information to node status for 6-year survival in breast cancer. Br J Cancer 1998;77:492–496.

131. Gillesby BE, Zacharewski TR. pS2 (TFF1) levels in human breast cancer tumor samples: correlation with clinical and histological prognostic markers. Breast Cancer Res Treat 1999;56:253–265.

132. Fuqua SA, Oesterreich S, Hilsenbeck SG, et al. Heat shock proteins and drug resistance. Breast Cancer Res Treat 1994;32:67–71.

133. Chamness GC. Estrogen-inducible heat shock protein hsp27 predicts recurrence in node-negative breast cancer. Proc Am Assoc Cancer Res 1989; 30:252.

134. Ciocca DR, Clark GM, Tandon AK, et al. Heat shock protein hsp70 in patients with axillary lymph node-negative breast cancer: Prognostic implications. J Natl Cancer Inst 1993;85:570–574.

135. Tetu B, Brisson J, Landry J, et al. Prognostic significance of heat-shock protein-27 in node-positive breast carcinoma: an immunohistochemical study. Breast Cancer Res Treat 1995;36:93–97.

136. Oesterreich S, Hilsenbeck SG, Ciocca DR, et al. The small heat shock protein HSP27 is not an independent prognostic marker in axillary lymph node-negative breast cancer patients. Clin Cancer Res 1996;2:1199–1206.

137. Ciocca DR, Green S, Elledge RM, et al. Heat shock proteins hsp27 and hsp70: lack of correlation with response to tamoxifen and clinical course of disease in estrogen receptor-positive metastatic breast cancer (a Southwest Oncology Group Study). Clin Cancer Res 1998;4:1263–1266.

138. Seymour L, Bezwoda WR, Meyer K. Tumor factors predicting for prognosis in metastatic breast cancer. The presence of P24 predicts full response to treatment and duration of survival. Cancer 1990;66: 2390–2394.

139. Ciocca DR, Fuqua SA, Lock-Lim S, et al. Response to human breast cancer cells to heat shock and chemotherapeutic drugs. Cancer Res 1992;52:3648–3654.

140. Rochefort H, Chalbos D, Cunat S, et al. Estrogen regulated proteases and antiproteases in ovarian and breast cancer cells. J Steroid Biochem Mol Biol 2001;76:119–124.

141. Thorpe SM, Rocheford H, Garcia M, et al. Association between high concentration of M52,000 cathepsin D and poor prognosis in primary breast cancer. Cancer Res 1989;49:6008–6014.

142. Tandon AK, Clark GM, Chamness GC, et al. Cathepsin D and prognosis in breast cancer. N Engl J Med 1990;322:297–302.

143. Kute TE, Shao ZM, Sugg NK, et al. Cathepsin D as a prognostic indicator for node-negative breast cancer patients using both immunoassays and enzymatic assays. Cancer Res 1992;52:5198–5203.

144. Henry JA, McCarthy AL, Angus B, et al. Prognostic significance of the estrogen regulated protein, cathepsin D, in breast cancer. An immunohistochemical study. Cancer 1990;65:265–271.

145. Visscher DW, Sarkar F, LoRusso P, et al. Immunohistologic evaluation on invasion-associated proteases in breast carcinoma. Mod Pathol 1993;6: 302–306.

146. Jensen EV, Block GE, Smith S, et al. Estrogen receptors and breast cancer response to adrenalectomy. J Natl Cancer Inst Monogr 1971;34:55–70.

147. McGuire WL, Carbone P, Vollmer EP. Estrogen Receptors in Human Breast Cancer. New York: Raven Press, 1995.

148. Cole M, Jones CT, Todd ID. A new anti-oestrogenic agent in late breast cancer. An early clinical appraisal of ICI46474. Br J Cancer 1971;25: 270–275.

149. Ward HW. Anti-oestrogen therapy for breast cancer: a trial of tamoxifen at two dose levels. Br Med J 1973;1:13–14.

150. Greene GL, Nolan C, Engler JP, et al. Monoclonal antibodies to human estrogen receptor. Proc Natl Acad Sci USA 1980;77:5115–5119.

151. Harvey JM, Clark GM, Osborne CK, et al. Estrogen receptor status by immunohistochemistry is superior to the ligand-binding assay for predicting response to adjuvant endocrine therapy in breast cancer. J Clin Oncol 1999;17:1474–1481.

152. Esteban JM, Battifora H, Warsi Z, et al. Quantitation of estrogen receptors on paraffin-embedded tumors by image analysis. Mod Pathol 1991;4:53–57.

153. King WJ, DeSombre ER, Jensen EV, et al. Comparison of immunocytochemical and steroid binding assays for estrogen receptor in human breast tumors. Cancer Res 1985;45:293–304.

154. Wilbur DC, Willis J, Mooney RA, et al. Estrogen and progesterone detection in archival formalin-fixed paraffin embedded tissue from breast carcinoma: a comparison of immunocytochemistry with dextran coated charcoal assay. Mod Pathol 1992;5:79–84.

155. Lemieux P, Fuqua S. The role of the estrogen receptor in tumor progression. J Steroid Biochem Mol Biol 1996;56:87–91.

156. Clemons M, Danson S, Howell A. Tamoxifen ("Nolvadex"): a review. Cancer Treat Rev 2002;28: 165–180.

157. de Cremoux P, Tran-Perennou C, Brockdorff B, et al. Validation of real-time RT-PCR for analysis of human breast cancer cell lines resistant or sensitive to treatment with antiestrogens. Endocr Relat Cancer 2003;10:409–418.

158. Gruvberger S, Ringner M, Chen Y, et al. Estrogen receptor status in breast cancer is associated with remarkably distinct gene expression patterns. Cancer Res 2001;15;61:5979–5984.

159. Frasor J, Danes JM, Komm B, et al. Profiling of estrogen up- and down-regulated gene expression in human breast cancer cells: insights into gene networks and pathways underlying estrogenic control of proliferation and cell phenotype. Endocrinology 2003;144:4562–4574.

160. Pusztai L, Ayers M, Stec J, et al. Gene expression profiles obtained from fine-needle aspirations of breast cancer reliably identify routine prognostic markers and reveal large-scale molecular differences between estrogen-negative and estrogen-positive tumors. Clin Cancer Res 2003;9:2406–2415.

161. Omoto Y, Hayashi S. A study of estrogen signaling using DNA microarray in human breast cancer. Breast Cancer 2002;9:308–311.

162. Hayashi SI. Prediction of hormone sensitivity by DNA microarray. Biomed Pharmacother 2004;58: 1–9.

163. Perou CM, Sorlie T, Eisen MB, et al. Molecular portraits of human breast tumours. Nature 2000;406: 747–752.

164. Hortobagyi GN. Treatment of breast cancer. N Engl J Med 1998;339:974–984.

165. Hortobagyi GN. Endocrine treatment of breast cancer. In: Becker KL, ed. Principles and Practice of Endocrinology and Metabolism. Philadelphia: Lippincott Williams & Wilkins, 2001:2039–2046.

166. Allred DC, Bryant J, Land S, et al. Estrogen receptor expression as a predictive marker of the effectiveness of tamoxifen in the treatment of DCIS: findings from NSABP Protocol B-24. Breast Cancer Res Treat 2002;76(suppl 1):S36(abst. 30).

167. Hortobagyi GN. Treatment of breast cancer. N Engl J Med 1998;339:974–984.

168. Osborne CK, Zhao H, Fuqua SA. Selective estrogen receptor modulators: structure, function, and clinical use. J Clin Oncol 2000;18:3172–3186.

169. Osborne CK, Pippen J, Jones SE, et al. Double-blind, randomized trial comparing the efficacy and tolerability of fulvestrant versus anastrozole in postmenopausal women with advanced breast cancer progressing on prior endocrine therapy: results of a North American trial. J Clin Oncol 2002;20:3386–3395.

170. Vergote I, Faslodex 0020 and 0021 Investigators. Fulvestrant versus anastrozole as second-line treatment of advanced breast cancer in postmenopausal women. Eur J Cancer 2002;38(suppl 6):S57–S58.

171. Pritchard KI. GnRH analogues and ovarian ablation: their integration in the adjuvant strategy. Recent Results Cancer Res 1998;152:285–297.

172. Nicholson RI, Walker KJ. GN-RH agonists in breast and gynaecologic cancer treatment. J Steroid Biochem 1989;33:801–804.

173. Buzdar AU, Robertson JF, Eiermann W, et al. An overview of the pharmacology and pharmacokinetics of the newer generation aromatase inhibitors anastrozole, letrozole, and exemestane. Cancer 2002;95: 2006–2016.

174. Buzdar A. An overview of the use of non-steroidal aromatase inhibitors in the treatment of breast cancer. Eur J Cancer 2000;36(suppl 4):S82–S84.

175. Goss PE, Strasser K. Aromatase inhibitors in the treatment and prevention of breast cancer. J Clin Oncol 2001;19:881–894.

176. Lonning PE, Bajetta E, Murray R, et al. Activity of exemestane in metastatic breast cancer after failure of nonsteroidal aromatase inhibitors: a phase II trial. J Clin Oncol 2000;18:2234–2244.

177. Howell A, Robertson JF, Quaresma AJ, et al. Fulvestrant, formerly ICI 182,780, is as effective as anastrozole in postmenopausal women with advanced breast cancer progressing after prior endocrine treatment. J Clin Oncol 2002;20:3396–3403.

178. Osborne CK, Pippen J, Jones SE, et al. Double-blind, randomized trial comparing the efficacy and tolerability of fulvestrant versus anastrozole in postmenopausal women with advanced breast cancer progressing on prior endocrine therapy: results of a North American trial. J Clin Oncol 2002;20:3386–3395.

179. Hortobagyi GN. Future directions in the endocrine therapy of breast cancer. Breast Cancer Res Treat 2003;80(suppl 1):S37–S39.

180. Osipo C, Gajdos C, Liu H, et al. Paradoxical action of fulvestrant in estradiol-induced regression of tamoxifen-stimulated breast cancer [see comment]. J Natl Cancer Inst 2003;95:1597–1608.

181. Liu H, Lee ES, Gajdos C, et al. Apoptotic action of 17beta-estradiol in raloxifene-resistant MCF-7 cells in vitro and in vivo [see comment]. J Natl Cancer Inst 2003;95:1586–1597.

182. Klijn JG, Blamey RW, Boccardo F, et al. Combined tamoxifen and luteinizing hormone-releasing hormone (LHRH) agonist versus LHRH agonist alone in premenopausal advanced breast cancer: a meta-analysis of four randomized trials. J Clin Oncol 2001;19:343–353.

183. Osborne CK, Boldt D, Clark G. Effects of tamoxifen on human breast cancer cell cycle kinetics. Cancer Res 1983;43:3583.

184. Sweeney K, Musgrove E, Watts C. Cyclins and breast cancer. In: Dickson RB, Lippman ME, eds. Mammary Tumor Cell Cycle, Differentiation and Metastasis. Boston: Kluwer Academic Publishers, 1996:141.

185. Zafonte B, Hulit J, Amanatullah D, et al. Cell-cycle dysregulation in breast cancer: breast cancer therapies targeting the cell cycle. Front Biosci 2000;5: D938–D961.

186. Ciocca DR, Elledge R. Molecular markers for predicting response to tamoxifen in breast cancer patients. Endocrine 2000;13:1–10.

187. Scorilas A, Yotis J, Pateras C, et al. Predictive value of c-erbB-2 and cathepsin-D for Greek breast cancer patients using univariate and multivariate analysis. Clin Cancer Res 1999;5:815–821.

188. Ellis MJ, Coop A, Singh B, et al. Letrozole is more effective neoadjuvant endocrine therapy than tamoxifen for ErbB-1- and/or ErbB-2-positive, estrogen receptor-positive primary breast cancer: evidence from a phase III randomized trial. J Clin Oncol 2001;19:3808–3816.

189. Dowsett M, Harper-Wynne C, Boeddinghaus I, et al. HER-2 amplification impedes the antiproliferative effects of hormone therapy in estrogen receptor-positive primary breast cancer. Cancer Res 2001;61: 8452–8458.

190. Early Breast Cancer Trialists' Collaborative Group. Tamoxifen for early breast cancer: an overview of the randomised trials. Lancet 1998;351:1451–1467.

191. Goss PE, Ingle JN, Martino S, et al. A randomized trial of letrozole in postmenopausal women after five years of tamoxifen therapy for early-stage breast cancer. N Engl J Med 2003;349:1793–1802.

192. Coombes RC, Hall E, Gibson LJ, et al. A randomized trial of exemestane after two to three years of tamoxifen therapy in postmenopausal women with primary breast cancer. N Engl J Med 2004;350: 1081–1092.

193. Boccardo F, Rubagotti A, Amoroso D, et al. Anastrozole appears to be superior to tamoxifen in women already

receiving adjuvant tamoxifen treatment. Breast Cancer Res Treat 2003;82(suppl 1):S6–S7(abst. 3).

194. Jones AL, Powles TJ, Law M, et al. Adjuvant aminoglutethimide for postmenopausal patients with primary breast cancer: analysis at 8 years. J Clin Oncol 1992;10:1547–1552.

195. Boccardo F, Rubagotti A, Amoroso D, et al. Italian Breast Cancer Adjuvant Chemo-Hormone Therapy Cooperative Group trials. GROCTA trials. Recent Results Cancer Res 1998;152:453–470.

196. Talley RW, Segaloff A, Gregory EJ, et al. Adjuvant therapy of breast cancer with megestrol acetate. Breast Cancer Res Treat 1983;3:323(abst. 81).

197. Chlebowski RT, Collyar DE, Somerfield MR, et al. American Society of Clinical Oncology Technology assessment on breast cancer risk reduction strategies: tamoxifen and raloxifene. J Clin Oncol 1999;17:1939–1955.

198. Cuzick J. Point of view. Continuation of the International Cancer Intervention Study (IBIS). Eur J Cancer 1998;34:1647–1648.

199. Veronesi U. Prevention of breast cancer with tamoxifen: the Italian study in hysterectomized women. The Breast 1995;4:267–272.

200. Veronesi U, Maisonneuve P, Sacchini V, et al. Tamoxifen for breast cancer among hysterectomised women. Lancet 2002;359:1122–1124.

201. Cuzick J, Powles T, Veronesi U, et al. Overview of the main outcomes in breast-cancer prevention trials. Lancet 2003;361:296–300.

202. Ross PJ, Powles TJ. Results and implications of the Royal Marsden and other tamoxifen chemoprevention trials. Clin Breast Cancer 2001;2:33–36.

203. Powles T, Eeles R, Ashley S, et al. Interim analysis of the incidence of breast cancer in the Royal Marsden Hospital tamoxifen randomised chemoprevention trial. Lancet 1998;352:98–101.

204. Cuzick J, Forbes J, Edwards R, et al. First results from the International Breast Cancer Intervention Study (IBIS-I): a randomised prevention trial. Lancet 2002;360:817–824.

205. Cummings SR, Eckert S, Krueger KA, et al. The effect of raloxifene on risk of breast cancer in postmenopausal women: results from the MORE randomized trial. Multiple Outcomes of Raloxifene Evaluation. JAMA 1999;281:2189–2197.

CHAPTER

16

HER-2/*neu* Gene and Protein in Breast Cancer

Fatima Cardoso, MD,[1] Virginie Durbecq, MD,[1] Christos Sotiriou, MD,[1] and Jeffrey S. Ross, MD[2]

[1]Department of Medical Oncology, Jules Bordet Institute, Brussels, Belgium
[2]Department of Pathology and Laboratory Medicine, Albany Medical College, Albany, New York and Division of Molecular Medicine, Millennium Pharmaceuticals, Inc., Cambridge, Massachusetts

Biologic Background

The HER-2/*neu* (c-*erb*B2) (*HER-2*) gene is localized to chromosome 17q and encodes a transmembrane tyrosine kinase receptor protein that is a member of the erbB or HER family (FIG. 16.1).[1] This family of receptors is involved in cell–cell and cell–stroma communication, primarily through a process known as signal transduction, in which external growth factors, or ligands, affect the transcription of various genes by phosphorylating or dephosphorylating a series of transmembrane proteins and intracellular signaling intermediates, many of which possess enzymatic activity. Signal propagation occurs as the enzymatic activity of one protein turns on the enzymatic activity of the next protein in the pathway.[2] Major pathways involved in signal transduction, including the Ras/MAPK pathway, the PI3K/Akt pathway, the JAK/STAT pathway, and the phospholipase C-gamma (PLC-γ) pathway, ultimately affect cell proliferation, survival, motility, and adhesion.

Receptor activation requires three variables, a ligand, a receptor, and a dimerization partner.[3] After a ligand binds to a receptor, that receptor must interact with another receptor of any identical or related structure in a process known as dimerization to trigger phosphorylation and activate signaling cascades. Therefore,

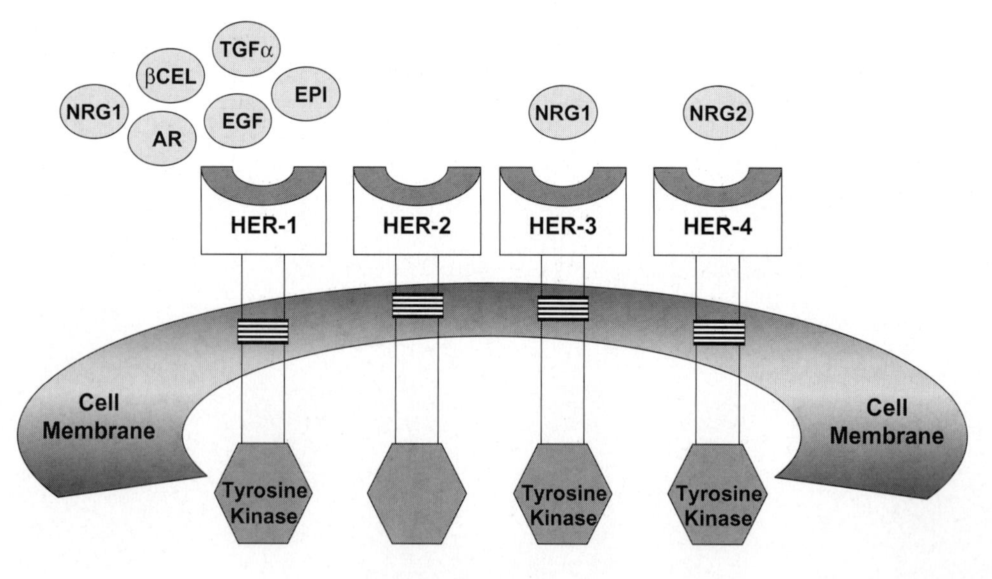

FIGURE 16.1 The HER (*erb*) gene family. Note that HER-2/*neu* has no known ligands and that HER-3 has no intrinsic tyrosine kinase activity.

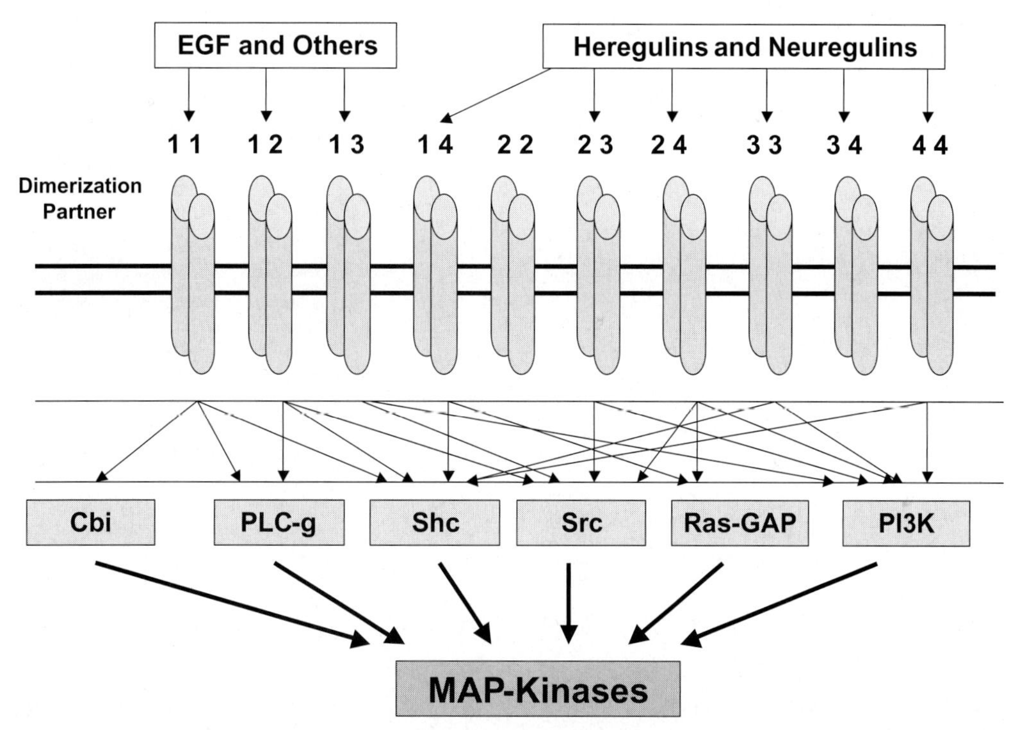

FIGURE 16.2 Dimerization and downstream signaling of the HER (EGFR) family. Major pathways involved in signal transduction, including the Ras, MAPK, PI3K/Akt, JAK/STAT, and the PLC-γ pathway, ultimately affect cell proliferation, cell survival, motility, and adhesion.

after ligand binding to a HER family member, the receptor can dimerize with various members of the family (epidermal growth factor receptor, HER-2, HER-3, or HER-4). It may dimerize with a like member of the family (homodimerization), or it may dimerize with a different member of the family (heterodimerization) (**FIG. 16.2**). The specific tyrosine residues on the intracellular portion of the HER-2 receptor that are phosphorylated, and hence the signaling pathways that are activated, depend on the ligand and dimerization partner. The wide variety of ligands and intracellular cross-talk with other pathways allow for significant diversity in signaling. Although no known ligand for the HER-2/*neu* receptor has been identified, it is the preferred dimerization partner of the other family members. HER-2 heterodimers are more stable[4] and their signaling is more potent[5] than receptor combinations without HER-2. *HER-2* gene amplification and/or protein overexpression has been identified in 10–34% of invasive breast cancers.[1]

Clinical Testing Methods

Both morphology based and molecular based techniques have been used to measure HER-2 status in breast cancer clinical samples (**Table 16.1**).[6–85] Of the 80 studies considering 26,309 patients listed in **Table 16.2**, 72 studies (90%) and 24,314 cases (92%) found that either *HER-2* gene amplification or HER-2 (p185 neu) protein overexpression predicted breast cancer outcome on either univariate or multivariate analysis. In 51 of 72 studies (71%) that featured multivariate analysis of

Table 16.1 Methods of Detection of HER-2/*neu* Status in Breast Cancer

Method	Target	FDA Approved	Slide Based
IHC	Protein	Yes*	Yes
FISH	Gene	Yes*	Yes
CISH	Gene	No	Yes
Southern blot	Gene	No	No
RT-PCR	mRNA	No	No
Microarray TP	mRNA	No	No
Tumor ELISA	Protein	No	No
Serum ELISA	Protein	Yes†	No

*For prognosis and prediction of response and eligibility to receive trastuzumab therapy.

†For monitoring response of breast cancer to treatment.

RT-PCR, reverse transcriptase-polymerase chain reaction; TP, transcriptional profiling.

Table 16.2 HER-2/*neu* Status and Prognosis in Breast Cancer

Study/ Year	First Author/ Reference	No. of Cases	Specimen Type	Method(s) of Analysis	Univariate Significance	Multivariate Significance	Comment
1 1987	Slamon 6	189	Frozen	Southern blot	Yes	Yes	HER-2 amplification predicted overall survival and time to relapse
2 1988	Berger 7	51	Frozen paraffin	Southern blot, IHC	Yes	—	HER-2 protein by IHC correlated with node status and tumor grade
3 1988	van de Vijver 8	189	Paraffin	IHC	No	No	Correlated with size, comedocarcinoma
4 1989	Wright 10	185	Paraffin	IHC	Yes	Yes	Overexpression correlated with high grade, negative ER, not size or node status
5 1990	Heintz 9	50	Frozen	Southern analysis	No	No	Mitoses, ER/PR negative
6 1990	Tsuda 11	176	Paraffin	Southern analysis	Yes	No	Impact on prognosis absorbed by grade
7 1990	Borg 12	300	Fresh	Western	Yes	No	Gene amplification predicted relapse and death in node negative on univariate only
8 1990	Paik 13	292	Paraffin	IHC	Yes	Yes	HER-2 protein overexpression second only to nodal status in predicting outcome
9 1991	Battifora 14	245	Paraffin	IHC	Yes	Yes	Significant only for low grade, low stage cases
10 1991	Kallionemi 15	319	Paraffin	IHC	Yes	Yes	Independent predictor in node negative and positive
11 1991	Gullick 16	483	Paraffin	IHC	Yes	Yes	Overexpression correlated with grade, but not size, ER or nodal status
12 1991	Clark 17	362	Fresh	Slot blot	No	No	Amplification marginally predictive in node positive only
13 1991	Lovekin 18	782	Paraffin	IHC	Yes	Yes	HER-2 overexpression predicted outcome in early and advanced cases; grade more significant predictor
14 1991	McCann 19	314	Paraffin	IHC	Yes	Yes	Predicted outcome in node negative and node positive
15 1991	Dykens 20	178	Paraffin	IHC	Yes	Yes	Overexpression predicted shortened survival in node-negative but not in node-positive patients
16 1991	Rilke 21	1210	Paraffin	IHC	Yes	No	Protein overexpression predicted outcome in node positive only
17 1991	Winstanley 22	465	Paraffin	IHC	Yes	Yes	HER-2 protein staining independently predicted survival
18 1991	O'Reily 23	172	Paraffin	IHC	Yes	Yes	HER-2 protein overexpression predicted outcome in node-positive but not in node-negative disease
19 1991	Patterson 24	115	Paraffin	Slot blot	Yes	Yes	Gene amplification predicted disease-free interval in node-negative patients

Table 16.2 HER-2/*neu* Status and Prognosis in Breast Cancer (*continued*)

Study/ Year	First Author/ Reference	No. of Cases	Specimen Type	Method(s) of Analysis	Univariate Significance	Multivariate Significance	Comment
20 1992	Toikkanen 25	209	Paraffin	IHC	Yes	Yes	Protein overexpression predicted shortened survival in node-positive patients
21 1992	Molina 26	301	Paraffin	IHC	Yes	Yes	Both methods of protein levels predicted worse overall survival
22 1992	Noguchi 27	234	Paraffin	IHC	Yes	No	Only nodal status correlated independently
23 1992	Allred 28	613	Paraffin	IHC	Yes	—	Significant for node negative low risk only
24 1992	Babiak 29	78	Paraffin	Slot blot	Yes	—	Only when combined with aneuploid status
25 1992	Tiwari 30	61	Frozen	Southern	Yes	—	HER-2 amplification associated with nodal metastasis
26 1992	Gusterson 31	1506	Paraffin	IHC	Yes	Yes	Predicted outcome in node positive only
27 1993	Bianchi 32	230	Paraffin	IHC	Yes	No	Only diffuse intense staining correlated
28 1993	Press 33	210	Paraffin	IHC	Yes	Yes	Predicts relapse in node negative
29 1993	Seshadri 34	1056	Fresh	Slot blot	Yes	Yes	Independent for both node negative and positive
30 1993	Descotes 35	298	Frozen	Slot blot	Yes	Yes	With PR-negative node status
31 1994	Giai 36	159	Paraffin	Western blot	Yes	Yes	Independent predictor in node negative
32 1994	Muss 37	442	Paraffin	IHC	Yes	—	Predicts response to chemotherapy
33 1994	Tetu 38	888	Paraffin	IHC	Yes	Yes	Survival in node positive, only membranous pattern correlates
34 1994	Hartmann 39	340	Paraffin	IHC	Yes	No	Predicts lymph node status
35 1994	Jaquemeier 40	81	Paraffin	IHC	No	No	No correlation with therapy response
36 1994	Marks 41	230	Paraffin	IHC	Yes	Yes	p53 and node status also independent
37 1994	Rosen 42	440	Paraffin	IHC	No	No	Medullar carcinoma is negative
38 1995	Quenel 43	942	Paraffin	IHC	Yes	Yes	HER-2 predicted relapse-free and metastasis-free survival
39 1996	Sundblad 44	271	Paraffin	IHC	Yes	Yes	Node positive, HER-2/*neu* negative, CEA positive, and *bcl-2* positive had the best prognosis
40 1996	O'Malley 45	107	Paraffin	IHC	Yes	Yes	Both p53 and HER-2/*neu* staining independently predicted outcome
41 1996	Hieken 46	100	Paraffin; cytosol	IHC	No	No	HER-2 protein by ELISA, not IHC predictive
42 1996	Xing 47	37	Paraffin	FISH	Yes	—	HER-2 amplification more predictive than nodal status in young patients

(*continued*)

Table 16.2 HER-2/*neu* Status and Prognosis in Breast Cancer (*continued*)

Study/ Year	First Author/ Reference	No. of Cases	Specimen Type	Method(s) of Analysis	Univariate Significance	Multivariate Significance	Comment
43 1997	Dittadi 48	115	Cytosol	ELISA	Yes	Yes	HER-2 protein levels predicted disease-free survival
44 1997	Fernandez-Acenero 49	112	Paraffin	IHC	Yes	No	Only TNM stage independent predictor
45 1997	Eissa 50	100	Cytosol	ELISA	Yes	Yes	HER-2 strongest predictor of recurrence in node negative
46 1997	Charpin 51	148	Frozen	IHC	Yes	Yes	Overall and disease-free survival independent of nodal status
47 1997	Press 52	324	Paraffin	FISH	Yes	Yes	HER-2/*neu* gene amplification predicts recurrence and death
48 1998	Ross 53	224	Paraffin	FISH	Yes	Yes	Three-tiered amplification system
49 1998	Depowski 54	145	Paraffin	FISH	Yes	Yes	HER-2 amplification predicted death independent of nodal status
50 1998	Querzoli 55	164	Paraffin	IHC	Yes	Yes	HER-2 overexpression independently predicted recurrence and survival
51 1998	Andrulis 56	580	Fresh	Southern	Yes	Yes	HER-2 amplification independently predicted recurrence in node negative; impact more in adjuvant treated cases
52 1998	Sjogren 57	315	Paraffin	IHC	Yes	Yes	CB-11 antibody IHC predicted overall survival independent of age, node status, size, grade, hormone receptor status, S phase, p53 status, and adjuvant therapy
53 1999	Harbeck 58	112	Paraffin	FISH	Yes	Yes	HER-2 amplification by FISH and (plasminogen protease [uPA, PAI-1]) levels were both independent predictors in node-negative patients
54 1999	Scorilas 59	136	Fresh	Southern	Yes	No	Cathepsin D was an independent predictor, but HER-2 was not
55 1999	Rudolph 60	356	Paraffin	IHC	Yes	No	HER-2 overexpression not an independent predictor of prognosis
56 2000	Reed 61	385	Paraffin	IHC	No	No	The HER-2–positive rate was only 10% in the noncorrelating IHC study
57 2000	Pauletti 62	189	Paraffin	IHC	Yes	Yes	Amplification detected by FISH was consistently predictive of outcome, but IHC testing was not
58 2000	Kakar 63	117	Paraffin	IHC	Yes	—	IHC 3+ positive and FISH positive cases correlated with outcome. IHC 2+ positive cases did not
59 2000	Agrup 64	110	Paraffin	IHC	Yes	Yes	HER-2 overexpression predicted survival in node-positive disease in young women independent of adjuvant therapy

Table 16.2 HER-2/*neu* Status and Prognosis in Breast Cancer (*continued*)

Study/ Year	First Author/ Reference	No. of Cases	Specimen Type	Method(s) of Analysis	Univariate Significance	Multivariate Significance	Comment
60 2000	Umekita 65	159	Paraffin	IHC	Yes	No	Coexpression of TGF-α and EGFR had worst prognosis
61 2000	Pawlowski 66	365	Fresh	RT-PCR	No	No	HER-2 mRNA levels correlated with grade and ER/PR status but not survival; increased HER-4 levels were a favorable prognostic finding
62 2000	Volpi 67	286	Paraffin	IHC	Yes	No	HER-2 status predictive of short term prognosis only and not an independent predictor
63 2000	Carr 68	190	Paraffin	FISH	Yes	Yes	HER-2 amplification predicted recurrence independent of ER and LN status
64 2000	Ferrero-Pous 69	488	Fresh	ELISA	Yes	Yes	HER-2 overexpression also predicted reduced response to tamoxifen
65 2000	Platt-Higgins 70	349	Paraffin	IHC	Yes	No	S1004A protein marker was an independent predictor. HER-2 expression was significant on univariate analysis only
66 2001	Eppenberger-Castori 71	1123	Fresh	ELISA	Yes	Yes	HER-2 levels predicted survival independent of hormonal or cytotoxic therapy
67 2001	Jukkola 72	650	Fresh	Southern	Yes	Yes	HER-2–positive patients also had reduced responses to both hormonal and cytotoxic therapies
68 2001	Gaci 73	100	Fresh	ELISA	Yes	Yes	HER-2 and cathepsin D expression were both independent predictors of OS
69 2001	Rudolph 74	261	Paraffin	IHC	Yes	Yes	Cell cycling ratio was most predictive factor in multivariate analysis
70 2001	Beenken 75	90	Paraffin	IHC	Yes	Yes	HER-2 and p53 coexpression had the worst prognosis
71 2001	Pinto 76	306	Paraffin	IHC	Yes	Yes	HER-2–positive ER-positive patients had reduced response rate to tamoxifen
72 2001	Riou 77	172	Fresh	Southern	Yes	Yes	HER-2 amplification independently predicted survival; no correlation with survival for IHC-positive cases with no gene amplification
73 2001	Horita 78	76	Paraffin	IHC	Yes	Yes	HER-2 expression correlated with PCNA expression
74 2002	Suo 79	100	Paraffin	IHC	Yes	Yes	Co-expression of HER-2 with EGFR (HER-1) had worse DFS
75 2002	Ristimaki 80	1576	Paraffin	FISH	Yes	Yes	HER-2 amplification was an independent predictor of survival and correlated with Cox-2 expression
76 2002	Rosenthal 81	71	Paraffin	FISH	Yes	No	Invasive lobular cancer had lower HER-2 amplification rate (<10%) than invasive ductal cancer and reached near significance as an independent predictor of OS

(*continued*)

Table 16.2 HER-2/*neu* Status and Prognosis in Breast Cancer (*continued*)

Study/ Year	First Author/ Reference	No. of Cases	Specimen Type	Method(s) of Analysis	Univariate Significance	Multivariate Significance	Comment
77 2002	Tsutsui 82	698	Paraffin	IHC	Yes	Yes	Combined HER-2–positive and ER-negative had the worst outcome
78 2002	Spizzo 83	205	Paraffin	IHC	Yes	Yes	Both HER-2 and Ep-CAM expression were independent predictors of OS
79 2002	Kato 84	408	Paraffin	IHC	Yes	No	HER-2 status was an independent predictor in some but not all patient cohorts
80 2002	El-Amady 85	94	Fresh	ELISA	Yes	Yes	HER-2 protein level by ELISA most predictive factor for short term prognosis
Total		26,309					

CEA, carcinoembryonic antigen; DFS, disease-free survival; LN, lymph node; OS, overall survival; RT-PCR, reverse transcriptase-polymerase chain reaction.

outcome data, the adverse prognostic significance of *HER-2* gene amplification, messenger RNA (mRNA), or protein overexpression was independent of all other prognostic variables. Thirteen studies (16%) reported prognostic significance on univariate analysis only (in 8 studies multivariate analysis was not performed). Only eight studies (10%), encompassing 1995 patients (8%), showed no correlation between HER-2 status and outcome. Of these eight studies, 5 (63%) used immunohistochemistry (IHC) on paraffin-embedded tissues as the HER-2 protein detection technique, two (25%) used Southern analysis, and one (13%) used a reverse transcriptase-polymerase chain reaction technique. All eight studies that used the fluorescence in situ hybridization (FISH) technique showed univariate prognostic significance of gene amplification, and seven of these (83%) showed prognostic significance on multivariate analysis as well.

HER-2 Testing Techniques

IHC staining (**FIG. 16.3**) has been the predominant method used. Unlike most immunohistochemical assays, the assessment of HER-2 status is quantitative rather than qualitative, because HER-2 is expressed in all breast epithelial cells. To provide a meaningful interpretation of a HER-2/*neu* immunostaining, it was necessary to establish a relationship between the number of HER-2 receptors on a cell's surface and the distribution and intensity of the immunostaining. Using cell lines, it was possible to establish a standardized immunohistochemical procedure and scoring system in which cells containing less than 20,000 receptors would show no staining (0), cells containing approximately 100,000 receptors would show partial membrane staining with less than 10% of the cells showing complete membrane staining (1+), cells containing approximately 500,000 receptors would show light to moderate complete membrane staining in more than 10% of the cells (2+), and cells containing approximately 2,300,000 receptors would show strong complete membrane staining in more than 10% of the cells (3+). Studies have shown that when a standardized IHC assay is performed on specimens that are carefully fixed, processed, and embedded, there is excellent correlation between gene copy status and protein expression levels.[1,86–88] However, alterations can be significantly impacted by technical issues, especially in archival, fixed, paraffin-embedded tissues. Advantages of IHC testing (**Table 16.3**) include its wide availability, relatively low cost, easy preservation of stained slides, and use of a familiar routine microscope. Disadvantages of IHC include the impact of preanalytic issues, including storage, duration and type of fixation, intensity of antigen retrieval, type of antibody (polyclonal vs. monoclonal), nature of system control samples, and, most importantly, the difficulties in applying a subjective slide scoring system.

In a study of a large panel of antibodies, Press et al.[89] reported a significant variation in detection rates for HER-2 protein by IHC using a large tissue block containing multiple breast tumors. Problems with standardization in slide scoring have been recently highlighted

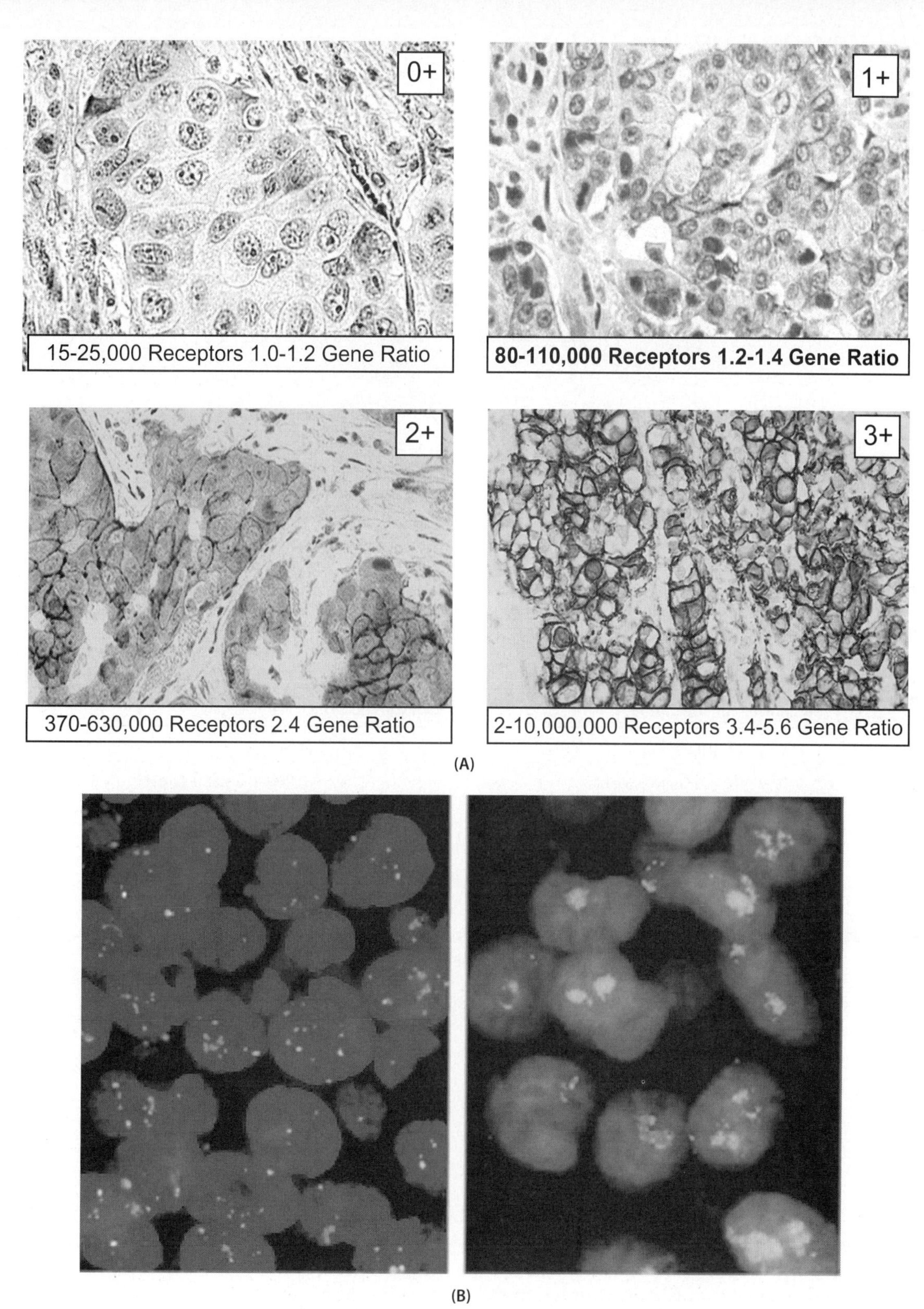

FIGURE 16.3 HER-2/*neu* testing in breast cancer. (A) HER-2/neu protein expression in infiltrating ductal breast cancer measured by IHC using the Herceptest™ Slide Scoring System. (Top, left) 0+ (negative) staining for HER-2/neu protein. This level of staining is typically associated with 15,000–25,000 surface receptor molecules per cell and HER-2/*neu* gene copy to chromosome 17 copy ratios measured by FISH of 1.0 to 1.2. (Top, right) 1+ staining associated with 80,000 to 110,000 receptors and gene ratio of 1.2 to 1.4. (Bottom, left) 2+ staining with membranous distribution, but no total cell encirclement associated with 370,000 to 630,000 receptors and gene ratio of 1.4 to 2.4. (Bottom, right) 3+ staining with diffuse positive membranous distribution, total cell encirclement, and "chicken wire" appearance associated with 2,000,000 to 10,000,000 receptors and gene ratio of 3.4 to 5.6. (peroxidase–anti-peroxidase with Herceptest™ antibody, ×200). (Receptor count and FISH gene ratio data provided by Dr. Kenneth Bloom, USLabs, Inc., Irvine, CA, USA.) (B) HER-2/*neu* gene amplification in infiltrating ductal breast cancer detected by FISH. (Left) HER-2/*neu* gene amplification demonstrated by the Abbott-Vysis Pathvysion™ method showing significant increase in HER-2/*neu* gene signals (red) compared with chromosome 17 signals (green) with a HER-2/*neu* gene ratio of 3.9. (Right) HER-2/*neu* gene amplification using the Ventana Inform™ method showing another breast cancer specimen with an absolute (raw) HER-2/*neu* gene copy number of 24. **See Plate 35 for color image.**

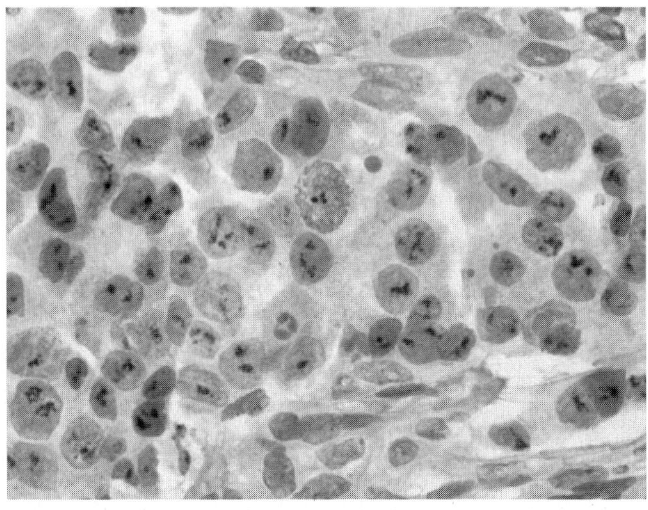

(C)

FIGURE 16.3 *(continued)* (C) HER-2/*neu* gene amplification in infiltrating breast cancer detected by CISH using anti–HER-2/*neu* probe and IHC with diaminobenzidine chromagen (SpotLight™ HER-2/*neu* probe, Zymed Corp., South San Francisco, CA, USA). **See Plate 36 for color image.**

Table 16.3 Advantages and Disadvantages of IHC and FISH-Based Testing for HER-2/*neu* in Breast Cancer

	IHC	FISH
Advantages	• Widely available • Rapid procedure • Light microscope based • HercepTest™ and Pathway™ FDA-approved assays for Herceptin eligibility selection	• DNA is a stable target • Standardized threshold for positivity • Built-in internal control • Low interlaboratory variability • High accuracy (sensitivity and specificity)
Disadvantages	• Variable antibody sensitivity and specificity • Highly impacted by tissue processing variables • Affected by antigen retrieval and reagent variability • Non–FDA-approved assays in routine use • Subjective scoring criteria • Low pathologist concordance and high interlaboratory variability	• Fluorescence microscope equipped with correct filter sets is required • Certain fixatives interfere with assay (noninformative result) • Limited community experience with tissue-based FISH • Difficulty in assessing aneuploid cases

in reference to the best method for using HER-2 status to predict response to the anti–HER-2 therapeutic monoclonal antibody, trastuzumab (Herceptin®).[90] Slide scoring can be improved by avoiding specimen edges, retraction artifacts, under- or over-fixation, cases with significant staining of benign elements, and tumor cells lacking a complete membranous staining pattern (the so-called chicken wire appearance).

Recent data presented by the National Surgical Adjuvant Breast Cancer Program (NSABP) has shown that certified laboratories, defined as those laboratories performing a high volume of HER-2 testing and demonstrating high concordance between IHC and FISH results, approach 98% interlaboratory concordance when tumors assessed as 3+ were reanalyzed by both IHC and FISH testing at the NSABP laboratory.[91] Because most of the submitting laboratories were reference laboratories that cannot control tissue fixation or storage, it has been suggested that preanalytical issues may not be the major cause of interlaboratory variability. Results from the United Kingdom National External Quality Assessment Scheme for IHC also suggested that the lack of reproducibility of HER-2 scoring between laboratories was not the result of tumor heterogeneity or differences in fixation or processing, but rather the result of how the scoring system was applied.[92] However, in a recent report, prolonged formalin fixation did not impact on the staining for cell cycle markers such as Ki-67 and p27 but did cause significant degradation in hormone receptor proteins and HER-2.[93]

The use of a quantitative image analysis system can reduce slide-scoring variability among pathologists, especially in 2+ cases.[94] When 130 HER-2 immunostained slides were reviewed by 10 pathologists and then were later reviewed with the aid of image analysis, the use of image analysis eliminated most of the interobserver variability, which was significant by routine microscopy (FIG. 16.4).[95] Two commercially available HER-2 IHC kits, the Dako Herceptest™ and the Ventana Pathway™, are approved by the US Food and Drug Administration (FDA) for determining eligibility of patients to receive trastuzumab.

Southern and slot blotting were the first gene-based *HER-2* detection methods used in breast cancer specimens. These methods can be significantly hampered when tumor cell DNA extracted from the primary carcinoma sample is diluted by DNA from benign breast tissue and inflammatory cells. The FISH technique (FIG. 16.3), which is morphology driven and like IHC can be automated, has the advantages of a more objective scoring system and the presence of a built-in internal control consisting of the two *HER-2* gene signals present in all non-neoplastic cells in the specimen. Disadvantages of FISH testing include the higher cost of each test, longer time required for slide scoring, requirement of a fluorescent microscope, the inability to preserve the slides for storage and review, and occasionally in identifying the invasive tumor cells. Of recent concern is the issue of tumor aneuploidy, especially polysomy of chromosome 17. Tumors that feature greater than four copies of chromosome 17 will, by definition, contain more than four copies of the *HER-2* gene and have been associated with overexpression of the HER-2 protein. Although there are no published data, it has been suggested that these patients lacking true amplification but featuring more than four gene copies

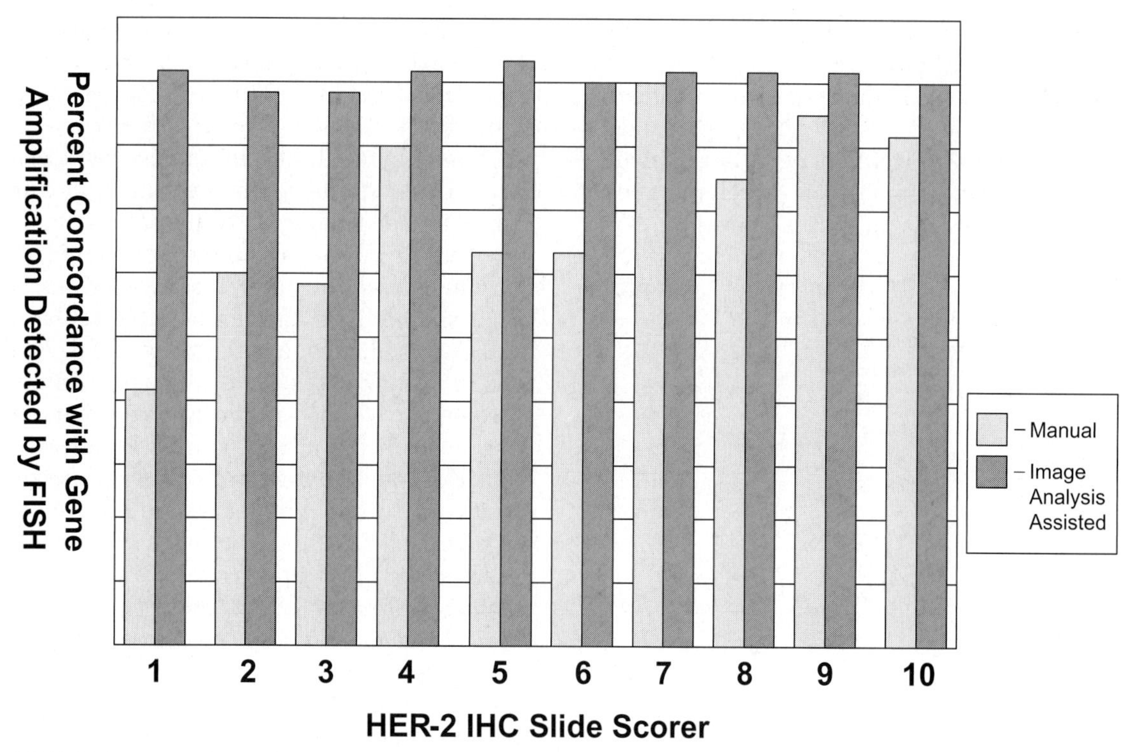

FIGURE 16.4 Image analysis and HER-2/*neu* slide scoring. Interobserver variability among 10 pathologists reviewing the same 130 HER-2/*neu* immunostained slides. Concordance with gene amplification status, assessed by FISH, is plotted on the *y*-axis. Slides assessed as 2+ or 3+ were called positive. Manual assessment showed significant interobserver variability, whereas assessment with the aid of image analysis showed little variability. Additionally, all pathologists improved their concordance with the aid of image analysis.

per cell on average are candidates for trastuzumab therapy, similar to the patients whose tumors are *HER-2* gene amplified. Two versions of the FISH assay are FDA approved: the Ventana Inform™ test, which measures only *HER-2* gene copies, and the Abbott-Vysis Pathvysion™ test, which includes a chromosome 17 probe in a dual color format. Published studies indicate that the two assays are highly correlative.[96] The advantages and disadvantages of IHC and FISH-based HER-2 testing are summarized in **Table 16.3**.

The chromogenic in situ hybridization (CISH) method (FIG. 16.3) features the advantages of both IHC (routine microscope, lower cost, familiarity) and FISH (built-in internal control, subjective scoring, the more robust DNA target) but to date is not FDA approved for selecting patient eligibility for trastuzumab treatment.[97,98] In a recent study using both FISH and CISH to assay 31 cases of infiltrating breast carcinoma with testing performed in laboratories at two institutions, identical results for both methods were found in 26 cases (84%).[99] The CISH and IHC detection methods can be combined to provide simultaneous evaluation of gene copy number and protein expression (FIG. 16.5), but such methods are experimental and have not yet been adopted in clinical practice.

The reverse transcriptase-polymerase chain reaction technique,[100,101] predominantly used to detect HER-2 mRNA in peripheral blood and bone marrow samples, has correlated more with gene amplification status than IHC levels[102] but failed to predict survival. However, it correlated well with the estrogen receptor (ER)/progesterone receptor (PR) status and tumor grade in one breast cancer outcome study of 365 patients.[67] With the advent of laser capture microscopy and the acceptance of reverse transcriptase-polymerase chain reaction as a routine and reproducible laboratory technique, its role in HER-2 status assessment may increase in the future.

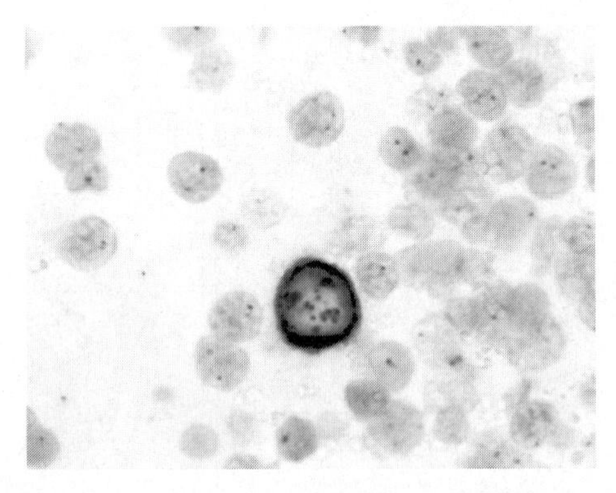

FIGURE 16.5 Simultaneous HER-2/*neu* gene amplification and protein overexpression determination. HER-2/*neu* combined gene and protein evaluation in pleural effusion by CISH (DAB = brown) and immunohistochemistry (Vector = blue), respectively. A single breast carcinoma cell with HER-2/*neu* gene amplification and protein overexpression is seen among nonneoplastic cells. **See Plate 37 for color image.**

The complementary DNA microarray based method (FIG. 16.6) of detecting HER-2 mRNA expression levels has recently achieved interest as an alternative method for measuring HER-2/*neu* status in breast cancer.[103–105] This method has the advantage of being able to assess downstream signaling of the HER-2 pathway at the same time that the level of HER-2 mRNA is measured. Other pathways relevant to HER-2, such as the ER pathway, can also be assessed simultaneously with this technique (see Chapter 15). In a recent study, the *HER-2* gene amplification status detected by FISH on 20 paraffin-embedded breast cancer core biopsy samples was correctly predicted in all cases by the quantification of the HER-2 mRNA levels obtained by expression profiling of mRNA extracted from paired fine-needle aspiration biopsies from the same patients.[103] Tissue microarrays have also been recently introduced into breast cancer research and have shown excellent correlation with FISH and IHC HER-2 results obtained from the donor tissue blocks used to produce the array.[106–112]

The enzyme-linked immunosorbent assay (ELISA) technique when performed on tumor cytosols made from fresh tissue samples avoids the potential antigen damage associated with fixation, embedding, and uncontrolled storage. In the six published studies listed in **Table 16.2**, ELISA-based measurements of HER-2 protein in tumor cytosols, mostly performed in Europe, have uniformly correlated with disease outcome.[48,50,69,71,73,85] In a recent study from Europe, the results of ELISA measurements on tumor extracts correlated closely with both *HER-2* gene amplification results detected by FISH and protein expression results measured by IHC.[113] However, the small size of breast cancers associated with expanded screening programs in the United States generally precludes tumor tissue ELISA methods because insufficient tumor tissue is available to produce a cytosol.

Using recombinant technologies, trastuzumab (Herceptin™), a monoclonal IgG1 class humanized murine antibody, was developed

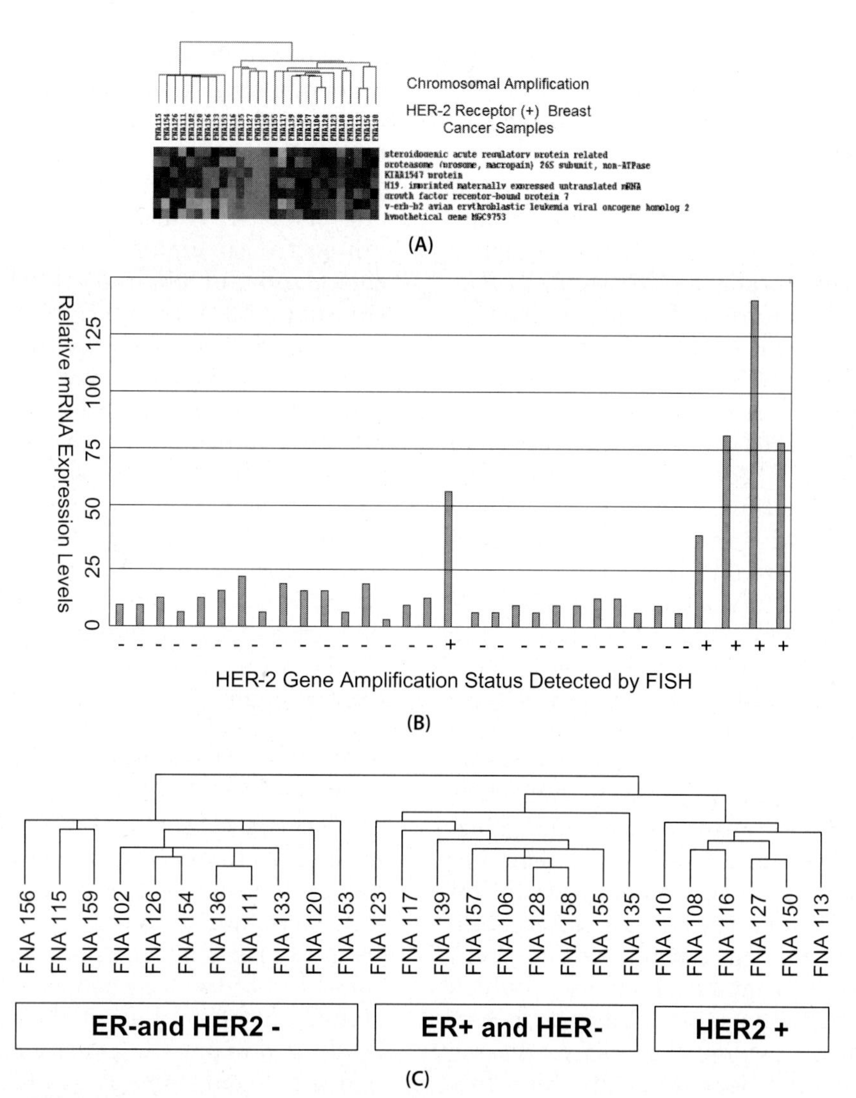

FIGURE 16.6 HER-2/*neu* mRNA detection by gene expression profiling. (A) Cluster analysis of transcriptional profile of a series of primary stages I–III invasive breast cancers that were sampled by fine-needle aspiration biopsy before the commencement of neoadjuvant chemotherapy. After mRNA extraction, mRNA expression profiling was performed using the Affymetrix u133 GeneChip™ oligonucleotide microarray system (Affymetrix Corp., Santa Clara, CA, USA). Heat map demonstrates a series of six genes at chromosomal locus 17q12 that were coamplified with HER-2/*neu* in a significant number of cases. (B) Comparison of HER-2/*neu* mRNA expression levels measured by transcriptional profiling on 32 fine-needle aspiration samples and HER-2/*neu* gene amplification status determined by FISH on the corresponding core biopsy specimen. Note the close correlation of gene expression with amplification status. (Original data from Ref. 117.) (C) Hierarchical clustering after transcriptional profiling of 26 infiltrating breast cancer specimens sampled by fine-needle aspiration using a 62 gene model (25 genes for determining ER status and 37 genes for determining HER-2/*neu* status). Note the clustering of the HER-2/*neu*–positive cases and their separation from the HER-2/*neu*–negative cases. The HER-2/*neu*–negative ER-positive and HER-2/*neu*–negative ER-negative tumors can also be separated by this technique. (Original data from Ref. 114.) **See Plate 38 for color image.**

by the Genentech Corporation (South San Francisco, CA, USA) to specifically bind the extracellular portion of HER-2. This antibody therapy was initially used in patients with advanced relapsed breast cancer that overexpressed the HER-2 protein.[114] Since its launch in 1998, trastuzumab has become an important therapeutic option for all patients with HER-2–positive metastatic breast cancer and is also being studied in the adjuvant and neoadjuvant settings in well-designed, large, prospective, clinical trials.[115–118] Trastuzumab has been approved since 1998 by the FDA for use as first- and second-line treatment for metastatic HER-2–positive breast cancer patients, both as single agent and in combination with paclitaxel. Using a clinical trial IHC assay to select patients for the phase III pivot trial, the addition of trastuzumab to chemotherapy (either anthracycline plus cyclophosphamide or taxane) was associated with a longer time to disease progression (median, 7.4 vs. 4.6 months; $P < 0.001$), a higher rate of objective response (50% vs. 32%; $P < 0.001$), a longer duration of response (median 9.1 vs. 6.1 months; $P < 0.001$), a lower rate of death at 1 year (22% vs. 33%; $P = 0.008$), longer survival (median survival 25.1 vs. 20.3 months; $P = 0.01$), and a 20% reduction in the risk of death.[119] Although not completely understood at this time, mechanisms believed to be associated with the antitumor effects of trastuzumab are listed in **Table 16.4**.

In the original pivot trial, class III or IV cardiac dysfunction occurred in 27% of the anthracycline and cyclophosphamide plus trastuzumab treated group compared with 8% of the group given an anthracycline and cyclophosphamide alone.[120] The exact pathophysiology of trastuzumab cardiac toxicity is currently under intense research and, most probably, involves multiple mechanisms. For instance, studies have demonstrated that HER-2 and HER-4 with its ligand heregulin are necessary for the normal development of the heart and that knockout mice, which lack the *HER-2* gene expression in their cardiac myocytes, develop progressive dilated cardiomyopathy.[121] Because of this high rate of cardiac toxicity, combination treatment with anthracycline-based chemotherapy should not be used, outside well-designed and strictly followed clinical trials. Furthermore, because of the long half-life of trastuzumab (approximately 28 days), anthracycline-based regimens must also be avoided during the 4- to 5-month period after interruption of trastuzumab administration. Nonanthracycline regimens are in the later stages of phase II and phase III trials, including protocols combining trastuzumab with taxanes, platinum compounds, or vinorelbine.[122] In a study of trastuzumab plus vinorelbine, the response rate was 84% in patients when the regimen was used as first-line therapy for metastatic disease.[123] High response rates were also seen for the trastuzumab–vinorelbine combination in women treated with second- or third-line therapy and among patients previously treated with anthracyclines and/or taxanes.[123] A recent report listed a time to progression of 17 months for patients with *HER-2* amplified metastatic breast cancer treated with the combination of docetaxel, carboplatin, and trastuzumab.[124]

Table 16.4 Mechanism of Action of Trastuzumab

- Antibody dependent cellular cytotoxicity
  - Stimulation of natural killer cells
- Immune system mediated
- Prevention of HER-2/*neu* receptor dimerization
- Downstream signaling is reduced or completely interrupted
- Direct cell cycle inhibitor
  - Induces p27^{kip1}cyclin dependent kinase inhibitor to slow G1 phase
- Antiangiogenesis
  - Reduces angiogenic factors: VEGF, TGF-α and PAI-1
- Proapoptotic
  - Inhibits Akt signaling

Modified from Nahata R, Esteva FJ. HER-2 targeted therapy: lessons learned and future directions. Clin Cancer Res 2003;9:5078–5084.

HER-2 Status and the Prediction of Response to Trastuzumab Therapy

The best method to identify patients who are most likely to respond to trastuzumab therapy has been a source of controversy. The original IHC technique used in the trastuzumab pivot trial was the clinical trial assay (CTA), which consisted of two antibodies: 4D5, the monoclonal antibody that is the actual antigen binding murine component of trastuzumab and is not commercially available, and CB-11, a monoclonal antibody directed toward the internal domain of the p185neu receptor, commercially available both as a research reagent and as an FDA-approved diagnostic test (Pathway™, Ventana Medical Systems, Tucson, AZ, USA). The original CTA was succeeded by the FDA-approved polyclonal HercepTest® (Dako Corporation, Glostrup, Denmark). There is good concordance between the CTA and HercepTest™, although 58 of 274 tumors (21%) that scored as positive with the CTA were scored as negative with HercepTest® and 59 of 274 tumors (22%) that scored as negative with the CTA were scored as positive with HercepTest®. (Package Insert, Herceptest®, Dako Corporation, Glosstrup, Denmark). After its FDA approval and launch, the HercepTest® assay was initially criticized for yielding false-positive results,[125,126] although better performance was ultimately achieved when the test was performed exactly according to the manufacturer's instructions. Although widely used in the United States,

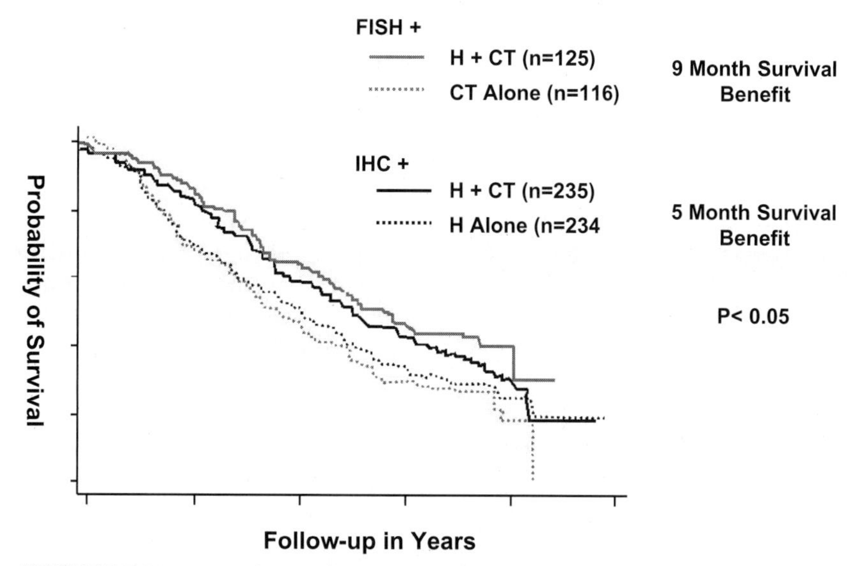

FIGURE 16.7 FISH vs. IHC and survival in trastuzumab single agent and combination chemotherapy for advanced metastatic breast cancer. Survival curves demonstrate a significant ($P < 0.05$) increase in median duration of survival for patients treated with either trastuzumab alone or trastuzumab plus chemotherapy in patients whose primary tumors were classified by FISH (Pathvysion™ test) for HER-2/*neu* gene amplification status vs. patients who were classified by IHC (Genentech Clinical Trial Assay with antibodies 4D5 and CB-11). (Data from Ref. 127.) **See Plate 39 for color image.**

the HercepTest® is not commonly chosen in Europe, where the monoclonal antibody CB-11 with antigen retrieval is the IHC protocol of choice in many institutions.

Concern over IHC accuracy using standard formalin-fixed paraffin-embedded tissue sections has encouraged the use of the FISH assay for its ability to predict trastuzumab response rates. Reports that FISH could outperform IHC in predicting trastuzumab response (FIG. 16.7)[127] and well-documented lower response rates of 2+ IHC staining versus 3+ staining tumors[128] resulted in an approach that uses either IHC as a primary screen with FISH testing of all 2+ cases or primary FISH-based testing.[129] In a recently published study where trastuzumab was used as a single agent, the response rates in 111 assessable patients with 3+ IHC staining was 35% and the response rates for 2+ cases was 0%; the response rates in patients with and without *HER-2* gene amplification detected by FISH were 34% and 7%, respectively.[128]

In another study of breast cancer treated with trastuzumab plus paclitaxel, in patients with HER-2/*neu* overexpressing tumors detected by IHC and confirmed by FISH-based testing, overall response rates ranged from 67% to 81% compared with 41% to 46% in patients with normal expression of HER2.[130] The CB-11 and TAB250 antibodies for IHC and FISH featured the strongest significance.[130]

Interestingly, although FISH-based testing is more expensive and not as widely available as IHC, a recent published review from New York and Italy suggested that FISH was actually the most cost-effective option.[131] Other recent studies have favored the FISH approach not only to confirm 2+ IHC cases but to also confirm 3+ and prevent the use of potentially toxic trastuzumab to patients with false-positive IHC results who are unlikely to benefit from this therapy.[132]

In summary, although the superiority of one method versus the other remains controversial,[133–136] most laboratories are either screening all cases with IHC and triaging selected cases for FISH testing or using FISH as the only method for HER-2 testing. For the laboratories that use IHC as the primary screen, the decision as to when to triage to FISH testing is also controversial. Some laboratories refer only their 2+ IHC cases, some triage 1+ and 2+, and others refer 1+, 2+, and any other cases where the HER-2 IHC results are not consistent with other disease parameters such as grade, stage, ploidy, S phase, and hormone receptor status or with clinical evolution of the disease in the specific patient.

Prediction of Response of Breast Cancer to Other Therapies

The best established correlation between HER-2 status and nontrastuzumab therapy response is the reported resistance of HER-2–positive patients to some types of hormonal therapy, namely tamoxifen.[137–140] Tumors that overexpress HER-2 are more likely to be ER negative and PR negative than tumors that do not show overexpression. In fact, when measured as continuous variables, the expression of HER-2 appears to be inversely related to the expression of ER and PR even in hormone receptor–positive tumors.[141] In some studies, HER-2–positive tumors were specifically resistant to tamoxifen therapy.[69,72,76,142–144] However, in one study, HER-2/*neu* status failed to predict tamoxifen resistance in ER-positive cases.[145] In another study, ER-positive/HER-2–positive tumors were not only resistant to tamoxifen, but treatment with this drug actually had an adverse impact compared with untreated patients.[146] However, to date this finding has not been confirmed by large intergroup studies in the United States.[147]

More recently, an important study by Ellis et al.,[148] conducted in the neoadjuvant setting, raised the hypothesis that ER-positive/HER-2–positive tumors may be the ones that derive the higher benefit from endocrine therapy using aromatase inhibitors, as compared with tamoxifen. HER-2 protein overexpression has also been linked to resistance, or at least to a lesser degree of response to CMF (cyclophosphamide + methotrexate + 5-fluorouracil) adjuvant chemotherapy.[36,146] However, and very importantly, patients who cannot be treated with anthracycline-based regimens should not be denied CMF treatment on the sole basis of their HER-2 status. The predictive value of HER-2 for response to taxane-based chemotherapy is still controversial with some studies,[149–151] suggesting resistance and others finding a near threefold increased sensitivity.[152] HER-2 overexpression has also been associated with enhanced response rates to anthracycline-containing chemotherapy regimens in some but not all studies.[37,90,153–157] A recent study of the response of locally advanced breast cancer to a neoadjuvant chemotherapy regimen consisting of 5-fluorouracil, doxorubicin, and cyclophosphamide found the pretreatment HER-2/*neu* status detected by IHC was not predictive of treatment response.[158] Finally, HER-2 immunostaining has successfully predicted local recurrence in patients receiving surgery and radiation.[158,159]

Because anthracyclines are topoisomerase inhibitors and topoisomerase IIα (TOP2A) is frequently coamplified with HER-2/*neu*, it has been suggested that HER-2/*neu* may be serving as a surrogate marker.[160] Several studies have consistently linked coexpression and coamplification of the *TOP2A* and *HER-2* genes with adverse prognosis and sensitivity to anthracycline drugs.[161–168] It has been suggested that *TOP2A* gene amplification and deletion can account for both relative chemosensitivity and resistance to topoisomerase II inhibitors and that *TOP2A* amplification/overexpression may occur in breast cancer independent from the *HER-2* status.[161–168] Furthermore, a recent study showed that about 4% of *TOP2A* aberrations can occur in HER-2–negative cases.[169]

In summary, despite being studied for more than a decade, HER-2 status has not yet achieved universal acceptance as a useful predictive factor for the selection of optimal systemic therapy other than trastuzumab in routine clinical practice. The main difficulties with the published literature is the lack of prospective studies needed to validate the proposed marker and the loss of reproducibility of the assessment methods across different laboratories. Only large, prospective, molecular-based trials will be able to overcome these limitations and improve the ability to individualize the selection of breast cancer treatment. Two examples of these trials, which are already ongoing, are the p53-EORTC trial (evaluation of p53 as predictive marker of response to taxanes) and the TOP trial (evaluation of *TOP2A* as predictive marker of response to anthracyclines), both conducted in the neoadjuvant setting.

Serum HER-2 Antigen Levels as a Tumor Marker

Circulating HER-2 receptor protein levels have successfully predicted the presence and progression of HER-2–positive breast cancer. Among the 22 published studies on 4088 patients, 16 studies (73%) involving 3458 patients (85%) reported a significant correlation of serum HER-2 protein levels with either disease recurrence, metastasis, or shortened survival.[170–185] Two studies involving 379 patients reported no significant association of serum levels with prognosis.[186,187] Of the 11 studies in which serum HER-2 protein levels were tested for their ability to predict response to therapy, 8 studies (73%) found that elevated serum HER-2 protein levels predicted therapy resistance,[180–182,185–188] whereas

3 additional studies did not demonstrate this association.[189–191] Serum HER-2 levels have correlated with decreased survival and absence of clinical response to hormonal therapy in ER-positive tumors in some[177,185] but not all[187] studies. Serum HER-2 protein measurements have successfully predicted resistance to high dose chemotherapy,[180–182] bone marrow transplantation,[181] and response to trastuzumab single agent and combination treatment for metastatic HER-2–positive disease.[192,193]

HER-2 Expression and Breast Pathology

HER-2 overexpression has been consistently associated with higher grade and extensive forms of ductal carcinoma in situ.[194–196] *HER-2* gene amplification occurs at a lower rate (less than 10%) and has been linked to an adverse outcome in invasive lobular carcinoma.[81] The frequency of *HER-2* gene amplification appears to be strongly correlated with tumor grade and ductal versus lobular status. Less than 3% of grade I invasive ductal carcinomas and classic lobular carcinomas feature amplification of the *HER-2* gene. Most cases of invasive lobular carcinoma that do display *HER-2* gene amplification are of the pleomorphic or high grade type.[81] HER-2 overexpression has been a consistent feature of both mammary and extramammary Paget disease.[197–199] Most studies that have compared the HER-2 status in paired primary and metastatic tumor tissues have found an overwhelming consistency of the patient's status regardless of the method of testing (IHC vs. FISH).[168,200–206]

In one study of node-positive tumors that were defined as biclonal by DNA ploidy profile, HER-2/*neu* status was determined by IHC in 17 primary tumors and their 82 axillary lymph node metastases.[205] Despite this apparent heterogeneity of the predominant clone measured by ploidy status, in each metastatic site, the HER-2 status was consistent between primary tumors and their corresponding metastases.[205] In another study of 107 paired primary tumor and distant metastatic lesions, 94% and 93% of the lesions had concordant HER-2 status when analyzed by IHC or FISH, respectively.[206] The authors concluded that support routine determination of HER-2 status on metastatic sites is not warranted, particularly when FISH results from the primary tumor are available.[206] *HER-2* amplification and overexpression has been associated with adverse outcome in some studies of male breast carcinoma[207–210] but not in others.[211–213] Finally, low level HER-2 overexpression has been identified in benign breast disease biopsies and associated with a near-significant increased risk of subsequent invasive breast cancer.[214] This finding must be validated in future larger studies using multiple techniques to assess HER-2/*neu* status. However, HER-2 overexpression is frequently detected in ductal carcinoma in situ, also suggesting that HER-2 signaling may play a role in early breast cancer development.

Interaction of HER-2 Expression with Other Prognosis Variables

HER-2 gene amplification and protein overexpression have been associated consistently with high tumor grade, DNA aneuploidy, high cell proliferation rate, negative assays for nuclear protein receptors for estrogen and progesterone, p53 mutation, *TOP2A* aberrations, and alterations in a variety of other molecular biomarkers of breast cancer invasiveness and metastasis.[215–218]

Phosphorylated HER-2 as a Potential Prognostic and Predictive Marker

It has been argued that one potential confounding aspect of the existing HER-2 tests is that they only detect gene amplification or protein overexpression, which does not necessarily reflect the functional activity of the receptor. The hypothesis is that for HER-2 to be truly important in the pathobiology of breast cancer, the receptor must be activated to exert its effects. A common feature of signal transduction through membrane bound receptor tyrosine kinases is autophosphorylation of the receptor. Autophosphorylation of HER-2 may therefore be used as a surrogate for active signaling. Monoclonal antibodies have been developed to detect autophosphorylated HER-2 by IHC.[219] In invasive breast cancer with HER-2 overexpression, the receptor appears to be activated only in a small subset of patients (12%).[219,220] Interestingly, the proportion of cases with phosphorylated HER-2 appears to be greater in ductal carcinoma in situ (58%).[221] In one large study of 800 cases of invasive breast cancer with HER-2 overexpression, only cases with phosphorylated HER-2 displayed adverse prognosis.[220] Cases with overexpressed but unphosphorylated receptor had a prognosis as favorable as non–HER-2 overexpressing cases, which supports the concept that phospho-HER-2 may be a more powerful prognostic marker than overall HER-2 protein overexpression. The role of phosphorylated HER-2 as a predictor to trastuzumab therapy is currently being investigated. A small, retrospective, currently unpublished study analyzed phospho-HER-2 in 89 patients treated with trastuzumab in the context of two compassionate use national programs, and the results suggest that phospho-HER-2 may be an important predictive marker of response to trastuzumab. Tissue specimens from the pivotal Herceptin studies are being reanalyzed and may help to answer this very important question.

Clinical Relevance and Current Clinical Testing Utility: Summary and Future Clinical Potential

Since the development of trastuzumab, the testing of newly diagnosed and metastatic breast cancer specimens for HER-2/*neu* has now achieved "standard of practice" status for the management of breast cancer. The discussion as to the best method to determine HER-2 status continues, with the FISH method gaining popularity because of recent evidence that, in comparison to IHC, it may more accurately predict clinical responses to trastuzumab-based therapies. Nevertheless, false FISH results may also occur,[222] and continuous research and improvement are needed in this area. With trastuzumab achieving excellent results in the treatment of HER-2–positive advanced disease and under extensive evaluation in major clinical trials for its potential efficacy when used earlier, the potential role(s) for HER-2 testing as predictor(s) of responses to other therapies being resolved by large prospective clinical outcome studies and the more convenient gene-based CISH technique "waiting in the wings," the story of HER-2 testing in breast cancer will certainly continue to unfold over the next several years.

References

1. Navolanic PM, Steelman LS, McCubrey JA. EGFR family signaling and its association with breast cancer development and resistance to chemotherapy. Int J Oncol 2003;22:237–252.
2. Karunagaran D, Tzahar E, Beerli RR, et al. ErbB-2 is a common auxiliary subunit of NDF and EGF receptors: implications for breast cancer. EMBO J 1996;15:254–264.
3. Yarden Y, Sliwkowski MX. Untangling the ErbB signalling network. Nat Rev Mol Cell Biol 2001;2:127–137.
4. Tzahar E, Waterman H, Chen X, et al. A hierarchical network of interreceptor interactions determines signal transduction by Neu differentiation factor/neuregulin and epidermal growth factor. Mol Cell Biol 1996;16:5276–5287.
5. Karunagaran D, Tzahar E, Beerli RR, et al. ErbB-2 is a common auxiliary subunit of NDF and EGF receptors: implications for breast cancer. EMBO J 1996;15:254–264.
6. Slamon DJ, Clark GM, Wong SG, et al. Human breast cancer: Correlation of relapse and survival with amplification of the Her-2/neu oncogene. Science 1987;235:177–182.
7. Berger MS, Locher GW, Saurer S, et al. Correlation of c-erb B2 gene amplification and protein expression in human breast carcinoma with nodal status and nuclear grading. Cancer Res 1988;48:1238–1243.
8. van de Vivjer MJ, Peterse JL, Mooi WJ, et al. Neu-protein overexpression in breast cancer. N Engl J Med 1988;319:1239–1245.
9. Heintz NH, Leslie KO, Rogers LA, et al. Amplification of the c-erb B 2 oncogene in prognosis of breast adenocarcinoma. Arch Pathol Lab Med 1990;114:160–163.
10. Wright C, Angus B, Nicholson S, et al. Expression of c-erbB-2 oncoprotein: a prognostic indicator in human breast cancer. Cancer Res 1989;49:2087–2090.
11. Tsuda H, Hirohashi S, Shimosato Y, et al. Correlation between histologic grade of malignancy and copy number of c-erbB-2 gene in breast carcinoma. A retrospective analysis of 176 cases. Cancer 1990;65:1794–1800.
12. Borg A, Tandon AK, Sigurdsson H, et al. HER-2/neu amplification predicts poor survival in node-positive breast cancer. Cancer Res 1990;50:4332–4337.
13. Paik S, Hazan R, Fisher ER, et al. Pathologic findings from the national surgical adjuvant breast and bowel project: Prognostic significance of erb B2 protein overexpression in primary breast cancer. J Clin Oncol 1990;8:103–112.
14. Battifora H, Gaffey M, Esteban J, et al. Immunohistochemical assay of neu/c-erb B-2 oncogene product in paraffin-embedded tissues in early breast cancer: retrospective follow-up study of 245 stage I and II cases. Mod Pathol 1991;4:466–474.
15. Kallioniemi OP, Holli K, Visakorpi T, et al. Association of c-erb B2 protein over-expression with high rate of cell proliferation, increased risk of visceral metastasis and poor long-term survival in breast cancer. Int J Cancer 1991;49:650–655.
16. Gullick WJ, Love SB, Wright C, et al. c-erbB-2 protein overexpression in breast cancer is a risk factor in patients with involved and uninvolved lymph nodes. Br J Cancer 1991;63:434–438.
17. Clark GM, McGuire WL. Follow-up study of HER-2/neu amplification in primary breast cancer. Cancer Res 1991;51:944–948.
18. Lovekin C, Ellis IO, Locker A, et al. c-erb B2 oncoprotein expression in primary and advanced breast cancer. Br J Cancer 1991;63:439–443.
19. McCann AH, DeDervan TA, O'Regan M, et al. Prognostic significance of C-erb B2 and estrogen receptor

status in human breast cancer. Cancer Res 1991;51: 3296–3303.

20. Dykens R, Corbett IP, Henry J, et al. Long term survival in breast cancer related to overexpression of the C-erb B2 oncoprotein: an immunohistochemical study using monoclonal antibody NCL-CB11. J Pathol 1991;163:105–110.
21. Rilke F, Colnaghi MI, Cascinelli N, et al. Prognostic significance of HER-2/neu expression in breast cancer and its relationship to other prognostic factors. Int J Cancer 1991;49:44–49.
22. Winstanley J, Cooke T, Murray GD, et al. The long term prognostic significance of C-erb B2 in primary breast cancer. Br J Cancer 1991;63:447–450.
23. O'Reilly SM, Barnes DM, Camplejohn RS, et al. The relationship between C-erb B2 expression, and s-phase fraction in prognosis in breast cancer. Br J Cancer 1991;63:444–446.
24. Paterson MC, Dietrich KD, Danyluk J, et al. Correlation between C erb B2 amplification and risk of recurrent disease in node-negative breast cancer. Cancer Res 1991;51:556–567.
25. Toikkanen S, Helin H, Isola J, et al. Prognostic significance of Her-2 oncoprotein expression in breast cancer: a 30-year follow up. J Clin Oncol 1992;10: 1044–1048.
26. Molina R, Ciocca DR, Candon AK, et al. Expression of HER-2/neu oncoprotein in breast cancer: a comparison of immunohistochemical and western blot techniques. Anticancer Res 1992;12:1965–1991.
27. Noguchi M, Koyasaki M, Ohta N, et al. c-erb B-2 oncoprotein expression versus internal mammary lymph node metastases as additional prognostic factors in patients with axillary lymph node-positive breast cancer. Cancer 1992;69:2953–2960.
28. Allred DC, Clark GM, Tandon AK, et al. HER-2/neu node-negative breast cancer: prognostic significance of overexpression influenced by the presence of in-situ carcinoma. J Clin Oncol 1992;10:599–605.
29. Babiak J, Hugh J, Poppeme S. Significance of c-erb B-2 amplification in DNA aneuploidy. Analysis in 78 patients with node-negative breast cancer. Cancer 1992;70:770–776.
30. Tiwari RK, Borgen PI, Wong GY, et al. HER-2/neu amplification and overexpression in primary human breast cancer is associated with early metastasis. Anticancer Res 1992;12:419–426.
31. Gusterson BA, Gelber RD, Goldhirsch A, et al. Prognostic importance of C-erb B2 expression in breast cancer. J Clin Oncol 1992;10:1049–1056.
32. Bianchi S, Paglierani M, Zampi G, et al. Prognostic significance of C-erb B2 expression in node negative breast cancer. Br J Cancer 1993;67:625–629.
33. Press MF, Pike MC, Chazin VR, et al. Her-2/neu expression in node-negative breast cancer: direct tissue quantification by computerized image analysis and association of overexpression with increased risk of recurrent disease. Cancer Res 1993;53:4960–4970.
34. Seshadri R, Firgaira FA, Horsfall DJ, et al. Clinical significance of Her-2/neu oncogene amplification in primary breast cancer. J Clin Oncol 1993;11: 1936–1942.
35. Descotes F, Pavy J-J, Adessi GL. Human breast cancer: Correlation study between Her-2/neu amplification and prognostic factors in an unselected population. Anticancer Res 1993;13:119–124.
36. Giai M, Roagna R, Ponzone R, et al. Prognostic and predictive relevance of C-erb B2 and ras expression in node positive and negative breast cancer. Anticancer Res 1994;14:1441–1450.
37. Muss HB, Thor AD, Berry DA, et al. Cerb-B2 expression and response to adjuvant therapy in women with node-positive early breast cancer. N Engl J Med 1994;330:1260–1266.
38. Tetu B, Brisson J. Prognostic significance of Her-2/neu oncogene expression in node-positive breast cancer. The influence of the pattern of immunostaining and adjuvant therapy. Cancer 1994;73:2359–2365.
39. Hartmann LC, Ingle JN, Wold LE, et al. Prognostic value of CerbB2 overexpression in axillary lymph node positive breast cancer. Results from a randomized adjuvant treatment protocol. Cancer 1994;74: 2956–2963.
40. Jacquemeier J, Penault-Llorca P, Viens P, et al. Breast cancer response to adjuvant chemotherapy in correlation with erb B2 and p53 expression. Anticancer Res 1994;14:2773–2778.
41. Marks JR, Humphrey PA, Wu K, at al. Overexpression of p53 and Her 2/neu proteins as prognostic markers in early stage breast cancer. Ann Surg 1994;219:332–341.
42. Rosen PP, Lesser ML, Arroyo CD, et al. Immunohistochemical detection of Her-2/neu expression in patients with axillary lymph node-negative breast carcinoma. A study of epidemiologic risk factors, histologic features and prognosis. Cancer 1995;75:1320–1326.
43. Quenel N, Wafflart J, Bonichon F, et al. The prognostic value of c-erbB2 in primary breast carcinomas: a study on 942 cases. Breast Cancer Res Treat 1995; 35:283–291.
44. Sundblad AS, Pellicer EM, Ricci L. Carcinoembryonic expression in stages I and II breast cancer; its relationship with clinicopathologic factors. Hum Pathol 1996;27;297–300.
45. O'Malley FP, Saad Z, Kerkvliet N, et al. The predictive power of semiquantitative immunohistochemical assessment of p53 and C-erbB2 in lymph node-negative breast cancer. Hum Pathol 1996;27:955–963.
46. Hieken TJ, Mehta RR, Shilkaitis A, et al. Her-2/neu and p53 expression in breast cancer: valid prognostic markers when assessed by direct immunoassay, but not by immunochemistry. Proc Annu Meet Am Soc Clin Oncol 1996;15:A113.
47. Xing W-R, Gilchrist KW, Harris CP, et al. FISH detection of HER-2/neu oncogene amplification in early onset breast cancer. Breast Cancer Res Treat 1996; 39:203–212.
48. Dittadi R, Brazzale A, Pappagallo G, et al. ErbB2 assay in breast cancer: possibly improved clinical information using a quantitative method. Anticancer Res 1997;17:1245–1247.
49. Fernandez Acenero MJ, Farina Gonzalez J, Arangoncillo Ballesteros P. Immunohistochemical expression of p53 and c-erbB-2 in breast carcinoma: relation with epidemiologic factors, histologic features and prognosis. Gen Diagn Pathol 1997;142: 289–296.
50. Eissa S, Khalifa A, el-Gharib A, et al. Multivariate analysis of DNA ploidy, p53, c-erbB-2 proteins, EGFR, and steroid hormone receptors for short-term prognosis in breast cancer. Anticancer Res 1997;17: 3091–3097.

51. Charpin C, Garcia S, Bouvier C, et al. c-erbB-2 oncoprotein detected by automated quantitative immunocytochemistry in breast carcinomas correlates with patients' overall and disease-free survival. Br J Cancer 1997;75:1667–1673.
52. Press MJ, Bernstein L, Thomas PA, et al. Her-2/neu gene amplification characterized by fluorescence in situ hybridization: poor prognosis in node negative breast carcinomas. J Clin Oncol 1997;15:2894–2904.
53. Ross JS, Muraca PJ, Jaffe D, et al. Multivariate analysis of prognostic factors in lymph node negative breast cancer. Mod Pathol 1998;11:26A.
54. Depowski PL, Brien TP, Sheehan CE, et al. Prognostic significance of p34cdc2 cyclin-dependent kinase and MIB1 overexpression, and HER-2/neu gene amplification detected by fluorescence in situ hybridization in breast cancer. Am J Clin Pathol 1999;112: 459–469.
55. Querzoli P, Albonico G, Ferretti S, et al. Modulation of biomarkers in minimal breast carcinoma. Cancer 1998;83:89–97.
56. Andrulis IL, Bull SB, Blackstein ME, et al. Neu/erbB-2 amplification identifies a poor-prognosis group of women with node-negative breast cancer. J Clin Oncol 1998;16:1340–1349.
57. Sjogren S, Inganas M, Lindgren A, et al. Prognostic and predictive value of c-erbB-2 overexpression in primary breast cancer, alone and in combination with other prognostic markers. J Clin Oncol 1998;16: 462–469.
58. Harbeck N, Ross JS, Yurdseven S, et al. HER-2/neu gene amplification by fluorescence in situ hybridization allows risk-group assessment in node-negative breast cancer. Int J Oncol 1999;14:663–671.
59. Scorilas A, Yotis J, Pateras C, et al. Predictive value of c-erbB-2 and cathepsin-D for Greek breast cancer patients using univariate and multivariate analysis. Clin Cancer Res 1999;5:815–821.
60. Rudolph P, Olsson H, Bonatz G, et al. Correlation between p53, c-erbB-2, and topoisomerase II alpha expression, DNA ploidy, hormonal receptor status and proliferation in 356 node-negative breast carcinomas: prognostic implications. J Pathol 1999;187: 207–216.
61. Reed W, Hannisdal E, Boehler PJ, et al. The prognostic value of p53 and c-erb B-2 immunostaining is overrated for patients with lymph node negative breast carcinoma: a multivariate analysis of prognostic factors in 613 patients with a follow-up of 14–30 years. Cancer 2000;88:804–813.
62. Pauletti G, Dandekar S, Rong H, et al. Assessment of methods for tissue-based detection of the HER-2/neu alteration in human breast cancer: a direct comparison of fluorescence in situ hybridization and immunohistochemistry. J Clin Oncol 2000;18:3651–3664.
63. Kakar S, Puangsuvan N, Stevens JM, et al. HER-2/neu assessment in breast cancer by immunohistochemistry and fluorescence in situ hybridization: comparison of results and correlation with survival. Mol Diagn 2000;5:199–207.
64. Agrup M, Stal O, Olsen K, et al. C-erbB-2 overexpression and survival in early onset breast cancer. Breast Cancer Res Treat 2000;63:23–29.
65. Umekita Y, Ohi Y, Sagara Y, et al. Co-expression of epidermal growth factor receptor and transforming growth factor-alpha predicts worse prognosis in breast-cancer patients. Int J Cancer 2000;89: 484–487.
66. Pawlowski V, Revillion F, Hebbar M, et al. Prognostic value of the type I growth factor receptors in a large series of human primary breast cancers quantified with a real-time reverse transcription-polymerase chain reaction assay. Clin Cancer Res 2000;6:4217– 4225.
67. Volpi A, De Paola F, Nanni O, et al. Prognostic significance of biologic markers in node-negative breast cancer patients: a prospective study. Breast Cancer Res Treat 2000;63:181–192.
68. Carr JA, Havstad S, Zarbo RJ, et al. The association of HER-2/neu amplification with breast cancer recurrence. Arch Surg 2000;135:1469–1474.
69. Ferrero-Pous M, Hacene K, Bouchet C, et al. Relationship between c-erbB-2 and other tumor characteristics in breast cancer prognosis. Clin Cancer Res 2000;6:4745–4754.
70. Platt-Higgins AM, Renshaw CA, West CR, et al. Comparison of the metastasis-inducing protein S100A4 (p9ka) with other prognostic markers in human breast cancer. Int J Cancer 2000;89:198–208.
71. Eppenberger-Castori S, Kueng W, Benz C, et al. Prognostic and predictive significance of ErbB-2 breast tumor levels measured by enzyme immunoassay. J Clin Oncol 2001;19:645–656.
72. Jukkola A, Bloigu R, Soini Y, et al. c-erbB-2 positivity is a factor for poor prognosis in breast cancer and poor response to hormonal or chemotherapy treatment in advanced disease. Eur J Cancer 2001;37: 347–354.
73. Gaci Z, Bouin-Pineau MH, Gaci M, et al. Prognostic impact of cathepsin D and c-erbB-2 oncoprotein in a subgroup of node-negative breast cancer patients with low histological grade tumors. Int J Oncol 2001;18:793–800.
74. Rudolph P, Alm P, Olsson H, et al. Concurrent overexpression of p53 and c-erbB-2 correlates with accelerated cycling and concomitant poor prognosis in node-negative breast cancer. Hum Pathol 2001;32: 311–319.
75. Beenken SW, Grizzle WE, Crowe DR, et al. Molecular biomarkers for breast cancer prognosis: coexpression of c-erbB-2 and p53. Ann Surg 2001;233: 630–638.
76. Pinto AE, Andre S, Pereira T, et al. C-erbB-2 oncoprotein overexpression identifies a subgroup of estrogen receptor positive (ER+) breast cancer patients with poor prognosis. Ann Oncol 2001;12: 525–533.
77. Riou G, Mathieu MC, Barrois M, et al. c-erbB-2 (HER-2/neu) gene amplification is a better indicator of poor prognosis than protein over-expression in operable breast-cancer patients. Int J Cancer 2001; 95:266–270.
78. Horita K, Yamaguchi A, Hirose K, et al. Prognostic factors affecting disease-free survival rate following surgical resection of primary breast cancer. Eur J Histochem 2001;45:73–84.
79. Suo Z, Risberg B, Kalsson MG, et al. EGFR family expression in breast carcinomas. c-erbB-2 and c-erbB-4 receptors have different effects on survival. J Pathol 2002;196:17–25.
80. Ristimaki A, Sivula A, Lundin J, et al. Prognostic significance of elevated cyclooxygenase-2 expression in breast cancer. Cancer Res 2002;62:632–635.

81. Rosenthal SI, Depowski PL, Sheehan CE, et al. Comparison of HER-2/neu oncogene amplification detected by fluorescence in situ hybridization in lobular and ductal breast cancer. Appl Immunohistochem Mol Morphol 2002;10:40–46.
82. Tsutsui S, Ohno S, Murakami S, et al. Prognostic value of c-erbB2 expression in breast cancer. J Surg Oncol 2002;79:216–223.
83. Spizzo G, Obrist P, Ensinger C, et al. Prognostic significance of Ep-CAM and Her-2/neu overexpression in invasive breast cancer. Int J Cancer 2002;98: 883–888.
84. Kato T, Kameoka S, Kimura T, et al. C-erbB-2 and PCNA as prognostic indicators of long-term survival in breast cancer. Anticancer Res 2002;22:1097–1103.
85. el-Ahmady O, el-Salahy E, Mahmoud M, et al. Multivariate analysis of bcl-2, apoptosis, P53 and HER-2/neu in breast cancer: a short-term follow-up. Anticancer Res 2002;22:2493–2499.
86. Schnitt SJ, Jacobs TW. Current status of HER2 testing: caught between a rock and a hard place. Am J Clin Pathol 2001;116:806–810.
87. Hayes DF, Thor AD. c-erbB-2 in breast cancer: development of a clinically useful marker. Semin Oncol 2002;29:231–245.
88. Masood S, Bui MM. Prognostic and predictive value of HER2/neu oncogene in breast cancer. Microsc Res Tech 2002;59:102–108.
89. Press MF, Hung G, Godolphin W, et al. Sensitivity of HER-2/neu antibodies in archival tissue samples: potential source of error in immunohistochemical studies of oncogene expression. Cancer Res 1994;54: 2771–2777.
90. Paik S, Bryant J, Tan-Chiu E, et al. Real-world performance of HER2 testing—national Surgical Adjuvant Breast and Bowel Project experience. J Natl Cancer Inst 2002;94:852–854.
91. Paik S, Tan-chui E, Bryan J, et al. Successful quality assurance program for HER-2 testing in the NSAPB trial for Herceptin. Breast Cancer Res Treat 2002; 76(suppl 1):S31.
92. Rhodes A, Jasani B, Anderson E, et al. Evaluation of HER-2/neu immunohistochemical assay sensitivity and scoring on formalin-fixed and paraffin-processed cell lines and breast tumors: a comparative study involving results from laboratories in 21 countries. Am J Clin Pathol 2002;118:408–417.
93. Arber DA. Effect of prolonged formalin fixation on the immunohistochemical reactivity of breast markers. Appl Immunohistochem Mol Morphol 2002;10: 183–186.
94. Wang S, Saboorian MH, Frenkel EP, et al. Assessment of HER-2/neu status in breast cancer. Automated Cellular Imaging System (ACIS)-assisted quantitation of immunohistochemical assay achieves high accuracy in comparison with fluorescence in situ hybridization assay as the standard. Am J Clin Pathol 2001;116:495–503.
95. Bloom KJ, Torre-Bueno J, Press M, et al. Comparison of HER-2/neu analysis using FISH and IHC when HercepTest is scored using conventional microscopy and image analysis. Breast Cancer Res Treat 2000; 64:99.
96. Wang S, Saboorian MH, Frenkel E, et al. Laboratory assessment of the status of Her-2/neu protein and oncogene in breast cancer specimens: comparison of immunohistochemistry assay with fluorescence in situ hybridization assays. J Clin Pathol 2000;53: 374–381.
97. Tanner M, Gancberg D, Di Leo A, et al. Chromogenic in situ hybridization: a practical alternative for fluorescence in situ hybridization to detect HER-2/neu oncogene amplification in archival breast cancer samples. Am J Pathol 2000;157:1467–1472.
98. Zhao J, Wu R, Au A, et al. Determination of HER2 gene amplification by chromogenic in situ hybridization (CISH) in archival breast carcinoma. Mod Pathol 2002;15:657–665.
99. Gupta D, Middleton LP, Whitaker MJ, et al. Comparison of fluorescence and chromogenic in situ hybridization for detection of HER-2/neu oncogene in breast cancer. Am J Clin Pathol 2003;119:381–387.
100. Pawlowski V, Revillion F, Hornez L, et al. A real-time one-step reverse transcriptase-polymerase chain reaction method to quantify c-erbB-2 expression in human breast cancer. Cancer Detect Prev 2000;24: 212–223.
101. Bieche I, Onody P, Laurendeau I, et al. Real-time reverse transcription-PCR assay for future management of ERBB2-based clinical applications. Clin Chem 1999;45:1148–1156.
102. Tubbs RR, Pettay JD, Roche PC, et al. Discrepancies in clinical laboratory testing of eligibility for trastuzumab therapy: apparent immunohistochemical false-positives do not get the message. J Clin Oncol 2001;19:2714–2721.
103. Pusztai L, Ayers M, Stec J, et al. Gene expression profiles obtained from fine-needle aspirations of breast cancer reliably identify routine prognostic markers and reveal large-scale molecular differences between estrogen-negative and estrogen-positive tumors. Clin Cancer Res 2003;9:2406–2415.
104. Mackay A, Jones C, Dexter T, et al. cDNA microarray analysis of genes associated with ERBB2 (HER2/neu) overexpression in human mammary luminal epithelial cells. Oncogene 2003;22:2680–2688.
105. Kauraniemi P, Barlund M, Monni O, et al. New amplified and highly expressed genes discovered in the ERBB2 amplicon in breast cancer by cDNA microarrays. Cancer Res 2001;61:8235–8240.
106. Kononen J, Bubendorf L, Kallioniemi A, et al. Tissue microarrays for high-throughput molecular profiling of tumor specimens. Nat Med 1998;4:844–847.
107. Camp RL, Charette LA, Rimm DL. Validation of tissue microarray technology in breast carcinoma. Lab Invest 2000;80:1943–1949.
108. Andersen CL, Hostetter G, Grigoryan A, et al. Improved procedure for fluorescence in situ hybridization on tissue microarrays. Cytometry 2001; 45:83–86.
109. Gancberg D, Di Leo A, Rouas G, et al. Reliability of the tissue microarray based FISH for evaluation of the HER-2 oncogene in breast carcinoma. J Clin Pathol 2002;55:315–317.
110. O'Grady A, Flahavan CM, Kay EW, et al. HER-2 analysis in tissue microarrays of archival human breast cancer: comparison of immunohistochemistry and fluorescence in situ hybridization. Appl Immunohistochem Mol Morphol 2003;11:177–182.
111. Zhang D, Salto-Tellez M, Do E, et al. Evaluation of HER-2/neu oncogene status in breast tumors on tissue microarrays. Hum Pathol 2003;34:362–368.
112. Camp RL, Dolled-Filhart M, King BL, et al. Quantitative analysis of breast cancer tissue microarrays

shows that both high and normal levels of HER2 expression are associated with poor outcome. Cancer Res 2003;63:1445–1448.

113. Muller V, Thomssen C, Karakas C, et al. Quantitative assessment of HER-2/neu protein concentration in breast cancer by enzyme-linked immunosorbent assay. Int J Biol Markers 2003;18:13–20.

114. Huston JS, George AJ. Engineered antibodies take center stage. Hum Antibodies 2001;10:127–142.

115. Hortobagyi GN. Overview of treatment results with trastuzumab (Herceptin) in metastatic breast cancer. Semin Oncol 2001;28:43–47.

116. McKeage K, Perry CM. Trastuzumab: a review of its use in the treatment of metastatic breast cancer overexpressing HER2. Drugs 2002;62:209–243.

117. Shawver LK, Slamon D, Ullrich A. Smart drugs: tyrosine kinase inhibitors in cancer therapy. Cancer Cell 2002;1:117–123.

118. Ligibel JA, Winer EP. Trastuzumab/chemotherapy combinations in metastatic breast cancer. Semin Oncol 2002;29:38–43.

119. Slamon DJ, Leyland-Jones B, Shak S, et al. Use of chemotherapy plus a monoclonal antibody against HER2 for metastatic breast cancer that overexpresses HER2. N Engl J Med 2001;344:783–792.

120. Schneider JW, Chang AY, Garratt A. Trastuzumab cardiotoxicity: Speculations regarding pathophysiology and targets for further study. Semin Oncol 2002;29(3 suppl 11):22–28.

121. Ozcelik C, Erdmann B, Pilz B, et al. Conditional mutation of the ErbB2 (HER2) receptor in cardiomyocytes leads to dilated cardiomyopathy. Proc Natl Acad Sci USA 2002;99:8880–8885.

122. Nahta R, Esteva FJ. HER-2-targeted therapy: lessons learned and future directions. Clin Cancer Res 2003;9:5078–5084.

123. Burstein HJ, Kuter I, Campos SM, et al. Clinical activity of trastuzumab and vinorelbine in women with HER2-overexpressing metastatic breast cancer. J Clin Oncol 2001;19:2722–2730.

124. Slamon DJ, Patel R, Northfelt R, et al. Phase II pilot study of Herceptin combined with taxotere and carboplatin in metastatic breast cancer patients overexpressing the HER-2/neu proto-oncogene: a pilot study of the UCLA network. Proc Am Soc Clin Oncol 2001;20:193.

125. Jacobs TW, Gown AM, Yaziji H, et al. Specificity of HercepTest in determining HER-2/neu status of breast cancers using the United States Food and Drug Administration-approved scoring system. J Clin Oncol 1999;17:1983–1987.

126. Roche PC, Ingle JN. Increased HER2 with U.S. Food and Drug Administration-approved antibody. J Clin Oncol 1999;17:434.

127. Mass RD, Press MF, Anderson S, et al. Improved survival benefit from Herceptin (trastuzumab) in patients selected by fluorescence in situ hybridization (FISH). Proc ASCO 2001;21:85.

128. Vogel CL, Cobleigh MA, Tripathy D, et al. Efficacy and safety of trastuzumab as a single agent in first-line treatment of HER2-overexpressing metastatic breast cancer. J Clin Oncol 2002;20:719–726.

129. Ogura H, Akiyama F, Kasumi F, et al. Evaluation of HER-2 status in breast carcinoma by fluorescence in situ hybridization and immunohistochemistry. Breast Cancer 2003;10:234–240.

130. Seidman AD, Fornier MN, Esteva FJ, et al. Weekly trastuzumab and paclitaxel therapy for metastatic breast cancer with analysis of efficacy by HER2 immunophenotype and gene amplification. J Clin Oncol 2001;19:2587–2595.

131. Fornier M, Risio M, Van Poznak C, et al. HER-2 testing and correlation with efficacy in trastuzumab therapy. Oncology 2003;16:1340–1358.

132. Hammock L, Lewis M, Phillips C, et al. Strong HER-2/neu protein overexpression by immunohistochemistry often does not predict oncogene amplification by fluorescence in situ hybridization. Hum Pathol 2003;34:1043–1047.

133. Nichols DW, Wolff DJ, Self S, et al. A testing algorithm for determination of HER2 status in patients with breast cancer. Ann Clin Lab Sci 2002;32:3–11.

134. Schnitt SJ, Jacobs TW. Current status of HER2 testing: caught between a rock and a hard place. Am J Clin Pathol 2001;116:806–810.

135. Yaziji H, Gown AM. Testing for HER-2/neu in breast cancer: is fluorescence in situ hybridization superior in predicting outcome? Adv Anat Pathol 2002;9: 338–344.

136. Press MF, Slamon DJ, Flom KJ. Evaluation of HER-2/neu gene amplification and overexpression: comparison of frequently used assay methods in a molecularly characterized cohort of breast cancer specimens. J Clin Oncol 2002;20:3095–3105.

137. Dowsett M. Overexpression of HER-2 as a resistance mechanism to hormonal therapy for breast cancer. Endocr Relat Cancer 2001;8:191–195.

138. Muss HB. Role of adjuvant endocrine therapy in early-stage breast cancer. Semin Oncol 2001;28:313–321.

139. Schmid P, Wischnewsky MB, Sezer O, et al. Prediction of response to hormonal treatment in metastatic breast cancer. Oncology 2002;63:309–316.

140. Nunes RA, Harris LN. The HER2 extracellular domain as a prognostic and predictive factor in breast cancer. Clin Breast Cancer 2002;3:125–135.

141. Konecny G, Pauletti G, Pegram M, et al. Quantitative association between HER-2/neu and steroid hormone receptors in hormone receptor-positive primary breast cancer. J Natl Cancer Inst 2003;95: 142–153.

142. Sjogren S, Inganas M, Lindgren A, et al. Prognostic and predictive value of c-erbB-2 overexpression in primary breast cancer, alone and in combination with other prognostic markers. J Clin Oncol 1998;16: 462–469.

143. Carlomagno C, Perrone F, Gallo C, et al. CerbB2 overexpression decreases the benefit of adjuvant tamoxifen in early-stage breast cancer without ancillary lymph node metastases. J Clin Oncol 1996;14: 2702–2708.

144. Burke HB, Hoang A, Iglehart JD, et al. Predicting response to adjuvant and radiation therapy in patients with early stage breast carcinoma. Cancer 1998;82:874–877.

145. Elledge RM, Green S, Ciocca D, et al. HER-2 expression and response to tamoxifen in estrogen receptor-positive breast cancer: a southwest oncology group study. Clin Cancer Res 1998;4:7–12.

146. Bianco AR, De Laurentiis M, Carlomagno C, et al. 20 year update of the Naples Gun trial of adjuvant breast cancer therapy: evidence of interaction between c-erb-B2 expression and tamoxifen efficacy. Proc ASCO 1998;17:97a.

147. Ravdin PM, Green S, Albain V, et al. Initial report of the SWOG biological correlative study of c-ERBB2 expression as a predictor of outcome in a trial comparing adjuvant CAF with tamoxifen alone. Proc ASCO 1998;17:97a.

148. Ellis MJ, Coop A, Singh B, et al. Letrozole is more effective neoadjuvant endocrine therapy than tamoxifen for ErbB-1- and/or ErbB-2-positive, estrogen receptor-positive primary breast cancer: evidence from a phase III randomized trial. J Clin Oncol 2001;19:3808–3816.

149. Berns EM, Foekens JA, van Staveren IL, et al. Oncogene amplification and prognosis in breast cancer: Relationship with systemic treatment. Gene 1995; 159:11–18.

150. Sparano JA. Taxanes for breast cancer: an evidence-based review of randomized phase II and phase III trials. Clin Breast Cancer 2000;1:32–40.

151. Yu D. Mechanisms of ErbB2-mediated paclitaxel resistance and trastuzumab-mediated paclitaxel sensitization in ErbB2-overexpressing breast cancers. Semin Oncol 2001;28(5 suppl 16):12–17.

152. Baselga J, Seidman AD, Rosen PP, et al. HER-2 overexpression and paclitaxel sensitivity in breast cancer: therapeutic implications. Oncology 1997;11: 43–48.

153. Kim R, Tanabe K, Uchida Y. The role of HER-2 oncoprotein in drug-sensitivity in breast cancer. Oncol Rep 2002;9:3–9.

154. Hamilton A, Larsimont D, Paridaens R, et al. A study of the value of p53, HER2, and Bcl-2 in the prediction of response to doxorubicin and paclitaxel as single agents in metastatic breast cancer: a companion study to EORTC 10923. Clin Breast Cancer 2000;1:233–240.

155. Di Leo A, Larsimont D, Gancberg D, et al. HER-2 and topoisomerase II alpha as predictive markers in a population of node-positive breast cancer patients randomly treated with adjuvant CMF or epirubicin plus cyclophosphamide. Ann Oncol 2001; 12:1081–1089.

156. Petit T, Borel C, Ghnassia JP, et al. Chemotherapy response of breast cancer depends on HER-2 status and anthracycline dose intensity in the neoadjuvant setting. Clin Cancer Res 2001;7:1577–1581.

157. Harris LN, Yang L, Liotcheva V, et al. Induction of topoisomerase II activity after ErbB2 activation is associated with a differential response to breast cancer chemotherapy. Clin Cancer Res 2001;7:1497–1504.

158. Zhang F, Yang Y, Smith T, et al. Correlation between HER-2 expression and response to neoadjuvant chemotherapy with 5-fluorouracil, doxorubicin, and cyclophosphamide in patients with breast carcinoma. Cancer 2003;97:1758–1765.

159. Haffty BG, Brown F, Carter D, et al. Evaluation of HER-2/neu oncoprotein expression as a prognostic indicator of local recurrence in conservatively treated breast cancer: a case-control study. Int J Radiat Oncol Biol Phys 1996;35:751–757.

160. Cardoso F, Bernard-Marty C, Rouas G, et al. Is HER-2 a true or a surrogate predictive marker (PM) of response to anthracycline-based chemotherapy (CT) in metastatic breast cancer (MBC) patients? Proc Am Soc Clin Oncol 2002;21:302b(abst. 3027).

161. Depowski PL, Rosenthal SI, Brien TP, et al. Topoisomerase II alpha expression in breast cancer: correlation with outcome variables. Mod Pathol 2000;13: 542–547.

162. Jarvinen TA, Tanner M, Rantanen V, et al. Amplification and deletion of topoisomerase II alpha associate with ErbB-2 amplification and affect sensitivity to topoisomerase II inhibitor doxorubicin in breast cancer. Am J Pathol 2000;156:839–847.

163. Tanner M, Jarvinen P, Isola J. Amplification of HER-2/neu and topoisomerase II alpha in primary and metastatic breast cancer. Cancer Res 2001;61: 5345–5348.

164. Harris LN, Yang L, Liotcheva V, et al. Induction of topoisomerase II activity after ErbB2 activation is associated with a differential response to breast cancer chemotherapy. Clin Cancer Res 2001;7:1497–1504.

165. Di Leo A, Gancberg D, Larsimont D, et al. HER-2 amplification and topoisomerase II alpha gene aberrations as predictive markers in node-positive breast cancer patients randomly treated either with an anthracycline-based therapy or with cyclophosphamide, methotrexate, and 5-fluorouracil. Clin Cancer Res 2002;8:1107–1116.

166. Coon JS, Marcus E, Gupta-Burt S, et al. Amplification and overexpression of topoisomerase II alpha predict response to anthracycline-based therapy in locally advanced breast cancer. Clin Cancer Res 2002;8:1061–1067.

167. Cardoso F, Durbecq V, Larsimont D, et al. Correlation between complete response to anthracycline-based chemotherapy and topoisomerase II-a gene amplification and protein overexpression in locally advanced/metastatic breast cancer. Int J Oncol 2004; 24:201–209.

168. Cardoso F, Larsimont D, Di Leo A, et al. Evaluation of Her2, p53, bcl-2, topoisomerase II-α, heat shock proteins 27 and 70 in primary breast cancer and metastatic ipsilateral axillary lymph nodes. Ann Oncol 2001;12:615–620.

169. Knoop A, Knudsen H, Balslev E, et al. Topoisomerase II alpha (Top2A) alteration as a predictive marker for epirubicin sensitivity in 805 high risk breast cancer patients. A randomised DBCG trial (DBCG89D). Eur J Cancer 2003;1:abst. 674.

170. Isola JJ, Holli K, Oksa H, et al. Elevated erb B-2 oncoprotein levels in preoperative and follow-up serum samples defined in aggressive course in patients with breast cancer. Cancer 1994;73:652–658.

171. Andersen TI, Paus E, Nesland JM, et al. Detection of C-erb B2 related proteins in sera from breast cancer patients. Relationship to ERBB2 gene amplification and Cerb B2 protein overexpression in tumour. Acta Oncol 1995;34:499–504.

172. Willsher PC, Beaver J, Pinder S, et al. Prognostic significance of serum CerbB2 protein in breast cancer patients. Breast Cancer Res Treat 1996;40: 251–255.

173. Fehm T, Maimonis P, Wetz S, et al. Influence of circulating C-erbB-2 serum protein on response to adjuvant chemotherapy in node-positive breast cancer patients. Breast Cancer Res Treat 1997;43:87–95.

174. Mansour OA, Zekri AR, Harvey J, et al. Tissue in serum CerbB2 and tissue EGFR in breast carcinoma: Three years follow-up. Anticancer Res 1997; 17:3101–3106.

175. Disis ML, Pupa SM, Gralow JR, et al. High-titer HER-2/neu protein-specific antibody can be detected in patients with early-stage breast cancer. J Clin Oncol 1997;15:3363–3367.

176. Krainer M, Brodowicz T, Zeillinger R, et al. Tissue expression and serum levels of HER-2/neu in patients with breast cancer. Oncology 1997;54:475–481.

177. Burke HB, Hoang A, Iglehart JD, et al. Predicting response to adjuvant and radiation therapy in patients with early stage breast carcinoma. Cancer 1998;82:874–877.

178. Fehm T, Maimonis P, Katalinic A, et al. The prognostic significance of c-erbB-2 serum protein in metastatic breast cancer. Oncology 1998;55:33–38.

179. Mehta RR, McDermott JH, Hieken TJ, et al. Plasma c-erbB-2 levels in breast cancer patients: prognostic significance in predicting response to chemotherapy. J Clin Oncol 1998;16:2409–2416.

180. Colomer R, Montero S, Lluch A, et al. Circulating HER2 extracellular domain and resistance to chemotherapy in advanced breast cancer. Clin Cancer Res 2000;6:2356–2362.

181. Bewick M, Conlon M, Gerard S, et al. HER-2 expression is a prognostic factor in patients with metastatic breast cancer treated with a combination of high-dose cyclophosphamide, mitoxantrone, paclitaxel and autologous blood stem cell support. Bone Marrow Transplant 2001;27:847–853.

182. Harris LN, Liotcheva V, Broadwater G, et al. Comparison of methods of measuring HER-2 in metastatic breast cancer patients treated with high-dose chemotherapy. J Clin Oncol 2001;19:1698–1706.

183. Ali SM, Leitzel K, Chinchilli VM, et al. Relationship of serum HER-2/neu and serum CA 15–3 in patients with metastatic breast cancer. Clin Chem 2002;48: 1314–1320.

184. Classen S, Kopp R, Possinger K, et al. Clinical relevance of soluble c-erbB-2 for patients with metastatic breast cancer predicting the response to second-line hormone or chemotherapy. Tumour Biol 2002;23:70–75.

185. Lipton A, Ali SM, Leitzel K, et al. Elevated serum Her-2/neu level predicts decreased response to hormone therapy in metastatic breast cancer. J Clin Oncol 2002;20:1467–1472.

186. Kandl H, Seymour L, Bezwoda WR. Soluble c-erb B-2 fragments in serum correlates with disease stage and predicts more shortened survival in patients with early-stage and advanced breast cancer. Br J Cancer 1994;70:739–742.

187. Volas GH, Leitzel K, Teramoto Y, et al. Serial serum C-erbB-2 levels in patients with breast carcinoma. Cancer 1996;78:267–272.

188. Willsher PC, Beaver J, Pinder S, et al. Prognostic significance of serum CerbB2 protein in breast cancer patients. Breast Cancer Res Treat 1996;40: 251–255.

189. Revillion F, Hebbar M, Boneterre J, et al. Plasma CerbB2 concentrations in relation to chemotherapy in breast cancer patients. Eur J Cancer 1996;32a: 231–234.

190. Volas GH, Leitzel K, Teramoto Y, et al. Serial serum C-erbB-2 levels in patients with breast carcinoma. Cancer 1996;78:267–272.

191. Mehta RR, McDermott JH, Hieken TJ, et al. Plasma c-erbB-2 levels in breast cancer patients: prognostic significance in predicting response to chemotherapy. J Clin Oncol 1998;16:2409–2416.

192. Nunes RA, Harris LN. The HER2 extracellular domain as a prognostic and predictive factor in breast cancer. Clin Breast Cancer 2002;3:125–135.

193. Slamon D, Pegram M. Rationale for trastuzumab (Herceptin) in adjuvant breast cancer trials. Semin Oncol 2001;28:13–19.

194. Bose S, Lesser ML, Norton L, et al. Immunophenotype of intraductal carcinoma. Arch Pathol Lab Med 1996;120:81–85.

195. Moreno A, Lloveras B, Figueras A, et al. Ductal carcinoma in-situ of the breast: Correlation between histologic classification and biologic markers. Mod Pathol 1997;10:1088–1092.

196. Mack L, Kerkzelit N, Doig G, et al. Relationship of a new histological categorization of ductal carcinoma in-situ of the breast with size and the immunohistochemical expression of p53, C-erb B2, bcl2 and ki-67. Hum Pathol 1997;28:974–979.

197. Wolber RA, DuPuis BA, Wick MR. Expression of cerb B2 oncoprotein in mammary and extramammary Paget's disease. Am J Clin Pathol 1991;96:243–247.

198. Fu W, Lobocki CA, Silberberg BK. Molecular markers in Paget disease of the breast. J Surg Oncol 2001;77:171–178.

199. Hanna W, Alowami S, Malik A. The role of HER-2/neu oncogene and vimentin filaments in the production of the Paget's phenotype. Breast J 2003; 9:485–490.

200. Masood S, Bui MM. Assessment of Her-2/neu overexpression in primary breast cancers and their metastatic lesions: an immunohistochemical study. Ann Clin Lab Sci 2000;30:259–265.

201. Dittadi R, Zancan M, Perasole A, et al. Evaluation of HER-2/neu in serum and tissue of primary and metastatic breast cancer patients using an automated enzyme immunoassay. Int J Biol Markers 2001;16:255–261.

202. Simon R, Nocito A, Hubscher T, et al. Patterns of her-2/neu amplification and overexpression in primary and metastatic breast cancer. J Natl Cancer Inst 2001;93:1141–1146.

203. Vincent-Salomon A, Jouve M, Genin P, et al. HER2 status in patients with breast carcinoma is not modified selectively by preoperative chemotherapy and is stable during the metastatic process. Cancer 2002;94:2169–2173.

204. Xu R, Perle MA, Inghirami G, et al. Amplification of Her-2/neu gene in Her-2/neu-overexpressing and -nonexpressing breast carcinomas and their synchronous benign, premalignant, and metastatic lesions detected by FISH in archival material. Mod Pathol 2002;15:116–124.

205. Symmans WF, Liu J, Knowles DM, et al. Breast cancer heterogeneity: evaluation of clonality in primary and metastatic lesions. Hum Pathol 1995;26:210–216.

206. Gancberg D, Di Leo A, Cardoso F, et al. Comparison of HER-2 status between primary breast cancer and corresponding distant metastatic sites. Ann Oncol 2002;13:1036–1043.

207. Gattuso P, Reddy VB, Green LK, et al. Prognostic factors for carcinoma of the male breast. Int J Surg Pathol 1995;2:199–206.

208. Joshi MG, Lee AK, Loda M, et al. Male breast carcinoma: an evaluation of prognostic factors contributing to a poorer outcome. Cancer 1996;77:490–498.

209. Pich A, Margaria E, Chiusa L. Oncogenes and male breast carcinoma: c-erbB-2 and p53 coexpression predicts a poor survival. J Clin Oncol 2000;18:2948–2956.

210. Wang-Rodriguez J, Cross K, Gallagher S, et al. Male breast carcinoma: correlation of ER, PR, Ki-67, Her2-Neu, and p53 with treatment and survival, a study of 65 cases. Mod Pathol 2002;15:853–861.

211. Rayson D, Erlichman C, Suman VJ, et al. Molecular markers in male breast carcinoma. Cancer 1998;83:1947–1955.

212. Shpitz B, Bomstein Y, Sternberg A, et al. Angiogenesis, p53, and c-erbB-2 immunoreactivity and clinicopathological features in male breast cancer. J Surg Oncol 2000;75:252–257.

213. Bloom KJ, Govil H, Gattuso P, et al. Status of HER-2 in male and female breast carcinoma. Am J Surg 2001;182:389–392.

214. Stark A, Hulka BS, Joens S, et al. HER-2/neu amplification in benign breast disease and the risk of subsequent breast cancer. J Clin Oncol 2000;18:267–274.

215. Eccles SA. The role of c-erbB-2/HER2/neu in breast cancer progression and metastasis. J Mamm Gland Biol Neoplasia 2001;6:393–406.

216. Piccart M, Lohrisch C, Di Leo A, et al. The predictive value of HER2 in breast cancer. Oncology 2001;61(suppl 2):73–82.

217. Yarden Y. Biology of HER2 and its importance in breast cancer. Oncology 2001;61(suppl 2):1–13.

218. Zemzoum I, Kates RE, Ross JS, et al. Invasion factors uPA/PAI–1 and HER2 status provide independent and complementary information on patient outcome in node-negative breast cancer. J Clin Oncol 2003;21:1022–1028.

219. DiGiovanna MP, Stern DF. Activation state-specific monoclonal antibody detects tyrosine phosphorylated p185neu/erbB-2 in a subset of human breast tumors overexpressing this receptor. Cancer Res 1995;55:1946–1955.

220. Thor AD, Liu S, Edgerton S, et al. Activation (tyrosine phosphorylation) of ErbB-2 (HER-2/neu): a study of incidence and correlation with outcome in breast cancer. J Clin Oncol 2000;18:3230–3239.

221. DiGiovanna MP, Chu P, Davison TL, et al. Active signaling by HER-2/neu in a subpopulation of HER-2/neu-overexpressing ductal carcinoma in situ: clinicopathological correlates. Cancer Res 2002;62:6667–6673.

222. Roche PC, Suman VJ, Jenkins RB, et al. Concordance between local and central laboratory HER2 testing in the breast intergroup trial N9831. J Natl Cancer Inst 2002;94:855–857.

CHAPTER 17

Epidermal Growth Factor Receptor and Other Growth Factors and Receptors

Joan Albanell, MD[1] and Jeffrey S. Ross, MD[2,3]

[1]*Department of Medical Oncology, ICMHO & IDIBAPS, Hospital Clinic, Barcelona, Spain*
[2]*Department of Pathology and Laboratory Medicine, Albany Medical College, Albany, New York*
[3]*Division of Molecular Medicine, Millennium Pharmaceuticals, Inc., Cambridge, Massachusetts*

Breast cancers are characterized by the frequent dysregulation of growth factors and their associated transmembrane receptors. The ErbB (HER) signaling family is the most well-characterized growth factor receptor system in breast cancer. The ErbB signaling serves as a successful model of translational research because of the key importance that it currently has in routine clinical practice, yet it is still an area of continued active research.[1–3] The insulin-like growth factor (IGF) signaling family is also emerging as a key element in breast cancer progression and in the cellular response to anticancer agents.[4] Because of the current clinical potential of the ErbB and IGF families and considering that agents targeting these systems are ready to or under current testing in clinical trials, we review these two families in this chapter. There is a large number of additional growth factor families[5] that play diverse roles in human breast cancers, including the fibroblast growth factor family,[6] the platelet-derived growth factor (PDGF) family,[7] the Wnt growth factor receptor family,[8] the transforming growth factor (TGF)-β family, the met receptor,[9] cytokines such as interleukin-6,[10] and chemokine receptors,[11] to name a few. Despite their biologic importance in both the normal and malignant breast, the clinical implications (i.e., for breast cancer diagnosis, prognosis, prediction of response, or as targets) of these systems are as yet poorly characterized, and their review is beyond the scope of this chapter (for a review see Refs. 12 and 13).

The ErbB Signaling Family

Biologic Background

The ErbB signaling family (also known as epidermal growth factor receptor [EGFR] signaling family) comprises four receptors, HER-1 (ErbB1/EGFR), HER-2 (ErbB2), HER-3 (ErbB3), and HER-4 (ErbB4), and a large family of ligands (growth factors) (FIG. 17.1).[2] Because of the particular importance of HER-2 as a prognostic and predictive factor as well as a therapeutic target in breast cancer, this receptor has a full chapter in this book (see Chapter 16)[1,14,15]. In addition to HER-2, the other members of the family (EGFR, HER-3, and HER-4 and their ligands) are also dysregulated in many human breast cancers[16] and represent attractive molecules for predictive medicine and as targets for breast cancer treatment.[17–19]

The ErbB receptors have three distinct domains: an extracellular domain, which recognizes and binds the ligands that result in receptor activation; a hydrophobic transmembrane region; and a cytoplasmic intracellular domain, which has enzymatic tyrosine kinase activity and is able to phosphorylate tyrosine residues on specific adaptor proteins[2] (FIG. 17.2). The carboxy-terminal tail of the cytoplasmic domain contains critical tyrosine residues and receptor regulatory motifs. EGFR, HER-2, and HER-4 have intrinsic tyrosine kinase activity, whereas HER-3 is a kinase defective receptor. Although these receptors have a high degree of homology in the tyrosine kinase domains, the

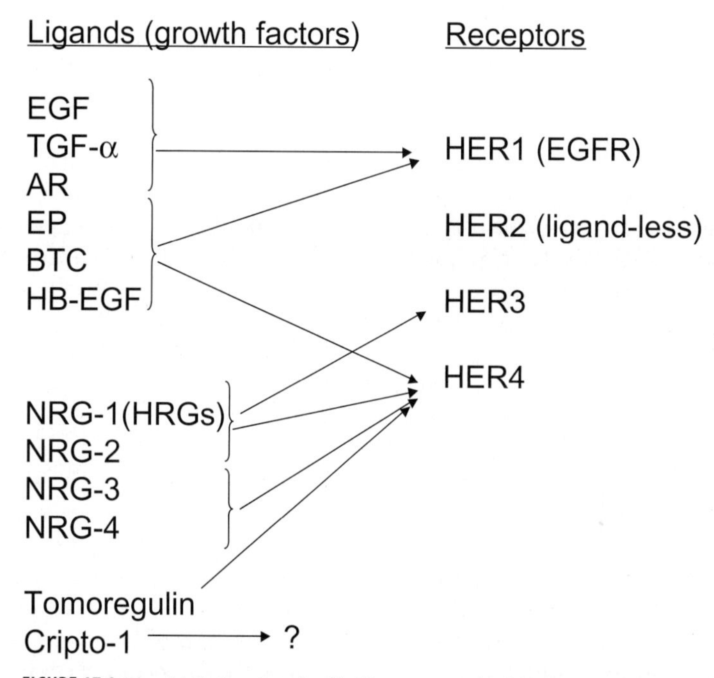

FIGURE 17.1 The ErbB signaling family. There are multiple ligands and four receptors in the family. Arrows indicate the ability of each ligand to bind to individual receptors of the family.

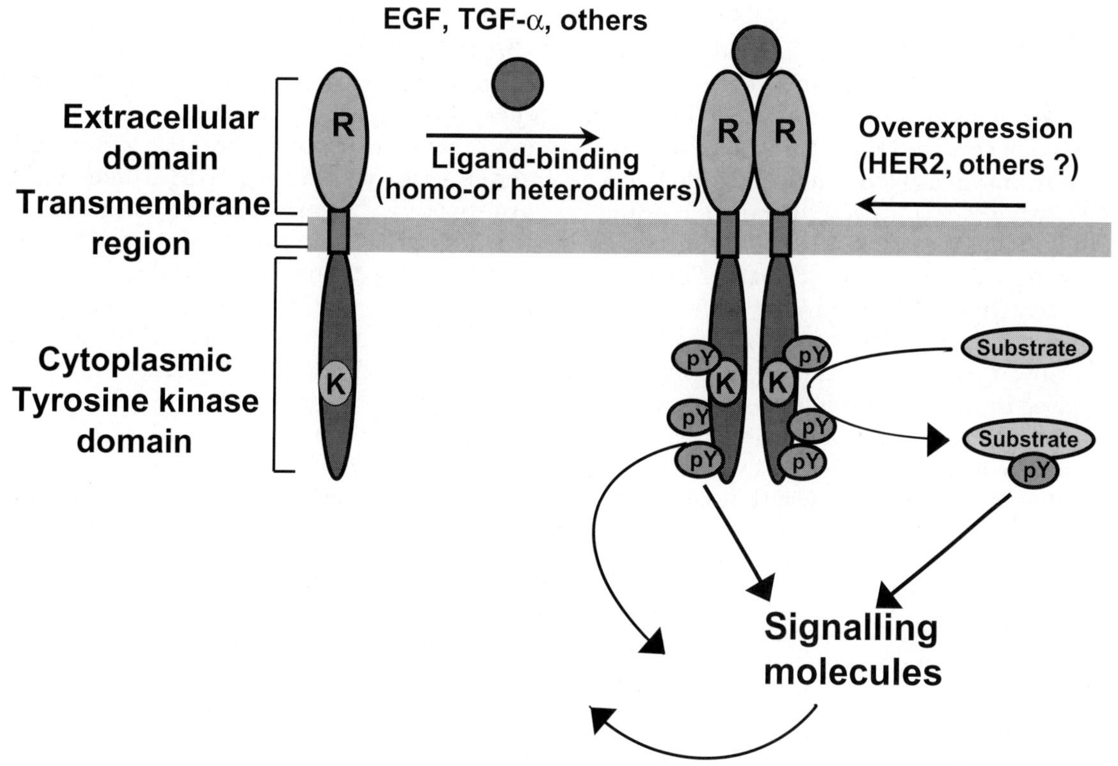

FIGURE 17.2 The ErbB receptors have an extracellular ligand-binding domain, a transmembrane region, and an intracellular cytoplasmic domain with tyrosine kinase activity. The receptors are inactive in their monomeric state. Upon binding of ligand, dimerization between identical (homodimerization) or different receptors (heterodimerization) occurs.

extracellular domains are less conserved. The differences in the extracellular domains allow different specificities in the ability of individual receptors to recognize and bind ligands.

The binding of a ligand to the extracellular domain is the key event that triggers the activation of the ErbB receptors. These ligands belong to the EGF family of peptide growth factors.[20,21] The known members of this family include EGF, TGF-α, amphiregulin, heparin binding EGF, betacellulin, epiregulin, tomoregulin, and the neuregulin (NRG) subfamily. The NRG subfamily is composed of four genes, NGR-1 to NGR-4. The NGR-1 family includes (but is not limited to) the rat neu differentiation factor and the heregulins. An EGF subfamily that includes cripto-1 and criptin has been also described.

The ligand–receptor interactions within the family are summarized in FIGURE 17.1. In essence, EGF, amphiregulin, and TGF-α only bind to the EGFR; NRGs/heregulins only bind to HER-3 and HER-4; epiregulin, betacellulin, and heparin binding-EGF can bind to EGFR as well as HER-3/4. HER-2 is the only ligand orphan receptor.[2,21] Typically, ligands bind and activate the EGFR via the so-called EGF-like motif. These ligands are synthesized as transmembrane molecules that can release their extracellular domains containing the EGF motif through a specialized type of limited proteolysis that is known as ectodomain shedding. It is widely accepted that ectodomain shedding regulates the availability of soluble EGFR ligands that act in autocrine or paracrine fashion, activating EGFRs in the very EGFR ligand-producing cell or in proximal cells, respectively. Further studies on the mechanisms implicated in sending the signals that activate these receptors may offer additional insight in the complex biology of this family. Along this line, it has been recently shown that when proteolytic shedding is prevented, the transmembrane form of the TGF-α interacts with, but does not activate, the EGFR.[22] Thus, shedding seems to control not only the availability of the soluble form of the growth factor (TGF-α) but also the activity of the transmembrane form.

The binding of a ligand to the extracellular domain promotes receptor dimerization, resulting in activation of the tyrosine kinase domain and autophosphorylation of the cytoplasmic domain. Receptor dimerization can take place with an identical ErbB receptor as dimerization partner (homodimerization) or another member of the ErbB receptor family (heterodimerization) (FIG. 17.2). Dimerized receptors undergo phosphorylation of specific tyrosines, which in turn results in the recruitment of downstream signaling proteins. Despite being a ligand-orphan receptor, HER2 is the preferred coreceptor for EGFR, HER-3, and HER-4. Conversely, HER-3 binds a number of ligands, but because it has a defective tyrosine kinase, HER-3 requires heterodimerization with another ErbB member to be activated.[2,23,24]

The preference of HER2 for heterodimerization within the ErbB receptor family explains how HER2 signals in the absence of a cognate ligand.[25] The formation of HER2-based heterodimers relies initially on the binding of a ligand to the HER-2 receptor partner. The heterodimers between HER-2 and the other ErbB receptors have relatively high ligand affinity, potent signaling activity, and are synergistic for cell transformation.[26–28] These features may be related to the ability of HER-2 to decelerate the rate of ligand dissociation that would imply a prolonged signaling by all ErbB family ligands. The best example of the ability of HER-2 to transactivate signaling initiated by ligands that bind to other ErbB receptors is that the HER-2/HER-3 heterodimers exhibit an extremely potent mitogenic activity, whereas HER-3 homodimers are inactive.[29] Overall, HER-2 is proposed as the master coordinator of the signaling network that functions as a shared coreceptor for ErbB ligands rather than as a receptor that mediates the action of a specific (unidentified) ligand. In addition to heterodimerization, it is thought that when human HER2 is overexpressed, the receptor also forms active dimers or oligomers of HER-2 (homodimerization) in the absence of ligand (see Chapter 16). To add further complexity to ErbB signaling, transactivation of ErbB receptors may be also related, directly or indirectly, by G protein–coupled receptors, integrin receptors, cytokine receptors, or other receptor tyrosine kinases.[30]

Phosphorylated tyrosine residues of the cytoplasmic domain (as a consequence of receptor dimerization) serve as docking sites of various signal transducers and adaptor molecules. These events are followed by the activation of diverse signaling pathways, such as Ras-Raf-MAPK, PI3K-Akt, PLC-γ1, Src,

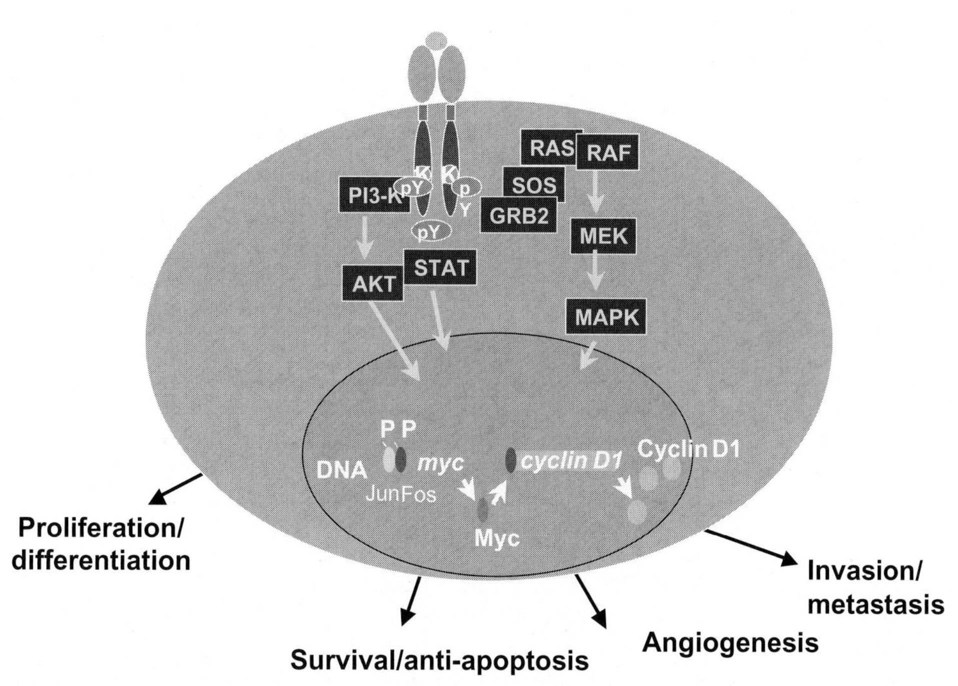

FIGURE 17.3 Signaling transduction pathways activated by ErbB receptors.

STATs, and others[2] (FIG. 17.3). Ultimately, the ErbB signaling family regulates cell proliferation, survival, angiogenesis, and invasion[2,23,24] (FIG. 17.3). Each ligand/receptor complex may activate different signaling pathways that elicit specific cellular responses, resulting in an enormous signaling diversity.[2] The physiologic functions of the ErbB signaling network are the regulation of normal cell growth and differentiation by means of their role in mesenchyme–epithelial cross-talk and in the interactions between neurons and muscle, glia, and Schwan cells. However, the overexpression of certain receptors of the family or their ligands confer an oncogenic role to them.[23]

Significance in Human Breast Cancer

The ErbB receptors have an important role in the development of the normal breast[31] and are commonly dysregulated in premalignant and malignant breast diseases.[16] In fact, most members behave as oncogenes when they are overexpressed or mutated.[32] A series of observations suggested that the autocrine pathway constituted by the EGFR and TGF-α have an important role in breast cancer (FIG. 17.2). EGFR and receptor ligand expression are found in breast carcinoma cell lines and in primary breast carcinoma. High levels of EGFR are generally associated to an enhanced level of EGFR messenger RNA or protein expression, whereas gene amplification is thought to be uncommon.[33] Three mutated forms of the EGFR, named EGFRvI, vII, and vIII, have been also identified and are the result of three different deletions of the extracellular domain of the receptor. EGFRvIII is a tumor-specific, ligand-independent, constitutively active variant of the EGFR that has been involved in breast cancer progression in preclinical models and is frequently detected in human breast cancers.[34,35] Its significance is under study.

EGFR and Prognosis in Breast Cancer

An overview of 28 published studies[36–63] that evaluated the impact of EGFR expression on breast cancer prognosis in 7384 patients is shown in **Table 17.1**. Of the 27 studies that performed univariate statistical analysis, 19 (70%) reported that EGFR expression detected either by radioligand binding assay or immunohistochemistry correlated with either disease-free or overall survival in various groups of breast cancer patients, including both node-negative and node-positive cohorts. Of the 26 studies listed in **Table 17.1** that performed multivariate analysis, only 7 (27%) found that EGFR expression was an independent predictor of breast cancer outcome. Thus, based on these results, EGFR expression testing for patient management in breast cancer has not achieved widespread use at the current time (see below). A significant number of

Table 17.1 Summary of Selected Studies of EGFR Expression and Breast Cancer Prognosis

Year	Author	Reference	No. of Patients	Specimen Type	Disease Outcome Statistical Correlation: Univariate Significance	Disease Outcome Statistical Correlation: Multivariate Independent Significance	Comment
1985	Sainsbury	36	108	Radio ligand binding	—	—	EGFR overexpression correlated with high tumor grade, negative ER status but not with nodal status
1987	Macias	37	72	Radio ligand binding	Yes	—	EGFR status correlated with ploidy and cell proliferation rate and disease relapse
1987	Sainsbury	38	135	Radio ligand binding	Yes	Yes	EGFR status was the most important predictor of early recurrence and disease related death
1989	Grimaux	39	68	Radio ligand binding	Yes	Yes	Studied LN+ patients only, EGFR status was an independent predictor of survival
1990	Spyratos	40	109	Radio ligand binding	Yes	Yes	EGFR was the only independent predictor of survival in patients who did not receive adjuvant therapy
1990	Nicholson	41	231	Radio ligand binding	Yes	Yes	EGFR status was independent of size, grade and stage in predicting outcome of LN– patients
1991	Toi	42	135	Radio ligand binding	Yes	No	EGFR inversely correlated with ER and predicted outcome in LN+, but not in LN– patients
1992	Bolla	43	303	Radio ligand binding	No	No	Only PR was an independent predictor of prognosis
1992	Gasparini	44	165	IHC	Yes	No	EGFR status predicted relapse free survival on univariate analysis only
1993	Koenders	45	459	Radio ligand binding	Yes	No	EGFR was a predictor of overall and disease-free survival
1994	Noguchi	46	93	IHC	Yes	No	Only LN status was an independent predictor of survival
1994	Fox	47	370	Radio ligand binding	Yes	No	EGFR status predicted relapse free survival, but not overall survival on univariate analysis
1994	Klijn	48	214	Radio ligand binding	Yes	No	EGFR prediction of prognosis significant at 5 years follow-up is lost at 10 years
1994	Bolla	49	229	Radio ligand binding	No	No	Tumor grade and stage were the only significant prognostic factors
1995	Newby	50	88	IHC and Radio ligand binding	Yes	No	EGFR was a significant predictor on univariate analysis when measured by IHC but not when measured by the radioligand binding assay
1995	Pirinen	51	213	IHC	No	No	Tumor size, PR, and node status were independent predictors
1995	Gohring	52	244	IHC	No	No	This study also did not correlate EGFR status with ER status
1996	Seshadri	53	345	Radio ligand binding	Yes	Yes	EGFR overexpression also correlated with ER– status but did not predict prognosis in ER– patients
1996	Hawkins	54	215	Radio ligand binding	No	No	Only tumor grade and ER status were significant prognostic factors

Table 17.1 Summary of Selected Studies of EGFR Expression and Breast Cancer Prognosis (*continued*)

					Disease Outcome Statistical Correlation		
Year	Author	Reference	No. of Patients	Specimen Type	Univariate Significance	Multivariate Independent Significance	Comment
1997	Eissa	55	100	Radio ligand binding	No	No	Only HER-2/neu and p53 status predicted disease outcome
1997	Torregrosa	56	825	Radio ligand binding	Yes	Yes	HER-2/neu status was the most significant variable and EGFR independently predicted overall survival
1997	Schroeder	57	111	IHC	Yes	No	EGFR status not an independent predictor at 5 years follow-up
1999	Neskovic-Konstantinovic	58	106	Radio ligand binding	Yes	No	EGFR was a powerful predictor of antiestrogen therapy response but a weak prognostic factor
2000	Umekita	59	173	IHC	Yes	No	EGFR was an independent predictor of survival when combined with TGF-α expression status
2000	Pawlowski	60	365	RT-PCR	Yes	No	Only size and ER/PR status were independent variables
2001	Gaci	61	99	Radio ligand binding	No	No	Tumor grade, cathepsin D, and HER-2/neu status were the only significant factors related to prognosis
2001	Ferrero	62	780	Radio ligand binding	No	No	Only tumor size, grade, and node status were independent predictors
2002	Tsutsui	63	1029	IHC	Yes	Yes	Combining EGFR+ with ER− formed a subgroup with worst prognosis that was an independent predictor of outcome in node+, node−, and combined groups

IHC, immunohistochemistry; LN, lymph node.

the studies described the consistent inverse relationship between EGFR expression and estrogen receptor (ER) expression. For example, one recent study reported that EGFR overexpression was an independent predictor of prognosis only when it was linked to ER-negative status.[63]

In a comprehensive review of studies that assessed EGFR expression in breast cancer, rates of expression ranged from 15% to 91%, with a median percentage of 48%.[64] Both relapse-free survival and total survival appeared to be significantly shorter for EGFR-positive as compared with EGFR-negative tumors.[65,66] A literature review pointed at EGFR as a modest prognostic indicator in breast cancer. In that review, just over half of the published studies reported a relationship between increased EGFR expression and poor survival.[65] Interestingly, EGFR appeared to be a better prognosticator for short-term (2-year) than for long-term (5-year) survival, suggesting the possibility that the EGFR may drive the rapid proliferation of a subset of tumor cells. Definitive studies in well-defined subsets of patients are needed to clarify the role of the EGFR. Along this line, a recent study in 82 patients with locally advanced breast cancer treated at MD Anderson with neoadjuvant FAC chemotherapy reported that patients with immunohistochemically scored EGFR-positive disease had a significantly worse disease-free survival rate (9-year rates: 43% vs. 58%) and overall survival rate (9-year rates: 43% versus 60%) compared with those with EGFR-negative disease.[67] Collectively, the field of EGFR as a prognostic factor in breast cancer warrants further studies in homogeneous patient populations and using accepted methods of measurement.

EGFR and Other Prognostic Variables

High levels of EGFR have been also correlated to advanced clinical stage, high proliferation rates, and failure to respond to endocrine treatment.[64,68] Furthermore, there is an inverse correlation between EGFR and ER expression.[64] The reason for the high expression of EGFR in ER-negative breast cancer cells may be a selective stimulation of a transcription-enhancing element located in the first intron of the EGFR gene in ER-negative cells.[69,70] Increased EGFR expression and activation also evolves during treatment with antihormones, limiting their efficacy and promoting resistance.[71,72] Elevated levels of EGFR and TGF-α were found on breast tumor specimens at the time of relapse or progression to hormone treatment compared with the basal expression levels, thus suggesting a causal relationship between the development of hormone resistance and the acquisition of EGFR/TGF-α expression.[71] Expression of EGFR has been also linked to decreased response to chemotherapy.[73] All these observations are of scientific interest and suggest important roads of further research but are yet insufficiently powered to have implications in routine clinical practice.

EGFR and Benign Breast Disease, Breast Cancer Histology, and Other Conditions EGFR expression has been linked to a variety of breast cancer subtypes and other noncarcinomatous tumors. Overexpression of EGFR has been associated with ductal carcinoma in situ in a similar pattern as HER-2 (see Chapter 16).[74,75] EGFR expression in ductal carcinoma in situ is found in lesions with high proliferation rates and high grade histology.[74,75] EGFR expression is highest in moderate to poorly differentiated solid and tubular infiltrating ductal carcinomas.[76] In well-differentiated ductal carcinoma and ductal carcinomas with favorable prognosis, including papillary, mucinous, and medullary carcinomas, EGFR expression levels are lower than in solid infiltrating carcinoma.[77] Similarly, although EGFR overexpression has been linked to adverse outcome in infiltrating lobular carcinoma, the frequency of EGFR overexpression in lobular lesions is significantly lower than that seen in ductal tumors.[78] Similar to the HER-2 protein, overexpression of EGFR has been described in Paget's disease of the nipple, but this observation has not reached the consistency and level of predictive capability that HER-2 status has achieved in this association.[79] Similar to HER-2, EGFR expression has not been consistently linked to disease outcome in male breast cancer.[80] Although EGFR expression has been described in breast sarcomas, including phylloides tumors, its role in the pathogenesis and progression of these tumors is not well characterized.[81] Finally, EGFR expression has been consistently observed in benign typical and atypical proliferative lesions of the breast, which has led to the concept of targeting this tyrosine kinase receptor pathway in breast cancer chemoprevention trials.[82]

EGFR and Its Relationship with Other EGFR (HER) Family Members in Breast Cancer Unlike the EGFR, HER-2 has an established role in breast cancer management. HER-2 is overexpressed, most commonly by gene amplification, in approximately 20–25% of human breast cancers, and the adverse clinical implications of such overexpression are discussed in detail in Chapter 16.[1,83] Ligand-dependent HER-2 activation (i.e., heterodimerization in the absence of HER2 overexpression) may be more frequent in breast cancer compared with rate of specific HER2 gene amplification and HER2 protein overexpression. HER-3 is overexpressed in about 20% of breast cancers, and the frequent coexpression of HER-2 and HER-3 suggests a role for the heterodimer in carcinogenesis.[29] Increased HER-2 expression has been also linked to endocrine treatment resistance and to a poor prognosis.[71] In contrast to the oncogenic role of HER-1–3, HER-4 has been shown to mediate heregulin-dependent antiproliferative responses in human breast cancer cell lines.[84,85] Expression of HER-4 in human breast tumors appears to be associated with differentiation rather than with proliferation and an improved prognosis,[85] although not all studies support this notion.[86] The clinical role of NRG expression in breast cancer is yet unclear.[16,21]

Given the rich cross-talk in the ErbB signaling family, studies assessing the expression of multiple members of the ErbB family (both receptors and ligands) are needed to assess the significance of ligand/receptor expression and of potential homo/heterodimers in human breast cancers. A recent immunohistochemical study analyzed the expression and prognostic significance of the four EGFR receptors.[85] In that series, overexpression of EGFR, HER-2, or HER-3 predicted for a significantly worse survival, whereas HER-4 predicted for a better survival. Overexpression of one or more of the following receptors, EGFR, HER-2, or HER-3, was reported in approximately 40% of tumors.[85] A trend toward a poorer prognosis in tumors co-overexpressing EGFR/HER-2 was found.[85] Another study supported the view that co-overexpression of EGFR and HER2 confers a worse prognosis than overexpression of only one of the receptors.[87] On the other hand, concomitant overexpression of HER-4 and other ErbB receptors was very rare,[85] although other investigators did not observe the same finding.[88] Additional studies of both EGFR activating ligands and EGFR receptor expression are emerging[88] and should help to establish that cross-talk that may occur in clinical samples.

EGFR and Downstream Signaling Despite extensive studies of EGFR expression, our understanding of EGFR dependent pathways in vivo in human tumors and on our ability to identify tumors that are truly EGFR driven is yet limited. A significant relationship between expression of EGFR and downstream molecules (i.e., activated MAPK (mitogen-activated protein kinase) has been reported in various tumor types, such as head and neck, breast, and gastric carcinomas.[89–92] However, some specimens have high levels of activated MAPK without high levels of EGFR, thus suggesting that other receptors, or Ras mutations, are activating MAPK. The reverse is also true. In a preliminary analysis, a significant relationship between expression levels of total EGFR was found with TGF-α, phosphorylated EGFR, phosphorylated MAPK, and Ki-67 but not with phosphorylated Akt in human breast cancer specimens.[90] Hopefully, additional studies and the use of transcriptional and proteomic profiling of tumor biopsies[13] will further aid in the identification of tumors and pathways that are EGFR dependent and that are likely to respond to EGFR inhibition.

Clinical Testing Methods

As discussed above, there is not enough evidence to use expression of EGFR, HER-3, HER-4, or receptor ligands such as TGF-α as prognostic or predictive factors to be incorporated in clinical practice. However, it is possible that the true role of EGFR or other members of the family is currently underestimated because the published studies lack standardized methods of testing and interpretation (i.e., cutoffs) and were generally performed in heterogeneous patient populations.[64,65] Nonetheless, the current interest in developing EGFR targeted compounds for breast cancer therapy, as well as in many other cancers, has strengthened the need to develop reliable methods to assess expression of EGFR and other relevant markers. In trials performed to date, the response rates to EGFR targeted therapies across multiple tumors tested has ranged from 0% to approximately 20% (see below).[93,94] This figure reflects the need to improve our understanding of which tumors are truly dependent on the activation of EGFR and, consequently, which patients are most likely to benefit from EGFR targeted therapies.[95] Recently, two independent studies of patients with non-small cell lung cancer co-discovered specific point mutations in the EGFR gene that was associated with activation of tyrosinase kinase activity and selective response to the anti-EGFR small molecule drug gefitinib (Iressa).[96,97] The EGFR is obviously the most studied putative predictor of response. EGFR may be measured at the DNA, RNA, or protein levels, each by different methods. However, for practical reasons and because high expression of EGFR is thought to be related to posttranslational modifications, immunohistochemistry has become the preferred method of study (FIG. 17.4). However, there are no published data to support a relationship between the level of expression of EGFR and response to EGFR targeted compounds.[98]

An alternative, and perhaps better, method to study the potential prognostic or predictive

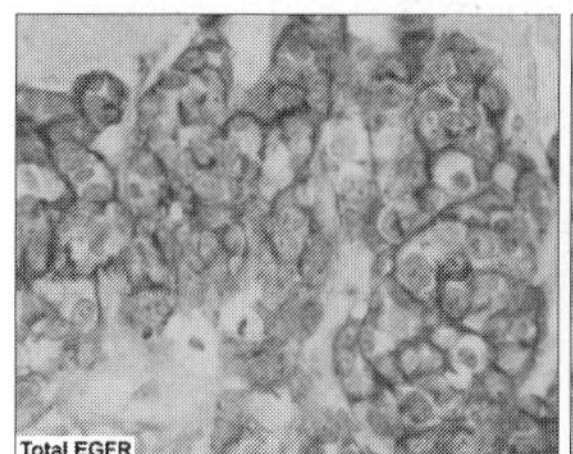

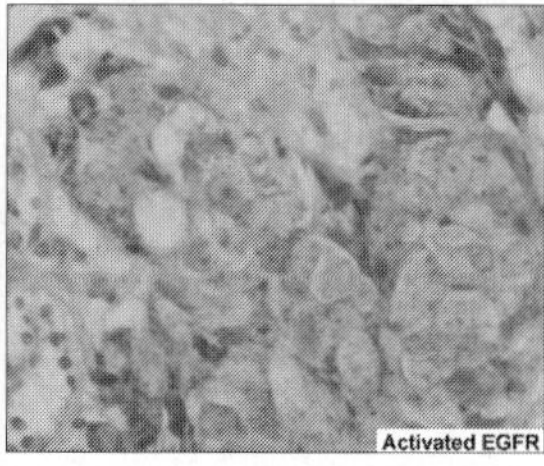

FIGURE 17.4 Example of human breast cancer specimens that expresses high levels of total EGFR (left) and activated EGFR (right). To detect the activated phosphorylated form of the EGFR, an antibody that recognizes the ligand-activated and phosphorylated form of the EGFR and does not recognize other tyrosine phosphorylated proteins was used (see Refs. 60 and 62 for details). Note the intense membranous staining in most tumor cells for total EGFR (left). Activated EGFR (right) had a both membranous and cytoplasmic staining, with some clustered areas and different level of expression among tumor cells. This pattern of staining for the activated form of the receptor is similar to the one observed in cancer cell lines. (Courtesy Dr. F. Rojo, Department of Pathology, Hospital Vall d'Hebron, Barcelona, Spain.) **See Plate 40 for color image.**

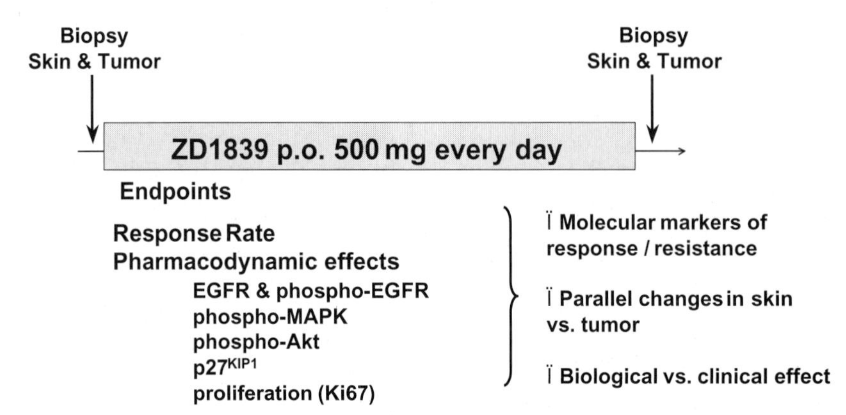

FIGURE 17.5 Design of a phase II and pharmacodynamic trial of the EGFR tyrosine kinase inhibitor gefitinib in advanced breast cancer patients previously treated with chemotherapy. (Adapted from Ref. 90.)

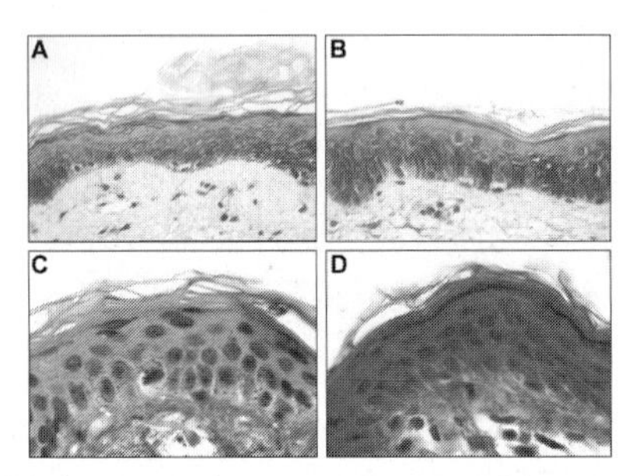

FIGURE 17.6 Pharmacodynamic assessments of the EGFR tyrosine kinase inhibitor gefitinib in skin biopsies. (A) The expression of total EGFR is evident in epidermal keratinocytes, mainly in the basal and suprabasal layers in a pretherapy skin biopsy. (B) In a biopsy obtained at day 28 of oral treatment with gefitinib, no changes in total EGFR expression were noted. (C) Activated EGFR is found in epidermal keratinocytes, mainly in the basal and suprabasal layers in a pretherapy skin biopsy. Activated EGFR was seen in all the EGFR-positive cell types present in the skin. The pattern of staining for the activated EGFR was granular and mainly cytoplasmic, suggesting that the activated receptor is internalized and located in intracellular vesicles in vivo in human skin cells. (D) In a biopsy obtained at day 28 of oral treatment with gefitinib, activated EGFR was undetected, indicating that this compound inhibits its target in vivo. (Adapted from Ref. 101.) **See Plate 41 for color image.**

value of the EGFR is the immunohistochemical assessment of activated phosphorylated EGFR, by using antibodies that recognize only the active forms of the receptor, instead of methods that assess the levels of total EGFR (that may be not active) (FIGS. 17.4–17.6).[65,99–101] Notably, a considerable proportion of breast cancer tumors express total EGFR, but the activated form of the receptor is frequently undetected by the currently used immunohistochemical methods. Emerging data suggest that the expression of phosphorylated forms of a receptor may offer additional prognostic information to assessments of total receptor levels.[102] In addition, expression of phosphorylated EGFR may also reflect the type of activating point mutation associated with drug response in lung cancer.[96,97] The data of activated EGFR as a predictor of response to EGFR targeted agents is still limited. A challenging observation is that in post-treatment biopsies, the activation of EGFR was inhibited in both responding and nonresponding patients treated with the anti-EGFR antibody EMD-72000 or in patients treated with the EGFR tyrosine kinase inhibitor gefitinib.[90,103] This observation indicates that, in contrast to a subset of patients with lung cancer featuring a specific activating mutation of the EGFR gene, activated receptor inhibition may not be sufficient for clinical response, thus suggesting the potential role of downstream molecules in defining the sensitivity to EGFR inhibitors. Diverse studies with a variety of EGFR targeted agents have supported the use of other downstream surrogates such as MAPK, Akt, p27, cyclin D, and c-*fos* as an additional measure of EGFR pathway inhibition.[89,100,104,105] The expression of EGF ligands or HER-3 and HER-4 has been also assessed by various methods, and there is an urgent need for standardization to clarify their significance in breast cancers and potentially use them in the clinic for prognostic or predictive medicine in the future.

Targeting ErbB Signaling Family Members for Breast Cancer Treatment

Based on the considerations discussed above, EGFR and HER2 were selected as targets for cancer therapy[14,17,106–108] (**Table 17.2**). HER-3 and HER-4 were considered less appropriate specific targets because HER-3 is a kinase deficient receptor and HER-4 might have

Table 17.2 Rationale to Investigate EGFR Tyrosine Kinase Inhibitors in Breast Cancer Patients

- Expression of EGFR correlates with a worse prognosis and ER-negative tumors.
- EGFR is linked to resistance to chemotherapy and hormonal therapy.
- The tyrosine kinase activity of the receptor is key for the activation of the receptor.
- EGFR tyrosine kinase inhibitors are active in cultured and xenografted human breast cancer cells.
- Selective EGFR tyrosine kinase inhibitors are generally well tolerated in patients.

Table 17.3 ErbB Receptor Inhibitors in Clinical Development

Selective EGFR Inhibitors	Dual or Pan-ErbB Inhibitors
Gefitinib (reversible)	GW 572016 (EGFR/HER2/reversible)
Erlotinib (reversible)	CI-1033 (pan-ErbB/irreversible)
EKB-569 (irreversible)	

antiproliferative actions. Targeting the ligands is, in principle, of more limited value because of the extensive number of them.

Early studies demonstrated that monoclonal antibodies directed at EGFR and HER2 receptors inhibit breast cancer growth in laboratory model systems. Later on it was shown that trastuzumab (Herceptin), a recombinant humanized anti-HER2 antibody, is effective for the treatment of women with advanced breast cancers that express high levels of HER2. Trastuzumab is the only anti-HER2 antibody that has been approved for breast cancer therapy for use as a single agent or in combination with chemotherapy[109] (see Chapter 16). However, trastuzumab activity relies on the presence of HER2 overexpression, whereas it is not active against tumors that express moderate or normal levels of HER2. Although overt HER2 overexpression occurs only in about 25% of human breast tumors, other members of the ErbB receptor family are commonly dysregulated in many breast cancers. Therefore, agents targeting other EGFR family members or even monoclonal antibodies, such as 2C4,[110,111] that can disrupt HER2 association with other ErbB receptors, have been produced and might have broader clinical applications than trastuzumab.[111,112]

An initial study with the anti-EGFR monoclonal antibody C225 (cetuximab; Erbitux) plus paclitaxel in breast cancer was halted because of folliculitis.[113] Additional breast cancer studies with C225 or other anti-EGFR antibodies in clinical development would be of clear interest.[3] In addition to monoclonal antibodies, additional strategies including the use of specific tyrosine kinase inhibitors have also shown promising activity in preclinical models and are being explored in the clinic.[93,114] Initial development of ErbB receptor tyrosine kinase inhibitors was based on data that revealed that the tyrosine kinase activity of the EGFR was required for many of the biochemical responses induced by this receptor. A potential advantage of ErbB receptor tyrosine kinase small molecule inhibitors over monoclonal antibodies is the ability to enter into the cell and to act on intracellular active forms of the receptor, such as EGFRvIII,[34] or truncated HER2 receptor ($p95^{HER2}$).[114–118] Quinazoline compounds, such as PD153035[119] and AG1478, represent the first of a class of competitive ATP inhibitors that are highly selective for the EGFR kinase. Since then, a series of agents with different degrees of specificity toward single or multiple ErbB receptors has been produced. The inhibition of the receptor tyrosine kinase can be also reversible or irreversible. Because of the increasing complexity of the compounds available, a classification of these agents has been recently proposed[112] (**Table 17.3**). A variety of these small molecule inhibitors are in clinical trials for breast cancer.

Specific EGFR Tyrosine Kinase Inhibitors

Gefitinib (ZD1839; Iressa) and erlotinib (OSI-774; Tarceva) are two orally active, highly specific, and reversible EGFR tyrosine kinase inhibitors that are being investigated in patients with breast cancer. These agents inhibit EGFR tyrosine kinase at concentrations that do not affect other kinases tested.[104] Notably, these agents are able to inhibit in vivo HER-2 phosphorylation and the growth of HER-2-overexpressing breast cancer cells. This is possibly due to sequestration of HER-2 and HER-3 receptors in an inactive heterodimer configuration with the EGFR.[120–122] This class of compounds also has antitumor activity against human breast cancer xenografts. In addition to the antitumor effect, studies performed in tumors excised at the end of treatment of xenografted mice have shown inhibition of EGFR phosphorylation, MAPK phosphorylation, and reduced proliferation rates.[123] Interestingly, Akt activation was also reduced but to a lesser extent than EGFR or MAPK. As discussed later, a similar finding was found in patients with breast cancer treated with gefitinib.[90] In addition to the single agent activity of these agents, preclinical studies have demonstrated that combined treatments with chemotherapeutic agents, antihormonal drugs, or biologic agents (i.e., trastuzumab, mTOR inhibitors, and others) frequently result in enhanced antitumor effects. Another specific EGFR is EKB-569.[85] This agent differs from gefitinib and erlotinib in that is an irreversible inhibitor of the EGFR and is currently in a less advanced stage of clinical development.

The early clinical development of these agents incorporated pharmacodynamic studies (FIGS. 17.5 and 17.6). Pharmacodynamic studies may be useful to define the biologic response to the treatment, to discover biomarkers that predict response, and to define potential mechanisms of therapeutic resistance. The early pharmacodynamic studies were performed in serial (pretreatment and on-therapy) skin biopsies.[101] Skin was the selected tissue for pharmacodynamic studies due to its well-characterized EGFR dependence. In most patients treated with gefitinib, inhibition of EGFR activation was achieved in vivo during treatment (FIG. 17.6). The expression of activated MAPK was also reduced and keratinocyte proliferation decreased, whereas concomitantly there was an increase in the CDK inhibitor p27. All these biologic effects were profound at doses well below the one associated with unacceptable toxicity. In another study, erlotinib also resulted in inhibition of phosphorylated EGFR and p27 up-regulation in skin biopsies.[125] Collectively, these studies were useful to demonstrate that the orally administered drugs can elicit a full range of biologic effects associated with effective EGFR inhibition in vivo in a surrogate tissue and provided basis to use optimal biologic dose instead of maximal tolerated doses for phase II/III trials.

Several studies of gefitinib and erlotinib in breast cancer have been reported recently in abstract form. All the phase II gefitinib studies have selected a dose of 500 mg/day. Pooling the data presented at meetings in the four gefitinib trials reported to date, a total of 167 patients were treated. In this population, the overall clinical benefit rate was 10.7% (4 patients with partial responses + 14 patients with prolonged stable disease).[90,126–128] Similar results have been reported with erlotinib. An initial phase II study included 69 patients with previously treated locally advanced or metastatic breast cancer that received the recommended erlotinib dose of 150 mg/day. One partial response was seen in a heavily pretreated patient, and three patients experienced stable disease.[129] Further studies in less heavily pretreated patients and in combination with other therapies are underway. In another pilot and pharmacodynamic trial with erlotinib, the biologic effect was demonstrated by decreased thickness in the stratum corneum and reduced Ki-67 in the epidermis but not in tumor tissue. However, 9 of 10 tumors were EGFR negative. In this heavily pretreated group, there was no evidence of tumor response, which may be due to the absence of EGFR expression.[130] Gefitinib and erlotinib are generally well tolerated. The more common side effects occur in skin and in the gastrointestinal tract, whereas the typical toxicities of conventional chemotherapy are mostly lacking.[114,131]

The low rates of clinical benefit establish that in unselected populations of women with advanced or metastatic breast cancer, the single agent activity of these anti-EGFR agents at the doses tested is anecdotal, although definite partial responses were seen. How should we interpret these results? A simplistic view could be that the level of clinical activity is below the conventional threshold considered for further development. However, open questions and preliminary results suggest that these agents merit a deeper analysis. Two concepts emerge from the available studies that need to be taken into account for further trial design. First, biologic activity is achieved in EGFR-positive breast tumor samples, as assessed by a gefitinib-induced loss/decrease of phosphorylated EGFR and decrease in phospho-MAPK (FIG. 17.5). In the same specimens, effects on phospho-Akt and Ki67 were not significant.[90] The demonstration of a biologic effect, although not sufficient to elicit a clinical benefit in most patients, may offer a scientific rationale for further combination strategies. Based on this evidence and on the lack of inhibition of Akt, combined treatments of EGFR inhibitors and agents targeting the PI3K/Akt pathway are justified. Second, selection of patients based on clinical or biologic criteria may identify patients with a higher chance of responding. A study suggested that the subset of ER-positive breast cancer patients with clinically acquired tamoxifen resistance may derive a considerable clinical benefit from gefitinib. In particular, 60% of patients who had ER-positive/tamoxifen-resistant disease had clinical benefit (one partial response and five stable disease more than 6 months out of nine patients).[128] In a recent analysis of determinants of response to gefitinib in breast cancer, the tumors of the patients that responded all expressed HER-2 only at low levels and exhibited a weak activation of MAPK and Akt. The clinical benefit rate to gefitinib in patients who had all three features was 20%.[132]

It is conceivable that as we learn more on the clinical and molecular characteristics that identify patients likely to respond to anti-EGFR drugs including the wider use of the lung cancer-based test for gefitinib responder activating mutations, we may eventually identify a subset of breast cancer patients with a high likelihood of benefit.[94,133] It has been also speculated that in addition to inhibiting the EGFR, these kinase inhibitors may affect also other cellular targets. In addition to the use of these drugs as single agents, many clinical studies are addressing the role of combining them with hormonal agents, biologic agents (i.e., trastuzumab, bevazucimab, etc.), or chemotherapy.

Dual or Pan-ErbB Receptor Tyrosine Kinase Inhibitors Many human breast cancer express multiple members of the ErbB receptor signaling family, a fact that points to the interest of the clinical testing of inhibitors of both EGFR and HER-2 (dual inhibitors) or inhibitors of the four receptors (pan-ErbB inhibitors) of the family, compared with inhibitors of single receptors (**Table 17.4**). Furthermore, increased cellular HER-2/HER-3 signaling can overcome selective EGFR inhibition. Novel anti-HER-2 antibodies such as 2C4 (Omnitarg), which prevents HER-2 heterodimers, are also of interest. In support of the potential benefit of targeting more than one receptor, for example, treatment with gefitinib plus the anti-HER-2 antibody trastuzumab induced a greater apoptotic effect than either inhibitor alone in HER-2-overexpressing human breast cancer cells.[122]

GW572016 is a novel reversible inhibitor of the tyrosine kinase activity of both EGFR and HER-2.[134,135] In a phase I pharmacodynamic study in patients with advanced cancers, oral treatment resulted in intratumor inhibition of activated phospho-Akt. This was associated with tumor cell apoptosis and regression of metastasis. Moreover, inhibition of intratumor expression of activated MAPK and cyclin D protein predicted clinical response to GW572016 (e.g., partial response, stable disease). In general, lack of biologic response was associated with disease progression. Notably, two trastuzumab-resistant breast cancer patients had objective responses.[136] A phase II study is ongoing in patients with disease progression after one or two prior treatments with trastuzumab alone or in a combination chemotherapy regimen.[137] Another agent of interest is CI-1033, an orally available irreversible pan-erbB receptor tyrosine kinase inhibitor that effectively blocks signal transduction through all four members of the erbB family.[138] Preclinical data have shown that CI-1033 is active against a variety of human tumors in mouse xenograft models, including breast carcinomas. CI-1033 has been shown to have an acceptable side effect profile at potentially therapeutic dose levels and demonstrates evidence of target biomarker modulation.[139–142] Antitumor activity has also been observed, including one partial clinical response and stable disease in over 30% of patients, one patient with heavily pretreated breast cancer. Further clinical studies are needed to clarify whether dual or pan-ErbB tyrosine kinase inhibitors such as GW572016 or CI-1033 will translate into benefits over specific EGFR inhibitors and whether they have potentially enhanced adverse events.

An additional strategy is the use of inhibitors that affect ErbB receptors and also the tyrosine kinase activity of other transmembrane receptors such as vascular endothelial growth factor receptor (VEGFR; e.g., the agent AEE788[143]) or PDGF receptor (e.g., the agent ZD6474[144]). Additional studies with anti-EGFR or anti-HER-2 antibodies are also eagerly

Table 17.4 Rationale to Investigate Dual/Pan-ErbB Tyrosine Kinase Inhibitors in Breast Cancer Patients

- ErbB heterodimers are more potent than ErbB homodimers.
- Coexpression of different ErbB receptors in human breast cancer specimens is common.
- Combined treatments of specific anti-EGFR agents and anti-HER2 agents result in greater antitumor activity than when each agent is used alone.

awaited.[110,111,114] Another field of research is the potential use of EGFR-targeted agents in premalignant lesions such as ductal carcinoma in situ.[145] It is anticipated that the ongoing and planned clinical studies with ErbB receptor targeted agents will shed light on the potential use of these agents, alone or combined with other treatments, for patients with breast cancer.

Other Growth Factors and Breast Cancer

IGF System

The IGF system comprises the insulin-like growth factor receptors (IGF-IR and IGF-IIR), two ligands (IGF-I and II), IGF-binding proteins (IGFBPs), and IGFBP proteases.[4] The best characterized receptor of the family is IGF-IR, a transmembrane tyrosine kinase receptor, closely related to the insulin receptor, which is activated by IGF ligands. The binding of its ligands (IGF-I, IGF-II) to the extracellular domains of IGF-IR activates its intracellular tyrosine kinase domain, resulting in autophosphorylation of the receptor. Activated IGF-IR phosphorylates its substrates and initiates proliferative and antiapoptotic signal transduction pathways that involve PI3K/Akt and Ras/Raf/MAPK.

Preclinical evidence indicates that the IGF-IR plays an important role in cancer. This receptor is important for oncogenic transformation and for cell survival. Genetic approaches leading to down-regulation of activation of the IGF-IR function resulted in inhibition of tumor formation. In addition to its role in proliferation of cancer cells, the IGF-IR protects cells from apoptosis caused by growth factor deprivation, anchorage independence, or cytotoxic drug treatment. IGF-IR also confers an enhanced metastatic potential to malignant cells. As far as breast cancer is concerned, diverse lines of evidence indicate that this receptor plays an important role in this tumor[4,145,146] and IGFs exert proliferative and antiapoptotic effects in many breast cancer cell lines[147,148]; targeted disruption of the IGF-I receptor, by using antireceptor antibodies or antisense to this receptor, limits breast cancer proliferation[149]; the IGF-IR and its ligands are expressed in many human breast tumors; and high levels of circulating IGF-I predict increased breast cancer risk in premenopausal women.[150] IGF-IR is frequently coexpressed with ERs,[151] and estrogens stimulate the expression of the IGF-IR. In primary breast tumors, the IGF-IR may be overexpressed and hyperphosphorylated, and this has been correlated with radioresistance and tumor recurrence.[146] However, a link between IGF-IR expression and disease-free or overall survival in breast cancer has not been yet established. The IGF-IIR lacks tyrosine kinase activity, and its role in breast cancer is less characterized than for IGF-IR.[4]

Recent data indicate a critical role of IGF-IR signaling in the response to ErbB receptor targeted agents.[152] In human cancer cells, an increased level of IGF-IR signaling adversely interfered with trastuzumab action on cell growth.[153] Conversely, the addition of IGFBP-3, which decreased IGF-IR signaling, restored the ability of trastuzumab to suppress growth. Based on these findings, it was proposed that strategies that target IGF-IR signaling may prevent or delay development of resistance to trastuzumab. A link between IGF-IR signaling in the modulation of response has been shown also for the EGFR targeted agents C225 and gefitinib.[154,155] Searching for interactions between both receptor systems (ErbB receptors and IGF-IR) will not be an easy task, because diverse study models have revealed different hierarchical cross-regulations between ErbB receptors and the IGF-IR.[30,156,157] However, a preliminary assessment revealed that IGF-IR is commonly coexpressed in HER-2/3-positive/fluorescence in situ hybridization–positive breast cancer.[158] Some degree of IGF-IR staining was seen in all the cases analyzed, but the cases varied with regard to the distribution, pattern, and intensity of staining. In a group of trastuzumab-naive patients, a patchy membranous IGF-IR staining was associated with less favorable response to trastuzumab/paclitaxel or trastuzumab/vinorelbine. This particular type of staining may represent clustered receptor complexes, leading to activated IGF-IR and relative resistance to trastuzumab.[158] Further studies are warranted to assess IGF-IR staining pattern in relation to response to trastuzumab-containing therapy.

Different strategies have been devised to interrupt IGF-IR activation, and some of them are ready for clinical testing.[159] In a similar fashion to EGFR targeted agents, the two classes of compounds in more advanced stage of development are monoclonal antibodies and tyrosine kinase inhibitors. Recently, a potent antagonistic anti-IGF-IR antibody (EM164) that binds to the extracellular domains of IGF-IR and inhibits receptor activation has been reported. EM164 inhibits the serum- and IGF-I–stimulated proliferation of diverse human cancer cell lines in vitro.[143,160] Activity in pancreatic cancer xenografts has been also noted. A humanized version of the EM164 antibody has been developed to target the IGF-IR function in cancer cells for human clinical trials. Another promising strategy to inhibit the function of IGF-IR is the development of selective and potent small molecular weight inhibitors of the kinase activity of the IGF-IR. Novel compounds that abolish receptor autophosphorylation and the consequent activation of the downstream target, Akt, have been recently reported. In agreement with their biochemical effects, these compounds also abrogate the ability of IGF-I to promote survival and anchorage-independent growth and have also activity in in vivo models.[161] Data on the clinical activity of IGF-IR targeted agents is expected in the near future, either as a single agent or in combination with other available treatments, in patients with breast cancer.

TGFs

TGF-α expression, an activating ligand for EGFR, has been associated with disease recurrence and adverse prognosis in breast cancer[162,163] and may mediate its effects through activation of the ER pathway.[164] TGF-β is a regulatory peptide that in addition to a role as a cell growth mediator is a potent stimulator of fibroblasts and extracellular matrix production.[165] The potential role(s) of TGF-β in breast cancer development, progression, and response to antiestrogen therapy are not completely understood at this time.[5,166–168] TGF-β expression has been linked to stromal proliferations in breast tissues[169] but has not been linked as a prognostic factor for epithelial breast malignancy.

PDGF

PDGF has been linked to the desmoplastic stromal response in breast cancer[170] and has been identified as a prognostic factor for the disease.[171] PDGF has become a target for therapy for some types of cancer using novel anti-tyrosine kinase small molecule drugs such as imatinib (Gleevec).[172]

Fibroblast Growth Factor

The fibroblast growth factor family, including the related *int-2* and *HST-1* genes, has been linked to breast cancer prognosis in some studies[173–175] but not in others.[176] Amplification studies of both *int-1* and its neighbor *HST-1* often include amplification of cyclin D1, which has led to the use of the term the 11q13 amplicon. In a comparative study of oncogene amplification and prognosis in breast cancer, *int-2* amplification was associated with an increased risk of breast cancer relapse, especially in node-negative patients and patients with positive ER and PR assays but failed to achieve independent status on multivariate analysis.[177] Serum monitoring of fibroblast growth factor levels have not been successful biomarkers of disease outcome.[178]

VEGF and Angiogenesis

VEGF, the most potent endothelial cell mitogen and a regulator of vascular permeability, and its various receptors have been extensively studied in breast cancer and associated with adverse prognosis in some studies[179–182] but not in others.[183–185] More recently, VEGF expression in breast cancer has been significantly associated with relapse-free survival, overall survival, or both.[186] Elevated VEGF levels may identify early stage breast cancers with a higher likelihood of recurrence or death and poor prognosis patients who benefit minimally from conventional adjuvant therapy but who may benefit from validated anti-VEGF treatments.[186,187] The number of microvessels in the richest vascular area of invasive breast cancer has been an inconsistent predictor of prognosis in breast cancer predominantly due to differences in the microvessel density measurement technique (see Chapter 19).[188,189] Tumor VEGF expression may be more reliable than microvessel density measurements as a predictor of angiogenesis and adverse prognosis.[190] To date, antiangiogenesis therapies, including small molecules, ribozymes, and antibodies, have failed to achieve significant efficacy for the treatment of metastatic breast cancer. A detailed consideration of VEGF expression, anti-VEGF therapies, and angiogenesis in breast is provided in Chapter 19.

Acknowledgment

Supported in part by grant SAF 2003-8181 from the Spanish Ministry of Health and Technology to J. A.

References

1. Ross JS, Fletcher JA, Linette GP, et al. The Her-2/neu gene and protein in breast cancer 2003: biomarker and target of therapy. Oncologist 2003;8:307–325.
2. Yarden Y, Sliwkowski MX. Untangling the ErbB signalling network. Nat Rev Mol Cell Biol 2001;2:127–137.
3. Mendelsohn J. Targeting the epidermal growth factor receptor for cancer therapy. J Clin Oncol 2002;20:1–13.
4. Sachdev D, Yee D. The IGF system and breast cancer. Endocr Relat Cancer 2001;8:197–209.
5. Dumont N, Arteaga CL. Transforming growth factor-beta and breast cancer: tumor promoting effects of transforming growth factor-beta. Breast Cancer Res 2000;2:125–132.
6. Ruohola JK, Viitanen TP, Valve EM, et al. Enhanced invasion and tumor growth of fibroblast growth factor 8b-overexpressing MCF-7 human breast cancer cells. Cancer Res 2001;61:4229–4237.
7. Yi B, Williams PJ, Niewolna M, et al. Tumor-derived platelet-derived growth factor-BB plays a critical role in osteosclerotic bone metastasis in an animal model of human breast cancer. Cancer Res 2002;62:917–923.
8. Wong SC, Lo SF, Lee KC, et al. Expression of frizzled-related protein and Wnt-signalling molecules in invasive human breast tumours. J Pathol 2002;196:145–153.
9. Kang JY, Dolled-Filhart M, Ocal IT, et al. Tissue microarray analysis of hepatocyte growth factor/Met pathway components reveals a role for Met, matriptase, and hepatocyte growth factor activator inhibitor 1 in the progression of node-negative breast cancer. Cancer Res 2003;63:1101–1105.
10. Badache A, Hynes NE. Interleukin 6 inhibits proliferation and, in cooperation with an epidermal growth factor receptor autocrine loop, increases migration of T47D breast cancer cells. Cancer Res 2001;61:383–391.
11. Muller A, Homey B, Soto H, et al. Involvement of chemokine receptors in breast cancer metastasis. Nature 2001;410:50–56.
12. Dickson R, Lippman M. Autocrine and paracrine growth factors in the normal and neoplastic breast. In: Harris J, Lippman M, Morrow M, et al., ed. Disease of the Breast. Philadelphia: Lippincott Williams & Wilkins, 2000:303–317.
13. Ross J, Linette G, Stec J, et al. Breast cancer biomarkers and molecular medicine. Expert Rev Mol Diagn 2003;3:573–585.
14. Slamon DJ, Clark GM, Wong SG, et al. Human breast cancer: correlation of relapse and survival with amplification of the HER-2/neu oncogene. Science 1987;235:177–182.
15. Press MF, Bernstein L, Thomas PA, et al. HER-2/neu gene amplification characterized by fluorescence in situ hybridization: poor prognosis in node-negative breast carcinomas. J Clin Oncol 1997;15:2894–2904.
16. Salomon DS, Brandt R, Ciardiello F, et al. Epidermal growth factor-related peptides and their receptors in human malignancies. Crit Rev Oncol Hematol 1995;19:183–232.
17. Albanell J, Baselga J. The ErbB receptors as targets for breast cancer therapy. J Mamm Gland Biol Neoplasia 1999;4:337–351.
18. Normanno N, Bianco C, De Luca A, et al. Target-based agents against ErbB receptors and their ligands: a novel approach to cancer treatment. Endocr Relat Cancer 2003;10:1–21.
19. Normanno N, Ciardiello F, Brandt R, et al. Epidermal growth factor-related peptides in the pathogenesis of human breast cancer. Breast Cancer Res Treat 1994;29:11–27.
20. Yarden Y. Biology of HER2 and its importance in breast cancer. Oncology 2001;61(suppl 2):1–13.
21. Normanno N, Bianco C, De Luca A, et al. The role of EGF-related peptides in tumor growth. Front Biosci 2001;6:D685–D707.
22. Borrell-Pages M, Rojo F, Albanell J, et al. TACE is required for the activation of the EGFR by TGF-alpha in tumors. EMBO J 2003;22:1114–1124.
23. Olayioye MA, Neve RM, Lane HA, et al. The ErbB signaling network: receptor heterodimerization in development and cancer. EMBO J 2000;19:3159–3167.
24. Lemmon MA, Schlessinger J. Regulation of signal transduction and signal diversity by receptor oligomerization. Trends Biochem Sci 1994;19:459–463.
25. Graus-Porta D, Beerli RR, Daly JM, et al. ErbB-2, the preferred heterodimerization partner of all ErbB receptors, is a mediator of lateral signaling. EMBO J 1997;16:1647–1655.
26. Pinkas-Kramarski R, Soussan L, Waterman H, et al. Diversification of Neu differentiation factor and epidermal growth factor signaling by combinatorial receptor interactions. EMBO J 1996;15:2452–2467.
27. Pinkas-Kramarski R, Alroy I, Yarden Y. ErbB receptors and EGF-like ligands: cell lineage determination and oncogenesis through combinatorial signaling. J Mamm Gland Biol Neoplasia 1997;2:97–107.
28. Lenferink AE, Pinkas-Kramarski R, van de Poll ML, et al. Differential endocytic routing of homo- and hetero-dimeric ErbB tyrosine kinases confers signaling superiority to receptor heterodimers. EMBO J 1998;17:3385–3397.
29. Alimandi M, Romano A, Curia MC, et al. Cooperative signaling of ErbB3 and ErbB2 in neoplastic transformation and human mammary carcinomas. Oncogene 1995;10:1813–1821.
30. Lee AV, Schiff R, Cui X, et al. New mechanisms of signal transduction inhibitor action: receptor tyrosine kinase down-regulation and blockade of signal transactivation. Clin Cancer Res 2003;9(1 Pt 2):516S–523S.
31. Stern DF. ErbBs in mammary development. Exp Cell Res 2003;284:89–98.
32. Di Marco E, Pierce JH, Aaronson SA, et al. Mechanisms by which EGF receptor and TGF alpha contribute to malignant transformation. Nat Immun Cell Growth Regul 1990;9:209–221.
33. Davidson N, Gelmann E. Epidermal growth factor receptor gene expression in estrogen receptor-positive and negative human breast cancer cell lines. Mol Endocrinol 1987;1:216–223.

34. Ge H, Gong X, Tang CK. Evidence of high incidence of EGFRvIII expression and coexpression with EGFR in human invasive breast cancer by laser capture microdissection and immunohistochemical analysis. Int J Cancer 2002;98:357–361.
35. Wikstrand CJ, Hale LP, Batra SK, et al. Monoclonal antibodies against EGFRvIII are tumor specific and react with breast and lung carcinomas and malignant gliomas. Cancer Res 1995;55:3140–3148.
36. Sainsbury JR, Malcolm AJ, Appleton DR, et al. Presence of epidermal growth factor receptor as an indicator of poor prognosis in patients with breast cancer. J Clin Pathol. 1985;38:1225–1228.
37. Macias A, Azavedo E, Hagerstrom T, et al. Prognostic significance of the receptor for epidermal growth factor in human mammary carcinomas. Anticancer Res 1987;7:459–464.
38. Sainsbury JR, Farndon JR, Needham GK, et al. Epidermal-growth-factor receptor status as predictor of early recurrence of and death from breast cancer. Lancet 1987;1:1398–1402.
39. Grimaux M, Romain S, Remvikos Y, et al. Prognostic value of epidermal growth factor receptor in node-positive breast cancer. Breast Cancer Res Treat 1989;14:77–90.
40. Spyratos F, Delarue JC, Andrieu C, et al. Epidermal growth factor receptors and prognosis in primary breast cancer. Breast Cancer Res Treat 1990;17: 83–89.
41. Nicholson S, Wright C, Sainsbury JR, et al. Epidermal growth factor receptor (EGFR) as a marker for poor prognosis in node-negative breast cancer patients: neu and tamoxifen failure. J Steroid Biochem Mol Biol 1990;37:811–814.
42. Toi M, Osaki A, Yamada H, et al. Epidermal growth factor receptor expression as a prognostic indicator in breast cancer. Eur J Cancer 1991;27:977–980.
43. Bolla M, Chedin M, Colonna M, et al. Prognostic value of epidermal growth factor receptor in a series of 303 breast cancers. Eur J Cancer 1992;28A: 1052–1054.
44. Gasparini G, Gullick WJ, Bevilacqua P, et al. Human breast cancer: prognostic significance of the c-erbB-2 oncoprotein compared with epidermal growth factor receptor, DNA ploidy, and conventional pathologic features. J Clin Oncol 1992;10:686–695.
45. Koenders PG, Beex LV, Kienhuis CB, et al. Epidermal growth factor receptor and prognosis in human breast cancer: a prospective study. Breast Cancer Res Treat 1993;25:21–27.
46. Noguchi M, Mizukami Y, Kinoshita K, et al. The prognostic significance of epidermal growth factor receptor expression in breast cancer. Surg Today 1994;24:889–894.
47. Fox SB, Smith K, Hollyer J, et al. The epidermal growth factor receptor as a prognostic marker: results of 370 patients and review of 3009 patients. Breast Cancer Res Treat 1994;29:41–49.
48. Klijn JG, Look MP, Portengen H, et al. The prognostic value of epidermal growth factor receptor (EGF-R) in primary breast cancer: results of a 10 year follow-up study. Breast Cancer Res Treat 1994;29:73–83.
49. Bolla M, Chedin M, Colonna M, et al. Lack of prognostic value of epidermal growth factor receptor in a series of 229 T1/T2, N0/N1 breast cancers, with well defined prognostic parameters. Breast Cancer Res Treat 1994;29:265–270.
50. Newby JC, A'Hern RP, Leek RD, et al. Immunohistochemical assay for epidermal growth factor receptor on paraffin-embedded sections: validation against ligand-binding assay and clinical relevance in breast cancer. Br J Cancer 1995;71:1237–1242.
51. Pirinen R, Lipponen P, Syrjanen K. Expression of epidermal growth factor receptor (EGFR) in breast cancer as related to clinical, prognostic and cytometric factors. Anticancer Res 1995;15:2835–2840.
52. Gohring UJ, Ahr A, Scharl A, et al. Immunohistochemical detection of epidermal growth factor receptor lacks prognostic significance for breast carcinoma. J Soc Gynecol Invest 1995;2:653–659.
53. Seshadri R, McLeay WR, Horsfall DJ, et al. Prospective study of the prognostic significance of epidermal growth factor receptor in primary breast cancer. Int J Cancer 1996;69:23–27.
54. Hawkins RA, Tesdale AL, Killen ME, et al. Prospective evaluation of prognostic factors in operable breast cancer. Br J Cancer 1996;74:1469–1478.
55. Eissa S, Khalifa A, el-Gharib A, et al. Multivariate analysis of DNA ploidy, p53, c-erbB-2 proteins, EGFR, and steroid hormone receptors for short-term prognosis in breast cancer. Anticancer Res 1997;17: 3091–3097.
56. Torregrosa D, Bolufer P, Lluch A, et al. Prognostic significance of c-erbB-2/neu amplification and epidermal growth factor receptor (EGFR) in primary breast cancer and their relation to estradiol receptor (ER) status. Clin Chim Acta 1997;262:99–119.
57. Schroeder W, Biesterfeld S, Zillessen S, et al. Epidermal growth factor receptor-immunohistochemical detection and clinical significance for treatment of primary breast cancer. Anticancer Res 1997;17: 2799–2802.
58. Neskovic-Konstantinovic Z, Nikolic-Vukosavljevic D, Brankovic-Magic M, et al. Expression of epidermal growth factor receptor in breast cancer, from early stages to advanced disease. J Exp Clin Cancer Res 1999;18:347–355.
59. Umekita Y, Ohi Y, Sagara Y, et al. Co-expression of epidermal growth factor receptor and transforming growth factor-alpha predicts worse prognosis in breast-cancer patients. Int J Cancer 2000;89: 484–487.
60. Pawlowski V, Revillion F, Hebbar M, et al. Prognostic value of the type I growth factor receptors in a large series of human primary breast cancers quantified with a real-time reverse transcription-polymerase chain reaction assay. Clin Cancer Res 2000;6: 4217–4225.
61. Gaci Z, Bouin-Pineau MH, Gaci M, et al. Prognostic impact of cathepsin D and c-erbB-2 oncoprotein in a subgroup of node-negative breast cancer patients with low histological grade tumors. Int J Oncol 2001;18:793–800.
62. Ferrero JM, Ramaioli A, Largillier R, et al. Epidermal growth factor receptor expression in 780 breast cancer patients: a reappraisal of the prognostic value based on an eight-year median follow-up. Ann Oncol 2001;12:841–846.
63. Tsutsui S, Ohno S, Murakami S, et al. Prognostic value of epidermal growth factor receptor (EGFR) and its relationship to the estrogen receptor status in 1029 patients with breast cancer. Breast Cancer Res Treat 2002;71:67–75.
64. Klijn JG, Berns PM, Schmitz PI, et al. The clinical significance of epidermal growth factor receptor

(EGF-R) in human breast cancer: a review on 5232 patients. Endocr Rev 1992;13:3–17.

65. Nicholson RI, Gee JM, Harper ME. EGFR and cancer prognosis. Eur J Cancer 2001;37(suppl 4): S9–S15.

66. Harris AL, Nicholson S, Sainsbury JR, et al. Epidermal growth factor receptors in breast cancer: association with early relapse and death, poor response to hormones and interactions with neu. J Steroid Biochem 1989;34:123–131.

67. Buchholz T, Ang K, Tu X, et al. Epidermal growth factor receptor expression correlates with poor survival in breast cancer patients treated with doxorubicin-based neoadjuvant chemotherapy. Presented at the San Antonio Breast Cancer Symposium 2003, abstract 511 (http://www.sabcs.org/).

68. Nicholson S, Sainsbury JR, Halcrow P, et al. Expression of epidermal growth factor receptors associated with lack of response to endocrine therapy in recurrent breast cancer. Lancet 1989;1:182–185.

69. McInerney JM, Wilson MA, Strand KJ, et al. A strong intronic enhancer element of the EGFR gene is preferentially active in high EGFR expressing breast cancer cells. J Cell Biochem 2001;80:538–549.

70. Chrysogelos SA, Yarden RI, Lauber AH, et al. Mechanisms of EGF receptor regulation in breast cancer cells. Breast Cancer Res Treat 1994;31:227–236.

71. Nicholson RI, Gee JM, Knowlden J, et al. The biology of antihormone failure in breast cancer. Breast Cancer Res Treat 2003;80(suppl 1):S29–S34, discussion S35.

72. Johnston SR, Head J, Pancholi S, et al. Integration of signal transduction inhibitors with endocrine therapy: an approach to overcoming hormone resistance in breast cancer. Clin Cancer Res 2003; 9(1 Pt 2):524S–532S.

73. Tsutsui S, Kataoka A, Ohno S, et al. Prognostic and predictive value of epidermal growth factor receptor in recurrent breast cancer. Clin Cancer Res 2002;8: 3454–3460.

74. Stanta G, Bonin S, Losi L, et al. Molecular characterization of intraductal breast carcinomas. Virchows Arch 1998;432:107–111.

75. Lebeau A, Unholzer A, Amann G, et al. EGFR, HER-2/neu, cyclin D1, p21 and p53 in correlation to cell proliferation and steroid hormone receptor status in ductal carcinoma in situ of the breast. Breast Cancer Res Treat 2003;79:187–198.

76. Tsutsui S, Ohno S, Murakami S, et al. Histological classification of invasive ductal carcinoma and the biological parameters in breast cancer. Breast Cancer 2003;10:149–152.

77. Martinazzi M, Crivelli F, Zampatti C, et al. Epidermal growth factor receptor immunohistochemistry in different histological types of infiltrating breast carcinoma. J Clin Pathol 1993;46:1009–1010.

78. Sainsbury JR, Nicholson S, Angus B, et al. Epidermal growth factor receptor status of histological subtypes of breast cancer. Br J Cancer 1988;58:458–460.

79. Schelfhout VR, Coene ED, Delaey B, et al. Pathogenesis of Paget's disease: epidermal heregulin-alpha, motility factor, and the HER receptor family. J Natl Cancer Inst 2000;92:622–628.

80. Willsher PC, Leach IH, Ellis IO, et al. Male breast cancer: pathological and immunohistochemical features. Anticancer Res 1997;17:2335–2338.

81. Suo Z, Nesland JM. Phyllodes tumor of the breast: EGFR family expression and relation to clinicopathological features. Ultrastruct Pathol 2000;24: 371–381.

82. Kimler BF, Fabian CJ, Wallace DD. Breast cancer chemoprevention trials using the fine-needle aspiration model. J Cell Biochem 2000;34(suppl):7–12.

83. Ross JS, Fletcher JA. The HER-2/neu oncogene: prognostic factor, predictive factor and target for therapy. Semin Cancer Biol 1999;9:125–138.

84. Sartor CI, Zhou H, Kozlowska E, et al. Her4 mediates ligand-dependent antiproliferative and differentiation responses in human breast cancer cells. Mol Cell Biol 2001;21:4265–4275.

85. Witton CJ, Reeves JR, Going JJ, et al. Expression of the HER1–4 family of receptor tyrosine kinases in breast cancer. J Pathol 2003;200:290–297.

86. Lodge AJ, Anderson JJ, Gullick WJ, et al. Type 1 growth factor receptor expression in node positive breast cancer: adverse prognostic significance of c-erbB-4. J Clin Pathol 2003;56:300–304.

87. Tsutsui S, Ohno S, Murakami S, et al. Prognostic value of the combination of epidermal growth factor receptor and c-erbB-2 in breast cancer. Surgery 2003;133:219–221.

88. Hudelist G, Singer CF, Manavi M, et al. Co-expression of ErbB-family members in human breast cancer: Her-2/neu is the preferred dimerization candidate in nodal-positive tumors. Breast Cancer Res Treat 2003; 80:353–361.

89. Albanell J, Codony-Servat J, Rojo F, et al. Activated extracellular signal-regulated kinases: association with epidermal growth factor receptor/transforming growth factor alpha expression in head and neck squamous carcinoma and inhibition by anti-epidermal growth factor receptor treatments. Cancer Res 2001;61:6500–6510.

90. Baselga J, Albanell J, Ruiz A, et al. Phase II and tumor pharmacodynamic study of gefinitib (ZD1839) in patients with advanced breast cancer. Proc Am Soc Clin Oncol 2003;22:7(abst 24).

91. Gee JM, Robertson JF, Ellis IO, et al. Phosphorylation of ERK1/2 mitogen-activated protein kinase is associated with poor response to anti-hormonal therapy and decreased patient survival in clinical breast cancer. Int J Cancer 2001;95:247–254.

92. Rojo F, Tabernero J, Van Cutsem E, et al. Pharmacodynamic studies of tumor biopsy specimens from patients with advanced gastric carcinoma undergoing treatment with gefinitib (ZD1839). Proc Am Soc Clin Oncol 2003;22:191 (abstr 764).

93. Arteaga CL. Inhibiting tyrosine kinases: successes and limitations. Cancer Biol Ther 2003;2(4 suppl 1): S79–S83.

94. Arteaga CL. EGF receptor as a therapeutic target: patient selection and mechanisms of resistance to receptor-targeted drugs. J Clin Oncol 2003; 21(23 suppl):289s–291s.

95. Arteaga CL. Epidermal growth factor receptor dependence in human tumors: more than just expression? Oncologist 2002;7(suppl 4):31–39.

96. Paez JG, Janne PA, Lee JC, et al. EGFR mutations in lung cancer: Correlation with clinical response to gefitinib therapy. Science 2004;304:1497–1500.

97. Lynch TJ, Bell DW, Sordella R, et al. Activating mutations in the epidermal growth factor receptor underlying responsiveness of non-small-cell lung cancer to Gefitinib. N Engl J Med 2004;350: 2129–2139.

98. Campiglio M, Locatelli A, Olgiati C, et al. Inhibition of proliferation and induction of apoptosis in breast cancer cells by the epidermal growth factor receptor (EGFR) tyrosine kinase inhibitor ZD1839 ('Iressa') is

independent of EGFR expression level. J Cell Physiol 2004;198:259–268.

99. Campos-Gonzalez R, Glenney JR. Immunodetection of the ligand-activated receptor for the epidermal growth factor. Growth Factors 1991;4:305–316.

100. Albanell J, Rojo F, Baselga J. Pharmacodynamic studies with the epidermal growth factor receptor tyrosine kinase inhibitor ZD1839. Semin Oncol 2001;28(5 suppl 16):56–66.

101. Albanell J, Rojo F, Averbuch S, et al. Pharmacodynamic studies of the epidermal growth factor receptor inhibitor ZD1839 in skin from cancer patients: histopathologic and molecular consequences of receptor inhibition. J Clin Oncol 2002;20:110–124.

102. Thor AD, Liu S, Edgerton S, et al. Activation (tyrosine phosphorylation) of ErbB-2 (HER-2/neu): a study of incidence and correlation with outcome in breast cancer. J Clin Oncol 2000;18:3230–3239.

103. Tabernero J, Rojo F, Jimenez E, et al. A phase I PK and serial tumor and skin pharmacodynamic (PD) study of weekly (q1w), every 2-week (q2w) or every 3-week (q3w) 1-hour (h) infusion EMD72000, a humanized monoclonal anti-epidermal growth factor receptor (EGFR) antibody, in patients (pt) with advanced tumors. Proc ASCO 2003;22(abst. 770).

104. Wakeling AE, Guy SP, Woodburn JR, et al. ZD1839 (Iressa): an orally active inhibitor of epidermal growth factor signaling with potential for cancer therapy. Cancer Res 2002;62:5749–5754.

105. Moulder S, Yakes M, Bianco R, et al. Small molecule EGF receptor tyrosine kinase inhibitor ZD1839 (IRESSA) inhibits HER2/Neu (erb-2) overexpressing breast tumor cells. Proc Am Soc Clin Oncol 2001; 20:8A.

106. Mendelsohn J, Baselga J. The EGF receptor family as targets for cancer therapy. Oncogene 2000;19: 6550–6565.

107. Nahta R, Hortobagyi GN, Esteva FJ. Growth factor receptors in breast cancer: potential for therapeutic intervention. Oncologist 2003;8:5–17.

108. Atalay G, Cardoso F, Awada A, et al. Novel therapeutic strategies targeting the epidermal growth factor receptor (EGFR) family and its downstream effectors in breast cancer. Ann Oncol 2003;14:1346–1363.

109. Slamon DJ, Leyland-Jones B, Shak S, et al. Use of chemotherapy plus a monoclonal antibody against HER2 for metastatic breast cancer that overexpresses HER2. N Engl J Med 2001;344:783–792.

110. Agus D, Gordon M, Taylor C, et al. Clinical activity in a phase I trial of HER-2-targeted rhuMAb 2C4 (pertuzumab) in patients with advanced solid malignancies (AST). Proc ASCO 2003;22(abst. 771).

111. Albanell J, Codony J, Rovira A, et al. Mechanism of action of anti-HER2 monoclonal antibodies: scientific update on trastuzumab and 2C4. Adv Exp Med Biol 2003;532:253–268.

112. Agus D, Akita R, Fox W, et al. Targeting ligand-activated ErbB2 signaling inhibits breast and prostate tumor growth. Cancer Cell 2002;2:127.

113. Clinical trial to evaluate antitumor activity of epidermal growth factor in breast cancer. Oncology (Huntingt) 1996;10:709.

114. Mendelsohn J, Baselga J. Status of epidermal growth factor receptor antagonists in the biology and treatment of cancer. J Clin Oncol 2003;21:2787–2799.

115. Codony-Servat J, Albanell J, Lopez-Talavera JC, et al. Cleavage of the HER2 ectodomain is a pervanadate-activable process that is inhibited by the tissue inhibitor of metalloproteases-1 in breast cancer cells. Cancer Res 1999;59:1196–1201.

116. Albanell J, Molina MA, Codony-Servat J, et al. The production of cleaved intracellular HER2 fragment is inducible in breast cancer cells and this fragment is phosphorylated in breast tumors (meeting abstract). Proc Am Soc Clin Oncol 2000;19:2458A.

117. Molina MA, Saez R, Ramsey EE, et al. NH(2)-terminal truncated HER-2 protein but not full-length receptor is associated with nodal metastasis in human breast cancer. Clin.Cancer Res 2002;8:347–353.

118. Molina MA, Codony-Servat J, Albanell J, et al. Trastuzumab (herceptin), a humanized anti-Her2 receptor monoclonal antibody, inhibits basal and activated Her2 ectodomain cleavage in breast cancer cells. Cancer Res 2001;61:4744–4749.

119. Bos M, Mendelsohn J, Kim YM, et al. PD153035, a tyrosine kinase inhibitor, prevents epidermal growth factor receptor activation and inhibits growth of cancer cells in a receptor number-dependent manner. Clin Cancer Res 1997;3:2099–2106.

120. Anido J, Matar P, Albanell J, et al. ZD1839, a specific epidermal growth factor receptor (egfr) tyrosine kinase inhibitor, induces the formation of inactive egfr/her2 and egfr/her3 heterodimers and prevents heregulin signaling in her2-overexpressing breast cancer cells. Clin Cancer Res 2003;9:1274–1283.

121. Moasser MM, Basso A, Averbuch SD, et al. The tyrosine kinase inhibitor ZD1839 ("Iressa") inhibits HER2-driven signaling and suppresses the growth of HER2-overexpressing tumor cells. Cancer Res 2001; 61:7184–7188.

122. Normanno N, Campiglio M, De LA, et al. Cooperative inhibitory effect of ZD1839 (Iressa) in combination with trastuzumab (Herceptin) on human breast cancer cell growth. Ann Oncol 2002;13:65–72.

123. Rojo F, Albanell J, Anido J, et al. Dose dependent pharmacodynamic effects of ZD1839 ('Iressa') correlate with tumor growth inhibition in BT-474 breast cancer xenografts. Proc Am Soc Clin Oncol 2002; 21(abstr 3893).

124. Wissner A, Brawner Floyd MB, Rabindran SK, et al. Syntheses and EGFR and HER-2 kinase inhibitory activities of 4-anilinoquinoline-3-carbonitriles: analogues of three important 4-anilinoquinazolines currently undergoing clinical evaluation as therapeutic antitumor agents. Bioorg Med Chem Lett 2002;12: 2893–2897.

125. Malik SN, Siu LL, Rowinsky EK, et al. Pharmacodynamic evaluation of the epidermal growth factor receptor inhibitor OSI-774 in human epidermis of cancer patients. Clin Cancer Res 2003;9:2478–2486.

126. Albain KS, Elledge R, Gradishar W, et al. Open-label, phase II, multicenter trial of ZD1839 ('Iressa') in patients with advanced breast cancer. Breast Cancer Res Treat 2002;(suppl)(abst. 20).

127. von Minckwitz G, Jonat W, Beckmann M, et al. A multicenter phase II trial to evaluate gefitinib (Iressa, ZD1839) in patients with metastatic breast cancer after previous chemotherapy treatment. Eur J Cancer 2003;1(suppl):S133.

128. Robertson J, Gutteridge E, Cheung K, et al. Gefitinib (ZD1839) is active in acquired tamoxifen (TAM)-resistant oestrogen receptor (ER)-positive and ER-negative breast cancer: Results from a phase II study. Proc Am Soc Clin Oncol 2003;22:7(abst. 23).

129. Winer E, Cobleigh M, Dickler M, et al. Phase II multicenter study to evaluate the efficacy and safety of Tarceva (erlotinib, OSI-774) in women with previously treated locally advanced or metastatic breast cancer. Breast Cancer Res Treat 2002;(suppl):(abst. 445).

130. Tan A, Yang X, Berman A, et al. Evaluation of epidermal growth factor receptor signaling in tumor and skin biopsies after treatment with OSI-774 in patients with metastatic breast cancer. Proc Am Soc Clin Oncol 2003;22:(abst. 784).

131. Albanell J, Gascon P. Pharmacodynamics and tolerabilities of EGFR-targeted compounds. Signal 2003; 4:7–13.

132. Christine A, Iacobuzio-Donahue, Albain K, et al. Determinants of response to gefitinib (Iressa®, ZD1839) in patients with advanced breast cancer. Clin Cancer Res 2003;9(suppl):(abst. 96).

133. Perez-Soler R, Ling Y-H, Lia M, et al. Molecular mechanisms of resistance to the HER1/EGFR tyrosine kinase inhibitor erlotinib HCl in human cancer cell lines. Proc Am Soc Clin Oncol 2003;22:190.

134. Xia W, Mullin RJ, Keith BR, et al. Anti-tumor activity of GW572016: a dual tyrosine kinase inhibitor blocks EGF activation of EGFR/erbB2 and downstream Erk1/2 and AKT pathways. Oncogene 2002; 21:6255–6263.

135. Mullin RJ, Keith BR, Murray DM, et al. Combination therapy with the dual EGFR-ERBB-2 tyrosine kinase inhibitor GW572016. Proc Am Soc Clin Oncol 2003;22:242(abst. 970).

136. Spector N, Raefsky E, Hurwitz H, et al. Safety, clinical efficacy, and biologic assessments from EGF10004: A randomized phase IB study of GW572016 for patients with metastatic carcinomas overexpressing EGFR or erbB2. Proc Am Soc Clin Oncol 2003;22:93(abst. 772).

137. Kaplan EH, Jones CM, Berger MS. A phase II, open-label, multicenter study of GW572016 in patients with trastuzumab refractory metastatic breast cancer. Proc Am Soc Clin Oncol 2003;22:245(abst. 981).

138. Slichenmyer WJ, Elliott WL, Fry DW. CI-1033, a pan-erbB tyrosine kinase inhibitor. Semin Oncol 2001;5(suppl 16):80–85.

139. Zinner RG, Donato NJ, Nemunaitis JJ, et al. Biomarker modulation in tumor and skin biopsy samples from patients with solid tumors following treatment with the pan-erbB tyrosine kinase inhibitor, CI-1033. Proc Am Soc Clin Oncol 2003;21:(abst. 58).

140. Rinehart JJ, Wilding G, Willson J, et al. A phase I clinical and pharmacokinetic (PK)/food effect study of oral CI-1033, a pan-erb B tyrosine kinase inhibitor, in patients with advanced solid tumors. Proc Am Soc Clin Oncol 2003;22:205(abst. 821).

141. Nemunaitis JJ, Eiseman I, Cunningham C, et al. A phase 1 trial of CI-1033, a pan-erbB tyrosine kinase inhibitor, given daily for 14 days every 3 weeks, in patients with advanced solid tumors. Proc Am Soc Clin Oncol 2003;22:243(abst. 974).

142. Rowinsky EK, Garrison M, Lorusso P, et al. Administration of CI-1033, an irreversible pan-erbB tyrosine kinase (TK) inhibitor is feasible on a 7-day-on/7-day-off schedule: a phase I, pharmacokinetic (PK), and food effect study. Proc Am Soc Clin Oncol 2003; 22:201(abst. 807).

143. Traxler P, Allegrini PR, Brandt R, et al. Preclinical activity of AEE788, a potent new inhibitor of the ErbB and VEGF receptor tyrosine kinases. In vivo profile of AEE788. Clin Cancer Res 2003;9(suppl):(abst. 87).

144. Ciardiello F, Caputo R, Damiano V, et al. Antitumor effects of ZD6474, a small molecule vascular endothelial growth factor receptor tyrosine kinase inhibitor, with additional activity against epidermal growth factor receptor tyrosine kinase. Clin Cancer Res 2003;9:1546–1556.

145. Chan KC, Knox WF, Gee JM, et al. Effect of epidermal growth factor receptor tyrosine kinase inhibition on epithelial proliferation in normal and premalignant breast. Cancer Res 2002;62:122–128.

146. Surmacz E. Function of the IGF-I receptor in breast cancer. J Mamm Gland Biol Neoplasia 2000;5: 95–105.

147. Chan TW, Pollak M, Huynh H. Inhibition of insulin-like growth factor signaling pathways in mammary gland by pure antiestrogen ICI 182,780. Clin Cancer Res 2001;7:2545–2554.

148. Bartucci M, Morelli C, Mauro L, et al. Differential insulin-like growth factor I receptor signaling and function in estrogen receptor (ER)-positive MCF-7 and ER-negative MDA-MB-231 breast cancer cells. Cancer Res 2001;61:6747–6754.

149. Arteaga CL, Osborne CK. Growth inhibition of human breast cancer cells in vitro with an antibody against the type I somatomedin receptor. Cancer Res 1989;49:6237–6241.

150. Hankinson SE, Willett WC, Colditz GA, et al. Circulating concentrations of insulin-like growth factor-I and risk of breast cancer. Lancet 1998;351:1393–1396.

151. Gee J, Rubini M, Robertson J, et al. Type 1 insulin-like growth factor receptor expression and activation in clinical breast cancer. San Antonio Breast Cancer Symposium, 2003 (abst. 429) (http://www.sabcs.org/).

152. Albanell J, Baselga J. Unraveling resistance to trastuzumab (Herceptin): insulin-like growth factor-I receptor, a new suspect. J Natl Cancer Inst 2001; 93:1830–1832.

153. Lu Y, Zi X, Zhao Y, et al. Insulin-like growth factor-I receptor signaling and resistance to trastuzumab (Herceptin). J Natl Cancer Inst 2001;93:1852–1857.

154. Liu B, Fang M, Lu Y, et al. Fibroblast growth factor and insulin-like growth factor differentially modulate the apoptosis and G1 arrest induced by anti-epidermal growth factor receptor monoclonal antibody. Oncogene 2001;20:1913–1922.

155. Nicholson R, Jones H, Gee J, et al. Insulin-like growth factor–1 receptor signalling and acquired resistance to gefitinib (Iressa, ZD1839) in MCF–7 human breast cancer cells. San Antonio Breast Cancer Symposium, 2003 (abst. 1011) (http://www.sabcs.org/).

156. Balana ME, Labriola L, Salatino M, et al. Activation of ErbB-2 via a hierarchical interaction between ErbB-2 and type I insulin-like growth factor receptor in mammary tumor cells. Oncogene 2001;20: 34–47.

157. Ram TG, Dilts CA, Dziubinski ML, et al. Insulin-like growth factor and epidermal growth factor independence in human mammary carcinoma cells with c-erbB-2 gene amplification and progressively elevated levels of tyrosine-phosphorylated p185erbB-2. Mol Carcinog 1996;15:227–238.

158. Harris L, Witkiewicz A, Freidman P, et al. Response to herceptin and chemotherapy in herceptin-naive patients with HER2 3+/FISH+ breast cancer is modified by pattern of expression of insulin-like growth factor-I receptor (IGF-IR) but not epidermal growth factor receptor (EGFR). San

Antonio Breast Cancer Symposium, 2003 (abst. 316) (http://www.sabcs.org/).

159. Surmacz E. Growth factor receptors as therapeutic targets: strategies to inhibit the insulin-like growth factor I receptor. Oncogene 2003;22:6589–6597.

160. Maloney EK, McLaughlin JL, Dagdigian NE, et al. An anti-insulin-like growth factor I receptor antibody that is a potent inhibitor of cancer cell proliferation. Cancer Res 2003;63:5073–5083.

161. Hofmann F, Kung AL, Mitsiades CS, et al. In vivo anti-tumor and anti-angiogenic activity of novel IGF-IR kinase inhibitors. Clin Cancer Res 2003;9(suppl): (abst. 165).

162. Castellani R, Visscher EW, Wykes S, et al. Interaction of transforming growth factor-alpha and epidermal growth factor receptor in breast carcinoma. Cancer 1994;73:344–349.

163. Umekita Y, Ohi Y, Sagara Y, et al. Co-expression of epidermal growth factor receptor and transforming growth factor-alpha predicts worse prognosis in breast-cancer patients. Int J Cancer 2000;89: 484–487.

164. Yarden RI, Wilson MA, Chrysogelos SA. Estrogen suppression of EGFR expression in breast cancer cells: a possible mechanism to modulate growth. J Cell Biochem 2001;36(suppl):232–246.

165. Dumont N, Arteaga CL. Transforming growth factor-beta and breast cancer: Tumor promoting effects of transforming growth factor-beta. Breast Cancer Res 2000;2:125–132.

166. Walker RA. Transforming growth factor beta and its receptors: their role in breast cancer. Histopathology 2000;36:178–180.

167. Reiss M, Barcellos-Hoff MH. Transforming growth factor-beta in breast cancer: a working hypothesis. Breast Cancer Res Treat 1997;45:81–95.

168. Berns EMJJ, Klijn JMG, van Staverin IL, et al. Sporadic amplification of the insulin-like growth factor receptor I gene in human breast tumors. Cancer Res 1992;52:1036–1039.

169. McCune BK, Mullin BR, Flanders KC, et al. Localization of transforming growth factor-isotypes in lesions of the human breast. Hum Pathol 1992; 23:13–20.

170. Shao ZM, Nguyen M, Barsky SH. Human breast carcinoma desmoplasia is PDGF initiated. Oncogene 2000;19:4337–4345.

171. Seymour L, Bezwoda WR. Positive immunostaining for platelet derived growth factor (PDGF) is an adverse prognostic factor in patients with advanced breast cancer. Breast Cancer Res Treat 1994;32: 229–233.

172. Rubin BP, Schuetze SM, Eary JF, et al. Molecular targeting of platelet-derived growth factor B by imatinib mesylate in a patient with metastatic dermatofibrosarcoma. J Clin Oncol 2002;20:3586–3591.

173. Yiangou C, Gomm JJ, Coope RC, et al. Fibroblast growth factor 2 in breast cancer: occurrence and prognostic significance. Br J Cancer 1997;75:28–33.

174. Blanckaert VD, Hebbar M, Louchez MM, et al. Basic fibroblast growth factor receptors and their prognostic value in human breast cancer. Clin Cancer Res 1998;4:2939–2947.

175. Faridi A, Rudlowski C, Biesterfeld S, et al. Long-term follow-up and prognostic significance of angiogenic basic fibroblast growth factor (bFGF) expression in patients with breast cancer. Pathol Res Pract 2002;198:1–5.

176. Smith K, Fox SB, Whitehouse R, et al. Upregulation of basic fibroblast growth factor in breast carcinoma and its relationship to vascular density, oestrogen receptor, epidermal growth factor receptor and survival. Ann Oncol 1999;10:707–713.

177. Berns EMJJ, Foekens JA, van Staveren IL, et al. Oncogene amplification and prognosis in breast cancer: Relationship with systemic treatment. Gene 1995;159:11–18.

178. Yiangou C, Gomm JJ, Coope RC, et al. Fibroblast growth factor 2 in breast cancer: occurrence and prognostic significance. Br J Cancer 1997;75:28–33.

179. Kinoshita J, Kitamura K, Kabashima A, et al. Clinical significance of vascular endothelial growth factor-C (VEGF-C) in breast cancer. Breast Cancer Res Treat 2001;66:159–164.

180. Linderholm BK, Lindahl T, Holmberg L, et al. The expression of vascular endothelial growth factor correlates with mutant p53 and poor prognosis in human breast cancer. Cancer Res 2001;61:2256–2260.

181. Foekens JA, Peters HA, Grebenchtchikov N, et al. High tumor levels of vascular endothelial growth factor predict poor response to systemic therapy in advanced breast cancer. Cancer Res 2001;61:5407–5414.

182. Manders P, Beex LV, Tjan-Heijnen VC, et al. The prognostic value of vascular endothelial growth factor in 574 node-negative breast cancer patients who did not receive adjuvant systemic therapy. Br J Cancer 2002;87:772–778.

183. Coradini D, Boracchi P, Daidone MG, et al. Contribution of vascular endothelial growth factor to the Nottingham prognostic index in node-negative breast cancer. Br J Cancer 2001;85:795–797.

184. De Paola F, Granato AM, Scarpi E, et al. Vascular endothelial growth factor and prognosis in patients with node-negative breast cancer. Int J Cancer 2002;98:228–233.

185. MacConmara M, O'Hanlon DM, Kiely MJ, et al. An evaluation of the prognostic significance of vascular endothelial growth factor in node positive primary breast carcinoma. Int J Oncol 2002;20:717–721.

186. Gasparini G. Prognostic value of vascular endothelial growth factor in breast cancer. Oncologist 2000; 5(suppl 1):37–44.

187. Toi M, Bando H, Kuroi K. The predictive value of angiogenesis for adjuvant therapy in breast cancer. Breast Cancer 2000;7:311–314.

188. Weidner N. Tumor angiogenesis: review of current applications in tumor prognostication. Semin Diagn Pathol 1993;10:302–313.

189. Siitonen SM, Haapasalo HK, Rintala IS, et al. Comparison of different immunohistochemical methods in the assessment of angiogenesis: lack of prognostic value in a group of 77 selected node-negative breast carcinomas. Mod Pathol 1995;8:745–752.

190. Callagy G, Dimitriadis E, Harmey J, et al. Immunohistochemical measurement of tumor vascular endothelial growth factor in breast cancer. A more reliable predictor of tumor stage than microvessel density or serum vascular endothelial growth factor. Appl Immunohistochem Mol Morphol 2000;8:104–109.

CHAPTER

18

Tumor-Associated Proteolytic Factors: Markers for Tumor Invasion and Metastasis

Nadia Harbeck, MD, PhD,[1] Viktor Magdolen, PhD,[1] Eleftherios Diamandis, MD, PhD,[2] Jeffrey S. Ross, MD,[3] Ronald E. Kates, PhD,[1] and Manfred Schmitt, PhD[1]

[1]Clinical Research Unit, Department of Obstetrics and Gynecology, Technical University of Munich, Munich, Germany
[2]Department of Pathology and Laboratory Medicine, Mount Sinai Hospital, and the Department of Laboratory Medicine and Pathobiology, University of Toronto, Toronto, Ontario, Canada
[3]Department of Pathology and Laboratory Medicine, Albany Medical College, Albany, New York, USA

Tumor-Associated Proteolytic Factors

Breast cancer is a potentially systemic disease already at the time of primary therapy. Those breast cancer patients who cannot be cured by optimal primary therapy will not die from their primary tumor but from distant metastases. A profound understanding of the complex processes involved in tumor cell invasion and metastasis is therefore of great clinical interest. An essential prerequisite for the invasive and metastatic capacity of cancer cells is their ability to attach to and degrade the extracellular matrix surrounding the primary tumor nest. Various extra- and intracellular proteases and their inhibitors are responsible for such tissue remodeling. The multifactorial role of tumor-associated protease systems in tissue degradation and remodeling, involving ovulation, embryogenesis, wound healing, neovascularization, atherosclerosis, tumor spread, and metastasis is well documented by now.[1–4] Invasiveness of solid malignant tumor cells into tissue, vessels, and the extracellular matrix (tumor stroma) depends on the regulated (limited) expression of proteases that dissociate cell–cell and/or cell–matrix interactions and degrade components of the surrounding extracellular matrix.

A variety of proteases has been associated with the clinical progression of breast cancer, including the matrix metalloproteases (MMPs), the lysosomal cysteine proteases, and the serine proteases. Serine proteases of the tissue kallikrein family, encompassing hK1 to hK15, and the plasminogen activator system, comprising urokinase plasminogen activator and plasma, act as key players in breast cancer progression, in concert with the other proteases and their inhibitors, including the MMPs and their tissue inhibitors of metalloproteases (TIMPs), and certain cysteine proteases, including the cathepsins B, H, L, cystatins, and stefins.[4,5]

Matrix Metalloproteases in Breast Cancer

The MMPs are a group of at least 25 zinc containing endopeptidase secreted as proenzymes with substantial sequence similarities that are inhibited by metallochelators and specific tissue inhibitors known as TIMPs.[6–8] The MMPs include the interstitial collagenases, gelatinases, stromelysins, and membrane-type MMPs and are involved in breast cancer initiation, invasion, and metastasis.[9–12] High levels of at least three MMPs (MMP-2, MMP-9, and MMP-11) have been found to correlate with poor disease outcome in breast cancer.[9–12] The multifunctional capacity of MMPs in cancer is shown by their function, but they also act as effectors of early stage tumorigenesis. MMPs and TIMPs affect tumor initiation and growth through loss of cell adhesion, evasion of apoptosis, and deregulation of cell division. MMPs are frequently overexpressed in cancer (e.g., cancer of the gastrointestinal and urogenital tract) and have been associated with an aggressive malignant phenotype and adverse prognosis. Synthetic MMP inhibitors affecting MMP proteolytic activity include collagen peptidomimetics and nonpeptidomimetic inhibitors of the MMP active site, and several members of this class of compounds are undergoing evaluation in clinical trials.[13]

MMP-2 (Gelatinase A)

MMP-2 degrades basement membranes, elastase and gelatin, which may facilitate stromal invasion and entry into blood vessels.[9–12] Stromal fibroblast production of MMP-2 in response to infiltration by breast carcinoma may be an indicator of an aggressive tumor.[14] Immunohistochemistry-based studies have linked overexpression of MMP-2 with adverse disease outcome.[15–17]

MMP-9 (Gelatinase B)

Studies of MMP-9 in breast cancer have been somewhat conflicting, with some studies showing an adverse outcome in cases with high MMP-9 expression[18] and others indicating that overexpression of MMP-9 indicated a favorable prognosis (FIG. 18.1).[19]

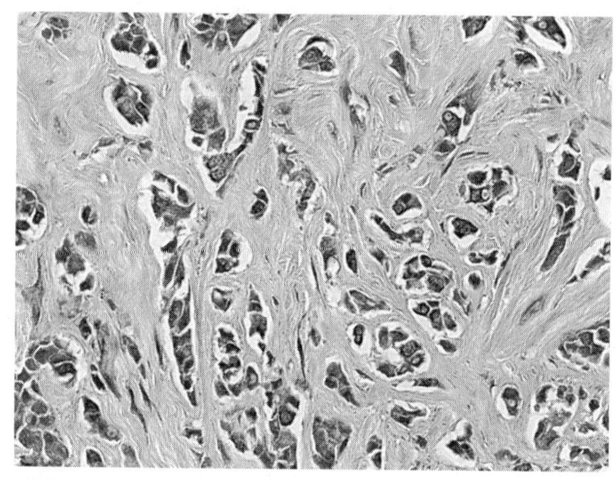

FIGURE 18.1 MMP9 overexpression in breast cancer. Although the association of high MMP expression with adverse prognosis in breast cancer has not been consistent, this patient with intense MMP immunostaining of the tumor cell cytoplasm of an infiltrating ductal carcinoma experienced early relapse and ultimately succumbed from her tumor (immunoperoxidase stain ×100). **See Plate 42 for color image.**

MMP-11 (Stromelysin-3)

The ability for the epithelial tumor cells to elicit the stromal cell elaboration of extracellular matrix–digesting enzymes may be a major pathway by which invasion of cancer is facilitated. Stromelysin-3 expression has been associated with poor prognosis in breast cancer.[20,21]

Continued interest in MMP and TIMP expression in breast cancer is being fueled, in part, by clinical studies using synthetic MMP inhibitor.[22] However, in breast cancer the status of MMP or TIMP expression still has not been linked to the response to MMP inhibitors or to other specific breast cancer therapies.

Lysosomal Proteases in Breast Cancer

Lysosomal cysteine proteases (e.g., cathepsins B, H, and L) play an important role in cancer spread and metastasis as well. Determination of their antigen levels in tumor tissue extracts can provide useful clinical information on disease-free or overall survival regarding patients with malignancies of the breast, liver, pancreas, colon lung, colorectum, brain, head and neck, and skin.[22–27] Nontoxic drugs and antibodies that are able to inhibit the proteolytic activity of tumor-associated cysteine-type cathepsins in vivo are under development.[28]

Cathepsin D is an estrogen-regulated lysosomal aspartate protease believed to facilitate cancer cell migration and promote stromal invasion by the digestion of basement membrane, matrix, and connective tissue.[29] Numerous studies in the early 1990s using an immunoassay approach on fresh breast tumor cytosolic preparations showed that elevated cathepsin D levels are an independent predictor of survival in breast cancer.[30–32] Attempts to convert the assay to an immunohistochemistry-based format have not been successful.[33,34] Thus, given the limitations of the current assay format, interest in using cathepsin D assessment as a prognostic factor and guide to therapy in breast cancer has all but disappeared in the United States, although it is still performed in Europe.

Novel Tissue Kallikreins in Breast Cancer

The optimal management of breast cancer patients involves a multidisciplinary approach, including the use of tumor biomarkers. Traditional prognostic and/or predictive factors in breast cancer include tumor size, grade, nodal status, and steroid hormone receptor (estrogen and progesterone receptors) status.[35] A number of tumor biologic factors that relate to tumor aggressiveness or metastatic potential, including markers of angiogenesis and proliferation, growth factor receptors, cell cycle regulators, and proteases (e.g., uPA, cathepsin B and L, MMPs), have been discovered.

The proteases of the tissue kallikrein family encompass 15 trypsin-like, chymotrypsin-like serine proteases[36] clustered together on chromosome 19q13.4. In the past, most of the information on the (patho-)physiologic role and function of the tissue kallikreins was only available for the well-known tissue kallikreins hK1 to hK3. The collective efforts of a number of independent researchers, including Diamandis and coworkers,[36–39] led to the expansion of the human tissue kallikrein gene family from 3 to 15 members. Of the "classic" tissue kallikreins hK1, hK2, and hK3, pancreatic kallikrein (hK1) is the only tissue kallikrein that acts like plasma kallikrein (on chromosome 4), that is, it releases lysyl-bradykinin (kallidin) from LMW-kininogen; hK2 is known as glandular kallikrein. hK3 is better known as prostate-specific antigen (PSA) and, under physiologic conditions, plays an important role in liquefaction of the seminal plasma clot after ejaculation. In prostate cancer, hK3 is widely used for cancer diagnosis, staging, and for monitoring therapy response.

The pathophysiologic functions of the newly discovered tissue kallikreins hK4 to hK15 are not yet resolved in detail. The de novo expression or up-regulation of some of these kallikreins in different types of hormone-regulated cancer, especially ovarian, breast, and prostate cancers, indicates that these proteases are part of tumor cell associated proteolytic cascades aimed at increasing the invasive potential of the tumor cells. Most tissue kallikrein genes are subject to steroid hormone regulation.[36] Tissue kallikreins are supposed to act as mediators of tumor invasion and metastasis through degradation of the basement membrane and the extracellular matrix[40] or by activation of hormones, growth factors, and other (pro-)enzymes (e.g., pro-uPA) by hK2 or hK4.[41] Furthermore, tissue kallikreins are considered to be potential tumor suppressors.[42,43]

Most tissue kallikreins represent prospective biomarkers for the early detection and/or prognosis and monitoring of cancer at the messenger RNA (mRNA) and/or protein level. Some of the tissue kallikreins (e.g., hK5 and hK8) have the potential to serve as true tumor markers in breast and ovarian cancer.[44,45] Several of the tissue kallikreins have prognostic or predictive value not only in breast cancer (**Table 18.1**) but also in ovarian, prostate, and testicular cancers.[36,46] The expression of KLK5 and KLK14 mRNA in breast tumors is indicative of a poor patient prognosis,[47] whereas higher levels of KLK9, KLK13, and KLK15 mRNA and hK3 protein forecast a favorable disease outcome.[48–52] Elevated levels of hK3 and hK10 protein in breast carcinomas are significantly correlated with poor response to tamoxifen therapy.[53,54] Interestingly, elevated levels of PSA in breast cancer patients

Table 18.1 Clinical Relevance of Several Tissue Kallikreins in Breast Cancer

Type	Name When First Cloned	Source of Material	Method Used	Prognostic Value	Reference
3	PSA	cytosol	ELISA	Favorable	53–55
5	KLK-L2, human stratum corneum tryptic enzyme (HSCTE)	mRNA	Q-RT-PCR	Unfavorable	47
7	Human stratum corneum chymotryptic enzyme (HSCCE)	mRNA	RT-PCR	Unfavorable	56
9	KLK-L3	mRNA	Q-RT-PCR	Favorable	50
10	Normal epithelial cell-specific 1 gene (NES1)	cytosol	ELISA	Predictive	57
13	KLK-L4	mRNA	Q-RT-PCR	Favorable	48
14	KLK-L6	mRNA cytosols	Q-RT-PCR ELISA	Unfavorable	58, 59
15	KLK15	mRNA	Q-RT-PCR	Favorable	49

Q-RT-PCR, quantitative reverse transcriptase-polymerase chain reaction.

have been associated with a favorable prognosis,[51,52] although one report has described a resistance to antiestrogen therapy with tamoxifen in patients with high PSA levels.[53] The human stratum corneum chymotryptic enzyme (tissue kallikrein 7) has been associated with an adverse prognosis in breast cancer.[55] In addition, increased levels of KLK10 and KLK14 have been associated with adverse outcome or therapy resistance.[56,57] The diagnostic and prognostic value of kallikreins in various forms of cancer has been recently reviewed.[36,38,39,58,59]

Plasminogen Activation System in Breast Cancer

Consistent with the key role of the plasminogen activation system in cancer progression and metastasis (FIG. 18.2), various groups have shown independently that high antigen levels of the same urokinase-type plasminogen activator (uPA), its inhibitor type 1 (PAI-1) or the uPA receptor (uPAR) in primary cancers are associated with poor clinical outcome. Elevation of these biomarkers in the tumor tissue of cancer patients indicates an increased risk of the patients to experience early disease recurrence (metastasis) and shorter survival.

uPA, PAI-1, and uPAR are key players in a proteolytic cascade involved in physiologic and pathophysiologic degradation and remodeling of the extracellular matrix (FIG. 18.2). uPA, when bound to its cellular receptor uPAR, efficiently converts plasminogen into the broad spectrum serine protease plasmin; its action on plasminogen is controlled by the serine protease inhibitors PAI-1. Plasmin disintegrates various components of the connective tissue and, in malignancy, the tumor stroma surrounding the tumor cell nests. Interaction of PAI-1 with the uPA–uPAR complex induces internalization of the ternary complex uPAR–uPA–PAI-1 via help of transmembrane receptors of the low density lipoprotein receptor family, which subsequently results in intracellular degradation of uPA and PAI-1, whereas uPAR is recycled to the cell surface. In this way, the proteolytic activity is efficiently reorganized on the cell surface, enabling pericellular proteolysis. PAI-2, another uPA inhibitor, also forms a ternary cell surface-associated complex with uPA and uPAR (uPAR–uPA–PAI-2), but this complex is not internalized; in contrast, PAI-2 is degraded after being bound to uPAR–uPA.[60]

Besides its proteolytic potential, the uPA system in concert with other proteins exerts several other important biologic effects, including chemotaxis, migration/invasion, adhesion, proliferation, and angiogenesis (for reviews see Refs. 2, 3, 61–63). Binding of uPA to cell surface associated uPAR leads to activation of various intracellular signaling molecules such as tyrosine and serine protein kinases[61]; uPAR interacts with many other proteins such as members of the mannose receptor family and the extracellular matrix protein vitronectin as well as certain types of integrins,[64]

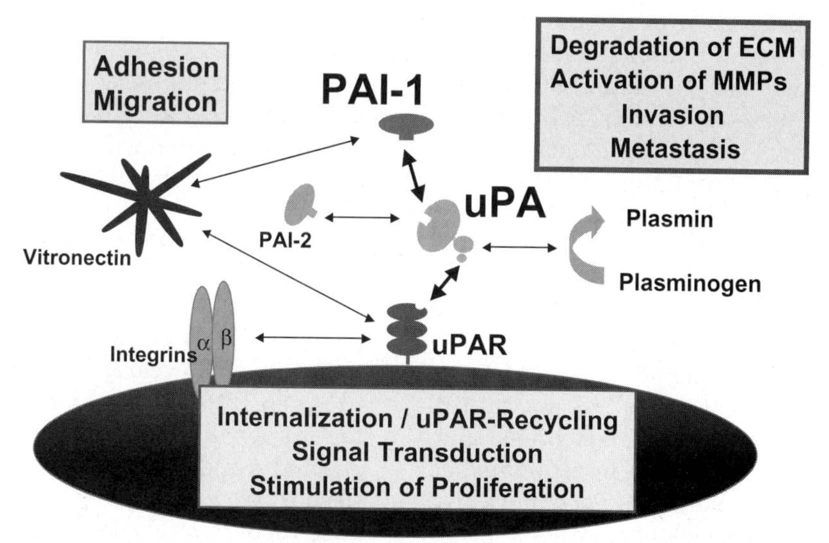

FIGURE 18.2 Multifunctional role of uPA and PAI-1 in tumor growth, invasion, and metastasis. In addition to pericellular proteolysis, many other important tumor biologic processes, such as cell migration, adhesion, intracellular signaling, and proliferation, are induced or modulated via multiple interactions of several components of the uPA system with adhesion molecules, receptors, and extracellular matrix proteins. (From Ref 4).

thereby modulating cell adhesion and migration.[3] Interestingly, PAI-1 modulates these uPAR-mediated processes by competing with uPAR for binding to the extracellular matrix protein vitronectin and to integrins. Thus, the biologic role of PAI-1 goes beyond that of a mere protease inhibitor.[65] PAI-2 lacks a cleavable signal peptide and thus is mainly present intracellularly. Only a small amount of PAI-2 (approximately 20%) is secreted. In contrast to PAI-1, no other functions in addition to inhibition of urokinase have been attributed to extracellular PAI-2. A well-balanced production and cellular assembly of cellular and pericellular uPA, uPAR, and PAI-1 is the prerequisite for efficient focal proteolysis, angiogenesis, cell adhesion, and migration and hence tumor cell invasion and metastasis.[4,66]

Clinical Relevance of uPA and PAI-1

Stand-Alone Prognostic Value of uPA and PAI-1

In 1988, Duffy and coworkers[67] were the first to show that measuring enzymatic activity of uPA provides prognostic information in primary breast cancer patients. This initial report was then strengthened by our finding that not only the enzymatic activity in the tumor tissue but to an even greater extent the uPA antigen level in the tumor tissue are of prognostic relevance.[68] In 1990, our group was the first to report that, in addition to uPA, its inhibitor PAI-1 is also of significant prognostic impact in both node-positive and node-negative breast cancer patients. Patients with high tumor antigen levels of either factor have a significantly worse survival probability than patients with low levels of these antigens.[69–85] At that time, the clinical finding that high levels of a protease inhibitor would indicate poor prognosis seemed surprising, but this finding has since been explained by thorough basic research on the concerted interaction of uPA and PAI-1 in tumor invasion and metastasis.[4,66] The strong prognostic impact of both uPA and the inhibitor PAI-1 on disease-free and overall survival in patients with primary breast cancer has since been confirmed in all published studies using immunometric biochemical assays (**Table 18.2**). The lack of contradictory evidence on the prognostic impact of uPA and PAI-1 in breast cancer is quite unique for any tumor biologic factor and is remarkable considering the variety of demographic conditions covered by the studies (Europe, United States, and Japan).

uPA and PAI-1 are predictors for distant metastasis but not for locoregional disease recurrence in breast cancer.[78] Both factors are only weakly correlated with other traditional prognostic factors,[86] and the prognostic impact of uPA and PAI-1 is independent of these other prognostic factors. With regard to risk group assessment, the particular combination of both factors, uPA/PAI-1 (both low vs. either or both factors high), is superior to either factor taken alone or to established prognostic factors such as tumor size, grade, and steroid hormone receptor, or menopausal status.[87] uPA/PAI-1 thus enables optimal distinction between high risk and low risk patients[82,83] and allows risk group assessment even within patient risk groups defined by established prognostic factors. Moreover, uPA/PAI-1 outperforms other tumor biologic factors such as cathepsins B, D, and L, p53, S phase, MIB1, or DNA ploidy with regard to prognostic importance.[87] Interestingly, HER-2 status and uPA/PAI-1 reveal independent but complementary information in breast cancer.[88,89] In node-negative disease, HER-2 status as assessed either by gene amplification or by protein overexpression does not add significant information to risk group selection determined by uPA/PAI-1.[89]

Prognostic Impact of uPA and PAI-1

In breast cancer, serine protease uPA and its inhibitor PAI-1 were the first novel tumor biologic factors to be validated at the highest level of evidence (LOE I) regarding their clinical utility. For this, the published data on the prognostic impact of uPA and PAI-1 in primary breast cancer were validated in 2002 by a pooled analysis performed by the Receptor and Biomarker Group (RBG) of the European Organization for Research and Treatment of Cancer (EORTC). This pooled analysis comprised 18 data sets provided by EORTC RBG members with 8377 primary breast cancer patients and a median follow-up of 6.5 years.[90,91] Next to nodal status, uPA and PAI-1 were the strongest prognostic factors for disease-free and overall survival. High uPA and/or PAI-1 levels in primary tumor tissue more than doubled a patient's risk for disease recurrence or death from breast cancer. Within the clinically relevant subgroup of node-negative patients without any adjuvant systemic therapy (n = 3362), uPA and PAI-1 were

Table 18.2 Individual Key Studies Supporting the Prognostic Impact of uPA and PAI-1 in Primary Breast Cancer

Author (year)	Country	Assays* (method of tissue extraction†)	Patients (N0)	Follow-Up (median, months)	Reference
Duffy (1988)‡	Ireland	Activity^{ih} (Cyt)	52 (25)	17	67
Jänicke (1991)	Germany	ELISA (TX)			70
Jänicke (1993)	Germany	ELISA^{ADI} (TX)	247 (101)	30	71
Foekens (1994)	Netherlands	ELISA^{ADI} (Cyt)	657 (273)	48	72
Grøhndahl-H. (1995)	Denmark	ELISA^{Mo} (Cyt)	505 (193)	54	73
Fernö (1996)‡	Sweden	LIA^{San} (Cyt)	688 (265)	42	74
Eppenberger (1998)	Switzerland	ELISA^{ADI} (Cyt)	305 (159)	37	75
Kim (1998)	Japan	ELISA^{Bio} (Cyt)	130 (130)	53	76
Kute (1998)	USA	ELISA^{Mo} (Cyt)	168 (168)	58	77
Knoop (1998)	Denmark	ELISA^{Mo} (TX)	429 (178)	61	78
Bouchet (1999)	France	ELISA^{ADI} (Cyt)	499 (233)	72	79
Foekens (2000)	Netherlands	ELISA^{ADI} (Cyt)	2780 (1405)	88	80
Jänicke (2001)	Germany	ELISA^{ADI} (TX)	556 (556)	32	81
Konecny (2001)	USA/Germany	ELISA^{ADI} (TX)	587 (283)	26	82
Harbeck (2002)	Germany	ELISA^{ADI} (TX)	761 (269)	60	83
Cufer (2003)	Slovenia	ELISA^{ADI} (TX)	460 (214)	33	84
Hansen (2003)	Denmark	ELISA^{ih} (TX)	228 (101)	150	85

Data drawn from Manders, Cancer 2004; Manders, Cancer Res 2004; Meijer-van Gelder, Cancer Res 2004; Zemzoum, J Clin Oncol 2003; Look, Thromb Haemost 2003.

*Assays: ADI (American Diagnostica Inc., Greenwich, CT, USA). Mo (Monozyme, Horsholm, Denmark). San (Sangtec, Bromma, Sweden). Bio (Biopool, Umea, Sweden), ih (in-house assay).

†Tissue extraction: Cyt (cytosol), TX (Triton-X-100, i.e., detergens extraction).

‡Only determination of uPA.

the strongest prognostic factors. This pooled analysis cannot be used to determine absolute survival expectations based on uPA/PAI-1, because the analysis comprises patients treated between 1970 and 1990 and—unlike the Chemo N_0 trial[81]—may therefore not be representative of modern interdisciplinary treatment strategies and thus modern survival expectations.

The prognostic impact of uPA and PAI-1 was also validated by the prospective randomized multicenter Chemo N_0 trial in node-negative breast cancer performed between 1993 and 1998 in 13 German academic centers and 1 center in Slovenia.[81] In this study, uPA and PAI-1 were prospectively determined in detergent extracts of primary tumor tissue using commercially available enzyme-linked immunosorbent assays (ELISAs) (American Diagnostica Inc., Stamford, CT, USA). Patients with low uPA/PAI-1 levels were observed and not treated, whereas high-uPA/PAI-1 patients were randomized to adjuvant CMF or observation only. The first scheduled interim analysis ($n = 556$) of the Chemo N_0 trial validated the independent prognostic impact of uPA/PAI-1 with regard to disease-free survival and confirmed the previously optimized cutoff values for uPA and PAI-1 used to discriminate between low and high uPA and PAI-1.[47,55] Even after a short median follow-up time of 32 months, a considerable and statistically significant benefit (per protocol analysis) from adjuvant CMF chemotherapy was observed in high risk patients. A second scheduled interim analysis of the trial comprising 647 patients and a longer follow-up (median, 50 months) confirmed the prognostic impact of uPA/PAI-1, not only with regard to disease-free but also to overall survival. It showed that node-negative breast cancer patients with low levels of uPA and PAI-1 have an excellent prognosis, with an approximately 95% 5-year overall survival rate without any adjuvant systemic therapy.[92]

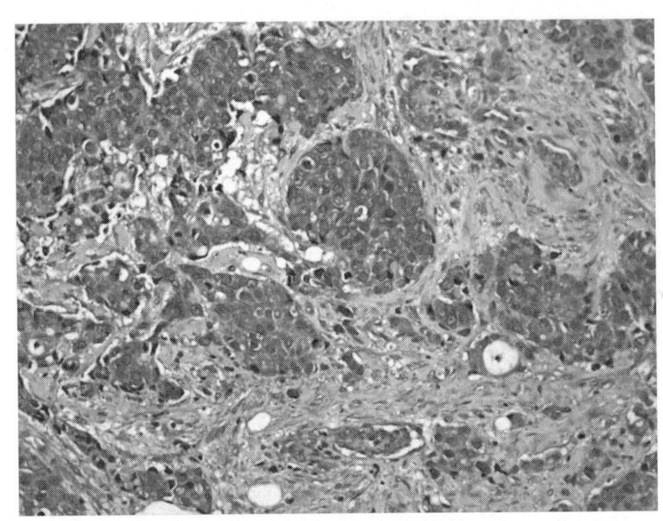

FIGURE 18.3 uPA expression detected by immunohistochemistry in primary breast cancer. Similar to MMP assessment, the stromal expression of tissue proteases can interfere with the accurate slide scoring of tumor-associated proteases. However, in this case of a moderately differentiated adenocarcinoma with intense expression of uPA the stromal staining is relatively scant compared to the tumor cell cytoplasm. This tumor progressed rapidly to systemic metastatic disease and was associated with disease-related death (immunoperoxidase staining ×100). **See Plate 43 for color image.**

Clinical Testing Methods for uPA and PAI-1

Uses and Limitations of Current Test Formats

Antigen levels of uPA and PAI-1 are determined by standardized immunometric assays (ELISA) in extracts of the primary tumor tissue. For this, ELISAs are commercially available and robust enough for clinical routine use. International quality assurance of the kits is guaranteed, and kits have been assessed by the EORTC RBG.[35,93] Extraction of tumor tissue in the presence of the non-ionic detergent Triton X-100 is recommended, because such a detergent-based extraction method yields about twice as much uPA antigen and provides a considerably better assessment of disease-free patient survival than uPA measured in detergent-free tumor extracts (cytosol fraction). For PAI-1, no such difference between extraction methods was found.[94] Extracts prepared from as little as 100 μg tumor tissue corresponding to about 1 μg protein suffice for routine ELISA testing. Therefore, the commercially available ELISAs can also be applied to extracts prepared from core biopsy specimens or cryostat sections. Good concordance between uPA/PAI-1 determination in core-needle biopsies and in surgically removed tumor tissue was recently shown.[95]

Future Test Formats in Development

At present, no clinically relevant data have been published applying immunohistochemistry or other techniques for determination of uPA and PAI-1 protein expression in breast carcinoma tissue. The use of archived material enabling more widely available determination techniques such as immunohistochemistry is unfortunately hampered by expression of the factors in both tumor tissue and surrounding stroma, thus making reliable scoring rather difficult (FIG. 18.3). Polymerase chain reaction techniques are currently under investigation. The ELISA procedure using primary tumor tissue extracts is therefore the only test format that can currently be recommended to determine uPA and PAI-1 for clinical routine use. It should also be used as the gold standard when evaluating other determination techniques.

Current Clinical Utility of uPA and PAI-1

Therapy Selection: To Treat or Not to Treat

To date, the combination uPA/PAI-1 (both low vs. uPA and/or PAI-1 high) has its greatest clinical relevance in node-negative breast cancer, where adequate risk group assessment is particularly important, considering the rising percentage of node-negative patients and the increasing issue of overtreatment in potentially cured women. Strictly following current international guidelines based on established risk factors alone such as the St. Gallen Consensus,[96] more than 90% of node-negative patients would be candidates for adjuvant systemic therapy, even though only 30% of node-negative patients will eventually relapse. The low risk patient group identified by uPA/PAI-1, comprising about 50% of node-negative breast cancer patients, is substantially larger than that characterized only by St. Gallen criteria and hence is much closer to the actual 70% of node-negative patients cured by locoregional treatment alone. The Chemo N_0 trial showed that node-negative patients with low uPA *and* PAI-1 have a rather low risk of disease recurrence and an estimated 5-year survival of more than 95%, even without receiving any adjuvant systemic therapy.[81,93] These low risk patients may thus be candidates for being spared the necessity of adjuvant chemotherapy. In contrast, high uPA/PAI-1 node-negative patients have a considerably increased risk of disease recurrence comparable with that of patients with three or more involved axillary lymph nodes.[81,93] Hence, these patients have to be considered high risk despite having no lymph node involvement, and for these high risk patients adjuvant systemic therapy, particularly chemotherapy, is clearly indicated. The current Working Group for Gynecological Oncology breast cancer guidelines in Germany recommend the use of uPA and PAI-1 for treatment decision making in node-negative breast cancer (www.ago-online.de).

Predictors of Therapy Response

In primary breast cancer, the Chemo N_0 trial showed a substantial benefit from adjuvant chemotherapy (six courses of CMF, days 1 + 8) in node-negative patients classified as high risk according to uPA/PAI-1. This observed treatment benefit is consistent with retrospective data suggesting a benefit from adjuvant systemic therapy in high risk patients according to uPA/PAI-1.[97,98] In a collective study comprising 3424 primary breast cancer patients treated in two different breast cancer centers (Technical University of Munich and Erasmus Medical Center Rotterdam) and a median follow-up of 83 months, a predictive impact of uPA/PAI-1 regarding response to adjuvant chemotherapy was observed.[99] For this retrospective analysis, uPA and PAI-1 had been quantified by ELISA in primary tumor tissue extracts; cutoff levels determining "high" versus "low" uPA/PAI-1 were defined consistently with the design of the Chemo N_0 trial. This predictive impact was seen as a significant interaction between chemotherapy and uPA/PAI-1 in all patients ($P < 0.003$; relative risk, 0.68; 95% confidence interval, 0.53–0.88) and also in patient subgroups according to nodal involvement. The enhanced benefit is seen over and above the significant impact of chemotherapy and/or endocrine therapy in all patients, and it is quite substantial in patients with more than three involved lymph nodes. Interestingly, patients with 0–3 involved axillary lymph nodes and low uPA/PAI-1 only seem to have a moderate benefit from adjuvant chemotherapy (Hazard Ratio [HR] = 0.72 [95% Confidence Limits = 0.53–0.98]). This result does not necessarily imply that node-negative patients with low uPA/PAI-1 benefit from adjuvant chemotherapy, but unfortunately the issue cannot be decided by this retrospective study because of the small number of node-negative patients treated with adjuvant systemic therapy in this group.[97] Further evaluation of this key question using the EORTC-RBG pooled analysis data is currently in progress. Recently, a predictive impact regarding response to adjuvant chemotherapy was also shown in patients with high tumor levels of uPA–PAI-1 complex.[99]

No statistically significant interaction was found between endocrine therapy and uPA/PAI-1 status; there was no significant difference in benefit from adjuvant endocrine therapy between patients with high uPA/PAI-1 and those with low uPA/PAI-1.[97] So far, the available data suggest that all primary breast cancer patients with high uPA/PAI-1—regardless of their nodal status—should receive adjuvant chemotherapy. There is no evidence for a subgroup defined by uPA and PAI-1 for which adjuvant endocrine therapy should not be given. However, in patients where the decision between endocrine and chemoendocrine therapy is clinically open, uPA and PAI-1 could provide decisive evidence.

Although prospective randomized studies are considered to be the gold standard for evaluating therapy response, such studies are often unfeasible or unethical to perform. Hence, as shown here for uPA/PAI-1 in the adjuvant setting, retrospective analyses may render valuable information on therapy response if they are sufficiently powered for multivariate models using interaction terms, provided that the factors analyzed had not originally been used for therapy decision making.

In the neoadjuvant setting, Pierga and coworkers[100] looked at the correlation between PAI-1 levels determined in initial core biopsies taken before and in tumor tissue taken after primary chemotherapy in a small collective of 69 patients. Although PAI-1 levels did correlate before and after primary chemotherapy, no correlation was found between initial PAI-1 antigen levels and clinical response to primary anthracycline-containing chemotherapy. In their article, the authors did not report on pathologic response or further course of the disease.

In metastatic breast cancer, retrospective studies have suggested that high uPA or PAI-1 tumor levels at primary therapy are associated with poor response to later palliative endocrine therapy.[101,102] Recently, a lower response rate to first-line endocrine therapy was also demonstrated for patients with high uPA–PAI-1 complex levels in their primary tumor tissue.[103]

Taken together, the findings regarding therapy response from the adjuvant and the palliative setting (**Table 18.3**) are quite consistent with the tumor biology of uPA and PAI-1: High levels of uPA and/or PAI-1 do reflect an aggressive tumor phenotype that may be overcome or suppressed by early systemic therapy in the adjuvant setting but may be far too advanced for response to palliative therapy at a later stage.[83,84,97,99–103]

Current Clinical Trials Using uPA and PAI-1

Looking at the prognostic and predictive impact of uPA/PAI-1 in primary breast cancer presented above, an interesting and still open clinical question remains the optimal systemic therapy for high risk patients according to high uPA/PAI-1, particularly for node-negative ones. Currently, two clinical trials seek to answer this question.

First, as a follow-up to the Chemo N_0 trial, a European prospective multicenter therapy trial in node-negative breast cancer (NNBC-3), supported by the German Working Group for Gynecological Oncology and the EORTC RBG, was initiated in October 2002. This trial not only compares established prognostic factors (i.e., age, grade, steroid hormone receptor status, tumor size) and uPA/PAI-1 for risk group stratification but also evaluates the optimal chemotherapy for high risk node-negative patients. In conjunction with uPA/PAI-1, tumor grade is also taken into account for patient stratification because grade was the strongest prognostic factor in the Chemo N_0 trial next to uPA/PAI-1. Before study entry, centers choose a criterion for stratifying node-negative patients: either according to uPA/PAI-1 or according to an algorithm defined using only established prognostic factors. High risk node-negative patients (as determined by the center's chosen criterion) are randomized either to anthracycline-containing adjuvant chemotherapy (6× $FE_{100}C$) or for a sequential anthracycline-docetaxel

Table 18.3 Studies Evaluating the *Predictive* Impact of uPA and PAI-1 in Breast Cancer

Therapeutic Setting	Treatment	Therapy Response If uPA/PAI-1 High	Patients	Study Design	Reference
Palliative	Endocrine therapy (tamoxifen)	Poor	*n* = 40	Retrospective	101
Palliative	Endocrine therapy (tamoxifen)	Poor*	*n* = 235	Retrospective	102
Palliative	Endocrine therapy (tamoxifen)	Poor*	*n* = 170	Retrospective	103
Neoadjuvant	Chemotherapy (FEC)	No correlation	*n* = 69	Prospective	100
Adjuvant	Chemotherapy (CMF)	Good	*n* = 556	Prospective	81
Adjuvant	Systemic therapy (chemo- or endocrine therapy)	Good*	*n* = 761	Retrospective	83
Adjuvant	Chemotherapy	Good*	*n* = 3424	Retrospective	97
Adjuvant	Endocrine therapy	Good	*n* = 460	Retrospective	84
Adjuvant	Chemotherapy	Good*	*n* = 1119	Retrospective	99

*Statistically significant prediction of therapy response.

regimen (3× $FE_{100}C$ → 3× $docetaxel_{100}$). For patients considered low risk, endocrine therapy (but not chemotherapy) is recommended provided that the steroid hormone receptor status is positive.

Second, prospective uPA and PAI-1 determinations are being performed in a randomized adjuvant therapy trial in Germany (ADEBAR) evaluating the added benefit from an anthracycline–docetaxel sequence versus an anthracycline-containing polychemotherapy in primary breast cancer patients with more than three involved lymph nodes. The prospective uPA and PAI-1 determinations will help to resolve the question of whether high risk patients according to uPA/PAI-1 do benefit from the addition of a taxane (i.e., docetaxel) to standard anthracycline chemotherapy.

Future Clinical Potential

The convincing experimental and clinical data demonstrating a key role for uPA and PAI-1 in tumor cell invasion and metastasis render these two factors promising targets for tumor therapy.[104,105] A first compound, a synthetic serine protease inhibitor (WX-UK1), targeting enzymatically active uPA, has already entered a phase I/II clinical cancer trial in ovarian, gastric, pancreatic, brain, and head and neck cancer patients (Wilex AG, press release, October 14, 2002, www.wilex.com). A trial with WX-UK1 in combination with the 5-FU prodrug Xeloda (capecitabine) in advanced breast cancer has begun as well. Other novel therapeutics are receptor ligand analogues designed to interfere with the cellular uPAR–uPA interaction, which are currently under investigation in preclinical tumor models.[3]

Taking the data on uPA/PAI-1 and response to adjuvant chemotherapy into account, chemotherapy seems to be an ideal complement to these novel therapeutic agents interfering with the uPA–uPAR system. Results from two prospective therapy trials discussed above (NNBC-3, ADEBAR) will not only help to define the most promising chemotherapy for high risk patients according to high uPA/PAI-1, but will also identify the most promising "conventional" chemotherapeutic regimen that may then be combined with the novel targeted therapeutics.

References

1. Andreasen PA, Egelund R, Petersen HH. The plasminogen activation system in tumor growth, invasion, and metastasis. Cell Mol Life Sci 2000;57: 25–40.
2. Reuning U, Magdolen V, Wilhelm O, et al. Multifunctional potential of the plasminogen activation system in tumor invasion and metastasis (review). Int J Oncol 1998;13:893–906.
3. Reuning U, Sperl S, Kopitz C, et al. Urokinase-type plasminogen activator (uPA) and its receptor (uPAR): development of antagonists of uPA/uPAR interaction and their effects in vitro and in vivo. Curr Pharm Des 2003;9:1529–1543.
4. Schmitt M, Harbeck N, Thomssen C, et al. Clinical impact of the plasminogen activation system in tumor invasion and metastasis: prognostic relevance and target for therapy. Thromb Haemost 1997;78: 285–296.
5. Schmitt M, Magdolen V, Reuning U. Theme Issue: update on the role of the fibrinolysis/plasminogen activation system in a cellular context. Thromb Haemost 2003;89:596–598.
6. Egeblad M, Werb Z. New functions for the matrix metalloproteinases in cancer progression. Nat Rev Cancer 2002;2:161–174.
7. Hojilla CV, Mohammed FF, Khokha R. Matrix metalloproteinases and their tissue inhibitors direct cell fate during cancer development. Br J Cancer 2003; 89:1817–1821.
8. Visse R, Nagase H. Matrix metalloproteinases and tissue inhibitors of metalloproteinases: structure, function, and biochemistry. Circ Res 2003;92: 827–839.
9. Egeblad M, Werb Z. New functions for the matrix metalloproteinases in cancer progression. Nat Rev Cancer 2002;2:161–174.
10. Brinckerhoff CE, Matrisian LM. Matrix metalloproteinases: a tail of a frog that became a prince. Nat Rev Mol Cell Biol 2002;3:207–214.
11. McCawley LJ, Matrisian LM. Matrix metalloproteinases: multifunctional contributors to tumor progression. Mol Med Today 2000;6:149–156.
12. Benaud C, Dickson RB, Thompson EW. Roles of the matrix metalloproteinases in mammary gland development and cancer. Breast Cancer Res Treat 1998; 50:97–116.
13. Coussens LM, Fingleton B, Matrisian LM. Matrix metalloproteinase inhibitors and cancer: trials and tribulations. Science 2002;295:2387–2392.
14. Porter-Jordan K, Hoyhtya M, Barnes R, et al. Prognostic value of the level of matrix metalloprotease-2 in the fibroblasts surrounding infiltrating ductal carcinoma of the breast. Breast Cancer Res Treat 1992; 23:149.
15. Talvensaari-Mattila A, Paakko P, Hoyhtya M, et al. Matrix metalloproteinase-2 immunoreactive protein: a marker of aggressiveness in breast carcinoma. Cancer 1998;83:1153–1162.
16. Talvensaari-Mattila A, Paakko P, Turpeenniemi-Hujanen T. MMP-2 positivity and age less than 40 years increases the risk for recurrence in premenopausal patients with node-positive breast carcinoma. Breast Cancer Res Treat 1999;58: 287–293.

17. Talvensaari-Mattila A, Paakko P, Blanco-Sequeiros G, et al. Matrix metalloproteinase-2 (MMP-2) is associated with the risk for a relapse in postmenopausal patients with node-positive breast carcinoma treated with antiestrogen adjuvant therapy. Breast Cancer Res Treat 2001;65:55–61.
18. Pacheco MM, Nishimoto IN, Mourao Neto M, et al. Prognostic significance of the combined expression of matrix metalloproteinase-9, urokinase type plasminogen activator and its receptor in breast cancer as measured by Northern blot analysis. Int J Biol Markers 2001;16:62–68.
19. Scorilas A, Karameris A, Arnogiannaki N, et al. Overexpression of matrix-metalloproteinase-9 in human breast cancer: a potential favourable indicator in node-negative patients. Br J Cancer 2001;84: 1488–1496.
20. Chenard MP, O'Siorain L, Shering S, et al. High levels of stromelysin-3 correlate with poor prognosis in patients with breast carcinoma. Int J Cancer 1996; 69:448–451.
21. McCarthy K, Maguire T, McGreal G, et al. High levels of tissue inhibitor of metalloproteinase-1 predict poor outcome in patients with breast cancer. Int J Cancer 1999;84:44–48.
22. Rasmussen HS, McCann PP. Matrix metalloproteinase inhibition as a novel anticancer strategy: a review with special focus on batimastat and marimastat. Pharmacol Ther 1997;75:69–75.
23. Cavallo-Medved D, Sloane BF. Cell-surface cathepsin B: understanding its functional significance. Curr Top Dev Biol 2003;54:313–341.
24. Kos J, Werle B, Lah T, et al. Cysteine proteinases and their inhibitors in extracellular fluids: markers for diagnosis and prognosis in cancer. Int J Biol Markers 2000;15:84–89.
25. Kos J, Lah TT. Cysteine proteinases and their endogenous inhibitors: target proteins for prognosis, diagnosis and therapy in cancer (review). Oncol Rep 1998;5:1349–1361.
26. Podgorski I, Sloane BF. Cathepsin B and its role(s) in cancer progression. Biochem Soc Symp 2003;70: 263–276.
27. Roshy S, Sloane BF, Moin K. Pericellular cathepsin B and malignant progression. Cancer Metastasis Rev 2003;22:271–286.
28. Premzl A, Zavasnik-Bergant V, Turk V, et al. Intracellular and extracellular cathepsin B facilitate invasion of MCF-10A neoT cells through reconstituted extracellular matrix in vitro. Exp Cell Res 2003;283: 206–214.
29. Schwartzberg LS. Clinical experience with edrecolomab: a monoclonal antibody therapy for colorectal carcinoma. Crit Rev Oncol Hematol 2001;40: 17–24.
30. Rochefort H, Chalbos D, Cunat S, et al. Estrogen regulated proteases and antiproteases in ovarian and breast cancer cells. J Steroid Biochem Mol Biol 2001;76:119–124.
31. Thorpe SM, Rocheford H, Garcia M, et al. Association between high concentration of M52,000 cathepsin D and poor prognosis in primary breast cancer. Cancer Res 1989;49:6008–6014.
32. Tandon AK, Clark GM, Chamness GC, et al. Cathepsin D and prognosis in breast cancer. N Engl J Med 1990;322:297–302.
33. Kute TE, Shao ZM, Sugg NK, et al. Cathepsin D as a prognostic indicator for node-negative breast cancer patients using both immunoassays and enzymatic assays. Cancer Res 1992;52:5198–5203.
34. Henry JA, McCarthy AL, Angus B, et al. Prognostic significance of the estrogen regulated protein, cathepsin D, in breast cancer. An immunohistochemical study. Cancer 1990;65:265–271.
35. Sweep FC, Fritsche HA, Gion M, et al. Considerations on development, validation, application, and quality control of immuno(metric) biomarker assays in clinical cancer research: an EORTC-NCI working group report. Int J Oncol 2003;23:1715–1726.
36. Borgono CA, Michael IP, Diamandis EP. Human tissue kallikreins: physiologic roles and applications in cancer. Mol Cancer Res 2004;2:257–280.
37. Diamandis EP, Yousef GM, Luo LY, et al. The new human kallikrein gene family: implications in carcinogenesis. Trends Endocrinol Metab 2000;11:54–60.
38. Diamandis EP, Yousef GM. Human tissue kallikrein gene family: a rich source of novel disease biomarkers. Expert Rev Mol Diagn 2001;1:182–190.
39. Diamandis EP, Yousef GM. Human tissue kallikreins: a family of new cancer biomarkers. Clin Chem 2002;48:1198–1205.
40. Webber MM, Waghray A, Bello D. Prostate-specific antigen, a serine protease, facilitates human prostate cancer cell invasion. Clin Cancer Res 1995;1: 1089–1094.
41. Takayama TK, McMullen BA, Nelson PS, et al. Characterization of hK4 (prostase), a prostate-specific serine protease: activation of the precursor of prostate specific antigen (pro-PSA) and single-chain urokinase-type plasminogen activator and degradation of prostatic acid phosphatase. Biochemistry 2001;40:15341–15348.
42. Goyal J, Smith KM, Cowan JM, et al. The role for NES1 serine protease as a novel tumor suppressor. Cancer Res 1998;58:4782–4786.
43. Wolf WC, Evans DM, Chao L, et al. A synthetic tissue kallikrein inhibitor suppresses cancer cell invasiveness. Am J Pathol 2001;159:1797–1805.
44. Kishi T, Grass L, Soosaipillai A, et al. Human kallikrein 8, a novel biomarker for ovarian carcinoma. Cancer Res 2003;63:2771–2774.
45. Yousef GM, Polymeris ME, Grass L, et al. Human kallikrein 5: a potential novel serum biomarker for breast and ovarian cancer. Cancer Res 2003;63: 3958–3965.
46. Yousef GM, Diamandis EP. Expanded human tissue kallikrein family—a novel panel of cancer biomarkers. Tumour Biol 2002;23:185–192.
47. Yousef GM, Scorilas A, Kyriakopoulou LG, et al. Human kallikrein gene 5 (KLK5) expression by quantitative PCR: an independent indicator of poor prognosis in breast cancer. Clin Chem 2002;48:1241–1250.
48. Chang A, Yousef GM, Scorilas A, et al. Human kallikrein gene 13 (KLK13) expression by quantitative RT-PCR: an independent indicator of favourable prognosis in breast cancer. Br J Cancer 2002;86: 1457–1464.
49. Yousef GM, Scorilas A, Magklara A, et al. The androgen-regulated gene human kallikrein 15 (KLK15) is an independent and favourable prognostic marker for breast cancer. Br J Cancer 2002;87: 1294–1300.
50. Yousef GM, Scorilas A, Nakamura T, et al. The prognostic value of the human kallikrein gene 9 (KLK9) in breast cancer. Breast Cancer Res Treat 2003;78: 149–158.

51. Yu H, Giai M, Diamandis EP, et al. Prostate-specific antigen is a new favorable prognostic indicator for women with breast cancer. Cancer Res 1995;55: 2104–2110.
52. Yu H, Levesque MA, Clark GM, et al. Prognostic value of prostate-specific antigen for women with breast cancer: a large United States cohort study. Clin Cancer Res 1998;4:1489–1497.
53. Foekens JA, Diamandis EP, Yu H, et al. Expression of prostate-specific antigen (PSA) correlates with poor response to tamoxifen therapy in recurrent breast cancer. Br J Cancer 1999;79:888–894.
54. Luo LY, Diamandis EP, Look MP, et al. Higher expression of human kallikrein 10 in breast cancer tissue predicts tamoxifen resistance. Br J Cancer 2002;86:1790–1796.
55. Talieri M, Diamandis EP, Gourgiotis D, et al. Expression analysis of the human kallikrein 7 (KLK7) in breast tumors: a new potential biomarker for prognosis of breast carcinoma. Thromb Haemost 2004;91:180–186.
56. Borgono CA, Grass L, Soosaipillai A, et al. Human kallikrein 14: a new potential biomarker for ovarian and breast cancer. Cancer Res 2003;63:9032–9041.
57. Yousef GM, Borgoño CA, Scorilas A, et al. Quantitative analysis of human kallikrein gene 14 expression in breast tumours indicates association with poor prognosis. Br J Cancer 2002;87:1287–1293.
58. Yousef GM, Diamandis EP. The new human tissue kallikrein gene family: structure, function, and association to disease. Endocr Rev 2001;22:184–204.
59. Yousef GM, Diamandis EP. Human tissue kallikreins: a new enzymatic cascade pathway? Biol Chem 2002; 383:1045–1057.
60. Conese M, Blasi F. The urokinase/urokinase-receptor system and cancer invasion. Baillieres Clin Haematol 1995;8:365–389.
61. Blasi F, Carmeliet P. uPAR: a versatile signalling orchestrator. Nat Rev Mol Cell Biol 2002;3:932–943.
62. Preissner KT, Kanse SM, May AE. Urokinase receptor: a molecular organizer in cellular communication. Curr Opin Cell Biol 2000;12:621–628.
63. Rosenberg S. The urokinase-type plasminogen activator system in cancer and other pathological conditions: introduction and perspective. Curr Pharm Des 2003;9:4p.
64. Rosenberg S. New developments in the urokinase-type plasminogen activator system. Expert Opin Ther Targets 2001;5:711–722.
65. Stefansson S, McMahon GA, Petitclerc E, et al. Plasminogen activator inhibitor-1 in tumor growth, angiogenesis and vascular remodeling. Curr Pharm Des 2003;9:1545–1564.
66. Andreasen PA, Kjoller L, Christensen L, et al. The urokinase-type plasminogen activator system in cancer metastasis: a review. Int J Cancer 1997;72:1–22.
67. Duffy MJ, O'Grady P, Devaney D, et al. Urokinase-plasminogen activator, a marker for aggressive breast carcinomas. Preliminary report. Cancer 1988; 62:531–533.
68. Janicke F, Schmitt M, Ulm K, et al. Urokinase-type plasminogen activator antigen and early relapse in breast cancer. Lancet 1989;2:1049.
69. Jänicke F, Schmitt M, Hafter R, et al. Urokinase-type plasminogen activator (u-PA) antigen is a predictor of early relapse in breast cancer. Fibrinolysis 1990;4:69–78.
70. Janicke F, Schmitt M, Graeff H. Clinical relevance of the urokinase-type and tissue-type plasminogen activators and of their type 1 inhibitor in breast cancer. Semin Thromb Hemost 1991;17:303–312.
71. Janicke F, Schmitt M, Pache L, et al. Urokinase (uPA) and its inhibitor PAI-1 are strong and independent prognostic factors in node-negative breast cancer. Breast Cancer Res Treat 1993;24:195–208.
72. Foekens JA, Schmitt M, van Putten WL, et al. Plasminogen activator inhibitor-1 and prognosis in primary breast cancer J Clin Oncol 1994;12:1648–1658.
73. Grondahl-Hansen J, Peters HA, van Putten WL, et al. Prognostic significance of the receptor for urokinase plasminogen activator in breast cancer. Clin Cancer Res 1995;1:1079–1087.
74. Ferno M, Bendahl PO, Borg A, et al. Urokinase plasminogen activator, a strong independent prognostic factor in breast cancer, analysed in steroid receptor cytosols with a luminometric immunoassay. Eur J Cancer 1996;32A:793–801.
75. Eppenberger U, Kueng W, Schlaeppi JM, et al. Markers of tumor angiogenesis and proteolysis independently define high- and low-risk subsets of node-negative breast cancer patients. J Clin Oncol 1998; 16:3129–3136.
76. Kim SJ, Shiba E, Kobayashi T, et al. Prognostic impact of urokinase-type plasminogen activator (PA), PA inhibitor type-1, and tissue-type PA antigen levels in node-negative breast cancer: a prospective study on multicenter basis. Clin Cancer Res 1998;4:177–182.
77. Kute TE, Grondahl-Hansen J, Shao SM, et al. Low cathepsin D and low plasminogen activator type 1 inhibitor in tumor cytosols defines a group of node negative breast cancer patients with low risk of recurrence. Breast Cancer Res Treat 1998;47:9–16.
78. Knoop A, Andreasen PA, Andersen JA, et al. Prognostic significance of urokinase-type plasminogen activator and plasminogen activator inhibitor-1 in primary breast cancer. Br J Cancer 1998;77:932–940.
79. Bouchet C, Hacene K, Martin PM, et al. Dissemination risk index based on plasminogen activator system components in primary breast cancer. J Clin Oncol 1999;17:3048–3057.
80. Foekens JA, Peters HA, Look MP, et al. The urokinase system of plasminogen activation and prognosis in 2780 breast cancer patients. Cancer Res 2000;60:636–643.
81. Janicke F, Prechtl A, Thomssen C, et al. Randomized adjuvant chemotherapy trial in high-risk, lymph node-negative breast cancer patients identified by urokinase-type plasminogen activator and plasminogen activator inhibitor type 1. J Natl Cancer Inst 2001;93:913–920.
82. Konecny G, Untch M, Arboleda J, et al. Her-2/neu and urokinase-type plasminogen activator and its inhibitor in breast cancer. Clin Cancer Res 2001;7: 2448–2457.
83. Harbeck N, Kates RE, Schmitt M, et al. Clinical relevance of invasion factors urokinase-type plasminogen activator and plasminogen activator inhibitor type 1 for individualized therapy decisions in primary breast cancer is greatest when used in combination. J Clin Oncol 2002;20:1000–1007.
84. Cufer T, Borstnar S, Vrhovec I. Prognostic and predictive value of the urokinase-type plasminogen activator (uPA) and its inhibitors PAI-1 and PAI-2 in operable breast cancer Int J Biol Markers 2003;18: 106–115.

85. Hansen S, Overgaard J, Rose C, et al. Independent prognostic value of angiogenesis and the level of plasminogen activator inhibitor type 1 in breast cancer patients. Br J Cancer 2003;88:102–108.

86. Harbeck N, Thomssen C, Berger U, et al. Invasion marker PAI-1 remains a strong prognostic factor after long-term follow-up both for primary breast cancer and following first relapse. Breast Cancer Res Treat 1999;54:147–157.

87. Harbeck N, Dettmar P, Thomssen C, et al. Risk-group discrimination in node-negative breast cancer using invasion and proliferation markers: 6-year median follow-up. Br J Cancer 1999;80:419–426.

88. Konecny G, Untch M, Pihan A, et al. Association of urokinase-type plasminogen activator and its inhibitor with disease progression and prognosis in ovarian cancer. Clin Cancer Res 2001;7:1743–1749.

89. Zemzoum I, Kates RE, Ross JS, et al. Invasion factors uPA/PAI-1 and HER2 status provide independent and complementary information on patient outcome in node-negative breast cancer. J Clin Oncol 2003;21:1022–1028.

90. Look M, van Putten W, Duffy M, et al. Pooled analysis of prognostic impact of uPA and PAI-1 in breast cancer patients. Thromb Haemost 2003;90:538–548.

91. Look MP, van Putten WL, Duffy MJ, et al. Pooled analysis of prognostic impact of urokinase-type plasminogen activator and its inhibitor PAI-1 in 8377 breast cancer patients. J Natl Cancer Inst 2002;94: 116–128.

92. Harbeck N, Meisner C, Prechtl A, et al. Level-I evidence for prognostic and predictive impact of uPA and PAI-1 in node-negative breast cancer provided by second scheduled analysis of multicenter Chemo-N0 therapy trial. Breast Cancer Res Treat 2001; 69:213.

93. Sweep CG, Geurts-Moespot J, Grebenschikov N, et al. External quality assessment of trans-European multicentre antigen determinations (enzyme-linked immunosorbent assay) of urokinase-type plasminogen activator (uPA) and its type 1 inhibitor (PAI-1) in human breast cancer tissue extracts. Br J Cancer 1998;78:1434–1441.

94. Janicke F, Pache L, Schmitt M, et al. Both the cytosols and detergent extracts of breast cancer tissues are suited to evaluate the prognostic impact of the urokinase-type plasminogen activator and its inhibitor, plasminogen activator inhibitor type 1. Cancer Res 1994;54:2527–2530.

95. Abraha RS, Thomssen C, Harbeck N, et al. Micromethod for determination of uPA and PAI-1 from preoperative core-needle biopsies in breast cancer. Breast Cancer Res Treat 2003;82:S144.

96. Goldhirsch A, Wood WC, Gelber RD, et al. Meeting highlights: updated international expert consensus on the primary therapy of early breast cancer. Clin Oncol 2003;21:3357–3365.

97. Harbeck N, Kates RE, Look MP, et al. Enhanced benefit from adjuvant chemotherapy in breast cancer patients classified high-risk according to urokinase-type plasminogen activator (uPA) and plasminogen activator inhibitor type 1 (n = 3424). Cancer Res 2002;62:4617–4622.

98. Harbeck N, Alt U, Krüger A, et al. Prognostic impact of proteolytic factors (uPA, PAI-1, cathepsins B, D, L) in primary breast cancer reflects effects of adjuvant systemic therapy. Clin Cancer Res 2001;7:2757–2764.

99. Manders P, Tjan-Heijnen VC, Span PN, et al. Predictive impact of urokinase-type plasminogen activator: plasminogen activator inhibitor type-1 complex on the efficacy of adjuvant systemic therapy in primary breast cancer. Cancer Res 2004;64:659–664.

100. Pierga JY, Laine-Bidron C, Beuzeboc P, et al. Plasminogen activator inhibitor-1 (PAI-1) is not related to response to neoadjuvant chemotherapy in breast cancer. Br J Cancer 1997;76:537–540.

101. Jänicke F, Thomssen C, Pache et al. Urokinase (uPA) and PAI-1 as selection criteria for adjuvant chemotherapy in axillary node-negative breast cancer patients. In: Schmitt M, Graeff H, Jänicke F, eds. Prospects in Diagnosis and Treatment of Cancer. Netherlands: Elsevier Science, 1994:207–218.

102. Foekens JA, Look MP, Peters HA, et al. Urokinase-type plasminogen activator and its inhibitor PAI-1: predictors of poor response to tamoxifen therapy in recurrent breast cancer. J Natl Cancer Inst 1995;87: 751–756.

103. Manders P, Tjan-Heijnen VC, Span PN, et al. The complex between urokinase-type plasminogen activator (uPA) and its type-1 inhibitor (PAI-I) independently predicts response to first-line endocrine therapy in advanced breast cancer. Thromb Haemost 2004;9:514–521.

104. Sperl S, Mueller MM, Wilhelm OG, et al. The uPA/uPA receptor system as a target for tumor therapy. Drug News Perspect 2001;14:401–411.

105. Muehlenweg B, Sperl S, Magdolen V, et al. Interference with the urokinase plasminogen activator system: a promising therapy concept for solid tumors. Exp Opin Biol Ther 2001;1:683–691.

CHAPTER

19

Breast Cancer and Angiogenesis

Manfred Schmitt, PhD,[1] Nadia Harbeck MD, PhD,[1] and Jeffrey S. Ross, MD[2]

[1]Clinical Research Unit, Department of Obstetrics and Gynecology, Technical University of Munich, Germany
[2]Department of Pathology and Laboratory Medicine, Albany Medical College, Albany, New York, USA

Angiogenesis (from the Greek words *angeion* = vessel and *genesis* = birth) is a normal event during fetal growth and development whereby blood vessels are formed *de novo* in the embryo from endothelial cell precursor cells (angioblasts), a process called *vasculogenesis,* which creates the major network of blood vessels, also in the adult. At the outset of this process, the primary capillary plexus is formed, which is then remodeled by the sprouting and branching of new vessels from preexisting ones.[1–3] Angiogenesis is limited in the adult and occurs, for instance, during corpus luteum development after ovulation and in physiologic repair processes such as wound healing and during the menstrual cycle.[4] Normal turnover of endothelial cells in the adult vasculature is relatively scarce.[5] Vascular endothelial cells divide about once every 3 years on average. However, and only when the situation requires it, angiogenesis can stimulate them to divide more rapidly.[3] Still, angiogenesis ensures that developing or healing tissues receive a sufficient supply of nutrients. Normal angiogenesis requires induction of quiescent endothelial cells in a monolayer to divide and to spread to the vascular network, but only to the extent needed by the growing tissue. Thereby, several regulatory events influence the angiogenic process, encompassing effectors of cell–cell and cell–matrix interaction and hemodynamics. The molecules and mechanical forces that mediate and coordinate normal angiogenesis are numerous,[1] including soluble mediators (vascular endothelial growth factor [VEGF], Ang1 and 2, acidic and basic fibroblast growth factor [FGF], platelet-derived growth factor [PDGF], transforming growth factor-α and -β, tumor necrosis factor-α, epidermal growth factor, granulocyte and granulocyte-macrophage colony-stimulating factor, angiogenin, angiotropin, tissue factor, factor V, prostaglandin, nicotinamide, monobutyrin) and cell membrane-bound proteins ($\alpha_v\beta_3$-integrin, $\alpha_v\beta_5$-integrin, $\alpha_5\beta_1$-integrin, VE-cadherin, Eph-4B/ephrin-B2, ephrin-A1, Eph-2A).

Normal angiogenesis depends on the coordination of several independent processes as described by Hughes et al.[6]: (1) Removal of pericytes from the endothelium and destabilization of the vessel by angiopoietin-2, (2) VEGF-induced hyperpermeability permits local extravasation of proteases and matrix components from the bloodstream, (3) proliferation and migration of endothelial cells through extracellular matrix and formation of new blood vessels, (4) proliferation and migration of mesenchymal cells along the new vessel and differentiation into mature pericytes, (5) endothelial cell quiescence, and (6) strengthening of cell–cell contacts and formation of new extracellular matrix to stabilize the new blood vessel. Thus, soluble factors (e.g., growth factors, proteolytic factors tissue and urokinase plasminogen activator [uPA], uPA receptor, plasminogen activator inhibitor type 1 [PAI-1], and matrix metalloproteinases) secreted from different sometimes distant cells, transmembrane receptor proteins (e.g., $\alpha_v\beta_3$-integrin, VE-cadherin, ephrin-2B) binding to extracellular matrix components, and hemodynamic forces all act together to coordinate normal angiogenesis. Excess of angiogenesis, for instance, is associated with rheumatoid arthritis, obesity, psoriasis, diabetic neuropathy, endometriosis, and cancer; insufficient angiogenesis is implicated in stroke, coronary heart disease, atherosclerosis, gestational disorders, osteoporosis, and delayed wound healing.[7,8] It is widely held that during the neoplastic process, an imbalance of proangiogenic and angiogenesis-inhibitory factors enables the spread of the malignant disease by initiating and supporting tumor angiogenesis.[9]

Tumor Angiogenesis

Widely disseminated cancer is characterized by spread of tumor cells to distant loci in the body, the formation of metastases via local infiltration of tumor cells from the primary tumor, intravasation into lymphatic and blood vessels, circulation through the bloodstream, and extravasation to normal tissues elsewhere.[3] Hence, it is the ability of tumor cells to escape from the primary site and to spread to other tissues and organs that causes many forms of cancer to become a life-threatening disease. Though several features distinguish tumor cells from nontransformed cells, they both require an ample supply of oxygen and nutrients and an effective way to remove wastes to maintain normal homeostatic metabolic processes and cell survival. Tumor cells can induce their own new network of blood vessels from the existing normal vasculature in a process that mimics normal angiogenesis.[1,3,10,11] As a result, tumor cells induce the formation of their own blood supply from the aberrant deployment of the normal preexisting vascular tree to supply the tumor tissue with oxygen and nutrients, which affords tumor cells the ability to survive and disseminate to other parts of the body.

The mechanisms by which developing tumors switch to an angiogenic phenotype are still not known in detail.[12] Before the late 1960s, cancer researchers and physicians were under the assumption that the blood reached tumors simply because the existing blood vessels dilated. It was Judah Folkman, in 1971, who proposed the then novel theory that tumors lay dormant yet viable, unable to grow beyond 2 to 3 mm^3 in size in the absence of neovascularization.[10,11] In essence, Folkman postulated that a tumor remains in a dormant state unable to grow in size beyond a few millimeters in the absence of a putative angiogenic factor to ensure blood supply of the tumor tissue cells. Such an angiogenic factor is eventually used by the tumor to initiate the shift of nearby vascular endothelial cells from a resting state into one resulting in the formation of new blood vessels to vascularize the tumor bed.[1,13] Even after all these years, Folkman's model is as relevant today as it was novel in 1971, and considerable evidence has amounted that tumor growth, invasion, and metastasis are associated with angiogenesis.[9,14–19]

Since that time, he and other cancer researchers have identified and characterized a number of substances that either promote (angiogenic factors) or stop (angiostatic substances) tumor angiogenesis. Angiogenic (VEGF, FGF, Heparinase, Ang 2, interleukin-8, matrix metalloproteinase-2) and angiostatic factors (interferons, interleukins, tissue inhibitors of matrix metalloproteinases) may be released by tumor cells, endothelial cells, mast cells, fibroblasts, or macrophages[1] or are derived from blood proteins (e.g., angiostatin from plasminogen)[20] or extracellular matrix components (e.g., endostatin from collagen XVIII).[21]

Resulting from this, the local balance between angiogenic and angiostatic factors determines whether the tumor will attract blood vessels or not and thereby connect with the bloodstream, grow large, and metastasize.[12,22] Tumor angiogenesis is also implicated in tumor metastasis; for shedding of tumor cells from the primary tumor, a full network of blood vessels is advantageous because shedding of tumor cells into the bloodstream is essential for hematogenous metastasis. In general, cancer patients with higher densities of blood vessels in their primary tumors are more likely to develop metastasis,[23] and the extent of tumor cell shedding during surgery is related to vascular density.[24]

Counting microvessels after immunostaining in a microscopic field of a tumor tissue section is an established methodology of angiogenesis quantification and gives an estimate of the net result of phases of angiogenesis and of the angioregression a tumor went through.[25,26] Quantification of angiogenesis or tumor vascularization is of prognostic and/or predictive value in oncology,[26] and the aim of the College of American Pathologists[27] and the International Consensus Board on the Methodology and Criteria of Evaluation of Angiogenesis Quantification in Solid Human Tumors[26,28] is to integrate methodological novelties for guidelines for the estimation of the amount of blood vessels in a solid tumor and for the estimation of ongoing angiogenesis. Qualitative aspects of tumor blood vessels and the tumor vasculature other than angiogenesis are also addressed. Interestingly, in a second report of the International Consensus Board in 2002, it was stated that the highly appreciated opinion of Judah Folkman[9] that tumors would need angiogenesis to grow, invade, and metastasize is not applicable to all tumors. Based on recent histomorphological data, the updated view is that all tumors do need blood vessels, but these vessels are not necessarily derived by sprouting of new blood vessels from preexisting ones as in angiogenesis (the so-called angiogenesis-independent growth pattern or vasculogenic mimicry).[29–38] Also, in these consensus guidelines, it was argued that high vessel density does not necessarily predict susceptibility of cancer patients for antiangiogenic therapy.[26]

Angiogenesis Inhibitors in Clinical Development to Treat Cancer

Excessive tumor angiogenesis contributes to cancer growth and metastasis, and angiogenesis can be a major complication of nonneoplastic diseases such as diabetes mellitus. According to a recent report by Featherstone and Griffiths in 2002,[7] The Angiogenesis Foundation claimed that in western nations at least 184 million patients with some kind of angiogenesis-related disease could benefit from some form of antiangiogenic therapy. This is also reflected by heavy investments of the pharmaceutical industry in this challenging sector of medical research, mainly for the treatment of cancer.[39–46] However, ongoing clinical cancer trials reveal that antiangiogenic treatment with single antiangiogenic molecules is more challenging than initially anticipated.[43,45] It is possible that cancer therapy with a single antiangiogenic agent alone may not meet the requirements to fight the multitude of angiogenic factors produced by the cancer cells and their neighboring stromal cells. This is not a surprise, considering that the formation of new blood vessels involves the interaction of various angiogenic molecules[8] that are not specific for the neoplastic process but are produced by all tissues as a general response to hypoxia.[47]

Under normal physiologic circumstances, angiogenesis is controlled through the equilibrium of proangiogenic and antiangiogenic factors, a balance that is disrupted in cancerous tissues.[46] In addition, the recently acquired expanded knowledge of the (patho)physiologic network of angiogenic factors has encouraged the development of novel angiogenesis inhibitors targeting well-defined cellular or soluble effector molecules.[7,40,41,43–46,48–52] However, the devising and testing of angiogenesis-directed reagents in preclinical tumor models or in cancer patients continues to pose problems with the traditional cancer trial design.[13,43,45] Of note, the classic drug development strategy, featuring preclinical testing and phase I through phase III clinical trials, of a typical anticancer drug that has been used for cytotoxic drugs is based on the traditional concept that the cytotoxic therapeutic should lead to shrinking of the tumor. This is in contrast to the new angiogenesis inhibitors in clinical development whose action is mainly cytostatic and causes a deceleration of tumor growth and weakening of tumor spread and metastasis.[45] Randomized cancer therapy trials will be necessary to further demonstrate the clinical utility of the antiangiogenic therapeutics.[13,45] For this purpose, tumor types and clinical end points must be selected with care. Recent evidence with the use of bevacizumab (Avastin™) in both third-line and first-line treatment for metastatic colorectal cancer (see below) suggests that it may not be sufficient or appropriate to perform typical trials for angiogenesis inhibitors in diseases where survival with current therapy is of short duration (e.g., advanced ovarian, lung, colorectal, or breast cancer) or in patients who have been subject of multiple other therapeutic regimens first. In the end, randomized adjuvant studies dispensing these therapeutics to untreated patients with an estimated high-risk of disease recurrence (metastases) might be essential to demonstrate efficacy.

Different kinds of angiogenesis inhibitors have been developed to target vascular endothelial cells in tumor tissue with the intention to block tumor angiogenesis, tumor growth, and tumor cell dissemination. Cristofanilli et al.[45] grouped tumor-associated modulators of angiogenesis into four categories on the basis of their biologic action (**Table 19.1**): (1) modulators of proteolytic enzymes, (2) inhibitors of endothelial cell proliferation and/or survival, (3) upstream modulators, and (4) molecules with nonspecific or unknown mechanism of action. Within these categories, they listed important target

Table 19.1 Classification of Tumor-Associated Modulators of Angiogenesis

Category
Category 1: Modulators of proteolytic enzymes • Urokinase-type plasminogen system • Matrix metalloprotease system
Category 2: Inhibitors of endothelial-cell proliferation and/or survival • Plasminogen • Collagen XVIII • VEGF- and PDGF-receptor tyrosine kinases • Integrins • Aminopeptidase N
Category 3: Upstream modulators • EGFR- and HER-2–associated tyrosine kinases • RAS
Category 4: Molecules with nonspecific or unknown mechanism of action • Calcium • COX-2 • p53 • NF-κB • Cellular hypoxia

Modified from Ref. 45.

molecules to which or from which antiangiogenic drugs have been generated: the urokinase-type plasminogen system and the matrix metalloprotease system (category 1); plasminogen, collagen XVIII, VEGF and PDGF receptor tyrosine kinases, integrins, and aminopeptidase N (category 2); epidermal growth factor receptor (EGFR)– and HER-2–associated tyrosine kinases, and RAS (category 3); and calcium, COX-2, p53, NF-κB, and hypoxic cells (category 4).

Kerbel and Folkman[43] classified the angiogenic inhibitors on their mode of action: direct angiogenesis inhibitors, which prevent endothelial cells from responding to angiogenic factors, and indirect angiogenesis inhibitors, which inhibit tumor-associated proangiogenic proteins or designed to block endothelial cell receptor expression. A summary of antiangiogenesis drugs is provided in **Table 19.2**.[43–48] The list of direct angiogenesis inhibitors includes a variety of different classes of inhibitors, including angiostatin, bevacizumab (Avastin™), arresten, canstatin, combretastatin, endostatin, NM-3, thrombospondin, tumstatin, Vitaxin, and 2-methoxyestradiol.[43] Indirect angiogenesis inhibitors include drugs that target growth factor receptor kinases (EGFR, HER-2, VEGF receptor, PDGF) and the interferon-α receptor. A number of recent

Table 19.2 Classification of Antiangiogenesis Drugs

Class
Inhibitors of proteolysis and matrix digestion • BMS-275291 • Dalteparin (Fragmin®) • Suramin
Direct endothelial cell antagonists • 2-Methoxyestradiol (2-ME) • CC-5013 (thalidomide analogue) • Combretastatin A4 phosphate • LY317615 (protein kinase C beta inhibitor) • Soy isoflavone (Genistein; soy protein isolate) • Thalidomide
Inhibitors of endothelial activation • AE-941 (Neovastat™; GW786034) • Anti-VEGF antibody (bevacizumab [Avastin™]) • Interferon-alpha • PTK787/ZK 222584 • VEGF-Trap • ZD6474
Inhibitors of endothelial survival and integrin signaling • EMD 121974 • Anti-Anb3 integrin antibody (Medi-522; Vitaxin™)
Nonspecific angiogenesis inhibitors • Carboxyamidotriazole (CAI) • Celecoxib (Celebrex®) • Halofuginone hydrobromide (Tempostatin™) • Interleukin-12 • Rofecoxib (VIOXX®) • Bortezomib (Velcade®)

reviews focused on the clinical development of angiogenesis inhibitors in cancer therapy have emphasized the reasons for failure of these agents in clinical trials.[43–49] The list of prominent antiangiogenesis drugs that failed to achieve its planned end point in clinical trials includes the synthetic VEGF receptor tyrosine kinase inhibitors (e.g., SU5416, SU6668), ribozymes (Angiozyme), and protease inhibitors marimastat, prinomastat, and BAY 12-9566, designed to inhibit tumor-associated matrix metalloproteases. These drugs were shown to effectively inhibit further growth of tumors, but tumor regressions were rare. These results reflect the challenges in the design of antiangiogenesis clinical trials and the critical need for surrogate markers of efficacy and their implication for guiding long-term therapy and the use of angiogenesis inhibitors in combination with other therapeutics.[43–45] Naturally occurring inhibitors of endothelial cell proliferation (endostatin, angiostatin, thrombospondin-1, and TNP-40) are currently

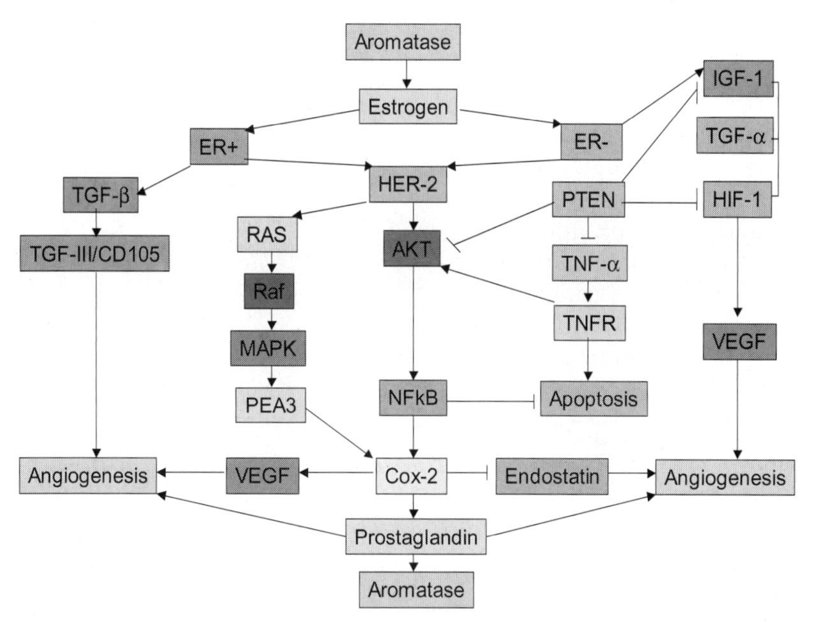

FIGURE 19.1 Angiogenic pathways in breast cancer. →, activation; ⊣, inhibition.

in early clinical testing.[23,40,41,43–46,53] The $\alpha_v\beta_3$-integrin is present on both tumor cells and endothelial cells of newly developed blood vessels and is considered to be a promising target for tumor therapy.[54] Other therapeutics in early phase clinical trials are Vitaxin and Medi-522, which are monoclonal antibodies targeting the $\alpha_v\beta_3$-integrin, and Cilengitide, a small $\alpha_v\beta_3/\alpha_v\beta_5$-integrin antagonist (cyclic RGD peptide).[44] Antiangiogenic inhibitors (e.g., Combretastatin, ZD6126) causing depolymerization of endothelial cell tubulin are also undergoing clinical cancer therapy trials. In essence, clinical trials with these novel therapeutics performed in recent years showed that some but not all of these angiogenesis inhibitors can be administered safely to patients. What remains is the question of how to define accurately the extent of target inhibition and the effective dose of angiogenesis inhibitors to be given to the cancer patient for advanced clinical testing, because some of the reagents tested showed antibiologic activity at rather low concentrations. For this, noninvasive imaging studies applying magnetic resonance imaging, computer tomography, ultrasonography, radionucleotide tracers, and smart contrast agents (fluorescent molecular beacons) could have a key role in assessing the efficacy of treatment.[45,54–63]

Differences between cancer chemotherapy directly aiming to target tumor cell proliferation and antiangiogenic therapy are based on the fact that angiogenesis inhibitors target dividing endothelial cells and not necessarily the tumor cells. Thus, antiangiogenic cancer therapy may prove useful in combination with chemotherapy. Avastin™ (bevacizumab), developed by Genentech/Roche, is a therapeutic humanized antibody that inhibits angiogenic signaling by binding circulating VEGF ligands to prevent its interaction with the receptors VEGFR1 or VEGFR2 (**FIG. 19.1**) and was designed to interfere with the blood supply to a tumor to impair tumor growth and tumor cell dissemination. VEGF acts via two receptors, VEGFR1 and VEGFR2, which are expressed on the vascular endothelium. VEGFR expression may be increased in response to hypoxia, oncogenes, and cytokines.[64–67] Expression of VEGFR is associated with poor prognosis.[68–73] In June 2003, Avastin™ was granted fast track status by the US Food and Drug Administration (FDA) for the treatment of first-line colorectal cancer. On February 26, 2004, Avastin™ was approved by the FDA to be used in combination with 5-fluorouracil based chemotherapy as a treatment for patients with first-line or previously untreated metastatic colorectal cancer. This decision was based on a randomized trial that demonstrated that the addition of bevacizumab to standard combination chemotherapy significantly prolonged survival of patients with metastatic colorectal cancer compared with chemotherapy alone. Avastin™ is the first FDA-approved therapy designed to inhibit tumor angiogenesis and has also shown evidence of biologic activity in patients with metastatic clear cell carcinoma of the kidney, non–small cell lung cancer, and in anthracycline-refractory and taxane-refractory metastatic breast cancer.[74–81] Thus, after a period of declining interest, antiangiogenic therapy has regained considerable interest as a novel approach to cancer therapy, and one can foresee that the combination of antiangiogenic therapeutics with other types of cancer therapeutics will open new vistas for treatment of patients afflicted with solid malignant tumors.

Overview of Angiogenesis-Related Cancer Studies

As of late spring 2004, more than 19,000 publications were listed in PubMed under the key word angiogenesis, of which approximately 58% are focused on cancer/tumor-related topics (**Table 19.3**). Approximately 14% of these publications concern breast cancer, including 84% that cover human research topics or clinical trials. Interestingly, only 84 publications related to angiogenesis are listed under clinical trials, of which only 7 are found under the topic *randomized controlled clinical trials*. Almost half of the breast cancer–related publications are engaged in angiogenesis and therapy-related breast cancer research, whereas only 29% of the publications center on breast cancer prognosis and angiogenesis.

The first reports using the term angiogenesis in PubMed were published in 1952. In 1968, Greenblatt and Shubi[82] described tumor angiogenesis in the hamster (FIG. 19.2). Since around 1971–1972, especially triggered by the pioneering work of Folkman and colleagues,[10,11,83–86] some 900 articles were published within the next 20 years. About 10 years ago, interest in tumor angiogenesis was stimulated even more by the idea of targeting the tumor vasculature

Table 19.3 List of Publications Under the Heading of the Key Word Angiogenesis and Cancer-Related Subheadings, Focusing on Breast Cancer in PubMed

No. of Publications	Key Words
18,999	Angiogenesis
11,106	Angiogenesis + (cancer/tumor/tumour)
1502	Angiogenesis + (cancer/tumor/tumour) + (breast/mammary)
1267	Angiogenesis + (cancer/tumor/tumour) + (breast/mammary) + human
490	Angiogenesis + (cancer/tumor/tumour) + (breast/mammary) + mouse/mice/murine
300	Angiogenesis + (cancer/tumor/tumour) + (breast/mammary) + review
84	Angiogenesis + (cancer/tumor/tumour) + (breast/mammary) + clinical trial
7	Angiogenesis + (cancer/tumor/tumour) + (breast/mammary) + randomized controlled clinical trial
711	Angiogenesis + (cancer/tumor/tumour) + (breast/mammary) + therapy/therapeutic
440	Angiogenesis + (cancer/tumor/tumour) + (breast/mammary) + prognosis/prognostic
272	Angiogenesis + (cancer/tumor/tumour) + (breast/mammary) + response
104	Angiogenesis + (cancer/tumor/tumour) + (breast/mammary) + anti-angiogenic/ anti-angiogenesis
98	Angiogenesis + (cancer/tumor/tumour) + (breast/mammary) + prediction/predictive

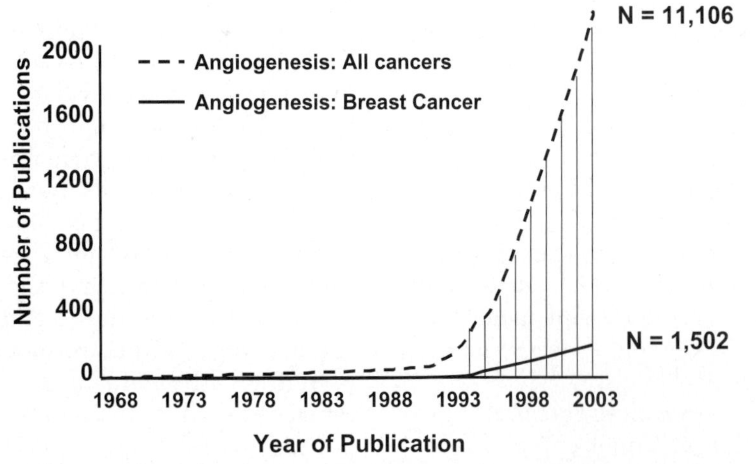

FIGURE 19.2 Angiogenesis and (breast) cancer: publication record listed in PubMed.

with antiangiogenesis agents to inhibit tumor growth and metastasis. Since then, the number of publications has gone up steadily. Exactly 3158 publications dealing with the topic of angiogenesis were published in PubMed in 2003, of which 1917 were addressing cancer/angiogenesis-related topics. Of these, 191 publications (10%) were connected with human breast cancer.

Prognostic Value of Tumor Vascularity in Breast Cancer

Breast cancer is a heterogeneous disease, and a major challenge for clinicians, physician–scientists, and basic cancer researchers lies in assessing the individual patient's risk profile at the time of diagnosis and in improving disease outcome.[87,88] Yet in contrast to many other solid malignancies, breast cancer is a potentially curable disease because of certain clinicobiologic features of cancer tissue and the availability of effective treatment modalities.[87] The identification of breast cancer patients at high risk of disease recurrence and the selection of appropriate systemic adjuvant therapy are based on both prognostic and predictive factors. Nonetheless, there are only a few accepted prognostic and predictive factors that include disease-related characteristics such as tumor size and axillary lymph node status, histologic grade, and some biologic tumor features like the steroid hormone receptors estrogen receptor (ER) and progesterone receptor, the oncogene *HER-2*, and the proteolytic factors uPA and its inhibitor PAI-1 (see Chapter 10).[89–96] A variety of new breast cancer tissue markers related to tumor cell proliferation, apoptosis, angiogenesis, tumor cell invasion, or metastasis is under investigation as prognostic factors and/or predictive factors.[80,81]

Similarly to other solid malignant tumors, tumor cell invasion and metastasis of breast cancer depend on vascularization and angiogenesis of the tumor tissue and the surrounding extracellular matrix. For more than 10 years, microvessel density (MVD) count and surrogate markers of tumor vascularity/angiogenesis proved to be of prognostic value, predicting disease recurrence and early death both in node-negative and node-positive patients (FIG. 19.3). Quantitative assessment of MVD is still the most commonly used technique to determine intratumoral angiogenesis in breast cancer and its clinical value as a prognostic factor.[23] In 1991–1993, Weidner first described MVD by using immunohistochemical processing of fixed breast and prostate cancer tissue specimens by staining endothelial cells of blood vessels for factor VIII.[97–99] Subsequently, additional blood vessel antigens (e.g., CD31, CD34, CD105, PECAM-1, or integrin $\alpha_v\beta_3$, collagen) were stained by antibodies for improved quantitative MVD (**Table 19.4**). **Table 19.4** lists a series of 30 individual studies of the prognostic significance of angiogenesis in both lymph node–negative

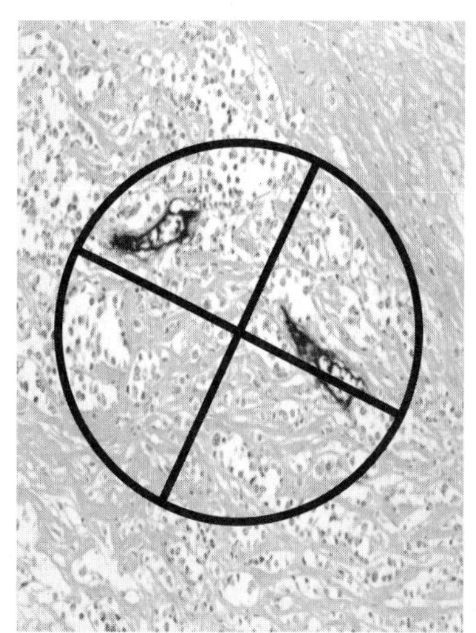
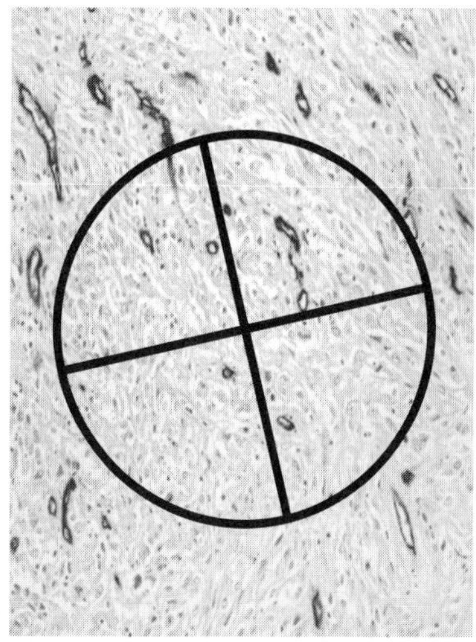

FIGURE 19.3 Measuring microvessel density. Examples of Chalkley counts in hot spots with low (left), intermediate (center), and high (right) numbers of vessel profiles. The ocular containing the Chalkley grid is rotated until the maximal number of immunostained vessels in the three most vascular areas of the tumors are contacted by the Chalkley graticule. The Chalkley grid area is 0.196 mm^2; ×250. (Immunoperoxidase, Anti-CD31 ×200) (From Ref. 122.) **See Plate 44 for color image.**

Table 19.4 Selected Studies of Microvessel Density and Prognosis in Breast Cancer

Year	Author	Reference	No. of Patients	Marker Determined by IHC	Disease Outcome Statistical Correlation		Comment
					Univariate	Multivariate	
1991	Weidner	97	49	Factor VIII	Yes	Yes	Factor VIII antigen marker for endothelial cells. MVD independently predicted local and systemic metastasis.
1992	Bosari	102	88	Factor VIII	Yes	Yes	MVD and vascular invasion were independent predictors.
1992	Hall	103	87	Factor VIII	No	No	No prognostic significance found.
1992	Weidner	98	165	Factor VIII	Yes	Yes	MVD was an independent predictor of overall and disease-free survival.
1993	Toi	104	125	Factor VIII	Yes	Yes	MVD was an independent prognostic factor in both LN+ and LN– groups.
1994	Fox	105	109	CD31	Yes	Yes	High MVD counts also predicted adverse prognosis in ER+ group.
1995	Axelsson	106	386	Factor VIII	No	No	Only ER, grade, and p53 status were of prognostic significance.
1995	Bevilacqua	107	211	CD31	Yes	Yes	MVD predicted overall survival in LN– patients and correlated with other prognosis variables.
1995	Costello	108	87	Factor VIII	No	No	LN– patients only. No correlation with ER, HER-2, p53, or overall survival.
1995	Fox	109	211	Factor VIII	Yes	Yes	MVD predicted survival only when the Chalkley point counting method was used.
1995	Siitonen	110	95	Factor VIII CD31, CD34	No	No	LN– patients only. Two MVD measurement systems. MVD correlated with ER and HER-2 status but did not predict prognosis.
1995	Toi	111	328	Factor VIII	Yes	Yes	MVD was an independent predictor.
1996	Heimann	112	167	CD34	Yes	Yes	MVD measured by image analysis was an independent prognostic factor in LN– patients.
1996	Morphopoulos	113	160	Factor VIII	No	No	No prognostic significance of MVD in infiltrating lobular carcinoma.
1996	Simpson	114	178	Factor VIII	Yes	Yes	MVD predicted overall survival when computer image analysis determined total endothelial area in LN– patients.
1997	Marinho	115	45	Factor VIII	No	No	MVD was correlated only with patient age and not with other prognostic variables or survival.
1997	Martin	116	197	FVIIIRAg CD31, CD34	Yes	No	Close association for all three antigens determined between MVD and survival.
1999	Kato	117	109	Factor VIII	No	No	20-year follow-up. Average, central, and highest microvessel count data were used.

Table 19.4 Selected Studies of Microvessel Density and Prognosis in Breast Cancer (*continued*)

Year	Author	Reference	No. of Patients	Marker Determined by IHC	Disease Outcome Statistical Correlation		Comment
					Univariate	Multivariate	
1999	Kumar	118	106	CD34, CD105	Yes	Yes	MVD predicted survival only when CD105 staining was used to detect endothelial cells.
1999	Viens	119	135	CD31	Yes	Yes	MVD predicted survival in patients treated with adjuvant chemotherapy.
2000	Fridman	120	216	Factor VIII	No	No	No correlation of MVD with prognosis.
2000	Guidi	121	110	Factor VIII	Yes	Yes	LN+ patients only. Only MVD measured in LNs was an independent predictor of survival.
2000	Hansen	122	836	Factor VIII	Yes	Yes	Using the Chalkley counting system, MVD was an independent prognostic factor.
2001	Kato	123	260	Factor VIII	Yes	Yes	Average microvessel count was an independent prognostic factor.
2002	Arora	124	36	CD31	Yes	Yes	MVD independently predicted overall and disease-free survival.
2002	Guidi	125	577	Factor VIII	No	No	MVD did not predict response to adjuvant CAF chemotherapy.
2002	Gunel	126	42	Factor VIII	No	No	MVD reached near significance in predicting resistance to both anthracycline and CMF adjuvant chemotherapy regimens.
2002	Offersen	127	455	CD31	Yes	Yes	Chalkley measurements of MVD predicted survival in both LN− and LN+ patients but was an independent factor in LN+ patients only.
2003	Hansen	128	228	CD34	Yes	Yes	MVD using the Chalkley system independently predicted survival.
2003	Olewniczak	129	226	CD31	Yes	Yes	MVD predicted relapse in LN− patients.

LN−, LN+, lymph node negative, lymph node positive.

and lymph node–positive breast cancer.[97,98,102–129] In 2001, Gasparini[23] reviewed the prognostic value of MVD and listed 44 studies concerned with the prognostic value of MVD in breast cancer; in 35 studies a prognostic relevance of MVD was demonstrated, whereas 9 studies showed no prognostic relevance. The explanation for the lack of achievement of uniform prognostic significance for MVD may depend, in part, on the variance in endothelial detection techniques, methods of microvessel counting, different uses of image analysis hardware and software, and different methods of vessel scoring and data analysis.

Of note, the most consistent results obtained in the MVD studies of breast cancer have been achieved when the Chalkley method of vessel counting has been used (FIG. 19.3).[109,122,127] With the Chalkley technique, the three most vascular areas (hot spots) with the highest number of microvessel profiles are subjectively selected from each tumor section and measured for MVD by a 25-point Chalkley eyepiece graticule applied to each hot spot area.

In an updated extended report, Uzzan et al.[100] reviewed in 2004 all of the then available published studies describing the potential clinical value of MVD and performed a meta-analysis on the basis of these published data. The authors listed 87 studies of which 43 studies with 8936 patients qualified for the meta-analysis, linking MVD to disease-free and/or overall survival. Although only in 22 (4,779 patients) of the 43 studies could a prognostic value for MVD be shown, in their meta-analysis the authors calculated a significant but only weak prognostic power of MVD for breast cancer prognosis. It is worth mentioning that MVD is not considered a predictor of treatment efficacy, because it is not a measure of the angiogenic dependence of a tumor. Consequently, in breast cancer MVD is not recommended for stratifying patients to be enrolled in clinical therapy trials.[101]

Prognostic Significance of Angiogenic Factor VEGF in Breast Cancer

Several angiogenic factors have been associated with prognosis in various kinds of solid malignant tumors. In breast cancer, most of the studies addressing the clinical relevance of angiogenic factors to predict the course of the disease have centered on VEGF and associated factors.[46] Preclinical and clinical studies support the notion that human breast cancer is an angiogenesis-dependent disease and that VEGF plays a key role in this process.[40] A variety of cells, but usually not endothelial cells themselves, secretes VEGF, which is a heparin-binding glycoprotein secreted as a homodimer of 45 kDa.[130,131] VEGF, also known as VEGF-A, increases vascular permeability and was therefore initially termed vascular permeability factor.[132] VEGF initiates vasodilatation, partly through stimulation of nitric oxide synthase in endothelial cells[133]; stimulates cell migration; and inhibits apoptosis.[133,134] Several splice variants of VEGF-A are known, with VEGF[165] being the most predominant protein that, via its heparin-binding domain, binds to extracellular matrix and is involved in binding to heparin sulfate and to VEGF receptors. Other members of the VEGF family have been cloned, including VEGF-B, -C, and -D.[135] VEGF-A transcription is augmented in response to hypoxia and by activated oncogenes.[136–138] Hypoxia inducible factor-1α (HIF-1α) and -2α (HIF-2α) are degraded by proteasomes in normoxia but stabilized in hypoxia. VEGF transcription in normoxia is activated by H-*ras* and other oncogenes and several transmembrane tyrosine kinases such as EGFR and HER-2. These pathways together account for a marked up-regulation of VEGF-A in tumors compared with normal tissues and are often of prognostic importance.[139–142] VEGF interacts with three different receptors: VEGFR1 (flt-1), VEGFR2 (KDR/flk-1), and VEGFR3 (flt-4),[143,144] which display common properties of multiple IgG-like extracellular domains and tyrosine kinase activity. Endothelial cells express VEGF receptors. VEGFR1 and VEGFR2 are up-regulated on tumor cells and the proliferating endothelium to a certain extent by hypoxia but also in response to VEGF-A itself. The VEGF-C/VEGFR3 pathway is also important for lymphatic proliferation. Platelets release VEGF upon activation and are a major source of VEGF present in tumors.[145]

Toi and Gasparini were the first to develop methods (immunohistochemistry [IHC], enzyme-linked immunosorbent assay [ELISA]) for the quantitative assessment of VEGF and its isoforms in tumor tissue extracts and serum/plasma. These workers also explored the potential of VEGF to serve as a prognostic marker in breast cancer. VEGF is emerging as a powerful endothelial cell mitogen in breast cancer with significant prognostic impact for disease-free and overall survival.[40,53,140,146] This is particularly true for early stage breast cancer patients who are at risk if displaying elevated VEGF tumor tissue levels.[147] As can be seen from the publications listed in **Table 19.5**, a significant number of studies have implicated high levels of VEGF in patient serum or tumor protein extracts as an adverse prognostic factor for both node-negative and node-positive disease.[91,148–181] Such patients are unlikely to benefit from standard adjuvant therapy but might be candidates for novel antiangiogenesis therapy[53,147] (FIG. 19.4). For this targeted therapy, however, VEGF has to be determined in the tumor tissue by IHC or ELISA but not in plasma or serum. Collection and storage of fresh (frozen) tumor tissue is therefore mandatory and a prerequisite for testing.[46]

Other Angiogenic Factors with Clinical Impact in Breast Cancer

Relf et al.[151] determined the expression of seven angiogenic molecules in primary human breast cancers, and the relationship of expression to

Table 19.5 VEGF and Prognosis in Breast Cancer

Year	Author	Reference	No. of Patients	Source	Method	Prognostic Relevance	Comment
1994	Toi	148	103	Fixed tissue	IHC	Yes	VEGF expression predicted relapse-free survival
1995	Toi	149	328	Fixed tissue	IHC	Yes	VEGF status was a significant prognostic indicator for disease-free survival
1997	Gasparini	140	260	Tumor tissue cytosol	ELISA	Yes	VEGF expression independently predicted relapse-free and overall survival
1997	Obermair	150	120	Tumor tissue cytosol	ELISA	No	VEGF correlated with MVD; MVD was an independent prognostic factor, but VEGF level was not
1997	Relf	151	64	Tissue	RNAse protection assay	Yes	VEGF, ER, and EGFR expression all predicted prognosis on univariate analysis
1997	Toi	152	260	Tumor tissue cytosol	ELISA	Yes	VEGF was found to be a statistically significant prognostic factor for both relapse-free and overall survival
1998	Eppenberger	153	305	Tumor tissue cytosol	ELISA	Yes	Tumor VEGF level is an important prognostic parameter among several markers of tumor angiogenesis and proteolysis
1998	Linderholm	154	525	Tumor tissue cytosol	ELISA	Yes	VEGF was an independent prognostic factor in LN– patients
1999	Gasparini	155	301	Tumor tissue cytosol	ELISA	Yes	VEGF levels were significantly predictive for disease-free and overall survival
2000	Adams	156	201	Plasma Serum	ELISA	No	Plasma and serum VEGF levels did not predict prognosis
2000	Linderholm	157	362	Tumor tissue cytosol	ELISA	Yes	VEGF levels predicted overall and disease-free survival in LN+ patients
2000	Linderholm	158	833	Tumor tissue cytosol	ELISA	Yes	VEGF is a significant predictor of survival
2001	Gasparini	159	168	Tumor tissue cytosol	ELISA	Yes	Determination of VEGF gave significant prognostic information for node-negative breast cancer patients
2001	Jing	160	70	Fixed tissue	IHC	Yes	VEGF status is a significant prognostic indicator
2001	Kinoshita	161	98	Fixed tissue	IHC	Yes	VEGF predicted disease-free survival
2001	Linderholm	162	224	Tumor tissue cytosol	ELISA	Yes	Elevated VEGF content was statistically significantly correlated with a poor outcome of the patients
2002	Mac-Conmara	163	108	Fixed tissue	IHC	No	No prognostic significance for VEGF expression
2002	Manders	164	574	Tumor tissue cytosol	ELISA	Yes	VEGF levels predicted recurrence in LN– patients who did not receive adjuvant chemotherapy

(continued)

Table 19.5 VEGF and Prognosis in Breast Cancer (*continued*)

Year	Author	Reference	No. of Patients	Source	Method	Prognostic Relevance	Comment
2002	Toi	165	110	Serum tissue	ELISA Immunoblot	Yes	High ratio of serum/tissue VEGF expression was an independent prognostic factor
2002	Wen	166	116	Tumor tissue cytosol	ELISA	Yes	VEGF levels were of prognostic value for both relapse-free and overall survival
2002	Wu	167	125	Plasma	ELISA	Yes	Plasma VEGF independently predicted recurrence and overall survival
2003	Berns	91	160	Tumor tissue cytosol	ELISA	Yes	Combined VEGF/p53 status independently predicted prognosis in patients treated with tamoxifen only
2003	Coradini	168	212	Tumor tissue cytosol	ELISA	Yes	The findings indicate that tumor progression, activated or sustained by high VEGF levels, may be counteracted by tamoxifen in high ER (LBA) cancers
2003	Coskun	169	61	Serum	ELISA	Yes	Serum VEGF levels were higher in patients with metastatic disease than in those with nonmetastatic disease
2003	de Paola	170	242	Fixed tissue	IHC	No	Study did not show a prognostic relevance of VEGF expression in patients with node-negative breast cancer
2003	Linderholm	171	1307	Tumor tissue cytosol	ELISA	Yes	VEGF levels independently predicted survival in ER+ patients treated with adjuvant hormonal therapy
2003	Li	172	81	Fixed tissue	IHC	Yes	Patients with low expression of VEGF had longer disease-free and overall survival
2003	Lim	173	128	Fixed tissue	IHC	Yes	Results indicate that elevated VEGF expression is more common in breast cancer patients with an unfavorable outcome
2003	Ludovini	174	228	Fixed tissue	IHC	No	VEGF expression was not of prognostic relevance
2003	Nakamura	175	105	Fixed tissue	IHC	Yes	VEGF-D was an independent predictor of disease-free survival and correlated with LN and HER-2 status
2003	Nakamura	176	123	Fixed tissue	IHC	Yes	Positive VEGF-C was associated with both disease-free and overall survival
2003	Nishimura	177	187	Plasma	ELISA	Yes	High plasma VEGF predicted disease recurrence
2003	Zhukova	172	83	Fixed tissue	IHC	No	VEGF expression was not a significant prognostic factor, but expression of VEGF receptors Flt-1 and Flk-1 were predictive
2004	Dales	178	905	Fixed tissue	IHC	No	CD105 and Tie-2/Tek expression were independent predictors, VEGFR1 and VEGFR2 levels were not significant

Table 19.5 VEGF and Prognosis in Breast Cancer (*continued*)

Year	Author	Reference	No. of Patients	Source	Method	Prognostic Relevance	Comment
2004	Konecny	179	611	Tumor tissue cytosol	ELISA	Yes	The positive association between HER-2/*neu* and VEGF expression implicates VEGF in the aggressive phenotype exhibited by HER-2/*neu* overexpression and supports the use of combination therapies
2004	Linderholm	180	656	Tumor tissue cytosol	ELISA	Yes	Combined survival analyses regarding c-*erb*B2 and VEGF yielded additional prognostic information
2004	Zhao	181	59	Serum	ELISA	Yes	Findings suggest a new angiogenesis balance is formed in the patients after surgery and such a resultant balance may be beneficial for the prognosis of breast cancer

LN–, lymph node negative, LN+, lymph node positive.

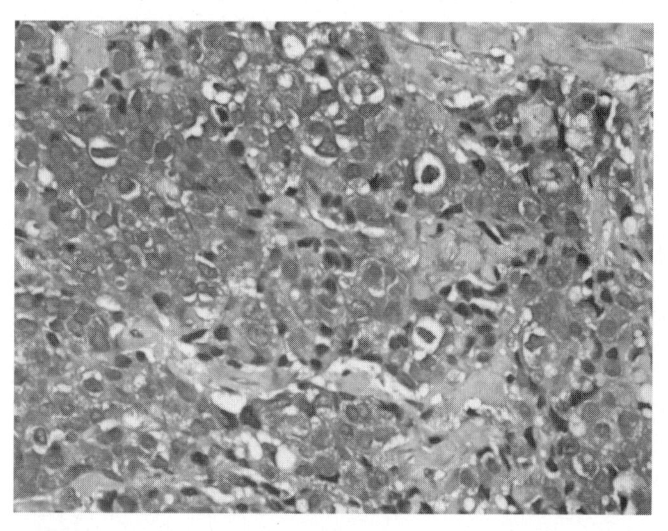

FIGURE 19.4 VEGF expression in breast cancer. Demonstration of high VEGF expression by immunohistochemistry using the Sigma Corp. antibody to VEGF. The image shows a high VEGF expression level in a case Stage II breast cancer which failed adjuvant chemotherapy and relapsed less than 2 years after diagnosis. **See Plate 45 for color image.**

ER status and MVD was also examined. VEGF and its four isoforms (121, 165, 189, and 206), transforming growth factor-β_1, pleiotrophin, acidic and basic FGF, placental growth factor, and thymidine phosphorylase (also known as platelet-derived endothelial cell growth factor) were quantitated by RNase protection analysis or measured by ELISA (β-FGF).[151] All tumors expressed at least six different vascular growth factors. VEGF was most abundant, and the transcript for $VEGF_{121}$ predominated. Thymidine phosphorylase and transforming growth factor-β_1 were also highly expressed. Expression of most of the angiogenic factors did not correlate with that of ER or MVD, except for thymidine phosphorylase. High VEGF expression correlated with poor prognosis in univariate analysis ($P = 0.03$), as did the ER.

Thymidine phosphorylase is involved in degradation of the pyrimidine nucleosides through phosphorolysis and also a converting enzyme of the prodrug 5′-deoxy-5-fluorouridine to 5-fluorouracil in tumors.[23,182] It is mainly expressed by host cells, especially monocytes/macrophages, and not by the tumor cells. Thymidine phosphorylase most probably is an indirect angiogenic factor, although its mechanism of action has not yet been clarified.[46] Expression of thymidine phosphorylase in breast cancer is associated with a favorable outcome,[152,183–186] also in the subgroups of breast cancer patients treated with adjuvant chemotherapy.[23,187,188]

Hypoxia regulates and influences the level of thymidine phosphorylase, demonstrating that tumor microenvironmental factors can result in the specific up-regulation of an angiogenic enzyme that can also activate 5-fluorouracil prodrugs and therefore may serve as a potential therapeutic target.[189] Tumor tissue hypoxia initiates a cascade of molecular pathways that include angiogenesis, glycolysis, and alterations in microenvironmental pH and is defined as the reduction of oxygen supply below the normal level for a tissue such that cellular mechanisms are compromised or an adaptive response occurs.[190,191] Regions of low oxygen tension are common findings in malignant tumors and are associated with increased frequency of tumor invasion and metastasis.[192,193] The cellular response to hypoxia is modulated by transcription factors known as HIFs, of which the best characterized is HIF-1. HIF-1 is differentially expressed under normoxic and hypoxic conditions.[194–197] Both low levels of oxygen, perhaps via reduced activity of a 2-oxoglutarate dependent oxygenase, and a number of tumor-specific genetic alterations synergistically act to induce the HIF system.[191]

Targets of HIF-1 mediated transcription are the angiogenic factor VEGF and a variety of genes encoding essential pathways in systemic, local, or intracellular homeostasis encompassing erythropoietin, inducible nitric oxide synthase, hemoxygenase-1, transferrin, and several glycolytic enzymes.[190,193] By interaction of HIF-1 with these target genes, HIF-1 can govern several metabolic pathways relevant for tumor progression, which may be considered as adaptive responses to tissue injury, thereby supporting tumor growth and metastasis. HIF-1α is overexpressed in a large number of human tumors, and its overexpression correlates with poor prognosis and treatment failure.[198] Four recent breast cancer studies have demonstrated the potential utility of HIF-1 to predict breast cancer prognosis. Gruber et al.[199] found that HIF-1α is expressed in most patients with node-positive breast cancer and that it can serve as a prognostic marker for an unfavorable outcome in breast cancer patients with T1/T2 tumors and positive axillary lymph nodes. In another study by Bos et al.,[200] increased levels of HIF-1α determined by IHC assessment were associated independently with shortened survival in patients with lymph node–negative breast carcinoma. The study by Schindl et al.[201] also showed that HIF-1α is an independent prognostic factor for an unfavorable prognosis in patients with lymph node–positive breast cancer. Taken together, these results indicate that patients with advanced stage breast cancers might profit from future therapies targeting HIF-1α. Interestingly, Cayre et al.[202] found a strong positive association between HIF-1α and sHIF-1α, sHIF-1α and aHIF, and an inverse correlation between HIF-1α/sHIF-1α and aHIF. In their series of breast cancer patients, aHIF, and not HIF-1α transcript, showed up as a marker of poor prognosis.

A potential link between up-regulation of HIF-1 in hypoxia and maintenance of pH balance are transmembrane carbonic anhydrases, which catalyze the reversible hydration of carbon dioxide to carbonic acid.[190] Particularly, carbonic anhydrases IX and XII are strongly up-regulated by hypoxia, also in breast cancer tissue.[203–205] Carbonic anhydrase IX expression, assessed by IHC by Chia et al.,[205] was associated with poor prognosis in patients with invasive breast carcinoma. Span et al.[203] determined the carbonic anhydrase IX messenger RNA expression level in breast cancer tissues, and their results indicate that patients with low carbonic anhydrase IX levels benefit more from adjuvant treatment than do patients with high levels. Watson et al.[204] found that carbonic anhydrase IX expression in breast cancer is influenced by factors related to differentiation and hypoxia and that carbonic anhydrase XII expression is associated with a better prognosis in invasive breast carcinoma patients.

It is of interest to note that in microenvironments where oxygen is scarce and glucose consumption high, such as within the growing tumor, a metabolic shift occurs to maintain local homeostasis.[190] The primary pathway for the efficient generation of ATP in eukaryotic cells in normoxia involves oxidative phosphorylation, but in tumors, a metabolic shift will happen in hypoxia under the regulation of HIF-1, leading to down-regulation of electron transport chain enzymes and affecting transcription of enzymes controlling the glycolytic pathway.[206] Key glycolytic enzymes are aldolase A, pyruvate kinase M, and phosphoglycerate kinase 1. In tumors, phosphoglycerate kinase 1 also functions as a reductase that reduces disulfide bonds in the serine proteinase plasmin. Disulfide bonds in secreted proteins are considered to be inert because of the oxidizing nature of the extracellular milieu.[207–212] As a result, reduction of plasmin initiates proteolytic cleavage in the kringle 5 domain and release of the tumor blood vessel inhibitor angiostatin. Phosphoglycerate

Table 19.6 Advantages and Disadvantages of Targeting Angiogenesis in the Treatment of Breast Cancer

Advantages

- Activated endothelium in tumor vessels express accessible targets for therapy.
- Tumor vascular endothelium is molecularly different from normal vessel endothelium allowing for the protection of the normal vasculature from untoward side effects of antiangiogenic drugs.
- Antiangiogenic drugs show synergy with targeted therapeutics and cytotoxic drugs in preclinical models.
- Preclinical data suggest that tumor vascular endothelium does not develop resistance to antiangiogenic therapies.

Disadvantages

- Tumor vessels are molecularly heterogeneous, which could potentially interfere with antiangiogenesis targeted therapies.
- Well-established and advanced primary tumors and metastatic lesions may produce abundant levels of vessel-stimulating substances overwhelming the potential efficacy of current antiangiogenesis agents.
- Interference with normal blood vessel growth may cause delayed wound healing in postsurgical patients, block revascularization in sites of ischemic injury, and cause cardiovascular complications including hypertension, stroke, and vascular thrombosis.
- The lack of validated serum or tissue based markers of angiogenesis and surrogate biomarkers of response to antiangiogenic therapies.

Modified from Ref. 225.

Table 19.7 Selected Antiangiogenesis Therapy in Breast Cancer

Agent	Source	Target	Clinical Trial Status	Comment
Endostatin	Entremed	Endothelial cell growth and apoptosis	Phase I completed	Further testing on hold due to lack of proven efficacy
Marimastat	British Biotech	Matrix metalloproteases	Phase III trial completed	Development stopped due to lack of efficacy and prominent toxicity
Bevacizumab	Genentech/Roche	VEGF	Phase I/II trial failed to achieve end point	Further trials in progress with combinations with cytotoxic drugs
SU5416	Sugen	VEGF RTK (flk-1)	Phase II/III trials	Pending results of later stage trials
Thalidomide	Celgene	Not certain; possible tumor necrosis factor-α modulation	Phase II trial failed to achieve end point	Development stopped due to lack of efficacy
Gefitinib, Erlotinib, Cetuximab	Astra-Zeneca/ Genentech-OSI/ Imclone	EFGR	Phase II/III trials ongoing in combination with cytotoxic drugs	Pending results of later stage trials
Celecoxib	Searle	COX-2	Phase II/III trials in combination with cytotoxic drugs and aromatase inhibitors	Pending results of later stage
Bortezomib	Millennium	Proteasome	Phase I	In early stage clinical trials in combination with docetaxel and trastuzumab

kinase 1 thus participates in the angiogenic process, not only as a glycolytic enzyme but also as a disulfide reductase. Elevated levels of phosphoglycerate kinase 1 in the blood of patients with lung adenocarcinomas[213] were significantly correlated with poor outcome. Breast cancer studies demonstrating the clinical utility of phosphoglycerate kinase 1 for breast cancer have not been published so far.

Several studies have shown a relationship between angiogenesis and cyclooxygenase (COX)-2 expression.[214] COX-2 regulates the rapid production of high levels of prostaglandins during inflammation and induces proangiogenic factors such as VEGF, nitric oxide synthase, interleukin-6 and interleukin-8, and the angiopoietin 1 and 2 receptors.[215] The prognostic impact of COX-2 in breast cancer was reviewed recently by Denkert et al.,[216] including a survey of eight different immunohistochemical studies that investigated the expression of COX-2 in a total of 2392 primary breast carcinomas. Of these, 40% were found to be COX-2 positive. In general, overexpression of COX-2 has been associated with poor prognosis. These investigations will provide the basis for further evaluation of a possible clinical benefit of COX-2 inhibitors in breast cancer therapy. Also, DNA methylation of the COX-2 promoter region may be one of the mechanisms by which tumor cells regulate COX-2 expression.[217,218]

A new autocrine–paracrine signaling pathway within both angiogenesis and breast cancer progression was brought recently to our attention by Menendez et al.[219] CYR61 (CNN1) is a member of the cysteine-rich 61/connective tissue growth factor/nephroblastoma (CYR61/CTFG/NOV) family of growth regulators. CNN is a proangiogenic factor that mediates diverse roles in development, cell proliferation, and tumorigenesis. Elevated levels of CYR61 in breast cancer are associated with more advanced disease. The exact mechanisms by which CYR61 promotes an aggressive breast cancer phenotype are still unknown, and large-scale prognostic studies in breast cancer have not been published to date.

Tumor angiogenesis is associated with extracellular degradation and remodeling, involving different proteolytic systems such as the matrix metalloprotease system, the cysteine protease system, and, most importantly for breast cancer prognosis and therapy response prediction, the serine protease system of plasminogen activation, involving the two key molecules uPA and its inhibitor PAI-1. To review the prominent roles of uPA and PAI-1 in breast cancer as prognostic and predictive factors and also as potential targets for therapy, see Chapter 18. Recently, several reviews have been published linking the plasminogen activation system to tumor angiogenesis and poor breast cancer patient outcome.[128,220–224] Because uPA- and PAI-1–positive tumors are considered to follow different and independent pathways to angiogenesis than classic direct microvessel generation, drugs targeting these proteases may have synergistic therapeutic benefits with other direct antiangiogenesis agents.[220]

Antiangiogenic Therapeutic Strategies in Breast Cancer

Angiogenesis has been a target for the treatment of breast cancer for more than 10 years.[39,40,44,52,225] The potential advantages and disadvantages of developing antiangiogenic drugs for breast cancer patients has recently been reviewed by Morabito et al.[40] and summarized in **Table 19.6**. A summary of the results and status of antiangiogenesis clinical trials in breast cancer is shown in **Table 19.7**. Several antiangiogenic agents are of particular significance.

Endostatin, an inhibitor of endothelial cell proliferation, has completed two phase I trials that included some patients with metastatic breast cancer showing no significant toxicity and no apparent clinical benefit.[226]

Marimastat, an oral anti–matrix metalloprotease, failed to achieve its time to progression end point compared with placebo in a randomized phase III trial when used as first-line therapy for metastatic breast cancer.[227] This agent also featured significant dose-dependent musculoskeletal toxicity.[228]

Bevacizumab, the recently approved humanized monoclonal antibody targeting the VEGF, is now being used in first-line treatment of colorectal cancer. In breast cancer, bevacizumab showed promise in phase I and II trials, producing clearcut objective responses in advanced chemotherapy treated metastatic breast cancer. However, bevacizumab combined with capecitabine failed to achieve a survival benefit in a phase III randomized trial comparing it with capecitabine alone in patients with advanced refractory breast cancer, although the combination produced a significant increase in overall response rate.[79]

The promising effect on response rate in this trial and the concept that earlier use of antiangiogenic agents may enhance their efficacy has led to the continuation of clinical trials of bevacizumab in combination with paclitaxel for early breast cancer patients.

SU5416 is a small molecule inhibitor of the VEGF activating flk-1 tyrosine kinase receptor. Early phase I testing has included several breast cancer patients, who experienced several partial responses and disease stabilization.[229] Further testing of this compound is currently in progress.

Thalidomide, originally a sedative and antiemetic, was withdrawn from the market about 40 years ago because its administration during pregnancy was associated with birth defects, mostly stunted fetal extremity development (phocomelia). More recently, research demonstrated its activity in the management of leprosy that resulted in its approval by the FDA for this indication. The drug has been widely used off-label for the treatment of multiple myeloma. Thalidomide is believed to be an antiangiogenic agent, but the mechanism of action is not currently well understood. Clinical trials of thalidomide in the treatment of breast cancer failed to demonstrate objective activity as a single agent in multiple institutions.[230,231] Nonetheless, current ongoing clinical trials are testing thalidomide in combinations with both hormonal therapies and cytotoxic agents.

Of the *COX-2 inhibitors,* celecoxib has been the most widely tested in breast cancer clinical trials. By targeting the arachidonic acid–prostaglandins pathway, investigators have sought to reduce the prostaglandins-mediated stimulation of tumor angiogenesis. A neoadjuvant study of celecoxib in combination with 5-fluorouracil/epirubin/cyclophosphamide (FEC) versus FEC alone showed promising increase in pathologic complete response rates in patients with locally advanced breast cancer.[232] Other preliminary studies have suggested that celecoxib may potentiate the activity of aromatase inhibitors such as exemestane.[233] Based on these results, COX-2 inhibition continues to be the target of several major ongoing clinical trials. There is particular interest in the evaluation of celecoxib and other COX-2-inhibitors in chemoprevention studies.

Small molecule *EGFR inhibitors* have been associated with antiangiogenic properties in preclinical testing (see Chapter 17). Although well tolerated, agents such as gefitinib (ZD1839, Iressa™) have produced only minor responses in patients with advanced metastatic breast cancer.[234] Nonetheless, clinical trials are ongoing to test these agents in combinations with cytotoxic drugs and novel hormonal-based therapies.

HER-2 targeted agents are under study. *HER-2* overexpression or amplification is associated with enhanced VEGF expression, and treatment of *HER-2*–positive tumors with the monoclonal antibody, trastuzumab, results in down-regulation of VEGF. Preclinical experiments suggest that VEGF expression is partly responsible for the aggressive phenotype associated with *HER-2* overexpressing tumors. There is currently much interest and ongoing trials combining trastuzumab and anti-VEGF strategies.

ER inhibitors are another agent of particular interest. It has been theorized that antiestrogen drugs such as tamoxifen may benefit ER-negative patients by their antiangiogenic properties (see Chapter 15). Although clinical trials with existing antiestrogens and selective ER modulators have failed to support this hypothesis, novel antiestrogens with preclinical evidence of antiangiogenic effects are beginning to enter phase I clinical trials.

The *proteasome inhibitor,* bortezomib (PS-341, Velcade™), has shown antiangiogenic properties in preclinical models (see Chapter 24).[235,236] Bortezomib is currently being tested in multiple phase I clinical trials for the treatment of metastatic breast cancer in combination with docetaxel and also with trastuzumab in patients with *HER-2/neu*-positive disease.[237,238]

Summary and Conclusions

It is now widely accepted that angiogenesis is a fundamental process for the establishment, local progression, and dissemination of breast cancer. Biomarkers testing angiogenesis, including MVD and VEGF expression levels, continue to show promise as significant predictors of overall disease outcome. Although clinical trials with antiangiogenesis drugs in breast cancer have been disappointing, many investigators are encouraged by the events surrounding the testing of bevacizumab in colorectal cancer where an initial failure in third-line use was followed by an impressive result in first-line use, including a major survival benefit and a fast-track FDA approval for the drug in this indication. This has led to continuing strategy modifications for the timing of and selection of therapeutic combinations for the treatment of breast cancer. In the next several years, we will learn the outcome of these trials and whether antiangiogenic drugs will become cornerstones in the treatment of this disease.

References

1. Papetti M, Herman IM. Mechanisms of normal and tumor-derived angiogenesis. Am J Physiol Cell Physiol 2002;282:C947–C970.
2. Risau W. Mechanisms of angiogenesis. Nature 1997; 386:671–674.
3. Kleinsmith LJ, Kerrigan D, Kelly J. Understanding angiogenesis. http://press2.nci.nih gov/sciencebehind/angiogenesis/angio01_htm, 2004.
4. Klagsbrun M. Regulators of angiogenesis: stimulators, inhibitors, and extracellular matrix. J Cell Biochem 1991;47:199–200.
5. Denekamp J. Endothelial cell proliferation as a novel approach to targeting tumour therapy. Br J Cancer 1982;45:136–139.
6. Hughes S, Yang H, Chan-Ling T. Vascularization of the human fetal retina: roles of vasculogenesis and angiogenesis. Invest Ophthalmol Vis Sci 2000;41: 1217–1228.
7. Featherstone J, Griffiths S. From the analyst's couch. Drugs that target angiogenesis. Nat Rev Drug Discov 2002;1:413–414.
8. Carmeliet P. Angiogenesis in health and disease. Nat Med 2003;9:653–660.
9. Folkman J. What is the evidence that tumors are angiogenesis dependent? J Natl Cancer Inst 1990;82: 4–6.
10. Folkman J. Tumor angiogenesis: therapeutic implications. N Engl J Med 1971;285:1182–1186.
11. Folkman J, Merler E, Abernathy C, et al. Isolation of a tumor factor responsible for angiogenesis. J Exp Med 1971;133:275–288.
12. Folkman J. Tumor angiogenesis. Mol Basis Cancer 1995;206–232.
13. Pluda JM. Tumor-associated angiogenesis: mechanisms, clinical implications, and therapeutic strategies. Semin Oncol 1997;24:203–218.
14. Folkman J, Klagsbrun M. Angiogenic factors. Science 1987;235:442–447.
15. Folkman J. Angiogenesis in cancer, vascular, rheumatoid and other disease. Nat Med 1995;1: 27–31.
16. Folkman J, Browder T, Palmblad J. Angiogenesis research: guidelines for translation to clinical application. Thromb Haemost 2001;86:23–33.
17. Folkman J. Role of angiogenesis in tumor growth and metastasis. Semin Oncol 2002;29:15–18.
18. Folkman J. Fundamental concepts of the angiogenic process. Curr Mol Med 2003;3:643–651.
19. Folkman J. Angiogenesis and apoptosis. Semin Cancer Biol 2003;13:159–167.
20. Cao Y, Ji RW, Davidson D, et al. Kringle domains of human angiostatin. Characterization of the antiproliferative activity on endothelial cells. J Biol Chem 1996;271:29461–29467.
21. O'Reilly MS, Boehm T, Shing Y, et al. Endostatin: an endogenous inhibitor of angiogenesis and tumor growth. Cell 1997;88:277–285.
22. Saaristo A, Karpanen T, Alitalo K. Mechanisms of angiogenesis and their use in the inhibition of tumor growth and metastasis. Oncogene 2000;19:6122–6129.
23. Gasparini G. Clinical significance of determination of surrogate markers of angiogenesis in breast cancer. Crit Rev Oncol Hematol 2001;37:97–114.
24. McCulloch P, Choy A, Martin L. Association between tumour angiogenesis and tumour cell shedding into effluent venous blood during breast cancer surgery. Lancet 1995;346:1334–1335.
25. Vermeulen PB, Libura M, Libura J, et al. Influence of investigator experience and microscopic field size on microvessel density in node-negative breast carcinoma. Breast Cancer Res Treat 1997;42:165–172.
26. Vermeulen PB, Gasparini G, Fox SB, et al. Second international consensus on the methodology and criteria of evaluation of angiogenesis quantification in solid human tumours. Eur J Cancer 2002;38:1564–1579.
27. Fitzgibbons PL, Page DL, Weaver D, et al. Prognostic factors in breast cancer. College of American Pathologists Consensus Statement 1999. Arch Pathol Lab Med 2000;124:966–978.
28. Vermeulen PB, Gasparini G, Fox SB, et al. Quantification of angiogenesis in solid human tumours: an international consensus on the methodology and criteria of evaluation. Eur J Cancer 1996;32A:2474–2484.
29. Vermeulen PB, Colpaert C, Salgado R, et al. Liver metastases from colorectal adenocarcinomas grow in three patterns with different angiogenesis and desmoplasia. J Pathol 2001;195:336–342.
30. Patan S, Munn LL, Jain RK. Intussusceptive microvascular growth in a human colon adenocarcinoma xenograft: a novel mechanism of tumor angiogenesis. Microvasc Res 1996;51:260–272.
31. McDonald DM, Munn L, Jain RK. Vasculogenic mimicry: how convincing, how novel, and how significant? Am J Pathol 2000;156:383–388.
32. Mattern J, Koomagi R, Volm M. Biological characterization of subgroups of squamous cell lung carcinomas. Clin Cancer Res 1999;5:1459–1463.
33. Maniotis AJ, Folberg R, Hess A, et al. Vascular channel formation by human melanoma cells in vivo and in vitro: vasculogenic mimicry. Am J Pathol 1999; 155:739–752.
34. Holash J, Wiegand SJ, Yancopoulos GD. New model of tumor angiogenesis: dynamic balance between vessel regression and growth mediated by angiopoietins and VEGF. Oncogene 1999;18:5356–5362.
35. Gunsilius E, Duba HC, Petzer AL, et al. Evidence from a leukaemia model for maintenance of vascular endothelium by bone-marrow-derived endothelial cells. Lancet 2000;355:1688–1691.
36. Chang YS, di Tomaso E, McDonald DM, et al. Mosaic blood vessels in tumors: frequency of cancer cells in contact with flowing blood. Proc Natl Acad Sci USA 2000;97:14608–14613.
37. Asahara T, Takahashi T, Masuda H, et al. VEGF contributes to postnatal neovascularization by mobilizing bone marrow-derived endothelial progenitor cells. EMBO J 1999;18:3964–3972.

38. Asahara T, Masuda H, Takahashi T, et al. Bone marrow origin of endothelial progenitor cells responsible for postnatal vasculogenesis in physiological and pathological neovascularization. Circ Res 1999;85:221–228.
39. Ranieri G, Gasparini G. Angiogenesis and angiogenesis inhibitors: a new potential anticancer therapeutic strategy. Curr Drug Targets Immune Endocr Metabol Disord 2001;1:241–253.
40. Morabito A, Sarmiento R, Bonginelli P, et al. Antiangiogenic strategies, compounds, and early clinical results in breast cancer. Crit Rev Oncol Hematol 2004;49:91–107.
41. Longo R, Sarmiento R, Fanelli M, et al. Anti-angiogenic therapy: rationale, challenges and clinical studies. Angiogenesis 2002;5:237–256.
42. Kerbel RS, Yu J, Tran J, et al. Possible mechanisms of acquired resistance to anti-angiogenic drugs: implications for the use of combination therapy approaches. Cancer Metast Rev 2001;20:79–86.
43. Kerbel R, Folkman J. Clinical translation of angiogenesis inhibitors. Nat Rev Cancer 2002;2:727–739.
44. Eskens FA. Angiogenesis inhibitors in clinical development; where are we now and where are we going? Br J Cancer 2004;90:1–7.
45. Cristofanilli M, Charnsangavej C, Hortobagyi GN. Angiogenesis modulation in cancer research: novel clinical approaches. Nat Rev Drug Discov 2002;1:415–426.
46. Bamias A, Dimopoulos MA. Angiogenesis in human cancer: implications in cancer therapy. Eur J Intern Med 2003;14:459–469.
47. Miller KD. Issues and challenges for antiangiogenic therapies. Breast Cancer Res Treat 2002;75(suppl 1):S45–S50.
48. Sridhar SS, Shepherd FA. Targeting angiogenesis: a review of angiogenesis inhibitors in the treatment of lung cancer. Lung Cancer 2003;42(suppl 1):S81–S91.
49. Manley PW, Bold G, Bruggen J, et al. Advances in the structural biology, design and clinical development of VEGF-R kinase inhibitors for the treatment of angiogenesis. Biochim Biophys Acta 2004;1697:17–27.
50. Lo S, Johnston SR. Novel systemic therapies for breast cancer. Surg Oncol 2003;12:277–287.
51. Kerbel RS. Antiangiogenic drugs and current strategies for the treatment of lung cancer. Semin Oncol 2004;31:54–60.
52. Atiqur RM, Toi M. Anti-angiogenic therapy in breast cancer. Biomed Pharmacother 2003;57:463–470.
53. Morabito A, Magnani E, Gion M, et al. Prognostic and predictive indicators in operable breast cancer. Clin Breast Cancer 2003;3:381–390.
54. Shimaoka M, Springer TA. Therapeutic antagonists and conformational regulation of integrin function. Nat Rev Drug Discov 2003;2:703–716.
55. Kircher MF, Weissleder R, Josephson L. A dual fluorochrome probe for imaging proteases. Bioconjug Chem 2004;15:242–248.
56. Perez JM, Josephson L, Weissleder R. Use of magnetic nanoparticles as nanosensors to probe for molecular interactions. Chem Biochem 2004;5:261–264.
57. Harisinghani MG, Barentsz J, Hahn PF, et al. Noninvasive detection of clinically occult lymph-node metastases in prostate cancer. N Engl J Med 2003;348:2491–2499.
58. Bremer C, Ntziachristos V, Weissleder R. Optical-based molecular imaging: contrast agents and potential medical applications. Eur Radiol 2003;13:231–243.
59. Rudin M, Weissleder R. Molecular imaging in drug discovery and development. Nat Rev Drug Discov 2003;2:123–131.
60. Weissleder R, Ntziachristos V. Shedding light onto live molecular targets. Nat Med 2003;9:123–128.
61. Bremer C, Tung CH, Weissleder R. In vivo molecular target assessment of matrix metalloproteinase inhibition. Nat Med 2001;7:743–748.
62. Weissleder R. A clearer vision for in vivo imaging. Nat Biotechnol 2001;19:316–317.
63. Law B, Curino A, Bugge TH, et al. Design, synthesis, and characterization of urokinase plasminogen-activator-sensitive near-infrared reporter. Chem Biol 2004;11:99–106.
64. Brekken RA, Thorpe PE. Vascular endothelial growth factor and vascular targeting of solid tumors. Anticancer Res 2001;21:4221–4229.
65. Ruddell A, Mezquita P, Brandvold KA, et al. B lymphocyte-specific c-Myc expression stimulates early and functional expansion of the vasculature and lymphatics during lymphomagenesis. Am J Pathol 2003;163:2233–2245.
66. Lohela M, Saaristo A, Veikkola T, et al. Lymphangiogenic growth factors, receptors and therapies. Thromb Haemost 2003;90:167–184.
67. Vacca A, Ria R, Ribatti D, et al. A paracrine loop in the vascular endothelial growth factor pathway triggers tumor angiogenesis and growth in multiple myeloma. Haematologica 2003;88:176–185.
68. Harris M. Monoclonal antibodies as therapeutic agents for cancer. Lancet Oncol 2004;5:292–302.
69. Rosen LS. Clinical experience with angiogenesis signaling inhibitors: focus on vascular endothelial growth factor (VEGF) blockers. Cancer Control 2002;9:36–44.
70. Arinaga M, Noguchi T, Takeno S, et al. Clinical significance of vascular endothelial growth factor C and vascular endothelial growth factor receptor 3 in patients with nonsmall cell lung carcinoma. Cancer 2003;97:457–464.
71. Dales JP, Garcia S, Bonnier P, et al. Prognostic significance of VEGF receptors, VEGFR-1 (Flt-1) and VEGFR-2 (KDR/Flk-1) in breast carcinoma. Ann Pathol 2003;23:297–305.
72. Yokoyama Y, Charnock-Jones DS, Licence D, et al. Expression of vascular endothelial growth factor (VEGF)-D and its receptor, VEGF receptor 3, as a prognostic factor in endometrial carcinoma. Clin Cancer Res 2003;9:1361–1369.
73. Kato H, Yoshikawa M, Miyazaki T, et al. Expression of vascular endothelial growth factor (VEGF) and its receptors (Flt-1 and Flk-1) in esophageal squamous cell carcinoma. Anticancer Res 2002;22:3977–3984.
74. Ramaswamy B, Shapiro CL. Phase II trial of bevacizumab in combination with docetaxel in women with advanced breast cancer. Clin Breast Cancer 2003;4:292–294.
75. Miller KD. Recent translational research: antiangiogenic therapy for breast cancer—where do we stand? Breast Cancer Res 2004;6:128–132.
76. Kaklamani V, O'regan RM. New targeted therapies in breast cancer. Semin Oncol 2004;31:20–25.
77. De Vore R, Fehrenbacher L, Herbst R, et al. A randomised phase II trial comparing rhumab

(recombinant humanised monoclonal antibody to vascular-endothelial growth factor) plus carboplatin/paclitaxel (CP) to CP alone in patients with stage 3B/4 NSCLC. Proc Am Soc Clin Oncol 2000;19: 485(abstr).

78. Novotny W, Holmgren E, Griffing S, et al. Identification of squamous histology and central, cavitary tumours as possible risk factors for pulmonary (PH) haemorrhage in patients with advanced NSCLC receiving bevacizumab. Proc Am Soc Clin Oncol 2001; 20:330(abstr).
79. Miller K, Rugo H, Cobleigh M, et al. Phase III trial of capecitabine plus bevacizumab versus capecitabine alone in women with metastatic breast cancer previously treated with anthracycline and a taxane. Breast Cancer Res Treat 2002;76:abstr 36.
80. Sweep FC, Fritsche HA, Gion M, et al. Considerations on development, validation, application, and quality control of immuno(metric) biomarker assays in clinical cancer research: an EORTC-NCI working group report. Int J Oncol 2003;23:1715–1726.
81. Schrohl AS, Holten-Andersen M, Sweep F, et al. Tumor markers: from laboratory to clinical utility. Mol Cell Proteomics 2003;2:378–387.
82. Greenblatt M, Shubi P. Tumor angiogenesis: transfilter diffusion studies in the hamster by the transparent chamber technique. J Natl Cancer Inst 1968; 41:111–124.
83. Brem S, Cotran R, Folkman J. Tumor angiogenesis: a quantitative method for histologic grading. J Natl Cancer Inst 1972;48:347–356.
84. Cavallo T, Sade R, Folkman J, et al. Tumor angiogenesis. Rapid induction of endothelial mitoses demonstrated by autoradiography. J Cell Biol 1972; 54:408–420.
85. Folkman J. Anti-angiogenesis: new concept for therapy of solid tumors. Ann Surg 1972;175:409–416.
86. Holleb AI, Folkman J. Tumor angiogenesis. CA Cancer J Clin 1972;22:226–229.
87. Daidone MG, Paradiso A, Gion M, et al. Biomolecular features of clinical relevance in breast cancer. Eur J Nucl Med Mol Imaging 2004;31(Suppl 1):S3–S14.
88. Coradini D, Daidone MG. Biomolecular prognostic factors in breast cancer. Curr Opin Obstet Gynecol 2004;16:49–55.
89. Schrohl AS, Holten-Andersen MN, Peters HA, et al. Tumor tissue levels of tissue inhibitor of metalloproteinase–1 as a prognostic marker in primary breast cancer. Clin Cancer Res 2004;10:2289–2298.
90. Span PN, Waanders E, Manders P, et al. Mammaglobin is associated with low-grade, steroid receptor-positive breast tumors from postmenopausal patients, and has independent prognostic value for relapse-free survival time. J Clin Oncol 2004;22: 691–698.
91. Berns EM, Klijn JG, Look MP, et al. Combined vascular endothelial growth factor and TP53 status predicts poor response to tamoxifen therapy in estrogen receptor-positive advanced breast cancer. Clin Cancer Res 2003;9:1253–1258.
92. Look M, van Putten W, Duffy M, et al. Pooled analysis of prognostic impact of uPA and PAI-1 in breast cancer patients. Thromb Haemost 2003;90: 538–548.
93. Look MP, van Putten WL, Duffy MJ, et al. Pooled analysis of prognostic impact of urokinase-type plasminogen activator and its inhibitor PAI-1 in 8377 breast cancer patients. J Natl Cancer Inst 2002;94: 116–128.
94. Dorssers LC, Van der Flier S, Brinkman A, et al. Tamoxifen resistance in breast cancer: elucidating mechanisms. Drugs 2001;61:1721–1733.
95. Foekens JA, Romain S, Look MP, et al. Thymidine kinase and thymidylate synthase in advanced breast cancer: response to tamoxifen and chemotherapy. Cancer Res 2001;61:1421–1425.
96. Van der Flier S, van der Kwast TH, Claassen CJ, et al. Immunohistochemical study of the BCAR1/p130Cas protein in non-malignant and malignant human breast tissue. Int J Biol Markers 2001;16: 172–178.
97. Weidner N, Semple JP, Welch WR, et al. Tumor angiogenesis and metastasis—correlation in invasive breast carcinoma. N Engl J Med 1991;324:1–8.
98. Weidner N, Folkman J, Pozza F, et al. Tumor angiogenesis: a new significant and independent prognostic indicator in early-stage breast carcinoma. J Natl Cancer Inst 1992;84:1875–1887.
99. Weidner N, Carroll PR, Flax J, et al. Tumor angiogenesis correlates with metastasis in invasive prostate carcinoma. Am J Pathol 1993;143:401–409.
100. Uzzan B, Nicolas P, Cucherat M, et al. Microvessel density as a prognostic factor in women with breast cancer: a systematic review of the literature and meta-analysis. Cancer Res 2004;64:2941–2955.
101. Hlatky L, Hahnfeldt P, Folkman J. Clinical application of antiangiogenic therapy: microvessel density, what it does and doesn't tell us. J Natl Cancer Inst 2002;94:883–893.
102. Bosari S, Lee AK, DeLellis RA, et al. Microvessel quantitation and prognosis in invasive breast carcinoma. Hum Pathol 1992;23:755–761.
103. Hall NR, Fish DE, Hunt N, et al. Is the relationship between angiogenesis and metastasis in breast cancer real? Surg Oncol 1992;1:223–229.
104. Toi M, Kashitani J, Tominaga T. Tumor angiogenesis is an independent prognostic indicator in primary breast carcinoma. Int J Cancer 1993;55: 371–374.
105. Fox SB, Leek RD, Smith K, et al. Tumor angiogenesis in node-negative breast carcinomas—relationship with epidermal growth factor receptor, estrogen receptor, and survival. Breast Cancer Res Treat 1994; 29:109–116.
106. Axelsson K, Ljung BM, Moore DH, et al. Tumor angiogenesis as a prognostic assay for invasive ductal breast carcinoma. J Natl Cancer Inst 1995;87:997–1008.
107. Bevilacqua P, Barbareschi M, Verderio P, et al. Prognostic value of intratumoral microvessel density, a measure of tumor angiogenesis, in node-negative breast carcinoma—results of a multiparametric study. Breast Cancer Res Treat 1995;36:205–217.
108. Costello P, McCann A, Carney DN, et al. Prognostic significance of microvessel density in lymph node negative breast carcinoma. Hum Pathol 1995;26: 1181–1184.
109. Fox SB, Leek RD, Weekes MP, et al. Quantitation and prognostic value of breast cancer angiogenesis: comparison of microvessel density, Chalkley count, and computer image analysis. J Pathol 1995;177: 275–283.
110. Siitonen SM, Haapasalo HK, Rantala IS, et al. Comparison of different immunohistochemical methods in the assessment of angiogenesis: lack of prognostic value in a group of 77 selected node-negative breast carcinomas. Mod Pathol 1995;8: 745–752.

111. Toi M, Inada K, Suzuki H, et al. Tumor angiogenesis in breast cancer: its importance as a prognostic indicator and the association with vascular endothelial growth factor expression. Breast Cancer Res Treat 1995;36:193–204.
112. Heimann R, Ferguson D, Powers C, et al. Angiogenesis as a predictor of long-term survival for patients with node-negative breast cancer. J Natl Cancer Inst 1996;88:1764–1769.
113. Morphopoulos G, Pearson M, Ryder WD, et al. Tumour angiogenesis as a prognostic marker in infiltrating lobular carcinoma of the breast. J Pathol 1996;180:44–49.
114. Simpson JF, Ahn C, Battifora H, et al. Endothelial area as a prognostic indicator for invasive breast carcinoma. Cancer 1996;77:2077–2085.
115. Marinho A, Soares R, Ferro J, et al. Angiogenesis in breast cancer is related to age but not to other prognostic parameters. Pathol Res Pract 1997;193: 267–273.
116. Martin L, Green B, Renshaw C, et al. Examining the technique of angiogenesis assessment in invasive breast cancer. Br J Cancer 1997;76:1046–1054.
117. Kato T, Kimura T, Ishii N, et al. The methodology of quantitation of microvessel density and prognostic value of neovascularization associated with long-term survival in Japanese patients with breast cancer. Breast Cancer Res Treat 1999;53: 19–31.
118. Kumar S, Ghellal A, Li C, et al. Breast carcinoma: vascular density determined using CD105 antibody correlates with tumor prognosis. Cancer Res 1999; 59:856–861.
119. Viens P, Jacquemier J, Bardou VJ, et al. Association of angiogenesis and poor prognosis in node-positive patients receiving anthracycline-based adjuvant chemotherapy. Breast Cancer Res Treat 1999;54: 205–212.
120. Fridman V, Humblet C, Bonjean K, et al. Assessment of tumor angiogenesis in invasive breast carcinomas: absence of correlation with prognosis and pathological factors. Virchows Arch 2000;437:611–617.
121. Guidi AJ, Berry DA, Broadwater G, et al. Association of angiogenesis in lymph node metastases with outcome of breast cancer. J Natl Cancer Inst 2000;92: 486–492.
122. Hansen S, Grabau DA, Sorensen FB, et al. The prognostic value of angiogenesis by Chalkley counting in a confirmatory study design on 836 breast cancer patients. Clin Cancer Res 2000;6:139–146.
123. Kato T, Kameoka S, Kimura T, et al. Angiogenesis and blood vessel invasion as prognostic indicators for node-negative breast cancer. Breast Cancer Res Treat 2001;65:203–215.
124. Arora R, Joshi K, Nijhawan R, et al. Angiogenesis as an independent prognostic indicator in node-negative breast cancer. Anal Quant Cytol Histol 2002;24:228–233.
125. Guidi AJ, Berry DA, Broadwater G, et al. Association of angiogenesis and disease outcome in node-positive breast cancer patients treated with adjuvant cyclophosphamide, doxorubicin, and fluorouracil: a Cancer and Leukemia Group B correlative science study from protocols 8541/8869. J Clin Oncol 2002; 20:732–742.
126. Gunel N, Akcali Z, Coskun U, et al. Prognostic importance of tumor angiogenesis in breast carcinoma with adjuvant chemotherapy. Pathol Res Pract 2002; 198:7–12.
127. Offersen BV, Sorensen FB, Yilmaz M, et al. Chalkley estimates of angiogenesis in early breast cancer—relevance to prognosis. Acta Oncol 2002;41: 695–703.
128. Hansen S, Overgaard J, Rose C, et al. Independent prognostic value of angiogenesis and the level of plasminogen activator inhibitor type 1 in breast cancer patients. Br J Cancer 2003;88:102–108.
129. Olewniczak S, Chosia M, Kolodziej B, et al. Angiogenesis as determined by computerised image analysis and the risk of early relapse in women with invasive ductal breast carcinoma. Pol J Pathol 2003; 54:53–59.
130. Houck KA, Ferrara N, Winer J, et al. The vascular endothelial growth factor family: identification of a fourth molecular species and characterization of alternative splicing of RNA. Mol Endocrinol 1991;5: 1806–1814.
131. Park JE, Keller GA, Ferrara N. The vascular endothelial growth factor (VEGF) isoforms: differential deposition into the subepithelial extracellular matrix and bioactivity of extracellular matrix-bound VEGF. Mol Biol Cell 1993;4:1317–1326.
132. Dvorak HF, Nagy JA, Feng D, et al. Vascular permeability factor/vascular endothelial growth factor and the significance of microvascular hyperpermeability in angiogenesis. Curr Top Microbiol Immunol 1999; 237:97–132.
133. Yang R, Thomas GR, Bunting S, et al. Effects of vascular endothelial growth factor on hemodynamics and cardiac performance. J Cardiovasc Pharmacol 1996;27:838–844.
134. Alon T, Hemo I, Itin A, et al. Vascular endothelial growth factor acts as a survival factor for newly formed retinal vessels and has implications for retinopathy of prematurity. Nat Med 1995;1:1024–1028.
135. Carmeliet P, Collen D. Molecular basis of angiogenesis. Role of VEGF and VE-cadherin. Ann N Y Acad Sci 2000;902:249–262.
136. Komatsu DE, Hadjiargyrou M. Activation of the transcription factor HIF-1 and its target genes, VEGF, HO-1, iNOS, during fracture repair. Bone 2004;34:680–688.
137. Mizukami Y, Li J, Zhang X, et al. Hypoxia-inducible factor-1-independent regulation of vascular endothelial growth factor by hypoxia in colon cancer. Cancer Res 2004;64:1765–1772.
138. Choi KS, Bae MK, Jeong JW, et al. Hypoxia-induced angiogenesis during carcinogenesis. J Biochem Mol Biol 2003;36:120–127.
139. Arbiser JL, Moses MA, Fernandez CA, et al. Oncogenic H-ras stimulates tumor angiogenesis by two distinct pathways. Proc Natl Acad Sci USA 1997; 94:861–866.
140. Gasparini G, Toi M, Gion M, et al. Prognostic significance of vascular endothelial growth factor protein in node-negative breast carcinoma. J Natl Cancer Inst 1997;89:139–147.
141. Okada F, Rak JW, Croix BS, et al. Impact of oncogenes in tumor angiogenesis: mutant K-ras up-regulation of vascular endothelial growth factor/vascular permeability factor is necessary, but not sufficient for tumorigenicity of human colorectal carcinoma cells. Proc Natl Acad Sci USA 1998;95: 3609–3614.
142. Petit AM, Rak J, Hung MC, et al. Neutralizing antibodies against epidermal growth factor and

ErbB-2/neu receptor tyrosine kinases down-regulate vascular endothelial growth factor production by tumor cells in vitro and in vivo: angiogenic implications for signal transduction therapy of solid tumors. Am J Pathol 1997;151:1523–1530.
143. Shibuya M. Angiogenesis—vascular endothelial growth factor and its receptors. Hum Cell 1999;12: 17–24.
144. Shibuya M. Structure and function of VEGF/VEGF-receptor system involved in angiogenesis. Cell Struct Funct 2001;26:25–35.
145. Pinedo HM, Verheul HM, D'Amato RJ, et al. Involvement of platelets in tumour angiogenesis? Lancet 1998;352:1775–1777.
146. Nicosia RF. What is the role of vascular endothelial growth factor-related molecules in tumor angiogenesis? Am J Pathol 1998;153:11–16.
147. Gasparini G. Prognostic value of vascular endothelial growth factor in breast cancer. Oncologist 2000; 5(suppl 1):37–44.
148. Toi M, Hoshina S, Takayanagi T, et al. Association of vascular endothelial growth factor expression with tumor angiogenesis and with early relapse in primary breast cancer. Jpn J Cancer Res 1994;85: 1045–1049.
149. Toi M, Inada K, Suzuki H, et al. Tumor angiogenesis in breast cancer: its importance as a prognostic indicator and the association with vascular endothelial growth factor expression. Breast Cancer Res Treat 1995;36:193–204.
150. Obermair A, Bancher-Todesca D, Bilgi S, et al. Correlation of vascular endothelial growth factor expression and microvessel density in cervical intraepithelial neoplasia. J Natl Cancer Inst 1997;89: 1212–1217.
151. Relf M, LeJeune S, Scott PA, et al. Expression of the angiogenic factors vascular endothelial cell growth factor, acidic and basic fibroblast growth factor, tumor growth factor beta–1, platelet-derived endothelial cell growth factor, placenta growth factor, and pleiotrophin in human primary breast cancer and its relation to angiogenesis. Cancer Res 1997; 57:963–969.
152. Toi M, Gion M, Biganzoli E, et al. Co-determination of the angiogenic factors thymidine phosphorylase and vascular endothelial growth factor in node-negative breast cancer: prognostic implications. Angiogenesis 1997;1:71–83.
153. Eppenberger U, Kueng W, Schlaeppi JM, et al. Markers of tumor angiogenesis and proteolysis independently define high- and low-risk subsets of node-negative breast cancer patients. J Clin Oncol 1998;16:3129–3136.
154. Linderholm B, Tavelin B, Grankvist K, et al. Vascular endothelial growth factor is of high prognostic value in node-negative breast carcinoma. J Clin Oncol 1998;16:3121–3128.
155. Gasparini G, Toi M, Miceli R, et al. Clinical relevance of vascular endothelial growth factor and thymidine phosphorylase in patients with node-positive breast cancer treated with either adjuvant chemotherapy or hormone therapy. Cancer J Sci Am 1999;5:101–111.
156. Adams J, Carder PJ, Downey S, et al. Vascular endothelial growth factor (VEGF) in breast cancer: comparison of plasma, serum, and tissue VEGF and microvessel density and effects of tamoxifen. Cancer Res 2000;60:2898–2905.
157. Linderholm B, Lindh B, Tavelin B, et al. p53 and vascular-endothelial-growth-factor (VEGF) expression predicts outcome in 833 patients with primary breast carcinoma. Int J Cancer 2000;89:51–62.
158. Linderholm B, Grankvist K, Wilking N, et al. Correlation of vascular endothelial growth factor content with recurrences, survival, and first relapse site in primary node-positive breast carcinoma after adjuvant treatment. J Clin Oncol 2000;18:1423–1431.
159. Gasparini G, Toi M, Biganzoli E, et al. Thrombospondin-1 and -2 in node-negative breast cancer: correlation with angiogenic factors, p53, cathepsin D, hormone receptors and prognosis. Oncology 2001; 60:72–80.
160. Jing J, Zhao Y, Li H, et al. The prognostic value of vascular endothelial growth factor in breast cancer. Hua Xi Yi Ke Da Xue Xue Bao 2001;32:566–568. [Chinese]
161. Kinoshita J, Kitamura K, Kabashima A, et al. Clinical significance of vascular endothelial growth factor-C (VEGF-C) in breast cancer. Breast Cancer Res Treat 2001;66:159–164.
162. Linderholm BK, Lindahl T, Holmberg L, et al. The expression of vascular endothelial growth factor correlates with mutant p53 and poor prognosis in human breast cancer. Cancer Res 2001;61:2256–2260.
163. MacConmara M, O'Hanlon DM, Kiely MJ, et al. An evaluation of the prognostic significance of vascular endothelial growth factor in node positive primary breast carcinoma. Int J Oncol 2002;20:717–721.
164. Manders P, Beex LV, Tjan-Heijnen VC, et al. The prognostic value of vascular endothelial growth factor in 574 node-negative breast cancer patients who did not receive adjuvant systemic therapy. Br J Cancer 2002;87:772–778.
165. Toi M, Bando H, Ogawa T, et al. Significance of vascular endothelial growth factor (VEGF)/soluble VEGF receptor-1 relationship in breast cancer. Int J Cancer 2002;98:14–18.
166. Wen XY, Bai Y, Stewart AK. Adenovirus-mediated human endostatin gene delivery demonstrates strain-specific antitumor activity and acute dose-dependent toxicity in mice. Hum Gene Ther 2001; 12:347–358.
167. Wu Y, Saldana L, Chillar R, et al. Plasma vascular endothelial growth factor is useful in assessing progression of breast cancer post surgery and during adjuvant treatment. Int J Oncol 2002;20: 509–516.
168. Coradini D, Biganzoli E, Pellizzaro C, et al. Vascular endothelial growth factor in node-positive breast cancer patients treated with adjuvant tamoxifen. Br J Cancer 2003;89:268–270.
169. Coskun U, Gunel N, Sancak B, et al. Significance of serum vascular endothelial growth factor, insulin-like growth factor-I levels and nitric oxide activity in breast cancer patients. Breast 2003;12:104–110.
170. De Paola F, Granato AM, Scarpi E, et al. Vascular endothelial growth factor and prognosis in patients with node-negative breast cancer. Int J Cancer 2002;98:228–233.
171. Linderholm BK, Lindh B, Beckman L, et al. Prognostic correlation of basic fibroblast growth factor and vascular endothelial growth factor in 1307 primary breast cancers. Clin Breast Cancer 2003; 4:340–347.
172. Zhukova LG, Zhukov NV, Lichinitser MR. Expression of Flt-1 and Flk-1 receptors for vascular

endothelial growth factor on tumor cells as a new prognostic criterion for locally advanced breast cancer. Bull Exp Biol Med 2003;135:478–481.

173. Lim SC. Role of COX-2, VEGF and cyclin D1 in mammary infiltrating duct carcinoma. Oncol Rep 2003;10:1241–1249.

174. Ludovini V, Sidoni A, Pistola L, et al. Evaluation of the prognostic role of vascular endothelial growth factor and microvessel density in stages I and II breast cancer patients. Breast Cancer Res Treat 2003;81:159–168.

175. Nakamura Y, Yasuoka H, Tsujimoto M, et al. Prognostic significance of vascular endothelial growth factor D in breast carcinoma with long-term follow-up. Clin Cancer Res 2003;9:716–721.

176. Nakamura Y, Yasuoka H, Tsujimoto M, et al. Clinicopathological significance of vascular endothelial growth factor-C in breast carcinoma with long-term follow-up. Mod Pathol 2003;16:309–314.

177. Nishimura R, Nagao K, Miyayama H, et al. Higher plasma vascular endothelial growth factor levels correlate with menopause, overexpression of p53, and recurrence of breast cancer. Breast Cancer 2003;10: 120–128.

178. Dales JP, Garcia S, Carpentier S, et al. Prediction of metastasis risk (11 year follow-up) using VEGF-R1, VEGF-R2, Tie-2/Tek and CD105 expression in breast cancer (n = 905). Br J Cancer 2004;90: 1216–1221.

179. Konecny GE, Meng YG, Untch M, et al. Association between HER-2/neu and vascular endothelial growth factor expression predicts clinical outcome in primary breast cancer patients. Clin Cancer Res 2004;10:1706–1716.

180. Linderholm B, Andersson J, Lindh B, et al. Overexpression of c-erbB-2 is related to a higher expression of vascular endothelial growth factor (VEGF) and constitutes an independent prognostic factor in primary node-positive breast cancer after adjuvant systemic treatment. Eur J Cancer 2004; 40:33–42.

181. Zhao J, Yan F, Ju H, et al. Correlation between serum vascular endothelial growth factor and endostatin levels in patients with breast cancer. Cancer Lett 2004;204:87–95.

182. Kim R, Murakami S, Toge T. Effects of introduction of dThdPase cDNA on sensitivity to 5'-deoxy-5-fluorouridine and tumor angiogenesis. Int J Oncol 2003;22:835–841.

183. Harris AL, Zhang H, Moghaddam A, et al. Breast cancer angiogenesis—new approaches to therapy via antiangiogenesis, hypoxic activated drugs, and vascular targeting. Breast Cancer Res Treat 1996;38: 97–108.

184. Fox SB, Westwood M, Moghaddam A, et al. The angiogenic factor platelet-derived endothelial cell growth factor/thymidine phosphorylase is up-regulated in breast cancer epithelium and endothelium. Br J Cancer 1996;73:275–280.

185. Harris AL. Hypoxia—a key regulatory factor in tumour growth. Nat Rev Cancer 2002;2:38–47.

186. Yang Q, Barbareschi M, Mori I, et al. Prognostic value of thymidine phosphorylase expression in breast carcinoma. Int J Cancer 2002;97:512–517.

187. Tominaga T, Toi M, Ohashi Y, et al. Prognostic and predictive value of thymidine phosphorylase activity in early-stage breast cancer patients. Clin Breast Cancer 2002;3:55–64.

188. Fox SB, Engels K, Comley M, et al. Relationship of elevated tumour thymidine phosphorylase in node-positive breast carcinomas to the effects of adjuvant CMF. Ann Oncol 1997;8:271–275.

189. Griffiths L, Dachs GU, Bicknell R, et al. The influence of oxygen tension and pH on the expression of platelet-derived endothelial cell growth factor/ thymidine phosphorylase in human breast tumor cells grown in vitro and in vivo. Cancer Res 1997; 57:570–572.

190. Goonewardene TI, Sowter HM, Harris AL. Hypoxia-induced pathways in breast cancer. Microsc Res Tech 2002;59:41–48.

191. Acker T, Plate KH. Hypoxia and hypoxia inducible factors (HIF) as important regulators of tumor physiology. Cancer Treat Res 2004;117:219–248.

192. Hockel M, Vaupel P. Biological consequences of tumor hypoxia. Semin Oncol 2001;28:36–41.

193. Maxwell PH, Pugh CW, Ratcliffe PJ. The pVHL-hIF-1 system. A key mediator of oxygen homeostasis. Adv Exp Med Biol 2001;502:365–376.

194. Semenza GL. Hypoxia, clonal selection, and the role of HIF-1 in tumor progression. Crit Rev Biochem Mol Biol 2000;35:71–103.

195. Semenza GL. HIF-1 and tumor progression: pathophysiology and therapeutics. Trends Mol Med 2002; 8:S62–S67.

196. Semenza GL. Involvement of hypoxia-inducible factor 1 in human cancer. Intern Med 2002;41: 79–83.

197. Semenza GL. Targeting HIF-1 for cancer therapy. Nat Rev Cancer 2003;3:721–732.

198. Welsh SJ, Powis G. Hypoxia inducible factor as a cancer drug target. Curr Cancer Drug Targets 2003;3:391–405.

199. Gruber G, Greiner RH, Hlushchuk R, et al. Hypoxia-inducible factor 1 alpha in high-risk breast cancer: an independent prognostic parameter? Breast Cancer Res 2004;6:R191–R198.

200. Bos R, van der GP, Greijer AE, et al. Levels of hypoxia-inducible factor-1alpha independently predict prognosis in patients with lymph node negative breast carcinoma. Cancer 2003;97:1573–1581.

201. Schindl M, Schoppmann SF, Samonigg H, et al. Overexpression of hypoxia-inducible factor 1alpha is associated with an unfavorable prognosis in lymph node-positive breast cancer. Clin Cancer Res 2002; 8:1831–1837.

202. Cayre A, Rossignol F, Clottes E, et al. aHIF but not HIF-1alpha transcript is a poor prognostic marker in human breast cancer. Breast Cancer Res 2003;5: R223–R230.

203. Span PN, Bussink J, Manders P, et al. Carbonic anhydrase-9 expression levels and prognosis in human breast cancer: association with treatment outcome. Br J Cancer 2003;89:271–276.

204. Watson PH, Chia SK, Wykoff CC, et al. Carbonic anhydrase XII is a marker of good prognosis in invasive breast carcinoma. Br J Cancer 2003;88: 1065–1070.

205. Chia SK, Wykoff CC, Watson PH, et al. Prognostic significance of a novel hypoxia-regulated marker, carbonic anhydrase IX, in invasive breast carcinoma. J Clin Oncol 2001;19:3660–3668.

206. Weber G, Stubbs M, Morris HP. Metabolism of hepatomas of different growth rates in situ and during ischemia. Cancer Res 1971;31:2177–2183.

207. Daly EB, Wind T, Jiang XM, et al. Secretion of phosphoglycerate kinase from tumour cells is controlled by oxygen-sensing hydroxylases. Biochim Biophys Acta 2004;1691:17–22.

208. Hogg PJ. Biological regulation through protein disulfide bond cleavage. Redox Rep 2002;7:71–77.

209. Lay AJ, Hogg PJ. Measurement of reduction of disulfide bonds in plasmin by phosphoglycerate kinase. Methods Enzymol 2002;348:87–92.

210. Lay AJ, Jiang XM, Daly E, et al. Plasmin reduction by phosphoglycerate kinase is a thiol-independent process. J Biol Chem 2002;277:9062–9068.

211. Lay AJ, Jiang XM, Kisker O, et al. Phosphoglycerate kinase acts in tumour angiogenesis as a disulphide reductase. Nature 2000;408:869–873.

212. Geiger JH, Cnudde SE. What the structure of angiostatin may tell us about its mechanism of action. J Thromb Haemost 2004;2:23–34.

213. Chen G, Gharib TG, Wang H, et al. Protein profiles associated with survival in lung adenocarcinoma. Proc Natl Acad Sci USA 2003;100:13537–13542.

214. Costa C, Soares R, Reis-Filho JS, et al. Cyclo-oxygenase 2 expression is associated with angiogenesis and lymph node metastasis in human breast cancer. J Clin Pathol 2002;55:429–434.

215. Gasparini G, Longo R, Sarmiento R, et al. Inhibitors of cyclo-oxygenase 2: a new class of anticancer agents? Lancet Oncol 2003;4:605–615.

216. Denkert C, Winzer KJ, Hauptmann S. Prognostic impact of cyclooxygenase-2 in breast cancer. Clin Breast Cancer 2004;4:428–433.

217. Lee S, Hwang KS, Lee HJ, et al. Aberrant CpG island hypermethylation of multiple genes in colorectal neoplasia. Lab Invest 2004;84:884–893.

218. Fosslien E. Review: molecular pathology of cyclooxygenase-2 in cancer-induced angiogenesis. Ann Clin Lab Sci 2001;31:325–348.

219. Menendez JA, Mehmi I, Griggs DW, et al. The angiogenic factor CYR61 in breast cancer: molecular pathology and therapeutic perspectives. Endocr Relat Cancer 2003;10:141–152.

220. Fox SB, Taylor M, Grondahl-Hansen J, et al. Plasminogen activator inhibitor-1 as a measure of vascular remodelling in breast cancer. J Pathol 2001;195:236–243.

221. Pepper MS. Lymphangiogenesis and tumor metastasis: more questions than answers. Lymphology 2000;33:144–147.

222. Pepper MS. Extracellular proteolysis and angiogenesis. Thromb Haemost 2001;86:346–355.

223. Pepper MS. Role of the matrix metalloproteinase and plasminogen activator-plasmin systems in angiogenesis. Arterioscler Thromb Vasc Biol 2001;21:1104–1117.

224. Rakic JM, Maillard C, Jost M, et al. Role of plasminogen activator-plasmin system in tumor angiogenesis. Cell Mol Life Sci 2003;60:463–473.

225. Awada A, Cardoso F, Atalay G, et al. The pipeline of new anticancer agents for breast cancer treatment in 2003. Crit Rev Oncol Hematol 2003;48:45–63.

226. Eder JP Jr, Supko JG, Clark JW, et al. Phase I clinical trial of recombinant human endostatin administered as a short intravenous infusion repeated daily. J Clin Oncol 2002;20:3772–3784.

227. Sparano JA, Bernardo P, Gradishar WJ, et al. Randomized phase II trial of marimastat versus placebo in patients with metastatic breast cancer who have responding or stable disease after first-line chemotherpay: an Eastern Cooperative Oncology Trial (E2196). Proc Am Soc Clin Oncol 2002;21:44a.

228. Miller KD, Gradishar W, Schuchter L, et al. A randomized phase II pilot trial of adjuvant marimastat in patients with early-stage breast cancer. Ann Oncol 2002;13:1220–1224.

229. Stopeck A, Sheldon M, Vahedian M, et al. Results of a Phase I dose-escalating study of the antiangiogenic agent, SU5416, in patients with advanced malignancies. Clin Cancer Res 2002;8:2798–2805.

230. Baidas SM, Winer EP, Fleming GF, et al. Phase II evaluation of thalidomide in patients with metastatic breast cancer. J Clin Oncol 2000;18:2710–2717.

231. Eisen T, Boshoff C, Mak I, et al. Continuous low dose Thalidomide: a phase II study in advanced melanoma, renal cell, ovarian and breast cancer. Br J Cancer 2000;82:812–817.

232. Chow LW, Wong JL, Toi M. Celecoxib anti-aromatase neoadjuvant (CAAN) trial for locally advanced breast cancer: preliminary report. J Steroid Biochem Mol Biol 2003;86:443–447.

233. Dirix ly, Ignacio J, Ng S, et al. Final results of an open-label multi-center, controlled study of exemestane +/– celecoxib in post-menopausal women with advanced breast cancer. Proc Am Assoc Cancer Res 2003;42:abstract 77.

234. Baselga J. Combined anti-EGF receptor and anti-HER2 receptor therapy in breast cancer: a promising strategy ready for clinical testing. Ann Oncol 2002;13:8–9.

235. Adams J. Potential for proteasome inhibition in the treatment of cancer. Drug Discov Today 2003;8:307–315.

236. Ohta T, Fukuda M. Ubiquitin and breast cancer. Oncogene 2004;23:2079–2088.

237. Orlowski RZ, Dees EC. The role of the ubiquitination-proteasome pathway in breast cancer: applying drugs that affect the ubiquitin-proteasome pathway to the therapy of breast cancer. Breast Cancer Res 2003;5:1–7.

238. Voorhees PM, Dees EC, O'Neil B, Orlowski RZ. The proteasome as a target for cancer therapy. Clin Cancer Res 2003;9:6316–6325.

CHAPTER 20

Cell Adhesion Markers

Heinz Höfler, MD, and Birgit Luber, MD

Institut fur Allgemeine Pathologie und Pathologische Anatomie, Technische Universität München, Munich, Germany

Breast cancer is one of the most frequent malignancies worldwide. Molecular prognostic markers include the estrogen (ER) and progesterone receptors, proliferation markers, and HER-2/*neu*, which belongs to the epidermal growth factor family of transmembrane receptor tyrosine kinases.[1,2] During breast cancer development, normal breast epithelium is transformed into invasive cancer cells. An important feature of tumor invasiveness is the modulation of the capacity of tumor cells to adhere to other cells or to the extracellular matrix. E-cadherin, CD44, and integrins mediate cell–cell and cell–matrix interactions and abnormal expression or function of these molecules is therefore implicated in tumor progression.

E-cadherin and the associated catenins form a cell adhesion complex that plays an important role in breast carcinogenesis. Alterations of components of the adherens junctions range from abnormal expression levels and altered subcellular localization to mutations within the E-cadherin gene. The prognostic value of E-cadherin in breast cancer has been studied intensively during recent years. An increasing number of reports point to a possible association between alterations in E-cadherin expression and function and poor prognosis, discussed below.

CD44 is a family of transmembrane glycoproteins that are encoded by one single gene.[3–5] CD44 transcripts are characterized by alternative splicing of up to 10 exons. The variant CD44 isoforms differ in their size, function, and glycosylation.[6] CD44 is composed of an extracellular domain, a transmembrane region, and a cytoplasmic domain, which binds to the cytoskeleton through ankyrin, and members of the ezrin, radixin, and moesin complex.[7–11] CD44 binds to hyaluronan,[12,13] which is present in the extracellular matrix, and to glycosaminoglycans, fibronectin, collagen, laminin, and osteopontin.[14–18] In experimental systems, CD44 has been shown to play a role in tumor progression and metastasis.[19] In normal breast epithelium, CD44 is not expressed.[14,20] In contrast, in breast tumors variant CD44 isoforms are highly expressed in neoplastic cells.[20–22] The prognostic relevance of the presence of different CD44 isoforms for patient survival and chemotherapy response in breast cancer is discussed controversially in different studies and therefore needs to be investigated further.[14]

Integrins mediate the interaction of tumor cells with the extracellular matrix. A decrease of the expression level of the $\alpha_2\beta_1$ integrin is frequently observed in breast cancers.[23] Down-regulation of $\alpha_2\beta_1$ integrin expression has been found to correlate with the stage of differentiation.[24] Moreover, $\alpha_6\beta_1$ and $\alpha_6\beta_4$ integrins have been shown to be implicated in breast cancer progression.[25,26] The prognostic significance of these adhesive receptors in mammary carcinogenesis remains to be established.

Biologic Background

E-cadherin is a cell–cell adhesion molecule that consists of an extracellular domain comprising five subdomains, a transmembrane region and a cytoplasmic domain, which associates with catenins.[27] β-Catenin or plakoglobin (γ-catenin) directly interacts with E-cadherin and simultaneously binds to α-catenin, which links the E-cadherin/catenin complex to the actin cytoskeleton.[27,28] Another catenin, p120^{ctn}, associates with E-cadherin and is involved in strengthening adherens junctions.[29] Besides playing an important role in cell adhesion, β-catenin functions also in the Wnt signaling pathway. β-Catenin forms a complex with members of the TCF/LEF1 transcription factor family.[30–32] This transcription factor complex activates target genes such as c-*myc*[33] and cyclin D1.[34,35] Mutations in β-catenin were found in colon cancer and malignant melanoma that reveal the oncogenic function of the molecule.[36–38] These mutations result in stabilization and nuclear localization of β-catenin. The human E-cadherin gene is located on chromosome 16q22.1,[39-41] spans a region of 100 kbp, and contains 16 exons. Loss of heterozygosity at 16q is one of the most frequent events in breast carcinomas.[42]

In experimental tumor cell systems, a growth, invasion, and tumor suppressor role of E-cadherin was demonstrated.[43,44] E-cadherin expression reduced cell proliferation in mammary carcinoma cell lines and mice tumors obtained after xenotransplantation.[45-47] Epidemiologic studies demonstrated a correlation between loss of E-cadherin or catenin expression and invasive tumor growth in a variety of tumors,[48–52] including breast carcinomas.[53,54] In breast cancer, partial or complete loss of E-cadherin expression correlates with metastatic behavior and poor prognosis.[54–58] Complete loss of E-cadherin expression is frequently observed in infiltrating lobular breast carcinomas.[59–63] Simultaneous loss of E-cadherin expression and catenins is also observed in lobular breast cancer.[64]

E-cadherin can be regulated by a variety of transcription factors, including snail and sip1.[65–67] It can also be silenced by DNA hypermethylation.[68,69] Beside transcriptional E-cadherin regulation, somatic E-cadherin mutations have been described in breast,[70–72] gastric,[73–77] gynecologic,[78] and thyroid carcinomas.[49,79] E-cadherin mutations are an infrequent event, except for breast and gastric carcinoma. Somatic E-cadherin mutations in the E-cadherin gene *CDH1* were detected in about 56% of lobular breast carcinomas, and in most cases (more than 90%), the wild-type allele was lost.[71,72] But no mutations in the E-cadherin gene were detected in ductal breast carcinomas.[71,80] Most mutations in lobular breast carcinomas are nonsense or frameshift mutations[49] that lead to truncated E-cadherin variants in 86% of cases.[49] No hotspot regions for mutations were found in the E-cadherin gene in lobular breast cancer. In gastric cancer, E-cadherin mutations were identified in 50% of diffuse type gastric carcinomas, whereas no mutations were detected in intestinal type gastric carcinoma.[73,81] Splice-site mutations and in-frame deletions were the predominant mutations in gastric cancer, and most of the mutations were located in exon 8 or 9.[81] This region is therefore considered as a hot spot region in gastric carcinoma. In breast and gastric cancer, E-cadherin mutations are detected only in the histologic subtypes lobular breast cancer and diffuse type gastric carcinoma, both of which are characterized by scattered tumor cell growth. These data suggest that E-cadherin mutations might be the reason for the scattered tumor cell morphology. E-cadherin inactivation by mutation is an early event in lobular breast cancer formation, because loss of E-cadherin expression associated with the presence of mutations is already found in early noninvasive lobular carcinoma in situ.[61]

In families with diffuse-type gastric carcinoma, E-cadherin mutations are found.[82–85] In one case, a patient with an E-cadherin mutation developed breast cancer in addition to a diffuse type gastric carcinoma.[84] In an attempt to identify germline mutations in breast carcinomas, patients with bilateral lobular breast carcinomas were investigated but only sporadic nongermline mutations were found.[72] In familial breast cancer patients, no germline E-cadherin alterations were detected in 19 patients.[86]

Reduced E-cadherin expression can also be associated with abnormal expression of P-cadherin, N-cadherin, or cadherin-11,[87,88] a phenomenon called cadherin switch. The loss of E-cadherin induces an epithelial to mesenchymal transition of the cellular phenotype.[89] N-cadherin expression is believed to promote the interaction of invasive breast cancer cells with mammary stromal cells. Expression of cadherin-11 is preferentially found in mesenchymal tissue, but has also been detected in breast carcinoma cell lines.[90] Tumor cell motility and invasiveness may be promoted by interaction between cadherin-11–positive tumor cells and surrounding stromal cells. Recently, a correlation between expression of cadherin-11 and enhanced metastatic potential of mammary carcinoma cells has been shown.[90]

Clinical Relevance

E-Cadherin as a Potential Prognostic Factor

The prognostic significance of E-cadherin has been investigated in breast cancer in a variety of studies during the last several years. A number of recent reports points to a possible association between abnormal E-cadherin expression and poor prognosis.

An enzyme-linked immunosorbent assay study of E-cadherin in human breast cancers revealed that patients with tumors containing low levels of E-cadherin had a significantly shorter disease-free interval than patients with high levels ($P = 0.041$).[91] The prognostic significance of E-cadherin for disease-free interval was also found in node-negative patients and in patients with tumors smaller than or equal to 2 cm. Loss of E-cadherin expression in human breast cancers was associated with increased metastatic potential.

The expression pattern of E-cadherin has been investigated by immunohistochemistry in 120 breast carcinomas, and the clinical outcome was ascertained for 108 patients.[92] Complete loss of E-cadherin expression was found in 19% of infiltrating ductal carcinomas (IDCs) and 64% of infiltrating lobular carcinomas (ILCs). Loss of heterozygosity was detected in 46% of IDCs and 89% of ILCs. In the ILCs, loss of heterozygosity was associated with complete loss of membranous E-cadherin expression and appearance of cytoplasmic E-cadherin staining. This association was not observed in IDCs. In a multivariate analysis, loss of E-cadherin staining was shown to be a significant independent risk factor for a poorer disease-free survival ($P = 0.019$), especially in node-negative patients ($P = 0.029$). The authors concluded that different mechanisms seemed to be involved in the altered E-cadherin expression observed in different subtypes of breast carcinomas. Loss of E-cadherin was suggested in this report as an independent prognostic marker for disease recurrence, especially in node-negative breast cancer patients, irrespective of the histologic type.

E-cadherin expression was analyzed immunohistochemically in frozen sections of 362 breast carcinomas using a monoclonal antibody (HECD-1).[58] Reduced E-cadherin expression was confirmed by messenger RNA in situ hybridization. The percentage of tumors with reduced or lost E-cadherin expression increased significantly from pure intraductal carcinomas (20%) through invasive ductal (IDCs, 52%) to recurrent carcinomas (64%, $P = 0.004$). ILCs and IDCs differed from each other in their E-cadherin expression, because none of the ILCs retained normal E-cadherin expression in contrast to 48% of the IDCs, which expressed it. In 259 primary IDCs, reduced E-cadherin expression was associated with high histologic grade ($P < 0.001$), negative ER status ($P = 0.042$), and marginally with axillary node involvement ($P = 0.063$). In a subset of 109 primary IDC patients with clinical follow-up, reduced E-cadherin expression was associated with shortened disease-free survival ($P = 0.027$). In a multivariate regression analysis, progesterone receptor status ($P = 0.018$) and E-cadherin expression ($P = 0.072$) were identified as independent predictors of disease-free survival. This study provides clinical evidence that loss of normal E-cadherin expression is an indicator of increased invasiveness and dedifferentiation in breast carcinoma. E-cadherin was also suggested as a potentially important prognostic factor in primary IDCs.

In 174 breast cancer patients, E-cadherin expression was investigated immunohistochemically using the monoclonal antibody HECD-1.[93] In 66% of the breast cancers, E-cadherin expression was reduced. Significant correlations were found between E-cadherin expression and tumor type ($P \leq 0.001$), ER status ($P = 0.026$), and histologic grade ($P = 0.031$). Expression of E-cadherin was not found to be associated with recurrence, distant metastases, lymph node stage, vascular invasion, primary tumor size, prognostic group, or survival in this study. Therefore, the conclusion from this study is that E-cadherin may have minimal prognostic value.

The expression of E-cadherin and P-cadherin was immunohistochemically investigated in 51 breast cancer patients, including 29 node-negative and 22 node-positive tumors.[94] A significant reduction of expression of both cadherins was found in node-positive tumors in comparison with node-negative tumors. By multivariate analysis, both cadherins were found to be independent prognostic markers.

In contrast to reports that describe an association of low E-cadherin expression levels and poor survival, in a study of 149 grade I infiltrating ductal breast carcinomas, high E-cadherin expression correlated with poor survival, tumor size, and nodal status.[95] High E-cadherin scores and tumor size were both significant predictors of survival in an univariate analysis.

The conclusion from the above-mentioned data is that abnormal E-cadherin expression in breast cancer has been described as a valuable prognostic marker in some studies, but a number of other reports do not confirm these observations. The prognostic significance of

E-cadherin was also investigated in combination with other factors with prognostic relevance.

In 55 breast cancer patients, expression of E-cadherin and its associated catenins as well as β1 and α2 integrins was evaluated by immunohistochemistry.[96] A significant correlation between loss of E-cadherin expression and loss of β/γ-catenin immunostaining was found. In 20% of cases, E-cadherin staining was negative. α- and β/γ-catenin expression was negative in 24% and 22% of cases, respectively. The intensity of E-cadherin expression showed a significant correlation with histologic grade ($P = 0.002$) and tumor type ($P < 0.001$). Lobular carcinomas showed frequent loss of expression of E-cadherin and catenin. Expression of α-catenin was associated with grade ($P = 0.008$) and with ER status ($P = 0.006$), whereas β/γ-catenin expression was not associated with other known prognostic factors. Forty-nine percent and 42% of cases showed no membrane immunostaining with β1 and α2 integrin, respectively, and coordinated loss of β1 and α2 integrin expression was found. Both β1 and α2 integrin expression were associated with histologic grade ($P = 0.003$ and $P = 0.031$, respectively) and β1 expression with tumor type ($P = 0.010$). None of the variables examined showed a statistically significant association with tumor size or lymph node stage or with overall survival, although a trend was seen ($P = 0.087$) toward poorer survival of patients with tumors with absent or weak expression of β1 integrin.

The expression of the tyrosine kinase receptor c-*met* was evaluated immunohistochemically in 69 invasive breast carcinomas and correlated with E-cadherin and β-catenin expression.[97] c-*met* expression was observed in 58% of cases and was associated with the lobular type of breast carcinomas ($P = 0.012$), low histologic grade ductal carcinomas ($P = 0.05$), favorable prognostic and predictive factors such as ER and progesterone receptor immunohistochemical expression, and negative c-*erb*B2 expression ($P = 0.05$, $P = 0.014$ and $P = 0.03$, respectively). c-*met* expression was not associated with lymph node status, tumor size, and stage of the disease. c-*met* immunoreactivity correlated significantly with favorable patient survival ($P = 0.028$). A statistically significant correlation was found between c-*met* expression and abnormal β-catenin staining ($P = 0.025$), suggesting possible involvement of c-*met* in the down-regulation of components of the E-cadherin–catenin adhesion complex, possibly through tyrosine phosphorylation of β-catenin.

In another study, it was investigated whether immunohistochemical expression of E-cadherin and β-catenin in conjunction with CD44 correlated with the clinical evolution and prognosis of breast cancer.[98] One hundred forty-two breast tissue samples, including normal breast, benign lesions, in situ, and invasive carcinomas, were investigated. E-cadherin and β-catenin were strongly expressed by luminal and basal cells in normal glands, benign proliferative, and early neoplastic intraductal lesions. In contrast, CD44 was expressed exclusively by myoepithelial cells in normal breast, whereas different isoform expression patterns were observed in premalignant and malignant lesions. E-cadherin and β-catenin expression was lost simultaneously in in situ and invasive lobular carcinomas. In contrast, in ductal lesions, the differential loss of the molecules was associated with poorer differentiation, irrespective of CD44 immunophenotype. Reduced E-cadherin ($P = 0.003$), β-catenin ($P = 0.03$), and increased CD44v4 ($P = 0.005$) and v7 ($P = 0.007$) expression were significantly correlated with positive lymph node status. Reduction of E-cadherin immunostaining and loss of CD44v6 expression correlated with poor survival.

Expression of E-cadherin has been examined in 187 primary breast carcinomas by immunohistochemistry.[99] In ILCs, complete loss of membrane staining was observed, whereas IDCs retained some level of expression. In IDCs, E-cadherin expression was not related to tumor grade, but there was a significant association between reduced membrane levels of E-cadherin and the presence of lymph node metastasis and a highly significant correlation between the presence of cytoplasmic E-cadherin and metastasis. A significant relationship was also demonstrated between reduced E-cadherin reactivity and expression of the epidermal growth factor receptor.

E-Cadherin Expression as a Diagnostic Test

In most cases, the HECD-1 antibody is used for the immunohistochemical detection of E-cadherin. Mutation-specific antibodies have been generated that recognize mutant E-cadherin lacking exon 8[100] or exon 9.[101] Although truncating mutations are predominantly found in breast cancer, an in-frame deletion of exon 9 has also been found in the breast cancer cell line MPE600.[102,103]

E-Cadherin as a Therapy Target

Mutant E-cadherin variants could serve as a target for individualized cancer therapy.[104] As a cancer-associated mutant cell surface molecule, E-cadherin is an attractive candidate to target tumor cells. Monoclonal antibodies against E-cadherin lacking exon 8 or 9 are available.[101,104] Mutation-specific antibodies could also be linked to toxins and drugs or could be radiolabeled to enable selective and effective tumor cell killing.

A monoclonal antibody directed against a mutant E-cadherin lacking exon 9 was conjugated with the high linear energy transfer α-emitter 213Bi and tested for its binding specificity in subcutaneous and intraperitoneal nude mice tumor models.[105] After intratumoral application in subcutaneous tumors expressing mutant E-cadherin, the 213Bi-labeled antibody was specifically retained at the injection site as shown by autoradiography. After injection into the peritoneal cavity, uptake in small intraperitoneal tumor nodules expressing mutant E-cadherin was 17-fold higher than in tumor nodules expressing wild-type E-cadherin. Seventy-eight percent of the total activity in the ascites fluid was bound to free tumor cells expressing mutant E-cadherin, whereas in control cells binding was only 18%. The selective binding of the 213Bi-labeled, mutation-specific, monoclonal antibody directed against E-cadherin lacking exon 9 suggests a possible usefulness of α-radioimmunotherapy of disseminated tumors after locoregional application. Locoregional application of the above-mentioned tumor-specific antibody conjugated with the highly cytotoxic α-emitter 213Bi was used for treatment of intraperitoneal tumor cell spread in a nude mouse model.[106] Therapeutic efficacy of the 213Bi-conjugated monoclonal antibody together with low bone marrow toxicity support the locoregional therapy for the subgroup of patients expressing E-cadherin lacking exon 9.

Because expression of E-cadherin is often down-regulated in cancer cells, restoration of E-cadherin expression was suggested to lead to differentiation and anti-invasive behavior of tumor cells.[107] Several drugs have been described to up-regulate the function of the E-cadherin catenin complex, at least in vitro: insulin-like growth factor I, tamoxifen, Taxol, retinoic acid, and progesterone.[108–112]

Clinical Testing Methods

For the immunohistochemical detection of E-cadherin expression level, the HECD-1 antibody is currently used most frequently (FIG. 20.1).

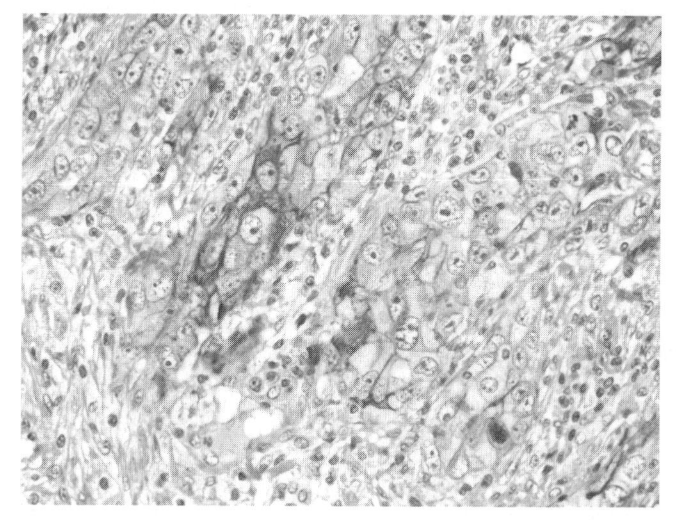

FIGURE 20.1 E-cadherin expression in breast cancer. In this photomicrograph significant loss of membranous E-cadherin immunostaining is seen in the poorly differentiated infiltrating ductal adenocarcibnoma that recurred after surgery and progressed with widespread metastatic disease (immunoperoxidase using the HECD-1 antibody ×200). **See Plate 46 for color image.**

Mutation-specific antibodies that have been generated to recognize mutant E-cadherin lacking exon 8[100] or exon 9[101] are also useful for mutation screening.

Current Clinical Utility

As discussed above, the use of E-cadherin expression as a prognostic marker in breast cancer is currently under evaluation.

Future Clinical Potential

Understanding of the mechanisms that underlie abnormal E-cadherin expression and function may also lead to the development of new therapeutic approaches, based on small molecule drugs that interfere with abnormal signaling triggered by mutant E-cadherin.

References

1. Esteva FJ, Sahin AA, Cristofanilli M, et al. Molecular prognostic factors for breast cancer metastasis and survival. Semin Radiat Oncol 2002;12:319–328.
2. Zhou BP, Hung MC. Dysregulation of cellular signaling by HER2/neu in breast cancer. Semin Oncol 2003;30:38–48.
3. Screaton GR, Bell MV, Jackson DG, et al. Genomic structure of DNA encoding the lymphocyte homing receptor CD44 reveals at least 12 alternatively spliced exons. Proc Natl Acad Sci USA 1992;89:12160–12164.
4. Ponta H, Wainwright D, Herrlich P. The CD44 protein family. Int J Biochem Cell Biol 1998;30:299–305.
5. Ponta H, Sherman L, Herrlich PA. CD44: from adhesion molecules to signalling regulators. Nat Rev Mol Cell Biol 2003;4:33–45.
6. Rudzki Z, LeDuy L, Jothy S. Changes in CD44 expression during carcinogenesis of the mouse colon. Exp Mol Pathol 1997;64:114–125.
7. Bourguignon LY, Lokeshwar VB, He J, et al. A CD44-like endothelial cell transmembrane glycoprotein (GP116) interacts with extracellular matrix and ankyrin. Mol Cell Biol 1992;12:4464–4471.
8. Lokeshwar VB, Fregien N, Bourguignon LY. Ankyrin-binding domain of CD44(GP85) is required for the expression of hyaluronic acid-mediated adhesion function. J Cell Biol 1994;126:1099–1109.
9. Legg JW, Isacke CM. Identification and functional analysis of the ezrin-binding site in the hyaluronan receptor, CD44. Curr Biol 1998;8:705–708.
10. Tsukita S, Oishi K, Sato N, et al. ERM family members as molecular linkers between the cell surface glycoprotein CD44 and actin-based cytoskeletons. J Cell Biol 1994;126:391–401.
11. Yonemura S, Hirao M, Doi Y, et al. Ezrin/radixin/moesin (ERM) proteins bind to a positively charged amino acid cluster in the juxta-membrane cytoplasmic domain of CD44, CD43, and ICAM-2. J Cell Biol 1998;140:885–895.
12. Aruffo A, Stamenkovic I, Melnick M, et al. CD44 is the principal cell surface receptor for hyaluronate. Cell 1990;61:1303–1313.
13. Peach RJ, Hollenbaugh D, Stamenkovic I, et al. Identification of hyaluronic acid binding sites in the extracellular domain of CD44. J Cell Biol 1993;122:257–264.
14. Herrera-Gayol A, Jothy S. Adhesion proteins in the biology of breast cancer: contribution of CD44. Exp Mol Pathol 1999;66:149–156.
15. Sleeman JP, Kondo K, Moll J, et al. Variant exons v6 and v7 together expand the repertoire of glycosaminoglycans bound by CD44. J Biol Chem 1997;272:31837–31844.
16. Jalkanen S, Jalkanen M. Lymphocyte CD44 binds the COOH-terminal heparin-binding domain of fibronectin. J Cell Biol 1992;116:817–825.
17. Ishii S, Ford R, Thomas P, et al. CD44 participates in the adhesion of human colorectal carcinoma cells to laminin and type IV collagen. Surg Oncol 1993;2:255–264.
18. Weber GF, Ashkar S, Glimcher MJ, et al. Receptor-ligand interaction between CD44 and osteopontin (Eta-1). Science 1996;271:509–512.
19. Culty M, Shizari M, Thompson EW, et al. Binding and degradation of hyaluronan by human breast cancer cell lines expressing different forms of CD44: correlation with invasive potential. J Cell Physiol 1994;160:275–286.
20. Bankfalvi A, Terpe HJ, Breukelmann D, et al. Gains and losses of CD44 expression during breast carcinogenesis and tumour progression. Histopathology 1998;33:107–116.
21. Kaufmann M, Heider KH, Sinn HP, et al. CD44 variant exon epitopes in primary breast cancer and length of survival. Lancet 1995;345:615–619.
22. Sinn HP, Heider KH, Skroch-Angel P, et al. Human mammary carcinomas express homologues of rat metastasis-associated variants of CD44. Breast Cancer Res Treat 1995;36:307–313.
23. Alford D, Pitha-Rowe P, Taylor-Papadimitriou J. Adhesion molecules in breast cancer: role of alpha 2 beta 1 integrin. Biochem Soc Symp 1998;63:245–259.
24. Zutter MM, Sun H, Santoro SA. Altered integrin expression and the malignant phenotype: the contribution of multiple integrated integrin receptors. J Mamm Gland Biol Neoplasia 1998;3:191–200.
25. Shaw LM. Integrin function in breast carcinoma progression. J Mamm Gland Biol Neoplasia 1999;4:367–376.
26. Mercurio AM, Bachelder RE, Chung J, et al. Integrin laminin receptors and breast carcinoma progression. J Mamm Gland Biol Neoplasia 2001;6:299–309.
27. Gumbiner BM. Regulation of cadherin adhesive activity. J Cell Biol 2000;148:399–404.
28. Kemler R. From cadherins to catenins: cytoplasmic protein interactions and regulation of cell adhesion. Trends Genet 1993;9:317–321.
29. Thoreson MA, Anastasiadis PZ, Daniel JM, et al. Selective uncoupling of p120(ctn) from E-cadherin disrupts strong adhesion. J Cell Biol 2000;148:189–202.
30. Behrens J, von Kries JP, Kuhl M, et al. Functional interaction of beta-catenin with the transcription factor LEF-1. Nature 1996;382:638–642.
31. Huber O, Korn R, McLaughlin J, et al. Nuclear localization of beta-catenin by interaction with transcription factor LEF-1. Mech Dev 1996;59:3–10.

32. Molenaar M, van de Wetering M, Oosterwegel M, et al. XTcf-3 transcription factor mediates beta-catenin-induced axis formation in *Xenopus* embryos. Cell 1996;86:391–399.
33. He TC, Sparks AB, Rago C, et al. Identification of c-MYC as a target of the APC pathway. Science 1998; 281:1509–1512.
34. Shtutman M, Zhurinsky J, Simcha I, et al. The cyclin D1 gene is a target of the beta-catenin/LEF-1 pathway. Proc Natl Acad Sci USA 1999;96:5522–5527.
35. Tetsu O, McCormick F. Beta-catenin regulates expression of cyclin D1 in colon carcinoma cells. Nature 1999;398:422–426.
36. Morin PJ, Sparks AB, Korinek V, et al. Activation of beta-catenin-Tcf signaling in colon cancer by mutations in beta-catenin or APC. Science 1997;275: 1787–1790.
37. Ilyas M, Tomlinson IP, Rowan A, et al. Beta-catenin mutations in cell lines established from human colorectal cancers. Proc Natl Acad Sci USA 1997;94: 10330–10334.
38. Rubinfeld B, Robbins P, El-Gamil M, et al. Stabilization of beta-catenin by genetic defects in melanoma cell lines. Science 1997;275:1790–1792.
39. Mansouri A, Spurr N, Goodfellow PN, et al. Characterization and chromosomal localization of the gene encoding the human cell adhesion molecule uvomorulin. Differentiation 1988;38:67–71.
40. Natt E, Magenis RE, Zimmer J, et al. Regional assignment of the human loci for uvomorulin (UVO) and chymotrypsinogen B (CTRB) with the help of two overlapping deletions on the long arm of chromosome 16. Cytogenet Cell Genet 50:145–148.
41. Berx G, Staes K, van Hengel J, et al. Cloning and characterization of the human invasion suppressor gene E-cadherin (CDH1). Genomics 1995;26:281–289.
42. Cleton-Jansen AM, Moerland EW, Kuipers-Dijkshoorn NJ, et al. At least two different regions are involved in allelic imbalance on chromosome arm 16q in breast cancer. Genes Chromosomes Cancer 1994;9:101–107.
43. Frixen UH, Behrens J, Sachs M, et al. E-cadherin-mediated cell-cell adhesion prevents invasiveness of human carcinoma cells. J Cell Biol 1991;113: 173–185.
44. Vleminckx K, Vakaet L, Jr, Mareel M, et al. Genetic manipulation of E-cadherin expression by epithelial tumor cells reveals an invasion suppressor role. Cell 1991;66:107–119.
45. Meiners S, Brinkmann V, Naundorf H, et al. Role of morphogenetic factors in metastasis of mammary carcinoma cells. Oncogene 1998;16:9–20.
46. St. Croix B, Sheehan C, Rak JW, et al. E-Cadherin-dependent growth suppression is mediated by the cyclin-dependent kinase inhibitor p27(KIP1). J Cell Biol 1998;142:557–571.
47. Kremer M, Quintanilla-Martinez L, Fuchs M, et al. Influence of tumor-associated E-cadherin mutations on tumorigenicity and metastasis. Carcinogenesis 2003;24:1879–1886.
48. Hirohashi S. Inactivation of the E-cadherin-mediated cell adhesion system in human cancers. Am J Pathol 1998;153:333–339.
49. Berx G, Becker KF, Hofler H, et al. Mutations of the human E-cadherin (CDH1) gene. Hum Mutat 1998; 12:226–237.
50. Birchmeier W, Behrens J. Cadherin expression in carcinomas: role in the formation of cell junctions and the prevention of invasiveness. Biochim Biophys Acta 1994;1198:11–26.
51. Van Aken E, De Wever O, Correia da Rocha AS, et al. Defective E-cadherin/catenin complexes in human cancer. Virchows Arch 2001;439:725–751.
52. Shino Y, Watanabe A, Yamada Y, et al. Clinicopathologic evaluation of immunohistochemical E-cadherin expression in human gastric carcinomas. Cancer 1995;76:2193–2201.
53. Zschiesche W, Schonborn I, Behrens J, et al. Expression of E-cadherin and catenins in invasive mammary carcinomas. Anticancer Res 1997;17:561–567.
54. Bukholm IK, Nesland JM, Karesen R, et al. E-cadherin and alpha-, beta-, and gamma-catenin protein expression in relation to metastasis in human breast carcinoma. J Pathol 1998;185:262–266.
55. Heimann R, Lan F, McBride R, et al. Separating favorable from unfavorable prognostic markers in breast cancer: the role of E-cadherin. Cancer Res 2000;60:298–304.
56. Hunt NC, Douglas-Jones AG, Jasani B, et al. Loss of E-cadherin expression associated with lymph node metastases in small breast carcinomas. Virchows Arch 1997;430:285–289.
57. Oka H, Shiozaki H, Kobayashi K, et al. Expression of E-cadherin cell adhesion molecules in human breast cancer tissues and its relationship to metastasis. Cancer Res 1993;53:1696–1701.
58. Siitonen SM, Kononen JT, Helin HJ, et al. Reduced E-cadherin expression is associated with invasiveness and unfavorable prognosis in breast cancer. Am J Clin Pathol 1996;105:394–402.
59. Gamallo C, Palacios J, Suarez A, et al. Correlation of E-cadherin expression with differentiation grade and histological type in breast carcinoma. Am J Pathol 1993;142:987–993.
60. Moll R, Mitze M, Frixen UH, et al. Differential loss of E-cadherin expression in infiltrating ductal and lobular breast carcinomas. Am J Pathol 1993;143: 1731–1742.
61. Vos CB, Cleton-Jansen AM, Berx G, et al. E-cadherin inactivation in lobular carcinoma in situ of the breast: an early event in tumorigenesis. Br J Cancer 1997;76:1131–1133.
62. Huiping C, Sigurgeirsdottir JR, Jonasson JG, et al. Chromosome alterations and E-cadherin gene mutations in human lobular breast cancer. Br J Cancer 1999;81:1103–1110.
63. Droufakou S, Deshmane V, Roylance R, et al. Multiple ways of silencing E-cadherin gene expression in lobular carcinoma of the breast. Int J Cancer 2001; 92:404–408.
64. De Leeuw WJ, Berx G, Vos CB, et al. Simultaneous loss of E-cadherin and catenins in invasive lobular breast cancer and lobular carcinoma in situ. J Pathol 1997;183:404–411.
65. Batlle E, Sancho E, Franci C, et al. The transcription factor snail is a repressor of E-cadherin gene expression in epithelial tumour cells. Nat Cell Biol 2000; 2:84–89.
66. Cano A, Perez-Moreno MA, Rodrigo I, et al. The transcription factor snail controls epithelial-mesenchymal transitions by repressing E-cadherin expression. Nat Cell Biol 2000;2:76–83.
67. Comijn J, Berx G, Vermassen P, et al. The two-handed E box binding zinc finger protein SIP1 downregulates

E-cadherin and induces invasion. Mol Cell 2001;7: 1267–1278.

68. Graff JR, Herman JG, Lapidus RG, et al. E-cadherin expression is silenced by DNA hypermethylation in human breast and prostate carcinomas. Cancer Res 1995;55:5195–5199.
69. Kanai Y, Ushijima S, Hui AM, et al. The E-cadherin gene is silenced by CpG methylation in human hepatocellular carcinomas. Int J Cancer 1997;71: 355–359.
70. Kanai Y, Oda T, Tsuda H, et al. Point mutation of the E-cadherin gene in invasive lobular carcinoma of the breast. Jpn J Cancer Res 1994;85:1035–1039.
71. Berx G, Cleton-Jansen AM, Nollet F, et al. E-cadherin is a tumour/invasion suppressor gene mutated in human lobular breast cancers. EMBO J 1995;14:6107–6115.
72. Berx G, Cleton-Jansen AM, Strumane K, et al. E-cadherin is inactivated in a majority of invasive human lobular breast cancers by truncation mutations throughout its extracellular domain. Oncogene 1996;13:1919–1925.
73. Becker KF, Atkinson MJ, Reich U, et al. E-cadherin gene mutations provide clues to diffuse type gastric carcinomas. Cancer Res 1994;54:3845–3852.
74. Muta H, Noguchi M, Kanai Y, et al. E-cadherin gene mutations in signet ring cell carcinoma of the stomach. Jpn J Cancer Res 1996;87:843–848.
75. Tamura G, Sakata K, Nishizuka S, et al. Inactivation of the E-cadherin gene in primary gastric carcinomas and gastric carcinoma cell lines. Jpn J Cancer Res 1996;87:1153–1159.
76. Machado JC, Soares P, Carneiro F, et al. E-cadherin gene mutations provide a genetic basis for the phenotypic divergence of mixed gastric carcinomas. Lab Invest 1999;79:459–465.
77. Oda T, Kanai Y, Oyama T, et al. E-cadherin gene mutations in human gastric carcinoma cell lines. Proc Natl Acad Sci USA 1994;91:1858–1862.
78. Risinger JI, Berchuck A, Kohler MF, et al. Mutations of the E-cadherin gene in human gynecologic cancers. Nat Genet 1994;7:98–102.
79. Soares P, Berx G, van Roy F, et al. E-cadherin gene alterations are rare events in thyroid tumors. Int J Cancer 1997;70:32–38.
80. Kashiwaba M, Tamura G, Suzuki Y, et al. Epithelial-cadherin gene is not mutated in ductal carcinomas of the breast. Jpn J Cancer Res 1995;86:1054–1059.
81. Becker KF, Atkinson MJ, Reich U, et al. Exon skipping in the E-cadherin gene transcript in metastatic human gastric carcinomas. Hum Mol Genet 1993;2: 803–804.
82. Gayther SA, Gorringe KL, Ramus SJ, et al. Identification of germ-line E-cadherin mutations in gastric cancer families of European origin. Cancer Res 1998; 58:4086–4089.
83. Guilford P, Hopkins J, Harraway J, et al. E-cadherin germline mutations in familial gastric cancer. Nature 1998;392:402–405.
84. Keller G, Vogelsang H, Becker I, et al. Diffuse type gastric and lobular breast carcinoma in a familial gastric cancer patient with an E-cadherin germline mutation. Am J Pathol 1999;155:337–342.
85. Richards FM, McKee SA, Rajpar MH, et al. Germline E-cadherin gene (CDH1) mutations predispose to familial gastric cancer and colorectal cancer. Hum Mol Genet 1999;8:607–610.
86. Salahshor S, Haixin L, Huo H, et al. Low frequency of E-cadherin alterations in familial breast cancer. Breast Cancer Res 2001;3:199–207.
87. Hazan RB, Kang L, Whooley BP, et al. N-cadherin promotes adhesion between invasive breast cancer cells and the stroma. Cell Adhes Commun 1997;4: 399–411.
88. Nieman MT, Prudoff RS, Johnson KR, et al. N-cadherin promotes motility in human breast cancer cells regardless of their E-cadherin expression. J Cell Biol 1999;147:631–644.
89. Thiery JP. Epithelial-mesenchymal transitions in tumour progression. Nat Rev Cancer 2002;2: 442–454.
90. Pishvaian MJ, Feltes CM, Thompson P, et al. Cadherin-11 is expressed in invasive breast cancer cell lines. Cancer Res 1999;59:947–952.
91. Maguire TM, Shering SG, McDermott EW, et al. Assay of E-cadherin by ELISA in human breast cancers. Eur J Cancer 1997;33:404–408.
92. Asgeirsson KS, Tryggvad L, Olafsdottir K, et al. Altered expression of E-cadherin in breast cancer patterns, mechanisms and clinical significance. Eur J Cancer 2000;36:1098–1106.
93. Parker C, Rampaul RS, Pinder SE, et al. E-cadherin as a prognostic indicator in primary breast cancer. Br J Cancer 2001;85:1958–1963.
94. Madhavan M, Srinivas P, Abraham E, et al. Cadherins as predictive markers of nodal metastasis in breast cancer. Mod Pathol 2001;14:423–427.
95. Tan DS, Potts HW, Leong AC, et al. The biological and prognostic significance of cell polarity and E-cadherin in grade I infiltrating ductal carcinoma of the breast. J Pathol 1999;189:20–27.
96. Gonzalez MA, Pinder SE, Wencyk PM, et al. An immunohistochemical examination of the expression of E-cadherin, alpha- and beta/gamma-catenins, and alpha2- and beta1-integrins in invasive breast cancer. J Pathol 1999;187:523–529.
97. Nakopoulou L, Gakiopoulou H, Keramopoulos A, et al. c-met tyrosine kinase receptor expression is associated with abnormal beta-catenin expression and favourable prognostic factors in invasive breast carcinoma. Histopathology 2000;36:313–325.
98. Bankfalvi A, Terpe HJ, Breukelmann D, et al. Immunophenotypic and prognostic analysis of E-cadherin and beta-catenin expression during breast carcinogenesis and tumour progression: a comparative study with CD44. Histopathology 1999;34: 25–34.
99. Jones JL, Royall JE, Walker RA. E-cadherin relates to EGFR expression and lymph node metastasis in primary breast carcinoma. Br J Cancer 1996;74: 1237–1241.
100. Becker KF, Kremmer E, Eulitz M, et al. Functional allelic loss detected at the protein level in archival human tumours using allele-specific E-cadherin monoclonal antibodies. J Pathol 2002;197:567–574.
101. Becker KF, Kremmer E, Eulitz M, et al. Analysis of E-cadherin in diffuse-type gastric cancer using a mutation-specific monoclonal antibody. Am J Pathol 1999;155:1803–1809.
102. Hiraguri S, Godfrey T, Nakamura H, et al. Mechanisms of inactivation of E-cadherin in breast cancer cell lines. Cancer Res 1998;58:1972–1977.
103. van de Wetering M, Barker N, Harkes IC, et al. Mutant E-cadherin breast cancer cells do not display

constitutive Wnt signaling. Cancer Res 2001;61: 278–284.

104. Becker KF, Hofler H. Mutant cell surface receptors as targets for individualized cancer diagnosis and therapy. Curr Cancer Drug Targets 2001;1:121–128.

105. Senekowitsch-Schmidtke R, Schuhmacher C, Becker KF, et al. Highly specific tumor binding of a 213Bi-labeled monoclonal antibody against mutant E-cadherin suggests its usefulness for locoregional alpha-radioimmunotherapy of diffuse-type gastric cancer. Cancer Res 2001;61:2804–2808.

106. Huber R, Seidl C, Schmid E, et al. Locoregional alpha-radioimmunotherapy of intraperitoneal tumor cell dissemination using a tumor-specific monoclonal antibody. Clin Cancer Res 2003;9:3922S–3928S.

107. Wijnhoven BP, Dinjens WN, Pignatelli M. E-cadherin-catenin cell-cell adhesion complex and human cancer. Br J Surg 2000;87:992–1005.

108. Matarrese P, Giandomenico V, Fiorucci G, et al. Antiproliferative activity of interferon alpha and retinoic acid in SiHa carcinoma cells: the role of cell adhesion. Int J Cancer 1998;76:531–540.

109. Fujimoto J, Ichigo S, Hori M, et al. Progestins and danazol effect on cell-to-cell adhesion, and E-cadherin and alpha- and beta-catenin mRNA expressions. J Steroid Biochem Mol Biol 1996;57:275–282.

110. Bracke ME, Charlier C, Bruyneel EA, et al. Tamoxifen restores the E-cadherin function in human breast cancer MCF-7/6 cells and suppresses their invasive phenotype. Cancer Res 1994;54:4607–4609.

111. Vermeulen SJ, Bruyneel EA, van Roy FM, et al. Activation of the E-cadherin/catenin complex in human MCF-7 breast cancer cells by all-trans-retinoic acid. Br J Cancer 1995;72:1447–1453.

112. Guvakova MA, Surmacz E. Overexpressed IGF-I receptors reduce estrogen growth requirements, enhance survival, and promote E-cadherin-mediated cell-cell adhesion in human breast cancer cells. Exp Cell Res 1997;231:149–162.

CHAPTER
21

Programmed Cell Death and Breast Cancer

Xiao-Ming Yin, MD, PhD,[1] and Gerald P. Linette, MD, PhD[2]

[1]Department of Pathology, University of Pittsburgh School of Medicine, Pittsburgh, Pennsylvania
[2]Division of Oncology, Washington University School of Medicine, St. Louis, Missouri

Fundamentals of Programmed Cell Death

Apoptosis Overview

Programmed cell death (PCD) plays a prominent role in normal breast development and the pathogenesis of breast cancer. Developmental biologists have clearly shown the influence of hormonal changes on cell proliferation that accompany mammary development during puberty as well as the changes that accompany pregnancy.[1–3] In this chapter we summarize recent developments in PCD and breast cancer research with emphasis on cell survival pathways and potential targets that could be exploited for new therapeutics. In addition, resistance mechanisms that impact pathways of PCD in breast cancer cells will be highlighted to underscore the fundamental issues of cancer biology. Recent discoveries and insights suggest that we are now poised to made fundamental advances in the development of new targeted therapies for advanced breast cancer.

PCD: The Essentials

Apoptosis refers to the morphologic (cytologic) features that typically accompany PCD.[4] The classic features of apoptosis usually noted include chromatin condensation and cell shrinkage with plasma membrane blebbing. Later, DNA fragmentation (induced by endogenous endonucleases) and exposure of phosphatidylserine on the outer cell membrane as a signal for rapid engulfment were noted as prominent features. Subsequent discoveries detailed the participation of caspases as effector cysteine proteases active in the final common pathway of PCD.[5,6]

It has become apparent that PCD is an evolutionarily conserved and genetically defined pathway that teleologically serves the embryo, developing organism, and adult to control "normal" cell deaths within tissues of each organ system at defined stages of development. Detailed investigation of model organisms such as *Drosophila, Caenorhabditis elegans,* and rodents reveal that interruption of normal cell death can lead to cancer and autoimmunity, whereas inadvertent premature cell death often leads to infertility, immunodeficiency, and neurodegenerative disorders. The t(14;18) translocation that accompanies approximately 85% of low grade human follicular lymphoma and juxtaposes Bcl-2 with the immunoglobulin heavy chain enhancer was the pivotal discovery that allowed investigators to appreciate the consequences of disrupted cell deaths that result in the accumulation of B cells that are otherwise destined to die. The overexpression of Bcl-2 prevents the physiologic PCD that often accompanies the acquired genetic changes (mutations) that can accumulate during B-cell differentiation. Avoidance of PCD in the context of a proliferative advantage for any cell results in cancer.[7] In contrast to other typical proto-oncogenes (such as c-*myc* and H-*ras*) that serve to promote cell cycle progression, Bcl-2 and related family members serve as "regulators of cell death" dedicated to establishing the control points with PCD pathways.[8]

This observation prompted an intense interest into the fundamental problem of metazoan cell death; the past decade has witnessed remarkable progress in defining the genetic elements and pathologic conditions

Table 21.1 The Mammalian Caspase Family

Group	Category	Human Caspase	Mouse Caspases	Prodomain
I	Proinflammation	1, 4, 5	1, 11	Long, containing the CARD domain
II	Effector	3, 6, 7, 14	3, 6, 7, 14	Short, no interaction domain
III	initiator	2, 9, 8, 10	2, 9, 8, 12	Long, containing the CARD (2, 9, 12) or DED domain (8, 10)

relevant to PCD. Parallel observations in *C. elegans* that identified the *ced* (cell death-3, -4, -9 members) genes as essential for controlling PCD were pivotal to the notion that this process is genetically defined and conserved in evolution.[9] Several critical concepts that have since emerged facilitate the discussion about pathologic processes related to cancer. First, multiple gene families exist that serve to regulate PCD pathways. The relevant gene families elucidated in detail both structurally and functionally include caspase members, inhibitor of apoptosis protein (IAP) members, tumor necrosis factor (TNF) receptors/ligands members, and Bcl-2 members. Clearly, other well-characterized molecules participate in the process of cell survival and death but cannot be formally assigned membership into one of the four major families. Second, two major death pathways have been defined: the extrinsic pathway, which is initiated by the death receptor family proteins, and the intrinsic pathway, which is centered on the mitochondria. Cross-talk between the two pathways has also been established. Third, death signaling compartmentalization at or within critical subcellular locales has emerged as an unexpected theme. Mitochondria, endoplasmic reticulum, nuclear membranes, and outer/inner cell membranes serve as "anatomic control points" by virtue of the subcellular localization of the various pro- and antiapoptotic family members within each cell lineage.

The Caspase Family

The central executioner molecules for apoptosis are a group of cysteine proteases, called caspases,[5] for which the prototype nematode homolog is CED-3. All caspases contain an active site QACRG pentapeptide. The molecule is synthesized in precursor form and processed to form an active enzyme in response to death signals. The activated proteases have unique substrate specificity on aspartic acid (P1). The substrates of caspases include caspases themselves and other cellular proteins. The activation of one caspase by another constitutes the caspase cascade that not only amplifies the process but also transmits the signals from one compartment to another within cells.

There are currently 11 caspases in human and 10 caspases in mouse[10] (**Table 21.1**). Mice do not have caspase-4 and -5 but have the functional homolog, caspase-11, which is not found in human. In addition, only human has caspase-10, which is very similar to caspase-8. Both are located on chromosome 2 adjacent to each other, suggesting a recent gene duplication event. In contrast, only functional transcript of caspase-12 is found in mice, and most humans do not encode a functional caspase-12. Finally, caspase-13 was found to be a bovine complementary DNA contaminant in a human library and is thus no longer considered to be present in either human or mouse genome.

Based on the relative substrate specificity, structure, and function, caspases can be categorized into three subgroups[5,11,12] (**Table 21.1**). Group I caspases are mainly involved in inflammation response, including caspase-1 and caspase-4 and -5, or their murine homolog caspase-11. Group II members serve as downstream "effector" caspases. Group III members are upstream "initiator" caspases, including caspase-2, -9, -8, and -10, and serve mainly to activate the downstream group II effector caspases, including caspase-3, -6, and -7. Structurally, group I and III caspases have a long prodomain, which often include some sort of protein–protein interaction domain, such as the CARD domain (caspase-1, -4, -5, -11, -2, and -9) or the DED domain (caspase-8 and -10). This feature is consistent with processing mechanism used in a multiple protein complex (see below). Murine caspase-12 is particularly involved in endoplasmic reticulum stress induced apoptosis. It has a long CARD-containing prodomain and thus may be an initiator caspase. However, how it is activated is not clear, although some reports suggest that it can be activated by calpain through cleavage.[13] On the other hand, caspase-14 has the typical short prodomain of the effector caspases. It can be cleaved by caspase-8 and thus likely belongs to the effector caspase.[14]

The Activation of Caspases Multiple mechanisms are used to activate different caspases. It seems that activation of the initiator caspases and the inflammation caspases often needs a large protein assembly, such as the death-inducing signaling complex (DISC) for caspase-8 and -10, the apoptosome for caspase-9, and the inflammasome for caspase-1 and -5. In contrast, cleavage of effector caspases by the initiator caspases is sufficient to activate the former, and there is no need for any large molecular complex. Once activated, however, effector caspases may cleave the initiator caspases in a feedback loop, resulting in further enhancement of the caspase activity. Thus, caspase activation often results in a cascade and multiple feedback amplifications.[15] There are also multiple endogenous and exogenous inhibitors that can block caspase activation at the various steps involved in caspase activation or just simply inhibit the activity of the activated caspases. Below is a brief discussion of the activation mechanisms of some of the better-studied caspases.

Activation of Caspase-8 Caspase-8 is generally activated after the death receptor ligation, although it may also be activated through other mechanisms. Recent studies have shown that caspase-8 would have to be activated through dimerization and cleavage of the zymogen is not necessary for the activation.[16] It is generally thought that when the death receptors are ligated, they recruit either directly or indirectly the adaptor molecule FADD (see below), which then recruits caspase-8. FADD interacts with the receptor through its DD domain and interacts with caspase-8 through its DED domain. The actual stoichiometry has not been worked out in terms of how many FADD molecules are associated with each trimeric receptor complex and how many caspase-8 molecules are recruited into each receptor–FADD complex. However, it seems that the induced proximity of the multiple caspase-8 molecules in the complex leads to dimerization that is required for the activation of this initiator caspase.[16,17]

Activation of caspase-8 could be thus inhibited by a dominant negative FADD, which lacks part of the DED domain.[18] In addition, a cellular caspase inhibitor, FLIP, which shares sequence homology with caspase-8, can also inhibit its activation.[19] There are two forms of FLIP: The long form has the DED domain that can interact with FADD but does not have the catalytic capability of caspase-8, whereas the short form has the DED domain only but no catalytic domain. Both can thus serve as a competitive inhibitor for caspase-8 by binding to FADD. Furthermore, it has been reported that caspase-8 activation could be inhibited by cIAP1 and cIAP2 in the presence of TRAF-2; all are under the control of NF-κB activation.[20] It is not clear how this may work because it is known that IAP has no direct inhibitory effects on caspase-8 based on in vitro analysis[21] (see below). Finally, CrmA, a cowpox virus encoded protein belonging to the serpin family, can bind to the activated caspase-8 and effectively inhibit its activity.[22] A mammalian counterpart of CrmA is likely present but has not been reported yet.

Activation of Caspase-9 Caspase-9 exits as a monomer in solution.[23] It is mainly activated in the apoptosome, which is composed of Apaf-1, cytochrome *c*, and dATP. Apaf-1 is likely present in a monomer in normal cells. When the mitochondria are activated by death signals, cytochrome *c* is released, which binds to Apaf-1 and in the presence of dATP or ATP causes a conformation change and oligomerization of Apaf-1.[24] It is believed that caspase-9 is subsequently recruited to the Apaf-1 complex through its CARD domain and activated. Monomer caspase-9 has no activities. It is proposed that caspase-9 would have to form dimers or even higher order oligomers to be active.[23,25] There would be only one active site for each dimer.[25] However, exactly how this may occur in the apoptosome is not known, which has a sevenfold symmetry. One thing that does seem clear is that cleavage of caspase-9 is not required for its activation.[26] In fact, cleavage of caspase-9 at Asp315 by autocatalysis will result in the exposure of the XIAP-binding motif at the N-terminus of the small subunit, rendering its subsequent inhibition by XIAP. XIAP belongs to the IAPs. As discussed in more detail in the following section, caspase-9 activation and the activity of activated caspase-9 can be inhibited by IAPs, such as XIAP. However, the downstream effector caspases, which is activated by caspase-9, could lead to further cleavage of caspase-9 at Asp330, which will remove the XIAP binding motif and may weaken the inhibition by XIAP.[27]

Activation of the Effector Caspase-3 and -7 It is well documented that cleavage of the proenzyme is required for the activation of these caspases. Structure analysis indicates that they may first form homodimers under normal conditions, which, however, have no activities. Cleavage by the initiator caspase-8, -9, or -10 or granzyme B, which generates the large and small subunits, will yield a productive structure element, called L2′ loop, which is important for the formation of the catalytic active site[23,28] (stage I). The activated effector caspases now possess a weak activity that can autocatalyze itself to remove the short N-terminal prodomain (stage II). This process leads to a fully activated caspase. For caspase-3, these two steps generate the p20/p12 complex and then p19/p12 and p17/p12 complex, respectively.[29–31] Note the series of cleavage eventually results in a function unit of heterotetramer derived from two procaspases, and there are two active sites per function unit.

IAPs can bind to the fully activated effector caspases through its linker sequence immediately preceding the BIR2 domain to suppress their activities.[32,33] Furthermore, IAPs can also suppress the stage II activation of these caspases.[21,34] In the case of caspase-3, binding of XIAP will arrest it at the stage of p20/p12 during its activation. This feature can be detected through Western blot analysis and is a good in vivo indication of IAP-mediated inhibition.[34] Finally, in addition to the IAP molecules, p35 is another baculovirus-encoded protein that could also inhibit the catalytic activities of the effector caspase.[35]

Activation of the Proinflammatory Caspase-1 and -5 Caspase-1 was originally discovered as the enzyme processing the precursor interleukin (IL)-1β to its matured form, thus the name IL-1 converting enzyme or ICE. Because the sequence and catalytic mechanisms (cysteine protease) of ICE is conserved with other caspases, it is renamed as caspase-1. However, although overexpression of caspase-1 can induce cell death, its normal function is not to kill cells but rather to process IL-1β. Thus, this caspase is considered to play important roles in inflammation, as are caspase-4 and caspase-5.

A potent proinflammatory stimulus is bacterial lipopolysaccharide, which binds to the toll-like receptor-4 and can activate caspase-1 and -5, resulting in the processing of IL-1β. Recent studies have shown that a critical molecule, NALP1, can be activated by toll-like receptor-4 signals and then recruit caspase-5 through the interaction of the CARD domains present in both molecules. In addition, NALP1 can recruit another molecule, ASC, through the PYD domain shared by the two molecules. ASC contains an additional CARD domain, which can interact with the CARD domain of caspase-1 and thus recruit it to the complex. This complex, dubbed as inflammasome, is considered to be crucial to the activation of caspase-1 and caspase-5, which then are able to process pro-IL-1β near the inner surface of the cytoplasmic membranes.[36] Although the actual mechanism of activation is not precisely defined, it is thought that the proximity of these recruited procaspases in the complex will activate them in a similar way as caspase-8 or caspase-9 is activated in the DISC or apoptosome.

NALP1 belongs to the large NALP family with 13 additional members. NALP2 through NALP14 do not encode the CARD domain and thus do not interact with caspase-5. However, they can promote pro-IL-1β processing and thus are able to activate caspase-1. Further studies using NALP2 and NALP3 found that they could interact with a novel protein called Cardinal.[37] This interaction is mediated by the NACHT domain of the NALP molecule with the FIIND domain of Cardinal. Interestingly, Cardinal also contains a CARD domain, which can interact with the CARD domain of caspase-1 but not that of caspase-4, -5, or -9. Meanwhile, NALP2 or NALP3 can still interact with ASC, which recruits another caspase-1 molecule to the complex. It is likely that in this type of inflammasome, caspase-1 can be activated through the proximity provided by the scaffold protein, NALP2 or NALP3.

The Function of Caspases Caspases participate in multiple cellular functions, including pivotal roles in execution of PCD as well as important roles in inflammation and cellular differentiation. There are more than 60 proteins identified as caspase substrates, representing a broad range of protein families involved in different cellular functions,[6] such as ICAD (inhibitor for caspase activated DNAase) and PARP (poly ADP-(ribosyl) polymerase), among others. Many of the known substrates are associated with the loss of normal cell functions and constitute the phenotypic presentation typical of apoptotic cells, such as cytoskeleton change, cell body shrinkage, externalization of membrane phosphatidylserine, nucleus condensation, and nucleus fragmentation. One of the best studied examples is ICAD, which can be cleaved by caspase-3, followed by the release of its binding partner, CAD.[38] CAD is then able to digest DNA at the junction of nucleosomes, creating the 180-bp repeats of the DNA ladder, the best-defined apoptosis feature.[39] Cleavage of various cellular substrates can ensure the death of cells.

Some of the caspase substrates, however, are not related to cell death, such as the proinflammatory cytokines IL-1 and IL-18, which are processed by caspase-1, -5 and -11, and IL-16, which can be processed by caspase-3 (reviewed in Ref. 6). As mentioned above, caspase-1, -4, and -5 are normally not involved in killing cells but are involved in the inflammatory response. Genetic mutations have been defined in the components of the inflammasome, such as NALP3, which lead to increased activation of caspase-1 and therefore increased production of IL-1β as seen in the Muckle-Wells autoinflammatory disorder.[37]

Notably, caspase activation is also seen in developmentally regulated PCD. Inhibition of cell death by blocking caspase activation often lead to developmental abnormalities, as seen in caspase-3, -8, and -9 knockout mice (reviewed in Ref. 6). In these examples, caspase activation and subsequent PCD is apparently a normal regulated process.

The IAP Family Proteins and Their Inhibition of Caspases

The IAP Molecules Can Inhibit the Activation of Caspases IAP represent a unique family of proteins that were originally identified in baculovirus-infected cells in a report published in 1993.[40] Cellular orthologues were later identified and shown to act primarily through inhibition of caspase-3, -7, and -9 to regulate cell death; no other caspases appear to be directly affected by IAP members.[21] To date, eight human IAP members have been identified: NAIP, c-IAP1, c-IAP2, XIAP, survivin, apollon (BRUCE), ML-IAP (Livin), and ILP-2.[10] As mentioned, IAP molecules are well conserved through evolution and have been found also in virus, yeast, *C. elegans,* and *Drosophila.*[21] It should be pointed out that not all IAPs have been shown to be able to directly suppress caspases, such as NAIP and some IAPs normally do not participate in apoptosis regulation, such as survivin (see below).

Two unique domains are characteristic features of this family: the zinc-binding baculovirus IAP repeat (BIR) domain and the RING domain. At least one (up to three) BIR domain(s) is present in each family member, and the BIR domain is required and sufficient for IAP function to repress caspase activation. The function of the RING domain may be related to the ability of IAPs to target to the ubiquitin pathway. The major mechanism for the IAP molecules to suppress caspase activation is a direct interaction with the caspase, as best illustrated in the study of XIAP. A minor mechanism, as illustrated by cIAP1 and cIAP2, is that these two IAPs could bind to TRAF1 or TRAF2, adaptor molecules recruited to TNF-R1 complex after TNF binding. The complex is then able to inhibit caspase-8 activation through a yet to be determined mechanism.[20] Most studies on the suppression of caspase by IAPs, however, are based on the direct interaction of XIAP with the caspases, and it is clear that XIAP suppresses the downstream effector caspase-3 and -7 and the initiator caspase-9 using different mechanisms. It is possible that cIAP1 and cIAP2 can also work by a similar mechanism.

XIAP preferentially binds to the large subunit of the activated caspase-3 or -7 much more strongly than to the zymogen.[33,41] For example, the initial cleavage of caspase-3 at Asp175 results in the p20 large subunit that can bind to XIAP, which inhibits the subsequent further processing of the p20 to p19 and p17.[33,41] The initial cleavage of caspase-3 or -7 after dimerization provides the necessary conformation for the activated site to be productive, which also allows the binding of XIAP.[23,33,42] XIAP is thus also able to bind to the fully matured activated caspase-3 or -7 and suppress their activities. In these cases, XIAP uses a linker sequence immediately preceding the BIR2 domain to bind to the active site of caspase dimers, which prevents the substrate entry and catalysis.[32,33]

Inhibition of caspase-9 by XIAP can be accomplished by two mechanisms. XIAP can suppress caspase-9 activation by binding to the inactive caspase-9 monomer using its

BIR3 domain. Structure analysis has found that this interaction will prevent caspase-9 dimerization or further oligomerization in the apoptosome, which would activate the caspase.[23] In addition, XIAP can also inhibit the activity of processed caspase-9 by using its BIR3 domain.[26] There are two XIAP-binding sites on caspase-9. The binding interface present in the caspase-9 monomer is quite large and continuous,[23] whereas the binding site of the activated caspase-9 is created as a result of caspase-9 cleavage and is located at the N-terminus of the nascent small subunit.[26,43] The first four amino acids (ATPV) at this N-terminal constitute the binding motif for the BIR3 domain of XIAP.

There Are Cellular Mechanisms to Antagonize IAP Functions to Promote Apoptosis The IAP-mediated caspase suppression could be removed by several mechanisms. One major mechanism is the release of Smac/DIABO[26,43–45] or HtrA2/Omi[46,47] from the mitochondria, which can be induced by prodeath Bcl-2 family protein, such as Bid. Smac/DIABLO can competitively bind to IAPs (XIAP, cIAP1, and cIAP2) to liberate the bound caspases.[26,43] The dimerized form of Smac/DIABLO is required to bind to the BIR2 domain, whereas both the monomer and dimer forms can bind to the BIR3 domain. Smac/DIABLO is synthesized in the cytosol and imported to the mitochondria, where it is matured by removing the 55 amino acid N-terminal mitochondria targeting sequence. This processing reveals a new N-terminal sequence (AVPI), which is the perfect binding motif for the BIR3 domain. The interaction of matured Smac/DIABLO dimers with both the BIR2 and BIR3 domains of XIAP molecules will prevent the latter from interacting with caspases.[26,43,48,49] This can be of functional significance. For example, Fas activation induced hepatocyte apoptosis is mitochondria dependent, because it relies on this mechanism to relieve XIAP inhibition.[34] This mechanism is apparently conserved through evolution, as several prodeath molecules in *Drosophila* (i.e., Reaper, Grim, and Hid) contain the conserved tetrapeptide domain and work in the same way as Smac/DIABLO to remove the inhibition of the *Drosophila* IAP molecules (DIAP 1 and DIAP2).[26,48]

HtrA2/Omi can also reverse IAP suppression by binding to it in a similar mechanism.[46,47] The 50 kDa HtrA2/Omi precursor is processed in the mitochondria, and its matured N-terminus contains the conserved tetrapeptide binding motif (AVPS). When released from the mitochondria, the matured HtrA2/Omi can bind to XIAP in much the same way as Smac/DIABLO. This redundancy may explain why mice deficient in Smac/DIABLO do not present any obvious phenotype and have a normal apoptosis response.[50] However, the prodeath activity of HtrA2/Omi is not limited to this function, because it also possesses a serine protease activity that can independently contribute to cell death.[47]

The second major mechanism that may remove XIAP inhibition is proteolytic cleavage and/or ubiquitination-mediated degradation. XIAP could be cleaved by caspase-8 and/or caspase-3 as found in the in vitro studies.[41] It is possible that cleavage of XIAP is an event more important for death receptor pathway, where activation of caspase-8 is an early event and is not subject to XIAP suppression. Cleavage of XIAP has also been observed in a mouse model of liver injury using anti-Fas antibody.[34] Alternatively, effector caspase activity would have to be accumulated to exceed the ability of XIAP to inhibit, thus making the cleavage of XIAP likely a later event. Furthermore, XIAP and cIAPs have been reported to possess the ubiquitin ligase activity and can be ubiquitinated. Such a mechanism of removing IAP inhibition by degrading the protein may be important in certain type of cell death, such as that of mouse thymocytes induced by dexamethasone or VP-16.[51] Finally, in *Drosophila,* binding of DIAP1 to the caspase leads to the cleavage of DIAP1 by the activate caspase, revealing the N-terminal Asn residue and causing ubiquitination mediated degradation of DIAP1 and associated caspases through the N-end rule pathway.[52] Degradation of DIAP1 is particularly observed when Hid, Reaper, or Grim is expressed and displaces the bound caspases, thus contributing to cell death induced by these molecules.[52]

Inhibition of IAP Activity May Sensitize Tumors to Apoptosis The increased expression of IAP is one apparent mechanism used by tumors to avoid chemotherapy-induced PCD. When the tumor cells are in active proliferation, cell death mechanisms are often engaged, for example, in the case of c-*myc* activation.[7] This consideration led to the initial belief that survivin, because it is highly expressed in tumor cells but not in terminally differentiated adult

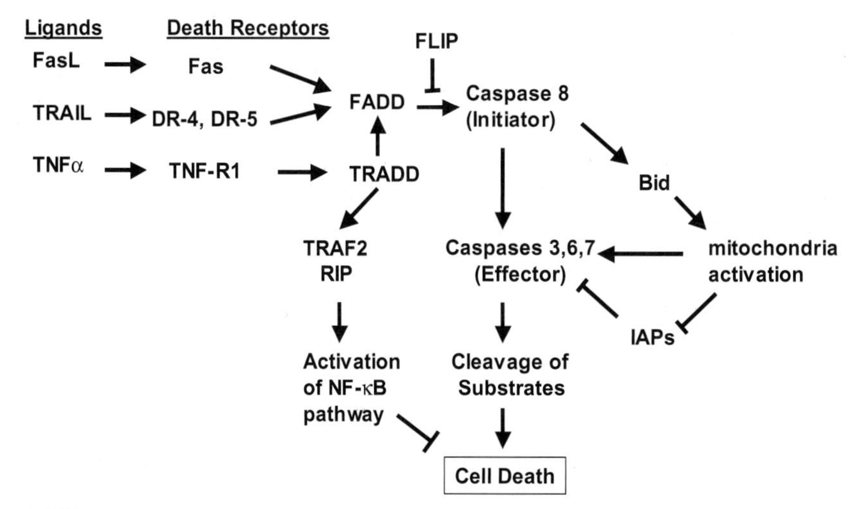

FIGURE 21.1 The death receptor apoptosis pathway. The extrinsic pathway is initiated by ligation of the death receptors (Fas, TRAIL, TNF-R1) by the corresponding ligand. A caspase mediated cell death is the primary mode of apoptosis. Cross-talk to the intrinsic pathway is mediated by Bid, which is proteolytically activated by caspase-8 (after death receptor cross-linking) and serves to activate the mitochondria pathway upon engagement with Bax or Bak. This cross-talk is essential for death receptor induced apoptosis in some tissues, such as the liver, where the inhibition of caspases by the IAP family proteins is significant. A protective pathway mediated by the transcriptional activity of NF-κB is also activated after TNF-R1 engagement in many cell types. See the text for additional details.

tissues, was normally involved in suppressing apoptosis of tumor cells.[53] Down-regulation of survivin in a number of cell lines leads to cell death.[54] However, the subsequent gene deletion study indicated that the normal cellular function of survivin is coordination of events like microtubule organization and cytokinesis.[55] This function is well conserved from yeast to mammalian cells. Survivin deletion causes embryonic lethality with disrupted microtubule formation.[55] This study demonstrates that IAP molecules may not normally participate in regulating cell death. However, when their expression is unduly increased in tumor cells, they may begin to perform functions that are normally not associated with them, such as inhibition of apoptosis. This perhaps is the case for survivin as well as ML-IAP/Livin, which is highly expressed in melanoma cell lines.[56] This observation is instructive to the development of novel cancer therapy, because the use of small molecules of IAP inhibitors, such as those based on the mechanism of Smac, should sensitize the tumor cells to cell death induced by the activation of caspases.

The Death Receptor Family and the Extrinsic Apoptosis Pathway

The Death Receptor Family This family of receptor proteins is also known as the TNF receptor superfamily, which has approximately 30 members sharing sequence homology in the cysteine-rich extracellular domains. Not all members are associated with PCD. Those members that are associated with PCD share additional sequence homology at the intracellular death domain. These include TNF-receptor 1 (TNF-R1, CD120a), Fas (CD95, Apo1), TNF-related apoptosis-inducing ligand (TRAIL)-receptor 1 (DR-4), and TRAIL-receptor 2 (DR-5). The corresponding ligands are TNF-α, FasL, and TRAIL, respectively[57–60] (FIG. 21.1). Other receptors, such as DR-3 and DR-6, may also be involved in cell death, but their ligands have not been defined yet.[60]

The Extrinsic Death Pathway The extrinsic pathway is initiated through oligomerization of the death receptors by their ligands or agonists presented in the extracellular space (FIG. 21.1). Ligation of Fas by either its ligand, FasL, or its agonistic antibody triggers the homotrimeric association of the receptor. The clustering of the death domain in the intracellular portion of the receptors recruits the adapter molecule, FADD/Mort1, which recruits an initiator caspase to cause its oligomerization.[61] These molecules form the DISC, which induces the activation of caspase-8, which in turn activates downstream effector caspase-3 and other substrates. Another initiator caspase, caspase-10, which shares a high degree of homology with caspase-8, can also be recruited to the DISC, acting in a similar way as caspase-8.[62]

TRAIL has several receptors. Its full functional receptors, DR-4 and DR-5 (TRAIL-R1 and TRAIL-2), when engaged, initiate a similar signaling cascade recruiting FADD and initiator caspase-8 or caspase-10 to the receptor complex.[63] Thus, TRAIL is a potent inducer of PCD. Interestingly, TRAIL also binds to the so-called decoy receptors, DcR1 and DcR2 (TRAIL-R3 and TRAIL-R4), which are defective in their intracellular domains and thus have effectively lost the ability to transmit death signals. The initial finding that most normal tissue, but not many tumor cells, express these decoy receptors[59] and that TRAIL has a tumoricidal activity without affecting the normal tissues of the mouse, including the liver,[64,65] led to the consideration that TRAIL could serve as a specific therapeutic agent for cancer therapy. However, different preparations of recombinant human TRAIL proteins may possess different toxicity to hepatocytes.[66–68] It seems that the zinc content, which is required to stabilize the trimer status of the protein, is important for its nontoxicity to human hepatocytes.[67] The nontoxicity of TRAIL to most, if not all, normal cells may not be simply attributed to the coexistence of the decoy receptor but also to a number of other factors, including the expression of regulatory proteins like IAPs, Smac, Bcl-xL, and FLIP.[59] Interestingly, many of them (XIAP, Bcl-xL, and FLIP) could be up-regulated by the NF-κB pathway, which TRAIL can activate (see below).

Activation of the TNF-R1 by TNF-α leads to a much more complex response. The engagement of the receptors results in the trimerization of the receptor and recruitment of TRADD via interactions of death domains present on both molecules.[69] Association of TRADD with FADD is followed by a similar reaction seen in the activation of Fas receptors.[69] However, recently it has been found that the TNF-R1–TRADD complex and TRADD–FADD–caspase-8 complex are not in the same compartment, raising the possibility that TRADD may have to dissociate from TNF-R1 before it binds to FADD to activate caspase-8.[70,71] How the TNF-R1–TRADD complex is converted to TRADD–FADD–caspase-8 complex is thus still unknown. In addition, TNF-R1 engagement could also activate effector caspases by other means or cause cell death by caspase-independent mechanisms. TNF-α/TNF-R1 interaction activates a MAPKKK, ASK-1, through TRADD, and TRAF2, which eventually leads to the activation of JNK and p38 MAPK, contributing to the death.[72,73]

Perhaps the most important pleiotropic effect of the death receptor engagement is the activation of the NF-κB pathway (see Chapter 24), most notably demonstrated in TNF-α/TNF-R1 system, but has also been reported in TRAIL receptor and Fas activation (FIG. 21.1). In the best studied case of TNF-R1,[58] TRADD recruits TRAF2 and RIP, which in turn recruit and activate the IKK complex. The activation of IKK complex, which includes the IκB kinase 1 (IKK1 or IKKα), IκB kinase 2 (IKK2 or IKKβ), and the regulatory subunit IKKγ leads to the phosphorylation of IκB and its degradation, which is followed by the release of the NF-κB p50 and p65 dimers. The nuclear translocation of NF-κB is followed by its transcriptional activation of multiple protective genes, leading to the inhibition of the death effects caused by the ligation of the very same receptor. TRAIL mediated NF-κB activation may use the similar mechanisms involving TRAF2.[59,74] However, Fas-mediated NF-κB activation is likely through different mechanisms. For example, in hepatocytes, this is usually observed in vitro under certain culture conditions but not observed in living animals.[75] It was found that NF-κB could be activated by Akt kinase, which was in turn activated and phosphorylated by PI-3 kinase in response to Fas engagement.[75] Activation of NF-κB by TNF-α in most cases and by Fas agonist or TRAIL in some cases inhibits cell death signaling simultaneously activated in the same cells by these ligands. Thus transcription inhibitors, or NF-κB specific inhibitors, are required for apoptosis to occur.

Cross-Talk Between the Death Receptor and the Mitochondria Pathway Ligation of the death receptors can also lead to the activation of the mitochondria pathway (FIG. 21.1). The BH3-only prodeath Bcl-2 family protein, Bid, is a good substrate of caspase-8.[76–78] When proteolytically cleaved by caspase-8 at Asp59, the C-terminal part of Bid could be further post-translationally modified by N-myristoylation and translocated to the mitochondria to initiate the mitochondria pathway.[79] Thus, Bid can bridge the extrinsic pathway to the intrinsic pathway and provide a way to amplify the death signals. Interestingly, an earlier study recognized that some lymphoid cells could be grouped into the so-called type I cells, where anti-Fas induced apoptosis could not be blocked by Bcl-2 or Bcl-x_L, and type II cells, where the anti-Fas induced death could be suppressed by these molecules.[80] Notably, both DISC formation and caspase-8 activation are weak in the type II cells, and thus the mitochondria pathway is required for the amplification of the death signals, which can be blocked by Bcl-2 or Bcl-x_L. It should be noted that cross-talk mediated by Bid in hepatocytes is actually indispensable for liver injury caused by Fas activation.[34,81]

Why is the mitochondrial pathway required in death receptor mediated apoptosis in certain types of cells despite the fact that initiator caspase-8 could directly activate downstream effector caspase-3? By analyzing the activation of caspase-3 in an in vivo model of anti-Fas induced liver injury that requires the participation of Bid-mediated mitochondrial activation, Li et al.[34] found that in the absence of Bid, and thus in the absence of mitochondrial activation, in vivo engagement of Fas receptors on hepatocytes could not lead to a full processing of caspase-3 and was accompanied by a very low caspase activity. This was most likely due to the inhibition by the IAP molecules as well as to a constitutively low caspase-8 activity. In wild-type animals, Bid could still be cleaved, and tBid induced the mitochondrial release of Smac, which antagonized the effects of XIAP. In addition, XIAP was also cleaved in a Bid-dependent fashion. These effects, together with the cytochrome *c* directed caspase activation, led to a full recovery of caspase-3 activity. Similarly, Sun et al.[82] found a similar mechanism in the type II lymphoid cells, which showed that Bcl-2 and Bcl-x_L suppressed the full processing of caspase-3 by inhibiting Smac release and thus maintaining the inhibitory effects of XIAP.

In conclusion, the cross-talk between the extrinsic pathway and the mitochondria pathway removes the inhibition posted by IAP molecules, and it seems that the molecular decision as to which pathway to use after death receptor engagement seems to be dependent on the stoichiometry among activated Caspase 8, IAPs, and Bid.

Bcl-2 Gene Family and the Mitochondrial Apoptosis Pathway

The Bcl-2 Family Proteins Are Evolutionarily Conserved Bcl-2 was originally cloned at the t(14;18) translocation breakpoint common to human follicular B-cell lymphoma, and at that time it was found to be a unique protein with no known function.[83–85] Astute observations made in several laboratories suggested that Bcl-2 serves to promote cell survival and thus functions as an antiapoptotic molecule.[86,87] The initial experiments using growth factor deprivation of cytokine dependent murine cell lines showed that Bcl-2 could maintain cell viability in the absence of IL-3.[86,87] These findings were quickly extended and validated using transgenic animals that overexpressed Bcl-2 in the B-cell compartment.[88] Moreover, Bcl-2-Ig transgenic animals developed aggressive monoclonal B-cell lymphoma over time, recapitulating the transformation of low grade follicular lymphoma often seen in patients with t(14;18) non-Hodgkin lymphoma.[88,89] Transgenic overexpression of Bcl-2 in other lineages beyond B lymphocytes supports the central thesis that repression of cell death is oncogenic but typically requires second cooperating events that usually promote cell cycle progression to ensure a proliferative advantage.[90] Based on these essential findings, the notion that cancer is a function of increased cell proliferation and decreased cell death was adopted.[91]

Currently, there are over 20 family members that are broadly classified into three major subgroups based on structure and function[10] (FIG. 21.2): antiapoptotic (Bcl-2 prototype), proapoptotic (Bax prototype), and the BH3-only domain proapoptotic (BID prototype), which are sensors for the peripheral death signals and are able to activate the "multidomain" executioner molecules, Bax or Bak.[92–95] Mutational analysis combined with traditional gain of function and loss of function experimental approaches have validated the classification of most family members studied in detail. Importantly, experimental overexpression of various antiapoptotic family members promotes solid tumor formation in other organ systems.[96–100]

Notably, this family of proteins is evolutionarily conserved. A number of viruses encode Bcl-2 homologs, including most, if not all, gamma herpesviruses.[101] Most of these viral homologs are antiapoptotic, probably because viruses need to keep the infected cells alive for

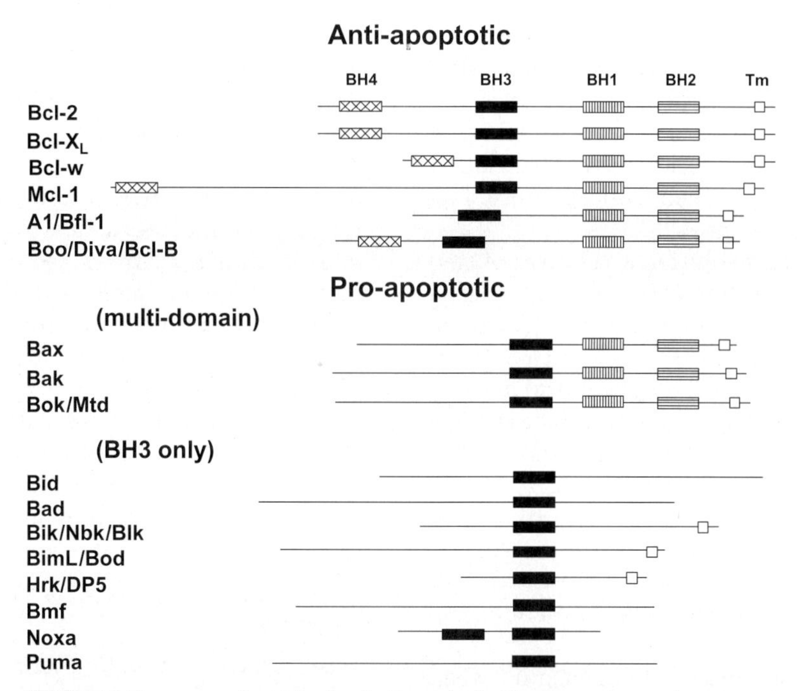

FIGURE 21.2 The mammalian Bcl-2 family. The Bcl-2 family of proteins can be grouped into the antiapoptotic and proapoptotic members based on function or into the multi-domain members and BH3-only members based on structure. Four BH (Bcl-2 homology) domains are indicated: the BH1 domain (box with vertical lines), the BH2 domain (box with horizontal lines), the BH3 domain (closed box), and the BH4 domain (box with crossed lines). In addition, the transmembrane domain (Tm) is also shown (open box). For those members with different spliced forms (Bcl-2, Bcl-xL, Bax, Bim, Bcl-G, etc.), only the long form (most common) is indicated. In addition, proteins with BH3-like domains (Nip3, Nix, Map1, p193, etc.) are not included in the scheme. See the text for more details.

the latent and persistent infection.[101,102] The nematode *C. elegans* has its sequence and functional homologs for a death antagonist, CED-9[103] and a BH3-only death agonist, EGL-1.[104] On the other hand, only prodeath homologs (dBorg-1/Drob-1/Debcl/ and dBorg-2/Buffy) have been described in the *Drosophila*.[105]

The Structure of Bcl-2 Family Proteins and Protein–Protein Interactions One of the key features of the Bcl-2 family proteins is that members share sequence homology in four domains, the BH1, 2, 3, and 4 domains, although not all members have all the domains[106–109] (FIG. 21.2). Mutagenesis studies have revealed that these domains are important for the function as well as for protein interactions among the family members. The BH1 and BH2 domains are necessary for the death repression function of the antiapoptotic family members, whereas the BH3 domain is required for the death promotion function of the proapoptotic family members.[95,110] In addition, the BH4 domain, which is present mainly in the antiapoptosis molecules, is also important for the death-inhibition functions.[108,109]

The Bcl-2 family proteins can interact with each other through homo- and heterodimerization. Several interaction patterns can be defined. The most common pattern is the interaction between the antiapoptotic and proapoptotic members, such as Bcl-2 versus Bax[111] or Bid.[112] Such protein–protein interactions serve as a rheostat control of the death program.[113] Interestingly, not all antideath molecules can interact with all prodeath molecules, and there are considerable differences in the affinity of these interactions.[114,115] This type of selectivity suggests that only a specific subset of Bcl-2 family proteins is involved in certain tissues and for certain death stimuli.

The second type of interaction occurs between two prodeath members, usually one “BH3-only” molecule and one “multidomain” molecule, such as Bid to Bax[112] or Bak,[116] MAP-1 to Bax,[117] and Bim$_S$ or BimAS to Bax.[118] This could be important for the activation of the multidomain executioner molecules, Bax or Bak, by the BH3-only sensor molecules.[116,119] The third type of interaction is multimerization of the same molecule. This has been observed in both antideath molecules, such as Bcl-2 or Bcl-x_L,[106,120] and prodeath molecules, such as Bax, Bak, and MAP-1.[116,117,121,122] Oligomerization of Bax or Bak is considered essential for their role as a mitochondrial channel for releasing mitochondrial apoptotic factor, such as cytochrome *c*.[116,122]

The crystal and solution structures of several Bcl-2 family proteins, Bcl-xL, Bcl-2, Bcl-w, Bax, and Bid, have been defined.[123–127] One of the common structural features is that these proteins are composed of α helices and assume an overall similar conformation. The arrangement of α helices is similar to that of the membrane translocation domain of bacterial toxins, in particular diphtheria toxin and the colicins, and suggests that the Bcl-2 family proteins may have pore forming capability, which serves to regulate mitochondrial permeability.

One major structural difference between the multidomain proteins, Bcl-2, Bcl-x_L, and Bax, and the BH3-only protein, Bid, is that the BH1, BH2, and BH3 domains form a hydrophobic pocket in the former, which is missing in Bid. This suggests that the BH3 domain of Bid or any other BH3-only protein may function as a "donor" in its interaction with the multidomain proteins, whose hydrophobic pocket can serve as an "acceptor."[112,128] One interesting structural feature of Bax is that its transmembrane domain is actually occupying its own hydrophobic pocket.[125] It is known that Bax needs to be activated for its proapoptotic function through conformation change.[119,129] Upon such a change, the transmembrane domain of Bax may be released, thus freeing the hydrophobic pocket for interaction with other Bcl-2 family proteins, such as Bid, and/or exposing the BH3 domain to exercise the prodeath function. The later possibility is further confirmed in Bid, which has assumed an overall conserved structure before activation.[124] However, activation of Bid by proteolysis can cause the exposure of its BH3 domain for its killing function.[124] In contrast, the antiapoptotic molecules usually have their BH3 domain buried, which may explain why they do not promote PCD under most conditions. It can be postulated that if their BH3 domain is ever exposed, these molecules may become proapoptotic. It may very likely be the case that if Bcl-2 or Bcl-x_L is cleaved by caspase-3 to remove the N-termini, they are endowed with the apoptotic activity.[130,131] Another mechanism of Bcl-2 being converted from an antideath to a prodeath molecule is via the interaction with Nur-77.[132] When activated by proper death stimuli, Nur-77 expression can be up-regulated and translocate from nucleus to the mitochondria, where it interacts with Bcl-2. Such an interaction can cause a conformation change of Bcl-2, exposing its BH3 domain and converting it to a prodeath molecule, which induces mitochondrial release of cytochrome *c*.[132] It seems that the killing induced by Nur-77 is thus completely dependent on Bcl-2. Thus, this observation suggests that increased expression of Bcl-2 may not always be associated with survival, and depending on the cellular context, Bcl-2 can sometimes promote cell death.

Control of Apoptosis by the Bcl-2 Family Proteins

The Intrinsic Death Pathway The mitochondrial apoptosis pathway or the intrinsic pathway is involved in many physiologic cell deaths such as the response to DNA damage or deprivation of hormone or growth factor (FIG. 21.3) (reviewed in Refs. 110, 133, and 134). In this scenario, the death signals are transmitted to the mitochondria, which release a number of apoptogenic factors, including cytochrome *c*, Smac/DIABLO, HtrA2/Omi, apoptosis inducing factor, and endonuclease G. Cytochrome *c* activates Apaf-1 in the presence of dATP, which in turn activates pro-caspase-9.[135] Activated caspase-9 can then cleave downstream effector caspases. Smac and HtrA2 can promote caspase activation by antagonizing the inhibitory effect of IAPs, whereas apoptosis inducing factor and endonuclease G work independently of caspases and directly induce nuclear DNA degradation.[110,133,134] It has to be pointed out that death signals transmitted through the mitochondria pathway not only activate the effector caspases but also promote mitochondrial dysfunction through mitochondrial depolarization, increased permeability transition, and generation of reactive oxygen species.[133,136]

Mitochondria Are the Major Targets of Bcl-2 Family Proteins The mitochondria death pathway is mainly regulated by the Bcl-2 family proteins. It is inhibited by the death antagonists (Bcl-2, Bcl-x_L, etc.) but promoted by the death agonists (Bax, Bak, Bid, Bad, etc.). The multidomain prodeath proteins, Bax or Bak, are responsible for the induction of

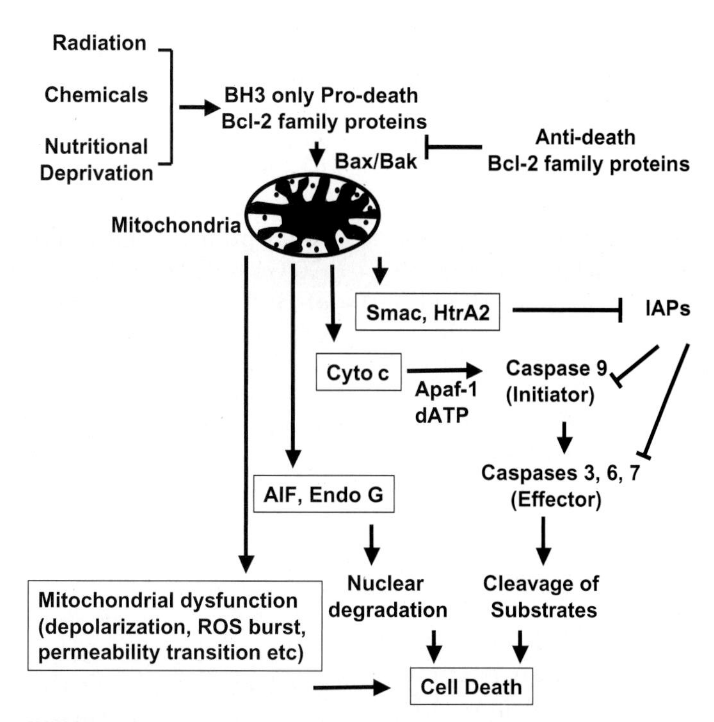

FIGURE 21.3 The mitochondria apoptosis pathway. This pathway is centered on the mitochondrion, which serves to integrate a multitude of survival and death signals. The BH3-only proapoptotic family members are the sentinels for a variety of death signals. Remarkably, distinct signals preferentially activate different BH3-only proteins as discussed in the text. The prototype BH3-only Bid translocates to the mitochondria, activating the multidomain prodeath molecule, Bax and/or Bak, and induces the release of a plethora of apoptogenic factors, which initiates both caspase-dependent (cytochrome *c*, Smac, HtrA2) and caspase-independent (apoptosis inducing factor, endonuclease G) cell death. In addition, mitochondria dysfunction (inhibition of respiration, alteration of ion gradients, increase ROS) also contributes to tumor cell death. The stoichiometry of various anti- and proapoptotic molecules functions as a rheostat to control decisions of life or death.

mitochondria apoptosis. Deletion of both Bax and Bak, but not one of them, renders the cell completely resistant to all major mitochondrial death signals, including DNA damage agents, growth factor deprivation, and endoplasmic reticulum stress, and to the extrinsic pathway signals mediated by Bid.[92–94,137] Mice deficient in both Bax and Bak also have defects in developmentally regulated apoptosis.[137] It should be pointed out, however, that Bax and Bak may not be completely redundant in serving their functions. Bax is usually localized in the cytosol in healthy cells and translocated to the mitochondria in response to death stimuli. In contrast, Bak resides in the outer mitochondrial membrane. This feature may contribute to the differential stimulation of Bax and Bak in certain cases.

It now seems clear that Bax and Bak are activated by the BH3-only molecules, which are activated by the peripheral death signals.[92–94] Over the past several years, several studies have found that each BH3-only molecule is preferentially activated by a distinct set of death stimuli. For example, Bid is mainly activated by proteases, such as caspase-8 in response to death receptor activation.[76–78] Cleaved Bid translocates from the cytosol to the mitochondria. On the other hand, cytokine withdrawal can cause dephosphorylation of Bad, releasing it from the association with 14-3-3 and promoting its mitochondria translocation.[138] In addition, DNA damage can transcriptionally activate PUMA or Noxa, which then translocates to the mitochondria.[139–141]

The activated BH3 molecules activate Bax/Bak either directly or indirectly.[142] Bid is the type of BH3-only molecule that can directly interact with Bax and Bak.[112,116,119] Bid can induce Bax insertion into the mitochondrial membranes with subsequent oligomerization[119,122] and can also induce Bak oligomerization.[116,142] Bad is the prototype BH3-only molecule that may indirectly activate Bax or Bak. Bad functions by inhibiting the antideath molecule, such as Bcl-x_L/Bcl-2, to remove their suppression on Bax or Bak.[93,95,142] This is also true for the *C. elegans* BH3-only molecule, EGL-1, which binds competitively to the antideath molecule, Ced-9, so that Ced-4 is released from the binding with Ced-9 to activate the caspase, Ced-3.[143] The antiapoptotic Bcl-2 family members thus can inhibit the intrinsic death pathway by inhibiting either type of prodeath molecules.[93–95]

The Bcl-2 Family Proteins Can Also Regulate Apoptosis Through Endoplasmic Reticulum It is now apparent that certain Bcl-2 family members, such as Bcl-2, Bax, and Bak, can localize to the endoplasmic reticulum and regulate the Ca^{2+} balance in endoplasmic reticulum ($[Ca^{2+}]_{ER}$). Bcl-2 can reduce the $[Ca^{2+}]_{ER}$ pool, and thus the endoplasmic reticulum will release less Ca^{2+} when it is stimulated by certain death stimulus, such as H_2O_2 or ceramides.[144–146] Bax and Bak can cause an opposing effect and increase the $[Ca^{2+}]_{ER}$ pool to promote more Ca^{2+} release from the endoplasmic reticulum upon death stimulation.[147–149] The increased uptake of calcium by the mitochondria contributes to cell death in the setting of Bax or Bak overexpression or, conversely, Bcl-2 deficiency. It is interesting to note that although some death stimuli, such as those mediated by the BH3-only molecules, are mainly dependent on the mitochondrial pathway, others are more dependent on the endoplasmic reticulum pathway, such as H_2O_2 or ceramides.[147] There is yet a third group of death stimuli that can activate cell death through both mitochondria and endoplasmic reticulum pathways, such as staurosporine, etoposide, and brefeldin-A.[147]

Bcl-2 Family Proteins Are Also Able to Regulate Cell Proliferation An intriguing finding is that Bcl-2 family members can also regulate cell proliferation. An early study on T-cell activation found that T cells from Bcl-2 transgenic mice proliferated much more slowly than the wild-type cells, whereas T cells from Bcl-2 deficient mice entered into cell cycle more rapidly than the wild-type cells.[150] A detailed kinetic analysis found that Bcl-2 had the antiproliferation effect by delaying the cell cycle transition from G0/G1 to S phase.[150] Others have since confirmed this finding and found that Bcl-x_L had similar effects.[151] Moreover, the antiproliferation effects of Bcl-2/Bcl-x_L could be also observed in a delayed reentry of serum deprived fibroblasts into cell cycle[151,152] or in reentry of quiescent hepatocytes into cell cycle after partial hepatectomy.[153] Interestingly, it was found that the prodeath molecules Bad[152] and Bax[154] could promote cell proliferation by enhancing G0/G1 to S phase transition, which in the case of Bad works through a direct antagonistic protein–protein interaction with Bcl-x_L.[151,152]

The precise mechanism of how these molecules might regulate cell proliferation is not known. However, it seems that at least some of the biochemical activities of the Bcl-2 family proteins are shared in the regulation of both apoptosis and proliferation, because all important structural features required for apoptosis regulation are also required for proliferation.[151,155] Thus, the functional manifestation (influenced by the Bcl-2 family) would be dependent on whether the cells are stimulated with a death signal or a proliferation signal. In the context of certain malignancies such as breast cancer, it is important to consider this unique property with potential effects on two fundamental processes such as proliferation and apoptosis. As discussed later, the functional consequences of Bcl-2 overexpression need to be better understood and how it ultimately impacts prognosis in breast cancer.

Apoptosis in Human Breast Cancer: The Paradox

The initial studies in primary human breast cancer focused on Bcl-2 expression as measured with conventional immunohistochemistry (IHC). The initial observations published in 1994 concluded that increased Bcl-2 expression in primary (stage I) tumors correlated with improved disease-free and overall survival in univariate analysis.[156,157] Interestingly, Bcl-2 expression correlated closely with estrogen receptor (ER) positivity and a low S phase growth fraction. Moreover, several reports suggested a correlation between Bcl-2 expression and sensitivity to hormonal therapy with improved outcome.[158,159] This was an unexpected finding because Bcl-2 was strongly implicated in the pathogenesis of follicular B-cell lymphoma and nominated as the founding member of a new category of protooncogene (regulators of cell death).[91] This discovery was further supported by transgenic animal models that recapitulated much of the disease process and advanced the thesis that repression of cell death is a primary oncogenic event in lymphoid malignancy.[88,89] The initial surveys performed looking at Bcl-2 protein expression in various tissues reinforced the notion that Bcl-2 serves to repress PCD in other lineages at critical developmental periods.[160,161] Bcl-2 is present in approximately 60–80% of primary breast cancers as monitored by IHC using a monoclonal antibody. Additional studies found that primary breast neoplasms with increased in situ apoptosis are more often ER-negative/Bcl-2–negative high-grade tumors and associated with a poorer survival.[162] A priori, many would assume that higher rates of in situ apoptosis (implying a smaller growth fraction) would correlate with an improved prognosis; however, the opposite was found and thus the initial paradox. A decade later, most studies find that high rates of apoptosis and undetectable Bcl-2 expression are poor prognostic indicators and are associated with an increased growth fraction that lack expression of ER.[163]

It is well accepted that overexpression of Bcl-2 (or other antiapoptotic family members such as Bcl-x) renders most cancer cells less sensitive to chemotherapeutic agents.[164] However, the initial observation suggested that Bcl-2 expression in breast cancer was associated with improved outcome. How can we explain this?

Most studies have focused on the regulation of apoptosis by Bcl-2 in promoting tumor formation. The general thesis is that inhibition of apoptosis will lead to tumor development by allowing cells to survive that normally die upon acquisition of certain mutations, such as c-*myc*.[91] This is indeed the case in several well-characterized murine models dependent on c-*myc* overexpression. For example, double transgenic mice expressing both c-*myc* and Bcl-2 in B lymphocytes develop a much more aggressive lymphoma[88,89] compared with the single transgenics. Overexpression of c-*myc*/Bcl-x_L or SV40-T-Ag/Bcl-x_L in pancreatic β-islet cells also resulted in rapid growth of angiogenic invasive tumors that were not seen in either single transgenic mouse strain.[165,166] In both cases, Bcl-2 or Bcl-x_L was thought to inhibit the increased apoptosis often associated with the c-*myc* oncogene and thus allow for an increased accumulation of cancer cells with no apparent effect on the cell cycle transit (doubling) time.

Interestingly, this may not seem to be the case for other tumor models. For example, overexpression of either SV40-T-Ag[167,168] or c-*myc*[169] in the liver induced hepatic adenoma and carcinoma. However, overexpression of Bcl-2 with c-*myc*[169] did not accelerate further tumor development. Contrary to expectations, tumor growth in the hepatic carcinoma model was retarded. The tumor incidence was decreased, accompanied with reduced cell death, and, interestingly, reduced cell proliferation. The authors concluded that the antiproliferation effects of Bcl-2 were dominant during the emergence of neoplastic foci.[169] This observation was later confirmed in Bcl-2/transforming growth factor-α double transgenic mice; the development of liver cancer stimulated by the overexpression of transforming growth factor-α was clearly delayed and reduced by the coexpression of Bcl-2 in the mice.[170] Thus, it seems that Bcl-2 overexpression in the liver functions mainly as a regulator of cell proliferation in the context of tumor development and has a pronounced dominant effect. Similar effects of Bcl-2 on tumor cell growth rates have been documented in murine breast cancer models driven by large T antigens, suggesting that this cell cycle effect can occur in other tissues beyond the liver.

Studies have also pointed out that high expression of Bcl-2 may actually promote cell death.[171,172] In addition, the orphan receptor Nur-77 can bind Bcl-2 and induce a conformational change to expose the BH3 domain, thus converting Bcl-2 into a killer molecule.[132] These studies demonstrate that Bcl-2 can promote cell death in certain cellular environments. This observation together with the growth inhibitory effects may explain, in part, why primary Bcl-2–positive breast cancer is associated with better outcomes.

Human Breast Cancer: Bcl-2 as a Prognostic Marker

The initial series of studies focused on Bcl-2 as a prognostic marker in early stage (stage I–II) breast cancer. From 1994 to 2004, a series of reports were published that evaluated the prognostic value of Bcl-2 as a biomarker in early stage breast cancer. A list of 18 published studies (**Table 21.2**) is presented evaluating a total of 2767 patients.[156,157,173–188] Of the 18 studies, 13 studies (72%) found a significant correlation of Bcl-2 expression with improved disease-free or overall survival. All the reports except one[181] were exclusively female

Table 21.2 Published Series That Evaluated the Prognostic Value of Bcl-2 as a Biomarker in Early Stage Breast Cancer

Year	Author and Reference	No. of Cases	Method	Regression-Free Survival		Overall Survival		Comment
				Univariate	Multivariate	Univariate	Multivariate	
1994	Joensuu[157]	174	IHC			Yes		46% cases positive for Bcl-2; Bcl-2 expression associated with low mitotic count
1994	Silvestrini[156]	283	IHC	Yes	No	Yes	No	Bcl-2 expression correlated with lack of p53 staining
1995	Hellemans[173]	251	IHC	Yes	No	No	No	75% cases positive for Bcl-2; survival advantage for node positive cases only
1995	Lipponen[174]	202	IHC	Yes	No	Yes	No	Intensity of bcl-2 expression was inversely related to tumour grade, S-phase fraction, and mitotic index
1996	Barbareschi[175]	178	IHC	No				62% cases positive for Bcl-2; Bcl-2 expression associated wit small tumor size, nonductal morphology, low tumor grade, ER positivity, and p53 negativity
1996	van Slooten[176]	202	IHC	Yes				Premenopausal women only; 32% cases strongly positive for Bcl-2
1997	Kapranos[177]	90	IHC	Yes				
1997	Krajewski[178]	135	IHC	No		No		All cases studies were p53 IHC positive
1997	Lee[179]	91	IHC			Yes	No	53% cases positive for Bcl-2; ER/PR expression correlate with bcl-2 positivity
1998	Charpin[180]	170	IHC	Yes	No	No		
1998	Veronese[182]	236	IHC	Yes	No	Yes	No	Bax expression was associated with high grade tumors but was not prognostic; 67% cases IHC positive for Bcl-2
1998	Pich[181]	34 (male)	IHC			No	No	82% cases positive for Bcl-2; no association with ER/PR expression
1999	Holmqvist[183]	165	IHC	No		No		Bcl-2 immunoreactivity correlated with ER/PR positivity and was inversely correlated with p53 expression
1999	Le[184]	175	IHC			Yes	Yes	63% cases IHC positive for Bcl-2
2001	McLaughlin[185]	108	IHC	Yes		Yes		
2002	Bukholm[186]	126	IHC			Yes	No	69% cases positive for Bcl-2; 58% cases positive for Bcl-x but no correlation with survival was noted
2003	O'Driscoll[187]	106	Reverse transcriptase-polymerase chain reaction	Yes	Yes	Yes	No	Bax, survivin, and bag-1 lack prognostic significance in this study; 78% cases reverse transcriptase-polymerase chain reaction positive for Bcl-2
2004	Linjawi[188]	75	IHC	No	No	No	No	76% cases positive for Bcl-2
		2767						

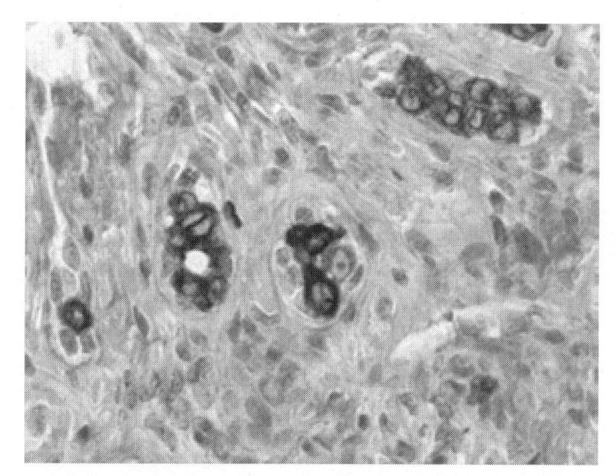

FIGURE 21.4 Bcl-2 expression in breast cancer. In this photomicrograph, Bcl-2 is overexpressed in an infiltrating ductal breast cancer that relapsed after primary treatment and progressed to chemotherapy-resistant metastatic disease (immunoperoxidase with anti-Bcl-2 antibody).

patients; Pich et al.[181] studied 34 cases of male breast cancer and found no significant correlation between Bcl-2 expression and outcome. With the exception of one report, the remaining 17 studies used IHC analysis using monoclonal antibodies to quantify Bcl-2 expression (FIG. 21.4).[187] It should be noted that conflicting reports may be due to variations in scoring criteria and the lack of standardization. Some authors judge Bcl-2 expression based on 5% or greater of cells staining positive, whereas other investigators use 20% or even 30% positivity to score Bcl-2 expression. One study used reverse transcriptase-polymerase chain reaction as the method to detect Bcl-2 expression using total RNA isolated from snap-frozen primary breast tumors that were carefully dissected at the time of resection; however, the authors did not state if any attempts were made to validate their findings with IHC or other methodologies.[187] Amplified *Bcl-2* was detected by ethidium bromide staining of agarose gels in a nonquantitative manner using conventional techniques as stated by the authors.

Other Bcl-2 Family Members as Prognostic Markers

Additional prognostic information in primary breast cancer for other Bcl-2 family members is limited. Six studies evaluated Bax expression in primary (female) breast tumors, and all studies found no correlation with outcome.[182,186–190] Three reports studied Bcl-x as a prognostic marker in breast cancer. The first report found 18 of 42 tumors (43%) IHC positive for Bcl-x expression that correlated with higher tumor grade and inversely with Bcl-2 expression; Bcl-x expression levels failed to correlate with ER status, tumor size, and p53 status, but a trend toward decreased survival was noted in patients with Bcl-x–positive tumors.[191] A second study evaluating Bcl2/Bcl-x ratio by IHC-based image cytometry revealed that a ratio of 1 or greater was associated with improved disease-free survival; in contrast, the Bcl-2/Bax ratio was not predictive of disease-free survival or overall survival.[192] In the third report, 84 of 116 primary breast tumors (72%) were positive for Bcl-x expression by IHC and confirmed positive by Western blot analysis using monoclonal antibodies specific for Bcl-x_L. Interestingly, Bcl-x expression correlated significantly with ER/progesterone receptor (PR) positivity; furthermore, improved disease-free survival ($P = 0.027$) was associated with Bcl-x expression by univariate analysis.

A survey of Bcl-2 family members within human breast cancer cell lines and primary tumors revealed prominent expression of Bcl-2, Bcl-x, Mcl-1, Bax, and Bak by IHC and confirmed by Western blot analysis.[193] Interestingly, Bax and Bak were expressed in all cell lines and primary tumors, whereas the remaining members were present in some but not all breast cell lines ($n = 11$) and primary tumors ($n = 20$). At this time, however, prognostic information related to these additional family members has not been published.

IAP, Caspase, and Death Receptor Family Members as Prognostic Factors

Within the family of IAP proteins, survivin has been studied as a prognostic factor in breast cancer by two groups. In the initial report, IHC analysis (using the 8E2 monoclonal antibody) revealed expression of survivin in 118 of 167 primary breast tumors; survivin was not found in adjacent normal tissue.[194] Interestingly, survivin expression was strongly correlated with Bcl-2 expression ($P = 0.005$) and reduced apoptotic index ($P < 0.0001$). However, survivin was not found to be an independent prognostic variable by univariate analysis.[194] In the second study, 293 cases of primary breast carcinoma were evaluated by IHC using a polyclonal antisera against survivin.[195] Specific staining for survivin was observed for 176 tumors (60%) in the study, and the authors further characterized the pattern of staining as nuclear ($n = 91$), cytoplasmic ($n = 37$), or both ($n = 48$). In multivariate analysis, nuclear survivin expression (greater than 20% cells positive) was a favorable prognostic variable for regression-free and overall survival ($P = 0.01$).[195] It has been suggested that the 8E2 monoclonal antibody recognizes only the cytoplasmic form of survivin and may account for the discrepancy between the two published studies.

Preliminary information regarding caspase-3 was published by the Burnham Institute group evaluating 56 cases of primary intraductal malignancy.[190,196] All tumors examined expressed caspase-3 at relatively higher levels compared with adjacent mammary epithelium; however, it was not clear if the caspase-3 was functional (or activated). In addition, the prognostic significance of caspase-3 (over)expression was not reported, and thus the significance of this finding remains unclear. At this time, there are no other reports evaluating the prognostic value of other caspases in human breast cancer.

The Fas death receptor has been postulated to play a role in breast cancer pathogenesis. The initial report showed that exposure of cancer cells to cytotoxic chemotherapy agents (such as cis-platinum and doxorubicin) could cause the transcriptional (and posttranscriptional) up-regulation of Fas and render cells susceptible to apoptotic cell death after engagement by monoclonal anti-Fas antibodies or soluble FasL (Fas ligand).[197] Follow-up studies in human breast cancer suggest that coregulation of Fas and FasL by tumor cells modulates the response to cytotoxic chemotherapy.[198] Evaluation of 167 early (stage I–II) breast cancer patients reveals that loss of Fas and up-regulation of FasL is related to poor clinical outcome in women that receive adjuvant epirubicin and cyclophosphamide chemotherapy.[199] Both Fas and FasL expression were studied by IHC using affinity purified polyclonal antisera and confirmed with monoclonal antibodies specific for Fas and FasL and, later, immunoprecipitation experiments using the monoclonal antibody reagents to confirm expression on a subset of tumor samples. In the published report, the following distribution of Fas/FasL was found: Fas−/FasL− (13%), Fas+/FasL− (41%), Fas−/FasL+ (29%), and Fas+/FasL+ (17%). FasL expression correlated with HER-2 overexpression ($P = 0.03$), but Fas/FasL expression did not correlate with menopausal status, nuclear grade, or ER/PR status. The median follow-up of the patient cohort was 69 months (range, 16–97 months), and the disease-free survival rate was 72%, with overall survival reported as 78%. Univariate (Cox model) analysis revealed that nuclear grade, stage, lack of Fas, and expression of FasL were significant predictors of disease-free and overall survival. In multivariate analysis, the relative risk was 8.50 ($P < 0.0001$) for disease-free survival in patients with Fas-negative tumors. The relative risk was 2.38 ($P = 0.01$) for disease-free survival in patients with FasL-positive tumors. In both early stage (I–IIa, $n = 126$) and more advanced stage (IIb, $n = 41$) patients, women with the Fas−/FasL+ phenotype has the poorest outcome, with inferior 5-year estimated disease-free and overall survival rates compared with patients with Fas+/FasL−, Fas+/FasL+, and Fas−/Fas− profiles. For example, the 5-year estimated survival rate for women with Fas−/FasL+ primary tumors was 14.9% (95% confidence interval, 2.3–36.5) compared with 68.1% (95% confidence interval, 28–88) for Fas+/FasL+ tumors and 100% for Fas+/FasL− tumors ($P < 0.0001$).[199] This observation supports the notion that death receptors such as Fas play an important role in executing tumor cell death after exposure with conventional cytotoxic agents; loss of critical cell surface death receptors (such as Fas) appears to promote drug resistance in breast cancer. Additional studies are needed to further define the role of other TNF death receptor family members in breast cancer.

Apoptosis Regulators and Response to Chemotherapy

Parton and colleagues[200] examined the dynamic changes of multiple apoptosis regulators in newly diagnosed patients treated with neoadjuvant combination chemotherapy. Of the 35 patients enrolled in the study, 11 had a documented complete pathologic response, 20 had a partial response, and the remaining 3 assessable patients had either stable disease or progression. Core biopsies were obtained pretreatment and at 24 hours after treatment. IHC analysis was performed for Bcl-2, Bax, caspase-3, caspase-6, DFF40, and XIAP; in addition, the proliferative index (Ki-67 staining) and the apoptotic index (TUNEL staining) was recorded for each case. A strong correlation was noted between Ki-67 and the TUNEL staining in pretreatment samples ($r = 0.64$, $P < 0.0001$); however, no correlation was found for any apoptotic regulator and the apoptotic index among the pretreatment samples. At 24 hours, significant changes were observed for Bax ($P = 0.05$) and DFF40 ($P = 0.02$), with increases in both biomarkers compared with pretreatment levels. Caspase-6 showed a nonsignificant ($P = 0.06$) increase at 24 hours, whereas Bcl-2, XIAP, and caspase-3 remained unchanged. Among the biomarkers tested, only elevated Ki-67 was associated with improved clinical outcome ($r = 0.42$, $P = 0.02$).[200] Interestingly, this correlation between elevated Ki-67 and clinical response has been noted previously by other investigators.[162] In summary, an increase in Bax and DFF40 is noted after treatment (at 24 hours) and often accompanies the increased apoptotic index and fall in Ki-67 staining that is seen early during the course of treatment. Additional studies are needed to identify potential biomarkers of response and clinical outcome after initial treatment to develop optimal treatment options for patients with locally advanced or metastatic breast cancer.

Novel Proapoptotic Therapies

Exciting progress has been made in the development of new therapies directed at several targets within the cell death signaling pathways.[201] An appreciation of the central role of the Bcl-2 gene family and its impact on chemotherapy resistance prompted initial efforts to identify inhibitors specific for Bcl-2. Significant advances further clarified the central role of the mitochondrion in controlling apoptotic signaling events and provided the conceptual starting point to identify novel small molecules that selectively perturb mitochondrial function. Finally, definition of the TNF-related death receptor family has led to a search for potential agonists that serve to cross-link and selectively activate death receptors expressed on the surface of various tumor cells. In this section, the development of three novel therapies is discussed: the antisense oligonucleotide specific for Bcl-2 (G3139), the novel mitochondriotoxic small molecule (F16) that targets HER-2 overexpressing breast cancers, and, finally, the naturally occurring ligand for death receptor-4 and -5 that initiate caspase-8 activation leading to apoptosis (TRAIL).

Initial studies by Reed and colleagues[202] revealed the potential utility of antisense oligonucleotides as an approach to reduce endogenous Bcl-2 messenger RNA as a strategy to sensitize tumor cells for apoptosis. Phosphorothioate oligonucleotides (18-mer complimentary to the first six codons of Bcl-2) are nuclease-resistant derivatives that optimize stability and allow drug delivery into the cytoplasm. A variety of human cancer cell lines (including breast cancer) has been shown to be sensitive to the effects of down regulation of Bcl-2 using the antisense oligonucleotides.[203] Mayer and associates have performed the most comprehensive studies in breast cancer with the Bcl-2 antisense compound, G3139 (oblimersen sodium, Genta). The initial studies used MCF-7 (Bcl-2 hi, ER positive) and MDA435/LCC6 (Bcl-2 lo, ER negative) as a model system to evaluate G3139 alone or in combination with doxorubicin.[204–207] Ribonuclease protection assays and Western blot analysis confirmed the down-regulation of Bcl-2 that occurs at 48–72 hours after G3139 exposure. Bcl-2 protein levels are reduced 80–90% upon G3139 exposure compared with approximately 20% reduction seen with a control antisense oligonucleotide. Alterations of other Bcl-2 family members (Bcl-x, Bax, Mcl-1, Bik, Bak, Bad, Bcl-w) was insignificant, consistent with specific inhibition of Bcl-2 mediated by G3139. Morphologic evidence of cell death along with detection of increased caspase-3 activity in cell cultures confirm the potent effects of G3139 on both breast cancer cell lines. These observations suggest that G3139 can equally sensitize tumor cells that express either high or low levels of endogenous Bcl-2. Confirmatory data using the MDA432/LCC6 xenograft model shows synergistic activity of G3139 in the presence of lipodoxorubicin, and thus a strong preclinical rationale is presented for breast cancer development.[205] A phase I/II clinical trial has been initiated to evaluate the safety and efficacy of G3139 in combination with docetaxel and doxorubicin in patients with advanced breast cancer.[207]

F16 is a novel small molecule identified in a screen of a chemical library of 16,000 compounds selected for inhibition of HER-2/*neu* overexpressing immortalized murine mammary epithelial cell line referred to as EpH4.[208] F16 is a lipophilic cation that selectively accumulates in the mitochondrial matrix and perturbs mitochondrial function culminating in cell cycle arrest and ultrastructural changes consistent with PCD. Interestingly, F16 promotes apoptosis in cell lines driven by HER-2/*neu*, H-*ras*, and, to a lesser degree β-catenin but not c-*myc*. Moreover, F16 is active on 8 cell lines from a panel of 10 well-characterized human breast cancer cell lines. The authors made the striking observation that F16 sensitivity correlates with higher resting mitochondrial membrane potential, a frequent defect noted in cancer cells. Furthermore, F16 inhibits downstream signaling of MAP kinase pathway that emanates from proximal HER-2/*neu* activation. Additional experiments reveal that EpH4 HER-2/*neu* cells that overexpress Bcl-2 remains susceptible to F16 induced PCD despite the finding that cytochrome *c* release and caspase-3 activation are both completely inhibited.[209] Also, the addition of the pan-caspase inhibitor z-VAD had minimal to no effect on the rate of cell death of EpH4 HER-2/*neu* Bcl-2+ cells after F16 exposure, suggesting that necrosis was the primary mode of death under these conditions. In support of this proposal, maintenance of the cellular ATP pool or protection against an increase in ROI had a partial effect in F16 mediated death in EpH4 HER-2/*neu* Bcl-2+ cells. In summary, F16 and related delocalized lipophilic compounds have unique properties of mitochondrial localization and induction of apoptotic or necrotic forms of cell death depending on the genetic background of the tumor cell. This family of compounds provides the initial evidence that selective mitochondriotoxic agents can induce cell death irrespective of the genetic lesions that accumulate with tumor progression.

TRAIL was identified originally as a soluble factor that mediates the induction of cell death upon binding to cognate cell surface receptors.[61] DR4 and DR5 are the primary cell surface receptors that bind TRAIL and signal to upstream caspases; however, DcR1, DcR2, and osteoprotegrin serve as decoy receptors that fail to activate initiator caspases. Interestingly, TRAIL deficient mice are more susceptible to spontaneous tumor formation compared with wild-type animals, implying an important role for TRAIL in immune surveillance.[210] Initial preclinical data using either recombinant TRAIL or monoclonal antibodies directed against DR4 or DR5 support the hypothesis that activation of proximal initiator caspase activity can promote tumor cell death.[65,211,212] It has been noted that exposure of tumor cells can actually up-regulate the cell surface expression of DR4 and DR5, providing one mechanism responsible for drug synergy with TRAIL agonists.[213,214] Several reports conclude that doxorubicin is the most effective chemotherapeutic sensitizing agent for TRAIL in breast cancer. Reports evaluating potential mutations in either DR4 or DR5 in breast cancer metastasis conclude that loss of function mutations likely play a role in disease progression. For instance, seven mutations (in the coding region) were identified among the 34 primary tumors from patients with known metastases in contrast to no mutations found in the remaining 23 patients with no known metastases.[215] A comprehensive study evaluating anti-DR5 monoclonal antibody given with additional agents such as doxorubicin, paclitaxel, or radiation showed that optimal combinations and schedules can be determined using xenograft models for breast cancer.[216] Anti-DR5 monoclonal antibody alone did have clear antitumor activity (in four of nine breast cancer cell lines tested) consistent with earlier reports using similar xenograft models. Interestingly, the anti-DR5 monoclonal antibody showed synergy with doxorubicin in four of eight cell lines tested, whereas no synergy was evident with paclitaxel in any cell line tested. Notably, the optimal combination was shown to be anti-DR5 monoclonal antibody plus doxorubicin followed by radiation (3 Gy total dose).[217] Similar observations with recombinant TRAIL or DR specific monoclonal antibodies have been published in other tumor histologies, supporting the general strategy of using death receptor signaling as a mode to trigger apoptosis in cancer cells.[64,65] Clinical trials using DR specific monoclonal antibodies or recombinant TRAIL as therapeutic agents in breast cancer are eagerly anticipated.

Summary

Exciting progress in the identification and characterization of the molecules involved in cell death signaling provides unique targets for drug discovery programs. However, the complexities of the four major gene families discussed in this chapter—Bcl-2, caspase, IAP, and death receptor—underscore the challenges facing cancer cell biologists and clinical scientists. There are two major challenges. First, a more detailed understanding of the expression and function of the four major gene families in human breast cancer is critical toward a better understanding of drug resistance mechanisms. Second, improved biomarkers are urgently needed to identify individual patients that could be selected for the appropriate combination therapy. Despite these challenges, new targeted therapies are in development for breast cancer and should appear in the clinic in the next several years.

References

1. Wiseman BS, Werb Z. Stromal effects on mammary gland development and breast cancer. Science 2002; 296:1046–1049.
2. Clarkson RW, Wayland MT, Lee J, et al. Gene expression profiling of mammary gland development reveals putative roles for death receptors and immune mediators in post-lactational regression. Breast Cancer Res 2004;6:R92–R109.
3. Lund LR, Romer J, Thomasset N, et al. Two distinct phases of apoptosis in mammary gland involution: proteinase-independent and -dependent pathways. Development 1996;122:181–193.
4. Kerr JF, Wyllie AH, Currie AR. Apoptosis: a basic biological phenomenon with wide-ranging implications in tissue kinetics. Br J Cancer 1972;26: 239–257.
5. Thornberry NA, Lazebnik Y. Caspases: enemies within. Science 1998;281:1312–1316.
6. Chang HY, Yang X. Proteases for cell suicide: functions and regulation of caspases. Microbiol Mol Biol Rev 2000;64:821–846.
7. Evan G, Littlewood T. A matter of life and cell death. Science 1998;281:1317–1322.
8. Korsmeyer SJ. Bcl-2: an antidote to programmed cell death. Cancer Surv 1992;15:105–118.
9. Ellis HM, Horvitz HR. Genetic control of programmed cell death in the nematode *C. elegans*. Cell 1986;44:817–829.
10. Reed JC, Doctor K, Rojas A, et al. Comparative analysis of apoptosis and inflammation genes of mice and humans. Genome Res 2003;13:1376–1388.
11. Thornberry NA, Rano TA, Peterson EP, et al. A combinatorial approach defines specificities of members of the caspase family and granzyme B. Functional relationships established for key mediators of apoptosis. J Biol Chem 1997;272:17907–17911.
12. Garcia-Calvo M, Peterson EP, Leiting B, et al. Inhibition of human caspases by peptide-based and macromolecular inhibitors. J Biol Chem 1998;273: 32608–32613.
13. Nakagawa T, Yuan J. Cross-talk between two cysteine protease families. Activation of caspase-12 by calpain in apoptosis. J Cell Biol 2000;150: 887–894.
14. Van de Craen M, Van Loo G, Pype S, et al. Identification of a new caspase homologue: caspase-14. Cell Death Differ 1998;5:838–846.
15. Slee EA, Harte MT, Kluck RM, et al. Ordering the cytochrome c-initiated caspase cascade: hierarchical activation of caspases-2, -3, -6, -7, -8, and -10 in a caspase-9-dependent manner. J Cell Biol 1999; 144:281–292.
16. Boatright KM, Renatus M, Scott FL, et al. A unified model for apical caspase activation. Mol Cell 2003; 11:529–541.
17. Salvesen GS, Dixit VM. Caspase activation: the induced-proximity model. Proc Natl Acad Sci USA 1999;96:10964–10967.
18. Chinnaiyan AM, Tepper CG, Seldin MF, et al. FADD/MORT1 is a common mediator of CD95 (Fas/APO-1) and tumor necrosis factor receptor-induced apoptosis. J Biol Chem 1996;271:4961-4965.
19. Tschopp J, Irmler M, Thome M. Inhibition of fas death signals by FLIPs. Curr Opin Immunol 1998; 10:552–558.
20. Wang CY, Mayo MW, Korneluk RG, et al. NF-kappaB antiapoptosis: induction of TRAF1 and TRAF2 and cIAP1 and c-IAP2 to suppress caspase-8 activation. Science 1998;281:1680–1683.
21. Deveraux QL, Reed JC. IAP family proteins-suppressors of apoptosis. Genes Dev 1999;13:239–252.
22. Zhou Q, Snipas S, Orth K, et al. Target protease specificity of the viral serpin CrmA. Analysis of five caspases. J Biol Chem 1997;272:7797–7800.
23. Shiozaki EN, Chai J, Rigotti DJ, et al. Mechanism of XIAP-mediated inhibition of caspase-9. Mol Cell 2003;11:519–527.
24. Budihardjo I, Oliver H, Lutter M, et al. Biochemical pathways of caspase activation during apoptosis. Annu Rev Cell Dev Biol 1999;15:269–290.
25. Renatus M, Stennicke HR, Scott FL, et al. Dimer formation drives the activation of the cell death protease caspase 9. Proc Natl Acad Sci USA 2001;98: 14250–14255.
26. Srinivasula SM, Hegde R, Saleh A, et al. A conserved XIAP-interaction motif in caspase-9 and Smac/DIABLO regulates caspase activity and apoptosis. Nature 2001;410:112–116.
27. Shi Y. A conserved tetrapeptide motif: potentiating apoptosis through IAP-binding. Cell Death Differ 2002;9:93–95.
28. Chai J, Wu Q, Shiozaki E, Srinivasula SM, et al. Crystal structure of a procaspase-7 zymogen: mechanisms of activation and substrate binding. Cell 2001; 107:399–407.
29. Martin SJ, Amarante-Mendes GP, Shi L, et al. The cytotoxic cell protease granzyme B initiates apoptosis in a cell-free system by proteolytic processing and activation of the ICE/CED-3 family protease, CPP32, via a novel two-step mechanism. EMBO J 1996;15: 2407–2416.
30. Fernandes-Alnemri T, Armstrong RC, Krebs J, et al. In vitro activation of CPP32 and Mch3 by Mch4, a novel human apoptotic cysteine protease containing two FADD-like domains. Proc Natl Acad Sci USA 1996;93:7464–7469.
31. Stennicke HR, Jurgensmeier JM, Shin H, et al. Procaspase-3 is a major physiologic target of caspase-8. J Biol Chem 1998;273:27084–27090.
32. Huang Y, Park YC, Rich RL, et al. Structural basis of caspase inhibition by XIAP: differential roles of the linker versus the BIR domain. Cell 2001;104: 781–790.
33. Riedl SJ, Renatus M, Schwarzenbacher R, et al. Structural basis for the inhibition of caspase-3 by XIAP. Cell 2001;104:791–800.

34. Li S, Zhao Y, He X, et al. Relief of extrinsic pathway inhibition by the bid-dependent mitochondrial release of Smac in Fas-mediated hepatocyte apoptosis. J Biol Chem 2002;277:26912–26920.
35. Riedl SJ, Renatus M, Snipas SJ, et al. Mechanism-based inactivation of caspases by the apoptotic suppressor p35. Biochemistry 2001;40:13274–13280.
36. Tschopp J, Martinon F, Burns K. NALPs: a novel protein family involved in inflammation. Nat Rev Mol Cell Biol 2003;4:95–104.
37. Agostini L, Martinon F, Burns K, et al. NALP3 forms an IL-1beta-processing inflammasome with increased activity in Muckle-Wells autoinflammatory disorder. Immunity 2004;20:319–325.
38. Sakahira H, Enari M, Nagata S. Cleavage of CAD inhibitor in CAD activation and DNA degradation during apoptosis [see comments]. Nature 1998;391: 96–99.
39. Wyllie AH. Glucocorticoid-induced thymocyte apoptosis is associated with endogenous endonuclease activation. Nature 1980;284:555–556.
40. Crook NE, Clem RJ, Miller LK. An apoptosis-inhibiting baculovirus gene with a zinc finger-like motif. J Virol 1993;67:2168–2174.
41. Deveraux QL, Leo E, Stennicke HR, et al. Cleavage of human inhibitor of apoptosis protein XIAP results in fragments with distinct specificities for caspases. EMBO J 1999;18:5242–5251.
42. Chai J, Shiozaki E, Srinivasula SM, et al. Structural basis of caspase-7 inhibition by XIAP. Cell 2001;104: 769–780.
43. Chai J, Du C, Wu JW, et al. Structural and biochemical basis of apoptotic activation by Smac/DIABLO. Nature 2000;406:855–862.
44. Du C, Fang M, Li Y, et al. Smac, a mitochondrial protein that promotes cytochrome c-dependent caspase activation by eliminating IAP inhibition. Cell 2000; 102:33–42.
45. Verhagen AM, Ekert PG, Pakusch M, et al. Identification of DIABLO, a mammalian protein that promotes apoptosis by binding to and antagonizing IAP proteins. Cell 2000;102:43–53.
46. Suzuki Y, Imai Y, Nakayama H, et al. A serine protease, htra2, is released from the mitochondria and interacts with xiap, inducing cell death. Mol Cell 2001;8:613–621.
47. Wolf BB, Green DR. Apoptosis: letting slip the dogs of war. Curr Biol 2002;12:R177–R179.
48. Wu G, Chai J, Suber TL, et al. Structural basis of IAP recognition by Smac/DIABLO. Nature 2000;408: 1008–1012.
49. Huang Y, Rich RL, Myszka DG, et al. Requirement of both the second and third BIR domains for the relief of X-linked inhibitor of apoptosis protein (XIAP)-mediated caspase inhibition by Smac. J Biol Chem 2003;278:49517–49522.
50. Okada H, Suh WK, Jin J, et al. Generation and characterization of Smac/DIABLO-deficient mice. Mol Cell Biol 2002;22:3509–3517.
51. Yang Y, Fang S, Jensen JP, et al. Ubiquitin protein ligase activity of IAPs and their degradation in proteasomes in response to apoptotic stimuli. Science 2000;288:874–877.
52. Ditzel M, Wilson R, Tenev T, et al. Degradation of DIAP1 by the N-end rule pathway is essential for regulating apoptosis. Nat Cell Biol 2003;5: 467–473.
53. Ambrosini G, Adida C, Altieri DC. A novel anti-apoptosis gene, survivin, expressed in cancer and lymphoma. Nat Med 1997;3:917–921.
54. Ambrosini G, Adida C, Sirugo G, et al. Induction of apoptosis and inhibition of cell proliferation by survivin gene targeting. J Biol Chem 1998;273: 11177–11182.
55. Uren AG, Wong L, Pakusch M, et al. Survivin and the inner centromere protein INCENP show similar cell-cycle localization and gene knockout phenotype. Curr Biol 2000;10:1319–1328.
56. Vucic D, Stennicke HR, Pisabarro MT, et al. ML-IAP, a novel inhibitor of apoptosis that is preferentially expressed in human melanomas. Curr Biol 2000;10: 1359–1366.
57. Peter ME, Krammer PH. The CD95(APO-1/Fas) DISC and beyond. Cell Death Differ 2003;10:26–35.
58. Wajant H, Pfizenmaier K, Scheurich P. Tumor necrosis factor signaling. Cell Death Differ 2003;10: 45–65.
59. LeBlanc HN, Ashkenazi A. Apo2L/TRAIL and its death and decoy receptors. Cell Death Differ 2003; 10:66–75.
60. Bridgham JT, Wilder JA, Hollocher H, et al. All in the family: evolutionary and functional relationships among death receptors. Cell Death Differ 2003;10: 19–25.
61. Ashkenazi A, Dixit VM. Death receptors: signaling and modulation. Science 1998;281:1305–1308.
62. Kischkel FC, Lawrence DA, Tinel A, et al. Death receptor recruitment of endogenous caspase-10 and apoptosis initiation in the absence of caspase-8. J Biol Chem 2001;276:46639–46646.
63. Suliman A, Lam A, Datta R, et al. Intracellular mechanisms of TRAIL: apoptosis through mitochondrial-dependent and -independent pathways. Oncogene 2001;20:2122–2133.
64. Walczak H, Miller RE, Ariail K, et al. Tumoricidal activity of tumor necrosis factor-related apoptosis-inducing ligand in vivo. Nat Med 1999;5:157–163.
65. Ashkenazi A, Pai RC, Fong S, et al. Safety and antitumor activity of recombinant soluble Apo2 ligand. J Clin Invest 1999;104:155–162.
66. Jo M, Kim TH, Seol DW, et al. Apoptosis induced in normal human hepatocytes by tumor necrosis factor-related apoptosis-inducing ligand. Nat Med 2000;6: 564–567.
67. Lawrence D, Shahrokh Z, Marsters S, et al. Differential hepatocyte toxicity of recombinant Apo2L/TRAIL versions. Nat Med 2001;7:383–385.
68. Gores GJ, Kaufmann SH. Is TRAIL hepatotoxic? Hepatology 2001;34:3–6.
69. Hsu H, Xiong J, Goeddel DV. The TNF receptor 1-associated protein TRADD signals cell death and NF-kappa B activation. Cell 1995;81:495–504.
70. Harper N, Hughes M, MacFarlane M, et al. Fas-associated death domain protein and caspase-8 are not recruited to the tumor necrosis factor receptor 1 signaling complex during tumor necrosis factor-induced apoptosis. J Biol Chem 2003;278:25534–25541.
71. Micheau O, Tschopp J. Induction of TNF receptor I-mediated apoptosis via two sequential signaling complexes. Cell 2003;114:181–190.
72. Takeda K, Matsuzawa A, Nishitoh H, et al. Roles of MAPKKK ASK1 in stress-induced cell death. Cell Struct Funct 2003;28:23–29.

73. Deng Y, Ren X, Yang L, et al. A JNK-dependent pathway is required for TNF alpha-induced apoptosis. Cell 2003;115:61–70.
74. Kim YS, Schwabe RF, Qian T, et al. TRAIL-mediated apoptosis requires NF-kappaB inhibition and the mitochondrial permeability transition in human hepatoma cells. Hepatology 2002;36:1498–1508.
75. Hatano E, Bennett BL, Manning AM, et al. NF-kappaB stimulates inducible nitric oxide synthase to protect mouse hepatocytes from TNF-alpha- and Fas-mediated apoptosis. Gastroenterology 2001;120:1251–1262.
76. Li H, Zhu H, Xu CJ, et al. Cleavage of BID by caspase 8 mediates the mitochondrial damage in the Fas pathway of apoptosis. Cell 1998;94:491–501.
77. Luo X, Budihardjo I, Zou H, et al. Bid, a Bcl2 interacting protein, mediates cytochrome c release from mitochondria in response to activation of cell surface death receptors. Cell 1998;94:481–490.
78. Gross A, Yin XM, Wang K, et al. Caspase cleaved BID targets mitochondria and is required for cytochrome c release, while BCL-XL prevents this release but not tumor necrosis factor-R1/Fas death. J Biol Chem 1999;274:1156–1163.
79. Zha J, Weiler S, Oh KJ, et al. Posttranslational N-myristoylation of BID as a molecular switch for targeting mitochondria and apoptosis. Science 2000;290:1761–1765.
80. Scaffidi C, Fulda S, Srinivasan A, et al. Two CD95 (APO-1/Fas) signaling pathways. EMBO J 1998;17:1675–1687.
81. Yin XM, Wang K, Gross A, et al. Bid-deficient mice are resistant to Fas-induced hepatocellular apoptosis. Nature 1999;400:886–891.
82. Sun XM, Bratton SB, Butterworth M, et al. Bcl-2 and Bcl-xL inhibit CD95-mediated apoptosis by preventing mitochondrial release of Smac/DIABLO and subsequent inactivation of X-linked inhibitor-of-apoptosis protein. J Biol Chem 2002;277:11345–11351.
83. Tsujimoto Y, Finger LR, Yunis J, et al. Cloning of the chromosome breakpoint of neoplastic B cells with the t(14;18) chromosome translocation. Science 1984;226:1097–1099.
84. Cleary ML, Sklar J. Nucleotide sequence of a t(14;18) chromosomal breakpoint in follicular lymphoma and demonstration of a breakpoint-cluster region near a transcriptionally active locus on chromosome 18. Proc Natl Acad Sci USA 1985;82:7439–7443.
85. Bakhshi A, Jensen JP, Goldman P, et al. Cloning the chromosomal breakpoint of t(14;18) human lymphomas: clustering around JH on chromosome 14 and near a transcriptional unit on 18. Cell 1985;41:899–906.
86. Vaux DL, Cory S, Adams JM. Bcl-2 gene promotes haemopoietic cell survival and cooperates with c-myc to immortalize pre-B cells. Nature 1988;335:440–442.
87. Nunez G, Seto M, Seremetis S, et al. Deregulated Bcl-2 gene expression selectively prolongs survival of growth factor-deprived hemopoietic cell lines. J Immunol 1990;144:3602–3610.
88. McDonnell TJ, Korsmeyer SJ. Progression from lymphoid hyperplasia to high-grade malignant lymphoma in mice transgenic for the t(14; 18). Nature 1991;349:254–256.
89. Strasser A, Harris AW, Bath ML, et al. Novel primitive lymphoid tumours induced in transgenic mice by cooperation between myc and bcl-2. Nature 1990;348:331–333.
90. Linette GP, Hess JL, Sentman CL, et al. Peripheral T-cell lymphoma in lckpr-bcl-2 transgenic mice. Blood 1995;86:1255–1260.
91. Korsmeyer SJ. Bcl-2 initiates a new category of oncogenes: regulators of cell death. Blood 1992;80:879–886.
92. Wei MC, Zong WX, Cheng EH, et al. Proapoptotic BAX and BAK: a requisite gateway to mitochondrial dysfunction and death. Science 2001;292:727–730.
93. Zong WX, Ditsworth D, Bauer DE, et al. BH3-only proteins that bind pro-survival Bcl-2 family members fail to induce apoptosis in the absence of Bax and Bak. Genes Dev 2001;15:1481–1486.
94. Cheng EH, Wei MC, Weiler S, et al. BCL-2, BCL-X(L) sequester BH3 domain-only molecules preventing BAX- and BAK-mediated mitochondrial apoptosis. Mol Cell 2001;8:705–711.
95. Puthalakath H, Strasser A. Keeping killers on a tight leash: transcriptional and post-translational control of the pro-apoptotic activity of BH3-only proteins. Cell Death Differ 2002;9:505–512.
96. Murphy KL, Kittrell FS, Gay JP, et al. Bcl-2 expression delays mammary tumor development in dimethylbenz(a)anthracene-treated transgenic mice. Oncogene 1999;18:6597–6604.
97. Furth PA, Bar-Peled U, Li M, et al. Loss of anti-mitotic effects of Bcl-2 with retention of anti-apoptotic activity during tumor progression in a mouse model. Oncogene 1999;18:6589–6596.
98. Bruckheimer EM, Brisbay S, Johnson DJ, et al. Bcl-2 accelerates multistep prostate carcinogenesis in vivo. Oncogene 2000;19:5251–5258.
99. Fedorov LM, Tyrsin OY, Papadopoulos T, et al. Bcl-2 determines susceptibility to induction of lung cancer by oncogenic CRaf. Cancer Res 2002;62:6297–6303.
100. Pena JC, Rudin CM, Thompson CB. A Bcl-xL transgene promotes malignant conversion of chemically initiated skin papillomas. Cancer Res 1998;58:2111–2116.
101. Hardwick JM. Viral interference with apoptosis. Semin Cell Dev Biol 1998;9:339–349.
102. Gangappa S, van Dyk LF, Jewett TJ, et al. Identification of the in vivo role of a viral bcl-2. J Exp Med 2002;195:931–940.
103. Hengartner MO, Horvitz HR. *C. elegans* cell survival gene ced-9 encodes a functional homolog of the mammalian proto-oncogene bcl-2. Cell 1994;76:665–676.
104. Conradt B, Horvitz HR. The *C. elegans* protein EGL-1 is required for programmed cell death and interacts with the Bcl-2-like protein CED-9. Cell 1998;93:519–529.
105. Vernooy SY, Copeland J, Ghaboosi N, et al. Cell death regulation in *Drosophila:* conservation of mechanism and unique insights. J Cell Biol 2000;150:F69–F76.
106. Yin XM, Oltvai ZN, Korsmeyer SJ. BH1 and BH2 domains of Bcl-2 are required for inhibition of apoptosis and heterodimerization with Bax [see comments]. Nature 1994;369:321–323.
107. Chittenden T, Flemington C, Houghton AB, et al. A conserved domain in Bak, distinct from BH1 and BH2, mediates cell death and protein binding functions. EMBO J 1995;14:5589–5596.
108. Hunter JJ, Bond BL, Parslow TG. Functional dissection of the human Bc12 protein: sequence

requirements for inhibition of apoptosis. Mol Cell Biol 1996;16:877–883.

109. Huang DC, Adams JM, Cory S. The conserved N-terminal BH4 domain of Bcl-2 homologues is essential for inhibition of apoptosis and interaction with CED-4. EMBO J 1998;17:1029–1039.

110. Gross A, McDonnell JM, Korsmeyer SJ. BCL-2 family members and the mitochondria in apoptosis. Genes Dev 1999;13:1899–1911.

111. Oltvai ZN, Milliman CL, Korsmeyer SJ. Bcl-2 heterodimerizes in vivo with a conserved homolog, Bax, that accelerates programmed cell death. Cell 1993; 74:609–619.

112. Wang K, Yin XM, Chao DT, et al. BID: a novel BH3 domain-only death agonist. Genes Dev 1996;10: 2859–2869.

113. Oltvai ZN, Korsmeyer SJ. Checkpoints of dueling dimers foil death wishes [comment]. Cell 1994;79: 189–192.

114. Ke N, Godzik A, Reed JC. Bcl-B, a novel Bcl-2 family member that differentially binds and regulates Bax and Bak. J Biol Chem 2001;276:12481–12484.

115. Hsu SY, Kaipia A, McGee E, et al. Bok is a pro-apoptotic Bcl-2 protein with restricted expression in reproductive tissues and heterodimerizes with selective anti-apoptotic Bcl-2 family members. Proc Natl Acad Sci USA 1997;94:12401–12406.

116. Wei MC, Lindsten T, Mootha VK, et al. tBID, a membrane-targeted death ligand, oligomerizes BAK to release cytochrome c. Genes Dev 2000;14: 2060–2071.

117. Tan KO, Tan KM, Chan SL, et al. MAP-1, a novel proapoptotic protein containing a BH3-like motif that associates with Bax through its Bcl-2 homology domains. J Biol Chem 2001;276:2802–2807.

118. Marani M, Tenev T, Hancock D, et al. Identification of novel isoforms of the BH3 domain protein Bim which directly activate Bax to trigger apoptosis. Mol Cell Biol 2002;22:3577–3589.

119. Desagher S, Osen-Sand A, Nichols A, et al. Bid-induced conformational change of Bax is responsible for mitochondrial cytochrome c release during apoptosis. J Cell Biol 1999;144:891–901.

120. Hanada M, Aime-Sempe C, Sato T, et al. Structure-function analysis of Bcl-2 protein. Identification of conserved domains important for homodimerization with Bcl-2 and heterodimerization with Bax. J Biol Chem 1995;270:11962–11969.

121. Gross A, Jockel J, Wei MC, et al. Enforced dimerization of BAX results in its translocation, mitochondrial dysfunction and apoptosis. EMBO J 1998;17: 3878–3885.

122. Eskes R, Desagher S, Antonsson B, et al. Bid induces the oligomerization and insertion of Bax into the outer mitochondrial membrane. Mol Cell Biol 2000; 20:929–935.

123. Muchmore SW, Sattler M, Liang H, et al. X-ray and NMR structure of human Bcl-xL, an inhibitor of programmed cell death. Nature 1996;381: 335–341.

124. McDonnell JM, Fushman D, Milliman CL, et al. Solution structure of the proapoptotic molecule BID: a structural basis for apoptotic agonist and antagonists. Cell 1999;96:625–634.

125. Suzuki M, Youle RJ, Tjandra N. Structure of Bax: coregulation of dimer formation and intracellular localization. Cell 2000;103:645–654.

126. Petros AM, Medek A, Nettesheim DG, et al. Solution structure of the antiapoptotic protein bcl-2. Proc Natl Acad Sci USA 2001;98:3012–3017.

127. Hinds MG, Lackmann M, Skea GL, et al. The structure of Bcl-w reveals a role for the C-terminal residues in modulating biological activity. EMBO J 2003;22:1497–1507.

128. Sattler M, Liang H, Nettesheim D, et al. Structure of Bcl-xL-Bak peptide complex: recognition between regulators of apoptosis. Science 1997;275: 983–986.

129. Hsu YT, Youle RJ. Bax in murine thymus is a soluble monomeric protein that displays differential detergent-induced conformations. J Biol Chem 1998; 273:10777–10783.

130. Cheng EH, Kirsch DG, Clem RJ, et al. Conversion of Bcl-2 to a Bax-like death effector by caspases. Science 1997;278:1966–1968.

131. Clem RJ, Cheng EH, Karp CL, et al. Modulation of cell death by Bcl-XL through caspase interaction. Proc Natl Acad Sci USA 1998;95:554–559.

132. Lin B, Kolluri SK, Lin F, et al. Conversion of Bcl-2 from protector to killer by interaction with nuclear orphan receptor Nur77/TR3. Cell 2004;116: 527–540.

133. Kroemer G, Reed JC. Mitochondrial control of cell death. Nat Med 2000;6:513–519.

134. Newmeyer DD, Ferguson-Miller S. Mitochondria: releasing power for life and unleashing the machineries of death. Cell 2003;112:481–490.

135. Li P, Nijhawan D, Budihardjo I, et al. Cytochrome c and dATP-dependent formation of Apaf-1/caspase-9 complex initiates an apoptotic protease cascade. Cell 1997;91:479–489.

136. Lemasters JJ, Qian T, He L, et al. Role of mitochondrial inner membrane permeabilization in necrotic cell death, apoptosis, and autophagy. Antioxid Redox Signal 2002;4:769–781.

137. Lindsten T, Ross AJ, King A, et al. The combined functions of proapoptotic Bcl-2 family members Bak and Bax are essential for normal development of multiple tissues. Mol Cell 2000;6:1389–1399.

138. Zha J, Harada H, Yang E, et al. Serine phosphorylation of death agonist BAD in response to survival factor results in binding to 14-3-3 not BCL-X(L). Cell 1996;87:619–628.

139. Oda E, Ohki R, Murasawa H, et al. Noxa, a BH3-only member of the Bcl-2 family and candidate mediator of p53-induced apoptosis. Science 2000;288:1053–1058.

140. Yu J, Zhang L, Hwang PM, et al. PUMA induces the rapid apoptosis of colorectal cancer cells. Mol Cell 2001;7:673–682.

141. Nakano K, Vousden KH. PUMA, a novel proapoptotic gene, is induced by p53. Mol Cell 2001;7: 683–694.

142. Letai A, Bassik MC, Walensky LD, et al. Distinct BH3 domains either sensitize or activate mitochondrial apoptosis, serving as prototype cancer therapeutics. Cancer Cell 2002;2:183–192.

143. Chen F, Hersh BM, Conradt B, et al. Translocation of *C. elegans* CED-4 to nuclear membranes during programmed cell death. Science 2000;287:1485–1489.

144. Foyouzi-Youssefi R, Arnaudeau S, Borner C, et al. Bcl-2 decreases the free Ca2+ concentration within the endoplasmic reticulum. Proc Natl Acad Sci USA 2000;97:5723–5728.

145. Pinton P, Ferrari D, Magalhaes P, et al. Reduced loading of intracellular Ca(2+) stores and downregulation of capacitative Ca(2+) influx in Bcl-2-overexpressing cells. J Cell Biol 2000;148:857–862.

146. He H, Lam M, McCormick TS, Distelhorst CW. Maintenance of calcium homeostasis in the endoplasmic reticulum by Bcl-2. J Cell Biol 1997;138:1219–1228.

147. Scorrano L, Oakes SA, Opferman JT, et al. BAX and BAK regulation of endoplasmic reticulum Ca2+: a control point for apoptosis. Science 2003;300:135–139.

148. Zong WX, Li C, Hatzivassiliou G, et al. Bax and Bak can localize to the endoplasmic reticulum to initiate apoptosis. J Cell Biol 2003;162:59–69.

149. Nutt LK, Pataer A, Pahler J, et al. Bax and Bak promote apoptosis by modulating endoplasmic reticular and mitochondrial Ca2+ stores. J Biol Chem 2002;277:9219–9225.

150. Linette GP, Li Y, Roth K, et al. Cross talk between cell death and cell cycle progression: BCL-2 regulates NFAT-mediated activation. Proc Natl Acad Sci USA 1996;93:9545–9552.

151. Janumyan YM, Sansam CG, Chattopadhyay A, et al. Bcl-xL/Bcl-2 coordinately regulates apoptosis, cell cycle arrest and cell cycle entry. EMBO J 2003;22:5459–5470.

152. Chattopadhyay A, Chiang CW, Yang E. BAD/BCL-[X(L)] heterodimerization leads to bypass of G0/G1 arrest. Oncogene 2001;20:4507–4518.

153. Vail ME, Chaisson ML, Thompson J, et al. Bcl-2 expression delays hepatocyte cell cycle progression during liver regeneration. Oncogene 2002;21:1548–1555.

154. Knudson CM, Johnson GM, Lin Y, et al. Bax accelerates tumorigenesis in p53-deficient mice. Cancer Res 2001;61:659–665.

155. Cheng N, Janumyan YM, Didion L, et al. Bcl-2 inhibition of T-cell proliferation is related to prolonged T-cell survival. Oncogene 2004;23:3770–3780.

156. Silvestrini R, Veneroni S, Daidone MG, et al. The Bcl-2 protein: a prognostic indicator strongly related to p53 protein in lymph node-negative breast cancer patients. J Natl Cancer Inst 1994;86:499–504.

157. Joensuu H, Pylkkanen L, Toikkanen S. Bcl-2 protein expression and long-term survival in breast cancer. Am J Pathol 1994;145:1191–1198.

158. Elledge RM, Green S, Howes L, et al. bcl-2, p53, and response to tamoxifen in estrogen receptor-positive metastatic breast cancer: a Southwest Oncology Group study. J Clin Oncol 1997;15:1916–1922.

159. Silvestrini R, Benini E, Veneroni S, et al. p53 and bcl-2 expression correlates with clinical outcome in a series of node-positive breast cancer patients. J Clin Oncol 1996;14:1604–1610.

160. Hockenbery DM, Zutter M, Hickey W, et al. BCL2 protein is topographically restricted in tissues characterized by apoptotic cell death. Proc Natl Acad Sci USA 1991;88:6961–6965.

161. Novack DV, Korsmeyer SJ. Bcl-2 protein expression during murine development. Am J Pathol 1994;145:61–73.

162. Parton M, Dowsett M, Smith I. Studies of apoptosis in breast cancer. BMJ 2001;322:1528–1532.

163. Fitzgibbons PL, Page DL, Weaver D, et al. Prognostic factors in breast cancer. College of American Pathologists Consensus Statement 1999. Arch Pathol Lab Med 2000;124:966–978.

164. Reed JC. Mechanisms of Bcl-2 family protein function and dysfunction in health and disease. Behring Inst Mitt 1996:72–100.

165. Naik P, Karrim J, Hanahan D. The rise and fall of apoptosis during multistage tumorigenesis: downmodulation contributes to tumor progression from angiogenic progenitors. Genes Dev 1996;10:2105–2116.

166. Pelengaris S, Khan M, Evan GI. Suppression of Myc-induced apoptosis in beta cells exposes multiple oncogenic properties of Myc and triggers carcinogenic progression. Cell 2002;109:321–334.

167. Messing A, Chen HY, Palmiter RD, et al. Peripheral neuropathies, hepatocellular carcinomas and islet cell adenomas in transgenic mice. Nature 1985;316:461–463.

168. Araki K, Hino O, Miyazaki J, et al. Development of two types of hepatocellular carcinoma in transgenic mice carrying the SV40 large T-antigen gene. Carcinogenesis 1991;12:2059–2062.

169. de La Coste A, Mignon A, Fabre M, et al. Paradoxical inhibition of c-myc-induced carcinogenesis by Bcl-2 in transgenic mice. Cancer Res 1999;59:5017–5022.

170. Vail ME, Pierce RH, Fausto N. Bcl-2 delays and alters hepatic carcinogenesis induced by transforming growth factor alpha. Cancer Res 2001;61:594–601.

171. Uhlmann EJ, Subramanian T, Vater CA, et al. A potent cell death activity associated with transient high level expression of BCL-2. J Biol Chem 1998;273:17926–17932.

172. Subramanian T, Chinnadurai G. Pro-apoptotic activity of transiently expressed BCL-2 occurs independent of BAX and BAK. J Cell Biochem 2003;89:1102–1114.

173. Hellemans P, van Dam PA, Weyler J, et al. Prognostic value of bcl-2 expression in invasive breast cancer. Br J Cancer 1995;72:354–360.

174. Lipponen P, Pietilainen T, Kosma VM, et al. Apoptosis suppressing protein bcl-2 is expressed in well-differentiated breast carcinomas with favourable prognosis. J Pathol 1995;177:49–55.

175. Barbareschi M, Caffo O, Veronese S, et al. Bcl-2 and p53 expression in node-negative breast carcinoma: a study with long-term follow-up. Hum Pathol 1996;27:1149–1155.

176. van Slooten HJ, Clahsen PC, van Dierendonck JH, et al. Expression of Bcl-2 in node-negative breast cancer is associated with various prognostic factors, but does not predict response to one course of perioperative chemotherapy. Br J Cancer 1996;74:78–85.

177. Kapranos N, Karaiosifidi H, Valavanis C, et al. Prognostic significance of apoptosis related proteins Bcl-2 and Bax in node-negative breast cancer patients. Anticancer Res 1997;17:2499–2505.

178. Krajewski S, Thor AD, Edgerton SM, et al. Analysis of Bax and Bcl-2 expression in p53-immunopositive breast cancers. Clin Cancer Res 1997;3:199–208.

179. Lee AK, Loda M, Mackarem G, et al. Lymph node negative invasive breast carcinoma 1 centimeter or less in size (T1a,bNOMO): clinicopathologic features and outcome. Cancer 1997;79:761–771.

180. Charpin C, Garcia S, Bonnier P, et al. bcl-2 automated and quantitative immunocytochemical assays in breast carcinomas: correlation with 10-year follow-up. J Clin Oncol 1998;16:2025–2031.

181. Pich A, Margaria E, Chiusa L. Bcl-2 expression in male breast carcinoma. Virchows Arch 1998;433: 229–235.
182. Veronese S, Mauri FA, Caffo O, et al. Bax immunohistochemical expression in breast carcinoma: a study with long term follow-up. Int J Cancer 1998; 79:13–18.
183. Holmqvist P, Lundstrom M, Stal O. Apoptosis and Bcl-2 expression in relation to age, tumor characteristics and prognosis in breast cancer. South-East Sweden Breast Cancer Group. Int J Biol Markers 1999;14:84–91.
184. Le MG, Mathieu MC, Douc-Rasy S, et al. c-myc, p53 and bcl-2, apoptosis-related genes in infiltrating breast carcinomas: evidence of a link between bcl-2 protein over-expression and a lower risk of metastasis and death in operable patients. Int J Cancer 1999;84:562–567.
185. McLaughlin R, O'Hanlon D, McHale T, et al. Prognostic implications of p53 and bcl-2 expression in 108 women with stage two breast cancer. Ir J Med Sci 2001;170:11–13.
186. Bukholm IR, Bukholm G, Nesland JM. Reduced expression of both Bax and Bcl-2 is independently associated with lymph node metastasis in human breast carcinomas. APMIS 2002;110:214–220.
187. O'Driscoll L, Linehan R, M Kennedy S, et al. Lack of prognostic significance of survivin, survivin-deltaEx3, survivin-2B, galectin-3, bag-1, bax-alpha and MRP-1 mRNAs in breast cancer. Cancer Lett 2003;201:225–236.
188. Linjawi A, Kontogiannea M, Halwani F, et al. Prognostic significance of p53, bcl-2, and Bax expression in early breast cancer. J Am Coll Surg 2004; 198:83–90.
189. Dimitrakakis C, Konstadoulakis M, Messaris E, et al. Molecular markers in breast cancer: can we use c-erbB-2, p53, bcl-2 and bax gene expression as prognostic factors? Breast 2002;11:279–285.
190. Krajewski S, Krajewska M, Turner BC, et al. Prognostic significance of apoptosis regulators in breast cancer. Endocr Relat Cancer 1999;6:29–40.
191. Olopade OI, Adeyanju MO, Safa AR, et al. Overexpression of BCL-x protein in primary breast cancer is associated with high tumor grade and nodal metastases. Cancer J Sci Am 1997;3:230–237.
192. Schiller AB, Clark WS, Cotsonis G, et al. Image cytometric bcl-2:bax and bcl-2:bcl-x ratios in invasive breast carcinoma: correlation with prognosis. Cytometry 2002;50:203–209.
193. Zapata JM, Krajewska M, Krajewski S, et al. Expression of multiple apoptosis-regulatory genes in human breast cancer cell lines and primary tumors. Breast Cancer Res Treat 1998;47:129–140.
194. Tanaka K, Iwamoto S, Gon G, et al. Expression of survivin and its relationship to loss of apoptosis in breast carcinomas. Clin Cancer Res 2000;6: 127–134.
195. Kennedy SM, O'Driscoll L, Purcell R, et al. Prognostic importance of survivin in breast cancer. Br J Cancer 2003;88:1077–1083.
196. Vakkala M, Paakko P, Soini Y. Expression of caspases 3, 6 and 8 is increased in parallel with apoptosis and histological aggressiveness of the breast lesion. Br J Cancer 1999;81:592–599.
197. Micheau O, Solary E, Hammann A, et al. Sensitization of cancer cells treated with cytotoxic drugs to fas-mediated cytotoxicity. J Natl Cancer Inst 1997; 89:783–789.
198. Mottolese M, Buglioni S, Bracalenti C, et al. Prognostic relevance of altered Fas (CD95)-system in human breast cancer. Int J Cancer 2000;89:127–132.
199. Botti C, Buglioni S, Benevolo M, et al. Altered expression of FAS system is related to adverse clinical outcome in stage I-II breast cancer patients treated with adjuvant anthracycline-based chemotherapy. Clin Cancer Res 2004;10:1360–1365.
200. Parton M, Krajewski S, Smith I, et al. Coordinate expression of apoptosis-associated proteins in human breast cancer before and during chemotherapy. Clin Cancer Res 2002;8:2100–2108.
201. Danial NN, Korsmeyer SJ. Cell death: critical control points. Cell 2004;116:205–219.
202. Reed JC, Stein C, Subasinghe C, et al. Antisense-mediated inhibition of BCL2 protooncogene expression and leukemic cell growth and survival: comparisons of phosphodiester and phosphorothioate oligodeoxynucleotides. Cancer Res 1990;50:6565–6570.
203. Klasa RJ, Gillum AM, Klem RE, et al. Oblimersen Bcl-2 antisense: facilitating apoptosis in anticancer treatment. Antisense Nucleic Acid Drug Dev 2002; 12:193–213.
204. Lopes de Menezes DE, Hudon N, McIntosh N, et al. Molecular and pharmacokinetic properties associated with the therapeutics of bcl-2 antisense oligonucleotide G3139 combined with free and liposomal doxorubicin. Clin Cancer Res 2000;6: 2891–2902.
205. Lopes de Menezes DE, Hu Y, Mayer LD. Combined treatment of Bcl-2 antisense oligodeoxynucleotides (G3139), p-glycoprotein inhibitor (PSC833), and sterically stabilized liposomal doxorubicin suppresses growth of drug-resistant growth of drug-resistant breast cancer in severely combined immunodeficient mice. J Exp Ther Oncol 2003;3:72–82.
206. Lopes D, Mayer LD. Pharmacokinetics of Bcl-2 antisense oligonucleotide (G3139) combined with doxorubicin in SCID mice bearing human breast cancer solid tumor xenografts. Cancer Chemother Pharmacol 2002;49:57–68.
207. Nahta R, Esteva FJ. Bcl-2 antisense oligonucleotides: a potential novel strategy for the treatment of breast cancer. Semin Oncol 2003;30(5 suppl 16):143–149.
208. Fantin VR, Berardi MJ, Scorrano L, et al. A novel mitochondriotoxic small molecule that selectively inhibits tumor cell growth. Cancer Cell 2002;2:29–42.
209. Fantin VR, Leder P. F16, a mitochondriotoxic compound, triggers apoptosis or necrosis depending on the genetic background of the target carcinoma cell. Cancer Res 2004;64:329–336.
210. Cretney E, Takeda K, Yagita H, et al. Increased susceptibility to tumor initiation and metastasis in TNF-related apoptosis-inducing ligand-deficient mice. J Immunol 2002;168:1356–1361.
211. Ichikawa K, Liu W, Zhao L, et al. Tumoricidal activity of a novel anti-human DR5 monoclonal antibody without hepatocyte cytotoxicity. Nat Med 2001;7: 954–960.
212. Chuntharapai A, Dodge K, Grimmer K, et al. Isotype-dependent inhibition of tumor growth in vivo by monoclonal antibodies to death receptor 4. J Immunol 2001;166:4891–4898.

213. Gibson SB, Oyer R, Spalding AC, et al. Increased expression of death receptors 4 and 5 synergizes the apoptosis response to combined treatment with etoposide and TRAIL. Mol Cell Biol 2000;20:205–212.

214. Keane MM, Ettenberg SA, Nau MM, et al. Chemotherapy augments TRAIL-induced apoptosis in breast cell lines. Cancer Res 1999;59:734–741.

215. Shin MS, Kim HS, Lee SH, et al. Mutations of tumor necrosis factor-related apoptosis-inducing ligand receptor 1 (TRAIL-R1) and receptor 2 (TRAIL-R2) genes in metastatic breast cancers. Cancer Res 2001;61:4942–4946.

216. Singh TR, Shankar S, Chen X, et al. Synergistic interactions of chemotherapeutic drugs and tumor necrosis factor-related apoptosis-inducing ligand/Apo-2 ligand on apoptosis and on regression of breast carcinoma in vivo. Cancer Res 2003;63:5390–5400.

217. Buchsbaum DJ, Zhou T, Grizzle WE, et al. Antitumor efficacy of TRA-8 anti-DR5 monoclonal antibody alone or in combination with chemotherapy and/or radiation therapy in a human breast cancer model. Clin Cancer Res 2003;9(10 Pt 1):3731–3741.

CHAPTER 22

Oncogenes

Andre M. Oliveira, MD, and Jonathan A. Fletcher, MD

Department of Pathology, Brigham and Women's Hospital, Boston, MA

History and Definition

Oncogenes are constitutively activated forms of genes that play key roles in cell functions such as adhesion, proliferation, and survival. There are various mechanisms by which normal genes are activated to give rise to oncogenes in breast cancer. These mechanisms include genomic amplification, gene rearrangement, intragenic mutation, and transcriptional deregulation. Many oncogenes have been implicated in subsets of breast cancer, and the clinical relevance of these genes has surged in recent years due to the burgeoning opportunities for targeted therapies.

Oncogenes were initially discovered through studies of DNA and RNA viruses.[1,2] The classic work of Payton Rous, performed almost a century ago, identified a cell-free agent that induced sarcomas in chickens. This groundbreaking discovery led, some 70 years later, to a series of investigations that revealed the first oncogene–v-*src* (from *v*iral and *sar*coma, respectively).[3] v-*src* is a retrovirus containing a mutant constitutively activated form of the protein tyrosine kinase gene, *SRC*.[3,4] Subsequent to the identification of v-*src,* more than 50 additional oncogenes were discovered through analyses of retroviruses, including *PDGF*β, *jun, myc, fos,* and H-*ras*. Oncogenes were also discovered through studies of DNA viruses, including the SV-40 T antigen, the human papilloma viruses genes *E6* and *E7,* and the adenovirus *E1A* and *E1B* genes.[2] However, in contrast to the retroviral v-oncogenes, there are no normal cellular counterparts for DNA virus oncogenes.

Besides retrovirus studies, other approaches have been used to identify oncogenes in breast carcinomas and in other human tumors. In cell transformation assays or in vitro transfection assays,[5] fragments of human tumor DNA are transfected into NIH3T3 mouse fibroblasts and morphologic changes consistent with cell transformation, such as loss of contact inhibition, are observed. Transformed colonies are then selected and screened for human oncogenic sequences responsible for the cell transformation observed in vitro; several genes of the *ras* family were discovered using this approach.[6] Another extensively used discovery method is based on the observation of recurrent cytogenetic abnormalities in cancer. Several oncogenes that were previously discovered by studying transforming retroviruses were later shown to be targeted in chromosomal translocations, including *abl*[7] and *myc*.[8] Other cytogenetic abnormalities observed in cancer cells, known as homogeneously staining chromosomal regions and double minute chromosomes, led to the discovery of oncogenic gene amplification.[9]

Functional Classes of Oncoproteins

Most oncoproteins can be grouped into classes, depending on the primary biologic function of each protein. These include transcriptional regulators, tyrosine and serine-threonine kinases, GTPases and guanine nucleotide exchange factors, and cell cycle regulators. Although many oncoproteins can be assigned to one or the other of these functional groups, such classifications are not entirely "clean" in that some oncoproteins have several functions and some fall outside of these classes (**Table 22.1**). Transcription regulators include oncoproteins that bind directly to DNA and oncoproteins that complex with DNA binding proteins and serve as coregulators of gene expression. Transcriptional regulators that bind DNA directly are known as transcription factors, and these include several essential domains, of which activation protein domains bind to coregulator proteins, and DNA-binding domains, such as leucine zippers,

Table 22.1 Oncogenes in Breast Cancer

Gene	Chromosome Location	Function	Mechanism
c-myc	8q24	Transcription factor	Overexpression/amplification
AIB1	20q12	Transcriptional coactivator	Overexpression/amplification
EGFR (HER-1)	7p12	Tyrosine kinase	Overexpression/amplification
ERBB2 (HER-2)	17q21	Tyrosine kinase	Overexpression/amplification
FGFR1	8p11	Tyrosine kinase	Overexpression/amplification
FGFR2	10q26	Tyrosine kinase	Overexpression/amplification
SRC	20q12–13	Tyrosine kinase	Overexpression/amplification
FGF2	4q25-q27	Tyrosine kinase ligand	Overexpression/amplification
FGF4	11q13	Tyrosine kinase ligand	Overexpression/amplification
Neuregulin	8p12	Tyrosine kinase ligand	Translocation
PSK1	17q23	Serine-theronine kinase	Overexpression
AKT	14q32/19q13	Serine-theronine kinase	Overexpression
Cyclin D1	11q13	Cell cycle regulator	Overexpression/amplification
Cyclin E1	19q13.1	Cell cycle regulator	Overexpression
Hras	11p15	GTPase	Activating mutations
TGF-α	2p13	Growth factor	Overexpression
Bcl-2	18q21	Inhibitor of apoptosis	Overexpression
MDM2	12q14	Regulator of p53	Overexpression
Cathepsin D	11p15	Lysosomal protease	Overexpression
EMS1	11q13	Actin binding protein	Overexpression/amplification
β-Catenin	3p22	Cell adhesion protein	Overexpression
CD44	11pter-p13	Cell adhesion protein	Overexpression

zinc fingers, and helix-loop-helixes, bind to consensus nucleotide sequences in the promoter region of the genes whose expression is regulated. Transcription factor oncogenes in breast carcinoma are exemplified by *myc,* which is highly amplified in a subset of breast carcinomas and which contains both helix-loop-helix and leucine zipper domains.[10–12]

Tyrosine kinases comprise a large group of proteins that regulate cell signaling pathways by tyrosine phosphorylation of substrates such as STATs, AKT, and mitogen activated protein kinase (MAPK). Protein tyrosine kinases are broadly divided into receptor and nonreceptor forms, the former having transmembrane and extracellular (ligand-binding) domains and the latter having only the cytoplasmic domain. Various tyrosine kinase proteins play oncogenic roles in breast carcinoma, most notably the four members of the epidermal growth factor receptor (EGFR) family: EGFR, ERBB2, ERBB3, and ERBB4 (see Chapter 16).

Serine-threonine kinases, which are also known as "dual-specificity" kinases, have functions similar to the tyrosine kinases, but they phosphorylate serine and threonine residues, rather than tyrosine, in substrate proteins. The cyclic AMP dependent protein kinase A is one serine-threonine kinase that contributes to breast tumorigenesis, protein kinase A overexpression having been associated with malignant transformation of the breast epithelium and poor prognosis in breast cancer.[13] Tyrosine and serine-threonine kinases can function jointly in the highly coordinated phosphorylation of a given substrate protein (e.g., MAPK), resulting in selective activation of specific protein functions by differential phosphorylation of tyrosine, serine, and threonine residues.

GTPase GTP-binding proteins participate in cell signaling as molecular switches that convert proteins from active to inactive state. This is accomplished by processing GTP to GDP, the GTP-bound forms of the target proteins being activated, whereas the GDP-bound forms of the same proteins are inactive. GTPase proteins play key roles in regulating activity for the various RAS-mediated signaling pathways that play key roles in breast tumorigenesis. Approximately 5% of breast cancers contain activating mutations of the H-*ras* GTPases that stabilize H-*ras* in its activated GTP-bound form, resulting in constitutive activation of the H-*ras* GTPase function.[14]

Guanine nucleotide exchange factors have also been implicated in breast cancer tumorigenesis.[15] Guanine nucleotide exchange factors complex with the inactive GDP-bound forms of GTPases, thereby mediating the release of GDP and restoring GTPase activity by enabling reassociation with GTP. Several components of the cell cycle regulatory machinery (see Chapter 11) have been implicated as either oncogenes or tumor suppressor genes in breast cancer. For example, cyclin D1—a key regulator of the G1/S cell cycle transition—is overexpressed in a substantial number of breast carcinomas (discussed below).[16,17] Survival proteins also function as oncogenes by inhibiting apoptosis or otherwise preventing normal cell death mechanisms. The classic example is the antiapoptotic protein Bcl-2, which was initially discovered by cloning the chromosomal translocation (14;18) in B-cell lymphomas[18–20]; Bcl-2 overexpression has been demonstrated in many breast cancers and correlates with hormone receptor status and survival.[21–23]

Mechanisms of Oncogene Activation

Direct mutational activation of oncogenes in breast cancer generally occurs by one of the following three major mechanisms: (1) point mutations or other sequence changes in the gene coding sequence or promoter region, (2) gene amplification, and (3) gene rearrangement, usually by chromosomal translocations. Indirect activation of oncogenes can occur through upstream activation events or when normal regulatory mechanisms are suppressed, as seen following inactivation of the *p53* or *PTEN* tumor suppressor genes (Chapter 23).

Point mutations in the coding sequence of an oncogene often result in constitutional activation of its protein product. Because these activation events behave in a dominant fashion, they confer a survival advantage and are retained by the malignant clones as the cancer progresses. For example, *ras* family point mutations are responsible for abnormal and uncontrolled activation of the ras signaling pathways, resulting in cell proliferation.[24] Gene amplification, as it is observed with HER-2/*neu* (FIG. 22.1) and *CCND1* (the gene that encodes cyclin D1) is also selectively retained by malignant clones because the resultant overexpression of these genes confers survival and proliferative advantages.

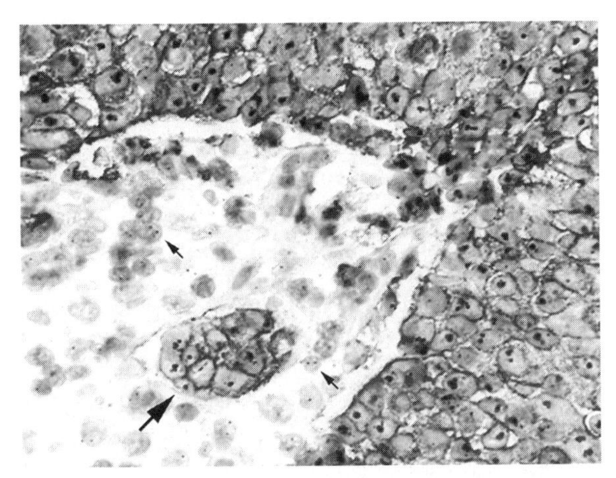

FIGURE 22.1 HER-2 amplification and overexpression demonstrated by HER-2 gene chromogenic in situ hybridization in brown and HER-2 protein immunohistochemistry in blue. Cells in this invasive ductal carcinoma show HER-2 gene clusters with high level amplification accompanied by strong membranous HER-2 expression (large arrow). Cells with small normal HER-2 gene signals lack HER-2 overexpression and are probably non-neoplastic cells (small arrows). **See Plate 47 for color image.**

Chromosomal rearrangements participate in oncogenesis by bringing together two different chromosome regions and thereby placing a quiescent oncogene from one of the regions under the transcriptional influence of a highly active promoter from a strongly expressed gene in the other chromosome region. This results in overexpression of the oncogene and of its encoded protein product. Some chromosomal rearrangements also disrupt the coding sequence of the oncogene, fusing it to part of another gene and creating a fusion oncogene with novel and constitutively activated function. The classic example of a fusion oncogene is *BCR-ABL,* which is created by the chromosomal translocation t(9;22)(q34;q11) in chronic myelogenous leukemia. Although most known fusion oncogenes do result from chromosomal translocations, they can also result from any other type of chromosomal rearrangement—including inversion, insertion, or deletion—where two different chromosome regions are juxtaposed. Cytogenetic karyotyping studies have played a key role in identifying novel fusion oncogenes, because such studies are uniquely effective in identifying the balanced chromosomal rearrangements that target most fusion oncogenes. Such studies have been extremely useful in discovery of fusion oncogenes among leukemias, lymphomas, and sarcomas, where the karyotypes are noncomplex and where translocations and other balanced chromosomal rearrangements can be recognized readily. However, fusion oncogene discovery has been hampered, in most carcinomas, by the profound complexity of the karyotypes, which confounds attempts to identify recurring chromosomal translocations. Therefore, only a few fusion oncogenes have been identified in breast carcinoma, although doubtless others remain to be discovered. Neuregulin is targeted by translocation breakpoints in a subset of breast cancer cell lines, and these rearrangements might result in autocrine/paracrine activation of ERBB family proteins.[25] Another fusion oncogene results from the chromosomal translocation t(12;15)(p13;q25) in secretory carcinomas, which are an uncommon type of breast cancer. This translocation fuses the transcription factor gene *ETV6* (*TEL*) to the protein tyrosine kinase gene *NTRK3*.[26,27]

Oncogenic Cooperative Networks

As more is learned about the complex oncogenic circuitry in breast cancer, it is increasingly evident that the various oncogenic signaling pathways are interconnected, sometimes in unexpected ways. Using the EGFR family of receptor tyrosine kinases (also discussed in Chapters 16 and 17) as a jumping off point, the discussion below touches upon the complex biology of breast cancer, in which major oncogenic networks involving cell signaling, cell cycle regulation, and apoptosis are integrated to serve the purpose of cell transformation (FIG. 22.2).

Tyrosine Kinase Signaling: EGFR Family as a Model

Receptor tyrosine kinases regulate critical aspects of cellular behavior, including cell growth, differentiation, and death, which contribute to breast tumorigenesis. As an example, the EGFR family members—ERBB1

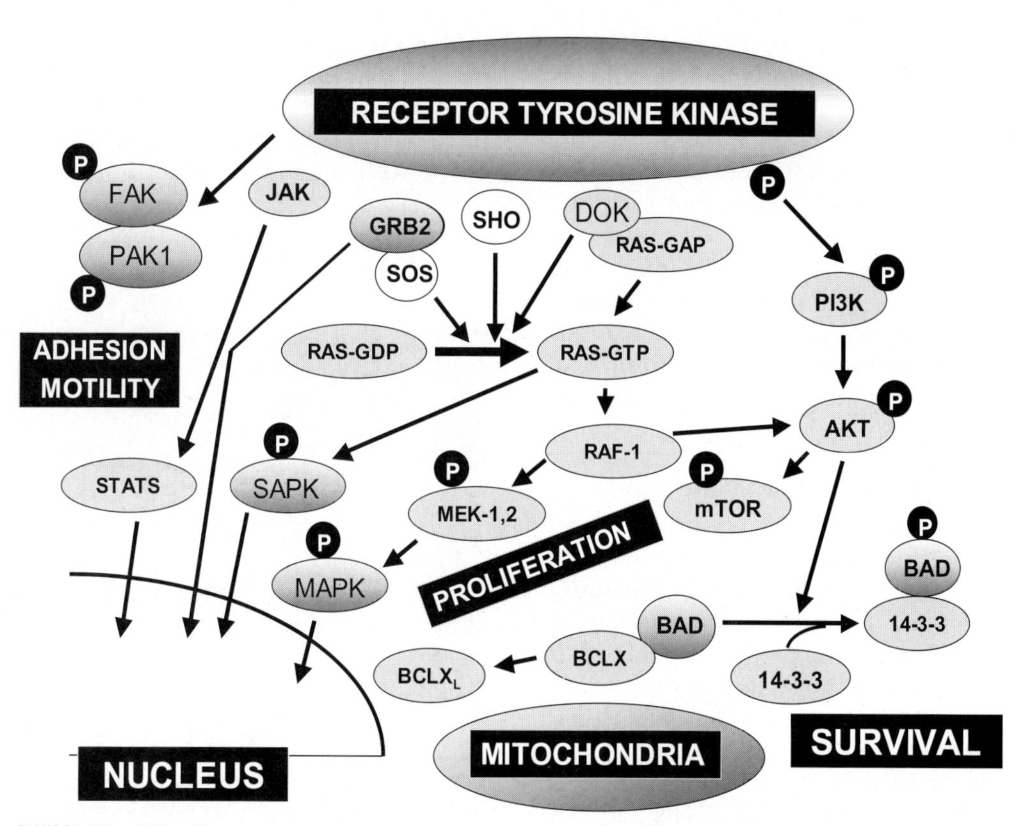

FIGURE 22.2 Cell signaling schematic of some of the essential downstream pathways activated by oncogenic receptor tyrosine kinase proteins, such as EGFR and HER-2.

(EGFR), ERBB2 (HER-2/*neu*), ERBB3, and ERBB4—have roles in the normal development of the mammary gland, including duct and lobule formation.[28] In a normal cell, the receptor tyrosine kinase proteins are maintained in a state of fine equilibrium by their constant alternation between monomeric inactive and dimeric active conformations, which depend on the presence of stimulatory and inhibitory growth factor signals. For example, EGFR family activation is dependent on various stimulatory growth factors, including EGF, amphiregulin, epiregulin, neuregulins, transforming growth factor-α, and others.[29] In addition, EGFR family proteins can be activated by other signaling pathways, including those mediated by G protein coupled receptors and by cytokines.[30] Oncogenic activation of EGFR family proteins, whether by overexpression, activation point mutations, or autocrine mechanisms, disrupts the delicate equilibrium in the above-mentioned pathways. In the case of ERBB2 in breast cancer, there is a quantitative change in the concentration of the receptor, which results in a stochastically favorable condition for increased ERBB2 homodimerization (binding to another ERBB2 protein) and heterodimerization (binding to another EGFR family protein, such as EGFR or ERBB3). Homodimerization and heterodimerization induces activation of the tyrosine kinase functions in the bound proteins, resulting in cross-phosphorylation of critical tyrosine residues on each of the bound proteins. The phosphorylated tyrosines serve as binding sites for adaptor, scaffold, relay, and anchoring proteins that trigger the activation of signaling pathways, such as the RAS MAPK cell proliferation pathway and the phosphatidylinositol-3 kinase (PI-3K)/AKT cell survival pathway, among others.

Serine-Threonine Kinases: The PI-3K/AKT Pathway

Among the cell signaling pathways activated by receptor tyrosine kinase proteins, the PI-3K/AKT pathway seems to be particularly active in breast cancers that overexpress ERBB2.[31–34] This results, in part, from HER-2/*neu* ERBB2 heterodimerization and cross-phosphorylation of the kinase deficient EGFR family member, ERBB3. ERBB3, once phosphorylated by ERBB2, recruits and activates the p85 regulatory subunit of PI-3K.[35] PI-3K activation in turn leads to AKT pathway activation by generating 3′-phosphorylated phosphatidylinositols. In addition, other kinases stimulated by PI-3K mediate the phosphorylation of the critical AKT threonine 308 and serine 473 residues. Activated AKT then mediates many important cellular events, including survival, cell cycle control, and differentiation.[36]

Integration of Cell Signaling with Cell Cycle Regulation

The downstream consequences of ERBB2 and PI-3K/AKT activation include various effects on proteins involved in cell cycle regulation. For example, activation of these pathways can induce increased expression of cyclin D1, which is a key stimulatory protein of the G1/S transition (see Chapter 13). Cyclin D1 overexpression has been associated with poor prognosis in breast cancer,[37] and in some breast cancers cyclin D1 overexpression results from amplification of the chromosome band 11q13 region, which contains the cyclin D1 gene (*CCND1*). However, chromosome 11q13 amplification is found in no more than 15% of breast cancers.[38,39] It is likely, therefore, that alternate mechanisms of cyclin D1 overexpression (FIG. 22.3), such as transcriptional up-regulation resulting from ERBB2 and PI-3K/AKT pathway activation, might be more frequent than cyclin D1 genomic amplification in breast cancer. The biologic pathways that connect AKT activation and cyclin D1 overexpression have been partly characterized, and these involve AKT-mediated inhibition of glycogen synthase kinase-3, which normally inhibits cyclin D1 gene transcription through regulation

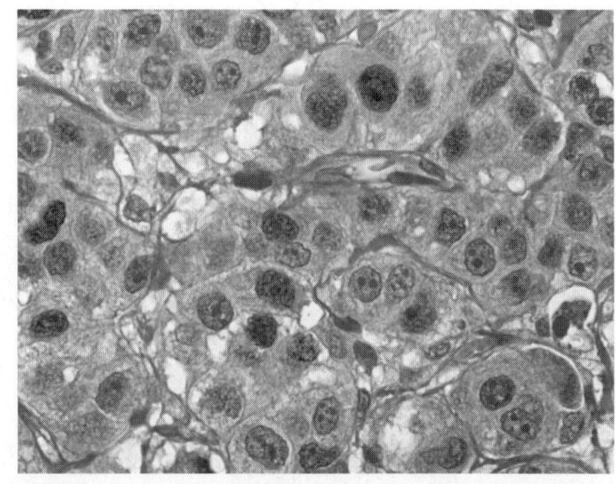

FIGURE 22.3 Cyclin D1 overexpression in a high grade infiltrating ductal carcinoma. Dysregulated cyclin D1 expression may be associated both with lack of cell cycle control and a dominant growth promoting effect. Image courtesy of Dr. Jeffrey S. Ross (immunoperoxidase ×400). **See Plate 48 for color image.**

of β-catenin signaling pathways.[36] However, the relationships between ERBB2 oncogenic activation and cyclin D1 overexpression are indeed complex and involve activation of pathways beyond the PI-3K/AKT axis. For example, Lenferink et al.[40] showed that ERBB2 overexpressing cells use both MAPK and PI-3K/AKT signaling pathways to accomplish cyclin D1 overexpression. In addition, although ERBB2 inhibitors have been shown to block cell cycle progression by disrupting the PI-3K/AKT pathways,[40] cyclin D1 knockout mice are resistant to oncogenic activation of both ERBB2 and RAS.[41] These observations show that multiple upstream pathways can function—either in a coordinated manner or as parallel "alternative" pathways—to accomplish the activation of critical downstream targets, such as cyclin D1. It is also clear that each upstream pathway regulates multiple downstream targets, whose collective activation (or inhibition) determines the overall biologic cell response. For example, PI-3K/AKT pathway activation, in addition to the abovementioned role in cyclin D1 overexpression, inhibits activation of the forkhead family of transcription factors, which results in an antiapoptotic signal and which also inhibits the cell cycle inhibitor p27.[42] Decreased p27 expression stimulates cell proliferation and is a marker of poor prognosis in breast cancer.[43–45]

Bcl-2, *myc*, and Survival Signaling

Bcl-2 is an antiapoptotic protein that is overexpressed in as many as 50% of breast cancers (see Chapter 21).[21–23] Bcl-2 overexpression has been associated with worse prognosis in various types of cancer. Surprisingly however, high levels of Bcl-2 expression have been associated with favorable outcome in patients with breast cancer.[46] The biologic basis for this seeming discrepancy remains to be determined, but it is possible that Bcl-2 is not the key survival protein in breast cancer or that differential expression and interactions among different members of the Bcl-2 family is the final determinant of cell survival.[47] Indeed, the expression of other apoptosis-related proteins have had unanticipated survival correlates in breast cancer. For example, decreased expression of the proapoptotic molecule Bax was associated with worse clinical outcome.[48] Bcl-2 expression is also under the influence estrogens, and a direct correlation between Bcl-2 and estrogen receptor expression has been observed.[46] In addition, Bcl-2 expression seems to cooperate with the *myc* (see below) pathway to induce mammary tumorigenesis.[49–52]

The c-*myc* Gene

The c-*myc* proto-oncogene located on chromosome 8 encodes for a 439 amino acid nuclear binding protein that directly stimulates cell division and participates in most aspects of cellular function, including replication, metabolism, differentiation, and apoptosis.[52,53] The c-*myc* gene is amplified in approximately 16% of breast cancer cases[54] and in most outcome-based studies is associated with decreased disease-free patient survival.[54,55] Some studies have failed to link c-*myc* status with prognosis.[56] Interestingly, c-*myc* amplification may be seen early in breast cancer specimens, including in situ tumors, but may not persist in the same patient when the lymph node metastasis tumoral tissue is analyzed.[57] Amplification of c-*myc* may result from upstream activation of HER-2/*neu*.[58]

The H-*ras* Gene

Of the three *ras* signal transduction encoding genes, the H-*ras* gene has been consistently associated with breast cancer progression.[59] Unlike K-*ras* and N-*ras* where point mutations are the most common cause of gene malfunction resulting in abnormal levels of activated ras p21 protein, in breast cancer mutations are rarely observed and loss of heterozygosity is far more frequently observed.[59,60] Loss of heterozygosity has been associated with progression and adverse outcome in breast cancer.[61,62] Immunohistochemistry-based studies of p21 protein overexpression have been less consistent, with publications of both noncorrelating studies[63] and reports of favorable prognostic impact.[64]

Other Breast Cancer Oncogenes

Measurements of the c-*fos* and c-*jun* regulators of the activating protein-1 complex and c-*myb* have successfully predicted breast cancer recurrence, response to hormonal therapy, and survival.[65–69]

Mouse Models

Murine breast cancer models are increasingly important in the elucidation of oncogenic networks. Such models are particularly important when the issue under question requires analysis of complex developmental and cancer biology relationships. In addition, murine breast cancer models serve as a starting point for identifying and tailoring novel therapeutic avenues.

The advantages in using mice for breast cancer models include their close evolutionary and developmental relationships with humans, their rapid breeding, and their ability to withstand the genetic perturbations that are required in various breast cancer models. There are essentially two types of engineered mice: transgenic and knockout. In the former, one or more genes are introduced into fertilized mice eggs and the phenotypic consequences are observed (e.g., predisposition to tumor development). The genes can be altered forms of known genes, such as oncogenes harboring specific activating mutations and/or driven to high levels of expression by juxtaposition with an active promoter region. A classic example of the use of transgenic mice in understanding breast cancer biology came from the studies performed by Leder and colleagues in the 1980s.[12] These authors observed that the introduction of a strongly expressed human *MYC* gene was followed by development of breast carcinoma in mice.[12] Subsequently, other transgenic mice have been shown to have increased predisposition for breast cancer,[70–72] including those expressing tumor growth factor (*TGF*), HER-2/*neu*, *PRAD1* (cyclin D1), *RAS*, and others. In knockout mice, target gene or genes are deleted and the phenotypic consequences are observed. Knockout mice models have been particularly useful in the biologic characterization of tumor suppressor genes (see Chapter 23).

Oncogenic Pathways as Therapeutic Targets

As outlined above, the constitutive activation of one or more oncogenic pathways (FIG. 22.2) is probably characteristic of all human breast cancers. Given the therapeutic opportunities for oncogene inhibition, it is essential to develop efficient predictive assays that can identify the clinically relevant oncogenic pathways in patients with breast cancer. In the coming years, it will be particularly important to screen for the constitutively activated targets—as are found in subsets of human breast cancer—that can be inhibited to clinical advantage by the burgeoning array of targeted therapies. The available evidence suggests that both the observation of oncogene activation and the specific mechanism(s) of activation will be important to clinical decision making. For example, oncogene activation can result from intrinsic mutations that alter the structure of the associated oncoprotein or from overexpression of the normal protein resulting from genomic amplification, transcriptional up-regulation, or impaired proteolysis. The success of a given targeted therapy option can depend on the activation mechanisms of the drug target and also on the interactions between the drug target and other oncogenic pathways in the cell (see Chapter 14). Well-known breast cancer targeted therapies include the use of partial estrogen receptor agonists (e.g., tamoxifen) in estrogen receptor–positive tumors (see Chapter 15) and the humanized monoclonal antibody trastuzumab (Herceptin®) in HER-2/*neu* overexpressing tumors.[73] Even in these seemingly straightforward examples, the mechanisms of target inhibition can be surprisingly complex. In the case of trastuzumab, the drug does not simply "block" the HER-2/ neu target protein but rather triggers tumor cell destruction in conjunction with antibody-dependent host cell cytotoxicity mechanisms.[73] In addition, trastuzumab may restore basal levels of HER-2/*neu* expression by increasing the internalization and processing of the overexpressed receptor.[73]

Given the diversity of oncogenic pathways in breast cancer, it can be assumed that different combinations of targeted therapies will ultimately be tailored to each patient, just as has been the case in targeting trastuzumab to that subset of patients whose breast cancers betray the biologic evidence of HER-2/*neu* dependence. Tyrosine kinase inhibitors have been prominent among the newer generations of targeted therapies, and there is undoubtedly reason for optimism, given that many cancers feature oncogenic kinase signaling. Agents that target EGFR,[74] downstream effectors of the PI-3K/AKT pathway,[74] or critical cell cycle regulators, such as the cyclin D1 inhibitor flavopiridol, are some of the new promising therapeutic possibilities for treating breast cancer.[75,76]

References

1. Bishop JM. The discovery of proto-oncogenes. FASEB J 1996;10:362–364.
2. Vecchio G. Oncogenes of DNA and RNA tumor viruses and the origin of cellular oncogenes. Hist Philos Life Sci 1993;15:59–74.
3. Stehelin D, Varmus HE, Bishop JM, et al. DNA related to the transforming gene(s) of avian sarcoma viruses is present in normal avian DNA. Nature 1976;260:170–173.
4. Irby RB, Yeatman TJ. Role of Src expression and activation in human cancer. Oncogene 2000;19:5636–5642.
5. Graham FL, van der Eb AJ. A new technique for the assay of infectivity of human adenovirus 5 DNA. Virology 1973;52:456–467.
6. Parada LF, Tabin CJ, Shih C, et al. Human EJ bladder carcinoma oncogene is homologue of Harvey sarcoma virus ras gene. Nature 1982;297:474–478.
7. de Klein A, van Kessel AG, Grosveld G, et al. A cellular oncogene is translocated to the Philadelphia chromosome in chronic myelocytic leukaemia. Nature 1982;300:765–767.
8. Taub R, Kirsch I, Morton C, et al. Translocation of the c-myc gene into the immunoglobulin heavy chain locus in human Burkitt lymphoma and murine plasmacytoma cells. Proc Natl Acad Sci USA 1982;79: 7837–7841.
9. Schwab M. Oncogene amplification in solid tumors. Semin Cancer Biol 1999;9:319–325.
10. Bonilla M, Ramirez M, Lopez-Cueto J, et al. In vivo amplification and rearrangement of c-myc oncogene in human breast tumors. J Natl Cancer Inst 1988;80: 665–671.
11. Liao DJ, Dickson RB. c-Myc in breast cancer. Endocr Relat Cancer 2000;7:143–164.
12. Stewart TA, Pattengale PK, Leder P. Spontaneous mammary adenocarcinomas in transgenic mice that carry and express MTV/myc fusion genes. Cell 1984; 38:627–637.
13. Miller WR. Regulatory subunits of PKA and breast cancer. Ann N Y Acad Sci 2002;968:37–48.
14. Clark GJ, Der CJ. Aberrant function of the Ras signal transduction pathway in human breast cancer. Breast Cancer Res Treat 1995;35:133–144.
15. Gotoh T, Cai D, Tian X, et al. p130Cas regulates the activity of AND-34, a novel Ral, Rap1, and R-Ras guanine nucleotide exchange factor. J Biol Chem 2000;275:30118–30123.
16. Weinstat-Saslow D, Merino MJ, Manrow RE, et al. Overexpression of cyclin D mRNA distinguishes invasive and in situ breast carcinomas from nonmalignant lesions. Nat Med 1995;1:1257–1260.
17. Oyama T, Kashiwabara K, Yoshimoto K, et al. Frequent overexpression of the cyclin D1 oncogene in invasive lobular carcinoma of the breast. Cancer Res 1998; 58:2876–2880.
18. Cleary ML, Smith SD, Sklar J. Cloning and structural analysis of cDNAs for bcl-2 and a hybrid bcl-2/immunoglobulin transcript resulting from the t(14;18) translocation. Cell 1986;47:19–28.
19. Cleary ML, Sklar J. Nucleotide sequence of a t(14;18) chromosomal breakpoint in follicular lymphoma and demonstration of a breakpoint-cluster region near a transcriptionally active locus on chromosome 18. Proc Natl Acad Sci USA 1985;82: 7439–7443.
20. Tsujimoto Y, Gorham J, Cossman J, et al. The t(14;18) chromosome translocations involved in B-cell neoplasms result from mistakes in VDJ joining. Science 1985;229:1390–1393.
21. Silvestrini R, Veneroni S, Daidone MG, et al. The Bcl-2 protein: a prognostic indicator strongly related to p53 protein in lymph node-negative breast cancer patients. J Natl Cancer Inst 1994;86:499–504.
22. Bhargava V, Kell DL, van de Rijn M, Warnke RA. Bcl-2 immunoreactivity in breast carcinoma correlates with hormone receptor positivity. Am J Pathol 1994;145:535–540.
23. Joensuu H, Pylkkanen L, Toikkanen S. Bcl-2 protein expression and long-term survival in breast cancer. Am J Pathol 1994;145:1191–1198.
24. Downward J. Targeting RAS signalling pathways in cancer therapy. Nat Rev Cancer 2003;3:11–22.
25. Adelaide J, Huang HE, Murati A, et al. A recurrent chromosome translocation breakpoint in breast and pancreatic cancer cell lines targets the neuregulin/NRG1 gene. Genes Chromosomes Cancer 2003;37: 333–345.
26. Tognon C, Knezevich SR, Huntsman D, et al. Expression of the ETV6-NTRK3 gene fusion as a primary event in human secretory breast carcinoma. Cancer Cell 2002;2:367–376.
27. Euhus DM, Timmons CF, Tomlinson GE. ETV6-NTRK3-Trk-ing the primary event in human secretory breast cancer. Cancer Cell 2002;2:347–348.
28. Stern DF. Tyrosine kinase signalling in breast cancer: ErbB family receptor tyrosine kinases. Breast Cancer Res 2000;2:176–183.
29. Salomon DS, Brandt R, Ciardiello F, et al. Epidermal growth factor-related peptides and their receptors in human malignancies. Crit Rev Oncol Hematol 1995; 19:183–232.
30. Fischer OM, Hart S, Gschwind A, et al. EGFR signal transactivation in cancer cells. Biochem Soc Trans 2003;31:1203–1208.
31. Nicholson KM, Streuli CH, Anderson NG. Autocrine signalling through erbB receptors promotes constitutive activation of protein kinase B/Akt in breast cancer cell lines. Breast Cancer Res Treat 2003;81: 117–128.
32. Bacus SS, Altomare DA, Lyass L, et al. AKT2 is frequently upregulated in HER-2/neu-positive breast cancers and may contribute to tumor aggressiveness by enhancing cell survival. Oncogene 2002;21:3532–3540.
33. Zhou BP, Liao Y, Xia W, et al. Cytoplasmic localization of p21Cip1/WAF1 by Akt-induced phosphorylation in HER-2/neu-overexpressing cells. Nat Cell Biol 2001;3:245–252.
34. Perez-Tenorio G, Stal O. Activation of AKT/PKB in breast cancer predicts a worse outcome among endocrine treated patients. Br J Cancer 2002;86: 540–545.
35. Holbro T, Beerli RR, Maurer F, et al. The ErbB2/ErbB3 heterodimer functions as an oncogenic unit: ErbB2 requires ErbB3 to drive breast tumor cell proliferation. Proc Natl Acad Sci USA 2003;100:8933–8938.
36. Blume-Jensen P, Hunter T. Oncogenic kinase signalling. Nature 2001;411:355–365.

37. Sutherland RL, Musgrove EA. Cyclins and breast cancer. J Mamm Gland Biol Neoplasia 2004;9:95–104.
38. Ormandy CJ, Musgrove EA, Hui R, et al. Cyclin D1, EMS1 and 11q13 amplification in breast cancer. Breast Cancer Res Treat 2003;78:323–335.
39. Rennstam K, Baldetorp B, Kytola S, et al. Chromosomal rearrangements and oncogene amplification precede aneuploidization in the genetic evolution of breast cancer. Cancer Res 2001;61:1214–1219.
40. Lenferink AE, Busse D, Flanagan WM, et al. ErbB2/neu kinase modulates cellular p27(Kip1) and cyclin D1 through multiple signaling pathways. Cancer Res 2001;61:6583–6591.
41. Yu Q, Geng Y, Sicinski P. Specific protection against breast cancers by cyclin D1 ablation. Nature 2001; 411:1017–1021.
42. Medema RH, Kops GJ, Bos JL, et al. AFX-like Forkhead transcription factors mediate cell-cycle regulation by Ras and PKB through p27kip1. Nature 2000; 404:782–787.
43. Porter PL, Malone KE, Heagerty PJ, et al. Expression of cell-cycle regulators p27Kip1 and cyclin E, alone and in combination, correlate with survival in young breast cancer patients. Nat Med 1997;3: 222–225.
44. Tan P, Cady B, Wanner M, et al. The cell cycle inhibitor p27 is an independent prognostic marker in small (T1a,b) invasive breast carcinomas. Cancer Res 1997;57:1259–1263.
45. Hulit J, Lee RJ, Russell RG, et al. ErbB-2-induced mammary tumor growth: the role of cyclin D1 and p27Kip1. Biochem Pharmacol 2002;64:827–836.
46. Schorr K, Li M, Krajewski S, et al. Bcl-2 gene family and related proteins in mammary gland involution and breast cancer. J Mamm Gland Biol Neoplasia 1999;4:153–164.
47. Reed JC. Balancing cell life and death: bax, apoptosis, and breast cancer. J Clin Invest 1996;97:2403–2404.
48. Krajewski S, Blomqvist C, Franssila K, et al. Reduced expression of proapoptotic gene BAX is associated with poor response rates to combination chemotherapy and shorter survival in women with metastatic breast adenocarcinoma. Cancer Res 1995; 55:4471–4478.
49. Jager R, Herzer U, Schenkel J, et al. Overexpression of Bcl-2 inhibits alveolar cell apoptosis during involution and accelerates c-myc-induced tumorigenesis of the mammary gland in transgenic mice. Oncogene 1997;15:1787–1795.
50. Sierra A, Castellsague X, Escobedo A, et al. Synergistic cooperation between c-Myc and Bcl-2 in lymph node progression of T1 human breast carcinomas. Breast Cancer Res Treat 1999;54:39–45.
51. Sierra A, Castellsague X, Escobedo A, et al. Bcl-2 with loss of apoptosis allows accumulation of genetic alterations: a pathway to metastatic progression in human breast cancer. Int J Cancer 2000;89: 142–147.
52. Aulmann S, Bentz M, Sinn HP. C-myc oncogene amplification in ductal carcinoma in situ of the breast. Breast Cancer Res Treat 2002;74:25–31.
53. Liao DJ, Dickson RB. c-Myc in breast cancer. Endocr Relat Cancer 2000;7:143–164.
54. Deming SL, Nass SJ, Dickson RB, et al. C-myc amplification in breast cancer: a meta-analysis of its occurrence and prognostic relevance. Br J Cancer 2000;83:1688–1695.
55. Mizukami Y, Nonomura A, Takizawa T, et al. N-myc protein expression in human breast carcinoma: prognostic implications. Anticancer Res 1995;15:2899–2905.
56. Pietilainen T, Lipponen P, Aaltomaa S, et al. Expression of c-myc proteins in breast cancer as related to established prognostic factors and survival. Anticancer Res 1995;15:959–964.
57. Watson PH, Safneck JR, Le K, et al. Relationship of c-myc amplification to progression of breast cancer from in-situ to invasive tumor and lymph node metastasis. J Nat Cancer Inst 1993;85:902–907.
58. Hynes NE, Lane HA. Myc and mammary cancer: Myc is a downstream effector of the ErbB2 receptor tyrosine kinase. J Mamm Gland Biol Neoplasia 2001;6: 141–150.
59. Rochlitz CF, Scott GK, Dodson JM, et al. Incidence of activating ras oncogene mutations associated with primary and metastatic human breast cancer. Cancer Res 1989;49:357–360.
60. Schondorf T, Andrack A, Niederacher D, et al. H-ras gene amplification or mutation is not common in human primary breast cancer. Oncol Rep 1999;6: 1029–1033.
61. Cline MJ, Battifora H, Yokota J, et al. Proto-oncogene abnormalities in breast cancer: Correlations with anatomic features and clinical course of disease. J Clin Oncol 1987;5:999–1006.
62. Fromowitz FB, Viola MV, Chao S, et al. ras p21 expression in the progression of breast cancer. Hum Pathol 1987;18:1268–1275.
63. Mizukami Y, Nonomura A, Noguchi M, et al. Immunohistochemical study of oncogene product ras p21, c-myc and growth factor EGF in breast carcinomas. Anticancer Res 1991;11:1485–1494.
64. Gohring UJ, Schondorf T, Kiecker VR, et al. Immunohistochemical detection of H-ras protooncoprotein p21 indicates favorable prognosis in node-negative breast cancer patients. Tumour Biol 1999; 20:173–183.
65. Bland KI, Konstadoulakis MM, Vezeridis MP, et al. Oncogene protein coexpression. Value of HAras, c-myc, c-fos, and p53 as prognostic discriminants for breast carcinoma. Ann Surg 1995;221:706–720.
66. Guerin M, Sheng ZM, Andrieu N, et al. Strong association between c-myb and oestrogen-receptor expression in human breast cancer. Oncogene 1990;5: 131–135.
67. Gee JM, Ellis IO, Robertson JF, et al. Immunocytochemical localization of Fos protein in human breast cancers and its relationship to a series of prognostic markers and response to endocrine therapy. Int J Cancer 1995;64:269–273.
68. Guerin M, Sheng ZM, Andrieu N, et al. Strong association between c-myb and oestrogen-receptor expression in human breast cancer. Oncogene 1990;5: 131–135.
69. Gee JM, Barroso AF, Ellis IO, et al. Biological and clinical associations of c-jun activation in human breast cancer. Int J Cancer 2000;89:177–186.
70. Hennighausen L. Mouse models for breast cancer. Breast Cancer Res 2000;2:2–7.
71. Cardiff RD, Anver MR, Gusterson BA, et al. The mammary pathology of genetically engineered mice: the consensus report and recommendations

from the Annapolis meeting. Oncogene 2000;19: 968–988.

72. Hennighausen L. Mouse models for breast cancer. Oncogene 2000;19:966–967.

73. Ross JS, Fletcher JA, Linette GP, et al. The Her-2/neu gene and protein in breast cancer 2003: biomarker and target of therapy. Oncologist 2003;8: 307–325.

74. Averbuch S, Kcenler M, Morris C, et al. Therapeutic potential of tyrosine kinase inhibitors in breast cancer. Cancer Invest 2003;21:782–791.

75. Wu K, Wang C, D'Amico M, et al. Flavopiridol and trastuzumab synergistically inhibit proliferation of breast cancer cells: association with selective cooperative inhibition of cyclin D1-dependent kinase and Akt signaling pathways. Mol Cancer Ther 2002;1: 695–706.

76. Nahta R, Trent S, Yang C, et al. Epidermal growth factor receptor expression is a candidate target of the synergistic combination of trastuzumab and flavopiridol in breast cancer. Cancer Res 2003;63: 3626–3631.

CHAPTER 23

p53 and Tumor Suppressor Genes in Breast Cancer

Andre M. Oliveira, MD, Jeffrey S. Ross, MD and Jonathan A. Fletcher, MD

Department of Pathology, Brigham and Women's Hospital, Boston, Massachusetts

History and Definition

Tumor suppressors are proteins whose normal function is inhibitory to the transformed state and whose inactivation is therefore advantageous for tumor cell growth and survival. In reality, it can be difficult to categorize some proteins as "tumor suppressor" versus "oncoprotein" because many proteins display varied effects on cell behavior, depending on the level of protein expression, the cell context in which the protein is expressed, and the exact structure of the expressed protein. In addition, some proteins have domains that inhibit cell growth and survival and others that promote those same properties. At the most basic level, tumor suppressors can be defined as the group of proteins that generally suppress tumor survival and proliferation in cell culture or in vivo models.

The concept of tumor suppressors was first validated by experiments in somatic cell hybrids, showing that fusions between cancer cells and normal cells resulted in loss of the malignant phenotype.[1–3] Notably, the somatic cell hybrids with loss of the malignant phenotype could be maintained in cell culture and would occasionally reacquire the malignant phenotype, coincident with deletion of certain chromosome regions from the hybrid cells.[4,5] These critical chromosome regions were subsequently shown to contain tumor suppressors whose inactivation is responsible for the malignant phenotype.

A seminal advance in the identification and understanding of tumor suppressor genes came from studies of the pediatric tumor, retinoblastoma. The retinoblastoma model was important in the development of Knudson's "two hit" hypothesis of tumor suppression, in which the sequential inactivation of two gene alleles might result in the development of retinoblastoma (FIG. 23.1).[6] In the case of retinoblastoma, and subsequently validated in familial breast cancer predisposition by *RB1*, p53, BRCA1, and BRCA2, one inactive allele is inherited—and is present in that inactive form in all the person's cells—whereas the remaining allele is inactivated somatically in the tumor cells to fully abrogate the tumor suppressor function. This hypothesis was also validated in nonfamilial retinoblastomas, where both *RB1* alleles are inactivated somatically and only within the tumor cell population.[7] These studies also highlighted that in both sporadic and familial tumors, the first "hit" is often a point mutation in one of the tumor suppressor gene alleles, whereas the second hit is often caused by loss of the homologous chromosome or chromosome locus, known as loss of heterozygosity.[8] However, as discussed below, the mechanisms of tumor suppressor inactivation are myriad and can include loss of expression by methylation mediated transcriptional silencing or by increased proteolysis.

Tumor suppressor genes, in breast cancer and in other tumors, have been discovered by various methods. In the case of familial breast cancer genes, causal tumor suppressor genes have been identified by linkage analysis in which inheritance of the cancer predisposition is localized to a particular chromosomal region. The *p53, BRCA1,* and *BRCA2* breast cancer tumor suppressors were identified in this manner. Loss of heterozygosity studies are based on the identification of losses of DNA markers in certain chromosomal regions in the cancer cells that are preserved in normal cells. By determining a possible critical gene or genes that are deleted in this area, a putative tumor suppressor gene may be identified. Comparative genomic hybridization is another technique used to identify regions of chromosome gains or losses, therefore opening possibilities for the identification of oncogenes (e.g., chromosomal gains might represent gene amplification, see Chapter 22) and tumor suppressor genes.

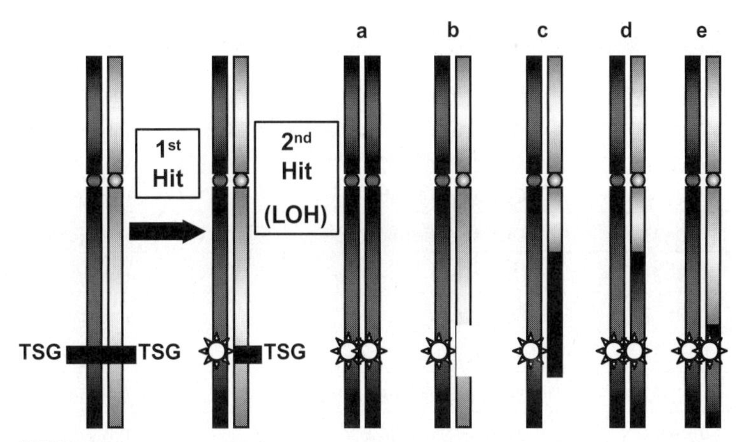

FIGURE 23.1 Knudson's model for the inactivation of a tumour suppressor gene (TSG). The first hit is usually a mutation in the DNA sequence of the gene (a small deletion or base substitution; star). This mutation can be transmitted through the germline, giving rise to an inherited form of cancer. The second hit (loss of heterozygosity, LOH) is often a gross chromosomal mechanism that occurs at higher rates in somatic cells and that leads to hemi- or homozygosity of the chromosome region containing the mutation. This includes nondisjunctional loss with reduplication of the chromosome carrying the mutated TSG (a), subchromosomal deletion (b), unbalanced translocation (c), and mitotic recombination (d and e). (From Reference 6.)

Mechanisms of Tumor Suppressor Inactivation

Tumor suppressor inactivation is accomplished by mechanisms that vary from tumor to tumor and depending on which gene is involved. Intragenic mutations can abrogate selected aspects of the tumor suppressor function, as in the case of missense mutations involving critical domains, or can abrogate the entire protein function, as in the case of nonsense mutations that prevent synthesis of a full-length protein product. Intragenic mutations can involve the tumor suppressor selectively or, as the case of microsatellite instability associated with mismatch repair defects, can be a manifestation of genome-wide genetic instability. Chromosomal deletions typically result in physical loss, and hence complete inactivation, of a tumor suppressor allele. Various mechanisms accomplish down-regulation of tumor suppressor expression by perturbations involving the tumor suppressor promoter region. Such mechanisms can include inappropriate expression of transcriptional repressors that inhibit tumor suppressor expression, loss of transcriptional activators that normally maintain constitutive tumor suppressor expression, or silencing of tumor suppressors by promoter methylation. Tumor suppressors are also inactivated via interactions at the protein level. One example is the inhibition of p53 function by ubiquitination, resulting in increased p53 proteolysis; another example is the deregulation of tumor suppressor proteins by perturbation of critical phosphorylation sites. The complexity of these inactivation mechanisms underscores the fact that tumor suppressors can be inactivated, partially or totally, by myriad mechanisms. Therefore, the classic two-hit model of tumor suppressor inactivation, in which an inactivating mutation is followed by physical deletion of the remaining allele, does not necessarily apply to most tumor suppressor events in breast cancer.

Functional Classes of Tumor Suppressor Genes

Tumor suppressor genes participate in a variety of critical and highly conserved cell functions: regulation of the cell cycle and apoptosis, differentiation, surveillance of genomic integrity and repair of DNA errors, signal transduction, and cell adhesion. Conceptually, tumor suppressor functions can be separated into two major categories, depending on whether the functions are viewed as having *gatekeeper* or *caretaker* roles.[9] Gatekeeper tumor suppressors are those that directly inhibit tumor growth or promote tumor death. Inactivation of these genes contributes directly to cancer formation and progression. Gatekeeper tumor suppressors often show biallelic inactivation, as in the examples of Rb and p53 aberrations in retinoblastoma and Li-Fraumeni syndromes, respectively, where one allele is inactivated in the germline and the second is somatically inactivated in the tumor

Table 23.1 Tumor Suppression Genes in Breast Cancer

Gene	Location	Function	Mechanism
TP53 (p53)	17p13	DNA repair, cell cycle, apoptosis, and angiogenesis regulator	Intragenic mutation, deletion
RB1	13q14	Cell cycle inhibitor	Intragenic mutation, deletion
PTEN	10q23	Dual specific phosphatase	Intragenic mutation, deletion
BRCA1	17q21	DNA repair	Intragenic mutation, deletion
BRCA2	13q12	DNA repair	Intragenic mutation, deletion
ATM	11q22	DNA repair	Intragenic mutation, deletion
STK11 (LKB1)	19p13	Serine-theronine kinase	Deletion
CDKN1B (p27kip1)	12p13	Cell cycle inhibitor	Proteolytic degradation, relocalization
CDKN2A (p16INK4)	9p21	Cell cycle inhibitor	Deletion, methylation
SERPINR5 (Maspin)	18q21	Serine protease inhibitor	Decreased expression
IGFII-R	6q26	Growth factor receptor	Deletion
CDH1 (E-cadherin)	16q22	Cell adhesion molecule	Methylation
RARbeta2	3p24	Retinoic acid receptor	Intragenic mutation, deletion, methylation
MLH1	3p21	Mismatch repair	Methylation
MSH2	2p22	Mismatch repair	Inactivating mutation
APC	5q21	Inhibitor of b-catenin transcription	Deletion, methylation

cells. Caretaker tumor suppressors are genes whose loss of function is not directly responsible for tumor development. Rather, their inactivation results in increased occurrence of genomic mutations, which in turn inactivate gatekeeper tumor suppressors as well as activate breast cancer oncogenes. Well-studied examples of caretakers include the DNA repair genes *MLH1* and *MSH2*, whose inactivation occurs in the hereditary nonpolyposis colorectal cancer syndrome and in the breast cancer predisposing syndrome, Muir-Torre. Although it is useful to conceptualize tumor suppressors as gatekeepers and caretakers, these distinctions should not be taken to an extreme. Some tumor suppressors (e.g., BRCA1 and BRCA2) belong to both camps, and the functions of even the best-studied tumor suppressors are still only partly understood (**Table 23.1**).

p53

Studies of SV40-transformed cells led to the discovery of the p53 protein in the late 1970s.[8,10,11] Initially regarded as an oncogene because of its transforming effects in NIH3T3 mouse fibroblasts, *p53* was later reclassified as a tumor suppressor gene when it was determined that only defective mutant forms of the gene had transforming properties.[12,13]

The *p53* gene is located on chromosome band 17p13 and encodes a 53-kDa multifunctional transcription factor that regulates the expression of genes involved in cell cycle control, apoptosis, DNA repair, and angiogenesis.[14] p53 activity can be enhanced by various mechanisms,[13] such as the activation of protein kinases ATM and CHK2 after DNA damage (FIG. 23.2) induced by ionizing radiation. Other mechanisms for increased p53 activity include ATM/CHK2 independent pathways activated by stress signals—such as ultraviolet radiation, cytotoxic drugs, and protein kinase inhibitors—and the upstream effects of the p14^{ARF} cell cycle regulator (FIG. 23.2). One of the widely studied p53 functions involves control of the G1/S cell cycle transition. p53 enhances transcription of the cell cycle inhibitor p21$^{WAF/CIP}$ and hence exerts a negative effect on the cyclin dependent kinase complex 4/6 (CDK 4/6). Inhibition of CDK 4/6 then blocks cell cycle progression by activating the *RB1* tumor suppressor and by inhibiting expression of cyclin D1, an oncogene that is commonly amplified in breast cancer (see

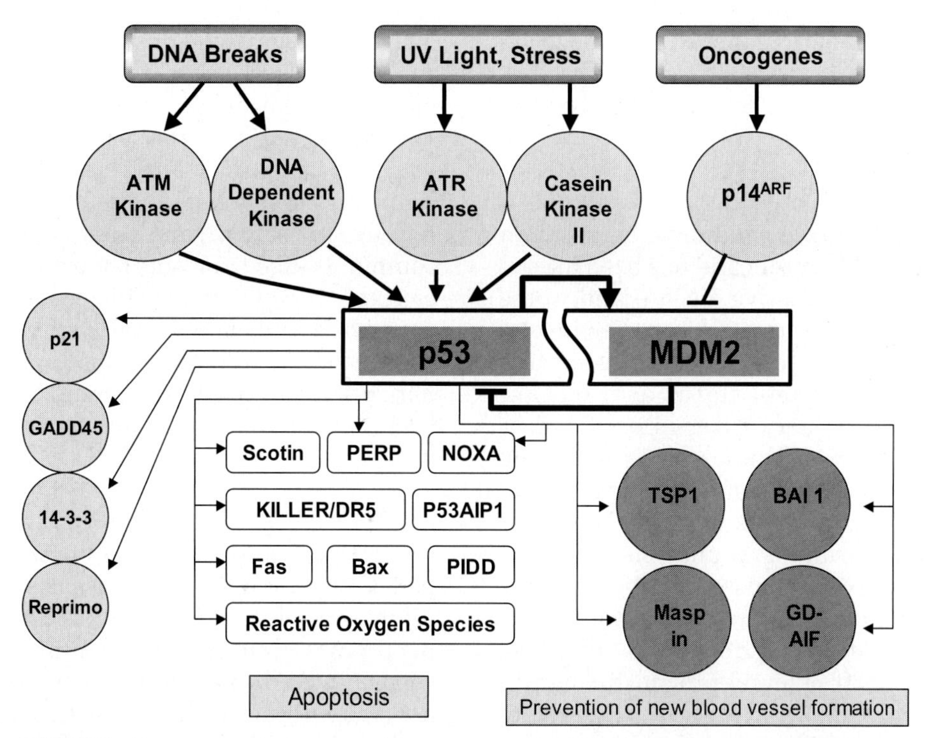

FIGURE 23.2 The p53 network. Activation of the network can result from DNA damage, ultraviolet light, or oncogene interactions that modify p53 and its negative regulator, MDM2. Activated p53 then activates the expression of target genes by binding to their regulatory regions. These genes—which include cell cycle inhibitors and inducers of apoptosis—slow down the development of tumors. In addition, p53 activation results in a variety of other effects, including maintenance of genetic stability and induction of cellular differentiation. (From Reference 14.)

Chapters 13 and 22). p53 can also regulate G2/M transition, which is accomplished by inhibiting Cdc2 and stimulating the transcription of cell cycle inhibitory genes *reprimo, B99,* and *mcg10*.[15] Additional cell cycle inhibitors that are stimulated by p53 include GADD45 and 14-3-3α.[15]

p53 also promotes apoptosis by activating the expression of several proapoptotic proteins, including BAX, APAF1, PUMA, p53AIP1, NOXA, and PIDD.[16] Likewise, DNA repair mechanisms can be regulated by p53, as in the p53-mediated transcriptional activation of the DNA repair genes *GADD45* and *p53R2*. p53 regulates expression of other genes relevant in breast tumorigenesis, including the mammary serine protease inhibitor maspin,[17] which has antiangiogenic, antimetastatic, and antiinvasiveness properties.

The multifunctional nature of p53 places this transcription factor at the epicenter of critical regulatory networks controlling cell fate after stress signals (**FIG. 23.2**). In this capacity, p53 prevents cell cycle progression and facilitates DNA repair but will also include apoptosis when the damage is irreparable.

p53 is deleted or mutated in up to 50% of all cancers, and it is estimated that between 15% and 71% of sporadic breast cancers harbor *p53* mutations.[18] Among the histologic subtypes of breast cancer, the highest frequencies of *p53* mutations are found in medullary breast carcinomas and lowest frequencies are observed in papillary and mucinous carcinoma.[16,19] *p53* mutations are more common in ductal carcinomas than in lobular carcinomas, and they occur in approximately 40% of high grade ductal carcinomas in situ and in 5% of intermediate grade ductal carcinomas in situ.[19] p53 mutations are rare or nonexistent in low grade ductal carcinomas in situ[20] and are not found in ductal hyperplasias.[21] *p53* mutations are more common in breast cancers associated with *BRCA1* and *BRCA2* germline mutations (see Chapters 4 and 25). Interestingly, the distribution of these mutations within the *p53* gene differs from nonsyndromic breast cancers.[22–25] This observation suggests that the genomic instability caused by BRCA mutations may influence the type and distribution of *p53* mutations in breast cancer.[23] Germline *p53* mutations are found in most patients with Li-Fraumeni syndrome (see

Chapter 4), a genetic cancer predisposition syndrome characterized by increased risk of sarcomas, brain tumors, leukemias, and breast carcinomas.[26,27]

p53 and Breast Cancer Prognosis

The *p53* mutation rate is lower in breast than in other epithelial cancers and has been associated with more aggressive disease and worse overall survival.[15] A series of 29 studies involving 9793 patients testing the association of *p53* gene and protein status as a marker of prognosis in breast cancer is provided in **Table 23.2**.[28–56] Seven studies (24%) including 1910 patients (20%) found no association of *p53* status with prognosis. Six studies (21%) that featured multivariate data analysis including 2822 patients (30%) found a significant prognostic impact on univariate but not multivariate analyses. Fifteen studies (54%) involving 4814 patients (50%) found prognostic significance on both univariate and multivariate analyses. Immunohistochemistry (IHC) evaluations with and without image analysis assisted slide scoring using either the DO-1 or the PAb1801 antibodies yielded variable associations of p53 stabilized mutant protein nuclear staining with outcome in breast cancer. Of the seven negative outcome studies in **Table 23.2**, 6 (86%) used an IHC approach for determining p53 status. The high number of IHC false-positive and false-negative results (compared with gene sequencing) precludes reliable use of IHC as an indicator for *p53* gene mutation in human breast cancer.[48] A recent meta-analyses confirmed these observations.[57]

The prognostic significance of *p53* status in breast cancer has been impacted by the accuracy of IHC (**FIG. 23.3**) versus molecular methods, such as single strand conformation polymorphism detection, polymerase chain reaction amplification followed by direct sequencing (polymerase chain reaction) with direct sequencing, microarray based testing using a p53 DNA sequencing chip (microarray), the ligase chain reaction testing, and the yeast colony functional assay (**Table 23.3**).[15] As mentioned above, the IHC method is generally unreliable even when the preferred DO-1 antibody is used[58] and molecular detection methods are favored by most investigators for both pre-clinical and clinical research studies.[59,60]

Site-specific mutations in the *p53* gene have been linked to differing impact on prognosis in breast cancer.[60] For example, there is a strong association between mutations in the L2/L3 loop and shorter survival or poor response to treatment.[60]

In addition to its association with overall tumor progression,[61] the acquisition of *p53* mutations is associated with resistance to doxorubicin, paclitaxel, and other cytotoxic agents commonly used for the treatment of breast cancer.[62–64] *p53* polymorphisms have also been associated with phenotypic features of breast carcinoma.[64] Interestingly, *p53* mutations are not restricted to the epithelial cells and have been detected in the breast cancer stroma as well.[65]

p53 and *MDM2*

The *MDM2* gene encodes a protein that binds to p53, thereby reducing its cell cycle progression inhibitory role.[66] *MDM2* amplification appears to be a rare event in breast cancer.[67,68] Multiple reports have linked mdm2 overexpression to adverse outcome in patients with node-negative and node-positive breast cancer[69–71]; however, one report found no correlation.[72]

p53 Cooperative Networks: ATM and CHK2

Various tumor suppressor and oncogene pathways interface with p53, and these pathways can be perturbed as an alternate to p53 mutation in breast cancer. Ataxia-telangiectasia provides one example in which p53 mechanisms are abrogated indirectly. Ataxia-telangiectasia is an autosomal recessive cancer syndrome caused by inactivating mutations of the *ATM* gene and characterized by cerebellar ataxia, oculocutaneous telangiectasias, immunodeficiency, and increased risk for the development of hematologic malignancies and breast carcinomas.[73,74] The ATM protein participates in DNA repair processes by phosphorylating and activating p53.[75–78] Patients heterozygous for *ATM* mutations may be at increased risk for the development of breast carcinoma.[79] In addition, loss of heterozygosity at the *ATM* gene (chromosome 11q22-23) region is common in sporadic breast carcinomas.[80–82] The CHK2 protein is another tumor suppressor that phosphorylates and activates p53.[14,83] *CHK2* gene mutations are found in Li-Fraumeni syndrome individuals who lack germline *p53* mutations[84] and therefore serve as alternate mechanisms of constitutive p53 inactivation in this syndrome. However, *CHK2* mutations are infrequent in other familial breast cancer syndromes and in sporadic breast cancers.[85]

Table 23.2 Summary of Selected Studies on the Correlation of the *p53* Tumor Suppressor Gene with Prognosis in Breast Cancer

Study Year	First Author/ Reference	No. of Cases	Specimen Type	Method(s) of Analysis	Univariate Significance	Multivariate Significance	Comment
1992	Thor (28)	304	Paraffin	IHC	Yes	Yes	p53 status was an independent predictor more in sporadic than familial breast cancer
1994	Cunningham (29)	247	Paraffin	IHC	Yes	—	DO-1 antibody 16% positive; significant prognosis only for DNA diploid tumors
1994	Caleffi (30)	192	Frozen	Direct sequencing	No	No	Associates with ER status; 22% mutations
1994	Marks (31)	230	Paraffin	IHC	Yes	Yes	PAb1801 antibody
1995	Beck (32)	462	Paraffin	IHC	Yes	No	Correlates with ER, proliferation (MIB-1), grade, DO-1 antibody
1995	Borg (33)	205	Cytosol	Immunoluminometric assay	Yes	Yes	30% cutoff for protein over-expression; associates with ER-negative, high proliferation rate
1995	Bland (34)	85	Paraffin	IHC	No	No	PAb1801 antibody; RAS/FOS status did not correlate; small sample
1995	Peyrat (35)	353	Plasma	ELISA	Yes	Yes	Associates with ER negative, significant survival differences at 5 years; 12% positive rate
1996	Katoh (36)	125	Paraffin	IHC	No	No	p53 status correlated with ER/PR, but did not predict prognosis
1998	Clahsen (38)	441	Paraffin	IHC	No	No	p53 did not predict prognosis; ER, Ki-67 status and patient age were significant predictors; p53 status did predict response to adjuvant therapy
1998	Rozan (39)	329	Paraffin	IHC	No	No	Neither p53 nor HER-2/neu predicted response to chemotherapy or radiotherapy
1998	Peyrat (40)	634	Fresh	ELISA	Yes	Yes	Only tumor grade was independent; p53 and uPA status were not
1998	Levesque (41)	998	Fresh	ELISA	Yes	Yes	p53 levels independently predicted relapse and disease-related death
1998	Jansen (42)	345	Paraffin	IHC	Yes	Yes	p53 prognostic significance dependent on the bcl-2 expression status
1999	Harbeck (43)	125	Paraffin	IHC	No	No	PAI-1 levels were independent predictors in node-negative patients

(continued)

Table 23.2 Summary of Selected Studies on the Correlation of the *p53* Tumor Suppressor Gene with Prognosis in Breast Cancer (*continued*)

Study Year	First Author/ Reference	No. of Cases	Specimen Type	Method(s) of Analysis	Univariate Significance	Multivariate Significance	Comment
1999	Broet (44)	1245	Fresh	Immunoluminometric assay	Yes	No	p53 status not independent predictor when uPA status was known
2000	Soong (37)	375	Fresh	SSCP	Yes	Yes	p53 mutation predicted poor survival independent of lymph node status
2000	Reed (45)	613	Paraffin	IHC	No	No	Only tumor size and grade were significant prognostic factors
2000	Berns (46)	243	Fresh	Direct sequencing	Yes	Yes	p53 mutation independently predicted outcome in tamoxifen-treated patients
2000	Ferrero (47)	297	Fresh	Immunoluminometric assay	Yes	No	Only tumor size and ER status were independent predictors in node-negative patients
2000	Bottini (48)	143	Paraffin	IHC	Yes	Yes	p53 status predicted response to chemotherapy independent of HER-2/neu, ER, PR, and Ki-67 expression
2001	Geisler (49)	90	Paraffin	IHC	Yes	No	p53 status predicted prognosis in patients treated with anthracycline-based chemotherapy regimens
2001	Montero (50)	458	Fresh	Immunoblotting	Yes	Yes	p53 status predicted outcome in operable cases
2001	Tsutui (51)	514	Frozen	IHC	Yes	Yes	p53 IHC predicted disease-free survival
2001	Shao (53)	46	Blood	PCR sequencing	Yes	Yes	p53 mutations in blood DNA in breast cancer patients independently predicted prognosis
2002	O'Hanlon (52)	105	Paraffin	IHC	Yes	Yes	p53 but not p21 status predicted prognosis in node-positive patients
2002	Kato (54)	420	Paraffin	IHC	Yes	Yes	p53 status predicted relapse-free and overall survival
2002	El-Ahmady (55)	94	Fresh	ELISA	Yes	No	HER-2/neu was independent predictor
2002	Chen (56)	75	Paraffin	IHC	Yes	Yes	p53 and HER-2/neu were independent predictors, but bcl-2 was not

ELISA, enzyme-linked immunosorbant assay; ER, estrogen receptor; PCR, polymerase chain reaction; PR, progesterone receptor; SSCP, single strand conformation polymorphism; uPA, urokinase plasminogen activator.

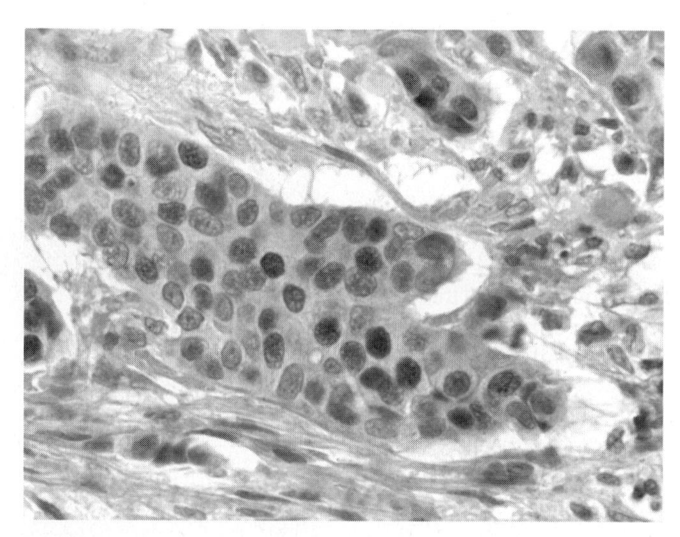

FIGURE 23.3 p53 expression determined by IHC. p53 nuclear staining has been linked to mutation in the *p53* gene. Wild-type p53 protein is expressed for an extremely short portion of the cell cycle generally, rendering it undetectable by IHC methods. The production of mutant p53 protein is associated with protein stabilization, allowing detection by IHC techniques. However, frame shift and missense mutations of the p53 gene may not be associated with sufficient mutant p53 protein production for IHC detection (false-negative IHC result), and on occasion, wild-type p53 protein may accumulate and be immunodetectable by IHC (false-positive IHC result). It is estimated that IHC is between 80% and 85% accurate in the prediction of *p53* gene status (see text). (Peroxidase-antiperoxidase using the DO-1 antibody, Dako Corp., Carpenteria, CA, ×200.)

Table 23.3 Methods of Detection of p53 Status in Breast Cancer

Method	Sample Type	Current Clinical Use	Advantages	Disadvantages	Comment
IHC	Paraffin	Yes	Rapid Inexpensive Paraffin ready	Low accuracy False positives False negatives	In clinical use for management of bladder cancer Mutations that do not produce a stabilized protein are missed Posttranslational modifications are missed by most routine antibodies
SSCP	Paraffin fresh tissue	Yes	More accurate than IHC	5–8% false-negative rate	Mutations that do not produce a stabilized protein are missed Post-translational modifications are missed
PCR + direct sequencing	Paraffin fresh tissue	No	More accurate than SSCP	2–4% false-negative rate	Mutations that do not produce a stabilized protein are missed Posttranslational modifications are missed
Genomic microarray	Paraffin fresh tissue	No	Sequencing chip High throughput	2–3% false-negative rate	Mutations that do not produce a stabilized protein are missed Posttranslational modifications are missed
Ligase chain reaction	Paraffin fresh tissue	No	Highest accuracy	Unfamiliar technique Not commercialized	Frame shift and missense mutations are detected Posttranslational modifications can be detected
Yeast reporter functional assay	Fresh tissue blood	Yes	Functional assay High specificity	Requires fresh sample Type and location of mutation not detected	In clinical use for germline detection of Li-Fraumeni syndrome Research use

PCR, polymerase chain reaction; SSCP, single strand conformation polymorphism.

RB1 and Cell Cycle Control

The retinoblastoma gene *RB1* is a prototype tumor suppressor gene in breast cancer. *RB1* encodes a 110-kDa protein that, among other cellular functions, plays a pivotal role in cell cycle checkpoint control, particularly in the transition from G1 to S phase.[86,87] RB1 alternates between hypophosphorylated (active) and hyperphosphorylated (inactive) conformations, depending on phosphorylation events from upstream signaling pathways. Active RB1 blocks cell cycle progression by binding and inactivating the transcription factor E2F, which controls the expression of several target genes necessary for entry into the S phase. During cell cycle progression, RB1 is gradually phosphorylated and inactivated by cyclin dependent kinase complexes (e.g., cyclin D–cdk4/6 and cyclin E–cdk2), causing RB1 to unbind from E2F (see Chapter 13). Constitutive RB1 inactivation can be demonstrated in approximately 20% of breast cancers,[88] resulting from intragenic mutation, chromosomal deletion, transcriptional silencing, and/or functional inactivation.[88–90] Several studies have addressed the frequency and prognostic value of RB1 inactivation in breast carcinomas. Conflicting results were reported for correlations between RB1 expression and known adverse prognostic factors; however, there was no correlation between RB1 expression and relapse-free or overall survival.[91–94]

Cowden Syndrome and *PTEN*

Cowden syndrome, also known as multiple hamartoma syndrome, is an autosomal dominant disorder characterized by skin trichilemmomas and predisposition to breast, thyroid, and endometrial neoplasms.[95] The causal gene in Cowden syndrome is the well-known tumor suppressor *PTEN,* which codes for a dual specificity phosphatase protein (FIG. 23.4). One of the key functions of the PTEN protein is inhibition of the oncogenic AKT/PI3-K signaling pathway, which is accomplished by dephosphorylating phosphatidylinositol-3,4,5-triphosphate. Beyond its etiologic role in Cowden syndrome breast cancers, *PTEN* inactivation also participates in the pathogenesis of sporadic breast cancers. Although somatic intragenic *PTEN* point mutations are rare in nonsyndromic breast cancers,[96] deletions of the gene are common.[97] The deletions, together with epigenetic events such as promoter methylation, presumably account for the loss of PTEN protein expression seen in many breast cancers.[98] It is intriguing that certain mechanisms of PTEN derangement are not associated with increased risk of breast cancer. For example, Bannayan-Ruvalcaba-Riley syndrome is caused by inactivating mutations of *PTEN* and is characterized by macrocephaly, multiple lipomas, and hemangiomas but not by increased risk of breast cancer. Whereas most Cowden syndrome *PTEN* mutations involve the phosphatase domain, the Bannayan-Ruvalcaba-Riley syndrome mutations involve other aspects of the PTEN protein.[99] Likewise, the Proteus and the Proteus-like syndromes result from *PTEN* mutations outside the phosphatase domain and do not seem to be associated with increased risk of breast cancer.[95] These findings suggest that perturbations of the PTEN phosphatase function are critical in the predisposition to breast cancer.

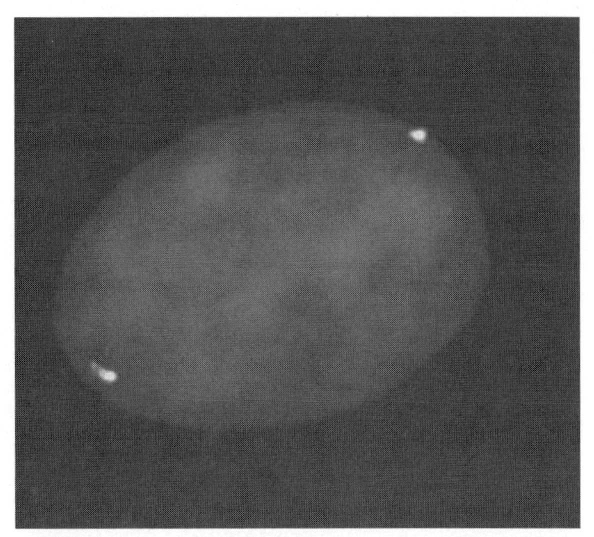

FIGURE 23.4 Fluorescence in situ hybridization (FISH) evaluation showing deletion of a large region on the chromosome 8 short arm (8p12, green; 8p21, red; 8 centromere, yellow), manifested by deletion of a green–red probe pair from one copy of chromosome 8 (top). Chromosome 8p deletions are found in approximately 50% of primary breast cancers and breast cancer cell lines, but the causative tumor suppressor gene(s) have not yet been identified. **See Plate 49 for color image.**

mTOR and Targeting the PTEN/AKT/PI3-K Pathway in Breast Cancer

mTOR, the mammalian target of rapamycin, is a downstream effector of the PI3K/AKT (protein kinase B) signaling pathway that mediates cell survival and proliferation and has become a target for therapy of breast cancer.[100] By targeting mTOR, the immunosuppressant and antiproliferative agent rapamycin inhibits signals required for cell cycle progression, cell growth, and proliferation.[100] Both rapamycin and novel rapamycin analogues are highly specific inhibitors of mTOR and block the actions of downstream signaling elements, which result in cell cycle arrest in the G1 phase.[100] Rapamycin and its analogues also prevent CDK activation, inhibit retinoblastoma protein phosphorylation, and accelerate the turnover of cyclin D1. In preclinical studies and early human clinical trials, rapamycin and rapamycin analogues have demonstrated growth inhibitory effects in breast cancer,[100] possibly by inhibiting the proliferative responses mediated by the epidermal growth factor receptor, the insulin growth factor receptor, and the estrogen receptor. Although pre-clinical testing of the mTOR inhibitor CCI-779 has shown promise for the treatment of metastatic breast cancer, current phase II clinical trials have not, to date, been associated with major clinical responses to the drug.[100]

Peutz-Jeghers Syndrome and LKB1/STK11 Serine-Threonine Kinase Signaling

Peutz-Jeghers syndrome is an autosomal recessive disorder characterized by mucocutaneous melanin pigmentation, gastrointestinal polyposis, and predisposition to gastrointestinal malignancies. Peutz-Jeghers syndrome patients are also predisposed to a variety of extraintestinal tumors, including breast and cervical cancers.[101] Inactivating mutations in the tumor suppressor gene *LKB1* have been identified in most Peutz-Jeghers syndrome patients.[102] *LKB1* encodes for a serine-threonine kinase that exhibits a high degree of homology to the *Xenopus* serine-threonine kinase XEEK1.[103] Several functions have been attributed to LKB1, including roles in normal embryonic development, a cofactor role in p53-mediated apoptosis, and regulation of cell proliferation by cell cycle checkpoint mechanisms. In addition, LKB1 interacts with the chromatin remodeling protein BRG161. Somatic mutations of *LKB1* have not been identified in sporadic breast cancer,[104,105] but low expression levels have been reported,[106] suggesting epigenetic mechanisms for LKB1 inactivation. Reduced LKB1 expression has been associated with worse clinical outcome and with adverse prognostic variables, including higher histologic grade, larger tumor size, lack of progesterone receptor expression, and presence of lymph node metastases.[106]

nm23

Localized to chromosome 17q, *nm23* belongs to a large family of structurally and functionally conserved proteins that exhibit nucleoside diphosphate kinase activity and bind DNA.[107] Although the exact mechanism of metastasis suppression associated with *nm23* expression is not completely understood, this gene is believed to function by regulating downstream signal transduction associated with an as yet unconfirmed receptor. *nm23* expression has predicted a favorable outcome in some studies of breast cancer patients[108–110] but not in others.[111,112]

p16

The *p16* (INK4A) tumor suppressor gene CDK inhibitor inhibits cell growth at the G1/S checkpoint of the cell cycle in concert with Rb, p14, and p15.[113–115] *p16* expression may be lost by a process of mutation (less common) or CpG island hypermethylation of the gene promoter (most common).[113–115]

Maspin

Maspin is a novel serine protease inhibitor related to the serpin family with a tumor suppressing function possibly associated with poor prognosis in breast cancer.[116–118] Maspin messenger RNA detection by reverse transcriptase-polymerase chain reaction has also been used to detect micrometastases and minimal residual disease in breast cancer.[119] Maspin expression has been correlated with breast cancer prognosis, with maspin nuclear staining significantly associated with good prognostic factors and cytoplasmic staining associated with poor prognostic markers.[120]

Other Tumor Suppressor Mechanisms: *APC* and Mismatch Repair Defects

In various tumors, including colorectal cancer and desmoid tumors, the WNT1/β-catenin signaling pathway is activated by loss-of-function mutations involving the *APC* tumor suppressor gene or by gain-of-function mutations in β-catenin. The WNT pathway might also be relevant in breast tumorigenesis, as suggested by the increased predisposition for breast cancer in *APC*-deficient $Apc^{min+/-}$ mice[121] and the role of *Wnt1* as a mammary oncogene in mice.[122] WNT and *APC* mutations are infrequent in human breast cancers, but overexpression of the β-catenin oncogene has been associated with a worse prognosis.[123] Recent studies have shown *APC* gene promoter methylation and mutations in a substantial number of breast carcinomas, but the significance of these findings remain to be elucidated.[124,125]

As discussed above, DNA repair genes are a subgroup of tumor suppressor genes that are viewed as caretaker genes. These include *BRCA1, BRCA2, MLH1,* and *MSH2* (see Chapters 4, 25, and 26). Cadherins also play key tumor suppressor roles in breast cancer (see Chapter 20).

Uncharacterized Tumor Suppressor Mechanisms

Although a diverse group of tumor suppressor genes have been implicated in breast cancer, it is likely that a much larger number remain to be identified. The main evidence for the presence of such genes is found in the many nonrandom chromosomal deletions that have been identified in breast cancer. For example, deletion of various regions on the chromosome 8 short arm have been identified in at least 50% of primary breast cancers and breast cancer cell lines (**FIG. 23.4**).[126,127] These recurring chromosomal deletions presumably mark the location of yet uncharacterized tumor suppressors, and although several candidate genes have been implicated, the major targets of the 8p deletions have not yet been identified. Progress in the identification of tumor suppressor genes has been hampered by the difficulty in identifying highly informative breast cancers, such as those with very small deletions that implicate single genes, or small numbers of genes, as the deletion target. This challenge should be overcome, in part, by the increasing availability of high throughput genomic assays. In one high throughput approach, the breast cancer DNA is hybridized against a microarray of bacterial artificial chromosome clones, with each clone containing an insert of 75 to 250 kilobases of contiguous human genomic DNA. This approach, known as bacterial artificial chromosome array comparative genomic hybridization, enables a genome-wide high density evaluation of breast cancer DNA gains and deletions.[128] The same goal can be achieved by analyzing breast cancer DNAs with high throughput single nucleotide polymorphism assays. Single nucleotide polymorphism assays can interrogate breast cancer DNA gains and deletions at very high levels of resolution and can therefore detect small deletions that affect only a single gene.[129,130] In the coming years, it is likely that increasing numbers of novel breast tumor suppressors will be identified using these powerful high throughput screens.

Mouse Models

Various mouse models of breast carcinoma have been developed by transgenic knockout of the same tumor suppressor genes responsible for familial or syndromic breast cancer in humans. Among the tumor suppressors whose targeted inactivation results in murine breast cancer are *p53, BRCA1, BRCA2, ATM, PTEN.* and *APC*.[131] However, the penetrance and phenotype of these tumor suppressor aberrations depends on the mouse background in which they are created. For example, breast cancers develop frequently in BALB/c;*p53*+/− mice,[132] whereas other *p53* null or p53+/− mice rarely develop breast carcinomas despite their predisposition to other types of cancers. These observations suggest that some mouse genetic backgrounds confer resistance to the development of mammary neoplasias, perhaps because they lack critical cofactors that are requisite for p53 associated mammary tumorigenesis. Mice heterozygous for *Apc* tumor suppressor inactivation ($Apc^{Min/+}$) are predisposed to both breast and colorectal carcinomas,[132] especially after exposure to carcinogenic compounds such as ethylnitrosourea.[133] Mice with germline *BRCA1* homozygous inactivation die early in embryonic development,[131,134] but the Cre-LoxP method can be used to restrict BRCA1 inactivation to mammary epithelium, resulting in development of breast cancer by age 10 months.[134] Mice with germline *BRCA2* homozygous inactivation are also embryonic lethals, whereas mice expressing a hypomorphic *BRCA2* allele develop lymphomas but not mammary tumors.[131,132] Similarly, mice with germline ATM homozygous mutants develop thymic lymphomas but not breast cancer, and mice with ATM heterozygous inactivation are phenotypically normal.[131,135,136] Mice with germline PTEN homozygous inactivation are embryonic lethals, but heterozygous mutants exhibit an increased rate of tumor development, including mammary tumors.[137]

References

1. Harris H. The analysis of malignancy by cell fusion: the position in 1988. Cancer Res 1988;48:3302–3306.
2. Ephrussi B, Davidson RL, Weiss MC, et al. Malignancy of somatic cell hybrids. Nature 1969;224: 1314–1316.
3. Stanbridge EJ, Der CJ, Doersen CJ, et al. Human cell hybrids: analysis of transformation and tumorigenicity. Science 1982;215:252–259.
4. Saxon PJ, Srivatsan ES, Stanbridge EJ. Introduction of human chromosome 11 via microcell transfer controls tumorigenic expression of HeLa cells. EMBO J 1986;5:3461–3466.
5. Weissman BE, Saxon PJ, Pasquale SR, et al. Introduction of a normal human chromosome 11 into a Wilms' tumor cell line controls its tumorigenic expression. Science 1987;236:175–180.
6. Knudson AG Jr. Mutation and cancer: statistical study of retinoblastoma. Proc Natl Acad Sci USA 1971;68:820–823.
7. Cavenee WK, Drencema NA, Gyja TP, et al. Expression of recessive alleles by chromosomal mechanisms in retinoblastoma. Nature 1983;305:779–784.
8. Fearon ER. Tumor-suppressor genes. In: Scriver CR, Beaudet AL, Sly WS, et al., eds. The Metabolic and Molecular Bases of Inherited Disease. New York: McGraw-Hill, 2004.
9. Yang Y, Li CC, Weissman AM. Regulating the p53 system through ubiquitination. Oncogene 2004;23: 2096–2106.
10. Kinzler KW, Vogelstein B. Familial cancer syndromes: the role of caretakers and gatekeepers. In: Scriver CR, Beaudet AL, Sly WS, et al., eds. The Metabolic and Molecular Bases of Inherited Disease. New York: McGraw-Hill, 2004.
11. Linzer DI, Levine AJ. Characterization of a 54K dalton cellular SV40 tumor antigen present in SV40-transformed cells and uninfected embryonal carcinoma cells. Cell 1979;17:43–52.
12. Eliyahu D, Raz A, Gruss P, et al. Participation of p53 cellular tumour antigen in transformation of normal embryonic cells. Nature 1984;312:646–649.
13. Finlay CA, Hinds PW, Levine AJ. The p53 proto-oncogene can act as a suppressor of transformation. Cell 1989;57:1083–1093.
14. Vogelstein B, Lane D, Levine AJ. Surfing the p53 network. Nature 2000;408:307–310.
15. Taylor WR, Stark GR. Regulation of the G2/M transition by p53. Oncogene 2001;20:1803–1815.
16. Gasco M, Shami S, Crook T. The p53 pathway in breast cancer. Breast Cancer Res 2002;4:70–76.
17. Maass N, Hojo T, Zhang M, et al. Maspin—a novel protease inhibitor with tumor-suppressing activity in breast cancer. Acta Oncol 2000;39:931–934.
18. Borresen-Dale AL. TP53 and breast cancer. Hum Mutat 2003;21:292–300.
19. Couch FJ, Weber BL. Breast cancer. In: Scriver CR, Beaudet AL, Sly WS, et al., eds. The Metabolic and Molecular Bases of Inherited Disease. New York: McGraw-Hill, 2004.
20. Done SJ, Eskandarian S, Bull S, et al. p53 missense mutations in microdissected high-grade ductal carcinoma in situ of the breast. J Natl Cancer Inst 2001;93:700–704.
21. Done SJ, Arneson NC, Ozcelik H, et al. p53 mutations in mammary ductal carcinoma in situ but not in epithelial hyperplasias. Cancer Res 1998;58: 785–789.
22. Phillips KA, Nichol K, Ozcelik H, et al. Frequency of p53 mutations in breast carcinomas from Ashkenazi Jewish carriers of BRCA1 mutations. J Natl Cancer Inst 1999;91:469–473.
23. Greenblatt MS, Chappuis PO, Bond JP, et al. TP53 mutations in breast cancer associated with BRCA1 or BRCA2 germ-line mutations: distinctive spectrum and structural distribution. Cancer Res 2001;61: 4092–4097.
24. Smith PD, Crossland S, Parker G, et al. Novel p53 mutants selected in BRCA-associated tumours which dissociate transformation suppression from other wild-type p53 functions. Oncogene 1999;18: 2451–2459.
25. Greenblatt MS, Chappuis PO, Bond JP, et al. TP53 mutations in breast cancer associated with BRCA1 or BRCA2 germ-line mutations: distinctive spectrum and structural distribution. Cancer Res 2001;61: 4092–4097.
26. Varley JM. Germline TP53 mutations and Li-Fraumeni syndrome. Hum Mutat 2003;21:313–320.
27. Frebourg T, Abel A, Bonaiti-Pellie C, et al. Li-Fraumeni syndrome: update, new data and guidelines for clinical management. Bull Cancer 2001;88: 581–587.
28. Thor AD, Moore DH, Edgerton SM, et al. p53 tumor suppressor gene: an independent marker of prognosis in breast cancers. J Natl Cancer Inst 1992;84: 845–855.
29. Cunningham JM, Ingle JN, Jung SH, et al. p53 gene expression in node positive breast cancer: Relationship to DNA ploidy and prognosis. J Nat Cancer Inst 1994;86:1871–1873.
30. Caleffi M, Teague MW, Jensen RA, et al. p53 gene mutations and steroid receptor status in breast carcinoma. Clinical pathologic correlations and prognostic assessment. Cancer 1994;73:2147–2156.
31. Marks JR, Humphrey PA, Wu K, et al. Overexpression of p53 and Her2/neu proteins as prognostic markers in early stage breast cancer. Ann Surg 1994; 219:332–341.
32. Beck T, Weller EE, Weikel W, et al. Usefulness of immunohistochemical staining for p53 in the prognosis of breast carcinomas: correlation with established prognosis parameters and with the proliferation marker, MIB-1. Gynecol Oncol 1995;57:96–104.
33. Borg A, Lennerstand J, Stenmark-Askmalm M, et al. Prognostic significance of p53 overexpression in primary breast cancer; a novel luminometric immunoassay applicable on steroid receptor cytosols. Br J Cancer 1995;71:1013–1017.
34. Bland KI, Konstadoulakis MM, Vezeridis MP, et al. Oncogene protein co-expression. Value of Ha-ras, c-myc, c-fos, and p53 as prognostic discriminants for breast carcinoma. Ann Surg 1995;221:706–718.

35. Peyrat J-P, Bonneterre J, Lubin R, et al. Prognostic significance of circulating P53 antibodies in patients undergoing surgery for local regional breast cancer. Lancet 1995;345:621–622.
36. Katoh A, Breier S, Stemmler N, et al. p53 protein expression in human breast carcinoma: lack of prognostic potential for recurrence of the disease. Anticancer Res 1996;16:1301–1304.
37. Soong R, Iacopetta BJ, Harvey JM, et al. Detection of p53 gene mutation by rapid PCR-SSCP and its association with poor survival in breast cancer. Clin Cancer Res 2000;6:443–451.
38. Clahsen PC, van de Velde CJ, Duval C, et al. p53 protein accumulation and response to adjuvant chemotherapy in premenopausal women with node-negative early breast cancer. J Clin Oncol 1998;16: 470–479.
39. Rozan S, Vincent-Salomon A, Zafrani B, et al. No significant predictive value of c-erbB-2 or p53 expression regarding sensitivity to primary chemotherapy or radiotherapy in breast cancer. Int J Cancer 1998; 79:27–33.
40. Peyrat JP, Vanlemmens L, Fournier J, et al. Prognostic value of p53 and urokinase-type plasminogen activator in node-negative human breast cancers. Clin Cancer Res 1998;4:189–196.
41. Levesque MA, Yu H, Clark GM, et al. Enzyme-linked immunoabsorbent assay-detected p53 protein accumulation: a prognostic factor in a large breast cancer cohort. J Clin Oncol 1998;16:2641–2650.
42. Jansen RL, Joosten-Achjanie SR, Volovics A, et al. Relevance of the expression of bcl-2 in combination with p53 as a prognostic factor in breast cancer. Anticancer Res 1998;18:4455–4462.
43. Harbeck N, Dettmar P, Thomssen C, et al. Risk-group discrimination in node-negative breast cancer using invasion and proliferation markers: 6-year median follow-up. Br J Cancer 1999;80:419–426.
44. Broet P, Spyratos F, Romain S, et al. Prognostic value of uPA and p53 accumulation measured by quantitative biochemical assays in 1245 primary breast cancer patients: a multicentre study. Br J Cancer 1999;80:536–545.
45. Reed W, Hannisdal E, Boehler PJ, et al. The prognostic value of p53 and c-erb B-2 immunostaining is overrated for patients with lymph node negative breast carcinoma: a multivariate analysis of prognostic factors in 613 patients with a follow-up of 14–30 years. Cancer 2000;88:804–813.
46. Berns EM, Foekens JA, Vossen R, et al. Complete sequencing of TP53 predicts poor response to systemic therapy of advanced breast cancer. Cancer Res 2000; 60:2155–2162.
47. Ferrero JM, Ramaioli A, Formento JL, et al. p53 determination alongside classical prognostic factors in node-negative breast cancer: an evaluation at more than 10-year follow-up. Ann Oncol 2000;11: 393–397.
48. Bottini A, Berruti A, Bersiga A, et al. p53 but not bcl-2 immunostaining is predictive of poor clinical complete response to primary chemotherapy in breast cancer patients. Clin Cancer Res 2000;6: 2751–2758.
49. Geisler S, Lonning PE, Aas T, et al. Influence of TP53 gene alterations and c-erbB-2 expression on the response to treatment with doxorubicin in locally advanced breast cancer. Cancer Res 2001;61:2505–2512.
50. Montero S, Guzman C, Vargas C, et al. Prognostic value of cytosolic p53 protein in breast cancer. Tumour Biol 2001;22:337–344.
51. Tsutsui S, Ohno S, Murakam S, et al. Prognostic value of p53 protein expression in breast cancer; an immunohistochemical analysis of frozen sections in 514 Japanese women. Breast Cancer 2001;8: 194–201.
52. O'Hanlon DM, Kiely M, MacConmara M, et al. An immunohistochemical study of p21 and p53 expression in primary node-positive breast carcinoma. Eur J Surg Oncol 2002;28:103–107.
53. Shao ZM, Wu J, Shen ZZ, et al. p53 mutation in plasma DNA and its prognostic value in breast cancer patients. Clin Cancer Res 2001;7:2222–2227.
54. Kato T, Kameoka S, Kimura T, et al. p53, mitosis, apoptosis and necrosis as prognostic indicators of long term survival in breast cancer. Anticancer Res 2002;22:1105–1112.
55. El-Ahmady O, el-Salahy E, Mahmoud M, et al. Multivariate analysis of bcl-2, apoptosis, p53, and HER-2/neu in breast cancer: a short-term follow-up. Anticancer Res 2002;22:2493–2499.
56. Chen HH, Su WC, Guo HR, et al. p53 and c-erbB-2 but not bcl-2 are predictive of metastasis-free survival in breast cancer patients receiving post-mastectomy adjuvant radiotherapy in Taiwan. Jpn J Clin Oncol 2002;32:332–339.
57. Pharoah PD, Day NE, Caldas C. Somatic mutations in the p53 gene and prognosis in breast cancer: a meta-analysis. Br J Cancer 1999;80:1968–1973.
58. Lohmann D, Ruhri C, Schmitt M, et al. Accumulation of p53 protein as an indicator for p53 gene mutation in breast cancer. Diagn Mol Pathol 1993;2: 36–41.
59. Bhargava V, Thor A, Deng G, et al. The association of p53 immunopositivity with tumor proliferation and other prognostic indicators in breast cancer. Mod Pathol 1994;7:361–368.
60. Soussi T, Beroud C. Assessing TP53 status in human tumours to evaluate clinical outcome. Nat Rev Cancer 2001;1:233–240.
61. Norberg T, Klaar S, Karf G, et al. Increased p53 mutation frequency during tumor progression—results from a breast cancer cohort. Cancer Res 2001;61: 8317–8321.
62. Geisler S, Lonning PE, Aas T, et al. Influence of TP53 gene alterations and c-erbB-2 expression on the response to treatment with doxorubicin in locally advanced breast cancer. Cancer Res 2001;61: 2505–2512.
63. Kandioler-Eckersberger D, Ludwig C, Rudas M, et al. TP53 mutation and p53 overexpression for prediction of response to neoadjuvant treatment in breast cancer patients. Clin Cancer Res 2000;6:50–56.
64. Borresen-Dale AL. TP53 and breast cancer. Hum Mutat 2003;21:292–300.
65. Kurose K, Gilley K, Matsumoto S, et al. Frequent somatic mutations in PTEN and TP53 are mutually exclusive in the stroma of breast carcinomas. Nat Genet 2002;32:355–357.
66. Quesnel B, Preudhomme C, Fournier J, et al. MDM2 gene amplification in human breast cancer. Eur J Cancer 1994;30A:982–984.
67. McCann AH, Kirley A, Carney DN, et al. Amplification of the MDM2 gene in human breast cancer and its association with MDM2 and p53 protein status. Br J Cancer 1995;71:981–985.

68. Bueso-Ramos CE, Manshouri T, Haidar MA, et al. Abnormal expression of MDM-2 in breast carcinomas. Breast Cancer Res Treat 1996;37:179–188.
69. Jiang M, Shao ZM, Wu J, et al. p21/waf1/cip1 and mdm-2 expression in breast carcinoma patients as related to prognosis. Int J Cancer 1997;74:529–534.
70. Mathoulin-Portier MP, Viens P, Cowen D, et al. Prognostic value of simultaneous expression of p21 and mdm2 in breast carcinomas treated by adjuvant chemotherapy with anthracycline. Oncol Rep 2000; 7:675–680.
71. Cuny M, Kramar A, Courjal F, et al. Relating genotype and phenotype in breast cancer: an analysis of the prognostic significance of amplification at eight different genes or loci and of p53 mutations. Cancer Res 2000;60:1077–1083.
72. Bankfalvi A, Tory K, Kemper M, et al. Clinical relevance of immunohistochemical expression of p53-targeted gene products mdm-2, p21 and bcl-2 in breast carcinoma. Pathol Res Pract 2000;196: 489–501.
73. Becker-Catania SG, Gatti RA. Ataxia-telangiectasia. Adv Exp Med Biol 2001;495:191–198.
74. Stankovic T, Kidd AM, Sutcliffe A, et al. ATM mutations and phenotypes in ataxia-telangiectasia families in the British Isles: expression of mutant ATM and the risk of leukemia, lymphoma, and breast cancer. Am J Hum Genet 1998;62:334–345.
75. Shiloh Y. ATM and related protein kinases: safeguarding genome integrity. Nat Rev Cancer 2003; 3:155–168.
76. Canman CE, Lim DS, Cimprich KA, et al. Activation of the ATM kinase by ionizing radiation and phosphorylation of p53. Science 1998;281:1677–1679.
77. Waterman MJ, Stavridi ES, Waterman JL, et al. ATM-dependent activation of p53 involves dephosphorylation and association with 14-3-3 proteins. Nat Genet 1998;19:175–178.
78. Banin S, Moyal L, Shieh S, et al. Enhanced phosphorylation of p53 by ATM in response to DNA damage. Science 1998;281:1674–1677.
79. Dork T, Bendix R, Bremer M, et al. Spectrum of ATM gene mutations in a hospital-based series of unselected breast cancer patients. Cancer Res 2001; 61:7608–7615.
80. Hampton GM, Mannermaa A, Winquist R, et al. Loss of heterozygosity in sporadic human breast carcinoma: a common region between 11q22 and 11q23.3. Cancer Res 1994;54:4586–4589.
81. Lu Y, Condie A, Bennett JD, et al. Disruption of the ATM gene in breast cancer. Cancer Genet Cytogenet 2001;126:97–101.
82. Launonen V, Laake K, Huusko P, et al. European multicenter study on LOH of APOC3 at 11q23 in 766 breast cancer patients: relation to clinical variables. Breast Cancer Somatic Genetics Consortium. Br J Cancer 1999;80:879–882.
83. Carr AM. Cell cycle. Piecing together the p53 puzzle. Science 2000;287:1765–1766.
84. Bell DW, Varley JM, Szydlo TE, et al. Heterozygous germ line hCHK2 mutations in Li-Fraumeni syndrome. Science 1999;286:2528–2531.
85. Ingvarsson S, Sigbjornsdottir BI, Huiping C, et al. Mutation analysis of the CHK2 gene in breast carcinoma and other cancers. Breast Cancer Res 2002; 4:R4.
86. Chau BN, Wang JY. Coordinated regulation of life and death by RB. Nat Rev Cancer 2003;3:130–138.
87. Classon M, Harlow E. The retinoblastoma tumour suppressor in development and cancer. Nat Rev Cancer 2002;2:910–917.
88. Fung YK, T'Ang A. The role of the retinoblastoma gene in breast cancer development. Cancer Treat Res 1992;61:59–68.
89. T'Ang A, Varley JM, Chakraborty S, et al. Structural rearrangement of the retinoblastoma gene in human breast carcinoma. Science 1988;242:263–266.
90. Lee EY, To H, Shew JY, et al. Inactivation of the retinoblastoma susceptibility gene in human breast cancers. Science 1988;241:218–221.
91. Ceccarelli C, Santini D, Chieco P, et al. Retinoblastoma (RB1) gene product expression in breast carcinoma. Correlation with Ki-67 growth fraction and biopathological profile. J Clin Pathol 1998;51: 818–824.
92. Berns EM, de Klein A, van Putten WL, et al. Association between RB-1 gene alterations and factors of favourable prognosis in human breast cancer, without effect on survival. Int J Cancer 1995;64: 140–145.
93. Sawan A, Randall B, Angus B, et al. Retinoblastoma and p53 gene expression related to relapse and survival in human breast cancer: an immunohistochemical study. J Pathol 1992;168:23–28.
94. Pietilainen T, Lipponen P, Aaltomaa S, et al. Expression of retinoblastoma gene protein (Rb) in breast cancer as related to established prognostic factors and survival. Eur J Cancer 1995;31A:329–333.
95. Eng C. PTEN: one gene, many syndromes. Hum Mutat 2003;22:183–198.
96. Petrocelli T, Slingerland JM. PTEN deficiency: a role in mammary carcinogenesis. Breast Cancer Res 2001;3:356–360.
97. Feilotter HE, Coulon V, McVeigh JL, et al. Analysis of the 10q23 chromosomal region and the PTEN gene in human sporadic breast carcinoma. Br J Cancer 1999;79:718–723.
98. Perren A, Weng LP, Boag AH, et al. Immunohistochemical evidence of loss of PTEN expression in primary ductal adenocarcinomas of the breast. Am J Pathol 1999;155:1253–1260.
99. Marsh DJ, Coulon V, Lunetta KL, et al. Mutation spectrum and genotype-phenotype analyses in Cowden disease and Bannayan-Zonana syndrome, two hamartoma syndromes with germline PTEN mutation. Hum Mol Genet 1998;7:507–515.
100. Mita MM, Mita A, Rowinsky EK. Mammalian target of rapamycin: a new molecular target for breast cancer. Clin Breast Cancer 2003;4:126–137.
101. Hemminki A. The molecular basis and clinical aspects of Peutz-Jeghers syndrome. Cell Mol Life Sci 1999;55:735–750.
102. Hemminki A, Markie D, Tomlinson I, et al. A serine/threonine kinase gene defective in Peutz-Jeghers syndrome. Nature 1998;391:184–187.
103. Yoo LI, Chung DC, Yuan J. LKB1—a master tumour suppressor of the small intestine and beyond. Nat Rev Cancer 2002;2:529–535.
104. Forster LF, Defres S, Goudie DR, et al. An investigation of the Peutz-Jeghers gene (LKB1) in sporadic breast and colon cancers. J Clin Pathol 2000;53: 791–793.
105. Bignell GR, Barfoot R, Seal S, et al. Low frequency of somatic mutations in the LKB1/Peutz-Jeghers syndrome gene in sporadic breast cancer. Cancer Res 1998;58:1384–1386.

106. Shen Z, Wen XF, Lan F, et al. The tumor suppressor gene LKB1 is associated with prognosis in human breast carcinoma. Clin Cancer Res 2002;8:2085–2090.

107. Postel EH. NM23-NDP kinase. Int J Biochem Cell Biol 1998;30:1291–1295.

108. Duenas-Gonzalez A, Abad-Hernandez MM, Garcia-Mata J, et al. Analysis of nm23-H1 expression in breast cancer. Correlation with p53 expression and clinicopathologic findings. Cancer Lett 1996;101: 137–142.

109. Mao H, Liu H, Fu X, et al. Loss of nm23 expression predicts distal metastases and poorer survival for breast cancer. Int J Oncol 2001;18:587–591.

110. Terasaki-Fukuzawa Y, Kijima H, Suto A, et al. Decreased nm23 expression, but not Ki-67 labeling index, is significantly correlated with lymph node metastasis of breast invasive ductal carcinoma. Int J Mol Med 2002;9:25–29.

111. Gohring UJ, Eustermann I, Becker M, et al. Lack of prognostic significance of nm23 expression in human primary breast cancer. Oncol Rep 2002;9:1205–1208.

112. Belev B, Aleric I, Vrbanec D, et al. Nm23 gene product expression in invasive breast cancer—immunohistochemical analysis and clinicopathological correlation. Acta Oncol 2002;41:355–361.

113. Lee MH, Yang HY. Negative regulators of cyclin-dependent kinases and their roles in cancers. Cell Mol Life Sci 2001;58:1907–1922.

114. Sherr CJ. The INK4a/ARF network in tumour suppression. Nat Rev Mol Cell Biol 2001;2:731–737.

115. Esteller M. CpG island hypermethylation and tumor suppressor genes: a booming present, a brighter future. Oncogene 2002;21:5427–5440.

116. Maass N, Teffner M, Rosel F, et al. Decline in the expression of the serine proteinase inhibitor maspin is associated with tumour progression in ductal carcinomas of the breast. J Pathol 2001;195:321–326.

117. Umekita Y, Ohi Y, Sagara Y, et al. Expression of maspin predicts poor prognosis in breast-cancer patients. Int J Cancer 2002;100:452–455.

118. Ukemita Y, Yoshida H. Expression of maspin is upregulated during the progression of mammary ductal carcinoma. Histopathol 2003;42:541–545.

119. Corradini P, Voena C, Astolfi M, et al. Maspin and mammaglobin genes are specific markers for RT-PCR detection of minimal residual disease in patients with breast cancer. Ann Oncol 2001;12: 1693–1698.

120. Mohsin SK, Zhang M, Clark GM, et al. Maspin expression in invasive breast cancer: association with other prognostic factors. J Pathol 2003;199: 432–435.

121. Su LK, Kinzler KW, Vogelstein B, et al. Multiple intestinal neoplasia caused by a mutation in the murine homolog of the APC gene. Science 1992;256: 668–670.

122. Brown AM. Wnt signaling in breast cancer: have we come full circle? Breast Cancer Res 2001;3: 351–355.

123. Lin SY, Xia W, Wang JC, et al. Beta-catenin, a novel prognostic marker for breast cancer: its roles in cyclin D1 expression and cancer progression. Proc Natl Acad Sci USA 2000;97:4262–4266.

124. Jin Z, Tamura G, Tsuchiya T, et al. Adenomatous polyposis coli (APC) gene promoter hypermethylation in primary breast cancers. Br J Cancer 2001; 85:69–73.

125. Furuuchi K, Tada M, Yamada H, et al. Somatic mutations of the APC gene in primary breast cancers. Am J Pathol 2000;156:1997–2005.

126. Seitz S, Rohde K, Bender E, et al. Deletion mapping and linkage analysis provide strong indication for the involvement of the human chromosome region 8p12-p22 in breast carcinogenesis. Br J Cancer 1997;76:983–991.

127. Kerangueven F, Allione F, Noguchi T, et al. Patterns of loss of heterozygosity at loci from chromosome arm 13q suggests a possible involvement of BRCA2 in sporadic breast tumors. Genes Chromosomes Cancer 1995;13:291–294.

128. Heiskanen M, Kononen J, Barlund M, et al. CGH, cDNA and tissue microarray analyses implicate FGFR2 amplification in a small subset of breast tumors. Anal Cell Pathol 2001;22:229–234.

129. Wang ZC, Lin M, Wei LJ, et al. Loss of heterozygosity and its correlation with expression profiles in subclasses of invasive breast cancers. Cancer Res 2004;64:64–71.

130. Lindblad-Toh K, Tanenbaum DM, Daly MJ, et al. Loss-of-heterozygosity analysis of small-cell lung carcinomas using single-nucleotide polymorphism arrays. Nat Biotechnol 2000;18:1001–1005.

131. Deng CX, Brodie SG. Knockout mouse models and mammary tumorigenesis. Semin Cancer Biol 2001; 11:387–394.

132. Kuperwasser C, Hurlbut GD, Kittrell FS, et al. Development of spontaneous mammary tumors in BALB/c p53 heterozygous mice. A model for Li-Fraumeni syndrome. Am J Pathol 2000;157:2151–2159.

133. Kohlhepp RL, Hegge LF, Nett JE, et al. ROSA26 mice carry a modifier of Min-induced mammary and intestinal tumor development. Mamm Genome 2000;11:1058–1062.

134. Xu X, Wagner KU, Larson D, et al. Conditional mutation of Brca1 in mammary epithelial cells results in blunted ductal morphogenesis and tumour formation. Nat Genet 1999;22:37–43.

135. Barlow C, Hirotsune S, Paylor R, et al. Atm-deficient mice: a paradigm of ataxia telangiectasia. Cell 1996; 86:159–171.

136. Xu Y, Ashley T, Brainerd EE, et al. Targeted disruption of ATM leads to growth retardation, chromosomal fragmentation during meiosis, immune defects, and thymic lymphoma. Genes Dev 1996;10:2411–2422.

137. Stambolic V, Tsao MS, Macpherson D, et al. High incidence of breast and endometrial neoplasia resembling human Cowden syndrome in pten+/− mice. Cancer Res 2000;60:3605–3611.

CHAPTER 24

The Ubiquitin–Proteasome Pathway and Breast Cancer

Robert Z. Orlowski

Department of Medicine, Division of Hematology/Oncology, and the Lineberger Comprehensive Cancer Center, University of North Carolina at Chapel Hill, Chapel Hill, North Carolina

The ubiquitin–proteasome pathway (UPP) is used by eukaryotic cells as the major mechanism responsible for intracellular protein degradation. UPP activity is essential for many fundamental cellular processes, including timely degradation of cyclins, cyclin dependent kinases, and cyclin dependent kinase inhibitors as part of mitosis,[1] removal of misfolded or damaged and oxidized proteins that would otherwise accumulate throughout the cell's life cycle, turnover and processing of transcription factors and other short-lived regulatory proteins,[2–6] processing of antigens for presentation in association with major histocompatibility class I molecules,[7] and angiogenesis.[8] As such, it plays a central role in maintaining homeostasis in both normal mammary epithelial cells and mammary carcinoma cells. Inactivating mutations introduced into many UPP constituent proteins in vitro are fatal,[2–6] and of those that are not, none has as yet been noted that contribute to mammary carcinogenesis. However, many genetic lesions have been described that either alter the interactions of proteins with the UPP or alter the function of some UPP components and thereby contribute to the pathophysiology of breast cancer.

With the greater understanding of the functions of the UPP that has recently been gained, several of the pharmacologic agents currently in clinical use have been found to work, at least in part, by impacting on UPP function. An appreciation of the molecular complexity of this pathway and its role in carcinogenesis has led to the identification of new drugs, and also potential new drug targets, that may impact on therapy in the future. This chapter reviews the role of the UPP in some of the more important biologic processes involved in breast cancer and the known UPP associated lesions that contribute to carcinogenesis. Current therapeutics in use or under investigation for patients with breast cancer are also reviewed, as well as the future potential of targeting other UPP components as part of breast cancer therapy.

Biologic Background

Functional Components of the UPP

There is evidence that some proteins, such as ornithine decarboxylase[9] and 3-hydroxy-3-methylglutaryl coenzyme A reductase,[10] are degraded in a ubiquitin independent manner, and this seems to be the case for oxidized proteins as well.[11] Turnover of these proteins may occur in part through the 20S multicatalytic proteinase complex, which forms the core proteolytic particle of the 26S complex that is active in the UPP. Also, there seems to be a contribution from the calpain family of proteases in degrading many proteins important in tumorigenesis, such as p53[12,13] and c-Myc.[12,14] The most important mechanism for protein turnover in all cells, however, is unquestionably the UPP.[2–6] Proteins that are targeted for degradation are labeled with a chain of ubiquitin (Ub) molecules in a process that begins with ATP-dependent activation of ubiquitin (FIG. 24.1). An E1 ubiquitin activating enzyme is responsible for this action and forms a high energy thiolester bond between Ub and a cysteine residue of E1. Ubiquitin is then transferred to a cysteine residue of an E2 ubiquitin conjugating enzyme (Ubc), maintaining the high energy thiolester bond. In some cases it appears that the E2 can by itself then transfer

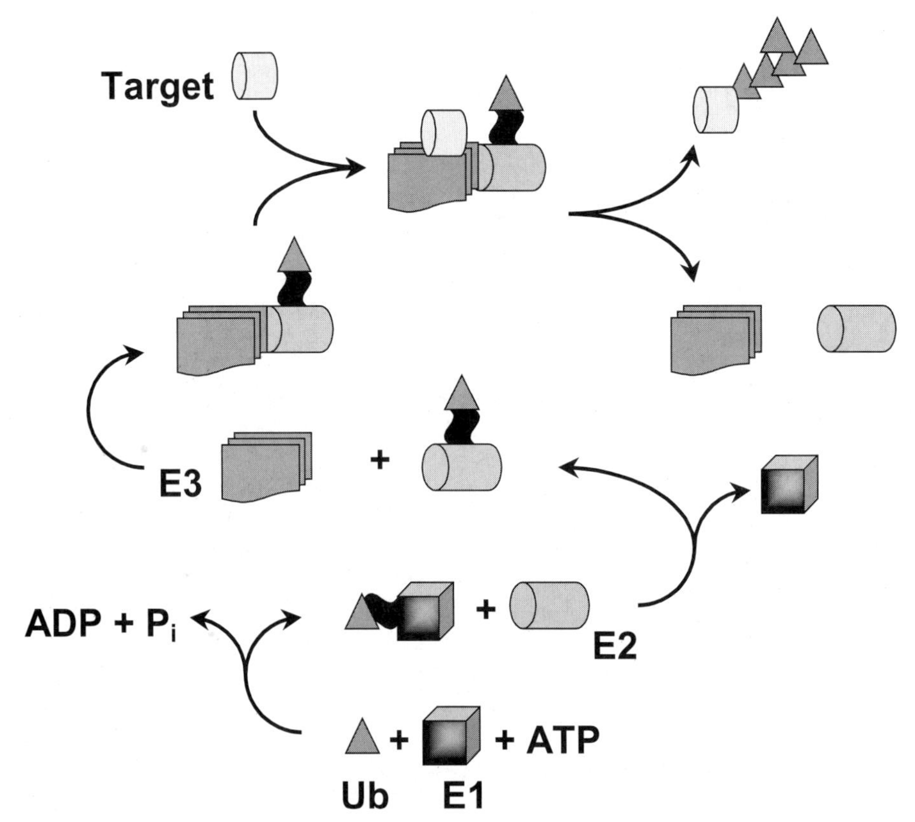

FIGURE 24.1 The ubiquitin conjugation pathway. Ubiquitin, a highly conserved 76-amino acid protein, is conjugated to many proteins whose fate is to undergo degradation. This occurs through the sequential action of an E1 Ub-activating enzyme in an ATP-dependent process, followed by contributions from E2 Ub conjugating and E3 Ub ligase proteins, as described in the text. The ultimate product is a protein labeled with one or more polyubiquitin chains, which becomes a substrate for the 26S proteasome as described in FIGURE 24.3. (From Ref. 86.)

Ub to a target protein, but more commonly this occurs in conjunction with an E3 ubiquitin ligase. Here too there is some heterogeneity, because some E3s act as scaffold proteins, bringing together the target protein and the E2–Ub complex, and it is the E2 that transfers the Ub molecule. In other cases the E3 itself is the catalytic agent, and Ub is first transferred from the E2–Ub complex to the E3, and only then to the target protein. For all these cases, an isopeptide bond is formed between the ϵ-amino group of a lysine residue of the target protein and the first ubiquitin. After this initial modification, several additional cycles of this process then occur, leading to formation of a chain of ubiquitin molecules, which is commonly used as a signal for proteolytic degradation. It should be noted that there is a growing body of literature indicating that monoubiquitination is also an important event, but unlike polyubiquitination it does not seem to have a role in proteolysis. Monoubiquitination is used by cells as a form of posttranslational modification that may be important in a variety of processes, including control of DNA damage repair mechanisms involving the Fanconi anemia pathway[15] and receptor endocytosis and trafficking.[16]

The specificity of protein turnover is provided in part by the E3 enzymes, and these also appear to be one important rate limiting step in proteolysis.[17,18] Though there is only one E1 enzyme and a handful of E2s, it seems likely that there are several hundred E3 enzymes. Of the three classes of E3s identified to date, one is the SCF family of ligases that contain S phase kinase associated protein 1 (Skp1), Cullin-1 (Cul1), regulator of cullins 1 (Roc1), and an F-box protein such as Skp2, with the latter being involved in turnover of $p27^{Kip1}$.[19] These ligases use a series of components to bridge the gap between the target protein and the Ub–E2 complex. HECT-domain containing E3s, so named because they have sequences that are homologous to the E6-associated protein (E6-AP) carboxy-terminus, are the second class. They include the prototype papillomavirus E6-AP that is involved in p53 degradation[20,21] and usually bring together the target and Ub–E2 using domains from a single E3 protein. Finally, RING-finger domain-containing proteins are the third class, with examples such as breast cancer growth suppressor protein 1 (BRCA1)[22] and murine double minute 2 (Mdm2),[23] which also use different domains of the same polypeptide to juxtapose the target with an Ub–E2 complex. In that p53 can be targeted for ubiquitination and ultimate degradation by both E6-AP and Mdm2, it is clear that some proteins may be substrates for several types of E3s. Each E3 is itself responsible for a small subset of protein clients and thus provides specificity to the protein degradatory pathway.

Once a target protein is polyubiquitinated, this polypeptide chain is recognized by constituents of the 19S cap structure present at each end of the mature 26S complex (FIG. 24.2).[2–6] Deubiquitinating enzymes in the cap cleave off this tag, and the ubiquitin molecules are recycled for later reuse (FIG. 24.3). Proteins are then unwound in an ATP-dependent process and enter a central channel in the proteasome, where they come into contact with the catalytic proteases encoded by some of the β subunits of the inner two proteasome rings. Up to five different activities have been described,[24] including a chymotrypsin-like activity that cleaves after hydrophobic amino acids such as phenylalanine and leucine, a trypsin-like activity that cleaves after basic amino acids including lysine and arginine, a peptidyl-glutamyl-peptide hydrolyzing activity that cleaves after acidic residues such as aspartate and glutamate, and a branched chain amino acid preferring activity and a small neutral amino acid preferring activity. Together, these activities are capable of hydrolyzing proteins into oligopeptides, though the chymotrypsin-like activity appears to be the most important rate limiting step in this process. Once generated, oligopeptides then exit the proteasome and are subject to further hydrolysis into their constituent amino acids by endopeptidases and aminopeptidases.

UPP Function in the Biology of Breast Cancer

Up to 80% or more of cellular proteins are degraded through the UPP,[25] many of which play important roles in the biology of normal and neoplastic mammary cells. To provide an example of one of these pathways, the role of the UPP in IκBα degradation is described. The interested reader is referred to other recent reviews of the role of UPP mediated proteolysis in pathways important to breast cancer biology, such as the turnover of the epidermal growth factor receptor[26] and control of p27^{Kip1} expression levels.[27]

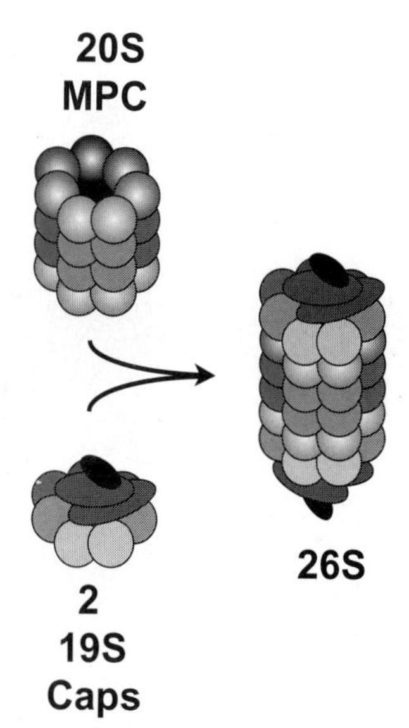

FIGURE 24.2 Structure of the 26S proteasome. The 20S multicatalytic proteinase complex (MPC), which forms the core of the 26S proteasome, is composed of four rings stacked one on top of another around a central pore or channel. Each of the two outer rings, depicted in light gray, contains seven nonidentical α subunits that are predominantly structural, whereas the two inner rings, shown in dark gray, contain seven nonidentical β subunits. These latter subunits encode the proteases, which are key to the proteasome's role in intracellular protein degradation. In the absence of additional subunits the 20S proteasome appears to function in the turnover of some short-lived regulatory proteins, unfolded, and damaged or misfolded proteins. To form the 26S proteasome that is active in ubiquitin dependent, and possibly some ubiquitin independent, proteolytic processes, two 19S cap structures, also called PA700 activators, are added to the 20S complex. Within these caps are proteins that function in recognition of polyubiquitinated sequences, deubiquitinating enzymes that remove these ubiquitin tags for later reuse, and other activities that unwind the protein target that is fed into the central cavity of the proteasome. Some cells also contain an 11S cap structure, or PA28 activator, whose addition at either end of the 20S core forms the so-called immunoproteasome, which plays a role in antigen processing (not shown). (From Ref. 86.)

IκBα The nuclear factor kappa-B (NF-κB) is a ubiquitous transcription factor that plays important roles in a variety of processes, including cell proliferation, angiogenesis, metastasis, and suppression of apoptosis.[28] Its ability to interfere with programmed cell death by enhancing transcription of targets such as the Bcl-2 homologs A1/Bfl-1, Bcl-x_L, and X-linked

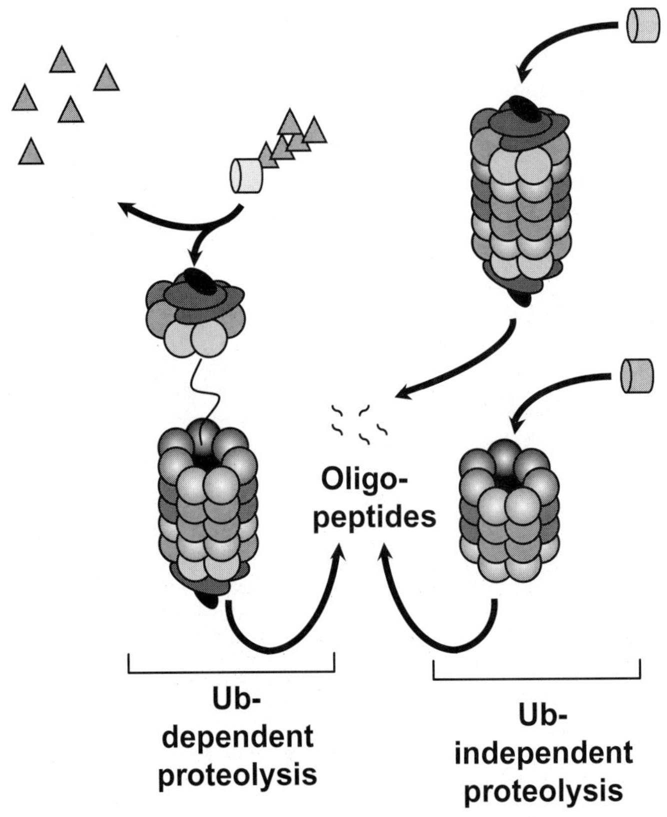

FIGURE 24.3 Proteasome mediated proteolysis. Polyubiquitinated proteins are substrates for proteolysis through the 26S proteasome (left), as described in the text, resulting in the degradation of targets into oligopeptides that are 3 to 24 amino acids in length. The 19S cap has been moved upward, away from the proteasome to show entry of the target protein into the core of the MPC. Once generated, oligopeptides are then generally further processed into their constituent amino acids by the activity of endopeptidases and aminopeptidases. In some situations (right), the 26S proteasome seems also able to degrade proteins in an ubiquitin independent fashion, and the 20S multicatalytic proteinase complex also is involved in some such processes. (From Ref. 86.)

inhibitors of apoptosis is important in the transformation process of tumors that have an increased incidence of activation of NF-κB, which includes breast cancer.[29] In addition, further NF-κB activation is induced by several therapies that are commonly used in treatment of breast cancer patients, including radiation therapy and chemotherapeutics such as anthracyclines, alkylating agents, and taxanes. This leads to the phenomenon of inducible chemoresistance, whereby some of these therapeutic interventions are limited in their own ability to activate apoptosis by the stimulation of this NF-κB mediated survival program. Thus, inhibition of NF-κB activation has potential both as a therapy in and of itself and as a strategy for chemosensitization.

Under unstimulated conditions, NF-κB is present in the cytoplasm, typically as a heterodimer of p50 and p65 (RelA) subunits, bound to the inhibitory protein IκBα, which masks the p65 nuclear localization signal. Upon stimulation, such as by tumor necrosis factor, radiation, or chemotherapy, the upstream IκB kinase (IKK) complex, which is often itself formed from a heterodimer of two subunits, IKKα and IKKβ, is activated.[30] IKK activation results in the phosphorylation of IκBα at two serine residues, Ser-32 and -36,[31] whose mutation to alanine residues that cannot be phosphorylated stabilizes IκBα. Once these serine residues are modified, IκBα becomes a substrate for an ubiquitin ligase complex containing the E2 Ubc4, Cul1, Roc1, Fbw1, and Skp1, known as E3(SCF^{Fbw1}). Activity of this E3 complex results in the polyubiquitination of phospho-Ser-32/36-IκBα at two lysine residues, Lys-21 and -22. Polyubiquitinated IκBα is then recognized by proteins that comprise the cap structure of the 26S

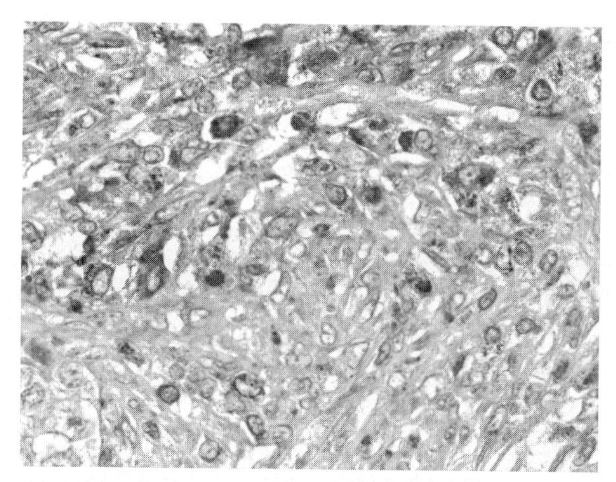

FIGURE 24.4 NF-κB expression in breast cancer. Nuclear and cytoplasmic staining of NF-κB in a case of high grade ER-infiltrating ductal breast cancer that relapsed and progressed after intensive chemotherapy using a Phospho-NF-κB p65 (Ser536) antibody that detects NF-κB p65 only when phosphorylated at Ser536. (Cell Signaling Technologies, Inc., Beverly, MA) (Immunoperoxidase ×200). **See Plate 50 for color image.**

proteasome, and the ubiquitin moieties are removed, followed by proteasome mediated degradation of IκBα. With the nuclear localization signal of p65 exposed, NF-κB is then freed to translocate to the nucleus (FIG. 24.4), where it activates transcription of target genes.

UPP Dysfunction and Breast Tumorigenesis

The molecular mechanisms of several events that are important in breast cancer pathogenesis in part involve function of components of the UPP. Also, many mutations have been described and at least partially characterized that impact directly or indirectly on some aspect of UPP function and may contribute to tumorigenesis (**Table 24.1**). Among the latter are lesions such as amplification or overexpression of Cul-4A that may promote cell cycle progression,[32] as may mutation of human cell division control protein (Cdc) 4,[33] whereas loss of extracellular signal regulated kinase 7 (ERK7), which enhances destruction of estrogen receptor α (ER-α), may lead to breast cancer progression.[34] Of those pathogenetic mechanisms affecting UPP components that have been better characterized, seem to play definite roles in tumorigenesis, and are discussed in more detail include BRCA1 and 2, Efp, HER-2/*neu* (*HER-2*), Mdm2, and Skp2.

BRCA1 and 2 Mutation of the breast and ovarian cancer susceptibility genes *BRCA1* and *BRCA2* predisposes both women and men to hereditary breast cancer as well as women to familial ovarian cancer (see Chapters 4 and 25).[35] Several functions have been ascribed to these large proteins, including transcriptional regulation in association with partners such as p53 and RNA polymerase II and a role in DNA repair.[36] The latter might account for the increased sensitivity of BRCA mutant cells to DNA damaging agents and radiation and, if correlated with increased genomic instability, could provide a link to an increased risk for malignancies. One of the motifs incorporated into BRCA1 is a RING finger domain, suggesting that BRCAs may have a role as E3 ubiquitin ligases. Although BRCA1 was indeed shown to be a weak ligase on its own, in combination with a partner protein, BRCA1-associated RING domain protein 1 (BARD-1), also a weak E3 in its own right, the ligase activity of the complex is enhanced, as is the stability of the two proteins.[37,38] Mutations in the BRCA RING finger motif that are associated with breast cancer abolish this ubiquitin ligase activity, further supporting the importance of this function. How this activity as an E3 may contribute to breast carcinogenesis, however, has not yet been established.

More recently, a large holoenzyme complex was identified containing BRCA1, BRCA2, BARD-1, RAD-51, and BRCC-36 and -45 proteins.[39] Depletion of the latter two by small interfering RNAs increased sensitivity to radiation and defects at the G_2/M checkpoint. Furthermore, aberrant expression of BRCC-36 was shown to occur in some sporadic breast cancers, suggesting that this complex may have a role to play in tumorigenesis outside of familial cases. It is also notable that nuclear localization signals aid in the appropriate nuclear expression of BRCA1, but these in and of themselves are not sufficient, and BARD-1 is in fact necessary as well to mask the BRCA nuclear export signal. Thus, mutations of BARD-1 may result in phenotypes that are similar to those of BRCA1 by virtue of the improper subcellular location of BRCA1 and consistent with this possibility BARD-1 is a known tumor suppressor.

Efp Estrogen-responsive finger protein (Efp) was originally isolated with a genomic binding site cloning technique intended to identify downstream targets of the ER.[40] Efp was noted to be expressed in estrogen target tissues, including the mammary glands and uterus, and also in breast cancer tissue. This protein was determined to contain several interesting

Table 24.1 Molecular Lesions in Breast Carcinogenesis and the UPP

Gene	Changes in Breast Cancer	Impact on UPP	Possible Role in Carcinogenesis	Reference
α6β1 integrin	Increased in many breast tumors, perhaps especially in metastatic ones	Induces proteasome-dependent cleavage of the erbB2 cytoplasmic domain	May act as a suppressor gene to regulate tumor growth	135
BAG-1	Overexpressed in many breast cancers	May bind ubiquitin and interact with HSC and HSP-70, leading to accumulation of HSP-90 client proteins	Protects cells from a variety of apoptotic stimuli; mechanism unknown	136, 137
BRCA1, 2 (with BARD-1)	Mutations increase breast cancer risk	Ubiquitin protein ligase	May impact on DNA damage repair mechanisms; impact of loss of ligase function unknown	
BRCC36	Aberrant expression in sporadic tumors	Part of the BRCC E3 Ub-protein ligase with BRCAs	Depletion increases radiation sensitivity and G_2/M checkpoint defects	39
cbl-b	Mutations not described	Increases phosphorylation, ubiquitination, and degradation of EGFR	Mutation could lead to unopposed EGFR signaling	138
CK2	Overexpressed in many human breast cancers	Phosphorylates β-catenin, inhibiting its proteasome mediated degradation	Could lead to increased expression of cyclin D1, which is a target of β-catenin	139
		Promotes IκBα degradation, activating NF-κB	Supports transformation of breast cancer cells, as well as survival and angiogenesis	140
CUL-4A	Amplified and/or overexpressed in breast cancer	Involved in degradation of cell cycle regulators	May facilitate cell cycle progression	32
EDD	Allelic instability common in breast cancer	HECT domain E3 ubiquitin protein ligase	Unknown	141
Efp	A target gene for ER	E3 ligase for 14-3-3σ	May eliminate 14-3-3σ mediated G_2 arrest	142
ER-α	Expressed in many breast cancers	Enhances the stability of IRS-1 and -2	May contribute to breast cancer growth, and decrease apoptosis	143
ERK7	Enhances destruction of ER-α	Targets ER-α for ubiquitination	Regulator of estrogen responsiveness	34
hCdc4	Mutated in a breast cancer-derived cell line	Part of the E3 responsible for cyclin E turnover	Overexpression of cyclin E may facilitate cell cycle progression	33
HER-2/*neu*	Overexpressed in many patients with breast cancer	Impairs expression or function of LMP-2, -10, and PA28α, β	May decrease MHC class I surface antigen expression, aiding tumor subversion of immune surveillance	52
		Up-regulates Cdk6 and cyclins D1 and E; enhances degradation and relocalization of $p27^{Kip1}$	Drives cell cycle progression	117
Hypoxia	Present in many tumors, especially in the central areas	Induces proteasome mediated degradation of ER-α	May contribute to antiestrogen resistance and outgrowth of estrogen independence	144
IGF-1	Signaling increased in many breast cancers	Increases expression of Skp2, leading to decreased in $p27^{Kip1}$	Antagonizes effects of trastuzumab, contributing to chemoresistance	111

Table 24.1 Molecular Lesions in Breast Carcinogenesis and the UPP (*continued*)

Gene	Changes in Breast Cancer	Impact on UPP	Possible Role in Carcinogenesis	Reference
Mdm2	Expression increased in some breast tumors	E3 ligase for p53	Interferes with the G_1/S checkpoint and genomic stability by abrogating p53 function	See text
NEDD8	Abnormalities not described	Role in ER-α polyubiquitination	Disruption impairs efficacy of antiestrogens; possible resistance mechanism	125
p44S10	Increased gene copy number and expression in MCF-7 cells	Sequence homology to a 26S proteasome subunit	Unknown role	144
p62	Increased in malignant breast tissue	Binds ubiquitin	Mechanism unknown	145
PSMB3	Coexpressed with erb-B2 in breast cancers	Proteasome subunit β-type 3	Unknown	146
Rad6	Up-regulated in metastatic mammary tumor cell lines	Ub conjugating enzyme, interacts with p53	Possible role in maintaining genomic integrity	147, 148
RNF11	Overexpressed in some breast cancers	Interacts with Smurf2, and both become ubiquitinated	Smurf2 represses TGF-β signaling; RNF11 may restore TGF-β responsiveness	149
SGK-1	Up-regulation associated with cell survival	Normally rapidly degraded through the UPP	GR-mediated transcriptional activation stabilizes protein levels	150
SHP-1	Expression decreased or abolished in some ER-negative cell lines	Promotes JAK and TYK2 kinase degradation	If present at normal levels, suppresses tumor cell growth	151
Skp2	Present at higher levels in ER-negative breast tumors	Part of the E3 ligase involved in $p27^{Kip1}$ degradation	May facilitate cell cycle progression; possible role in antiestrogen resistance	67
TGF-β1	Tumor suppressor; receptor may be inactivated; role in osteolytic metastases	Induces proteasome-mediated degradation of ER-α	Unknown	152

EGFR, epidermal growth factor receptor; IRS, insulin receptor substrate; MHC, major histocompatibility complex; TGF, transforming growth factor.

motifs, including a RING finger B-box coiled-coil of the type used by one of the three classes of E3 ubiquitin protein ligases. Further studies showed that Efp acted as the E3 for 14-3-3σ,[41] a protein that normally sequesters Cdc2 in the cytoplasm and therefore slows or prevents cell cycle progression.[42] Reduced levels of 14-3-3σ due to either posttranslational mechanisms[43] or promoter hypermethylation[44–46] have been reported in breast cancer. When Efp was overexpressed in MCF-7 cells, which are commonly used as a model of estrogen dependent breast cancer, they could proliferate even in ovariectomized mice.[41] Conversely, treatment of xenografts with antisense oligonucleotides to Efp slowed tumor growth, and loss of Efp in this setting led to accumulation of 14-3-3σ. Thus, Efp seems to represent a protein that is important in the growth of estrogen responsive tumors, allowing them to progress more rapidly through the cell cycle. Moreover, overexpression of Efp could be a step leading to the development of estrogen independence in tumor cells and possibly in resistance to antiestrogen therapy. Antisense oligonucleotides against Efp or other agents targeting this E3 ligase would appear to represent a rational therapeutic modality possibly for all patients with ER-positive tumors, and especially for those with Efp overexpressing lesions.

HER-2/*neu* The human epidermal growth factor receptor-2 (*HER-2*) oncogene is overexpressed in 20–30% of patients with breast cancer and in some studies has been associated with rapid disease progression, poor prognosis, and chemoresistance (see Chapter 16).[47,48] HER-2 can be targeted using the monoclonal antibody trastuzumab (Herceptin®),[49] whose mechanism of action in part involves effects on the UPP, described later, but this transmembrane protein has also been a target for immunotherapies such as vaccines.[50] After studies showed that the rat c-*neu* gene affected expression of components of the antigen-processing machinery,[51] the effects of HER-2 overexpression were also studied. HER-2 reduced levels of surface major histocompatibility complex class I antigens in fibroblasts, associated with impaired expression of the peptide transporter that functions in the antigen processing pathway.[52] Furthermore, decreased expression of the low molecular weight protein (LMP)-2 and LMP-10 proteasome subunits was noted, as well as decreased levels of the proteasome activators PA28α and PA28β. In that the PA activators enhance proteasome function, whereas LMP-2 and -10 are expressed in the so-called immunoproteasome and may contribute to the formation of more antigenic peptides,[53] these changes might enhance the ability of HER-2 expressing tumors to evade immune surveillance. Consistent with this possibility, these changes were associated with decreased sensitivity of HER-2 overexpressing cells to cytotoxic T-cell mediated lysis.[52] Thus, inhibition of HER-2 might be a rational treatment strategy not only in its effect on downstream signaling, such as through the growth and survival-associated p44/42 mitogen activated protein kinase (MAPK), but also by increasing the ability of these cells to serve as targets for immune based therapies.

Mdm2 The Mdm2 protein is best known as the E3 ubiquitin ligase responsible for ubiquitination and eventual proteasome mediated degradation of p53 and normally functions in a negative feedback loop because it is itself induced by p53 (see Chapter 23).[54] Mdm2 has also recently been described to have another role in breast tumorigenesis arising from the estrogen receptor. Glucocorticoid receptor (GR) activation often opposes functions mediated by the ER, and in mammary tissue GR stimulation is antiproliferative.[55] ER activation with agonists such as 17β-estradiol, diethylstilbestrol, and genistein, but not antagonists such as tamoxifen or raloxifene, was found to lead to ubiquitination and subsequent degradation of the GR.[56] This occurred by activation of both p53 and Mdm2, and the latter, in addition to functioning as an E3 for p53, also is the ubiquitin ligase for GR. Thus, in ER-positive tumors, growth inhibitory signals that might normally act as a control on tumor growth from activation of the GR were dampened by enhanced GR degradation. Because the increased levels of p53 would tend to decrease tumorigenicity given this oncogene's activity as a tumor suppressor, this may provide further selective pressure for suppression of p53 function.

Mutational inactivation of p53 is a well-recognized event in many solid tumors, including breast cancer.[57] Given the role of p53 in sensing damage to DNA and arresting cells to allow for either subsequent DNA repair or apoptosis, removal of p53 decreases the ability of cells to arrest at the G_1/S cell cycle checkpoint, increases genomic instability, and also inhibits sensitivity to a variety of chemotherapeutics, such as anthracyclines and radiation, that normally induce p53 levels. Another mechanism by which p53 can be abrogated, however, is by increasing expression levels of the Mdm2 protein. Overexpression of Mdm2 has been reported to occur in breast cancer[58,59] and resulted in increased p53 ubiquitination followed by proteasome mediated degradation, leading to decreased overall levels of p53.[60] This would be expected to have a similar impact on the pathogenesis of breast cancer as deletion of p53, though because some active p53 may persist it could be an intermediate phenotype, with some residual ability to induce p53 due to chemotherapy, for example. As for Efp, antisense oligonucleotides to Mdm2 might be of benefit in such patients, because they should down-regulate Mdm2 and restore normal levels of p53 protein and its related activities.[61] Perhaps arguing against this possibility, however, are recent studies showing that overexpression of ligase dead Mdm2 mutants did not reactivate p53 transcriptional transactivation,[62] had no effect on p53-induced apoptosis after DNA damage, and indeed promoted growth of one of the two cell lines studied.

Skp2 The cyclin dependent kinase inhibitor $p27^{Kip1}$ is an important part of the G_1/S cell cycle checkpoint and functions by inhibiting cyclin dependent kinase (Cdk) 2 activity, whereas earlier in G_1 it aids in assembling the cyclin D–Cdk4/6 complex (see Chapter 13).[63] Expression levels of $p27^{Kip1}$ have been noted to be decreased in many tumor samples, and this decrease is a negative prognostic factor in breast cancer[27] as well as in several other malignancies, including colon and prostate cancer and mantle cell non-Hodgkin lymphoma.[64] Estrogen activity is an important factor in growth and development of both normal mammary tissue and breast carcinomas and also impacts on events at G_1.[65] A link between these has recently been described in that 17β-estradiol treatment of MCF-7 cells was shown to reduce p27 levels while inducing Skp2,[66] the F-box protein that forms part of the SCF complex responsible for $p27^{Kip1}$ ubiquitination. Overexpression of Skp2 prevented growth arrest induced by antiestrogen therapy and resulted in decreased levels of $p27^{Kip1}$, whereas antisense oligonucleotides to Skp2 and a dominant negative form of Cul-1 inhibited p27 turnover. Of note, a Skp2-independent mechanism was also identified, involving inhibition by 17β-estradiol of $p27^{Kip1}$ nuclear localization through changes in Ras/Raf-1/ERK signaling. Nonetheless, it seems likely from this and other work that down-regulation of p27 through enhanced expression and activity of Skp2 mediated by estrogens is an important step in breast tumorigenesis.

In addition to a role for 17β-estradiol induced events in breast cancer, Skp2 may in and of itself have an oncogenic role. Microarray analysis and immunohistochemistry has shown that patients with ER-negative tumors had higher levels of Skp2 than those patients who were ER positive.[67] Given the role of ER and Skp2 in enhancing cell cycle progression described above, this makes some teleologic sense. ER-negative tumors would lose the ability to down-regulate $p27^{Kip1}$ in response to estrogens, but overexpression of Skp2 would restore this mechanism. A subpopulation of patients with ER-negative lesions that overexpressed Skp2 was identified who had high grade tumors that were negative for HER-2 and had a basal-like phenotype. These tumors showed increased expression of proteins known to interact with Skp2, including cyclin A2, cyclin E1, Cdk2, and Cdc kinase subunit 1 (Cks1), as well as other markers of proliferation, including cyclin B1 and Cks2, whereas p27 levels were decreased. Expression of Skp2 was cell adhesion dependent in normal mammary epithelium but not in mammary carcinoma cells, and inhibition of Skp2 with antisense oligonucleotides decreased adhesion independent growth. Also, overexpression of Skp2 was noted to abolish the effects of antiestrogens. These studies strongly suggest that Skp2 overexpression is an important event in the development of at least a subset of breast carcinomas and support the possible utility of oligonucleotide based therapies for such patients.

It is also interesting to note that function of the Skp2-containing SCF complex may play an important role in another subset of breast tumors. D-type cyclins are important in the G_1/S transition, overexpression of cyclin D1 is tumorigenic in laboratory mammary carcinoma models, and tumors from up to 35% of breast cancer patients have increased levels of cyclin D1 (see Chapter 13).[68,69] In the MCF-7 cell line, as well as in some primary breast tumors, studies showed that expression levels of cyclin D1, cyclin D3, and $p21^{Cip1}$ were coordinately increased and that cyclin D3 could associate with the Skp2-containing SCF.[70,71] Elevated levels of $p21^{Cip1}$ would normally be expected to inhibit cell cycle progression, and p21 was indeed active as a cyclin dependent kinase inhibitor. The elevated levels of cyclin D3, however, may have bound p21 and decreased its ability to interact with cyclin D1–Cdk4/6 complexes, thereby promoting transition through the G_1/S checkpoint. Studies are ongoing to define the precise molecular lesion resulting in this defect, and it may not involve Skp2. Nonetheless, it is interesting to note that this SCF may contribute to carcinogenesis in pleiotropic ways, including through either increased or decreased function. This latter finding suggests that strategies that may augment Skp2 expression and/or SCF function could be useful in treating some patients, including those whose tumors overexpress cyclin D1 and cyclin D3.

Clinical Relevance

Testing of UPP Function

The multiple mechanisms by which function of components of the UPP can be impacted as part of the process of tumorigenesis suggest that a functional evaluation of the UPP may be useful clinically. One such assay for the chymotrypsin-like activity of the proteasome is already available and has been applied during the development of the proteasome inhibitor bortezomib (VELCADE®).[72] Making use of the proteolytic activity of the proteasome, the technique is based on the introduction of a peptide substrate, succinyl-leucyl-leucyl-valyl-tyrosyl-amino-methyl-coumarin (AMC), into cell lysates. Cleavage of this peptide by a chymotryptic activity would occur on the carboxy-terminus of the tyrosine residue, thereby releasing the fluorescent AMC moiety, which can be quantified spectrophotometrically. Although proteases other than the proteasome could exhibit a similar activity, their possible contributions can be evaluated and then subtracted by performing parallel experiments in the presence of specific proteasome inhibitors, such as lactacystin[73] or bortezomib.[74] Some of the other activities of the proteasome, such as the trypsin-like activity, can also be evaluated in an analogous fashion using peptides with AMC moieties whose amino acid sequences take advantage of the substrate preferences of each of the different proteasome activities.[72] This assay can be performed either on tumor tissue, if available, or in lysates prepared from whole blood or peripheral blood mononuclear cells, as was done in the initial phase I trials of bortezomib,[75,76] which demonstrated a dose-dependent inhibition of 20S function and helped to guide drug dosing. Though useful as a screening test of other drugs with potential as proteasome inhibitors, there is as yet no clear application of this technique in a diagnostic setting, because global abnormalities of these vital activities have not yet been described to contribute to tumorigenesis.

Of more immediate clinical utility is testing for some of the specific gene mutations detailed above, including deletion of *BRCA1* and *BRCA2*, which is currently available through standard methods such as isolation of genomic DNA followed by sequencing.[35] In other cases immunohistochemistry can be performed on tumor tissue to evaluate expression levels of the proteins in question, most notably HER-2,[77] whose overexpression suggests that the HER-2 directed monoclonal antibody trastuzumab may be a viable treatment option. Evaluation of many of the other lesions is also possible, such as the expression level of Mdm2 or p27^{Kip1}, and should become clinically relevant once therapeutics directed at these targets become available. The disadvantage of such testing is that it does not directly identify the responsible genetic lesion and therefore the most appropriate therapy. For example, decreased levels of p27 detected by immunohistochemistry could be the result of deletions or other mutations of the p27^{Kip1} coding sequence itself, or they could be due to overexpression of Skp2 resulting in enhanced degradation. Differentiation between these possibilities has clear therapeutic implications, because patients with the latter lesion would be candidates for inhibitors of Skp2 or the SCF of which it is a part, but this would not necessarily be the case for patients with the former abnormality. Thus, testing with a panel of antibodies in immunohistochemistry designed to elucidate not just the primary genetic lesions but also its mechanism would be just the beginning. Ultimately, a combination of both genomic as well as proteomic approaches will likely be necessary. Although many of these are technically feasible at this time on a research basis, it remains incumbent upon investigators in this field to develop individualized targeted treatment options for each of these patient subpopulations that would justify such testing on a routine basis. As a demonstration of the promise of such an approach, however, pharmacogenomic analysis of patients with multiple myeloma receiving the proteasome inhibitor bortezomib on a phase II trial[78] identified gene expression profiles that correlated with a response to therapy.[79] These are currently being evaluated prospectively in a phase III trial and, interestingly, included some components of the UPP.

Current Clinical Utility

Targeting Components of the UPP with Current Cancer Therapies

The role of the UPP in both the normal biology of mammary carcinoma cells and the many lesions that contribute to carcinogenesis and involve the UPP supports the possibility that this pathway should provide many rational targets for cancer therapy. Several drugs currently in use, as well as potential new drug candidates, work, at least in part, by impacting upon the UPP (**Table 24.2**). All-trans-retinoic acid, for example, appears to promote ubiquitination and degradation of Skp2 with

Table 24.2 Agents That Impact UPP Function and Their Therapeutic Implications in Breast Cancer

Agent	Mechanism of Action	Possible Application	Reference
Apigenin	Promotes degradation of HER-2/*neu* and decreased expression of HER-3	Antitumor agent for HER-2/*neu* overexpressing tumors	153
ATRA	Promotes ubiquitination and degradation of Skp2, accumulation of p27, and cyclin E-cdk2 inactivation	Could induce cell cycle arrest	80
Ciglitazone	Peroxisome proliferator-activated receptor γ agonist; enhances ubiquitination of cyclin D1 and ER-α	Antitumor agent for PPAR γ-containing tumors, particularly if ER positive	154
Estrogens	Down-regulate $p27^{Kip1}$ in part by stimulating Skp-2	Antiestrogens may restore $p27^{Kip1}$ levels	66
	Down-regulate GR levels due to induction of its E3 ligase Mdm2, inhibiting $p21^{Cip1}$ induction	Antiestrogens may restore $p21^{Cip1}$ levels	56
Antiestrogens	Down-regulate ER levels by stimulating its degradation	In use for ER-positive tumors	155
Geldanamycin	Binds GRP94, interfering with posttranslational modification of erbB-2, and inducing its proteasome mediated degradation	Antitumor agent, especially for HER-2/*neu* overexpressing tumors	96–98
Genistein	Inhibits the chymotrypsin-like activity of the proteasome	Possible chemosensitizer, in analogy with bortezomib (see text)	156
Heregulin β1	Induces HER-2 phosphorylation, polyubiquitination, and degradation	Antitumor agent for HER-2/*neu* overexpressing tumors	157
LAQ824	Histone deacetylase inhibitor that shifted HER-2 binding from HSP-90 to HSP-70, promoting HER-2 proteasomal degradation	Antitumor agent for HER-2/*neu* overexpressing tumors	112
Progestins	Induce MAPK mediated PR phosphorylation, leading to its degradation	Some may induce apoptosis in breast cancer models in vitro	158
Quinidine	Induces proteolysis of HDAC1 through the proteasome	Possible differentiating agent	159
Radicicol	Inhibitor of HSP-90; induces ER-α ubiquitination and degradation	Similar applications to geldanamycin (see text)	160
TCDD	Aryl hydrocarbon receptor ligand; induces proteasome mediated ER-α degradation	Possible antitumor agent for ER-α–positive tumors	161
Trastuzumab	Modifies phosphorylation of $p27^{Kip1}$, decreasing its susceptibility to proteasome mediated degradation, causing accumulation and G_1 arrest	Already in use for patients with HER-2/*neu* overexpressing tumors	110
WR1065	Cytoprotective aminothiol; activates JNK-mediated p53 phosphorylation, protecting p53 from proteasome mediated degradation	May represent a mechanism to stabilize p53, enhancing cell cycle arrest and chemo-/radiosensitivity	162

ATRA, all-trans-retinoic acid; PPAR, peroxisome proliferator activated receptor; TCDD, 2,3,7,8-tetrachlorodibenzo-p-dioxin.

resultant accumulation of $p27^{Kip1}$ and decreased activation of cyclin E-Cdk 2,[80] possibly contributing to differentiation and eventually apoptosis of carcinoma cells. Of those agents that are most clinically relevant at this time or under ongoing development, more detailed discussions of the mechanism of action and rationale for the use of antiestrogens, bortezomib, geldanamycin and its analogues, and trastuzumab are presented below.

Antiestrogens Estrogens have a prominent role to play in the biology of breast cancer and that in part involves the UPP.[26] Once

17β-estradiol enters cells and binds ER-α, this receptor then undergoes conformational changes that release it from a molecular chaperone, which may be heat shock protein (HSP) 90. After binding DNA and promoting gene transcription, ER-α is subjected to ubiquitination and proteasome mediated degradation. When pure estrogen antagonists were studied, however, they were found to induce a much more rapid and profound decrease in ER-α.[81] Binding of ER-α by estradiol led to a reduction of receptor content with a $t_{1/2}$ of 100 minutes, whereas binding with pure antagonists did so with a $t_{1/2}$ of only 12 minutes. Interestingly, the partial agonist 4-hydroxy-tamoxifen stabilized ER-α and induced accumulation in a different subcellular compartment than the other agents. The pure estrogen antagonist fulvestrant (Faslodex®), which has been approved for postmenopausal patients with ER-positive breast cancer who have progressed after other antiestrogen therapy,[82] also works in part by enhancing proteasome-dependent degradation of ER-α.[83]

Bortezomib Although many agents impact on the UPP as part of their mechanisms of action, the only drug currently in clinical use that was specifically designed for such activity is the proteasome inhibitor bortezomib. This dipeptide boronic acid derivative binds to the chymotrypsin-like activity of the proteasome with great specificity and potency[84] and has activity against a variety of malignancies in both in vitro and in vivo models, including breast cancer.[85] After encouraging results in both phase I[76] and II[78] studies in advanced multiple myeloma, this agent has been approved by the US Food and Drug Administration for patients with myeloma who have received at least two prior therapies and progressed on the last of these. It is also under active investigation in other malignancies, including solid tumors such as breast cancer.[86] Several lines of investigation suggest that this drug should have promise in the therapy of patients with breast cancer based on some of its known mechanisms of action.[87] By preventing the proteasome mediated degradation of the inhibitory protein IκB, proteasome inhibitors block activation of NF-κB, which is constitutively activated in a subset of breast cancers. This transcription factor has a role in a variety of processes, including cell proliferation through transactivation of cyclin D1, cell survival through induction of antiapoptotic proteins such as Bcl-x_L and X-linked inhibitors of apoptosis, and angiogenesis through effects on vascular endothelial growth factor.[28] Another target of proteasome inhibitors is the MAPK phosphatase 1, which blocks activation of p44/42 MAPK by dephosphorylation of the activated form of this kinase.[88] p44/42 also has roles in several tumorigenesis associated pathways, including proliferation through modulation of c-Myc, cell survival by phosphorylation of Bad, discussed subsequently, and angiogenesis.[89] Proteasome inhibitors also may have a role in chemo- and radiosensitization, because they can abrogate important survival mechanisms used by tumor cells.[86] Among these are inherited resistance through expression of Bcl-2 by inducing phosphorylation and cleavage of this antiapoptotic protein.[90,91] Acquired resistance induced by expression of the P-glycoprotein can also be overcome, because proteasome function is needed for maturation of this membrane pump[92,93] that extrudes xenobiotics, including chemotherapeutics, from tumor cells. Finally, inducible chemoresistance, which can occur due to activation of pathways such as NF-κB and p44/42 MAPK, can be repressed as described above.

In preclinical studies bortezomib as a single agent has some activity against breast cancer,[85] but the most exciting data have been in combination with other chemotherapeutics, consistent with the role of proteasome inhibitors as chemosensitizers. Several studies of proteasome inhibitor-based regimens are currently ongoing, including combinations with paclitaxel and anthracyclines. At our institution, pegylated liposomal doxorubicin (Doxil®) has been evaluated with bortezomib and encouraging results have been noted in several patients,[94] including a near complete response in one patient who was previously anthracycline naive, as well as minor responses or stable disease for prolonged periods in patients who have previously received doxorubicin. Further evaluation of at least some of these regimens in a phase II setting seems to be warranted.

Geldanamycin and Its Analogues Geldanamycin is a benzoquinoid ansamycin, analogues of which, including 17-allylamino-17-demethoxygeldanamycin, are now in clinical trials. This agent is a selective and specific inhibitor of HSP-90 and binds to the ATP binding site of this chaperone, preventing it from interacting with its client proteins.[95] These likely then instead form complexes with HSP-70, which promotes polyubiquitination and subsequent proteasome-mediated degradation. One of the known client proteins of HSP-90 relevant to breast cancer is the p185 HER-2 protein tyrosine kinase. After cells were exposed to geldanamycin, HER-2 rapidly became polyubiquitinated[96] and then underwent degradation. Given the role of HER-2 in breast carcinogenesis, it seems logical to expect a role for this class of drugs in treating patients with tumors that overexpress HER-2.[97,98]

Beyond the applications of geldanamycin and its analogues as single agents, however, it is possible that, due to an effect on other client proteins of HSP-90, they may have activity as chemotherapy and radiation therapy sensitizers. Phosphatidylinositol-3 kinase (PI3K) is one such client whose binding to HSP-90 is inhibited, leading to its proteasome mediated degradation with resultant decreased activation of the downstream AKT8 virus oncogene cellular homolog (Akt)/protein kinase B.[99] Of note, Akt activation in HER-2 overexpressing cells may also be decreased by geldanamycin through the serine/threonine protein phosphatase PP1.[100] Akt has an important role in promoting cell survival through effects on a number of signaling pathways,[101] but it occurs in part through phosphorylation of Bad.[102,103]

Raf-1 is another protein that normally binds to HSP-90, whose interaction with this chaperone is disrupted by geldanamycin.[104] In addition to its role in growth signaling through the p44/42 MAPK pathway, Raf-1 also impacts on Bad phosphorylation.[105] When Bad is hypophosphorylated, it cannot interact with 14-3-3 proteins that normally sequester it in the cytoplasm,[106] and instead Bad translocates to mitochondria, where it can form proapoptotic heterodimers with Bcl-2 family proteins. In the absence of Bad, these Bcl-2 proteins form antiapoptotic homodimers that contribute to tumor cell survival and resistance to both chemotherapy and radiation therapy. Several chemotherapeutics used in breast cancer therapy, including anthracyclines, activate p44/42 MAPK through diverse mechanisms, such as activation of upstream protein kinases[107,108] and downregulation of MAPK phosphatases.[109] Therefore, geldanamycin may be able to impact on several pathways that not only promote tumor cell growth but also tumor cell survival and chemotherapy resistance.

Trastuzumab Trastuzumab is a humanized monoclonal antibody directed against HER-2 that has activity in patients whose tumors express this oncogene protein (see Chapter 16). Treatment with trastuzumab or other antibodies directed against HER-2 has been noted to result in the induction of $p27^{Kip1}$ protein and G_1 cell cycle arrest, whereas reducing p27 expression with small interfering RNAs blocked these activities.[110] These changes were accompanied by decreased phosphorylation of p27 at threonine 187, but increased phosphorylation at serine 10. $p27^{Kip1}$ phosphorylation at Thr-187 is known to lead to its ubiquitination by the Skp2-containing SCF complex, followed by proteasome mediated degradation, and it is also possible that phosphorylation at Ser-10 may impact on p27 stability. Thus, at least one of the mechanisms by which trastuzumab acts is by modulating the sensitivity of p27 to UPP mediated degradation.

One of the limitations in the use of trastuzumab is the development of drug resistance, which has been reported to occur, at least in part, through insulin-like growth factor (IGF) signaling. This pathway is involved in induction of cyclins and cyclin-associated Cdk activity as well as down-regulation of $p27^{Kip1}$. IGF-I was recently noted to antagonize the trastuzumab mediated induction of p27, resulting in increased Cdk2 activity and decreased arrest at the G_1/S cell cycle checkpoint.[111] Decreased $p27^{Kip1}$ levels were associated with an induction of Skp2 and increased association between Skp2 and p27. IGF mediated depletion of $p27^{Kip1}$ appeared to occur in a manner dependent on activity of the PI3K pathway. The latter suggests that combination regimens of trastuzumab with a PI3K inhibitor might restore sensitivity in tumors that were previously resistant to this antibody therapy and also might sensitize them even when used initially, before resistant cell clones were allowed to emerge. Histone deacetylase (HDAC) inhibitors have also been reported to increase the efficacy of trastuzumab in vitro in association with induction of $p21^{Cip1}$ and $p27^{Kip1}$,[112] although it is not clear if this occurs through a mechanism involving the UPP. More clearly related to UPP function is the ability of HDAC inhibitors to induce acetylation of HSP-90. This inhibited binding of the chaperone to ATP and impaired its association with client proteins such as HER-2, Akt, and c-Raf-1, depleting their levels and possibly also contributing to chemosensitization as described previously.

Future Clinical Potential

The full potential of some of the drugs impacting on the UPP that are already in use or undergoing clinical testing has not yet been fully evaluated. Even more exciting, however, is the likelihood that some of the targets detailed in **Table 24.1** and the drugs listed in **Table 24.2** may soon find their way into our chemotherapeutic armamentarium to treat patients with breast cancer. Some discussion of the possibilities has already been presented, but special attention is focused below on Mdm2 and Skp2 as therapeutic targets. Brief mention is also made of two other potential pathways that are related to ubiquitination whose modulation could prove fruitful, including NEDDylation and SUMOylation, as well as the use of proteolysis targeting chimeric molecules.

Mdm2

Removal of the tumor suppressor role of p53 is a critical step in carcinogenesis for many solid tumors, including breast cancer.[57] Although this often occurs by mutational inactivation, overexpression of the p53 E3 ligase Mdm2 has been described in breast cancer,[59,113] and for those patients with this molecular lesion, inhibition of Mdm2 is a very rational treatment strategy. Increased levels of p53 would result in cell cycle arrest through transactivation of $p21^{Cip1}$, whereas increased transcription of proapoptotic proteins such as Bax would lead to programmed cell death. As described earlier, Mdm2 blockade could also augment GR mediated growth inhibition given the role of Mdm2 in ubiquitination of the glucocorticoid receptor.[56] Such intervention might also enhance tumor sensitivity to the variety of DNA damaging agents that work in part by stabilizing p53, such as anthracyclines, as well as radiation therapy, which in part uses this same mechanism of action. Moreover, because the same mechanisms would be intact provided that p53 retained wild-type function, inhibitors of Mdm2 could be anticipated to have similar applicability to patients even in the absence of overexpression of this ligase. Among the approaches that could prove fruitful in blocking Mdm2 include antisense technology to decrease protein expression[61] and inhibition of the nuclear export that is necessary for its function.[114] Given that Mdm2 also acts as the ligase for other proteins such as the histone acetyltransferase complex Tip60,[115] however, which may result in modulation of transcription of a variety of genes, further study into the consequences of such inhibitory agents on breast cancer biology is indicated.

Skp2

Among the most attractive targets within the UPP for therapy of breast cancer, as well as other malignancies, is Skp2, a component of the SCF E3 ligase responsible for $p27^{Kip1}$ turnover. The ability to inhibit Skp2, or other components of the SCF complex of which it is a part, could result in antitumor activity through several mechanisms. Because estrogens seem to work in part through induction of Skp2,[66] blockade of its function would provide another pathway by which to abrogate hormonal effects on breast cancer and could enhance the efficacy of antiestrogens in patients with ER-positive tumors. Overexpression of Skp2 itself has been implicated as an important event in a subset of breast carcinomas that are ER negative,[67] and inhibition of its function would be the most rational treatment strategy for this subpopulation. Patients with tumors that overexpress HER-2 would also be logical candidates for such an agent, because this oncogene induces degradation and relocalization of p27 as part of its mechanism of action.[116,117] Even in patients without these lesions, however, the accumulation of $p27^{Kip1}$ that would be induced by targeting Skp2 could result in cell cycle arrest at G_1/S and apoptosis.[118] Furthermore, because IGFs seem to mediate resistance to trastuzumab by inducing Skp2 and antagonizing the induction of p27,[111] Skp2 inhibition could restore sensitivity to trastuzumab in patients whose tumors express HER-2. Also, it is possible that initial use of a trastuzumab/Skp2 inhibitor regimen could enhance antitumor activity and prevent the emergence of drug resistant clones. As is the case for targeting Mdm2, however, a cautionary note must be sounded, in that the Skp2-containing SCF is also responsible, at least in part, for degradation of c-Myc.[119] Given the known role of this oncogene in breast tumorigenesis and progression,[120] its accumulation could result in promotion of tumor growth, though this impact could be abrogated by cell cycle arrest due to $p27^{Kip1}$ and $p21^{Cip1}$, the latter of which is also a Skp2 client.[121]

NEDDylation and SUMOylation

Although ubiquitination is the best characterized example of posttranslational modification of proteins through covalent attachment of other proteins, several comparable mechanisms have also been described. One of these is called NEDDylation, because it involves addition of NEDD8 (neural precursor cell expressed developmentally down-regulated) to a target protein.[18] The cascade leading to NEDDylation starts with the amyloid precursor protein binding protein APP-BP1/Uba3 complex, which activates NEDD8 using ATP to form a high energy thiol ester bond. Ubc12, an E2 conjugating enzyme, then transfers activated NEDD8 to the protein target, possibly without the aid of an E3-like protein. Members of the cullin family are one of the better characterized targets for NEDDylation, and such modification is necessary for protein polyubiquitination. Interruption of this process blocks ubiquitination of a variety of other proteins as well, including $p27^{Kip1}$,[122] IκBα,[123] and the p105 precursor of the p50 subunit of NF-κB.[124] Inhibitors of NEDDylation could therefore be very potent anticancer agents, given the previously described roles of just these three proteins in angiogenesis, cell cycle progression, chemotherapy resistance, and tumor cell survival. One recent study indicated, however, that a functional NEDDylation pathway was necessary for polyubiquitination and proteasome mediated degradation of ER-α.[125] Antiestrogens that normally induce ER-α proteolysis could not do so in cells with a disrupted NEDD8 pathway and also could not inhibit cell proliferation. This suggests that interruption of NEDDylation may be a mechanism of antiestrogen resistance in breast cancer. Moreover, these findings clearly point to the need for further study of such pathways to determine their full role in breast carcinogenesis before they are targeted therapeutically with unanticipated and possibly deleterious consequences.

A similar protein modification pathway in cells is that which leads to covalent attachment of the small ubiquitin-like modifier-1 (SUMO-1). SUMO activating enzymes also use ATP to activate this protein tag, after which it is transferred to the E2 Ubc9, and eventually to proteins through the intervention of the protein inhibitor of activated STAT ligase. Although the likely many roles of SUMOylation are far from being completely appreciated, some of the known activities do impact on protein turnover. Interestingly, SUMOylation of IκBα stabilizes its association with NF-κB by preventing its ubiquitination.[126] Thus, if SUMOylation of IκBα could be stimulated, this would provide another mechanism of inhibiting nuclear NF-κB translocation and enhancing chemosensitivity that could be more specific than through the use of proteasome inhibitors such as bortezomib.

Protacs

The antitumor efficacy of agents such as geldanamycin that enhance degradation of some proteins[97,98] and drugs such as bortezomib that inhibit protein turnover[86] show that therapeutic benefits may be obtained by both up-regulation and down-regulation of proteolysis. Although the multiple targets of such agents may help to reduce the emergence of drug-resistant cell lines, both of these approaches also have the potential for activating antiapoptotic signal transduction pathways that would promote the survival of cancer cells, thereby limiting their own efficacy. Geldanamycin, for example, induces a heat shock response through heat shock factor-1[127] that results in activation of HSP-72,[128] and proteasome inhibitors also activate HSP-72[129] through the heat shock factors.[130] Though HSP-72 likely contributes to apoptosis by acting as a chaperone to former HSP-90 client proteins that are displaced by geldanamycin, taking them to the proteasome for degradation, HSP-72 can also inhibit activity of the c-Jun-N-terminal kinase and thereby negatively regulate apoptosis.[131] Indeed, an antiapoptotic activity for HSP-72 has been documented for proteasome inhibitors[129] and antisense down-regulation of HSP-72 enhanced programmed cell death.[132] More specific targeting of proteins critical to tumor biology might avoid such pleiotropic effects, and one method that could be helpful in this regard is the use of Protacs, or proteolysis targeting chimeric molecules. This technique was originally applied to the normally stable protein methionine aminopeptidase-2 (MetAP-2).[133] Through the use of a phosphopeptide derived from IκBα that binds an SCF ubiquitin ligase complex linked to the MetAP-2 binding protein ovalicin, MetAP-2 was targeted for more rapid ubiquitination and proteasome mediated degradation. Subsequent studies showed that an estradiol-based Protac could similarly induce ER-α degradation in vitro, whereas a dihydroxytestosterone-based Protac promoted turnover of the androgen receptor in a cell line model.[134] Delivery of such molecules in vivo is certainly one possible obstacle, but this powerful technique could be applied to breast cancer therapy for those patients whose tumors depend on overexpression of certain key molecules, such as HER-2, ER-α, and Skp2.

Conclusions

Coordinated regulated control of proteolysis through the UPP is crucial to normal cellular homeostasis, and related activities such as monoubiquitination, NEDDylation, and SUMOylation are also proving to be critical regulatory mechanisms. For many years the complexity of these pathways was very much underappreciated, and whereas recently there has been a dramatic increase in interest and research focused on these areas, we are likely only to have begun to understand their contributions to carcinogenesis. Despite this, it is already clear that mutations that impact on the function of these systems play important roles in breast tumorigenesis. In a relatively short period of time, drugs have been either identified or specifically designed to target components of these pathways that offer the possibility of new treatment options to patients afflicted with breast cancer. Furthermore, as our understanding of these areas grows, it is rational to expect that more therapeutics will be developed that are targeted at the pathogenetic mechanisms underlying cancer and that their enhanced specificity will improve their therapeutic index. Agents in this class should definitely be able to make a significant contribution to our future armamentarium against breast cancer.

Acknowledgments

I acknowledge support from the Department of Defense Breast Cancer Research Program, BC991049, the National Cancer Institute, RO1 CA10227, the National Cancer Institute SPORE in Breast Cancer, and the Leukemia and Lymphoma Society, R6206-02. Support for the phase I trial of bortezomib and pegylated liposomal doxorubicin was provided by Millennium Pharmaceuticals, Inc. (Cambridge, MA) and from the General Clinical Research Centers program of the Division of Research Resources, National Institutes of Health (RR00046).

References

1. Hershko A. Roles of ubiquitin-mediated proteolysis in cell cycle control. Curr Opin Cell Biol 1997;9: 788–799.
2. Bush KT, Goldberg AL, Nigam SK. Proteasome inhibition leads to a heat-shock response, induction of endoplasmic reticulum chaperones, and thermotolerance. J Biol Chem 1997;272:9086–9092.
3. DeMartino GN, Slaughter CA. The proteasome, a novel protease regulated by multiple mechanisms. J Biol Chem 1999;274:22123–22126.
4. Tanahashi N, Kawahara H, Murakami Y, et al. The proteasome-dependent proteolytic system. Mol Biol Rep 1999;26:3–9.
5. Zwickl P, Baumeister W, Steven A. Disassembly lines: The proteasome and related ATPase-assisted proteases. Curr Opin Struct Biol 2000;10:242–250.
6. Ciechanover A, Orian A, Schwartz AL. Ubiquitin-mediated proteolysis: biological regulation via destruction. Bioessays 2000;22:442–451.
7. Kloetzel PM. Antigen processing by the proteasome. Nat Rev Mol Cell Biol 2001;2:179–187.
8. Oikawa T, Sasaki T, Nakamura M, et al. The proteasome is involved in angiogenesis. Biochem Biophys Res Commun 1998;246:243–248.
9. Murakami Y, Matsufuji S, Kameji T, et al. Ornithine decarboxylase is degraded by the 26S proteasome without ubiquitination. Nature 1992;360:597–599.
10. McGee TP, Cheng HH, Kumagai H, et al. Degradation of 3-hydroxy-3-methylglutaryl-CoA reductase in endoplasmic reticulum membranes is accelerated as a result of increased susceptibility to proteolysis. J Biol Chem 1996;271:25630–25638.
11. Davies KJ. Degradation of oxidized proteins by the 20S proteasome. Biochimie 2001;83:301–310.
12. Gonen H, Shkedy D, Barnoy S, et al. On the involvement of calpains in the degradation of the tumor suppressor protein p53. FEBS Lett 1997;406:17–22.
13. Pariat M, Carillo S, Molinari M, et al. Proteolysis by calpains: a possible contribution to degradation of p53. Mol Cell Biol 1997;17:2806–2815.
14. Small GW, Chou TY, Dang CV, et al. Evidence for involvement of calpain in c-Myc proteolysis in vivo. Arch Biochem Biophys 2002;400:151–161.
15. Gregory RC, Taniguchi T, D'Andrea AD. Regulation of the Fanconi anemia pathway by monoubiquitination. Semin Cancer Biol 2003;13:77–82.
16. Haglund K, Di Fiore PP, Dikic I. Distinct monoubiquitin signals in receptor endocytosis. Trends Biochem Sci 2003;28:598–603.
17. Glickman MH, Ciechanover A. The ubiquitin-proteasome proteolytic pathway: destruction for the sake of construction. Physiol Rev 2002;82:373–428.
18. Nalepa G, Wade Harper J. Therapeutic anti-cancer targets upstream of the proteasome. Cancer Treat Rev 2003;29(suppl 1):49–57.
19. Tsvetkov LM, Yeh KH, Lee SJ, et al. p27(Kip1) ubiquitination and degradation is regulated by the SCF(Skp2) complex through phosphorylated Thr187 in p27. Curr Biol 1999;9:661–664.
20. Scheffner M, Huibregtse JM, Vierstra RD, et al. The HPV-16 E6 and E6-AP complex functions as a ubiquitin-protein ligase in the ubiquitination of p53. Cell 1993;75:495–505.
21. Rolfe M, Beer-Romero P, Glass S, et al. Reconstitution of p53-ubiquitinylation reactions from purified

components: The role of human ubiquitin-conjugating enzyme UBC4 and E6-associated protein. Proc Natl Acad Sci USA 1995;92:3264–3268.

22. Hashizume R, Fukuda M, Maeda I, et al. The RING heterodimer BRCA1-BARD1 is a ubiquitin ligase inactivated by a breast cancer-derived mutation. J Biol Chem 2001;276:14537–14540.
23. Shmueli A, Oren M. Regulation of p53 by Mdm2: fate is in the numbers. Mol Cell 2004;13:4–5.
24. Orlowski M, Wilk S. Catalytic activities of the 20 S proteasome, a multicatalytic proteinase complex. Arch Biochem Biophys 2000;383:1–16.
25. Rock KL, Gramm C, Rothstein L, et al. Inhibitors of the proteasome block the degradation of most cell proteins and the generation of peptides presented on MHC class I molecules. Cell 1994;78:761–771.
26. Lipkowitz S. The role of the ubiquitination-proteasome pathway in breast cancer: ubiquitin mediated degradation of growth factor receptors in the pathogenesis and treatment of cancer. Breast Cancer Res 2003;5:8–15.
27. Alkarain A, Slingerland J. Deregulation of p27 by oncogenic signaling and its prognostic significance in breast cancer. Breast Cancer Res 2004;6:13–21.
28. Orlowski RZ, Baldwin AS. NF-kappaB as a therapeutic target in cancer. Trends Mol Med 2002;8:385–389.
29. Cao Y, Karin M. NF-kappaB in mammary gland development and breast cancer. J Mamm Gland Biol Neoplasia 2003;8:215–223.
30. Karin M, Yamamoto Y, Wang QM. The IKK NF-kappa B system: a treasure trove for drug development. Nat Rev Drug Discov 2004;3:17–26.
31. Hu X. Proteolytic signaling by TNFalpha: caspase activation and IkappaB degradation. Cytokine 2003;21:286–294.
32. Chen LC, Manjeshwar S, Lu Y, et al. The human homologue for the *Caenorhabditis elegans* cul-4 gene is amplified and overexpressed in primary breast cancers. Cancer Res 1998;58:3677–3683.
33. Strohmaier H, Spruck CH, Kaiser P, et al. Human F-box protein hCdc4 targets cyclin E for proteolysis and is mutated in a breast cancer cell line. Nature 2001;413:316–322.
34. Henrich LM, Smith JA, Kitt D, et al. Extracellular signal-regulated kinase 7, a regulator of hormone-dependent estrogen receptor destruction. Mol Cell Biol 2003;23:5979–5988.
35. Nicoletto MO, Donach M, De Nicolo A, et al. BRCA-1 and BRCA-2 mutations as prognostic factors in clinical practice and genetic counseling. Cancer Treat Rev 2001;27:295–304.
36. Powell SN, Kachnic LA. Roles of BRCA1 and BRCA2 in homologous recombination, DNA replication fidelity and the cellular response to ionizing radiation. Oncogene 2003;22:5784–5791.
37. Baer R, Ludwig T. The BRCA1/BARD1 heterodimer, a tumor suppressor complex with ubiquitin E3 ligase activity. Curr Opin Genet Dev 2002;12:86–91.
38. Brzovic PS, Keeffe JR, Nishikawa H, et al. Binding and recognition in the assembly of an active BRCA1/BARD1 ubiquitin-ligase complex. Proc Natl Acad Sci USA 2003;100:5646–5651.
39. Dong Y, Hakimi MA, Chen X, et al. Regulation of BRCC, a holoenzyme complex containing BRCA1 and BRCA2, by a signalosome-like subunit and its role in DNA repair. Mol Cell 2003;12:1087–1099.
40. Inoue S, Orimo A, Hosoi T, et al. Genomic binding-site cloning reveals an estrogen-responsive gene that encodes a RING finger protein. Proc Natl Acad Sci USA 1993;90:11117–11121.
41. Urano T, Saito T, Tsukui T, et al. Efp targets 14-3-3 sigma for proteolysis and promotes breast tumour growth. Nature 2002;417:871–875.
42. Laronga C, Yang HY, Neal C, et al. Association of the cyclin-dependent kinases and 14-3-3 sigma negatively regulates cell cycle progression. J Biol Chem 2000;275:23106–23112.
43. Vercoutter-Edouart AS, Lemoine J, Le Bourhis X, et al. Proteomic analysis reveals that 14-3-3 sigma is down-regulated in human breast cancer cells. Cancer Res 2001;61:76–80.
44. Ferguson AT, Evron E, Umbricht CB, et al. High frequency of hypermethylation at the 14-3-3 sigma locus leads to gene silencing in breast cancer. Proc Natl Acad Sci USA 2000;97:6049–6054.
45. Iwata N, Yamamoto H, Sasaki S, et al. Frequent hypermethylation of CpG islands and loss of expression of the 14–3–3 sigma gene in human hepatocellular carcinoma. Oncogene 2000;19:5298–5302.
46. Umbricht CB, Evron E, Gabrielson E, et al. Hypermethylation of 14-3-3 sigma (stratifin) is an early event in breast cancer. Oncogene 2001;20:3348–3353.
47. Zhou BP, Hung MC. Dysregulation of cellular signaling by HER2/neu in breast cancer. Semin Oncol 2003;30:38–48.
48. Menard S, Pupa SM, Campiglio M, et al. Biologic and therapeutic role of HER2 in cancer. Oncogene 2003;22:6570–6578.
49. Spigel DR, Burstein HJ. Trastuzumab regimens for HER2-overexpressing metastatic breast cancer. Clin Breast Cancer 2003;4:329–337, discussion 338–339.
50. Ko BK, Kawano K, Murray JL, et al. Clinical studies of vaccines targeting breast cancer. Clin Cancer Res 2003;9:3222–3234.
51. Lollini PL, Nicoletti G, Landuzzi L, et al. Down regulation of major histocompatibility complex class I expression in mammary carcinoma of HER–2/*neu* transgenic mice. Int J Cancer 1998;77:937–941.
52. Herrmann F, Lehr HA, Drexler I, et al. HER-2/*neu*-mediated regulation of components of the MHC class I antigen-processing pathway. Cancer Res 2004;64:215–220.
53. Kruger E, Kuckelkorn U, Sijts A, et al. The components of the proteasome system and their role in MHC class I antigen processing. Rev Physiol Biochem Pharmacol 2003;148:81–104.
54. Vargas DA, Takahashi S, Ronai Z. Mdm2: a regulator of cell growth and death. Adv Cancer Res 2003;89:1–34.
55. Picard D, Bunone G, Liu JW, et al. Steroid-independent activation of steroid receptors in mammalian and yeast cells and in breast cancer. Biochem Soc Trans 1997;25:597–602.
56. Kinyamu HK, Archer TK. Estrogen receptor-dependent proteasomal degradation of the glucocorticoid receptor is coupled to an increase in Mdm2 protein expression. Mol Cell Biol 2003;23:5867–5881.
57. Borresen-Dale AL. TP53 and breast cancer. Hum Mutat 2003;21:292–300.
58. Sheikh MS, Shao ZM, Hussain A, et al. The p53-binding protein Mdm2 gene is differentially expressed in human breast carcinoma. Cancer Res 1993;53:3226–3228.

59. Quesnel B, Preudhomme C, Fournier J, et al. Mdm2 gene amplification in human breast cancer. Eur J Cancer 1994;30A:982–984.
60. McCann AH, Kirley A, Carney DN, et al. Amplification of the Mdm2 gene in human breast cancer and its association with Mdm2 and p53 protein status. Br J Cancer 1995;71:981–985.
61. Wang H, Nan L, Yu D, et al. Antisense anti-Mdm2 oligonucleotides as a novel therapeutic approach to human breast cancer: *in vitro* and *in vivo* activities and mechanisms. Clin Cancer Res 2001;7: 3613–3624.
62. Swaroop M, Sun Y. Mdm2 ligase dead mutants did not act in a dominant negative manner to re-activate p53, but promoted tumor cell growth. Anticancer Res 2003;23:3167–3174.
63. Moller MB. P27 in cell cycle control and cancer. Leuk Lymph 2000;39:19–27.
64. Nho RS, Sheaff RJ. p27kip1 contributions to cancer. Progr Cell Cycle Res 2003;5:249–259.
65. Doisneau-Sixou SF, Sergio CM, Carroll JS, et al. Estrogen and anti-estrogen regulation of cell cycle progression in breast cancer cells. Endocr Relat Cancer 2003;10:179–186.
66. Foster JS, Fernando RI, Ishida N, et al. Estrogens down-regulate p27Kip1 in breast cancer cells through Skp2 and through nuclear export mediated by the ERK pathway. J Biol Chem 2003;278:41355–41366.
67. Signoretti S, Di Marcotullio L, Richardson A, et al. Oncogenic role of the ubiquitin ligase subunit Skp2 in human breast cancer. J Clin Invest 2002;110: 633–641.
68. Steeg PS, Zhou Q. Cyclins and breast cancer. Breast Cancer Res Treat 1998;52:17–28.
69. Zhou Q, Hopp T, Fuqua SA, et al. Cyclin D1 in breast pre-malignancy and early breast cancer: Implications for prevention and treatment. Cancer Lett 2001;162:3–17.
70. Russell A, Hendley J, Germain D. Inhibitory effect of p21 in MCF-7 cells is overcome by its coordinated stabilization with D-type cyclins. Oncogene 1999; 18:6454–6459.
71. Russell A, Thompson MA, Hendley J, et al. Cyclin D1 and D3 associate with the SCF complex and are coordinately elevated in breast cancer. Oncogene 1999;18:1983–1991.
72. Lightcap ES, McCormack TA, Pien CS, et al. Proteasome inhibition measurements: clinical application. Clin Chem 2000;46:673–683.
73. Fenteany G, Schreiber SL. Lactacystin, proteasome function, and cell fate. J Biol Chem 1998;273:8545–8548.
74. Adams J, Behnke M, Chen S, et al. Potent and selective inhibitors of the proteasome: dipeptidyl boronic acids. Bioorg Med Chem Lett 1998;8: 333–338.
75. Aghajanian C, Soignet S, Dizon DS, et al. A phase I trial of the novel proteasome inhibitor PS341 in advanced solid tumor malignancies. Clin Cancer Res 2002;8:2505–2511.
76. Orlowski RZ, Stinchcombe TE, Mitchell BS, et al. Phase I trial of the proteasome inhibitor PS-341 in patients with refractory hematologic malignancies. J Clin Oncol 2002;20:4420–4427.
77. Bartlett J, Mallon E, Cooke T. The clinical evaluation of HER-2 status: which test to use? J Pathol 2003;199:411–417.
78. Richardson PG, Barlogie B, Berenson J, et al. A phase 2 study of bortezomib in relapsed, refractory myeloma. N Engl J Med 2003;348:2609–2617.
79. Richardson PG, Barlogie B, Berenson J, et al. Prognostic factors for response parameters and overall survival in patients with multiple myeloma following treatment with bortezomib. Blood 2003;102: 446a(abst. 1629).
80. Dow R, Hendley J, Pirkmaier A, et al. Retinoic acid-mediated growth arrest requires ubiquitylation and degradation of the F-box protein Skp2. J Biol Chem 2001;276:45945–45951.
81. Marsaud V, Gougelet A, Maillard S, et al. Various phosphorylation pathways, depending on agonist and antagonist binding to endogenous estrogen receptor alpha (ERalpha), differentially affect ERalpha extractability, proteasome-mediated stability, and transcriptional activity in human breast cancer cells. Mol Endocrinol 2003;17:2013–2027.
82. Robertson JF. ICI 182,780 (Fulvestrant)—the first oestrogen receptor down-regulator. Current clinical data. Br J Cancer 2001;85(suppl 2):11–14.
83. Preisler-Mashek MT, Solodin N, Stark BL, et al. Ligand-specific regulation of proteasome-mediated proteolysis of estrogen receptor-alpha. Am J Physiol Endocrinol Metab 2002;282:E891–E898.
84. Adams J, Palombella VJ, Sausville EA, et al. Proteasome inhibitors: a novel class of potent and effective antitumor agents. Cancer Res 1999;59:2615–2622.
85. Teicher BA, Ara G, Herbst R, et al. The proteasome inhibitor PS-341 in cancer therapy. Clin Cancer Res 1999;5:2638–2645.
86. Voorhees PM, Dees EC, O'Neil B, et al. The proteasome as a target for cancer therapy. Clin Cancer Res 2003;9:6316–6325.
87. Orlowski RZ, Dees EC. The role of the ubiquitination-proteasome pathway in breast cancer: applying drugs that affect the ubiquitin-proteasome pathway to the therapy of breast cancer. Breast Cancer Res 2003;5:1–7.
88. Orlowski RZ, Small GW, Shi YY. Evidence that inhibition of p44/42 mitogen-activated protein kinase signaling is a factor in proteasome inhibitor-mediated apoptosis. J Biol Chem 2002;277:27864–27871.
89. Hilger RA, Scheulen ME, Strumberg D. The Ras-Raf-MEK-ERK pathway in the treatment of cancer. Onkologie 2002;25:511–518.
90. Zhang XM, Lin H, Chen C, et al. Inhibition of ubiquitin-proteasome pathway activates a caspase-3-like protease and induces Bcl-2 cleavage in human M-07e leukaemic cells. Biochem J 1999;340(Pt 1): 127–133.
91. Chadebech P, Brichese L, Baldin V, et al. Phosphorylation and proteasome-dependent degradation of Bcl–2 in mitotic-arrested cells after microtubule damage. Biochem Biophys Res Commun 1999;262:823–827.
92. Loo TW, Clarke DM. Superfolding of the partially unfolded core-glycosylated intermediate of human P-glycoprotein into the mature enzyme is promoted by substrate-induced transmembrane domain interactions. J Biol Chem 1998;273:14671–14674.
93. Loo TW, Clarke DM. The human multidrug resistance P-glycoprotein is inactive when its maturation is inhibited: potential for a role in cancer chemotherapy. FASEB J 1999;13:1724–1732.
94. Dees EC, O'Neil B, Humes E, et al. Phase I trial of the proteasome inhibitor bortezomib in combination

with pegylated liposomal doxorubicin in patients with refractory solid tumors. Proc Am Soc Clin Oncol 2003;22:217(abst. 868).

95. Neckers L, Schulte TW, Mimnaugh E. Geldanamycin as a potential anti-cancer agent: its molecular target and biochemical activity. Invest New Drugs 1999;17: 361–373.
96. Mimnaugh EG, Chavany C, Neckers L. Polyubiquitination and proteasomal degradation of the p185c-erbB-2 receptor protein-tyrosine kinase induced by geldanamycin. J Biol Chem 1996;271: 22796–22801.
97. Maloney A, Workman P. HSP-90 as a new therapeutic target for cancer therapy: the story unfolds. Exp Opin Biol Ther 2002;2:3–24.
98. Neckers L. HSP-90 inhibitors as novel cancer chemotherapeutic agents. Trends Mol Med 2002;8: S55–S61.
99. Fujita N, Sato S, Ishida A, et al. Involvement of HSP-90 in signaling and stability of 3-phosphoinositide-dependent kinase-1. J Biol Chem 2002;277:10346–10353.
100. Xu W, Yuan X, Jung YJ, et al. The heat shock protein 90 inhibitor geldanamycin and the ErbB inhibitor ZD1839 promote rapid PP1 phosphatase-dependent inactivation of Akt in ErbB2 overexpressing breast cancer cells. Cancer Res 2003;63:7777–7784.
101. Franke TF, Hornik CP, Segev L, et al. PI3K/Akt and apoptosis: size matters. Oncogene 2003;22: 8983–8998.
102. Datta SR, Dudek H, Tao X, et al. Akt phosphorylation of BAD couples survival signals to the cell-intrinsic death machinery. Cell 1997;91:231–241.
103. del Peso L, Gonzalez-Garcia M, Page C, et al. Interleukin-3-induced phosphorylation of BAD through the protein kinase Akt. Science 1997;278: 687–689.
104. Schulte TW, Blagosklonny MV, Ingui C, et al. Disruption of the Raf-1-HSP90 molecular complex results in destabilization of Raf-1 and loss of Raf-1-Ras association. J Biol Chem 1995;270:24585–24588.
105. Fang X, Yu S, Eder A, et al. Regulation of BAD phosphorylation at serine 112 by the Ras-mitogen-activated protein kinase pathway. Oncogene 1999; 18:6635–6640.
106. Tzivion G, Shen YH, Zhu J. 14-3-3 proteins; bringing new definitions to scaffolding. Oncogene 2001;20: 6331–6338.
107. Zhu W, Zou Y, Aikawa R, et al. MAPK superfamily plays an important role in daunomycin-induced apoptosis of cardiac myocytes. Circulation 1999;100: 2100–2107.
108. Mas VM, Hernandez H, Plo I, et al. Protein kinase Czeta mediated Raf-1/extracellular-regulated kinase activation by daunorubicin. Blood 2003;101: 1543–1550.
109. Small GW, Somasundaram S, Moore DT, et al. Repression of mitogen-activated protein kinase (MAPK) phosphatase-1 by anthracyclines contributes to their anti-apoptotic activation of p44/42-MAPK. J Pharmacol Exp Ther 2003;307:861–869.
110. Le XF, Claret FX, Lammayot A, et al. The role of cyclin-dependent kinase inhibitor p27Kip1 in anti-HER2 antibody-induced G1 cell cycle arrest and tumor growth inhibition. J Biol Chem 2003;278: 23441–23450.
111. Lu Y, Zi X, Pollak M. Molecular mechanisms underlying IGF-I-induced attenuation of the growth-inhibitory activity of trastuzumab (Herceptin) on SKBR3 breast cancer cells. Int J Cancer 2004; 108:334–341.
112. Fuino L, Bali P, Wittmann S, et al. Histone deacetylase inhibitor LAQ824 down-regulates HER-2 and sensitizes human breast cancer cells to trastuzumab, taxotere, gemcitabine, and epothilone B. Mol Cancer Ther 2003;2:971–984.
113. Bueso-Ramos CE, Manshouri T, Haidar MA, et al. Abnormal expression of Mdm-2 in breast carcinomas. Breast Cancer Res Treat 1996;37:179–188.
114. Freedman DA, Levine AJ. Nuclear export is required for degradation of endogenous p53 by Mdm2 and human papillomavirus E6. Mol Cell Biol 1998;18: 7288–7293.
115. Legube G, Linares LK, Lemercier C, et al. Tip60 is targeted to proteasome-mediated degradation by Mdm2 and accumulates after UV irradiation. EMBO J 2002;21:1704–1712.
116. Lane HA, Beuvink I, Motoyama AB, et al. ErbB2 potentiates breast tumor proliferation through modulation of p27(Kip1)-Cdk2 complex formation: Receptor overexpression does not determine growth dependency. Mol Cell Biol 2000;20:3210–3223.
117. Timms JF, White SL, O'Hare MJ, et al. Effects of ErbB-2 overexpression on mitogenic signalling and cell cycle progression in human breast luminal epithelial cells. Oncogene 2002;21:6573–6586.
118. Philipp-Staheli J, Payne SR, Kemp CJ. p27(Kip1): regulation and function of a haploinsufficient tumor suppressor and its misregulation in cancer. Exp Cell Res 2001;264:148–168.
119. Kim SY, Herbst A, Tworkowski KA, et al. Skp2 regulates Myc protein stability and activity. Mol Cell 2003;11:1177–1188.
120. Liao DJ, Dickson RB. c-Myc in breast cancer. Endocr Relat Cancer 2000;7:143–164.
121. Yu ZK, Gervais JL, Zhang H. Human Cul-1 associates with the SKP1/SKP2 complex and regulates p21(CIP1/WAF1) and cyclin D proteins. Proc Natl Acad Sci USA 1998;95:11324–11329.
122. Podust VN, Brownell JE, Gladysheva TB, et al. A NEDD8 conjugation pathway is essential for proteolytic targeting of p27Kip1 by ubiquitination. Proc Natl Acad Sci USA 2000;97:4579–4584.
123. Tanaka K, Kawakami T, Tateishi K, et al. Control of IkappaBalpha proteolysis by the ubiquitin-proteasome pathway. Biochimie 2001;83:351–356.
124. Amir RE, Iwai K, Ciechanover A. The NEDD8 pathway is essential for SCF(beta-TrCP)-mediated ubiquitination and processing of the NF-kappa B precursor p105. J Biol Chem 2002;277:23253–23259.
125. Fan M, Bigsby RM, Nephew KP. The NEDD8 pathway is required for proteasome-mediated degradation of human estrogen receptor (ER)-alpha and essential for the antiproliferative activity of ICI 182,780 in ERalpha-positive breast cancer cells. Mol Endocrinol 2003;17:356–365.
126. Desterro JM, Rodriguez MS, Hay RT. SUMO-1 modification of IkappaBalpha inhibits NF-kappaB activation. Mol Cell 1998;2:233–239.
127. Zou J, Guo Y, Guettouche T, et al. Repression of heat shock transcription factor HSF1 activation by HSP-90 (HSP-90 complex) that forms a stress-sensitive complex with HSF1. Cell 1998;94:471–480.
128. Bagatell R, Paine-Murrieta GD, Taylor CW, et al. Induction of a heat shock factor 1-dependent stress response alters the cytotoxic activity of

HSP-90-binding agents. Clin Cancer Res 2000;6: 3312–3318.

129. Meriin AB, Gabai VL, Yaglom J, et al. Proteasome inhibitors activate stress kinases and induce HSP–72. Diverse effects on apoptosis. J Biol Chem 1998;273:6373–6379.

130. Kawazoe Y, Nakai A, Tanabe M, et al. Proteasome inhibition leads to the activation of all members of the heat-shock-factor family. Eur J Biol Chem 1998;255: 356–362.

131. Gabai VL, Meriin AB, Yaglom JA, et al. Role of HSP-70 in regulation of stress-kinase JNK: implications in apoptosis and aging. FEBS Lett 1998;438:1–4.

132. Robertson JD, Datta K, Biswal SS, et al. Heat-shock protein 70 antisense oligomers enhance proteasome inhibitor-induced apoptosis. Biochem J 1999;344 (Pt 2):477–485.

133. Sakamoto KM, Kim KB, Kumagai A, et al. Protacs: chimeric molecules that target proteins to the Skp1-Cullin-F box complex for ubiquitination and degradation. Proc Natl Acad Sci USA 2001;98:8554–8559.

134. Sakamoto KM, Kim KB, Verma R, et al. Development of Protacs to target cancer-promoting proteins for ubiquitination and degradation. Mol Cell Proteomics 2003;2:1350–1358.

135. Shimizu H, Seiki T, Asada M, et al. Alpha6beta1 integrin induces proteasome-mediated cleavage of erbB2 in breast cancer cells. Oncogene 2003;22: 831–839.

136. Townsend PA, Cutress RI, Sharp A, et al. BAG-1 prevents stress-induced long-term growth inhibition in breast cancer cells via a chaperone-dependent pathway. Cancer Res 2003;63:4150–4157.

137. Doong H, Rizzo K, Fang S, et al. CAIR-1/BAG-3 abrogates heat shock protein-70 chaperone complex-mediated protein degradation: accumulation of poly-ubiquitinated HSP–90 client proteins. J Biol Chem 2003;278:28490–28500.

138. Ettenberg SA, Rubinstein YR, Banerjee P, et al. Cbl-b inhibits EGF-receptor-induced apoptosis by enhancing ubiquitination and degradation of activated receptors. Mol Cell Biol Res Commun 1999;2: 111–118.

139. Song DH, Dominguez I, Mizuno J, et al. CK2 phosphorylation of the armadillo repeat region of beta-catenin potentiates Wnt signaling. J Biol Chem 2003;278:24018–24025.

140. Romieu-Mourez R, Landesman-Bollag E, Seldin DC, et al. Protein kinase CK2 promotes aberrant activation of nuclear factor-kappaB, transformed phenotype, and survival of breast cancer cells. Cancer Res 2002;62:6770–6778.

141. Clancy JL, Henderson MJ, Russell AJ, et al. EDD, the human orthologue of the hyperplastic discs tumour suppressor gene, is amplified and overexpressed in cancer. Oncogene 2003;22:5070–5081.

142. Horie K, Urano T, Ikeda K, et al. Estrogen-responsive RING finger protein controls breast cancer growth. J Steroid Biochem Mol Biol 2003;85: 101–104.

143. Morelli C, Garofalo C, Bartucci M, et al. Estrogen receptor-alpha regulates the degradation of insulin receptor substrates 1 and 2 in breast cancer cells. Oncogene 2003;22:4007–4016.

144. Ren S, Smith MJ, Louro ID, et al. The p44S10 locus, encoding a subunit of the proteasome regulatory particle, is amplified during progression of cutaneous malignant melanoma. Oncogene 2000;19: 1419–1427.

145. Thompson HG, Harris JW, Wold BJ, et al. p62 overexpression in breast tumors and regulation by prostate-derived Ets factor in breast cancer cells. Oncogene 2003;22:2322–2333.

146. Dressman MA, Baras A, Malinowski R, et al. Gene expression profiling detects gene amplification and differentiates tumor types in breast cancer. Cancer Res 2003;63:2194–2199.

147. Shekhar MP, Lyakhovich A, Visscher DW, et al. Rad6 overexpression induces multinucleation, centrosome amplification, abnormal mitosis, aneuploidy, and transformation. Cancer Res 2002;62:2115–2124.

148. Lyakhovich A, Shekhar MP. Supramolecular complex formation between Rad6 and proteins of the p53 pathway during DNA damage-induced response. Mol Cell Biol 2003;23:2463–2475.

149. Subramaniam V, Li H, Wong M, et al. The RING-H2 protein RNF11 is overexpressed in breast cancer and is a target of Smurf2 E3 ligase. Br J Cancer 2003; 89:1538–1544.

150. Brickley DR, Mikosz CA, Hagan CR, et al. Ubiquitin modification of serum and glucocorticoid-induced protein kinase-1. J Biol Chem 2002;277:43064–43070.

151. Wu C, Guan Q, Wang Y, et al. SHP-1 suppresses cancer cell growth by promoting degradation of JAK kinases. J Cell Biol Chem 2003;90:1026–1037.

152. Petrel TA, Brueggemeier RW. Increased proteasome-dependent degradation of estrogen receptor-alpha by TGF-beta1 in breast cancer cell lines. J Cell Biol Chem 2003;88:181–190.

153. Way TD, Kao MC, Lin JK. Apigenin induces apoptosis through proteasomal degradation of HER2/*neu* in HER2/*neu*-overexpressing breast cancer cells via the phosphatidylinositol 3-kinase/Akt-dependent pathway. J Biol Chem 2004;279:4479–4489.

154. Qin C, Burghardt R, Smith R, et al. Peroxisome proliferator-activated receptor gamma agonists induce proteasome-dependent degradation of cyclin D1 and estrogen receptor alpha in MCF-7 breast cancer cells. Cancer Res 2003;63:958–964.

155. El Khissiin A, Leclercq G. Implication of proteasome in estrogen receptor degradation. FEBS Lett 1999; 448:160–166.

156. Kazi A, Daniel KG, Smith DM, et al. Inhibition of the proteasome activity, a novel mechanism associated with the tumor cell apoptosis-inducing ability of genistein. Biochem Pharmacol 2003;66: 965–976.

157. Magnifico A, Tagliabue E, Ardini E, et al. Heregulin beta1 induces the down regulation and the ubiquitin-proteasome degradation pathway of p185HER2 oncoprotein. FEBS Lett 1998;422:129–131.

158. Lange CA, Shen T, Horwitz KB. Phosphorylation of human progesterone receptors at serine-294 by mitogen-activated protein kinase signals their degradation by the 26S proteasome. Proc Natl Acad Sci USA 2000;97:1032–1037.

159. Zhou Q, Melkoumian ZK, Lucktong A, et al. Rapid induction of histone hyperacetylation and cellular differentiation in human breast tumor cell lines following degradation of histone deacetylase-1. J Biol Chem 2000;275:35256–35263.

160. Lee MO, Kim EO, Kwon HJ, et al. Radicicol represses the transcriptional function of the estrogen receptor by suppressing the stabilization of the receptor by heat shock protein 90. Mol Cell Endocrinol 2002;188:47–54.

161. Wormke M, Stoner M, Saville B, et al. The aryl hydrocarbon receptor mediates degradation of estrogen receptor alpha through activation of proteasomes. Mol Cell Biol 2003;23:1843–1855.

162. Pluquet O, North S, Bhoumik A, et al. The cytoprotective aminothiol WR1065 activates p53 through a non-genotoxic signaling pathway involving c-Jun N-terminal kinase. J Biol Chem 2003;278:11879–11887.

CHAPTER

25

DNA Repair and the Maintenance of Genomic Stability in Breast Cancer

Jennifer E. Quinn, PhD, Richard D. Kennedy, MB, and D. Paul Harkin, PhD

Cancer Research Centre, Queens University Belfast, University Floor, Belfast City Hospital, Lisburn Road, Belfast BT9 7AB

In this chapter we focus on the *BRCA1* tumor suppressor gene (TSG) and its role in the maintenance of genomic stability through its multiple cellular functions in transcription, DNA damage response and DNA repair, cell cycle regulation, and ubiquitination. The mechanism through which inactivation of *BRCA1* leads to genetic instability and ultimately the development of breast and ovarian tumors is discussed. The use of *BRCA1* as a potential predictive marker of response to chemotherapy is also addressed. We conclude with a brief consideration of the *BRCA2* gene and other genes associated with susceptibility to DNA damage and genomic instability in breast cancer.

The molecular biology of breast cancer is complex, with multiple factors contributing to its development. These include the ability of a breast tumor to acquire growth signal autonomy, to evade apoptosis, to become insensitive to antigrowth signals, to develop sustained angiogenesis, to replicate limitlessly, and to invade tissue and undergo metastasis.[1] The underlying genetic changes by which these characteristics are acquired include alterations in the three major classes of genes that have been implicated in tumor initiation: oncogenes, DNA repair genes, and TSGs.

TSGs (see Chapter 23) are negative regulators of cellular growth and function to prevent cells that bear damaged DNA from proliferating, either temporarily until damage is repaired or permanently by pushing the damaged cell down an irreversible apoptotic pathway.[2] TSGs can be inactivated through multiple point mutations or loss of heterozygosity (LOH), resulting in genomic instability. Typically, both copies of TSG must be inactivated to reveal the malignant phenotype.[3] In hereditary forms of cancer, a single mutated copy of a TSG is inherited through the germline with a second mutational event resulting in inactivation of the remaining wild-type allele in a target somatic cell. Well-characterized TSGs that have been identified over the last two decades include *p53, Rb, WT1, ATM, BRCA1,* and *BRCA2* (see Chapter 4).

Breast cancer is a leading cause of mortality among women in the Western Hemisphere, with highest frequencies being observed in North America and the countries of Northern Europe. It is currently estimated that one in eight women in these countries will be diagnosed with breast carcinoma. The disease is multifactorial, and some of the risk factors are well established, including advancing age (>50 years), reproductive factors (nulliparity, early menarche, late menopause), family history, and previous benign disease. However, there are other less well-established factors that may also increase risk, such as poor diet, excessive alcohol consumption, lack of exercise, exposure to ionizing radiation, and taking exogenous hormones. Of these risk factors, a family history of the disease has been identified as a particularly significant risk factor for developing breast cancer.

Approximately 5–10% of all breast cancers are hereditary (familial) in origin, whereas the remaining 90–95% of cases are described as sporadic breast cancers. Hereditary breast cancers are characterized by early age at onset (<40 years), bilateral disease, a high number of first-degree relatives affected, and a strong association with ovarian cancer. These patients appear to have a genetic predisposition to developing breast and/or ovarian cancer. Approximately 80% of hereditary breast cancers are accounted for by mutations in the breast cancer

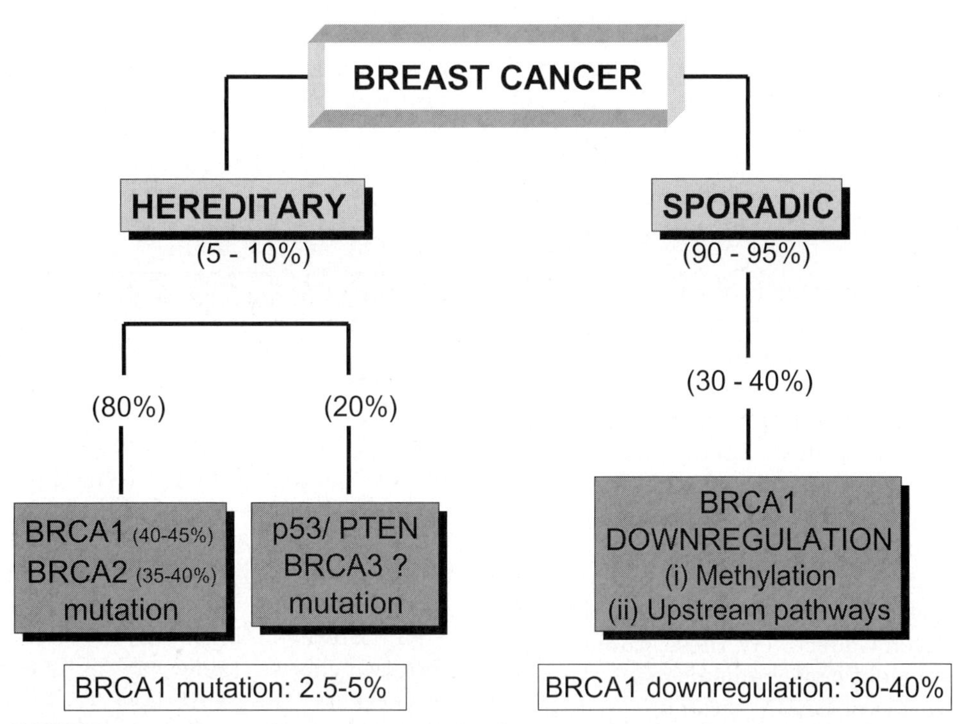

FIGURE 25.1 Of the 5 to 10% of breast cancer cases that develop on a hereditary basis, approximately 80% are accounted for by mutations in the breast cancer susceptibility genes, BRCA1 (40–45%) and BRCA2 (35–40%). The remaining 20% of familial breast cancer may be due to mutations in other tumour suppressor genes such as p53, PTEN and ATM in addition to other as yet undiscovered BRCA genes (see Chapter 4).

susceptibility genes, *BRCA1* (40–45%) and *BRCA2* (35–40%). It is proposed that some of the remaining 20% of familial breast cancer may be due to mutations in other TSGs, such as *p53, PTEN,* and *ATM* in addition to other as yet undiscovered BRCA genes (FIG. 25.1).[4]

The initial breakthrough in identifying genes that confer susceptibility to familial breast cancer came with the identification of a susceptibility locus on 17q21 by genetic linkage analysis.[5] The first breast cancer susceptibility gene, named *BRCA1,* was subsequently cloned and found to be mutated in families with early onset breast cancer.[6] Mutations within the *BRCA1* coding region are thought to account for approximately 45% of inherited breast cancers and approximately 90% of families with increased incidence of both early onset breast and ovarian cancer.[6] *BRCA1* mutated tumors are characterized by poor differentiation, high rates of proliferation, estrogen and progestogen receptor negativity, p53 mutation, and HER-2 negativity.

Carriers of *BRCA1* germline mutations have a lifetime risk of 82% and 54% of developing breast or ovarian cancer, respectively, by age 80.[7] Further analyses of these disease predisposing germline mutations indicate that 90% of mutations result in the premature truncation of the BRCA1 protein. Among these mutations are missense mutations that result in frameshift or nonsense mutations.[6,8–10] Unlike other TSGs such as p53, there are no mutational hot spots and no clear connection between phenotype, type of mutation, or its location within the gene. However, BRCA1 founder mutations have been identified in a number of select population groupings. The main example of this occurs in the Ashkenazi Jewish population, where approximately 1.25% of women carry the 185del AG mutation, making the overall penetrance of this mutation 38% by age 40.[11,12] In addition, founder mutations that predispose to breast and ovarian cancer have been observed in the Dutch (2804delAA) and Swedish populations (3175ins5).[13,14] These findings have led to genetic testing for specific mutations in these families; however, the identification of over 200 distinct mutations to date has made predictive genetic testing difficult in the general population.

It was also speculated that BRCA1 might play a role in the development of the sporadic form of breast and ovarian cancer due to the detection of frequent LOH in the BRCA1 region in these tumors. However, unlike other TSGs such as p53, there have been few reports of BRCA1 mutations in sporadic ovarian cancer and none reported for sporadic breast cancer, leading to some speculation over the role, if any, of BRCA1 in the sporadic form of the disease.

It was demonstrated that BRCA1 messenger RNA and protein expression levels are reduced in most invasive breast tumors[15,16] and in sporadic ovarian cancers.[17] In addition, it was found that BRCA1 protein expression was significantly reduced in about 34.3% of patients with early sporadic breast cancer.[18] Taken together, these findings suggest that abrogation of BRCA1 function by decreased expression is probably a critical step in the development of the sporadic form of breast and ovarian cancer. Collectively, these studies suggest that epigenetic mechanisms may result in the inactivation of BRCA1 protein expression in sporadic disease. Such events may include hypermethylation of the BRCA1 CpG rich promoter, aberrant subcellular localization of BRCA1, and defects in the upstream molecular pathways that modulate BRCA1 function.

BRCA1 Expression Pattern

Studies of Brca1 during development in mice provide clues about the distribution and function of the human BRCA1 protein. Brca1 expression is most abundant in tissues undergoing rapid proliferation and differentiation such as the thymus, testis, breast, and ovaries. In the mammary gland, BRCA1 is expressed in an epithelial-specific fashion consistent with the fact that germline alterations in BRCA1 ultimately result in malignant transformation of this cell type in the breast. In addition, Brca1 is found at highest levels in terminal end buds as they differentiate into the ductal epithelium during puberty.[19] Expression has also been found to increase sharply during pregnancy induced mammary gland development and to decrease after parturition in a pattern consistent with its regulation by pregnancy associated hormonal changes. A link between BRCA1 and estrogen regulated signaling pathways has also been inferred. Evidence for this has come from a recent study that demonstrates that BRCA1 can inhibit signaling by the ligand activated estrogen receptor (ER) and suggests that BRCA1 may be a negative regulator of ER mediated proliferation.[20–22] Collectively, these reports may also help to explain why *BRCA1* mutated tumors manifest specifically in the breast and ovary, although extensive research is still required to conclusively address this issue.

BRCA1 and Tumor Suppression

BRCA1 deficiency results in spontaneous chromosomal instability and allows cells to undergo neoplastic transformation. An acquired *BRCA1* germline mutation is sufficient to cause cancer predisposition. However, LOH analyses indicates that the second allele is constitutively lost in these predisposed individuals. Thus, the *BRCA1* gene conforms to Knudson's two hit hypothesis of a classic TSG.[3] However, TSGs can be further divided into two classes, known as "gatekeepers" and "caretakers."[23] A gatekeeper TSG such as p53 directly regulates cellular proliferation by inhibiting growth or promoting cell death. The inactivation of this type of gene is defined as the rate limiting step in tumor initiation. The disruption of a TSG that causes genomic instability increasing the frequency of alterations in other genes is known as a caretaker. Inactivation of a caretaker TSG is not a rate limiting step in cancer progression but rather destabilizes the genome, which results in the accumulation of mutations in all genes, including gatekeepers like p53, which then leads to tumor formation. We now discuss further the role of BRCA1 as a caretaker of the genome.

BRCA1 Is a Multifunctional Protein

BRCA1 is a large gene, spanning approximately 81 kb of genomic DNA, and is comprised of 24 exons (22 of which are coding) and displays a predominant messenger RNA transcript size of 7.8 kb. The BRCA1 open reading frame encodes a nuclear phosphoprotein comprising 1863 amino acids with a predicted molecular weight of 220 kDa. BRCA1 is a multifunctional protein that has been implicated in modulating the cellular response to DNA damage, the regulation of transcription, cell cycle checkpoint control, and ubiquitination (FIG. 25.2). However, the exact mechanism through which BRCA1 inactivation directly leads to tumorigenesis remains to be determined. Cellular

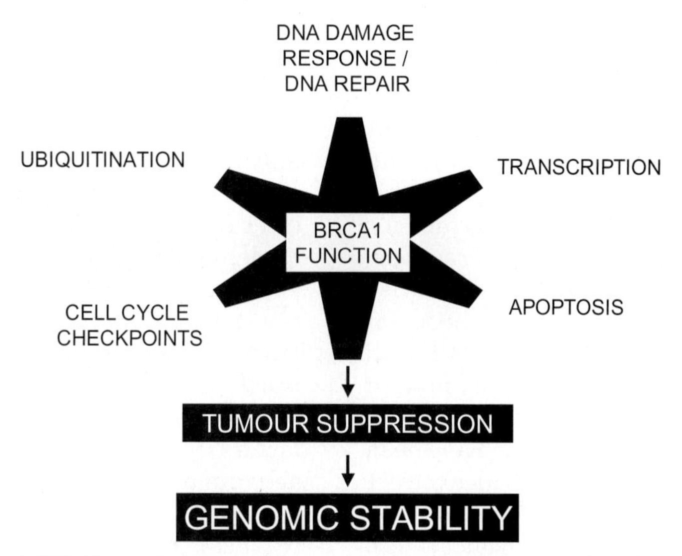

FIGURE 25.2 BRCA1 is a multifunctional protein that has been implicated in modulating the cellular response to DNA damage, the regulation of transcription, cell cycle checkpoint control and ubiquitination. Through these proposed mechanisms, BRCA1 plays a major role in tumor suppression and genomic stability.

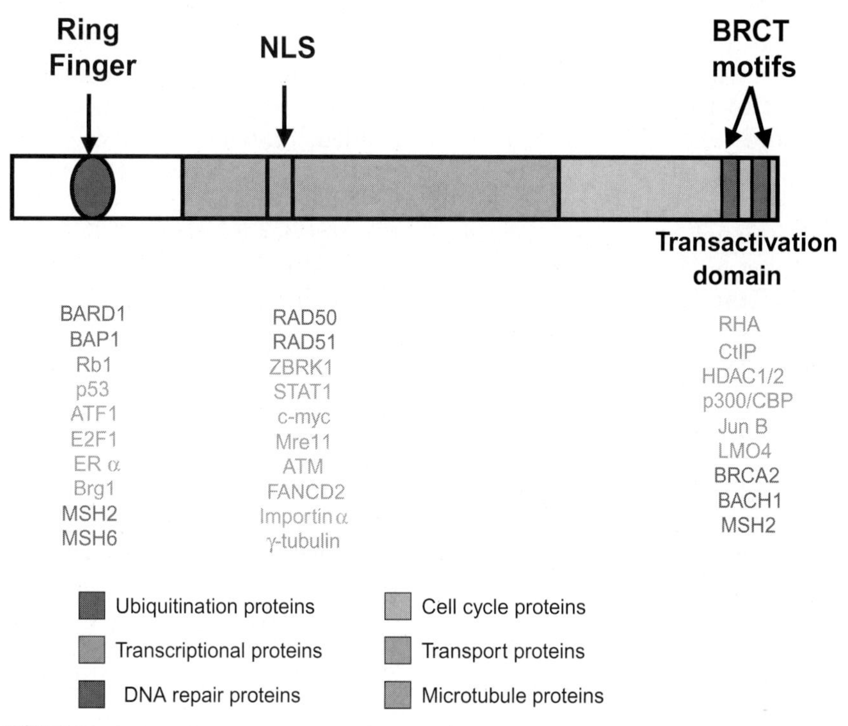

FIGURE 25.3 The specific domains within the BRCA1 protein appear to regulate specific groups of BRCA1 interacting proteins. The respective interacting proteins are listed by color-coded classes and linked to the four BRCA1 domains: 1) the ring finger domain, 2) the two central nuclear localization signals (NLS), 3) the two tandem BRCA1 carboxy terminus (BRCT) motifs and 4) the transactivation domain. **See Plate 51 for color image.**

pathways regulated by BRCA1 were initially inferred from the identification of specific domains within the BRCA1 protein and through the identification of numerous BRCA1 interacting proteins, several of which are referred to throughout the text (FIG. 25.3).

There are four known recognizable domains within BRCA1: 1) a RING finger at the amino terminus, 2) two central nuclear localization signals, and 3) two tandem BRCA1 C-terminal (BRCT) motifs embedded within a 4) potent acidic carboxy transactivation domain. The RING finger (or zinc finger) motif, located at the N-terminus between amino acids 20–68, consists of an interleaved structure in which two ions of zinc are coordinated by eight conserved amino acids (seven cysteines and one histidine). It has been proposed that this RING domain may serve as an interface for DNA recognition, protein–protein interactions, or in targeted protein proteolysis, otherwise known as ubiquitination. This has been inferred from the interaction of this motif with proteins involved in ubiquitination and protein stability including BARD1[24,25] and BAP1.[26]

Two nuclear localization signals have been identified within the large central exon 11 (amino acids 503–508 and 606–615), and these are important for BRCA1 translocation into the nucleus where BRCA1 is predominantly localized.[27] Two previously undescribed BRCT motifs have been identified at amino acids 1699–1736 and 1818–1855.[28] These motifs are now recognized in an expanding family of proteins, many of which have roles in DNA damage response and in cell cycle checkpoint control.[29] Of interest are the interacting proteins that can bind to the BRCT motifs of BRCA1. These include the DNA repair proteins BRCA2 and MSH2; the cell cycle regulatory proteins Rb and p53; and proteins involved in transcriptional regulation such as RNA helicase A, (part of the core transcriptional machinery), HDAC1 and HDAC2 (part of the histone deacetylase complex), and the transcriptional corepressor, CtIP (FIG. 25.3).

In addition, the C-terminal domain of BRCA1 is highly acidic, a feature that is characteristic of many transcription factors. Indeed the C-terminus has been shown to be a potent transactivator in Gal4 fusion assays.[30] Numerous other studies have now firmly established a role for BRCA1 in transcriptional regulation. Extensive investigation into the various functions of BRCA1 and how its loss or inactivation leads to breast and ovarian tumorigenesis are currently ongoing. The most significant developments to date are discussed further, with particular emphasis on the role of BRCA1 in DNA damage response and DNA repair.

BRCA1 and Transcription

To date, there is a substantial volume of literature to suggest a central role for BRCA1 in transcriptional regulation. In addition to possessing an acidic C-terminal domain, BRCA1 associates with the RNA pol II holoenzyme complex through an association with RNA helicase A.[31] This suggests that BRCA1 may be a component of the core transcriptional machinery. BRCA1 has also been implicated in the modification of chromatin through an interaction with the SWI/SNF chromatin remodeling complex; with regulators of histone acetylation and deacetylation; with COBRA, a regulator of chromatin decondensation; and DNA helicases such as BLM and BACH1.[32] Furthermore, BRCA1 interacts with a range of sequence-specific DNA binding transcription factors (p53, c-*myc*, Rb, ERα, NF-κB) to transactivate a variety of genes (p21, p27, 14-3-3σ, GADD45, DDB2, XPC, IRF7).[33] Reference to these transcriptional targets are made throughout the remainder of this chapter.

BRCA1 and Ubiquitination

Ubiquitin tagging of proteins labels them for degradation by the proteasome. However, ubiquitin hydrolases degrade ubiquitin and prevent tagged proteins from being degraded.[34] Evidence for a role for BRCA1 in this process was derived from two proteins (BARD1 and BAP1) that bind to BRCA1 through its N-terminal RING finger motif.[35] Interestingly, these interactions can be abrogated by germline mutations within the BRCA1 RING finger domain.

BARD1 also contains a RING domain and in complex with BRCA1 has potent ubiquitin E3 ligase activity, suggesting a role for this complex in ubiquitination.[25] In addition, BARD1 has been found to colocalize with BRCA1 in nuclear foci containing proliferating cell nuclear antigen (PCNA), suggesting a potential role for this interaction in DNA repair. Collectively, these observations suggest that the BRCA1–BARD1 complex may degrade proteins at a site of DNA damage and initiate repair mechanisms.

BRCA1 also interacts with BAP1, a carboxy terminal hydrolase and a new member of the ubiquitin C-terminal hydrolase family.[26] It is therefore possible that depending on its protein binding partner, BRCA1 may be involved in both targeting proteins for destruction (BRCA1–BARD1) or enhancing target protein stability by preventing ubiquitination (BRCA1–BAP1). In other words, these protein complexes may be effectors and/or regulators of BRCA1 mediated tumor suppression.

Although targets of BRCA1 mediated ubiquitination have yet to be identified, one possible candidate is the FANCD2, Fanconi anemia protein D2,[36] that is found to be altered in a disease characterized by chromosomal instability and cancer predisposition. After DNA damage, FANCD2 is found to relocalize to sites of DNA damage induced nuclear foci with BRCA1; significantly, this depends on monoubiquitination of FANCD2, although it remains to be investigated whether the BRCA1–BARD1 complex or the BRCA1–BAP1 complex may be involved in either the ubiquitination or deubiquitination, respectively, of FANCD2. A detailed consideration of ubiquitination and proteasome inhibition is presented in Chapter 24.

Sensing and Responding to DNA Damage

Accurate transmission of genetic information from one cell to its daughter cells requires extreme accuracy in DNA replication, precise chromosomal segregation, and the ability to survive spontaneous and induced DNA damage. To achieve this, cells have evolved surveillance mechanisms that detect DNA damage and coordinate repair and cell cycle arrest.[37]

This genomic surveillance is essential to prevent the accumulation of DNA damage, which leads to dysregulated growth and an increase in genomic instability that results in a progressive cycle of further genomic aberrations and advancing malignancy. These occurrences signify the failure of DNA damage response mechanisms that normally detect DNA lesions and evoke appropriate repair mechanisms. During evolution, mammalian cells have acquired multiple repair pathways reflecting the many different types of DNA damage that can occur.

DNA sequence modifications can occur during normal DNA replication processes. However, DNA damage also arises from endogenous agents such as oxygen free radicals that are released during normal cellular metabolism and from exogenous environmental factors such as ultraviolet (UV) or ionizing radiation. Ionizing radiation results in DNA double stranded breaks (DSBs) that is one of the most disruptive form of DNA damage.

The DSB response involves a series of steps that includes detection by as yet unknown sensor proteins that recognize the site of DNA damage.[38] These sensors activate components, known as transducers, which convey the signal to downstream effector proteins (FIG. 25.4). In the case of DSBs, ATM is the primary transducer and BRCA1 appears to function as an effector. ATM is the product of the ataxia telangiectasia gene, a member of the phosphatidylinositol-3

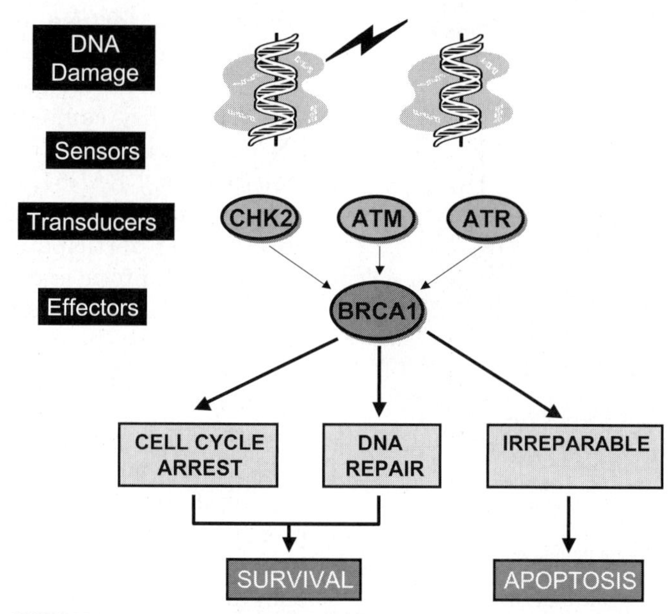

FIGURE 25.4 DNA damage leads to double strand breaks in DNA. The DSB response involves a series of steps that includes detection by as yet unknown sensor proteins that recognize the site of DNA damage, activate components known as transducers which, in turn, convey the signal to downstream effector proteins. ATM is the primary transducer after DSB and BRCA1 appears to function as an effector regulating cell survival and apoptosis.

kinase family or phosphatidylinositol-like protein kinases. ATM mutation carriers are characterized by immunodeficiency, premature aging, sensitivity to ionizing radiation, and a predisposition to cancer, including breast cancer.[39]

Substantial evidence exists to support a role for BRCA1 in mediating the cellular response to DNA damage and, in particular, the response to DSBs. BRCA1 becomes hyperphosphorylated in response to various DNA damaging agents, and these phosphorylation events are mediated by various kinases, including ATM, ATR and chk2. BRCA1 is phosphorylated on multiple serine residues after different types of DNA damage. For example, in response to ionizing radiation, ATM phosphorylates BRCA1 on serines 1387, 1423, and 1457 and chk2 phosphorylates BRCA1 on serine 988 (FIG. 25.5). However, ATR (the ATM related protein kinase) phosphorylates BRCA1 on serine 1423 upon exposure to hydroxyurea and UV.[40]

The functional consequences of these different phosphorylation events are now being uncovered. For example, phosphorylation of BRCA1 on serine 1423 by ATM after gamma irradiation results in activation of the G2/M checkpoint. In contrast, ATM mediated phosphorylation of BRCA1 on serine 1387 has been shown to be required for the BRCA1 mediated intra S phase checkpoint after irradiation (FIG. 25.6). It is clear, therefore, that BRCA1 is likely to regulate different cellular responses depending on the nature of the DNA damage and the specific kinase involved.[40]

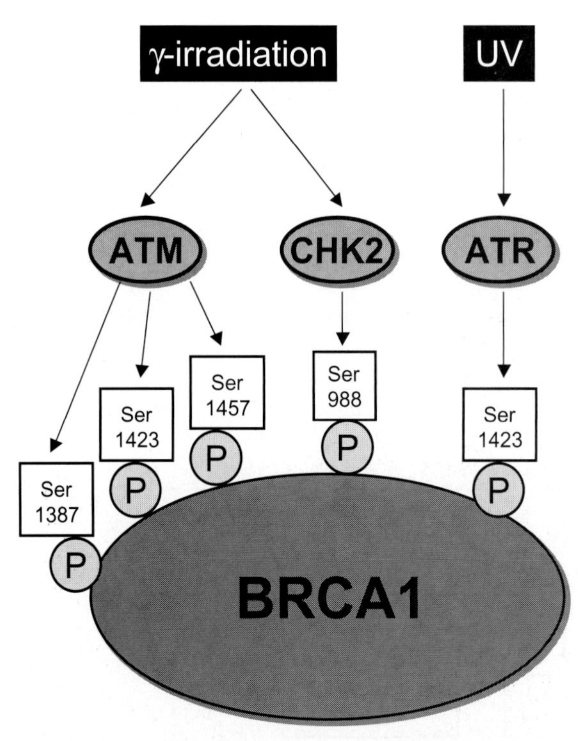

FIGURE 25.5 In support of its proposed role as mediator of the cellular response to DNA damage, BRCA1 becomes hyperphosphorylated in response to various DNA damaging agents and these phosphorylation events are mediated by various kinases including ATM, ATR and chk2. As shown in the diagram, in response to ionizing radiation ATM phosphorylates BRCA1 on serines 1387, 1423 and 1457 and chk2 phosphorylates BRCA1 on serine 988. However, ATR (the ATM related protein kinase) phosphorylates BRCA1 on serine 1423 upon exposure to hydroxyurea and UV (40).

BRCA1 and DNA Repair

A substantial body of evidence also exists to suggest that, in addition to modulating the DNA damage response, BRCA1 also functions at the actual site of DNA damage and initiates specific DNA repair mechanisms. Initial evidence for this was obtained from the observation that after DNA damage, hyperphosphorylated BRCA1 relocates to sites of damage.[41] To date, BRCA1 has been in implicated in three major repair pathways: homologous recombination (HR), nonhomologous end joining (NHEJ), and nucleotide excision repair (NER).

BRCA1 and HR

HR is an error free mechanism that repairs DSBs. This type of repair predominates during the S and G2 phases of the cell cycle when sister chromatids (undamaged homologous strands) are available for use as a template through which the damaged strand can be repaired.

The initial evidence to suggest a role for BRCA1 in HR came from the report that BRCA1 colocalizes with RAD51 in mitotic and meiotic cells.[42] RAD51, the human homolog of bacterial RecA, is essential for efficient HR and in yeast participates in this repair pathway during mitotic and meiotic recombination events. Its main function in HR is to catalyze strand exchange by binding to single stranded DNA that invades and aligns the homologous DNA strand.[43] RAD51 is localized to chromosome 15q15,[44] and in breast tumors, high rates of LOH have been detected in this region, suggesting its involvement in the development of breast cancer.[45] Additionally, the mammalian tissue expression pattern of RAD51 is very similar to that of BRCA1, being highest in the thymus, testis, and ovaries.[46]

BRCA1–RAD51 S phase foci were shown to disperse after treatment with a range of

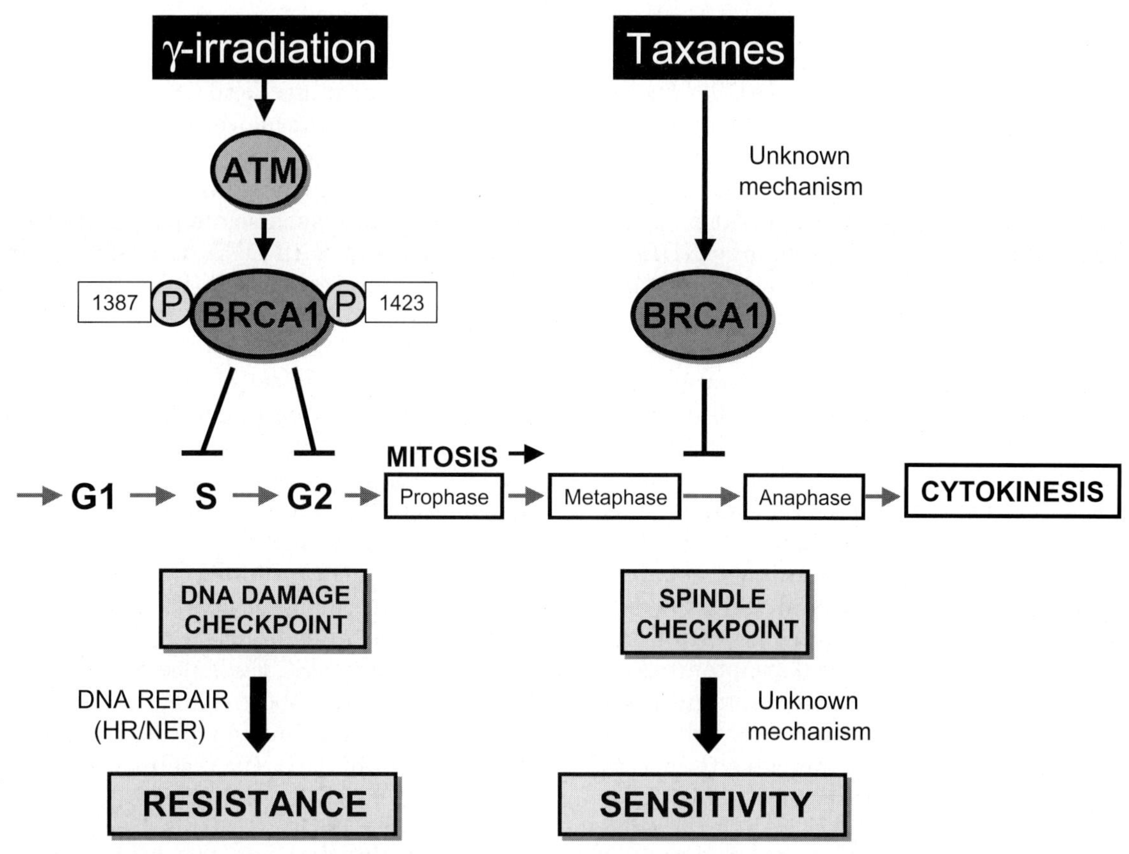

FIGURE 25.6 Following γ irradiation, ATM dependent phosphorylation of BRCA1 on serine 1423 by ATM results in activation of the G2/M cell cycle checkpoint. In contrast, ATM mediated phosphorylation of BRCA1 on serine 1387 is required for the BRCA1 mediated intra S phase checkpoint following γ irradiation. The mechanism by which taxanes impact BRCA1 mediated cell cycle inhibition during mitosis is currently unknown.

DNA damaging agents, including hydroxyurea, UV, mitomycin C, and gamma irradiation. Subsequently, it was demonstrated that upon DNA damage, BRCA1 and RAD51 relocate to sites of stalled replication forks marked by PCNA. The recruitment of these proteins to these sites suggests that BRCA1 acts as a sensor of DNA damage, relocating to sites of damage to initiate repair.[41] Defects in such a process could lead to chromosomal abnormalities and increased cancer risk.

BRCA1 also interacts with the second breast cancer susceptibility gene, *BRCA2,* which is also involved in HR through an interaction with RAD51. Like BRCA1, BRCA2 defective cell lines and *Brca2* deficient mice display similar sensitivities to DNA damaging agents. BRCA2 depleted cell lines also exhibit defects in DSB repair by HR, and BRCA2 colocalizes to nuclear foci during S phase and after DNA damage. In addition, BRCA2 also relocates to replication sites containing PCNA after DNA damage.[47]

Moynahan et al.[48] provided the first direct evidence to suggest a role for BRCA1 in controlling HR. *Brca1* deficient mouse embryonic stem cells were reported to have impaired repair of DSBs and gross chromosomal rearrangements, typical of chromosome instability disorders such as Bloom syndrome, Nijmegen breakage syndrome, and ataxia telangiectasia. In addition, it was discovered that although these cells had consistently lower levels of HR, an alternative DSB repair mechanism known as NHEJ was intact and slightly increased, suggesting that in BRCA1 deficient cells, the alternative NHEJ pathway substitutes for HR.

BRCA1 and NHEJ

Unlike the conservative process of HR, NHEJ is a repair mechanism wherein broken ends of DNA are directly rejoined. However, nucleotide alterations are permitted at these sites of rejoining, leading to an error prone method of repair.

BRCA1 has also been implicated in NHEJ based on the observation that it has been identified as a component of the RAD50/Mre11/NBS1 complex, involved in mediating NHEJ repair. The effects of null mutations of RAD50 or Mre11 are severe, and these cells sustain high levels of chromosome aberrations and have a reduced capacity for repair by NHEJ.[49] Patients with mutations in Mre11 display similar clinical symptoms to the cancer prone syndrome ataxia telangiectasia, emphasizing the importance of the NHEJ repair process in cancer development. In normal human fibroblasts, Mre11 localizes to DNA breaks within 30 minutes of exposure to gamma irradiation, prompting speculation that the NHEJ complex may function as a sensor of DNA damage, which relocates to the lesion site to initiate repair.[50] *NBS* is the gene mutated in Nijmegen breakage syndrome.[51] Interestingly, patients with these mutations also exhibit clinical symptoms similar to those involved in ataxia telangiectasia patients. These observations are consistent with a recent report that indicates that ATM phosphorylates *NBS* in response to ionizing radiation.[39] Confocal microscopy has revealed that BRCA1 colocalizes with RAD50/Mre11/NBS1 in response to ionizing radiation, an effect that is abrogated in the BRCA1 mutant HCC1937 cells,[52] suggesting that BRCA1 is a critical component of the NHEJ pathway.

In summary, BRCA1 can regulate both HR and NHEJ pathway to repair DSBs, but it remains unclear as to how a choice is made between these two repair mechanisms. It has been suggested that in BRCA1 deficient cells, when HR is compromised cells are repaired by the more erroneous NHEJ pathway. It is therefore possible that the chromosomal instability observed in BRCA1 deficient tumors may arise because of the incorrect rerouting of repair along the NHEJ pathway.[36]

BRCA1 and NER

NER is comprised of two related repair events: global genome repair (GGR), which removes lesions from the whole genome, and transcription coupled repair (TCR), which repairs lesions preferentially from the transcribed strand of expressed genes. Two different reports have suggested a role for BRCA1 in NER. The first report demonstrated that GGR occurred with greater efficiency when BRCA1 was overexpressed in the presence of functional p53. In agreement with this, a second report using antisense oligonucleotides to BRCA1 revealed the impaired removal of UV induced cyclobutane pyrimidine dimers and a 6-4 photoproduct from DNA.[53] Furthermore, BRCA1 can transcriptionally activate genes that are known to be involved in NER, including *GADD45, DDB2,* and *XPC*.[54] It was also demonstrated that upon treatment with DNA damaging agents (UVC, adriamycin, or cisplatin), BRCA1 could up-regulate the *XPE* and *DDB2* genes in the presence of p53.[55]

An important effect of genomic lesions is to block transcription of genes on a damaged DNA strand. TCR has evolved to ensure high priority repair of these types of lesions by removing the stalled RNA polymerase. This type of repair is characterized by the rapid repair of lesions in transcriptionally active DNA as compared with nontranscribed DNA. BRCA1 has been implicated in the TCR of oxidative induced DNA damage. Evidence for this came from a report by Abbott et al.,[56] who demonstrated that reintroduction of BRCA1 into the BRCA1 mutant HCC1937 cell line rescued their radiosensitive phenotype, an effect that was correlated with a 15-fold increase in the rate of TCR in these cells. In addition, BRCA1 has been found to reside within a large DNA repair complex called the BRCA1 associated genome surveillance complex (BASC)[57] that contains several mismatch repair genes that are required for TCR. It is postulated that the involvement of BRCA1 in TCR may be mediated by the ability of BRCA1 to bind to the RNA pol II holoenzyme complex through an interaction with RNA helicase A, a core element of the RNA pol II complex.[31]

BASC

It is proposed that BRCA1 may act as a scaffold protein that is a component of different protein complexes that are formed or dismantled in response to different types of DNA damage. As described above, BRCA1 is a component of a large multisubunit protein complex (>2 MD) called BASC.[57] This complex is comprised of at least 15 proteins, including various DNA damage responsive and DNA repair proteins, such as ATM, the NHEJ complex (Rad50/Mre11/Nbs1), BLM (Bloom's helicase); the mismatch repair proteins, MLH1, MSH2, and MSH6; and DNA replication factor C, which loads PCNA onto DNA. Interestingly, all the members of this complex have roles in the recognition of abnormal DNA structures and damaged DNA. BASC may therefore serve as a general sensor of DNA damage as well as marking sites of DNA lesions and activating DNA repair mechanisms. In summary, impaired function of BRCA1 may result in an inactive BASC complex that could render cells more prone to genomic instability.

Surprisingly the DNA repair protein RAD51 was not found in BASC. However, more recently, RAD51 has been identified along with BRCA1 and BRCA2 in another holoenzyme complex termed BRCC (BRCA1 and BRCA2 containing complex). In addition to RAD51 and BRCA2, the BRCC complex also contains two novel proteins, termed BRCC36 and BRCC45. Of particular importance was the observation that depletion of BRCC36 and BRCC45 by siRNA resulted in increased sensitivity to ionizing radiation and a defective G2/M checkpoint.[58] Collectively, all these reports indicate that BRCA1 is involved in multiple repair pathways that participate in the maintenance of genomic integrity; however, further investigation into the exact role played by BRCA1 in these repair mechanisms is required.

BRCA1 and Cell Cycle Checkpoint Control

Cell cycle checkpoints are surveillance systems that interrupt cell cycle progression in mammalian cells when damage to the genome or spindle is detected or when cells have failed to complete an event in the cell cycle.[59] The control of cell cycle checkpoints have been shown to be associated with the maintenance of genomic fidelity. Defects in the regulation of cell cycle progression may result in genomic instability, including gene mutation, amplification, or chromosomal aberration, which are associated with malignant transformation and tumorigenesis.[60,61] The mammalian cell cycle is composed of four phases: the G1 gap phase that precedes DNA replication, the S phase where DNA synthesis occurs, the G2 gap phase where the cells are preparing for mitosis, and the M phase where cell division occurs.[59] DNA damage checkpoints are mechanisms that detect damaged DNA and generate signals that can arrest cells at the G1/S transition, the S phase, or the G2/M transition of the cell cycle so that DNA damage incurred can be repaired. Arrest at the G1/S or S phase checkpoint prevents damaged DNA from being replicated, and arrest at the G2/M checkpoint avoids the segregation of damaged chromosomes. In addition to these DNA damage checkpoints, mitosis is monitored by a spindle checkpoint at the metaphase–anaphase boundary that prevents the unequal segregation of chromosomes. The role of BRCA1 in the regulation of the DNA damage and spindle checkpoints is discussed.

BRCA1 has been implicated in this G1/S checkpoint through p53 dependent and independent transactivation of p21, a cdk inhibitor that functions to negatively regulate the cyclin D–cdk4/6 and cyclin E–cdk2 complexes (see Chapter 13). This prevents phosphorylation of Rb and release of E2F such that the cell remains in G1/S. BRCA1 can bind p53 (amino acids 224–500 of BRCA1) and enhance transcription of p21, leading to a G1 arrest.[62,63] However, BRCA1 can also activate p21 transcription in a p53 independent manner.[64] In addition, it was noted that adenoviral mediated expression of BRCA1 led to dephosphorylation of Rb and a decrease in cyclin dependent kinase activity, an effect that is dependent on p21.[65]

More recently, $p27^{Kip1}$ has been identified as a BRCA1 transcriptional target. This cdk inhibitor specifically inhibits the cyclin D–cdk4/6 complex and in doing so can activate the G1/S checkpoint. Therefore, $p27^{Kip1}$ represents another mechanism by which BRCA1 may regulate this checkpoint.[66]

BRCA1 has also been implicated in the regulation of the intra–S phase checkpoint. It has been demonstrated that BRCA1 regulates this checkpoint in response to γ irradiation, an effect that is dependent on ATM mediated phosphorylation on serine 1387.[67]

The role of BRCA1 in the G2/M checkpoint has been more firmly established. As discussed earlier, ATM can phosphorylate BRCA1 in response to γ irradiation on serine 1423 to induce G2/M arrest. However, further evidence was derived from the report that mouse embryonic fibroblasts containing a targeted deletion of the *Brca1* gene exhibited a defective G2/M checkpoint in response to gamma irradiation, suggesting that BRCA1 specifically regulates this checkpoint after exposure to DNA damage.[68] Furthermore, the BRCA1 mutated HCC1937 cells failed to activate the G2/M checkpoint upon treatment with the DNA damaging agents, bleomycin and etoposide. However, when BRCA1 was reconstituted into these cells, G2/M checkpoint activation was restored in response to these chemotherapeutic agents.[69]

The cyclin B–cdc2 kinase complex is the principal target of G2/M checkpoint inhibitory signals because its activity is essential for progression through the G2/M phase of the cell cycle. The cyclin B–cdc2 complex is regulated at multiple levels in response to a variety of cellular signals. Decreased expression of cyclin B and inhibitory phosphorylation of cdc2 are thought to be the major mechanisms by which this complex is inactivated, thereby leading to G2/M arrest.

BRCA1 can up-regulate the expression of the G2/M checkpoint genes *14-3-3σ*[70] and *GADD45*[71] to induce G2/M arrest.[72] These two genes prevent the nuclear formation of the cyclin B–cdc2 complex and prevent progression from G2 into mitosis. *14-3-3σ* can effect this by sequestering the entire complex in the cytoplasm, whereas *GADD45* can bind to and sequester the activity of cdc2 kinase. In addition, BRCA1 can transcriptionally repress cyclin B to activate the G2/M checkpoint.[65] BRCA1 is also required for the activation of the DNA damage responsive Chk1 kinase in response to gamma irradiation.[73] Chk1 is responsible for both phosphorylating and inactivating cdc25C (an activator of the cyclin B–cdc2 complex). Furthermore, BRCA1 can also modulate the activity of the Wee1 kinase which hyperphosphorylates cdc2 on Tyr 15, leading to inactivation of the cyclin B–cdc2 complex.[73]

The formation of the cyclin B–cdc2 complex promotes entry into mitosis and activates a cascade of events that results in mitotic spindle formation, the segregation of chromosomes, and the initiation of cytokinesis. The mitotic spindle assembly checkpoint arrests cells in metaphase and prevents the onset of anaphase until these processes have been accurately accomplished.

Although the S and G2 checkpoints are activated in response to DNA damage, it has been proposed that BRCA1 may also regulate a mitotic checkpoint that is unresponsive to DNA damage, independent of ATM, and monitors cells as they exit from the G2 checkpoint and before the onset of mitosis. The main evidence for this is the fact that mouse embryonic stem cells deficient in *brca1* exhibit extensive genetic instability characterized by chromosomal rearrangements and multiple functional centrosomes leading to aneuploidy, a hallmark of a defective spindle checkpoint.

Weaver et al.[74] demonstrated that mice conditionally mutant for *brca1* also display aneuploidy and contained supernumerary functional centrosomes. Intriguingly, this pattern of chromosome loss and gain resembles that of human *BRCA1* mutated tumors, and indeed loss of *BRCA1* expression in sporadic breast cancers is associated with chromosomal instability.[75,76]

In vitro experiments support these in vivo findings. Larson et al.[77] demonstrated that overexpression of a BRCA1 dominant negative increased growth rate and abrogated the G2/M checkpoint in response to the microtubule inhibitor, colchicine. In addition, it has been observed that BRCA1 is required for the decatenation and therefore accurate separation of sister chromatids at anaphase.[78] In addition, inducible BRCA1 expression in the presence of the antimicrotubule agents, paclitaxel and vincristine, resulted in G2/M arrest followed by apoptosis.[72] Closer examination of this G2/M checkpoint activation revealed a 32% increase in cells arrested in the mitotic phase of the cell cycle when BRCA1 was induced compared with noninduced in the presence of paclitaxel (**FIG. 25.6**). Inducible expression of GADD45 exhibited a similar mitotic arrest in the presence of paclitaxel, suggesting GADD45 as a potential effector of BRCA1 and paclitaxel-induced mitotic arrest.[72]

In support of these observations, the defective G2/M checkpoint observed in the BRCA1 mutant HCC1937 cells in response to paclitaxel was restored after reconstitution of these cells with wild-type exogenous BRCA1.[69] Further evidence that BRCA1 may regulate the accurate segregation of chromosomes before the onset of mitosis came from the observation that BRCA1 is localized to the centrosome through an interaction with gamma tubulin.[79] Loss of BRCA1 might therefore result in centrosomal abnormalities, a major cause of chromosomal instability.[79]

In summary, these reports clearly demonstrate that BRCA1 is a major player in the regulation of the cell cycle based on its ability to modulate the G1/S, S, G2, and mitotic checkpoints and is strongly supportive of a role for BRCA1 in genomic surveillance (**FIG. 25.6**).

BRCA1 and Apoptosis

Typically, if DNA damage is too great to repair, apoptosis occurs. The mechanism underlying the switch from cell cycle checkpoints to repair pathways or apoptotic pathways is not known. In addition to regulating DNA damage response, cell cycle checkpoints and DNA repair pathways, BRCA1 has also been shown to regulate apoptotic pathways.

An initial publication to support a role for BRCA1 as a negative regulator of cell growth came from Thompson et al.,[80] who demonstrated that BRCA1 is expressed at higher levels in normal mammary cells compared with breast cancer cells and demonstrated that inhibition of BRCA1 expression with antisense oligonucleotides accelerated the growth of both normal and malignant breast epithelial cells. It was also demonstrated that transfection of wild-type BRCA1 into breast and ovarian cancer cell lines resulted in growth inhibition.[81]

BRCA1 has also been shown to induce apoptosis, suggesting that the lack or decreased levels of BRCA1 expression in breast or ovarian cancers may be responsible for the increased resistance of these cells to apoptosis.[82] Subsequently, inducible expression of BRCA1 was demonstrated to mediate apoptosis, an effect that was shown to be dependent on activation of the JNK kinase pathway. This effect was coupled with up-regulation of *GADD45,* a target gene that was identified in this study by oligonucleotide array based expression profiling.[71] Evidence to support a role for *GADD45* in mediating JNK activation was derived from the observation that *GADD45* could associate with MEKK4, an upstream regulator of the JNK pathway. Further delineation of this apoptotic pathway identified other proteins involved, including H-ras, MEKK4, JNK, Fas, Fas ligand, and caspase 9 activation, and demonstrated that a dominant negative C-terminally truncated BRCA1 protein diminished the apoptotic response to growth factor withdrawal.[83]

More recent reports indicate that BRCA1 mediates apoptosis in response to different forms of cellular stress. Evidence for this comes from the observation that BRCA1 regulates interferon-gamma mediated apoptosis, an effect that was correlated with synergistic induction of IRF7.[84] It has also been shown that increased expression of BRCA1 results in increased sensitivity to the antimicrotubule agents, paclitaxel and vinorelbine.[72] In accordance with this, loss of BRCA1 by antisense strategies or mutation results in increased resistance to paclitaxel.[85] More recently, BRCA1 has been identified as a differential modulator of chemotherapy induced apoptosis. In essence, BRCA1 can regulate both pro- and antiapoptotic pathways, depending on the type of chemotherapeutic agent. Specifically, BRCA1 induced dramatic resistance to a range of DNA damaging agents, including bleomycin (a radiomimetic), etoposide (a topoisomerase II inhibitor), and cisplatin (a DNA cross-linker) (FIG. 25.6). Conversely, BRCA1 induced acute sensitivity to the spindle poisons, paclitaxel and vinorelbine, that exert their cytotoxicity by the polymerization and depolymerization of tubulin, respectively (FIG. 25.6).[69]

Clinical Implications

Various regimens for the treatment of breast cancer are used depending on the category of the disease (invasive/noninvasive), stage, grade, and overall health of the patient. Primary treatment of early breast cancer usually involves partial or total mastectomy, which permits removal of the axillary lymph nodes to determine adequate staging of the disease. Adjuvant therapy includes chemotherapy, radiotherapy, and hormonal therapy or a combination of these. Chemotherapy for breast cancer is typically anthracycline based and regimens often include combinations of the topoisomerase II inhibitor doxorubicin and the DNA cross-linker cyclophosphamide. More recently, it has been shown that patients treated with paclitaxel after standard chemotherapy had a longer disease-free survival,[86,87] leading to increased interest in using taxanes in combination with anthracyclins and cyclophosphamide.[88]

However, many breast cancer patients develop metastatic disease despite treatment, and this group of patients remain an enormous challenge in the management of the disease. Currently, the only predictive markers to guide that treatment of breast cancer in the clinic are ER and HER-2/*neu* status. For example, 70–80% of breast cancers are ER positive (they express the ER) and these receive the ER antagonist, tamoxifen, which has been shown to reduce disease recurrence by 40%. In addition, 25% of breast cancers overexpress the *erb*2/HER-2/*neu* receptor, which promotes aggressive disease with a poor prognosis. These patients are treated with trastuzumab (Herceptin^{TM}), a HER-2 antagonist, that has demonstrated 21% response rates when given as a single agent therapy.[89]

BRCA1 and the Prediction of Response to Chemotherapy

Currently, there are no predictive markers routinely used in the choice of chemotherapy for breast cancer patients (see Chapter 28). However, we and others have demonstrated through in vitro studies that BRCA1 may be an important molecular marker in predicting response to chemotherapeutic agents used in the treatment of breast cancer. Essentially, BRCA1 mutated tumors may respond well to DNA damaging based chemotherapy (bleomycin, etoposide, and cisplatin) (FIG. 25.7). There is evidence to support this in preclinical and clinical models.

Moynahan et al.[90] demonstrated that BRCA1 exon11 mutant embryonic stem cells were highly sensitive to the alkylating agent mitomycin C, and this was coupled with a defective HR repair pathway. In addition, antisense inhibition of BRCA1 caused increased cisplatin sensitivity,[91] and loss of BRCA1 in breast cancer cells using ribozyme technology displayed increased sensitivity to etoposide.[85] Furthermore, *Brca1* disrupted murine cell lines were found to be more sensitive to etoposide, doxorubicin, cisplatin, and carboplatin.[92] Several small clinical studies have also be conducted, and these have concluded that BRCA1 mutation carriers may have significantly increased sensitivity to DNA damaging based chemotherapy.[93,94]

In contrast, BRCA1 mutated tumors may be more resistant to taxane based chemotherapy (FIG. 25.7). Preclinical models include an overexpression model of BRCA1 that demonstrated increased sensitivity to antimicrotubule agents. In addition, inhibition of endogenous BRCA1 expression using messenger RNA specific ribozymes and siRNA technology in breast cancer cell lines caused resistance to vincristine and paclitaxel, respectively.[69,85] Furthermore, reconstitution of BRCA1 into HCC1937 cells conferred increased sensitivity to paclitaxel and vinorelbine.[69,95] Clinical studies assessing the role of BRCA1 in response to taxanes have been less well characterized and are therefore essential to further determine whether BRCA1 may be a predictive marker of response to this type of chemotherapy (FIG. 25.7).

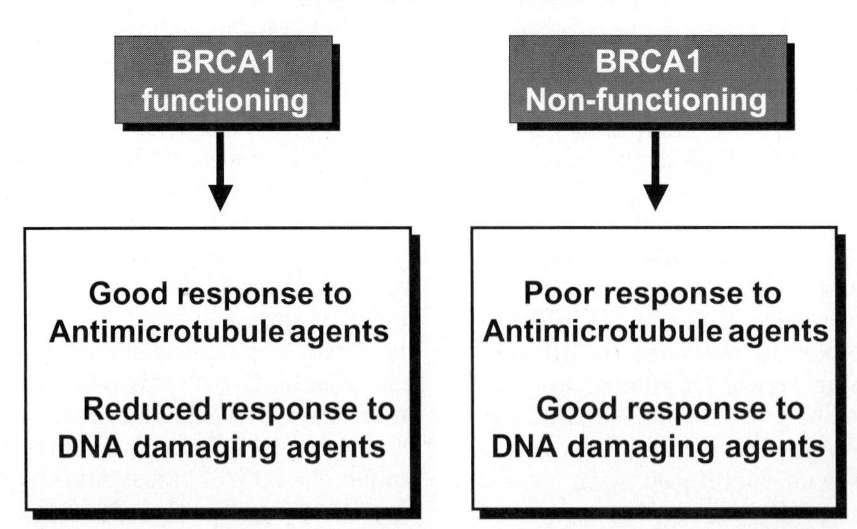

FIGURE 25.7 Recent evidence favors that BRCA1 may be an important molecular marker in predicting response to anti-tubulin and DNA damaging chemotherapies in the treatment of breast cancer. Preclinical and early clinical studies have found that *BRCA1* mutated tumors may respond well to DNA damaging based chemotherapy (bleomycin, etoposide and cisplatin).[90,91,92,93,94] In contrast, preclinical studies have shown that BRCA1 mutated tumors may be more resistant to microtubulin targeting drugs such as the taxanes (see text for details).[69,85,95]

Other Genes Associated with DNA Repair in Breast Cancer

As mentioned throughout this BRCA1-focused chapter on DNA repair and the maintenance of genomic stability in breast cancer, a number of other genes have been implicated as DNA damage response genes, genomic stability-related TSGs, and genes associated with breast cancer predisposition (see Chapter 4).

BRCA2 Similar to BRCA1, the BRCA2 protein containing 3418 amino acids also facilitates HR[96,97] but does not appear to be associated with either NHEJ or cell cycle checkpoint regulation.[97] As mentioned above, the BRCA2 protein complexed with the RAD51 protein is involved in DNA repair by HR. BRCA2 mutation is associated with both a reduction in HR and an increase in radiation sensitivity.[97,98] The clinical significance of germline BRCA2 mutation is discussed in Chapter 4.

Fanconi Anemia Associated Genes Fanconi anemia (FA) is a rare autosomal recessive disease characterized by chromosome instability and breast (and other) cancer susceptibility. As discussed above, genes associated with FA include both *BRCA1* and *BRCA2*. Multiple FA proteins form a complex required for activation of FANCD2, a downstream component of the FA pathway, in response to DNA cross-links. Active FANCD2 interacts with BRCA1 and is independently phosphorylated by ATM in response to ionizing radiation.[99] In addition, the FA proteins are interconnected with other nuclear and cytoplasmic factors, all related to cellular responses to carcinogenic stress and to both caretaker and gatekeeper functions.[99] FANCD2 is ubiquitinated in response to DNA damage and translocates to the nucleus where it colocalizes to foci containing BRCA1.[100]

p53 The so-called guardian of the genome, p53, is reviewed as a breast cancer predisposition gene in Chapter 4 and as a prognostic factor for the disease in Chapter 27. As discussed above and reviewed in the published literature, the p53 protein has a wide range of roles in the maintenance of genomic stability and in response to DNA damaging agents in breast cancer.[101]

Ataxia Telangiectasia Ataxia telangiectasia is a rare, autosomal, recessive disease characterized by neuronal degeneration, genomic instability, an increased risk of lymphoid malignancies, and, less commonly, breast cancer. The ataxia telangiectasia gene (*ATM*), which maps to chromosomal region 11q22-23, encodes a kinase related to the phosphatidylinositol-3 kinase and has been shown to function in DNA repair and cell cycle checkpoint control after DNA damage.[102] As discussed above, *ATM* is a caretaker TSG and is activated in response to DSBs. Ataxia telangiectasia heterozygotes have a slightly increased risk of breast cancer.

Conclusion

In summary, as our knowledge of the various functions associated with *BRCA1* and other DNA damage response genes increases, it is clear that it functions as a master regulator of multiple cellular stress response pathways. Crucially, it is intimately involved in the recognition and cellular response to DNA damage, in addition to other types of cellular stress such as disruption of the mitotic spindle. The mechanistic basis through which it achieves these diverse roles is likely to stem from its ability to interact with and modulate the activity of multiple partners, including proteins involved in transcriptional regulation and those directly involved at sites of DNA damage. The clinical implications of these functions relating to patient management are now emerging, and it is likely that the next few years will see an exhaustive examination of its potential as a predictive marker of response to chemotherapy (FIG. 25.7).

References

1. Hanahan D, Weinberg RA. The hallmarks of cancer. Cell 2000;100:57–70.
2. Lane TF, Deng C, Elson A, et al. Expression of Brca1 is associated with terminal differentiation of ectodermally and mesodermally derived tissues in mice. Genes Dev 1995;9:2712–2722.
3. Knudson AG. Mutation in cancer: statistical study of retinoblastoma. Proc Natl Acad Sci USA 1971;68: 820–823.
4. Rosen EM, Fan S, Pestell RG, et al. BRCA1 gene in breast cancer. J Cell Physiol 2003;196:19–41.
5. Hall JM, Lee MK, Newman B, et al. Linkage of early-onset familial breast cancer to chromosome 17q21. Science 1990;250:1684–1689.
6. Miki Y, Swensen J, Shattuck-Eidens D, et al. A strong candidate for the breast and ovarian cancer susceptibility gene BRCA1. Science 1994;266: 66–71.
7. King MC, Marks HH, Mandell JB. Breast and ovarian cancer risks due to inherited mutations in BRCA1 and BRCA2. Science 2003;302:643–646.

8. Futreal PA, Liu Q, Shattuck-Eidens D, et al. BRCA1 mutations in primary breast and ovarian carcinomas. Science 1994;266:120–122.
9. Castilla LH, Couch FJ, Erdos MR, et al. Mutations in the BRCA1 gene in families with early-onset breast and ovarian cancer. Nat Genet 1994;8:387–391.
10. Friedman LS, Szabo CI, Ostermeyer EA, et al. Novel inherited mutations and variable expressivity of BRCA1 alleles, including the founder mutation 185delAG in Ashkenazi Jewish families. Am J Hum Genet 1995;57:1284–1297.
11. FitzGerald MG, MacDonald DJ, Krainer M, et al. Germ-line BRCA1 mutations in Jewish and non-Jewish women with early-onset breast cancer. N Engl J Med 1996;334:143–149.
12. Struewing JP, Abeliovich D, Peretz T, et al. The carrier frequency of the BRCA1 185delAG mutation is approximately 1 percent in Ashkenazi Jewish individuals. Nat Genet 1995;11:198–200.
13. Peelen T, van Vliet M, Petrij-Bosch A, et al. A high proportion of novel mutations in BRCA1 with strong founder effects among Dutch and Belgian hereditary breast and ovarian cancer families. Am J Hum Genet 1997;60:1041–1049.
14. Einbeigi Z, Bergman A, Kindblom LG, et al. A founder mutation of the BRCA1 gene in Western Sweden associated with a high incidence of breast and ovarian cancer. Eur J Cancer 2001;37:1904–1909.
15. Thompson ME, Jensen RA, Obermiller PS, et al. Decreased expression of BRCA1 accelerates growth and is often present during sporadic breast cancer progression. Nat Genet 1995;9:444–450.
16. Wilson CA, Ramos L, Villasenor MR, et al. Localization of human BRCA1 and its loss in high grade, non-inherited breast carcinomas. Nat Genet 1999; 21:236–240.
17. Zheng W, Luo F, Lu JJ, et al. Reduction of BRCA1 expression in sporadic ovarian cancer. Gynecol Oncol 2000;76:294–300.
18. Yang Q, Sakurai T, Mori I, et al. Prognostic significance of BRCA1 expression in Japanese sporadic breast carcinomas. Cancer 2001;92:54–60.
19. Marquis ST, Rajan JV, Wynshaw-Boris A, et al. The developmental pattern of Brca1 expression implies a role in differentiation of the breast and other tissues. Nat Genet 1995;11:17–26.
20. Fan S, Wang J, Yuan R, et al. BRCA1 inhibition of estrogen receptor signaling in transfected cells. Science 1999;284:1354–1356.
21. Fan S, Ma YX, Wang C, et al. Role of direct interaction in BRCA1 inhibition of estrogen receptor activity. Oncogene 2001;20:77–87.
22. Fan S, Ma YX, Wang C, et al. p300 Modulates the BRCA1 inhibition of estrogen receptor activity. Cancer Res 2002;62:141–151.
23. Kinzler KW, Vogelstein B. Cancer-susceptibility genes. Gatekeepers and caretakers. Nature 1997; 386:761–763.
24. Wu LC, Wang ZW, Tsan JT, et al. Identification of a RING protein that can interact in vivo with the BRCA1 gene product. Nat Genet 1996;14: 430–440.
25. Hashizume R, Fukuda M, Maeda I, et al. The RING heterodimer BRCA1-BARD1 is a ubiquitin ligase inactivated by a breast cancer-derived mutation. J Biol Chem 2001;276:14537–14540.
26. Jensen DE, Proctor M, Marquis ST, et al. BAP1: a novel ubiquitin hydrolase which binds to the BRCA1 RING finger and enhances BRCA1-mediated cell growth suppression. Oncogene 1998;16:1097–1112.
27. Scully R, Ganesan S, Brown M, et al. Location of BRCA1 in human breast and ovarian cancer cells. Science 1996;272:123–126.
28. Koonin EV, Altschul SF, Bork P. BRCA1 protein products. Functional motifs. Nat Genet 1996;13:266–268.
29. Huyton T, Bates PA, Zhang X, et al. The BRCA1 C-terminal domain: structure and function. Mutat Res 2000;460:319–332.
30. Chapman MS, Verma IM. Transcriptional activation by BRCA1. Nature 1996;382:678–679.
31. Scully R, Anderson SF, Chao DM, et al. BRCA1 is a component of the RNA polymerase II holoenzyme. Proc Natl Acad Sci USA 1997;94:5605–5610.
32. Starita LM, Parvin JD. The multiple nuclear functions of BRCA1: transcription, ubiquitination and DNA repair. Curr Opin Cell Biol 2003;15:345–500.
33. Wang Q, Zhang H, Fishel R, et al. BRCA1 and cell signalling. Oncogene 2000;19:6152–6158.
34. Freemont PS. RING for destruction? Curr Biol 2000; 10:R84–R87.
35. Meza JE, Brzonic PS, King MC, et al. Mapping the functional domains of BRCA1: interaction of RING finger domains of BRCA1 and BARD1. J Biol Chem 1999;9:5659–5665.
36. Venkitaraman AR. Cancer susceptibility and the functions of BRCA1 and BRCA2. Cell 2002;108: 171–182.
37. Zhou B-BS, Elledge SJ. The DNA damage response: putting checkpoints in perspective. Nature 2000; 408:433–439.
38. Shiloh Y. ATM and ATR: networking cellular responses to DNA damage. Curr Opin Genet Dev 2001;1:71–77.
39. Khanna KK. Cancer risk and the ATM gene: a continuing debate. J Natl Cancer Inst 2000;92: 795–802.
40. Venkitaraman AR. Functions of BRCA1 and BRCA2 in the biological response to DNA damage. J Cell Sci 2001;114:3591–3598.
41. Scully R, Chen J, Ochs RL, et al. Dynamic changes of BRCA1 subnuclear location and phosphorylation state are initiated by DNA damage. Cell 1997;90: 425–435.
42. Scully R, Chen J, Plug A, et al. Association of BRCA1 with Rad51 in mitotic and meiotic cells. Cell 1997; 88:265–275.
43. Baumann P, Benson FE, West SC. Human Rad51 protein promotes ATP-dependent homologous pairing and strand transfer reactions in vitro. Cell 1996;87:757–766.
44. Kote-Jarai Z, Eeles RA. BRCA1, BRCA2 and their possible function in DNA damage response. Br J Cancer 1999;81:1099–1102.
45. Gonzalez R, Silva JM, Dominguez G, et al. Detection of loss of heterozygosity at RAD51, RAD52, RAD54 and BRCA1 and BRCA2 loci in breast cancer: pathological correlations. Br J Cancer 1999;81: 503–509.
46. Zhang H, Tombline G, Weber BL. BRCA1, BRCA2, and DNA damage response: collision or collusion? Cell 1998;92:433–436.
47. Chen Y, Farmer AA, Chen CF, et al. BRCA1 is a 220-kDa nuclear phosphoprotein that is expressed and phosphorylated in a cell cycle-dependent manner. Cancer Res 1996;56:3168–3172.

48. Moynahan ME, Chiu JW, Koller BH, et al. Brca1 controls homology-directed DNA repair. Mol Cell 1999;4:511–518.
49. Karran P. DNA double strand break repair in mammalian cells. Curr Opin Genet Dev 2000;10:144–150.
50. Dasika GK, Lin SC, Zhao S, et al. DNA damage-induced cell cycle checkpoints and DNA strand break repair in development and tumorigenesis. Oncogene 1999;18:7883–7899.
51. Digweed M, Reis A, Sperling K. Nijmegen breakage syndrome: consequences of defective DNA double strand break repair. Bioessays 1999;21:649–656.
52. Zhong Q, Chen CF, Li S, et al. Association of BRCA1 with the hRad50-hMre11-p95 complex and the DNA damage response. Science 1999;285:747–750.
53. Takimoto R, MacLachlan TK, Dicker DT, et al. BRCA1 transcriptionally regulates damaged DNA binding protein (DDB2) in the DNA repair response following UV-irradiation. Cancer Biol Ther 2002;1: 177–186.
54. Hartman AR, Ford JM. BRCA1 induces DNA damage recognition factors and enhances nucleotide excision repair. Nat Genet 2002;32:180–184.
55. MacLachlan TK, Takimoto R, El-Deiry WS. BRCA1 directs a selective p53-dependent transcriptional response towards growth arrest and DNA repair targets. Mol Cell Biol 2002;22:4280–4292.
56. Abbott DW, Thompson ME, Robinson-Benion C, et al. BRCA1 expression restores radiation resistance in BRCA1-defective cancer cells through enhancement of transcription-coupled DNA repair. J Biol Chem 1999;274:18808–18812.
57. Wang Y, Cortez D, Yazdi P, et al. BASC, a super complex of BRCA1-associated proteins involved in the recognition and repair of aberrant DNA structures. Genes Dev 2000;14:927–939.
58. Dong Y, Hakimi MA, Chen X, et al. Regulation of BRCC, a holoenzyme complex containing BRCA1 and BRCA2, by a signalosome-like subunit and its role in DNA repair. Mol Cell 2003;5:1087–1099.
59. Hartwell LH, Weinert TA. Checkpoints: controls that ensure the order of cell cycle events. Science 1989;246:629–634.
60. Hartwell LH, Kastan MB. Cell cycle control and cancer. Science 1994;266:1821–1828.
61. Elledge SJ. Cell cycle checkpoints: preventing an identity crisis. Science 1996;274:1664–1671.
62. Zhang H, Somasundaram K, Peng Y, et al. BRCA1 physically associates with p53 and stimulates its transcriptional activity. Oncogene 1998;16:1713–1721.
63. Ouchi T, Monteiro AN, August A, et al. BRCA1 regulates p53-dependent gene expression. Proc Natl Acad Sci USA 1998;95:2302–2306.
64. Somasundaram K, Zhang H, Zeng YX, et al. Arrest of the cell cycle by the tumour suppressor BRCA1 requires the CDK-inhibitor p21WAF1/CiP1. Nature 1997;389:187–190.
65. MacLachlan TK, Somasundaram K, Sgagias M, et al. BRCA1 effects on the cell cycle and the DNA damage response are linked to altered gene expression. J Biol Chem 2000;275:2777–2785.
66. Williamson EA, Dadmanesh F, Koeffler HP. BRCA1 transactivates the cyclin-dependent kinase inhibitor p27(Kip1). Oncogene 2002;21:3199–3206.
67. Xu B, O'Donnell AH, Kim ST, et al. Phosphorylation of serine 1387 in Brca1 is specifically required for the Atm-mediated S-phase checkpoint after ionizing irradiation. Cancer Res 2002;62:4588–4591.
68. Xu X, Weaver Z, Linke SP, et al. Centrosome amplification and a defective G2-M cell cycle checkpoint induce genetic instability in BRCA1 exon 11 isoform-deficient cells. Mol Cell 1999;3:389–395.
69. Quinn JE, Kennedy RD, Mullan PB, et al. BRCA1 functions as a differential modulator of chemotherapy induced apoptosis. Cancer Res 2003;63: 6221–6228.
70. Aprelikova O, Pace AJ, Fang B, et al. BRCA1 is a selective co-activator of 14-3-3 sigma gene transcription in mouse embryonic stem cells. J Biol Chem 2001;276:25647–25650.
71. Harkin DP, Bean JM, Miklos D, et al. Induction of GADD45 and JNK/SAPK-dependent apoptosis following inducible expression of BRCA1. Cell 1999;97: 575–586.
72. Mullan PB, Quinn JE, Gilmore PM, et al. BRCA1 and GADD45 mediated G2/M cell cycle arrest in response to antimicrotubule agents. Oncogene 2001; 20:6123–6131.
73. Yarden RI, Pardo-Reoyo S, Sgagias M, et al. BRCA1 regulates the G2/M checkpoint by activating Chk1 kinase upon DNA damage. Nat Genet 2002;30: 285–289.
74. Weaver Z, Montagna C, Xu X, et al. Mammary tumors in mice conditionally mutant for Brca1 exhibit gross genomic instability and centrosome amplification yet display a recurring distribution of genomic imbalances that is similar to human breast cancer. Oncogene 2002;21:5097–5107.
75. Miyoshi Y, Iwao K, Takahashi Y, et al. Acceleration of chromosomal instability by loss of BRCA1 expression and p53 abnormality in sporadic breast cancers. Cancer Lett 2000;159:211–216.
76. Staff S, Isola J, Tanner M. Haplo-insufficiency of BRCA1 in sporadic breast cancer. Cancer Res 2003; 63:4978–4983.
77. Larson JS, Tonkinson JL, Lai MT. A BRCA1 mutant alters G2-M cell cycle control in human mammary epithelial cells. Cancer Res 1997;57:3351–3355.
78. Deming PB, Cistuilli CA, Zhao H, et al. The human decatenation checkpoint. Proc Natl Acad Sci USA 1998;95:12983–12988.
79. Lingle WL, Barrett SL, Negron VC, et al. Centrosome amplification drives chromosomal instability in breast tumor development. Proc Natl Acad Sci USA 2002;99:1978–1983.
80. Thompson ME, Jensen RA, Obermiller PS, et al. Decreased expression of BRCA1 accelerates growth and is often present during sporadic breast cancer progression. Nat Genet 1995;9:444–450.
81. Holt JT, Thompson ME, Szabo C, et al. Growth retardation and tumour inhibition by BRCA1. Nat Genet 1996;12:298–302.
82. Shao N, Chai YL, Shyam E, et al. Induction of apoptosis by the tumor suppressor protein BRCA1. Oncogene 1996;13:1–7.
83. Thangaraju M, Kaufmann SH, Couch FJ. BRCA1 facilitates stress induced apoptosis in breast and ovarian cancer cell lines. J Biol Chem 2000;275: 33487–33496.
84. Andrews HN, Mullan PB, McWilliams S, et al. BRCA1 regulates the interferon gamma-mediated apoptotic response. J Biol Chem 2002;277:26225–26232.

85. La Farge S, Sylvain V, Ferrara M, et al. Inhibition of BRCA1 leads to increased chemoresistance to microtubule interfering agents, an effect that involves the JNK pathway. Oncogene 2001;20:6597–6606.

86. Henderson IC, Berry DA, Demetri GD, et al. Improved outcomes from adding sequential paclitaxel but not from escalating doxorubicin dose in an adjuvant chemotherapy regimen for patients with node-positive primary breast cancer. J Clin Oncol 2003; 21:976–983.

87. Nabholtz JM, Falkson C, Campos D, et al. Docetaxel and doxorubicin compared with doxorubicin and cyclophosphamide as first-line chemotherapy for metastatic breast cancer: results of a randomized, multicenter, phase III trial. J Clin Oncol 2003;21: 968–975.

88. Kennedy RD, Quinn JE, Johnston PG, et al. BRCA1: mechanisms of inactivation and implications for management of patients. Lancet 2002;360:1007–1014.

89. Pegram MD, Pauletti G, Slamon DJ. HER-2/neu as a predictive marker of response to breast cancer therapy. Breast Cancer Res Treat 1998;52:65–77.

90. Moynahan ME, Cui TY, Jasin M. Homology directed DNA repair, mitomycin C resistance and chromosome stability is restored with correction of Brca1 mutation. Cancer Res 2001;61:4842–4850.

91. Husain A, Venkatraman ES, Spriggs DR. BRCA1 upregulation is associated with repair-mediated resistance to cis-diamminedichloroplatinum (II). Cancer Res 1998;58:1120–1123.

92. Brodie SG, Xu X, Qiao W, et al. Multiple genetic changes are associated with mammary tumorigenesis in Brca1 conditional knockout mice. Oncogene 2001;20:7514–7523.

93. Chappius PO, Goffin J, Wong N. et al. A significant response to neoadjuvant chemotherapy in BRCA1/2 related breast cancer. J Med Genet 2002;398: 608–610.

94. Goffin JR, Chappius P, Begin LR, et al. Impact of germline mutation and overexpression of p53 on prognosis and response to treatment following breast carcinoma: 10 year followup data. Cancer 2003;979:2187–2195.

95. Tassone P, Tagliaferri P, Perricelli A, et al. BRCA1 expression modulates chemosensitivity of BRCA1-defective HCC1937 human breast cancer cells. Br J Cancer 2003;888:1285–1291.

96. Moynahan ME, Pierce AJ, Jasin M. BRCA2 is required for homology-directed repair of chromosomal breaks. Mol Cell 2001;7:263–272.

97. Xia F, Taghian DG, DeFrank JS, et al. Deficiency of human BRCA2 leads to impaired homologous recombination but maintains normal nonhomologous end joining. Proc Natl Acad Sci USA 2001;98: 8644–8649.

98. Chen CF, Chen PL, Zhong Q, et al. Expression of BRC repeats in breast cancer cells disrupts the BRCA2-Rad51 complex and leads to radiation hypersensitivity and loss of G(2)/M checkpoint control. J Biol Chem 1999;274:32931–32935.

99. Bogliolo M, Cabre O, Callen E, et al. The Fanconi anaemia genome stability and tumour suppressor network. Mutagenesis 2002;17:529–538.

100. Yamashita T, Nakahata T. Current knowledge on the pathophysiology of Fanconi anemia: from genes to phenotypes. Int J Hematol 2001;74:33–41.

101. Liu MC, Gelmann EP. P53 gene mutations: case study of a clinical marker for solid tumors. Semin Oncol 2002;29:246–257.

102. Boultwood J. Ataxia telangiectasia gene mutations in leukaemia and lymphoma. J Clin Pathol 2001; 54:512–516.

CHAPTER

26

Biologic and Clinical Significance of DNA Gene Methylation and Epigenetic Events in Breast Cancer

Manel Esteller, MD, PhD, and Esteban Ballestar, PhD

Cancer Epigenetics Laboratory, Molecular Pathology Program, Spanish National Cancer Center (CNIO), Madrid, Spain

Role of Aberrant DNA Methylation and Epigenetics in Normal Cells, Cancer Formation, and Progression

Overview of the Epigenetic Regulation in a Normal Cell

The inheritance of information based on gene expression levels independent of the underlying DNA nucleotide sequence is known as epigenetics, as opposed to genetics, which refers to information transmitted on the basis of gene sequence. The main epigenetic modification in humans is methylation of the nucleotide cytosine (C) when it precedes a guanine (G), forming the dinucleotide CpG. It is thought that about 6–8% of all cytosines are methylated in normal human DNA. The distribution of CpGs dinucleotides in the human genome is not uniform. There are CpG rich regions called CpG islands. These are usually unmethylated in all normal tissues and span the 5′ end of genes in the regulatory regions that determine gene expression. If transcription factors are available and the island remains in an unmethylated state with open chromatin configuration associated with hyperacetylated histones, transcription will occur (FIG. 26.1).[1–3] This is the case for all housekeeping genes. Although this is the general rule, certain CpG islands exist that are normally methylated, for example, one allele of imprinted genes and genes from one inactivated X-chromosome in women. DNA methylation also has a normal protective role in repressing parasitic DNA sequences acquired through human evolution and in inducing latency in several viral pathogens.

A Profound but Specific Disturbance of the DNA Methylation and Chromatin Patterns Occurs in a Cancer Cell

The epigenetic homeostasis of the cell undergoes a dramatic transformation in cancer: transcriptional silencing of tumor suppressor genes by CpG island hypermethylation and histone deacetylation, global genomic hypomethylation, and genetic defects in chromatin related genes.

Silencing of tumor suppressor genes is due to dense hypermethylation of the CpG island located in the regulatory promoter region (FIG. 26.1). Not every gene is methylated and inactivated in every tumor type. We and others have shown the exquisite profile of hypermethylation that occurs in human tumors.[4,5] The portrait of CpG island hypermethylation lesions can be so specific that using blind samples the tumor type can be specifically identified by its methylation pattern.[6] CpG island methylation associated silencing affects all cellular pathways: *p16*^{INK4a} leading to avoidance of senescence, *p14*^{ARF} leading to *p53* degradation, *APC* in aerodigestive tumors, E-cadherin in breast tumors (see Chapter 20), DNA repair genes such as *hMLH1* in microsatellite instability, or O6-methylguanine DNA methyltransferase *MGMT* leading to K-*ras* and *p53* mutations, or even *BRCA1* causing double strand breaks. This silencing also affects hormonal receptors, such as the estrogen (ER), progesterone (PR), and retinoic receptors, and many other targets (*DAPK, p73, VHL, GSTP1, LKB1/STK11, p15*^{INK4b}*, SOCS-1, RASSF1*).[1–3]

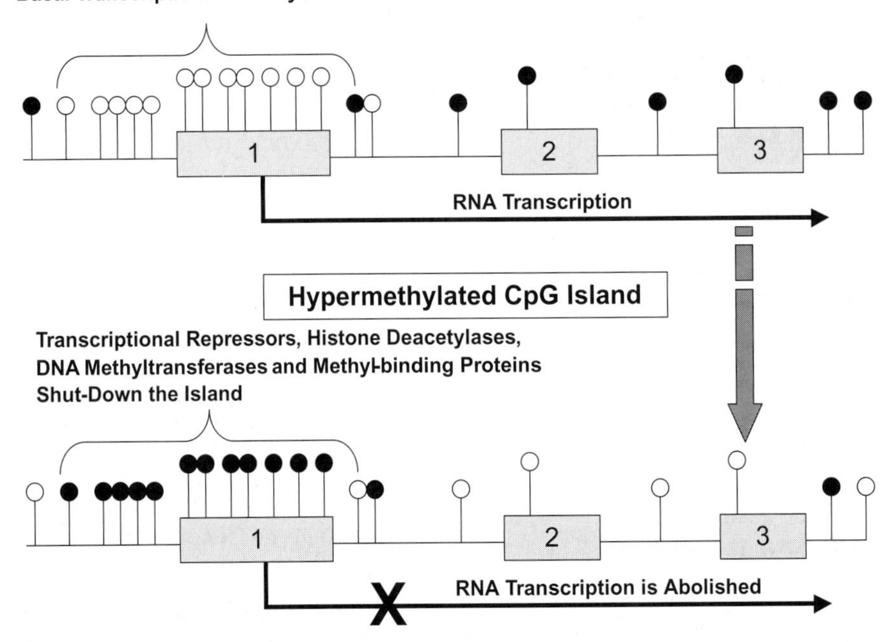

FIGURE 26.1 Comparison of an unmethylated and methylated CpG island of a prototype tumor suppressor gene. If transcription factors are available and the island remains in an unmethylated state with open chromatin configuration-associated with hyperacetylated histones, transcription will occur. The transcription of tumor suppressor genes can be abolished by dense hypermethylation of the CpG island located in the regulatory-promoter region under the influence of transcriptional repressors, histone deacetylases, DNA methyl transferases and methyl-binding proteins.

Although CpG islands become hypermethylated, the genome of the cancer cell undergoes a dramatic global hypomethylation, with 20–60% less genomic 5mC accomplished from the hypomethylation of the "body" of genes and repetitive DNA sequences.[4,7] Global DNA hypomethylation may contribute to carcinogenesis through the induction of chromosomal instability, reactivation of transposable elements, and loss of imprinting.

A third area deals with the existence of genetic defects in methyl-chromatin related genes in cancer cells and other human pathologies. Candidates are the DNA methyltransferases, *DNMT1* ("maintenance" enzyme), *DNMT3a,* and *DNMT3b* ("de novo" enzymes) that are up-regulated in human neoplasms. A germline genetic defect in *DNMT3b* causes the immunodeficiency disease ICF syndrome. Methyl-binding proteins (MBD1, MBD2, MBD3, MBD4, and MeCP2), responsible to recruit histone deacetylases (HDACs) and other chromatin factors to methylated CpGs, are also gene mutation candidates. *MBD4* mutations occur in unstable tumors, and germline mutations of *MeCP2* cause the mental retardation Rett syndrome. The chromatin universe offers more candidates: HDACs, which through deacetylation of the histone tails "pack" the CpG islands to create transcriptional repressive domains; histone acetyltransferases, activators of transcription thanks to their histone acetylase activity ("unpacking" of CpG islands); and the chromatin remodeling factors *SWI/SNF* and polycomb, which rearrange promoter region chromatin and nucleosomes to facilitate gene expression. A few examples exist, such as truncations of *hSNF5/INI1* in pediatric cancers and somatic inactivation of the histone acetyltransferases *p300* in several tumors and *CBP* in families with the Rubinstein-Taybi syndrome.

The list of genes altered by epigenetic mechanisms is increasing at a rapid rate with new putative methylation markers in breast cancer derived from candidate gene "guessing"[2] or genomic approaches[8–12] added to the published literature on a regular basis. In this

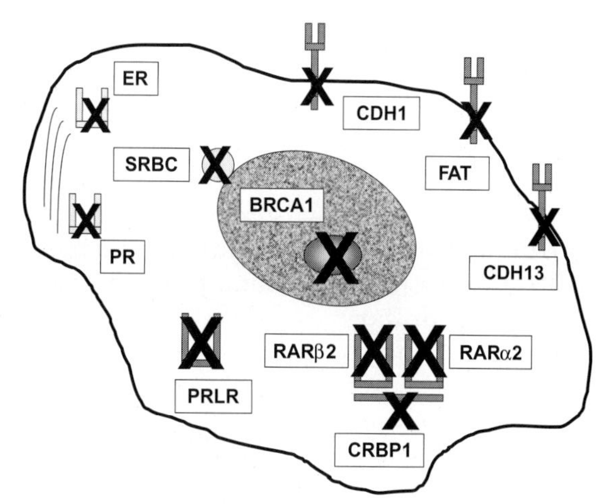

FIGURE 26.2 Tumor suppressor genes with methylation-mediated silencing that are relevant for breast cancer tumorigenesis in the estrogen and progesterone receptor genes (*ER* and *PR*) (Chapter 15), *BRCA1* (Chapters 4, 25 and 27), the prolactin hormone receptor (*PRLR*), retinoic acid receptor α2 (*RARα2*), retinoic acid receptor β2 (*RARβ2*), cellular retinol binding protein 1 (*CRBP1*), E-cadherin (*CDH1*) (Chapter 20), protocadherin (*FAT*), cadherin superfamily members (*CDH13*), and the serum deprivation response factor (SDR)-related gene product that binds to c-kinase (*SRBC*).

chapter we focus on those tumor suppressor genes with methylation mediated silencing that are clearly relevant for breast cancer tumorigenesis (FIG. 26.2).

BRCA1 Is Inactivated by Promoter Hypermethylation in Breast Tumors

An unexpected finding in the molecular genetics of human cancer was the absence of somatic mutations in the breast cancer susceptibility gene *BRCA1* in sporadic cases of breast carcinoma. Since the gene's cloning, germline mutations in *BRCA1* have been found in hereditary cases of breast and ovarian cancers (see Chapters 4 and 25).[13,14] In fact, germline alterations in *BRCA1* have been estimated to be responsible for about 40% of familial breast cancer.[14,15] However, despite an extensive search, the *BRCA1* gene has not been shown to be mutated in any cases of truly sporadic breast cancer. These findings challenged the concept of the role of *BRCA1* as a tumor suppressor gene in the nonhereditary forms of breast tumors that constitute 90% of the total cases. However, two lines of evidence continued to support *BRCA1* loss of function as an important contributor to breast tumorigenesis in the nonfamilial cases. First, a high rate of loss of heterozygosity (LOH), an allelic deletion that usually pinpoints the presence of a tumor suppressor gene, has been observed at the *BRCA1* locus in approximately one half of sporadic breast carcinomas.[13,14] Second, the BRCA1 transcript and protein are often decreased or lost in sporadic breast carcinomas.[15,16] However, it remained uncertain which mechanisms, apart from LOH, were behind the above-mentioned loss of function of *BRCA1* in the nonfamilial breast and ovarian tumors. The subsequent observation of promoter hypermethylation of BRCA1 resolved this dilemma.

The first reports suggested that epithelial breast cancer cell lines displayed methylation at the regulatory regions of the *BRCA1* gene related to down-regulation of its expression.[17,18] After these observations, studies indicated that BRCA1 hypermethylation was also present in primary sporadic breast neoplasms.[19–21] However, the establishment of a definitive relationship among promoter methylation, *BRCA1* gene expression, incidence of LOH, pathologic subtype, or clinical translation was an urgent requirement. This objective was addressed by studying a comprehensive cohort of more than 200 primary breast tumors, breast cancer cell lines, and breast xenografts (human breast tumors grown in immunodeficient mice).[22] These results demonstrated that the inactivation of *BRCA1* by promoter hypermethylation is a common event in sporadic breast cancer, affecting 25–30% of cases. The CpG island hypermethylation of *BRCA1* was also commonly associated with LOH. Interestingly, it is particularly frequent in the medullary and mucinous subtypes of breast carcinoma that the same subtypes are overrepresented in the BRCA1 families.[22] These findings were also later confirmed by several others laboratories.[23,24]

Ensuing studies have revealed new facts about the basic mechanisms and biologic implications of BRCA1 methylation. The detailed analysis of the regulation of the CpG island of BRCA1 has demonstrated that the establishment of gene silencing is also related to the binding of methyl-CpG binding proteins to those methylated CpGs[25] and changes in histone acetylation and chromatin structure.[26] It is also known that other genes encoding proteins that are partners of BRCA1 can also have methylation associated silencing in breast cancer, such as SRBC,[27] and can contribute to the shut-down of the BRCA1 pathway in breast tumorigenesis. It has been clearly demonstrated that, unlike *BRCA1, BRCA2* does not undergo promoter hypermethylation.[28,29] However, there are alternative epigenetic mechanisms in the absence of DNA methylation that can also lead to *BRCA2* loss of function in breast cancer,[30] discussed at the end of this chapter.

There are two clinically important issues relevant to *BRCA1* status. First, the expression microarray patterns of the familial breast tumors related to *BRCA1* and *BRCA2* families and those of sporadic cases yield three different profiles.[31] There is one exception: If a breast tumor has a germline mutation in *BRCA1* or a somatic hypermethylation of *BRCA1,* it will have a similar portrait of gene expression.[31] Thus, the inactivation of *BRCA1* by genetic or epigenetic means leads to the same kind of alterations in breast physiology. The second clinically relevant point also relates to the interplay between mutation and methylation. According to Knudson's "two-hit" hypothesis in familial cancer syndromes, an individual inherits one mutant copy of the tumor suppressor gene and the second copy is lost by genomic deletion. *BRCA1* generally conforms to this model, but not always: there are inherited *BRCA1* breast tumors that retain the second allele. The explanation for this phenomenon was that this conserved nonmutated allele of *BRCA1* becomes methylated in breast tumors.[4,3] All these data support a predominant and specific role for BRCA1 CpG island promoter methylation as the main cause of BRCA1 loss of function in breast cancer.

Hormonal Status in Breast Cancer Depends on the Methylation Status of the Relevant Hormone Receptors

Hormone responsiveness is a critical determinant of breast cancer progression and management, and the response to endocrine therapy is highly correlated with the ER and PR status of tumor cells (see Chapter 15). Thus, key areas of study in breast cancer are those mechanisms that regulate ER and PR expression in normal and malignant breast tissues. One third of all breast cancers lack ER and PR, which is associated with less differentiated tumors and poorer clinical outcome. In addition, approximately one half of ER-positive tumors lack PR protein, and patients with this phenotype are less likely to respond to hormonal therapies than those whose tumors express both receptors. Although a variety of possible mechanisms might account for the loss of ER and PR gene expression, including structural changes within each gene such as deletions or polymorphisms, it is promoter methylation that is predominantly responsible.[2,33]

Initial studies in cultured human breast cancer cell lines showed that the absence of ER gene expression in ER-negative cells was associated with extensive methylation of the CpG island located in its regulatory region,[34] and treatment with pharmacologic agents that demethylate the DNA, and thus the ER gene CpG island, resulted in the restoration of a functional ER protein.[35] The methylation related silencing of ER is removed if an inhibitor HDAC is added to the demethylating drug.[9,36] In primary mammary carcinomas, it has been found that ER is hypermethylated in around 25–45% of tumors with good correlation with the loss of gene expression and clinical progression.[37,38] The PR receptor gene also undergoes methylation associated silencing in breast carcinomas with a similar frequency to that of ER,[37] but the picture is more complicated because PR expression is induced by ER. Overall, these data demonstrate that methylation of the ER and PR gene CpG islands is associated with the lack of ER and PR gene expression in a significant fraction of human breast cancers.

Additional hormonal receptors are also disrupted by CpG island promoter hypermethylation in breast malignancies. The prolactin hormone receptor[39] is methylated in breast cancer cell lines and in approximately 30% of primary tumors in association with the loss of prolactin hormone receptor expression. Demethylating drugs restore prolactin hormone receptor expression.[12] In this regard, prolactin and its receptor are increasingly being recognized as playing important roles in breast tumorigenesis.[39,40] Finally, another cellular network extremely important for breast homeostasis compromised by aberrant DNA methylation is the retinoid response. The main target of epigenetic silencing, the retinoid response is the gene encoding the retinoic acid receptor β2. Retinoic acid receptor β2 is found to be methylated and repressed in cultured breast cancer cells[36,41–44] and primary tumors.[44,45] Two additional genes from this pathway also undergo methylation-associated silencing in breast tumorigenesis: the retinoic acid receptor α2,[46] and the cellular retinol binding protein 1.[47] Of potential interest in breast cancer epidemiology related to the incidence of the disease, it is worth noting that high dietary intakes of retinoids "prevented" the hypermethylation of the retinoic acid receptor β2 and the cellular retinol binding protein 1 genes.[47]

Disruption of Cell Adhesion in Breast Cancer by Hypermethylation of the E-Cadherin and Related Genes

E-cadherin (CDH1), a cell adhesion protein that plays a key role in maintaining the epithelial phenotype (see Chapter 20), is regarded as an invasion suppressor gene in light of accumulating evidence from in vitro experiments and clinical observations. Decreased CDH1 expression is common in poorly differentiated advanced stage carcinomas, but the mechanism by which CDH1 was silenced remained unclear.[48] Except for the case of lobular breast cancer where somatic gene mutation in CDH1 seems to play an important role,[49] E-cadherin is another gene where the basis of its inactivation is promoter hypermethylation. CDH1 displays methylation associated silencing in breast cancer cell lines[49,50] and primary tumors,[49–51] where the greatest decreases are found in the most aggressive tumors.[38] Furthermore, treatment of E-cadherin–negative carcinoma cells with the demethylating agent 5-azacytidine results in reexpression of the gene and reversion of scattered spindle-shaped cells to cells with epithelial morphology.[49] Thus, demethylating drugs can induce epithelial morphogenesis with acquisition of the homophilic cell–cell adhesive property.

Epigenetics and metastasis are not static processes. Indeed, whereas initial loss of CDH1 may promote invasion, reexpression may facilitate cell survival within metastatic deposits. The mechanisms underlying such plasticity are unclear. One possibility is that breast cancer cells can cycle between different states of CDH1 promoter methylation and gene expression.[52] In cultured tumor cells, the methylation patterns may shift in relation to the tumor microenvironment. After invasion in vitro, which favors diminished E-cadherin expression, the density of promoter methylation markedly increased. When these cells are cultured as spheroids, which requires homotypic cell adhesion, promoter methylation decreases dramatically, and E-cadherin is reexpressed. This observation becomes more complicated when the methylation status of other members of the cadherin superfamily, such as CDH13[53] and the protocadherin FAT,[11] are included. Finally, there exists another level of epigenetic regulation of CDH1 expression in normal breast epithelial and cancer cells mediated by numerous members of the chromatin–histone network, discussed in the next sections.

Further Chromatin Mechanisms of Epigenetic Silencing in Breast Cancer Cells: Methyl-Binding Proteins, Histones, and Chromatin Remodeling Factors

Intense research efforts are currently underway to understand the mechanisms of epigenetic silencing of tumor suppressor genes in cancer. The study of these processes not only helps to understand the epigenetic pathway of cancer, but it is also a source of novel therapies.

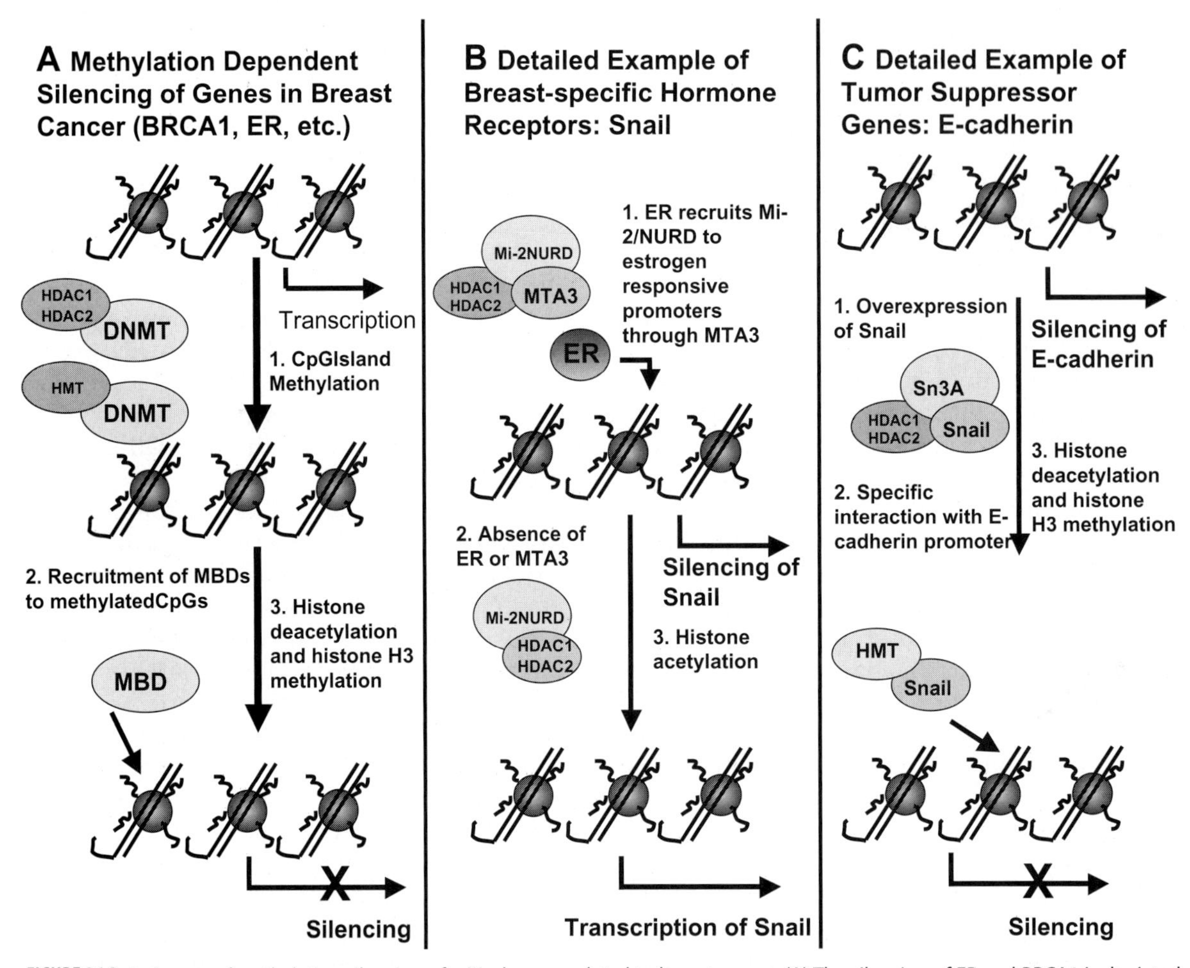

FIGURE 26.3 Pathways of methylation silencing of critical genes related to breast cancer. (A) The silencing of ER and BRCA1 is depicted highlighting the roles of histone deacetylation and histone methylation. (B) The activation of the ER pathway is associated with silencing of the Snail zinc finger transcription factor which plays a role in repressing E-cadherin transcription and the function of E-cadherin as a cell adhesion molecule and tumor suppressor gene. (C) The relationship between the silencing of Snail and the downstream silencing of E-cadherin is shown.

It can be anticipated that general chromatin mechanisms will be involved in translating the information encoded by hypermethylation of the CpG islands and that the specific effects in breast tissue will arise from the set of genes that are silenced in this tumor type. In the case of breast cancer, different types of chromatin mediated regulation can be distinguished: general chromatin mechanisms involving methylation dependent inactivation of genes and breast-specific mechanisms that comprise hormone receptors and specific nuclear factors that interact with tumor suppressor genes (FIG. 26.3).

Methylation-Dependent Silencing of Genes in Breast Cancer

As explained above, hypermethylation of CpG islands and subsequent inactivation of tumor suppressor genes is a general epigenetic mechanism in cancer. One important conclusion from the massive and comprehensive screening of the methylation status of many genes was that the profile of methylation is predominantly tumor type specific. The chromatin associated mechanisms that link DNA methylation and transcriptional silencing are now starting to be uncovered. Methylation of *BRCA1* and other genes is coupled with transcriptional silencing through the action of multiprotein complexes that contain both methylated DNA binding subunits and histone modifying and nucleosome remodeling activities. The methylated DNA binding subunits of these complexes are members of a family of proteins that share a methyl-CpG binding domain (MBD).[54,55] Among the remaining subunits of these complexes, it is remarkable for the presence of histone deacetylases, histone methyltransferases, and, in some cases, nucleosome remodeling subunits that modify the chromatin structure to generate a transcriptionally silenced state. Chromatin immunoprecipitation (ChIP) assays are used to investigate the presence of MBD proteins in methylated promoters. These assays consist in the immunoprecipitation of previously formaldehyde fixed and sonicated chromatin with specific antibodies against a DNA-binding protein. Isolation of immunoprecipitated DNA is followed by polymerase chain reaction amplification with specific primers. In the particular case of the methylated *BRCA1* gene in breast cancer, the presence of MBD2 in its promoter has recently been demonstrated by ChIP analysis.[12] Association of MBD2 to the methylated BRCA1 promoter was shown to be associated with histone deacetylation and histone H3 K9 methylation, chromatin states that are associated with transcriptional repression.

Because the presence of MBD proteins is a general feature of methylated genes in cancer, the use of ChIP assays with antibodies against different MBD proteins has recently been used to identify methylated genes in breast cancer. The rationale of this approach is the following: ChIP will allow the isolation of MBD-bound sequences that, consistent with the properties of the MBD proteins, should be methylated. By hybridizing the DNA isolated in ChIP assays from breast cancer cells to hybridize a microarray containing CpG island sequences, it is possible to discover genes that are methylated in breast cancer.[12]

Searching for Breast-Specific Chromatin Mechanisms of Regulation: Hormone Receptors

Another important issue concerns the identification of the specific complexes that mediate the silencing of different hormone receptors. For instance, the ER is a key regulator of proliferation and differentiation in mammary epithelia and represents a crucial prognostic indicator and therapeutic target in breast cancer. Estrogens have long been known to have mitogenic functions in breast cancer cell lines and in breast tumors. Approximately 30% of breast carcinomas lack ER expression. Presumably, these breast cancers become estrogen independent through genetic alterations that bypass the requirement for ER dependent stimulation of cell proliferation. Mechanistically, ER induces changes in gene expression through direct gene activation and also through the biologic functions of target loci.

It has been reported that ER-positive breast cancer cells exhibit activation of the heregulin/HER-2 pathway that leads to disruption of estradiol responsiveness and thus contributes to progression of tumors to more invasive phenotypes. This process is effected by heregulin-β_1 mediated overexpression of metastatic-associated protein 1 (MTA1), a component of HDAC and nucleosome–remodeling complexes.[56] MTA1 and HDACs act as potent corepressors of ER element driven transcription. Overexpression of MTA1 in breast cancer cells is accompanied by enhancement of the ability of cells to invade and to grow in an anchorage-independent manner.

Moreover, the product of the human *MTA3* gene is an estrogen-dependent component of the Mi-2/NuRD transcriptional corepressor in breast epithelial cells.[57] The Mi-2/NuRD complex contains an MBD protein, namely MBD3, as an integral subunit. This complex is recruited to methylated genes through its association with MBD2 or, alternatively, can be recruited to other targets through interaction with some sequence-specific DNA binding proteins.[58,59] MTA3 constitutes a key component of an estrogen-dependent pathway, regulating growth and differentiation. The absence of ER or of MTA3 leads to aberrant expression of the transcriptional repressor Snail, a master regulator of epithelial to mesenchymal transitions. Aberrant Snail expression results in loss of expression of the cell adhesion molecule E-cadherin, which, as described above, is an event associated with changes in epithelial architecture and invasive growth. These results establish mechanistic links between the ER status and invasive growth of breast cancers.

Tumor Suppressor Genes and Chromatin Connections

The study of the mechanisms by which tumor suppressor genes exert their effects is bringing chromatin-associated mechanisms into sharp focus. Most of the genes described here interact with different histone modifying activities; an important example of a tumor suppressor gene that is a key element in cancer development and progression is again the E-cadherin gene (*CDH1*). Its inactivation occurs not only through instability at its chromosomal locus and mutations, but also through epigenetic mechanisms such as promoter hypermethylation and transcriptional silencing. In skin cancer, it has been recently demonstrated that Snail mediates silencing of E-cadherin through the recruitment of the Sin3A/HDAC 1/2 complex.[60]

Another example of the chromatin involvement in tumor suppressor activities comes from the breast cancer metastasis suppressor 1 (BRMS1), whose gene product suppresses metastasis of multiple human and murine cancer cells without inhibiting tumorigenicity. It has been recently demonstrated that BRMS1 interacts with retinoblastoma binding protein 1 and at least seven members of the mSin3 HDAC complex.[61] Presumably, histone deacetylation of BRMS1 targets will contribute to the malignant phenotype.

A final example of the chromatin involvement in tumor suppressor activities is provided by EMSY, a novel nuclear factor that binds *BRCA2*. Although the mechanistic implications of the BRCA2–EMSY interactions remain to be defined, EMSY seems to participate in chromatin remodeling during double strand DNA break repair. Interestingly, aberrant overexpression of this factor may contribute to tumorigenesis in a subset of sporadic breast cancers.[30]

Using Promoter Hypermethylation for Translational Purposes in Breast Cancer Patients

To realize the potential for the molecular understanding of the cancer cell to be translated into improved diagnosis and treatment for cancer patients, the epigenetic profiles described in this chapter can be exploited in three ways (FIG. 26.4).

First, new lines of treatments can be based on epigenetics. Unlike genetic changes in cancer, epigenetic changes are potentially reversible. In cultured breast cancer cell lines, it has been possible to reexpress genes silenced by methylation using demethylating agents such as 5-aza-2-deoxycytidine.[62] These compounds have been previously used in the clinic, but the doses administered were quite toxic. Interestingly, the dose can be reduced by adding inhibitors of histone deacetylases, such as phenyl butyrate, and several clinical trials in phase I and II are underway to test this strategy.

Second, methylation can be used as a biomarker of breast cancer cells. Because the presence of CpG island hypermethylation of the tumor suppressor genes described in this

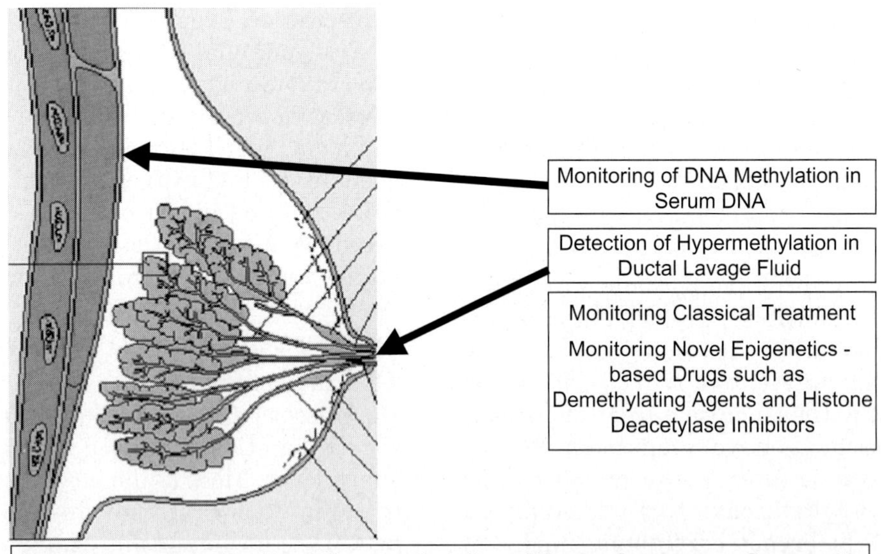

FIGURE 26.4 The potential clinical applications of the study of epigenetic profiles of breast cancer are shown in this illustration. There are three possible routes of exploitation of these observations: 1) the development of new lines of treatments based on epigenetics; 2) the use of gene methylation as a diagnostic biomarker of breast cancer in blood, body fluids, and nipple aspirates (Chapters 5 and 27); and 3) the use of gene promoter hypermethylation as a prognostic factor and predictor of response to therapy.

chapter is specific for transformed cells,[1–3] and a particular profile of methylation exists for breast cancer cells,[11,51] methylation can be used as an indicator for the presence of breast malignancy and may also be used as a tool to detect cancer cells in multiple biologic fluids (see Chapter 5), such as ductal lavage fluids,[63] or even to monitor hypermethylated promoter loci in serum DNA from cancer patients.[64]

Third, gene promoter hypermethylation can be used as a prognostic/predictor factor. Methylation is not only a "black and white" marker but also a qualitative one. There is now compelling evidence that methylation-associated silencing of the DNA repair *MGMT* in tumors indicates which patients are going to be sensitive to chemotherapy with alkylating agents (see Chapter 27).[65] Similar scenarios can now be outlined using the methylation status of the ER, PR, prolactin, or retinoid receptors for breast cancer.

In summary, the disruption of the DNA methylation and epigenetic machinery and their patterns is a major hallmark of breast cancer. Much is still unknown, but the unfolding scenario shows great promise for a better understanding of cancer biology and for the improvement in the management of human breast tumors.

References

1. Jones PA, Laird PW. Cancer epigenetics comes of age Nat Genet 1999;21:163–167.
2. Esteller M. CpG island hypermethylation and tumor suppressor genes: a booming present, a brighter future. Oncogene 2002;21:5427–5440.
3. Herman JG, Baylin SB. Gene silencing in cancer in association with promoter hypermethylation. N Engl J Med 2003;349:2042–2054.
4. Esteller M, Corn PG, Baylin SB, et al. A gene hypermethylation profile of human cancer. Cancer Res 2001;61:3225–3229.
5. Costello JF, Fruhwald MC, Smiraglia DJ, et al. Aberrant CpG-island methylation has non-random and tumour-type-specific patterns. Nat Genet 2000;24: 132–138.
6. Paz MF, Fraga MF, Avila S, et al. A systematic profile of DNA methylation in human cancer cell lines. Cancer Res 2003;63:1114–1121.
7. Goelz SE, Vogelstein B, Hamilton SR. Hypomethylation of DNA from benign and malignant human colon neoplasms. Science 1985;228:187–190.
8. Huang TH, Laux DE, Hamlin BC, et al. Identification of DNA methylation markers for human breast carcinomas using the methylation-sensitive restriction fingerprinting technique. Cancer Res 1997;57: 1030–1034.
9. Yang X, Phillips DL, Ferguson AT, et al. Synergistic activation of functional estrogen receptor (ER)-alpha by DNA methyltransferase and histone deacetylase inhibition in human ER-alpha-negative breast cancer cells. Cancer Res 2001;61:7025–7029.
10. Shi H, Yan PS, Chen CM, et al. Expressed CpG island sequence tag microarray for dual screening of DNA hypermethylation and gene silencing in cancer cells. Cancer Res 2002;62:3214–3220.
11. Paz MF, Wei S, Cigudosa JC, et al. Genetic unmasking of epigenetically silenced tumor suppressor genes in colon cancer cells deficient in DNA methyltransferases. Hum Mol Genet 2003;12:2209–2219.
12. Ballestar E, Paz MF, Valle L, et al. Methyl-CpG binding proteins identify novel sites of epigenetic inactivation in human cancer. EMBO J 2003;22:1–11.
13. Gayther SA, Pharoah PD, Ponder BA. The genetics of inherited breast cancer. J Mamm Gland Biol Neoplasia 1998;3:365–376.
14. Wooster R, Weber BL. Breast and ovarian cancer. N Engl J Med 2003;348:2339–2347.
15. Thompson ME, Jensen RA, Obermiller PS, et al. Decreased expression of BRCA1 accelerates growth and is often present during sporadic breast cancer progression. Nat Genet 1995;9:444–450.
16. Wilson CA, Ramos L, Villasenor MR, et al. Localization of human BRCA1 and its loss in high-grade, non-inherited breast carcinomas. Nat Genet 1999;21: 236–240.
17. Rice JC, Massey-Brown KS, Futscher BW. Aberrant methylation of the BRCA1 CpG island promoter is associated with decreased BRCA1 mRNA in sporadic breast cancer cells. Oncogene 1998;17: 1807–1812.
18. Mancini DN, Rodenhiser DI, Ainsworth PJ, et al. CpG methylation within the 5′ regulatory region of the BRCA1 gene is tumor specific and includes a putative CREB binding site. Oncogene 1998;16:1161–1169.
19. Dobrovic A, Simpfendorfer D. Methylation of the BRCA1 gene in sporadic breast cancer. Cancer Res 1997;57:3347–3350.
20. Magdinier F, Ribieras S, Lenoir GM, et al. Down-regulation of BRCA1 in human sporadic breast cancer; analysis of DNA methylation patterns of the putative promoter region. Oncogene 1998;17:3169–3176.
21. Catteau A, Harris WH, Xu CF, et al. Methylation of the BRCA1 promoter region in sporadic breast and ovarian cancer: correlation with disease characteristics. Oncogene 1999;18:1957–1965.
22. Esteller M, Silva JM, Dominguez G, et al. Promoter hypermethylation and BRCA1 inactivation in sporadic breast and ovarian tumors. J Natl Cancer Inst 2000;92:564–569.
23. Bianco T, Chenevix-Trench G, Walsh DC. Tumour-specific distribution of BRCA1 promoter region methylation supports a pathogenetic role in breast and ovarian cancer. Carcinogenesis 2000;21:147–151.
24. Rice JC, Ozcelik H, Maxeiner P, et al. Methylation of the BRCA1 promoter is associated with decreased BRCA1 mRNA levels in clinical breast cancer specimens. Carcinogenesis 2000;21:1761–1765.
25. Fraga MF, Ballestar E, Montoya G, et al. The affinity of different MBD proteins for a specific methylated locus depends on their intrinsic binding properties. Nucleic Acids Res 2003;31:1765–1774.
26. Rice JC, Futscher BW. Transcriptional repression of BRCA1 by aberrant cytosine methylation, histone hypoacetylation and chromatin condensation of the BRCA1 promoter. Nucleic Acids Res 2000;28: 3233–3239.

27. Xu XL, Wu LC, Du F, et al. Inactivation of human SRBC, located within the 11p15.5-p15.4 tumor suppressor region, in breast and lung cancers. Cancer Res 2001;61:7943–7949.
28. Collins N, Wooster R, Stratton MR. Absence of methylation of CpG dinucleotides within the promoter of the breast cancer susceptibility gene BRCA2 in normal tissues and in breast and ovarian cancers. Br J Cancer 1997;76:1150–1156.
29. Hilton JL, Geisler JP, Rathe JA, et al. Inactivation of BRCA1 and BRCA2 in ovarian cancer. J Natl Cancer Inst 2002;94:1396–1406.
30. Hughes-Davies L, Huntsman D, Ruas M, et al. EMSY links the BRCA2 pathway to sporadic breast and ovarian cancer. Cell 2003;115:523–535.
31. Hedenfalk I, Duggan D, Chen Y, et al. Gene-expression profiles in hereditary breast cancer. N Engl J Med 2001;344:539–548.
32. Geisler JP, Rathe JA, Manahan KJ, et al. Methylation: a second hit in the two-hit hypothesis. Eur J Gynaecol Oncol 2003;24:361.
33. Ferguson AT, Lapidus RG, Davidson NE. The regulation of estrogen receptor expression and function in human breast cancer. Cancer Treat Res 1998;94: 255–278.
34. Ottaviano YL, Issa JP, Parl FF, et al. Methylation of the estrogen receptor gene CpG island marks loss of estrogen receptor expression in human breast cancer cells. Cancer Res 1994;54:2552–2555.
35. Ferguson AT, Lapidus RG, Baylin SB, et al. Demethylation of the estrogen receptor gene in estrogen receptor-negative breast cancer cells can reactivate estrogen receptor gene expression. Cancer Res 1995; 55:2279–2283.
36. Bovenzi V, Momparler RL. Antineoplastic action of 5-aza-2′-deoxycytidine and histone deacetylase inhibitor and their effect on the expression of retinoic acid receptor beta and estrogen receptor alpha genes in breast carcinoma cells. Cancer Chemother Pharmacol 2001;48:71–76.
37. Lapidus RG, Ferguson AT, Ottaviano YL, et al. Methylation of estrogen and progesterone receptor gene 5′ CpG islands correlates with lack of estrogen and progesterone receptor gene expression in breast tumors. Clin Cancer Res 1996;2:805–810.
38. Nass SJ, Herman JG, Gabrielson E, et al. Aberrant methylation of the estrogen receptor and E-cadherin 5′ CpG islands increases with malignant progression in human breast cancer. Cancer Res 2000;60: 4346–4348.
39. Goffin V, Binart N, Touraine P, et al. Prolactin: the new biology of an old hormone. Annu Rev Physiol 2002;64:47–67.
40. Clevenger CV, Furth PA, Hankinson SE, et al. The role of prolactin in mammary carcinoma. Endocr Rev 2003;24:1–27.
41. Bovenzi V, Le NL, Cote S, et al. DNA methylation of retinoic acid receptor beta in breast cancer and possible therapeutic role of 5-aza-2′-deoxycytidine. Anticancer Drugs 1999;10;471–476.
42. Arapshian A, Kuppumbatti YS, Mira-y-Lopez R. Methylation of conserved CpG sites neighboring the beta retinoic acid response element may mediate retinoic acid receptor beta gene silencing in MCF-7 breast cancer cells. Oncogene 2000;19:4066–4070.
43. Sirchia SM, Ferguson AT, Sironi E, et al. Evidence of epigenetic changes affecting the chromatin state of the retinoic acid receptor beta2 promoter in breast cancer cells. Oncogene 2000;19:1556–1563.
44. Widschwendter M, Berger J, Hermann M, et al. Methylation and silencing of the retinoic acid receptor-beta2 gene in breast cancer. J Natl Cancer Inst 2000;92:826–832.
45. Esteller M, Fraga MF, Paz MF, et al. Cancer epigenetics and methylation. Science 2002;297:1807–1808.
46. Farias EF, Arapshian A, Bleiweiss IJ, et al. Retinoic acid receptor alpha2 is a growth suppressor epigenetically silenced in MCF-7 human breast cancer cells. Cell Growth Differ 2002;13:335–341.
47. Esteller M, Guo M, Moreno V, et al. Hypermethylation-associated inactivation of the cellular retinol-binding-protein 1 gene in human cancer. Cancer Res 2002;62: 5902–5905.
48. Berx G, Van Roy F. The E-cadherin/catenin complex: an important gatekeeper in breast cancer tumorigenesis and malignant progression. Breast Cancer Res 2001;3:289–293.
49. Graff JR, Herman JG, Lapidus RG, et al. E-cadherin expression is silenced by DNA hypermethylation in human breast and prostate carcinomas. Cancer Res 1995;55:5195–5199.
50. Yoshiura K, Kanai Y, Ochiai A, et al. Silencing of the E-cadherin invasion-suppressor gene by CpG methylation in human carcinomas. Proc Natl Acad Sci USA 1995;92:7416–7419.
51. Esteller M, Fraga MF, Guo M, et al. DNA methylation patterns in hereditary human cancers mimic sporadic tumorigenesis. Hum Mol Genet 2001;10: 3001–3007.
52. Graff JR, Gabrielson E, Fujii H, et al. Methylation patterns of the E-cadherin 5′ CpG island are unstable and reflect the dynamic, heterogeneous loss of E-cadherin expression during metastatic progression. J Biol Chem 2000;275:2727–2732.
53. Toyooka KO, Toyooka S, Virmani AK, et al. Loss of expression and aberrant methylation of the CDH13 (H-cadherin) gene in breast and lung carcinomas. Cancer Res 2001;61:4556–4560.
54. Hendrich B, Bird A. Identification and characterization of a family of mammalian methyl-CpG binding proteins. Mol Cell Biol 1998;18:6538–6547.
55. Ballestar E, Wolffe AP. Methyl-CpG-binding proteins: targeting specific gene repression. Eur J Biochem 2001;268:1–6.
56. Mazumdar A, Wang RA, Mishra SK, et al. Transcriptional repression of oestrogen receptor by metastasis-associated protein 1 corepressor. Nat Cell Biol 2001; 3:30–37.
57. Fujita N, Jaye DL, Kajita M, et al. MTA3, a Mi-2/NuRD complex subunit, regulates an invasive growth pathway in breast cancer. Cell 2003;113:207–219.
58. Hendrich B, Guy J, Ramsahoye B, et al. Closely related proteins MBD2 and MBD3 play distinctive but interacting roles in mouse development. Genes Dev 2001;15:710–723.
59. Feng Q, Zhang Y. The MeCP1 complex represses transcription through preferential binding, remodeling, and deacetylating methylated nucleosomes. Genes Dev 2001;15:827–832.
60. Peinado H, Ballestar E, Esteller M, et al. The transcription factor Snail mediates E-cadherin repression by the recruitment of the Sin3A/histone deacetylase 1/2 complex. Mol Cell Biol 2004;24:316–319.

61. Meehan WJ, Samant RS, Hopper JE, et al. Breast cancer metastasis suppressor 1 (BRMS1) forms complexes with retinoblastoma-binding protein 1 (RBP1) and the mSin3 histone deacetylase complex and represses transcription. J Biol Chem 2004;279: 1562–1569.

62. Villar-Garea A, Esteller M. DNA demethylating agents and chromatin-remodelling drugs: which, how and why? Curr Drug Metab 2003;4:11–31.

63. Evron E, Dooley WC, Umbricht CB, et al. Detection of breast cancer cells in ductal lavage fluid by methylation-specific PCR. Lancet 2001;357: 1335–1336.

64. Esteller M, Sanchez-Cespedes M, Rosell R, et al. Detection of aberrant promoter hypermethylation of tumor suppressor genes in serum DNA from non-small cell lung cancer patients. Cancer Res 1999;59: 67–70.

65. Esteller M, Herman JG. Generating mutations but providing chemosensitivity: the role of $O^{(6)}$-methylguanine DNA methyltransferase in human cancer. Oncogene 2004;23:1–8.

CHAPTER 27

DNA Methylation in Breast Cancer as Diagnostic Markers and Targets for Cancer Therapy

Sabine Maier, MD, PhD,[1] Nadia Harbeck, MD, PhD,[2] Manfred Schmitt, PhD,[2] John A. Foekens, PhD,[3] and John W. M. Martens, PhD[3]

[1]Department of Biomedical Research and Development and Technology Development, Epigenomics AG, Berlin, Germany
[2]Department of Obstetrics and Gynecology, Klinikum rechts der Isar, Technical University of Munich, Munich, Germany
[3]Department of Medical Oncology, Erasmus MC, Rotterdam, The Netherlands

General Introduction to DNA Methylation

Methylation on the fifth carbon of cytosine nucleotides can be considered the fifth base in the mammalian genome (FIG. 27.1)[1,2]; it contains important biologic information that is not included in the human genome as we currently know it.[3,4] DNA methylation does not alter the nucleotide sequence[5] but adds potentially reversible information on top of the genetically inherited genome.[6,7] Therefore, it has been termed epigenetic information. Even though epigenetic information is not inheritable from one individual to another, it is often passed on within an organism from a precursor cell to its daughter cells, thereby maintaining biologic knowledge that is gathered along the road of differentiation and/or development. With few exceptions, in humans and other mammals DNA methylation occurs predominantly on cytosine residues that precede a guanine residue.[2] These sites have therefore traditionally been referred to as CpG sites (CpGs). Methylation on CpGs in general occurs on both strands in the regulatory regions of genes and in highly repetitive DNA sequences. The bulk of the genomic DNA methylation (70%) is found on extragenic highly repetitive regions that comprise 30–35% of the genome. These regions include centromeric regions, transposons (Line-1 elements) and retroviral sequences, and junk DNA with unknown function. In the coding regions, CpG sites are underrepresented and occur at a frequency of 1 per 100 bp. In contrast, regulatory regions of genes contain CpGs at expected frequency (1 per 8 bp), indicating selection against CpGs outside the regulatory regions. The CpG dinucleotides are often clustered together in so-called CpG islands that are rich in CpG sites.[8] CpG islands are frequently found in the promoter region of genes or the first five exon regions of genes.[8–10] There are approximately 25,000 CpG islands in the human genome, and 50–60% of the genes contain at least one CpG island. Through methylation of the CpG islands, gene activation can directly be suppressed by masking binding sites for transcription factors and indirectly by recruitment of repressor complexes that remodel the local chromatin structure in the regulatory region of genes (FIG. 27.2 and discussion on methyl binding proteins, below). The result of either of these mechanisms is that the interaction of the promoter with the transcriptional machinery is prevented.[11] In normal adult tissues, unlike in cancer, CpG islands are generally nonmethylated, with the exception of transcriptionally silent genes on the inactive X chromosome,[12] maternally and paternally imprinted genes,[13] and certain genes involved in tissue specific expression.[14]

In this chapter, we discuss the natural function of DNA methylation, the enzymes catalyzing the reaction, and the mechanisms causing DNA methylation. Furthermore, methods to study DNA methylation are reviewed. We also discuss aberrant DNA methylation that is observed in cancer and its possible role in the malignant behavior of

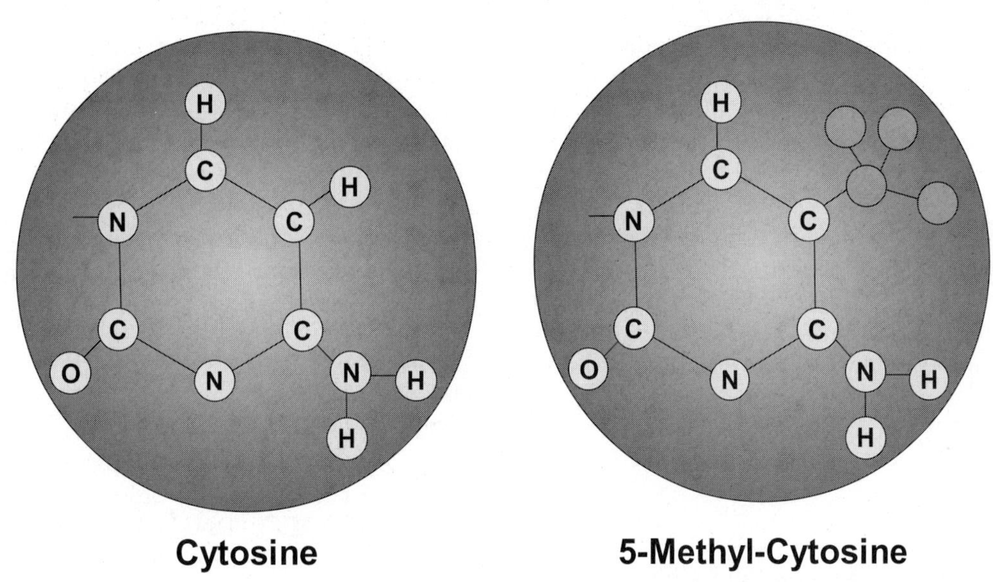

FIGURE 27.1 The fifth base in the genome. Methylation of the carbon 5 position is the epigenetic DNA modification in the mammalian genome that contributes to cancer. (Courtesy Epigenomics, www.epigenomics.com.)

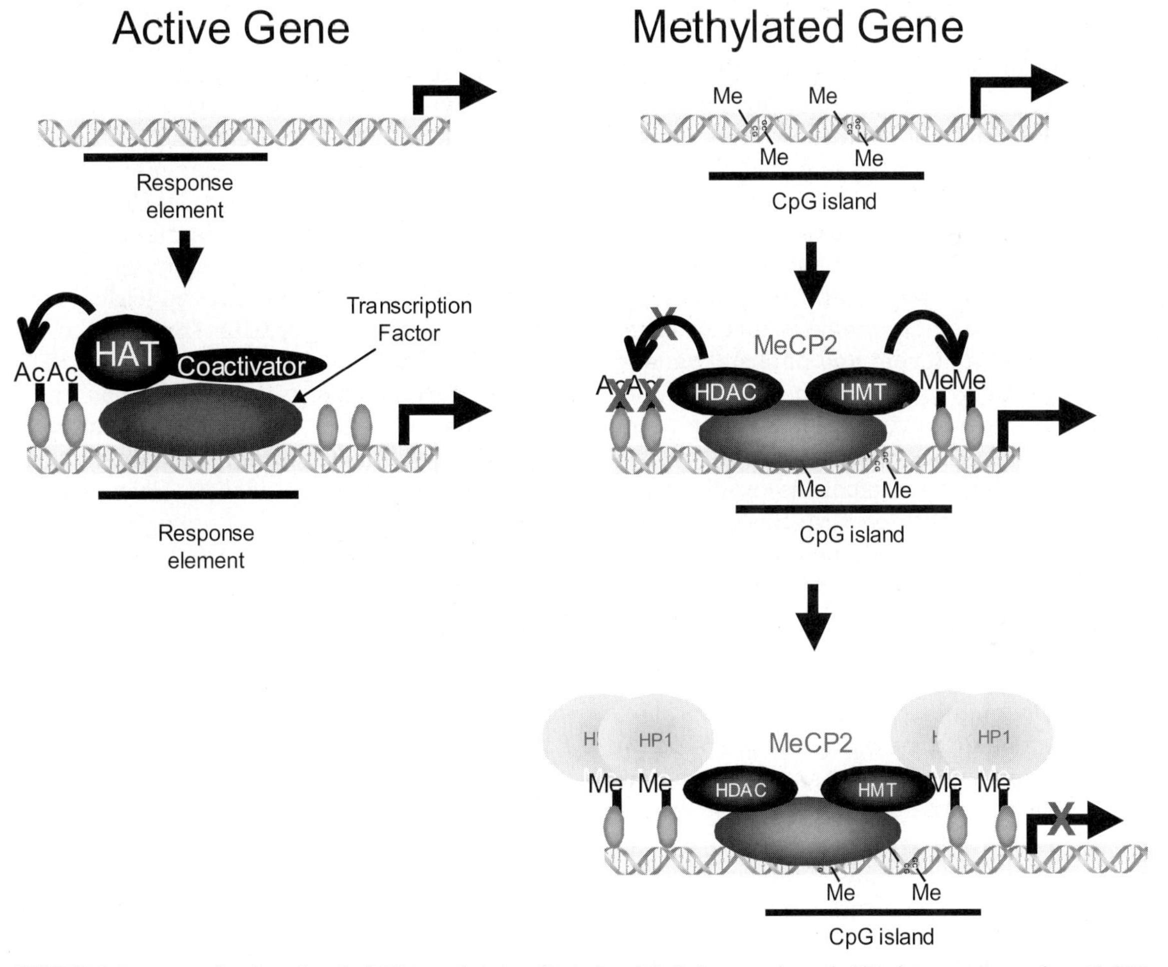

FIGURE 27.2 Promoter silencing after CpG DNA methylation. Methylated CpG sites recruit methyl binding proteins such as MeCP2. These subsequently recruit repressor complexes, including histone deacetylases and histone H3-K9/27 methyltransferases. K9/27 methylated histone H3 subsequently recruits HP1, leading to packing of this genomic region into heterochromatin.

cancer. The focus of this chapter is on breast cancer; however, when appropriate other types of cancers are discussed as well. Finally, the clinical impact of DNA methylation as a diagnostic tools and as therapeutic targets in the near future is addressed. For the impact of DNA methylation on DNA repair and genomic instability in relation to the BCRA1 gene, we refer to Chapter 25. For significance of promoter methylation on cell adhesion molecule expression, see Chapter 20. Finally, for additional views on clinical aspects of DNA methylation in breast cancer, readers are referred to Chapter 26.

DNA Methylation in Normal Development

Events that are aberrantly induced in cancer in general have a normal function, and so does DNA methylation. The normal function of DNA methylation, particularly promoter DNA methylation, has been a debate for many years, a debate that is still ongoing. Still, there are a couple of clearcut functions for DNA methylation that are relatively well understood: genomic imprinting and X chromosome inactivation. Methylation of a major part of one X chromosome in women is considered necessary to have equal gene dosage of genes of the X chromosome in cells from men and women. Genomic imprinting controls for equal contribution of maternal and paternal genetic material in the zygote. A second specific function of DNA methylation is the silencing of repetitive sequences (Alu repeats, satellite DNA, and LINE-1 transposons) and foreign genomic DNA (integrated viral RNA and DNA genomes).[15] It is generally believed that DNA methylation in these regions protects the cell against the unwanted foreign genetic information being expressed.[16] In addition, more recent work suggests that DNA methylation of repetitive sequence is necessary to protect the genome against unwanted recombination between homologous regions in the genome, giving rise to translocations and other rearrangements leading to undesired chromosomal instability.[17] Finally, for a limited number of genes, tissue specific DNA methylation occurs, but our understanding of this is still too limited to understand its significance. The role of DNA methylation in normal differentiation, in addition to specific processes such as genomic imprinting and X chromosome inactivation, is poorly understood. However, evidence is accumulating that DNA methylation plays a pivotal role in the normal differentiation of many tissues. This is concluded from the presence of tissue-specific DNA methylation of certain genes and reactivation of these tissue specific genes in cells that lack the enzymes crucial for maintaining the epigenetic DNA methylation mark (see below).

DNA Methyltransferases, the Enzymes That Perform DNA Methylation

DNA methyltransferases (DNMT) are the catalytic enzymes that put the epigenetic methyl group on the DNA.[18] At this moment, three DNMTs (DNMT1, DNMT3a, and DNMT3b) have been characterized that can catalyze this step.[19,20] All three enzymes are crucial for normal mammalian development, underscoring the pivotal role of DNA methylation in mammalian biology.[21,22] DNMT1 is the enzyme responsible for the maintenance of DNA methylation. It recognizes hemimethylated DNA and copies the epigenetic mark on the newly synthesized DNA strand during DNA replication. DNMT1 can thus be considered the Xerox machine of the epigenetic methylation footprints on the DNA.[23] DMNT1 expression in most cancer types is elevated, most likely, not to limit optimal copying of epigenetic tags during continuous replication of the cancer genome, implicating that to a certain extent DNA methylation needs to be maintained during tumor growth. DMNT3a and 3b are considered the enzymes responsible for de novo DNA methylation. Hemimethylated DNA is not the target of these enzymes. It is thought that the DNMT3 methylating enzymes are directed to the DNA by other proteins. DNMT3s have an ADD domain that is also found in ATRX, a member of the SNF2/SWI2 protein family that is a component of ATP-dependent chromatin remodeling complex, suggesting this enzyme could be part of this complex. A second region in the protein, the PWWP domain, is also essential for the chromatin targeting of the enzymes.[24] The functional significance of PWWP mediated chromatin targeting is illustrated by the fact that a missense mutation in this domain of human DNMT3b causes the ICF syndrome.[25] Besides immunodeficiency, the ICF syndrome is characterized by loss of methylation in satellite DNA and pericentromeric genomic instability. Silencing DNMT1 together with

DNMT3b greatly increased the expression of several methylated genes and resulted in demethylation of methylated loci and selected repetitive sequences. The study confirms that DNMTs control the cellular DNA methylation status.[26]

Induction of DNA Methylation (Upstream Events)

A general picture is starting to emerge how DNA methylation is induced. This current view is mainly based on knowledge from nonmammalian species, including fungi (*Neurospora crassa*), yeast (*Saccharomyces pombe* and *Saccharomyces cerivisea*), and plant species (*Arabidopsis thalia*).[27–29] At least two processes can act upstream of DNA methylation. The first one, which is currently considered as being directly upstream of DNA methylation, is histone H3 lysine 9/27 double/triple methylation.[30,31] This enzymatic activity, catalyzed by histone methylation enzymes such as EZH2, is the epigenetic hallmark that initiates long-term gene silencing through heterochromatin formation. Knocking out the single histone H3-K9/27 methylating enzyme in *S. pombe* causes loss of DNA methylation and reactivation of the mating locus. A second mechanism connected to DNA methylation is posttranscriptional RNA mediated gene silencing (PTGS). In *A. thalia,* it was revealed that key components of the RNAi silencing machinery (RISC complex) such as Argonaute, Dicer, and RNA-dependent RNA polymerase (Rdp1) responsible for PTGS are crucial for silencing of centromeric repeats.[32] In this process, the PTGS using the RNAi silencing machinery acts upstream of histone H3-K9/27 methylation. That PTGS acts upstream of histone H3-K9/27 methylation has also been observed in *S. pombe* at the centromere-homologous repeat present at the silent mating type region, suggesting that this mechanism might be general. Currently, it is unclear whether these components are required for all types of silencing leading to DNA methylation or even play a role in mammalian cells. However, for genomic imprinting of the *igf2* and the *igf2r* locus and for X chromosome inactivation, RNA intermediates are essential. During X chromosome inactivation, the Xist and Tsix noncoding RNAs are crucial for the inactivation of the complete X chromosome. Similarly, in the *igf2r* locus the antisense noncoding *Air* mRNA is required for DNA methylation of the *igf2r* promoter as well as for other nearby located transcripts.[33] From this one might speculate that micro RNA-induced histone H3-K9/27 methylation precedes DNA methylation in cancer. Whether the PTGS/RNAi pathway and/or histone methylation are indeed required for CpG methylation in tumors remains to be seen.

Gene Silencing Through DNA Methylation (Downstream Events)

There are currently two ways in which actual gene silencing is established as soon as the DNA is methylated. First is the direct effect of CpG methylation. The methyl group in the DNA excludes transcription factors from binding the canonical binding site. This, however, can only explain silencing of genes that are controlled by transcription factors that contain a CpG site in their recognition site, e.g., Sp1. It is therefore currently considered that the second route, which involves the binding of proteins that specifically recognize methylated CpG sites, is a more general way to silence gene expression and repetitive chromatin. In the genome there are currently five proteins (MeCP2, MBD1 to 4) identified that display the capacity to perform this function.[34] When bound to the methylated CpGs, they act as transcriptional repressors by recruiting chromatin remodeling complex that include histone deacetylases (HDACs)[35] and histone H3-K9/27 methylating enzymes.[36] K9-methylated histone H3 recruits HP1 to the chromatin, resulting in chromatin condensation and leading to a promoter region that is refractory to gene transcription (FIG. 27.2).

DNA Methylation of Malignant Tissue

Evidence is accumulating that cancer, in addition to being a genetic disease, is also an epigenetic disease. Oncogene activation[37,38] and loss of suppressor gene function[11] can occur via DNA methylation. It is currently clear that DNA methylation contributes to malignancy; however, the fact that DNA methylation contributes to cancer has long been viewed with considerable skepticism. From several independent studies, however, it is evident that the level and the kind of DNA methylation in tumor tissue is different from that of normal tissue.[11] In fact, during oncogenesis two different changes in DNA methylation occur. On the one hand, in many cancer types the entire genome is hypomethylated, with 20–60% less genomic 5-methylcytosine compared with its normal counterpart. Thus, the levels of DNA methylation is globally decreased, termed global hypomethylation.[5,39] Global hypomethylation is thought to occur in regions of the genome that are normally methylated, such as highly repetitive sequences, foreign genetic material, and centromeric regions. It has been suggested that unmethylated homologous sequences can engage in recombination during repair, leading to genomic instability characterized by translocation and loss of heterozygosity. In instances, loss of methylation in single copy genes such as growth hormone, c-*myc*, *MAGE* genes, and chorionic gonadotropin has been observed in cancer. Also in breast cancer, hypomethylation of repetitive sequences has been reported as well as loss of methylation of single copy genes, that is, *pS2*, *PTH*, *SNCG*, and *uPA*. Hypomethylation of single copy genes probably activates genes that are normally silent in a particular tissue. (For specific details on hypomethylation, see Ref. 40.) On the other hand, opposite to the genome-wide hypomethylation stands increased DNA methylation of CpG islands present in the regulatory regions of genes, referred to as CpG hypermethylation. Hypermethylation is often found in promoter areas of suppressor genes and growth regulatory genes.[41,42] In addition to its active role in repression of gene transcription, CpG methylation may also play a passive role in promoting point mutations[34] as suggested for p53.[43,44] Thus, it appears that DNA methylation in the cancer genome, besides a global loss, shifts from repetitive intergenic regions to specific promoter regions.

Hypermethylation in Cancer

Since first introduced by Knudson, it has been firmly established that many tumor suppressor genes that predispose to hereditary cancer obtain their second hit due to chromosomal loss or local deletions, allelic recombination, or point mutations. Locus specific DNA methylation of the promoter of the tumor suppressor (see Chapter 23) has been recently added (FIG. 27.3, **Table 27.1**). In hereditary breast cancer (see Chapter 4), in some cases the second allele of the *BRCA1* locus is silenced through

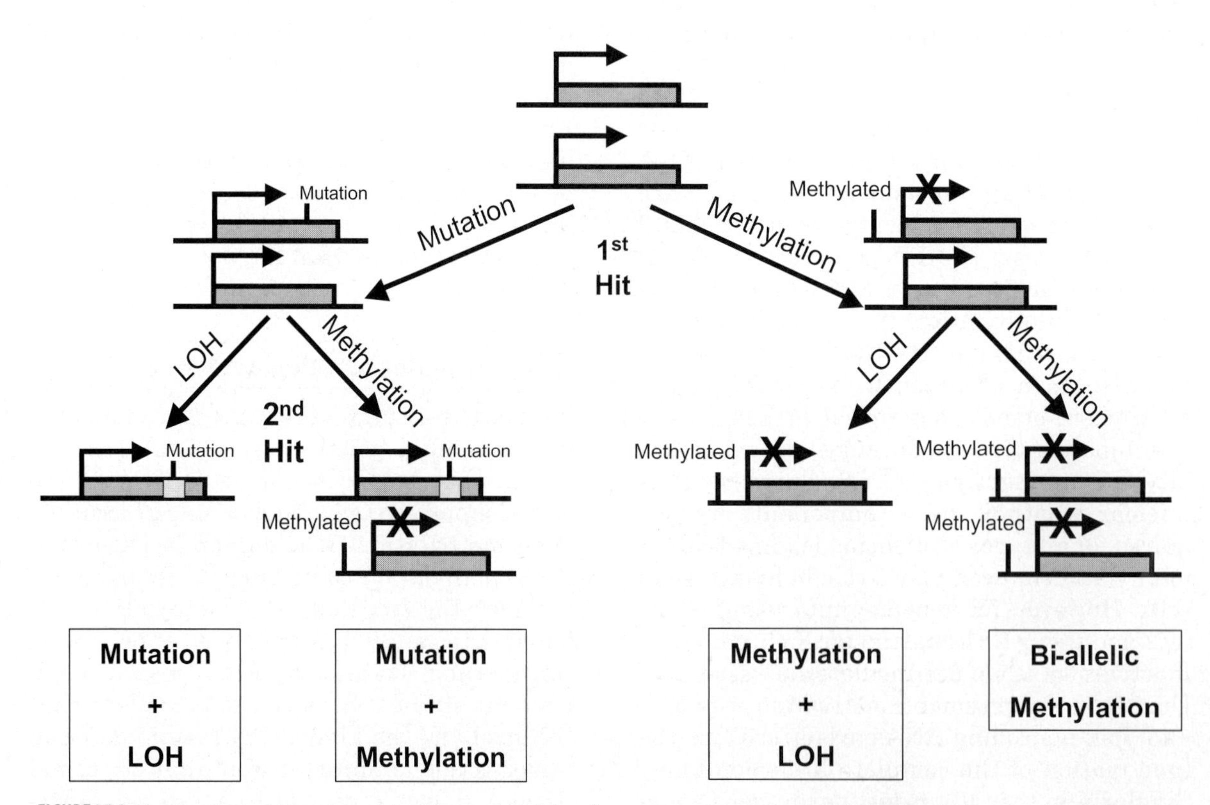

FIGURE 27.3 Knudson's two-hit model revisited. Two active alleles of a tumor suppressor gene, indicated by the boxes at the top, undergo sequential inactivation through mutation, loss of heterozygosity (LOH), promoter DNA methylation, or a combination of either of these. (From Ref. 8.)

Table 27.1 Tumor Suppressor Genes Methylated in Corresponding Human Cancers

Gene	Disease
LKB1	Hamartomatous colorectal cancer/ Peutz-Jeghers syndrome 1[127]
hMLH1	Colorectal cancer (HNPCC)[128,129]
BRCA1	Hereditary breast cancer[130,131]
CDH1	Diffuse gastric cancer[132]
VHL	Clear cell renal carcinoma[133]
Rb1	Retinoblastoma[134]
APC	Colorectal cancer (FAP)[135]

DNA methylation of the *BRCA1* promoter region. In addition, promoter methylation of the tumor suppressor genes is often observed in corresponding sporadic tumors. In colorectal cancer, the *hMLH1* locus is responsible for the microsatellite instability phenotype (MIN+) in hereditary nonpolyposis colon cancer, whereas in sporadic colon cancer, promoter DNA methylation of *hMLH1* almost exclusively explains the MIN+ phenotype.[45] The definite proof for the role of locus specific DNA methylation is at this moment difficult to address in mice models because in contrast to genetic knockout procedures, locus specific epigenetic silencing through DNA methylation is not yet feasible. In addition, a "chicken and egg" discussion is still ongoing of whether DNA methylation is the initial event leading to silencing of the tumor suppressor or whether upstream pathways (e.g., those described above) eventually leading to DNA methylation are causative. Recent work by the group of Vogelstein is in this respect intriguing.[46] In the cell line HCT116 in which the *p16* gene normally is silenced through DNA methylation, this group showed that in the absence of functional DNA methylating enzymes, the locus becomes silenced through histone H3-K9/27 methylation. This work, although performed in vitro, indicates that a normally upstream-acting silencing mechanism might play an upstream role in promoter DNA methylation of p16 in colon cancer.

Besides the inactivation of the initiator of cancer, additional hits are required for a full-blown cancer. Currently, it is argued that a tumor requires at least six capacities to grow independent of its surrounding tissue.[47] Unlimited replicative potential, self-sufficiency in growth factors, insensitivity to growth inhibitors, resistance to the apoptotic program, control of angiogenesis, and invasive or metastatic capacity are the processes that a tumor needs to survive. Because many of the genes involved in one or more of these pathways (i.e., *p16, Rb-1, p53, uPA, TIMP3, SYK, TGFBR2, 14-3-3, RASSF1*) show altered promoter methylation in many tumor types, this further supports that changes in promoter methylation can provide additional potential mechanisms that accelerate cancer progression.[48–50] In addition, when patients receive chemotherapy, radiotherapy, and/or endocrine therapy, altered methylation of genes that influences the response to these treatments (*ERα*,[51] *MGTG*,[52] *FANCF*[53]) also contributes to cancer progression. In conclusion, the current data imply that DNA methylation is not just a surrogate marker for cancer development and cancer progression but a critical component of the process itself.

What Causes Altered Methylation in Cancer?

The general current idea is that DNA methylation does not occur out of the blue or at random in cancer; however, definite proof is certainly lacking and the mechanisms responsible for DNA methylation in cancer, if not at random, are completely unknown. However, during aging increased DNA methylation does occur in various tissues,[54] suggesting that the epigenetic integrity is not completely maintained during normal life span. Current models suggest[55] that the DNA methylation status of a particular site is not static but is in a continuous active equilibrium between methylation and demethylation. Because hypermethylation and hypomethylation both occur frequently in cancer, this implies that the normal equilibrium in cancer must be disturbed. Furthermore, because the methylation pattern is tumor type and subtype specific, specific forces must be underlying the patterns that are observed. Aberrant or altered induction of DNA methylation might cause the specific patterns observed. Whether the micro-RNA machinery or the histone methylating enzymes, which lie upstream in normal silencing events leading to DNA methylation, play a role will be one of the exciting things to uncover in the near future. Another hypothesis is that transcription factors protect regulatory promoters against DNA methylation. Down-regulation of these transcription factors in cancer resulting in lack of protection of the regulatory regions might cause the balance to flip toward DNA methylation of a particular locus. Finally, loss of demethylation, the other side of the equilibrium, is another mechanism that could cause altered DNA methylation status in cancer. Although this pathway has high potential in being disrupted in cancer, little is known about the specific enzymes as well as the pathways that mediate this step.

Reduction of DNMT Activity and Cancer

That alteration of the levels of DNA methylation through inhibition of DNA methylating enzymes contributes to cancer is now well established.[40] For instance, APC mutated mice crossed with mice that have only 10% DMNT1 activity (DMNT1 chip/−) additionally treated with low doses of 5-azacytidine showed a significant reduction of the number of colon polyps compared with APC mutated mice that had wild-type DMNT1 activity and did not receive inhibitor treatment.[56] Also *mlh*−/− mice, with similarly reduced DNA methylation activity, developed intestinal tumors more slowly. These experiments clearly illustrate that DNA methylating enzymes presumably causing promoter hypermethylation are required during certain important steps of cancer progression.[57] These tumor types would probably benefit from increased DMNT activity that is often seen in cancer. On the other hand, *mlh*−/− mice, with reduced DNA methylation activity, developed B- and T-cell lymphomas more frequently.[57] In another mouse model,[17] which has a p53 and an Nf1 mutation on one allele of chromosome 11 and which specifically required loss of heterozygosity of the remaining allele for sarcoma development, hypomethylation enhanced tumor development. Similarly, earlier experiments with mice fed on a methyl deficient diet showed that these mice developed liver tumors. Because loss of promoter DNA methylation of oncogenes was concordant, the authors suggested activation of oncogenes was causing the liver cancer.[37]

The problem with most of these studies is that the inhibition of DMNTs by chemical agents or even by genetic modification affects all types of DNA methylation in the genome performed by the methylating enzymes. Still, they prove that imbalance of the genomic DNA methylation status contributes to malignancy. Hypomethylation has been connected to genomic instability. This was also reinforced by the observation of loss of heterozygosity in the p53/Nf1 mouse model described above. However, in human cell lines, combined deletion of DMNT1 and 3b did not result in any detectable gain or loss of chromosomes even though repetitive sequences, the imprinted igf*2* locus, and the regulatory region of the tumor suppressor gene p16 in this cell line were to a high degree demethylated. Thus, whether lowering the DNA methylation status stimulates or inhibits cancer is currently unpredictable or may be tumor type specific. Therefore, although DNMT inhibitors bare a high potential to be applied for cancer therapy, the effect of altering the DNA methylation state in a particular cancer type needs to be thoroughly investigated, and thus one must be cautious with the application of DMNT inhibitors as anticancer agents. Still, the possibility to revert methylation by the use of inhibitors of DNA methylation might be exploited in the clinic.[7] Using these agents, epigenetically silenced genes could also be reactivated for therapeutic benefit.[58] The reversibility of methylation, however, may allow the tumor to escape the different challenges it faces during the different steps in tumor progression (e.g., hypoxia, matrix degradation, migration, survival, challenges by the immune system, and homing into distant locations, etc.). These challenges may also include different treatment regimens such as radiation, chemotherapy, and endocrine therapy. Even though experimental evidence supports the fact that epigenetic modification allows biology to adopt more efficiently to the short-term environmental change than do genetic changes,[59] a role for this phenomenon in cancer progression has not been well documented.

DNA Methylation in Breast Cancer

In breast cancer, a large number of recent studies have unequivocally shown that aberrant promoter methylation of many genes occurs often and at a high frequency, including a variety of (candidate) tumor suppressor genes and genes that regulate the cell cycle in breast cancer.[60] Among many other genes are *BRCA1, p16INK4a/CDKN2/MTS-1, MDG1, HIC-1, ZAC, LKB1, maspin, NM23–H1, p53, TES, APC, SYK, HIN-1, hSRBC, RIZ1, Cx26, TMS1, DUTT1, TSLC1, FHIT, E-cadherin/CDH1, H-cadherin/CDH13, SLIT2, RASSF1A, LOT1, glypican 3, cyclin D2, 14-3-3σ,* and *14-3-3ζ*.[42,61]

Similarly, hypermethylation of promoter regions of genes coding for transcription factors has frequently been observed, including those for the steroid and nuclear hormone receptors estrogen receptor alpha (ER-α), progesterone receptor, thyroid hormone receptors (*TRβ*)1, retinoic acid receptor α2, and retinoic acid receptor β2.[42,61] Furthermore, methylation of a variety of other genes with diverse functions seems to play a role in breast cancer promotion and progression. These genes, among many others, include those of the protease urokinase plasminogen activator,[62] the protease inhibitor TIMP-3, the death-associated protein kinase, the mismatch repair genes *hMSH2* and *hMLH1,* gelsolin, glypican-3/GPC3, TMS1,[42,61] and many other genes predominantly identified by a high throughput approach (i.e., *CAV-1, CYP27B1, HSPA2, TIF-1, CpG12F10, MP2A3, MP3D7*).[63–65] Also in breast cancer, besides DNA hypermethylation of CpG islands in promoter areas of genes, DNA hypomethylation occurs frequently in the same tumors and has been linked to elevated mutation rates and genomic instability.[66,67] Hypomethylation of the estrogen-regulated *PS2* gene[68] has been observed in breast cancer cells (see Chapter 15). In breast tumor tissues, the *γ-globin* gene, the *MUC1* gene, the *HLA-DRα* gene, the *p16INK4a/CDKN2/MTS-1* gene, and the 11p15.5 chromosomal domain were found to be hypomethylated as well. Global hypomethylation has also been observed in breast tumor tissues[69] and was suggested to be associated with tumor progression[70] and poor prognostic features.[71] Thus, also in breast cancer, altered DNA methylation of many genes and of intragenic regions is observed that could be exploited for diagnostic purposes.

Detection of DNA Methylation

Over the last years, the area of DNA methylation detection has been moving from a hypothesis driven analysis of single candidate genes via the analysis of multiple candidates in high throughput to the ultimate goal of accurate epigenome-wide analysis of all CpG sites in a particular sample or a set of samples. Traditionally, DNA methylation was measured using methylation sensitive restriction enzymes (e.g., *Hpa* I). Differentially digested fragments, initially detected by Southern blot using promoter specific probes but later by polymerase chain reaction (PCR), revealed whether the studied regions in the genome were methylated. More recently, several methods based on the use of methylation sensitive restriction enzymes have been developed that take a broader look at the genome and allow unbiased discovery of differentially methylated sites between groups of samples. These methods include arbitrarily primed PCR, methylated CpG island amplification, differential methylation hybridization,[64] and restriction landmark genomic scanning.[72] Differential methylation hybridization uses a digest with methylation sensitive enzymes followed by a selective amplification of the undigested sites and hybridization to glass arrays containing over 7000 CpG islands. Thus, thousands of CpGs sites differentially methylated between two samples can be detected. To illustrate the power, differential methylation hybridization revealed that on average 83 (range, 15–207) of the 7000 CpGs are differentially methylated in breast tumors compared with normal breast tissue.[64] In addition, it revealed in these breast tumor specimen hypomethylation of several repetitive sequences and sequences containing satellite DNA. Restriction landmark genomic scanning uses, instead of an array, two-dimensional electrophoresis to visualize the differentially methylated CpG sites. This method, although tedious, is probably the most complete and unbiased method to detect differentially methylated DNA in a sample.[72] Because these methods allow an unbiased look at the genome, it can be anticipated that the described assays will revolutionize our knowledge about DNA methylation in normal and malignant physiology in the near future.

Another major leap forward occurred with the ability to convert epigenetic DNA methylation mark into a genetic DNA change by bisulfite treatment of genomic DNA.[73] During such a treatment, unmethylated cytosines are converted to uracil residues, whereas methylated cytosines remain unaffected. Turning the epigenetic code into a genetic one allows general molecular biologic tools, such as sequencing, PCR, allele-specific oligodetection, representation difference analysis, differential display, single nucleotide primer extension, or single stranded conformational polymorphism, to be applied to detect epigenetic modified DNA. Thus, with human genome in hand,[3,4] all methylated cytosine can theoretically be uncovered. An epigenomic sequencing project has been initiated by the Human Epigenome Consortium, a public/private collaboration that aims to identify and catalogue methylation variable positions in the human genome (www.epigenome.org). The development of the methylation specific PCR technique,[73,74] which generates a positive signal irrespective of the presence of contaminating normal cells, has revolutionized DNA methylation research in the cancer field, and the methylation specific PCR technique is currently the most commonly used way to measure altered DNA methylation levels. The methylation specific PCR technique exploits the nucleotide changes obtained after bisulfite treatment. Primers are designed that specifically amplify the methylated and unmethylated allele separately. Gel electrophoreses of the PCR products allows detection of the DNA methylation status of a particular locus in a particular sample.

More recently, the bisulfite conversion methodology has been extended to quantitative PCR techniques, COBRA[75] and Methylight.[76] COBRA uses a PCR on bisulfite-treated DNA. After that, restriction enzyme digestion is used to discriminate between the methylated and unmethylated state of the original genomic DNA. Relative amounts of digested and undigested PCR product are linearly quantitative across a wide spectrum of DNA methylation levels. Methylight is a regular quantitative real-time PCR on bisulfite converted DNA that uses fluorescent oligonucleotides differentiating between the methylated and unmethylated status of original genomic DNA. For detection in minute clinical samples or in body fluids, such as serum, sputum, and urine, that contain diluted tumor DNA, COBRA and Methylight are suitable and can also be reliably applied to DNA from microdissected paraffin-embedded material. The specific and real-time PCR detection of methylated DNA was recently further improved by including specific oligonucleotide blockers. The assay, termed HeavyMethyl, detects as little as 30 and 60 pg of methylated DNA and does not amplify contaminating unmethylated DNA.[77] Most of the currently used techniques are highly labor intensive and cannot be automated and used to analyze many genes in parallel. Considerable progress in DNA microarray technologies has led to new possibilities in the methylation research, allowing high throughput approaches for the quantification of CpG island methylation incidence at each CpG site.[63,78,79] The quantitative DNA methylation status of many candidate genes can now be analyzed simultaneously in a large number of tumor samples, and thus candidates identified with genome-wide methods can easily be validated in larger populations. Thus, the microarray-based methods represent the missing link between genome-wide screening approaches and single gene assays.

Clinical Application of DNA Methylation

With the knowledge of DNA methylation in human cancers accumulating, this information should now be transferred from the bench to application in the clinic. It is anticipated that one of the most important applications will be the use of DNA methylation markers as diagnostic tools, in particular in oncology. Several characteristics of DNA methylation make it an ideal diagnostic tool. The most prominent features of DNA methylation that are highly valuable considering a diagnostic application are the stability of DNA and thus the applicability to paraffin-embedded tissues, the possibility to detect methylation signals on free DNA in body fluids (e.g., for screening applications), and the ease of quantification due to the binary nature of the signal. Further clinical benefit may arise from the use of DNA methylation for therapeutic applications, either as a tool to identify new potential drug targets or as a therapeutic target by itself (e.g., agents targeting the methylation machinery; see below). In the following paragraphs we give a short overview of each of the areas.

Cancer Diagnosis and Detection Tumor Specimen and Body Fluids

One of the most relevant parameters defining the prognosis of a cancer patient is the stage at which the cancer is diagnosed. Large differences in survival and cure rates are observed depending on the stage at diagnosis. The strongest evidence that screening reduces mortality from a particular cancer comes from breast, colorectal, and cervical cancer.[80] To be suitable for cancer screening, a marker has to have certain properties: Ideally, the marker is cancer specific and is present in a large number of cancers. Furthermore, the marker should occur early in carcinogenesis, thus increasing the chances to detect early lesions. Being one of the earliest and most frequent events in tumorigenesis, DNA methylation promises to be a very powerful tool for cancer screening. Furthermore, a screening test is ideally performed on body fluids. Because of their immense importance, such noninvasive tests are considered the Holy Grail of cancer research.[81] Noninvasive tests are easy to integrate into screening programs, are inexpensive, and allow screening of large populations without exposing the screened individuals to additional risks, while keeping the hurdles low which prevent many individuals from getting regular check-ups. There is ample evidence that body fluids such as serum, plasma, and urine can be used to detect DNA methylation markers, which have previously been identified to be differentially methylated between tumor and normal. Several publications demonstrate that concentration of free DNA is elevated in the serum and plasma of cancer patients.[82–84] These elevated levels generally decrease after treatment, in particular in patients in whom treatment has been successful.[85] Some of this DNA, in some cases even most of it,[86] is derived from tumor cells.

Many groups have demonstrated that the aberrantly methylated genes, which are present in the original tumor, can also be detected on the free DNA shed into the serum or plasma. The first cancer for which detection of DNA methylation in serum was reported was non–small cell lung carcinomas. Esteller et al.[87] detected methylation in 73% of the serum samples from individuals with methylation present in the tumor. Several studies in lung cancer probably fueled the search for specific kinds of promoter DNA methylation in sera from patients with other kinds of cancer.[88–90] For example, Zou et al.[91] detected p16 hypermethylation in the sera of 70% of the cases with colorectal cancer with aberrant methylation in the corresponding tumor tissue. No methylated p16 sequences were detected in the sera of the other colorectal cancer cases without these changes in the tumor, in patients with adenomatous polyps, or in 10 healthy control subjects. Another group detected methylated APC sequences in preoperative serum and plasma samples: Of 99 patients analyzed, 96 had methylated APC in their tumors. In 47% of the 96 patients, methylated APC was also detected in serum and plasma.[92]

Other body fluids have also proven to be a useful specimen for cancer screening. For example, GSTP1 methylation was detected in 77% of simple voided urine samples from patients with prostate cancer. Serum samples from these patients proved to be positive for the methylation signal only in 72% of cases.[93] In postprostatic massage urine, even higher levels of methylated GSTP1 were identified.[93,94] Urine was also shown to be useful for detection of methylated sequences originating from bladder cancer.[95] When screening for markers, urine is generally conceived to be more specific for cancers of the urogenital system, thus putting less constraints onto the specificity of the biomarkers. However, there has been at least one study reporting that pieces of free DNA are able to pass the kidney barrier.[96] If this is confirmed by other studies, it needs to be taken into account for the development for biomarkers based on nucleic acid fragments from urine. Moreover, this could potentially allow screening for an even broader spectrum of malignancies in urine samples. Finally, tumor specific methylated sequences have also been identified in sputum,[97] saliva,[98] ductal lavage fluid,[99] and semen.[93]

Considering the requirements for implementing a biomarker as a screening tool into clinical routine, DNA methylation has several advantages compared with other biomarkers. In most cases, the detection assay would screen for a positive signal, that is, the presence of hypermethylation. This allows the development of very sensitive assays. In contrast, detecting the loss of expression or a genomic loss would be very difficult due to the excess of messenger RNA (mRNA)/DNA present from normal cells. The DNA containing the methylation information is far more stable than, for example, RNA or proteins. Thus, DNA can be isolated from almost all body fluids, and even paraffin-embedded material, without having to put into place extremely strict sample handling procedures, as would be the case if an RNA or protein marker would be used. Being highly compatible with transport of samples, DNA methylation would enable widespread use of screening tests, with the analysis taking place in more specialized central laboratories. Apart from true screening applications, DNA methylation tests may also be used for monitoring purposes, for instance, for earlier detection of tumor recurrences, or for screening of bone marrow samples for minimal residual disease. Furthermore, DNA methylation is potentially suitable to improve detection of tumor cells in biopsies and lymph nodes. Considering these fundamental facts, and therefore the feasibility of such an approach, one can anticipate that DNA methylation based early detection methods will very soon enter clinical trials for thorough testing of the applicability to large populations of nonsymptomatic people. There is hope that this application of methylation will lead to a major improvement of survival and much decreased suffering of millions of cancer patients.

Cancer Classification, Prognosis, and Prediction

Currently, tumors are described mainly by their organ of origin, the extension of disease (tumor size, involvement of lymph nodes, and distant metastasis to other organs), histology, and grade of differentiation. With very few exceptions in breast cancer and hematologic malignancies, the molecular background is not taken into account for patient management. However, an increasing number of studies have become available that show the large influence of particular molecular alterations on prognosis and clinical outcome. Initially, most reports focused on mRNA expression patterns (see Chapter 6).[100] However, this technique is still poorly compatible with paraffin-embedded specimens, thus leaving large retrospective collections of tissues with valuable follow-up untapped. Furthermore, mRNA expression markers have some severe disadvantages when being developed into a final application, including the stability of mRNA itself and the rapid changes in expression levels occurring under hypoxic conditions in the tumor (e.g., during tumor surgery). DNA methylation patterns are at least as informative and possibly an even richer source of information. Several studies have identified DNA methylation patterns that are associated with a particular cancer subtype.[50,78,101] Furthermore, DNA methylation seems to determine new subclasses in some cancers.

The CIMP phenotype (an acronym for "CpG island methylator phenotype") was first described by Jean-Pierre Issa in colon cancer and has subsequently been verified in other cancers as well. This phenotype seems to be associated with distinct genetic and histologic features.[102] More importantly, DNA methylation is not only useful to define already known tumor classes, DNA methylation markers have also been shown to be independent prognostic predictors for different tumor types.[92,103–105] Even though the number of validated prognostic markers are limited, initial studies suggest that DNA methylation has the potential to be used as prognostic and predictive markers. A good overview on the currently available knowledge regarding the prognostic value of DNA methylation is given by Fruhwald in a recent review.[106]

DNA methylation also promises to be useful as a predictive marker in cancer. The methylation of certain genes such as several members of the family of DNA repair genes is associated with differential response to drugs.[107,108] Two mechanisms have been proposed for the involvement of these genes in drug response. A subset, when active, has the potential to repair DNA damages set by cancer cytotoxic agents, thus contributing to potential resistance. Others function to recognize DNA damage and act as signals, leading to subsequent apoptosis and thus contributing to increased sensitivity. Many of these genes, if not all, are regulated by DNA methylation, either in promoter or gene body sequences. Promoter methylation of O6-methylguanine-DNMT has been shown to be responsible for decreased expression and response to alkylating agents.[109] Similarly, the same group found that promoter hypermethylation of this gene was associated with increase in overall- and progression-free survival in patients with diffuse large B-cell lymphoma.[52] Because of the experimental design, however, it cannot be excluded that the marker acts as a prognostic rather than a predictive marker. Further evidence that O6-methylguanine-DNMT is a predictive marker comes from studies in cell lines, which acquire resistance to *N*-methyl-*N*-nitrosourea by reactivation of previously silent O6-methylguanine-DNMT gene by demethylation of nonpromoter sequences[107] or by repression of hMSH6 by DNA methylation of its promoter. Christmann et al.[110] showed that in melanoma cells, gene body hypermethylation is associated with decreased expression of the gene and altered sensitivity to alkylating agents.

The mismatch repair gene *hMLH1* is a promising candidate for prediction of response to DNA damaging agents, in particular platinum compounds (see Chapter 25). Methylation of the promoter of this gene is predictive of resistance to these drugs. Resensitization in ovarian cancer xenografts can be achieved by treating them with decitabine, a demethylating agent. This leads to demethylation of the *hMLH1* gene promoter, an increase in MLH1 expression, and subsequently to restoration of cisplatinum sensitivity. Interestingly, the effect is specific for DNA-damaging agents, and no effect is seen regarding the sensitivity to taxol.[108] Experiments and clinical trials to translate these results into a clinical application are currently underway at the University of Glasgow.

Another mismatch repair pathway, the disruption of which leads to increased sensitivity of ovarian tumors, is the Fanconi anemia–BRCA pathway.[53] DNA methylation seems to be involved in regulation of at least one gene upstream in this pathway, *FANCF*. There is evidence from cell line experiments that DNA methylation is also relevant for another frequently used class of drugs, the antimetabolites. Worm et al.[111] showed in a study in breast cancer cells that methotrexate sensitivity is determined by two proteins, reduced folate carrier and MDR1, both of which are regulated by DNA methylation. Consistently, treatment with 5-aza-2′-deoxycytidine is able to modulate responsiveness to methotrexate. Few studies have shown the involvement of DNA methylation in endocrine resistance in breast cancer. In one study, methylation of ER-α has been involved in silencing of the ER and in abolishing response to antihormonal agents in breast cancer cells. Reexpression of ER was induced in ER-α–negative human breast cancer cells by 5-aza-2′-deoxycytidine. The reexpression is associated with partial demethylation of the ER CpG island and restores estrogen responsiveness.[51] We recently screened 117 candidate genes for methylation patterns associated with endocrine responsiveness. Among them, we identified at least five candidate genes, the methylation of which was strongly associated with response to tamoxifen. A panel of five genes can be used to predict failure to endocrine treatment in advanced breast cancer (Martens et al., manuscript in preparation).

On a technical level, DNA methylation offers several advantages over gene expression or proteomic approaches. Most importantly, the stability of the DNA and the possibility to recover the signal from archived material opens up large archives of retrospective collections with the follow-up required for marker development. Large-scale analysis of these archived specimens, which are not accessible anymore to any other genomic method, will greatly enhance our knowledge on biomarkers and reveal the large power of DNA methylation. Naturally, applicability to paraffin-embedded tissues is equally important for the design of the final assay, which may contain only a handful of genes but which has to be as simple as possible for integration into daily routine. Furthermore, the technical properties of DNA methylation address one of the key issues concerning mRNA expression and proteomics: the need of references. Whereas it is challenging in the best case to design proper controls for standardization of mRNA expression levels between different tissues and different measurements, DNA methylation can easily be compared with absolute reference points (completely methylated or unmethylated sequences). Tissue-based tests in general often struggle with tissue heterogeneity, for example, the "contamination" with stromal cells or lymphocytes. This can constitute a significant problem for a marker for which the tumor differs only quantitatively from normal (e.g., overexpressed genes), whereas it is excellently solved when using a marker, which is qualitatively different in tumor cells such as DNA methylation markers.

Taken together with advantages of a methylation testing platform as such, we see a large opportunity for methylation tests to complement or even replace expression microarrays in many of those applications that are most intensively discussed today. Perhaps the clearest advantage of methylation technology in this application may be the ability to analyze very small amounts of dissected material from paraffin samples that were previously analyzed by conventional pathology. This allows seamless integration of a modern molecular methodology with current clinical practice.

Therapeutic Applications (Methylation Therapy)

DNA methylation patterns contain highly useful information regarding a differentiation state of a cell that can be exploited to develop a new generation of highly sensitive and specific diagnostic tools. Moreover, with epigenetic silencing of genes being a fundamental mechanism in tumorigenesis, therapeutic strategies reversing the silencing and leading to reexpression of these genes also seem to hold great promises. In the following sections, we describe several approaches currently under investigation in the clinical or preclinical phase.

DNA Methylation Machinery as a Therapy Target DNA methylation in promoter regions can be targeted by inhibiting DNMTs. The resulting reversal of hypermethylation in these regulatory regions has been shown to lead to reexpression of several genes that either suppress tumor growth or reestablish sensitivity to other anticancer drugs.[112,113] The first DNMT inhibitors that have been synthesized are 5-aza-cytidine and its derivative 5-aza-2-deoxy-cytidine (decitabine).[114] As pyrimidine analogues, both agents get incorporated into genomic DNA and form a covalent complex with DNMT1, leading to its inactivation.[115] Because DNMT1 is mainly responsible for maintenance of DNA methylation, the newly synthesized DNA strands remain unmethylated. Decitabine has been in clinical trials for the treatment of several hematopoietic diseases where it has differentiation inducing effects.[116] Decitabine has been shown to restore sensitivity to other chemotherapeutic drugs in a xenograft model by reinducing the expression of *hMLH1,*[108] a mismatch repair gene crucial for response to DNA damaging drugs such as carboplatin and epirubicin. Hence, strong synergies with conventional cytotoxic drugs can be expected, and decitabine may prove highly valuable in dealing with a priori or acquired drug resistance in many cancers. A different approach to inhibit the DNA methylation machinery has been taken by Methylgene, a company currently running phase I and II clinical trials with MG98, an antisense oligonucleotide against DNMT1. Treatment of human tumors with MG98 reduces the levels of DNMT1 protein and induces reexpression of several genes, which were silenced in the respective tumors, including the p16 tumor suppressor gene.[117] The major concern with gene-unspecific demethylation is that global demethylation may lead to reactivation of silenced proviral sequences, oncogenes, or imprinted genes with tumor growth promoting effects. Inhibition of methyltransferases that have mainly been shown to be responsible for de novo methylation, such as DNMT3A and DNMT3B, can be expected not to be associated with the danger of demethylating these elements but may lack the strong demethylating effect of DNMT1 inhibition.

Chromatin as a Therapy Target Another highly promising target for epigenetic therapy are histone modifications. Histone modifications influence gene expression by remodeling of chromatin to active or inactive states.[118] DNA methylation and histone modifications influence each other in a variety of ways.[119–121] The histone modification that has received most attention regarding its systematic manipulation is (de-)acetylation. Deacetylation of lysine residues, predominantly in histones H3 and H4, has been demonstrated to be associated with silencing of genes. Several families of HDACs and histone acetyl transferases have been identified so far. For therapeutic purposes, HDACs can be inhibited by several compound classes: short-chain fatty acids (e.g., butyrate), hydroxamic acids (e.g., trichostatin A, suberoylanilide hydroxamic acid [SAHA], oxamflatin), cyclic tetrapeptides (e.g., trapoxin A, apicidin), and benzamides (e.g., MS-275).[122] In human tumor cells, HDAC inhibitors have been shown to induce growth arrest, cellular differentiation, and apoptosis.[119,122] Interestingly, only about 2% of the genes were observed to change their expression levels after treatment with HDAC inhibitors, among them the tumor suppressor and cell cycle regulator p21.[122,123] As it might have been anticipated due to the intensive interplay among HDACs and the DNA methylation machinery, combination of inhibition of DNMTs and HDAC inhibitions yields synergistic effects in reexpression of silenced genes.[124] Following these highly promising preclinical studies, several HDAC inhibitors are currently undergoing early phases of clinical trials (SAHA, depsipeptide, MS-27-275). Crystallographic studies have revealed the structure of the catalytic site of HDACs, and competition with substrates for the catalytic center has been identified as the mechanism of action of hydroxamic acid HDAC inhibitors such as TSA and SAHA.[124,125] These findings are expected to greatly enhance the development of more specific HDAC inhibitors. This is of particular importance because most naturally occurring HDAC inhibitors are thought to have several other effects apart their HDAC inhibition. Again, as with unspecific demethylating agents, the question arises how to target the effects to specific genes the reexpression of which is needed to restore apoptotic or tumor growth suppressing pathways or sensitivity to other anticancer agents.

Targeting Specific Sites Sangamo Biosciences has undertaken a promising approach that may well overcome the problem of unspecific activation of genes by both DNA demethylating agents and HDAC inhibitors. To direct a specific activity (acetyl transferase, HDAC, K9 methyl transferase, etc.) to a particular site in the genome, the most common DNA binding motif found in nature, the Cys2-His2 zinc-finger DNA binding domain, is used as a scaffold to construct factors with the desired specificity for a particular DNA binding site. Depending on whether reexpression of a gene or its down-regulation is the therapeutic goal, a functional domain that acts as a transcriptional activator or repressor can be coupled to the DNA binding domain of the artificial peptide. Among the most interesting functional domains are factors targeting the chromatin surrounding the transcription factor binding site, for example, domains with HDAC or histone acetyl transferases activity or domains influencing K9 methylation. Because of its gene specificity, this approach is expected to control gene function much more efficiently than unspecific approaches and can be assumed to be associated with less side effects.[126]

Concluding Remarks

The altered DNA methylation status in breast cancer compared with normal breast epithelia should be exploited for clinical benefit. In this chapter, we reviewed that DNA methylation can be used as markers for cancer classification, for cancer detection in tissue biopsies and body fluids, for determination of prognosis, and for therapy response prediction. In addition, DNA methylation may be targeted by broad range or specific therapies. Although most of the studies discussed are still in a preclinical phase, the considerable progress made during recent years suggests that it is just a matter of time before DNA methylation markers as well as sophisticated DNA methylation-based targeted therapies are applied in the clinic.

References

1. Bird AP. DNA methylation and the frequency of CpG in animal DNA. Nucleic Acids Res 1980;8: 1499–1504.
2. Bird AP. CpG-rich islands and the function of DNA methylation. Nature 1986;321:209–213.
3. Lander ES, Linton LM, Birren B, et al. Initial sequencing and analysis of the human genome. Nature 2001;409:860–921.
4. Venter JC, Adams MD, Myers EW, et al. The sequence of the human genome. Science 2001;291: 1304–1351.
5. Feinberg AP, Vogelstein B. Hypomethylation distinguishes genes of some human cancers from their normal counterparts. Nature 1983;301:89–92.
6. Baylin SB, Esteller M, Rountree MR, et al. Aberrant patterns of DNA methylation, chromatin formation and gene expression in cancer. Hum Mol Genet 2001;10:687–692.
7. Widschwendter M, Jones PA. The potential prognostic, predictive, and therapeutic values of DNA methylation in cancer. Clin Cancer Res 2002;8: 17–21.
8. Jones PA, Laird PW. Cancer epigenetics comes of age. Nat Genet 1999;21:163–167.
9. Cross SH, Bird AP. CpG islands and genes. Curr Opin Genet Dev 1995;5:309–314.
10. Bird A. DNA methylation patterns and epigenetic memory. Genes Dev 2002;16:6–21.
11. Jones PA, Baylin SB. The fundamental role of epigenetic events in cancer. Nat Rev Genet 2002;3: 415–428.
12. Heard E, Avner P. Role play in X-inactivation. Hum Mol Genet 1994;3:1481–1485.
13. Tremblay KD, Saam JR, Ingram RS, et al. A paternal-specific methylation imprint marks the alleles of the mouse H19 gene. Nat Genet 1995;9:407–413.
14. Ehrlich M. Expression of various genes is controlled by DNA methylation during mammalian development. J Cell Biochem 2003;88:899–910.
15. Clark SJ, Melki J. DNA methylation and gene silencing in cancer: which is the guilty party? Oncogene 2002;21:5380–5387.
16. Bestor TH, Tycko B. Creation of genomic methylation patterns. Nat Genet 1996;12:363–367.
17. Eden A, Gaudet F, Waghmare A, et al. Chromosomal instability and tumors promoted by DNA hypomethylation. Science 2003;300:455.
18. Robertson KD. DNA methylation, methyltransferases, and cancer. Oncogene 2001;20:3139–3155.
19. Bestor T, Laudano A, Mattaliano R, et al. Cloning and sequencing of a cDNA encoding DNA methyltransferase of mouse cells. The carboxyl-terminal domain of the mammalian enzymes is related to bacterial restriction methyltransferases. J Mol Biol 1988;203:971–983.
20. Okano M, Xie S, Li E. Cloning and characterization of a family of novel mammalian DNA (cytosine-5) methyltransferases. Nat Genet 1998;19:219–220.
21. Li E, Bestor TH, Jaenisch R. Targeted mutation of the DNA methyltransferase gene results in embryonic lethality. Cell 1992;69:915–926.
22. Okano M, Bell DW, Haber DA, et al. DNA methyltransferases Dnmt3a and Dnmt3b are essential for de novo methylation and mammalian development. Cell 1999;99:247–257.
23. Martienssen RA, Colot V. DNA methylation and epigenetic inheritance in plants and filamentous fungi. Science 2001;293:1070–1074.
24. Ge YZ, Pu MT, Gowher H, et al. Chromatin targeting of de novo DNA methyltransferases by the PWWP domain. J Biol Chem 2004;279:25447–25454.
25. Xu GL, Bestor TH, Bourc'his D, et al. Chromosome instability and immunodeficiency syndrome caused by mutations in a DNA methyltransferase gene. Nature 1999;402:187–191.

26. Leu YW, Rahmatpanah F, Shi H, et al. Double RNA interference of DNMT3b and DNMT1 enhances DNA demethylation and gene reactivation. Cancer Res 2003;63:6110–6115.
27. Allshire R. RNAi and heterochromatin—a hushed-up affair. Science 2002;297:1818–1819.
28. Nelson P, Kiriakidou M, Sharma A, et al. The microRNA world: small is mighty. Trends Biochem Sci 2003;28:534–540.
29. Jenuwein T. An RNA-guided pathway for the epigenome. Science 2002;297:2215–2218.
30. Tamaru H, Zhang X, McMillen D, et al. Trimethylated lysine 9 of histone H3 is a mark for DNA methylation in *Neurospora crassa*. Nat Genet 2003; 34:75–79.
31. Jackson JP, Lindroth AM, Cao X, et al. Control of CpNpG DNA methylation by the KRYPTONITE histone H3 methyltransferase. Nature 2002;416: 556–560.
32. Zilberman D, Cao X, Jacobsen SE. ARGONAUTE4 control of locus-specific siRNA accumulation and DNA and histone methylation. Science 2003;299: 716–719.
33. Sleutels F, Zwart R, Barlow DP. The non-coding Air RNA is required for silencing autosomal imprinted genes. Nature 2002;415:810–813.
34. Ordway JM, Curran T. Methylation matters: modeling a manageable genome. Cell Growth Differ 2002; 13:149–162.
35. Jones PL, Veenstra GJ, Wade PA, et al. Methylated DNA and MeCP2 recruit histone deacetylase to repress transcription. Nat Genet 1998;19:187–191.
36. Fuks F, Hurd PJ, Wolf D, et al. The methyl-CpG-binding protein MeCP2 links DNA methylation to histone methylation. J Biol Chem 2003;278: 4035–4040.
37. Simile MM, Pascale R, De Miglio MR, et al. Correlation between S-adenosyl-L-methionine content and production of c-myc, c-Ha-ras, and c-Ki-ras mRNA transcripts in the early stages of rat liver carcinogenesis. Cancer Lett 1994;79:9–16.
38. Watt PM, Kumar R, Kees UR. Promoter demethylation accompanies reactivation of the HOX11 proto-oncogene in leukemia. Genes Chromosomes Cancer 2000;29:371–377.
39. Goelz SE, Vogelstein B, Hamilton SR, et al. Hypomethylation of DNA from benign and malignant human colon neoplasms. Science 1985;228:187–190.
40. Ehrlich M. DNA methylation in cancer: too much, but also too little. Oncogene 2002;21:5400–5413.
41. Esteller M, Fraga MF, Paz MF, et al. Cancer epigenetics and methylation. Science 2002;297: 1807–1808.
42. Esteller M. Epigenetic lesions causing genetic lesions in human cancer promoter hypermethylation of DNA repair genes. Eur J Cancer 2000;36: 2294–2300.
43. Greenblatt MS, Bennett WP, Hollstein M, et al. Mutations in the p53 tumor suppressor gene: clues to cancer etiology and molecular pathogenesis. Cancer Res 1994;54:4855–4878.
44. Denissenko MF, Chen JX, Tang MS, et al. Cytosine methylation determines hot spots of DNA damage in the human P53 gene. Proc Natl Acad Sci USA 1997;94:3893–3898.
45. Baylin SB, Herman JG. DNA hypermethylation in tumorigenesis: epigenetics joins genetics. Trends Genet 2000;16:168–174.
46. Bachman KE, Park BH, Rhee I, et al. Histone modifications and silencing prior to DNA methylation of a tumor suppressor gene. Cancer Cell 2003; 3:89–95.
47. Hanahan D, Weinberg RA. The hallmarks of cancer. Cell 2000;100:57–70.
48. Rainier S, Feinberg AP. Genomic imprinting, DNA methylation, and cancer. J Natl Cancer Inst 1994;86: 753–759.
49. Esteller M, Corn PG, Baylin SB, et al. A gene hypermethylation profile of human cancer. Cancer Res 2001;61:3225–3229.
50. Toyota M, Kopecky KJ, Toyota MO, et al. Methylation profiling in acute myeloid leukemia. Blood 2001;97:2823–2829.
51. Yang X, Phillips DL, Ferguson AT, et al. Synergistic activation of functional estrogen receptor (ER)-alpha by DNA methyltransferase and histone deacetylase inhibition in human ER-alpha-negative breast cancer cells. Cancer Res 2001;61:7025–7029.
52. Esteller M, Gaidano G, Goodman SN, et al. Hypermethylation of the DNA repair gene O(6)-methylguanine DNA methyltransferase and survival of patients with diffuse large B-cell lymphoma. J Natl Cancer Inst 2002;94:26–32.
53. Taniguchi T, Tischkowitz M, Ameziane N, et al. Disruption of the Fanconi anemia-BRCA pathway in cisplatin-sensitive ovarian tumors. Nat Med 2003; 9:568–574.
54. Issa JP, Ottaviano YL, Celano P, et al. Methylation of the oestrogen receptor CpG island links ageing and neoplasia in human colon. Nat Genet 1994;7: 536–540.
55. Riggs AD. X chromosome inactivation, differentiation, and DNA methylation revisited, with a tribute to Susumu Ohno. Cytogenet Genome Res 2002; 99:17–24.
56. Laird PW, Jackson-Grusby L, Fazeli A, et al. Suppression of intestinal neoplasia by DNA hypomethylation. Cell 1995;81:197–205.
57. Gaudet F, Hodgson JG, Eden A, et al. Induction of tumors in mice by genomic hypomethylation. Science 2003;300:489–492.
58. Cheng JC, Matsen CB, Gonzales FA, et al. Inhibition of DNA methylation and reactivation of silenced genes by zebularine. J Natl Cancer Inst 2003;95: 399–409.
59. Rutherford SL, Henikoff S. Quantitative epigenetics. Nat Genet 2003;33:6–8.
60. Yang X, Yan L, Davidson NE. DNA methylation in breast cancer. Endocr Relat Cancer 2001;8:115–127.
61. Widschwendter M, Jones PA. DNA methylation and breast carcinogenesis. Oncogene 2002;21: 5462–5482.
62. Xing RH, Rabbani SA. Transcriptional regulation of urokinase (uPA) gene expression in breast cancer cells: role of DNA methylation. Int J Cancer 1999; 81:443–450.
63. Huang TH, Laux DE, Hamlin BC, et al. Identification of DNA methylation markers for human breast carcinomas using the methylation-sensitive restriction fingerprinting technique. Cancer Res 1997;57: 1030–1034.
64. Huang TH, Perry MR, Laux DE. Methylation profiling of CpG islands in human breast cancer cells. Hum Mol Genet 1999;8:459–470.
65. Yan PS, Perry MR, Laux DE, et al. CpG island arrays: an application toward deciphering epigenetic

signatures of breast cancer. Clin Cancer Res 2000; 6:1432–1438.

66. Almeida A, Kokalj-Vokac N, Lefrancois D, et al. Hypomethylation of classical satellite DNA and chromosome instability in lymphoblastoid cell lines. Hum Genet 1993;91:538–546.

67. Chen RZ, Pettersson U, Beard C, et al. DNA hypomethylation leads to elevated mutation rates. Nature 1998;395:89–93.

68. Martin V, Ribieras S, Song-Wang X, et al. Genomic sequencing indicates a correlation between DNA hypomethylation in the 5′ region of the pS2 gene and its expression in human breast cancer cell lines. Gene 1995;157:261–264.

69. Narayan A, Ji W, Zhang XY, et al. Hypomethylation of pericentromeric DNA in breast adenocarcinomas. Int J Cancer 1998;77:833–838.

70. Bernardino J, Roux C, Almeida A, et al. DNA hypomethylation in breast cancer: an independent parameter of tumor progression? Cancer Genet Cytogenet 1997;97:83–89.

71. Soares J, Pinto AE, Cunha CV, et al. Global DNA hypomethylation in breast carcinoma: correlation with prognostic factors and tumor progression. Cancer 1999;85:112–118.

72. Plass C, Yu F, Yu L, et al. Restriction landmark genome scanning for aberrant methylation in primary refractory and relapsed acute myeloid leukemia; involvement of the WIT-1 gene. Oncogene 1999;18:3159–3165.

73. Frommer M, McDonald LE, Millar DS, et al. A genomic sequencing protocol that yields a positive display of 5-methylcytosine residues in individual DNA strands. Proc Natl Acad Sci USA 1992;89: 1827–1831.

74. Herman JG, Graff JR, Myohanen S, et al. Methylation-specific PCR: a novel PCR assay for methylation status of CpG islands. Proc Natl Acad Sci USA 1996;93: 9821–9826.

75. Xiong Z, Laird PW. COBRA: a sensitive and quantitative DNA methylation assay. Nucleic Acids Res 1997;25:2532–2534.

76. Trinh BN, Long TI, Laird PW. DNA methylation analysis by MethyLight technology. Methods 2001; 25:456–462.

77. Cottrell SE, Distler J, Goodman NS, et al. A real-time PCR assay for DNA-methylation using methylation-specific blockers. Nucleic Acids Res 2004;32:e10.

78. Adorjan P, Distler J, Lipscher E, et al. Tumour class prediction and discovery by microarray-based DNA methylation analysis. Nucleic Acids Res 2002; 30:e21.

79. Shi H, Maier S, Nimmrich I, et al. Oligonucleotide-based microarray for DNA methylation analysis: principles and applications. J Cell Biochem 2003;88: 138–143.

80. US Preventive Task Force. Guide to Clinical Preventive Services. Alexandria, VA: International Medical Publishing, 1996.

81. Ransohoff DF. Cancer: developing molecular biomarkers for cancer. Science 2003;299:1679–1680.

82. Cottrell SE, Laird PW. Sensitive detection of DNA methylation. Ann NY Acad Sci 2003;983:120–130.

83. Sozzi G, Conte D, Mariani L, et al. Analysis of circulating tumor DNA in plasma at diagnosis and during follow-up of lung cancer patients. Cancer Res 2001; 61:4675–4678.

84. Jahr S, Hentze H, Englisch S, et al. DNA fragments in the blood plasma of cancer patients: quantitations and evidence for their origin from apoptotic and necrotic cells. Cancer Res 2001;61:1659–1665.

85. Leon SA, Shapiro B, Sklaroff DM, et al. Free DNA in the serum of cancer patients and the effect of therapy. Cancer Res 1977;37:646–650.

86. Chen XQ, Stroun M, Magnenat JL, et al. Microsatellite alterations in plasma DNA of small cell lung cancer patients. Nat Med 1996;2:1033–1035.

87. Esteller M, Sanchez-Cespedes M, Rosell R, et al. Detection of aberrant promoter hypermethylation of tumor suppressor genes in serum DNA from non-small cell lung cancer patients. Cancer Res 1999; 59:67–70.

88. Esteller M. Relevance of DNA methylation in the management of cancer. Lancet Oncol 2003;4: 351–358.

89. Grady WM, Rajput A, Lutterbaugh JD, et al. Detection of aberrantly methylated hMLH1 promoter DNA in the serum of patients with microsatellite unstable colon cancer. Cancer Res 2001;61:900–902.

90. Lo YM, Wong IH, Zhang J, et al. Detection of aberrant p16 methylation in the plasma and serum of liver cancer patients. Cancer Res 1999;59:71–73.

91. Zou HZ, Yu BM, Wang ZW, et al. Detection of aberrant p16 methylation in the serum of colorectal cancer patients. Clin Cancer Res 2002;8:188–191.

92. Usadel H, Brabender J, Danenberg KD, et al. Quantitative adenomatous polyposis coli promoter methylation analysis in tumor tissue, serum, and plasma DNA of patients with lung cancer. Cancer Res 2002; 62:371–375.

93. Goessl C, Muller M, Heicappell R, et al. DNA-based detection of prostate cancer in blood, urine, and ejaculates. Ann N Y Acad Sci 2001;945:51–58.

94. Cairns P, Esteller M, Herman JG, et al. Molecular detection of prostate cancer in urine by GSTP1 hypermethylation. Clin Cancer Res 2001;7:2727–2730.

95. Chan MW, Chan LW, Tang NL, et al. Hypermethylation of multiple genes in tumor tissues and voided urine in urinary bladder cancer patients. Clin Cancer Res 2002;8:464–470.

96. Botezatu I, Serdyuk O, Potapova G, et al. Genetic analysis of DNA excreted in urine: a new approach for detecting specific genomic DNA sequences from cells dying in an organism. Clin Chem 2000;46: 1078–1084.

97. Palmisano WA, Divine KK, Saccomanno G, et al. Predicting lung cancer by detecting aberrant promoter methylation in sputum. Cancer Res 2000;60: 5954–5958.

98. Rosas SL, Koch W, da Costa Carvalho MG, et al. Promoter hypermethylation patterns of p16, O6-methylguanine-DNA-methyltransferase, and death-associated protein kinase in tumors and saliva of head and neck cancer patients. Cancer Res 2001; 61:939–942.

99. Evron E, Dooley WC, Umbricht CB, et al. Detection of breast cancer cells in ductal lavage fluid by methylation-specific PCR. Lancet 2001;357: 1335–1336.

100. van't Veer LJ, Dai H, van de Vijver MJ, et al. Gene expression profiling predicts clinical outcome of breast cancer. Nature 2002;415:530–536.

101. Shi H, Yan PS, Chen CM, et al. Expressed CpG island sequence tag microarray for dual screening of

DNA hypermethylation and gene silencing in cancer cells. Cancer Res 2002;62:3214–3220.

102. Toyota M, Ahuja N, Suzuki H, et al. Aberrant methylation in gastric cancer associated with the CpG island methylator phenotype. Cancer Res 1999; 59:5438–5442.

103. Maruyama R, Toyooka S, Toyooka KO, et al. Aberrant promoter methylation profile of prostate cancers and its relationship to clinicopathological features. Clin Cancer Res 2002;8:514–519.

104. Brabender J, Usadel H, Metzger R, et al. Quantitative O(6)-methylguanine DNA methyltransferase methylation analysis in curatively resected non-small cell lung cancer: associations with clinical outcome. Clin Cancer Res 2003;9:223–227.

105. Kawakami K, Brabender J, Lord RV, et al. Hypermethylated APC DNA in plasma and prognosis of patients with esophageal adenocarcinoma. J Natl Cancer Inst 2000;92:1805–1811.

106. Fruhwald MC. DNA methylation patterns in cancer: novel prognostic indicators? Am J Pharmacogenomics 2003;3:245–260.

107. Bearzatto A, Szadkowski M, Macpherson P, et al. Epigenetic regulation of the MGMT and hMSH6 DNA repair genes in cells resistant to methylating agents. Cancer Res 2000;60:3262–3270.

108. Plumb JA, Strathdee G, Sludden J, et al. Reversal of drug resistance in human tumor xenografts by 2′-deoxy–5-azacytidine-induced demethylation of the hMLH1 gene promoter. Cancer Res 2000;60: 6039–6044.

109. Esteller M, Garcia-Foncillas J, Andion E, et al. Inactivation of the DNA-repair gene MGMT and the clinical response of gliomas to alkylating agents. N Engl J Med 2000;343:1350–1354.

110. Christmann M, Pick M, Lage H, et al. Acquired resistance of melanoma cells to the antineoplastic agent fotemustine is caused by reactivation of the DNA repair gene MGMT. Int J Cancer 2001;92: 123–129.

111. Worm J, Kirkin AF, Dzhandzhugazyan KN, et al. Methylation-dependent silencing of the reduced folate carrier gene in inherently methotrexate-resistant human breast cancer cells. J Biol Chem 2001;276:39990–40000.

112. Kopelovich L, Crowell JA, Fay JR. The epigenome as a target for cancer chemoprevention. J Natl Cancer Inst. 2003;95:1747–1757.

113. Strathdee G, Brown R. Aberrant DNA methylation in cancer: potential clinical interventions. Expert Rev Mol Med 2002;2002:1–17.

114. Sorm F, Piskala A, Cihak A, et al. 5-Azacytidine, a new, highly effective cancerostatic. Experientia 1964;20:202–203.

115. Bouchard J, Momparler RL. Incorporation of 5-Aza-2′-deoxycytidine–5′-triphosphate into DNA. Interactions with mammalian DNA polymerase alpha and DNA methylase. Mol Pharmacol 1983;24:109–114.

116. Pinto A, Zagonel V. 5-Aza-2′-deoxycytidine (Decitabine) and 5-azacytidine in the treatment of acute myeloid leukemias and myelodysplastic syndromes: past, present and future trends. Leukemia 1993; 7(suppl 1):51–60.

117. Reid GK, Besterman JM, MacLeod AR. Selective inhibition of DNA methyltransferase enzymes as a novel strategy for cancer treatment. Curr Opin Mol Ther 2002;4:130–137.

118. Tyler JK, Kadonaga JT. The "dark side" of chromatin remodeling: repressive effects on transcription. Cell 1999;99:443–446.

119. Cervoni N, Szyf M. Demethylase activity is directed by histone acetylation. J Biol Chem 2001;276: 40778–40787.

120. Fuks F, Burgers WA, Brehm A, et al. DNA methyltransferase Dnmt1 associates with histone deacetylase activity. Nat Genet 2000;24:88–91.

121. Robertson KD, et al. DNMT1 forms a complex with Rb, E2F1 and HDAC1 and represses transcription from E2F-responsive promoters. Nat Genet 2000;25: 338–342.

122. Marks PA, Richon VM, Breslow R, et al. Histone deacetylase inhibitors as new cancer drugs. Curr Opin Oncol 2001;13:477–483.

123. Van Lint C, Emiliani S, Verdin E. The expression of a small fraction of cellular genes is changed in response to histone hyperacetylation. Gene Exp 1996; 5:245–253.

124. Cameron EE, Bachman KE, Myohanen S, et al. Synergy of demethylation and histone deacetylase inhibition in the re-expression of genes silenced in cancer. Nat Genet 1999;21:103–107.

125. Finnin MS, Donigian JR, Cohen A. Structures of a histone deacetylase homologue bound to the TSA and SAHA inhibitors. Nature 1999;401: 188–193.

126. Reik A, Gregory PD, Urnov FD. Biotechnologies and therapeutics: chromatin as a target. Curr Opin Genet Dev 2002;12:233–242.

127. Esteller M, Avizienyte E, Corn PG, et al. Epigenetic inactivation of LKB1 in primary tumors associated with the Peutz-Jeghers syndrome. Oncogene 2000; 19:164–168.

128. Wheeler JM, Loukola A, Aaltonen LA, et al. The role of hypermethylation of the hMLH1 promoter region in HNPCC versus MSI+ sporadic colorectal cancers. J Med Genet 2000;37:588–592.

129. Kuismanen SA, Holmberg MT, Salovaara R, et al. Genetic and epigenetic modification of MLH1 accounts for a major share of microsatellite-unstable colorectal cancers. Am J Pathol 2000;156:1773–1779.

130. Esteller M, Silva JM, Dominguez G, et al. Promoter hypermethylation and BRCA1 inactivation in sporadic breast and ovarian tumors. J Natl Cancer Inst 2000;92:564–569.

131. Dobrovic A, Simpfendorfer D. Methylation of the BRCA1 gene in sporadic breast cancer. Cancer Res 1997;57:3347–3350.

132. Grady WM, Willis J, Guilford PJ, et al. Methylation of the CDH1 promoter as the second genetic hit in hereditary diffuse gastric cancer. Nat Genet 2000; 26:16–17.

133. Prowse AH, Webster AR, Richards FM, et al. Somatic inactivation of the VHL gene in Von Hippel-Lindau disease tumors. Am J Hum Genet 1997;60: 765–771.

134. Stirzaker C, Millar DS, Paul CL, et al. Extensive DNA methylation spanning the Rb promoter in retinoblastoma tumors. Cancer Res 1997;57: 2229–2237.

135. Esteller M, Sparks A, Toyota M, et al. Analysis of adenomatous polyposis coli promoter hypermethylation in human cancer. Cancer Res 2000;60:4366–4371.

GURE 2.1: HER-2 testing in breast cancer / FISH (top), IHC (middle), and CISH ottom).

GURE 2.2 (far right panels): Types of nomic microarrays showing fluores-nt (top) and radioactive (bottom) tection.

PLATE 1: Fig. 2.1

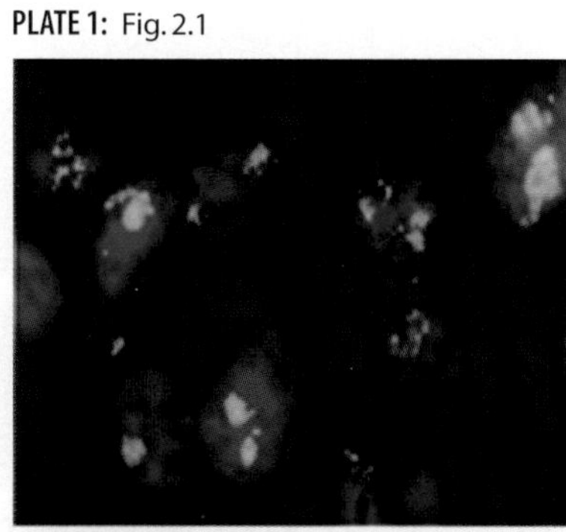

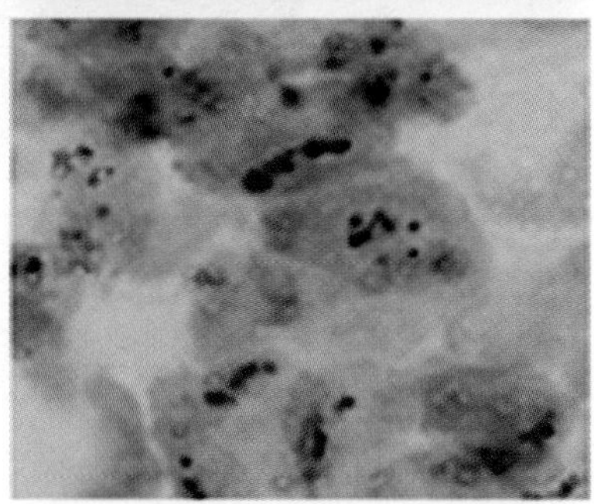

PLATE 2: Fig. 2.2

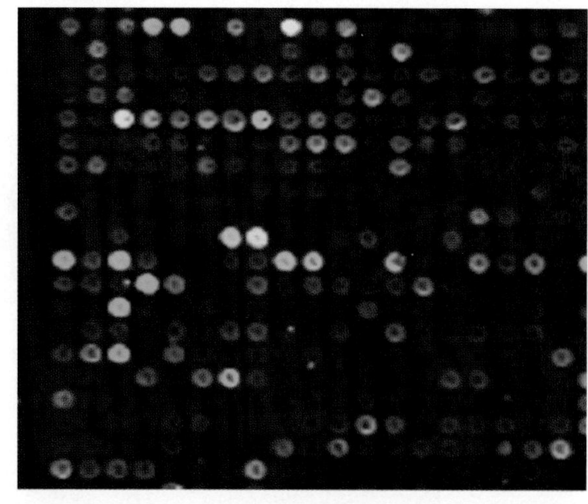

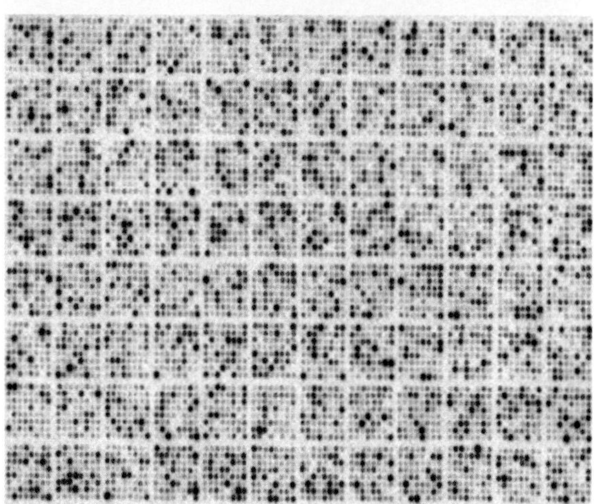

.ATE 3: Fig. 2.3

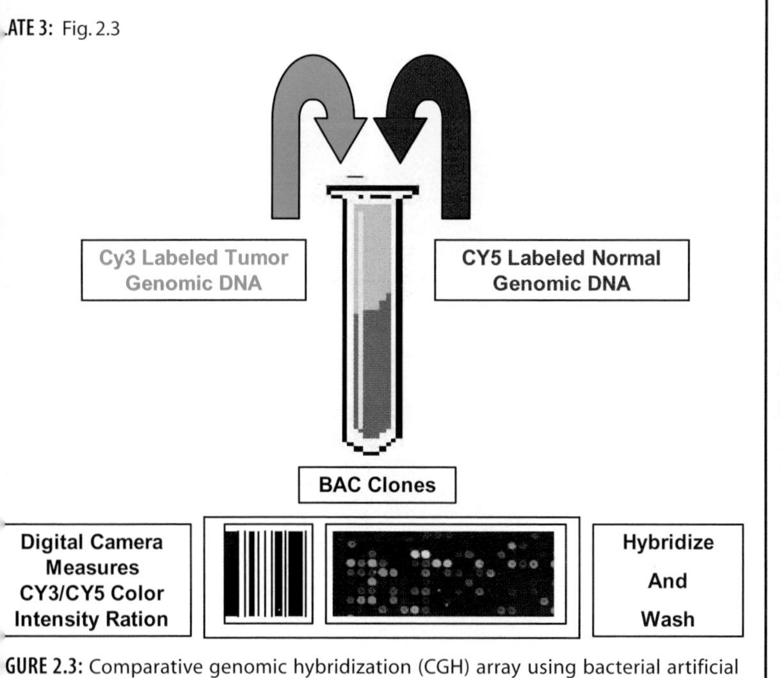

GURE 2.3: Comparative genomic hybridization (CGH) array using bacterial artificial ıromosomes.

PLATE 4: Fig. 2.4

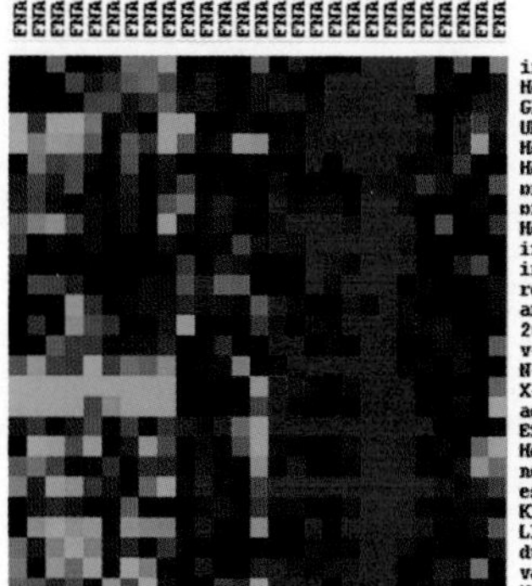

FIGURE 2.4: Hierarchical clustering of gene expression in breast cancer after microarray profiling.

PLATE 5: Fig. 2.5

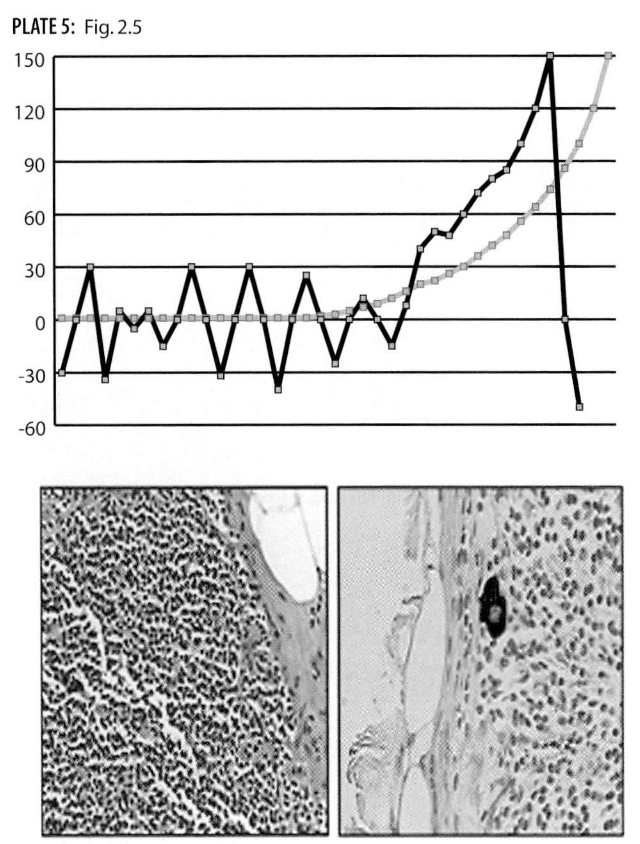

FIGURE 2.5: Micrometastasis detection in lymph node using RT-PCR (top) and IHC (bottom).

PLATE 7: Fig. 5.3

A B C D E F G H

FIGURE 5.3: Chromosomal aneusomies detected by FISH in breast cancer cells obtained by ductal lavage.

PLATE 6: Fig. 5.2

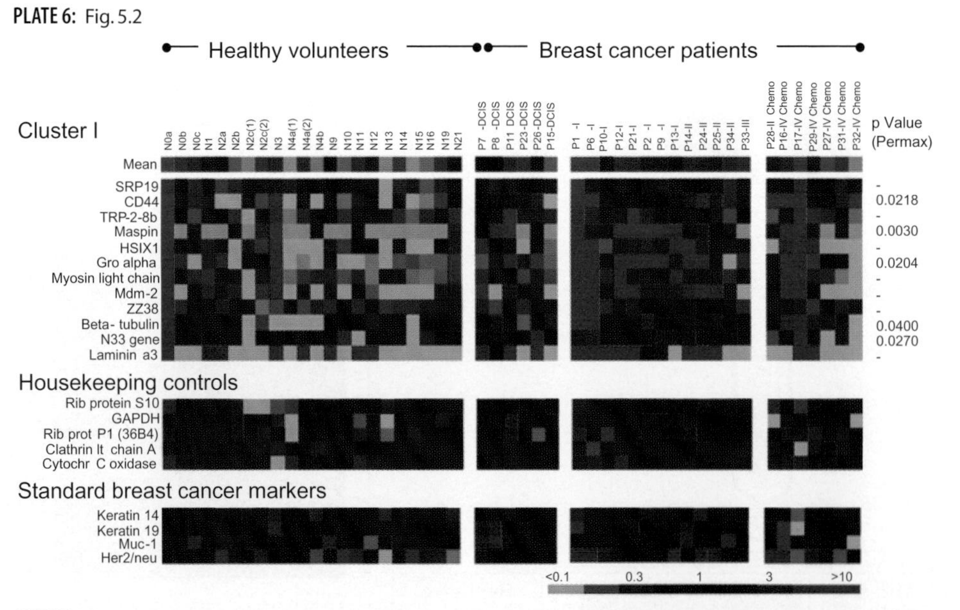

FIGURE 5.2: Circulating tumor cell detection in blood by gene expression pattern using microarray profiling.

IGURE 6.1: Sequential steps in a typical microarray nalysis using Affymetrix GeneChip™ system.

PLATE 8: Fig. 6.1

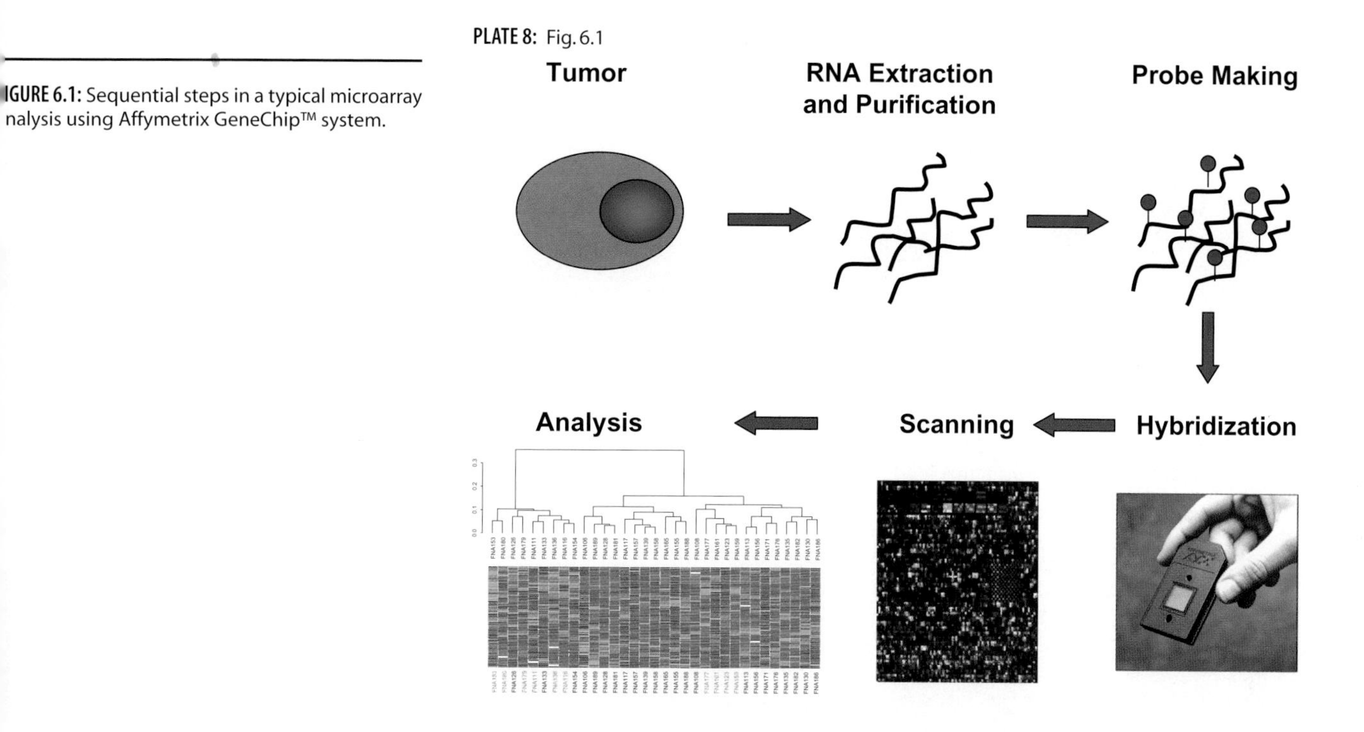

GURE 6.2: Dendrogram analysis of transcriptional profile of breast cancer subsets organized ccording to ER status.

PLATE 9: Fig. 6.2

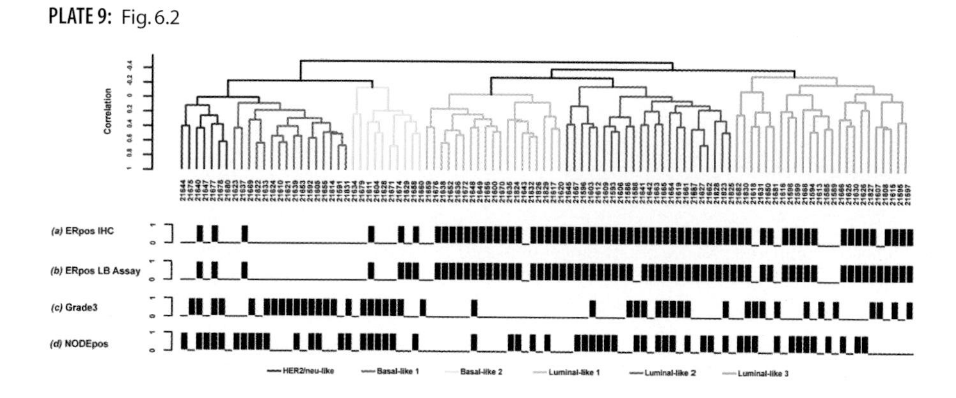

PLATE 10: Fig. 7.2

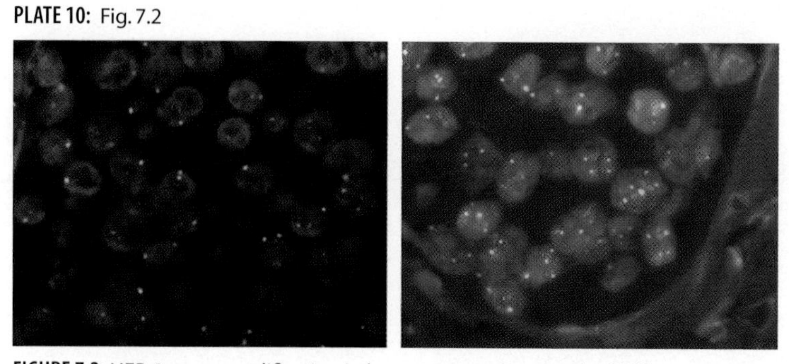

FIGURE 7.2: HER-2 gene amplification in breast cancer detected by FISH.

PLATE 11: Fig. 7.4

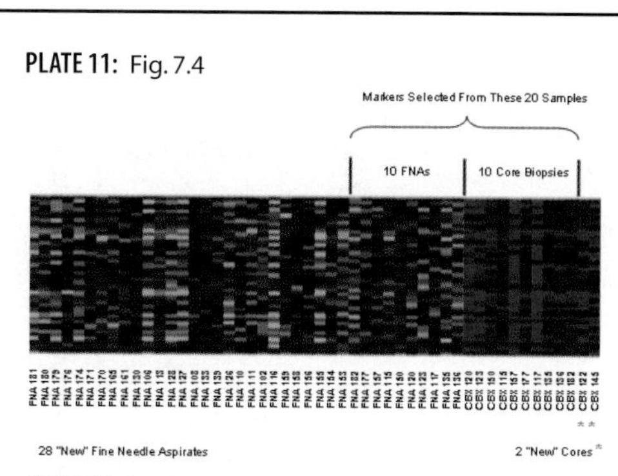

FIGURE 7.4: Comparison of breast cancer gene expression profiles in matched fine needle aspiration and core needle biopsies.

PLATE 12: Fig. 8.1

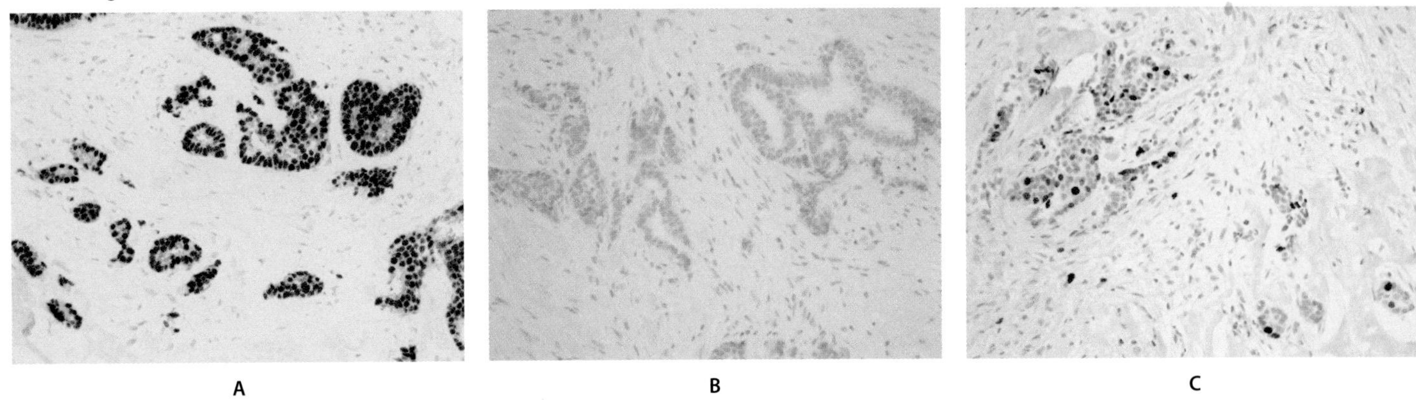

FIGURE 8.1: Well-differentiated invasive breast cancer that is (A) ER-positive, (B) HER-2 negative, and (C) has an intermediate proliferation rate on Ki-67 staining.

PLATE 13: Fig. 8.2

FIGURE 8.2: Poorly differentiated invasive breast cancer that is (A) ER-negative, (B) HER-2 negative, and (C) has a high cell proliferation rate on Ki-67 staining.

PLATE 14: Fig. 8.3

PLATE 15: Fig. 8.4

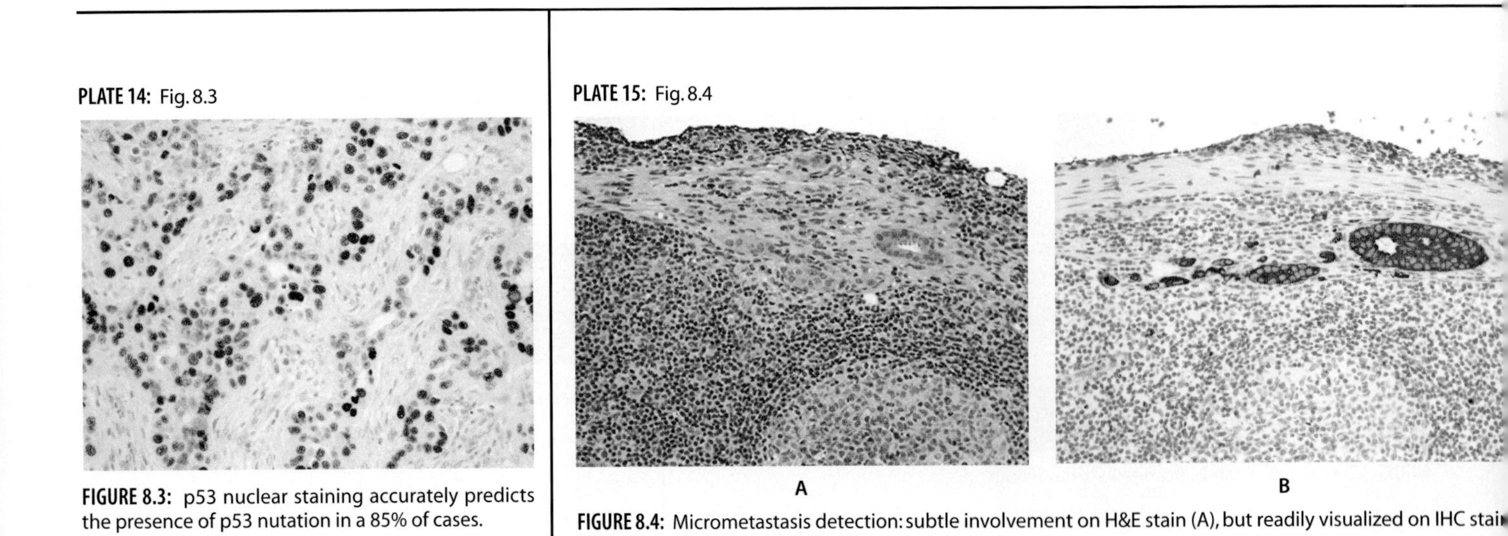

FIGURE 8.3: p53 nuclear staining accurately predicts the presence of p53 nutation in a 85% of cases.

FIGURE 8.4: Micrometastasis detection: subtle involvement on H&E stain (A), but readily visualized on IHC stain (B) using antibodies to cytokeratins.

GURE 8.5: Well-differentiated intra-
uctal carcinoma on H&E staining
\) with strong ER expression by
IC (B) which decreases as the
sion extends into the lobule.

PLATE 16: Fig. 8.5

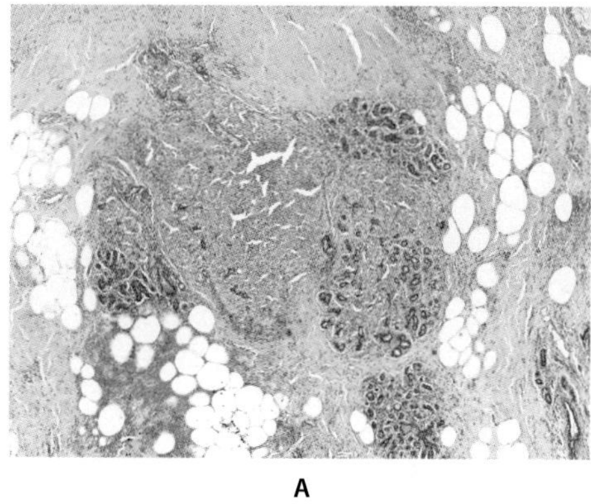

A

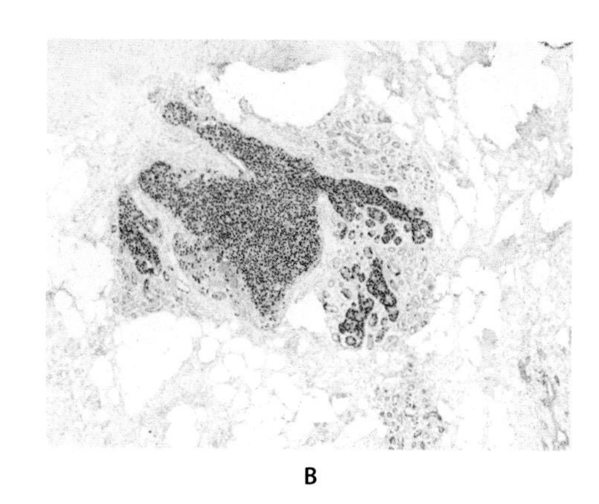

B

GURE 8.6: Invasive ductal carci-
oma with absence of basement
embrane structures on H&E stain-
g (A) containing myofibroblasts
the stroma on IHC staining for
nooth muscle action (B).

PLATE 17: Fig. 8.6

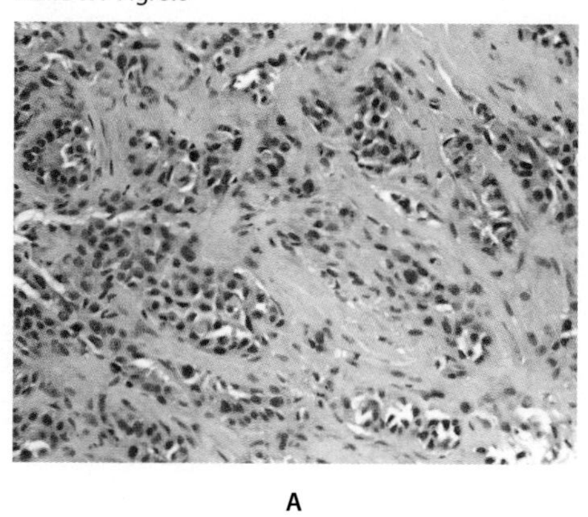

A

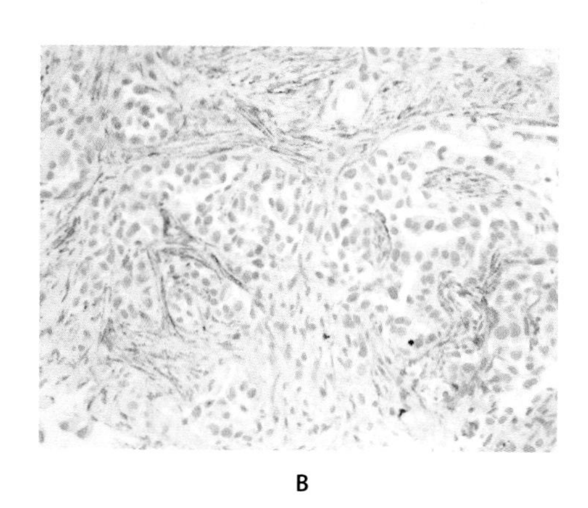

B

GURE 8.7: Classic single cell rows in
infiltrating lobular carcinoma
H&E stain (A). Complete loss of
cadherin staining (B).

PLATE 18: Fig. 8.7

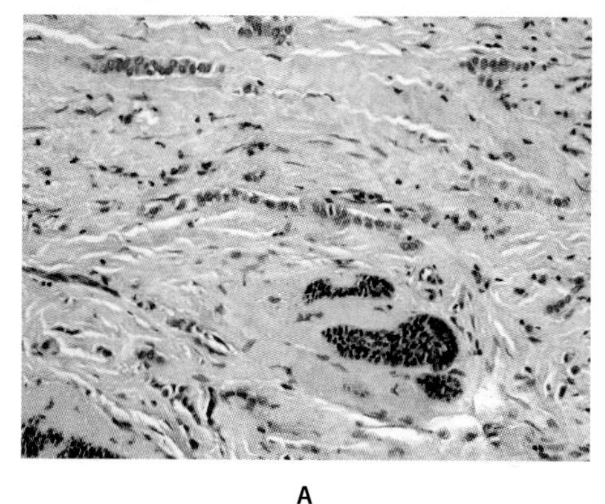

A

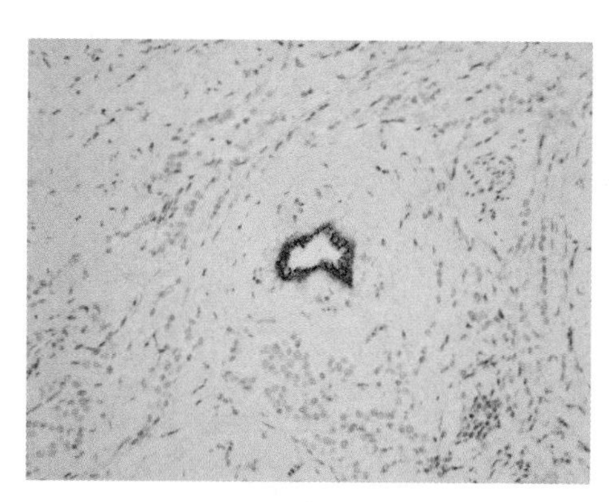

B

FIGURE 8.8: An invasive ductal carcinoma simulating the lobular carcinoma invasion pattern (A). The carcinoma is ER-positive (B), HER-2 positive (C), and expresses E-cadherin in a typical membranous pattern (D).

PLATE 19: Fig. 8.8

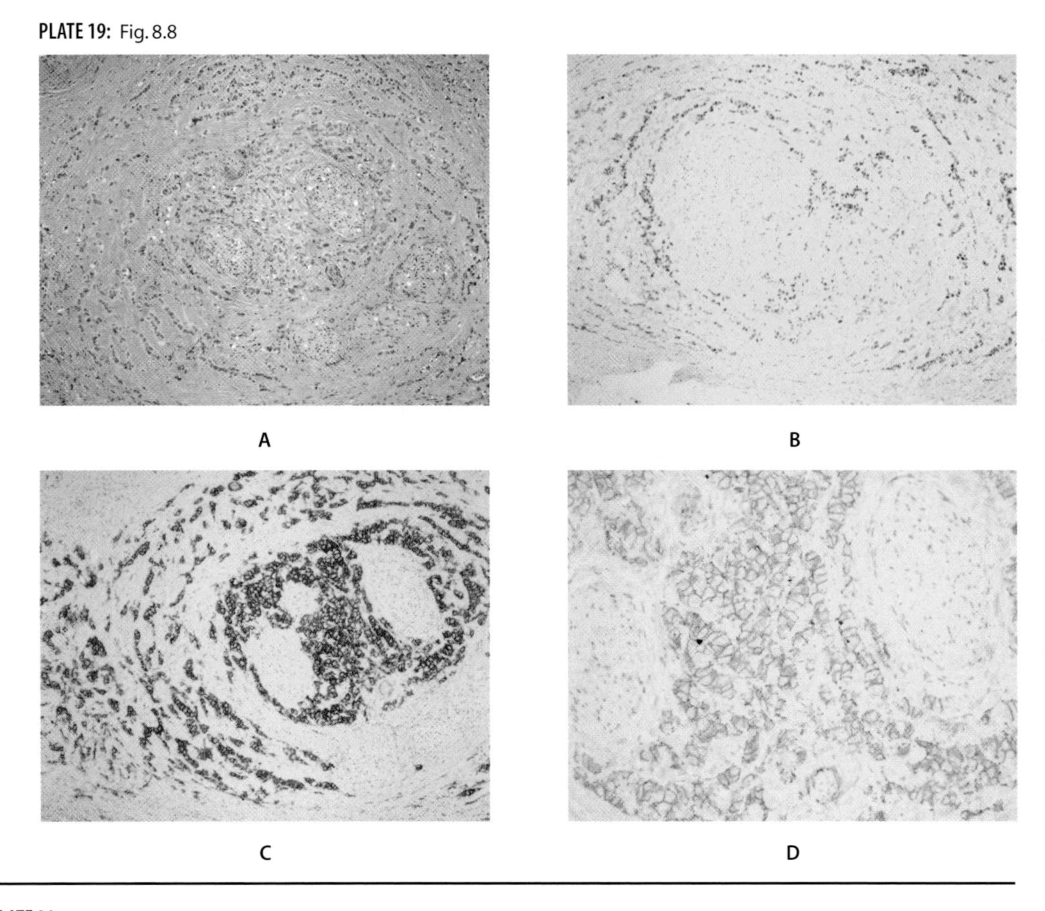

FIGURE 8.9: Poorly-differentiated carcinoma with lack of cytokeratins 5/6 expression (A). The lesion does stain positively for cytokeratins 8/18 (B).

PLATE 20: Fig. 8.9

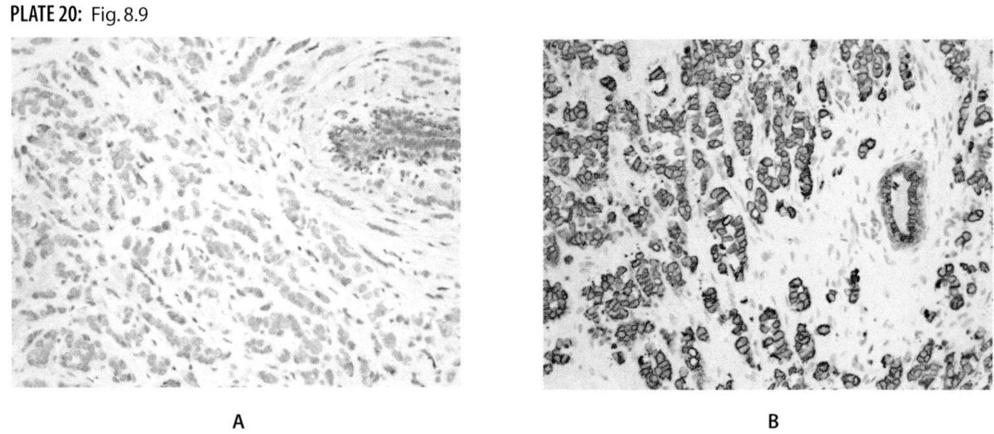

PLATE 21: Fig. 9.1

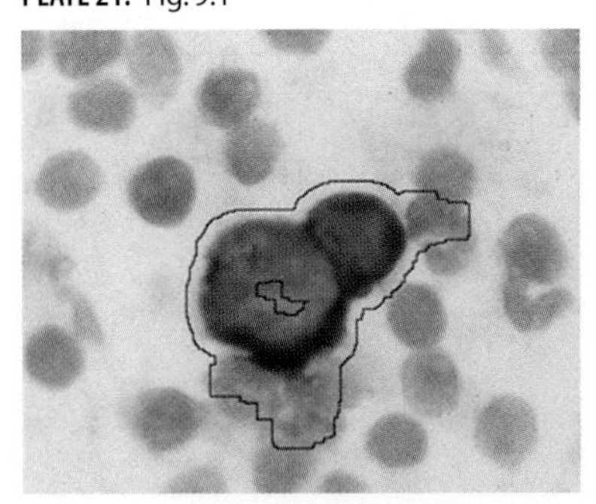

FIGURE 9.1: Disseminated tumor cell detected in bone marrow by cytokeratin immunostaining.

PLATE 22: Fig. 9.2

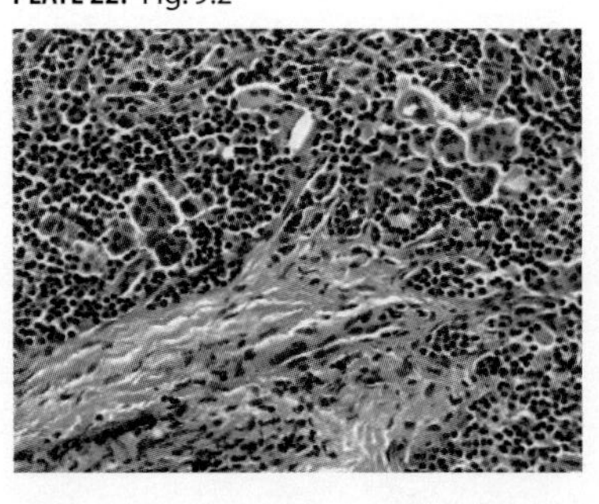

A

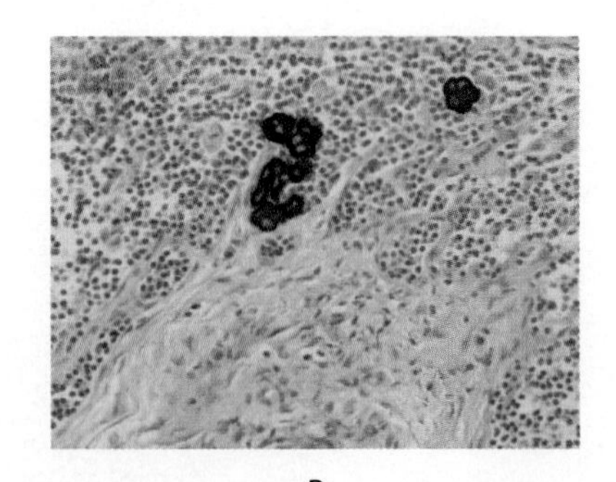

B

FIGURE 9.2: Micrometastasis detection in a lymph node using H&E staining (A) and cytokeratin immunostaining (B).

FIGURE 10.1: Traditional morphologic features associated with adverse prognosis in breast cancer.

PLATE 23: Fig. 10.1

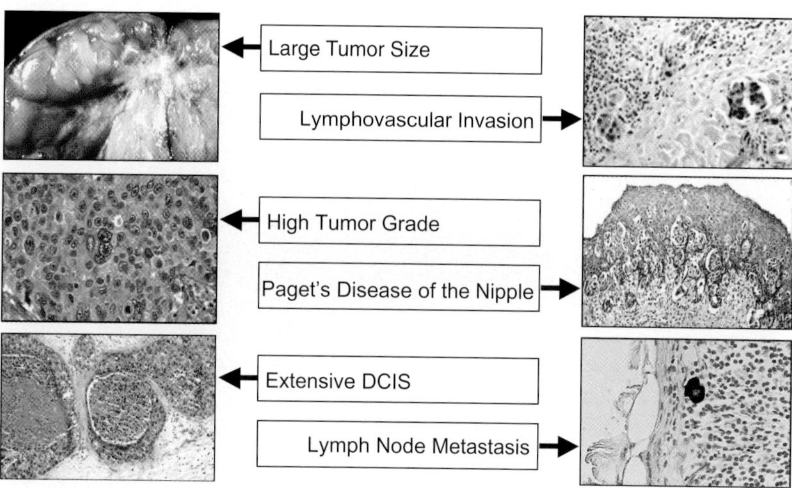

FIGURE 10.2: Currently used ancillary tests indicating adverse prognosis in breast cancer.

PLATE 24: Fig. 10.2

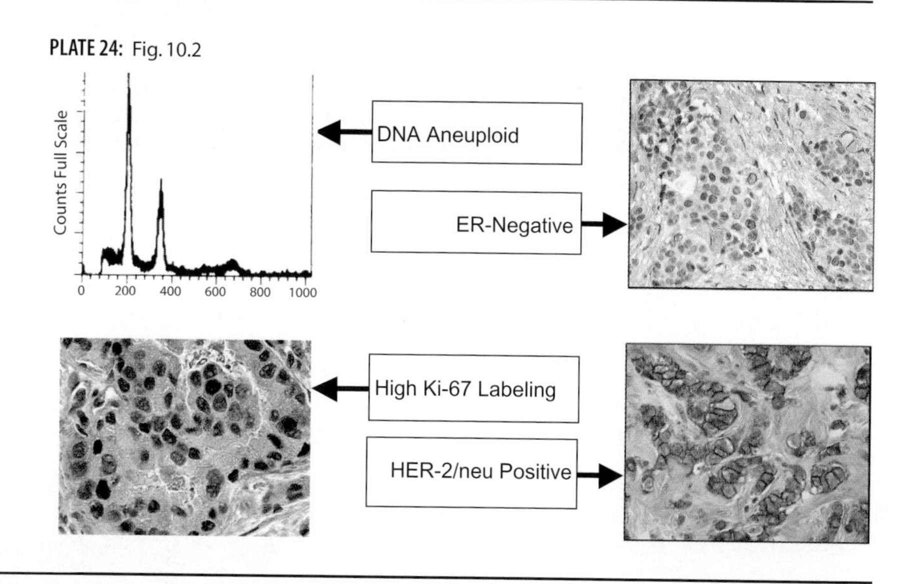

URE 11.2: Metaphase multicolor H (MFISH) analysis of whole omosomes in a breast cancer ng combinatorial colors.

PLATE 25: Fig. 11.2

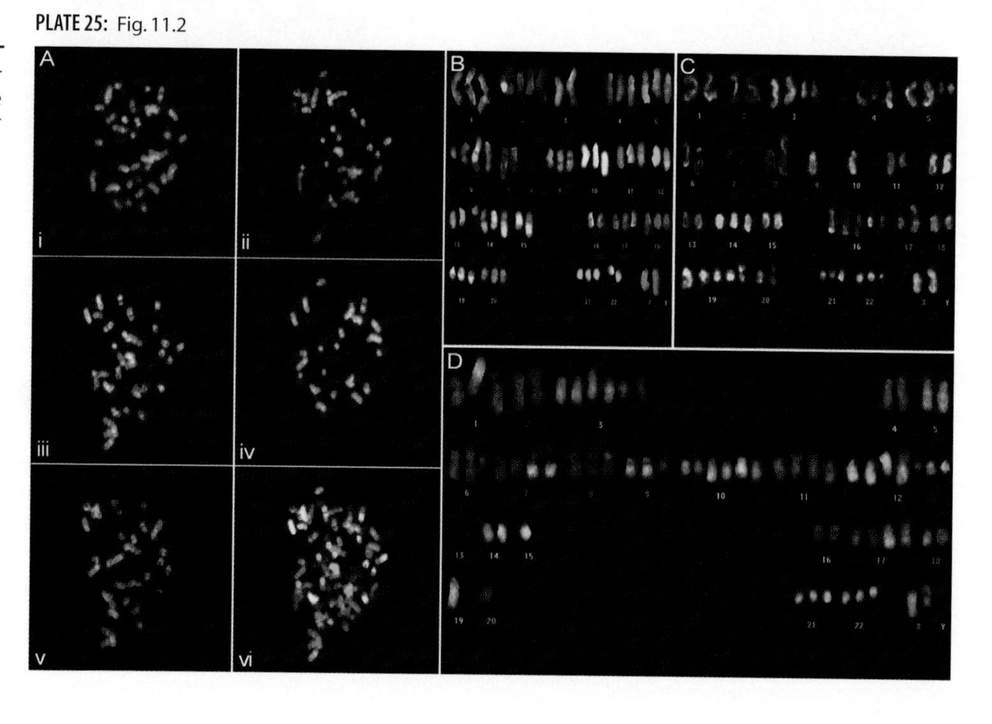

PLATE 26: Fig. 11.3

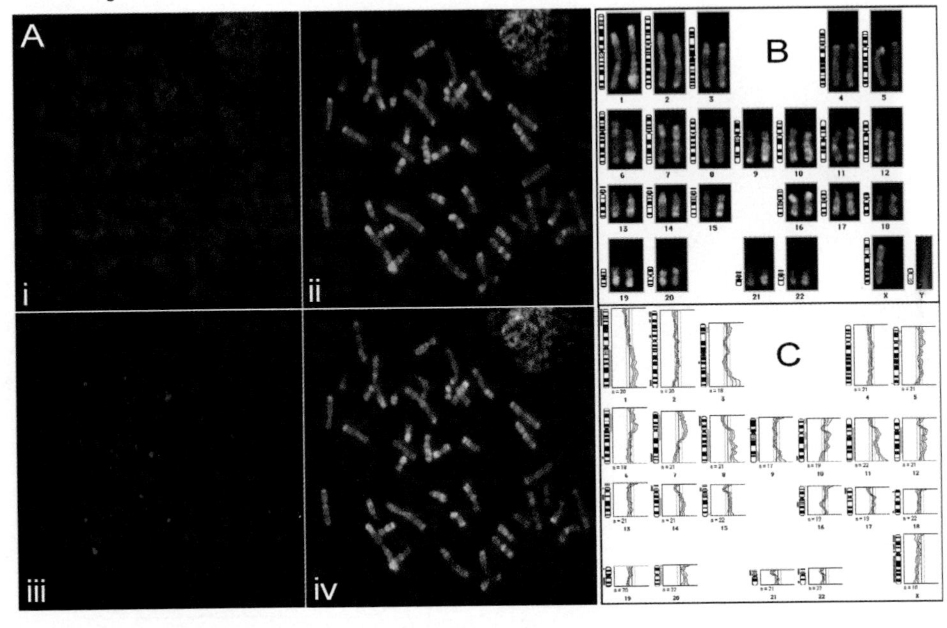

FIGURE 11.3: Spectral metaphase tumor cell karyotyping (SKY).

PLATE 27: Fig. 13.2

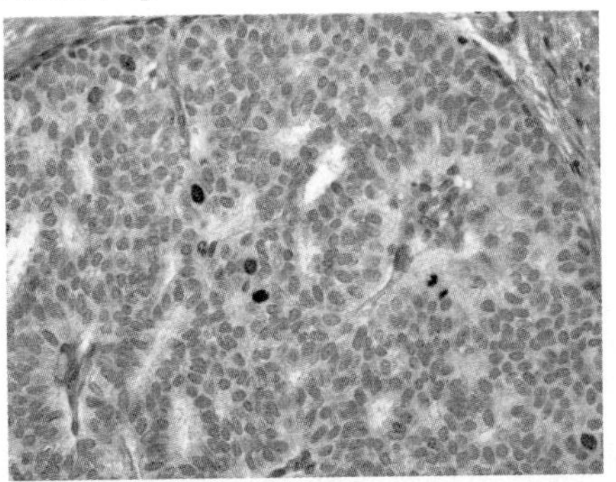

FIGURE 13.2: Well-differentiated breast cancer with low Ki-67 labelling on IHC staining.

PLATE 28: Fig. 13.4

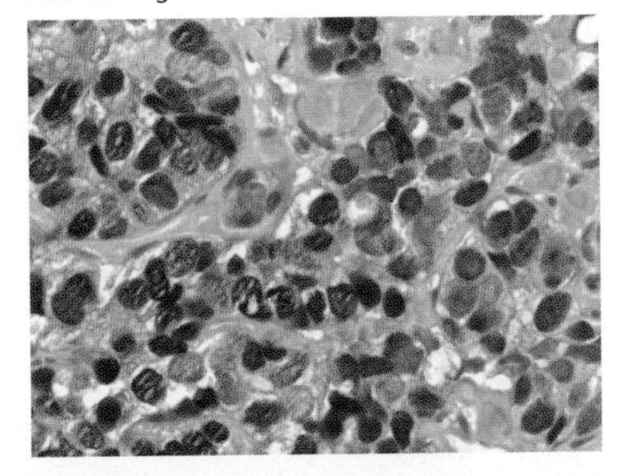

FIGURE 13.4: P27 expression in breast cancer which is associated with a favorable prognosis.

PLATE 29: Fig. 13.6

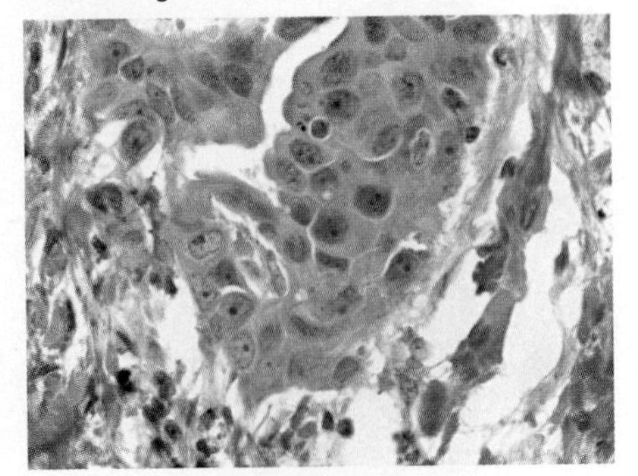

FIGURE 13.6: Low Skp-2 expression which generally correlates inversely with the expression p27 status.

PLATE 30: Fig. 14.1

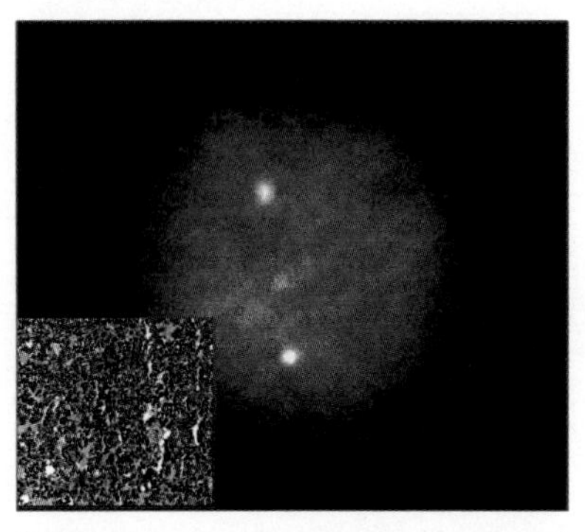

FIGURE 14.1: *Bcr-abl* translocation detected by FISH (single yellow signal) in patient with chronic mylegenous leukemia on bone marrow biopsy (inset).

PLATE 31: Fig. 14.2, B and C

B

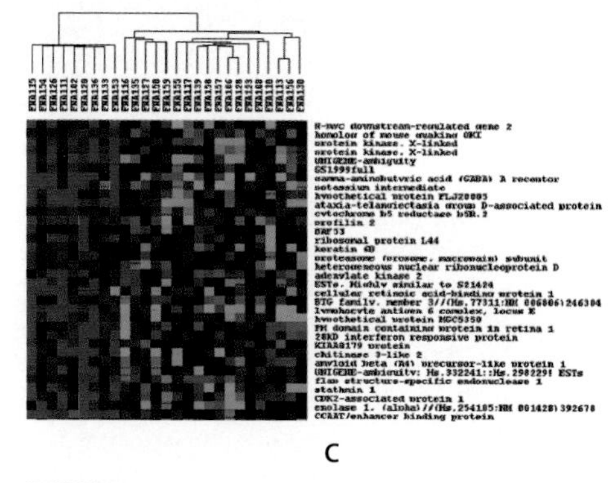

C

FIGURE 14.2, B and C: Cluster analysis of gene expression in ER positive (B) and ER negative (C) breast cancer.

PLATE 32: Fig. 15.2

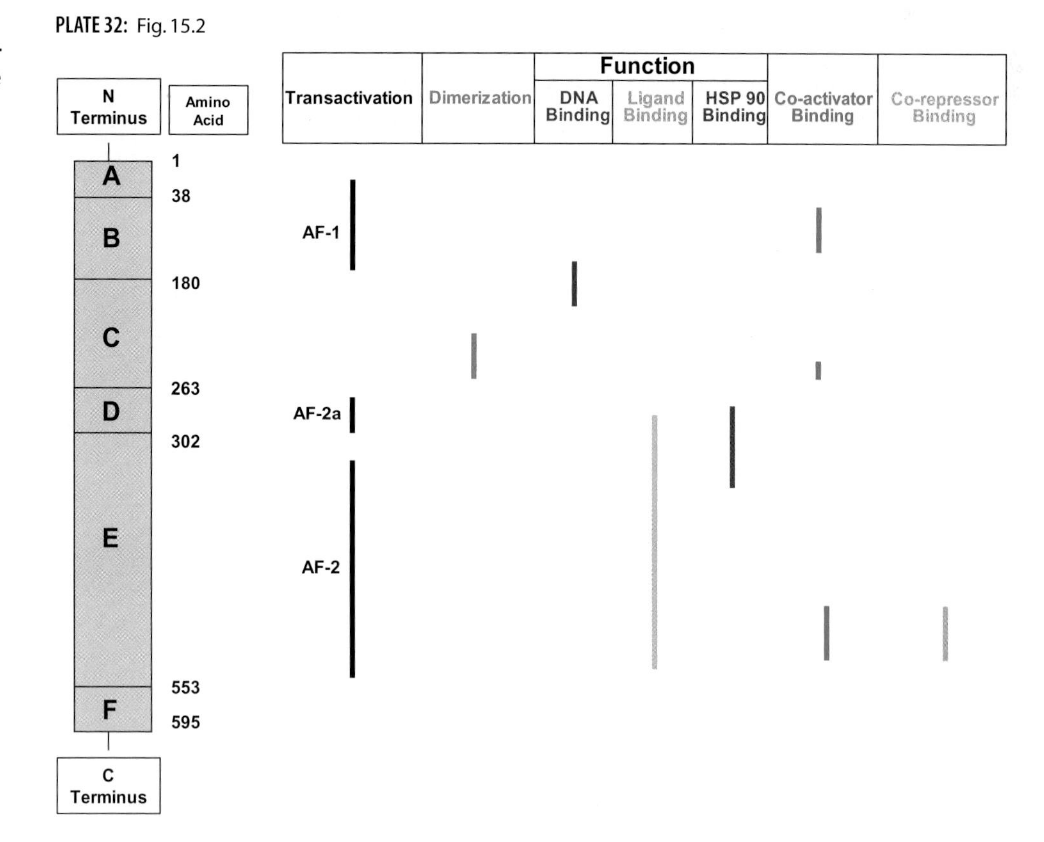

GURE 15.2: Structure of the -α gene.

PLATE 33: Fig. 15.3

FIGURE 15.3: The biochemical competitive binding assay for measuring ER status.

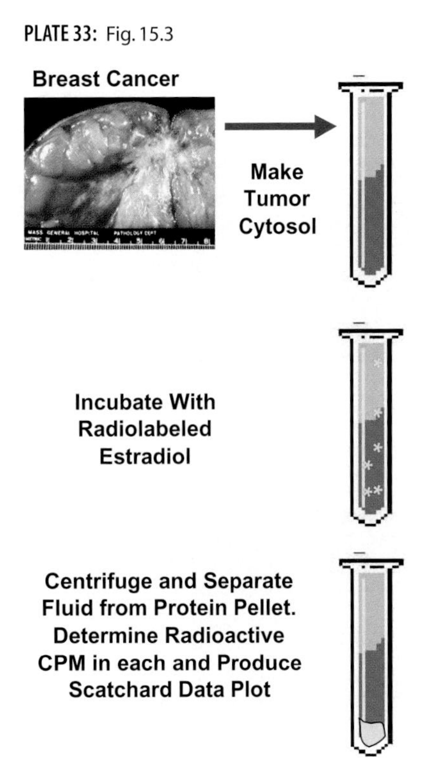

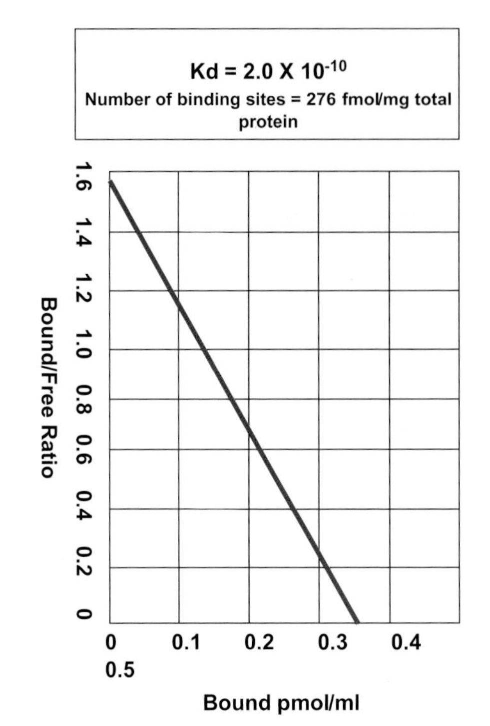

PLATE 34: Fig. 15.4

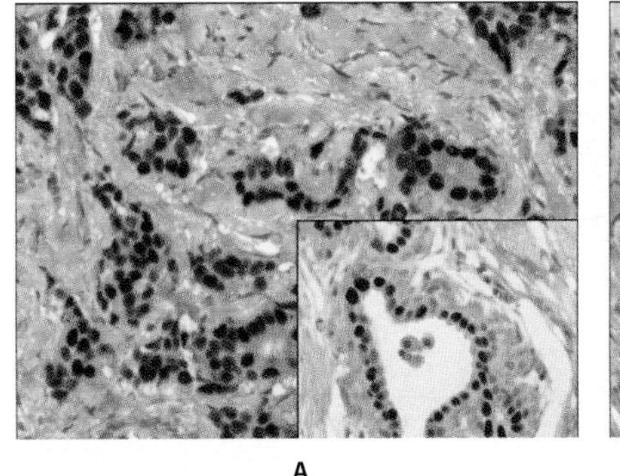

A

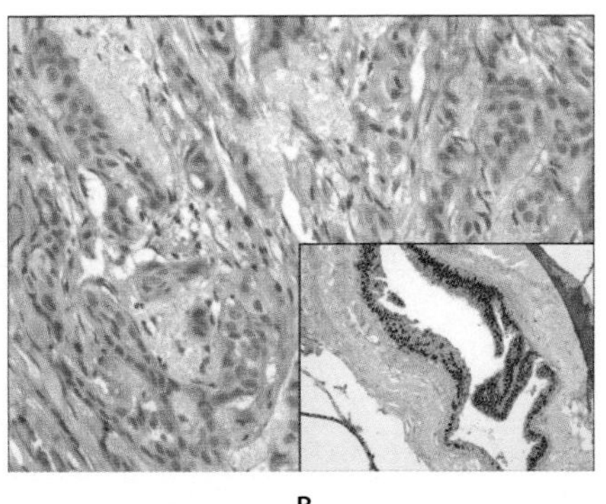

B

FIGURE 15.4: Positive (A) and negative (B) ER expression in breast cancer using IHC (control in inset).

PLATE 37: Fig. 16.5

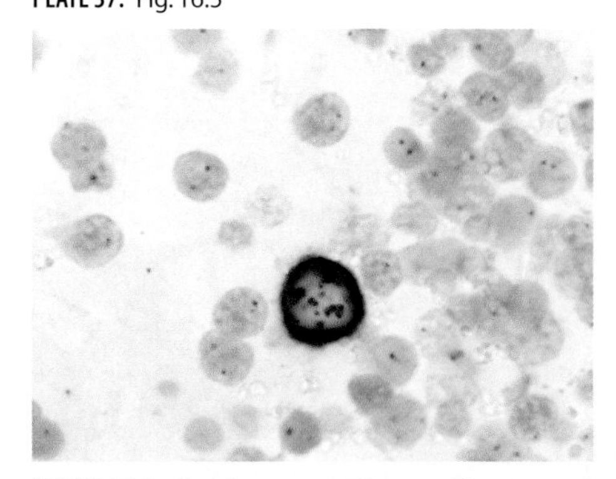

FIGURE 16.5: Simultaneous HER-2 amplification and overexpression detected by CISH and IHC in a single breast cancer cell.

PLATE 35: Fig. 16.3

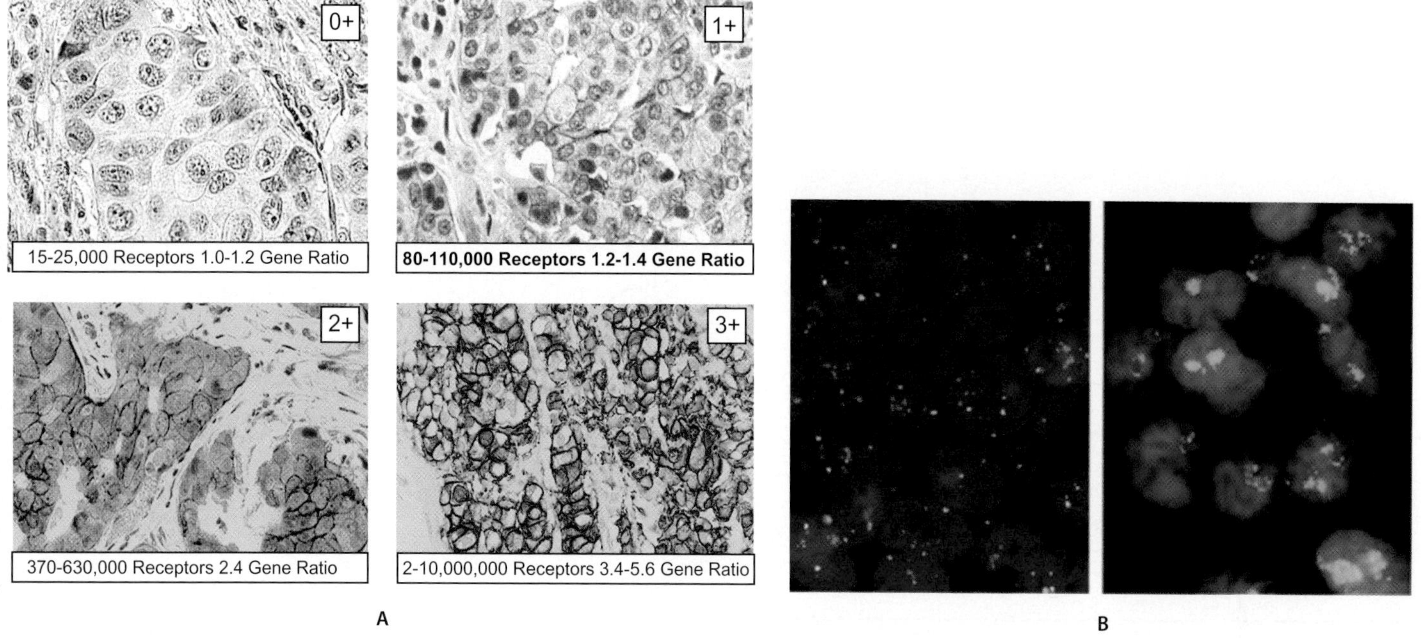

FIGURE 16.3: (A) HER-2 status detected by IHC showing 0+, 1+, 2+, and 3+ categories of expression. (B) HER-2 gene amplification detected by FISH showing the Pathyvision™ (left) and Inform™ (right) techniques.

PLATE 36: Fig. 16.3C

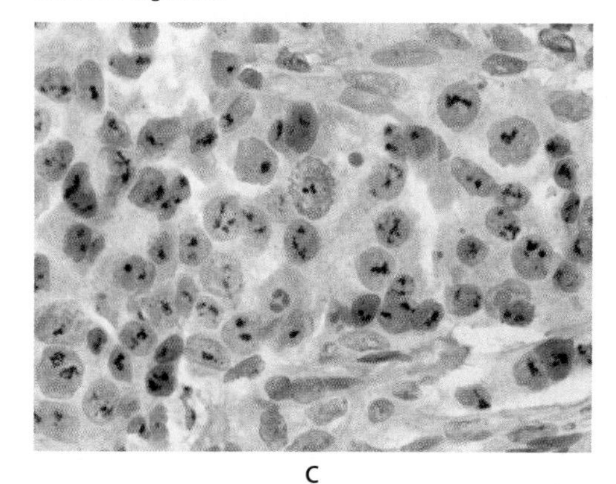

C

FIGURE 16.3: (C) HER-2 amplification detected by CISH.

PLATE 38: Fig. 16.6A

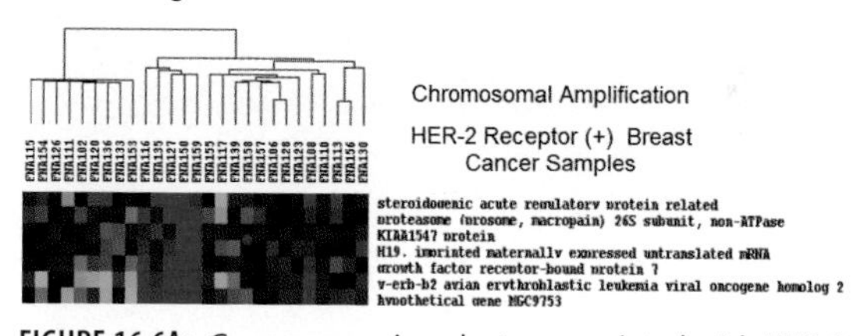

FIGURE 16.6A: Gene expression cluster associated with HER-2 amplification.

PLATE 39: Fig. 16.7

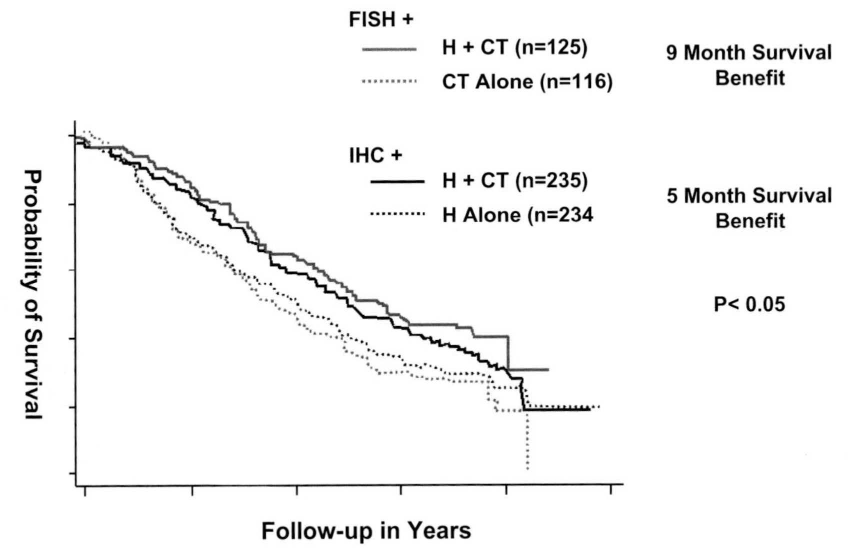

FIGURE 16.7: FISH is a significantly more accurate predictor of trastuzumab response than IHC.

PLATE 40: Fig. 17.4

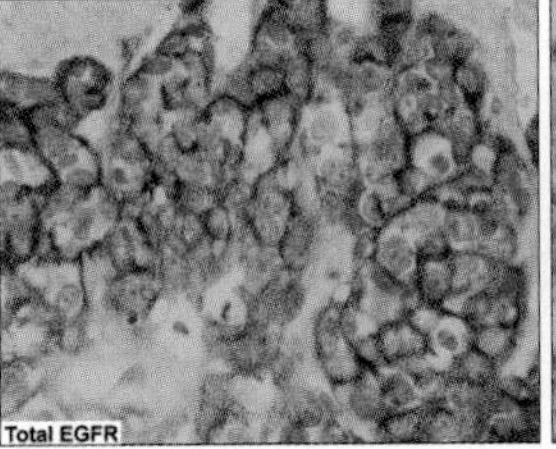

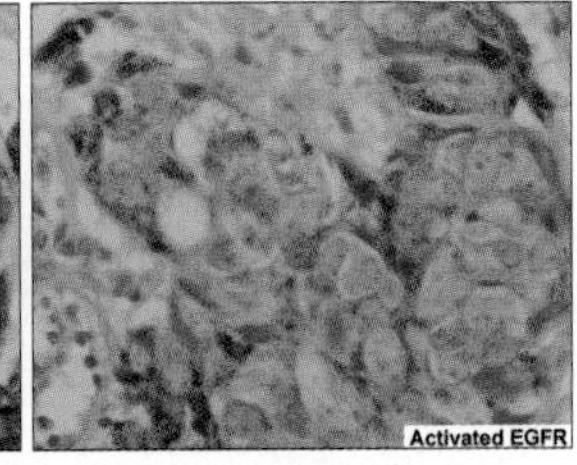

FIGURE 17.4: Expression of EGFR (left) and activated EGFR (right) in breast cancer.

PLATE 41: Fig. 17.6

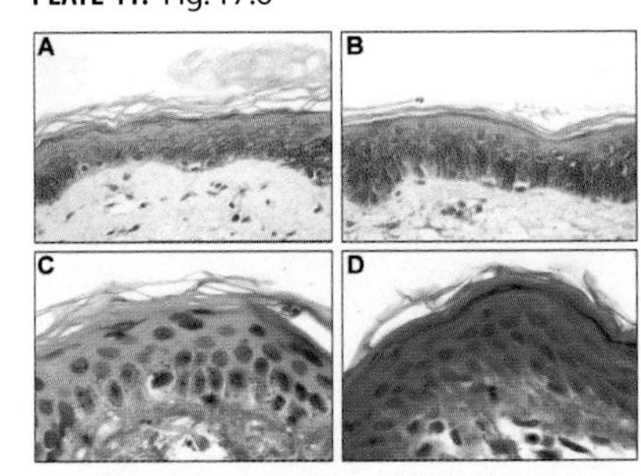

FIGURE 17.6: Pharmacodynamic assessment of anti-EGFR targeted therapy (gefitinib) using skin biopsies and EGFR.

PLATE 42: Fig. 18.1

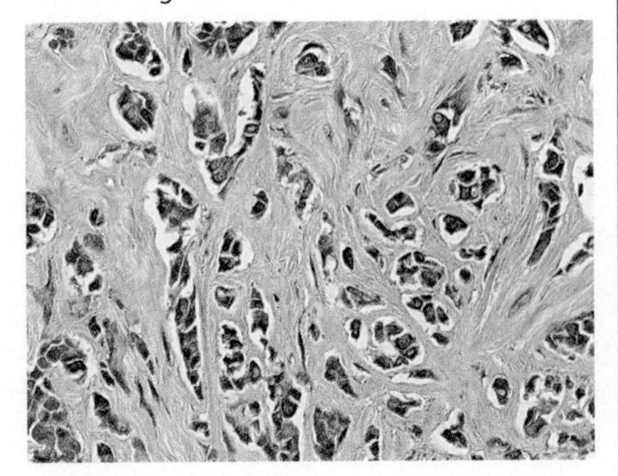

FIGURE 18.1: MMP-9 expression in breast cancer.

PLATE 43: Fig. 18.3

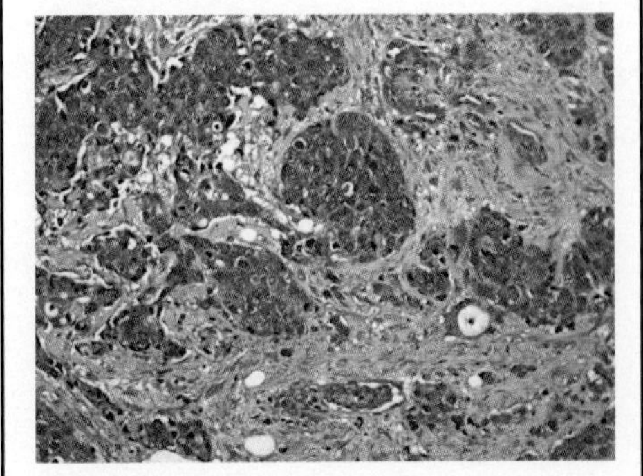

FIGURE 18.3: uPA expression in breast cancer.

PLATE 45: Fig. 19.4

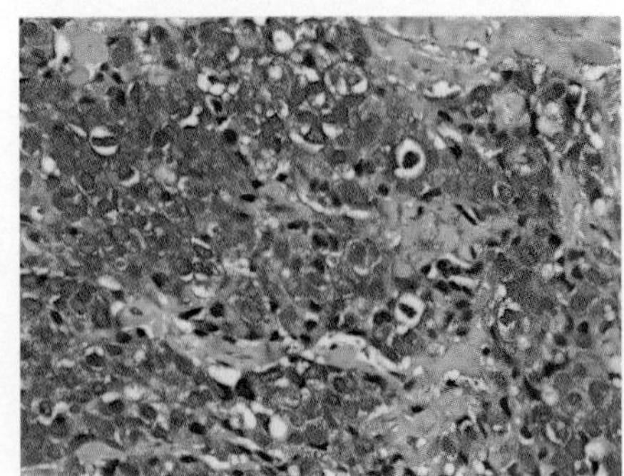

FIGURE 19.4: High VEGF expression in breast cancer.

PLATE 44: Fig. 19.3

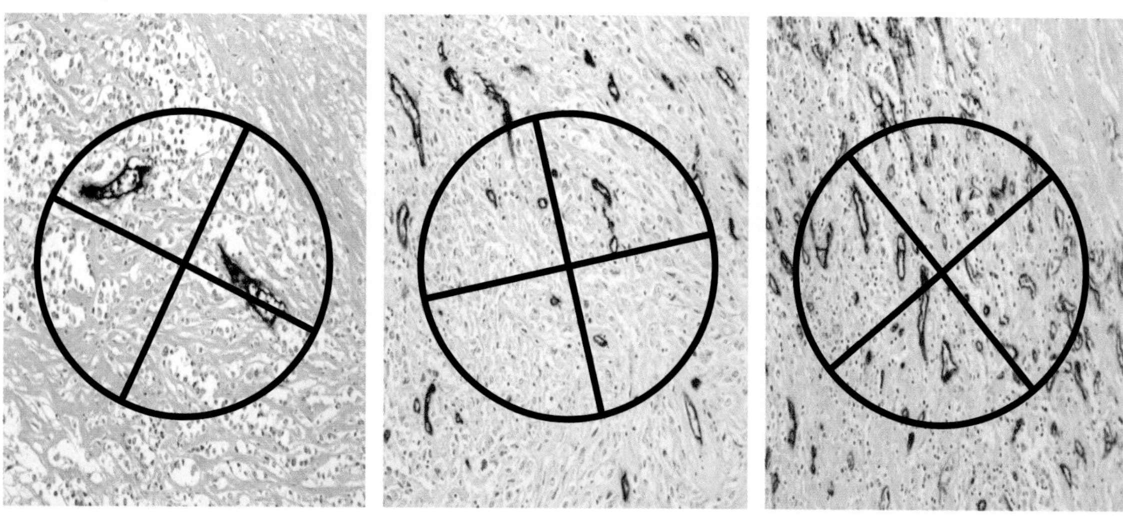

FIGURE 19.3: Measurement of tumor microvessel density (MVD) using the Chalkley grid system in low MVD (left), intermediate MVD (center) and high MVD (right) tumors.

ATE 46: Fig. 20.1

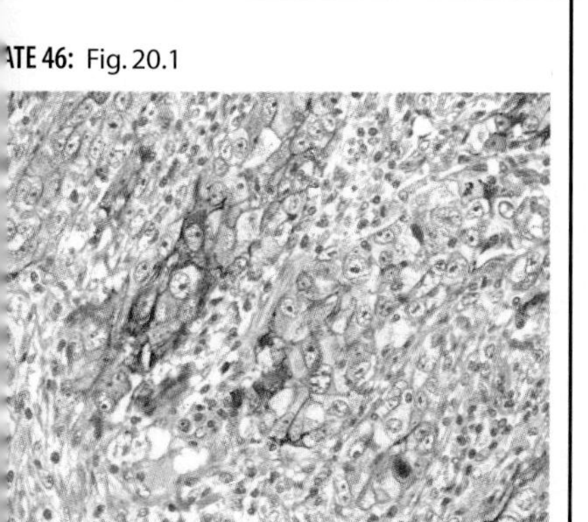

URE 20.1: E-cadherin expression loss in a poorly ferentiated infiltrating ductal carcinoma.

PLATE 47: Fig. 22.1

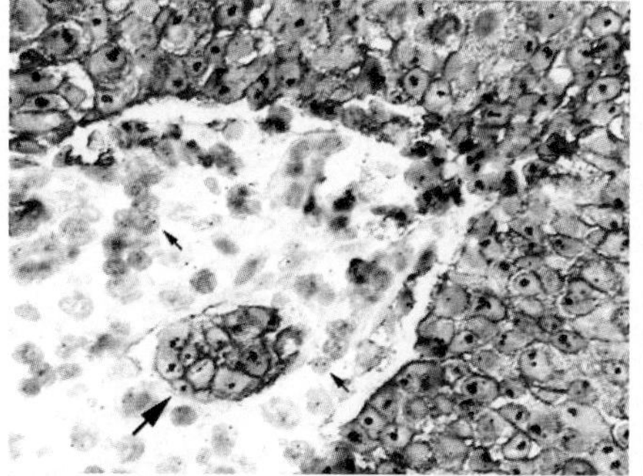

FIGURE 22.1: Simultaneous HER-2 amplification and overexpression detected by CISH and IHC in an infiltrating ductal adenocarcinoma.

PLATE 48: Fig. 22.3

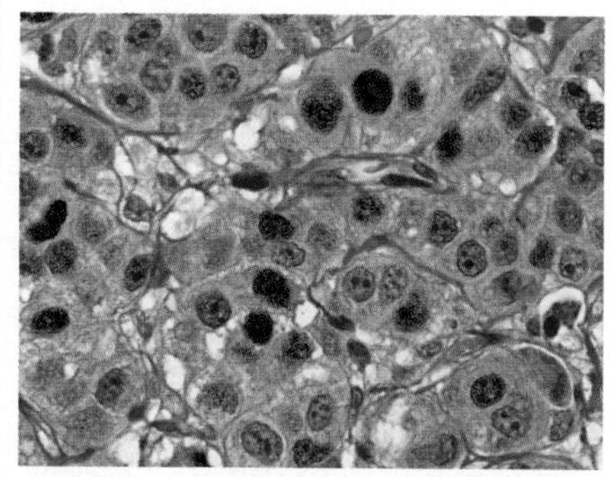

FIGURE 22.3: Overexpression of cyclin D in a high grade breast cancer.

PLATE 49: Fig. 23.4

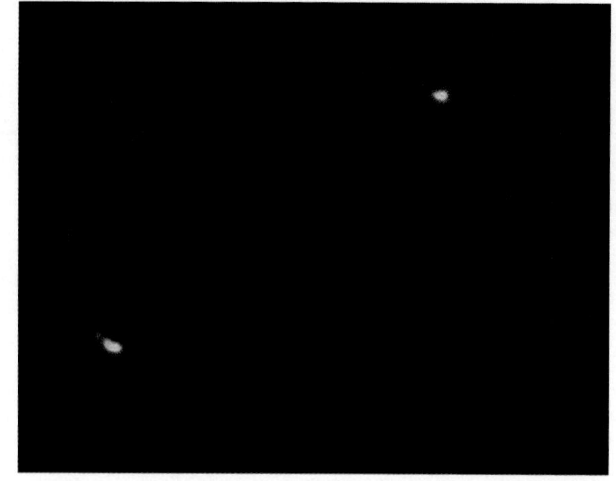

FIGURE 23.4: Chromosome 8p deletion in breast cancer detected by FISH.

PLATE 50: Fig. 24.4

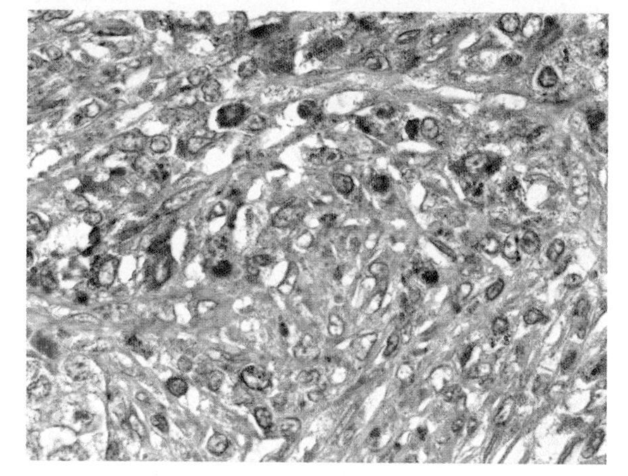

FIGURE 24.4: NF-κB overexpression in breast cancer with nuclear and cytoplasmic staining.

FIGURE 25.3: The domains of the BRCA1 gene.

PLATE 51: Fig. 25.3

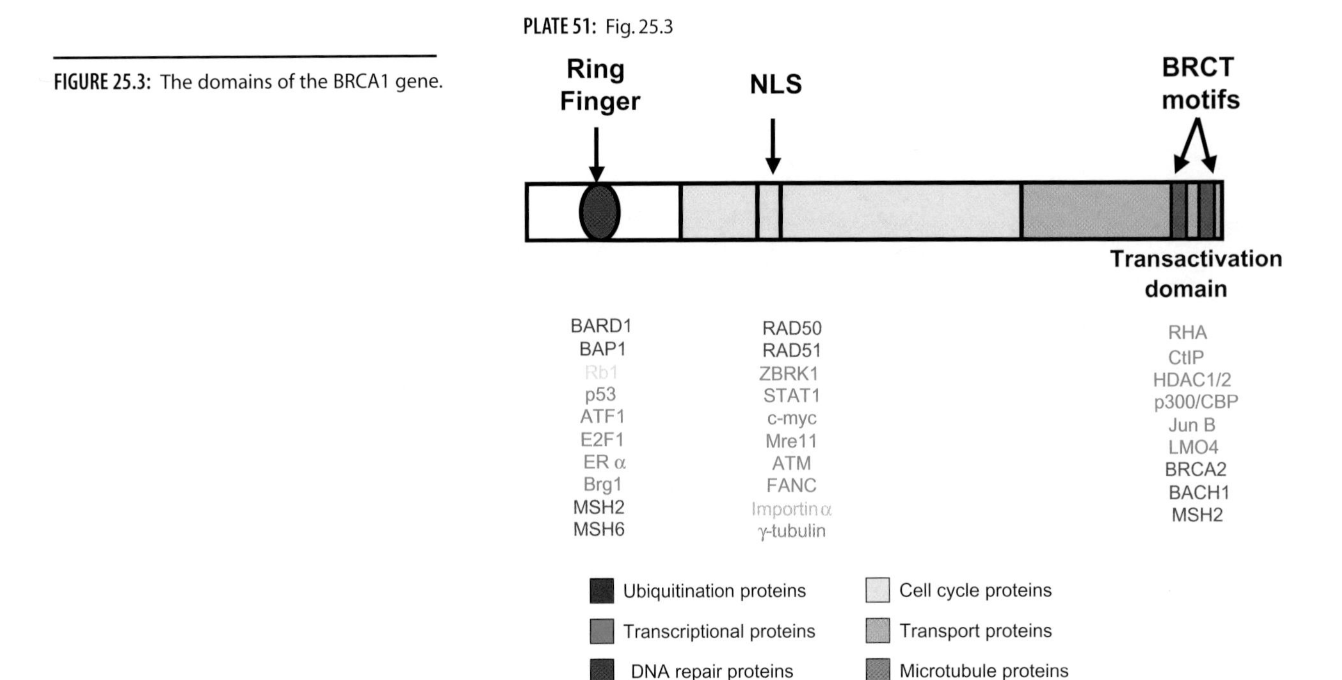

PLATE 52: Fig. 28.1

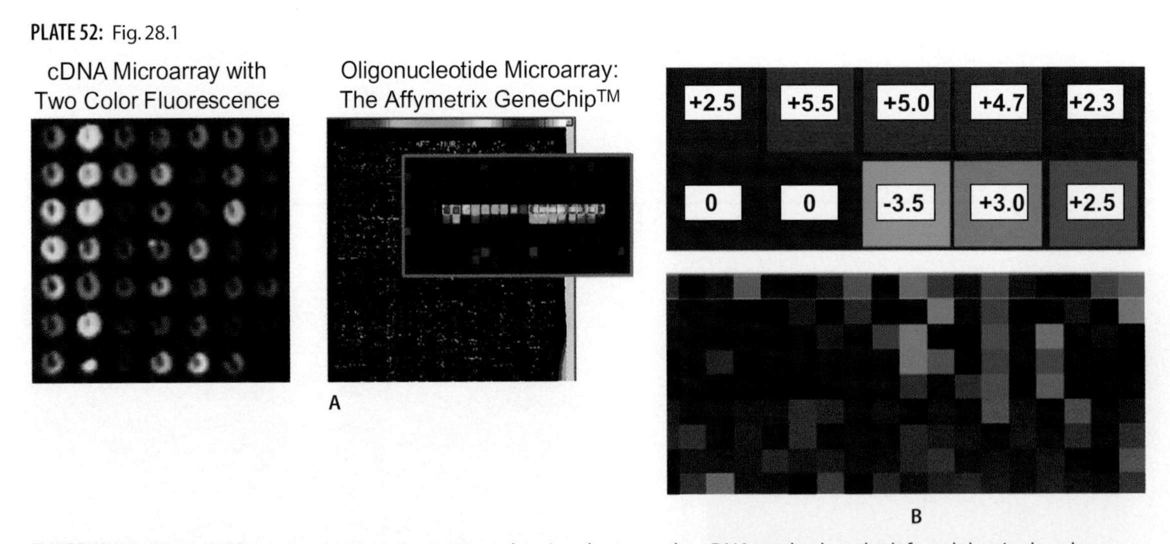

FIGURE 28.1: (A) Types of fluorescent microarray systems showing the two-color cDNA method on the left and the single color oligonucleotide method on the right. (B) Gene expression profiles are typical represented by "heat maps" which, by convention, show up-regulated genes in red, down-regulated genes in green and genes at similar expression levels to the background in black.

PLATE 53: Fig. 28.3

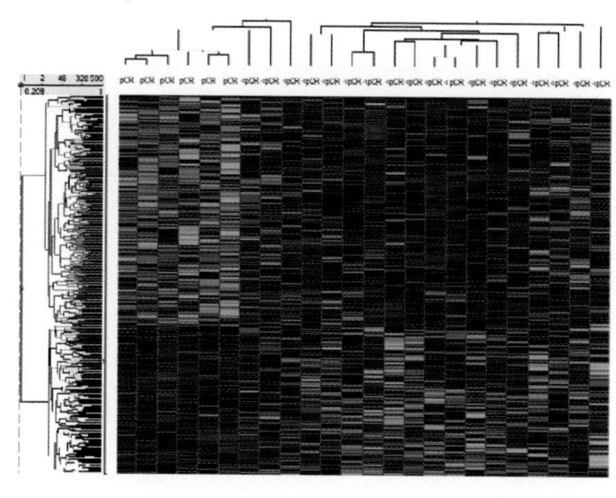

FIGURE 28.3: Heat map demonstrating genes that are up and down-regulated in a series of patients treated with the T/FAC regimen in the neo-adjuvant setting using pathologic complete response (PCR) as the end-point.

PLATE 54: Fig. 29.4

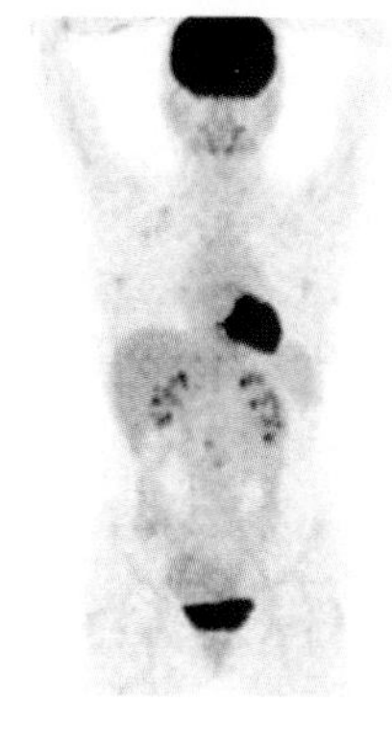

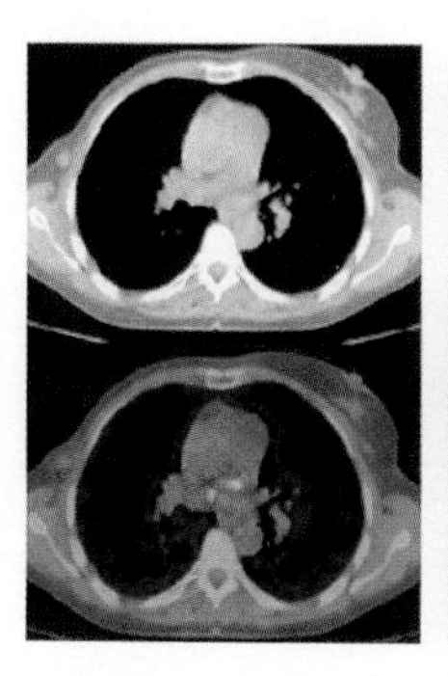

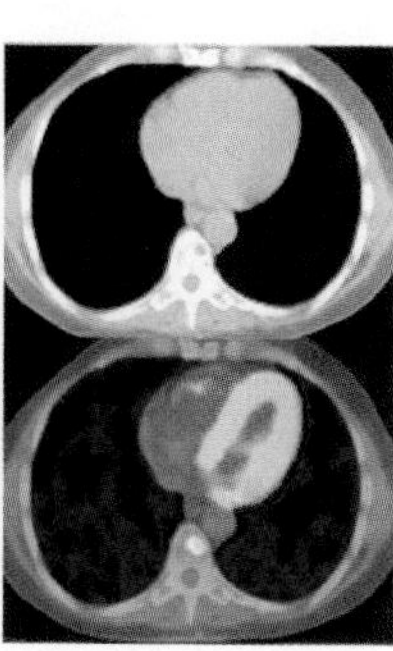

FIGURE 29.4: Detection of breast cancer recurrence using FDG PET scanning.

CHAPTER

28

Pharmacogenetics, Pharmacogenomics, and Predicting Response to Therapy

Lajos Pusztai, MD, PhD,[1] James Stec, BS,[2] Mark Ayers, BS,[2] Jeffrey S. Ross MD,[2,3] Peter Wagner, MD,[1] Roman Rouzier, MD,[1] Fraser Symmans, MD,[1] and Gabriel N. Hortobagyi, MD[1]

[1]The University of Texas M. D. Anderson Cancer Center, Houston, Texas
[2]Millennium Pharmaceuticals Inc., Cambridge, Massachusetts
[3]Albany Medical College, Albany, New York

Biologic Background

Since the introduction of systemic therapy for breast cancer, a great variability in response to therapy among cancers of different individuals has been apparent. In the late 1970s, standardized clinical criteria to assess tumor response to therapy were published to aid the communication of clinical trial results.[1] The response criteria were recently revised, but they continue to be based on a four-way categorization of response including complete response, partial response, stable disease, and progressive disease based on clinical tumor measurements before and during therapy continues.[2] These categories reflect the observation that clinical response to therapy is a continuum. In general, objective tumor shrinkage after treatment is more commonly observed earlier in the course of the disease than at later phases; the more prior treatment an individual received, the less likely that subsequent therapies would produce a response. This unfortunate phenomenon is documented in clinical trials that invariably report higher tumor response rates for any drug used as first-line treatment compared with second-, third-, and fourth-line use of the same drug. These clinical observations are consistent with the hypothesis that most human breast cancers contain a mixed population of cells, some sensitive to a given therapy and some not. During the course of the disease, the treatment resistant cells gradually acquire dominance.

The biologic mechanisms of resistance to any therapeutic agent are not well understood, although working hypotheses abound. At the cell population level, kinetic models of tumor resistance have been popular.[3,4] These models postulate that cytotoxic drugs produce units of log kill of neoplastic cells with each course of treatment. The classic models also imply that the likelihood of drug-resistant cells being present in a tumor before initiation of therapy is a function of the number of cells in the tumor, their mutation rate, and their proliferative status. It has been suggested that the growth fraction, the proportion of actively proliferating cells in the tumor, is a function of tumor size and that proliferating cells are more sensitive to many cytotoxic agents than resting cells. According to the Gompertzian growth kinetics, smaller tumors contain the highest fraction of proliferating cells and therefore may be most sensitive to cytotoxic therapy.[5]

We proposed a complementary model of clinical drug sensitivity.[6] This model assumes that tumors are formed of heterogeneous cell populations and, in terms of drug sensitivity, three distinct cell types may be present at any time during the course of the disease: physiologically drug resistant, drug sensitive, and pathologically drug resistant. Physiologic drug resistance is a cellular state when sensitivity of breast cancer cells to chemotherapy is similar to that of the corresponding normal breast epithelial cells. We suggested that increased sensitivity to cytotoxic drugs is an acquired trait of some neoplastic cells that is caused by loss of physiologic stress response and repair pathways during the neoplastic evolution. The common clinical pathologic observation that many invasive breast cancer cells are more sensitive to chemotherapy than the adjacent normal breast epithelial cells or noninvasive neoplastic cells from which they originate supports this hypothesis.[7] With further accumulation of genetic or epigenetic damage in the cells, pathologic drug resistance can develop due to impairment in apoptotic mechanisms, aberrant activation of survival pathways, alteration in drug targets, or activation of drug elimination mechanisms. Clinical response to therapy may be determined by the relative contribution of these different cell populations to the total tumor mass.

At the cellular level, a beneficial response to therapy may be described either as cytostatic or cytotoxic response. A cytostatic response is frequently thought of as proliferation arrest leading to stagnation of tumor growth. A cytotoxic response, on the other hand, leads to programmed cell death and may result in tumor shrinkage. It is quite plausible that both types of responses may occur after treatment with anticancer drugs. These biologic events may develop sequentially in the same cell manifesting first as cell cycle arrest followed by cell death or may affect distinct cell populations, some responding with growth arrest and others undergoing cell death after treatment.

At the subcellular molecular level, response to cytotoxic therapy is a complex molecular event that involves simultaneous activation of drug elimination, damage repair, and survival pathways along with the activation of programmed cell death.[8] The ultimate cellular outcome of the toxic insult, death versus survival, may depend on the balance of the prodeath and prosurvival signals.[9] Whether cell death or recovery ensues after exposure to a drug may be influenced by the activity of drug efflux pumps that remove toxic compounds from cells, the abundance of drug target within the cell, the magnitude of cellular damage caused by the insult, and the efficiency of physiologic repair pathways.

It should also be noted that at the level of the organism, several factors independent of the biology of the neoplastic cells also influence the efficacy of antitumor therapy. Foremost among these factors are the interindividual differences in activity of drug metabolism enzymes. Any systemically administered drug is subject to elimination and drug transformation processes that take place at several organs. The liver and the kidneys are the most important organs that eliminate and metabolize anticancer drugs. A large number of enzymes participate in the inactivation and elimination of drugs. The activity of these enzymes is a genetically inherited trait that may vary from individual to individual and from human population to population.

In summary, response to anticancer therapy can be defined at different organizational levels as clinical, cellular, or molecular response to treatment. At each level, response is determined by a complex interaction of events. Because of this complexity, it is unlikely that using only one or a few biologic markers, such as the proliferation rate of the tumor or the presence or absence of a single molecule in the cancer, will provide a robust enough predictor of clinical response to treatment. Indeed, very few single gene markers (estrogen receptor [ER] and HER-2 receptor) are of clinical value to predict response to therapy.[10] This motivates research to apply high throughput genomic and proteomic techniques to search for multigene markers that could yield clinically relevant predictive information.

Clinical Relevance of Predicting Response to Therapy

The clinical importance of predicting who will or will not respond to a particular therapy is intuitively obvious. If physicians could predict who will respond to a particular treatment, such a test could be used to select patients who benefit from therapy and minimize unnecessary toxicity for those who are unlikely to benefit. However, the practical development of response prediction tests is challenging. What level of prediction accuracy would be clinically useful is uncertain. Also, would a test that was developed to predict response to a given treatment in previously untreated patients also predict sufficiently accurately response in the second- and third-line treatment setting? There are also theoretical limits to the accuracy of any response predictor that measures the characteristics of the cancer only. It is likely that a combination of factors, some residing in the tumor itself and others in the host (such as rates of drug metabolism), determines the response to treatment together.

It is interesting to recall that ER and progesterone receptor (PR) (see Chapter 15) and

HER-2 amplification/overexpression (see Chapter 16) in breast cancer are primarily useful because they have high negative predictive values. Patients who test negative for these receptors have a 5% or less chance to respond to antiestrogens or to the anti–HER-2 antibody trastuzumab. On the other hand, the positive predictive value of these tests is no greater than 30–50%.[11] Therefore, patients whose tumors are positive for these receptors only have a 50% or less chance to respond to treatment. These examples show that a test with a 95% accuracy to rule out response may be clinically useful, even if its positive predictive value is limited. Whether such accurate tests for any chemotherapy agent can be developed remains to be seen. It is worth remembering that antiestrogens and trastuzumab are highly targeted agents with variable target expression across cancers in contrast to the currently used chemotherapy drugs that aim at universal cellular targets expressed in all cells.

Could predictive tests of lesser accuracy be clinically useful? By definition, a test is clinically useful if its results contribute to clinical decision making and influence treatment recommendations. What predictive accuracy is needed from a test is influenced to a large extent by the clinical setting in which it is applied.[12] Different levels of predictive accuracy may be required for treatment decisions that concern curative therapy compared with palliative treatment. For example, let's assume that a putative chemotherapy response prediction test has a 60% positive predictive value (60% chance to respond if test positive) and an 80% negative predictive value (20% chance of response if test negative), whether this test would be useful or not depends on the availability and success of alternative treatments options, the frequency and severity of adverse effects of the drug, and the risk associated with exposure to ineffective therapy (i.e., rapid disease progression with life-threatening complications). When applied in the palliative setting, especially when alternative treatment options are limited and generally ineffective, the use of this test may be of limited value. Patients and physicians may want to try a drug even if the test is negative and the expected response rate is only around 20%, particularly if side effects are uncommon and tolerable. In a situation like this, the usefulness of the test increases as its negative predictive value improves.

On the other hand, a test with the same characteristics could be quite useful if applied to try to select the best regimen from multiple available choices, particularly if the right choice leads to improved survival. A possible clinical scenario for this is selecting adjuvant chemotherapy for women with newly diagnosed stage I–III breast cancer. There are multiple adjuvant treatment choices for these patients, and the right choice may make the difference between cure and relapse. Assuming that pathologic complete response to preoperative chemotherapy may be used as an early surrogate for long-term disease-free survival, a test that could predict a 60% chance of pathologic complete response to a *particular* regimen over any other could be quite useful. This test could maximize the chances of cure for a subset of women by helping to select the best available neoadjuvant/adjuvant therapy for them. Unselected use of any of the currently best neoadjuvant chemotherapy regimens results in pathologic complete response rates of around 30% in the general patient population.[13]

Current Clinical Response Prediction Tests

The most recent guideline of the American Society of Clinical Oncology for the use of tumor markers only recommends routine testing for ER, PR, and HER-2 receptors in breast cancer.[13] The ER and PR expression is usually determined by immunohistochemistry in formaldehyde-fixed paraffin-embedded tissues. The test is primarily used to select patients for endocrine therapy. The current gold standard to determine HER-2 amplification is fluorescent in situ hybridization (see Chapter 16). For cost containment and simplicity, many laboratories use immunohistochemistry as an initial screening tool and consider patients with 0 or 1+ staining HER-2 normal and those with 3+ staining HER-2 amplified and perform fluorescent in situ hybridization only on cases that are immunohistochemistry indeterminate or 2+. HER-2 amplification is determined in order to select patients for trastuzumab therapy.[14]

There are no widely accepted clinically useful single gene markers of response to any chemotherapy in breast cancer. Several markers have been proposed (**Table 28.1**), but none has been validated or reproduced

Table 28.1 Predicting Response to Chemotherapy in Breast Cancer

I. Specific markers with accepted link to specific drug response
  A. ER/PR status and hormonal-based therapies
  B. HER-2/*neu* status and trastuzumab response
II. Traditional markers not universally accepted as predictive of drug response
  A. High tumor grade
  B. ER-negative status
  C. High S phase fraction
  D. High Ki-67 labeling index
  E. HER-2/*neu*–positive and anthracycline response
  F. p53 mutation status
  G. MDR (p-glycoprotein) status
  H. Topoisomerase IIα and anthracycline response
  I. Bcl-2 and apoptosis resistance
III. Emerging strategies for predicting drug response
  A. Pharmacogenomics on tumor tissue by transcriptional profiling
  B. Pharmacogenetics on germline tissues (blood) by direct sequencing
  C. Proteomics on serum by mass spectroscopy

without controversy. In general, high nuclear grade and ER-negative status are two clinicopathologic features that have been associated with better response to chemotherapy in most neoadjuvant treatment studies.[15,16] Similarly, tumors with higher proliferation rate assessed by S phase fraction or Ki-67 expression tend to show higher response rates to chemotherapy.[17,18] However, none of these clinicopathologic features are regimen specific and therefore are not useful to select any particular chemotherapy regimen over the other. Furthermore, their negative predictive value is not high enough to recommend against potentially curative adjuvant/neoadjuvant chemotherapy in the "low response" subset of women.

Among the numerous potential single gene markers of response, HER-2 has been examined extensively but failed to show a robust predictive value for response to preoperative anthracycline-based therapy.[19–21] Similarly, a large and controversial literature exists on the value of p53, topoisomerase IIα, bcl-2, and p-glycoprotein expression in predicting response to therapy.[22–30] Unfortunately, none of these markers demonstrated high enough positive or negative predictive values to be useful in treatment selection for individual patients. This is partly due to fragmented research efforts that consist of numerous small studies each with limited statistical power that precludes full assessment of the predictive value of these factors. The methodologies to detect any particular marker or biologic process (such as proliferative activity) are also not standardized. Definitions for test-positive and test-negative cases also differ from study to study. No prospective studies were conducted with any of these markers to validate their predictive value. Apart from methodologic and clinical trial design issues, the limited utility of individual molecules to predict response to chemotherapy may also be due to incomplete understanding of the biological function of these putative markers. It is clear that biologically important molecules act in concert and form complex, often redundant, interactive pathways.[31] Therefore, individual components may only contribute minimal information on the activity of the whole system or pathway, and this could limit their predictive value.

Molecular Tools for Pharmacogenomic and Pharmacogenetic Research

Pharmacogenomics refers to the science of examining the cancer genome with high throughput techniques to gain insight into mechanisms of response and resistance to therapy and to identify single gene or multigene predictors of response. There are three broad methodologic approaches to pharmacogenomics. One approach examines genetic alterations of the tumor DNA itself, including mutations, translocations, gene amplification/deletion, and hypermethylation. The second approach is to sequence the DNA of the patient's germline, often using peripheral blood mononuclear cells, to search for single nucleotide polymorphisms (SNPs) that might be predictive of a drug class or individual drug response. This approach is often referred to as pharmacogenetics (**Table 28.2**). A third approach is to study the consequences of the genetic alterations in the tumor cells at the transcriptional level by surveying the expression of tens of thousands of messenger RNA (mRNA) species simultaneously in the cancer tissue. This latter approach is called transcriptional profiling.

Table 28.2 Pharmacogenetics and Pharmacogenomics

	Pharmacogenetics	Pharmacogenomics
Parameter	Germline DNA variations	mRNA expression in cancer tissue
Platform	DNA sequencing SNP discovery Genotyping	Transcriptional profiling
Result	SNPs and complex polymorphisms	Gene expression profiles Biologic pathway activation
Prediction of efficacy	++ (Restricted classes of drugs)	++++
Prediction of toxicity	++++	++ (Limited capability)

Comparative Genomic Hybridization (CGH)

Gene amplifications and deletions are frequently seen in cancer cells and probably represent one of the most common mechanisms that lead to deranged gene expression in cancer. CGH is a technology that measures the proportional prevalence of chromosomal regions between normal and cancer tissue.[32] CGH arrays contain DNA fragments aligned according to their position on normal human chromosomes.[33] Arrays that survey all chromosomes with various levels of genomic resolution can be constructed and used as probes to simultaneously hybridize with differentially labeled DNA from normal (reference sample) and cancer (test sample). Customarily, the DNA is labeled with two distinct fluorochromes, and the different colors correspond to the reference and test samples. The data are reported as fluorescent intensity ratios for each chromosomal region on the array, which in turn reflects DNA copy number differences between the normal and cancer tissues. This method for detection of imbalances in chromosomal segments is quite sensitive and accurate. With the availability of the essentially complete sequence of the human genome, the chromosomal segments represented on a CGH array can be mapped to individual genes that reside on the segments. CGH technology has limitations: It only can detect over- or under-represented chromosomal segments compared with a normal reference tissue. It is unable to detect chromosomal translocations without copy number change or gene inversions. CGH is also not sensitive enough to detect point mutations or polymorphisms in particular genes. Also, hyper- or hypomethylation of regulatory regions of genes may alter gene expression unrelated to any amplification or deletion.

Profiling for CpG Island Methylation

Epigenetic changes that involve nonpermanent and noninherited alterations in DNA can lead to changes in gene expression. The major known mechanism of epigenetic regulation of gene expression in mammalian cells is DNA methylation.[34] DNA methylation occurs at the 5′position of cytosine within the dinucleotide sequence CpG. The human genome contains numerous CpG reach regions commonly referred to as CpG islands. These regions are often located within the 5′prime regulatory end of genes, and hypermethylation of the CpG islands leads to silencing of gene expression. It is hypothesized that methylation induced inactivation of tumor suppressor genes contributes to malignant transformation.[35] Methods for detection of DNA methylation include methylation Southern blotting, methylation specific polymerase chain reaction (PCR), bisulfite genomic sequencing, and restriction landmark genomic screening.[34,36] Most of these methods use methylation sensitive restriction enzymes (*Bsf*UI, *Hpa* II, *Hha* I). These enzymes are unable to cut methylated DNA, and by hybridizing gene specific probes to the digested DNA samples, it is possible to identify variation in DNA fragment size due to methylation compared with normal unmethylated controls.

Analysis of Human SNPs

The exact nucleotide sequence of any particular human gene shows variations from individual to individual even in the absence of any larger scale structural abnormality of the chromosomes. Each gene is encoded in two alleles that are located on each of the chromosome pairs that human cells have. Sometimes, the exact sequence of these two alleles is different due to single nucleotide base pair changes. A single nucleotide change may be categorized as a mutation if it occurs in less than 1% of the population or as a SNP if it occurs in more than 1% of the population. Single nucleotide changes in genes in cancer cells compared with the sequence of the same gene in normal tissues of the patient are also considered as mutations. Missense sequence variation occurs when the change in a nucleotide results in a change in the amino acids sequence of the protein. Nonsense variation occurs when the change results in premature termination of the protein. Frame shift mutation occurs when there is a nucleotide insertion or deletion that leads to a shift in the codon reading frame (a codon describes the three nucleotides that together code for a single amino acid) that yields a completely different new sequence of codons. Sequence changes that lead to no amino acid alterations, due to the redundancy in the genetic code, are called silent variations. Naturally, any of the nonsilent polymorphisms or mutations may cause functional alterations in the biologic properties of the proteins.[37] Single nucleotide changes in genes that are important in drug metabolism and cellular response to therapy can cause functional alterations that may contribute to the interindividual differences in response and toxicity to treatment.

Genome-wide screening for mutations and SNPs is possible by using oligonucleotide DNA arrays.[38,39] Oligonucleotide arrays contain 20–60 base pair long DNA sequences that are bound to the surface of glass slides or silicon chips and serve as molecular probes. Tens of thousands of distinct probe sets may be printed to the surface of a single chip or a small set of slides.[40] Individual probe sets represent perfect match sequences and a large number of known mutations and polymorphisms of any gene of interest. Multiplex quantitative real-time reverse transcription (RT)-PCR (TaqMan™) is also used to detect known SNPs, and screening for hundreds of polymorphisms can be performed rapidly.[41] Detection of polymorphisms is based on the design of two TaqMan PCR probes, specific for the wild-type allele and the polymorphic allele. Each of the two probes is labeled with a different fluorescent tag. The binding efficiency of the wild-type TaqMan probe to the mutant allele and vice versa is low because of the mismatch within the TaqMan probe and the target sequence. All the above high throughput methods for SNP analysis are limited by their inability to screen for previously unknown sequence variations. These unknown sequence variations are currently identified by direct sequencing, which is expensive and labor intensive.

Greater than one million genetic markers known as SNPs have recently become available for genotyping and phenotyping studies.[42–45] Novel genotyping strategies are emerging on a regular basis using a variety of techniques, including oligonucleotide genomic arrays,[42,43] gel and flow cytometry, classic gene sequencing, mass spectroscopy,[44] and microarray or gene chips,[45] all designed to increase the rate of data generation and analysis. In 1999, 10 companies and the Wellcome Trust formed a consortium to discover clinically relevant SNPs. To date, over 3 M SNPs have been deposited into the SNP Consortium's public database (http://www.snp.cshl.org). A high resolution SNP map will expedite the identification of genes for complex diseases such as asthma, diabetes mellitus, atherosclerosis, and psychiatric disorders. In oncology, the SNP technology has focused on detecting the predisposition for cancer, predicting of toxic responses to drugs, and selecting the best individual and combinations of anticancer drugs.

For cancer predisposition testing, SNP genotyping and gene sequencing have uncovered a variety of familiar cancer predisposition syndromes based on single and multiple gene variants.[45–48] Genotyping has been introduced widely for the detection of familial cancers of the breast, ovary, colon, melanoma, and multiple endocrine neoplasia.[47,49–51]

Prediction of drug toxicity is another focus. One of the earliest applications of SNP genotyping in cancer management was the discovery of variations in drug metabolism associated with genomic variations in drug metabolizing enzymes such as the cytochrome system.[52–54] Other detoxification pathways associated with SNP-based variations in drug metabolism have been described,[53–55] but compared with the cytochrome gene system, these markers have not yet achieved significant clinical utility. The potential clinical value of the pharmacogenetics approach for predicting drug toxicity will be uncovered as more candidate polymorphisms can be discovered.

The application of genotyping strategies to predict anticancer drug efficacy has recently emerged in a variety of clinical settings.[46,56–59] Genotype resistance testing of human immunodeficiency virus isolates has demonstrable clinical utility and provides a way to assist therapeutic decision making in patients whose human immunodeficiency

Table 28.3 Comparison of DNA Microarrays

	cDNA	Oligonucleotide	GeneChip™
Substrate	Spotted onto nylon membrane or glass slide	30-75mer oligonucleotides	25mer oligonucleotides
Probes per array	100 to 5000	10,000 to 22,000	500,000
Probes per transcript	1	1 to 6	1–20
Commercial sources	Clontech, Agilent (Incyte), Stanford	Agilent, Amersham, Mergen, Clontech, etc.	Affymetrix

virus RNA levels are rising.[60] In colorectal cancer, pretreatment genotyping on peripheral blood samples is currently being tested to select therapy based on the prediction of resistance associated with certain genetic polymorphisms.[59]

Transcriptional Profiling

In contrast to the previously described approaches that examined variations in DNA structure, transcriptional profiling surveys the expression of genes by quantifying mRNA. Three high throughput methods are commonly used to simultaneously measure the expression of a large number of genes in clinical specimens (**Table 28.3**): two distinct DNA microarray platforms, complementary DNA (cDNA) and oligonucleotide arrays, and multiplex quantitative real-time RT-PCR.[61,62] cDNA arrays contain DNA sequences of variable lengths (600–2000 nucleotides) where each cDNA fragment corresponds to a particular mRNA and functions as a molecular probe. Several thousand cDNA probes can be spotted onto the surface of a single glass slide or a special nylon membrane to make a high density DNA array. Each "spot" contains distinct probes and tests for a particular mRNA species. Using sequence verified cDNA clones, robotic printing, either fluorescent or radioisotope-based signal detection, and usually either glass slide or nylon membrane hybridization surfaces (**FIG. 28.1**), this technique has generated a wealth of new information in subtyping leukemia/lymphoma, solid tumor classification, pharmacogenomics, and drug and biomarker target discovery. Oligonucleotide arrays differ from cDNA arrays in the length of the probes they contain. Oligoarrays are populated by small probes of uniform length (40–70 nucleotides). There are subtle differences between the performances of these different types of arrays. In general, cDNA arrays are considered to be more sensitive for low abundance mRNA but are also more prone to cross-hybridization, which may lead to less specific signal.

The third method of gene expression profiling includes real-time RT-PCR. Real-time PCR is based on the quantitation of a fluorescent reporter generated during the PCR process.[63] This signal increases in direct proportion to the amount of PCR product that in turn reflects the abundance of the mRNA target that is measured. Real-time PCR offers a wider dynamic range of detection compared with DNA microarrays. It can also be optimized to detect mRNA fragments recovered from formaldehyde-fixed paraffin-embedded tissues. Current real-time RT-PCR systems can measure several dozen to a few hundred genes simultaneously, which is substantially less than the comprehensive profiling that DNA microarrays can provide. Transcriptional profiling carried out by hybridizing labeled probes derived from cancer specimens to oligo- or cDNAs arrays gives a picture of the transcriptome, the complete mRNA content of the specimen (**Table 28.3**). Probes may be directly hybridized to the array or may be hybridized in the presence of a reference sample that is labeled different from the disease sample. Reference samples can include cell lines, normal tissues, and cancer precursor lesions. **Table 28.4** lists the major types of arrays and their associated fluorescent, radioisotope, or mass spectroscopy reporter systems in current use or in development for oncology molecular diagnostics. Each of these techniques has in common the ability to generate hundreds of thousands of data points requiring sophisticated and complex information systems necessary for accurate and useful data analysis.

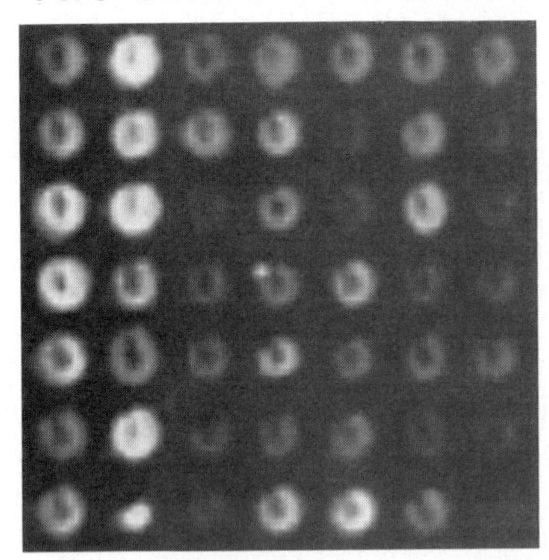

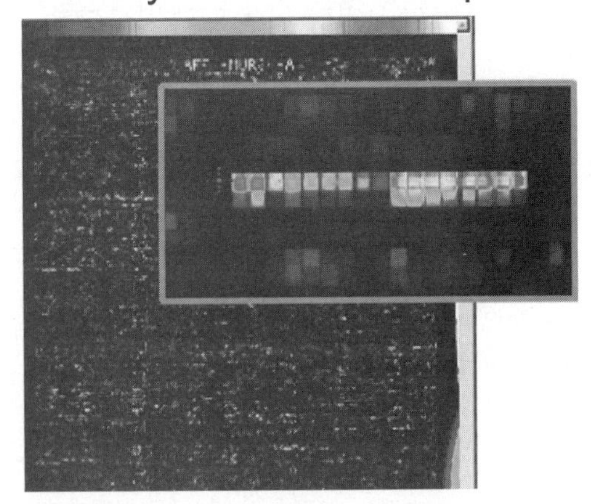

(A)

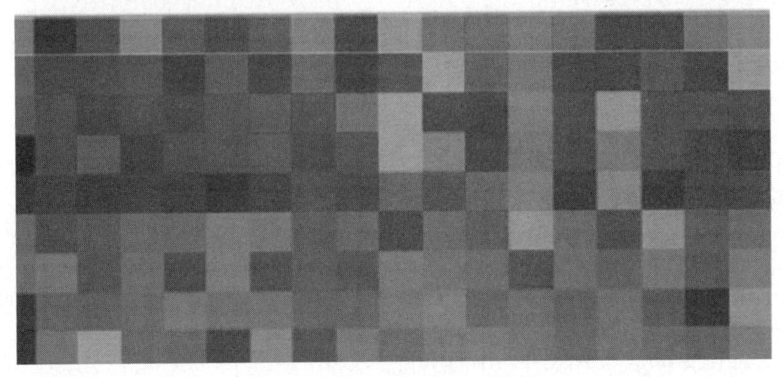

(B)

FIGURE 28.1 (A) Types of microarrays. The two most commonly used types of genomic microarray chips. Left: A dual-label microarray uses competitive hybridization in which cDNA derived from two separate mRNA sources is hybridized to the same microarray with one cDNA labeled with green (Cy3) dye and the other cDNA labeled with red (Cy5) dye. The cDNA or oligonucleotides representing different genes are immobilized on the glass slide with two corresponding measurements, one for each dye, generated from each individual spot. An appropriate scanning or imaging reader that provides the fluorescence intensity for Cy3 and Cy5 for each target can reveal the resulting hybridization pattern. Right: The Affymetrix GeneChip™ System (Affymetrix Corp., Santa Clara, CA) features a single color oligonucleotide microarray where each gene is represented by multiple probe pairs. The sequences of the probes are specific for a particular gene and the probe pairing design facilitates identification and subtraction of non-specific hybridization and background signals. Hybridization signals are developed using avidin-biotin reaction and a fluorochrome. (B) Heat map representation of gene expression data. The image demonstrates a standard data expression form for gene expression results. Genes expression at levels greater than median are displayed and deepening red color and genes expressed below baseline are shown in deepening green color. **See Plate 52 for color image.**

Table 28.4 Types of Microarrays in Oncology Molecular Diagnostics

Feature	cDNA Microarray (spotted arrays)	Oligonucleotide Microarrays	CGH–BAC/YAC Microarrays	Tissue Microarrays
Target type and technology	Double stranded cDNA Made from tumor mRNA >100 nucleotides PCR products from cDNA clones Competitive hybridization	16–24 bp oligonucleotides in situ synthesis or printing Up to 240,000 oligos/chip Probe redundancy decreases false positives	DNA fragment arrays Bacterial/yeast artificial chromosomes (BAC/YAC) 100 bp to 100 kb 5000 spots 1-Mb resolution	Paraffin section arrays Frozen section arrays
Types of array surfaces	Glass slides Silicon Nylon membranes	Glass slides Silicon Nylon membranes Gels Beads		Glass slides
Signal detection systems	Fluorophores Radioisotopes Typically dual color	Fluorophores Typically single color	Fluorophores	Fluorophores Chromagens Radioisotopes
Technical limitations	Printing inaccuracies False hybridizations	Greater density but limited by cross-hybridization errors	Limited resolution	Tumor heterogeneity Formalin-based protein and RNA degradation
Automation capability	Robot printing Laser scanning	Robot printing Laser scanning	Fluorescence microscopy Digital imaging	Image analysis
Clinical applications	Leukemia/lymphoma classification Solid tumor classification Carcinoma of unknown primary site Drug selection	Gene resequencing SNP discovery Tumor classification SNP prediction of toxicity Drug selection	High resolution cytogenetics for gains and losses Tumor classification New primary tumor vs. metastasis	Currently used for research purposes only
Drug discovery and development applications	Drug target selection Pharmacogenomics discovery Biomarker discovery	Drug target selection Pharmacogenomics discovery Biomarker discovery	Drug target discovery	Drug target validation Pharmacogenomic target validation Biomarker validation

Statistical Analysis of High Dimensional Transcriptional Profiling Data

Analysis of gene expression data is complex. It starts with log transformation, normalization, and scaling of the raw data so that cross-specimen comparisons can be performed.[64,65] PCR based data are commonly normalized to housekeeping genes where DNA microarray data are more often normalized to the median intensity of all gene expression on the array. When normalization is performed using housekeeping genes (i.e., β-actin, GAPDH), it is assumed that these ubiquitously expressed genes are also similarly expressed in different tumor specimens. This assumption unfortunately is not always correct because substantial variation in the expression of housekeeping genes have been observed across different specimens.[64,65] Normalization to the median assumes that when a large number of genes are considered across a large number of specimens, most genes should be expressed to a similar extent; therefore, the ratio of median gene expression of all genes should approach 1 across the specimens.

A frequent goal of pharmacogenomic studies that use gene expression profiling is to identify genes that are differentially expressed between responders and nonresponders. This can be accomplished by using various methods, including Student's two sample t-test for normally distributed gene expression data or by nonparametric tests for not normally distributed data.[66,67] Significance levels are often adjusted to account for the consequences of multiple testing to limit false discovery rates (FDRs). Controlling the FDR, which is the proportion of genes identified as significantly different when in fact they are not, is critical. Several methods have been developed for controlling FDR at specified levels.[68–70] One commonly used method proceeds by treating the distribution of *P* values computed for all genes as a mixture of two distributions, one from truly differentially expressed genes and one from the nondifferentially expressed genes.[71] Computing the parameters for these distributions and specifying a *P* value threshold for declaring significance allows for FDR estimation. One can select a subjectively acceptable FDR threshold and find a *P* value threshold that will lead to that prespecified FDR. Results of this type of analysis are often presented as a list of genes ranked by significance level. In the context of pharmacogenomics, these gene lists may provide clues toward the biology of drug resistance and sensitivity. Differentially expressed genes identified through gene expression profiling may also be used as markers of response or resistance. However, even the best single gene predictor, selected from several thousands of genes, may not be as accurate as the combination of multiple genes.

The next level of analysis of gene expression focuses on the identification of multiple genes that in combination predict response more accurately than any single gene could. The data are usually first explored with unsupervised class discovery techniques that use information from all genes to reveal molecular classes of breast cancer.[72,73] Hierarchical clustering, multidimensional scaling, and neural networks are often used for this purpose. However, these natural molecular subgroups may or may not correspond to groups of cases that respond or not to a particular therapy. A potentially more useful way to identify patients who respond to therapy is to take cases with known outcome and train machine learning algorithms on these cases to identify gene expression patterns that separate the responders from the nonresponders. This second approach is referred to as supervised learning.[74,75]

A commonly used strategy to develop and assess a learning based predictor is to randomly divide cases with known outcome into two groups: the *training set,* used to develop the predictor and the second group, the *validation set,* that is used to estimate its predictive accuracy in independent cases.[76] The first step in developing the predictor is often the identification of differentially expressed genes between the two outcome groups in the training set. Next, these informative genes are combined with machine learning algorithms such as support vector machine, K-nearest neighbor, weighted voting methods, or others to generate an outcome predictor test. The definition of a gene expression profile based predictive test therefore includes a particular class prediction algorithm and the unique set of genes that are fed into the predictive algorithm to generate a binary "yes or no" classification system.

It is customary to estimate the predictive accuracy of classifiers in leave-out cross-validation and random label permutation tests. Cross-validation within the training set may be used to determine the best classifier algorithm, the optimal number of genes to be included in the classifier, and to estimate the expected predictive accuracy. During this process a subset of cases is omitted from the training set, discriminating genes are selected again, and a class predictor is constructed from the remainder of the cases. This process is repeated many times with different sets of cases held out on each occasion, and the predictive accuracies are averaged to yield estimated misclassification error rates for each classifier. Typically, the classifier requiring the least number of genes to achieve an acceptable rate of correct classification is considered to be the best and applied to the entire training data to generate the final predictor. However, chance could produce a certain number of correct predictions.

To assess whether the classification error rates of a predictor differ significantly from what chance alone could produce, random label permutation test is performed.[77] During this process, the outcome label of each case (i.e., responder vs. nonresponder) is randomly changed. Hundreds of such randomly permutated data sets are created, and the classifier generated from the true data is applied to the randomly permutated data. The observed error rate for the true data set is compared with the distribution of the error rates observed with the randomly permutated data sets to calculate a permutation *P* value. A model that results in significant correct prediction is taken for further validation on independent data to determine its true predictive accuracy.

Clinical Applications of Pharmacogenomics in Breast Cancer

All the molecular analytic methods described above have been successfully applied to human tissue specimens in the context of cancer. CGH and transcriptional profiling have been performed on human breast cancer specimens. However, currently all these techniques are only used within the context of clinical research, and none is sufficiently confirmed and validated to be used in routine clinical care.

Unsupervised clustering of breast cancer specimens consistently separated tumors into ER-positive and ER-negative clusters, indicating that ER status has a major impact on the transcriptome of breast cancer (FIG. 28.2).[78–80]

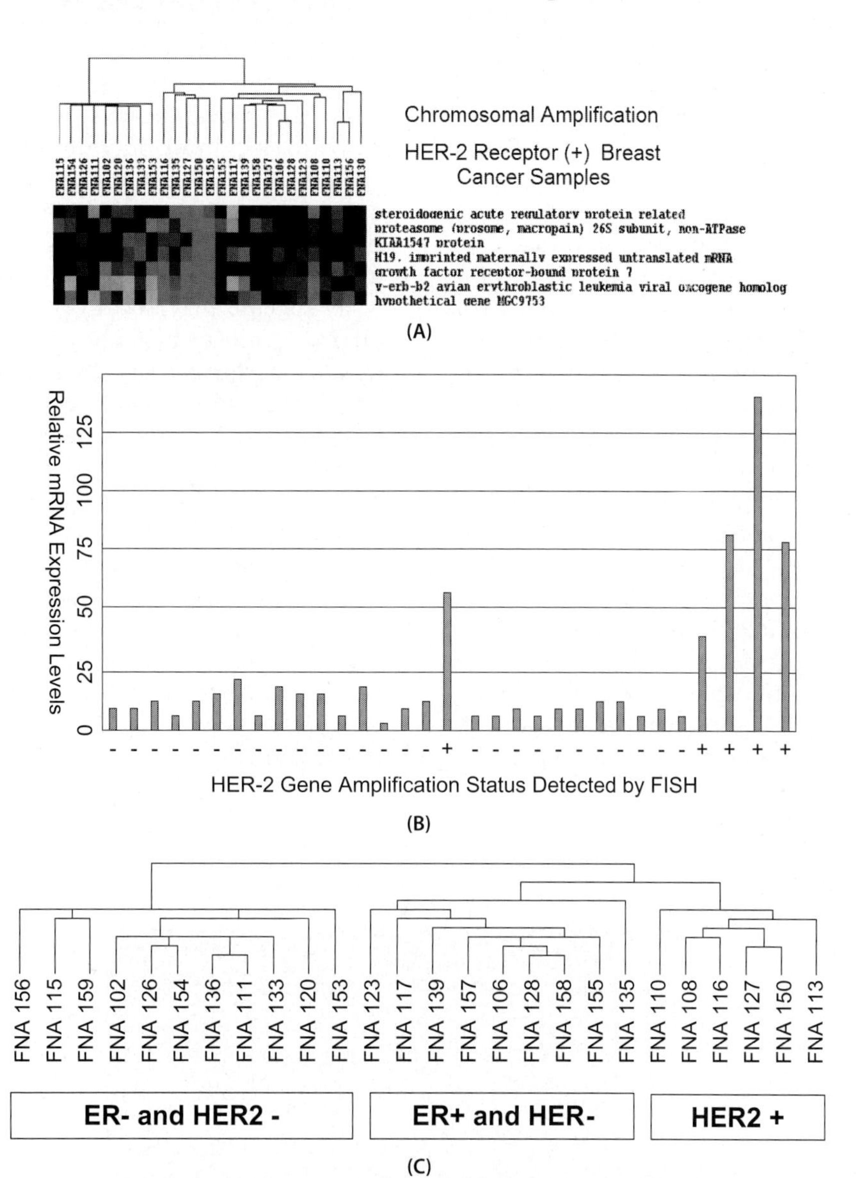

FIGURE 28.2 (A) Hierarchical clustering of gene expression results. Cluster analysis of a series of primary stages I to III invasive breast cancers using five informative genes that separate HER-2/*neu* amplified cancers from HER-2 normal cases. (B) Comparison of HER-2/*neu* mRNA expression levels measured by Affymetrix gene chip transcriptional profiling on 32 fine-needle aspiration samples and HER-2/*neu* gene amplification status determined by fluorescent in situ hybridization on the corresponding core biopsy specimens. Note the close correlation of gene expression with amplification status. (C) Hierarchical clustering after transcriptional profiling of 26 infiltrating breast cancer specimens sampled by fine-needle aspiration using a 62 gene model (25 genes for determining ER status and 37 genes for determining HER-2/*neu* status). Note the clustering of the HER-2/*neu*–positive cases and their separation from the HER-2/*neu*–negative cases. The HER-2/*neu*–negative ER-positive and HER-2/*neu*–negative ER-negative tumors can also be separated by this technique.

Analysis of gene expression profiles can also distinguish sporadic breast cancers from *BRCA* mutant cases.[81] Transcriptional profiles also revealed previously unrecognized molecular subgroups within existing histological categories of cancer. For example, breast cancer may be grouped into luminal or basal epithelial cell types.[78] What makes these attempts clinically relevant is that several investigators also reported distinct clinical outcomes for the different molecular subgroups.[82,83] Gene expression profiles have been shown to predict survival of patients with node-negative breast cancer, and this gene expression profile-based outcome prediction appears to outperform existing prognostic classification schemes.[84,85] An RT-PCR-based multigene prognostic score has also been developed, and its prognostic value was evaluated independently on RNA from paraffin-embedded tissues. The prognostic score could separate patients with higher risk for relapse after 5 years of adjuvant tamoxifen hormonal therapy from those with lesser risk.[86,87]

It was also reported that gene expression signatures can be identified that separate patients with good clinical response to preoperative single agent docetaxel from those with minimal or no response.[88] A multigene predictor of complete pathologic response to preoperative sequential paclitaxel and 5-fluorouracil, doxorubicin, cyclophosphamide (FAC) chemotherapy in breast cancer has also been described (FIG. 28.3).[89] If these observations on predicting response to chemotherapy are confirmed, these represent successful application of microarrays technology to a clinical challenge that could not be addressed adequately with previously existing technologies.

Clinical Applications of Pharmacogenetics in Breast and Other Cancers

Pharmacogenetics refers to the science of examining the *host* (patient's germline) genome with high throughput techniques to identify the causes of interindividual differences in response to treatment and toxicity from a given drug. Commonly applied tools of pharmacogenetics are SNP analysis, direct sequencing, and functional assessment of enzymatic activity of specific enzymes involved in drug metabolism. As described above, SNP analysis is usually performed on DNA obtained from peripheral blood nucleated cells. It is estimated that around 1.5 million SNPs are distributed throughout the human genome and that at least 60,000 of these are located within protein coding regions within exons of genes.[90]

There are several examples of genetic polymorphisms in enzymes that metabolize chemotherapy drugs and lead to increased toxicity from standard doses of these drugs. In the context of breast cancer treatment, the most relevant is dihydropyrimidine dehydrogenase (DPD) deficiency that can be

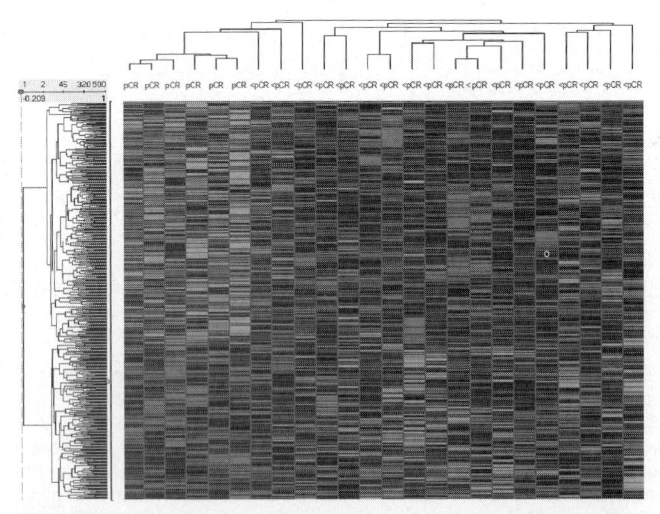

FIGURE 28.3 Distinct gene expression profiles are associated with complete pathologic response to neoadjuvant paclitaxel/FAC chemotherapy. Supervised clustering of the top 500 SNR markers associated with pathologic response from 24 training samples is shown. All the pathologic complete responders (pCR) cluster together and are separated from the samples that had incomplete pathologic response (<pCR). **See Plate 53 for color image.**

manifested with severe potentially life-threatening toxicity after administration of 5-fluorouracil.[91] DPD is responsible for the conversion of 5-fluorouracil into an inactive metabolite, 5-fluorodihydrouracil. As DPD activity is reduced, less 5-fluorouracil is catabolyzed and, not surprisingly, DPD deficiency clinically presents as 5-fluorouracil overdose with severe mucositis, diarrhea, and neutropenia after a standard dose of therapy. A large number of mutations in the DPD gene has been described that can lead to reduced enzymatic activity.[92] The gene is very large, and currently the most expeditious screening tool for DPD deficiency remains detection of elevated levels of uracil in the plasma or urine or radioenzymatic assay of DPD activity in cytosol extract.[93,94]

Other known polymorphisms that influence drug toxicity, but somewhat less relevant for breast cancer treatment, include polymorphisms of thiopurine methyltransferase (TPMT) and uridine diphosphate glucuronosyltransferase (UGT1A1). TPMT catalyzes methylation of azathiopirine, mercaptopurine, and thioguanine, commonly used in the treatment of leukemia, into less toxic compounds.[95,96] TPMT enzymatic activity is highly variable among individuals and is an autosomal inherited trait. Three alleles, TPMT*2, TPMT*3A, and TPMT*3C, account for 95% of intermediate or low enzyme activity. TPMT*2 is caused by a single nucleotide transversion (G238C) leading to an amino acid substitution Ala>Pro in the protein. TPMT*3A contains two distinct transversions (G460A, A719G) with corresponding amino acid changes, and TPMT*3C contains 1 transversion (A719G).[97–99] Patients with low TPMT activity due to these single nucleotide changes are predisposed to severe hematopoietic toxicity from conventional doses of these drugs.[100]

An important determinant of irinotecan-induced diarrhea is UGT1A1 activity. The diarrhea is caused by an active metabolite of irinotecan, SN-38. This toxic metabolite is inactivated through glucuronidation by UGT1A1. The promoter region of the UGT1A1 gene normally contains five to six cytosine-adenine dinucleotide repeats. Some individuals have more dinucleotide repeats, and this leads to decreased transcriptional activity of the gene that results in less UGT1A1 protein. Patients with seven repeats have a fourfold relative risk of experiencing severe SN-38 induced diarrhea compared with patients with a lesser number of repeats.[101] There are other examples of clinically relevant genetic polymorphisms. The activation of codeine to morphine depends on the catalytic function of CYP2D6, a P-450 enzyme. Mutations in CYP2D6 can lead to impaired conversion of codeine to morphine, and individuals with this form of CYP2D6 polymorphism do not receive pain relief from codeine-containing medications.[102,103] These elegant observations fuel great interest in systematic exploration of SNP in the human genome.[104] The expectation is that the discovery and functional annotation of polymorphisms in critical drug metabolizing enzymes and other molecules that play a role in drug elimination or response will yield clinically useful tests to adjust and individualize the doses of existing drugs to attain maximum benefit with the least toxicity.

Future Clinical Potential and Challenges

There are several unique potential strengths of gene expression profiling based outcome predictions. Several distinct predictors could be run on the same single comprehensive gene expression data to generate prognostic prediction, screen for *BRCA* mutations, and identify best chemotherapy regimen. Different sets of genes could be used for different prediction purposes. The expression of clinically important single gene markers such as ER and HER-2 status can also be reliably measured by several mRNA based high throughput technologies. Comprehensive gene expression profiling could evolve into an "all-in-one" diagnostic test (FIG. 28.4). An important and unique feature of learning algorithm-based multigene predictors is that the predictive accuracy of these tests is expected to improve as the classifiers are trained on increasingly larger numbers of cases.[105]

The potential of gene expression profiling as a novel tool to improve on existing diagnostic, prognostic, and predictive tests is very exciting. However, several challenges need to be addressed before routine clinical application could be considered. Some of these relate to the technology itself; others concern clinical utility. A very important challenge is standardization. Currently, multiple microarray platforms exist that use distinct sets of genes and different hybridization and signal detection methods. Some arrays contain cDNAs of variable lengths; others contain small oligonucleotide

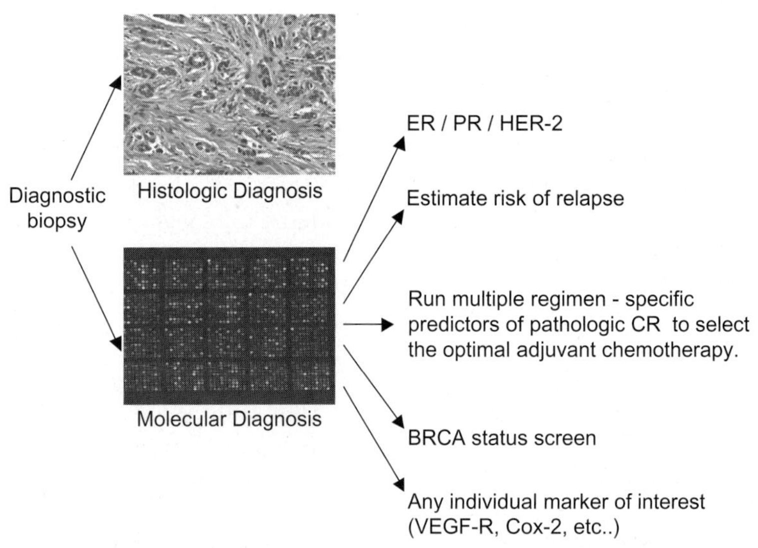

FIGURE 28.4 A potential future for breast cancer diagnostics: "all-in-one" molecular testing with gene expression profiling.

sequences. The same gene may be represented by different sequences in different arrays. Investigators who use competitive hybridization between fluorescein-labeled clinical samples and a control sample invariably use different controls from laboratory to laboratory. Not surprisingly, gene marker sets generated by one group often differ significantly from marker sets generated for the same purpose by others, and cross-platform application of predictors leads to substantial decrease in predictive accuracy.[106,107] The type of tissue sampling method also influences the profiling results. Transcriptional profiles are a composite of mRNA contributed by all tissue components of the sample. Fine-needle aspirations, core needle biopsies, or surgically resected specimens each yield somewhat different transcriptional profiles from the same tumor.[108] However, there is no single best profiling platform or tissue sampling method. This underscores the importance of collaborative efforts to create large uniform gene expression databases across various laboratories using standard operating procedures and a common platform if this technology is to be used for clinical diagnostic purposes.

Various DNA microarrays and other pharmacogenomic techniques have demonstrated in several "proof-of-principle" experiments that these methods can predict clinical outcomes, including outcomes that cannot currently be predicted with other methods. However, the clinical development of multigene predictors is a process similar to the development of new therapeutic agents. The feasibility and reproducibility of the method needs to be optimized, the accuracy of the test has to be estimated, and finally the clinical utility needs to be proved. These steps could only be accomplished in a series of marker discovery and validation studies analogous to therapeutic phase I, II, and III clinical trials. In this context, almost all published studies represent early phase I/II pharmacogenomic marker discovery studies.

There is also substantial interest in developing new drugs parallel with genomic response predictor tests. One could foresee extended phase II therapeutic trials of novel agents where microarray profiling, in the absence of single gene predictors of response, is used to identify patients who benefit from the new drug. Definitive phase III treatment trials could then be conducted with relatively modest sample sizes but with potentially large screened populations to demonstrate benefit for these prospectively selected patients. This approach to drug development currently requires substantial investment in pioneering an unprecedented novel diagnostic method with uncertain regulatory implications. The US Food and Drug Administration has issued preliminary guidelines on new drug applications that are accompanied with genomic data to facilitate the development of pharmacogenomic tests in 2003 (http://www.fda.gov/cder/guidance/5900dft.pdf). If the past history of laboratory technologies may serve as precedent, then gene expression-based pharmacogenomic diagnostic products have a bright future. All major technical innovations in laboratory methods, including the discovery of monoclonal antibodies, flow cytometry, PCR, and in situ hybridization, have found their niche in diagnostic medicine. It is likely that it is only a matter of time before gene expression profiling and other pharmacogenomic and pharmacogenetic methods also become routine clinical tests to fill an unmet diagnostic niche.

References

1. WHO Handbook for Reporting Results of Cancer Treatment. Geneva: World Health Organization Offset Publication No. 48, 1979.
2. Therasse P, Arbuck SG, Eisenhauer EA, et al. New guidelines to evaluate the response to treatment in solid tumors. European Organization for Research and Treatment of Cancer, National Cancer Institute of the United States, National Cancer Institute of Canada. J Natl Cancer Inst 2000;92:205–216.
3. Skipper HE. Historic milestones in cancer biology: a few that are important to cancer treatment. Semin Oncol 1979;6:506–514.
4. Goldie JH. Drug resistance. In: Perry MC, ed. The Chemotherapy Source Book. Baltimore: Williams & Wilkins, 1996:63–69.
5. Norton L, Simon R. Tumor size sensitivity to therapy, and design of treatment schedules. Cancer Treat Rep 1977;61:1307–1317.
6. Pusztai L, Siddik ZH, Mills GB, et al. Physiologic and pathologic drug resistance in ovarian carcinoma: a hypothesis based on a clonal progression model. Acta Oncol 1998;37:629–640.
7. Kennedy S, Merino MJ, Swain SM, et al. The effects of hormonal and chemotherapy on tumoral and nonneoplastic breast tissue. Hum Pathol 1990;21: 192–198.
8. Burns TF, el-Deiry WS. Cell death signaling in malignancy. Cancer Treat Res 2003;115:319–343.
9. Reed JC. Dysregulation of apoptosis in cancer. J Clin Oncol 1999;17:2941–2953.
10. Hortobagyi GN, Hayes D, Pusztai L. Integrating newer science into breast cancer prognosis and treatment. Molecular predictors and profiles. Proc Am Soc Clin Oncol 2002;22:191–202.
11. Ross JS, Linette GP, Stec J, et al. Breast cancer biomarkers and molecular medicine. Exp Rev Mol Diagn 2003;3:573–585.
12. Hayes DF, Trock B, Harris AL. Assessing the clinical impact of prognostic factors: When is "statistically significant" clinically useful? Breast Cancer Res Treat 1998;52:305–319.
13. Bast RC Jr, Ravdin P, Hayes D, et al. 2000 Update of recommendations for the use of tumor markers in breast and colorectal cancer: clinical practice guidelines of the American Society of Clinical Oncology. J Clin Oncol 2001;19:1865–1878.
14. Ross JS, Fletcher JA, Bloom KP, et al. HER-2/*neu* testing in breast cancer. Am J Clin Pathol 2004; 120(suppl 1):S50–S71.
15. Abu-Farsakh H, Sneige N, Atkinson EN, et al. Pathologic predictors of tumor response to preoperative chemotherapy in locally advanced breast carcinoma. Breast J 1995;1:96–101.
16. Colleoni M, Minchella I, Mazzarol G, et al. Response to primary chemotherapy in breast cancer patients with tumors not expressing estrogen and progesterone receptors. Ann Oncol 2000;11:1057–1059.
17. Chang J, Powles TJ, Allred DC, et al. Biologic markers as predictors of clinical outcome from systemic therapy for primary operable breast cancer. J Clin Oncol 1999;17:3058–3063.
18. Ellis P, Smith I, Ashley S, et al. Clinical prognostic and predictive factors for primary chemotherapy in operable breast cancer. J Clin Oncol 1998;16: 107–114.
19. Rozan S, Vincent-Salomon A, Zafrani B, et al. No significant predictive value of c-erbB-2 or p53 expression regarding sensitivity to primary chemotherapy or radiotherapy in breast cancer. Int J Cancer 1998;79:27–33.
20. Petit T, Borel C, Ghnassia JP, et al. Chemotherapy response of breast cancer depends on HER-2 status and anthracycline dose intensity in the neoadjuvant setting. Clin Cancer Res 2001;7:1577–1581.
21. Zhang F, Yang Y, Smith T, et al. Correlation between HER2 expression and response to neoadjuvant FAC chemotherapy in breast cancer. Cancer 2003;97: 1758–1765.
22. Chevillard S, Lebeau J, Pouillart P, et al. Biological and clinical significance of concurrent p53 gene alterations, MDR1 gene expression, and S-phase fraction analyses in breast cancer patients treated with primary chemotherapy or radiotherapy. Clin Cancer Res 1997;3:2471–2478.
23. Archer SG, Eliopoulos A, Spandidos D, et al. Expression of ras p21, p53 and c-erbB-2 in advanced breast cancer and response to first line hormonal therapy. Br J Cancer 1995;72:1259–1266.
24. Makris A, Powles TJ, Dowsett M, et al. p53 Protein overexpression and chemosensitivity in breast cancer. Lancet 1995;345:1181–1182.
25. Linn SC, Pinedo HM, van Ark-Otte J, et al. Expression of drug resistance proteins in breast cancer, in relation to chemotherapy. Int J Cancer 1997;71: 787–795.
26. Formenti SC, Dunnington G, Uzieli B, et al. Original p53 status predicts for pathological response in locally advanced breast cancer patients treated preoperatively with continuous infusion 5-fluorouracil and radiation therapy. Int J Radiat Oncol Biol Phys 1997;39:1059–1068.
27. Kandioler-Eckersberger D, Ludwig C, Rudas M, et al. TP53 mutation and p53 overexpression for prediction of response to neoadjuvant treatment in breast cancer patients. Clin Cancer Res 2000;6: 50–56.
28. Verrelle P, Meissonnier F, Fonck Y, et al. Clinical relevance of immunohistochemical detection of multidrug resistant P-glycoprotein in breast carcinoma. J Natl Cancer Inst 1991;83:111–116.
29. Ro J, Sahin A, Ro JY, et al. Immunohistochemical analysis of P-glycoprotein expression correlated with chemotherapy resistance in locally advanced breast cancer. Hum Pathol 1990;21:787–791.
30. Chevillard S, Pouillart P, Beldjord C, et al. Sequential assessment of multidrug resistance phenotype

and measurement of S-phase fraction as predictive markers of breast cancer response to neoadjuvant chemotherapy. Cancer 1996;77:292–300.

31. Hanahan D, Weinberg RA. The hallmark of cancer. Cell 2000;100:57–70.
32. Jeuken JW, Sprenger SH, Wesseling P. Comparative genomic hybridization: practical guidelines. Diagn Mol Pathol 2002;11:193–203.
33. Pollack JR, Perou CM, Alizadeh AA, et al. Genome-wide analysis of DNA copy-number changes using cDNA microarrays. Nat Genet 1999;23:41–46.
34. Robertson KD, Jones PA. DNA methylation: past, present and future directions. Carcinogenesis 2000; 21:461–467.
35. Huang THM, Perry MR, Laux DE. Methylation profiling of CpG islands in human breast cancer cells. Hum Mol Genet 1999;8:459–470.
36. Yan PS, Chen C-M, Shi H, et al. Applications of CpG island microarrays for high-throughput analysis of DNA methylation. J Nutr 2002;132:2430S–2434S.
37. Ng PC, Henikoff S. Accounting for human polymorphisms predicted to affect protein function. Genome Res 2002;12:436–446.
38. Wang DG, Fan J-B, Siao C-J, et al. Large-scale identification, mapping and genotyping of single-nucleotide polymorphisms in the human genome. Science 1998;280:1077–1082.
39. Kwok P-Y. High-throughput genotyping assay approaches. Pharmacogenomics 2000;1:95–100.
40. Wang E, Adams S. Discrimination of genetic polymorphism using DNA chips. ASHI Q 2001;24:45–48.
41. Huber M, Mundlein A, Dornstauder E, et al. Accessing single nucleotide polymorphisms in genomic DNA by direct multiplex polymerase chain reaction amplification on oligonucleotide microarrays. Anal Biochem 2002;303:25–33.
42. Taylor JG, Choi EH, Foster CB, et al. Using genetic variation to study human disease. Trends Mol Med 2001;7:507–512.
43. Carlson CS, Newman TL, Nickerson DA. SNPing in the human genome. Curr Opin Chem Biol 2001;5: 78–85.
44. Griffin TJ, Smith LM. Single-nucleotide polymorphism analysis by MALDI TOF mass spectrometry. Trends Biotechnol 2000;18:77–84.
45. Kallioniemi OP. Biochip technologies in cancer research. Ann Med 2001;33:142–147.
46. Relling MV, Dervieux T. Pharmacogenetics and cancer therapy. Nat Rev Cancer 2001;1:99–108.
47. Borg A. Molecular and pathological characterization of inherited breast cancer. Semin Cancer Biol 2001;11:375–385.
48. Renegar G, Rieser P, Manasco P. Pharmacogenetics: the Rx perspective. Exp Rev Mol Diagn 2001;1: 255–263.
49. Jass JR. Familial colorectal cancer: pathology and molecular characteristics. Lancet Oncol 2000;1: 220–226.
50. Weber W, Estoppey J, Stoll H. Familial cancer diagnosis. Anticancer Res 2001;21:3631–3635.
51. Calzone KA, Biesecker BB. Genetic testing for cancer predisposition. Cancer Nurs 2002;25:15–25.
52. Ingelman-Sundberg M. Genetic susceptibility to adverse effects of drugs and environmental toxicants. The role of the CYP family of enzymes. Mutat Res 2001;482:11–19.
53. Ingelman-Sundberg M. Pharmacogenetics: an opportunity for a safer more efficient pharmacotherapy. J Intern Med 2001;250:186–200.
54. Steimer W, Potter JM. Pharmacogenetic screening and therapeutic drugs. Clin Chim Acta 2002;315: 137–155.
55. Roses AD. Pharmacogenetics. Hum Mol Genet 2001;10:2261–2267.
56. Diasio RB, Johnson MR. The role of pharmacogenetics and pharmacogenomics in cancer chemotherapy with 5-fluorouracil. Pharmacology 2000;61: 199–203.
57. Linder MW, Valdes R Jr. Genetic mechanisms for variability in drug response and toxicity. J Anal Toxicol 2001;25:405–413.
58. Danesi R, De Braud F, Fogli S, et al. Pharmacogenetic determinants of anti-cancer drug activity and toxicity. Trends Pharmacol Sci 2001;22:420–426.
59. Pullarkat ST, Stoehlmacher J, Ghaderi V, et al. Thymidylate synthase gene polymorphism determines response and toxicity of 5-FU chemotherapy. Pharmacogenomics J 2001;1:65–70.
60. Durant J, Clevenbergh P, Halfon P, et al. Drug-resistance genotyping in HIV-1 therapy: the VIRADAPT randomised controlled trial. Lancet 1999;353:2195–2199.
61. Lockhart DJ, Dong H, Byrne MC, et al. Expression monitoring of hybridization to high-density oligonucleotide arrays. Nature Biotech 1996;14:1675–1680.
62. Brown PO, Botstein D. Exploring the new world of the genome with DNA microarrays. Nat Genet 1999;21:33–37.
63. Walker SJ, Worst TJ, Vrana KE. Semiquantitative real-time PCR for analysis of mRNA levels. Methods Mol Med 2003;79:211–227.
64. Yang YH, Dudoit S, Luu P, et al. Normalization of cDNA microarray data: a robust composite method addressing single and multiple slide systematic variation. Nucleic Acids Res 2002;30:e15.
65. Quackenbush J. Microarray data normalization and transformation. Nat Genet 2002;32s:496–501.
66. Cui X, Churchill GA. Statistical tests for differential expression in cDNA microarray experiments. Genome Biol 2003;4:210.
67. Hatfield GW, Hung S, Baldi P. Differential analysis of DNA microarray gene expression data. Molecular Microbiology 2003;47:871–877.
68. Benjamini Y, Hochberg Y. Controlling the false discovery rate: a practical and powerful approach to multiple testing. J Royal Stat Soc (B) 1995;57: 289–300.
69. Wittes J, Friedman HP. Searching for evidence of altered gene expression: a comment on statistical analysis of microarray data. J Natl Cancer Inst 1999;91:400–401.
70. Baggerly KA, Coombes KR, Hess KR, et al. Identifying differentially expressed genes in cDNA microarray experiments. J Comput Biol 2001;8:639–659.
71. Pounds S, Morris SW. Estimating the occurrence of false positive and false negatives in microarray studies by approximating and partitioning the empirical distribution of p-values. Bioinformatics 2003;19: 1236–1242.
72. Golub TR, Slonim DK. Tamayo P, et al. Molecular classification of cancer: class discovery and class prediction by gene expression monitoring. Science 1999;286:531–536.

73. Eisen MB, Spellman PT, Brown PO, et al. Cluster analysis and display of genome-wide expression patterns. Proc Natl Acad Sci USA 1998;95:14863–14868.
74. Radmacher MD, McShane LM, Simon R. A paradigm for class prediction using gene expression profiles. J Comput Biol 2002;9:505–511.
75. Ringner M, Peterson C, Khan J. Analyzing array data using supervised methods. Pharmacogenomics 2002;3:403–415.
76. Simon R, Radmacher MD, Dobbin K. Design of studies using DNA microarrays. Genet Epidemiol 2002; 23:21–36.
77. Pan W. On the use of permutation in and the performance of a class of nonparametric methods to detect differential gene expression. Bioinformatics 2003;19: 1333–1340.
78. Perou CM, Serlie T, Eisen MB, et al. Molecular portraits of human breast tumours. Nature 2000;406: 747–752.
79. Pusztai L, Ayers M, Stec J, et al. Gene expression profiles obtained from single passage fine needle aspirations (FNA) of breast cancer reliably identify prognostic/predictive markers such as estrogen (ER) and HER-2 receptor status and reveal large scale molecular differences between ER-negative and ER-positive tumors. Clin Cancer Res 2003;9: 2406–2415.
80. Gruvberger S, Ringner M, Chen Y, et al. Estrogen receptor status in breast cancer is associated with remarkably distinct gene expression patterns. Cancer Res 2001;61:5979–5984.
81. Hedenfalk I, Duggan D, Chen Y, et al. Gene-expression profiles in hereditary breast cancer. N Engl J Med 2001;344:539–548.
82. Sorlie T, Tibshirani R, Parker J, et al. Repeated observation of breast tumor subtypes in independent gene expression data sets. Proc Natl Acad Sci USA 2003;100:8418–8423.
83. Sotiriou C, Neo SY, McShane LM, et al. Breast cancer classification and prognosis based on gene expression profiles from a population-based study. Proc Natl Acad Sci USA 2003;100:10393–10398.
84. van't Veer LJ, Dai H, van de Vijver, et al. Gene expression profiling predicts clinical outcome of breast cancer. Nature 2001;415:530–536.
85. van de Vijver MJ, Yudong DH, van't Veer LJ, et al. A gene-expression signature as a predictor of survival in breast cancer. N Engl J Med 2002;347: 1999–2009.
86. Esteban J, Baker J, Cronin M, et al. Tumor gene expression and prognosis in breast cancer: multi-gene RT-PCR assay of paraffin-embedded tissue [abstract]. Proc Am Soc Clin Onc 2003;23:3416.
87. Paik S, Shak S, Tang G, et al. Multi-gene RT-PCR assay for predicting recurrence in node negative breast cancer patients – NSABP studies B-20 and B-14. Breast Cancer Res Treat 2003;82(suppl 1): S10(abst. 16).
88. Chang JC, Wooten EC, Tsimelzon A, et al. Gene expression profiling for the prediction of therapeutic response to docetaxel in patients with breast cancer. Lancet 2003;362:362–369.
89. Ayers M, Symmans WF, Stec J, et al. Gene expression profiling of fine needle aspirations of breast cancer identifies genes associated with complete pathologic response to neoadjuvant taxol/FAC chemotherapy. J Clin Oncol 2004;22:2284–2289.
90. Sachidanandam R, Weissman D, Schmidt SC, et al. A map of human genome sequence variation containing 1.42 million single nucleotide polymorphisms. Nature 2001;409:928–933.
91. Harris BE, Carpenter JT, Diasio RB. Severe 5-fluorouracil toxicity secondary to dihydropyrimidine dehydrogenase deficiency: a potentially more common pharmacogenetic syndrome. Cancer 1991;68: 499–501.
92. Mattison LK, Johnson MR, Diasio RB. A comparative analysis of translated dihydropyrimidine dehydrogenase (DPD) cDNA; conservation of functional domains and relevance to genetic polymorphisms. Pharmacogenetics 2002;12:133–144.
93. Yokota H, Fernandez-Salguero P, Furuya H, et al. cDNA cloning and chromosome mapping of human dihydropyrimidine dehydrogenase, an enzyme associated with 5-fluorouracil toxicity and congenital thymine uraciluria. J Biol Chem 1994;269:23192–23196.
94. Johnson MR, Yan J, Shao L, et al. Semi-automated radioassay for determination of dihydropyrimidine dehydrogenase (DPD) activity. Screening cancer patients for DPD deficiency, a condition associated with 5-fluorouracil toxicity. J Chromatogr B Biomed Sci Appl 1997;696:183–191.
95. McLeod HL, Krynetski E, Relling MV, et al. Genetic polymorphism of thiopurine methyltransferase and its clinical relevance for childhood acute lymphoblastic leukemia. Leukemia 2000;14:567–572.
96. Weinshilboum RM, Sladek SL. Mercaptopurine pharmacogenetics: monogenic inheritance of erythrocyte thiopurine methyltransferase activity. Am J Hum Genet 1980;32:651–662.
97. Krynetski EY, Schuetz JD, Galpin AJ, et al. A single point mutation leading to loss of catalytic activity in human thiopurine S-methyltransferase. Proc Natl Acad Sci USA 1995;92:949–953.
98. Tai HL, Krynetski EY, Yates CR, et al. Thiopurine S-methyltransferase deficiency: two nucleotide transitions define the most prevalent mutant allele associated with loss of catalytic activity in Caucasians. Am J Hum Genet 1996;58:694–702.
99. Loennechen T, Yates CR, Fessing MY, et al. Isolation of a human thiopurine S-methyltransferase (TPMT) complementary DNA with a single nucleotide transition A719G (TPMT*3C) and its association with loss of TPMT protein and catalytic activity in humans. Clin Pharmacol Ther 1998;64: 46–51.
100. Lennard L, Van Loon JA, Weinshilboum RM. Pharmacogenetics of acute azathioprine toxicity: relationship to thiopurine methyltransferase genetic polymorphism. Clin Pharmacol Ther 1989;46: 149–154.
101. Ando Y, Saka H, Ando M, et al. Polymorphisms of UDP-glucuronosyltransferase gene and irinotecan toxicity: a pharmacogenetic analysis. Cancer Res 2000;60:6921–6926.
102. Sindrup SH, Brosen K. The pharmacogenetics of codeine hypoalgesia. Pharmacogenetics 1995;5: 335–346.
103. Poulsen L, Arendt-Nielsen L, Brosen K, et al. The hypoalgesic effect of tramadol in relation to CYP2D6. Clin Pharmacol Ther 1996;60:636–644.

104. The International SNP Map Working Group. A map of human genome sequence variation containing 1.42 million single nucleotide polymorphisms. Nature 2001;409:928–933.

105. Mukherjee S, Tamayo P, Rogers S, et al. Estimating dataset size requirements for classifying DNA microarray data. J Comput Biol 2003;10:119–142.

106. Kuo WP, Jenssen T-K, Butte AJ, et al. Analysis of matched mRNA measurements from two different microarray technologies. Bioinformatics 2002;18: 405–412.

107. Yuen T, Wurmbach E, Pfeffer RL, et al. Accuracy and calibration of commercial oligonucleotide and custom of cDNA microarrays. Nucleic Acids Res 2002; 30:1–9.

108. Symmans WF, Ayers M, Clark EA, et al. Total RNA yield and microarray gene expression profiles from fine needle aspiration and core needle biopsy samples of breast cancer. Cancer 2003;97: 2960–2971.

CHAPTER 29

Mammography and Molecular Imaging

Hernan I. Vargas, MD, M. Perla Vargas, MD, and Iraj Khalkhali, MD

Departments of Surgery, Pathology and Radiology, Harbor UCLA Medical Center, Torrance, California

Mammography

Mammography is the only method with proven benefit for screening and early detection of breast cancer. The relative risk of death from breast cancer is decreased by 25–30% with systematic use of screening mammography.[1–8] However, some controversy exists regarding these large population based studies.[9]

The false-negative rate of diagnostic mammography ranges from 10% to 20%.[10–12] Rosenberg et al.[10] studied the impact of ethnicity, age, breast density, and hormone use on the sensitivity of screening mammography. They reported that sensitivity varied significantly in relation to these attributes. Sensitivity was 54% in women younger than 40 years, 77% in women aged 40 to 49 years, 78% in women aged 50 to 64 years, and 81% in women older than 64 years. Sensitivity was 68% for dense breasts and 85% for nondense breasts. Sensitivity was most markedly reduced with the combination of dense breasts and the use of estrogen replacement therapy. This, in turn, may lead to significant numbers of missed cancers.[10]

Low specificity and positive predictive value are other drawbacks to mammography. These limitations are illustrated by data reporting an estimated 75–80% of all breast biopsies performed for mammographic findings resulting in benign outcomes.[10–14] Clearly, better methods are needed to characterize the benign or malignant nature of breast lesions.

Introduction to Molecular Imaging

During most of last century, radical mastectomy was the treatment of choice for patients with breast cancer. This was based on Halsted's concept of sequential, orderly, and progressive dissemination of disease. However, investigators from the National Surgical Adjuvant Breast Involved Project questioned the validity of this concept.[15] Within the last four decades, results from multiple, randomized, prospective studies have led to a new understanding of breast cancer biology. Systemic progression of breast cancer may occur early and independent of lymph node involvement. As our understanding of the natural history and biology of breast cancer evolved, the treatment of this disease has also changed.

Current treatment of breast cancer is based primarily on anatomic principles of the extent of disease, a belief supported by the current staging system that reinforces anatomic and histologic tumor characteristics of the primary tumor, regional lymphatics, and identification of systemic metastases. Imaging is an essential component in the multidisciplinary approach to the care of patients with cancer. Conventional imaging modalities, such as x-rays, computed tomography (CT), magnetic resonance imaging (MRI), and sonography, identify anatomic abnormalities in tissue. In general, these architectural changes are not apparent until a tumor has reached a threshold of approximately 1 cm (10^6 cells).

The knowledge base acquired in the post-genomic era permits targeting of cellular and molecular events. A better understanding of the impact of gene/protein function is the next crucial step in the development of therapy for cancer. New insights into the interactions between tumor and surrounding stroma, tumor angiogenesis and the role of receptors, and regulatory proteins in tumor growth have provided a better understanding of the development and progression of cancer and have influenced the creation of new therapeutic approaches to cancer. Thus, noninvasive visualization of specific molecular targets, pathways, and physiologic effects in vivo is highly desirable. In the last few decades, a transition from imaging purely anatomic details to imaging on a molecular level has taken place. Molecular imaging has been defined as the noninvasive, quantitative, and repetitive imaging of targeted macromolecules and biologic processes.[16,17] This discipline, in the strictest characterization, provides an optical representation and measurement of certain endogenous molecular biologic events. Molecular imaging has its roots in nuclear medicine. Biologic tracers are injected, providing specific imaging of defined targets. Tracer techniques extensively used for in vitro studies in the laboratory allow direct visualization of physiologic and biochemical processes. The use of specific tracers for given biologic processes is remarkable and promises a high degree of sensitivity and specificity.[17,18]

Instrumentation and Techniques in Biologic and Molecular Imaging

In vivo imaging of biologic and molecular events has developed as an extension of methods commonly used in the laboratory, such as immunohistochemistry and in situ hybridization. The field has rapidly evolved and grown in part because of the increasing pressure for translational research and the needs of small animal research. These recent developments are just now finding their way into the clinical arena.

A variety of techniques are available for molecular imaging. In considering these techniques, we must keep in mind the following key points:

1. Spatial resolution, a measure of the detail of the image, is the minimum distance between two points in space detectable by an imaging method.
2. Temporal resolution, a measure of the time interval needed to record an image.
3. Depth, the ability of energy waves to penetrate through tissue.
4. Sensitivity, the ability to detect a molecular probe relative to background.
5. Energy expended for image generation.
6. Availability of injectable/biocompatible molecular probes.

Imaging representation may be planar or tomographic. Advantages of planar imaging are its speed and its ability to generate small concise data sets. Tomographic imaging allows volume representation but requires longer acquisition times and higher energy expenditures. It provides for the display of internal structures and functional information.

Radionuclide Imaging

Nuclear medicine procedures are based on the detection of gamma rays from an energy source, usually in combination with a targeting agent. Such probes include antibodies or ligands to surface receptors, substrates for intracellular enzymes, probes for intracellular receptors, antisense oligonucleotide probes for targeting messenger RNA, and genes transferred into the cell. The energy source is a radioisotope attached to a substrate molecule, antibody, or protein. This is known as a radiopharmaceutical. Scintillation crystals (NaI) in an array detect gamma radiation and convert it to visible light. An advantage of radionuclide imaging techniques is the ability to label a wide range of molecules with an isotope. Two types of imaging based on the radionuclide used are available.

Single photon emission produces lower energy gamma rays, generated during the radioactive decay process. Most frequently used radionuclides for single photon emission are ^{99m}Tc, ^{131}I, ^{123}I, ^{133}Xe, ^{201}Tl, ^{67}Ga, and ^{111}In.

The simplest acquisition protocol is the planar image. With planar imaging, the detector array is stationary over the patient and acquires data only from one angle. The image created with this type of acquisition is similar to an x-ray radiograph. The principal disadvantage of planar imaging is the low contrast generated due to the overlap of activity that occurs when a three-dimensional structure is compressed onto a single image. A lead collimator, drilled with small gaps is used to project an image of the radionuclide distribution onto the scintillation crystal by allowing passage of selected gamma rays from only one particular direction and absorbing all others.

With single photon emission CT (SPECT) (FIG. 29.1), rotation of the gamma camera results in the acquisition of views of the tracer distribution at a variety of angles. Generation of slices and a three-dimensional view of the radiotracer distribution are therefore possible through image reconstruction. The process of collimation in SPECT decreases the detection efficiency by approximately 10^{-4} times the emitted photons. Spatial resolution of 1–2 cm is possible with current SPECT systems.

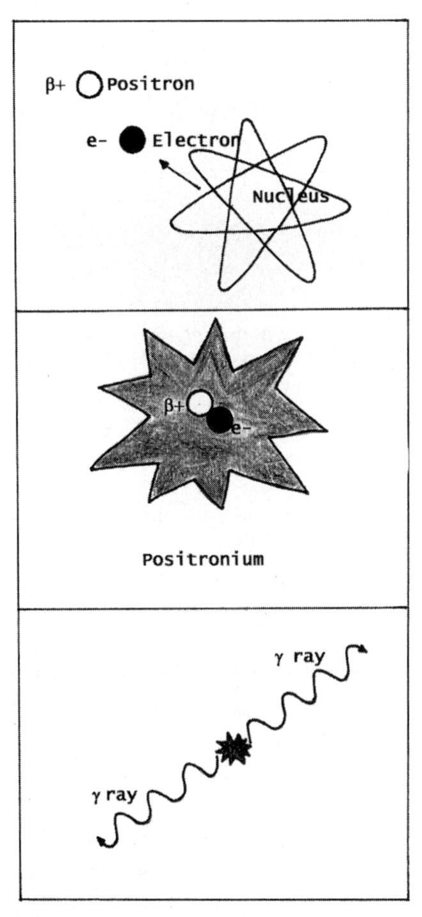

FIGURE 29.1 Single photon emission.

The second type of imaging based on the radionuclide used is positron emission tomography (PET). Certain radioisotopes with excess protons decay by positron emission. In this process, a proton in the nucleus is transformed into a neutron and a positron. This positron scatters through tissue, rapidly losing kinetic energy and making it possible to combine with an electron. With this encounter, an annihilation reaction takes place, releasing energy in the form of two 511 −keV gamma rays (FIG. 29.1). This energy is emitted as two gamma rays 180 degrees apart, facilitating coincidence detection and electronic collimation with higher efficiency. For this reason, PET is approximately 10- to 15-fold more sensitive than SPECT. Most clinical PET scanners have a spatial resolution of 6–8 mm, but newer generation investigational systems have spatial resolutions of 1–2 mm.

Positron-emitting isotopes most commonly used are ^{18}F, ^{15}O, ^{13}N, and ^{11}C and are produced in a cyclotron. These isotopes are unique because of their ability to be linked to biologic probes and can substitute for naturally occurring atoms in organic molecules. Because of the high spatial resolution and sensitivity of detection systems, minute amounts of these isotopes are sufficient to provide imaging and measurements of biologic processes using small quantities of biotracer (approximately 10^{-11} M). A disadvantage of radionuclide modalities is the poor definition of anatomic details. The combination of PET/CT and MRI/PET aids in overcoming this difficulty.

MRI

The principle that magnetic dipoles align themselves when placed into a magnetic field is used in MRI. A radiofrequency coil within the MRI scanner generates a radiofrequency pulse, which alters the alignment of the dipoles. Dipoles in different physicochemical environments have different relaxation times and thus generate different MRI signals. MRI has a high spatial resolution (25–100 μm), providing exquisite anatomic detail. It has the potential of offering anatomic, functional, and biochemical information simultaneously.

The optimal use of MRI for breast cancer has not been adequately defined.[19,20] MRI must be performed with the administration of intravenous contrast.[21,22] Gadolinium, a commonly used MRI contrast agent, is a strong ferromagnetic element that when injected alters the local magnetic field, thus enhancing the MR image. Dynamic imaging with sequential imaging of the breast over several minutes is important for identification of breast tumors, which have a characteristic early uptake and washout phase. This is thought to be caused by the characteristics of tumor vasculature and angiogenesis.[23] Biologic and molecular tracers for detection with MRI have been reported. Gadolinium based avidin-biotin has been used to target cell surface receptors for imaging with MRI.[24]

Nuclear MR spectroscopy provides spectroscopic data from a defined volume in a precisely localized anatomic location (voxel).[25] Studies of tumor extracts and cells have demonstrated higher levels of phosphomonoesther and nucleoside triphosphate.[26] Phosphocholine and total choline content correlate with malignant transformation and progression.

Optical Imaging

Optical imaging techniques are in current laboratory use for in vitro and ex vivo studies.[27]

Optical techniques rely on fluorescence or bioluminescence for imaging. Using fluorescence, the energy from an external source is reemitted at a longer lower energy wavelength. Bioluminescence uses light generated from a chemical reaction without excitation. A significant problem yet to be overcome by optical methods of imaging is tissue penetration using visible or infrared light. Hemoglobin is the primary absorber of visible light, with water and lipids the principal absorbers of infrared light. Tissue penetration is currently limited to 1–2 cm.[17,18] A strategy proposed to overcome this problem is the use of light in the near infrared spectrum. Hemoglobin, water, and lipids have their lowest absorption levels in the near infrared region (650–900 nm). The development of tomography coupled with fluorescent imaging is more promising and may improve on tissue penetration of 5–6 cm.

Preclinical and Clinical Experience

Scintimammography with ^{99m}Tc Sestamibi

The first report of radiopharmaceutical concentration in breast cancer was in 1946, when increased concentration of ^{32}P was documented in an ulcerated carcinoma of the breast.[28] The most commonly used radiopharmaceutical for breast imaging is ^{99m}Tc sestamibi. In 1987, Muller et al.[29] reported ^{99m}Tc sestamibi was taken up by tumor. The exact mechanism of cellular uptake by cancer cells is still under investigation.[30] ^{99m}Tc sestamibi is driven into the cell by a negative transmembrane potential and is sequestered within the cytoplasm and mitochondria.[31] ^{99m}Tc sestamibi is a substrate of the transmembrane p-glycoprotein protein[32] that is present in the cells overexpressing the *MDR1* (multiple drug resistance) gene. P-glycoprotein (protein PG170) acts as a protective pump by extruding out of the tumor cells a wide range of molecules, including ^{99m}Tc sestamibi and a number of chemotherapy agents.

The accuracy of diagnosis of breast cancer with ^{99m}Tc sestamibi ranges from 67% to 90% in single institution clinical trials. The results of a multicenter clinical trial[33] in more than 40 North American institutions and involving 673 female patients have been reported. Using double blinded readers and having 286 patients with palpable abnormalities and 387 patients with nonpalpable abnormalities, the overall diagnostic sensitivity and specificity were 80% and 81%, respectively. Focal uptake of ^{99m}Tc sestamibi is depicted in FIGURE 29.2. The sensitivity for lesions less than 10 mm is rather low (72%). An advantage of ^{99m}Tc sestamibi scintimammography in contrast to mammography is that it is not affected by the density of the breast tissue. Caveats of scintimammography with ^{99m}Tc sestamibi are the increased uptake of this radiotracer in a number of benign breast diseases such as acute mastitis, juvenile fibroadenomas, and proliferative breast disease.

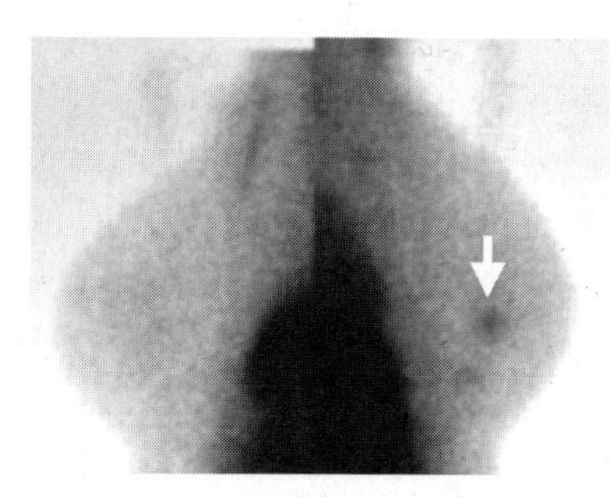

FIGURE 29.2 Breast cancer imaging with ^{99m}Tc sestamibi scintimammography. Arrowhead points to focal area of uptake of a biopsy proven infiltrating ductal carcinoma.

Staging of the regional lymphatics with ^{99m}Tc sestamibi has been studied. Taillefer et al.[34] reported on 100 consecutive patients with breast cancer. The sensitivity in detecting metastatic axillary lymph nodes was 79.2% and the specificity was 84.6% (FIG. 29.2). The positive and negative predictive values were 82.6% and 81.5%, respectively. Similar results were reported by Tolmos et al.[35]

Neoadjuvant chemotherapy is used in patients with locally advanced breast cancer with two objectives: to improve the rate of breast conservation surgery and to assess chemosensitivity of the tumor in vivo. It takes several weeks or months to determine whether a tumor response is occurring using conventional clinical assessment and imaging.

Scintimammography can provide functional imaging for evaluating the susceptibility of a breast cancer to chemotherapeutic agents. Ciarmielo et al.[36] studied 39 patients with stage III disease who underwent ^{99m}Tc sestamibi scintimammography before neoadjuvant therapy. Poor response to primary chemotherapy was seen in 88% of patients with rapid tumor clearance of ^{99m}Tc sestamibi, whereas 64% of cases with complete response had prolonged retention of ^{99m}Tc sestamibi. Rapid tumor clearance of ^{99m}Tc sestamibi predicts lack of tumor response to neoadjuvant chemotherapy, likely related to the presence of a multidrug resistance phenotype and function of p-glycoprotein.

2-Deoxy-2,2-[F-18] Fluoro-D-Glucose PET

The seminal work of Warburg and Dickens[37] established that tumors have increased glucose metabolism in comparison with normal tissues. The glucose analogue FDG is taken up at an increased rate by tumor cells. Once FDG is phosphorylated in the cell, FDG-6-phosphate cannot be metabolized and remains trapped intracellularly.

Clinicians and investigators have studied PET using ^{18}F-FDG as a method for delineating the anatomic extent of breast cancer.[38] Several clinical trials studying patients with breast lesions reported overall sensitivity and specificity ranging from 80% to 100% and 75% to 100%, respectively, in the diagnosis of breast cancer.[39–45] However, several caveats require mention. Avril et al.[45] reported that the sensitivity for detecting tumors smaller than 1 cm was 57% compared with 91% for tumors larger than 1 cm. Similarly, there is a decreased sensitivity in the detection of well-differentiated cancers.[45]

Accurate staging of breast cancer patients is important for optimal management. In a study assessing the extent of breast cancer to the regional lymphatics, Greco et al.[46] reported a sensitivity of 94% and a specificity of 86%. Similarly, Schirrmeister et al.[47] reported a sensitivity of 79% and a specificity of 92%. However, the sensitivity in detection of axillary metastasis in patients with small T1 tumors decreased significantly to 33%. A recently published multicenter trial of 360 patients studied ^{18}F-FDG PET for axillary staging.[48] The authors concluded that ^{18}F-FDG PET has moderate accuracy for detecting axillary metastasis but often fails to detect axillae with small and few nodal metastases. Although highly predictive for nodal tumor involvement when multiple intense foci of tracer uptake are identified, ^{18}F-FDG PET is not routinely recommended for axillary staging of patients with newly diagnosed breast cancer.[48]

^{18}F-FDG PET appears to be more sensitive than conventional imaging in the detection of systemic disease (FIG. 29.3).[49,50] Cook et al.[51] showed in a study of 23 breast cancer patients that ^{18}F-FDG PET detected more lesions than bone scintigraphy. This is significant for patients with purely lytic lesions where the bone scan has lower sensitivity.

The role of PET in diagnosis and staging of breast cancer is evolving. PET has a theoretical advantage over mammography, ultrasound, and MRI. In a single whole body examination, PET may provide not only accurate characterization of primary tumors and the staging of axillary and mediastinal lymph node involvement but also the detection of distant metastases.

Significant progress has been achieved with the development of combined molecular and anatomic imaging.[52,53] The simultaneous acquisition of PET and CT is ideal for fusion of biologic and anatomic images (FIG. 29.4). The combination of PET and CT is advantageous in the planning of radiation therapy and the monitoring of surgical and medical therapy. Wahl et al.[54] documented tumor regression by PET/CT in patients receiving chemotherapy. They reported decreases in standard substrate uptake between the 8th and 21st day after the first cycle of chemotherapy. No such decline was observed in nonresponders. Similar observations have been reported by Bassa et al.[55] This early experience suggests that monitoring cancer treatment with ^{18}F-FDG PET is useful in the prediction of responses early in treatment.

Mortimer et al.[56] reported on a series of 40 patients who had ^{18}F-FDG PET after 7 to 10 days of beginning tamoxifen treatment. ^{18}F-FDG uptake increased in patients who responded to tamoxifen treatment. This was confirmed by Dehdashti et al.[57] This is referred to as a flare phenomenon. Nonresponders with estrogen receptor–positive breast cancer did not exhibit this phenomenon.

Cell Surface Receptor Imaging with Antibodies

The rationale for imaging with antibodies to cell surface antigens is the very high specificity for cell surface antigens. Approximately two thirds of infiltrating breast carcinomas express carcinoembryogenic antigen (CEA) on immunohistochemistry.[58] Immunoglobulin against CEA labeled with ^{131}I was reported as a means of imaging tumors that express CEA.[59] Subsequently, murine monoclonal antibodies against CEA were labeled with ^{99m}Tc.[60] In clinical studies, tumors expressing

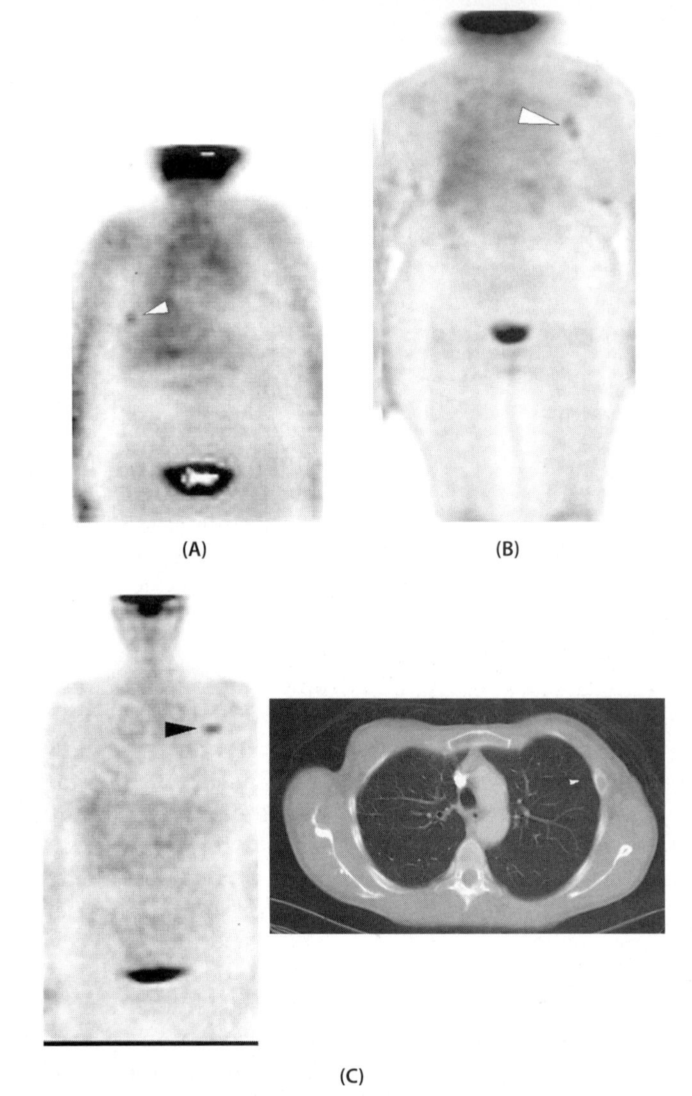

FIGURE 29.3 PET. (A) A 78-year-old woman with localized breast cancer in the right breast (white arrowhead). (B) A 60-year-old woman status/post modified radical mastectomy with axillary recurrence (white arrowhead). (C) A 58-year-old woman with systemic disease. Rib metastases seen in FDG scan (black arrowhead) and CT (white arrowhead).

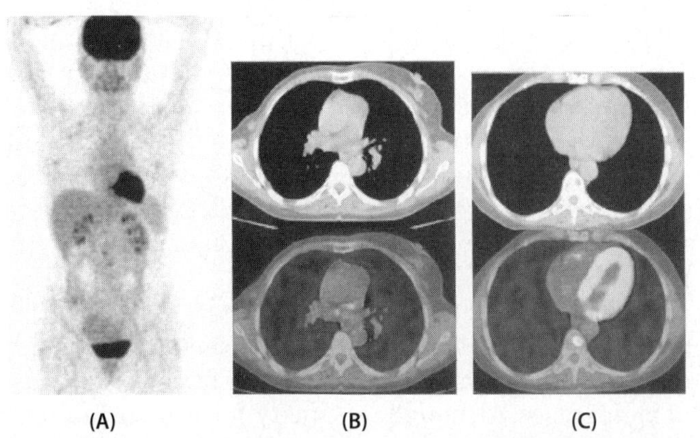

FIGURE 29.4 A 54-year-old woman with history of right sided breast cancer treated with operation and chemotherapy. Detection of suspected recurrent disease in the right axillary region with FDG uptake in an enlarged lymph node seen on maximal impulse projection scan (A) but more clearly on PET CT (B). Note the relatively low activity frequently seen in breast cancer. In addition, an unsuspected osteolytic metastasis in a vertebral body was identified (C), clearly depicted in the PET/CT images as a hypodense lesion with high FDG uptake. (From Wahl RL. PET and PET/CT imaging in breast cancer. In: von Schulthess GK, ed. Clinical Molecular Anatomic Imaging. Philadelphia: Lippincott, Williams and Wilkins, 2003:307.) **See Plate 54 for color image.**

CEA could be visualized with radionuclide imaging. Lind et al.[61] reported a sensitivity of 69% and specificity of 90% in the detection of cancer in patients with elevated CEA levels. This approach has also been used for the detection of axillary metastasis and skeletal metastasis in breast cancer.[62] Goldenberg et al.[63] reported a large prospective trial of a CEA antibody Fab' fragment labeled directly with ^{99m}Tc. The sensitivity of this test was 61% and the specificity was 91%. Tumor size significantly affected the detection rates. Two thirds of T1 lesions were missed by this study. Immunoscintigraphic detection with this labeled antibody is therefore related to tumor size.

Bakir et al.[64] reported on the successful imaging of HER-2 receptors with ^{124}I radiolabeled rat monoclonal antibody in a human breast cancer xenograft model. In 1992, De Santes et al.[65] documented a study of 10 cases that received the antibody against HER-2/neu labeled with ^{99m}Tc. Two cases revealed imaging of the primary tumor or axillary lymph nodes. A study of trastuzumab labeled with ^{111}In revealed that patients who suffered from trastuzumab-related cardiotoxicity showed uptake in the myocardium.[66] This suggests the presence of a cross-reactive antigen in the heart. In vitro and in vivo studies of a biotinylated HER-2/neu monoclonal antibody and avidin GdDTPA conjugate have been used for detection of HER-2/neu receptor expression.[24] However, a major drawback of this technique is the high likelihood of provoking a profound immune response and thus has not been tested in humans. Immunofluorescent labeling of HER2/neu receptors with semiconductor quantum dots has been recently reported in an in vitro assay. Quantum dots are fluorescent probes and may be important for cellular imaging and biomedical applications in the future.[67] These studies are proof of the principle that radiolabeled antibodies can be used for imaging of primary and metastatic breast cancer. Imaging of specific antibodies may provide therapeutic targets, such as the HER-2/neu receptor or estrogen receptors.

A caveat of antibody-based imaging is the low diffusion coefficients of large labeled molecules to the target site. Antibodies can also be internalized after their interaction with the cognate antigen, making longitudinal studies dependent on the antigen turnover rate. Given these limitations, investigators have been exploring the use of normal human cell surface receptors as imaging targets using smaller molecules.

Receptor Ligand Imaging

Receptor ligand imaging may be advantageous because of the specific binding affinity for certain tumors. This also offers the potential for use of therapeutic radionuclides for labeling of specific receptor ligands. Peptide targeted imaging may lead to peptide targeted therapy. Investigators have attempted to target both endogenously expressed receptors and those on engineered tumor cells that overexpress a receptor of choice.

Hull et al.[68] reported that in vitro measurements of estrogen receptors by immunohistochemistry do not discriminate between functional and nonfunctional receptors and only provide an estimate of hormone sensitivity. Imaging of estrogen receptors is possible using 16α-[F-18]-fluororestradiol-17β.[69,70] Measurement of the level of blockade of estrogen receptors with 16α-[F-18]-fluororestradiol-17β may provide an estimate of the likelihood of response to hormone intervention[71] and theoretically facilitate tailoring of the dose for an individual patient. Vijaykumar et al.[72] similarly reported on the use of ^{18}F-labeled compounds for imaging of the progesterone receptor.

Dolan et al.[73] conducted a meta-analysis of 14 studies using somatostatin analogues for the treatment of stage IV breast cancer. Octreotide therapy was associated with a short-lived tumor response of over 40% with few side effects. Imaging of somatostatin receptors has therefore been studied; van Eijck et al.[74] reported positive results in 75% of cases using PET.

The sigma-2 receptor is believed to be overexpressed in cancer. The proliferative status of breast cancer is a marker for prognosis. Thus, sigma-2 receptor expression correlates with the proliferative state of tumor cells. Mach et al.[75] reported on the high affinity for both sigma-1 and sigma-2 receptors with the radiotracer [$^{(18)}$F]*N*-(4′-fluorobenzyl)-4-(3-bromophenyl)acetamide, [$^{(18)}$F]1 in nude mice bearing tumor xenografts of the mouse mammary adenocarcinoma.

The radiolabeled DNA precursors [$^{(124)}$I]IUdR and [$^{(11)}$C] thymidine have also been used to measure the proliferative status of breast tumors with PET and SPECT. Shields et al.[76] used ^{18}F-fluorothymidine to image cell replication and proliferation of tumors in vivo. This tracer is retained in proliferating tissues through the enzyme thymidine kinase 1.

Imaging Gene Expression In Vivo

Conventional study of gene expression or protein function is conducted ex vivo in tissue samples or cell culture under artificial conditions. A goal of noninvasive imaging modalities is to monitor gene expression and its effect on the cellular environment in vivo. PET plays an important role in the postgenome era and will allow the imaging and measurement of biologically directed diagnosis and monitored therapeutic interventions at the molecular level.

In direct imaging, PET may be used to image the expression of endogenous genes using short single strands of ^{18}F labeled oligodeoxynucleotides. Small oligonucleotide sequences complementary to a segment of messenger RNA (mRNA) or DNA could potentially target any specific mRNA or DNA sequence. These techniques represent an extrapolation of in situ hybridization for use in live biologic systems: in vivo hybridization. In this context, imaging of specific mRNAs with radiolabeled antisense oligonucleotides produces direct images of specific molecular–genetic events.[77,78] Limitations of these techniques are the very low number of target molecules, limited tracer delivery and transfer across cellular and physiologic barriers, degradation by RNAase, and comparatively high background activity and low specificity of localization (low target/background ratios).

Indirect imaging (reporter gene imaging) is another way of imaging gene expression. Measuring the expression of successful gene transduction in target tissue or specific organs is essential during animal experiments of gene transfer and clinical trials of gene therapy in human subjects. Conventional methods used tissue and blood samples to measure the expression of the enzyme β-galactosidase. Tjuvajev et al.[79] described an innovative approach for imaging the expression of successful gene transduction. An in vivo combination of a marker gene (herpes simplex virus 1 thymidine kinase [HSV1-tk]) and a marker probe (radiolabeled 5-iodo-2′-fluoro-2′deoxy-1β-D-arabinofuranosyluracil) were used in an experimental animal model (**FIG. 29.5**).

Wild-type *HSV1-tk*[80] and a mutant *HSV1-tk* gene, *HSV1-sr*39tk,[81] are the reporter genes

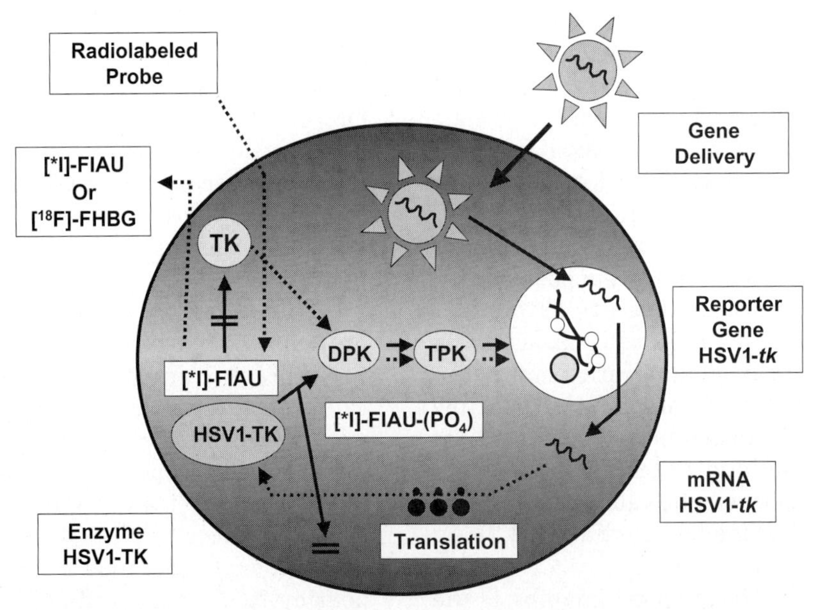

FIGURE 29.5 Indirect gene expression. Schematic for imaging herpes simplex virus 1 thymidine kinase reporter gene (HSV1-*tk*) expression with reporter probes ^{124}I- or ^{131}I-labeled FIAU and ^{18}F-labeled FHBG. The HSV1-*tk* gene complex is transfected into target cells by a vector. Inside the transfected cell, the HSV1-*tk* gene is transcribed to HSV1-*tk* mRNA and then translated on the ribosomes to a protein (enzyme), HSV1-TK. After administration of a radiolabeled probe and its transport into the cell, the probe is phosphorylated by HSV1-TK (gene product). The phosphorylated radiolabeled probe does not readily cross the cell membrane and is trapped within the cell. Thus, the magnitude of probe accumulation in the cell (i.e., the level of radioactivity) reflects the level of HSV1-TK enzyme activity and the level of HSV1-*tk* gene expression.

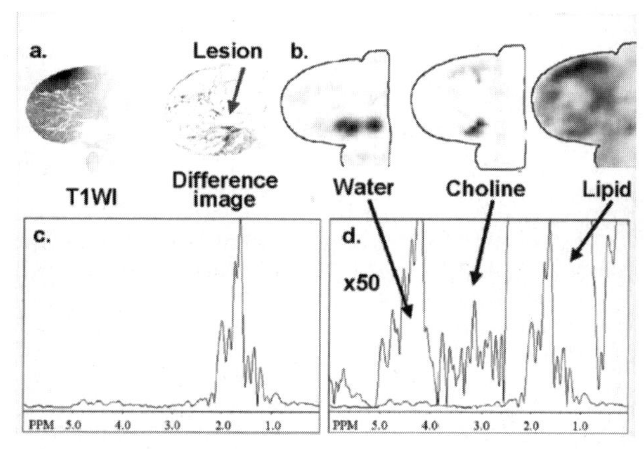

FIGURE 29.6 Spectroscopy: proton MRI and MR spectroscopy of a 41-year-old patient with infiltrating ductal carcinoma of the breast. (a) Postcontrast T1-weighted images of the breast lesion. (b) MR spectroscopy images of water, choline, and lipids. Representative spectrum of elevated choline signal to noise ratio (SNR = 10.6) within the lesion (c) and magnified (×50) region (d) demonstrates a detectable choline signal. (From Ref. 84.)

most commonly used in current molecular imaging studies using radiolabeled probes and PET. The reporter probe is an ^{18}F-labeled analogue of ganciclovir. If gene expression is present, the radiolabeled ganciclovir analogue is phosphorylated and trapped in the cell providing evidence of expression of the *HSV1-tk* gene detected by PET. Other approaches have used a dopamine receptor as the reporter gene and [F-18] fluoroethylspiperone as the reporter probe.

Imaging Apoptosis

Apoptosis (programmed cell death) is a normal event in cell life and is related to many disease processes. Phosphatidylserine, one component of cell membrane phospholipids, is normally confined to the inner leaflet of the plasma membrane but is exposed during apoptosis. A human conjugate of ^{99m}Tc-Annexin V binds to phosphatidylserine. Blankenberg et al.[82] reported on the correlation between ^{99m}Tc-Annexin V and apoptosis based on in vitro and in vivo experiments. This method may be useful in monitoring the efficacy of primary chemotherapy.

Imaging Biochemistry In Vivo: MR Spectroscopy

A study of approximately 200 fine-needle aspiration samples showed a choline peak that distinguishes between benign and malignant tumors with sensitivity and specificity of 95% and 96%, respectively.[83] Jacobs et al.[84] recently reported a pilot study of the successful imaging of the breast with proton MR spectroscopic imaging in 15 patients with breast cancer (FIG. 29.6).

The Future of Molecular Imaging

Molecular imaging is a reality at the preclinical level. By providing information on biology and function, molecular imaging techniques are likely to play an important role in the diagnosis, staging, and treatment of patients with cancer in the 21st century. Future applications of PET must use other tracers in addition to FDG to better characterize tumor biology and more effectively measure response to therapy.[85] This potential refinement in tumor characterization will help predict clinical behavior and tailor therapy to the individual tumor's biology. Novel approaches that combine mammography with PET or use a dedicated small high-resolution PET imaging system might further enhance the diagnostic accuracy of PET for primary breast tumors. The development of new tracers targeting specific biological properties of cancer cells is another important line of current research. Several new tracers are currently being evaluated for their clinical usefulness in cancer patients.

Acknowledgments

We are indebted to Katherine D. Gonzalez and Melissa Burla for critical review of this manuscript.

References

1. Shapiro S, Venet W, Strax P, et al. Ten- to fourteen-year effect of screening on breast cancer mortality. J Natl Cancer Inst 1982;69:349–355.
2. Andersson I, Aspegren K, Janzon L, et al. Mammographic screening and mortality from breast cancer: the Malmö mammographic screening trial. BMJ 1988;297:943–948.
3. Nyström L, Rutqvist LE, Wall S, et al. Breast cancer screening with mammography: overview of Swedish randomised trials. Lancet 1993;341:973–978.
4. Tabár L, Fagerberg CJ, Gad A, et al. Reduction in mortality from breast cancer after mass screening with mammography. Randomised trial from the Breast Cancer Screening Working Group of the Swedish National Board of Health and Welfare. Lancet 1985;1:829–832.
5. Tabàr L, Fagerberg G, Duffy SW, et al. Update of the Swedish two-county program of mammographic screening for breast cancer. Radiol Clin North Am 1992;30:187–210.
6. Tabar L, Fagerberg G, Duffy SW, et al. The Swedish two county trial of mammographic screening for breast cancer: recent results and calculation of benefit. J Epidemiol Community Health 1989;43:107–114.
7. Roberts MM, Alexander FE, Anderson TJ, et al. Edinburgh trial of screening for breast cancer: mortality at seven years. Lancet 1990;335:241–246.
8. Miller AB, To T, Baines CJ, et al. The Canadian National Breast Screening Study-1: breast cancer mortality after 11 to 16 years of follow-up. A randomized screening trial of mammography in women age 40 to 49 years. Ann Intern Med 2002;137(5 Part 1): 305–312.
9. Gotzsche PC, Olsen O. Correspondence. Authors' reply. Lancet 2000;355:752.
10. Rosenberg RD, Hunt WC, Williamson MR, et al. Effects of age, breast density, ethnicity, and estrogen replacement therapy on screening mammographic sensitivity and cancer stage at diagnosis: review of 183,134 screening mammograms in Albuquerque, New Mexico. Radiology 1998;209:511–518.
11. Kerlikowske K, Grady D, Barclay J, et al. Likelihood ratios for modern screening mammography. Risk of breast cancer based on age and mammographic interpretation. JAMA 1996;276:39–43.
12. Porter PL, El-Bastawissi AY, Mandelson MT, et al. Breast tumor characteristics as predictors of mammographic detection: comparison of interval- and screen-detected cancers. J Natl Cancer Inst 1999; 91:2020–2028.
13. Liberman L. Percutaneous image-guided core breast biopsy. Radiol Clin North Am 2002;40:483–500.
14. Elmore JG, Barton MB, Moceri VM, et al. Ten-year risk of false positive screening mammograms and clinical breast examinations. N Engl J Med 1998;338: 1089–1096.
15. Fisher B, Swamidoss S, Lee C. Detection and significance of occult axillary node metastasis in patients with invasive breast cancer. Cancer 1978;42: 2025–2031.
16. Choy G, Choyke P, Libutti SK. Current advances in molecular imaging: noninvasive in vivo bioluminescent and fluorescent optical imaging in cancer research. Mol Imaging 2003;2:303–312.
17. Herschman HR. Molecular imaging: looking at problems, seeing solutions. Gen Med 2003;302:605–608.
18. Massoud TF, Ghambhir SS. Molecular imaging in living subjects: seeing fundamental biological processes in a new light. Genes Dev 2003;17:545–580.
19. Harms SE, Flamig DP. Breast MRI. Clin Imag 2001; 25:227–246.
20. Schnall MD. Application of magnetic resonance imaging to early detection of breast cancer. Breast Cancer Res 2001;3:17–21.
21. Kaiser WA, Zeitler E. MR imaging of the breast: fast imaging sequences with and without Gd-DTPA. Radiology 1989;170:681–686.
22. Heywang SH, Wolf A, Pruss E, et al. MR imaging of the breast with Gd-DTPA: use and limitations. Radiology 1989;171:95–103.
23. Leach MO. Application of magnetic resonance imaging to angiogenesis in breast cancer. Breast Cancer Res 2001;3:22–27.
24. Artemov D, Mori N, Ravi R, et al. Magnetic resonance molecular imaging of the HER-2/neu receptor. Cancer Res 2003;63:2723–2727.
25. Hara T. 11C-choline and 2-deoxy-2-[18F]fluoro-D-glucose in tumor imaging with positron emission tomography. Mol Imaging Biol 2002;4:267–273.
26. Ronen SM, Leach MO. Imaging biochemistry: application to breast cancer. Breast Cancer Res 2001;3: 36–40.
27. Hage R, Galhanone PR, Zangaro RA, et al. Using the laser induced fluorescence spectroscopy in the differentiation between normal and neoplastic human breast tissue. Laser Med Sci 2003;18:171–176.
28. Low-Beer BV, Bell HG, McCorkle HJ. Measurement of radioactive phosphorus in breast tumors in situ: A possible diagnostic procedure. Radiology 1946;47: 492–493.
29. Muller ST, Guth-Tougelids B, Creutzig GH. Imaging of malignant tumors with Tc-99m MIBI SPECT. Eur J Nucl Med 1987;28:562–566.
30. Taillefer R. Clinical applications of 99mTc-sestamibi scintimammography. Semin Breast Dis 2002;5: 128–141.
31. Chiu ML, Kronange JF, Piwnica-Worms D. Effect of mitochondrial and plasma membrane potentials on accumulation of hexakis (2 methxy-isobutyl isonitrile) technetium in cultured mouse fibroblasts. J Nucl Med 1990;31:1646–1653.
32. Piwnica-Worms D, Chiu ML, Budding M, et al. Functional imaging of multidrug resistant P-glycoprotein with an organo technetium complex. Cancer Res 1993;53:977–984.
33. Khalkhali I, Baum JK, Villanueva-Meyer J, et al. (99m)Tc sestamibi breast imaging for the examination of patients with dense and fatty breasts: multicenter study. Radiology 2002;222:149–155.
34. Taillerfer R, Robidoux A, Lambert R, et al. Technetium-99m-sestamibi prone scintimammography to detect primary breast cancer and axillary lymph node involvement. J Nucl Med 1995;36:1758–1765.
35. Tolmos J, Khalkhali I, Vargas H, et al. Detection of axillary lymph node metastasis of breast carcinoma with Technetium-99m sestamibi scintimammography. Am Surg 1997;63:850–853.
36. Ciarmiello A, Del Vecchio S, Silvestro P, et al. Tumor clearance of technetium 99m-sestamibi as a predictor of response to neoadjuvant chemotherapy for locally advanced breast cancer. J Clin Oncol 1998;16: 1677–1683.
37. Warburg OH, Dickens F. The Metabolism of Tumours. New York: Richard R. Smith, 1931.

38. Wahl R, Cody RL, Hutchings GD, et al. Primary and metastatic breast carcinoma: Initial clinical evaluation with PET with radiolabeled glucose antigen 2-[F-18]-fluoro-2-deoxy-D-glucose. Radiology 1991; 179:765–770.
39. Nieweg OE, Kim EE, Wong WH, et al. Positron emission tomography with [fluorine-18]-deoxyglucose in the detection and staging of breast cancer. Cancer 1993;71:3920–3925.
40. Bruce DM, Evans NT, Keys SD, et al. Positron emission tomography: 2-deoxy-2-[18F]-fluoro-D-glucose uptake in locally advanced breast cancers. Eur J Surg Oncol 1995;21:280–283.
41. Tse NY, Koh CK, Kawkins RA, et al. The application of positron emission tomographic imaging with flurodeoxyglucose to the evaluation of breast disease. Ann Surg 1992;216:27–34.
42. Adler LP, Crowe JP, Al-Kaisi NK, et al. Evaluation of breast masses and axillary lymph nodes with [F-18]-2-deoxy-2-flouro-D-glucose PET. Radiology 1993;187: 743–750.
43. Hoh CK, Hawkins RA, Glaspy JA, et al. Cancer detection with whole-body PET using 2-[F-18]-fluoro-2-deoxy-D-glucose. J Comput Assist Tomogr 1993;17: 582–589.
44. Avril N, Dose J, Janicke F, et al. Assessment of axillary lymph node involvement in breast cancer patients with positron emission tomography using radiolabeled 2-[fluorine-18]-fluoro-2-deoxy-D-glucose. J Natl Cancer Inst 1996;88:1204–1209.
45. Avril N, Rose CA, Schelling M, et al. Breast imaging with positron emission tomography and [fluorine-18]-fluorodeoxyglucose: Use and limitations. J Clin Oncol 2000;18:3495–3502.
46. Greco M, Crippa F, Agresti R, et al. Axillary lymph node staging in breast cancer by 2-fluoro-2-deoxy-D-glucose-positron emission tomography: clinical evaluation and alternative management. J Natl Cancer Inst 2001;93:630–635.
47. Schirrmeister H, Kuhn T, Guhlmann A, et al. Fluorine-18 2-deoxy-2-fluoro-D-glucose PET in the preoperative staging of breast cancer: comparison with the standard staging procedures. Eur J Nucl Med 2001;28:351–358.
48. Wahl RL, Siegel BA, Coleman RE, et al. Prospective multicenter study of axillary nodal staging by positron emission tomography in breast cancer: a report of the staging breast cancer with PET Study Group. J Clin Oncol 2004;22:277–285.
49. Bohdiewicz PJ, Wong CY, Kondas D, et al. High predictive value of F-18 FDG PET patterns of the spine for metastases or benign lesions with good agreement between readers. Clin Nucl Med 2003;28:966–970.
50. Gallowitsch HJ, Kresnik E, Gasser J, et al. F-18 fluorodeoxyglucose positron-emission tomography in the diagnosis of tumor recurrence and metastases in the follow-up of patients with breast carcinoma: a comparison to conventional imaging. Invest Radiol 2003; 38:250–256.
51. Cook GJ, Houston S, Rubens R, et al. Detection of bone metastases in breast cancer by 18FDG PET: differing metabolic activity in osteoblastic and osteolytic lesions. J Clin Oncol 1998;16:3375–3379.
52. Cohade C, Osman M, Nakamoto Y, et al. Initial experience with oral contrast in PET/CT: phantom and clinical studies. J Nucl Med 2003;44:412–416.
53. Nakamoto Y, Tatsumi M, Cohade C, et al. Accuracy of image fusion of normal upper abdominal organs visualized with PET/CT. Eur J Nucl Med Mol Imaging 2003;30:597–602.
54. Wahl RL, Zasadny K, Helvie MA, et al. Metabolic monitoring of breast cancer chemohormonotherapy using positron emission tomography: initial evaluation. J Clin Oncol 1993;11:2101–2111.
55. Bassa P, Kim EE, Inoue T, et al. Evaluation of preoperative chemotherapy using PET with [fluorine-18]-fluorodeoxyglucose in breast cancer: a method for early therapy evaluation? J Clin Oncol 1995;13: 1470–1477.
56. Mortimer JE, Dehdashti F, Siegel BA, et al. Metabolic flare: indicator of hormone responsiveness in advanced breast cancer. J Clin Oncol 2001;19: 2797–2803.
57. Dehdashti F, Flanagan FL, Mortimer JE, et al. Positron emission tomographic assessment of "metabolic flare" to predict response of metastatic breast cancer to antiestrogen therapy. Eur J Nuc Med 1999; 26:51–56.
58. Kuhajda FP, Offutt LE, Mendelshon G, et al. The distribution of carcinoembryonic antigen in breast carcinoma. Diagnostic and prognostic implications. Cancer 1983;52:1257–1264.
59. Goldenberg DM, DeLand F, Kim E, et al. Use of radiolabeled antibodies to carcinoembryonic antigen for the detection and localization of diverse cancers by external photoscanning. N Engl J Med 1978;298: 1384–1386.
60. Baum RP, Lorenz M, Hertel A, et al. Successful immunoscintigraphic tumor detection with technetium99m marked monoclonal anti-CEA antibodies. Onkologie 1989;12:26–29.
61. Lind P, Smola MG, Lechner P, et al. The immunoscintigraphic use of Tc-99m-labelled monoclonal anti-CEA antibodies in patients with primary suspected recurrent breast cancer. Int J Cancer 1991;47:865–869.
62. Taylor JL, Taylor DN, Lowry C, et al. Radioimmunoscintigraphy of metastatic breast carcinoma. Eur J Surg Oncol 1992;18:57–63.
63. Goldenberg DM, Abdel-Nabi H, Sullivan C, et al. Carcinoembryonic antigen immunoscintigraphy complements mammography in the diagnosis of breast carcinoma. Cancer 2000;89:104–115.
64. Bakir MA, Eccles S, Babich JW, et al. c-erbB2 protein overexpression in breast cancer as a target for PET using iodine-124-labeled monoclonal antibodies. J Nucl Med 1992;33:2154–2160.
65. De Santes K, Slamon D, Anderson SK, et al. Radiolabeled antibody targeting of the HER-2/neu oncoprotein. Cancer Res 1992;52:1916–1923.
66. Behr TM, Wörmann B, Kaufmann CC, et al. Does external scintigraphy allow for predictions with respect to the toxicity and therapeutic efficacy of Herceptin therapy of HER2/neu expressing breast cancer? J Nucl Med 2002;43:295P (abst. 1189).
67. Wu X, Liu H, Liu J, et al. Immuno-fluorescent labeling of cancer marker Her2 and other cellular targets with semiconductor quantum dots. Nature Biotech 2003;21:41–46.
68. Hull DF, Clark GM, Osborne CK et al. Multiple estrogen receptor assays in human breast cancer. Cancer Res 1983;43:413–416.
69. Landvatter SW, Kiesewetter DO, Kilbourn MR, et al. (2R*, 3S*)-1-[18F]fluoro-2,3-bis(4-hydroxyphenyl) pentane [(18F]fluoronor-hexestrol), a positron-emitting estrogen that shows highly-selective,

receptor-mediated uptake by target tissues in vivo. Life Sci 1983;33:1933–1938.

70. Mintum MA, Welch MJ, Siegel BA, et al. Breast cancer: PET imaging of estrogen receptors. Radiology 1988;169:45–48.

71. McGuire AH, Dehdahshi F, Siegel BA, et al. Positron tomographic assessment of 16 alpha-[F-18}-fluoro-17-beta-estradiol uptake in metastatic breast carcinoma. J Nuc Med 1991;32:1526–1531.

72. Vijaykumar D, Mao W, Kirschbaum KS, et al. An efficient route for the preparation of a 21-fluoro progestin-16 alpha,17 alpha-dioxolane, a high-affinity ligand for PET imaging of the progesterone receptor. J Org Chem 2002;67:4904–4910.

73. Dolan JT, Miltenburg DM, Granchi TS, et al. Treatment of metastatic breast cancer with somatostatin analogues—a meta-analysis. Ann Surg Oncol 2001; 8:227–233.

74. van Eijck CH, Kwekkeboom DJ, Krenning EP. Somatostatin receptors and breast cancer. Q J Nucl Med 1998;42:18–25.

75. Mach RH, Huang Y, Buchheimer N, et al. [$^{(18)}$F]N-(4′-fluorobenzyl)-4-(3-bromophenyl) acetamide for imaging the sigma receptor status of tumors: comparison with [$^{(18)}$F]FDG, and [(125)I]IUDR. Nucl Med Biol 2001;28:451–458.

76. Shields AF, Grierson JR, Muzik O, et al. Kinetics of 3′-deoxy-3′-[F-18]fluorothymidine uptake and retention in dogs. Mol Imaging Biol 2002;4:83–89.

77. Dewanjee MK, Ghafouripour AK, Kapadvanjwala, et al. Noninvasive imaging of c-myc oncogene messenger RNA with indium-111-antisense probes in a mammary tumor-bearing mouse model. J Nucl Med 1994;6:1054–1063.

78. Tavitian B, et al. In vivo imaging of oligonucleotides with positron emission tomography. Nat Med 1998; 4:467–471.

79. Tjuvajev JG, et al. Imaging the expression of transfected genes in vivo. Cancer Res 1995;55:6126–6132.

80. Tjuvajev JG, et al. Imaging herpes virus thymidine kinase gene transfer and expression by positron emission tomography. Cancer Res 1998;58:4333–4341.

81. Gambhir SS, et al. A mutant herpes simplex virus type 1 thymidine kinase reporter gene shows improved sensitivity for imaging reporter gene expression with positron emission tomography. Proc Natl Acad Sci USA 2000;6:2785–2790.

82. Blankenberg FG, Katsikis PD, Tait JF, et al. Imaging of apoptosis (programmed cell death) with 99mTc annexin V. J Nucl Med 1999;40:184–191.

83. Mackinnon WB, Barry PA, Malycha PL, et al. Fine-needle biopsy specimens of benign breast lesions distinguished from invasive cancer ex vivo with proton MR spectroscopy. Radiology 1997;204:661–666.

84. Jacobs MA, Barker PB, Bottomley PA, et al. Proton magnetic resonance spectroscopic imaging of human breast cancer: a preliminary study. J Magn Reson Imaging 2004;19:68–75.

85. Blasberg R. PET Imaging of gene expression. Eur J Cancer 2002;38:2137–2146.

CHAPTER 30

Translation of Molecular Markers to Clinical Practice

Nadia Harbeck, MD, PhD, Christoph Thomssen, MD, and Ronald E. Kates, PhD

Clinical Needs in Breast Cancer

Breast cancer is a potentially curable carcinoma with high incidence in the Western world. Already at the time of primary therapy, the disease may be systemic, and local and systemic treatment modalities needs to be optimally combined to achieve cure for an individual patient. However, once a patient has developed overt distant metastases, breast cancer is no longer considered curable by therapies available today. Hence, the clinician is faced with the challenge to accurately assess an individual patient's risk situation from the start and to individualize locoregional and systemic treatment modalities accordingly.

Molecular markers as prognostic or predictive factors in breast cancer will most likely be used in the context of systemic therapy. In this setting, accurate risk group selection includes avoiding overtreatment for low-risk patients and administering more aggressive therapies to high-risk patients, where therapy side effects can be tolerated in view of potential therapy benefits. With regard to adjuvant systemic therapy, the goal is to identify those patients who are already cured by locoregional treatment alone. Because established prognostic factors and guidelines are not yet sufficient for accurate identification of this low-risk group, molecular markers are urgently needed. Moreover, in view of the increasing complexity of conventional systemic treatment options and the advent of tumor biologic therapies, molecular markers are also urgently needed for prediction of therapy response.

The following considerations mainly concern the use of prognostic or predictive markers in breast cancer. However, several of these markers may also be used for preclinical evaluation of markers as therapeutic targets. The important issue of which markers should be measured ("routinely" or under special circumstances) is ultimately related to the incremental improvement in the quality of the clinical decision reached using the information provided by the markers and is not adequately characterized by a simple minded quantity (such as the *P*-value in a study).

Prerequisites for Clinical Use of Novel Markers

A list of criteria to be fulfilled before transferring a new prognostic factor into routine clinical use was first put forward by McGuire[1] and Clark[2] (**Table 30.1**). Even though recent advances require a more generalized interpretation of these prerequisites, they can still be used as guidance for development of new prognostic and predictive markers:

1. A biologic hypothesis and model needs to underline the plausibility for the intended use of a new marker.
2. A simple, robust, and standardized determination method is a mandatory prerequisite for clinical routine use of a new marker. Quality assurance of the test system needs to be implemented.
3. Adequate statistical evaluation of a new factor needs to be planned prospectively. Ideally, evaluation of the marker should be the primary objective of at least one major study.
4. Correlations with established markers need to be evaluated to identify markers that actually add to the existing knowledge.
5. Optimized and validated cutoff values are sometimes the most appropriate way to express a single marker for clinical decision making. Because biologic processes are mostly continuous, advanced statistical models implementing continuous variables

Table 30.1 Criteria to be Fulfilled by New Markers Before Transfer into Clinical Practice

1. Biologic hypothesis
2. Simple, robust, and standardized determination method; quality assurance
3. Adequate statistical evaluation
4. Correlation to established markers
5. Cutoff values to distinguish low-risk and high risk patients
6. Independent validation of prognostic and/or predictive impact
7. Clinical relevance for clinical decision making

From Refs. 1 and 2.

for decision support may be a valid alternative in the future.

6. The prognostic and/or predictive impact of a new marker needs to be independently validated by well-conducted confirmatory studies. Multicenter studies are best able to show feasibility in day-to-day practice. Meta-analyses or pooled analyses may also be useful to validate a marker's proposed impact.
7. Clinical relevance for clinical decision making needs to be demonstrated. Ideally, a new marker can serve as a primary or secondary end point in a prospective randomized trial analyzing different therapeutic strategies according to marker expression.

Pitfalls When First Assessing Potential New Markers

Pretest Specimen Handling

In view of the increasing complexity of the biologic test systems and the variability of the results depending on pretest specimen handling, standardization, and quality control are absolutely mandatory for any new marker. For example, RNA-based techniques or activity measurements are rather sensitive to environmental influences before a sample is subjected to appropriate storage conditions. Hence, test results may be substantially impaired by the time period between sample retrieval and sample storage—a time period that is very variable in clinical routine. Moreover, immunohistochemistry results depend on tissue fixation. This factor may also vary between and even within laboratories. Consequently, pretest requirements need to be appropriately specified before clinical actions based on particular test results are recommended, especially if the test is sensitive to these conditions.

Patient Populations

When looking at a marker's impact on patient outcome, the patient population needs to be well characterized and as homogeneous as possible with regard to potentially biasing factors such as systemic therapy. Hence, a pure prognostic marker is best assessed in patients without adjuvant systemic therapy. Here, its impact on disease-free survival can be directly attributed to its potential to reflect disease aggressiveness. In patients who received adjuvant systemic therapy, a marker's impact on disease-free survival is influenced by its predictive capacity. For example, the prognostic impact of urokinase plasminogen activator (uPA) and plasminogen activator inhibitor type 1 (PAI-1) in primary breast cancer is strongest in patients without any adjuvant systemic therapy, whereas it is considerably diminished in patients who receive adjuvant systemic therapy.[3] However, this latter effect may be attributed to an enhanced benefit from adjuvant chemotherapy in high risk patients according to high uPA/PAI-1.[4] Moreover, the prognostic impact of HER-2 has been reported to differ between node-negative and node-positive patients.[5] This observation is not necessarily indicative of fundamentally different underlying tumor biology between node-negative and node-positive patients but could well be due to different therapeutic strategies used in these patient groups.

Data Analysis Techniques

Appropriate statistical evaluation, particularly realistic confidence estimates and the avoidance of undue optimism, has always been a concern and is likely to increase in importance in the future. One example is the need for corrections to confidence intervals (*P* values) arising from multiple testing or the use of optimal cutoffs.[6]

In contrast to individual markers, the data profile provided by array analyses contains many separate signals. As discussed below, the quality of risk assessment may depend on the data processing algorithm used to fuse this information. Hence, before test results can be supplied to clinicians, the process of standardization and validation must include not only laboratory techniques themselves, but also all subsequent data processing algorithms.

Clinical Relevance of New Markers

Even if all methodologic criteria as listed above are fulfilled, a new marker may not add any relevant information for clinical decision making. The determination of whether a new marker adds relevant information to that of "established" markers generally requires multivariate analysis, including the established markers. Moreover, particularly for a prognostic marker, but also for a predictive marker, a statistically significant impact may not necessarily be clinically relevant. The following scenarios may serve as examples.

Absolute Outcome Difference

A prognostic marker splits a patient population into a high risk and a low-risk group by showing a statistically significant difference with regard to outcome. However, this is by itself only clinically relevant if there is a substantial absolute difference between the two groups over a considerable length of follow-up. For example, statistically significant but only small differences at long-term follow-up are found between steroid hormone receptor positive and negative breast cancers.[7]

Risk Level in Low-Risk Group

Sometimes, a substantial and statistically significant difference in patient outcome according to marker expression is observed. However, if the level of risk of recurrence in the low-risk group is still rather high, no clinical consequences can be drawn due to this marker alone. As an example, assessment of tumor proliferation by MIB-1 immunohistochemistry or DNA flow cytometry provides statistically significant prognostic information without clinical impact.[8]

Size of Low-Risk Group

Again, a substantial and statistically significant separation between low-risk and high-risk group is observed. However, if the percentage of patients in the low-risk group is too small, the marker cannot be used as a sole decision criterion in the whole patient population. An example of this scenario is seen in risk classification based on tumor grade as used in the St. Gallen criteria.[9] The proportion of minimal-risk patients (node-negative with G1 tumors) is less than 10%. For most patients, relevant prognostic information is still lacking.

Subjective Value Systems

As a rule of thumb in clinical breast cancer management today, a less than 10% risk of recurrence and a less than 5% risk of death after 10 years is generally accepted to be low-risk. However, appropriate risk-to-benefit analysis, carefully weighing magnitude of risk of recurrence against potential benefits and side effects provided by a particular systemic therapy, is strongly influenced by the patient's subjective value system.[10]

In summary, clinically relevant risk group stratification by an ideal prognostic factor needs to demonstrate a substantial and statistically significant difference between a low-risk and a high-risk group that is sufficient in size for appropriate treatment recommendations. For example, in node-negative breast cancer, such a low-risk group should ideally comprise more than 50% of all node-negative patients, because about 70% of node-negative patients will be cured by locoregional therapy only and will thus not need any additional systemic therapy. The levels of tolerable risks depend on the potential benefits and side effects of the treatment options and involve individual patient considerations.

Evidence-Based Criteria for New Markers

Official guidelines for diagnosis and treatment of breast cancer use predefined and reproducible criteria to strengthen their recommendations. These criteria are based on the published scientific evidence available to support a specific recommendation. To enable easy transfer of evidence-based criteria into clinical practice, statements are then scored with regard to their clinical relevance. A number of different classification systems of levels of evidence and grades of recommendations that are applicable for molecular markers are currently used.

Tumor Marker Utility Grading System

To classify the increasing amount of new biologic markers for prognosis and prediction, a tumor marker utility grading system was proposed by American Society of Clinical Oncology members.[11] This system seeks to score the value of a new marker by evaluating and classifying the supporting scientific evidence using levels of evidence as in other formal classification systems (**Table 30.2**). In view of

Table 30.2 Levels of Evidence for Clinical Use of Tumor Markers

Level	Description
I	Single prospective study with high statistical power intended to test the clinical use of a specific marker (preferentially with therapeutic implication) or meta-analysis.
II	Prospective study with tumor marker analysis as secondary end point
III	Large retrospective studies
IV	Small retrospective studies, "matched pair" analyses, etc.
V	Small pilot studies, correlations to other known factors

From Ref. 11.

the supporting evidence, clinical utility of a new marker can then be scored on a scale from 0 ("marker has no utility at all") to +++ ("marker can be used as the sole criterion for clinical decision making").

Oxford Centre for Evidence-Based Medicine Levels of Evidence

In a more comprehensive manner, the Oxford Centre for Evidence-based Medicine published a system of levels of evidence, which is adapted for different clinical applications (e.g., diagnostic procedures, therapeutic interventions, economic decisions, side effects; http://www.cebm.net). This system is used by Cochrane institutions, particularly in establishing treatment guidelines. Similar evidence-based guidelines are used by the American Society of Clinical Oncology[12] or the Canadian Task Force on Preventive Health Care.

German AGO Breast Cancer Expert Panel

This group uses a combined system to evaluate prognostic and predictive markers in breast cancer in their S2 guidelines for treatment of early and advanced breast cancer (www.ago-online.de). Whereas evidence levels are given according to the Oxford classification, clinical recommendations of the expert panel are added to facilitate use of new markers in clinical practice (**Table 30.3**). We note that the determination of the advantage of a particular test to a given patient depends on what is already known about the patient and on the intended use of the marker in decision making.

In conclusion, published evidence is the backbone of most scoring and recommendation systems for new markers. However, expert recommendations for clinical use of new markers should not only look at published evidence

Table 30.3 AGO Breast Cancer Expert Panel Recommendation System for New Tests or Therapeutic Interventions in Breast Cancer

Grade	Recommendation
++	This test or therapeutic intervention has great advantage to the patients and can thus be recommended without hesitation and should be performed.
+	This test or therapeutic intervention could be advantageous to the patient and can thus be performed.
+/−	This test or therapeutic intervention has not yet been proven to be advantageous to the patient; it can be performed in individual cases. In view of the published literature, no clear recommendation can be given.
−	This test or therapeutic intervention may be of disadvantage to the patient and should thus not be performed.
−−	This test or therapeutic intervention is of disadvantage to the patient and should not be performed under all circumstances.

AGO (Arbeitsgemeinschaft Gynäkologische Onkologie), Working Group for Gynecological Oncology (German Cancer Society and German Ob&Gyn Society).

and address the question of potential clinical usefulness, but also consider standardization or quality control of the test system, which is not properly taken into account by thoroughly evidence-based systems. Yet clinicians will need this kind of expert judgment to decide on what markers to base their therapeutic decisions. In particular, with regard to molecular markers and their potential methodologic variability, technical considerations are just as important as the supporting evidence.

Marker Fusion: A Framework for Combining Molecular Staging, Advanced Statistical Models, and Decision Support

Data Fusion Paradigm

With the advent of array techniques for molecular staging, the concept of what constitutes a "marker" and questions of clinical translation and relevance require some generalization; additional challenges and demands on information processing will result, including the need to combine heterogeneous data sources within a complex information environment. In cancer or other diseases, the impact of one factor is sometimes so strong (and the number of possible treatments or interventions so limited) that decisions may be based on univariate survival analysis with respect to this one factor; however, the generic situation is multifactorial with more complex decision alternatives.

The interdisciplinary concept of data fusion offers a potentially useful paradigm to describe the flow of information beginning with marker and array measurements, feeding into estimation and application of an advanced statistical model for risk and treatment response assessment, and finally supporting clinical determination of an optimal therapy regimen for an individual patient.[13] A typical data fusion process involves several levels of integration. Each level reduces the volume of data and extracts information for the next level. The goal of data fusion is to optimize the quality of decisions, which are usually not produced until the top level is reached.

From the clinical point of view, "quality" refers to obtaining a clinically relevant result, that is, one that satisfies all the above criteria for individual markers (subject to suitable generalizations) and that leads to optimal individualized treatment. In the following, we refer to the data fusion process as it applies to translation markers to clinical practice as "marker fusion."

Reducing Dimensionality: Low-Level Marker Fusion

Familiar procedures that constitute relatively simple examples of low-level marker fusion are the assignment of histologic grade to a tumor[14] or the characterization of uPA levels in tumor tissue by enzyme-linked immunosorbent assay measurement.[15] Here, only a few measurements or "signals" are combined. However, as stated earlier, the data profile provided by array analyses contains many separate signals or components of information, some of which are highly correlated. For an individual patient, this profile may be viewed mathematically as a point in a "signal space" of high dimensionality. In this picture, the challenge for low-level data marker fusion is thus to reduce the dimensionality of the signal space by an appropriate mathematical algorithm. To justify this approach, one imagines that the set of candidate biologic processes is much smaller than the number of array signals and that each such biologic process leaves behind a unique signature identifiable as a signal pattern or cluster.

Procedures for reduction of dimensionality abound and include clustering algorithms (e.g., "K-means clustering"), unsupervised learning algorithms, and the like. Despite the bewildering arsenal of methodologies, universally accepted quality criteria for deciding which method is suitable for which application do not yet exist. In our view, quality criteria for unsupervised learning algorithms applied to array data deserve considerable attention in future research.

It may happen that determination of exactly one signal (e.g., amplified expression of a particular gene) is sufficient in principle to identify a relevant process. It would then be tempting to design a device to assign a marker based exclusively on this signal. However, as discussed above (see Pitfalls When First Assessing Potential New Markers), standardization and robustness of test results are important requirements. What happens if it is not technically feasible to manufacture a device with perfectly reproducible individual signals? Using an appropriate mathematical clustering procedure, it may still be possible to derive "composite marker" that offer sufficient standardization and remain robust even in the presence of imperfect measurements or partial information. This process depends on the presence of redundancy (e.g., internal correlations) between signals and also requires intelligent knowledge engineering techniques. Moreover, in the generic case, we cannot expect all relevant biologic processes to be marked by one gene expression signal. Hence, the output of low-level marker fusion will usually be a small set of generalized markers, some composite, that are thought to be associated with relevant biologic processes but are not yet necessarily associated with an outcome of interest.

Intelligent Risk Assessment: Second-Level Marker Fusion

Once such a reduced set of (generalized) markers has been obtained, the challenge for the second level of marker fusion is risk assessment (including both "prognostic" and "predictive applications"), that is, associating markers, treatment alternatives, and combinations thereof with probabilities of one or more outcomes with the goal of doing so as precisely and reproducibly as possible. Risk assessment is often discussed in terms of classification, but in the paradigm of data fusion it is often preferable to view the assessment of risk as a scoring problem, that is, one or more scores are to be assigned on the basis of the patient's individual available factors. Methods such as CART

("Classification and Regression Trees") that are usually thought of as classification techniques are in principle capable of associating scores with each individual.[16,17] As discussed below, the classification for therapy (decision support) is a distinct task that may subsequently be defined in terms of suitable decision criteria based on scores and is thus regarded as a higher level of marker fusion as discussed below.

As stronger biologic markers become available, and particularly in applications of array technologies, risk assessment will almost certainly require multivariate modeling including nonlinear effects and factor interactions. Intelligent knowledge engineering systems such as neural nets offer an attractive alternative compared with traditional models due to their flexibility, that is, their ability to represent and also identify these effects. In our view, there is no fundamental distinction between "knowledge engineering" and "statistical methods" in application to medical decision support,[18,19] and indeed a synthesis of these two approaches has become evident in the literature.[20–25]

In the following, we discuss the use of a trained neural network for risk assessment within a statistical approach[13,25–27] as an example of second-level marker fusion. In particular, we consider the problem of using markers not only to assess the risk of relapse in breast cancer, but also to differentiate among relapse sites that might call for different treatment approaches, which we refer to as "differential therapy benefit prediction." This problem is closely related to the statistical concept of competing risks.[28]

We now briefly sketch the mathematical structure of a solution to this problem using the learning capability of neural networks. Consider follow-up data containing a classification of first relapse in breast cancer into relapse "modes" with "B" for "bone," "D" for "distant not bone," and "L" for "local/regional relapse (not distant)," as illustrated in FIGURE 30.1. Patients are assumed to be observed at month t as either censored or relapsed. If relapsed, they are classified uniquely as B, D, or L in that order of precedence; for example, even if a patient is

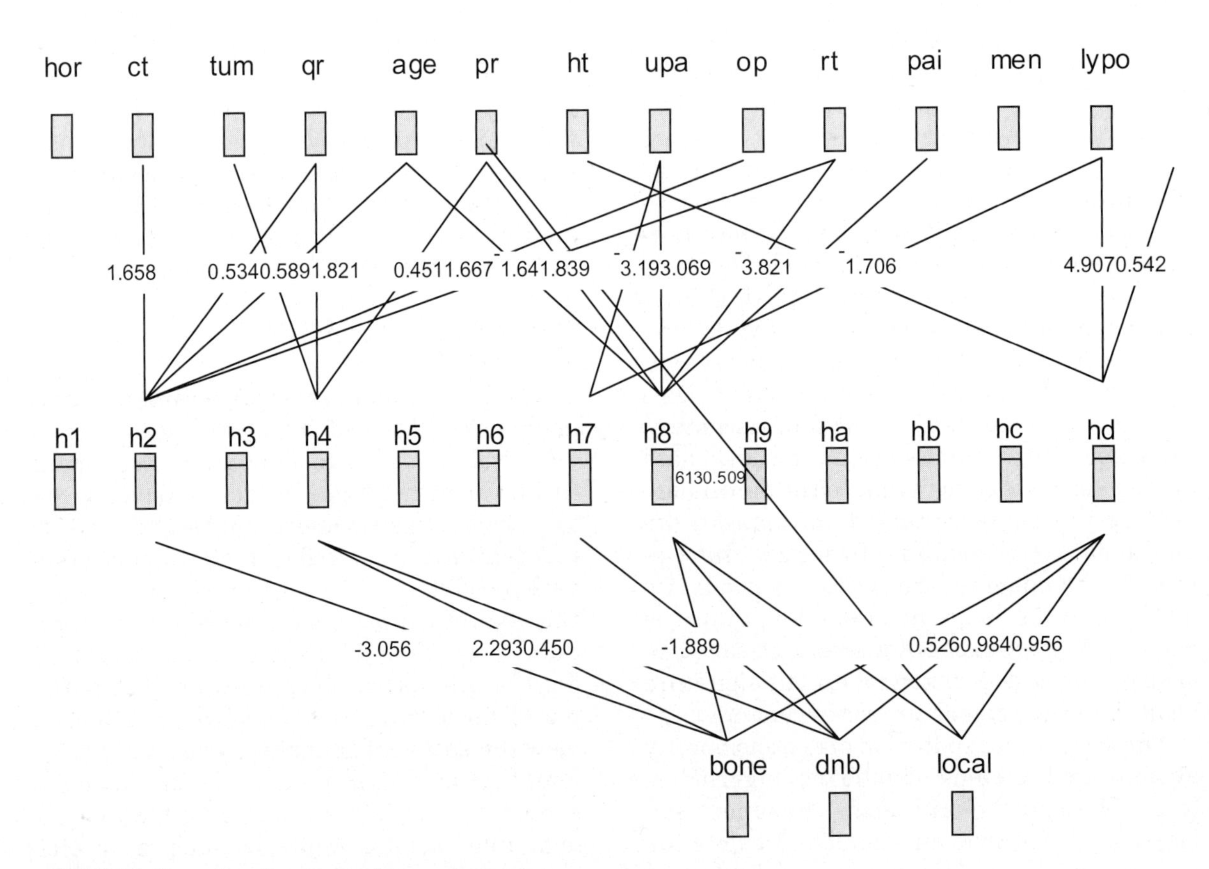

FIGURE 30.1 Illustration of competing risk neural network.

observed with both bone and other distant metastases in month t, then the category is B. Subsequent relapses are not included in this example; for example, if a patient has only local relapse at month t_1 but develops distant relapse at month t_2, then we code L at month t_1.

Suppose now that in addition to the follow-up data, risk factors and treatment variables denoted by x_j are recorded for the jth patient. It is then possible to train a neural network to learn the underlying relationship between risk factors/treatment variables and end points.[25,26] Advanced techniques for complexity reduction, which are critical for good generalization performance, are discussed in these references. An illustration of a neural network trained to represent competing risks is shown in FIGURE 30.1.

The output layer of the trained neural network (illustrated as the lower row of three nodes in FIG. 30.1) produces a risk profile for each patient, which can be thought of as a table of scores coding the risk of each mode of relapse. As discussed in Kates et al.,[25] additional output nodes can be associated with possible time varying (i.e., non-proportional) hazards, for example, a trend to early onset that could result from a particular constellation of risk factors. The proteases uPA and PAI-1 are known, for example, to be associated with early relapse risk[29] in breast cancer. Moreover, each risk depends in a known (and perhaps distinct) way not only on the risk factors but also on the treatment options. Hence, the representation of competing risks and/or time varying hazards by neural networks could facilitate targeting tumors with susceptibility to a particular mode of relapse or with a high probability of early relapse by particular adapted treatments.

From the risk profile, the statistical "hazard function" for each risk can be constructed as described in Kates et al.[25] Hence, this formulation of risk assessment by neural nets can be quantitatively compared with risk assessment by other statistical models. Incidentally, as should be expected whenever the impact of nonlinear interactions is limited, the scores of neural nets and of statistical models often turn out to be highly correlated and would thus lead to similar decisions when the inputs consist of clinical factors or even when they include some tumor biological factors. The real advantage of neural nets is expected when complex nonlinear factor interactions become dominant. In any case, whether neural networks or other methods of risk assessment are used, the result of this "level-2 marker fusion" is not yet a decision but a knowledge base (one or more risk scores) to *support* a decision.

Decision Support: High Level Data Fusion

In cancer, the designation of high-risk patient is sometimes used as a surrogate for the classification "likely to benefit from therapy," but such logical shortcuts are not conducive to translation of factors to clinical practice. For example, patients whose natural disease course puts them at relatively low-risk might still benefit from a particular therapy option. Conversely, there could be groups of patients with poor natural disease course who unfortunately have nothing to gain from a therapy option. In view of the broadening spectrum of new therapy options, including tumor biological therapy approaches such as Herceptin™,[30] prediction of response to specific therapy options will play an increasingly important role.

Prediction of response does not need to be associated with a single marker. If the multivariate risk assessment provided by level 2 marker fusion properly includes the influence of therapies, then it can in principle be used to support the decision process for clinical indications. An idealized systematic decision optimization process would consider a virtual study with an ensemble of virtual patients having the same markers as the patient in question, with an arm of the study for each treatment option. One can imagine following all patients in this virtual study for a long period, taking into account the probabilities of all subsequent events (including relapses, side effects of treatments, etc.) and their associated life quality impacts. (As explained earlier, subjective value systems could enter into these associated life quality impacts.) The treatment with the best impact on life quality would then correspond to the individual patient's preferred treatment.

A typical statistical framework for estimating the results of such a virtual study might use what is known as Monte Carlo simulation within a Markov model. The inputs to this model, known as "state transition probabilities," are obtained from the risk assessment (second) level (e.g., from a neural net).

Incidentally, the monetary cost of treatment was not considered in this line of reasoning. If included, we would not be discussing individualized decision support but rather optimal resource management. A good risk assessment knowledge basis (such as that provided by a trained neural network) can support both kinds of decision making, but distinguishing risk assessment (level-2 data fusion) from the decision process (higher-level data fusion) allows a clear specification of goals.

Conclusions

With the advent of array technologies for molecular staging, the issues in translation of markers to clinical practice are becoming increasingly complex. We have described a general framework for understanding the information processes in terms of several levels of marker fusion, beginning with algorithms for reduction of dimensionality (clustering, unsupervised learning), continuing with (differential) risk assessment (including the influence of therapy options, i.e., "prognosis" and "prediction"), and ending with systematic decision optimization.

At present, systematic decision optimization at the level of detail outlined above is neither used in routine clinical practice nor in regulatory approval of new assays, and many decisions remain in the realm of subjective judgment. However, given the increasing complexity of markers and treatments, systematic decision optimization could in the future play an important role in translation of markers to clinical practice.

Increasingly, quality assurance and standardization requirements will need to address not only laboratory measurement procedures per se but also the sequence of mathematical and algorithmic steps used to arrive at a clinical indication. This need does not imply that regulatory agencies should restrict marker fusion to a particular ostensibly "simple" algorithmic form (e.g., decision trees), because there is no evidence that one such method is universally superior to the others. An algorithm (e.g., for clustering) that can be shown to be functionally equivalent or superior in quality to an existing one (in the sense of providing indications with higher patient benefit) should ideally be permitted. However, the future may well pose some difficult issues both for physicians and for regulatory agencies at all levels of marker fusion, including definition of composite markers, performance of risk assessment, and the quality of therapy indications.

References

1. McGuire WL. Breast cancer prognostic factors: evaluation guidelines. J Natl Cancer Inst 1991;83:154–155.
2. Clark GM. Do we really need prognostic factors in breast cancer? Breast Cancer Res Treat 1994;30: 117–126.
3. Harbeck N, Kates R, Schmitt M. Clinical relevance of invasion factors uPA and PAI-1 for individualized therapy decisions in primary breast cancer is greatest when used in combination. J Clin Oncol 2002; 20:1000–1009.
4. Harbeck N, Kates RE, Look MP, et al. Enhanced benefit from adjuvant systemic chemotherapy in breast cancer patients classified high-risk according to uPA and PAI-1 (n=3,424). Cancer Res 2002;62: 4617–4622.
5. Yamauchi H, Stearns V, Hayes DF. When is a tumor marker ready for prime time? A case study of c-erbB-2 as a predictive factor in breast cancer. J Clin Oncol 2001;19:2334–2356.
6. Hilsenbeck SG, Clark GM. Practical p-value adjustment for optimally selected cutpoints. Stat Med 1996;15:103–112.
7. Zemzoum I, Kates RE, Ross JS, et al. Invasion factors uPA/PAI-1 and HER2 status provide independent and complementary information on patient outcome in node-negative breast cancer. J Clin Oncol 2003; 21:1022–1028.
8. Dettmar P, Harbeck N, Thomssen C, et al. Prognostic impact of the proliferation associated factors Ki-67 and S-phase in node-negative breast cancer using MIB1 immunostaining and flow cytofluorometric S-phase determination. Br J Cancer 1997;75:1525–1533.
9. Goldhirsch A, Wood WC, Gelber RD, et al. Meeting highlights: updated international expert consensus on the primary therapy of early breast cancer. J Clin Oncol 2002;21:3357–3365.
10. Ravdin PM, Siminoff IA, Harvey JA. Survey of breast cancer patients concerning their knowledge and expectations of adjuvant therapy. J Clin Oncol 1998; 16:515–521.
11. Hayes DF, Bast R, Desch CE, et al. A tumor marker utility grading system (TMUGS): a framework to evaluate clinical utility of tumor markers. J Natl Cancer Inst 1996;88:1456–1466.
12. ASCO. 1997 Update of recommendations for the use of tumor markers in breast and colorectal cancer. Adopted on November 7, 1997 by the American Society of Clinical Oncology. J Clin Oncol 1998;16: 793–795.
13. Kates R, Schmitt M, Harbeck N. Advanced statistical methods for the definition of new staging models. Recent Results Cancer Res 2003;162:101–113.
14. Elston CW, Ellis IO. Pathological prognostic factors in breast cancer. I. The value of histological grade in breast cancer: experience from a large study with long-term follow-up. Histopathology 1991;19:403–410.
15. Harbeck N, Dettmar P, Thomssen C, et al. Risk-group discrimination in node-negative breast cancer using invasion and proliferation markers: six-year median follow-up. Br J Cancer 1999;80:419–426.
16. Breiman L, Friedman J, Olshen R, et al. Classification and Regression Trees. New York: Chapman and Hall, 1984.
17. Danegger F. Improving prediction of tree-based models. In: Marx B, Friedl H, eds. Statistical Modeling.

Proceedings of the 13th International Workshop on Statistical Modeling. 1998.
18. Schumacher M, Rossner R, Vach W. Neural networks and logistic regression. Part I. Comput Stat Data Anal 1996;21:661.
19. Sauerbrei W, Blettner M, Schumacher M. The importance of basic statistical principles for the interpretation of epidemiological data. Onkologie 1997;20:455.
20. Ripley B. Statistical aspects of neural networks. In: Barndorff-Nielsen O, Jensen J, Kendall W, eds. Networks and Chaos—Statistical and Probabilistic Aspects. Chapman & Hall, 1993.
21. Liestol K, Anderson P, Anderson U. Survival analysis and neural nets. Stat Med 1994;13:1189.
22. Faraggi D, Simon R. A neural network model for survival data. Stat Med 1995;14:73.
23. Warner B, Misra M. Understanding neural networks as statistical tools. Am Stat 1996;50:284.
24. Biganzoli E, Boracchi P, Mariani L, et al. Feed forward neural networks for the analysis of censored survival data: a partial logistic regression approach. Stat Med 1998;17:1169–1186.
25. Kates R, Harbeck N, Ulm K, et al. Decision support in breast cancer: recent advances in prognostic and predictive techniques. In: Jain A, Jain A, Jain S, et al., eds. Artificial Intelligence Techniques in Breast Cancer Diagnosis & Prognosis. Singapore: World Scientific Publishing, 2000:55–95.
26. Harbeck N, Kates R, Ulm K, et al. Neural network analysis of follow-up data in primary breast cancer. Int J Biol Markers 2000;15:116–122.
27. Kates RE, Foekens JA, Look MP, et al. Generalization of competing-risk neural networks in breast cancer. Breast Cancer Res Treat 2001;69:267.
28. Kalbfleisch J, Prentice R. The Statistical Analysis of Failure Time Data. Wiley, 1980.
29. Schmitt M, Thomssen C, Ulm K, et al. Time-varying prognostic impact of tumor biological factors urokinase (uPA), PAI-1, and steroid hormone receptor status in primary breast cancer. Br J Cancer 1997; 76:306–311.
30. Slamon DJ, Leyland-Jones B, Shak S, et al. Use of chemotherapy plus a monoclonal antibody against HER2 for metastatic breast cancer that overexpresses HER2. N Engl J Med 2001;344:783–792.

CHAPTER 31

Regulatory Aspects and Implications of Molecular Testing in the United States

Robert G. Pietrusko, PhD

Senior Vice President, Worldwide Regulatory Affairs and Pharmacovigilence, Millennium Pharmaceuticals, Inc., Cambridge, Massachusetts

Based on increased understanding of underlying disease mechanisms, breast cancer is evolving into a group of diseases each with its own molecular basis. Mutations in otherwise important or at least benign biologic pathways may lead to expression of tumors. These molecular pathways need to be thoroughly explored to predict response to therapy. After treatment, it is also important to detect up-regulation or down-regulation of these molecular pathways with respect to survival, proliferation, and stress as well as to evaluate interactions with various cellular kinases that may activate resistance mechanisms. Molecular testing is taking on a more prominent role in the diagnosis, individualized treatment, response, prognosis, and predictive risk of developing breast cancer. Such testing is regulated in the United States in various ways depending on the type of test, where the test is performed, whether the test is supplied as a kit, and the relationship of the test to medications and their labeling.

Differing regulatory requirements for various tests has led to confusion. For example, certain genetic tests are being advertised directly to the public, whereas a new screening test, fluorescence in situ hybridization, used to identify women with breast cancer who could benefit from Herceptin® (trastuzumab) treatment, is tightly regulated and took several years to be included in the Herceptin® label (package insert, 2003). Furthermore, many new pharmacogenomic tests based on DNA (sequence variation analysis) or messenger RNA (expression or functional transcriptional profiling) will need to be considered as part of regulatory submissions in the near future.

The US Food and Drug Administration (FDA) is very interested and committed to understanding and evaluating these data in preparation for the regulatory review process. As a result, the Agency has initiated several workshops with industry to help create guidelines for when and how to submit pharmacogenomic data.[1,2] The Agency has called for voluntary submission of genomic data that will be evaluated separately from the regulatory review process. They have formed an internal Interdisciplinary Pharmacogenomic Review Group to evaluate these data across products and review divisions.

In Vitro Diagnostic Tests

In vitro diagnostic tests are regulated by the FDA under the 1976 Medical Device Amendments to the Federal Food, Drug and Cosmetic Act.[3] Within the Center for Devices and Radiological Health (CDRH), the Office of In Vitro Diagnostic Device Evaluation and Safety (OIVD) has responsibility for regulating these diagnostic devices. Medical devices are classified according to risk. Class I devices have the least risk and are subject to the fewest or general controls. Class II devices have intermediate risk and have a performance standard for control. Class III devices have the highest associated risk and have the most stringent requirements. This latter category includes implantable devices and life support devices. Until such time as a new device is found to be substantially equivalent to a previously marketed device, it is automatically placed in class III. A class III device that is not substantially equivalent to a marketed predicate device must have an FDA-approved premarket approval (PMA) application before it can be distributed commercially in interstate commerce.

Device products are either cleared as a 510(k) product or approved via PMA process. The 510(k) process refers to the section of the Food, Drug and Cosmetic Act that describes the regulatory requirements to market the device. This process is appropriate for a new version of a device that is substantially equivalent to a marketed predicate device. A predicate device may be a device that was legally marketed before the 1976 Device Amendments or another device that has already demonstrated substantial equivalence to one of those devices. The 510(k) submissions do not normally require clinical data and undergo a regulatory review process that averages 70 days.

The PMA approval is based on confirmation that the safety and effectiveness of the device is established by clinical investigation. Collecting sufficient information and preparing a PMA is costly and typically requires several years. The regulatory review process for a PMA (total time from submission to decision) also takes longer than that for a 510(k) and averages 12 months. Enactment in 2002 of the Medical Device User Fee and Modernization Act whereby device manufacturers are charged user fees to have their applications reviewed should bring down the average review times.[4,5] Specific to breast cancer treatment, the test for estrogen receptor positivity was approved as a PMA for commercial use in 1981 with subsequent 510(k) applications being approved.[6]

Clinical laboratories frequently develop and offer their own high complexity tests, known colloquially as "home brews." In 1997, the FDA issued the analyte specific reagent (ASR) regulation.[7] This regulation allows clinical laboratories to use ASRs as individual components or building blocks for developing genetic tests and other molecular assays if they are certified to perform high complexity testing by Clinical Laboratory Improvement Amendments (CLIAs). An ASR is the active ingredient of an in-house test. ASRs are specifically defined as "antibodies, both polyclonal and monoclonal, specific receptor proteins, ligands, nucleic acid sequences, and similar reagents which, through specific binding or chemical reaction with substances in a specimen, are intended for use in a diagnostic application for identification and quantification of an individual chemical substance or ligand in biological specimens."[8] The Healthcare Financing Administration regulates all laboratory testing performed in the United States through the CLIA program. CLIA requires a laboratory to meet certain criteria, including analytical validation but *not* clinical validity. The laboratories must report results from ASR-derived tests with the boilerplate disclaimer: "This test was developed and its performance characteristics determined by [laboratory name]. It has not been cleared or approved by the FDA." The manufacturers of the test are not required to seek FDA PMA for class I (low risk) ASRs but must register and list them with the FDA. To qualify for the regulatory exemption, the manufacturer cannot make analytical performance or clinical claims for the ASR nor can it provide clinical laboratories with instructions on how to use the ASR.[9] The product must be labeled as follows: "Analyte Specific Reagent. Analytical and performance characteristics are not established."

Recently, the FDA determined that a microarray test to identify polymorphisms related to drug metabolism could not be commercially distributed without an appropriate premarket determination from the FDA. The test was considered to be of substantial importance in preventing impairment of human health. Moreover, the Agency considered the test to be a unique, highly processed, multisignal device with unusual quality control features that required unique and sophisticated configurations of hardware and software to make accurate use of those features. After a premarket notification is submitted, the FDA said it would determine the device classification of the test and regulatory approval process.

One of the main issues facing the FDA is a lack of standardization in microarray technology, both in the use of different array platforms and the statistical approach used to analyze or mine the copious amounts of data generated.[10] There also is a lack of information regarding reproducibility with and between manufactured lots. This is compounded when multiple laboratories are involved. Studies are being done to examine the extent of variability.

Expression profiling has provided some encouraging data as a prognostic marker in breast cancer (see Chapter 6). For example, in 295 patients with stage I or stage II breast cancer, the probability of remaining free of metastasis was 51% in 180 patients with a "poor prognosis profile" versus 85% in 115 patients with a "good prognosis profile."[11,12] As these gene expression signature tests become available commercially, it will be important to observe how the FDA regulates such tests. If the FDA were to require laboratories doing such testing to submit these tests for premarket review, it could affect how quickly they could bring these tests on line and would ultimately impact physician and patient access to these new tests.

Pharmacogenomic microarray diagnostic screening tests that claim to be ASRs raise two main concerns for the FDA: whether the tests are intended to screen for serious medical conditions and whether they are presented as kits. The FDA has suggested that companies planning to market such products discuss the products with the Agency before marketing or much sooner in the development process. The FDA may rewrite the ASR rule using a risk stratification approach, that is, the risk or impact of the individual test result on the patient. One potential goal would be to obviate the need for dependency on predicate devices for the approval of microarrays, or other new molecular marker tests, for which no appropriate predicate device exists. Under this potential scenario, FDA approval under the 510(k) process would be possible if the sponsors were to have had ongoing and close discussions with the Agency. A rewrite of the ASR rule would provide an opportunity for public comment.

The CDRH published a draft guideline for multiplex tests for heritable DNA markers, mutations, and expression patterns in April 2003.[13] The guidance provided extensive information on analytical and clinical validation. According to the draft guidance,

> The measurement of expression changes, whether RNA or protein, will raise different validation and safety and effectiveness questions than the measurement of DNA changes or variations. Expression changes, in contrast to DNA changes, can be in response to a variety of factors. These may include simple individual to individual differences, time of day, and specific effect of a therapeutic treatment on a tissue. Results can vary markedly as a result of these factors. Tests to diagnose, predict, or select based on expression patterns may consequently be difficult to interpret. Sponsors of these tests should consider array physical design strategies, quality control (QC), reproducibility and readout/interpretation. To date, FDA has little experience with multiplex or array device submissions.

As mentioned above, only tests that are considered diagnostic devices that have a major impact on the health of the patient or those that are packaged and sold as kits require PMA of the FDA. Tests developed by laboratories and provided as a service do not receive premarket review beyond regulation of the reagents used. Certain manufacturers consider genetic tests to have met the criteria for this category. Testing for germline mutations in *BRACA1* and *BRACA2* is an important tool for predicting the risk of breast cancer and developing management strategies. Mutations in either of these genes confer a lifetime risk of breast cancer between 60% and 85%.[11] Testing is recommended for individuals with specific risk factors.[14,15] Direct to consumer advertisements for tests of "BRACA-1 and BRACA-2 genes" have caused major controversy.[16] The regulation of direct to consumer advertising of genetic testing is unclear because the tests would have the appropriate disclaimer that they have not been cleared or approved by the FDA. The former Secretary's Advisory Committee on Genetic Testing (replaced by the Secretary's Advisory Committee on Genetics, Health and Society) has suggested a risk-based regulatory approach by the FDA for these genetic tests with maximum transparency that would not impede new technology.

Use of Biomarkers in Drug Development

Biomarkers play an integral role in the discovery, development, and approval of new drug products. To illustrate these points, a recent report on a gene that may have an effect in breast cancer will serve as an example. β_1 integrin may be essential for development of mammary gland tumors.[17] The authors of the finding stated that this gene represents a target for biologic drug development. If a company were to consider developing a compound to block β_1 integrin, it would be important to have a high degree of process and analytical control of various molecular biomarkers early on in discovery process to avoid spurious results. Transcriptional profiling of gene pathways involved with β_1 would be evaluated for further understanding of all the potential effects as other pathways may have a synergistic or antagonistic effect. Assay development is important to determine the level of inhibition that would be needed to produce the effect in both "in vitro" and "in vivo" animal models as well as in later testing in humans. High throughput screening would be done for compounds that would block the effect of β_1 integrin and have appropriate pharmacokinetic characteristics. It is important to note that the FDA is not involved in the oversight of research at this stage. Use of molecular biomarkers, transcriptional profiling, and analytical tests and controls are done by companies under their own control. Research and development firms have to make critical decisions as to what compounds to select and when to move development forward to the next critical step. The results of how biomarkers are affected by the development compound or by in vivo tests including knock-out animals frequently play a major role in these decisions. The various points along the drug development pathway that incorporate biomarkers, including static (DNA) and dynamic (messenger RNA) transcriptional profiling, are outlined in FIGURE 31.1. Validation of assays for assessment of the molecular targeted effect is critical. The degree of documentation increases as the likelihood increases that the biomarker will be required to select patients for treatment or to predict benefit or possible risk. The intricacies of signaling pathways also highlight the need and complexity inherent in developing validated biomarkers.[18]

The FDA is becoming increasingly aware of the difficulties encountered in rapid drug development given the bludgeoning amount of new scientific data, including molecular targeting. In an effort to assist companies, the Agency will issue a guidance addressing translational medicine (going from "bench to patient bedside"). Dr. Janet Woodcock, FDA Deputy Commissioner for Operations, is evaluating these processes and is scheduled to prepare a position paper that will focus on improving "predictability" of drug development.

Pharmacodynamic effects based on biomarkers can be particularly useful for deciding drug dosing regimens in later phase clinical studies and for dosing regimens for toxicology studies. Exploratory biomarker data early on in the drug development process is not regulated by the FDA and does not have to be validated (i.e., including rigorous documentation and control of the data). However, the FDA

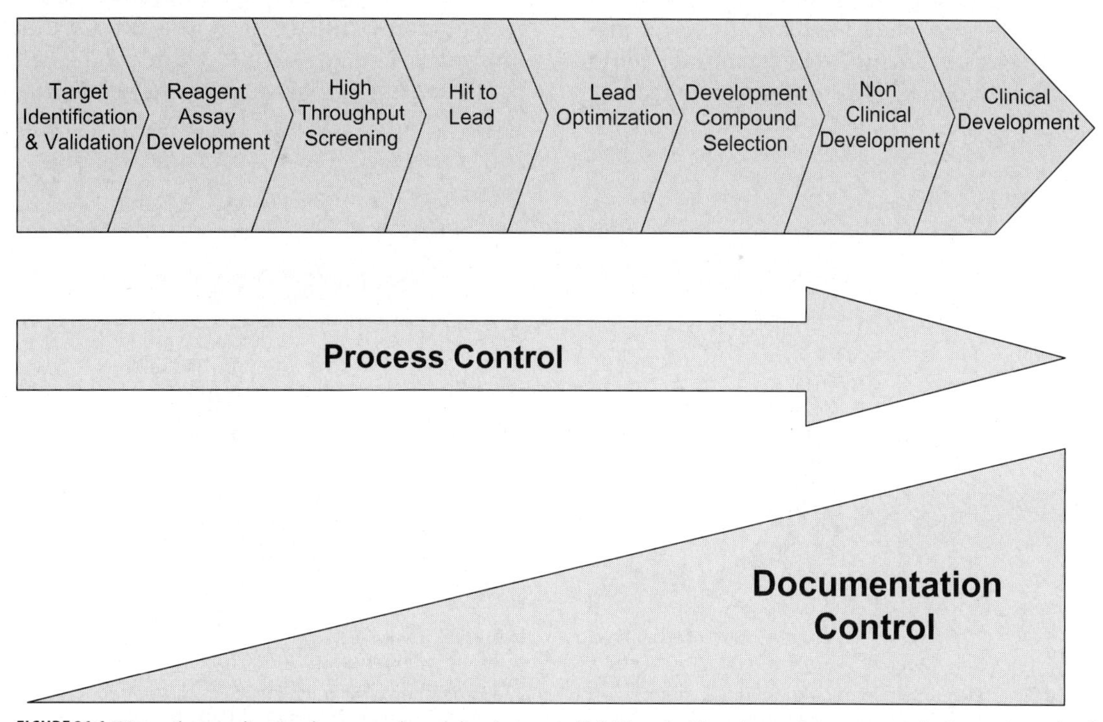

FIGURE 31.1 Biomarker application by research and development (R & D) stage. The picture demonstrates the increasing role of biomarkers and degree of documentation required as the candidate compounds pass through the earliest to the latest stages of drug discovery and development.

requires control and validation of pharmacodynamic biomarkers when used to determine pharmacokinetic/pharmacodynamic (PK/PD) relationships in humans. As an example, the degree and duration of proteasome inhibition associated with Velcade® (bortezomib) treatment was evaluated to define the dosing regimen. An "ex vivo" 20S proteasome activity assay was used for the toxicology studies and in early clinical trials. The goal of the dosing regimen was to achieve approximately 80% inhibition of proteasome activity with a return toward baseline to allow cells to recuperate from the effects. Based on these biomarker pharmacodynamic data, a twice a week dosing regimen followed by a 10-day rest period without dosing was selected.

Use of targeted therapies in combination with traditional chemotherapy is in the early development stage but will be critical to patient outcome. Establishment of appropriate timing and dosing for each component of the regimen may be critical to the overall response to the combination therapy. Biomarker-derived pharmacodynamic effects may be useful in evaluating dosing regimens for these combinations.

FIGURE 31.2 illustrates the "full circle" process for the use of molecular markers, including use of information obtained after approval of a product to provide feedback into the discovery process. Avastin™ (bevacizumab) could serve as an example. This product has recently been approved for front-line treatment of metastatic carcinoma of the colon or rectum (package insert, 2004). In 1992, results from a phase III trial indicated that Avastin™ did not meet the primary end point, improvement of progression-free survival, for treatment of relapsed metastatic breast cancer.[19] Further discovery efforts using molecular biomarker data may be useful in analyzing why this product worked in one form of cancer but not another. The data may also assist with discovery of insights into gene pathways for future development of products with potential targeted effects for breast cancer.

Full or "regular" marketing approval of oncology drug products requires substantial evidence of efficacy from adequate and well-controlled investigation. Efficacy should be demonstrated by clinical benefit or by an established surrogate end point for this benefit. Clinical benefit is defined as improved survival (i.e., the gold standard) or some other measure of patient well-being, such as reduction in the number of infections or prolongation of the time to new skeletal lesions (e.g., a skeletal-related event). Over the years, there have been very few "biomarker surrogates" that have been used for full approval. A high degree of stringency is required before a biomarker response can be substituted for a clinical outcome and is accepted for regulatory approval. A surrogate end point can be defined as follows: A laboratory measurement or a

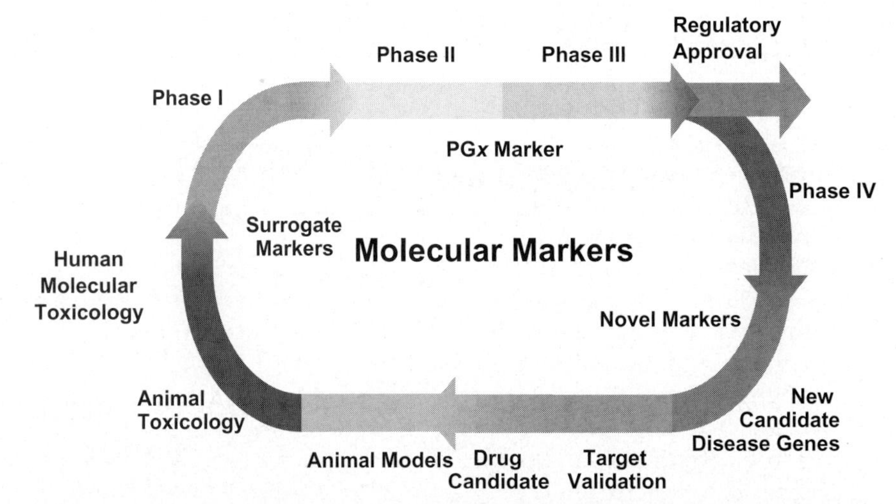

FIGURE 31.2 Integrated drug development. This schematic demonstrates how molecular markers can be applied to the early and late stages of drug development and after regulatory approval. Biomarkers have the potential to drive new information about an approved drug that will both further optimize its use and develop strategies to expand its indications.

physical sign, during a clinical trial, used as a substitute for a clinically meaningful end point that measures directly how a patient feels, functions, or survives. Changes induced by a therapy on a surrogate end point are expected to reflect changes in a clinically meaningful end point.[20] Reductions in cholesterol, blood glucose, and viral DNA and an increase in CD_4 counts have been accepted as surrogate markers for full approval of lipid-lowering, anti-diabetic, and anti-HIV products, respectively.[21] However, additional claims related to long-term benefits such as prevention of strokes or heart attacks require specific clinical data gathered over a prolonged period of time in many patients. For the treatment of advanced metastatic breast cancer, the surrogate end points of response rate and time to progression are considered adequate surrogates for clinical benefit associated with the use of hormonal therapy and therefore for regular approval. This is based on the favorable toxicity profile associated with hormonal drugs compared with conventional cytotoxic agents, and tamoxifen's well-accepted therapeutic role. Disease-free survival also can serve as the basis for regular approval in the adjuvant setting if a high proportion of symptomatic recurrences is present. Disease-free survival has supported applications for adjuvant therapy of breast cancer.[22] Submission of biomarker data as a component of a surrogate end point also has been used. For life-threatening conditions, compounds also can be approved under the accelerated approval process based on use of surrogates that are "reasonably likely to predict clinical benefit." A confirmatory study is to be ongoing at the time of submission of the new drug application (NDA).

Regarding biomarker surrogate end points for approval of oncology drug products, it should be noted that the FDA has not accepted changes in tumor biomarkers alone as a basis for any marketing approval.[22] Composite end points leading to approval may include evaluation of a tumor biomarker as part of an evaluation of disease progression.

Biomarkers can also be used to narrow the patient population to be treated in a clinical trial. The biomarker test and certain criteria are used to "enrich" the population, that is, make it more homogeneous regarding a specific aspect of the disease such as requiring that all patients' tumors be estrogen receptor positive. Use of enrichment techniques can markedly reduce the sample size of the study population and may also increase the degree of response in the evaluated population. However, the labeling claim for use of the product also would be limited to the population that was included in the clinical trial. For example, it is known that approximately 25–30% of women with breast cancer have the HER-2/neu receptor (see Chapter 16). Because Herceptin® is targeted for this receptor, all patients entered into the clinical trials with Herceptin® had to have a positive HER-2/neu test. Thus, the population was enriched based on this entrance criterion. As a result, the diagnostic test also had to be approved before Herceptin® could be approved because this test was required to screen for such patients in whom efficacy was demonstrated. The enriched study also limited the labeling claim to use in patients whose tumors express HER-2/neu.

FDA regulations establish different requirements for submission of specific data to (INDs), new NDA or biologic license application (BLA) submissions, and already approved drugs NDAs (or BLAs).[23] The FDA has provided definitions for biomarkers in their approach to defining what biomarker data need to be submitted to the Agency and at what stage of development in a draft guidance document.[2] Based on feedback from interested parties, these definitions could change.

Pharmacogenetic test: An assay intended to study interindividual variations in DNA sequence related to drug absorption and disposition (pharmacokinetics) or drug action (pharmacodynamics) including polymorphic variation in the genes that encode for functions of transporters, metabolizing enzymes, receptors, and other proteins.

Pharmacogenomic test: An assay intended to study interindividual variations in whole genome or candidate gene single nucleotide polymorphism maps, haplotype markers, and alterations in gene expression or inactivation that may be correlated with pharmacologic function and therapeutic response.

Biologic marker (biomarker): A characteristic that is objectively measured and evaluated as an indicator of normal biologic process, pathogenic processes, or pharmacologic responses to a therapeutic intervention.[24]

Valid biomarker: A biomarker that is measured in an analytical test system with well-established performance characteristics and for which there is an established scientific framework or body of evidence that elucidates the physiologic, toxicologic, pharmacologic, or clinical significance of the test results.

- *Known valid biomarker:* A biomarker that is measured in an analytical test system with well-established performance characteristics and for which there is widespread agreement in the medical or scientific community about the physiologic, toxicologic, pharmacologic, or clinical significance of the results. Known valid biomarker data must be submitted to INDs, new NDA submissions, and approved NDAs.
- *Probable valid biomarker:* A biomarker that is measured in an analytical test system with well-established performance characteristics and for which there is a scientific framework or body of evidence that appears to elucidate the physiologic, toxicologic, pharmacologic, or clinical significance of the test results. A probable valid biomarker may not have reached the status of a known valid biomarker for a variety of reasons, for example,
    a. The data elucidating its significance may have been generated within a single company and may not be available for public scientific scrutiny.
    b. The data elucidating its significance, although highly suggestive, may not be conclusive.
    c. Independent replication of the results may not have occurred. Probable valid biomarker data do not need to be submitted to INDs if the data are not used by the sponsor in decision making. For new NDA submissions, FDA recommends submission of these data in abbreviated reports. The data must be submitted to an approved NDA in the annual report as synopses or abbreviated reports.

Voluntary genomic data submission: The designation for pharmacogenomic data submitted voluntarily to the FDA. Data in this category would include results from pharmacogenomic tests that are not known or probable valid biomarkers. General exploratory or research data, such as gene expression screening, collection of sera or tissue samples, or pharmacogenomic tests that are not known or probable valid biomarkers, do not need to be submitted to an IND or an approved NDA. However, FDA recommends inclusion of such data in new NDA submissions as synopses.

Combination Products

Combination products within this definition can be considered "virtual" combination products that, in concept and labeling, must be used together to achieve the intended use, indication, or effect but are not packaged together. An example would be Herceptin® and the HER-2/neu expression test. Under FDA regulations, a biomarker test will need to be approved if mentioned specifically in the label of the drug product or if use of the product requires an established laboratory service capable of achieving the specific results. This has been applicable initially to breast cancer treatments. It has been extended to other oncology products as therapy becomes more targeted and likely to be effective only in tumors with certain features of overexpression or mutations. In addition to Herceptin®, another recent example is the approval of Erbitux™ (cetuximab), used in combination with irinotecan, for the treatment of epidermal growth factor (EGFR, HER-1) expressing, metastatic, colorectal carcinoma in patients who are refractory for irinotecan-based chemotherapy (package insert, 2004). Patients enrolled in the clinical studies were required to have immunohistochemical evidence of positive EGFR expression. Overexpression of EGFR (HER-1) has been detected in many human cancers, including colorectal, lung, breast, pancreatic, bladder, prostate, kidney, and head and neck. On the same day cetuximab was approved, the FDA approved a PMA application for EGFR pharm Dx kit by DakoCytomation. The PMA for the EGFR pharm Dx kit was submitted to FDA in parallel with the cetuximab submission. More recently, Avastin™ (bevacizumab) was approved for use, in combination with intravenous 5-fluorouracil based chemotherapy, for first-line treatment of patients with metastatic carcinoma of the colon or rectum (package insert 2004). Because the clinical trials did not require screening for vascular endothelial growth factor to enrich the population, this test is not required for selection of patients for Avastin™.

Nolvadex® (tamoxifen citrate) was initially approved in 1977. Although the importance of determining steroid hormone receptor status in the management of breast cancer is well recognized, the test is not listed in the labeling as being required for the use of tamoxifen. The labeling states that estrogen and progesterone values may help to predict whether adjuvant tamoxifen therapy is likely to be beneficial. The labeling further states that there are no data available regarding the effect of tamoxifen on breast cancer incidence in women with inherited mutations (*BRACA1, BRACA2*) (package insert, 2003).

A new Office of Combination Products has been formed within the Commissioner's Office of the FDA to facilitate the premarket review and approval process among the review divisions in the different centers of the FDA.[25] Guidance on intercenter consultations is being provided.

Further clarity is needed on how the primary center for review and approval, such as the Oncology Drug Products Division, works with OIVD within the CDRH to have the product, as well as the required test, approved and available simultaneously. The degree of complexity may be different if one company develops both the drug product and device, two companies collaborate to develop both, or if two companies act independently to develop the drug and device separately. A major problem could occur if the drug is ready for approval but the device is not ready for clearance at the same time. The recent approval of Erbitux and the EGFR expression test on the same day by two different companies is cause for cautious optimism. A process is needed to ensure that labeling for both drug product and the diagnostic test are updated and remain up to date, accurate, and mutually conforming.

References

1. Lesko LJ, Salerno RA, Spear BB, et al. Pharmacogenetics and pharmacogenomics in drug development and regulatory decision making: report of the first FDA-PWG-PhRMA-DruSafe workshop. J Clin Pharmacol 2003;43:342–358.
2. US Food and Drug Administration. Guidance for Industry—Pharmacogenomic Data Submissions. Rockville, MD: National Press Office, Nov. 2003.
3. Kahan JS. Medical Device Development. Waltham, MA: Parexel International Corporation, 2000.
4. Medical Device User Fee and Modernization Act of 2002 (H.R. 5651) Public Law 107–250 (107th Congress), Oct. 26, 2002.
5. Explanation of New Goal for Reducing Device Approval Times. http://www.fda.gov/cdrh/mdufma/goalpma.html, Feb. 2004.
6. US Food and Drug Administration, Center for Devices and Radiological Health. Draft: Premarketing Approval Review Criteria for Premarket Approval of Estrogen (ER) or Progesterone (PGR) Receptors In Vitro Diagnostic Devices Using Steroid Hormone Binding (SBA) with Dextran-Coated Charcoal (DCC) Separation, Histochemical Receptor Binding Assays, or Solid Phase Enzyme Immunoassay (EIA) Methodologies. Rockville, MD: National Press Office, 1997. http://www.fad.gov/cdrh/ode/odec1603.html, Feb. 2004.
7. 62 Federal Register 66260, Nov. 21, 1997.
8. 21 CFR 864.4020(a).
9. US Food and Drug Administration, Center for Devices and Radiological Health. Analyte Specific Reagents; Small Entity Compliance Guidance; Guidance for Industry. Rockville, MD: National Press Office, Feb. 26, 2003.
10. Hackett JL, Lesko LJ. Microarray data—the US FDA, industry and academia. Nat Biotechnol 2003; 21:742–743.
11. Wooster R, Weber BL. Breast and ovarian cancer. N Engl J Med 2003;348:2339–2347.
12. Van de Vijver MJ, He YD, van't Veer LJ, et al. A gene-expression signature as a predictor of survival in breast cancer. N Engl J Med 2002;347:1999–2009.
13. US Food and Drug Administration. Center for Devices and Radiological Health. Multiplex Tests for Heritable DNA Markers, Mutations and Expression Patterns; Draft Guidance for Industry and FDA Reviewers. Rockville, MD: National Press Office, April 21, 2003.
14. Claus EB, Schildkraut JM, Thompson WD, et al. The genetic attributable risk of breast and ovarian cancer. Cancer 1996;77:2318–2324.
15. Langston AA, Malone KE, Thompson JD, et al. BRACA1 mutations in a population-based sample of young women with breast cancer. N Engl J Med 1996; 334:137–142.
16. Gollust SE, Hull SC, Wilfond BS. Limitations of direct-to-consumer advertising for clinical genetic testing. JAMA 2002;288:1762–1767.
17. BBC News. Gene 'Switches Off' Breast Cancer, http://news.bbc.co.uk/1/hi/health/3236563.stm, Feb. 2004.
18. McClellan M. FDA commissioner says science is far from eliminating cancer as a threat. Cancer Lett 2004;30:1–3.
19. Phase III Trail with Avastin™ in Relapsed Metastatic Breast Cancer Does Not Meet Primary Endpoint. http://www.gene.com/gene/news/press-releases/display.do?method=detail&id=5427, Feb. 2004.
20. Temple RJ. A regulatory authority's opinion about surrogate endpoints. In: Nimmo WS, Tucker GT, eds. Clinical Measurement in Drug Evaluation. New York: John Wiley & Sons, 1995.
21. Lesko LJ, Atkinson AJ Jr. Use of biomarkers and surrogate endpoints in drug development and regulatory

decision making: criteria, validation, strategies. Annu Rev Pharmacol Toxicol 2001;41:347–366.

22. Johnson JR, Williams G, Pazdur R. End points and United States Food and Drug Administration approval of oncology drugs. J Clin Oncol 2003;21: 1404–1411.

23. Pisano DJ, Mantus D, eds. FDA Regulatory Affairs. Boca Raton, FL: CRC Press, 2004.

24. Biomarkers Definitions Working Group. Biomarkers and surrogate endpoints: preferred definitions and conceptual framework. Clin Pharmacol Ther 2001; 69:89–95.

25. US Food and Drug Administration, Office of Combination Products. Annual Report to Congress, Federal Food, Drug, and Cosmetic Act as amended by the Medical Device User Fee Act of 2002. Rockville, MD: National Press Office, Oct. 26, 2003.

INDEX

H

I

K

L

M

R

T

U

V

W

X